Lecture Notes in Computer Science 9453

Commenced Publication in 1973
Founding and Former Series Editors:
Gerhard Goos, Juris Hartmanis, and Jan van Leeuwen

More information about this series at http://www.springer.com/series/7410

Tetsu Iwata · Jung Hee Cheon (Eds.)

Advances in Cryptology – ASIACRYPT 2015

21st International Conference on the Theory
and Application of Cryptology and Information Security
Auckland, New Zealand, November 29 – December 3, 2015
Proceedings, Part II

Springer

Editors
Tetsu Iwata
Nagoya University
Nagoya
Japan

Jung Hee Cheon
Seoul National University
Seoul
Korea (Republic of)

ISSN 0302-9743   ISSN 1611-3349   (electronic)
Lecture Notes in Computer Science
ISBN 978-3-662-48799-0   ISBN 978-3-662-48800-3   (eBook)
DOI 10.1007/978-3-662-48800-3

Library of Congress Control Number: 2015953256

LNCS Sublibrary: SL4 – Security and Cryptology

Printed on acid-free paper

Springer-Verlag GmbH Berlin Heidelberg is part of Springer Science+Business Media
(www.springer.com)

Preface

ASIACRYPT 2015, the 21st Annual International Conference on Theory and Application of Cryptology and Information Security, was held on the city campus of the University of Auckland, New Zealand, from November 29 to December 3, 2015. The conference focused on all technical aspects of cryptology, and was sponsored by the International Association for Cryptologic Research (IACR).

The conference received 251 submissions from all over the world. The program included 64 papers selected from these submissions by a Program Committee (PC) comprising 43 leading experts of the field. In order to accommodate as many high-quality submissions as possible, the conference ran in two parallel sessions, and these two-volume proceedings contain the revised versions of the papers that were selected. The revised versions were not reviewed again and the authors are responsible for their contents.

The selection of the papers was made through the usual double-blind review process. Each submission was assigned three reviewers and submissions by PC members were assigned five reviewers. The selection process was assisted by a total of 339 external reviewers. Following the individual review phase, the selection process involved an extensive discussion phase.

This year, the conference featured three invited talks. Phillip Rogaway gave the 2015 IACR Distinguished Lecture on "The Moral Character of Cryptographic Work," Gilles Barthe gave a talk on "Computer-Aided Cryptography: Status and Perspectives," and Masayuki Abe spoke on "Structure-Preserving Cryptography." The proceedings contain the abstracts of these talks. The conference also featured a traditional rump session that contained short presentations on the latest research results of the field.

The best paper award was decided based on a vote by the PC members, and it was given to "Improved Security Proofs in Lattice-Based Cryptography: Using the Rényi Divergence Rather than the Statistical Distance" by Shi Bai, Adeline Langlois, Tancrède Lepoint, Damien Stehlé, and Ron Steinfeld. Two more papers, "Key-Recovery Attacks on ASASA" by Brice Minaud, Patrick Derbez, Pierre-Alain Fouque, and Pierre Karpman, and "The Tower Number Field Sieve" by Razvan Barbulescu, Pierrick Gaudry, and Thorsten Kleinjung, were solicited to submit full versions to the *Journal of Cryptology*.

ASIACRYPT 2015 was made possible by the contributions of many people. We would like to thank the authors for submitting their research results to the conference. We are deeply grateful to all the PC members and all the external reviewers for their hard work to determine the program of the conference. We sincerely thank Steven Galbraith, the general chair of the conference, and the members of the local Organizing Committee for handling all the organizational work of the conference. We also thank Nigel Smart for organizing and chairing the rump session.

We thank Shai Halevi for setting up and letting us use the IACR conference management software. Springer published the two-volume proceedings and made these available at the conference. We thank Alfred Hofmann, Anna Kramer, and their colleagues for handling the editorial process. Last but not least, we thank the speakers, session chairs, and all the participants for coming to Auckland and contributing to ASIACRYPT 2015.

December 2015   Tetsu Iwata
                                                                Jung Hee Cheon

ASIACRYPT 2015

The 21st Annual International Conference on Theory and Application of Cryptology and Information Security

Sponsored by the International Association for Cryptologic Research (IACR)

November 29–December 3, 2015, Auckland, New Zealand

General Chair

Steven Galbraith   University of Auckland, New Zealand

Program Co-chairs

Tetsu Iwata   Nagoya University, Japan
Jung Hee Cheon   Seoul National University, Korea

Program Committee

Daniel J. Bernstein   University of Illinois at Chicago, USA and Technische Universiteit Eindhoven, The Netherlands
Ignacio Cascudo   Aarhus University, Denmark
Chen-Mou Cheng   National Taiwan University, Taiwan
Sherman S.M. Chow   Chinese University of Hong Kong, Hong Kong, SAR China
Kai-Min Chung   Academia Sinica, Taiwan
Nico Döttling   Aarhus University, Denmark
Jens Groth   University College London, UK
Dawu Gu   Shanghai Jiaotong University, China
Dong-Guk Han   Kookmin University, Korea
Marc Joye   Technicolor, USA
Nathan Keller   Bar-Ilan University, Israel
Aggelos Kiayias   National and Kapodistrian University of Athens, Greece
Kaoru Kurosawa   Ibaraki University, Japan
Xuejia Lai   Shanghai Jiaotong University, China
Hyang-Sook Lee   Ewha Womans University, Korea
Jooyoung Lee   Sejong University, Korea
Soojoon Lee   Kyung Hee University, Korea
Arjen Lenstra   EPFL, Switzerland
Hemanta K. Maji   UCLA, USA
Alexander May   Ruhr University Bochum, Germany
Bart Mennink   KU Leuven, Belgium
Tatsuaki Okamoto   NTT Secure Platform Laboratories, Japan
Raphael C.-W. Phan   Multimedia University, Malaysia

Additional Reviewers

Jintai Ding
Itai Dinur
Christophe Doche
Ming Duan
Léo Ducas
Alina Dudeanu
Orr Dunkelman
Keita Emura
Martianus Frederic
  Ezerman
Xiong Fan
Antonio Faonio
Pooya Farshim
Sebastian Faust
Marc Fischlin
Eiichiro Fujisaki
Philippe Gaborit
Martin Gagné
Steven Galbraith
Nicolas Gama
Wei Gao
Peter Gaži
Essam Ghadafi
Hossein Ghodosi
Irene Giacomelli
Benedikt Gierlichs
Zheng Gong
Dov Gordon
Robert Granger
Sylvain Guilley
Jian Guo
Qian Guo
Zheng Guo
Divya Gupta
Florian Göpfert
Jaecheol Ha
Xue Haiyang
Keisuke Hakuta
Shuai Han
Neil Hanley
Malin Md Mokammel
  Haque
Yasufumi Hashimoto
Gottfried Herold
Javier Herranz
Shoichi Hirose
Viet Tung Hoang

Dennis Hofheinz
Justin Holmgren
Deukjo Hong
Wei-Chih Hong
Tao Huang
Yun-Ju Huang
Pavel Hubáček
Michael Hutter
Andreas Hülsing
Jung Yeon Hwang
Laurent Imbert
Sorina Ionica
Zahra Jafargholi
Tibor Jager
Jérémy Jean
Ik Rae Jeong
Hyungrok Jo
Thomas Johansson
Antoine Joux
Handan Kılınç
Taewon Kim
Alexandre Karlov
Pierre Karpman
Kenji Kashiwabara
Aniket Kate
Marcel Keller
Carmen Kempka
Dmitry Khovratovich
Dakshita Khurana
Jinsu Kim
Jongsung Kim
Min Kyu Kim
Sungwook Kim
Tae Hyun Kim
Taechan Kim
Taewan Kim
Paul Kirchner
Elena Kirshanova
Susumu Kiyoshima
Thorsten Kleinjung
Jessica Koch
Markulf Kohlweiss
Ilan Komargodski
Venkata Koppula
Ranjit Kumaresan
Po-Chun Kuo
Stefan Kölbl

Pascal Lafourcade
Russell W.F. Lai
Adeline Langlois
Martin M. Lauridsen
Changhoon Lee
Changmin Lee
Eunjeong Lee
Hyung Tae Lee
Juhee Lee
Tancrède Lepoint
Wen-Ding Li
Yang Li
Benoît Libert
Seongan Lim
Changlu Lin
Fuchun Lin
Tingting Lin
Wei-Kai Lin
Feng-Hao Liu
Junrong Liu
Shengli Liu
Ya Liu
Zhen Liu
Zhenhua Liu
Zhiqiang Liu
Satya Lokam
Carl Löndahl
Yu Long
Steve Lu
Yiyuan Luo
Atul Luykx
Vadim Lyubashevsky
Alex J. Malozemoff
Avradip Mandal
Giorgia Azzurra Marson
Luke Mather
Takahiro Matsuda
Christian Matt
Peihan Miao
Daniele Micciancio
Andrea Miele
Eric Miles
Kazuhiko Minematsu
Marine Minier
Takaaki Mizuki
Ameer Mohammed
Paweł Morawiecki

Daisuke Moriyama
Kirill Morozov
Nicky Mouha
Nadia El Mrabet
Pratyay Mukherjee
Yusuke Naito
Chanathip Namprempre
Mridul Nandi
María Naya-Plasencia
Khoa Nguyen
Ruben Niederhagen
Jesper Buus Nielsen
Ivica Nikolić
Svetla Nikova
Tobias Nilges
Ryo Nishimaki
Wakaha Ogata
Go Ohtake
Claudio Orlandi
Ilya Ozerov
Jiaxin Pan
Giorgos Panagiotakos
Omkant Pandey
Kostas Papagiannopoulos
Cheol-Min Park
Bryan Parno
Anat Paskin-Cherniavsky
Chris Peikert
Bo-Yuan Peng
Clément Pernet
Léo Perrin
Giuseppe Persiano
Thomas Peters
Christophe Petit
Albrecht Petzoldt
Thomas Peyrin
Le Trieu Phong
Cécile Pierrot
Bertram Poettering
Joop van de Pol
Antigoni Polychroniadou
Carla Ràfols
Yogachandran
  Rahulamathavan
Sergio Rajsbaum
Somindu C. Ramanna
Samuel Ranellucci

Vanishree Rao
Christian Rechberger
Oded Regev
Michał Ren
Oscar Reparaz
Reza Reyhanitabar
Vincent Rijmen
Matthieu Rivain
Vladimir Rožić
Saeed Sadeghian
Yusuke Sakai
Subhabrata Samajder
Simona Samardjiska
Katerina Samari
Alessandra Scafuro
Jacob C.N. Schuldt
Karn Seth
Yannick Seurin
Setareh Sharifian
Ji Sun Shin
Bo-Yeon Sim
Siang Meng Sim
Leonie Simpson
Shashank Singh
Arkadii Slinko
Mate Soos
Pierre-Jean Spaenlehauer
Martijn Stam
Ron Steinfeld
Christoph Striecks
Le Su
Koutarou Suzuki
Alan Szepieniec
Björn Tackmann
Katsuyuki Takashima
Syh-Yuan Tan
Qiang Tang
Christophe Tartary
Sidharth Telang
Isamu Teranishi
Stefano Tessaro
Ivan Tjuawinata
Daniel Tschudi
Yiannis Tselekounis
Yu-Hsiu Tung
Himanshu Tyagi
Aleksei Udovenko

Praveen Vadnala
Srinivas Vivek Venkatesh
Frederik Vercauteren
Damien Vergnaud
Gilles Villard
Dhinakaran
  Vinayagamurthy
Vanessa Vitse
Damian Vizár
Lei Wang
Qingju Wang
Weijia Wang
Bogdan Warinschi
Hoeteck Wee
Benjamin Wesolowski
Carolyn Whitnall
Daniel Wichs
Xiaodi Wu
Hong Xu
Sen Xu
Shota Yamada
Naoto Yanai
Bo-Yin Yang
Guomin Yang
Shang-Yi Yang
Masaya Yasuda
Takanori Yasuda
Kazuki Yoneyama
Taek-Young Youn
Ching-Hua Yu
Shih-Chun Yu
Yu Yu
Aaram Yun
Thomas Zacharias
Mark Zhandry
Bingsheng Zhang
Hui Zhang
Jiang Zhang
Liang Feng Zhang
Liting Zhang
Tao Zhang
Ye Zhang
Yongjun Zhao
Bo Zhu
Jens Zumbrägel

Organizing Committee

Chair

Steven Galbraith   University of Auckland, New Zealand

Local Committee Members

Peter Gutmann   Computer Science, University of Auckland, New Zealand
Hinne Hettema   ITS, University of Auckland, New Zealand
Giovanni Russello   Computer Science, University of Auckland, New Zealand
Arkadii Slinko   Mathematics, University of Auckland, New Zealand
Clark Thomborson   Computer Science, University of Auckland, New Zealand

Sponsors

The University of Auckland
Microsoft Research
Intel
STRATUS
Centre for Discrete Mathematics and Theoretical Computer Science

Invited Talks

Structure-Preserving Cryptography

Masayuki Abe

NTT Secure Platform Laboratories, NTT Corporation, Tokyo, Japan
abe.masayuki@lab.ntt.co.jp

Bilinear groups has been a common ground for building cryptographic schemes since its introduction in seminal works [3, 5, 6]. Not just being useful for directly designing schemes for their rich mathematical structure, they aim to modular construction of complex schemes from simpler building blocks that work over the same bilienar groups. Namely, given a description of blinear groups, several building blocks exchange group elements each other, and the security of the resulting scheme is proven based on the security of the underlying building blocks. Unfortunately, things are not that easy in reality. Building blocks often require grues that bridge incompatible interfaces or they have to be modified to work together and the security has to be re-proved.

Structure-preserving cryptography [2] is a paradigm for designing cryptographic schemes over bilinear groups. A cryptographic scheme is called structure preserving if its all public inputs and outputs consist of group elements of bilinear groups and the functional correctness can be verified only by computing group operations, testing group membership and evaluating pairing product equations. Due to the regulated interface, structure-preserving schemes are highly inter-operable as desired in modular constructions. In particular, combination of structure-preserving signatures and noninteractive proof system of [4] yields numerous applications that protect signers' or receivers' privacy. The required properties on the other hand make some important primitives such as pseudo-random functions and collision resistant shrinking commitments unavailable in the world of structure-preserving cryptography. Interestingly, however, the constraints on the verification of correctness aim to argue non-trivial lower bounds in some aspects of efficiency such as signature size in the structure-preserving signature schemes.

Since the first use of the term "structure-preserving" in [1] in 2010, intensive research has been done for the area. In this talk, we overview state of the art on several structure-preserving schemes including commitments and signatures with a careful look about underlying assumptions, known bounds, and impossibility results. We also show open questions and discuss promising directions for further research.

References

1. Abe, M., Fuchsbauer, G., Groth, J., Haralambiev, K., Ohkubo, M.: Structure-preserving signatures and commitments to group elements. In: Advances in Cryptology - CRYPTO 2010, 30th Annual Cryptology Conference, Santa Barbara, CA, USA, 15–19 August 2010. Proceedings, pp. 209–236 (2010)

2. Abe, M., Fuchsbauer, G., Groth, J., Haralambiev, K., Ohkubo, M.: Structure-preserving signatures and commitments to group elements. J. Cryptol. (2015). doi:http://dx.doi.org/10.1007/s00145-014-9196-7.
3. Boneh, D., Boyen, X.: Efficient selective-id secure identity-based encryption without random oracles. In: Advances in Cryptology - EUROCRYPT 2004, International Conference on the Theory and Applications of Cryptographic Techniques, Interlaken, Switzerland, 2–6 May 2004. Proceedings, pp. 223–238 (2004)
4. Groth, J., Sahai, A.: Efficient noninteractive proof systems for bilinear groups. SIAM J. Comput. **41**(5), 1193–1232 (2012)
5. Menezes, A., Okamoto, T., Vanstone, S.A.: Reducing elliptic curve logarithms to logarithms in a finite field. IEEE Trans. Inf. Theory **39**(5), 1639–1646 (1993)
6. Sakai, R., Kasahara, M.: ID based cryptosystems with pairing on elliptic curve. IACR Cryptology ePrint Archive 2003, vol. 54 (2003)

Computer-Aided Cryptography:
Status and Perspectives

Gilles Barthe

IMDEA Software Institute, Madrid, Spain

Computer-aided cryptography is an emerging discipline which advocates the use of computer tools for building and mechanically verifying the security of cryptographic constructions. Computer-aided cryptography builds on the code-based game-based approach to cryptographic proofs, and adopts a program verification approach to justify common patterns of reasoning, such as equivalence up to bad, lazy sampling, or simply program equivalence. Technically, tools like EasyCrypt use a program verification method based on probabilistic couplings for reasoning about the relationship between two probabilistic programs, and standard tools to reason about the probability of events in a single probabilistic program. The combination of these tools, together with general mechanisms to instantiate or combine proofs, can be used to verify many examples from the literature.

Recent developments in computer-aided cryptography have explored two different directions. On the one hand, several groups have developed fully automated techniques to analyze cryptographic constructions in the standard model or hardness assumptions in the generic group model. In turn, these tools have been used for synthesizing new cryptographic constructions. *Transformational* synthesis tools take as input a cryptographic construction, for instance a signature in Type I setting and outputs another construction, for instance a batch signature or a signature in Type III setting. In contrast, *generative* synthesis tools take as input some size constraints and output a list of secure cryptographic constructions, for instance padding-based encryption schemes, modes of operations, or tweakable blockciphers, meeting the size constraints. On the other hand, several groups are working on carrying security proofs to (assembly-level) implementations, building on advances in programming languages, notably verified compilers. These works open the possibility to reason formally about mitigations used by cryptography implementers and to deliver strong mathematical guarantees, in the style of provable security, for cryptographic code against more realistic adversaries.

For further background information, please consult: www.easycrypt.info.

The Moral Character of Cryptographic Work

Phillip Rogaway[1]

Department of Computer Science
University of California, Davis, USA

Abstract. Cryptography rearranges power: it configures who can do what, from what. This makes cryptography an inherently *political* tool, and it confers on the field an intrinsically *moral* dimension. The Snowden revelations motivate a reassessment of the political and moral positioning of cryptography. They lead one to ask if our inability to effectively address mass surveillance constitutes a failure of our field. I believe that it does. I call for a community-wide effort to develop more effective means to resist mass surveillance. I plea for a reinvention of our disciplinary culture to attend not only to puzzles and math, but, also, to the societal implications of our work.

Keywords: Cryptography · Democracy · Ethics · Mass surveillance · Privacy · Snowden revelations · Social responsibility

[1] Work on the paper and talk associated to this abstract has been supported by NSF Grant CNS 1228828. Many thanks to the NSF for their continuing support.

Contents – Part II

Contents – Part I

Signatures

Multiparty Computation I

Public Key Encryption

ABE and IBE

Zero-Knowledge

Multiparty Computation II

Attacks on ASASA

Key-Recovery Attacks on ASASA

Brice Minaud[1]([⊠]), Patrick Derbez[2], Pierre-Alain Fouque[1,3],
and Pierre Karpman[4,5]

[1] Université de Rennes 1, Rennes, France
brice.minaud@gmail.com
[2] SnT, University of Luxembourg, Luxembourg City, Luxembourg
patrick.derbez@uni.lu
[3] Institut Universitaire de France, Paris, France
pierre-alain.fouque@ens.fr
[4] Inria, Paris, France
[5] Nanyang Technological University, Singapore, Singapore
pierre.karpman@inria.fr

Abstract. The ASASA construction is a new design scheme introduced
at ASIACRYPT 2014 by Biryukov, Bouillaguet and Khovratovich. Its versatility was illustrated by building two public-key encryption schemes, a
secret-key scheme, as well as super S-box subcomponents of a white-box
scheme. However one of the two public-key cryptosystems was recently
broken at CRYPTO 2015 by Gilbert, Plût and Treger. As our main contribution, we propose a new algebraic key-recovery attack able to break at
once the secret-key scheme as well as the remaining public-key scheme, in
time complexity 2^{63} and 2^{39} respectively (the security parameter is 128
bits in both cases). Furthermore, we present a second attack of independent interest on the same public-key scheme, which heuristically reduces
its security to solving an LPN instance with tractable parameters. This
allows key recovery in time complexity 2^{56}. Finally, as a side result, we
outline a very efficient heuristic attack on the white-box scheme, which
breaks an instance claiming 64 bits of security under one minute on a
single desktop computer.

Keywords: ASASA · Algebraic cryptanalysis · Multivariate cryptography · LPN

1 Introduction

The idea of creating a public-key cryptosystem by obfuscating a secret-key cipher
was proposed by Diffie and Hellman in 1976, in the same seminal paper that
introduced the idea of public-key encryption [DH76]. While the RSA cryptosystem was introduced only a year later, creating a public-key scheme based on

P.Derbez—Supported by the CORE ACRYPT project from the Fond National de
la Recherche, Luxembourg.
P.Karpman—Partially supported by the Direction Générale de l'Armement and by
the Singapore National Research Foundation Fellowship 2012 (NRF-NRFF2012-06).

© International Association for Cryptologic Research 2015
T. Iwata and J.H. Cheon (Eds.): ASIACRYPT 2015, Part II, LNCS 9453, pp. 3–27, 2015.
DOI: 10.1007/978-3-662-48800-3_1

symmetric components has remained an open challenge to this day. The interest of this problem is not merely historical: beside increasing the variety of available public-key schemes, one can hope that a solution may help bridging the performance gap between public-key and secret-key cryptosystems, or at least offer new trade-offs in that regard.

Multivariate cryptography is one way to achieve this goal. This area of research dates back to the 1980's [MI88, FD86], and has been particularly active in the late 1990's and early 2000's [Pat95, Pat96, RP97, FJ03, ...]. Many of the proposed public-key cryptosystems build an encryption function from a structured, easily invertible polynomial, which is then scrambled by affine maps (or similarly simple transformations) applied to its input and output to produce the encryption function.

This approach might be aptly described as an ASA structure, which should be read as the composition of an affine map "A", a nonlinear transformation of low algebraic degree "S" (not necessarily made up of smaller S-boxes), and another affine layer "A". The secret key is the full description of the three maps A, S, A, which makes computing both ASA and $(ASA)^{-1}$ easy. The public key is the function ASA as a whole, which is described in a generic manner by providing the polynomial expression of each output bit in the input bits (or group of n bits if the scheme operates on $\mathbb{F}_{2^n}$). Thus the owner of the secret key is able to encrypt and decrypt at high speed, depending on the structure of S. The downside is slow public key operations, and a large key size.

The ASASA Construction. Historically, attempts to build public-key encryption schemes based on the above principle have been ill-fated [FJ03, BFP11, DGS07, DFSS07, WBDY98, ...][1]. However several new ideas to build multivariate schemes were recently introduced by Biryukov, Bouillaguet and Khovratovich at ASIACRYPT 2014 [BBK14]. The paradigm federating these ideas is the so-called ASASA structure: that is, combining two quadratic mappings S by interleaving random affine layers A. With quadratic S layers, the overall scheme has degree 4, so the polynomial description provided by the public key remains of reasonable size.

This is very similar to the 2R scheme by Patarin [PG97], which fell victim to several attacks [Bih00, DFKYZD99], including a powerful decomposition attack [DFKYZD99, FP06], later developed in a general context by Faugère *et al.* [FvzGP10, FP09a, FP09b]. The general course of this attack is to differentiate the encryption function, and observe that the resulting polynomials in the input bits live in a "small" space entirely determined by the first ASA layers. This essentially allows the scheme to be broken down into its two ASA sub-components, which are easily analyzed once isolated. A later attempt to circumvent this and other attacks by truncating the output of the cipher proved insecure against the same technique [FP06] — roughly speaking truncating does not prevent the derivative polynomials from living in too small a space.

---

[1] HFEv- seems to be an exception in this regard.

In order to thwart attacks including the decomposition technique, the authors of [BBK14] propose to go in the opposite direction: instead of truncating the cipher, a *perturbation* is added, consisting in new random polynomials of degree four added at fixed positions, prior to the last affine layer[2]. The idea is that these new random polynomials will be spread over the whole output of the cipher by the last affine layer. When differentiating, the "noise" introduced by the perturbation polynomials is intended to drown out the information about the first quadratic layer otherwise carried by the derivative polynomials, and thus to foil the decomposition attack.

Based on this idea, two public-key cryptosystems are proposed. One uses random quadratic expanding S-boxes as nonlinear components, while the other relies on the χ function, most famous for its use in the SHA-3 winner Keccak. However the first scheme was broken at Crypto 2015 by a decomposition attack [GPT15]: the number of perturbation polynomials turned out to be too small to prevent this approach. This leaves open the question of the robustness of the other cryptosystem, based on χ, to which we answer negatively.

Black-Box ASASA. Besides public-key cryptosystems, the authors of [BBK14] also propose a secret-key ("black-box") scheme based on the ASASA structure, showcasing its versatility. While the structure is the same, the context is entirely different. This black-box scheme is in fact the exact counterpart of the SASAS structure analyzed by Biryukov and Shamir [BS01]: it is a block cipher operating on 128-bit inputs; each affine layer is a random affine map on $\mathbb{Z}_2^{128}$, while the nonlinear layers are composed of 16 random 8-bit S-boxes. The secret key is the description of the three affine layers, together with the tables of all S-boxes.

In some sense, the "public key" is still the encryption function as a whole; however it is only accessible in a black-box way through known or chosen-plaintext or ciphertext attacks, as any standard secret-key scheme. A major difference however is that the encryption function can be easily distinguished from a random permutation because the constituent S-boxes have algebraic degree at most 7, and hence the whole function has degree at most 49; in particular, it sums up to zero over any cube of dimension 50. The security claim is that the secret key cannot be recovered, with a security parameter evaluated at 128 bits.

White-Box ASASA. The structure of the black-box scheme is also used as a basis for several white-box proposals. In that setting, a symmetric (black-box) ASASA cipher with small block (*e.g.* 16 bits) is used as a super S-box in a design with a larger block. A white-box user is given the super S-box as a table. The secret information consists in a much more compact description of the super S-box in terms of alternating linear and nonlinear layers. The security of the ASASA design is then expected to prevent a white-box user from recovering the secret information.

---

[2] A similar idea was used in [Din04].

1.1   Our Contribution

Algebraic Attack on the Secret-Key and χ-Based Public-Key Schemes.
Despite the difference in nature between the χ-based public-key scheme and the
black-box scheme, we present a new algebraic key-recovery attack able to break
both schemes at once. This attack does not rely on a decomposition technique.
Instead, it may be regarded as exploiting the relatively low degree of the encryption function, coupled with the low diffusion of nonlinear layers. Furthermore, in
the case of the public-key scheme, the attack applies regardless of the amount of
perturbation. Thus, contrary to the attack of [GPT15], there is no hope of patch-
ing the scheme by increasing the number of perturbation polynomials. As for the
secret-key scheme, our attack may be seen as a counterpart to the cryptanalysis
of SASAS in [BS01], and is structural in the same sense.

While the same attack applies to both schemes, their respective bottlenecks
for the time complexity come from different stages of the attack. For the χ
scheme, the time complexity is dominated by the need to compute the kernel
of a binary matrix of dimension 2^{13}, which can be evaluated to 2^{39} basic linear
operations[3]. As for the black-box scheme, the time complexity is dominated by
the need to encrypt 2^{63} chosen plaintexts, and the data complexity follows.

This attack actually only peels off the last linear layer of the scheme, reducing
ASASA to ASAS. In the case of the black-box scheme, the remaining layers can
be recovered in negligible time using Biryukov and Shamir's techniques [BS01].
In the case of the χ scheme, removing the remaining layers poses non-trivial
algorithmic challenges (such as how to efficiently recover quadratic polynomials
$A, B, C \in \mathbb{Z}_2[X_1, \ldots, X_n]/\langle X_i^2 - X_i \rangle$, given $A + B \cdot C$), and some of the algorithms we propose may be of independent interest. Nevertheless, in the end the
remaining layers are peeled off and the secret key is recovered in time complexity
negligible relative to the cost of removing the first layer.

LPN-Based Attack on the χ Scheme. As a second contribution, we present
an entirely different attack, dedicated to the χ public-key scheme. This attack
exploits the fact that each bit at the output of χ is "almost linear" in the input:
indeed the nonlinear component of each bit is a single product, which is equal to
zero with probability 3/4 over all inputs. Based on this property, we are able to
heuristically reduce the problem of breaking the scheme to an LPN-like instance
with easy-to-solve parameters. By LPN-like instance, we mean an instance of a
problem very close to the Learning Parity with Noise problem (LPN), on which
typical LPN-solving algorithms such as the Blum-Kalai-Wasserman algorithm
(BKW) [BKW03] are expected to immediately apply. The time complexity of
this approach is higher than the previous one, and can be evaluated at 2^{56} basic

---

[3] In practice, vector instructions operating on 128-bit inputs would mean that the
meaningful size of the matrix is $2^{13-7} = 2^6$, and in this context the number of basic
linear operations would be much lower. We also disregard asymptotic improvements
such as the Strassen or Coppersmith-Winograd algorithms and their variants. The
main point is that the time complexity is quite low — well within practical reach.

operations. However it showcases a different weakness of the χ scheme, providing a different insight into the security of ASASA constructions. In this regard, it is noteworthy that the security of another recent multivariate scheme, presented by Huang *et al.* at PKC'12 [HLY12], was also reduced to an easy instance of LWE [Reg05], which is an extension of LPN, in [AFF+14] [4].

Heuristic Attack on the White-Box Scheme. Finally as a side result, we describe a key-recovery attack on white-box ASASA. The attack technique is unrelated to the previous ones, and its motivation relies on heuristics rather than a theoretical model. On the other hand it is very effective on the smallest white-box instances of [BBK14] (with a security level of 64 bits), which we break under a minute on a laptop computer. Thus it seems that the security level offered by small-block ASASA is much lower than anticipated.

The same attack on white-box schemes was found independently by Dinur, Dunkelman, Kranz and Leander [DDKL15]. Their approach focuses on small-block ASASA instances, and is thus only applicable to the white-box scheme of [BBK14]. Section 5 of [DDKL15] is essentially the same attack as ours, minus some heuristic improvements (see [MDFK15]). On the other hand, the authors of [DDKL15] present other methods to attack small-block ASASA instances that are less reliant on heuristics, but as efficient as our heuristically improved variant, and thus provide a better theoretical basis for understanding small-block ASASA, as used in the white-box scheme of [BBK14].

1.2 Structure of the Article

Section 3 provides a brief description of the three ASASA schemes under attack. In Sect. 4, we present our main attack, as applied to the secret-key ("black-box") scheme. In particular, an overview of the attack is given in Sect. 4.1. The attack is then adapted to the χ public-key scheme in Sect. 5.1, while the LPN-based attack on the same scheme is presented in Sect. 5.2. Finally, our attack on the white-box scheme is presented in Sect. 6.

1.3 Implementation and Full Version

Due to space constraints, some subordinate algorithms and proofs were removed from the print version of this article. However none of the missing material is essential to understanding the attacks. The full version is available on ePrint [MDFK15]. It is also available at the following link, together with implementations of our attacks:

https://www.dropbox.com/sh/3glwc5x181fekre/AAASeG7D-CGKM2gLmr-UVBK9a

---

[4] On this topic, the authors of [BBK14] note that "the full application of LWE to multivariate cryptography is still to be explored in the future".

2 Notation and Preliminaries

The sign $\triangleq$ denotes an equality by definition. $|S|$ denotes the cardinality of a set S. The $\log()$ function denotes logarithm in base 2.

Binary Vectors. We write $\mathbb{Z}_2$ as a shorthand for $\mathbb{Z}/2\mathbb{Z}$. The set of n-bit vectors is denoted interchangeably by $\{0,1\}^n$ or $\mathbb{Z}_2^n$. However the vectors are always regarded as elements of $\mathbb{Z}_2^n$ with respect to addition $+$ and dot product $\langle\cdot|\cdot\rangle$. In particular, addition should be understood as bitwise XOR. The canonical basis of $\mathbb{Z}_2^n$ is denoted by $e_0, \dots, e_{n-1}$.

For any $v \in \{0,1\}^n$, v_i denotes the i-th coordinate of v. In this context, the index i is always computed modulo n, so $v_0 = v_n$ and so forth. Likewise, if F is a function mapping into $\{0,1\}^n$, F_i denotes the i-th bit of the output of F.

For $a \in \{0,1\}^n$, $\langle F|a\rangle$ is a shorthand for the function $x \mapsto \langle F(x)|a\rangle$.

For any $v \in \{0,1\}^n$, $\lfloor v \rfloor_k$ denotes the truncation $(v_0, \dots, v_{k-1})$ of v to its first k coordinates.

For any bit b, $\bar{b}$ stands for $b + 1$.

Derivative of a Binary Function. For $F : \{0,1\}^m \to \{0,1\}^n$ and $\delta \in \{0,1\}^m$, we define the derivative of F along δ as $\partial F/\partial\delta \triangleq x \mapsto F(x) + F(x+\delta)$. We write $\partial^d F/\partial v_0 \dots \partial v_{d-1} \triangleq \partial(\dots(\partial F/\partial v_0)\dots)/\partial v_{d-1}$ for the order-d derivative along $v_0, \dots, v_{d-1} \in \{0,1\}^m$. For convenience we may write F' instead of $\partial F/\partial v$ when v is clear from the context; likewise for F''.

The *degree* of F_i is its degree as an element of $\mathbb{F}_2[x_0, \dots, x_{m-1}]/\langle x_i^2 - x_i\rangle$ in the binary input variables. The degree of F is the maximum of the degrees of the F_i's.

Cube. A cube of dimension d in $\{0,1\}^n$ is simply an affine subspace of dimension d. The terminology comes from [DS09]. Note that summing a function F over a cube C of dimension d, i.e. computing $\sum_{c\in C} F(c)$, amounts to computing the value of an order-d differential of F at a certain point: it is equal to $\partial^d F/\partial v_0 \dots \partial v_{d-1}(a)$ for a, (v_i) such that $C = a + \mathrm{span}\{v_0, \dots, v_{d-1}\}$. In particular if F has degree d, then it sums up to zero over any cube of dimension $d + 1$.

Bias. For any probability $p \in [0,1]$, the *bias* of p is $|2p - 1|$. Note that the bias is sometimes defined as $|p - 1/2|$ in the literature. Our choice of definition makes the formulation of the Piling-up Lemma more convenient [Mat94]:

Lemma 1 (Piling-up Lemma). *For $X_1, \dots, X_n$ independent random binary variables with respective biases $b_1, \dots, b_n$, the bias of $X = \sum X_i$ is $b = \prod b_i$.*

Learning Parity with Noise (LPN). The LPN problem was introduced in [BKW03], and may be stated as follows: given $(A, As + e)$, find s, where:

- $s \in \mathbb{Z}_2^n$ is a uniformly random secret vector.
- $A \in \mathbb{Z}_2^{N \times n}$ is a uniformly random binary matrix.
- $e \in \mathbb{Z}_2^N$ is an *error* vector, whose coordinates are chosen according to a Bernoulli distribution with parameter p.

3   Description of ASASA schemes

3.1   Presentation and Notations

ASASA is a general design scheme for public or secret-key ciphers (or cipher components). An ASASA cipher is composed of 5 interleaved layers: the letter A represents an affine layer, and the letter S represents a nonlinear layer (not necessarily made up of smaller S-boxes). Thus the cipher may be pictured as:

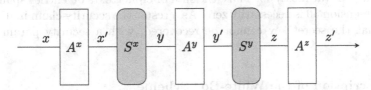

We borrow the notation of [GPT15] and write the encryption function F as:

$$F = A^z \circ S^y \circ A^y \circ S^x \circ A^x$$

Moreover, $x = (x_0, \ldots, x_{n-1})$ is used to denote the input of the cipher; x' is the output of the first affine layer A^x; and so on, as pictured above. The variables x'_i, y_i, etc., will often be viewed as polynomials over the input bits $(x_0, \ldots, x_{n-1})$. Similarly, F denotes the whole encryption function, while $F^y = S^x \circ A^x$ is the partial encryption function that maps the input x to the intermediate state y, and likewise $F^{x'} = A^x$, $F^{y'} = A^y \circ S^x \circ A^x$, etc.

One secret-key ("black-box") and two public-key ASASA ciphers are presented in [BBK14]. The secret-key and public-key variants are quite different in nature, even though our main attack applies to both. We now present in turn the black-box and white-box constructions and the public-key variant based on χ.

3.2   Description of the Black-Box Scheme

It is worth noting that the following ASASA scheme is the exact counterpart of the SASAS structure analyzed by Biryukov and Shamir [BS01], with swapped affine and S-box layers.

Black-box ASASA is a secret-key encryption scheme, parameterized by m, the size of the S-boxes and k, the number of S-boxes. Let $n = km$ be the number of bits of the scheme. The overall structure of the cipher follows the ASASA construction, with layers as follows:

- A^x, A^y, A^z are a random invertible affine mappings $\mathbb{Z}_2^n \to \mathbb{Z}_2^n$. Without loss of generality, the mappings can be considered purely linear, because the affine constant can be integrated into the preceding or following S-box layer. In the remainder we assume the mappings to be linear.
- S^x, S^y are S-box layers. Each S-box layer consists in the application of k parallel random invertible m-bit S-boxes.

All linear layers and all S-boxes are uniformly random among invertible elements, and independent from each other.

In the concrete instance of [BBK14], each S-box layer contains $k = 16$ S-boxes over $m = 8$ bits each, so that the scheme operates on blocks of $n = 128$ bits. The secret key consists in three n-bit matrices and $2k$ m-bit S-boxes, so the key size is $3 \cdot n^2 + 2k \cdot m2^m$-bit long. With the previous parameters this amounts to 14 KB.

It should be pointed out that the scheme is not IND-CPA secure. Indeed, an 8-bit invertible S-box has algebraic degree (at most) 7, so the overall scheme has algebraic degree (at most) 49. Thus, the sum of ciphertexts on entries spanning a cube of dimension 50 is necessarily zero. As a result the security claim in [BBK14] is only that the secret key cannot be recovered, with a security parameter of 128 bits.

3.3   Description of the White-Box Scheme

As an application of the symmetric ASASA scheme, Biryukov *et al.* propose its use as a basis for designing white-box block ciphers. In a nutshell, their idea is to use ASASA to create small ciphers of, say, 16-bit blocks and to use them as super S-boxes in *e.g.* a substitution-permutation network (SPN). Users of the cipher in the white-box model are given access to super S-boxes in the form a table, which allows them to encrypt and decrypt at will. Yet if the small ciphers used in building the super S-boxes are secure, one cannot efficiently recover their keys even when given access to their whole codebook, meaning that white-box users cannot extract a more compact description of the super S-boxes from their tables. This achieves *weak white-box security* as defined by Biryukov *et al.* [BBK14]:

Definition 1 (Key Equivalence [BBK14]). *Let* $E : \{0,1\}^\kappa \times \{0,1\}^n \to \{0,1\}^n$ *be a (symmetric) block cipher.* $\mathbb{E}(k)$ *is called the* equivalent key set *of* k *if for any* $k' \in \mathbb{E}(k)$ *one can efficiently compute* E' *such that* $\forall p\; E(k,p) = E'(k',p)$.

Definition 2 (Weak White-Box T-security [BBK14]). *Let* $E : \{0,1\}^\kappa \times \{0,1\}^n \to \{0,1\}^n$ *be a (symmetric) block cipher.* $\mathbb{W}(E)(k,\cdot)$ *is said to be a* T-secure weak white-box implementation *of* $E(k,\cdot)$ *if* $\forall p\; \mathbb{W}(E)(k,p) = E(k,p)$ *and if it is computationally expensive to find* $k' \in \mathbb{E}(k)$ *of length less than* T *bits when given full access to* $\mathbb{W}(E)(k,\cdot)$.

Example 1. If S_{16} is a secure cipher with 16-bit blocks, then the full codebook of $S_{16}(k,\cdot)$ as a table is a 2^{20}-secure weak white-box implementation of $S_{16}(k,\cdot)$.

For their instantiations, Biryukov *et al.* propose to use several super S-boxes of different sizes, among others:

- A 16-bit ASASA_{16} where the nonlinear permutations S are made of the parallel application of two 8-bit S-boxes, with conjectured security of 64 bits against key recovery.
- A 20-bit ASASA_{20} where the nonlinear permutations S are made of the parallel application of two 10-bit S-boxes, with conjectured security of 100 bits against key recovery.
- A 24-bit ASASA_{24} where the nonlinear permutations S are made of the parallel application of three 8-bit S-boxes, with conjectured security of 128 bits against key recovery.

3.4 Description of the χ-based Public-Key Scheme

The χ mapping was introduced by Daemen [Dae95] and later used for several cryptographic constructions, including the SHA-3 competition winner KECCAK. The mapping $\chi : \{0,1\}^n \to \{0,1\}^n$ is defined by:

$$\chi_i(a) = a_i + \overline{a_{i+1}} a_{i+2}$$

The χ-based ASASA scheme presented in [BBK14] is a public-key encryption scheme operating on 127-bit inputs, the odd size coming from the fact that χ is only invertible on inputs of odd length. The encryption function may be written as:

$$F = A^z \circ (P + \chi \circ A^y \circ \chi \circ A^x)$$

where:

- A^x, A^y, A^z are random invertible affine mappings $\mathbb{Z}_2^{127} \to \mathbb{Z}_2^{127}$. In the remainder we will decompose A^x as a linear map L^x followed by the addition of a constant C^x, and likewise for A^y, A^z.
- χ is as above.
- P is the *perturbation*. It is a mapping $\{0,1\}^{127} \to \{0,1\}^{127}$. For 24 output bits at a fixed position, it is equal to a random polynomial of degree 4. On the remaining 103 bits, it is equal to zero.

Since χ has degree only 2, the overall degree of the encryption function is 4. The public key of the scheme is the encryption function itself, given in the form of degree 4 polynomials in the input bits, for each output bit. The private key is the triplet of affine maps (A^x, A^y, A^z).

Due to the perturbation, the scheme is not actually invertible. To circumvent this, some redundancy is required in the plaintext, and the 24 bits of perturbation must be guessed during decryption. The correct guess is determined first by checking whether the resulting plaintext has the required redundancy, and second by recomputing the ciphertext from the tentative plaintext and checking that it matches. This is not relevant to our attack, and we refer the reader to [BBK14] for more information.

4   Structural Attack on Black-Box ASASA

Our goal in this section is to recover the secret key of the black-box ASASA scheme, in a chosen-plaintext model. For this purpose, we begin by peeling off the last linear layer, A^z. Once A^z is removed, we obtain an ASAS structure, which can be broken using Biryukov and Shamir's techniques [BS01] in negligible time. Thus the critical step is the first one.

4.1   Attack Overview

Before progressing further, it is important to observe that the secret key of the scheme is not uniquely defined. In particular, we are free to compose the input and output of any S-box with a linear mapping of our choosing, and use the result in place of the original S-box, as long as we modify the surrounding linear layers accordingly. Thus, S-boxes are essentially defined up to linear equivalence. When we claim to recover the secret key, this should be understood as recovering an equivalent secret key; that is, any secret key that results in an encryption function identical to the black-box instance under attack.

In particular, in order to remove the last linear layer of the scheme, it is enough to determine, for each S-box, the m-dimensional subspace corresponding to its image through the last linear layer. Indeed, we are free to pick any basis of this m-dimensional subspace, and assert that each element of this basis is equal to one bit at the output of the S-box. This will be correct, up to composing the output of the S-box with some invertible linear mapping, and composing the input of the last linear layer with the inverse mapping; which has no bearing on the encryption output.

Thus, peeling off A^z amounts to finding the image space of each S-box through A^z. For this purpose, we will look for linear masks $a, b \in \{0, 1\}^n$ over the output of the cipher, such that the two dot products $\langle F|a\rangle$ and $\langle F|b\rangle$ of the encryption function F along each mask are each equal to one bit at the output of the *same* S-box in the last nonlinear layer S^y. Let us denote the set of such pairs (a, b) by $\mathcal{S}$ (as in "solution").

In order to compute $\mathcal{S}$, the core property at play is that if masks a and b are as required, then the binary product $\langle F|a\rangle\langle F|b\rangle$ has degree only $(m-1)^2$ over the input variables of the cipher (meaning that $\langle F|a\rangle\langle F|b\rangle$ sums to zero over any cube of dimension $(m-1)^2 + 1$), whereas it has degree $2(m-1)^2$ in general.

We define the two linear masks a and b we are looking for as two vectors of binary unknowns. Then $f(a, b) = \langle F|a\rangle\langle F|b\rangle$ may be expressed as a quadratic polynomial over these unknowns, whose coefficients are $\langle F|e_i\rangle\langle F|e_j\rangle$ for (e_i) the canonical basis of $\mathbb{Z}_2^n$. Now, the fact that $f(a, b)$ sums to zero over some cube C gives us a quadratic condition on (a, b), whose coefficients are $\sum_{c \in C} \langle F(c)|e_i\rangle\langle F(c)|e_j\rangle$.

By computing $n(n-1)/2$ cubes of dimension $(m-1)^2 + 1$, we thus derive $n(n-1)/2$ quadratic conditions on (a, b). The resulting system can then be solved by relinearization. This yields the linear space K spanned by $\mathcal{S}$.

However we want to recover $\mathcal{S}$, rather its linear combinations K. Thus in a second step, we compute $\mathcal{S}$ as $\mathcal{S} = K \cap P$, where P is essentially the set of elements that stem from a single product of two masks a and b. While P is not a linear space, by guessing a few bits of the masks a, b, we can get many linear constraints on the elements of P satisfying these guesses, and intersect these linear constraints with K.

The first step may be regarded as the core of the attack, and it is also the computationally most expensive: essentially we need to encrypt plaintexts spanning $n(n-1)/2$ cubes of dimension $(m-1)^2 + 1$. We recall that in the actual black-box scheme of [BBK14], we have S-boxes over $m = 8$ bits, and the total block size is $n = 128$ bits, covered by $k = 16$ S-boxes, so the complexity is dominated by the computation of the encryption function over 2^{13} cubes of dimension 50, i.e. 2^{63} encryptions.

4.2   Description of the Attack

We use the notation of Sect. 3.1: let $F = A^z \circ S^y \circ A^y \circ S^x \circ A^x$ denote the encryption function. We are interested in linear masks $a \in \{0,1\}^n$ such that $\langle F|a \rangle$ depends only on the output of one S-box. Since $\langle F|a \rangle = \langle S^y \circ A^y \circ S^x \circ A^x|(A^z)^{\mathrm{T}}a \rangle$, this is equivalent to saying that the active bits of $(A^z)^{\mathrm{T}}a$ span a single S-box.

In fact we are searching for the set $\mathcal{S}$ of pairs of masks (a, b) such that $(A^z)^{\mathrm{T}}a$ and $(A^z)^{\mathrm{T}}b$ span the same single S-box. Formally, if we let $(e_0, \ldots, e_{n-1})$ be the canonical basis of $\mathbb{Z}_2^n$, and let $O_t = \mathrm{span}\{e_i : mt \leq i < m(t+1)\}$ be the span of the output of the t-th S-box, then:

$$\mathcal{S} = \{(a, b) \in \{0,1\}^n \times \{0,1\}^n : \exists t, (A^z)^{\mathrm{T}}a \in O_t \text{ and } (A^z)^{\mathrm{T}}b \in O_t\}$$

The core property exploited in the attack is that if (a, b) belongs to $\mathcal{S}$, then $\langle F|a \rangle \langle F|b \rangle$ has degree at most $(m-1)^2$, as shown by Lemma 2 below. On the other hand, if $(a, b) \notin \mathcal{S}$, then $\langle F|a \rangle \langle F|b \rangle$ is akin to the product of two independent random polynomials of degree $(m-1)^2$, and it reaches degree $2(m-1)^2$ with overwhelming probability.

Lemma 2. *Let G be an invertible mapping $\{0,1\}^m \to \{0,1\}^m$ for $m > 2$. For any two m-bit linear masks a and b, $H = \langle G|a \rangle \langle G|b \rangle$ has degree at most $m - 1$.*

Proof. It is clear that the degree cannot exceed m, since we depend on only m variables (and we live in $\mathbb{F}_2$). What we show is that it is less than $m-1$, as long as $m > 2$. If $a = 0$ or $b = 0$ or $a = b$, this is clear, so we can assume that a, b are linearly independent. Note that there is only one possible monomial of degree m, and its coefficient is equal to $\sum_{x \in \{0,1\}^m} H(x)$. So all we have to show is that this sum is zero.

Because G is invertible, $G(x)$ spans each value in $\{0,1\}^m$ once as x spans $\{0,1\}^m$. As a consequence, the pair $(\langle G|a \rangle, \langle G|b \rangle)$ takes each of its 4 possible values an equal number of times. In particular, it takes the value $(1, 1)$ exactly $1/4$ of the time. Hence $\langle G|a \rangle \langle G|b \rangle$ takes the value 1 exactly 2^{m-2} times, which is even for $m > 2$. Thus $\sum_{x \in \{0,1\}^m} H(x) = 0$ and we are done. $\square$

In the remainder, we regard two masks a and b as two sequences of n binary unknowns $(a_0, \ldots, a_{n-1})$ and $(b_0, \ldots, b_{n-1})$.

Step 1: Kernel Computation. If a, b are as desired, $\langle F|a\rangle\langle F|b\rangle$ has degree at most $(m-1)^2$. Hence the sum of this product over a cube of dimension $(m-1)^2+1$ is zero, as this amounts to an order-$(m-1)^2+1$ differential of a degree $(m-1)^2$ function. Let then C denote a random cube of dimension $(m-1)^2+1$ – that is, a random affine space of dimension $(m-1)^2+1$, over $\{0,1\}^n$. We have:

$$\sum_{c \in C} \langle F(c)|a\rangle\langle F(c)|b\rangle = \sum_{c \in C} \sum_{i<n} a_i F_i(c) \sum_{j<n} b_j F_j(c)$$

$$= \sum_{i,j<n} \Big(\sum_{c \in C} F_i(c)F_j(c)\Big) a_i b_j$$

$$= \sum_{i<j<n} \Big(\sum_{c \in C} F_i(c)F_j(c)\Big)(a_i b_j + a_j b_i)$$

To deduce the last line, notice that $\sum_{c\in C} F_i F_i = 0$ since F has degree less than $\dim C$. Since the equation above really only says something about $a_i b_j + a_j b_i$ rather than $a_i b_j$ (which is unavoidable, since the roles of a and b are symmetric), we define $E = \mathbb{Z}_2^{n(n-1)/2}$, see its canonical basis as $e_{i,j}$ for $i < j < n$, and define $\lambda(a,b) \in E$ by: $\lambda(a,b)_{i,j} = a_i b_j + a_j b_i$. By convention we set $\lambda_{j,i} = \lambda_{i,j}$ and $\lambda_{i,i} = 0$. The previous equations tells us that knowing only the $n(n-1)/2$ bits $\sum_{c \in C} F_i(c)F_j(c)$ yields a quadratic condition on (a,b), and more specifically a linear condition on $\lambda(a,b)$. Whence we proceed as follows:

Algorithm 1: GENERATECONDITION

Input: A random cube C of dimension $(m-1)^2+1$ over $\{0,1\}^n$
1 Let $sum = (0, \ldots, 0) \in E$
2 **for** $c \in C$ **do**
3   $\quad (x_0, \ldots, x_{n-1}) \leftarrow F(c)$
4   $\quad t \leftarrow (x_i x_j \text{ for } i < j < n) \in E$
5   $\quad sum = sum + t$
6 **return** sum

Let M be a binary matrix of size $(n^2/2) \times (n(n-1)/2)$, whose rows are separate outputs of Algorithm 1. Let K be the kernel of this matrix. Then for all $(a,b) \in \mathcal{S}$, $\lambda(a,b)$ is necessarily in K. Thus K contains the span of the $\lambda(a,b)$'s for $(a,b) \in \mathcal{S}$. Because M contains more than $n(n-1)/2$, with overwhelming probability K contains no other vector[5]. This is confirmed by our experiments.

---
[5] This point is the only reason we pick $n^2/2$ rows rather than only $n(n-1)/2$; but we may as easily choose $n(n-1)/2$ plus some small constant. In practice it we can just pick $n(n-1)/2$ rows, and add more as required until the kernel has the expected dimension $km(m-1)/2$.

Complexity Analysis. Overall, the dominant cost is to compute $2^{(m-1)^2+1}$ encryptions per cube, for $n^2/2$ cubes, which amounts to a total of $n^2 2^{(m-1)^2}$ encryptions. With the parameters of [BBK14], this is 2^{63} encryptions. In practice, we could limit ourselves to dimension-$(m-1)^2+1$ subcubes of a single dimension-$(m-1)^2+2$ cube, which would cost only $2^{(m-1)^2+2}$ encryptions. However we would still need to sum (pairwise bit products of) ciphertexts for each subcube, so while this approach would certainly be an improvement in practice, we believe it is cleaner to simply state the complexity as $n^2 2^{(m-1)^2}$ encryption equivalents.

Beside that, we also need to compute the kernel of a matrix of dimension $n(n-1)/2$, which incurs a cost of roughly $n^6/8$ basic linear operations. With the parameters of [BBK14], we need to invert a binary matrix of dimension 2^{13}, costing around 2^{39} (in practice, highly optimized) operations, so this is negligible compared to the required number of encryptions.

Step 2: Extracting Masks. Let:

$$P = \{\lambda \in E : \exists\, a, b \in \{0,1\}^n, \lambda = \lambda(a,b)\}$$

Clearly we have $\lambda(\mathcal{S}) \subseteq K \cap P$. In fact, we assume $\lambda(\mathcal{S}) = K \cap P$, which is confirmed by our experiments. We now want to compute $K \cap P$.

However we do not need to enumerate the whole intersection $K \cap P$ directly: for our purpose, it suffices to recover enough elements of $\lambda(\mathcal{S})$ such that the corresponding masks span the output space of all S-boxes. Indeed, recall that our end goal is merely to find the image of all k S-boxes through the last linear layer. Thus, in the remainder, we explain how to find a random element in $K \cap P$. Once we have found km linearly independent masks in this manner, we will be done.

The general idea to find a random element of $K \cap P$ is as follows. We begin by guessing the value of a few pairs (a_i, b_i). This yields linear constraints on the $\lambda_{i,j}$'s. As an example, if $(a_0, b_0) = (0,0)$, then $\forall i, \lambda_{0,i} = 0$. Because the constraints are linear and so is the space K, finding the elements of K satisfying the constraints only involves basic linear algebra. Thus, all we have to do is guess enough constraints to single out an element of $\mathcal{S}$ with constant probability, and recover that element as the one-dimensional subspace of K satisfying the constraints.

More precisely, assume we guess $2r$ bits of a, b as:

$$a_0, \dots, a_{r-1} = \alpha_0, \dots, \alpha_{r-1}$$
$$b_0, \dots, b_{r-1} = \beta_0, \dots, \beta_{r-1}$$

We view pairs (α_i, β_i) as elements of $\mathbb{Z}_2^2$. Assume there exists some linear dependency between the (α_i, β_i)'s: that is, for some $(\mu_i) \in \{0,1\}^r$:

$$\sum_{i=0}^{r-1} \mu_i(\alpha_i, \beta_i) = (0,0)$$

Then for all $j < n$, we have:

$$\sum_{i=0}^{r-1} \mu_i \lambda_{i,j} = b_j \sum_{i=0}^{r-1} \mu_i a_i + a_j \sum_{i=0}^{r-1} \mu_i b_i = 0 \qquad (1)$$

Now, since $\mathbb{Z}_2^2$ has dimension only 2, we can be sure that there exist $r - 2$ independent linear relations between the (α_i, β_i)'s, from which we deduce as above $(r - 2)n$ linear relations on the $\lambda_{i,j}$'s. In the full version of this article (see Sect. 1.3), we prove that at least $(r - 2)(n - r)$ of these relations are linearly independent.

Now, the cardinality of $\mathcal{S}$ is $k(2^m - 1)(2^m - 2) \approx k2^{2m}$. Hence if we choose $r = \lfloor \log(|\mathcal{S}|)/2 \rfloor \approx m + \frac{1}{2} \log k$, and randomly guess the values of (a_i, b_i) for $i < r$, then we can expect that with constant probability there exists exactly one element in $\mathcal{S}$ satisfying our guess. More precisely, each element has a probability (close to) $2^{-2\lfloor |\mathcal{S}|/2 \rfloor} \approx 2^{-|\mathcal{S}|}$ of fitting our guess of $2r$ bits, so this probability is close to $|\mathcal{S}|(|\mathcal{S}|^{-1}(1 - |\mathcal{S}|^{-1})^{|\mathcal{S}|-1}) \approx 1/e$. Thus, if we denote by T the subspace of E of vectors satisfying the linear constraints induced by our guess, with probability roughly $1/3$, $\lambda(\mathcal{S}) \cap T$ contains a single element.

On the other hand, K is generated by pairs of masks corresponding to distinct bits for each S-box in S^y. Hence $\dim K = km(m-1)/2 = n(m-1)/2$. As shown earlier, from our $2r$ guesses, we deduce (at least) $(r - 2)(n - r)$ linear conditions on the $(\lambda_{i,j})$'s, so codim $T \geq (r - 2)(n - r)$. Since we chose $r = m + \frac{1}{2} \log k$, this means:

$$\text{codim } T \geq \left(m - 2 + \frac{1}{2} \log k\right) \cdot \left(n - m - \frac{1}{2} \log k\right)$$
$$\dim K = (m - 1) \qquad \cdot (n/2)$$

Thus, having $\frac{1}{2} \log k \geq 1$, i.e. $k \geq 4$, and $m + \frac{1}{2} \log k \geq n/2$, which is easily the case with concrete parameters $m = 8$, $k = 16$, $n = 128$, we have codim $T \geq \dim K$, and so $K \cap T$ is not expected to contain any extra vector beside the span of $\lambda(\mathcal{S}) \cap T$. This is confirmed by our experiments.

In summary, if we pick $r = m + \frac{1}{2} \log k$ and randomly guess the first r pairs of bits (a_i, b_i), then with probability close to $1/e$, $K \cap T$ contains only a single vector, which belongs to $\lambda(\mathcal{S}) \cap T$ and in particular to $\lambda(\mathcal{S})$. In practice it may be worthwhile to guess a little less then $m + \frac{1}{2} \log k$ pairs to ensure $K \cap T$ is nonzero, then guess more as needed to single out a solution. Once we have a single element in $\lambda(\mathcal{S})$, it is easy to recover the two masks (a, b) it stems from[6].

In the end, we recover two masks (a, b) coming from the same S-box. If we repeat this process $n = km$ times on average, the masks we recover will span the output of each S-box (indeed we recover 2 masks each time, so n tries is more than enough with high probability). Furthermore, checking whether two masks belong to the same S-box is very cheap (for two masks a, b, we only need to check whether $\lambda(a, b)$ is in K), so we recover the output space of each S-box.

---

[6] It can be shown that λ is invertible except on its zero output, which is reached only when $a = 0$, $b = 0$ or $a = b$. An inversion algorithm is given in the full version of this article (cf. Sect. 1.3).

Complexity Analysis. In order to get a random element in $\mathcal{S}$, each guess of $2r$ bits yields roughly $1/3$ chance of recovering an element by intersecting linear spaces K and T. Since K has dimension $n(m-1)/2$, the complexity is roughly $(n(m-1)/2)^3$ per try, and we need 3 tries on average for one success. Then the process must be repeated n times. Thus the complexity may be evaluated to roughly $\frac{3}{8}n^4(m-1)^3$ basic linear operations. With the parameters of [BBK14], this amounts to 2^{36} linear operations, so this step is negligible compared to Step 1 (and quite practical besides).

Before closing this section, we note that our attack does not really depend on the randomness of the S-boxes or affine layers. All that is required of the S-boxes is that the degree of $z_i z_j$ vary depending on whether i and j belong to the same S-box. This makes the attack quite general, in the same sense as the structural attack of [BS01].

5   Attacks on the χ-based Public-Key Scheme

In this section, our goal is to recover the private key of the χ-based ASASA scheme, using only the public key. For this purpose, we peel off one layer at a time, starting with the last affine layer A^z. We actually propose two different ways to achieve this. The first attack is our main algebraic attack from Sect. 4, with some modifications to account for the peculiarity of χ and the presence of the perturbation. It is presented in Sect. 5.1. The second attack reduces the problem to an instance of LPN, and is presented in Sect. 5.2. Once the last affine layer has been removed with either attack, we move on to attacking the remaining layers in Sect. 5.3.

5.1   Algebraic Attack on the χ Scheme

The χ scheme can be attacked in exactly the same manner as the black-box scheme in Sect. 4. Using the notations of Sect. 3.1, we have:

$$z_i z_{i+1} = (y_i' + \overline{y_{i+1}'}y_{i+2}') \cdot (y_{i+1}' + \overline{y_{i+2}'}y_{i+3}')$$
$$= y_i'y_{i+1}' + y_i'\overline{y_{i+2}'}y_{i+3}'$$

Here the crucial point is that y_{i+2}' is shared by the only degree-4 term of both sides. Thus the degree of $z_i z_{i+1}$ is bounded by 6. Likewise, the degree of $z_{i+1}(z_i + z_{i+2}) = z_i z_{i+1} + z_{i+1}z_{i+2}$ is also bounded by 6, as the sum of two products of the previous form. On the other hand, any product of linear combinations $(\sum \alpha_i z_i)(\sum \beta_i z_i)$ not of the previous two forms does not share common y_i'''s in its higher-degree terms, so no simplification occurs, and the product reaches degree 8 with overwhelming probability.

As a result, we can proceed as in Sect. 4. Let $n = 127$ be the size of the scheme, $p = 24$ the number of perturbation polynomials. The positions of the p perturbation polynomials are not defined in the original paper; in the sequel we assume that they are next to each other. Other choices of positions increase the tedium of the attack rather than its difficulty. A brief discussion of random positions for perturbation polynomials is offered in the full version of this article (see Sect. 1.3). Due to the rotational symmetry of χ, the positions of the perturbed bits is only defined modulo rotational symmetry; for convenience, we assume that perturbed bits are at positions z_{n-p} to z_{n-1}.

The full attack presented below has been verified experimentally for small values of n.

Step 1: Kernel Computation. We fill the rows of an $n(n-1)/2 \times n(n-1)/2$ matrix with separate outputs of Algorithm 1, with the difference that the dimension of cubes in the algorithm is only 7 (instead of $(m-1)^2+1 = 50$ in the black-box case). Then we compute the kernel K of this matrix. Since $n(n-1)/2 \approx 2^{13}$ the complexity of this step is roughly 2^{39} basic linear operations.

Step 2: Extracting Masks. The second step is to intersect K with the set P of elements of the form $\lambda(a, b)$ to recover actual solutions (see Sect. 4, step 2). In Sect. 4 we were content with finding random elements of $K \cap P$. Now we want to find all of them. To do so, instead of guessing a few pairs (a_i, b_i) as earlier, we exhaust all possibilities for (a_0, b_0) then (a_1, b_1) and so forth along a tree-based search. For each branch, we stop when the dimension of K intersected with the linear constraints stemming from our guesses of (a_i, b_i)'s is reduced to 1. Each branch yields a solution $\lambda(a, b)$, from which the two masks a and b can be easily recovered.

Step 3: Sorting Masks. Let $a_i = ((L^z)^{\mathrm{T}})^{-1} e_i$ be the linear mask such that $z_i = \langle F|a_i \rangle$ (for the sake of clarity we first assume $C^z = 0$; this has no impact on the attack until step 4 in Sect. 5.3 where we will recover C^z). At this point we have recovered the set $\mathcal{S}$ of all (unordered) pairs of masks $\{a_i, a_{i+1}\}$ and $\{a_i, a_{i-1} + a_{i+1}\}$ for $i < n - p$, i.e. such that the corresponding z_i's are not perturbed. Now we want to distinguish masks $a_{i-1} + a_{i+1}$ from masks a_i. For each i such that z_{i-1}, z_i, z_{i+1} are not perturbed, this is easy enough, as a_i appears exactly three times among unordered pairs in $\mathcal{S}$: namely in the pairs $\{a_i, a_{i-1}\}$, $\{a_i, a_{i+2}\}$ and $\{a_i, a_{i-1} + a_{i+1}\}$; whereas masks of the form $a_{i-1} + a_{i+1}$ appear only once, in $\{a_{i-1} + a_{i+1}, a_i\}$.

Thus we have recovered every a_i for which z_{i-1}, z_i, z_{i+1} are not perturbed. Since perturbed bits are next to each other, we have recovered all unperturbed a_i's save the two a_i's on the outer edge of the perturbation, i.e. a_0 and a_{n-p-1}. We can also order all recovered a_i's simply by checking whether $\{a_i, a_{i+1}\}$ is in $\mathcal{S}$. In other words, we look at $\mathcal{S}$ as the set of edges of a graph whose vertices are the elements of pairs in $\mathcal{S}$; then the chain $(a_1, \ldots, a_{n-p-2})$ is simply the longest path in this graph. In fact we recover $(a_1, \ldots, a_{n-p-2})$, minus its direction: that is, so far, we cannot distinguish it from $(a_{n-p-2}, \ldots, a_1)$. If we look at the neighbours of the end points of the path, we also recover $\{a_0, a_0 + a_2\}$ and $\{a_{n-p-1}, a_{n-p-3} + a_{n-p-1}\}$. However we are not equipped to tell apart the members of each pair with only $\mathcal{S}$ at our disposal.

To find a_0 in $\{a_0, a_0 + a_2\}$ (and likewise a_{n-p-2} in $\{a_{n-p-1}, a_{n-p-3} + a_{n-p-1}\}$), a very efficient technique is to anticipate a little and use the distinguisher in Sect. 5.2. Namely, in short, we differentiate the encryption function F twice using two fixed random input differences $\delta_1 \neq \delta_2$, and check whether for a fraction $1/4$ of possible choices of (δ_1, δ_2), $\langle \partial^2 F / \partial \delta_1 \partial \delta_2 | x \rangle$ is equal to a constant with bias 2^{-4}: this property holds if and only if x is one of the a_i's. This only requires around 2^{16} encryptions for each choice of (δ_1, δ_2), and thus completes in negligible time. Another more self-contained approach is to move on to the next step (in Sect. 5.3), where the algorithm we use is executed separately on each recovered mask a_i, and fails for $a_0 + a_2$ but not a_1. However this would be slower in practice.

We assume either solution was chosen and we now know the whole ordered chain $(a_0, \ldots, a_{n-p-1})$ of masks corresponding to unperturbed bits. At this stage we are only missing the direction of the chain, i.e. we cannot distinguish $(a_0, \ldots, a_{n-p-1})$ from $(a_{n-p-1}, \ldots, a_0)$. This will be corrected at the next step.

As mentioned earlier, we propose two different techniques to recover the first linear layer of the χ scheme: one algebraic technique, and another based on LPN. We have now just completed the algebraic technique. In the next section we present the LPN-based technique. Afterwards we will move on to the remaining steps, which are common to both techniques, and fully break the cipher with the knowledge of $(a_0, \ldots, a_{n-p-1})$, in Sect. 5.3.

5.2   LPN-based attack on the χ scheme

We now present a different approach to remove the last linear layer of the χ scheme. This approach relies on the fact that each output bit of χ is almost linear, in the sense that the only nonlinear component is the product of two input bits. In particular this nonlinear component is zero with probability $3/4$. The idea is then to treat this nonlinear component as random noise. To achieve this we differentiate the encryption function F twice. So the first ASA layers of F'' yield a constant; then ASAS is a noisy constant due to the weak nonlinearity; and ASASA is a noisy constant accessed through A^z. This allows us to reduce the problem of recovering A^z to (a close variant of) an LPN instance with tractable parameters.

We now describe the attack in detail. First, pick two distinct random differences $\delta_1, \delta_2 \in \{0,1\}^n$. Then compute the order 2 differential of the encryption function along these two differences. That is, let $F'' = \partial F / \partial \delta_1 \partial \delta_2$. This second-order differential is constant at the output of $F^{y'} = A^y \circ \chi \circ A^x$, since χ has degree only two:

$$(F^{y'})''(x) \triangleq \partial F^{y'} / \partial \delta_1 \partial \delta_2 = C(\delta_1, \delta_2)$$

Now if we look at a single bit at the output of $F^z = \chi \circ F^{y'}$, we have:

$$(F^z)_i''(x) = (F^{y'})_i''(x) + \overline{F_{i+1}^{y'}} F_{i+2}^{y'}(x) + \overline{F_{i+1}^{y'}} F_{i+2}^{y'}(x + \delta_1)$$
$$+ \overline{F_{i+1}^{y'}} F_{i+2}^{y'}(x + \delta_2) + \overline{F_{i+1}^{y'}} F_{i+2}^{y'}(x + \delta_1 + \delta_2) \quad (2)$$

That is, a bit at the output of $(F^z)''$ still sums up to a constant, plus the sum of four bit products. If we look at each product as an independent random binary variable that is zero with probability $3/4$, i.e. bias 2^{-1}, then by the Piling-up Lemma (Lemma 1) the sum is equal to zero with bias 2^{-4}.

Experiments show that modeling the four products as independent is not quite accurate: a significant discrepancy is introduced by the fact that the four inputs of the products sum up to a constant. For the sake of clarity, we will disregard this for now and pretend that the four products are independent. We will come back to this issue later on.

Now a single linear layer remains between $(F^z)''$ and F''. Let $s_i \in \{0,1\}^n$ be the linear mask such that $\langle F|s_i\rangle = F_i^z$ (once again we assume $C^z = 0$, and postpone taking C^z into account until step 4 of the attack). Then $\langle F''|s_i\rangle$ is equal to a constant with bias 2^{-4}. Now let us compute N different outputs of F'' for some N to be determined later, which costs $4N$ calls to the encryption function F. Let us stack these N outputs in an $N \times n$ matrix A.

Then we know that $A \cdot s_i$ is either the all-zero or the all-one vector (depending on $(F^{y'})_i''$) plus a noise of bias 2^{-4}. Thus finding s_i is essentially an LPN problem with dimension $n = 127$ and bias 2^{-4} (i.e. noise $1/2 + 2^{-5}$). Of course this is not *quite* an LPN instance: A is not uniform, there are n solutions instead of one, and there is no output vector b (although we could isolate the last column of A and define it as the output vector). However in practice none of this should hinder the performance of a BKW algorithm [BKW03]. Thus we make the heuristic assumption that BKW performs here as it would on a standard LPN instance[7].

In the end, we recover the masks s_i such that $z_i = \langle F|s_i\rangle$. Before moving on to the next stage of the attack, we go back to the earlier independence assumption.

Dependency Between the Four Products. In the reasoning above, we have modeled the four bit products in Eq. 2 as independent binary random variables with bias 2^{-1}. That is, we assumed the four products would behave as:

$$\Pi = W_1 W_2 + X_1 X_2 + Y_1 Y_2 + Z_1 Z_2$$

where W_i, X_i, Y_i, Z_i are uniformly random *independent* binary variables. This yields an expectancy $\mathbb{E}[\Pi]$ with bias 2^{-4}. As noted above, this is not quite accurate, and we now provide a more precise model that matches with our experiments.

---

[7] To the best of our knowledge, we have yet to see an LPN-like problem with a matrix A on which BKW underperforms significantly compared to the uniform case, unless the problem was specifically crafted for this purpose. The existence of multiple solutions is also a notable difference in our case. However in a classic application of BKW with a fast Fourier transform at the end, this only means that the Fourier transform will output several solutions. Note that the dimension of the Fourier transform will be close to $127/3 \approx 42$ [LF06], and we have only $\approx 2^{14}$ solutions, so they are distinct on their last 42 bits with very high probability.

Since $F^{y'}$ has degree two, $(F^{y'})''$ is a constant, dependent only on δ_1 and δ_2. This implies that in the previous formula, we have $W_1 + X_1 + Y_1 + Z_1 = (F^{y'})''_{i+1}$ and $W_2 + X_2 + Y_2 + Z_2 = (F^{y'})''_{i+2}$. To capture this, we look at:

$$E(c_1, c_2) = \mathbb{E}[\Pi \mid W_1 + X_1 + Y_1 + Z_1 = c_1, W_2 + X_2 + Y_2 + Z_2 = c_2]$$

It turns out that $E(0,0)$ has a stronger bias, close to 2^{-3}; while perhaps surprisingly, $E(a, b)$ for $(a, b) \neq (0, 0)$ has bias zero, and is thus not suitable for our attack. Since G'' is essentially random, this means that our technique will work for only a fraction $1/4$ of output bits. However, once we have recovered these output bits, we can easily change δ_1, δ_2 to obtain a new value of G'' and start over to find new output bits.

After k iterations of the above process, a given bit at position $i \leq 127$ will have probability $(3/4)^k$ of remaining undiscovered. In order for all 103 unperturbed bits to be discovered with good probability, it is thus enough to perform $k = -\log(103)/\log(3/4) \approx 16$ iterations.

In the end we recover all linear masks a_i corresponding to unperturbed bits at the output of the second χ layer; i.e. $a_i = ((A^z)^{\mathrm{T}})^{-1} e_i$ for $0 \leq i < n-p$. The a_i's can then be ordered into a chain $(a_0, \ldots, a_{n-p-1})$ like in Sect. 5.1: neighbouring a_i's are characterized by the fact that $\langle F|a_i\rangle\langle F|a_{i+1}\rangle$ has degree 6. We postpone distinguishing between $(a_0, \ldots, a_{n-p-1})$ and $(a_{n-p-1}, \ldots, a_0)$ until Sect. 5.3.

Complexity Analysis. According to [LF06, Theorem 2], the number of samples needed to solve an LPN instance of dimension 127 and bias 2^{-4} is $N = 2^{44}$ (attained by setting $a = 3$ and $b = 43$). This requires $4N = 2^{46}$ encryptions. Moreover the dominant cost in the time complexity is to sort the 2^{44} samples a times, which requires roughly $3 \cdot 44 \cdot 2^{44} < 2^{52}$ basic operations. Finally, as noted above, we need to iterate the process 16 times to recover all unperturbed output bits with good probability, so our overall time complexity is increased to 2^{56} for BKW, and 2^{50} encryptions to gather samples (slightly less with a structure sharing some plaintexts between the 16 iterations).

5.3 Peeling Off the Remaining ASAS layers

Using either the algebraic attack from Sect. 5.1 or the LPN-based attack from Sect. 5.2, we have recovered the ordered chain $(a_0, \ldots, a_{n-p-1})$ of linear masks such that $z_i = \langle F|a_i\rangle$. More exactly we have recovered either $(a_0, \ldots, a_{n-p-1})$ or $(a_{n-p-1}, \ldots, a_0)$. For simplicity assume we have recovered $(a_0, \ldots, a_{n-p-1})$. We will be able to distinguish between the two cases later on.

Essentially, this means we have peeled off the last affine layer A^z — or more accurately, its linear component, over the unperturbed bits. Note that we cannot hope to recover A^z over perturbed bits, as perturbed bits are by definition uniformly random polynomials of degree 4, and a linear combination of uniformly random polynomials of degree 4 is still a uniformly random polynomial of degree 4. In other words, the perturbation is essentially defined modulo affine equivalence.

We now move on to peeling off the remaining layers one by one. We point out once again that all steps below have been verified experimentally.

Step 4: from ASAS to ASA. The next layer we wish to peel off is a χ layer, which is entirely public. It may seem that applying χ^{-1} should be enough. The difficulty arises from the fact that we do not know the full output of χ, but only $n - p$ bits. Furthermore, if our goal was merely to decrypt some specific ciphertext, we could use other techniques, e.g. the fact that guessing one bit at the input of χ produces a cascade effect that allows recovery of all other input bits from output bits, regardless of the fact that the function has been truncated [Dae95]. However our goal is different: we want to recover the secret key, not just be able to decrypt messages. For this purpose we want to cleanly recover the input of χ in the form of degree 2 polynomials, for every unperturbed bit. We propose a technique to achieve this below.

From the previous step, we are in possession of $(a_0, \ldots, a_{n-p-1})$ as defined above. Since by definition $z_i = \langle F | a_i \rangle$, this means we know z_i for $0 \leq i < n - p$. Note that y_i' has degree only 2, and we know that $z_i = y_i' + \overline{y_{i+1}'} y_{i+2}'$. In order to reverse the χ layer, we set out to recover y_i', y_{i+1}', y_{i+2}' from knowledge of only z_i, by using the fact that y_i', y_{i+1}', y_{i+2}' are quadratic.

This reduces to the following problem: given $P = A + B \cdot C$, where A, B, C are degree-2 polynomials, recover A, B, C. A closer look reveals that this problem is not possible exactly as stated, because P can be equivalently written in four different ways as: $A + B \cdot C$, $A + B + B \cdot \overline{C}$, $A + C + \overline{B} \cdot C$, $\overline{A + B + C} + \overline{B} \cdot \overline{C}$. On the other hand, we assume that for uniformly random A, B, C, the probability that P may be written in some unrelated way, i.e. $P = C + D \cdot E$ for C, D, E distinct from the previous four cases, is overwhelmingly low. This situation has never occurred in our experiments. Thus our problem reduces to:

Problem 1. Given $P = A + B \cdot C$, where A, B, C are degree-2 polynomials, recover degree-2 polynomials A', B', C' such that $P = A' + B' \cdot C'$.

Our previous assumption says $A' \in \text{span}\{A, B, C, 1\}$; $B', C' \in \text{span}\{B, C, 1\}$. A straightforward approach to tackle this problem is to write B formally as a generic degree-2 polynomial with unknown coefficients. This gives us $k = 1 + n + n(n+1)/2 \approx n^2/2$ binary unknowns. Then we observe that $B \cdot P$ has degree only 4 (since $B^2 = B$). Each term of degree 5 in $B \cdot P$ must have a zero coefficient, and thus each term gives us a linear constraint on the unknown coefficients of B. Collecting the constraints takes up negligible time, at which point we have a $k \times k$ matrix whose kernel is $\text{span}\{B, C, 1\}$. This gives us a few possibilities for B', C', which we can filter by checking that $A' = P - B' \cdot C'$ has degree 2. The complexity of this approach boils down to inverting a k-dimensional binary matrix, which costs essentially 2^{3k} basic linear operations. In our case this amounts to 2^{39} basic linear operations. In the full version of this article (cf. Sect. 1.3), we present a more elaborate, but faster algorithm to solve Problem 1.

At this point, we have essentially removed the first two ASASA layers (assuming $C^z = 0$, but this actually has no impact up to this point). More work is required to fully recover the layers, and analyze the remaining ASA layers. However the core of the attack is over. A detailed description of the remaining steps to fully recover the remaining layers is provided in the full version of this article (see Sect. 1.3).

6 A Practical Attack on White-Box ASASA

In this section we show that the actual security of small-block ASASA ciphers is much lower than was estimated by Biryukov *et al*. We describe a procedure that attempts to recover the secret components of the structure, thus breaking the weak white-box security notion (Definition 2). Our algorithm relies rather heavily on heuristics, and evaluating its efficiency requires actual implementation. We focused on two instance, the 16-bit ASASA_{16} with claimed security of 64 bits and the 20-bit ASASA_{20} with claimed security of 100 bits. A straightforward implementation of our algorithm is able to recover the secret components of the 16-bit instance in under a minute and of the 20-bit instance in a few hours, when running on a standard PC. We recall that the source code is publicly available (see Sect. 1.3). For the remainder of the section, we implicitly use the 16-bit instance when describing the attack.

6.1 Attack Overview

Our general black-box attack from Sect. 4 does not apply, because the block size is too small to allow computing cubes of dimension 50. On the other hand, the small block size makes it possible to compute the distribution of output differences for a single input difference in very reasonable time. For instance, one can compute and store the entire difference distribution table (DDT) of a 16-bit cipher in under a second using just a standard PC.

Remark 1. Our attack makes use of the full codebook of the ciphers, which in general may be seen as a very strong requirement. This is however only natural in the case of attacking white-box implementations, as the user is actually *required* to be given the full codebook of the super S-boxes as part of the implementation.

From the results of Biryukov and Shamir [BS01], it is already enough to recover only one of the external affine (or linear) layers in order to break the security of ASASA. Indeed, this allows to reduce the cipher to either of ASAS or SASA, which can then be attacked in practical time using their method. Thus we focus on removing the first linear layer. In accordance with the opening remarks of Sect. 4.1, this amounts to finding the image space of each S-box through $(A^x)^{-1}$.

The general idea of the attack is to create an oracle able to recognize whether an input difference δ activates one or two S-boxes in the first S-box layer S^x. More accurately, we create a ranking function $\mathcal{F}$ such that $\mathcal{F}(\delta)$ is expected to be significantly higher if δ activates only one S-box rather than two. We propose two choices for $\mathcal{F}$.

Both choices begin by computing the entire output difference distribution $D(\delta)$ for the input difference δ, i.e. the row corresponding to δ in the DDT. Then the value of $\mathcal{F}(\delta)$ is computed from $D(\delta)$. Choices for $\mathcal{F}$ are heuristic, but experiments show they are quite efficient. We now present our two choices for $\mathcal{F}$.

Walsh Transform. The idea behind this version of the attack is quite intuitive. If δ activates only one S-box, then after the first SA layers, two inner states computed from any two plaintexts with input difference δ are equal on the output of the inactive S-box. Hence after the first ASA layers, they are equal along $2^8 - 1$ non-zero linear masks. Since these masks only traverse a single S-box layer before the output of the cipher, linear cryptanalysis [Mat94] tells us that we can expect some linear masks to be biased at the output of the cipher. On the other hand if both S-boxes are active in the first round, no such phenomenon occurs, and linear biases on the output differences are expected to be weaker.

In order to measure this difference, we propose to compute, for every output mask a, the value $f(a) = \left(\sum_{x \in \{0,1\}^{16}} \langle \partial F \partial \delta(x) | a \rangle \right) - 2^{15}$ (where the sum is computed in $\mathbb{Z}$). That is, $2^{-15} f(a)$ is the bias of the output differences $D(\delta)$ along mask a. The function f can be computed efficiently, since it is precisely the Walsh transform of the characteristic function of $D(\delta)$, and we can use a fast Fourier transform algorithm. Then as a ranking function $\mathcal{F}$ we simply choose $\max(f)$, i.e. the highest bias among all output masks.

Number of Collisions. It turns out that performing the Walsh transform is not truly necessary. Indeed, the number of collisions in $D(\delta)$ is higher when δ activates only 1 S-box; where by number of collisions we mean 2^{15} minus the number of distinct values in $D(\delta)$. This may be understood as a consequence of the fact that whenever δ activates a single S-box, only 2^7 output differences are possible after the first ASA layers; and depending on the properties of the active (random) S-box, the distribution between these differences may be quite uneven. Whereas if both S-boxes are active, 2^{15} differences are possible and the distribution is expected to be less skewed. Thus we pick as ranking function $\mathcal{F}$ the number of collisions in $D(\delta)$ in the previous sense.

Once we have chosen a ranking function $\mathcal{F}$, we simply compute the ranking of every possible input difference, sort the differences, and choose the highest 16 linearly independent differences according to our ranking. Our hope is that these differences only activate a single S-box. In a second step, we will group together differences that activate the same S-box. A more detailed description of the attack, together with a discussion of the results, is provided in the full version of this article (see Sect. 1.3).

7   Conclusion

We presented a new algebraic attack able to efficiently break both the χ-based public-key cryptosystem and the secret-key scheme of [BBK14]. In addition we proposed another attack that heuristically reduces the key-recovery problem on the χ scheme to an easy instance of LPN. In the case of the public-key scheme, both attacks go through regardless of the amount of perturbation. For both schemes, the attacks are quite structural (in the case of the black-box scheme, it is in fact structural in the sense of [BS01]), and seem difficult to patch. Finally, although the general attack on the black-box scheme does not carry over to the small-block instances used for white-bow designs, we also showed a very efficient dedicated attack on some of the small-block instances, casting a doubt on their general suitability for that purpose.

References

[AFF+14] Albrecht, M.R., Faugére, J.-C., Fitzpatrick, R., Perret, L., Todo, Y., Xagawa, K.: Practical cryptanalysis of a public-key encryption scheme based on new multivariate quadratic assumptions. In: Krawczyk, H. (ed.) PKC 2014. LNCS, vol. 8383, pp. 446–464. Springer, Heidelberg (2014)

[BBK14] Biryukov, A., Bouillaguet, C., Khovratovich, D.: Cryptographic schemes based on the ASASA structure: black-box, white-box, and public-key (extended abstract). In: Sarkar, P., Iwata, T. (eds.) ASIACRYPT 2014. LNCS, vol. 8873, pp. 63–84. Springer, Heidelberg (2014)

[BFP11] Bettale, L., Faugère, J.-C., Perret, L.: Cryptanalysis of multivariate and odd-characteristic HFE variants. In: Catalano, D., Fazio, N., Gennaro, R., Nicolosi, A. (eds.) PKC 2011. LNCS, vol. 6571, pp. 441–458. Springer, Heidelberg (2011)

[Bih00] Biham, E.: Cryptanalysis of Patarin's 2-round public key system with S Boxes (2R). In: Preneel, B. (ed.) EUROCRYPT 2000. LNCS, vol. 1807, pp. 408–416. Springer, Heidelberg (2000)

[BKW03] Blum, A., Kalai, A., Wasserman, H.: Noise-tolerant learning, the parity problem, and the statistical query model. J. ACM (JACM) 50(4), 506–519 (2003)

[BS01] Biryukov, A., Shamir, A.: Structural cryptanalysis of SASAS. In: Pfitzmann, B. (ed.) EUROCRYPT 2001. LNCS, vol. 2045, pp. 395–405. Springer, Heidelberg (2001)

[Dae95] Daemen, J.: Cipher and hash function design strategies based on linear and differential cryptanalysis. Ph.D. thesis, Katholieke Universiteit Leuven, Leuven, Belgium (1995)

[DDKL15] Dinur, I., Dunkelman, O., Kranz, T., Leander, G.: Decomposing the asasa block cipher construction. Cryptology ePrint Archive, Report 2015/507 (2015). http://eprint.iacr.org/2015/507/

[DFKYZD99] Ding-Feng, Y., Kwok-Yan, L., Zong-Duo, D.: Cryptanalysis of 2R schemes. In: Wiener, M. (ed.) CRYPTO 1999. LNCS, vol. 1666, pp. 315–325. Springer, Heidelberg (1999)

[DFSS07] Dubois, V., Fouque, P.-A., Shamir, A., Stern, J.: Practical cryptanalysis of SFLASH. In: Menezes, A. (ed.) CRYPTO 2007. LNCS, vol. 4622, pp. 1–12. Springer, Heidelberg (2007)

[DGS07] Dubois, V., Granboulan, L., Stern, J.: Cryptanalysis of HFE with internal perturbation. In: Okamoto, T., Wang, X. (eds.) PKC 2007. LNCS, vol. 4450, pp. 249–265. Springer, Heidelberg (2007)

[DH76] Diffie, W., Hellman, M.E.: Multiuser cryptographic techniques. In: AFIPS 1976 National Computer Conference, pp. 109–112. ACM (1976)

[Din04] Ding, J.: A new variant of the Matsumoto-Imai cryptosystem through perturbation. In: Bao, F., Deng, R., Zhou, J. (eds.) PKC 2004. LNCS, vol. 2947, pp. 305–318. Springer, Heidelberg (2004)

[DS09] Dinur, I., Shamir, A.: Cube attacks on tweakable black box polynomials. In: Joux, A. (ed.) EUROCRYPT 2009. LNCS, vol. 5479, pp. 278–299. Springer, Heidelberg (2009)

[FD86] Fell, H., Diffie, W.: Analysis of a public key approach based on polynomial substitution. In: Williams, H.C. (ed.) CRYPTO 1985. LNCS, vol. 218, pp. 340–349. Springer, Heidelberg (1986)

[FJ03] Faugère, J.-C., Joux, A.: Algebraic cryptanalysis of hidden field equation (HFE) cryptosystems Using Gröbner bases. In: Boneh, D. (ed.) CRYPTO 2003. LNCS, vol. 2729, pp. 44–60. Springer, Heidelberg (2003)

[FP06] Faugère, J.-C., Perret, L.: Cryptanalysis of $2R^{--}$ schemes. In: Dwork, C. (ed.) CRYPTO 2006. LNCS, vol. 4117, pp. 357–372. Springer, Heidelberg (2006)

[FP09a] Faugère, J.-C., Perret, L.: An efficient algorithm for decomposing multivariate polynomials and its applications to cryptography. J. Symbolic Computat. **44**(12), 1676–1689 (2009)

[FP09b] Faugère, J.C., Perret, L.: High order derivatives and decomposition of multivariate polynomials. In: ISSAC 2009: Proceedings of the 2009 International Symposium on Symbolic and Algebraic Computation, pp. 207–214. ACM (2009)

[FvzGP10] Faugère, J.C., von zur Gathen, J., Perret, L.: Decomposition of generic multivariate polynomials. In ISSAC 2010: Proceedings of the 2010 International Symposium on Symbolic and Algebraic Computation, pp. 131–137. ACM (2010). ISBN 0747-7171 (updated version)

[GPT15] Gilbert, H., Plût, J., Treger, J.: Key-recovery attack on the ASASA cryptosystem with expanding S-Boxes. In: Gennaro, R., Robshaw, M. (eds.) CRYPTO 2015. LNCS, vol. 9215, pp. 475–490. Springer, Heidelberg (2015)

[HLY12] Huang, Y.-J., Liu, F.-H., Yang, B.-Y.: Public-key cryptography from new multivariate quadratic assumptions. In: Fischlin, M., Buchmann, J., Manulis, M. (eds.) PKC 2012. LNCS, vol. 7293, pp. 190–205. Springer, Heidelberg (2012)

[LF06] Levieil, É., Fouque, P.-A.: An improved LPN algorithm. In: De Prisco, R., Yung, M. (eds.) SCN 2006. LNCS, vol. 4116, pp. 348–359. Springer, Heidelberg (2006)

[Mat94] Matsui, M.: Linear cryptanalysis method for DES cipher. In: Helleseth, T. (ed.) EUROCRYPT 1993. LNCS, vol. 765, pp. 386–397. Springer, Heidelberg (1994)

[MDFK15] Minaud, B., Derbez, P., Fouque, P.-A., Karpman, P.: Key-recovery attacks on ASASA. Cryptology ePrint Archive, Report 2015/516 (2015). http://eprint.iacr.org/2015/516/

[MI88] Matsumoto, T., Imai, H.: Public quadratic polynomial-tuples for efficient signature-verification and message-encryption. In: Günther, C.G. (ed.) EUROCRYPT 1988. LNCS, vol. 330, pp. 419–453. Springer, Heidelberg (1988)

[Pat95] Patarin, J.: Cryptanalysis of the Matsumoto and Imai public key scheme of Eurocrypt '88. In: Coppersmith, D. (ed.) CRYPTO 1995. LNCS, vol. 963, pp. 248–261. Springer, Heidelberg (1995)

[Pat96] Patarin, J.: Hidden fields equations (HFE) and isomorphisms of polynomials (IP): two new families of asymmetric algorithms. In: Maurer, U.M. (ed.) EUROCRYPT 1996. LNCS, vol. 1070, pp. 33–48. Springer, Heidelberg (1996)

[PG97] Patarin, J., Goubin, L.: Asymmetric cryptography with S-Boxes. In: Han, Y., Quing, S. (eds.) ICICS 1997. LNCS, vol. 1334, pp. 369–380. Springer, Heidelberg (1997)

[Reg05] Regev, O.: On lattices, learning with errors, random linear codes, and cryptography. In: STOC 2005, pp. 84–93. ACM Press (2005)

[RP97] Rijmen, V., Preneel, B.: A family of trapdoor ciphers. In: Biham, E. (ed.) FSE 1997. LNCS, vol. 1267, pp. 139–148. Springer, Heidelberg (1997)

[WBDY98] Wu, H., Bao, F., Deng, R.H., Ye, Q.-Z.: Cryptanalysis of Rijmen-Preneel trapdoor ciphers. In: Ohta, K., Pei, D. (eds.) ASIACRYPT 1998. LNCS, vol. 1514, pp. 126–132. Springer, Heidelberg (1998)

Number Field Sieve

The Tower Number Field Sieve

Razvan Barbulescu[1]([✉]), Pierrick Gaudry[2], and Thorsten Kleinjung[3]

[1] CNRS, Univ Paris 6 and Univ Paris 7, Paris, France
razvan.barbulescu@imj-prg.fr
[2] CNRS, Inria, University of Lorraine, Nancy, France
pierrick.gaudry@loria.fr
[3] Institute of Mathematics, Universität Leipzig, Leipzig, Germany
thorsten.kleinjung@epfl.ch

Abstract. The security of pairing-based crypto-systems relies on the difficulty to compute discrete logarithms in finite fields $\mathbb{F}_{p^n}$ where n is a small integer larger than 1. The state-of-art algorithm is the number field sieve (NFS) together with its many variants. When p has a special form (SNFS), as in many pairings constructions, NFS has a faster variant due to Joux and Pierrot. We present a new NFS variant for SNFS computations, which is better for some cryptographically relevant cases, according to a precise comparison of norm sizes. The new algorithm is an adaptation of Schirokauer's variant of NFS based on tower extensions, for which we give a middlebrow presentation.

Keywords: Discrete logarithm · Number field sieve · Pairings

1 Introduction

The discrete logarithm problem (DLP) in finite fields is a central topic in public key cryptography. The case of $\mathbb{F}_{p^n}$ where p is prime and n is a small integer greater than 1, albeit less studied than the prime case, is at the foundation of pairing-based cryptography. The number field sieve (NFS) started life as a factoring algorithm but was rapidly extended to compute discrete logarithms in $\mathbb{F}_p$ [19,20,33] and has today a large number of variants. In 2000 Schirokauer [34] proposed the tower number field sieve (TNFS), as the first variant of NFS to solve DLP in fields $\mathbb{F}_{p^n}$ with $n > 1$. When n is fixed and the field cardinality $Q = p^n$ tends to infinity, he showed that TNFS has the heuristic complexity $L_Q(1/3, \sqrt[3]{64/9})$, where

$$L_Q(\alpha, c) = \exp\left((c + o(1))(\log Q)^\alpha (\log\log Q)^{1-\alpha}\right).$$

Schirokauer explicitly suggested that his algorithm might be extended to arbitrary fields $\mathbb{F}_{p^n}$ with $p = L_{p^n}(\alpha, c)$ and $\alpha > 2/3$, while maintaining the same complexity. Another question that he raised was whether his algorithm could take advantage of a situation where the prime p has a special SNFS shape, namely if it can be written $p = P(u)$ for an integer $u \approx p^{1/d}$ and a polynomial

© International Association for Cryptologic Research 2015
T. Iwata and J.H. Cheon (Eds.): ASIACRYPT 2015, Part II, LNCS 9453, pp. 31–55, 2015.
DOI: 10.1007/978-3-662-48800-3_2

$P \in \mathbb{Z}[x]$ of degree d, with coefficients bounded by an absolute constant. By that time, even for prime fields the answer was not obvious.

In 2006 Joux, Lercier, Smart and Vercauteren [21] presented a new variant of NFS which applies to all finite fields $\mathbb{F}_{p^n}$ with $p = L_Q(\alpha, c)$ for some $\alpha \geq 1/3$ and $c > 0$, the JLSV algorithm. When $\alpha > 2/3$, their variant has complexity $L_Q(1/3, \sqrt[3]{64/9})$. The question of extending TNFS to arbitrary finite fields became obsolete, because, in case of a positive answer, it would have the same complexity as the JLSV algorithm.

In 2013 Joux and Pierrot [22] designed another variant of NFS which applies to non-prime fields $\mathbb{F}_{p^n}$ where p is an SNFS prime. Their algorithm has complexity $L_Q(1/3, \sqrt[3]{32/9})$, which is the same as that of Semaev's SNFS algorithm for prime fields [35]. It shows that the pairing-based crypto-systems which use primes of a special form are more vulnerable to NFS attacks than the general ones. With this SNFS algorithm, the second question of Schirokauer lost its appeal as well, because this is the complexity that one can expect if Schirokauer's algorithm can be adapted when p is an SNFS prime.

In 2014 Barbulescu, Gaudry, Guillevic and Morain improved the algorithm in [21] and set a record computation in a field $\mathbb{F}_{p^2}$ of 180 decimal digits. However, since their improvements do not apply to SNFS fields and since the algorithm of Joux and Pierrot was never implemented, it is important to find a practical algorithm for this case.

In this work, we wish to rehabilitate Schirokauer's TNFS algorithm. First, we show that indeed, the heuristic complexity carries over to the expected range of finite fields. In order to make this analysis, we restate the original TNFS with less technicalities than in the original presentation, taking advantage of tools that were invented later (virtual logarithms).

We also show that for extension fields based on SNFS primes, the complexity of TNFS drops as expected to $L_Q(1/3, \sqrt[3]{32/9})$.

Finally, going beyond the asymptotic formulae, we compute estimates that strongly suggest that TNFS is currently the most efficient algorithm for solving discrete logarithms in small degree extensions of SNFS prime fields, like the ones arising naturally in several pairing constructions.

Outline. After a brief description of Schirokauer's TNFS algorithm in Sect. 2, we present it with sufficiently many details to get a proper asymptotic analysis in Sect. 3. In Sect. 4, several variants are described and analyzed, in particular the SNFS variant. This is followed, in Sect. 5 by more precise estimates for cryptographically relevant sizes and comparisons with other methods. Further technicalities about TNFS are given in an appendix; these are mostly details that could be useful for an implementation but which do not change the complexities.

2   Overview of TNFS

To fix ideas, we consider the case of "large" characteristic, so that we target fields $\mathbb{F}_Q$ with $Q = p^n$ so that $p = L_Q(\alpha, c)$ for some constants $\alpha > 2/3$ and $c > 0$.

Pohlig and Hellman explained how to retrieve the discrete logarithm modulo the group order N from the value of the discrete logarithms modulo each prime factor ℓ of N. Furthermore, Pollard's rho algorithm allows to compute discrete logarithms for small primes. Hence it is enough to explain how to use NFS to compute discrete logarithms modulo prime factors ℓ of $\#\mathbb{F}_{p^n}^*$ larger that $L_{p^n}(1/3, c)$ for some $c > 0$.

A classical variant of the NFS algorithm, e.g. one of the variants used for factoring and DLP in prime fields, would involve two irreducible polynomials f and g in $\mathbb{Z}[x]$ which have a common irreducible factor of degree n modulo p. Here, in TNFS, we consider two polynomials f and g defined over a ring R which is of the form $R = \mathbb{Z}[t]/(h(t))$ for a monic irreducible polynomial h of degree n. We ask furthermore that h remains irreducible modulo p, so that there is a unique ideal $\mathfrak{p}$ above p in R. Finally, we require that f and g are irreducible over $\mathbb{Q}[t]/(h(t))$ and have a common root modulo $\mathfrak{p}$ in R.

In the rest of the article, we denote by K_f the number field K_f defined by f, and by K_g the one defined by g. Also we write $\mathbb{Q}(\iota)$ for the number field defined by h, so that K_f and K_g are as in the figure aside.

The conditions imposed on f, g and h are such that there exist two ring homomorphisms from $R[x]$ to $R/\mathfrak{p} = \mathbb{F}_{p^n}$, one going through $R[x]/f(x)$, and the other through $R[x]/g(x)$, and for any polynomial in $R[x]$, the resulting values in $\mathbb{F}_{p^n}$ coincide, so that we get a commutative diagram as in the classical NFS algorithm. In Fig. 1, we recall this diagram, where we have denoted by α_f (resp. α_g) a root of f (resp. of g) and by m the common root of f and g modulo $\mathfrak{p}$ in R. These notations will be used all along the article.

Among the constructions that we tried, the best one uses polynomials f and g with coefficients in $\mathbb{Z}$, so that K_f and K_g can also be seen as compositum of two fields. If one could find a construction where f and g have coefficients in R one might find a faster algorithm. In any case, it is interesting to consider f and

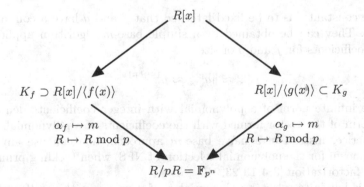

Fig. 1. Commutative diagram of TNFS for discrete logartihm in $\mathbb{F}_{p^n}$. In the classical case, $R = \mathbb{Z}$; here $R = \mathbb{Z}[\iota]$ is a subring of a number field of degree n where p is inert.

g as polynomials in $R[x]$, since this makes it easier to follow the analogy with the classical NFS.

Once this setting is done, the TNFS algorithm proceeds as usual. For many polynomials $a(\iota) - b(\iota)x$ in $R[x]$, we consider their two images in $R[x]/f(x)$ and $R[x]/g(x)$, and test them for smoothness as ideals. Each time the images are simultaneously smooth, we can write a relation: modulo the usual complications with principality defects and units that can be handled with the help of Schirokauer maps, it is possible to convert a relation into a linear relation between virtual logarithms of the factor base elements. Then follows a sparse linear algebra step to deduce the values of these virtual logarithms. And finally, the logarithm of an individual element of $\mathbb{F}_{p^n}$ can be computed using a descent step.

In the next section, we will enter into details, define more precisely the factor base elements and the associated smoothness notion, and estimate the size of the objects involved in the computation.

3 Detailed Description and Analysis

3.1 Polynomial Selection

In the overview of the previous section, nothing is said about the respective degrees of f and g. In fact, there is some freedom here, and we could in principle have balanced degrees and use for instance the algorithm of [20] or we can use a linear polynomial g, both methods leading to the same asymptotic complexity. The only difference comes in the individual logarithm stage. In order to keep the exposition short, we will only present this stage in the case where g is linear, but in practice one must take the one which minimizes the overall time.

To fix ideas, we take a linear polynomial g and a polynomial f with a degree of the form

$$\deg f = d = \delta \, (\log Q / \log \log Q)^{1/3},$$

where the constant δ is to be fixed later, so that f and g have a common root modulo p. They can be obtained by a simple base-m algorithm applied to p, yielding coefficients for f and g of size

$$\|f\|_\infty \approx \|g\|_\infty \approx p^{1/(d+1)},$$

where the infinite norm of a polynomial with integer coefficients denotes the infinite norm of the vector formed with the coefficients of a polynomial.

In practice, instead of a naïve base-m approach, one can use any of the methods known for the polynomial selection of NFS, when tackling prime fields or integer factorization [3,4,13,23,24].

What is left is to select a polynomial h of degree n with small coefficients which is irreducible modulo p. This is done by testing polynomials with small coefficients and, heuristically, we succeed after n trials, on average, because the proportion of irreducible polynomials modulo p is $\approx 1/n$. As we will explain later, rather than having the polynomial h with the smallest coefficients, we might prefer some polynomial with slightly larger coefficients but with the additional property that the Galois group of h is cyclic of order n. For this, we test polynomials in families with a cyclic Galois group; for example Foster [17] gives a list of such families when $\deg h = 2, 3, 4, 5$ or 6.

If one is interested in rigorous results and not in the most efficient polynomials, then one can give a proof of existence based on Corollary 10 given in the Appendix. Indeed, using cyclotomic fields one provably finds h with coefficients upper bounded by $(An^B \log(pn)^C)^n$ for some effective constants A, B and C.

3.2   Relation Collection

In the top of the diagram of Fig. 1 one usually takes $a - bx$ with $a, b \in R$. However, in the most general version of NFS one considers polynomials in $R[x]$ of arbitrary degrees; this is in particular necessary for the medium characteristic case [21]. In our study, we did not find any case where it was advantageous to consider polynomials of degree more than 1. Therefore we stick to the traditional (a, b)-pairs terminology for designating a linear polynomial $a(\iota) - b(\iota)x$ in $R[x]$ that we consider as a candidate for producing a relation.

Ideals of Degree 1. In our case, just like in the classical NFS, only ideals of degree 1 can occur in the factorizations of the elements in the number rings (except maybe for a finite number of ideals dividing the discriminants). This is, of course only true when thinking in the relative extensions; we formalize this in the following proposition that holds for f, but is also true for g if it happens to be non-linear.

Proposition 1. *Let $\mathbb{Q}(\iota)$ be a number field and let $\mathcal{O}_\iota$ be its ring of integers. Let f be a monic irreducible polynomial in $\mathcal{O}_\iota[x]$, and denote by α one of its roots. We denote by $K_f = \mathbb{Q}(\iota, \alpha)$ the corresponding extension field, and $\mathcal{O}_f$ its ring of integers.*

If $\mathfrak{q}$ is a prime ideal of $\mathcal{O}_\iota$ not dividing the index-ideal $[\mathcal{O}_f : \mathcal{O}_\iota[\alpha]]$, then the following statements hold.

(i) The prime ideals of $\mathcal{O}_f$ above $\mathfrak{q}$ are all the ideals of the form

$$\mathfrak{Q} = \langle \mathfrak{q}, T(\alpha) \rangle,$$

where $T(x)$ are the lifts to $\mathcal{O}_\iota[x]$ of the irreducible factors of f in $\mathcal{O}_\iota/\mathfrak{q}[x]$. Moreover $\deg \mathfrak{Q} = \deg T$.

(ii) If $a(t), b(t) \in \mathbb{Z}[t]$ are such that $\mathfrak{q}$ divides $\mathrm{N}_{K_f/\mathbb{Q}(\iota)}(a(\iota) - b(\iota)\alpha)$ and $a(\iota)\mathcal{O}_\iota + b(\iota)\mathcal{O}_\iota = \mathcal{O}_\iota$, then the unique ideal of $\mathcal{O}_f$ above $\mathfrak{q}$ which divides $a(\iota) - b(\iota)\alpha$ is $\mathfrak{Q} = \langle \mathfrak{q}, \alpha - r(\iota) \rangle$ with $r \equiv a(\iota)/b(\iota) \pmod{\mathfrak{q}}$.

Proof. (i) This is Proposition 2.3.9 of [14].

(ii) Let $\mathfrak{Q} = \langle \mathfrak{q}, T(\alpha) \rangle$ be a prime ideal of K above $\mathfrak{q}$ that divides $a(\iota) - b(\iota)\alpha$. If $\mathfrak{Q}$ divides $b(\iota)$ then it also divides $a(\iota)$, and therefore we have a contradiction with the condition $a(\iota)\mathcal{O}_\iota + b(\iota)\mathcal{O}_\iota = \mathcal{O}_\iota$. Therefore we can simplify $\mathrm{val}_\mathfrak{Q}(a(\iota) - b(\iota)\alpha)$ by dividing out by $b(\iota)$:

$$\mathrm{val}_\mathfrak{Q}(a(\iota) - b(\iota)\alpha) = \mathrm{val}_\mathfrak{Q}(b(\iota)) + \mathrm{val}_\mathfrak{Q}(a(\iota)/b(\iota) - \alpha) = \mathrm{val}_\mathfrak{Q}(\alpha - r(\iota)).$$

This expression is non-zero only when $\mathfrak{Q} = \langle \mathfrak{q}, \alpha - r(\iota) \rangle$, which proves the statement.

Note that the coprimality condition is similar to the one we have in the classical case, and the proportion of coprime pairs is

$$\prod_{\mathfrak{q} \text{ prime ideal in } \mathbb{Q}(\iota)} \left(1 - \frac{1}{N(\mathfrak{q})^2}\right) = \frac{1}{\zeta_{\mathbb{Q}(\iota)}(2)},$$

replacing $1/\zeta_\mathbb{Q}(2)$ in the classical variant.

Factor Base. The consequence of this result is that we keep only the degree 1 ideals in the factor bases for each side. With the same notations as above, and for a smoothness bound B, we define the factor base for f by

$$\mathcal{F}_f(B) = \left\{ \begin{array}{c} \text{prime ideals of } \mathcal{O}_f, \text{ coprime to } \mathrm{Disc}(K_f), \text{ of norm less than } B, \\ \text{whose inertia degree over } \mathbb{Q}(\iota) \text{ is one} \end{array} \right\}.$$

We define $\mathcal{F}_g(B)$ similarly; if g is linear this is just the set of prime ideals of $\mathcal{O}_\iota \cong \mathcal{O}_g$ of norm less than B. Prime ideals that divide the ideal-index $[\mathcal{O}_f : \mathcal{O}_\iota[\alpha]]$ are not covered by Proposition 1, and can still occur in the factorization of $(a(\iota) - b(\iota)\alpha)$. Moreover, since the index-ideal cannot be computed effectively, we consider together all the ideals above $\mathrm{Disc}(f)$ and above the leading coefficient of f. We denote them by $\mathcal{D}_f$ on the f-side, and $\mathcal{D}_g$ on the g-side. The cardinalities of these sets are bounded by a polynomial in $\log Q$. Since Proposition 1 cannot be used for detecting which elements of $\mathcal{D}_f$ divide $(a(\iota) - b(\iota)\alpha)$, we have to use general algorithms, and again, we refer to [14].

Finally, we join the two factor bases and these exceptional ideals in the global factor base defined by

$$\mathcal{F} = \mathcal{F}_f(B) \cup \mathcal{F}_g(B) \cup \mathcal{D}_f \cup \mathcal{D}_g.$$

We note that, as usual, the parameter B will be chosen of the form $B = L_Q(1/3, \beta)$, for a constant β to be fixed later.

By the prime ideal theorem, the number of prime ideals in $\mathbb{Q}(\iota)$ of norm less than B is $\frac{B}{\log B}(1 + o(1))$. Using Chebotarev's density theorem, the average number of roots of f (resp. g) modulo a random prime ideal $\mathfrak{q}$ is one. Hence the cardinality of the factor base is

$$\#\mathcal{F} = \frac{B}{\log B}(2 + o(1)),$$

which is similar to its value in the classical variant of NFS. As usual, in the complexity analysis, we approximate $\#\mathcal{F}$ by the quantity $L_Q(1/3, \beta)$, since polynomial-time factors are, in the end, hidden in the $o(1)$ added to the exponent constant.

Finding Doubly-smooth (a,b)-pairs. Among various choices for the shape of the $a(t)$ and $b(t)$ polynomials that we tried, the one giving the smallest norms is that where a and b are of maximal degree, $n-1$, and for which their coefficients are all of more or less the same size.

Let us denote by A a bound on these coefficients of $a(t)$ and $b(t)$. In the end, it will be chosen to be just large enough so that we get enough relations to get a full-rank system by browsing through all the possible coprime (a,b)-pairs of degree at most $n-1$ fitting this bound.

In order to estimate the probability that an (a,b)-pair gives a relation, the first step is to bound the size of the absolute norms on the f- and the g-side. The main tool is the following bound on the resultant.

Theorem 2 *[10, Thm 7]. If $f,g \in \mathbb{C}[c]$ have degree d_f and d_g, then*

$$|\operatorname{Res}(f,g)| \le \|f\|_\infty^{d_g} \|g\|_\infty^{d_f} (d_f+1)^{d_g/2}(d_g+1)^{d_f/2}.$$

We can now give the formula for the bound on the norm. We write it with the notations of the f-side, but it applies also to the g-side, after replacing the degree d by 1.

Theorem 3. *Let h and f be monic irreducible polynomials over $\mathbb{Z}$ of respective degrees n and d. Let K be the compositum of the number fields defined by h and f, and let ι and α_f be roots in K of h and f, respectively.*

Let $a(t)$ and $b(t)$ be two polynomials of degree less than n and with coefficients bounded by A. Then, the absolute norm of the element $a(\iota) - b(\iota)\alpha_f$ of K is bounded by

$$|\operatorname{N}_{K/\mathbb{Q}}(a(\iota) - b(\iota)\alpha_f)| < A^{nd} \|f\|_\infty^n \|h\|_\infty^{d(n-1)} C(n,d), \tag{1}$$

where $C(n,d) = (n+1)^{(3d+1)n/2}(d+1)^{3n/2}$.

Proof. We have $\operatorname{N}_{K/\mathbb{Q}} = \operatorname{N}_{\mathbb{Q}(\iota)/\mathbb{Q}} \circ \operatorname{N}_{K/\mathbb{Q}(\iota)}$ and, since f is monic, we get

$$\operatorname{N}_{K/\mathbb{Q}}(a(\iota) - b(\iota)\alpha_f) = \operatorname{N}_{\mathbb{Q}(\iota)/\mathbb{Q}}\Big(F(a,b)\Big),$$

where $F(a,b) = \sum_{i \in [0,d]} f_i a(t)^i b(t)^{d-i}$. The i-th term of this sum is a product of f_i and of d factors that are polynomials of degree less than n. Each term of the sum is therefore a polynomial of degree less than or equal to $d(n-1)$ with coefficients bounded by $\|f\|_\infty A^d n^d$. Therefore, we have

$$\|F(a,b)\|_\infty \le (d+1)\|f\|_\infty A^d n^d.$$

Finally, since h is monic, we have $\operatorname{N}_{\mathbb{Q}(\iota)/\mathbb{Q}}(F(a,b)) = \operatorname{Res}(h, F(a,b))$, and we can apply Theorem 2 to get the following upper bound:

$$\operatorname{N}_{\mathbb{Q}(\iota)/\mathbb{Q}}(F(a,b)) \le \|F(a,b)\|_\infty^n \|h\|_\infty^{d(n-1)}(n+1)^{d(n-1)/2}(d(n-1)+1)^{n/2}$$

$$< \|h\|_\infty^{d(n-1)} A^{nd}\|f\|_\infty^n (d+1)^{\frac{3}{2}n}(n+1)^{\frac{(3d+1)n}{2}}$$

If the polynomials f, g or h are not monic, the theorem does not apply, since the element $a(\iota) - b(\iota)\alpha_f$ is not an integer anymore. However, the denominators, that are powers of the primes dividing the leading coefficients are under control in term of smoothness (it suffices to add a few prime ideals in the factor bases). And in fact, the quantity based on resultants computed in the proof of the theorem is the one that is really used for smoothness testing. Therefore, the monic hypothesis is not a restriction, and is just there to avoid technicalities.

It remains to plug-in $\|h\| = O(1)$ and the bounds for $\|f\|_\infty$ and $\|g\|_\infty$ coming from our choice of polynomial selection and we get:

$$N_{K_f/\mathbb{Q}}(a - b\alpha_f) \le (A^{nd}\|f\|_\infty^n)^{1+o(1)} = (E^d Q^{1/(d+1)})^{1+o(1)}, \tag{2}$$

and

$$N_{K_g/\mathbb{Q}}(a - b\alpha_g) \le (A^n\|g\|_\infty^n)^{1+o(1)} = (E Q^{1/(d+1)})^{1+o(1)}, \tag{3}$$

where we have set $E = A^n$, so that the quantity of pairs that are tested is E^2, just like in the classical NFS analysis. It is to be noted that the contribution of $C(n,d)$ remains negligible. Indeed, it would reach a value of the form $L_Q(2/3)$, only when n gets larger than an expression of the form $(\log Q / \log\log Q)^{1/3}$, which is not the case, since we ask that p is larger than any expression of the form $L_Q(2/3)$. It is worth noticing that the expressions for the norms are the same as for the prime field case, where $Q = p$.

3.3   Writing and Solving Linear Equations

Mapping a factorization of ideals to a linear combination of logarithms is not immediate unless the ring is principal and there are no units other than ± 1; both things are highly unlikely since the fields K_f and K_g have large degrees over $\mathbb{Q}$. Therefore, we have to resort to the notion of virtual logarithms, just like in the classical case.

For this, it is easier to work with absolute extensions. Then, we can use the same strategy as in Sect. 4.3 of [21], that we summarize in the following theorem which can be applied to K_f and K_g.

Theorem 4 *([21, Section 4.3]). Let $K = \mathbb{Q}(\theta)$ be a number field and $\mathfrak{P}$ a non-ramified ideal of its ring of integers $\mathcal{O}_K$, with residual field isomorphic to $\mathbb{F}_{p^n}$ in which we fix a generator t. Let ℓ be a prime factor of $p^n - 1$ and let $U = \{x \in K \mid \forall \mathfrak{L}$ above ℓ, $\mathrm{val}_{\mathfrak{L}}(x) = 0\}$.*

We assume that there exists a Schirokauer function, i.e. an injective group homomorphism $\lambda = (\lambda_1, \ldots, \lambda_r) : (U/U^\ell, \cdot) \to (\mathbb{Z}/\ell\mathbb{Z}, +)^r$, where r is the unit rank of $\mathcal{O}_K$.

Assuming furthermore that ℓ neither divides the class number of K nor its discriminant, the following holds.

There exists a map $\log : \{$ideals of $\mathcal{O}_K$ coprime to $\mathfrak{P}\} \to \mathbb{Z}/\ell\mathbb{Z}$ and a map $\chi : \{1, \ldots, r\} \to \mathbb{Z}/\ell\mathbb{Z}$ called virtual logarithms, so that, for all $\phi \in \mathbb{Z}[x]$, such that $\phi(\theta)$ is in U and coprime to $\mathfrak{P}$, we have

$$\log_t \overline{\phi(\theta)}^{\mathfrak{P}} = \sum_{\mathfrak{Q} \ prime \ ideal} \mathrm{val}_{\mathfrak{Q}}(\phi(\theta)) \log \mathfrak{Q} + \sum_{j=1}^{r} \lambda_j(\phi(\theta)) \chi_j, \qquad (4)$$

where $\overline{\phi(\theta)}^{\mathfrak{P}}$ is the projection of $\phi(\theta)$ in the residual field $\mathbb{F}_{p^n}$ of $\mathfrak{P}$.

In [33], Schirokauer explained how to construct an explicitly and efficiently computable map λ as in the theorem and brought heuristics to support the assumptions. These heuristics and the fact that the other hypothesis of the theorem are expected to be true rely on the condition that ℓ is not too small. These are the main reasons why we asked that ℓ grows at least like $L_Q(1/3)$ in the beginning.

For each (a, b)-pair that gives two smooth ideals in K_f and K_g, the element $a(\iota) - b(\iota)\alpha_f$ can be expressed in the absolute representation of $K_f = \mathbb{Q}(\theta_f)$ by a polynomial form $\phi_f(\theta_f)$, and similarly $a(\iota) - b(\iota)\alpha_g$ can be written $\phi_g(\theta_g)$ in $K_g = \mathbb{Q}(\theta_g)$. We refer for instance to [14] for algorithms to manipulate relative extensions as absolute extensions. Then, applying Theorem 4 to ϕ_f in K_f and ϕ_g in K_g, we obtain two linear expressions that must be equal, since they both correspond to the logarithm of the same element in $\mathbb{F}_{p^n}$.

As a consequence, each relation is rewritten as a linear equation between the virtual logarithms of the elements of the factor base and the χ_j for each field. We make the now classical heuristic that collecting roughly the same number of relations as the size of the factor base (say, a polynomial factor times more), then the linear system obtained in such a manner has a kernel of dimension one. A vector of this kernel is computed using Wiedemann's algorithm [36] in a quasi-quadratic time $B^{2+o(1)}$. This gives the logarithms of all the ideals in the factor base.

3.4 Overall Complexity of the Main Phase

From the previous sections, we can now conclude about the complexity of the main steps of the algorithm. In fact, with our choice for the polynomial selection, and the kind of (a, b)-pairs that we test for smoothness, we have obtained exactly the same expressions for the sizes of the norms as in the usual NFS complexity analysis for prime fields, and in particular the same probability Prob that the product of the norms is smooth. Also, since the linear algebra step is also similar, the final complexity is the same: we have then to minimize $B^2 + E^2$ subject to the condition $E^2 \cdot \mathrm{Prob} \geq B^{1+o(1)}$, and we refer for example to Conjecture 11.2 of [13]. Hence, the optimal values of the parameters are $E = B = L_Q(1/3, \sqrt[3]{8/9})$ and $d = \sqrt[3]{3}(\frac{\log Q}{\log \log Q})^{1/3}$, and the heuristic complexity of the main phase of TNFS is $L_Q(1/3, \sqrt[3]{64/9})$.

3.5 Individual Logarithms

Let s be an element of $\mathbb{F}_{p^n}^*$ for which we want to compute the discrete logarithm. If s is very small, then it factors into ideals of the factor base, and its logarithm is easily retrieved. However, in general, this requires a 2-phase process that is not so trivial, although negligible compared to the other steps.

First, in what we call a smoothing phase, the element s is randomized and tested for B_1-smoothness with the ECM algorithm. The bound B_1 will be of the form $L_Q(2/3)$, so that the cost of the smoothing test is in $L_Q(1/3)$.

Thereafter, each prime ideal $\mathfrak{Q}$ which is not in the factor base is considered as a special-q and we search for a relation involving $\mathfrak{Q}$ and other smaller ideals. Continuing recursively, we get a special-q descent tree, from which the logarithm of s can be deduced.

Smoothing. The randomization is simple: we compute $z = s^e$ in $\mathbb{F}_{p^n}$ for random values e, and test z for smoothness. The logarithm of s is just the logarithm of z divided by e modulo ℓ.

To be more precise, the smoothness is not tested for the element z as an element of the finite field, but as the corresponding element in K_g. Indeed, in our construction, $z \in \mathbb{F}_{p^n}$ is represented by a polynomial of degree less than n with coefficients modulo p. Lifting these coefficients to integers, we obtain a polynomial which makes sense modulo $h(t)$, therefore an element of $\mathbb{Q}(\iota) = K_g$ (this is where we use that g is linear). As usual, to test the smoothness of z as an element of $\mathbb{Q}(\iota)$, we test the smoothness of its norm as an integer. Using again the estimate of Theorem 3, the size S of this norm is $Q^{1+o(1)}$.

The bound B_1 can then be optimized w.r.t. this only step, like in the classical NFS: if this is too small, the probability of being smooth is too small, while if it is too large, the cost of testing the smoothness by ECM is prohibitive. The analysis is the same as in [15] and gives a value $B_1 = L_Q(2/3, (\frac{1}{3})^{1/3})$; the corresponding cost for the smoothing phase is $L_Q(1/3, 3^{1/3})$.

After the smoothing phase, the logarithm of s has been rewritten in terms of the logarithms of small prime ideals of K_g for which the logarithm is already known, and some largish prime ideals of K_g, of norm bounded by B_1. The next step is to compute the logarithms of these largish ideals.

Descent by Special-q. As in NFS, the algorithm is recursive: if $\mathfrak{Q}$ is a prime ideal of degree one in K_f (respectively K_g), then we write $\log \mathfrak{Q}$ as a formal sum of virtual logs of ideals $\mathfrak{Q}'$ of K_f and K_g with norm less than $N(\mathfrak{Q})^c$, for a positive parameter $c < 1$. For this, we consider the lattice of (a, b)-pairs for which $\mathfrak{Q}$ divides the element $a - b\alpha_f$ (resp. $a - b\alpha_g$). A basis for this lattice can be constructed and LLL-reduced. Small combinations of these basis vectors are then formed and the norms of the corresponding (a, b) pairs are tested for $N(\mathfrak{Q})^c$-smoothness. We refer to Appendix 7.1 for the description of this special-q lattice technique, that is also used in practice during the collection of relations in the main stage. When a relation is found, this gives a new node in the descent tree, the children of it being the ideals of the relations that are still too large to be in the factor base. The total number of nodes is quasi-polynomial.

The cost of each step is determined by the size of $N(a(\iota) - \alpha_f b(\iota))$ (resp. $N(a(\iota) - \alpha_g b(\iota))$) which are tested during the computations. The matrix $M_{\mathfrak{Q}}$ of the basis of the lattice has determinant $\det M_{\mathfrak{Q}} = N(\mathfrak{Q})$, so a short vector in the LLL-reduced basis has coordinates of size $\approx N(\mathfrak{Q})^{1/(2n)}$. We make the heuristic assumption that all the vectors of the reduced basis, $(a^{(k)}, b^{(k)})$ for $k = 1, \ldots, 2n$, have coordinates of the same size. The pairs (a, b) tested for smoothness are linear combinations $(a, b) = \sum_{k=1}^{2n} i_k (a^{(k)}, b^{(k)})$ where i_k are rational integers with absolute value less than a parameter A', we set $E' = (A')^n$. By Theorem 3, the size of the norms tested for smoothness is

$$N_{K_f/\mathbb{Q}}(a - b\alpha_f) \le (\max(\|a\|_\infty, \|b\|_\infty)^{nd} \|f\|_\infty^n)^{1+o(1)} = (N(\mathfrak{Q})^{d/2}(E')^d Q^{1/d})^{1+o(1)},$$

$$N_{K_g/\mathbb{Q}}(a - b\alpha_g) \le (\max(\|a\|_\infty, \|b\|_\infty)^n \|g\|_\infty^n)^{1+o(1)} = (N(\mathfrak{Q})^{1/2} E' Q^{1/d})^{1+o(1)}.$$

These expressions coincide with the ones in the analogous stage of the classical variant (for example in Equation (7.11) in [5]) and we obtain a complexity of $L_Q(1/3, 1.1338...)$ which is the same as in the classical case [15]. We conclude that the overall complexity of individual logarithm is dominated by the $L_Q(1/3, 3^{1/3})$ complexity of the smoothing test.

4   Variants

Note on the Boundary Case. TNFS can be applied to the boundary case $p = L_Q(2/3, c_p)$, $c_p > 0$, where one obtains a complexity $L_Q(1/3, c)$. The constant c is strictly larger then $\sqrt[3]{64/9}$ as the factor $C(n, d)$ in Eq. (1) is not negligible any more. Yet, for some values of c_p, TNFS overcomes the method of [21], which was state-of-art until recently. Using the generalized Joux-Lercier method, the authors of [6,7] reduced the constant c to $(64/9)^{1/3} \approx 1.92$ and Pierrot [31] showed that a multiple fields variant allows to further reduce c to ≈ 1.90. Therefore, we do not reproduce here the tedious computations of the complexity in the boundary case.

The Case of Primes of Special Form (SNFS). Given a positive integer d, an integer p, not necessarily prime, is said to be a d-SNFS integer if it can be written as $p = P(u)$ for some integer $u \approx p^{1/d}$ and a polynomial $P \in \mathbb{Z}[x]$ such that $\|P\|_\infty$ is small (say, bounded by a constant). We remark that when a number is SNFS, then there can be several valid choices for d and P. This is typically the case for integers of the form $2^k + \varepsilon$, for tiny ε.

When solving DLP in fields $\mathbb{F}_{p^n}$ for d-SNFS primes p, we can follow the classical SNFS construction [27] and set $f(x) = P(x)$ and $g(x) = x - u$, which is possible since f and g share the root u modulo p.

When evaluating the sizes of the norms, Eq. (2) can be restated with $\|f\|_\infty = O(1)$, so we obtain the following bound:

$$N_{K_f/\mathbb{Q}}(a - b\alpha_f) N_{K_g/\mathbb{Q}}(a - b\alpha_g) \le (E^{d+1} Q^{1/d})^{1+o(1)}. \tag{5}$$

Following the analysis of Semaev [35], we obtain that if the degree d can be chosen to grow precisely as $d = \sqrt[3]{\frac{3}{2}} \left(\frac{\log Q}{\log \log Q} \right)^{1/3}$, then the overall complexity of SNFS is the same as that of factoring numbers from the Cunningham project, namely $L_Q \left(1/3, \sqrt[3]{\frac{32}{9}} \right)$.

Using Multiple Number Fields (MNFS). Given a choice of polynomials f and g selected as in the first step of TNFS, one can construct a large number of polynomials f_i which share with f and g the root m modulo p. The idea goes back to Coppersmith's variant of NFS for factorization [16] and has been used again in [8,28] and [31]. Let V be a parameter of size $L_Q(1/3, c_v)$ for some constant $c_v > 0$. For all $\mu(t)$ and $\nu(t) \in \mathbb{Z}[t]$ so that $\deg \mu, \deg \nu \le n - 1$ and $\|\mu\|_\infty, \|\nu\|_\infty \le V^{1/(2n)}$, we set

$$f_{\mu,\nu} = \mu(\iota)f + \nu(\iota)g, \tag{6}$$

keeping only those polynomials that are irreducible (most of them are, so we expect that the correcting factor on the bounds for $\|\mu\|_\infty$ and $\|\nu\|_\infty$ are only marginally adjusted). Let us denote by $K_{f_{\mu,\nu}}$ the number field generated by $f_{\mu,\nu}$ over $\mathbb{Q}(\iota)$, and call $\alpha_{\mu,\nu}$ a root of $f_{\mu,\nu}$ in its number field. For any pair (μ, ν) as above and (a, b) in the sieving domain, by Theorem 3 we have

$$N_{K_{\mu,\nu}}(a - \alpha_{\mu,\nu}b) \le A^{nd}(V^{1/(2n)}\|f\|_\infty)^n \|h\|_\infty^{nd} C(n,d) = (V^{1/2}E^d Q^{1/d})^{1+o(1)}. \tag{7}$$

In the multiple number field sieve a relation is given by a pair (a, b) in the sieving domain and a polynomial $f_{\mu,\nu}$ from the set constructed above so that $N_{K_g/\mathbb{Q}}(a - b\alpha_g)$ is B-smooth and $N_{K_{f_{\mu,\nu}}}(a - b\alpha_{\mu,\nu})$ is B/V-smooth. We use as factor base the set

$$\mathcal{F} = \left(\bigcup_{\mu,\nu} \mathcal{F}_{f_{\mu,\nu}}(B/V) \right) \bigcup \mathcal{F}_g(B).$$

We collect relations as in Coppersmith's modification: collect pairs (a, b) in the sieving domain and keep only those for which $N_{K_g/\mathbb{Q}}(a - \alpha_g b)$ is B-smooth. Then, for each surviving pair (a, b) we use ECM to collect polynomials $f_{\mu,\nu}$ such that $N_{K_{f_{\mu,\nu}}/\mathbb{Q}}(a - \alpha_{\mu,\nu}b)$ is B/V-smooth.

We choose parameter E so that the number of collected pairs exceeds $2B$, which is an upper bound on $\#\mathcal{F}$. The same considerations as in [16] allow us to find the optimal parameters: $V = L_Q(1/3, 1 - (\frac{\sqrt{13}-1}{3})^{1/3})$, $E = B = L_Q(1/3, (\frac{46+13\sqrt{13}}{108})^{1/3})$ and $d = \delta(\log Q/\log\log Q)^{1/3}$ where $\delta = (\frac{32-2\sqrt{13}}{9})^{1/3}$; the complexity of the multiple field variant of TNFS is $L_Q(1/3, (\frac{92+26\sqrt{13}}{27})^{1/3})$.

Automorphisms. Joux, Lercier, Smart and Vercauteren [21] proposed an improvement based on the field automorphisms of the number fields occurring in NFS. A recent preprint proves (a reformulation of) the following result:

Theorem 5 (Theorem 3.5(i) of [6]). *Let σ be a field automorphism of $K/\mathbb{Q}$. Assume that $\mathfrak{P}$ is a prime ideal of K above p such that $\sigma\mathfrak{P} = \mathfrak{P}$. Fix a prime ℓ dividing $N(\mathfrak{P}) - 1$, coprime to the class number and the discriminant of K. Fix a generator t of the residual field of $\mathfrak{P}$ and, for any prime ideal $\mathfrak{Q}$, denote by $\log\mathfrak{Q}$ the virtual logarithm of $\mathfrak{Q}$ with respect to t and a set of explicit generators so that $\gamma_{\sigma(\mathfrak{Q})} = \sigma(\gamma_\mathfrak{Q})$. Then, there exists a constant $\kappa \in [0, \mathrm{ord}(\sigma) - 1]$ such that for any $\mathfrak{Q}$ we have*

$$\log(\sigma\mathfrak{Q}) \equiv p^\kappa \log(\mathfrak{Q}) \mod \ell.$$

In Sect. 3.1 we noted that one might find ι so that $\mathbb{Q}(\iota)/\mathbb{Q}$ has n automorphisms over $\mathbb{Q}$. All these automorphisms can be used to speed-up computations, using the following result.

Corollary 6. *Let σ be an automorphism of $\mathbb{Q}(\iota)/\mathbb{Q}$ and call $\tilde{\sigma}$ the unique field automorphism of K_f such that $\tilde{\sigma}(\iota) = \sigma(\iota)$ and $\tilde{\sigma}(\alpha_f) = \alpha_f$. Assume that f has small coefficients so that virtual logarithms are defined using explicit generators. Then, there exists $\kappa \in [0, \mathrm{ord}(\sigma) - 1]$ such that, for all prime ideals $\mathfrak{Q}$ of K_f, we have*

$$\log(\tilde{\sigma}\mathfrak{Q}) \equiv p^\kappa \log \mathfrak{Q} \mod \ell.$$

Proof. The only non-trivial condition, $\tilde{\sigma}\mathfrak{P}_f = \mathfrak{P}_f$, is tested directly:

$$\tilde{\sigma}\mathfrak{P}_f = \tilde{\sigma}\langle p\mathbb{Z}[\iota], \alpha_f - m\rangle = \langle \tilde{\sigma}(p)\mathbb{Z}[\iota], \tilde{\sigma}(\alpha_f) - \tilde{\sigma}(m)\rangle = \langle p\mathbb{Z}[\iota], \alpha_f - m\rangle = \mathfrak{P}_f.$$

According to [7], automorphisms allow us to sieve n times faster and to speed-up the linear algebra stage by a factor n^2. Note that, contrary to the classical variant of NFS where automorphisms were available only for certain values of n, TNFS has no restrictions.

5 Comparison for Cryptographically Relevant Sizes

The complexity of NFS and its many variants is written in the form $C^{1+o(1)}$, which can hide large factors, and therefore we cannot decide which variant to implement based only on asymptotic complexity. We follow the methodology of [7, Section 4.4] and do a more precise comparison by evaluating the upper bound on the size of the integers which are tested for smoothness: the product of the norms with respect to the two sides. In particular, we make explicit the negligible terms of Eqs. (2) and (3) using Theorem 3.

5.1 The Case of General Primes

When p is not an SNFS number, we compare TNFS to the algorithm of Joux, Lercier, Smart and Vercauteren(JLSV) [21]. From Eqs. (2) and (3) we find a formula for the logarithm of the product of the norms in TNFS:

$$C_{\mathrm{TNFS}} = (d + 1)\log_2 E + \frac{2}{d+1}\log_2 Q = C_{\mathrm{NFS}},$$

where $d = \deg f$ can be chosen as desired (unlike the SNFS variant of the algorithm where d might be imposed by the shape of p). It is remarkable that this formula is the same as for NFS in the integer factorization case.

Since the JLSV algorithm comes with a variety of methods of polynomial selection, we cannot give a unified formula for the size of norms' product, so we use the minimum of the formulae in [7]. Therefore, in the following, when we say JLSV, this covers both variants explained in [21] as well as the Conjugation and Generalized Joux-Lercier methods. The choice of the parameter E depends on

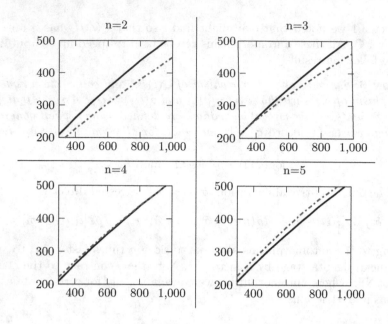

Fig. 2. Comparison of TNFS (in black) and the best variant of JLSV algorithm (in dashdotted blue). Vertical axis: bitlength of the norm's product; horizontal axis: bitlength of p^n (Color figure online).

the size of the norms, but for a first comparison we can use the default values of CADO-NFS [7, Table 2].

In Fig. 2 we compare TNFS to JLSV when p is a general prime (not SNFS), for a range $400 \leq \log_2 Q \leq 1000$. We conclude that in this range, when $n \geq 5$, TNFS is competitive and must be kept for an even more accurate comparison.

5.2   The Case of Primes of Special Shape (SNFS)

The Importance of the d Parameter. If we want to compute discrete logarithms in a field $\mathbb{F}_{p^n}$ such that p is d-SNFS for a parameter d, then the first question to ask is whether to use a general algorithm like TNFS and JLSV or a specialized variant of these two, namely the SNFS variant of TNFS that we denote STNFS or the Joux-Pierrot algorithm.

When $d = 6$ we can rely on a real-life example: Aoki et al. [2] factored a 1039-bit integer with SNFS, using sextic polynomials, i.e. $d = 6$. The current record, hold by Kleinjung et al. [26], was obtained with a MNFS variant and targeted d-SNFS integers for $d = 8$. Their computations were much faster than the evaluated time to factor a 1024-bit RSA modulus, so it is safe to say that SNFS is the best option when $\log_2 Q \approx 1024$ and $d = 6$ or when $d = 8$ for slightly larger targets. However, the value of d is fixed in most cases and can take very different values among curves used in pairing-based crypto-systems, going from $d = 2$ for MNT curves [29] to $d = 56$ in other constructions [18, Table 6.1],[30].

If the polynomial P such that $p = P(u)$ has a special shape, one can try to reduce the value of d using techniques from the Cunningham project records. On the one hand, if $P = T(x^a)$ with $T \in \mathbb{Z}[x]$ and $a \in \mathbb{N}$, we can also write $p = T(v)$ with $v = u^a$, so p is $(\deg T)$-SNFS. This technique can be used for example in the construction of Brezing-Weng [12, Section 3, item 3(b)] where $P(x) = \mu a^2 + \nu b^2$ for some small constants μ and ν and where $a, b \in \mathbb{Z}[x^5]$ have degree 5 and respectively 15; we replace P of degree 30 by a polynomial of degree 6.

On the other hand, a construction of Freeman, Scott and Teske [18, Construction 6.4] allows to divide the degree by 2. Indeed, in that case the polynomial P is almost a palindrome, in the sense that it can be written as $P(x) = \frac{1}{4} x^{(\deg P)/2} T(x - \frac{1}{x})$ with $T \in \mathbb{Z}[x]$. Then we select $f = T(x)$ and $g = ux - (u^2 - 1)$, which share the root $u - \frac{1}{u}$ modulo p and are so that $\|f\|_\infty = O(1)$ and $\|g\|_\infty = p^{1/\deg f}$.

Modeling. A good comparison requires a precise estimation of the norms. However, several factors in Eq. (1) can be negligible in some cases but can also be very large in the others:

$$\text{negligible factors} = C(n, d)\|f\|_\infty^n \|h\|_\infty^d.$$

The factor $C(n, d)$ is itself a bad estimation of the number of terms in the Sylvester discriminant, which can vary between 6 bits for $n = 2$ and $d = 3$ to 15 bits for $n = 5$ and $d = 6$. This determines us to restrict to $n \leq 5$ and $d \leq 6$. The factor $\|f\|_\infty^n$ equals 1 if $\|f\|_\infty = 1$ but can be as large as 2^{62} when $n = 12$ and $\|f\|_\infty = 36$. Hence, it is impossible to plot the size of the norms for all SNFS numbers, independently of the polynomial f.

For our modeling, we consider the case $\|f\|_\infty = \|h\|_\infty = 1$ and neglect the combinatorial factor $C(n, d)$ for small values of n and d. From Eq. (5) the dominant factor in the product of the norms for STNFS is

$$C_{\text{STNFS}}(n, d) = \log(E^{d+1}) + \log(Q^{1/d}).$$

Note again that this formula is the same as that of the complexity of the factoring variant of SNFS.

The product of the norms in the Joux-Pierrot algorithm is bounded by $(n + 1)^{2t}(\log n)^{nd} \, E^{2n(d+1)/t} \, Q^{(t-1)/(nd)}$ (discussion preceding Eq. (5) in [22]), and for the comparison we keep only the logarithm of most important factors:

$$C_{\text{JP}}(n, d, t) = \frac{2n}{t} \log(E^{d+1}) + \frac{t-1}{n} \log(Q^{1/d}).$$

Let us see two examples in which we tackle fields of about one kilobit, for which we use the approximation $\log_2 E = 30.4$, as in [2].

A First Example. We target a 1024-bit field $\mathbb{F}_{p^2}$ for a 6-SNFS prime p and we set the parameters equal to their value in the computation of Aoki et al. If one chooses to forget that p has a special shape and uses JLSV with conjugation method, then the product of the norms has bitsize ≈ 439. If instead one uses the special shape of p, the product of the norms for STNFS has bitsize $C_{\text{STNFS}}(n = 2, d = 6) \approx 386$, while the best parameters for the Joux-Pierrot algorithm yield $C_{\text{JP}}(n = 2, d = 6, t = 3) \approx 457$. A probabilistic experiment suggests that our model is quite precise, as the negligible factors do not add more than 6 bits.

Barreto-Naehrig. The elliptic curves proposed by Barreto and Naehrig [9] correspond to finite fields of parameters $n = 12$ and $d = 4$. We tackle a field of 1024-bit cardinality and we will use a value of E close to the one in the factorization record, i.e. $\log_2 E = 30.4$. If we forget that p is SNFS, then we can choose the value of d in TNFS and we find $C_{\text{TNFS}}(n = 12, d = 7) = 500$. If instead we use the special shape of p we obtain $C_{\text{STNFS}}(n = 12, d = 4) = 408$ and $C_{\text{JP}}(n = 12, d = 4, t = 12) = 539$.

In that case, the extension degree n (a.k.a. the embedding degree in the pairing context) is already pretty large, so that we are not at all in the nominal range of applicability of TNFS. As a consequence, our estimate for C_{TNFS} is way too optimistic, since the so-called negligible factors are no longer small. But in fact, it is not that bad: computing explicitly the norms for a sample of typical (a, b)'s of the appropriate size shows that the product of the norms for STNFS is 60 to 80 bits larger than the ideal model when $f = 36x^4 + 12x^3 + 16x^2 + 2x + 1$ and $h = x^{12} - x - 1$. Therefore, it might still be better than Joux-Pierrot.

There are however examples when the specialized algorithms do not apply.

Fact 7. *When $d = 2$, the JP and STNFS algorithms are not better than the general ones, i.e.*

$$C_{\text{JLSV}} \leq \min(C_{\text{JP}}, C_{\text{SNFS}}),$$

where C_{JLSV} is the complexity of the JLSV algorithm with conjugation method.

To see this, note first that the Joux-Pierrot algorithm keeps unchanged the stages of JLSV once finished the polynomial selection. In the Joux-Pierrot algorithm one constructs polynomials f and g such that $\deg(f) = nd$, $\deg(g) = n$, $\|f\|_\infty = O(1)$ and $\|g\|_\infty = Q^{1/(nd)}$. However, when $n = 2$, they have the same characteristics as the polynomials constructed by the Conjugation method, which applies to arbitrary primes.

Also note that the STNFS uses a polynomial g with coefficients of size $p^{1/d}$. When $d = 2$ the norm of the g-side has bitsize larger than $\frac{1}{2}\log_2 Q$, which is typical for algorithms of complexity $L_Q(1/2)$ and is larger than the norms considered in the JLSV algorithm in the range $\log_2 Q \leq 1000$ and $n \leq 5$.

Plots. Let us plot the modelled bitsize of the norms product for STNFS and Joux-Pierrot in the range which is currently feasible or might become in the near future: $400 \leq \log_2 Q \leq 1000$. Together with C_{STNFS} and C_{JP} (Joux-Pierrot), we also plot C_{NFS} which represents the bitsize of the product of the norms in NFS when factoring RSA numbers. We make separate graphs for each pair (n, d)

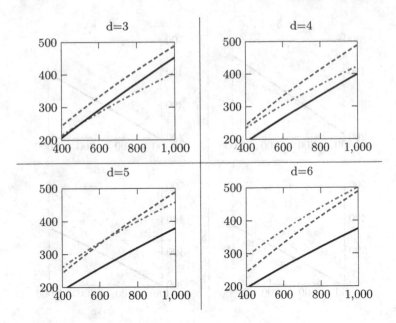

Fig. 3. Comparison of C_{NFS} (in dashed blue), C_{STNFS} (in black) and C_{JP} (in dasdotted red) in $\mathbb{F}_{p^n}$ with $n = 2$, for d-SNFS primes. Vertical axis: bitlength of the norm's product; horizontal axis: bitlength of p^n (Color figure online).

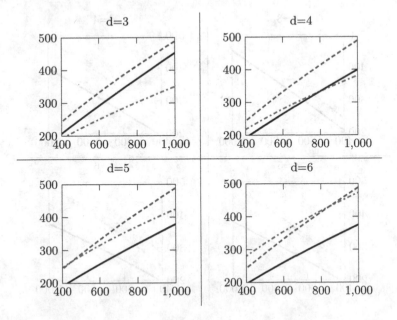

Fig. 4. Comparison of C_{NFS} (in dashed blue), C_{STNFS} (in black) and C_{JP} (in dashdotted red) in $\mathbb{F}_{p^n}$ with $n = 3$, for d-SNFS primes. Vertical axis: bitlength of the norm's product; horizontal axis: bitlength of p^n (Color figure online).

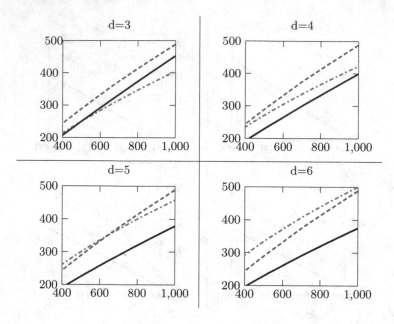

Fig. 5. Comparison of C_{NFS} (in dashed blue), C_{STNFS} (in black) and C_{JP} (in dashdotted red) in $\mathbb{F}_{p^n}$ with $n = 4$, for d-SNFS primes. Vertical axis: bitlength of the norm's product; horizontal axis: bitlength of p^n (Color figure online).

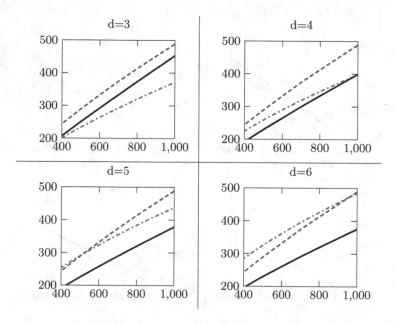

Fig. 6. Comparison of C_{NFS} (in dashed blue), C_{STNFS} (in black) and C_{JP} (in dashdotted red) in $\mathbb{F}_{p^n}$ with $n = 5$, for d-SNFS primes. Vertical axis: bitlength of the norm's product; horizontal axis: bitlength of p^n (Color figure online).

where n is the degree of the target field and d is the parameter such that p is d-SNFS, as those parameters are unique (in general) for each finite field: Fig. 3 ($n = 2$), Fig. 4 ($n = 3$), Fig. 5 ($n = 4$) and Fig. 6 ($n = 5$). Albeit the value of E depends on the size of the norms, in a first approximation we can use the formula $E = c \cdot L_Q(1/3, (4/9)^{1/3})$ where c is a constant chosen such that the formula fits the value of E in the example of Aoki et al.

We emphasize that our comparisons are imprecise since they are based only on the product of the norms. Nevertheless, one might make two remarks:

- when $d \geq 3$, the two algorithms specialized in fields of SNFS characteristic have smaller norms than those of NFS when factoring RSA numbers;
- when $d \geq 4$, STNFS is an important challenger for the Joux-Pierrot algorithm.

6   Cryptographic Consequences

The number field sieve algorithm is still far from being fully understood, in particular for extension fields that are so important for pairing-based cryptography. In the past few years, several improvements have been made in the asymptotic complexities in various scenarios, leading in particular to an $L(1/3, \sqrt[3]{32/9})$ complexity for small degree extensions of SNFS-prime fields, that are common in pairing-friendly constructions.

We have shown, that in this setting, an old NFS variant due to Schirokauer could compete and probably overcome the algorithm by Joux-Pierrot. We acknowledge that the comparison is not perfect since it is based on a model where the efficiency is directly linked to the size of product of the norms of the elements that have to be tested for smoothness. Still, in some cases, the difference is large enough (a few dozens of bits), so that we are confident that this should translate into a significant practical difference.

Of course, only a careful implementation of both algorithms could confirm this. Unfortunately, this goes way beyond the scope of this paper. As far as we know, Joux-Pierrot's algorithm has not been used so far for a record-setting computation, and Schirokauer's TNFS would require even more implementation work to handle the sieve in higher dimension. And since doing experiments with non-optimized implementations and small field sizes could lead to highly misleading conclusions, we preferred to keep this for future work.

7   Appendix: Technicalities

7.1   Special-q Sieving

In practice for prime fields the relation collection phase is split in subtasks following the so-called special-q sieving strategy. It is expected, but no so obvious, that this technique can be adapted to the case of TNFS.

The General Case. Given a prime ideal $\mathfrak{Q}$ of K_f (resp. of K_g), the special-q algorithm collects (most of) the coprime pairs $(a,b) \in \mathbb{Z}[\iota]^2$ which satisfy

- $a - b\alpha_f \equiv 0 \mod \mathfrak{Q}$;
- $N_{K_f/\mathbb{Q}}(a - b\alpha_f)/N_{K_f/\mathbb{Q}}(\mathfrak{Q})$ and $N_{K_g/\mathbb{Q}}(a - b\alpha_g)$ are B-smooth,

and which have coefficients bounded by $N_{K_f/\mathbb{Q}}(\mathfrak{Q})^{1/2n}I$ for a parameter I.

In the main lines, the sieving is done by Algorithm 1, where a key role is played by the lattice $L_{\mathfrak{Q}}$ of (a,b)-pairs such that $\mathfrak{Q} \mid a - b\alpha_f$:

$$L_{\mathfrak{Q}} = \left\{ (a_0, \ldots, a_{n-1}, b_0, \ldots, b_{n-1}) \in \mathbb{Z}^{2n} \mid \left(\sum_{k=0}^{n-1} a_k \iota^k \right) - \alpha_f \left(\sum_{k=0}^{n-1} b_k \iota^k \right) \equiv 0 \mod \mathfrak{Q} \right\}.$$

Algorithm 1. Special-q task

1: Compute an LLL-reduced basis of $L_{\mathfrak{Q}}$, $u^{(1)}, \ldots, u^{(2n)}$, and for each k define the pair $(a^{(k)}, b^{(k)})$ by $a^{(k)} = \sum_{i=0}^{n-1} u_i^{(k)} \iota^i$ and $b^{(k)} = \sum_{i=n}^{2n-1} u_i^{(k)} \iota^i$.
2: Initialize an array indexed by $(i_1, \ldots, i_{2n}) \in \prod_{k=1}^{2n} [-I, I]$ with the value of $\log_2 N_{K_f/\mathbb{Q}}(a - b\alpha_f)$ where

$$a = \sum_{k=1}^{2n} i_k a^{(k)} \text{ and } b = \sum_{k=1}^{2n} i_k b^{(k)}.$$

3: For each $\mathfrak{L}$ in $\mathcal{F}_f$ update the entries of the array such that $a - b\alpha_f \in \mathfrak{L}$.
4: Collect **yield**(f), the coprime pairs (a,b) associated to entries of the array with value less than a given threshold.
5: Repeat Steps 2-4 with f replaced by g, and collect **yield**(g).
6: **return yield**(f) $\bigcap$ **yield**(g)

In more detail, if $\mathfrak{Q} = \langle q, \alpha_f - \rho_{\mathfrak{Q}}(\iota) \rangle$ and $\mathfrak{q} = \langle q, \varphi_{\mathfrak{q}}(\iota) \rangle$, we can assume that $\varphi_{\mathfrak{q}}$ is monic and define the matrix

$$
M_{\mathfrak{Q}} = \begin{pmatrix}
q & & & & 0 \cdots \cdots 0 \\
& \ddots & & & \vdots \\
& & q & & \\
& & \boxed{\text{vector}(\varphi_{\mathfrak{q}})} & & \\
& & & \ddots & \vdots \\
& & \boxed{\text{vector}(\varphi_{\mathfrak{q}})} & 0 \cdots \cdots 0 \\
& \text{vector}(\rho_{\mathfrak{Q}}(\iota)) & & 1 \\
& \text{vector}(\rho_{\mathfrak{Q}}(\iota)\iota) & & & \ddots \\
& \vdots & & \\
& \text{vector}(\rho_{\mathfrak{Q}}(\iota)\iota^{n-1}) & & & & 1
\end{pmatrix}
$$

One can check that the rows of $M_{\mathfrak{Q}}$ form a basis of $L_{\mathfrak{Q}}$, and that $\det(L_{\mathfrak{Q}}) = q^{\deg(\varphi_{\mathfrak{q}})} = N_{\mathbb{Q}(\iota)/\mathbb{Q}}(\mathfrak{q}) = N_{K_f/\mathbb{Q}}(\mathfrak{Q})$ and $\dim L_{\mathfrak{Q}} = 2n$. Then, the coefficients of the shortest vector in an LLL-reduced basis have size about $N_{K_f/\mathbb{Q}}(\mathfrak{Q})^{1/(2n)}$.

We make the heuristic assumption that for a large proportion of ideals $\mathfrak{Q}$, all the vectors in the reduced basis have coefficients of this size. Then, the coefficients of the (a, b) pairs visited during Steps 3-4-5 of Algorithm 1 are approximatively equal to $I \, N_{K_f/\mathbb{Q}}(\mathfrak{Q})^{1/(2n)}$.

The critical part of Algorithm 1 is Step 4., where we need to solve a problem that Pollard [32] asked in the case $m = 2$.

Problem 1. Compute the intersection of a sub-lattice of $\mathbb{Z}^m$ with an interval product $\prod_{k=0}^{m-1} I_k$.

Since the dimension is fixed or small enough, we can use a generic lattice enumeration algorithm like the Kannan-Fincke-Pohst algorithm. In the case $m = 2$, Franke and Kleinjung [25, Appendix A] gave an elegant algorithm that proved very efficient in practice. Extending this algorithm to higher dimension is still an open problem.

The Particular Case of Gaussian Integers. When $h = x^2 + 1$, $\iota = i$ and we have a series of advantages. First of all, we have $\deg(h) = n = 2$, so the combinatorial overhead $C(n, d)$ in Theorem 3 is small. Secondly, the ring $\mathbb{Z}[i]$ is Euclidean, so that we can speed-up Step 1 of Algorithm 1.

Lemma 8. *Let q and r be two elements of $\mathbb{Z}[i]$ such that q is irreducible and r is not divisible by q. Assume that $\mathfrak{Q} = \langle q, \alpha_f - r \rangle$ is a prime ideal of K_f. Let $(u_j, v_j, d_j)_{j \geq 0}$ be the sequence of Bezout coefficients such that $u_j q + v_j r = d_j$, obtained during the Extended Euclidean Algorithm(EEA). Let $j \geq 0$ be an integer. For $k = 1, 2, 3, 4$ we set*

$$(a^{(1)}, b^{(1)}) = (d_j, v_j), \qquad (a^{(2)}, b^{(2)}) = (id_j, iv_j),$$
$$(a^{(3)}, b^{(3)}) = (d_{j+1}, v_{j+1}), \quad (a^{(4)}, b^{(4)}) = (id_{j+1}, iv_{j+1}),$$

and define $u^{(k)} = (\mathrm{Re}(a^{(k)}), \mathrm{Im}(a^{(k)}), \mathrm{Re}(b^{(k)}), \mathrm{Im}(b^{(k)}))$. Then the vectors $u^{(1)}$, $u^{(2)}$, $u^{(3)}$, $u^{(4)}$ form a basis of the lattice $L_{\mathfrak{Q}}$.

Proof. Note first that if two elements e_1, e_2 form a basis for a $\mathbb{Z}[i]$-module M, then the set $\{e_1, ie_1, e_2, ie_2\}$ is a basis of M seen as a $\mathbb{Z}$-module. We apply this fact to $M = \{(a, b) \in \mathbb{Z}[i] \times \mathbb{Z}[i] \mid a - br \equiv 0 \mod q\}$, so it is sufficient to show that (d_j, v_j) and (d_{j+1}, v_{j+1}) form a basis of M when seen as a $\mathbb{Z}[i]$-module.

By construction of the sequence $(u_j, v_j, d_j)_{j \geq 0}$, there exist invertible matrices $I_1, I_2, \ldots \in \mathrm{GL}(\mathbb{Z}[i], 2)$ so that, for all $j \geq 1$,

$$\begin{pmatrix} u_{j+1} & v_{j+1} & d_{j+1} \\ u_j & v_j & d_j \end{pmatrix} = I_j \begin{pmatrix} u_j & v_j & d_j \\ u_{j-1} & v_{j-1} & d_{j-1} \end{pmatrix}.$$

Therefore, for all j, the pairs (d_j, v_j) and (d_{j+1}, v_{j+1}) span the same $\mathbb{Z}[i]$-module. In particular, for $j = 0$, we have $(d_0, v_0) = (q, 0)$ and $(d_1, v_1) = (r, 1)$, which is a basis of M, so that any pair in the sequence spans M. Finally, a pair $(a, b) \in \mathbb{Z}[i] \times \mathbb{Z}[i]$ is in M if and only if the vector $u = (\mathrm{Re}(a), \mathrm{Im}(a), \mathrm{Re}(b), \mathrm{Im}(b))$ is in the lattice $L_{\mathfrak{Q}}$, which completes the proof.

We interrupt the execution of EEA at its middle point, i.e. for the least index j where $N_{\mathbb{Q}(i)/\mathbb{Q}}(d_j) < \sqrt{N_{\mathbb{Q}(i)/\mathbb{Q}}(q)}$. As in the classical variant of NFS, we make the heuristic that for all $k \in [1,4]$, we have $\|(a^{(k)}, b^{(k)})\|_\infty \approx \sqrt{|q|}$. Hence, we replaced Step 1 in Algorithm 1 by EEA in $\mathbb{Z}[i]$.

Another advantage of $\mathbb{Z}[i]$ is that we can easily deal with the roots of unity. Indeed, the roots of unity have a bad effect on the sieve since, for any pairs (a,b) found during the sieve, one will also find (ua, ub) for all roots of unity u. For a practical implementation one might prefer to choose h so that there are no roots of unity other than ± 1.

In the case $h = x^2 + 1$, we can impose that we have no duplicates due to roots of unity. For this, we modify Step 2 of Algorithm 1 so that the indices run in

$$(i_1, i_2, i_3, i_4) \in [0, I] \times [0, I] \times [-I, I] \times [-I, I]$$

instead of $[-I, I]^4$. By doing so we divide by four the number of pairs (a, b) sieved in the special-q task associated to $\mathfrak{Q}$. Indeed, if a pair (a, b) is written as $(a, b) = \sum_{k=1}^4 i_k(a^{(k)}, b^{(k)})$, then when we multiply (a, b) by roots of unity we use the following indices where exactly one of the pairs has $i_1, i_2 \geq 0$:

$$\begin{array}{ll}
(a, b) \leftrightarrow (i_1, i_2, i_3, i_4) & (-a, -b) \leftrightarrow (-i_1, -i_2, -i_3, -i_4) \\
(ia, ib) \leftrightarrow (-i_2, i_1, -i_4, i_3) & (-ia, -ib) \leftrightarrow (i_2, -i_1, i_4, -i_3).
\end{array}$$

7.2   Using a Cyclotomic Field for $\mathbb{Q}(\iota)$

Although we found no practical advantage for cyclotomic fields other than $\mathbb{Q}(i)$, they allow us to give a poof of existence for the polynomial h, as required in the TNFS construction of Sect. 3.1.

Theorem 9 (/1/, Prop. 3). *Assuming the Extended Riemann Hypothesis (ERH), there is a constant $c > 0$, such that for all $p, n \in \mathbb{N}$, p prime and $\gcd(n, p) = 1$, there exists a prime q such that $q \equiv 1 \pmod{n}$, $q < cn^4 \log(pn)^2$ and p is inert in the unique subfield K of $\mathbb{Q}(\zeta_q)$ with $[K : \mathbb{Q}] = n$.*

Corollary 10. *Under ERH, there exists a constant $c > 0$ such that, for any integer n and any prime $p > n$, there exists an effectively constructible polynomial $h \in \mathbb{Z}[x]$ such that:*

- *h is irreducible modulo p;*
- *$\|h\|_\infty < (2cn^4 \log(np)^2)^n$.*

Proof. Let c be the constant of the theorem above. Let q be a prime associated with p and n and let ζ_q be a primitive qth root of unity and η a Gaussian period:

$$\eta = \sum_{x \in \mathbb{F}_q^*/(\mathbb{F}_q^*)^{(q-1)/n}} \zeta_q^x.$$

If $r_1, \ldots, r_n$ are a set of representatives of $\mathbb{F}_q^*/(\mathbb{F}_q^*)^{(q-1)/n}$, then the conjugates of η are its images by the morphisms $\sigma_i : \zeta_q \mapsto \zeta_q^{r_i}$. Hence, the minimal polynomial of η over $\mathbb{Q}$ is

$$h = \prod_{i=0}^{n-1} (x - \sigma_i(\eta)).$$

For $k \in [0, n]$, a crude estimate of the kth coefficient of f is $\binom{n}{k}|\eta|^k$, which is further upper bounded by $2^n(q-1)^n$, and finally by $(2cn^4 \log(np)^2)^n$. The coefficients of h add a factor $\|h\|_\infty^{n(d-1)}$ in Eq. (1). It remains negligible, i.e. $L_Q(2/3)^{o(1)}$, when $n^2 = o(d)$ or equivalently when $p = L_Q(\alpha)$ for $\alpha > 5/6$.

7.3   The Waterloo Improvement

At the beginning of the individual logarithm stage, the smoothing step can be sped up in practice using the continued fraction method, also called "Waterloo improvement"[1]. It allows to replace the probability of an integer of size S to be smooth by the probability of two numbers of size $\sqrt{S}$ to be simultaneously smooth. This does not change the complexity, unless we make the $o(1)$ expression explicit, but has a measurable practical impact. A TNFS equivalent for the continued-fraction method is to LLL-reduce the lattice generated by the rows of the matrix

$$M(z) = \left(\begin{array}{ccc|ccc} p & & & & & \\ & \ddots & & & 0 & \\ & & p & & & \\ \hline z & & 1 & & & \\ \vdots & & & & \ddots & \\ \iota^{n-1}z & & & & & 1 \end{array} \right),$$

where z is a lift of the target element of the finite field, and $z, \ldots, \iota^{n-1}z$ are written by their coordinates as elements of $\mathbb{Q}(\iota)$. Since $\det M(z) = p^n = Q$, a short vector $(u_0, \ldots, u_{n-1}, v_0, \ldots, v_{n-1})$ has coordinates of size $\approx Q^{1/2n}$. The quotient u/v where $u = \sum_{k=0}^{n-1} u_k \iota^k$ and $v = \sum_{k=0}^{n-1} v_k \iota^k$ is an element of $\mathbb{Q}(\iota)$ that reduces to the same element of $\mathbb{F}_{p^n}$ as z. Therefore, instead of testing for smoothness the norm of z, of size $S = Q$, we test whether the norms of u and v, both of size $\sqrt{Q}$, are smooth.

References

1. Adleman, L.M., Lenstra, H.W.: Finding irreducible polynomials over finite fields. In: Proceedings of the Eighteenth Annual ACM Symposium on Theory of Computing, pp. 350–355. ACM (1986)

---

[1] The name, coined by Coppersmith, makes reference to the group who first used this technique [11].

2. Aoki, K., Franke, J., Kleinjung, T., Lenstra, A.K., Osvik, D.A.: A kilobit special number field sieve factorization. In: Kurosawa, K. (ed.) ASIACRYPT 2007. LNCS, vol. 4833, pp. 1–12. Springer, Heidelberg (2007)
3. Bai, S.: Polynomial selection for the number field sieve. Ph.D. thesis, Australian National University (2011)
4. Bai, S., Bouvier, C., Kruppa, A., Zimmermann, P.: Better polynomials for GNFS. Preprint (2014)
5. Barbulescu, R.: Algorithmes de logarithmes discrets dans les corps finis. Ph.D. thesis, Université de Lorraine (2013)
6. Barbulescu, R., Gaudry, P., Guillevic, A., Morain, F.: (Algebraic) improvements to the number field sieve for non-prime finite fields. Preprint http://hal.inria.fr/hal-01052449
7. Barbulescu, R., Gaudry, P., Guillevic, A., Morain, F.: Improving NFS for the discrete logarithm problem in non-prime finite fields. In: Oswald, E., Fischlin, M. (eds.) EUROCRYPT 2015. LNCS, vol. 9056, pp. 129–155. Springer, Heidelberg (2015)
8. Barbulescu, R., Pierrot, C.: The multiple number field sieve for medium- and high-characteristic finite fields. LMS J. Comput. Math. **17**, 230–246 (2014)
9. Barreto, P.S.L.M., Naehrig, M.: Pairing-friendly elliptic curves of prime order. In: Preneel, B., Tavares, S. (eds.) SAC 2005. LNCS, vol. 3897, pp. 319–331. Springer, Heidelberg (2006)
10. Bistritz, Y., Lifshitz, A.: Bounds for resultants of univariate and bivariate polynomials. Linear Algebra Appl. **432**(8), 1995–2005 (2010)
11. Blake, I.F., Fuji-Hara, R., Mullin, R.C., Vanstone, S.A.: Computing logarithms in finite fields of characteristic two. SIAM J. Algebraic Discrete Methods **5**(2), 276–285 (1984)
12. Brezing, F., Weng, A.: Elliptic curves suitable for pairing based cryptography. Des. Codes Crypt. **37**(1), 133–141 (2005)
13. Buhler, J.P., Lenstra Jr., H.W., Pomerance, C.: Factoring integers with the number field sieve. In: Lenstra, A.K., Lenstra Jr., H.W. (eds.) The Development of the Number Field Sieve. Lecture Notes in Mathematics, vol. 1554, pp. 50–94. Springer, Heidelberg (1993)
14. Cohen, H.: Advanced Topics in Computational Number Theory. Graduate Texts in Mathematics, vol. 193. Springer, New York (2000)
15. Commeine, A., Semaev, I.A.: An algorithm to solve the discrete logarithm problem with the number field sieve. In: Yung, M., Dodis, Y., Kiayias, A., Malkin, T. (eds.) PKC 2006. LNCS, vol. 3958, pp. 174–190. Springer, Heidelberg (2006)
16. Coppersmith, D.: Modifications to the number field sieve. J. Cryptol. **6**(3), 169–180 (1993)
17. Foster, K.: HT90 and "simplest" number fields. Illinois J. Math. **55**(4), 1621–1655 (2011)
18. Freeman, D., Scott, M., Teske, E.: A taxonomy of pairing-friendly elliptic curves. J. Cryptol. **23**(2), 224–280 (2010)
19. Gordon, D.M.: Discrete logarithms in GF(p) using the number field sieve. SIAM J. Discrete Math. **6**(1), 124–138 (1993)
20. Joux, A., Lercier, R.: The function field sieve is quite special. In: Fieker, C., Kohel, D.R. (eds.) ANTS 2002. LNCS, vol. 2369, pp. 431–445. Springer, Heidelberg (2002)
21. Joux, A., Lercier, R., Smart, N.P., Vercauteren, F.: The number field sieve in the medium prime case. In: Dwork, C. (ed.) CRYPTO 2006. LNCS, vol. 4117, pp. 326–344. Springer, Heidelberg (2006)

22. Joux, A., Pierrot, C.: The special number field sieve in $\mathbb{F}_{p^n}$. In: Cao, Z., Zhang, F. (eds.) Pairing 2013. LNCS, vol. 8365, pp. 45–61. Springer, Heidelberg (2014)
23. Kleinjung, T.: On polynomial selection for the general number field sieve. Math. Comput. **75**(256), 2037–2047 (2006)
24. Kleinjung, T.: Polynomial selection. Slides at CADO workshop (2008). http:// cado.gforge.inria.fr/workshop/slides/kleinjung.pdf
25. Kleinjung, T., Aoki, K., Franke, J., Lenstra, A.K., Thomé, E., Bos, J.W., Gaudry, P., Kruppa, A., Montgomery, P.L., Osvik, D.A., te Riele, H., Timofeev, A., Zimmermann, P.: Factorization of a 768-Bit RSA modulus. In: Rabin, T. (ed.) CRYPTO 2010. LNCS, vol. 6223, pp. 333–350. Springer, Heidelberg (2010)
26. Kleinjung, T., Bos, J.W., Lenstra, A.K.: Mersenne factorization factory. In: Sarkar, P., Iwata, T. (eds.) ASIACRYPT 2014. LNCS, vol. 8873, pp. 358–377. Springer, Heidelberg (2014)
27. Lenstra, A.K., Lenstra Jr., H.W., Manasse, M., Pollard, J.: The number field sieve. The Development of the Number Field Sieve. Lecture Notes in Mathematics, vol. 1554, pp. 11–42. Springer, Heidelberg (1993)
28. Matyukhin, D.V.: On asymptotic complexity of computing discrete logarithms over GF(p). Discrete Math. Appl. **13**(1), 27–50 (2003)
29. Miyaji, A., Nakabayashi, M., Takano, S.: New explicit conditions of elliptic curve traces for FR-reduction. IEICE Trans. Fundam. Electron. Commun. Comput. Sci. **84**(5), 1234–1243 (2001)
30. Murphy, A., Fitzpatrick, N.: Elliptic curves for pairing applications. Cryptology ePrint Archive, Report 2005/302 (2005). http://eprint.iacr.org/
31. Pierrot, C.: The multiple number field sieve with conjugation and generalized joux-lercier methods. In: Oswald, E., Fischlin, M. (eds.) EUROCRYPT 2015. LNCS, vol. 9056, pp. 156–170. Springer, Heidelberg (2015)
32. Pollard, J.M.: The lattice sieve. In: Lenstra, A.K., Lenstra, Jr., H.W.: The development of the number field sieve, vol. 1554 of Lecture Notes in Mathematics, pp. 43–49. Springer (1993)
33. Schirokauer, O.: Discrete logarithms and local units. Philos. Trans. Roy. Soc. London Ser. A **345**(1676), 409–423 (1993)
34. Schirokauer, O.: Using number fields to compute logarithms in finite fields. Math. Comp. **69**(231), 1267–1283 (2000)
35. Semaev, I.: Special prime numbers and discrete logs in finite prime fields. Math. Comp. **71**(237), 363–377 (2002)
36. Wiedemann, D.H.: Solving sparse linear equations over finite fields. IEEE Trans. Inform. Theory **32**(1), 54–62 (1986)

Hashes and MACs

On the Impact of Known-Key Attacks on Hash Functions

Bart Mennink[✉] and Bart Preneel

Department of Electrical Engineering, ESAT/COSIC,
KU Leuven and iMinds, Leuven, Belgium
{bart.mennink,bart.preneel}@esat.kuleuven.be

Abstract. Hash functions are often constructed based on permutations or blockciphers, and security proofs are typically done in the ideal permutation or cipher model. However, once these random primitives are instantiated, vulnerabilities of these instantiations may nullify the security. At ASIACRYPT 2007, Knudsen and Rijmen introduced known-key security of blockciphers, which gave rise to many distinguishing attacks on existing blockcipher constructions. In this work, we analyze the impact of such attacks on primitive-based hash functions. We present and formalize the weak cipher model, which captures the case a blockcipher has a certain weakness but is perfectly random otherwise. A specific instance of this model, considering the existence of sets of B queries whose XOR equals 0 at bit-positions C, where C is an index set, covers a wide range of known-key attacks in literature. We apply this instance to the PGV compression functions, as well as to the Grøstl (based on two permutations) and Shrimpton-Stam (based on three permutations) compression functions, and show that these designs do not seriously succumb to any differential known-key attack known to date.

Keywords: Hash functions · Known-key security · Knudsen-Rijmen · PGV · Grøstl · Shrimpton-Stam · Collision resistance · Preimage resistance

1 Introduction

Cryptographic hash functions are conventionally built on top of compression functions, and in turn on one or more blockciphers. Since the first appearance of such compression function $F(h, m) = DES_m(h)$ by Rabin [49] in the late 70s, many blockcipher-based functions appeared in the literature [23,25,29,30,40,43, 48,58]. These all enjoy security proofs in the ideal model, where the underlying ciphers are assum ed to behave ideally. Characteristic to these designs is that the key input to the cipher depends on the input to the compression function, and that the key scheduling needs to be sufficiently strong. For instance, Biryukov et al. [6] derived a related-key attack on AES and claimed that it invalidates the security of the Davies-Meyer compression function when the underlying primitive is instantiated with AES. A more recent approach to compression function design

© International Association for Cryptologic Research 2015
T. Iwata and J.H. Cheon (Eds.): ASIACRYPT 2015, Part II, LNCS 9453, pp. 59–84, 2015.
DOI: 10.1007/978-3-662-48800-3_3

is to base them on a limited number of permutations [8,41,42,51,57]. These permutations could be designed from scratch, or obtained by fixing a small set of keys and using a blockcipher for these keys only. Related- or chosen-key attacks on blockciphers do not help the adversary here, as the keys are fixed.

Known-Key Security of Blockciphers. While in the classical security models for blockciphers the key is secret and randomly drawn and the adversary's target is to distinguish the instantiation of the cipher from a random permutation (also known as (strong) pseudorandom permutation security), this notion does not apply if the key is known to the adversary. At ASIACRYPT 2007, Knudsen and Rijmen [27] introduced known-key security of blockciphers. Here, the key is presumed known, and the adversary succeeds in distinguishing if it identifies a structural property of the cipher. Andreeva et al. [1] proposed a way to formalize the known-key security of blockciphers based on the underlying primitives. The model is derived from the indifferentiability framework [37] and hence all composition results carry over. Intuitively: suppose some cryptosystem F is proven to achieve a certain level of security in the ideal permutation model, and consider F' to be F with the permutations replaced by independent blockcipher instantiations. Then, F' achieves the same level of security as F, up to the known-key indifferentiability bound of the underlying blockciphers.

In [1], several blockcipher constructions are proven to be known-key indifferentiable, such as the multiple Even-Mansour cipher and 14 rounds of balanced Feistel with random functions (using a result of Holenstein et al. [24]). For such ciphers, the above approach works well, although for Even-Mansour the composition is trivial (one essentially replaces an ideal permutation by an ideal permutation) and for Feistel with 14 rounds security is only guaranteed up to $2^{n/32}$ queries, where n is the state size of the cipher.

Known-Key Attacks on Blockciphers. Knudsen and Rijmen also demonstrated that the Feistel network on n bits with 7 rounds (called "Feistel$_7$") is *not* known-key indifferentiable [1,27]: an adversary can generically find $2^{n/2}$ plaintext/ciphertext tuples (m, c) and (m', c') satisfying $\mathsf{Ri}_{n/2}(m \oplus c \oplus m' \oplus c') = 0$ (where $\mathsf{Ri}_r(x)$ outputs the r rightmost bits of x). This result has lead to a wave of other known-key attacks on practical constructions, including generalized/extended variants of Feistel [1,27,47,53,56], reduced versions of AES or Rijndael [22,27,38,44,52], reduced variants of the blockciphers underlying SHA-2 and SHA-3 finalists BLAKE and Skein [2,7,31,34,60], and many more [3,11,12,14,17,18,28,33,46,47,54,55]. This paper will mostly be concerned with differential known-key attacks, including rebound- and boomerang-based attacks (the majority of above-mentioned attacks). We highlight two results that are among the best-known ones and that exemplify the idea of the other attacks. Gilbert and Peyrin [22] used the rebound technique [39] to derive a known-key attack on 8 rounds of AES (called "AES$_8$"). It starts from the middle, and results in a differential trail with four active words in the beginning, and four at the end. These active words are overlapping at two positions, hence one could consider this result as two tuples (m, c) and (m', c') satisfying $m \oplus c \oplus m' \oplus c' = 0$ at $10n/16$

bit-positions. The adversary has $2^{15} \leq 2^{n/8}$ degrees of freedom in the attack, and for any choice it results in such a tuple with a certain probability. (The bound of $2^{n/8}$ is used for simplicity later on.) The second attack we highlight is by Yu et al. [60], who employ the boomerang technique [59] to attack 36 rounds of the blockcipher Threefish-512 (called "Threefish$_{36}$") used in Skein. This attack results in four tuples $(m^1, c^1), \ldots, (m^4, c^4)$ satisfying $m^1 \oplus \cdots \oplus c^4 = 0$. The adversary has 2^n degrees of freedom, but any trial succeeds with probability approximately 2^{-454}. Therefore, the expected number of solutions is about $2^{n-454} \leq 2^{n/8}$. This attack is in fact a known-related-key attack, where a fixed difference in the key exists. For simplicity, we condone this, observing that an attack with *no* key difference must logically be harder.

In any of these cases, the traditional and commonly employed ideal cipher/permutation model falls short: results achieved in this model do not *necessarily* hold if the primitives are instantiated with Feistel$_7$, AES$_8$, Threefish$_{36}$, or any other known-key distinguishable cipher.

1.1   Our Contributions

In their seminal work, Knudsen and Rijmen state: "In some cases blockciphers are used with a key that is known to the adversary, and at least to a certain extent, the key is under the adversary's control. Our attacks are quite relevant to this case." We investigate this fundamental question whether known-key attacks invalidate the security of primitive-based hash functions, but we do so in a much more general way. At a high level, we present a model that goes beyond the traditional ideal cipher model as well as the principle of known-key attacks and that allows to generically analyze the impact of various weaknesses of blockciphers on various blockcipher- and permutation-based cryptosystems.

Model. A naive approach to analyzing the impact of known-key attacks would be to simply plug a certain blockcipher construction into a hash function and to analyze its security, but this would be a devious and complex combinatorial task: for a function based on r permutations, plugging Feistel$_7$ into it would lead to $7r$ underlying primitive calls. Note that proving security of the Feistel construction itself is already extraordinarily hard [16,24,32]. Instead, we model the blockciphers in such a way that they behave randomly, except that an adversary can exploit the particular relation. More formally, we pose a certain predicate Φ, and we draw blockciphers randomly from the set of all ciphers *that comply with predicate Φ*. Throughout, we refer to this model as the "weak cipher model (WCM)." It corresponds to the ideal cipher model if Φ is trivial.

We present an explicit description of a random weak cipher for the case where Φ implies for each key k the existence of A sets of B queries $\{(k, m^1, c^1), \ldots, (k, m^B, c^B)\}$ that comply with a certain condition φ. These ciphers are modeled to have three interfaces: forward queries, inverse queries, and predicate queries. Forward and inverse queries are as usual; on a predicate query, an adversary is given a set of B queries satisfying φ. Multiple technicalities are involved in this formalization. Most importantly, predicate Φ applies to tuples of queries, rather than single queries only, and some query responses may have a reduced entropy.

Above-mentioned known-key attacks are covered by our model if the condition φ states for some $C \subseteq \{1, \ldots, n\}$ that

$$\mathsf{Bits}_C \left(m^1 \oplus c^1 \oplus \cdots \oplus m^B \oplus c^B \right) = 0, \tag{1}$$

where $\mathsf{Bits}_C(x)$ outputs a string consisting of all bits of x whose index is in C. (In fact, our model is much more general: above-mentioned attacks aim to generate only *one* relation, while we allow an adversary to see multiple relations.) The value A usually depends on n and C is regularly a large subset. We consider B being a relatively small number (independent of n). For the above-mentioned attack on Feistel$_7$, $A = 2^{n/2}$, $B = 2$, and C corresponds to the rightmost $n/2$ bits. Similarly, the attacks on AES$_8$ (for $A = 2^{n/8}$, $B = 2$, and C a certain set of size $10n/16$) and Threefish$_{36}$ (for $A = 2^{n/8}$, $B = 4$, and $C = \{1, \ldots, n\}$) are covered, and so are almost all known differential (rebound- or boomerang-based) known-key attacks. We remark that, on the other hand, the predicate is not well-suited for integral-based known-key attacks: upon a predicate query an attacker would receive $B \approx 2^n$ queries.

The weak cipher model is similar to an approach followed by Bresson et al. [15] for the indifferentiability analysis of the SHA-3 candidate Shabal if the underlying blockcipher shows some non-random behavior, and by Bouillaguet et al. [13] to analyze the indifferentiability security of SIMD when the underlying compression function is distinguishable from a random function. However, in both approaches, the underlying biased primitives were relatively easy to model. For instance in [15] (using our terminology), predicate Φ is a relation that holds for single queries only, and not for combinations of queries. This considerably simplifies the analysis: one can derive a bias β to measure the distance between primitive responses and fully random responses, and consider oracle responses to be drawn from a set of size at least $2^{n-\beta}$, and the original indifferentiability analysis carries over with minor modifications. The predicate used in the analysis in [13], on the other hand, *does* apply to tuples of queries, but the model can simply be described using two sampling algorithms, and an adversary cannot hit a weak pair by accident (which *is* possible in our analysis). Liskov [35] used a similar approach to prove indifferentiability security of the zipper hash if the underlying compression function is invertible up to a certain degree. However, the analysis is significantly simpler, as this primitive can be perfectly modeled. We finally remark that Katz et al. [26] analyze the impact of related-key attacks on blockciphers to hash functions. However, in their model, the differences Δk, Δx, Δy are fixed, an ideal cipher is generated for half of the key space, and for the other half the cipher is adjusted as $\mathsf{E}_k(x, y) = \mathsf{E}_{k \oplus \Delta k}(x \oplus \Delta x) \oplus \Delta y$. This primitive can be easily modeled, but is also too generous to the attacker.

To our knowledge, this is the first attempt to formally analyze the effect of a wide class of blockcipher attacks on higher level cryptographic functions. Nonetheless, the weak cipher model is in essence still a model: we use an abstraction of the cryptanalytic known-key attacks in such a way that the ideal cipher

Table 1. Security results for the PGV, Grøstl, and Shrimpton-Stam compression functions in the weak cipher model. Ideal cipher/permutation model bounds match the ones of $B \geq 3$. All results are tight except for the case $(B = 1, |C| > n/2)$ for Shrimpton-Stam.

B	$|C|$	PGV		Grøstl		Shrimpton-Stam	
		collision	preimage	collision	preimage	collision	preimage
1	$\leq n/2$	$2^{(n-|C|)/2}$	$2^{n-|C|}$	$2^{(n-|C|)/4}$	$2^{(n-|C|)/2}$	$2^{(n-|C|)/2}$	$2^{n/2}$
	$> n/2$	$2^{(n-|C|)/2}$	$2^{n-|C|}$	$2^{(n-|C|)/4}$	$2^{(n-|C|)/2}$	$2^{(n-|C|)/2}$	$2^{n-|C|}$
2	$\leq n/2$	$2^{n/2}$	2^n	$2^{n/4}$	$2^{n/2}$	$2^{n/2}$	$2^{n/2}$
	$> n/2$	$2^{n-|C|}$	2^n	$2^{(n-|C|)/2}$	$2^{n/2}$	$2^{n-|C|}$	$2^{n/2}$
≥ 3	arbitrary	$2^{n/2}$	2^n	$2^{n/4}$	$2^{n/2}$	$2^{n/2}$	$2^{n/2}$

model can be relaxed to cope them. A further discussion on the accuracy of the model is given in Sect. 7.

Application to Blockcipher-Based Hash Functions. Preneel, Govaerts, and Vandewalle (PGV) [48] classified the 64 most basic ways of constructing a $2n$-to-n-bit compression function from a blockcipher with n-bit key and n-bit state, and claimed security of 12 of them. A formal security analysis of these functions in the ICM has been performed by Black et al. [9], and later by Duo and Li [19], Stam [58], and Black et al. [10]. In more detail, in the ICM these constructions achieve tight collision security up to about $2^{n/2}$ queries and preimage security up to about 2^n queries. Baecher et al. [4] recently showed that the 12 secure PGV functions can be divided into two classes, in such a way that if a primitive makes one function secure it makes the entire class secure.

As first application of our model, we consider the PGV compression functions in the WCM and derive collision and preimage bounds for general (A, B, C). A schematic summary of the results for various B and C is given in Table 1 (we remark that A is merely a technical parameter that has no influence on the results). We also show that the bounds are optimal, by providing matching attacks. Some of these attacks are similar to methods used in [27,53,56] to detect (near-)collisions in certain PGV modes of operations using known-key attacks.

Application to Permutation-Based Hash Functions. We also apply the WCM to permutation-based compression functions. This is particularly interesting for two reasons: (i) it allows us to understand the impact of distinguishers on permutations that are used in hash functions, and (ii) a blockcipher with a fixed and known key is a permutation and can be used as such. In more detail, we consider the Grøstl compression function [21] and the permutation-based equivalent of the Shrimpton-Stam compression function [57] (see also Fig. 4). In the IPM, the former is proven to achieve collision security up to $2^{n/4}$ queries, where n is the state size, and preimage security up to $2^{n/2}$ [20]. Rogaway and Steinberger [51] showed via an automated analysis that the latter function is collision and preimage resistant up to $2^{n/2}$ queries (asymptotically). This has been confirmed in the generalized work of Mennink and Preneel [41].

A summary of our findings for the Grøstl and Shrimpton-Stam compression functions in the WCM is given in Table 1. All results are tight, except for the case $(B = 1, |C| > n/2)$ for Shrimpton-Stam, for which we leave proving tightness as an open problem. We remark that the analysis for these schemes is much more demanding as multiple primitives are involved.

Impact. An application of our formalization to the PGV functions and various permutation-based functions shows that these achieve a comparable level of security in the ideal and weak cipher model for a spectrum of choices for (A, B, C). This result particularly implies that most relevant rebound-based (including [12,22,28,38,52,53,56]) and boomerang-based (including [2,7,31,54,60]) known-key attacks known to date do not invalidate the security of such functions, or only have a little effect. For instance, the above-discussed attack on Feistel$_7$ satisfies $B = 2$ and $|C| = n/2$ and it does not affect the security; similarly for Threefish$_{36}$ for which $B = 4$. The attack on AES$_8$ is covered for $B = 2$ and $|C| = 10n/16$, which demonstrates a slight security degradation to $2^{6n/16}$ for the PGV functions, but this may in part be due to our over-generosity to the adversary. We remark that, even though we focused on collision and preimage resistance, the techniques can be generalized to other security notions, such as near-collisions. This may entail differences in the security results.

We stress that these results do not mean that the analyzed functions are secure when the underlying permutations are instantiated with, say, Feistel$_7$ or Threefish$_{36}$: it only means that existing known-key attacks, or more general weaknesses such as relation (1), *alone* are not sufficient to invalidate the collision and preimage security of the construction. Indeed, more sophisticated attacks which are not yet covered by our application of the WCM may still invalidate the security of certain modes [6]. It remains a challenging open research problem to generalize the findings to underlying primitives that have multiple or different weaknesses.

1.2   Outline

In Sect. 2, we formally present the "weak cipher model," and in Sect. 3 we show how it relates to known-key attacks. We apply the model to the PGV functions in Sect. 4, to the Grøstl compression function in Sect. 5, and to Shrimpton-Stam in Sect. 6. We conclude this work in Sect. 7.

2   Weak Cipher Model

If X is a set, by $x \xleftarrow{\$} X$ we denote the uniformly random sampling of an element from X. By $X \xleftarrow{\cup} x$, we denote $X \leftarrow X \cup \{x\}$. For a bit string x, its bits are numbered $x = x_{|x|} \cdots x_2 x_1$. If $C \subseteq \{1, \ldots, |x|\}$, the function $\mathsf{Bits}_C(x)$ outputs a string consisting of all bits of x whose index is in C. Abusing notation, $\mathsf{Bits}_{\overline{C}}(x)$

always denotes the remaining bits (technically, $\overline{C} = \{1, \ldots, |x|\} \backslash C$). For $0 \leq r \leq |x|$, we consider $\mathsf{Ri}_r(x)$ that outputs the r rightmost bits of x. In other words, $\mathsf{Ri}_r(x) = \mathsf{Bits}_{\{1,\ldots,r\}}(x)$. For a function f, by $\mathsf{dom}(f)$ and $\mathsf{rng}(f)$ we denote its domain and range, respectively.

2.1 Security Model

For $\kappa \geq 0$ and $n \geq 1$, by $\mathrm{BC}(\kappa, n)$ we denote the set of all blockciphers with κ-bit key operating on n bits. If $\kappa = 0$, $\mathrm{BC}(n) := \mathrm{BC}(0, n)$ denotes the set of all n-bit permutations. If Φ is a predicate, by $\mathrm{BC}[\Phi](\kappa, n)$ we denote the subset of ciphers of $\mathrm{BC}(\kappa, n)$ that satisfy predicate Φ. For $\pi \in \mathrm{BC}[\Phi](\kappa, n)$, the input-output tuples are denoted (k, x, z), where $\pi(k, x) = \pi_k(x) = z$ and $\pi^{-1}(k, z) = \pi_k^{-1}(z) = x$. The key k is omitted in case $\kappa = 0$.

Let $\mathsf{F} : \{0, 1\}^s \rightarrow \{0, 1\}^n$ be a compressing function instantiated with $\ell \geq 1$ primitives from $\mathrm{BC}[\Phi](\kappa, n)$, for some predicate Φ. Throughout, we consider security of F in an idealized model: we consider an adversary $\mathcal{A}$ that is a probabilistic algorithm with oracle access to a randomly sampled primitive $\pi = (\pi_1, \ldots, \pi_\ell) \xleftarrow{\$} \mathrm{BC}[\Phi](\kappa, n)^\ell$. $\mathcal{A}$ is information-theoretic and its complexity is only measured by the number of queries made to its oracles. The adversary can make forward and inverse queries to its oracles, and these queries are stored in a query history $\mathcal{Q}$.

A collision-finding adversary $\mathcal{A}$ for F aims at finding two distinct inputs to F that compress to the same range value. In more detail, we say that $\mathcal{A}$ succeeds if it finds two distinct inputs X, X' such that $\mathsf{F}(X) = \mathsf{F}(X')$ and $\mathcal{Q}$ contains all queries required for these evaluations of F. We define by

$$\mathbf{Adv}_{\mathsf{F}}^{\mathrm{col}}(\mathcal{A}) = \mathbf{Pr}\left(\pi \xleftarrow{\$} \mathrm{BC}[\Phi](\kappa, n)^\ell, \ X, X' \leftarrow \mathcal{A}^\pi : X \neq X' \wedge \mathsf{F}(X) = \mathsf{F}(X')\right)$$

the probability that $\mathcal{A}$ succeeds in this. By $\mathbf{Adv}_{\mathsf{F}}^{\mathrm{col}}(q)$ we define the maximum collision advantage taken over all adversaries making q queries.

For preimage resistance, we focus on everywhere preimage resistance [50], which captures preimage security for every point of $\{0, 1\}^n$. Let $Z \in \{0, 1\}^n$ be any range value. Then, we say that $\mathcal{A}$ succeeds in finding a preimage if it obtains an input X such that $\mathsf{F}(X) = Z$ and $\mathcal{Q}$ contains all queries required for this evaluation of F. We define by

$$\mathbf{Adv}_{\mathsf{F}}^{\mathrm{epre}}(\mathcal{A}) = \max_{Z \in \{0,1\}^n} \mathbf{Pr}\left(\pi \xleftarrow{\$} \mathrm{BC}[\Phi](\kappa, n)^\ell, \ X \leftarrow \mathcal{A}^\pi(Z) : \mathsf{F}(X) = Z\right)$$

the probability that $\mathcal{A}$ succeeds, maximized over all possible choices for Z. By $\mathbf{Adv}_{\mathsf{F}}^{\mathrm{epre}}(q)$ we define the maximum (everywhere) preimage advantage taken over all adversaries making q queries.

If Φ is a trivial relation, we have $\mathrm{BC}[\Phi](\kappa, n) = \mathrm{BC}(\kappa, n)$, and the above definitions boil down to security in the ideal cipher model (ICM) if $\kappa > 0$ or the ideal permutation model (IPM) if $\kappa = 0$. On the other hand, if Φ is a non-trivial predicate, it strictly reduces the set $\mathrm{BC}(\kappa, n)$. In this case, we will refer to the model as the "weak cipher model (WCM)," for both $\kappa > 0$ and $\kappa = 0$. Very informally, this model still involves random ciphers/permutations, with the difference that an adversary may exploit a certain additional property. The modeling of a randomly drawn weak ciphers is much more delicate.

2.2   Random Weak Cipher

For a certain class of predicates, we discuss how to model a randomly drawn weak cipher π from $\mathrm{BC}[\Phi](\kappa, n)$. Let $A, B \in \mathbb{N}$. We will consider predicates that imply, *for every* $k \in \{0,1\}^{\kappa}$, the existence of A sets of B distinct queries $\{(x^1, z^1), \ldots, (x^B, z^B)\}$ that satisfy $\varphi_k(\{(x^1, z^1), \ldots, (x^B, z^B)\})$ for some condition φ depending on key k. The predicate is denoted $\Phi(A, B, \varphi)$. A is merely a technical parameter, and throughout we assume it is larger than q, the number of oracle calls an adversary can make. This definition of $\Phi(A, B, \varphi)$ is fairly general. Particularly, predicate B-sets may overlap and the condition φ can represent any function on the inputs. We note that Φ can be easily generalized to tuples of different length and/or to multiple types of conditions at the same time.

Traditionally, an adversary has only forward $\pi_k(x)$ and inverse $\pi_k^{-1}(z)$ query access. In order for the adversary to be able to exploit the weakness present in π, we give it additional access to π via a "predicate query" $\pi_k^{\Phi}(y)$: on input of $y \in \{1, \ldots, A\}$, the adversary obtains a B-set $\{(x^1, z^1), \ldots, (x^B, z^B)\}$ that satisfies $\varphi_k(\{(x^1, z^1), \ldots, (x^B, z^B)\})$.

A formal description of how to model $\pi \xleftarrow{\$} \mathrm{BC}[\Phi(A, B, \varphi)](\kappa, n)$ is given in Fig. 1. Here, for every $k \in \{0,1\}^{\kappa}$, P_k is an initially empty list of π_k-evaluations, where a regular forward/inverse query adds one element (x, z) to P_k and a π_k^{Φ}-query may add up to B elements. Additionally, P_k^{Φ} is an initially empty list of queries to π_k^{Φ}. We denote by $\Sigma_k(P_k, P_k^{\Phi}) \subseteq (\{0,1\}^n \times \{0,1\}^n)^B$ the set of all tuples $\{(x^1, z^1), \ldots, (x^B, z^B)\}$ such that

(i) $x^1, \ldots, x^B$ are pairwise distinct and $z^1, \ldots, z^B$ are pairwise distinct;
(ii) $\forall_{\ell=1}^{B} : x^{\ell} \in \mathrm{dom}(P_k) \implies z^{\ell} = P_k(x^{\ell})$ and $z^{\ell} \in \mathrm{rng}(P_k) \implies x^{\ell} = P_k^{-1}(z^{\ell})$;
(iii) $\varphi_k(\{(x^1, z^1), \ldots, (x^B, z^B)\})$ holds;
(iv) $\{(x^{p(1)}, z^{p(1)}), \ldots, (x^{p(B)}, z^{p(B)})\} \notin \mathrm{rng}(P_k^{\Phi})$ for any permutation p on $\{1, \ldots, B\}$.

For a new query $\pi_k^{\Phi}(y)$, the response is then randomly drawn from $\Sigma_k(P_k, P_k^{\Phi})$. Conditions (i–iii) are fairly self-evident; note particularly that an existing $(x, z) \in P_k$ may appear in multiple predicate queries. Condition (iv) assures that the drawing from $\Sigma_k(P_k, P_k^{\Phi})$ is not just an old predicate query or a reordering thereof. The usage of this set $\Sigma_k(P_k, P_k^{\Phi})$ allows for a uniform behavior of π_k^{Φ} for every k, and in general of $\pi \xleftarrow{\$} \mathrm{BC}[\Phi(A, B, \varphi)](\kappa, n)$, modulo the known existence of condition φ. This step is fundamental to our model and new compared with previous approaches of [13, 15, 35]. We remark that the model allows adversaries to make their queries at their own discretion, e.g., duplicate queries and regular queries after predicate queries are allowed.

procedure $\pi_k(x)$	procedure $\pi_k^{\Phi}(y)$
if $P_k(x) = \bot$:	if $P_k^{\Phi}(y) = \bot$:
$\quad z \xleftarrow{\$} \{0,1\}^n \backslash \mathrm{rng}(P_k)$	$\quad \{(x^1, z^1), \ldots, (x^B, z^B)\} \xleftarrow{\$} \bar{\Sigma}_k(P_k, P_k^{\Phi})$
$\quad P_k \xleftarrow{\cup} (x, z)$	$\quad$ for $\ell = 1, \ldots, B$:
end if	$\quad\quad$ if $(x^\ell, z^\ell) \notin P_k$:
return $P_k(x)$	$\quad\quad\quad P_k \xleftarrow{\cup} (x^\ell, z^\ell)$

procedure $\pi_k^{-1}(z)$	$\quad\quad$ end if
if $P_k^{-1}(z) = \bot$:	$\quad$ end for
$\quad x \xleftarrow{\$} \{0,1\}^n \backslash \mathrm{dom}(P_k)$	$\quad P_k^{\Phi} \xleftarrow{\cup} (y, \{(x^1, z^1), \ldots, (x^B, z^B)\})$
$\quad P_k \xleftarrow{\cup} (x, z)$	end if
end if	return $P_k^{\Phi}(y)$
return $P_k^{-1}(z)$	

Fig. 1. Random weak cipher π. An adversary has access to π, π^{-1}, and π^{Φ}.

2.3 Random Abortable Weak Cipher

Security analyses in the WCM are significantly more complex than in the ICM or IPM, which is in part because predicate queries may consist of older queries. This will particularly be an issue once collisions among queries are investigated. To suit the analysis for this case, we transform the WCM to an abortable weak cipher model (AWCM), which we denote as $\overline{\mathrm{BC}}[\Phi(A, B, \varphi)](\kappa, n)$. At a high-level, an abortable weak cipher responds to predicate queries with *new* query tuples only, and aborts once it turns out that an older query appears in a newer predicate query.

For any $k \in \{0,1\}^\kappa$ and partial P_k and P_k^{Φ}, define by $\bar{\Sigma}_k(P_k^{\Phi}) \subseteq (\{0,1\}^n \times \{0,1\}^n)^B$ the set of all tuples $\{(x^1, z^1), \ldots, (x^B, z^B)\}$ such that

(iii) $\varphi_k(\{(x^1, z^1), \ldots, (x^B, z^B)\})$ holds;

(iv) $\{(x^{p(1)}, z^{p(1)}), \ldots, (x^{p(B)}, z^{p(B)})\} \notin \mathrm{rng}(P_k^{\Phi})$ for any permutation p on $\{1, \ldots, B\}$.

$\bar{\Sigma}_k(P_k^{\Phi})$ differs from $\Sigma(P_k, P_k^{\Phi})$ in that conditions (i) and (ii) are omitted, and particularly: it is independent of P_k. A formal description of a random cipher $\bar{\pi} \xleftarrow{\$} \overline{\mathrm{BC}}[\Phi(A, B, \varphi)](\kappa, n)$ is given in Fig. 2. It deviates from Fig. 1 as follows: for every key k, $\bar{\pi}_k^{\Phi}$ responds randomly from $\bar{\Sigma}_k(P_k^{\Phi})$, and it aborts if the response violates one of the two skipped conditions of $\Sigma_k(P_k, P_k^{\Phi})$.

The next lemma shows that the WCM and AWCM are indistinguishable as long as the abortable weak cipher does not abort, approximately up to the birthday bound. Here, we assume that $\bar{\Sigma}_k(P_k^{\Phi})$ is always large enough.

Lemma 1. *Let $\bar{\pi} \xleftarrow{\$} \overline{\mathrm{BC}}[\Phi(A, B, \varphi^C)](\kappa, n)$. Consider an adversary that makes q queries to $\bar{\pi}$. Then,*

$$\mathbf{Pr}\left(\bar{\pi} \text{ sets abort}\right) \leq \frac{B^2 q(q+1)}{2^n - \frac{B! q 2^n}{|\bar{\Sigma}_k(\varnothing)|}}.$$

procedure $\bar{\pi}_k(x)$	**procedure** $\bar{\pi}_k^{\Phi}(y)$

if $P_k(x) = \bot$:
$\quad z \xleftarrow{\$} \{0,1\}^n \backslash \text{rng}(P_k)$
$\quad P_k \xleftarrow{\cup} (x,z)$
end if
return $P_k(x)$
	if $P_k^{\Phi}(y) = \bot$:
	$\quad \{(x^1,z^1),\ldots,(x^B,z^B)\} \xleftarrow{\$} \bar{\Sigma}_k(P_k^{\Phi})$
	$\quad$ for $\ell = 1,\ldots,B$:
	$\qquad$ if $x^\ell \in \text{dom}(P_k) \wedge z^\ell \neq P_k(x^\ell)$: abort
	$\qquad$ if $z^\ell \in \text{rng}(P_k) \wedge x^\ell \neq P_k^{-1}(z^\ell)$: abort
	$\qquad$ if $(x^\ell, z^\ell) \in \{(x^1,z^1),\ldots,(x^{\ell-1},z^{\ell-1})\}$: abort
	$\qquad$ if $(x^\ell, z^\ell) \notin P_k$:

procedure $\bar{\pi}_k^{-1}(z)$	

if $P_k^{-1}(z) = \bot$:
$\quad x \xleftarrow{\$} \{0,1\}^n \backslash \text{dom}(P_k)$
$\quad P_k \xleftarrow{\cup} (x,z)$
end if
return $P_k^{-1}(z)$
	$\qquad\quad P_k \xleftarrow{\cup} (x^\ell, z^\ell)$
	$\qquad$ end if
	$\quad$ end for
	$\quad P_k^{\Phi} \xleftarrow{\cup} (y, \{(x^1,z^1),\ldots,(x^B,z^B)\})$
	end if
	return $P_k^{\Phi}(y)$

Fig. 2. Random abortable weak cipher $\bar{\pi}$. An adversary has access to $\bar{\pi}, \bar{\pi}^{-1}$, and $\bar{\pi}^{\Phi}$.

Proof. Consider the ith query, for $i \in \{1,\ldots,q\}$, and assume it is a predicate query $\bar{\pi}_k^{\Phi}(y)$. We will consider the probability that this query makes $\bar{\pi}$ abort, provided it has not aborted so far. Prior to this i^{th} query, $|P_k| \leq B(i-1)$ and $|P_k^{\Phi}| \leq i$. Basic combinatorics shows that

$$|\bar{\Sigma}_k(P_k^{\Phi})| = |\bar{\Sigma}_k(\varnothing)| - B! \cdot |P_k^{\Phi}|,$$

where we use that $\bar{\pi}$ has not aborted so far. This i^{th} query aborts only if for some $\ell \in \{1,\ldots,B\}$, the value x^ℓ equals an element in $\text{dom}(P_k) \cup \{x^1,\ldots,x^{\ell-1}\}$ or the value z^ℓ equals an element in $\text{rng}(P_k) \cup \{z^1,\ldots,z^{\ell-1}\}$.

Define by $\bar{\Sigma}_k^{\text{abort}}(P_k^{\Phi})$ the set of all elements of $\bar{\Sigma}_k(P_k^{\Phi})$ that would lead to abort. We have $2B$ possible values to cause the abort (namely, $x^1,\ldots,z^B$), and it causes the abort if it equals an element in a set of size at most $|P_k| + B$. For any of these $2B(|P_k| + B)$ choices, the number of tuples in $\bar{\Sigma}_k(P_k^{\Phi})$ complying with this choice is at most $\frac{|\bar{\Sigma}_k(\varnothing)|}{2^n}$. Thus,

$$\mathbf{Pr}\left(\bar{\pi}^{\Phi}(y) \text{ sets abort}\right) = \frac{|\bar{\Sigma}_k^{\text{abort}}(P_k^{\Phi})|}{|\bar{\Sigma}_k(P_k^{\Phi})|} \leq \frac{2B(|P_k| + B) \cdot \frac{|\bar{\Sigma}_k(\varnothing)|}{2^n}}{|\bar{\Sigma}_k(\varnothing)| - B! \cdot |P_k^{\Phi}|} \leq \frac{2B^2 i}{2^n - \frac{B!q2^n}{|\bar{\Sigma}_k(\varnothing)|}}.$$

The proof is completed by summation over $i = 1,\ldots,q$. $\qquad\square$

3   Modeling Known-Key Attacks

We next apply the WCM to known-key attacks. For the sake of explanation, we first reconsider the Knudsen-Rijmen attack on Feistel$_7$ [27]. (A detailed description of the attack is also given in the full version of this paper.) Let $n \in \mathbb{N}$, and let $\pi := \pi_k$ be an instance of Feistel$_7$ with fixed key k. Knudsen and

Rijmen revealed four functions $f, f', g, g' : \{0,1\}^{n/2} \rightarrow \{0,1\}^n$ such that for all $y \in \{0,1\}^{n/2}$:

$$g(y) = \pi(f(y)) \text{ and } g'(y) = \pi(f'(y)),$$
$$\mathsf{Ri}_{n/2}\left(f(y) \oplus g(y)\right) = \mathsf{Ri}_{n/2}\left(f'(y) \oplus g'(y)\right). \tag{2}$$

These four functions depend on the cryptographic primitive underlying Feistel$_7$ in a complicated way. Therefore, we can safely assume that these functions behave sufficiently random, besides this particular relation (2), and that they are unknown to the adversary. f, f', g, g' are all injective and satisfy $f(y) \neq f'(y)$ and $g(y) \neq g'(y)$ for all y. On the other hand, collisions of the form $f(y) = f'(y')$ and $g(y) = g'(y')$ may occur.

Generically, the attack demonstrates that for key k there exist $2^{n/2}$ possibly overlapping sets of distinct queries $\{(x^1, z^1), (x^2, z^2)\}$ that satisfy $\mathsf{Ri}_{n/2}\left(x^1 \oplus z^1 \oplus x^2 \oplus z^2\right) = 0$. In other words, Feistel$_7$ meets predicate $\Phi(2^{n/2}, 2, \varphi^{\mathrm{Feistel}_7})$, where

$$\varphi_k^{\mathrm{Feistel}_7}\left(\{(x^1, z^1), (x^2, z^2)\}\right) \ : \ \mathsf{Ri}_{n/2}\left(x^1 \oplus z^1 \oplus x^2 \oplus z^2\right) = 0\,.$$

Here, we remark that the Knudsen-Rijmen attack works for *any* fixed but known key k, and that condition $\varphi_k^{\mathrm{Feistel}_7}$ is in fact independent of the key. In this work, we will consider a more general predicate $\Phi(A, B, \varphi^C)$ for $A, B \in \mathbb{N}$ and $C \subseteq \{1, \ldots, n\}$, where

$$\varphi_k^C\left(\{(x^1, z^1), \ldots, (x^B, z^B)\}\right) \ : \ \mathsf{Bits}_C\left(x^1 \oplus z^1 \oplus \cdots \oplus x^B \oplus z^B\right) = 0\,. \tag{3}$$

This generalized predicate considers the case of arbitrary but fixed and known keys, where the adversary can even choose the key every time it makes a predicate query. Note that also the attacks on AES$_8$ and Threefish$_{36}$ (see Sect. 1) are covered, as they satisfy $\Phi(2^{n/8}, 2, \varphi^C)$ for certain C of size $10n/16$ and $\Phi(2^{n/8}, 4, \varphi^{\{1,\ldots,n\}})$, respectively. In general, all rebound- or boomerang-based known-key attack in literature are covered by predicate $\Phi(A, B, \varphi^C)$ for some A, B, C. Here, B is always a value independent of n (usually 2 or 4) and C is regularly a large subset (of size at least $n/4$). Throughout, we consider A to be sufficiently large.

Basic Computations for AWCM

For the specific condition φ^C of (3), we derive a simpler bound on the probability that a primitive $\bar{\pi} \overset{\$}{\leftarrow} \overline{\mathrm{BC}}[\Phi(A, B, \varphi^C)](\kappa, n)$ aborts, along with some other elementary observations for $\bar{\pi}$. To this end, we define the notation "$[X]$," which equals 1 if X holds and 0 otherwise. For conciseness, we introduce the function $\delta_{B,C}[b]$ defined as

$$\delta_{B,C}[b] = 2^{|C|}[B = b] + [B > b]\,. \tag{4}$$

Lemma 2. *Let* $\bar{\pi} \xleftarrow{\$} \overline{BC}[\Phi(A, B, \varphi^C)](\kappa, n)$. *Consider an adversary that makes* $q \leq 2^{n-1}/B$ *queries to* $\bar{\pi}$. *Then,*

$$\mathbf{Pr}\left(\bar{\pi} \text{ sets abort}\right) \leq \frac{B^2 q(q+1)}{2^n - Bq}. \tag{5}$$

Let $k \in \{0,1\}^\kappa$ *and let* $Z, Z', Z'' \in \{0,1\}^n$. *Consider any new query* $\bar{\pi}_k^\Phi(y)$ *and assume it does not abort. Write the response as* $\{(x^1, z^1), \ldots, (x^B, z^B)\}$. *Then,*

(i) $\forall\, a \in \{1, \ldots, B\}:\ \mathbf{Pr}\left(x^a = Z\right), \mathbf{Pr}\left(z^a = Z\right) \leq \frac{1}{2^n - Bq}$;

(ii) $\forall\, a \in \{1, \ldots, B\}:\ \mathbf{Pr}\left(x^a \oplus z^a = Z\right) \leq \frac{\delta_{B,C}}{2^n - Bq}$;

(iii) $\forall\, \{a, b\} \subseteq \{1, \ldots, B\}:\ \mathbf{Pr}\left(x^a \oplus z^a = Z \wedge x^b \oplus z^b = Z'\right) \leq \frac{\delta_{B,C}[2]}{2^{2n} - Bq}$;

(iv) $\forall\, \{a, b\} \subseteq \{1, \ldots, B\}:$

$$\mathbf{Pr}\left(x^a = Z \wedge x^b = Z' \wedge x^a \oplus z^a \oplus x^b \oplus z^b = Z''\right) \leq \frac{\delta_{B,C}[2]}{2^{3n} - Bq}.$$

Proof. Recall from the proof of Lemma 1 that

$$|\bar{\Sigma}_k(P_k^\Phi)| = |\bar{\Sigma}_k(\varnothing)| - B!|P_k^\Phi|,$$

where $|P_k^\Phi| \leq q$. For the specific predicate analyzed in this lemma, $|\bar{\Sigma}_k(\varnothing)| = (2^n)^{2B-1} 2^{n-|C|}$. In the remainder, we regularly bound $B! \leq B \cdot (2^n)^{2B-2}$ for $B \geq 1$ or $B! \leq B \cdot (2^n)^{2B-4}$ for $B \geq 2$.

Probability of Abortion. The bound of (5) directly follows from Lemma 1, the above-mentioned size of $\bar{\Sigma}_k(\varnothing)$, and the bound on $B!$.

Part (i). Define by $\bar{\Sigma}_k^{(i)}(P_k^\Phi)$ the set of all elements of $\bar{\Sigma}_k(P_k^\Phi)$ that satisfy $x^a = Z$. Then, $|\bar{\Sigma}_k^{(i)}(P_k^\Phi)| \leq (2^n)^{2B-2} 2^{n-|C|}$, and

$$\mathbf{Pr}\left(x^a = Z\right) = \frac{|\bar{\Sigma}_k^{(i)}(P_k^\Phi)|}{|\bar{\Sigma}_k(P_k^\Phi)|} \leq \frac{1}{2^n - Bq}.$$

A similar analysis applies to the case $z^a = Z$.

Part (ii). Define by $\bar{\Sigma}_k^{(ii)}(P_k^\Phi)$ the set of all elements of $\bar{\Sigma}_k(P_k^\Phi)$ that satisfy $x^a \oplus z^a = Z$. We make a distinction between $B = 1$ and $B > 1$. In case $B > 1$, a similar reasoning as in (i) applies, and we have $|\bar{\Sigma}_k^{(ii)}(P_k^\Phi)| \leq (2^n)^{2B-2} 2^{n-|C|}$. On the other hand, if $B = 1$, we have $|\bar{\Sigma}_k^{(ii)}(P_k^\Phi)| = 0$ if $\mathsf{Bits}_C(Z) \neq 0$ and $|\bar{\Sigma}_k^{(ii)}(P_k^\Phi)| \leq 2^n$ if $\mathsf{Bits}_C(Z) = 0$. In any case,

$$|\bar{\Sigma}_k^{(ii)}(P_k^\Phi)| \leq (2^n)^{2B-2} 2^{n-|C|} \delta_{B,C}[1],$$

and

$$\mathbf{Pr}\left(x^a \oplus z^a = Z\right) = \frac{|\bar{\Sigma}_k^{(ii)}(P_k^\Phi)|}{|\bar{\Sigma}_k(P_k^\Phi)|} \leq \frac{\delta_{B,C}[1]}{2^n - Bq}.$$

Part (iii). This part only applies to $B > 1$; if $B = 1$ the probability equals 0 by construction. Define by $\bar{\Sigma}_k^{(iii)}(P_k^\Phi)$ the set of all elements of $\bar{\Sigma}_k(P_k^\Phi)$ that satisfy $x^a \oplus z^a = Z$ and $x^b \oplus z^b = Z'$. We make a distinction between $B = 2$ and $B > 2$. In case $B > 2$, a similar reasoning as in (i) and (ii) applies, and we have $|\bar{\Sigma}_k^{(iii)}(P_k^\Phi)| \leq (2^n)^{2B-3} 2^{n-|C|}$. On the other hand, if $B = 2$, we have $|\bar{\Sigma}_k^{(iii)}(P_k^\Phi)| = 0$ if $\mathsf{Bits}_C(Z \oplus Z') \neq 0$ and $|\bar{\Sigma}_k^{(iii)}(P_k^\Phi)| \leq (2^n)^2$ if $\mathsf{Bits}_C(Z \oplus Z') = 0$. In any case,

$$|\bar{\Sigma}_k^{(iii)}(P_k^\Phi)| \leq (2^n)^{2B-3} 2^{n-|C|} \delta_{B,C}[2],$$

and

$$\mathbf{Pr}\left(x^a \oplus z^a = Z \wedge x^b \oplus z^b = Z'\right) = \frac{|\bar{\Sigma}_k^{(iii)}(P_k^\Phi)|}{|\bar{\Sigma}_k(P_k^\Phi)|} \leq \frac{\delta_{B,C}[2]}{2^{2n} - Bq}.$$

Part (iv). The approach is fairly similar to case (iii). If $B = 1$ the probability is 0 by construction. Define by $\bar{\Sigma}_k^{(iv)}(P_k^\Phi)$ the set of all elements of $\bar{\Sigma}_k(P_k^\Phi)$ that satisfy $x^a = Z$, $x^b = Z'$, and $x^a \oplus z^a \oplus x^b \oplus z^b = Z''$. In case $B > 2$, we have $|\bar{\Sigma}_k^{(iv)}(P_k^\Phi)| \leq (2^n)^{2B-4} 2^{n-|C|}$. On the other hand, if $B = 2$, we have $|\bar{\Sigma}_k^{(iv)}(P_k^\Phi)| = 0$ if $\mathsf{Bits}_C(Z'') \neq 0$ and $|\bar{\Sigma}_k^{(iv)}(P_k^\Phi)| \leq 2^n$ if $\mathsf{Bits}_C(Z'') = 0$. In any case,

$$|\bar{\Sigma}_k^{(iv)}(P_k^\Phi)| \leq (2^n)^{2B-4} 2^{n-|C|} \delta_{B,C}[2],$$

and

$$\mathbf{Pr}\left(x^a = Z \wedge x^b = Z' \wedge x^a \oplus z^a \oplus x^b \oplus z^b = Z''\right) = \frac{|\bar{\Sigma}_k^{(iv)}(P_k^\Phi)|}{|\bar{\Sigma}_k(P_k^\Phi)|} \leq \frac{\delta_{B,C}[2]}{2^{3n} - Bq}.$$

$\square$

4 Application to PGV Compression Functions

We consider the 12 blockcipher-based compression functions from Preneel, Govaerts, and Vandewalle (PGV) [48]. In the ICM these constructions achieve tight collision security up to about $2^{n/2}$ queries and preimage security up to about 2^n queries [9,10,19,58]. The 12 constructions are depicted in Fig. 3. Here, we follow the ordering of [10], where PGV1, PGV2, and PGV5 are better known as the Matyas-Meyer-Oseas [36], Miyaguchi-Preneel, and Davies-Meyer [45] compression functions.

Baecher et al. [4] analyzed the 12 PGV constructions under ideal cipher reducibility, which at a high level covers the idea of two constructions being equally secure for the same underlying idealized blockcipher. They divide the PGV functions into two classes, in such a way that if some blockcipher makes one of the constructions secure, it makes all functions in the corresponding class secure. Applied to our WCM, the results of Baecher et al. imply the following:

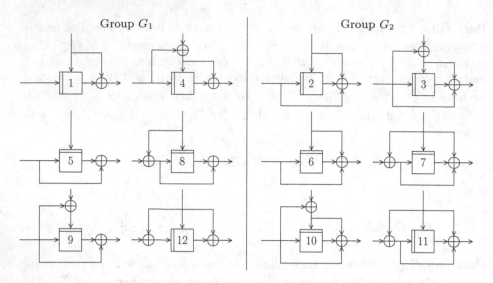

Fig. 3. The 12 PGV compression functions. When in iteration mode, the message comes in at the top. The groups G_1 and G_2 refer to Lemma 3.

Lemma 3 (Ideal Cipher Reducibility of PGV [4], Informal). *Let* $\pi \xleftarrow{\$}$ $\mathrm{BC}[\Phi](n, n)$ *for some predicate* Φ. *Let*

$$G_1 = \{1, 4, 5, 8, 9, 12\}, \ and \ G_2 = \{2, 3, 6, 7, 10, 11\}.$$

For any $\alpha \in \{1, 2\}$ *and* $i, j \in G_\alpha$, *PGVi and PGVj achieve the same level of collision and preimage security once instantiated with* π.

Baecher et al. also derive a reduction between the two classes, but this reduction requires a non-direct transformation on the ideal cipher π^1, making it unsuitable for our purposes. Thanks to Lemma 3, it suffices to only analyze PGV1 and PGV2 in the WCM: the bounds carry over to the other 10 PGV constructions. In Sect. 4.1 we analyze the collision security of these functions in the WCM. The preimage security is considered in Sect. 4.2.

4.1   Collision Security

Theorem 1. *Let* $n \in \mathbb{N}$. *Let* $\alpha \in \{1, 2\}$ *and consider* PGVα. *Suppose* $\pi \xleftarrow{\$}$ $\mathrm{BC}[\Phi(A, B, \varphi^C)](n, n)$. *Then, for* $q \leq 2^{n-1}/B$,

$$\mathbf{Adv}_{\mathrm{PGV}\alpha}^{\mathrm{col}}(q) \leq \frac{B^2 \delta_{B,C}[1] q^2}{2^n} + \binom{B}{2} \frac{2 \delta_{B,C}[2] q}{2^n} + \frac{4B^2 q^2}{2^n}.$$

---

[1] If π makes the PGV constructions from group G_1 secure, there is a transformation τ such that τ^π makes the constructions from G_2 secure, and vice versa.

Proof. We focus on PGV2. The analysis for PGV1 is a simplification due to the absence of the feed-forward of the key. We consider any adversary that has query access to $\pi \xleftarrow{\$} BC[\Phi(A, B, \varphi^C)](n, n)$ and makes q queries. As a first step, we move from π to $\bar{\pi} \xleftarrow{\$} \overline{BC}[\Phi(A, B, \varphi^C)](n, n)$. By Lemma 2, this costs us an additional term $\frac{B^2 q(q+1)}{2^n - Bq}$.

A collision for PGV2 would imply the existence of two distinct query pairs $(k, x, z), (k', x', z')$ such that $k \oplus x \oplus z = k' \oplus x' \oplus z'$. We consider the ith query $(i \in \{1, \ldots, q\})$ to be the first query to make this condition satisfied, and sum over $i = 1, \ldots, q$ at the end. For regular (forward or inverse) queries, the analysis of [9,10,58] mostly carries over. The analysis of predicate queries is a bit more technical.

Query $\bar{\pi}_k(x)$ or $\bar{\pi}_k^{-1}(z)$. The cases are the same by symmetry, and we consider $\bar{\pi}_k(x)$ only. Denote the response by z. There are at most $B(i - 1)$ possible (k', x', z'). As z is randomly drawn from a set of size at least $2^n - Bq$, it satisfies $z = k \oplus x \oplus k' \oplus x' \oplus z'$ with probability at most $\frac{B(i-1)}{2^n - Bq}$.

Query $\bar{\pi}_k^{\Phi}(y)$. Denote the query response by $\{(k, x^1, z^1), \ldots, (k, x^B, z^B)\}$. In case the B-set contributes only to (k, x, z), the same reasoning as for regular queries applies with the difference that any query of the B-set may be successful and that the bound of Lemma 2 part (ii) applies: $\frac{B^2 \delta_{B,C}[1](i-1)}{2^n - Bq}$.

Now, consider the case the predicate query contributes to both (k, x, z) and (k, x', z'). There are $\binom{B}{2}$ ways for the predicate query to contribute (or 0 if $B = 1$). By Lemma 2 part (iii), which considers the success probability for any such combination, the predicate query results in a collision with probability at most $\binom{B}{2} \frac{\delta_{B,C}[2] 2^n}{2^{2n} - Bq}$.

Conclusion. Taking the maximum of all success probabilities, the ith query is successful with probability at most $\frac{B^2 \delta_{B,C}[1](i-1)}{2^n - Bq} + \binom{B}{2} \frac{\delta_{B,C}[2] 2^n}{2^{2n} - Bq}$. Summation over $i = 1, \ldots, q$ gives

$$\mathbf{Adv}_{\text{PGV2}}^{\text{col}}(q) \leq \frac{B^2 \delta_{B,C}[1] q^2}{2(2^n - Bq)} + \binom{B}{2} \frac{\delta_{B,C}[2] q}{2^n - Bq} + \frac{B^2 q(q+1)}{2^n - Bq},$$

where the last part of the bound comes from the transition from WCM to AWCM. The proof is completed by using the fact that $2^n - Bq \geq 2^{n-1}$ for $Bq \leq 2^{n-1}$, and that $q + 1 \leq 2q$ for $q \geq 1$. □

We note that the bound gets worse for increasing values of B. This has a technical cause: predicate queries are counted equally expensive as regular queries, but result in up to B new query tuples. This leads to several factors of B in the bound. As this work is mainly concerned with differential known-key attacks for which B is regularly small, these factors are of no major influence.

The implications of the bound of Theorem 1 become more visible when considering particular choices of B and C.

(i) If $B = 1$, then $\mathbf{Adv}^{\mathrm{col}}_{\mathrm{PGV\alpha}}(q) \leq \frac{2^{|C|}q^2}{2^n} + \frac{4q^2}{2^n}$;

(ii) If $B = 2$, then $\mathbf{Adv}^{\mathrm{col}}_{\mathrm{PGV\alpha}}(q) \leq \frac{20q^2}{2^n} + \frac{4 \cdot 2^{|C|}q}{2^n}$;

(iii) If $B \geq 3$ (independent of n), then $\mathbf{Adv}^{\mathrm{col}}_{\mathrm{PGV\alpha}}(q) \leq \frac{5B^2q^2}{2^n} + \frac{B^2q}{2^n}$.

In other words, for $B = 2$ and C with $|C| \leq n/2$, or for $B \geq 3$ constant and C arbitrary, the PGV functions achieve the same $2^{n/2}$ collision security level as in the ICM. On the other hand, if $B = 1$, collisions can be found in about $2^{(n-|C|)/2}$ queries, and if $B = 2$ with $|C| > n/2$, in about $2^{n-|C|} < 2^{n/2}$ queries. See also Table 1.

Tightness

For the cases $B = 1$ and C arbitrary, and $B = 2$ and C arbitrary such that $|C| > n/2$, we derive generic attacks that demonstrate tightness of the bound of Theorem 1. Knudsen and Rijmen [27] and Sasaki et al. [53,56] already considered how to exploit a known-key pair for the underlying blockcipher to find a collision for the Matyas-Meyer-Oseas (PGV1) and/or Miyaguchi-Preneel (PGV2) compression functions. Their attacks correspond to our $B = 2$ case.

Proposition 1 ($B = 1$). Let $n \in \mathbb{N}$. Let $\alpha \in \{1, 2\}$ and consider PGVα. Suppose $\pi \xleftarrow{\$} \mathrm{BC}[\Phi(A, 1, \varphi^C)](n, n)$. Then, $\mathbf{Adv}^{\mathrm{col}}_{\mathrm{PGV\alpha}}(q) \geq \frac{q^2}{2^{n-|C|}}$.

Proof. We construct a collision-finding adversary $\mathcal{A}$ for PGV2. It fixes key $k = 0$, and makes predicate queries to π^Φ_k on input of distinct values y to obtain q queries (k, x_y, z_y) satisfying $\mathsf{Bits}_C(x_y \oplus z_y) = 0$. Any two such queries collide on the entire state, $k \oplus x_y \oplus z_y = k \oplus x_{y'} \oplus z_{y'}$, with probability at least $\frac{q^2}{2^{n-|C|}}$. The attack for PGV1 is the same as we have taken $k = 0$.   □

Proposition 2 ($B = 2$ and $|C| > n/2$). Let $n \in \mathbb{N}$. Let $\alpha \in \{1, 2\}$ and consider PGVα. Suppose $\pi \xleftarrow{\$} \mathrm{BC}[\Phi(A, 2, \varphi^C)](n, n)$. Then, $\mathbf{Adv}^{\mathrm{col}}_{\mathrm{PGV\alpha}}(q) \geq \frac{q}{2^{n-|C|}}$.

Proof. We construct a collision-finding adversary $\mathcal{A}$ for PGV2. It fixes key $k = 0$, and makes predicate queries to π^Φ_k on input of distinct values y to obtain q 2-sets $\{(k, x^1_y, z^1_y), (k, x^2_y, z^2_y)\}$ satisfying $\mathsf{Bits}_C(x^1_y \oplus z^1_y) = \mathsf{Bits}_C(x^2_y \oplus z^2_y)$. These two queries collide on the entire state, $k \oplus x^1_y \oplus z^1_y = k \oplus x^2_y \oplus z^2_y$, with probability at least $\frac{1}{2^{n-|C|}}$. If the adversary makes q predicate queries, we directly obtain our bound. The attack for PGV1 is the same as we have taken $k = 0$.   □

4.2   Preimage Security

Theorem 2. Let $n \in \mathbb{N}$. Let $\alpha \in \{1, 2\}$ and consider PGVα. Suppose $\pi \xleftarrow{\$} \mathrm{BC}[\Phi(A, B, \varphi^C)](n, n)$. Then, for $q \leq 2^{n-2}/B$,

$$\mathbf{Adv}^{\mathrm{epre}}_{\mathrm{PGV\alpha}}(q) \leq \left(\frac{2Bq}{2^n}\right)^B + \frac{2B^2 \delta_{B,C}[1]q}{2^n}.$$

The proof is given in Appendix A. It is much more involved than the one of Theorem 1, particularly as we cannot make use of abortable ciphers. Entering various choices of B and C shows that in the PGV functions remain mostly unaffected in the WCM if $B \geq 2$, and the same security level as in the ICM is achieved [9,10,58]. A slight security degradation appears for $B = 1$ as preimages can be found in about $2^{n-|C|}$. In the full version, we present a matching attack in the WCM.

5 Application to Grøstl Compression Function

We consider the provable security of the compression function mode of operation of Grøstl [21] (see also Fig. 4):

$$F_{\text{Grøstl}}(x_1, x_2) = x_2 \oplus \pi_1(x_1) \oplus \pi_2(x_1 \oplus x_2). \tag{6}$$

The Grøstl compression function is in fact designed to operate in a wide-pipe mode, and in the IPM, the function is proven collision secure up to about $2^{n/4}$ queries and preimage secure up to $2^{n/2}$ queries [20]. We consider the security of $F_{\text{Grøstl}}$ in the WCM, where $(\pi_1, \pi_2) \xleftarrow{\$} BC[\Phi(A, B, \varphi^C)](n)^2$. We remark that in this section we consider keyless primitives, hence $\kappa = 0$ and the k-input is dropped throughout. We furthermore note that finding collisions and preimages for $F_{\text{Grøstl}}$ is equivalent to finding them for

$$F'_{\text{Grøstl}}(x_1, x_2) = x_1 \oplus x_2 \oplus \pi_1(x_1) \oplus \pi_2(x_2), \tag{7}$$

as $F_{\text{Grøstl}}(x_1, x_2) = F'_{\text{Grøstl}}(x_1, x_1 \oplus x_2)$, and we will consider $F'_{\text{Grøstl}}$ throughout.

5.1 Collision Security

Theorem 3. *Let* $n \in \mathbb{N}$. *Suppose* $(\pi_1, \pi_2) \xleftarrow{\$} BC[\Phi(A, B, \varphi^C)](n)^2$. *Then, for* $q \leq 2^{n-1}/B$,

$$\mathbf{Adv}^{\text{col}}_{F'_{\text{Grøstl}}}(q) \leq \frac{B^4 \delta_{B,C}[1]q^4}{2^n} + \binom{B}{2}\frac{2\delta_{B,C}[2](q^2 + 2^{n/2-|C|}q)}{2^n} + \frac{B^2 q^2}{2 \cdot 2^{n/2}} + \frac{4B^2 q^2}{2^n}.$$

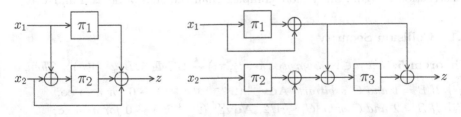

Fig. 4. Grøstl compression function (left) and Shrimpton-Stam (right).

The proof is given in the full version of the paper. If we enter particular choices of B and C into the bound, we find results comparable to the case of Sect. 4.1. In more detail, for $B = 2$ and C with $|C| \leq n/2$, or for $B \geq 3$ constant and C arbitrary, $\mathsf{F}_{\mathrm{Grøstl}}$ achieves the same $2^{n/4}$ collision security level as in the ICM [20]. If $B = 1$, the bound guarantees security up to about $2^{(n-|C|)/4}$, and if $B = 2$ with $|C| > n/2$, collisions can be found in about $2^{(n-|C|)/2}$ queries. See also Table 1. In the full version, we also show that the bound is optimal, by presenting tight attacks on $\mathsf{F}'_{\mathrm{Grøstl}}$ in the WCM.

5.2 Preimage Security

Theorem 4. Let $n \in \mathbb{N}$. Suppose $(\pi_1, \pi_2) \xleftarrow{\$} \mathrm{BC}[\Phi(A, B, \varphi^C)](n)^2$. Then, for $q \leq 2^{n-1}/B$,

$$\mathbf{Adv}^{\mathrm{epre}}_{\mathsf{F}'_{\mathrm{Grøstl}}}(q) \leq \frac{2B^2 \delta_{B,C}[1](q^2 + 2^{n/2-|C|}q)}{2^n} + \frac{Bq}{2^{n/2}} + \frac{4B^2 q^2}{2^n}.$$

The proof is given in the full version of the paper. As before, we find that $\mathsf{F}_{\mathrm{Grøstl}}$ remains unaffected in the WCM for most cases, the sole exception being $B = 1$ for which preimages can be found in about $2^{(n-|C|)/2}$. In the full version, we also show that the bound is optimal, by presenting a tight attack on $\mathsf{F}'_{\mathrm{Grøstl}}$ for $B = 1$ in the WCM.

6   Application to Shrimpton-Stam Compression Function

In this section, we consider the provable security of the Shrimpton-Stam compression function [57] (see also Fig. 4):

$$\mathsf{F}_{\mathrm{SS}}(x_1, x_2) = x_1 \oplus \pi_1(x_1) \oplus \pi_3(x_1 \oplus \pi_1(x_1) \oplus x_2 \oplus \pi_2(x_2)). \tag{8}$$

This function is proven asymptotically optimally collision and preimage secure up to $2^{n/2}$ queries in the IPM [41,51,57]. We consider the security of F_{SS} in the WCM, where $(\pi_1, \pi_2, \pi_3) \xleftarrow{\$} \mathrm{BC}[\Phi(A, B, \varphi^C)](n)^3$. (As in Sect. 5 we consider keyless functions, hence $\kappa = 0$ and the key inputs are dropped throughout.) Our findings readily apply to the generalization of F_{SS} of [41]. The analysis of this construction is significantly more complex than the ones of Sects. 4 and 5.

6.1 Collision Security

Theorem 5. Let $n \in \mathbb{N}$. Suppose $(\pi_1, \pi_2, \pi_3) \xleftarrow{\$} \mathrm{BC}[\Phi(A, B, \varphi^C)](n)^3$. Then,

(i) If $B = 1$ and C arbitrary, $\mathbf{Adv}^{\mathrm{col}}_{\mathsf{F}_{\mathrm{SS}}}(2^{(n-|C|)/2-n\varepsilon}) \to 0$ for $n \to \infty$;

(ii) If $B = 2$ and C with $|C| \leq n/2$, $\mathbf{Adv}^{\mathrm{col}}_{\mathsf{F}_{\mathrm{SS}}}(2^{n/2-n\varepsilon}) \to 0$ for $n \to \infty$;

(iii) If $B = 2$ and C with $|C| > n/2$, $\mathbf{Adv}^{\mathrm{col}}_{\mathsf{F}_{\mathrm{SS}}}(2^{n-|C|-n\varepsilon}) \to 0$ for $n \to \infty$;

(iv) If $B \geq 3$ (independent of n) and C arbitrary, $\mathbf{Adv}^{\mathrm{col}}_{\mathsf{F}_{\mathrm{SS}}}(2^{n/2-n\varepsilon}) \to 0$ for $n \to \infty$.

Due to the technicality of the proof, the results are expressed in asymptotic terms. The proof is given in the full version of the paper. For $B = 2$ and C with $|C| \leq n/2$, or for $B \geq 3$ constant and C arbitrary, F_{SS} achieves the same security level as in the IPM. On the other hand, if $B = 1$, or if $B = 2$ but $|C| > n/2$, Theorem 5 results in a worse bound. See also Table 1. In the full version, we also show that the bound is optimal, by presenting tight attacks on F_{SS} in the WCM.

6.2   Preimage Security

Theorem 6. *Let $n \in \mathbb{N}$. Suppose $(\pi_1, \pi_2, \pi_3) \xleftarrow{\$} BC[\Phi(A, B, \varphi^C)](n)^3$. Then,*

(i) *If $B = 1$ and C with $|C| \leq n/2$, $\mathbf{Adv}_{F_{SS}}^{epre}(2^{n/2-n\varepsilon}) \to 0$ for $n \to \infty$;*

(ii) *If $B = 1$ and C with $|C| > n/2$, $\mathbf{Adv}_{F_{SS}}^{epre}(2^{n-|C|-n\varepsilon}) \to 0$ for $n \to \infty$;*

(iii) *If $B \geq 2$ (independent of n) and C arbitrary, $\mathbf{Adv}_{F_{SS}}^{epre}(2^{n/2-n\varepsilon}) \to 0$ for $n \to \infty$.*

As for collision resistance, the results are expressed in asymptotic terms. The proof is given in the full version of the paper. The bounds match the ones in the IPM, except for the case of $B = 1$ and $|C| > n/2$. We leave it as an open problem to prove tightness of Theorem 6 part (ii).

7   Conclusions

Since their formal introduction by Knudsen and Rijmen at ASIACRYPT 2007 [27], numerous known-key attacks on blockciphers have appeared in literature. These attacks are often considered delicate, as it is not always clear to what extent they influence the security of cryptographic functions based on these known-key blockciphers. We presented the weak cipher model in order to investigate this impact. For a specific instance of this model, considering the existence of A sets of B queries that satisfy condition φ^C of (3), we proved that the PGV compression functions [48], the Grøstl compression function [21], and the Shrimpton-Stam compression function [57] remain mostly unaffected by the generalized weakness. Additionally, preimage security of the functions turned out to be significantly less susceptible to these types of weaknesses than collision security. The results can be readily generalized to other primitive-based functions, such as the double block length compression functions Tandem-DM, Abreast-DM, and Hirose's compression functions [23,30], and to the permutation-based sponge mode [5].

Our model is general enough to cover practically all differential known-key attacks in literature, such as latest results based on the rebound attack [12,22,28,38,52,53,56] and on the boomerang attack [2,7,31,54,60]. To our knowledge, our work provides the first attempt to formally analyze the effect of a wide class of cryptanalytic attacks from a modular and provable security point of view. It is a step in the direction of security beyond the ideal model, connecting practical attacks from cryptanalysis with ideal model provable security.

There is still a long way to go: in order to make the connection between the two fields, we abstracted known-key attacks to a certain degree. It remains a highly challenging open research problem to generalize our findings to multiple or different weaknesses, and to different permutation-based cryptographic functions. These generalizations include the analysis of known-key based constructions for more advanced conditions φ (such as arbitrary polynomials).

Acknowledgments. This work was supported in part by European Union's Horizon 2020 research and innovation programme under grant agreement No 644052 HECTOR and grant agreement No H2020-MSCA-ITN-2014-643161 ECRYPT-NET, and in part by the Research Council KU Leuven: GOA TENSE (GOA/11/007). Bart Mennink is a Postdoctoral Fellows of the Research Foundation – Flanders (FWO). The authors would like to thank the anonymous reviewers for their valuable help and feedback.

A   Proof of Theorem 2

We focus on PGV2. The analysis for PGV1 is a simplification due to the absence of the feed-forward of the key. We consider any adversary that has query access to $\pi \xleftarrow{\$} \mathrm{BC}[\Phi(A, B, \varphi^C)](n, n)$ and makes q queries. Let $Z \in \{0, 1\}^n$. A preimage for Z would imply the existence of a query (k, x, z) such that $x \oplus z = k \oplus Z$. We consider the ith query ($i \in \{1, \ldots, q\}$) to be the first query to make this condition satisfied, and sum over $i = 1, \ldots, q$ at the end. For regular (forward or inverse) queries, the analysis of [9,10,58] mostly carries over. The analysis of predicate queries is a more technical, particularly as we cannot make use of abortable ciphers.

Query $\pi_k(x)$ or $\pi_k^{-1}(z)$. The cases are the same by symmetry, and we consider $\pi_k(x)$ only. Denote the response by z. As z is randomly drawn from a set of size at least $2^n - Bq$, it satisfies $z = x \oplus k \oplus Z$ with probability at most $\frac{1}{2^n - Bq}$.

Query $\pi_k^\Phi(y)$. Denote the query response by $\{(k, x^1, z^1), \ldots, (k, x^B, z^B)\}$. If all tuples are old, the query cannot be successful as no earlier query was successful, and so we assume it contains at least one new tuple. The response is drawn uniformly at random from the set $\Sigma_k(P_k, P_k^\Phi)$. For $\ell = 0, \ldots, B$, denote by $\Sigma_k^\ell(P_k, P_k^\Phi)$ the subset of all responses that have ℓ new query tuples and $B - \ell$ old query tuples (which already appear in P_k). By construction,

$$\Sigma_k(P_k, P_k^\Phi) = \bigcup_{\ell=0}^{B} \Sigma_k^\ell(P_k, P_k^\Phi). \tag{9}$$

Define furthermore for $\ell = 1, \ldots, B$ by $\Sigma_k^{\ell,\mathrm{pre}}(P_k, P_k^\Phi)$ the subset of elements of $\Sigma_k^\ell(P_k, P_k^\Phi)$ for which one of the new query tuples satisfies $x \oplus z = k \oplus Z$ (recall that we have excluded the case of $\ell = 0$). The predicate query is successful with probability

$$\mathrm{Pr}\left(\pi_k^\Phi(y) \text{ sets } \mathrm{pre}(\mathcal{Q}_i)\right) = \sum_{\ell=1}^{B} \frac{|\Sigma_k^{\ell,\mathrm{pre}}(P_k, P_k^\Phi)|}{|\Sigma_k(P_k, P_k^\Phi)|}. \tag{10}$$

Using (9), we bound (10) as

$$\mathbf{Pr}\left(\pi_k^{\Phi}(y) \text{ sets } \mathsf{pre}(\mathcal{Q}_i)\right) \leq \frac{|\Sigma_k^{1,\mathrm{pre}}(P_k, P_k^{\Phi})|}{|\Sigma_k^{B}(P_k, P_k^{\Phi})|} + \sum_{\ell=2}^{B} \frac{|\Sigma_k^{\ell,\mathrm{pre}}(P_k, P_k^{\Phi})|}{|\Sigma_k^{\ell}(P_k, P_k^{\Phi})|}. \qquad (11)$$

The reason why $\ell = 1$ is treated differently, will become clear shortly.

We next bound all relevant sets. Here, for integers $a \geq b \geq 1$, we denote by $a^{\underline{b}} = \frac{a!}{(a-b)!}$ the falling factorial power. Starting with the numerators, for $\ell = 1$ we have

$$|\Sigma_k^{1,\mathrm{pre}}(P_k, P_k^{\Phi})| \leq B \cdot |P_k|^{\underline{B-1}} \cdot (2^n - |P_k|).$$

Indeed, we have B positions for the sole new query to appear and $|P_k|^{\underline{B-1}}$ choices for the old queries. For the new query, without loss of generality (k, x^B, z^B), it needs to satisfy $\mathsf{Bits}_C(x^B \oplus z^B) = \mathsf{Bits}_C(x^1 \oplus \cdots \oplus z^{B-1})$ and $x^B \oplus z^B = k \oplus Z$. We have $2^n - |P_k|$ possible choices for x^B, and any choice gives at most one possible z^B. We remark that $|\Sigma_k^{1,\mathrm{pre}}(P_k, P_k^{\Phi})|$ will probably be about a factor $2^{-|C|}$ less, as we should only count all possible solutions for the $B-1$ old queries that satisfy $\mathsf{Bits}_C(x^1 \oplus \cdots \oplus z^{B-1}) = \mathsf{Bits}_C(k \oplus Z)$. Deriving a tighter bound would be a cumbersome exercise, but fortunately there is no need to do so: the fraction of elements in $\Sigma_k(P_k, P_k^{\Phi})$ consisting of $B-1$ old tuples is already small enough for the case $B > 1$. This is the reason why we use a special treatment for the case of $\ell = 1$ in (11).

For $\ell \in \{2, \ldots, B\}$ we have

$$|\Sigma_k^{\ell,\mathrm{pre}}(P_k, P_k^{\Phi})| \leq \binom{B}{\ell} \cdot |P_k|^{\underline{B-\ell}} \cdot (2^n - |P_k|)^{\underline{\ell}} \cdot \ell \cdot (2^n - |P_k|)^{\underline{\ell-2}} \cdot 2^{n-|C|}.$$

Again, the first term comes from identifying at which positions the new queries appear and the second term comes from the selection of old queries. Next, we have $(2^n - |P_k|)^{\underline{\ell}}$ choices for the x-values and ℓ positions for the "winning query" to occur. For this particular winning query, the corresponding z-value is fixed by the equation $x \oplus z = k \oplus Z$. For the remaining $\ell - 1$ z-values, there are $(2^n - |P_k|)^{\underline{\ell-2}}$ possibilities to freely fix the first $\ell - 2$ of them, and the last one will be adapted to the predicate condition, and can take at most $2^{n-|C|}$ values.

Regarding the denominators, for $\ell \in \{1, \ldots, B\}$ we have

$$|\Sigma_k^{\ell}(P_k, P_k^{\Phi})| \geq \binom{B}{\ell} \cdot |P_k|^{\underline{B-\ell}} \cdot \left(\begin{array}{l} (2^n - |P_k|)^{\underline{\ell}} \cdot (2^n - |P_k|)^{\underline{\ell-1}} \cdot 2^{n-|C|} - \\ Bq \cdot (2^n - |P_k|)^{\underline{\ell-1}} \cdot (2^n - |P_k|)^{\underline{\ell-1}} \cdot 2^{n-|C|} \end{array} \right),$$

which can be seen as follows. As before, we have $\binom{B}{\ell}$ positions for the new queries to appear and $|P_k|^{\underline{B-\ell}}$ possible lists of old queries. Regarding the ℓ new queries, without loss of generality $(k, x^1, z^1), \ldots, (k, x^\ell, z^\ell)$, these need to satisfy $\mathsf{Bits}_C(x^1 \oplus \cdots \oplus z^\ell) = \mathsf{Bits}_C(x^{\ell+1} \oplus \cdots \oplus z^B)$. We first compute the number of choices for these new queries where z^ℓ is only used to adapt to this condition *and does not need to satisfy that it is fresh.* For this case, we have precisely $(2^n - |P_k|)^{\underline{\ell}} \cdot (2^n - |P_k|)^{\underline{\ell-1}}$ choices for $x^1, \ldots, z^{\ell-1}, x^\ell$, and $2^{n-|C|}$ possibilities for the adaption value z^ℓ.

Now, we subtract the cases where this adapted value happens to collide, either with an older value in $\mathsf{rng}(P_k)$ or with any of the new $z^1, \ldots, z^{\ell-1}$. Any of these choices would fix z^ℓ (in total at most $(|P_k| + \ell - 1)$ possibilities). Similarly to the analysis for $|\Sigma_k^{\ell,\mathrm{pre}}(P_k, P_k^\Phi)|$, where now x^ℓ will be used to be adapted to the predicate condition, there are at most

$$(|P_k| + \ell - 1) \cdot (2^n - |P_k|)^{\underline{\ell-1}} \cdot (2^n - |P_k|)^{\underline{\ell-1}} \cdot 2^{n-|C|}$$

choices for the fresh values. As $\ell \le B$, and additionally $|P_k| \le B(i-1) \le B(q-1)$ for the current query, we obtain our bound for $|\Sigma_k^\ell(P_k, P_k^\Phi)|$. The bound can be simplified to

$$|\Sigma_k^\ell(P_k, P_k^\Phi)| \ge \binom{B}{\ell} \cdot |P_k|^{\underline{B-\ell}} \cdot (2^n - |P_k|)^{\underline{\ell-1}} \cdot (2^n - |P_k|)^{\underline{\ell-1}} \cdot 2^{n-|C|} \cdot (2^n - 2Bq),$$

using that $\frac{(2^n - |P_k|)^{\underline{\ell}}}{(2^n - |P_k|)^{\underline{\ell-1}}} = 2^n - |P_k| - (\ell - 1) \ge 2^n - Bq$.

Plugging these bounds into (11), we find for the case $B = 1$:

$$\mathbf{Pr}\left(\pi_k^\Phi(y) \text{ sets } \mathsf{pre}(Q_i)\right) \le \frac{2^n - |P_k|}{2^{n-|C|} \cdot (2^n - 2q)} \le \frac{2^{|C|}}{2^n - 2q}.$$

For the case $B > 1$ the computation is a bit more elaborate:

$$\mathbf{Pr}\left(\pi_k^\Phi(y) \text{ sets } \mathsf{pre}(Q_i)\right) \le \frac{B \cdot (2^n - |P_k|)}{(2^n - |P_k|)^{\underline{B-1}} \cdot 2^{n-|C|} \cdot (2^n - 2Bq)} \cdot \frac{|P_k|^{\underline{B-1}}}{(2^n - |P_k|)^{\underline{B-1}}} +$$
$$\sum_{\ell=2}^{B} \frac{(2^n - |P_k|)^{\underline{\ell}} \cdot (2^n - |P_k|)^{\underline{\ell-2}}}{(2^n - |P_k|)^{\underline{\ell-1}} \cdot (2^n - |P_k|)^{\underline{\ell-1}}} \cdot \frac{\ell}{2^n - 2Bq}.$$

For the first fraction we use that $2^n - |P_k| \le (2^n - |P_k|)^{\underline{B-1}}$ as $B > 1$, and additionally that $|C| \le n$. For the falling factorial powers of the second fraction, we use that $|P_k|^{\underline{B-1}} \le (Bq)^{B-1}$ and $(2^n - |P_k|)^{\underline{B-1}} \ge (2^n - |P_k| - (B-1))^{B-1} \ge (2^n - 2Bq)^{B-1}$. For the fraction in the sum, we use that $\frac{(2^n - |P_k|)^{\underline{\ell}} \cdot (2^n - |P_k|)^{\underline{\ell-2}}}{(2^n - |P_k|)^{\underline{\ell-1}} \cdot (2^n - |P_k|)^{\underline{\ell-1}}} = \frac{2^n - |P_k| - (\ell-1)}{2^n - |P_k| - (\ell-2)} \le 1$. We obtain:

$$\mathbf{Pr}\left(\pi_k^\Phi(y) \text{ sets } \mathsf{pre}(Q_i)\right) \le \frac{B}{2^n - 2Bq} \cdot \frac{(Bq)^{B-1}}{(2^n - 2Bq)^{B-1}} + \sum_{\ell=2}^{B} \frac{\ell}{2^n - 2Bq}$$
$$\le \frac{B^B q^{B-1}}{(2^n - 2Bq)^B} + \frac{B^2}{2^n - 2Bq}.$$

Conclusion. Taking the maximum of all success probabilities, the ith query is successful with probability at most $\frac{B^B q^{B-1}}{(2^n - 2Bq)^B} + \frac{B^2 \delta_{B,C}[1]}{2^n - 2Bq}$. Summation over $i = 1, \ldots, q$ gives

$$\mathbf{Adv}^{\text{epre}}_{\text{PGV2}}(q) \leq \frac{B^B q^B}{(2^n - 2Bq)^B} + \frac{B^2 \delta_{B,C}[1]q}{2^n - 2Bq}.$$

The proof is completed by using the fact that $2^n - 2Bq \geq 2^{n-1}$ for $Bq \leq 2^{n-2}$.

References

1. Andreeva, E., Bogdanov, A., Mennink, B.: Towards understanding the known-key security of block ciphers. In: Moriai, S. (ed.) FSE 2013. LNCS, vol. 8424, pp. 348–366. Springer, Heidelberg (2014)
2. Aumasson, J.-P., Çalık, Ç., Meier, W., Özen, O., Phan, R.C.-W., Varıcı, K.: Improved cryptanalysis of skein. In: Matsui, M. (ed.) ASIACRYPT 2009. LNCS, vol. 5912, pp. 542–559. Springer, Heidelberg (2009)
3. Aumasson, J., Meier, W.: Zero-sum distinguishers for reduced Keccak- f and for the core functions of Luffa and Hamsi (2009)
4. Baecher, P., Farshim, P., Fischlin, M., Stam, M.: Ideal-cipher (Ir)reducibility for blockcipher-based hash functions. In: Johansson, T., Nguyen, P.Q. (eds.) EURO-CRYPT 2013. LNCS, vol. 7881, pp. 426–443. Springer, Heidelberg (2013)
5. Bertoni, G., Daemen, J., Peeters, M., Van Assche, G.: Sponge functions. In: ECRYPT Hash Function Workshop (2007)
6. Biryukov, A., Khovratovich, D., Nikolić, I.: Distinguisher and related-key attack on the full AES-256. In: Halevi, S. (ed.) CRYPTO 2009. LNCS, vol. 5677, pp. 231–249. Springer, Heidelberg (2009)
7. Biryukov, A., Nikolić, I., Roy, A.: Boomerang attacks on BLAKE-32. In: Joux, A. (ed.) FSE 2011. LNCS, vol. 6733, pp. 218–237. Springer, Heidelberg (2011)
8. Black, J.A., Cochran, M., Shrimpton, T.: On the impossibility of highly-efficient blockcipher-based hash functions. In: Cramer, R. (ed.) EUROCRYPT 2005. LNCS, vol. 3494, pp. 526–541. Springer, Heidelberg (2005)
9. Black, J., Rogaway, P., Shrimpton, T.: Black-box analysis of the block-cipher-based hash-function constructions from PGV. In: Yung, M. (ed.) CRYPTO 2002. LNCS, vol. 2442, pp. 320–335. Springer, Heidelberg (2002)
10. Black, J., Rogaway, P., Shrimpton, T., Stam, M.: An analysis of the blockcipher-based hash functions from PGV. J. Cryptology **23**(4), 519–545 (2010)
11. Blondeau, C., Peyrin, T., Wang, L.: Known-key distinguisher on full PRESENT. In: Gennaro, R., Robshaw, M. (eds.) CRYPTO 2015. LNCS, vol. 9215, pp. 455–474. Springer, Heidelberg (2015)
12. Bouillaguet, C., Dunkelman, O., Leurent, G., Fouque, P.A.: Attacks on hash functions based on generalized feistel: application to reduced-round *Lesamnta* and *SHAvite-3*$_{512}$. In: Biryukov, A., Gong, G., Stinson, D.R. (eds.) SAC 2010. LNCS, vol. 6544, pp. 18–35. Springer, Heidelberg (2011)
13. Bouillaguet, C., Fouque, P.-A., Leurent, G.: Security analysis of SIMD. In: Biryukov, A., Gong, G., Stinson, D.R. (eds.) SAC 2010. LNCS, vol. 6544, pp. 351–368. Springer, Heidelberg (2011)
14. Boura, C., Canteaut, A.: Zero-sum distinguishers for iterated permutations and application to KECCAK-f and Hamsi-256. In: Biryukov, A., Gong, G., Stinson, D.R. (eds.) SAC 2010. LNCS, vol. 6544, pp. 1–17. Springer, Heidelberg (2011)

15. Bresson, E., Canteaut, A., Chevallier-Mames, B., Clavier, C., Fuhr, T., Gouget, A., Icart, T., Misarsky, J.F., Naya-Plasencia, M., Paillier, P., Pornin, T., Reinhard, J., Thuillet, C., Videau, M.: Indifferentiability with distinguishers: why Shabal does not require ideal ciphers. Cryptology ePrint Archive, Report 2009/199 (2009)

16. Coron, J.-S., Patarin, J., Seurin, Y.: The random oracle model and the ideal cipher model are equivalent. In: Wagner, D. (ed.) CRYPTO 2008. LNCS, vol. 5157, pp. 1–20. Springer, Heidelberg (2008)

17. Dong, L., Wu, W., Wu, S., Zou, J.: Known-key distinguisher on round-reduced 3D block cipher. In: Jung, S., Yung, M. (eds.) WISA 2011. LNCS, vol. 7115, pp. 55–69. Springer, Heidelberg (2012)

18. Duan, M., Lai, X.: Improved zero-sum distinguisher for full round Keccak- f permutation. Chin. Sci. Bull. **57**(6), 694–697 (2012)

19. Duo, L., Li, C.: Improved collision and preimage resistance bounds on PGV schemes. Cryptology ePrint Archive, Report 2006/462 (2006)

20. Fouque, P.-A., Stern, J., Zimmer, S.: Cryptanalysis of tweaked versions of SMASH and reparation. In: Avanzi, R.M., Keliher, L., Sica, F. (eds.) SAC 2008. LNCS, vol. 5381, pp. 136–150. Springer, Heidelberg (2009)

21. Gauravaram, P., Knudsen, L.R., Matusiewicz, K., Mendel, F., Rechberger, C., Schläffer, M., Thomsen, S.: Grøstl - a SHA-3 candidate (2011). Submission to NIST's SHA-3 competition

22. Gilbert, H., Peyrin, T.: Super-Sbox cryptanalysis: improved attacks for AES-like permutations. In: Hong, S., Iwata, T. (eds.) FSE 2010. LNCS, vol. 6147, pp. 365–383. Springer, Heidelberg (2010)

23. Hirose, S.: Some plausible constructions of double-block-length hash functions. In: Robshaw, M. (ed.) FSE 2006. LNCS, vol. 4047, pp. 210–225. Springer, Heidelberg (2006)

24. Holenstein, T., Künzler, R., Tessaro, S.: The equivalence of the random oracle model and the ideal cipher model, revisited. In: Proceedings of ACM Symposium on Theory of Computing 2011, pp. 89–98. ACM, New York (2011)

25. Jetchev, D., Özen, O., Stam, M.: Collisions are not incidental: a compression function exploiting discrete geometry. In: Cramer, R. (ed.) TCC 2012. LNCS, vol. 7194, pp. 303–320. Springer, Heidelberg (2012)

26. Katz, J., Lucks, S., Thiruvengadam, A.: Hash functions from defective ideal ciphers. In: Nyberg, K. (ed.) CT-RSA 2015. LNCS, vol. 9048, pp. 273–290. Springer, Heidelberg (2015)

27. Knudsen, L.R., Rijmen, V.: Known-key distinguishers for some block ciphers. In: Kurosawa, K. (ed.) ASIACRYPT 2007. LNCS, vol. 4833, pp. 315–324. Springer, Heidelberg (2007)

28. Koyama, T., Sasaki, Y., Kunihiro, N.: Multi-differential cryptanalysis on reduced DM-PRESENT-80: collisions and other differential properties. In: Kwon, T., Lee, M.-K., Kwon, D. (eds.) ICISC 2012. LNCS, vol. 7839, pp. 352–367. Springer, Heidelberg (2013)

29. Kuwakado, H., Hirose, S.: Hashing mode using a lightweight blockcipher. In: Stam, M. (ed.) IMACC 2013. LNCS, vol. 8308, pp. 213–231. Springer, Heidelberg (2013)

30. Lai, X., Massey, J.L.: Hash function based on block ciphers. In: Rueppel, R.A. (ed.) Advances in Cryptology – EUROCRYPT 1992. LNCS, vol. 658, pp. 55–70. Springer, Heidelberg (1992)

31. Lamberger, M., Mendel, F.: Higher-order differential attack on reduced SHA-256. Cryptology ePrint Archive, Report 2011/037 (2011)

32. Lampe, R., Seurin, Y.: Security analysis of key-alternating feistel ciphers. In: Cid, C., Rechberger, C. (eds.) FSE 2014. LNCS, vol. 8540, pp. 243–264. Springer, Heidelberg (2015)

33. Lauridsen, M.M., Rechberger, C.: Linear distinguishers in the key-less setting: application to PRESENT. In: Leander, G. (ed.) FSE 2015. LNCS, vol. 9054, pp. 217–240. Springer, Heidelberg (2015)

34. Leurent, G., Roy, A.: Boomerang Attacks on Hash Function Using Auxiliary Differentials. In: Dunkelman, O. (ed.) CT-RSA 2012. LNCS, vol. 7178, pp. 215–230. Springer, Heidelberg (2012)

35. Liskov, M.: Constructing an ideal hash function from weak ideal compression functions. In: Biham, E., Youssef, A.M. (eds.) SAC 2006. LNCS, vol. 4356, pp. 358–375. Springer, Heidelberg (2007)

36. Matyas, S., Meyer, C., Oseas, J.: Generating strong one-way functions with cryptographic algorithm. IBM Techn. Disclosure Bull. **27**(10A), 5658–5659 (1985)

37. Maurer, U.M., Renner, R.S., Holenstein, C.: Indifferentiability, impossibility results on reductions, and applications to the random oracle methodology. In: Naor, M. (ed.) TCC 2004. LNCS, vol. 2951, pp. 21–39. Springer, Heidelberg (2004)

38. Mendel, F., Peyrin, T., Rechberger, C., Schläffer, M.: Improved cryptanalysis of the reduced Grøstl compression function, ECHO permutation and AES block cipher. In: Jacobson Jr, M.J., Rijmen, V., Safavi-Naini, R. (eds.) SAC 2009. LNCS, vol. 5867, pp. 16–35. Springer, Heidelberg (2009)

39. Mendel, F., Rechberger, C., Schläffer, M., Thomsen, S.S.: The rebound attack: cryptanalysis of reduced whirlpool and Grøstl. In: Dunkelman, O. (ed.) FSE 2009. LNCS, vol. 5665, pp. 260–276. Springer, Heidelberg (2009)

40. Mennink, B.: Optimal collision security in double block length hashing with single length key. In: Wang, X., Sako, K. (eds.) ASIACRYPT 2012. LNCS, vol. 7658, pp. 526–543. Springer, Heidelberg (2012)

41. Mennink, B., Preneel, B.: Hash functions based on three permutations: a generic security analysis. In: Safavi-Naini, R., Canetti, R. (eds.) CRYPTO 2012. LNCS, vol. 7417, pp. 330–347. Springer, Heidelberg (2012)

42. Mennink, B., Preneel, B.: Efficient parallelizable hashing using small non-compressing primitives. Int. J. Inf. Sec. (2015, to appear)

43. Meyer, C., Schilling, M.: Secure program load with manipulation detection code. In: Proceedings of Securicom, pp. 111–130 (1988)

44. Minier, M., Phan, R.C.-W., Pousse, B.: Distinguishers for ciphers and known key attack against rijndael with large blocks. In: Preneel, B. (ed.) AFRICACRYPT 2009. LNCS, vol. 5580, pp. 60–76. Springer, Heidelberg (2009)

45. Miyaguchi, S., Ohta, K., Iwata, M.: Confirmation that some hash functions are not collision free. In: Damgård, I.B. (ed.) EUROCRYPT 1990. LNCS, vol. 473, pp. 326–343. Springer, Heidelberg (1991)

46. NakaharaJr, J.: New impossible differential and known-key distinguishers for the 3D cipher. In: Bao, F., Weng, J. (eds.) ISPEC 2011. LNCS, vol. 6672, pp. 208–221. Springer, Heidelberg (2011)

47. Nikolić, I., Pieprzyk, J., Sokołowski, P., Steinfeld, R.: Known and chosen key differential distinguishers for block ciphers. In: Rhee, K.-H., Nyang, D.H. (eds.) ICISC 2010. LNCS, vol. 6829, pp. 29–48. Springer, Heidelberg (2011)

48. Preneel, B., Govaerts, R., Vandewalle, J.: Hash functions based on block ciphers: a synthetic approach. In: Stinson, D.R. (ed.) CRYPTO 1993. LNCS, vol. 773, pp. 368–378. Springer, Heidelberg (1994)

49. Rabin, M.: Digitalized signatures. In: Lipton, R., DeMillo, R. (eds.) Foundations of Secure Computation 1978, pp. 155–166. Academic Press, New York (1978)

50. Rogaway, P., Shrimpton, T.: Cryptographic hash-function basics: definitions, implications, and separations for preimage resistance, second-preimage resistance, and collision resistance. In: Roy, B., Meier, W. (eds.) FSE 2004. LNCS, vol. 3017, pp. 371–388. Springer, Heidelberg (2004)

51. Rogaway, P., Steinberger, J.P.: Constructing cryptographic hash functions from fixed-key blockciphers. In: Wagner, D. (ed.) CRYPTO 2008. LNCS, vol. 5157, pp. 433–450. Springer, Heidelberg (2008)

52. Sasaki, Y.: Known-key attacks on Rijndael with large blocks and strengthening *ShiftRow* parameter. In: Echizen, I., Kunihiro, N., Sasaki, R. (eds.) IWSEC 2010. LNCS, vol. 6434, pp. 301–315. Springer, Heidelberg (2010)

53. Sasaki, Y., Emami, S., Hong, D., Kumar, A.: Improved known-key distinguishers on Feistel-SP ciphers and application to Camellia. In: Susilo, W., Mu, Y., Seberry, J. (eds.) ACISP 2012. LNCS, vol. 7372, pp. 87–100. Springer, Heidelberg (2012)

54. Sasaki, Y., Wang, L.: Distinguishers beyond three rounds of the RIPEMD-128/-160 compression functions. In: Bao, F., Samarati, P., Zhou, J. (eds.) ACNS 2012. LNCS, vol. 7341, pp. 275–292. Springer, Heidelberg (2012)

55. Sasaki, Y., Wang, L., Takasaki, Y., Sakiyama, K., Ohta, K.: Boomerang distinguishers for full HAS-160 compression function. In: Hanaoka, G., Yamauchi, T. (eds.) IWSEC 2012. LNCS, vol. 7631, pp. 156–169. Springer, Heidelberg (2012)

56. Sasaki, Y., Yasuda, K.: Known-key distinguishers on 11-round feistel and collision attacks on its hashing modes. In: Joux, A. (ed.) FSE 2011. LNCS, vol. 6733, pp. 397–415. Springer, Heidelberg (2011)

57. Shrimpton, T., Stam, M.: Building a collision-resistant compression function from non-compressing primitives. In: Aceto, L., Damgård, I., Goldberg, L.A., Halldórsson, M.M., Ingólfsdóttir, A., Walukiewicz, I. (eds.) ICALP 2008, Part II. LNCS, vol. 5126, pp. 643–654. Springer, Heidelberg (2008)

58. Stam, M.: Blockcipher-based hashing revisited. In: Dunkelman, O. (ed.) FSE 2009. LNCS, vol. 5665, pp. 67–83. Springer, Heidelberg (2009)

59. Wagner, D.: The boomerang attack. In: Knudsen, L.R. (ed.) FSE 1999. LNCS, vol. 1636, pp. 156–170. Springer, Heidelberg (1999)

60. Yu, H., Chen, J., Wang, X.: The boomerang attacks on the round-reduced skein-512. In: Knudsen, L.R., Wu, H. (eds.) SAC 2012. LNCS, vol. 7707, pp. 287–303. Springer, Heidelberg (2013)

Generic Security of NMAC and HMAC with Input Whitening

Peter Gaži[1](✉), Krzysztof Pietrzak[1], and Stefano Tessaro[2]

[1] IST Austria, Klosterneuburg, Austria
{peter.gazi,pietrzak}@ist.ac.at
[2] UC Santa Barbara, Santa Barbara, USA
tessaro@cs.ucsb.edu

Abstract. HMAC and its variant NMAC are the most popular approaches to deriving a MAC (and more generally, a PRF) from a cryptographic hash function. Despite nearly two decades of research, their exact security still remains far from understood in many different contexts. Indeed, recent works have re-surfaced interest for *generic* attacks, i.e., attacks that treat the compression function of the underlying hash function as a black box.

Generic security can be proved in a model where the underlying compression function is modeled as a random function – yet, to date, the question of proving tight, non-trivial bounds on the generic security of HMAC/NMAC even as a PRF remains a challenging open question.

In this paper, we ask the question of whether a small modification to HMAC and NMAC can allow us to exactly characterize the security of the resulting constructions, while only incurring little penalty with respect to efficiency. To this end, we present simple variants of NMAC and HMAC, for which we prove tight bounds on the generic PRF security, expressed in terms of numbers of construction and compression function queries necessary to break the construction. All of our constructions are obtained via a (near) *black-box* modification of NMAC and HMAC, which can be interpreted as an initial step of key-dependent message pre-processing.

While our focus is on PRF security, a further attractive feature of our new constructions is that they clearly defeat all recent generic attacks against properties such as state recovery and universal forgery. These exploit properties of the so-called "functional graph" which are not directly accessible in our new constructions.

Keywords: Message authentication codes · HMAC · Generic attacks · Provable security

1 Introduction

This paper presents new variants of the HMAC/NMAC constructions of message authentication codes which enjoy *provable* security as a pseudorandom function (PRF) against generic distinguishing attacks, i.e., attacks which treat the compression function of the underlying hash function as a black-box. In particular,

© International Association for Cryptologic Research 2015
T. Iwata and J.H. Cheon (Eds.): ASIACRYPT 2015, Part II, LNCS 9453, pp. 85–109, 2015.
DOI: 10.1007/978-3-662-48800-3_4

we prove concrete *tight* bounds in terms of the number of queries to the construction *and* to the compression function necessary to distinguishing our construction from a random function. Our constructions are the first HMAC/NMAC variants to enjoy such a tight analysis, and we see this as an important stepping stone towards the understanding of the generic security of such constructions.

Hash-Based MACs. HMAC [3] is the most widely used approach to key a hash function H to obtain a PRF or a MAC. It computes the output on message M and a key K as

$$\mathsf{HMAC}(K, M) = H(K \oplus \mathsf{opad} \, \| \, H(K \oplus \mathsf{ipad} \, \| \, M)),$$

where $\mathsf{opad} \neq \mathsf{ipad}$ are constants.[1] Usually, H is a hash function like SHA-1, SHA-256 or MD5, in particular following the Merkle-Damgård paradigm [4,16]. That is, it extends a compression function $\mathsf{f} : \{0,1\}^c \times \{0,1\}^b \to \{0,1\}^c$ into a hash function $\mathsf{MD}_{\mathsf{IV}}^{\mathsf{f}}$ by first padding M into b-bit blocks $M[1], \ldots, M[\ell]$, and then producing the output $H(M) = S_\ell$, where

$$S_0 \leftarrow \mathsf{IV}, \quad S_i \leftarrow \mathsf{f}(S_{i-1} \, \| \, M[i]) \text{ for all } i = 1, \ldots, \ell. \tag{1}$$

starting with the c-bit initialization value IV. A cleaner yet slightly less practical variant of HMAC is NMAC, which instead outputs

$$\mathsf{NMAC}_{K_{\mathsf{in}}, K_{\mathsf{out}}}(M) = \mathsf{MD}_{K_{\mathsf{out}}}^{\mathsf{f}}(\mathsf{MD}_{K_{\mathsf{in}}}^{\mathsf{f}}(M)),$$

where $K_{\mathsf{in}}, K_{\mathsf{out}} \in \{0,1\}^c$ are key values.

Security of HMAC/NMAC. The security of both constructions has been studied extensively, both by obtaining security proofs and proposing attacks. On the former side, NMAC and HMAC were proven to be secure *pseudorandom functions* (PRFs) in the standard model [3], later also using weaker assumptions [2] and via a tight bound in the uniform setting [7]. However, as argued in [7], this standard-model bound might be overly pessimistic, covering also very unnatural constructions of the underlying compression function f (for example the one used in their tightness proof). The authors hence argue for the need of an analysis of the PRF security of HMAC in the so-called *ideal compression function model* where the compression function is modelled as an ideal random function and the adversary is allowed to query it. This model was previously used by Dodis *et al.* [6] to study *indifferentiability* of HMAC, which however only holds for certain key lengths.

This is also the model implicitly underlying many of the recently proposed attacks on hash-based MACs [5,10,15,17,19,20,22]. These attacks are termed *generic*, meaning they can be mounted for any underlying hash function as long as it follows the Merkle-Damgård (MD) paradigm. The complexity of such a generic attack is then expressed in the number of key-dependent queries to the construction (denoted q_C) as well as the number of queries to the underlying compression function (denoted q_f). These two classes of queries are also often referred to as *online* and *offline*, respectively.

---

[1] Some details such as padding and arbitrary key length are addressed in Sect. 2.

All iterated MACs are subject to the long-known Preneel and van Oorschot's attack [21] which implies a forgery (and hence also distinguishing) attack against HMAC/NMAC making $q_C = 2^{c/2}$ construction queries (consisting of constant-length messages) and no direct compression function queries (i.e., $q_f = 0$). This immediately raises two questions:

How does the security of HMAC and NMAC degrade (in terms of tolerable q_C) by increasing (1) the length ℓ of the messages and (2) the number q_f of compression-function evaluations?

The first question has been partially addressed in [7]. Their result[2] can be interpreted as giving tight bounds on the PRF security of NMAC against an attacker making q_C key-dependent construction queries (of length at most $\ell < 2^{c/3}$ b-bit blocks) but *no* queries to the compression function. They show that both constructions can only be distinguished from random function with advantage roughly $\epsilon(q_C, \ell) \approx \ell^{1+o(1)} q_C{}^2/2^c$, improving significantly on the bound $\epsilon(q_C, \ell) \approx \ell^2 q_C{}^2/2^c$ provable using standard folklore techniques. From our perspective, this bound can be read as a smooth trade-off: with increasing maximum allowed query length ℓ it tells us how many queries q_C can be tolerated for any acceptable upper bound on advantage.

Still, it is not clear how this trade-off changes when allowing extremely long messages ($\ell > 2^{c/3}$) and/or some queries to the compression function ($q_f > 0$). Note that while huge ℓ can be prevented by standards, in practical settings q_f is very likely to be much higher than q_C, as it represents cheap local (offline) computation of the attacker. We therefore focus on capturing the trade-off between q_C and q_f for values of q_C that do not allow to mount the attack from [21]. However, as we argue below, getting such a tight trade-off for NMAC/HMAC seems to be out of reach for now, we hence relax the problem by allowing for slight modifications to the vanilla NMAC/HMAC construction.

Our Contributions. We ask the following question here, and answer it positively:

Can we devise variants of HMAC/NMAC whose security provably degrades gracefully with an increasing number of compression function queries q_f, possibly retaining security for q_f being much larger than 2^c?

The main contribution of this paper is the introduction and analysis of a variant of NMAC (which we then adapt to the HMAC setting, as described below) which uses additional key material to "whiten" message blocks before being processed by the compression function. Concretely, our construction – termed WNMAC (for "whitened NMAC") uses an additional extra b-bit key K_w, and given a message M padded as $M[1], \ldots, M[\ell]$, operates as NMAC on input padded to blocks $M'[i] = M[i] \oplus K_b$, i.e., every message block is whitened with the *same* key (see also Fig. 1).

---

[2] Here we refer to Theorem 2 in [7] that formally considers a related construction NI in the standard model. However, its proof starts by a transition to the ideal-model analysis of a construction very closely related to NMAC, while disallowing compression-function queries.

The rationale behind WNMAC is two-fold. First, from the security viewpoint, the justification comes from the rich line of research on generic attacks on hash-based MACs. Most recent attacks [10,15,19,20] exploit the so-called "functional graph" of the compression function f, i.e., the graph capturing the structure of f when repeatedly invoked with its b-bit input fixed to some constant (say 0^b). Since our whitening denies the adversary the knowledge of b-bit inputs on which f is invoked during construction queries, intuitively it seems to be the right way to foil such attacks. Moreover, a recent work by Sasaki and Wang [22] suggests that keying *every* invocation of f is necessary in order to prevent suboptimal security against generic state recovery attacks. WNMAC arguably provides the simplest and most natural such keying. Second, from the practical perspective, WNMAC can be implemented on top of an existing implementation of NMAC, using it as a black-box.

PRF-Security of WNMAC. Our main result shows that WNMAC is a secure PRF; more precisely, no attacker making at most q_C construction queries (for messages padded into at most ℓ blocks) and q_f primitive queries can distinguish WNMAC from a random function, except with distinguishing advantage

$$\epsilon_{\mathsf{WNMAC}}(q_C, q_f, \ell) \leq \frac{q_f q_C}{2^{2c}} + 2 \cdot \frac{\ell q_C q_f}{2^{b+c}} + \frac{\ell q_C^2}{2^c} \cdot \left(d'(\ell) + \frac{64\ell^3}{2^c} + 1 \right).$$

Here, $d'(\ell)$ is the maximum, over all positive integers $\ell' \leq \ell$, of the number of positive divisors of ℓ', and grows very slowly, i.e., $d'(\ell) \approx \ell^{1/\ln\ln\ell}$. We also prove that this bound is essentially tight. Namely, we give an attack that achieves advantage roughly $q_C q_f / 2^{2c}$, showing the first term above to be necessary. Additionally, we know from [7] that the third term is tight for $\ell \leq 2^{c/3}$.

Note that in the case of $q_f = 0$, the bound matches exactly the bound from [7]. Moreover, observe that under the realistic assumption that $\ell < \min\{2^{c/3}, 2^{b-c}\}$, the bound simplifies to

$$\epsilon_{\mathsf{WNMAC}}(q_C, q_f, \ell) \leq 3\frac{q_f q_C}{2^{2c}} + (d'(\ell) + 2) \cdot \frac{\ell q_C^2}{2^c}.$$

Ignoring $d'(\ell)$ for simplicity, we see that we can tolerate up to $q_C \approx 2^{c/2}/\sqrt{\ell}$ construction queries and up to $q_f \approx 2^{1.5c}$ primitive queries. This corresponds to the security threshold ranging from 2^{192} f-queries for MD5 up to 2^{768} f-queries for SHA-512. The first term also clearly characterizes the complete trade-off curve between $q_C < 2^{c/2}/\sqrt{\ell}$ and q_f for any reasonable upper bound on the message length and acceptable distinguishing advantage.

Other Security Properties. Additionally, we also analyze the security level WNMAC achieves with respect to other security notions frequently considered in the attacks literature. By a series of reductions, we show that, roughly speaking, $\epsilon_{\mathsf{WNMAC}}$ also upper-bounds the adversary's advantage for *distinguishing-H* and *state recovery*. We believe that addressing these cryptanalytic notions also using the traditional toolbox of provable security is important and see this paper as taking the first step on that path.

Lifting to HMAC. We then move our attention from NMAC to HMAC and propose two analogous modifications to it. The first one, called WHMAC, is obtained from HMAC in the same way WNMAC is obtained from NMAC: by whitening the padded message blocks with an independent key, The second one, termed WHMAC$^+$, additionally processes a fresh key K^+ instead of the first block of the message. Both variants can be implemented given only black-box access to HMAC, and we prove that they maintain the same security level as WNMAC as long as the parameters b, c of f satisfy $b \gg 2c$ (for WHMAC) or $b \gg c$ (for WHMAC$^+$). Note that for existing hash functions, the former condition is satisfied for both MD5 and SHA-1, while the latter holds also for SHA-256 and SHA-512.

The Dual Construction. Motivated by the most restrictive term $q_C q_f / 2^{2c}$ in $\epsilon_{\mathsf{WNMAC}}$, the final construction we propose in this paper is a "dual" version of WNMAC denoted DWNMAC, that differs in the final, outer f-call. Instead of $f(K_2, s \parallel 0^{b-c})$ for a c-bit key K_2 and a c-bit state s padded with zeroes, the outer call in DWNMAC computes $f(s, K_2)$ for a longer, b-bit key. As expected, we prove that this tweak removes the need for the $q_C q_f / 2^{2c}$ term and replaces it by the strictly favourable term $q_C q_f / 2^{b+c}$, proving that the zero-padding in the outer call of WNMAC was actually responsible for the "bottle-neck" term in its security bound.

Our Techniques. In our information-theoretic analysis of WNMAC we employ the H-coefficient technique by Patarin [18], partially inheriting the notational framework from the recent analysis of keyed sponges by Gaži, Pietrzak, and Tessaro [8]. On a high level, the heart of our proof is a careful analysis of the probability that two sets intersect in the ideal experiment: (1) the set of adversarial queries to f, and (2) the set of inputs on which f is invoked when answering the adversary's queries to WNMAC. Obtaining a bound on the probability of this event then allows us to exclude it and use the result from [7] that considers $q_f = 0$, properly adapted to the WNMAC setting.

Related Work. As mentioned above, the motivation for our work partially stems from the recent line of work on generic attacks against iterated hash-based MACs [5,10,15,17,19,20,22]. While our security bound for WNMAC does not exclude attacks of the complexity (in terms of numbers of queries and message lengths) considered in these papers, the design of WNMAC was partially guided by the structure of these attacks and seems to prevent them. We find in particular the work [22] to be a good justification for investigating the security of WNMAC and related constructions. Iterated MAC that uses keying in every f-invocation was already considered by An and Bellare [1], their construction NI was later subject to analysis [7] that we adapt and reuse. One can see WNMAC as a conceptual simplification of NI where the key is simply used to whiten the b-bit input to the compression function. Finally, our dual construction considered in Sect. 5 bears resemblance to the Sandwich MAC analyzed by Yasuda [23], we believe that our methods could be easily adapted to cover this construction as well.

Perspective and Open Problems. We stress that the reader should not conclude from this work that NMAC and HMAC are necessarily less secure than the constructions proposed in this paper, specifically with respect to PRF security. In fact, we are not aware of any attacks showing a separation between the PRF security of our constructions and that of the original NMAC/HMAC constructions, finding one is an interesting open problem.

While obtaining a non-tight birthday-type bound for NMAC/HMAC is feasible (for most key-length values, a bound follow directly from the indifferentiability analysis of [6]), proving *tight* bounds in terms of compression function and construction queries on the generic PRF security of NMAC/HMAC is a challenging open problem, on which little progress has been made. The main challenge is to understand how partial information in form of f-queries can help the attacker to break security (i.e., distinguish) in settings with $q_C \ll 2^{c/2}/\sqrt{\ell}$, when the attack from [7] does not apply. This will require in particular developing a better understanding of the functional graph defined by queries to the function f. Some of its properties have been indeed exploited in existing generic attacks, but proving security appears to require a much deeper understanding: Most of the recent attacks, which are probably still not tight, do not come with rigorous proofs but instead rely on conjectures on the structure of these graphs [10]. The difficulty of this question for NMAC/HMAC is also well documented by the fact that even proving security of the whitened constructions presented in this paper required some novel tricks and considerable effort.

Similarly, it remains equally challenging to prove that for the properties considered by the recent HMAC/NMAC attacks (such as distinguishing-H, state recovery or various types of forgeries), the security of WNMAC/WHMAC is provably superior. Yet, we note that our construction invalidates direct application of all existing attacks, and hence we feel confident conjecturing that its security is much higher.

Black-box Instantiations. Throughout the paper we implicitly assume we can add a key to each b-bit input block, even though we aim for a black-box instantiation. For many MD-based hash functions, such fine-grained control of the input to the compression function is generally not possible via a black-box message pre-processing. Concretely, the functions from the SHA-family with 512-bit blocks only allow to effectively control (via alterations of the message) the first 447 bits of the last block, since the remaining 65 bits are reserved for the 64-bit length, and an additional 1-bit. Our analysis can be easily modified to take this into account. The resulting bound will change very little, and will result in the term $\ell q_C q_f/2^{b+c}$ being replaced by the term $(\ell-1+2^d) \cdot q_C \cdot q_f/2^{b+c}$, where d is the length of the non-controllable part of the input (for SHA-functions, $d = 65$). Note that since $d \ll b - c$, this will not affect the tightness of the bounds for concrete parameters.

2   Preliminaries

Basic Notation. We denote $[n] := \{1, \ldots, n\}$. Moreover, for a finite set $\mathcal{S}$ (e.g., $\mathcal{S} = \{0, 1\}$), we let $\mathcal{S}^n$, $\mathcal{S}^+$ and $\mathcal{S}^*$ be the sets of sequences of elements of $\mathcal{S}$ of length n, of arbitrary (but non-zero) length, and of arbitrary length, respectively (with ε denoting the empty sequence, as opposed to ϵ which is a small quantity). As a shorthand, let $\{0,1\}^{b*}$ denote $(\{0,1\}^b)^*$. We denote by $S[i]$ the i-th element of $S \in \mathcal{S}^n$ for all $i \in [n]$. Similarly, we denote by $S[i \ldots j]$, for every $1 \leq i \leq j \leq n$, the sub-sequence consisting of $S[i], S[i+1], \ldots, S[j]$, with the convention that $S[i \ldots i] = S[i]$. Moreover, we denote by $S \parallel S'$ the concatenation of two sequences in $\mathcal{S}^*$, and also, we let $S \mid T$ be the usual prefix-of relation: $S \mid T :\Leftrightarrow (\exists S' \in \mathcal{S}^* : S \parallel S' = T)$.

For an integer n, $d(n) = |\{i \in \mathbb{N} : i \mid n\}|$ is the number of its positive divisors and

$$d'(n) := \max_{n' \in \{1, \ldots, n\}} |\{d \in \mathbb{N} : d \mid n'\}| \approx n^{1/\ln \ln n}$$

is the maximum, over all positive integers $n' \leq n$, of the number of positive divisors of n'. More precisely, we have $\forall \varepsilon > 0 \; \exists n_0 \; \forall n > n_0 : d(n) < n^{(1+\varepsilon)/\ln \ln n}$ [11].

We also let $\mathcal{F}(\mathcal{D}, \mathcal{R})$ be the set of all functions from $\mathcal{D}$ to $\mathcal{R}$; and with a slight abuse of notation we sometimes write $\mathcal{F}(m, n)$ (resp. $\mathcal{F}(*, n)$) to denote the set of functions mapping m-bit strings to n-bit strings (resp. from $\{0,1\}^*$ to $\{0,1\}^n$). We denote by $x \xleftarrow{\$} \mathcal{X}$ the act of sampling x uniformly at random from $\mathcal{X}$. Finally, we denote the event that an adversary A, given access to an oracle O, outputs a value y, as $\mathsf{A}^\mathsf{O} \Rightarrow y$. To emphasize the random experiment considered, we sometimes denote the probability of an event A in a random experiment E by $\mathsf{P}^\mathsf{E}[A]$. Finally, the min-entropy $\mathsf{H}_\infty(X)$ of a random variable X with range $\mathcal{X}$ is defined as $- \log(\max_{x \in \mathcal{X}} \mathsf{P}_X(x))$.

Pseudorandom Functions. We consider *keyed* functions $\mathsf{F} : \mathcal{K} \times \mathcal{D} \to \mathcal{R}$ taking a κ-bit key (i.e., $\mathcal{K} = \{0,1\}^\kappa$), a message $M \in \mathcal{D}$ as input, and returning an output from $\mathcal{R}$. For a keyed function F under a key $k \in \mathcal{K}$ we often write $\mathsf{F}_k(\cdot)$ instead of $\mathsf{F}(k, \cdot)$. One often considers the security of F as a *pseudorandom function* (or PRF, for short) [9]. This is defined via the following advantage measure, involving an adversary A:

$$\mathsf{Adv}_\mathsf{F}^{\mathsf{prf}}(\mathsf{A}) := \left| \mathsf{P}\left[K \xleftarrow{\$} \{0,1\}^\kappa : \mathsf{A}^{\mathsf{F}_K} \Rightarrow 1 \right] - \mathsf{P}\left[f \xleftarrow{\$} \mathcal{F}(\mathcal{D}, \mathcal{R}) : \mathsf{A}^f \Rightarrow 1 \right] \right|.$$

Informally, we say that F is a PRF if this advantage is "negligible" for all "efficient" adversaries A.

PRFs in the Ideal Compression Function Model. For our analysis below, we are going to consider keyed constructions $\mathsf{C}[\mathsf{f}] : \{0,1\}^\kappa \times \mathcal{D} \to \mathcal{R}$ which make queries to a randomly chosen compression function $\mathsf{f} \xleftarrow{\$} \mathcal{F}(c+b, c)$ which can also be evaluated by the adversary (we sometimes write C^f instead of $\mathsf{C}[\mathsf{f}]$). For this case, we use the following notation to express the PRF advantage of A:

$$\mathsf{Adv}_{\mathsf{C}[\mathsf{f}]}^{\mathsf{prf}}(\mathsf{A}) := \left| \mathsf{P}\left[K \xleftarrow{\$} \{0,1\}^\kappa, \mathsf{f} \xleftarrow{\$} \mathcal{F}(c+b, c) : \mathsf{A}^{\mathsf{C}_K^\mathsf{f}, \mathsf{f}} \Rightarrow 1 \right] \right.$$
$$\left. - \mathsf{P}\left[R \xleftarrow{\$} \mathcal{F}(\mathcal{D}, \mathcal{R}), \mathsf{f} \xleftarrow{\$} \mathcal{F}(c+b, c) : \mathsf{A}^{\mathsf{R}, \mathsf{f}} \Rightarrow 1 \right] \right|.$$

We call A's queries to its first oracle *construction queries* (or C-queries) and its queries to the second oracle as *primitive queries* (or f-queries).

Note that the notion of PRF-security is identical to the notion of *distinguishing-R*, first defined in [13] and often used in the cryptanalytic literature on hash-based MACs.

Distinguishing-H. A further security notion defined in [13] is the so-called *distinguishing-H* security. Here, the goal of the adversary is to distinguish the hash-based MAC construction $C_K[f]$ using its underlying compression function f (say SHA-1) and a random key K, from the same construction $C_K[g]$ built on top of an independent random compression function g. In the ideal compression function model, where we model already the initial compression function f as ideal, this corresponds to distinguishing a pair of oracles $(C_K[f], f)$ from $(C_K[f], g)$. Formally,

$$\mathsf{Adv}_C^{\mathsf{dist\text{-}H}}(A) := \left| P\left[K \xleftarrow{\$} \{0,1\}^\kappa, f \xleftarrow{\$} \mathcal{F}(c+b,c) : A^{C_K^f, f} \Rightarrow 1 \right] \right.$$
$$\left. - P\left[K \xleftarrow{\$} \{0,1\}^\kappa, f, g \xleftarrow{\$} \mathcal{F}(c+b,c) : A^{C_K^f, g} \Rightarrow 1 \right] \right|.$$

State Recovery. An additional notion considered in the literature is security against *state recovery*. Since the definition of this notion needs to be tailored for the concrete construction it is applied to, we postpone the formal definition of security against state recovery to Sect. 3.10.

MACs and Unpredictability. It is well known that a good PRF also yields a good message-authentication code (MAC). A concrete security bound for unforgeability can be obtained from the PRF bound via a standard argument.

Iterated MACs. For a keyed function $f : \{0,1\}^c \times \{0,1\}^b \to \{0,1\}^c$ we denote with $\mathsf{Casc}^f : \{0,1\}^c \times \{0,1\}^{b*} \to \{0,1\}^c$ the cascade construction (also known as Merkle-Damgård) built from f as

$$\mathsf{Casc}^f(K, m_1\| \ldots \|m_\ell) := y_\ell \text{ where } y_0 := K \text{ and for } i \geq 1 : y_i := f(y_{i-1}, m_i),$$

in particular $\mathsf{Casc}^f(K, \varepsilon) := K$.

The construction $\mathsf{NMAC}^f : (\{0,1\}^c)^2 \times \{0,1\}^{b*} \to \{0,1\}^c$ is derived from Casc^f by adding an additional, independently keyed application of f at the end. It assumes that the domain sizes of f satisfy $b \geq c$ and the output of the cascade is padded with zeroes before the last f-call. Formally,

$$\mathsf{NMAC}^f((K_1, K_2), M) := f(K_2, \mathsf{Casc}^f(K_1, M)\|0^{b-c}).$$

Note that practical MD-based hash functions take as input arbitrary-length bit-strings and then pad them to a multiple of the block length, often including the message length in the so-called MD-strengthening. This padding then also appears in NMAC (and HMAC) but here we take the customary shortcut and our definition of NMAC above (resp. HMAC below) actually corresponds to the generalized constructions denoted as GNMAC (resp. GHMAC) in [2] where this step is also justified in detail.

HMAC^f is a practice-oriented version of NMAC^f, where the two keys (K_1, K_2) are derived from a single key $K \in \{0,1\}^b$ by xor-ing it with two fixed b-bit strings ipad and opad. In addition, the keys are not given through the key-input of the compression function f, but are prepended to the message instead. This allows for the usage of existing implementations of hash functions that contain a hard-coded initialization vector IV. Formally:

$$\mathsf{HMAC}^f(K, m) := \mathsf{Casc}^f(\mathsf{IV}, K_2\|\mathsf{Casc}^f(\mathsf{IV}, K_1\|m)\|\mathsf{fpad})$$
$$\text{where } (K_1, K_2) := (K \oplus \mathsf{ipad}, K \oplus \mathsf{opad})$$

and fpad is a fixed $(b-c)$-bit padding not affecting the security analysis. (Technically, [14] allows for arbitrary length of the key K: a key shorter than b bits is padded with zeroes before applying the xor transformations, a longer key is first hashed.)

3 The Whitened NMAC Construction

We now present our main construction called *Whitened NMAC* (or WNMAC for short). To that end, let us first consider a modification of the cascade construction Casc called *whitened cascade* and denoted WCasc. For a keyed function $f : \{0,1\}^c \times \{0,1\}^b \to \{0,1\}^c$ we denote with $\mathsf{WCasc}^f : (\{0,1\}^c \times \{0,1\}^b) \times \{0,1\}^{b*} \to \{0,1\}^c$ the whitened cascade construction built from f as

$$\mathsf{WCasc}^f((K_1, K_w), m_1\|\dots\|m_\ell) := y_\ell$$
$$\text{where } y_0 := K_1 \text{ and for } i \geq 1 : y_i := f(y_{i-1}, m_i \oplus K_w),$$

in particular $\mathsf{WCasc}^f((K_1, K_w), \varepsilon) := K_1$.

The construction WNMAC is derived from NMAC, the only difference being that the inner cascade Casc is replaced by the whitened cascade WCasc. More precisely,

$$\mathsf{WNMAC}^f((K_1, K_2, K_w), M) := f(K_2, \mathsf{WCasc}^f((K_1, K_w), M)\|0^{b-c}).$$

For a graphical depiction of WNMAC, see Fig. 1. We devote most of this section to the proof of the following theorem that quantifies the PRF-security of WNMAC.

Theorem 1 (PRF-Security of WNMAC). *Let* A *be an adversary making at most q_f queries to the compression function f and at most q_C construction queries, each of length at most ℓ b-bit blocks. Let $K = (K_1, K_2, K_w) \in \{0,1\}^c \times \{0,1\}^c \times \{0,1\}^b$ be a tuple of random keys. Then we have*

$$\mathsf{Adv}^{\mathsf{prf}}_{\mathsf{WNMAC}^f_K}(\mathsf{A}) \leq \frac{q_f q_C}{2^{2c}} + 2 \cdot \frac{\ell q_C q_f}{2^{b+c}} + \frac{\ell q_C{}^2}{2^c} \cdot \left(d'(\ell) + \frac{64\ell^3}{2^c} + 1 \right). \tag{2}$$

Note that as observed in Sect. 2, this also covers the so-called distinguishing-R security of WNMAC. Moreover, our analysis also implies security bounds for distinguishing-H and state recovery, as we discuss later.

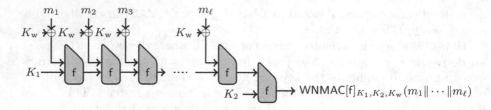

Fig. 1. The construction $\mathsf{WNMAC}[f]_{K_1,K_2,K_w}$.

3.1   Basic Notation, Message Trees and Repetition Patterns

Let us fix an adversary A. We assume that A is deterministic, it makes *exactly* q_f queries to f and q_C construction queries, and it never repeats the same query twice. All these assumptions are without loss of generality for an information-theoretic indistinguishability analysis, since an arbitrary (possibly randomized) adversary making at most this many queries can be transformed into one satisfying the above constraints and achieving advantage which is at least as large.

Let $\mathcal{Q}_C \subseteq (\{0,1\}^b)^*$ be any non-empty set of messages (later this will represent the set of A's C-queries). Based on it, we now introduce the *message tree* and its labeled version, which capture the inherent combinatorial structure of the messages $\mathcal{Q}_C$, as well as the internal values computed while these messages are processed by WCasc^f inside of WNMAC^f. The message tree $T(\mathcal{Q}_C) = (V, E)$ for $\mathcal{Q}_C$ is defined as follows:

- The vertex set is $V := \big\{ M' \in (\{0,1\}^b)^* : \exists M \in \mathcal{Q}_C : M' \mid M \big\}$, where is the prefix-of partial ordering of strings. In particular, note that the empty string ε is a vertex and that $\mathcal{Q}_C \subseteq V$.
- The set $E \subseteq V \times V$ of (directed) edges is

$$E := \big\{ (M, M') : \exists m \in \{0,1\}^b : M' = M \parallel m \big\}.$$

To simplify our exposition, we also define the following two mappings based on $T(\mathcal{Q}_C)$.

- The mapping $\pi(v) \colon V \backslash \{\varepsilon\} \to V$ returns the unique parent node of $v \in V \backslash \{\varepsilon\}$; i.e., the unique node u such that $(u, v) \in E$.
- The mapping $\mu(v) \colon V \setminus \{\varepsilon\} \to \{0,1\}^b$ returns the unique message block $m \in \{0,1\}^b$ such that $\pi(v) \parallel \mu(v) = v$ (intuitively, this will be the message block that is processed when "arriving" in vertex v).

Alternatively, with a slight abuse of notation we will also refer to the vertices in V as $v_1, \ldots, v_{|V|}$ which is an arbitrary ordering of them such that for all $1 \leq i, j \leq |V|$ it satisfies $v_i | v_j \Rightarrow i \leq j$. Note that one obtains such an ordering for example if one, intuitively speaking, processes the messages in $\mathcal{Q}_C$ block-wise and labels the vertices by their "first appearance": in particular $v_1 = \varepsilon$ is the tree root.

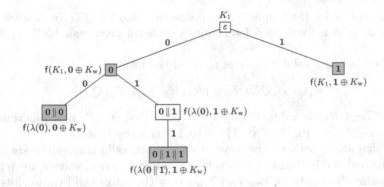

Fig. 2. Labeled message tree. Example of a labeled message tree $T_K^f(\mathcal{Q}_C)$ for four messages $\mathcal{Q}_C = \{0, 0 \,\|\, 0, 0 \,\|\, 1 \,\|\, 1, 1\}$, where $r = r^b$ for $r \in \{0, 1\}$. The gray vertices correspond to these four messages. Next to each vertex v and edge (u, v), we give the label $\lambda(v)$ and the value $\mu(v)$, respectively.

Additionally, for a mapping $f \colon \{0, 1\}^c \times \{0, 1\}^b \rightarrow \{0, 1\}^c$ and a key tuple $K = (K_1, K_2, K_w) \in \{0, 1\}^c \times \{0, 1\}^c \times \{0, 1\}^b$ we also consider an extended version of $T(\mathcal{Q}_C)$ which we call the *labeled message tree* and denote $T_K^f(\mathcal{Q}_C) = (V, E, \lambda)$, and which is defined as follows:

- The set of vertices V and edges E are defined exactly as for $T(\mathcal{Q}_C)$ above.
- The vertex-labeling function $\lambda \colon V \rightarrow \{0, 1\}^c$ is defined iteratively: $\lambda(\varepsilon) := K_1$ and for each non-root vertex $v \in V \setminus \{\varepsilon\}$ we put $\lambda(v) := f(\lambda(\pi(v)), \mu(v) \oplus K_w)$.

An example of a labeled message tree is given in Fig. 2. Note that each vertex label $\lambda(v)$ is exactly the output of the inner, whitened cascade $\mathsf{WCasc}_{K_1, K_w}^f(v)$ in WNMAC_K^f (recall that v is actually a message from $\{0, 1\}^{b*}$).

For any message tree $T(\mathcal{Q}_C) = (V, E)$, a *repetition pattern* is any equivalence relation ρ on V. For a labeled message tree $T_K^f(\mathcal{Q}_C) = (V, E, \lambda)$ we say that a repetition pattern ρ is *induced* by it if it satisfies

$$\forall u, v \in V : \lambda(u) = \lambda(v) \Leftrightarrow \rho(u, v).$$

3.2   Interactions and Transcripts

Let $\mathcal{QR}_C$ denote the set of q_C pairs (x, r) such that $x \in \{0, 1\}^{b*}$ is a construction query and $r \in \{0, 1\}^c$ is a potential response to it (what we mean by "potential" will be clear from below). Similarly let $\mathcal{QR}_f$ denote the set of q_f pairs (x, r) such that $x \in \{0, 1\}^c \times \{0, 1\}^b$ is an f-query and $r \in \{0, 1\}^c$ is a potential response to it. Let $\mathcal{Q}_C \subseteq \{0, 1\}^{b*}$ and $\mathcal{Q}_f \subseteq \{0, 1\}^c \times \{0, 1\}^b$ denote the sets of first coordinates (i.e., the queries) in $\mathcal{QR}_C$ and $\mathcal{QR}_f$, respectively; we have $|\mathcal{Q}_C| = q_C$ and $|\mathcal{Q}_f| = q_f$.

We call the pair of sets $(\mathcal{QR}_C, \mathcal{QR}_f)$ *valid* if the adversary A would indeed ask these queries throughout the experiment, assuming that each of her queries would be replied by the respective response in $\mathcal{QR}_C$ or $\mathcal{QR}_f$ (note that once a deterministic A is fixed, this determines whether a given pair $(\mathcal{QR}_C, \mathcal{QR}_f)$ is valid).

We then define a *valid transcript* to be of the form

$$\tau = \left(\mathcal{QR}_C, \mathcal{QR}_f, K = (K_1, K_2, K_w), T_K^f(\mathcal{Q}_C) \right),$$

where $(\mathcal{QR}_C, \mathcal{QR}_f)$ is valid, $f \colon \{0,1\}^c \times \{0,1\}^b \to \{0,1\}^c$ is a function and $K = (K_1, K_2, K_w) \in \{0,1\}^c \times \{0,1\}^c \times \{0,1\}^b$ is a key tuple.

We differentiate between the ways in which such valid transcripts are generated in the real and in the ideal worlds (or experiments), respectively, by defining corresponding distributions $\mathsf{T}_{\mathsf{real}}$ and $\mathsf{T}_{\mathsf{ideal}}$ over the set of valid transcripts:

Real World. The transcript $\mathsf{T}_{\mathsf{real}}$ for the adversary A is obtained by sampling $f \xleftarrow{\$} \mathcal{F}(c+b, c)$ and $K = (K_1, K_2, K_w) \leftarrow \{0,1\}^c \times \{0,1\}^c \times \{0,1\}^b$, and letting $\mathsf{T}_{\mathsf{real}}$ denote

$$\left(\mathcal{QR}_C = \{(M_i, Y_i)\}_{i=1}^{q_C}, \mathcal{QR}_f = \{(X_i, R_i)\}_{i=1}^{q_f}, K = (K_1, K_2, K_w), T_K^f(\mathcal{Q}_C) \right),$$

where we execute A, which asks construction queries $M_1, \ldots, M_{q_C}$ answered with $Y_i := \mathsf{WNMAC}[f]_K(M_i)$ for all $i \in [q_C]$; and f-queries $X_1, \ldots, X_{q_f}$ answered with $R_i := f(X_i)$ for all $i \in [q_f]$ (note that the C-queries and f-queries may in general be interleaved adaptively, depending on A). Finally, we let $T_K^f(\mathcal{Q}_C)$ be the labeled message tree corresponding to $\mathcal{Q}_C$, f and K.

Ideal World. The transcript $\mathsf{T}_{\mathsf{ideal}}$ for the adversary A is obtained similarly to the above, but here, together with the random function $f \xleftarrow{\$} \mathcal{F}(c+b, c)$ and the key tuple $K = (K_1, K_2, K_w) \leftarrow \{0,1\}^c \times \{0,1\}^c \times \{0,1\}^b$, we also sample q_C independent random values $Y_1, \ldots, Y_{q_C} \in \{0,1\}^r$. Then we let $\mathsf{T}_{\mathsf{ideal}}$ denote

$$\left(\mathcal{QR}_C = \{(M_i, Y_i)\}_{i=1}^{q_C}, \mathcal{QR}_f = \{(X_i, R_i)\}_{i=1}^{q_f}, K = (K_1, K_2, K_w), T_K^f(\mathcal{Q}_C) \right),$$

where we execute A, answer each its C-query M_i with Y_i for all $i \in [q_C]$ and each its f-query X_i with $R_i := f(X_i)$ for all $i \in [q_f]$. Then we let $T_K^f(\mathcal{Q}_C)$ be the labeled message tree corresponding to $\mathcal{Q}_C$, f and K.

Later we refer to the above two random experiments as real and ideal, respectively. Note that the range of $\mathsf{T}_{\mathsf{real}}$ is included in the range of $\mathsf{T}_{\mathsf{ideal}}$ by definition, and that the range of $\mathsf{T}_{\mathsf{ideal}}$ is easily seen to contain all valid transcripts.

3.3 The H-Coefficient Method

We upper-bound the advantage A in distinguishing $\mathsf{WNMAC}[f]_K$ for $f \xleftarrow{\$} \mathcal{F}(c+b, c)$ from a random function in terms of the statistical distance of the transcripts, i.e.,

$$\mathsf{Adv}_{\mathsf{WNMAC}}^{\mathsf{prf}}(\mathsf{A}) \leq \mathsf{SD}(\mathsf{T}_{\mathsf{real}}, \mathsf{T}_{\mathsf{ideal}}) = \frac{1}{2} \sum_\tau |\mathsf{P}\left[\mathsf{T}_{\mathsf{real}} = \tau\right] - \mathsf{P}\left[\mathsf{T}_{\mathsf{ideal}} = \tau\right]|, \quad (3)$$

where the sum is over all valid transcripts. This is because an adversary for T_{real} and T_{ideal}, whose optimal advantage is exactly $SD(T_{real}, T_{ideal})$, can always output the same decision bit as A, ignoring any extra information provided by the transcript.

We are going to use Patarin's H-coefficient method [18]. This means that we need to partition the set of valid transcripts into *good* transcripts GT and *bad* transcripts BT and then apply the following lemma.

Lemma 1 (The *H*-Coefficient Method [18]**).** *Let $\delta, \epsilon \in [0, 1]$ be such that:*

(a) $P[T_{ideal} \in BT] \leq \delta.$
(b) *For all* $\tau \in GT,$

$$\frac{P[T_{real} = \tau]}{P[T_{ideal} = \tau]} \geq 1 - \epsilon.$$

Then,
$$\mathsf{Adv}^{prf}_{WNMAC}(A) \leq SD(T_{real}, T_{ideal}) \leq \epsilon + \delta.$$

More verbally, we want a set of good transcripts GT such that with very high probability (i.e., $1 - \delta$) a generated transcript *in the ideal world* is going to be in this set, and moreover, for each such good transcript, the probabilities that it occurs in the real and in the ideal worlds are *roughly* the same, i.e., at most a multiplicative factor $1 - \epsilon$ apart.

3.4   Good and Bad Transcripts

Given a valid transcript τ we define the sets $\mathcal{L}_{in}, \mathcal{L}_{out} \subseteq \{0,1\}^c \times \{0,1\}^b$ as

$$\mathcal{L}_{in} := \{(\lambda(\pi(v)), \mu(v) \oplus K_w) : v \in V \setminus \{\varepsilon\}\}$$
$$\mathcal{L}_{out} := \{(K_2, \lambda(v) \| 0^{b-c}) : v \in \mathcal{Q}_C\},$$

and let $\mathcal{L} = \mathcal{L}_{in} \cup \mathcal{L}_{out}$. Intuitively, $\mathcal{L}$ represents the set of inputs on which f is evaluated while processing A's construction queries in the real experiment. This set is also well-defined in the ideal experiment by the above equations, and in both experiments it is determined by the transcript. We refer to $\mathcal{L}_{in}$ as the set of *inner* f-*invocations*, i.e., those invocations of f that were required to evaluate the inner, whitened cascade $WCasc^f$ in WNMAC; and similarly, $\mathcal{L}_{out}$ denotes the *outer invocations*.

If there is an intersection between the adversary's f-queries and the inputs in $\mathcal{L}_{in}$ (resp. $\mathcal{L}_{out}$), we call this an *inner (resp., outer) C-f-collision*. We then denote by C-f-coll$_{in}$ (resp., C-f-coll$_{out}$) the event that any inner (resp., outer) C-f-collision occurs. Formally,

$$\mathsf{C\text{-}f\text{-}coll}_{in} :\Leftrightarrow (\mathcal{Q}_f \cap \mathcal{L}_{in} \neq \emptyset) \quad \text{and} \quad \mathsf{C\text{-}f\text{-}coll}_{out} :\Leftrightarrow (\mathcal{Q}_f \cap \mathcal{L}_{out} \neq \emptyset)$$

and let C-f-coll := C-f-coll$_{in}$ $\cup$ C-f-coll$_{out}$. Furthermore, if the vertex labels $\lambda(M)$ collide for two messages $M, M' \in \mathcal{Q}_C$, we call this a C-collision and denote such an event by

$$\mathsf{C\text{-}coll} :\Leftrightarrow (\exists M, M' \in \mathcal{Q}_C : \lambda(M) = \lambda(M')).$$

Definition 1 (Good Transcripts). *Let*

$$\tau = \left(\mathcal{QR}_C, \mathcal{QR}_f, K = (K_1, K_2, K_w), T_K^f(\mathcal{Q}_C) = (V, E, \lambda)\right)$$

be a valid transcript. We say that the transcript is good *(and thus $\tau \in$ GT) if the following properties are true:*

(1) *The event* C-f-coll$_{out}$ *has not occurred.*
(2) *The event* C-coll *has not occurred.*
(3) *For any $v \in V$ we have $\lambda(v) \neq K_2$.*

We denote as GT the set of all good transcripts, and BT the set of all *bad* transcripts, i.e., transcripts which can possibly occur (i.e., they are in the range of T_{ideal}) and are not good. More specifically, we denote by BT_i the set of all bad transcripts that do not satisfy the i-th property in the definition of a good transcript above, hence we have $BT = \bigcup_{i=1}^{3} BT_i$.

3.5   Probability of a C-f-collision

In this section we upper-bound the probability of C-f-coll by considering inner and outer C-f-collisions separately.

Lemma 2. *We have* $\mathsf{P}^{ideal}[\text{C-f-coll}_{in}] \leq \ell q_C q_f / 2^{b+c}$.

Proof. We start by modifying the ideal experiment to obtain an experiment denoted ideal$'$ and the corresponding transcript distribution $T_{ideal'}$. The experiment ideal$'$ is given in Fig. 3. Clearly, ideal$'$ differs from the ideal experiment only in the way the vertex labeling function $\lambda(\cdot)$ is determined.

We now argue that $\mathsf{P}^{ideal}[\text{C-f-coll}_{in}] = \mathsf{P}^{ideal'}[\text{C-f-coll}_{in}]$. To see this, consider an intermediate experiment ideal$''$ that is defined exactly as ideal except that it uses a separate ideal compression function g to generate the vertex labels of the tree contained in the transcript, where g is completely independent of f queried by the adversary (i.e., the adversary queries f and the transcript contains $\mathcal{QR}_f$ and $T_K^g(\mathcal{Q}_C)$). It is now clear that $\mathsf{P}^{ideal}[\text{C-f-coll}_{in}] = \mathsf{P}^{ideal''}[\text{C-f-coll}_{in}]$ since as long as no inner C-f-collision happens, the experiments are identical.

The remaining equality $\mathsf{P}^{ideal''}[\text{C-f-coll}_{in}] = \mathsf{P}^{ideal'}[\text{C-f-coll}_{in}]$ follows from the definition of ideal$'$. It is easy to see that the distribution of vertex labels sampled in steps 2 and 3 of ideal$'$ and by labeling the tree $T_K^g(\mathcal{Q}_C)$ in ideal$''$ are the same. In both cases, repeated inputs to the compression function lead to consistent outputs, while fresh inputs lead to independent random outputs. The two experiments only differ in the order of sampling: ideal$''$ first samples g and then performs the labeling, while ideal$'$ starts by sampling the repetition pattern, and then chooses the actual labels correspondingly. The same distribution of vertex labels in these two experiments then implies the same probability of C-f-coll$_{in}$ occurring.

1. **The adversary asks its C-queries and f-queries and these are answered by independent random values.** Once the q_C queries in $\mathcal{Q}_C$ are fixed, they also determine the message tree $T(\mathcal{Q}_C)$ and mappings μ and π as defined in Section 3.1 (the labeling λ is so far undefined).

2. **Sample a repetition pattern ρ.** The equivalence relation ρ is determined indirectly by first iteratively defining a mapping $\hat{\rho} \colon V \to [|V|]$. Recall the vertex ordering $v_1, \ldots, v_{|V|}$ defined in Section 3.1. First, set $\hat{\rho}(v_1) := 1$. Then, for i taking values $2, \ldots, |V|$, determine $\hat{\rho}(v_i)$ as follows. If there exists $j \in [i-1]$ such that $\mu(v_j) = \mu(v_i)$ and $\hat{\rho}(\pi(v_j)) = \hat{\rho}(\pi(v_i))$ then let $\hat{\rho}(v_i) := \hat{\rho}(v_j)$ for the minimal such j. Otherwise let $z := \max_{j \in [i-1]}\{\hat{\rho}(v_j)\}$ and sample $\hat{\rho}(v_i)$ as

$$\hat{\rho}(v_i) := \begin{cases} 1 & \text{with probability } 2^{-c} \\ \vdots & \vdots \\ z & \text{with probability } 2^{-c} \\ z+1 & \text{with probability } 1 - z \cdot 2^{-c}. \end{cases}$$

Finally, for all $i, j \in [|V|]$ let $\rho(v_i, v_j) :\Leftrightarrow (\hat{\rho}(v_i) = \hat{\rho}(v_j))$.

3. **Sample a vertex labeling $\lambda(\cdot)$ according to ρ.** Namely, sample $|\rho|$ distinct uniformly random values $s_1, \ldots, s_{|\rho|} \in \{0,1\}^c$ where $|\rho|$ is the number of equivalence classes of ρ, and let $\lambda(v_i) := s_{\hat{\rho}(v_i)}$ for all $i \in [|V|]$. Also let $K_1 := \lambda(\varepsilon)$.

4. **Sample random keys $(K_2, K_w) \in \{0,1\}^c \times \{0,1\}^b$.**

Fig. 3. The random experiment ideal' for the proofs of Lemmas 2 and 3.

Finally, we upper-bound the probability $\mathsf{P}^{\mathsf{ideal}'}[\text{C-f-coll}_{\mathsf{in}}]$. Conditioned on the repetition pattern ρ taking some fixed value rp, in step 2, we have

$$\mathsf{P}^{\mathsf{ideal}'}[\text{C-f-coll}_{\mathsf{in}} \mid \rho = rp] \leq \sum_{v \in V \setminus \{\varepsilon\}} \mathsf{P}^{\mathsf{ideal}'}[(\lambda(\pi(v)), \mu(v) \oplus K_w) \in \mathcal{Q}_{\mathsf{f}} \mid \rho = rp]$$

$$= \sum_{v \in V \setminus \{\varepsilon\}} \mathsf{P}^{\mathsf{ideal}'}[(s_{\hat{\rho}(\pi(v))}, \mu(v) \oplus K_w) \in \mathcal{Q}_{\mathsf{f}} \mid \rho = rp]$$

$$= \sum_{v \in V \setminus \{\varepsilon\}} q_{\mathsf{f}}/2^{b+c} \leq \ell q_C q_{\mathsf{f}}/2^{b+c}$$

because the random variables s_i and K_w sampled in steps 3 and 4 are uniformly distributed and independent of $\mathcal{Q}_{\mathsf{f}}$. Since this bound holds conditioned on ρ being any fixed repetition pattern rp, it remains valid also without conditioning on it, hence concluding the proof. $\square$

We proceed by upper-bounding the probability of an outer C-f-collision.

Lemma 3. *We have*

$$\mathsf{P}^{\mathsf{ideal}}[\text{C-f-coll}_{\mathsf{out}}] \leq \frac{\ell q_C q_{\mathsf{f}}}{2^{b+c}} + \frac{q_C q_{\mathsf{f}}}{2^{2c}}.$$

Proof. Let us again consider the experiments ideal' and ideal'' defined in the proof of Lemma 2. We start by the simple observation that for any event A we have

$$\mathsf{P}^{\mathsf{ideal}}[A] = \mathsf{P}^{\mathsf{ideal}}[A \wedge \mathsf{C\text{-}f\text{-}coll_{in}}] + \mathsf{P}^{\mathsf{ideal}}[A \wedge \neg\mathsf{C\text{-}f\text{-}coll_{in}}]$$

$$\leq \frac{\ell q_C q_f}{2^{b+c}} + \mathsf{P}^{\mathsf{ideal}''}[A \wedge \neg\mathsf{C\text{-}f\text{-}coll_{in}}] \leq \frac{\ell q_C q_f}{2^{b+c}} + \mathsf{P}^{\mathsf{ideal}''}[A], \qquad (4)$$

which follows from Lemma 2 and the observation that ideal and ideal'' only differ if $\mathsf{C\text{-}f\text{-}coll_{in}}$ occurs.

Applying (4) to the event $\mathsf{C\text{-}f\text{-}coll_{out}}$ as A, it remains to bound the probability $\mathsf{P}^{\mathsf{ideal}''}[\mathsf{C\text{-}f\text{-}coll_{out}}]$; for this we observe that $\mathsf{P}^{\mathsf{ideal}''}[\mathsf{C\text{-}f\text{-}coll_{out}}] = \mathsf{P}^{\mathsf{ideal}'}[\mathsf{C\text{-}f\text{-}coll_{out}}]$ similarly as before: the repetition pattern ρ sampled in step 2 of ideal' has the same distribution as the repetition pattern induced by the tree $T_K^g(\mathcal{Q}_C)$ in ideal'', and this together with the sampling performed in step 3 results in the same distribution of vertex labels in ideal'' and ideal' and hence also in the same probability of $\mathsf{C\text{-}f\text{-}coll_{out}}$ in both experiments.

Finally, to upper-bound the probability $\mathsf{P}^{\mathsf{ideal}'}[\mathsf{C\text{-}f\text{-}coll_{out}}]$, again conditioned on the repetition pattern ρ sampled in step 2 taking some fixed value rp, we have

$$\mathsf{P}^{\mathsf{ideal}'}[\mathsf{C\text{-}f\text{-}coll_{out}} \mid \rho = rp] \leq \sum_{v \in \mathcal{Q}_C} \mathsf{P}^{\mathsf{ideal}'}\left[(K_2, \lambda(v) \,\|\, 0^{b-c}) \in \mathcal{Q}_f \mid \rho = rp\right]$$

$$\leq \sum_{v \in \mathcal{Q}_C} \mathsf{P}^{\mathsf{ideal}'}\left[(K_2, s_{\hat{\rho}(v)} \,\|\, 0^{b-c}) \in \mathcal{Q}_f \mid \rho = rp\right]$$

$$= \sum_{v \in \mathcal{Q}_C} q_f/2^{2c} \leq q_C q_f/2^{2c}$$

because the random variables s_i and K_2 sampled in steps 3 and 4 are uniformly distributed and independent of $\mathcal{Q}_f$. Since this bound holds conditioned on ρ being any fixed repetition pattern rp, it remains valid also without conditioning on it. $\qquad\square$

3.6   Probability of Repeated Outer Invocations

In this section we analyze the probability that any of the outer f-invocations in the ideal experiment will not be fresh, in particular we upper-bound both $\mathsf{P}[\mathsf{T_{ideal}} \in \mathsf{BT}_2]$ and $\mathsf{P}[\mathsf{T_{ideal}} \in \mathsf{BT}_3]$.

Lemma 4. *We have*

$$\mathsf{P}^{\mathsf{ideal}}[\mathsf{C\text{-}coll}] \leq \frac{\ell q_C q_f}{2^{b+c}} + \frac{\ell q_C^2}{2^c} \cdot \left(d'(\ell) + \frac{64\ell^3}{2^c}\right).$$

Proof. Applying (4) to the event C-coll, we have $\mathsf{P}^{\mathsf{ideal}}$ [C-coll] $\leq \ell q_C q_f / 2^{b+c} +$ $\mathsf{P}^{\mathsf{ideal}''}$ [C-coll]. Since the queries $\mathcal{Q}_C$ in the experiment ideal$''$ are chosen non-adaptively (with respect to the keys K_1, K_w and the function g used to later compute the tree labeling), we can obtain via a union bound that

$$\mathsf{P}^{\mathsf{ideal}''} \text{ [C-coll]} \leq q_C^2 \cdot \max_{\substack{M_1 \neq M_2 \\ |M_1|,|M_2| \leq \ell b}} \mathsf{P}^{\mathsf{g},K_1,K_w} \left[\mathsf{WCasc}^{\mathsf{g}}_{K_1,K_w}(M_1) = \mathsf{WCasc}^{\mathsf{g}}_{K_1,K_w}(M_2) \right].$$

Moreover, we have

$$\max_{\substack{M_1 \neq M_2 \\ |M_1|,|M_2| \leq \ell b}} \mathsf{P}^{\mathsf{g},K_1,K_w} \left[\mathsf{WCasc}^{\mathsf{g}}_{K_1,K_w}(M_1) = \mathsf{WCasc}^{\mathsf{g}}_{K_1,K_w}(M_2) \right]$$

$$= \max_{\substack{M_1 \neq M_2 \\ |M_1|,|M_2| \leq \ell b}} \sum_{\substack{K_1 \in \{0,1\}^c \\ K_w \in \{0,1\}^b}} \frac{1}{2^{c+b}} \cdot \mathsf{P}^{\mathsf{g}} \left[\mathsf{WCasc}^{\mathsf{g}}_{K_1,K_w}(M_1) = \mathsf{WCasc}^{\mathsf{g}}_{K_1,K_w}(M_2) \right]$$

$$\leq \sum_{\substack{K_1 \in \{0,1\}^c \\ K_w \in \{0,1\}^b}} \frac{1}{2^{c+b}} \cdot \max_{\substack{M_1 \neq M_2 \\ |M_1|,|M_2| \leq \ell b}} \mathsf{P}^{\mathsf{g}} \left[\mathsf{WCasc}^{\mathsf{g}}_{K_1,K_w}(M_1) = \mathsf{WCasc}^{\mathsf{g}}_{K_1,K_w}(M_2) \right]$$

$$= \sum_{\substack{K_1 \in \{0,1\}^c \\ K_w \in \{0,1\}^b}} \frac{1}{2^{c+b}} \cdot \max_{\substack{M_1 \neq M_2 \\ |M_1|,|M_2| \leq \ell b}} \mathsf{P}^{\mathsf{g}} \left[\mathsf{Casc}^{\mathsf{g}}_{K_1}(M_1 \oplus K_w) = \mathsf{Casc}^{\mathsf{g}}_{K_1}(M_2 \oplus K_w) \right]$$

$$= \underbrace{\sum_{\substack{K_1 \in \{0,1\}^c \\ K_w \in \{0,1\}^b}} \frac{1}{2^{c+b}} \cdot \max_{\substack{M_1 \neq M_2 \\ |M_1|,|M_2| \leq \ell b}} \mathsf{P}^{\mathsf{g}} \left[\mathsf{Casc}^{\mathsf{g}}_{K_1}(M_1) = \mathsf{Casc}^{\mathsf{g}}_{K_1}(M_2) \right]}_{\mathsf{CascColl}(\ell)},$$

where the notation $M_i \oplus K_w$ denotes XOR-ing the key K_w to each of the blocks of M_i.

The last maximization term above was already studied in the context of the construction NI2 in [7], where it was denoted as $\mathsf{CColl}(\ell)$, but we will refer to it as $\mathsf{CascColl}(\ell)$ to avoid confusion with the event C-coll considered here. It was shown in [7] that

$$\mathsf{CascColl}(\ell) \leq \frac{\ell \cdot d'(\ell)}{2^c} + \frac{64\ell^4}{2^{2c}}. \tag{5}$$

Putting all the above bounds together concludes the proof of Lemma 4.   □

Lemma 5. *We have*

$$\mathsf{P}^{\mathsf{ideal}} [\exists v \in V \colon \lambda(v) = K_2] \leq \frac{\ell q_C}{2^c}.$$

Proof. As is clear from the description of the ideal experiment, the key K_2 is chosen uniformly at random and independently of the rest of the experiment, in particular of the labels $\lambda(v)$. The lemma hence follows by a simple union bound over all ℓq_C vertices $v \in V$.   □

3.7   Good Transcripts and Putting Pieces Together

Let us consider a good transcript τ. First, since $\tau \notin \mathsf{BT}_1$, there is no overlap between the outer f-invocations and the f-queries issued by the adversary. Second, since $\tau \notin \mathsf{BT}_2$, there is also no repetition between the outer f-invocations themselves. Finally, since $\tau \notin \mathsf{BT}_3$, there is also no overlap between the outer f-invocations and the inner f-invocations (all the outer invocations contain K_2 as their first component). Altogether, this means that each outer f-invocation in real is fresh and hence its outcome can be seen as freshly uniformly sampled (since f is an ideal random function). Therefore, the distribution of these outcomes will be the same as in ideal, where they correspond to the independent random values Y_i. Hence, for all $\tau \in \mathsf{GT}$, we have

$$\frac{\mathsf{P}\left[\mathsf{T}_{\mathsf{real}} = \tau\right]}{\mathsf{P}\left[\mathsf{T}_{\mathsf{ideal}} = \tau\right]} = 1.$$

Plugging this into Lemma 1, together with the bounds from Lemmas 3, 4 and 5, we obtain

$$\mathsf{Adv}^{\mathsf{prf}}_{\mathsf{WNMAC}}(\mathsf{A}) \le \sum_{i=1}^{3} \mathsf{P}\left[\mathsf{T}_{\mathsf{ideal}} \in \mathsf{BT}_i\right]$$

$$\le \frac{q_{\mathsf{f}}q_{\mathsf{C}}}{2^{2c}} + 2 \cdot \frac{\ell q_{\mathsf{C}}q_{\mathsf{f}}}{2^{b+c}} + \frac{\ell q_{\mathsf{C}}^2}{2^c} \cdot \left(d'(\ell) + \frac{64\ell^3}{2^c}\right) + \frac{\ell q_{\mathsf{C}}}{2^c}$$

$$\le \frac{q_{\mathsf{f}}q_{\mathsf{C}}}{2^{2c}} + 2 \cdot \frac{\ell q_{\mathsf{C}}q_{\mathsf{f}}}{2^{b+c}} + \frac{\ell q_{\mathsf{C}}^2}{2^c} \cdot \left(d'(\ell) + \frac{64\ell^3}{2^c} + 1\right),$$

which concludes the proof of Theorem 1.   □

3.8   Tightness

We now argue that the $q_{\mathsf{C}}q_{\mathsf{f}}/2^{2c}$ term in our bound on the security of WNMAC as given in (2) is tight, by giving a matching attack (up to a linear factor $O(c)$). For most practical parameters, this will be the dominating term in (2), and thus for those parameters Theorem 1 gives a tight bound. Here we only describe an attack for the case where $q_{\mathsf{C}} = \Theta(c)$ is very small, and defer the general case to the full version.

The $q_{\mathsf{C}} = \Theta(c)$ Case. We must define an adversary $\mathsf{A}^{\mathcal{O},\mathsf{f}}$ who can distinguish the case where the first oracle $\mathcal{O}$ implements a random function R from the case where it implements $\mathsf{WNMAC}^{\mathsf{f}}((K_1, K_2, K_{\mathsf{w}}), \cdot)$ with random keys K_1, K_2, K_{w} using the random function $\mathsf{f} : \{0,1\}^{b+c} \to \{0,1\}^c$ which is given as the second oracle.

$\mathsf{A}^{\mathcal{O},\mathsf{f}}$ first picks $t := q_{\mathsf{f}}/2^c$ keys $\widetilde{K}_1, \ldots, \widetilde{K}_t$ arbitrarily, and then uses its q_{f} function queries to learn the outputs

$$\mathcal{Z}_i = \{\mathsf{f}(\widetilde{K}_i, x \| 0^{b-c}) \; : \; x \in \{0,1\}^c\}$$

for all the keys. When throwing 2^c balls randomly into 2^c bins, we expect a $1 - 1/e \approx 0.63$ fraction of the bins to be non-empty (and the value is strongly concentrated around this expectation). We can think of evaluating the random function $f(\widetilde{K}_i, \cdot \| 0^{b-c}) : \{0,1\}^c \to \{0,1\}^c$ as throwing 2^c balls (the inputs) to random bins (the outputs), and thus have $|\mathcal{Z}_i| \approx 0.63 \cdot 2^c$. Then $A^{\mathcal{O},f}$ queries $\mathcal{O}$ on $\Theta(c)$ random inputs, let $\mathcal{Q}_c$ denote the corresponding outputs. Now $A^{\mathcal{O},f}$ outputs 1 if and only if for some i we have $\mathcal{Q}_c \subset \mathcal{Z}_i$. If $\mathcal{O}(\cdot) = \mathsf{WNMAC}^f((K_1, K_2, K_w), \cdot) = f(K_2, \mathsf{WCasc}^f((K_1, K_w), \cdot) \| 0^{b-c})$ and moreover $K_2 = \widetilde{K}_i$ for some i – which happens with probability $t/2^c$ – then all the outputs of $\mathcal{O}(\cdot)$ are in the range of $f(\widetilde{K}_i, \cdot \| 0^{b-c})$ and thus $A^{\mathcal{O},f}$ outputs 1.

On the other hand, if $\mathcal{O}(\cdot)$ is a random function, then every single query will miss the set $\mathcal{Z}_i$ with constant probability 0.37. Using this, we get by a Chernoff bound (and the union bound over all t keys) that

$$P[\exists i : \mathcal{Q}_c \subset \mathcal{Z}_i] \le \frac{t}{2^{\Theta(q_c)}} .$$

Summing up we get for $q_C = \Theta(c)$ and $t = q_f/2^c$

$$\mathsf{Adv}^{\mathsf{prf}}_{\mathsf{WNMAC}}(A_{q_C,t}) \ge \left| \frac{t}{2^c} - \frac{t}{2^{\Theta(q_C)}} \right| \ge \frac{t}{2^{c-1}} \ge \frac{q_f}{2^{2c-1}} = \frac{q_f q_C}{2^{2c} \cdot \Theta(c)}$$

which matches our term $q_f q_C/2^{2c}$ from the lower bound up to a $\Theta(c)$ factor.

3.9   Distinguishing-H Security of WNMAC

The above results also imply a bound on the distinguishing-H security of WNMAC. To capture this, we first introduce the notion of distinguishing-C, which corresponds to PRF-security with the restriction that the distinguisher only uses construction queries.

Definition 2 (Distinguishing-C). *Let* $C[f]: \{0,1\}^\kappa \times \mathcal{D} \to \mathcal{R}$ *be a keyed construction making queries to a randomly chosen compression function* $f \xleftarrow{\$} \mathcal{F}(c+b, c)$. *The* distinguishing-C *advantage of an adversary* A *is defined as*

$$\mathsf{Adv}^{\mathsf{dist\text{-}C}}_{C[f]}(A) := \left| P\left[K \xleftarrow{\$} \{0,1\}^\kappa, f \xleftarrow{\$} \mathcal{F}(c+b, c) : A^{C^f_K} \Rightarrow 1 \right] \right.$$
$$\left. - P\left[R \xleftarrow{\$} \mathcal{F}(\mathcal{D}, \mathcal{R}) : A^R \Rightarrow 1 \right] \right|.$$

The notion of distinguishing-C is useful for bridging distinguishing-H and PRF-security, as the following lemma shows (we omit its simple proof).

Lemma 6. *For every adversary* A *asking* q_C *and* q_f *construction and primitive queries, respectively, there exists an adversary* A' *asking* q_C *queries to its single oracle such that*

$$\mathsf{Adv}^{\mathsf{dist\text{-}H}}_C(A) \le \mathsf{Adv}^{\mathsf{prf}}_{C[f]}(A) + \mathsf{Adv}^{\mathsf{dist\text{-}C}}_{C[f]}(A')$$

and

$$\mathsf{Adv}^{\mathsf{prf}}_{\mathsf{C[f]}}(\mathsf{A}) \leq \mathsf{Adv}^{\mathsf{dist\text{-}H}}_{\mathsf{C}}(\mathsf{A}) + \mathsf{Adv}^{\mathsf{dist\text{-}C}}_{\mathsf{C[f]}}(\mathsf{A}').$$

One can readily obtain a bound on the distinguishing-C security of WNMAC using Theorem 1 with $q_f = 0$.

Lemma 7 (Distinguishing-C Security of WNMAC). *Let A be an adversary making at most q_C construction queries, each of length at most ℓ b-bit blocks. Let $K = (K_1, K_2, K_w) \in \{0,1\}^c \times \{0,1\}^c \times \{0,1\}^b$ be a tuple of random keys. Then we have*

$$\mathsf{Adv}^{\mathsf{dist\text{-}C}}_{\mathsf{WNMAC}_K}(\mathsf{A}) \leq \frac{\ell q_C{}^2}{2^c} \cdot \left(d'(\ell) + \frac{64\ell^3}{2^c} + 1 \right).$$

By combining Theorem 1 and Lemmas 6 and 7, we get the following theorem.

Theorem 2 (Distinguishing-H Security of WNMAC). *Let A be an adversary making at most q_f queries to the compression function and at most q_C construction queries, each of length at most ℓ b-bit blocks. Let $K = (K_1, K_2, K_w) \in \{0,1\}^c \times \{0,1\}^c \times \{0,1\}^b$ be a tuple of random keys. Then we have*

$$\mathsf{Adv}^{\mathsf{dist\text{-}H}}_{\mathsf{WNMAC}_K}(\mathsf{A}) \leq \frac{q_f q_C}{2^{2c}} + 2 \cdot \frac{\ell q_C q_f}{2^{b+c}} + 2 \cdot \frac{\ell q_C{}^2}{2^c} \cdot \left(d'(\ell) + \frac{64\ell^3}{2^c} + 1 \right).$$

3.10 State Recovery for WNMAC

We now formally define the notion of security against state recovery for WNMAC. We consider the strong notion where the goal of the adversary is to output a pair (M, s) such that the state s occurs *at any point* during the evaluation of WCasc on M. Formally, we define $\mathsf{Adv}^{\mathsf{sr}}_{\mathsf{WNMAC[f]}}(\mathsf{A})$ to be

$$\mathsf{P}\left[K \xleftarrow{\$} \mathcal{K}, \mathsf{f} \xleftarrow{\$} \mathcal{F}, \mathsf{A}^{\mathsf{WNMAC}^{\mathsf{f}}_K, \mathsf{f}} \Rightarrow (M, s) : \right.$$

$$\left. \exists M' \in \{0,1\}^{b*} \text{ s.t. } M' | M \wedge \mathsf{WCasc}^{\mathsf{f}}_{K_1, K_w}(M') = s \right]$$

where $\mathcal{K} = \{0,1\}^c \times \{0,1\}^c \times \{0,1\}^b$, $K = (K_1, K_2, K_w)$ and $\mathcal{F} := \mathcal{F}(c + b, c)$.

Theorem 3 (State-Recovery Security of WNMAC). *Let A be an adversary making at most q_f queries to the compression function and at most q_C construction queries, each of length at most ℓ b-bit blocks. Let $K = (K_1, K_2, K_w) \in \{0,1\}^c \times \{0,1\}^c \times \{0,1\}^b$ be a tuple of random keys. Then we have*

$$\mathsf{Adv}^{\mathsf{sr}}_{\mathsf{WNMAC}^{\mathsf{f}}_K}(\mathsf{A}) \leq \frac{q_f q_C}{2^{2c}} + 2 \cdot \frac{\ell q_C q_f}{2^{b+c}} + 2 \cdot \frac{\ell q_C{}^2}{2^c} \cdot \left(d'(\ell) + \frac{64\ell^3}{2^c} + 2 \right).$$

Proof (sketch). First, we replace the compression function oracle f by an independent random function g completely unrelated to WNMAC^f. The error introduced by this is upper-bounded by Theorem 2 and now, compression-function queries are useless to the adversary, hence we can disregard them.

Let us denote by $\mathcal{E}$ the experiment where A interacts with WNMAC^f (without direct access to f). Consider an alternative experiment $\mathcal{E}'$ given in Fig. 4. As long as the key K_2 chosen in step 4 does not hit any of the internal states that occurred during the query evaluation, the experiment $\mathcal{E}'$ is identical to $\mathcal{E}$. Moreover, since K_2 is chosen independently at random, such a hit can only occur with probability at most $\ell q_C/2^c$. Since the vertex labels are only sampled after the adversary makes its guess for the state, the probability that the guess will be correct is at most $\ell/2^c$. □

---

1. **The adversary asks its C-queries.** For each of them, only the repetition pattern for the state values belonging to this query is sampled (as in the experiment ideal' in Figure 3) and the query is answered with a fresh random value, unless the outer f-invocation happens on a repeated value, in which case the query is answered consistently. After answering all queries, we have a complete repetition pattern ρ for all state values.
2. **Let A output its guess (M, s).**
3. **Sample a vertex labeling $\lambda(\cdot)$ according to ρ, let $K_1 := \lambda(\varepsilon)$.**
4. **Sample random keys $(K_2, K_w) \in \{0,1\}^c \times \{0,1\}^b$.**

Fig. 4. The random experiment $\mathcal{E}'$ for the proof of Theorem 3.

4 Whitening HMAC

HMAC is a "practice-oriented" variant of NMAC, see Sect. 2 for its definition. In this section we consider a "whitened" variant WHMAC of HMAC which is derived from HMAC in the same way as WNMAC was derived from NMAC, i.e., by XORing a random key K_w to every message block. We also consider a variant WHMAC$^+$ where the first message block is a fresh key $K^+ \in \{0,1\}^b$. More precisely,

$$\mathsf{WHMAC}_{K,K_w}[f](m) := f\left(K_2', \mathsf{WCasc}^f_{K_1',K_w}(m)\|\mathsf{fpad}\right)$$

where

$$K_1' := f(\mathsf{IV}, K \oplus \mathsf{ipad}) \quad \text{and} \quad K_2' := f(\mathsf{IV}, K \oplus \mathsf{opad}) \tag{6}$$

and fpad is some fixed padding; and

$$\mathsf{WHMAC}^+_{K,K_w,K^+}[f](m) := f\left(K_2', \mathsf{WCasc}^f_{K_1',K_w}(m)\|\mathsf{fpad}\right),$$

where this time

$$Z := f(\mathsf{IV}, K \oplus \mathsf{ipad}) \quad \text{and} \quad K_1' := f(Z, K^+) \quad \text{and} \quad K_2' := f(\mathsf{IV}, K \oplus \mathsf{opad})$$

and fpad is again some padding. Note that both variants, WHMAC and WHMAC$^+$, can be implemented given just black-box access to an implementation of HMAC.

The theorem below relates the security of WHMAC and WHMAC$^+$ to the security of WNMAC.

Theorem 4 (Relating Security of WHMAC to WNMAC). *Consider any* xxx $\in \{prf, dist\text{-}H, sr\}$. *Assume that for every adversary* A *making at most* q_f *queries to the compression function* f *and at most* q_C *construction queries, each of length at most* ℓ *b-bit blocks, we have*

$$\mathsf{Adv}^{\mathsf{xxx}}_{\mathsf{WNMAC}_{K_1, K_2, K_w}[f]}(A) \le \epsilon,$$

where here and below, $K_1, K_2 \in \{0,1\}^c$ *and* $K, K_w, K^+ \in \{0,1\}^b$ *are uniformly random keys. Then for every such adversary* A *we have*

$$\mathsf{Adv}^{\mathsf{xxx}}_{\mathsf{WHMAC}_{K, K_w}[f]}(A) \le \epsilon + 2^{-\frac{b-2c}{2}} \tag{7}$$

and

$$\mathsf{Adv}^{\mathsf{xxx}}_{\mathsf{WHMAC}^+_{K, K_w, K^+}[f]}(A) \le \epsilon + 2 \cdot 2^{-\frac{b-c}{2}} + 2^{-c}. \tag{8}$$

Proof. Intuitively, for WHMAC one can think of f as an extractor which extracts keys K_1', K_2' from K, and the bound then readily follows by the leftover hash lemma. For WNMAC$^+$ one can roughly think of K_1' and K_2' as being extracted from independent keys K^+ and K, respectively. For the latter it is thus sufficient that b (which is the length, and thus also the entropy of the uniform K and K^+) is sufficiently larger than c (the length of K_1', K_2'), whereas for the former we need b to be sufficiently larger than $2c$. We now give the details of the proof for WHMAC and postpone the treatment of WNMAC$^+$ to the full version.

In order to prove the bound (7) it is sufficient to show that the statistical distance between the transcripts (as seen by the adversary) when interacting with WNMAC or WHMAC is at most $2^{-\frac{b-2c}{2}}$. As the only difference between WNMAC and WHMAC is that we replace the uniform keys K_1, K_2 with keys K_1', K_2' derived according to (6), to bound the distance between the transcripts, it is sufficient to bound the distance between the random and derived keys. As K_1', K_2' are not independent of f, it is important to bound the distance when given f, concretely, we must show that

$$\mathsf{SD}\left((K_1', K_2', f), (K_1, K_2, f)\right) \le 2^{-\frac{b-2c}{2}}.$$

We will use the leftover hash lemma [12] which states that for any random variable $X \in \{0,1\}^m$ with min-entropy at least $H_\infty(X) \ge k$ and a hash function

$h : \{0,1\}^m \rightarrow \{0,1\}^\ell$ chosen from a family of pairwise independent hash functions we have (with U_ℓ being uniform over $\{0,1\}^\ell$)

$$\mathsf{SD}\left((h(X), h), (U_\ell, h)\right) \leq 2^{\frac{\ell - H_\infty(X)}{2}} \leq 2^{\frac{\ell - k}{2}}.$$

Since $\mathsf{f} : \{0,1\}^{b+c} \rightarrow \{0,1\}^c$ is uniformly random, also the function

$$\mathsf{f}'(K) = (\mathsf{f}(\mathsf{IV}, K \oplus \mathsf{ipad}), \mathsf{f}(\mathsf{IV}, K \oplus \mathsf{opad}))$$

is uniformly random, and thus also pairwise independent. Using $H_\infty(K) = H_\infty(K \oplus \mathsf{ipad}) = b$ and $(K_1', K_2') = \mathsf{f}'(K)$ we thus get

$$\mathsf{SD}\left((K_1', K_2', \mathsf{f}'), (K_1, K_2, \mathsf{f}')\right) = \mathsf{SD}\left((K_1', K_2', \mathsf{f}), (K_1, K_2, \mathsf{f})\right) \leq 2^{-\frac{b-2c}{2}}$$

as required. The first equality above holds as f defines all of f' and vice versa. $\square$

5   The Dual WNMAC Construction

Looking at the security bounds for WNMAC given in Sect. 3 from a distance, it seems that under reasonable assumptions the most restrictive term in the bounds is $q_\mathsf{f}q_\mathsf{C}/2^{2c}$. Intuitively speaking, the reason for this term is the outer f-call in WNMAC that only takes $2c$ bits of actual inputs and adds $b - c$ padding zeroes.

In an attempt to overcome this limitation, we propose a variant of the WNMAC construction that we call *Dual WNMAC* (DWNMAC). We prove the PRF-security of DWNMAC that goes beyond the restrictive term $q_\mathsf{f}q_\mathsf{C}/2^{2c}$ and our proof again extends also to distinguishing-H and state-recovery security. The price we pay for this improvement is a slight increase in the key length and the fact that DWNMAC cannot be implemented using only black-box access to NMAC. Similarly, if we apply the same modification to WHMAC, the resulting construction can no longer be implemented using black-box access to HMAC.

The construction DWNMAC is derived from WNMAC, the only difference being that the outer f-call is performed on the c-bit state and a b-bit key K_2. More precisely, for a key tuple $(K_1, K_2, K_\mathsf{w}) \in \{0,1\}^c \times \{0,1\}^b \times \{0,1\}^b$ and a message $M \in \{0,1\}^{b*}$, we define

$$\mathsf{DWNMAC}^\mathsf{f}((K_1, K_2, K_\mathsf{w}), M) := \mathsf{f}(\mathsf{WCasc}^\mathsf{f}_{K_1, K_\mathsf{w}}(M), K_2).$$

Note that DWNMAC is slightly similar to what we would obtain by whitening from the Sandwich MAC construction [23].

We now summarize the security of DWNMAC.

Theorem 5. (Security of DWNMAC). *Let* A *be an adversary making at most* q_f *queries to the compression function* f *and at most* q_C *construction queries, each of length at most* ℓ b-*bit blocks. Let* $K = (K_1, K_2, K_\mathsf{w}) \in \{0,1\}^c \times \{0,1\}^b \times \{0,1\}^b$ *be a tuple of random keys. Then we have*

$$\mathsf{Adv}^{\mathsf{xxx}}_{\mathsf{DWNMAC}^\mathsf{f}_K}(\mathsf{A}) \leq 3 \cdot \frac{\ell q_\mathsf{C} q_\mathsf{f}}{2^{b+c}} + 2 \cdot \frac{\ell q_\mathsf{C}{}^2}{2^c} \cdot \left(d'(\ell) + \frac{64\ell^3}{2^c} + 2\right)$$

for all $\mathsf{xxx} \in \{\mathsf{prf}, \mathsf{dist\text{-}H}, \mathsf{sr}\}$.

Proof (sketch). The proofs are analogous to the proofs for WNMAC given in Sect. 3, with the main modification needed in Lemma 3 where the probability of an outer C-f-collision can be upper-bounded by $q_C q_f / 2^{b+c}$. Roughly speaking, this is because the outer call in DWNMAC does not contain the 0^{b-c} padding and instead processes $b + c$ bits of input that are hard to predict for the attacker. □

Acknowledgments. We thank the anonymous reviewers for their helpful comments. Gaži and Pietrzak's work was partly funded by the European Research Council under an ERC Starting Grant (259668-PSPC). Tessaro's research was partially supported by NSF grant CNS-1423566 and by the Glen and Susanne Culler Chair.

References

1. An, J.H., Bellare, M.: Constructing VIL-MACs from FIL-MACs: message authentication under weakened assumptions. In: Wiener, M. (ed.) CRYPTO 1999. LNCS, vol. 1666, pp. 252–269. Springer, Heidelberg (1999)
2. Bellare, M.: New proofs for NMAC and HMAC: security without collision-resistance. In: Dwork, C. (ed.) CRYPTO 2006. LNCS, vol. 4117, pp. 602–619. Springer, Heidelberg (2006)
3. Bellare, M., Canetti, R., Krawczyk, H.: Keying hash functions for message authentication. In: Koblitz, N. (ed.) CRYPTO 1996. LNCS, vol. 1109, pp. 1–15. Springer, Heidelberg (1996)
4. Damgård, I.B.: A design principle for hash functions. In: Brassard, G. (ed.) CRYPTO 1989. LNCS, vol. 435, pp. 416–427. Springer, Heidelberg (1990)
5. Dinur, I., Leurent, G.: Improved generic attacks against hash-based MACs and HAIFA. In: Garay, J.A., Gennaro, R. (eds.) CRYPTO 2014, Part I. LNCS, vol. 8616, pp. 149–168. Springer, Heidelberg (2014)
6. Dodis, Y., Ristenpart, T., Steinberger, J., Tessaro, S.: To hash or not to hash again? (In)differentiability results for H^2 and HMAC. In: Safavi-Naini, R., Canetti, R. (eds.) CRYPTO 2012. LNCS, vol. 7417, pp. 348–366. Springer, Heidelberg (2012)
7. Gaži, P., Pietrzak, K., Rybár, M.: The exact PRF-security of NMAC and HMAC. In: Garay, J.A., Gennaro, R. (eds.) CRYPTO 2014, Part I. LNCS, vol. 8616, pp. 113–130. Springer, Heidelberg (2014)
8. Gaži, P., Pietrzak, K., Tessaro, S.: The exact PRF security of truncation: tight bounds for keyed sponges and truncated CBC. In: Gennaro, R., Robshaw, M.J.B. (eds.) CRYPTO 2015, Part I. LNCS, vol. 9215, pp. 368–387. Springer, Heidelberg (2015)
9. Goldreich, O., Goldwasser, S., Micali, S.: On the cryptographic applications of random functions. In: Blakely, G.R., Chaum, D. (eds.) CRYPTO 1984. LNCS, vol. 196, pp. 276–288. Springer, Heidelberg (1985)
10. Guo, J., Peyrin, T., Sasaki, Y., Wang, L.: Updates on generic attacks against HMAC and NMAC. In: Garay, J.A., Gennaro, R. (eds.) CRYPTO 2014, Part I. LNCS, vol. 8616, pp. 131–148. Springer, Heidelberg (2014)
11. Hardy, G.H., Wright, E.M.: An Introduction to the Theory of Numbers, 6th edn. Oxford University Press, Oxford (2008)
12. Håstad, J., Impagliazzo, R., Levin, L.A., Luby, M.: A pseudorandom generator from any one-way function. SIAM J. Comput. **28**(4), 1364–1396 (1999)

13. Kim, J.-S., Biryukov, A., Preneel, B., Hong, S.H.: On the security of HMAC and NMAC based on HAVAL, MD4, MD5, SHA-0 and SHA-1 (Extended abstract). In: De Prisco, R., Yung, M. (eds.) SCN 2006. LNCS, vol. 4116, pp. 242–256. Springer, Heidelberg (2006)

14. Krawczyk, H., Bellare, M., Canetti, R.: HMAC: keyed-hashing for message authentication. In: IETF Internet Request for Comments 2104, February 1997

15. Leurent, G., Peyrin, T., Wang, L.: New generic attacks against hash-based MACs. In: Sako, K., Sarkar, P. (eds.) ASIACRYPT 2013, Part II. LNCS, vol. 8270, pp. 1–20. Springer, Heidelberg (2013)

16. Merkle, R.C.: One way hash functions and DES. In: Brassard, G. (ed.) CRYPTO 1989. LNCS, vol. 435, pp. 428–446. Springer, Heidelberg (1990)

17. Naito, Y., Sasaki, Y., Wang, L., Yasuda, K.: Generic state-recovery and forgery attacks on ChopMD-MAC and on NMAC/HMAC. In: Sakiyama, K., Terada, M. (eds.) IWSEC 2013. LNCS, vol. 8231, pp. 83–98. Springer, Heidelberg (2013)

18. Patarin, J.: The "Coefficients H" technique. In: Avanzi, R.M., Keliher, L., Sica, F. (eds.) SAC 2008. LNCS, vol. 5381, pp. 328–345. Springer, Heidelberg (2009)

19. Peyrin, T., Sasaki, Y., Wang, L.: Generic related-key attacks for HMAC. In: Wang, X., Sako, K. (eds.) ASIACRYPT 2012. LNCS, vol. 7658, pp. 580–597. Springer, Heidelberg (2012)

20. Peyrin, T., Wang, L.: Generic universal forgery attack on iterative hash-based MACs. In: Nguyen, P.Q., Oswald, E. (eds.) EUROCRYPT 2014. LNCS, vol. 8441, pp. 147–164. Springer, Heidelberg (2014)

21. Preneel, B., van Oorschot, P.C.: MDx-MAC and building fast MACs from hash functions. In: Coppersmith, D. (ed.) CRYPTO 1995. LNCS, vol. 963, pp. 1–14. Springer, Heidelberg (1995)

22. Sasaki, Y., Wang, L.: Generic attacks on strengthened HMAC: n-bit secure HMAC requires key in all blocks. In: Abdalla, M., De Prisco, R. (eds.) SCN 2014. LNCS, vol. 8642, pp. 324–339. Springer, Heidelberg (2014)

23. Yasuda, K.: "Sandwich" Is indeed secure: how to authenticate a message with just one hashing. In: Pieprzyk, J., Ghodosi, H., Dawson, E. (eds.) ACISP 2007. LNCS, vol. 4586, pp. 355–369. Springer, Heidelberg (2007)

Symmetric Encryption

On the Optimality of Non-Linear Computations of Length-Preserving Encryption Schemes

Mridul Nandi[✉]

Indian Statistical Institute, Kolkata, India
mridul.nandi@gmail.com

Abstract. It is well known that three and four rounds of balanced Feistel cipher or Luby-Rackoff (LR) encryption for two blocks messages are pseudorandom permutation (PRP) and strong pseudorandom permutation (SPRP) respectively. A **block** is n-bit long for some positive integer n and a (possibly keyed) **block-function** is a nonlinear function mapping all blocks to themselves, e.g. blockcipher. XLS (eXtended Latin Square) encryption defined over two block inputs with three blockcipher calls was claimed to be SPRP. However, later Nandi showed that it is not a SPRP. Motivating with these observations, we consider the following questions in this paper: *What is the minimum number of invocations of block-functions required to achieve PRP or SPRP security over ℓ blocks inputs?* To answer this question, we consider all those length-preserving encryption schemes, called **linear encryption mode**, for which only nonlinear operations are block-functions. Here, we prove the following results for these encryption schemes:

1. At least 2ℓ (or $2\ell - 1$) invocations of block-functions are required to achieve SPRP (or PRP respectively). These bounds are also tight.
2. To achieve the above bound for PRP over $\ell > 1$ blocks, either we need at least two keys or it can not be *inverse-free* (i.e., need to apply the inverses of block-functions in the decryption). In particular, we show that a single-keyed inverse-free PRP needs 2ℓ invocations of block functions.
3. We show that 3-round LR using a single-keyed pseudorandom function (PRF) is PRP if we xor a block of input by a masking key.

Keywords: XLS · CMC · Luby-Rackoff · PRP · SPRP · Blockcipher

1 Introduction

BLOCK FUNCTION. For all symmetric key algorithms, domains (sometimes, also ranges) are desired to be sets of bit-strings of variable sizes. However, almost all known methodologies, known as **modes**, use one or more (usually keyed) functions defined over small and fixed lengths (e.g., blockcipher, compression function, permutations in sponge constructions etc.) in a black-box manner. We call a function from $I_n := \{0,1\}^n$ (elements of the set are called **blocks**) to itself a **block function**. Throughout the paper *we fix a positive integer n.*

© International Association for Cryptologic Research 2015
T. Iwata and J.H. Cheon (Eds.): ASIACRYPT 2015, Part II, LNCS 9453, pp. 113–133, 2015.
DOI: 10.1007/978-3-662-48800-3_5

A keyed blockcipher is a popular example of block function. Multiplying (as a field multiplication over I_n) an element by a secret key K can also be considered to be a block function as it maps a block input x to $K \cdot x \in I_n$. Outputs of a streamcipher with one block seed, can also be viewed as a sequence of execution of different block functions. In fact, any function mapping one block to multiple blocks can be viewed as a sequence of executions of block functions. Whereas, a function mapping multiple blocks to a single block can not be in general expressed through block functions. For example, compression function, or mapping (x, y) to $(x + K) \cdot (y + K)$ (known as pseudo dot-product) are not examples of block functions as they take more than one block as an input.

Length-Preserving Encryption. An encryption algorithm is called *length-preserving* if the the number of blocks in a plaintext and its corresponding ciphertext are same. A length-preserving encryption is called an **enciphering scheme**. In addition with the theoretical interest, an enciphering scheme has some practical applications. Among others, a popular application is disk-sector encryption addressed by the "IEEE Security in Storage" Work Group P1619. An enciphering scheme is said to be (S)PRP or (strong) pseudorandom permutation [34,35] if it is secure against adversaries making only plaintext queries (or both plaintext, ciphertext queries respectively). The building block keyed block function is assumed to be PRP or PRF (pseudorandom function [12]).

Linear Mode. In this paper we consider a wide class of enciphering schemes and pseudorandom functions based on linear mode. Informally, a **linear mode** (LM) is defined by an oracle algorithm which interacts with block functions (usually keyed) as oracles such that all inputs of the block functions are computed through some public linear functions (determined in the design) of the previous obtained responses. Finally, the output is also computed through a public linear function of all responses of block functions and the input.

This class is indeed a wide class of encryption algorithms. Most of the known symmetric key encryptions, e.g., Luby-Rackoff (LR) [23,28], Feistel type Encryption Schemes [6,17] CMC [15], EME [13,16] HCTR [9,51], TET [14], HEH [47] etc. are some examples of enciphering schemes based on linear mode. Almost all pseudorandom functions (e.g., CBC-MAC [5], PMAC [8], TMAC [22], OMAC [18], DAG-based constructions [20], a sub-class of affine domain extension or ADE [29] etc.) are also based on linear mode. Thus, the linear mode based keyed construction includes a wide class of symmetric key algorithms.

1.1 Brief Literature Survey

Now we briefly revisit the related results. Feistel structure is used to define different blockciphers e.g., Lucifer [50], DES etc. Later, Luby-Rackoff provides the PRP and SPRP security analysis of this type of ciphers and since then it is also popularly known as Luby-Rackoff (LR) cipher. In particular it was shown that three and four round LR cipher are PRP and SPRP secure respectively. Each round invokes exactly one block function. There are many results known for security analysis of different rounds of LR and for different forms of Feistel structures [6,28,39,40,48]. Many results are known for reducing the key-sizes (i.e. reusing the round keys [37,38,42,46]). Nandi [28] has characterized that all secure LR encryption schemes must have non-palindrome key-scheduling algorithms. Thus, we cannot use one single key.

XLS [43] is proposed to construct a generic encryption scheme which takes incomplete message blocks given that an encryption which can take complete message blocks. A particular instantiation of XLS invokes three block functions and claimed to be SPRP secure. However, the result is shown to be wrong [31] and some of implications (e.g., COPA [2] which uses XLS) are shown very recently [32]. Among all linear mode based length-preserving SPRP, the CMC and four-round Luby-Rackoff require only 2ℓ calls for encrypting ℓ blocks and others requires more (e.g., EME requires $2\ell+1$ calls etc.). Understanding optimality of SPRP and PRP, in terms of the number of blockcipher or block function calls, is our main motivation of this paper.

A class of authenticated encryption modes linear over the field was proposed by Jutla [21]. This class is more restricted than our linear mode as the linearity is considered over I_n instead of binary. In other words, only linear operation is bit-wise xor (without having any rotation or permutation of bit positions, multiplying by primitive element etc.). Jutla had shown that the number of invocations of blockcipher calls plus the number of masking keys should be about $\ell + O(\log_2 \ell)$.

1.2   Our Contribution

(1) Optimality in PRP and SPRP. Lear Bahack in his submission of the design called Julius [1] stated that $2\ell - 1$ blockcipher encryptions are required for achieving "simple linear mode" PRP over ℓ blocks. However, their result is still unpublished and so formalizing the issue and proving such a statement is yet to know. Moreover, no such claim is known for SPRP security. In this paper we provide a formal definition of linear mode in Sect. 3. In Sect. 4, we formally show that a linear mode based length-preserving PRP (or SPRP) over ℓ blocks must invoke block functions at least $2\ell - 1$ (or respectively, 2ℓ) times. This justifies why XLS or three rounds of Luby-Rackoff are not SPRP. This bound is tight as three and four-rounds LR, CMC (for arbitrary block messages) etc. achieve these bounds.

(2) Optimality in Single-key Inverse-Free PRP. Inverse-free encryptions [6,17,19,23] like LR cipher are useful in terms of implementation as we do not need to implement the inverse of the building-block for the combined implementation of encryption and decryption. In Sect. 5, we show that any linear-mode based inverse-free single key length-preserving PRP over ℓ blocks requires at least 2ℓ invocations (which is actually same for SPRP constructions). This shows that PRP and SPRP becomes equally costly for single-keyed inverse-free encryptions. Although all distinguishers of our paper are differential distinguishers, the PRP distinguisher for an inverse-free single key construction is different from the above SPRP attacks.

(3) Three-Round Single-PRF Based LR with a Masking is PRP. The above observation says that to achieve inverse-free double-block PRP with three invocations, we can use two independent PRF (e.g., the constructions in [28] are such examples). Two independent keyed PRF may be more costly than one keyed PRF due to key-scheduling or key set-up algorithms [10,44]. In the later part of the Sect. 5, we show that the single PRF based three round LR is indeed PRP if we simply mask one block of the input by a masking key.

Significance. Our above two distinguishing attacks provide a limitation on the performance of a (inverse-free) length-preserving encryption or pseudorandom function or permutation. This applies to a wide class of encryption algorithms including online encryption, authenticated encryption (without any nonce) etc. and so it has impact on designs and analysis in symmetric key cryptography.

Novelty of The Attack Idea. In [30] the minimum number of multiplications required to achieve Δ universal hash has been proposed. Like all other differential attacks (where zero differences are exploited), our PRP distinguisher and the ΔU attack from [30] basically finds zero differences in the input of non-linear functions for some executions. Basic intuition of our SPRP distinguishing attack is also similar to that of the distinguishing attack of XLS. However, to make all these applicable for general constructions, we need to find an appropriate difference in queries. For this, we adopt methodologies from linear algebra. The PRP distinguisher for single keyed inverse-free construction also exploits zero differential propagation. However, to achieve zero differential in one more block than expected (for a PRP distinguisher) is the tricky part of the attack. This essentially allows to achieve a PRP distinguisher even if we invoke one extra block function compared to usual PRP construction.

2   Preliminaries

A **block matrix** is a binary square matrix of size n. Let $\mathbb{M}_n(a, b)$ denote the set of all partitioned matrices $E_{a \times b}$ (of size $a \times b$ as a block partitioned matrix and of size $an \times bn$ as a binary matrix) whose $(i, j)^{\text{th}}$ entry, denoted $E[i, j]$, is a block-matrix for all $i \in [1..a] = \{1, \ldots, a\}$ and $j \in [1..b]$. The transpose of E, denoted E^{tr}, is applied as a binary matrix. Thus, $E^{tr}[i, j] = E[j, i]^{tr}$. Conventionally, any matrix $E_{a \times b}$ is written as the following block-wise partition matrices

$$
E = \begin{pmatrix} E[1,1] & E[1,2] & \cdots & E[1,b] \\ E[2,1] & E[2,2] & \cdots & E[2,b] \\ \vdots & \vdots & \vdots & \vdots \\ E[a,1] & E[a,2] & \cdots & E[a,b] \end{pmatrix} := \begin{pmatrix} E[1,*] \\ E[2,*] \\ \vdots \\ E[a,*] \end{pmatrix} := \begin{pmatrix} E[*,1] & E[*,2] & \cdots & E[*,b] \end{pmatrix}
$$

where $E[i, *]$ and $E[*, j]$ denote i^{th} block-row and j^{th} block-column respectively. For $1 \leq i \leq j \leq a$, we also write $E[i..j \; ; \; *]$ to mean the sub-matrix consisting of all rows in between i and j. We simply write $E[..j \; ; \; *]$ or $E[i.. \; ; \; *]$ to denote $E[1..j \; ; \; *]$ and $E[i..a \; ; \; *]$ respectively. Similar notation for columns are defined.

Definition 1. *A (square) matrix $E \in \mathbb{M}_n(a, a)$ is called* **(block-wise) strictly lower triangular** *if for all $1 \leq i \leq j \leq a$, $E[i, j] = 0$ (zero matrix).*

For all $x = (x_1, \dots, x_a) \in I_n^a$, we define a linear function mapping a blocks to b blocks as $E \cdot x = (y_1, \dots, y_b)$. Here, we consider x and y as binary column vectors (we follow this convention which should be understood from the context). So *the block matrix $E[i, j]$ represents the contribution of x_j to define y_i*. More formally,

$$y_i = E[i, 1] \cdot x_1 + E[i, 2] \cdot x_2 + \cdots + E[i, a] \cdot x_a, \quad 1 \leq i \leq b.$$

If E is a strictly lower triangular matrix then y_i is clearly functionally independent of $x_i, \dots, x_a$, $1 \leq i \leq a$. So if we associate y_i uniquely to each x_i (e.g., $y_i = \rho(x_i)$ for some function ρ) then the choice of the vectors x and y satisfying $E \cdot x = y$ becomes unique. This observation is useful while we define intermediate inputs and outputs of a black-box based construction.

2.1 Useful Properties of Matrices

It is well known that the maximum number of linearly independent (binary) rows and columns of a matrix $A \in \mathbb{M}_n(s, t)$ are same and this number is called rank of the matrix, denoted $\mathrm{rank}(A)$. So clearly we have $\mathrm{rank}(A) \leq \min\{ns, nt\}$. By using Gaussian elimination method, denoted $x = \mathsf{solve}(A, b)$, we can solve for some x (not necessarily unique) of the system of solvable linear equations $A \cdot x = b$. By convention, whenever a non-zero solution exists it returns a non-zero solution. Note that if $w^{tr} = \mathsf{solve}(A^{tr}, b^{tr})$ then $w \cdot A = b$ (by applying transpose). The following results are straightforward and so we skip the proofs.

Lemma 1. *Let $A \in \mathbb{M}_n(s, t)$ and $r := \mathrm{rank}(A)$.*

(1) If $r < ns$ (i.e. presence of row-dependency) then $\mathsf{solve}(A^{tr}, 0)$ returns a non-zero solution.

(2) Similarly for $r < nt$ (i.e. presence of column-dependency) $\mathsf{solve}(A, 0)$ returns a non-zero solution.

(3) Finally, let $r = nt$ (i.e., full column rank) and $b := A \cdot w$. Then, $\mathsf{solve}(A, b) = w$ (i.e., w is also the unique solution).

Lemma 2. *Suppose $A \in \mathbb{M}_n(s, s)$ is a non-singular matrix, i.e., $\mathrm{rank}(A) = ns$. Let $t < s$ and*

$$B = \begin{pmatrix} A[..t, *] & 0 \\ 0 & A[..t, *] \\ A[t+1.., *] & A[t+1.., *] \end{pmatrix}$$

where 0 denotes the zero matrix of appropriate size. Then, $\mathrm{rank}(B) = n(s+t)$ (i.e., full row-rank).

2.2   Security Definitions and Notation

In this section we quickly recall the security definitions of fixed length keyed constructions. One can also extend the definitions for variable length constructions.

PRF. We call an oracle algorithm $\mathcal{A}$ (t, q)-algorithm if it makes at most q queries and runs in time t. Let $\mathcal{K}$ be a key-space and $f : \mathcal{K} \times I_n^a \to I_n^b$ be a (keyed) function. We say that f is (q, t, ϵ)-PRF if for any (t, q)-algorithm $\mathcal{A}$ the **prf-distinguishing advantage**

$$\mathbf{Adv}_f^{\mathrm{prf}}(\mathcal{A}) := |\Pr[\mathcal{A}^{f_K} = 1; K \xleftarrow{\$} \mathcal{K}] - \Pr[\mathcal{A}^g = 1; g \xleftarrow{\$} \mathrm{Func}(a, b)]|$$

is at most ϵ where $\mathrm{Func}(a, b)$ denotes the set of all functions from I_n^a to I_n^b. We call randomly chosen g to be the (uniform) *random function*.

Notation. For notational simplicity, we skip the time parameter t which is irrelevant in this paper. We also simply write $\mathrm{Func} := \mathrm{Func}(1, 1)$ and Perm to mean the set of all functions and permutations over I_n.

(S)PRP. A keyed permutation g over I_n^a is a function $g : \mathcal{K} \times I_n^a \to I_n^a$ such that for all key $k \in \mathcal{K}$, $g_k := g(K, \cdot) \in \mathrm{Perm}(a)$ (the set of all permutations over I_n^a). We denote the uniformly chosen permutation by Π_a and call *uniform random permutation*. A keyed permutation g is called (q, ϵ)-PRP if for any q-algorithm $\mathcal{A}$ the **prp-distinguishing advantage**

$$\mathbf{Adv}_g^{\mathrm{prp}}(\mathcal{A}) := |\Pr[\mathcal{A}^{g_K(\cdot)} = 1; K \xleftarrow{\$} \mathcal{K}] - \Pr[\mathcal{A}^{\Pi_a} = 1]|$$

is at most ϵ. By PRF-PRP switching lemma [4,49], it is well known that $|\mathbf{Adv}_f^{\mathrm{prf}}(\mathcal{A}) - \mathbf{Adv}_f^{\mathrm{prp}}(\mathcal{A})| \leq \binom{q}{2}2^{-n}$. We define the **sprp-distinguishing advantage**

$$\mathbf{Adv}_f^{\mathrm{sprp}}(\mathcal{A}) := |\Pr[\mathcal{A}^{f_K, f_K^{-1}} = 1; K \xleftarrow{\$} \mathcal{K}] - \Pr[\mathcal{A}^{\Pi_a, \Pi_a^{-1}} = 1]|$$

and (q, ϵ)-SPRP.

2.3   Tools for Proving Security

Given a q-algorithm $\mathcal{A}$ interacting with an oracle $\mathcal{O}$ we denote the transcript $\tau(\mathcal{A}^{\mathcal{O}})$ by the random vector $((X_1, Y_1), \ldots, (X_q, Y_q))$ where $X_i \in I_n^a$ and $Y_i \in I_n^b$ are the ith query made by and response obtained by $\mathcal{A}$ respectively. The following theorem, known as coefficient-H technique [36,41] is very useful to show a construction is PRP or SPRP. It has also been adapted in [7,25]

Theorem 1 (Coefficient-H Technique). *Let* $f : \mathcal{K} \times I_n^a \to I_n^b$ *be a keyed function and* $\mathcal{V}_{\mathrm{bad}} \subseteq (I_n^a \times I_n^b)^q$. *Suppose*

1. *for all q-algorithm $\mathcal{B}$, $\Pr[\tau(\mathcal{B}^{\Gamma_{a,b}}) \in \mathcal{V}_{\mathrm{bad}}] \leq \epsilon_1$ and*
2. *for all $\tau = ((x_1, y_1), \ldots, (x_q, y_q)) \notin \mathcal{V}_{\mathrm{bad}}$,*

$$\Pr[f_K(x_1) = y_1, \ldots, f_K(x_q) = y_q; K \xleftarrow{\$} \mathcal{K}] \geq (1 - \epsilon_2) \times 2^{-nbq}.$$

Then, for all q-algorithm $\mathcal{A}$, $\mathbf{Adv}_f^{\mathrm{prf}}(\mathcal{A}) \leq \epsilon_1 + \epsilon_2$.

3   Linear Mode

3.1   Linear Query and Mode

A block matrix $U \in \mathbb{M}_n(\ell, a + \ell)$ is called (a, ℓ)-**query function** if $U[* \; ; \; a + 1..]$ is block-wise strictly lower triangular. Here ℓ represents the number of queries and a represents the number of blocks in the input. For any such *query function*, an input $X \in I_n^a$, (and a tuple of ℓ functions $\tilde{\rho} = (\rho_1, \ldots, \rho_\ell)$ over I_n), we can *uniquely define* or associate u and v, called **intermediate input and output vector** respectively, satisfying (1) $U \cdot \binom{X}{v} = u$ and (2) $\tilde{\rho}(u) := (\rho_1(u_1), \ldots, \rho_\ell(u_\ell)) = v$. This can be easily shown by recursive definitions of u_i's and v_i's. More precisely, u_i is uniquely determined by $v_1, \ldots, v_{i-1}$ and X (through the linear function) and v_i is uniquely determined by u_i through ρ_i, for all $1 \leq i \leq \ell$. Informally, a (a, b, ℓ)-linear mode is a mode which takes a blocks input and returns b blocks output based on executing block functions building blocks (see Fig. 1 for an illustration of a linear mode). Formally, (a, b, ℓ)-linear mode is defined by a block matrix $E \in \mathbb{M}_n(\ell + b, \ell + a)$ where $E[1..\ell \; ; \; *]$ is a (a, ℓ)-query function. For any ℓ-tuple of functions $\tilde{\rho} \in \mathrm{Func}^\ell$, the corresponding linear-mode function $E^{\tilde{\rho}} : I_n^a \to I_n^b$ is defined as $E^{\tilde{\rho}}(X) = Y$ where

$$E \cdot \binom{X}{v} = \binom{u}{Y}, \quad \tilde{\rho}(u) = v.$$

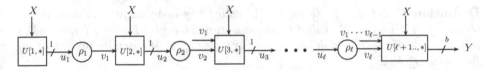

Fig. 1. Linear Mode: Here $U[i, *]$ means the ith block row which maps $(X, v_1, \ldots, v_{i-1}, 0^{\ell - i + 1})$ to u_i. Finally, $U[\ell + 1.. \; ; \; *]$ maps the input X and intermediate output vector v to the output Y consisiting of b blocks.

So v is the intermediate output vector associated to the input X and the final output $Y := E[\ell + 1.. \; ; \; *] \cdot \binom{X}{v}$, a linear function of v and X. Now we state a useful differential property of linear mode. Note that the functions of $\tilde{\rho}$ are non-linear and would be secret for the adversaries. So to obtain any information about the intermediate input and output, we only can equate intermediate outputs whenever two inputs collide for same function. Given any vectors x, x' of same size, we write Δx to mean $x \oplus x'$ and $\Delta_{a.b} x$ to mean $(x_a \oplus x_a', \ldots, x_b \oplus x_b')$. We simply write $\Delta_t x$ to mean $\Delta_{1..t} x$ (the first t elements of Δx) (Fig. 2).

Lemma 3. *Suppose* $E[..t \; ; \; *] \cdot X = E[..t \; ; \; *] \cdot X'$ *(i.e.,* $E[..t \; ; \; *] \cdot \Delta X = 0$*). Let* $E^{\tilde{\rho}}(X) = Y$, $E^{\tilde{\rho}}(X') = Y'$. *Let* v, v' *and* u, u' *denote intermediate outputs and inputs respectively associated with* X *and* X' *(for the function tuple* $\tilde{\rho}$*) respectively. Then,* $\Delta_t u = \Delta_t v = 0^t$ *and*

$$\Delta Y = E[\ell + 1.. \; ; \; ..a] \cdot \Delta X + E[\ell + 1.. \; ; \; a + t + 1..] \cdot \Delta v_{t+1..}.$$

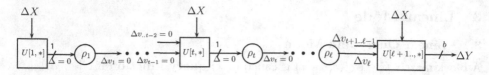

Fig. 2. Differential pattern of the linear mode: we choose ΔX such that the first t input differences of the ρ functions are zero. So the final difference ΔY can be expressed as the linear function of the rest of the differences $\Delta v_{t+1..}$ and ΔX.

Proof. Due to choice of X and X', by induction one can show that $(u_1, v_1) = (u_1', v_1'), \ldots (u_t, v_t) = (u_t', v_t')$ where u and u' denote the intermediate inputs associated with X and X' respectively (for the function tuple $\tilde{\rho}$). In other words, $\Delta_t u = \Delta_t v = 0^t$. Now, $Y = E[\ell+1..\,;\ a+1..] \cdot v + E[\ell+1..\,;\ ..a] \cdot X$ and similarly $Y' = E[\ell+1..\,;\ a+1..] \cdot v' + E[\ell+1..\,;\ ..a] \cdot X'$. The result is followed after we add these two equations and using that $\Delta_t v = 0^t$. $\square$

3.2 Keyed Constructions Based on Linear Mode

KEYED LINEAR MODE. Let $\mathcal{F} = \mathcal{F}_1 \times \cdots \times \mathcal{F}_f$ and k be a non-negative integer where $\mathcal{F}_i \subseteq$ Func. A key-space $\mathcal{K}$ for any keyed function is of the form $I_n^k \times \mathcal{F}$. We call $\mathcal{F}$ the function-key space and I_n^k masking-key space. Any function g is also written as g^{+1}.

Definition 2. *Let* $\mu : [1..\ell] \rightarrow [1..f]$, *called* key-assignment function, $\alpha := (\alpha_1, \ldots, \alpha_\ell) \in \{+1, -1\}^\ell$, *called* inverse-assignment tuple. *For any function-key* $\rho = (\rho_1, \ldots, \rho_f) \in \mathcal{F}$, *we define* $\rho_\mu^\alpha := (\rho_{\mu_1}^{\alpha_1}, \ldots, \rho_{\mu_\ell}^{\alpha_\ell})$. *We denote the set of all functions* ρ_μ^α *by* $\mathcal{F}_\mu^\alpha$.

Here we implicitly assume that whenever $\alpha_i = -1$, ρ_{μ_i} is a permutation. If $\alpha = 1^\ell$, we simply skip the notation α. In general, the presence of inverse call of building blocks may be required when we consider decryption of keyed function. For the encryption, or a keyed function where decryption is not defined, w.l.o.g. we may assume that $\alpha = 1^\ell$.

Definition 3. *A* (k, a, b) keyed linear mode *with key-space* $\mathcal{K}$, *key-assignment function* μ, *is a* $(a+k, b, \ell)$ *linear mode* E. *For each key* $\kappa := (L, \rho) \in \mathcal{K} := I_n^k \times \mathcal{F}$, *we define a keyed function* $E_\kappa(P) := E^{\rho_\mu}(L, P)$.

Keyed linear mode E is actually a linear mode with a part of the input is the masking key and function tuples are also derived by reusing some keyed block functions.

Example 1. Consider the simple variant of PMAC [8,45] defined over I_n^a (see Fig. 3 above). Let $(p_1, \ldots, p_a)$ be the input.

$$1 \leq i \leq a - 1, u_i = p_i \text{ and } u_a = p_a \oplus \left(\bigoplus_{i=1}^{a-1} v_i\right).$$

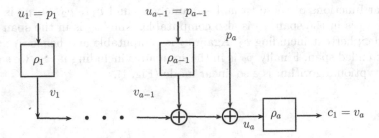

Fig. 3. The simplified structure of PMAC. The input is $(p_1, \ldots, p_a)$ and the output is c_1.

Finally the output is defined as $c_1 = v_a$. Here $\ell = a$ and $b = 1$. There is no masking key, i.e. $k = 0$ and $f = a$ (all function-keys are independently chosen). The key-assignment function μ is an identity function.

In a single function-key version of PMAC (with independent masking key), we have $f = 1 = k$. The $u_i = \alpha^i \cdot L \oplus p_i$ for $1 \leq i < a$ and $u_a = p_a \oplus (\bigoplus_{i=1}^{a-1} v_i) \oplus L$. Here the key-assignment function maps all indices to the key-index 1 (as there is only one choice of key).

Affine Domain Extension or ADE [29]. As defined in [29], affine domain extension over I_n^a is nothing but a $(a, 1, \ell)$-linear mode keyed function **E** such that the key-space is $\mathcal{K} = \mathcal{F} \subseteq \mathrm{Func}$, i.e., $f = 1$ (single function-key) and $k = 0$ (no masking key). Moreover, the final output is the response of the last oracle call, i.e. v_ℓ. Like PMAC, the key-assignment function for ADE maps all indices to the key-index 1. One can consider an injective padding rule and sequence of such constructions indexed by a to incorporate variable length inputs. CBC-MAC [5], PMAC [8,24,33], TMAC [22], OMAC [18,27], DAG-based constructions [20] etc. are some examples of ADE.

Length Preserving Linear Encryption Mode. A keyed linear mode E is called length-preserving (LP) encryption if E_κ is encryption scheme and $a = b$. In addition to these, we also assume that its decryption algorithm D is also a keyed linear mode which is indeed true for all known linear encryption modes. We first see an example below.

Example 2. As an example, consider Luby-Rackoff (LR) keyed function with three rounds using two random functions ρ_1, ρ_2, i.e. $f = 2$, $a = b = 2$ and $\ell = 3$ (three invocations of the underlying block functions). Consider the key-assignment function π with $\pi_1 = 1, \pi_2 = 1$ and $\pi_3 = 2$. So the function tuple after applying the key-assignment is (ρ_1, ρ_1, ρ_2). As there is no masking key, we have $k = 0$. So the key-space is Func^2. Given $(p_1, p_2) \in I_n^2$ we define

$$u_1 := p_1, v_1 = \rho_1(u_1), u_2 = v_1 + p_2, v_2 = \rho_1(u_2), u_3 = v_2 + p_1, v_3 = \rho_2(u_3).$$

Finally, the output is (c_1, c_2) where $c_1 := u_3$ and $c_2 = v_3 + u_2$. This is clearly decryptable. Consider u_i's, v_i's and p_i's as variables. The ciphertext provides two linear functions of these variables, namely u_3 and $v_3 + u_2$. So u_3 is in the span. As u_3 is in the span, v_3 is also computable. Thus u_2 is in the span of the extended ciphertext including v_3. Again v_2 is computable and hence $u_1 := p_1$ is in the extended span. Finally, p_2 is in the span after including v_1. So we see that that decryption algorithm is also linear mode (Fig. 4).

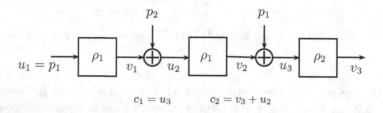

Fig. 4. LR with three round.

Decryption Algorithm of a Keyed Linear Encryption Mode. From the above example, it is clear that the intermediate input outputs for the building blocks would be same if we encrypt and then decrypt as we do in the correctness condition: $D_\kappa(E_\kappa(P)) = P$. Informally, if some input-output does not arise in the decryption then either this input-output is redundant in the encryption computation or the correctness condition does not hold (due to randomness of the output which has influence in the encryption but is not used in the decryption). We now describe the details of a length preserving linear encryption mode for which all invocations of block function calls are not redundant.

Definition 4 (Reordering of Vectors). *Let* $\alpha := (\alpha_1, \ldots, \alpha_\ell) \in \{1, -1\}^\ell$, *and* $\beta = (\beta_1, \ldots, \beta_\ell)$ *be a permutation over* $[1..\ell]$. *A pair of vectors* $(w, z) \in I_n^{2\ell}$ *is* (a, β)-*reordering of a pair of vectors* $(u, v) \in I_n^{2\ell}$ *if*

$$(w_i, z_i) = \begin{cases} (u_{\beta_i}, v_{\beta_i}) & \text{if } \alpha_i = 1, \\ (v_{\beta_i}, u_{\beta_i}) & \text{if } \alpha_i = -1. \end{cases}$$

Definition 5. *A* $(k + a, a, \ell)$-*linear mode* E *is called* linear-mode length-preserving encryption *with key-space* $\mathcal{K} := I_n^k \times \mathcal{F}$ *and key-assignment* π *if the corresponding decryption algorithm* D *is also a* $(k + a, a, \ell)$-*linear mode with* (1) *an inverse assignment-tuple* $\alpha := (\alpha_1, \ldots, \alpha_\ell) \in \{1, -1\}^\ell$ *and* (2) *key-assignment* $\pi' := \pi \circ \beta$ *where* $\beta = (\beta_1, \ldots, \beta_\ell)$ *is a permutation over* $[1..\ell]$. *Moreover,* $\forall P \in I_n^a, L \in I_n^k, \rho = (\rho_1, \ldots, \rho_f) \in \mathcal{F}$,

$$E \cdot \begin{pmatrix} L \\ P \\ v \end{pmatrix} = \begin{pmatrix} u \\ C \end{pmatrix}, \rho_{\pi_1}(u_1) = v_1, \ldots \rho_{\pi_\ell}(u_\ell) = v_\ell \text{ if and only if}$$

$$D \cdot \begin{pmatrix} L \\ C \\ z \end{pmatrix} = \begin{pmatrix} w \\ P \end{pmatrix}, \rho_{\pi_1'}^{\alpha_1}(w_1) = z_1, \ldots, \rho_{\pi_\ell'}^{\alpha_\ell}(w_\ell) = z_\ell$$

where (w, z) is (a, β)-reordering of (u, v).

The above definition implies that correctness condition of an encryption $D^{\rho_{\pi'}^\alpha}(L, E^\rho(L, P)) = P$. In addition with the correctness condition, the intermediate inputs and outputs for both encryption and decryption are simply reordered. In Example 2 (given above), we have $a = b = f = 2$, $\ell = 3$. For the decryption algorithm, we execute the function in the reverse order and so we set $\beta_1 = 3, \beta_2 = 2, \beta_1 = 3$. So the key-assignment function for the decryption is $\pi_1' = 2, \pi_2' = 2, \pi_3' = 1$. We do not need to apply inverse for the decryption (it is called inverse-free) and so inverse-assignment tuple is 1^3. So if $(u_1, v_1), (u_2, v_2)$ and (u_3, v_3) are the intermediate input-output pairs for encryption then $(u_2, v_3), (u_2, v_2)$ and (u_1, v_1) (reordering of the previous pairs) are the intermediate input-output pairs for decryption.

Examples. EME [16], ELmE [11], AEZ [1], CMC [15] (these follow Encrypt-Mix-Encrypt paradigm), Luby-Rackoff with $a = b = 2$, unbalanced Feistel [17, 48] etc. are some examples of length-preserving linear mode encryptions. HCBC1, HCBC2 [3], Modified-HCBC's, ELmD [1], MCBC [26], COPE [2] etc. are some examples of online computable length-preserving encryptions based on linear mode.

4   PRP and SPRP Distinguishing Attacks

Consider a length-preserving encryption scheme based on $(k+a, a, \ell)$ linear mode E. Now we show two main results in this section. Namely, we provide PRP and SPRP distinguishing attacks on the encryption scheme if $\ell \leq 2a - 2$. and $\ell \leq 2a - 1$ respectively. Thus, it gives lower bound on the number of invocations of building blocks for achieving PRP and SPRP security.

4.1   PRP Distinguishing Attack on E with $\ell = 2a - 2$

Let us assume $\ell = 2a - 2$. The attack can be trivially extended to all those constructions with $\ell < 2a - 2$. We recall that $E_L^{\tilde{\rho}}(P) = C$ if and only if

$$E \cdot \begin{pmatrix} L \\ P \\ v \end{pmatrix} = \begin{pmatrix} u \\ C \end{pmatrix}, \quad \tilde{\rho}(u) = v.$$

Distinguisher D_{prp} against $(k + a, a, 2a - 2)$-Linear mode E.

1. **step-1** (finding a suitable difference in a pair of plaintext queries): Let $d \in I_n^a$ be the non-zero solution of $\mathsf{solve}(E[..a - 1 \ ; \ k + 1..k + a], 0)$, i.e. $E[..a - 1 \ ; \ k + 1..k + a] \cdot d = 0$. Such a non-zero solution exists as the number of columns is more than that of rows (see lemma 1).

2. **step-2** (make the queries with the difference obtained in step-1): Now the distinguisher makes two queries 0^a and d and obtains corresponding responses $c = E_L^{\hat{\rho}}(0)$ and $c' = E_L^{\hat{\rho}}(d)$. Let

$$u_1, v_1, \ldots, u_{2a-2}, v_{2a-2}, \text{ and } u'_1, v'_1, \ldots, u'_{2a-2}, v'_{2a-2}$$

denote the intermediate inputs outputs for the two queries respectively. By lemma 2, we have $1 \leq i \leq a-1$, $u_i = u'_i, v_i = v'_i$ and

$$\Delta c = E[2a-1.. \ ; \ k+1..(a+k)] \cdot d + E[2a-1.. \ ; \ 2a+k..] \cdot \Delta v_{a..}$$

while it is interacting with the keyed construction.

3. **step-3** (find a nullifier of unknown intermediate values): As the matrix $E[2a-1.. \ ; \ 2a+k..]$ is a $\times (a-1)$ matrix, we find a non-zero binary vector $w \in \{0,1\}^{na}$ such that $w \cdot E[2a-1.. \ ; \ 2a+k..] = 0$. In particular, $w = \text{solve}(E[2a-1.. \ ; \ 2a+k..]^{\text{tr}}, 0)$.

4. **step-4** (the distinguisher event): If $w \cdot \Delta c = w \cdot E[2a-1.. \ ; \ k+1..(a+k)] \cdot d$ then it returns 1 (decision for the keyed construction), else returns 0 (decision for uniform random permutation).

The distinguishing advantage of the above distinguisher D is at least $1/2$ since for a random permutation $w \cdot \Delta c = w \cdot E[2a-1.. \ ; \ k+1..(a+k)] \cdot d$ with probability $1/2$ whereas we have seen this holds with probability one for the keyed construction. When $a = 2$, we know that LR with three rounds is PRP. This shows the bound is tight at least for $a = 2$.

A Generalized Distinguisher D_{prp}^{gen} Against $(k+a, a, \ell)$-Linear Mode E. Now we define a distinguisher against $(k+a, a, \ell)$-linear mode E assuming certain singularities in the sub-matrices.

Assumption: Suppose there exists an integer t such that

1. $\text{rank}(E[..t \ ; \ ..a]) < na$ and
2. $\text{rank}(E[\ell+1.. \ ; \ a+k+t+1..]) < na$.

Note the above assumption always holds for $t = a-1$ when $\ell \leq 2a-2$. However, if $\ell \geq 2a-1$, the both conditions not necessarily hold. Whenever the assumptions hold, we have the following similar distinguisher as mentioned before. This distinguisher would be used later on while describing SPRP distinguishers.

Distinguisher D_{prp}^{gen} Against $(k+a, a, \ell)$-linear Mode E.

1. **step-1.** Due to the assumptions, we can find d and w such that $E[..t \ ; \ ..a]) \cdot d = 0$ and $w \cdot E[\ell+1.. \ ; \ a+k+t+1..] = 0$.
2. **step-2.** Then we make two queries 0 and d and obtain responses c and c'.
3. **step-3.** The distinguisher returns 1 if $w \cdot \Delta c = w \cdot E[\ell+1.. \ ; \ k+1..(a+k)] \cdot d$, else 0.

4.2   SPRP Distinguishing Attack on E with $\ell = 2a - 1$

Now we show that if $\ell < 2a$ then we have a SPRP distinguisher. In other words, $2a$ many invocations is minimum to achieve SPRP and which is tight as it is achieved in CMC. The basic intuition of our attack is similar to that of XLS. However, to complete the attack for any linear-mode encryption we need to carefully set the queries and distinguishing event. Consider a length-preserving $(k, a, 2a - 1)$-encryption scheme based on $(k + a, a, 2a - 1)$-linear mode E. Let us denote the $(k + a, a, 2a - 1)$-linear mode for its decryption by D. We describe three distinguishers depending on cases.

Case 1: Rank($E[2a.. \; ; \; 2a + k..]$) $< na$. In this case, the two assumptions, mentioned above, hold for $t = a - 1$. So we can run the PRP-distinguisher D_{prp}^{gen}.

Case 2: Rank($D[..a \; ; \; k + 1..k + a]$) $< na$. In this case, the two assumptions also hold for $t = a$ for the decryption function. So we run our general PRP distinguisher D_{prp}^{gen} applied to the decryption function.

Case 3: Rank($D[..a \; ; \; k + 1..k + a]$) $= na$, rank($E[2a.. \; ; \; 2a + k..]$) $= na$.

Here we describe a SPRP distinguisher. Briefly, it works as follows. It first makes two queries as in step-2 (the first $a - 1$ intermediate input and outputs are identical for two encryption queries). Using the invertible property we can actually obtain all the differences of intermediate values. As the computation of decryption algorithm must use same internal input and outputs of the building blocks, we also know the differences of intermediate inputs and outputs if we decrypt the first two encryption queries. Now we find another decryption query for which the first a intermediate input and output differences with one of the first two queries are fixed. So we can nullify the unknown $a - 1$ differences and obtain a distinguishing event. The details are described below.

Distinguisher D_{sprp} Against $(k + a, a, 2a - 1)$-Linear Mode E.

1. **step-1** (make two queries with a certain difference, same as PRP distinguisher): Let $d \in I_n^a$ be the non-zero solution of $\mathsf{solve}(E[..a-1 \; ; \; k+1..k+a], 0)$, i.e. $E[..a - 1 \; ; \; k + 1..k + a] \cdot d = 0$. It makes two queries 0^a and d and obtains corresponding responses $c = E_L^{\tilde{\rho}}(0)$ and $c' = E_L^{\tilde{\rho}}(d)$.
   Let $u_1, v_1, \ldots, u_{2a-1}, v_{2a-1}$ and $u_1', v_1', \ldots, u_{2a-1}', v_{2a-1}'$ denote the intermediate inputs outputs for the two queries respectively. By lemm 3, we have $1 \le i \le a - 1$, $u_i = u_i', v_i = v_i'$ and

$$\Delta c = E[2a - 1.. \; ; \; k + 1..(a + k)] \cdot d + E[2a.. \; ; \; 2a + k..] \cdot \Delta v_{a..}$$

   while it is interacting with the keyed construction.
2. **step-2** (solve for Δu, Δv): Using the invertible property of $E[2a.. \; ; \; 2a + k..]$, we can actually solve $\Delta v_{a..}$ and hence $\Delta u_{a..}$. Thus, we know Δu and Δv. Suppose we make two (redundant) decryption queries c and c' (whose responses

must be 0 and d) and let $w_1, z_1, \ldots, w_{2a-1}, z_{2a-1}$ and $w'_1, z'_1, \ldots, w'_{2a-1}, z'_{2a-1}$ denote the intermediate inputs outputs for the two queries respectively. Then by the definition of decryption algorithm we also know Δw, Δz which are nothing but (β, π)-reordering of $(\Delta u, \Delta v)$.

3. **step-3** (find a difference for the final decryption query): Now we find a difference d' such that

$$D[..a \ ; \ k+1..k+a+1] \cdot \begin{pmatrix} d' \\ \Delta z_1 \end{pmatrix} = \begin{pmatrix} \Delta w_1 \\ 0^{a-1} \end{pmatrix}.$$

We can solve for a non-zero d'. This can be solved assuming that $\Delta w_1 \neq 0$ (see the remark below). Note that the matrix $D[..a \ ; \ k+1..k+a]$ is invertible. Now we make two decryption queries $\bar{c}$ and $\bar{c}' = \bar{c} + d'$. While we set two queries we should ensure that none of these have been obtained in the first two encryption queries (these are also called non-pointless or non-trivial queries). Let $\bar{w}_1, \bar{z}_1, \ldots, \bar{w}_{2a-1}, \bar{z}_{2a-1} \ \bar{w}'_1, \bar{z}'_1, \ldots, \bar{w}'_{2a-1}, \bar{z}'_{2a-1}$ denote the intermediate inputs outputs for these two queries respectively and let $\bar{p}$ and $\bar{p}'$ denote the corresponding responses. By choice of d' we know that $\bar{z}_1 = \bar{z}'_1$ and $\Delta \bar{z}_{2..a} = 0^{a-1}$.

4. **step-4** (find a nullifier of unknown intermediate values, same as PRP distinguisher): As $D[2a.. \ ; \ 2a+k..]$ is $a \times (a-1)$ matrix, we find a non-zero binary vector $w \in \{0,1\}^{nb}$ such that $w \cdot D[2a-1.., 2a+k..] = 0$.

5. **step-5** (the distinguisher event): If $w \cdot (\bar{p} \oplus \bar{p}') = w \cdot D[2a-1.. \ ; \ k+1..(a+k)] \cdot d'$ then it returns 1 (decision for the keyed construction), else returns 0 (decision for uniform random permutation).

Remark 1. In the above attack we assume that $\Delta w_1 \neq 0$ since otherwise we do not get a non-zero d'. Note that Δw_1 can be written as a function of c and c'. So for a random permutation, a function of c and c' become zero has low probability. So we may assume that the $\Delta w_1 \neq 0$.

5   Security Analysis of Inverse-Free Single Key Construction

5.1   PRP Attack of Single-Key Inverse-Free Constructions Without Masking

In the last section, we have seen that to obtain PRP, we need at least $2a - 1$ invocations and this is tight as three rounds of LR achieves this bound. Note that the three calls of the building block can not have same key. In [28], it is also shown that three rounds of LR-type rounds with same key building block can not be PRP. However, their result is applicable to a specific form of encryption schemes. Now, we generalize this result and show that any inverse-free single function-key (and no masking key) PRP requires at least $2a$ calls. In [28], there is a construction of inverse-free SPRP over two blocks invoking underlying function (single keyed) four times. So the bound is tight. Interestingly, the cost of PRP and SPRP become same when we want inverse-free single function-key constructions.

Consider a length-preserving encryption scheme based on $(a, a, 2a - 1)$-linear mode E. Let us denote the $(a, a, 2a-1)$-linear mode for its decryption by D. Since it is inverse-free the inverse-assignment for the decryption is $\beta = (1, 1, \ldots, 1)$. As it is based on single function-key, the key-assignment is a constant function, i.e., $\pi_i = \pi'_i = 1$. However, there exists a permutation β over $[1..2a - 1]$. such that w and z are π-reordering of u and v respectively where u, v denote the intermediate input and output, respectively for $E^\rho(P) = C$ and similarly w, z for $D^\rho(C) = P$. We first briefly describe how we can construct a PRP-distinguisher (as like SPRP). The attack is similar to SPRP but we can not make decryption queries. We see how we can manage even if we are not allowed to make decryption queries.

We make two encryption queries such that $\Delta_{a-1} u = \Delta_{a-1} v = 0^{a-1}$. This is possible as we have a many plaintext blocks. Assuming some invertible property, we can find out the whole differences Δu and Δv for these two queries. For these two queries, if we look at the decryption computation then the first inputs, say w_1, w'_1 and their corresponding output differences Δz_1 (not the exact outputs) for both decryption are known (as there is no masking key). So now we make two encryption queries with the the following restrictions on intermediate values $\overline{u}, \overline{v}, \overline{u}'$ and $\overline{v}'$: $\overline{u}_1 = w_1, \overline{u}'_1 = w'_1, \Delta_{2..a}\overline{u} = \Delta_{2..a}\overline{u}', \Delta_{2..a}\overline{v} = \Delta_{2..a}\overline{v}'$. As we have obtained differences for the first a inputs in a determined manner, we can nullify the remaining $a - 1$ intermediate differences and obtain a distinguishing event. The more details of the attack is given below depending on different cases. Note that the matrix $E \in \mathbb{M}_n(3a - 1, 3a - 1)$.

Distinguisher D_{prp} Against $(a, a, 2a - 1)$-Linear-Mode E (with Corresponding Decryption Mode D.

Case 1: Rank($E[2a.. ; 2a..]$) $<$ na. In this case, the two assumptions, mentioned before, hold for $t = a - 1$. So we have our general PRP distinguisher.

Case 2: Rank($E[1..a ; ..a]$) $<$ na. In this case, the two assumptions also hold for $t = a$. So we have our general PRP distinguisher.

Case 3: Rank($E[1..a ; ..a]$) $=$ na., rank($E[2a.. ; 2a..]$) $=$ na. Here we describe a PRP distinguisher which works similar to SPRP distinguisher and as described above.

1. **step-1** (make two queries with a certain difference, same as PRP distinguisher): Let $d \in I_n^a$ be the non-zero solution of $\mathsf{solve}(E[..a - 1 ; ..a], 0)$, i.e. $E[..a-1 ; ..a] \cdot d = 0$. It makes two queries 0^a and d and obtains corresponding responses $c = E^\rho(0)$ and $c' = E^\rho(d)$.

Let $u_1, v_1, \ldots, u_{2a-1}, v_{2a-1}$ and $u'_1, v'_1, \ldots, u'_{2a-1}, v'_{2a-1}$ denote the intermediate inputs outputs for the two queries respectively. By lemma 3, we have $1 \leq i \leq a-1$, $u_i = u'_i$, $v_i = v'_i$ and

$$\Delta c = E[2a.. \ ; \ ..a] \cdot d + E[2a.. \ ; \ 2a..] \cdot \Delta v_{a..}$$

while it is interacting with the keyed construction.

2. **step-2** (solve for $\Delta u, \Delta v$): Using the invertible property of $E[2a.. \ ; \ 2a..]$, we can actually solve $\Delta v_{a..}$ and hence $\Delta u_{a..}$. Thus, we know Δu and Δv. Now note that the first input of decryption D is only based on c and c'. Let β be the permutation corresponding to the reordering of intermediate input outputs for decryption. So the values of u_{β_1} and u'_{β_1} are known (as they depend only on c and c' due to no masking keys and inverse-free property). Moreover, we know Δv_{β_1}. Here we assume the difference Δu_{β_1} is non-zero, otherwise, we can have a different distinguishing event as zero difference can occur with low probability for random permutation.

3. **step-3** (find a difference for two more encryption queries): Now we find a solution p and p' such that

$$\begin{pmatrix} E[1,*] & 0 \\ 0 & E[1,*] \\ E[2..a,*] & E[2..a,*] \end{pmatrix} \cdot \begin{pmatrix} p \\ p' \end{pmatrix} = \begin{pmatrix} u_{\beta_1} \\ u'_{\beta_1} \\ 0 \end{pmatrix}.$$

This can be solved as it has full column rank (see Lemma 2). Now we make two encryption queries p and p' and obtain outputs $\bar{c}$ and $\bar{c}'$. Let $\bar{u}, \bar{v}, \bar{u}'$ and $\bar{v}'$ be the intermediate inputs and outputs for these two queries respectively. So $\bar{u}_1 = u_{\beta_1}, \bar{u}'_1 = u'_{\beta_1}, \Delta \bar{v}_1 = \Delta v_{\beta_1}$ and $\Delta_{2..a} \bar{u} = \Delta_{2..a} \bar{v} = 0^{a-1}$. Thus, the a block output difference $\Delta \bar{c}$ depends only on the $a-1$ blocks of the intermediate output difference $\Delta \bar{v}_{a+1..}$.

4. **step-4** (find a nullifier of unknown intermediate values, same as PRP distinguisher): As $E[2a.. \ ; \ 2a+1..]$ is $a \times (a-1)$ matrix, we find a non-zero binary vector $w \in \{0,1\}^{nb}$ such that $w \cdot E[2a.., 2a+1..] = 0$.

5. **step-5** (the distinguisher event): If $w \cdot (p \oplus d) = w \cdot D[2a.. \ ; \ ..a] \cdot d'$ then it returns 1 (decision for the keyed construction), else returns 0 (decision for uniform random permutation).

5.2 PRP Security of Single-Key Luby-Rackoff with Masking

Define one round Luby-Rackoff $LR^f(a, b) = (b \oplus f(a), a)$ where $a, b \in I_n$ and $f \in \text{Func}(a, a)$. In [28] it was shown that three rounds of some variants LR rounds with single function key is not PRP secure. In last section we have also generalized and showed that any encryption making three calls over two blocks input with key space $\mathcal{K} = \mathcal{F} = \text{Func}(a)$ is not PRP secure. However, we now show that a simple variant of LR with a masking key becomes PRP secure.

Definition 6. *For any* $f \in \text{Func}(a)$, $L \in I_n$, *we define (see the Fig. 5 below)*

$$LR_L^{f,3}(a, b) = LR^f(LR^f(LR^f(a+L, b))).$$

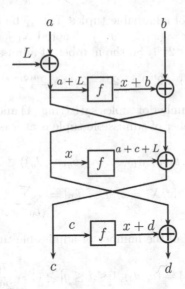

Fig. 5. LR-three rounds single function-key and one masking key.

Now we show that the above construction with key-space $\mathcal{K} = I_n \times$ Func is PRP. Note that we have constant key-assignment (i.e., we reuse the PRF for all invocations) and also inverse assignment tuple is 1^3. Let f denote the uniform random function on I_n. Given a tuple of elements $c = (c_1, \ldots, c_t)$ we say that the event $\mathsf{coll}(c)$ holds if there exists $i \neq j$ such that $c_i = c_j$. We define

$$\mathcal{V}_{bad} = \{((a_1, b_1, c_1, d_1), \ldots (a_q, b_q, c_q, d_q)) \in I_n^{4q} : \mathsf{coll}(c)\}.$$

It is easy to see that for random function Γ_2 and a q-algorithm $\mathcal{A}$,

$$\Pr[\tau(\mathcal{A}^{\Gamma_2}) \in \mathcal{V}_{bad}] \leq \binom{q}{2} 2^{-n}.$$

Now we show the high interpolation probability of the variant of 3 round LR construction.

Proposition 1. *For all $\tau = ((a_1, b_1, c_1, d_1), \ldots (a_q, b_q, c_q, d_q)) \notin \mathcal{V}_{bad}$, we have*

$$\Pr[\tau] := \Pr[LR_L^{f,3}(a_i, b_i) = (c_i, d_i), 1 \leq i \leq q] \geq (1 - \epsilon)2^{-2nq}$$

where $\epsilon = \frac{7q^2}{2^{n+1}}$.

Proof. We say that a tuple $(L_0, (x_i)_{1 \leq i \leq q})$ is *admissible* if

1. $L_0 \notin \{a_i + c_j; 1 \leq i, j \leq q\} \cup \{a_i + x_j; 1 \leq i, j \leq q\}$,
2. x_i's are distinct and $x_i \neq c_j$, $1 \leq i, j \leq q$ and
3. whenever $a_i = a_j$, we have $x_i + x_j = b_i + b_j$.

Let $\mathcal{A}$ denote the set of admissible tuples. Let q_1 be the number of distinct a_i's. The number of $(L_0, x = (x_1, \ldots, x_q))$, denoted $N_{1,3}$, satisfying only (1) and (3) is at least $(2^n - 2q^2) \times 2^{nq_1}$. So the number of admissible tuple is at least

$$(2^n - 2q^2) \times 2^{nq_1} - (2^n - 2q^2) \times 2^{n(q_1-1)}3q^2/2.$$

We mainly subtract the number of tuples satisfying (1) and (3) and not satisfying (2) from $N_{1,3}$. So the number of admissible tuple is at least $2^{n(q_1+1)}(1-\epsilon)$ where $\epsilon = \frac{7q^2}{2^{n+1}}$.

Now, for any $\tau = ((a_1, b_1, c_1, d_1), \ldots (a_q, b_q, c_q, d_q)) \notin \mathcal{V}_{bad}$ we have

$$\Pr[\tau] \geq \sum_{(L_0,x)\in\mathcal{A}} \Pr[\tau, X_i = x_i, L = L_0] = \sum_{(L_0,x)\in\mathcal{A}} 2^{-n(q_1+2q+1)}.$$

By using the lower bound of the number of admissible tuples we have

$$\Pr[\mathsf{LR}_L^{f,3}(a_i, b_i) = (c_i, d_i), 1 \leq i \leq q] \geq (1 - \frac{7q^2}{2^{n+1}})2^{-2nq}.$$

$\square$

Theorem 2. *For any q-adversary, the PRP advantage $\mathbf{Adv}_{\mathsf{LR}_L^{f,3}}^{\mathrm{prp}}$ against $\mathsf{LR}_L^{f,3}$ is at most $\frac{4q^2}{2^n}$.*

Proof. Armed with the above result and using Coefficient-H technique the theorem follows. $\square$

6   Conclusion

In this paper, we justify formally why we do not have any length-preserving PRP constructions more efficient than LR three rounds and length-preserving SPRP constructions more efficient than CMC or four round LR (in terms of the number of building block calls). We note that this optimality holds for all linear modes. We show that any such linear mode based constructions over ℓ blocks requires at leat $2\ell - 1$ blockcipher calls against chosen plaintext adversaries and at least 2ℓ blockcipher calls against chosen plaintext-ciphertext adversaries. This bounds are clearly tight as we know some constructions achieving the bound. Then we look into inverse-free single-key PRP constructions. Nandi has shown that three blockcipher call is no longer sufficient for LR-type constructions over two blocks (note that three call is sufficient using two independent PRF). We extend this result and show that any ℓ-block single-key inverse-free PRP must require 2ℓ calls like SPRP constructions. However, if we are allowed to use one masking key then we can have inverse-free PRP construction invoking only three blockcipher calls. We actually show that the three round LR using same keyed PRF is PRP if we mask a plaintext block by a masking key.

Acknowledgement. This work is supported by Centre of Excellence in Cryptology and R.C. Bose Center at Indian Statistical Institute, Kolkata. The author would like to thank Lear Bahack who found an error of the SPRP distinguisher in one of the sub-cases. Author is also grateful to Lear for pointing out the unpublished claim in the submission draft of authenticated encryption Julius. The author would also like to thank anonymous reviewers for their helpful comments.

References

1. CAESAR submissions (2014). http://competitions.cr.yp.to/caesar-submissions.html
2. Andreeva, E., Bogdanov, A., Luykx, A., Mennink, B., Tischhauser, E., Yasuda, K.: Parallelizable and authenticated online ciphers. In: Sako, K., Sarkar, P. (eds.) ASIACRYPT 2013, Part I. LNCS, vol. 8269, pp. 424–443. Springer, Heidelberg (2013)
3. Bellare, M., Boldyreva, A., Knudsen, L.R., Namprempre, C.: Online ciphers and the hash-CBC construction. In: Kilian, J. (ed.) CRYPTO 2001. LNCS, vol. 2139, p. 292. Springer, Heidelberg (2001)
4. Bellare, M., Rogaway, P.: The security of triple encryption and a framework for code-based game playing proofs. In: Vaudenay, S. (ed.) EUROCRYPT 2006. LNCS, vol. 4004, pp. 409–426. Springer, Heidelberg (2006)
5. Bellare, M., Kilian, J., Rogaway, P.: The security of cipher block chaining. In: Desmedt, Y.G. (ed.) CRYPTO 1994. LNCS, vol. 839, pp. 341–358. Springer, Heidelberg (1994)
6. Berger, T.P., Minier, M., Thomas, G.: Extended generalized feistel networks using matrix representation. In: Lange, T., Lauter, K., Lisoněk, P. (eds.) SAC 2013. LNCS, vol. 8282, pp. 289–305. Springer, Heidelberg (2014)
7. Bernstein, D.J.: A short proof of the unpredictability of cipher block chaining (2005)
8. Black, J.A., Rogaway, P.: A block-cipher mode of operation for parallelizable message authentication. In: Knudsen, L.R. (ed.) EUROCRYPT 2002. LNCS, vol. 2332, p. 384. Springer, Heidelberg (2002)
9. Chakraborty, D., Nandi, M.: An improved security bound for HCTR. In: Nyberg, K. (ed.) FSE 2008. LNCS, vol. 5086, pp. 289–302. Springer, Heidelberg (2008)
10. Daemen, J., Rijmen, V.: The Design of Rijndael: AES - The Advanced Encryption Standard. Information Security and Cryptography. Springer, Heidelberg (2002)
11. Datta, N., Nandi, M.: Misuse resistant parallel authenticated encryptions. IACR Cryptology ePrint Archive, vol. 2013, p. 767 (2013)
12. Goldreich, O., Goldwasser, S., Micali, S.: How to construct random functions. J. ACM **33**(4), 792–807 (1986)
13. Halevi, S.: EME*: extending EME to handle arbitrary-length messages with associated data. In: Canteaut, A., Viswanathan, K. (eds.) INDOCRYPT 2004. LNCS, vol. 3348, pp. 315–327. Springer, Heidelberg (2004)
14. Halevi, S.: Invertible universal hashing and the TET encryption mode. In: Menezes, A. (ed.) CRYPTO 2007. LNCS, vol. 4622, pp. 412–429. Springer, Heidelberg (2007)
15. Halevi, S., Rogaway, P.: A tweakable enciphering mode. In: Boneh, D. (ed.) CRYPTO 2003. LNCS, vol. 2729, pp. 482–499. Springer, Heidelberg (2003)
16. Halevi, S., Rogaway, P.: A parallelizable enciphering mode. In: Okamoto, T. (ed.) CT-RSA 2004. LNCS, vol. 2964, pp. 292–304. Springer, Heidelberg (2004)

17. Hoang, V.T., Rogaway, P.: On generalized feistel networks. In: Rabin, T. (ed.) CRYPTO 2010. LNCS, vol. 6223, pp. 613–630. Springer, Heidelberg (2010)
18. Iwata, T., Kurosawa, K.: OMAC: one-key CBC MAC. In: Johansson, T. (ed.) FSE 2003. LNCS, vol. 2887, pp. 129–153. Springer, Heidelberg (2003)
19. Iwata, T., Yasuda, K.: BTM: a single-key, inverse-cipher-free mode for deterministic authenticated encryption. In: Jacobson Jr, M.J., Rijmen, V., Safavi-Naini, R. (eds.) SAC 2009. LNCS, vol. 5867, pp. 313–330. Springer, Heidelberg (2009)
20. Jutla, C.S.: PRF domain extension using DAGs. In: Halevi, S., Rabin, T. (eds.) TCC 2006. LNCS, vol. 3876, pp. 561–580. Springer, Heidelberg (2006)
21. Jutla, C.S.: Lower bound on linear authenticated encryption. In: Matsui, M., Zuccherato, R.J. (eds.) SAC 2003. LNCS, vol. 3006. Springer, Heidelberg (2004)
22. Kurosawa, K., Iwata, T.: TMAC: two-key CBC MAC. In: Joye, M. (ed.) CT-RSA 2003. LNCS, vol. 2612, pp. 33–49. Springer, Heidelberg (2003)
23. Luby, M., Rackoff, C.: How to construct pseudo-random permutations from pseudorandom functions. In: Williams, H.C. (ed.) CRYPTO 1985. LNCS, vol. 218, pp. 447–447. Springer, Heidelberg (1986)
24. Minematsu, K., Matsushima, T.: New bounds for PMAC, TMAC, and XCBC. In: Biryukov, A. (ed.) FSE 2007. LNCS, vol. 4593, pp. 434–451. Springer, Heidelberg (2007)
25. Nandi, M.: A simple and unified method of proving indistinguishability. In: Barua, R., Lange, T. (eds.) INDOCRYPT 2006. LNCS, vol. 4329, pp. 317–334. Springer, Heidelberg (2006)
26. Nandi, M.: Two new efficient CCA-secure online ciphers: MHCBC and MCBC. In: Chowdhury, D.R., Rijmen, V., Das, A. (eds.) INDOCRYPT 2008. LNCS, vol. 5365, pp. 350–362. Springer, Heidelberg (2008)
27. Nandi, M.: Improved security analysis for OMAC as a pseudorandom function. J. Math. Cryptol. 3(2), 133–148 (2009)
28. Nandi, M.: The characterization of Luby-Rackoff and its optimum single-key variants. In: Gong, G., Gupta, K.C. (eds.) INDOCRYPT 2010. LNCS, vol. 6498, pp. 82–97. Springer, Heidelberg (2010)
29. Nandi, M.: A unified method for improving PRF bounds for a class of blockcipher based MACs. In: Hong, S., Iwata, T. (eds.) FSE 2010. LNCS, vol. 6147, pp. 212–229. Springer, Heidelberg (2010)
30. Nandi, M.: On the minimum number of multiplications necessary for universal hash functions. In: Cid, C., Rechberger, C. (eds.) FSE 2014. LNCS, vol. 8540, pp. 489–507. Springer, Heidelberg (2015)
31. Nandi, M.: XLS is not a strong pseudorandom permutation. In: Sarkar, P., Iwata, T. (eds.) ASIACRYPT 2014. LNCS, vol. 8873, pp. 478–490. Springer, Heidelberg (2014)
32. Nandi, M.: Revisiting security claims of XLS and COPA. In: IACR Cryptology ePrint Archive, vol. 2015, p. 444 (2015)
33. Nandi, M., Mandal, A.: Improved security analysis of PMAC. Cryptology ePrint Archive, Report 2007/031 (2007). http://eprint.iacr.org/
34. Naor, M., Reingold, O.: A pseudo-random encryption mode. www.wisdom. weizmann.ac.il/~naor
35. Naor, M., Reingold, O.: On the construction of pseudorandom permutations: Luby-Rackoff revisited. J. Cryptol. 12(1), 29–66 (1999)
36. Patarin, J.: Etude des Générateurs de Permutations Basés sur le Schéma du D.E.S. Ph. D. thèsis de Doctorat de l'Université de Paris 6 (1991)

37. Patarin, J.: New results on pseudorandom permutation generators based on the DES scheme. In: Feigenbaum, J. (ed.) CRYPTO 1991. LNCS, vol. 576, pp. 301–312. Springer, Heidelberg (1992)
38. Patarin, J.: How to construct pseudorandom and super pseudorandom permutations from one single pseudorandom function. In: Rueppel, R.A. (ed.) EURO-CRYPT 1992. LNCS, vol. 658, pp. 256–266. Springer, Heidelberg (1993)
39. Patarin, J.: Generic attacks on feistel schemes. In: Boyd, C. (ed.) ASIACRYPT 2001. LNCS, vol. 2248, pp. 222–238. Springer, Heidelberg (2001)
40. Patarin, J.: Security of random feistel schemes with 5 or more rounds. In: Franklin, M. (ed.) CRYPTO 2004. LNCS, vol. 3152, pp. 106–122. Springer, Heidelberg (2004)
41. Patarin, J.: The "Coefficients H" technique. In: Avanzi, R.M., Keliher, L., Sica, F. (eds.) SAC 2008. LNCS, vol. 5381, pp. 328–345. Springer, Heidelberg (2009)
42. Pieprzyk, J.P.: How to construct pseudorandom permutations from single pseudo-random functions. In: Damgård, I.B. (ed.) EUROCRYPT 1990. LNCS, vol. 473, pp. 140–150. Springer, Heidelberg (1991)
43. Ristenpart, T., Rogaway, P.: How to enrich the message space of a cipher. In: Biryukov, A. (ed.) FSE 2007. LNCS, vol. 4593, pp. 101–118. Springer, Heidelberg (2007)
44. Rivest, R.L., Robshaw, M.J.B,, Yin, Y.L.: RC6 as the AES. In: AES Candidate Conference, pp. 337–342 (2000)
45. Rogaway, P.: Efficient instantiations of tweakable blockciphers and refinements to modes OCB and PMAC. In: Lee, P.J. (ed.) ASIACRYPT 2004. LNCS, vol. 3329, pp. 16–31. Springer, Heidelberg (2004)
46. Sadeghiyan, B., Pieprzyk, J.P.: A construction of super pseudorandom permutations from a single pseudorandom function. In: Rueppel, R.A. (ed.) EUROCRYPT 1992. LNCS, vol. 658, pp. 267–284. Springer, Heidelberg (1993)
47. Sarkar, P.: Improving upon the TET mode of operation. In: Nam, K.-H., Rhee, G. (eds.) ICISC 2007. LNCS, vol. 4817, pp. 180–192. Springer, Heidelberg (2007)
48. Schneier, B., Kelsey, J.: Unbalanced feistel networks and block cipher design. In: Gollmann, D. (ed.) FSE 1996. LNCS, vol. 1039. Springer, Heidelberg (1996)
49. Shoup, V.: Sequences of games: a tool for taming complexity in security proofs. In: Cryptology ePrint Archive, Report 2004/332 (2004). http://eprint.iacr.org/
50. Sorkin, A.: Lucifer, a cryptographic algorithm. Cryptologia 8(1), 22–42 (1984)
51. Wang, P., Feng, D., Wu, W.: HCTR: a variable-input-length enciphering mode. In: Feng, D., Lin, D., Yung, M. (eds.) CISC 2005. LNCS, vol. 3822, pp. 175–188. Springer, Heidelberg (2005)

Beyond-Birthday-Bound Security
for Tweakable Even-Mansour Ciphers
with Linear Tweak and Key Mixing

Benoît Cogliati[1]([✉]) and Yannick Seurin[2]

[1] University of Versailles, Versailles, France
`benoitcogliati@hotmail.fr`
[2] ANSSI, Paris, France
`yannick.seurin@m4x.org`

Abstract. The iterated Even-Mansour construction defines a block cipher from a tuple of public n-bit permutations $(P_1, \ldots, P_r)$ by alternatively xoring some n-bit round key k_i, $i = 0, \ldots, r$, and applying permutation P_i to the state. The *tweakable* Even-Mansour construction generalizes the conventional Even-Mansour construction by replacing the n-bit round keys by n-bit strings derived from a master key *and a tweak*, thereby defining a tweakable block cipher. Constructions of this type have been previously analyzed, but they were either secure only up to the birthday bound, or they used a nonlinear mixing function of the key and the tweak (typically, multiplication of the key and the tweak seen as elements of some finite field) which might be costly to implement. In this paper, we tackle the question of whether it is possible to achieve beyond-birthday-bound security for such a construction by using only linear operations for mixing the key and the tweak into the state. We answer positively, describing a 4-round construction with a $2n$-bit master key and an n-bit tweak which is provably secure in the Random Permutation Model up to roughly $2^{2n/3}$ adversarial queries.

Keywords: Tweakable block cipher · Iterated Even-Mansour cipher · Key-alternating cipher · Beyond-birthday-bound security

1 Introduction

Background. A block cipher with key space $\mathcal{K}$ and message space $\mathcal{M}$ is a family of permutations of $\mathcal{M}$ indexed by the key $\mathbf{k} \in \mathcal{K}$. A *tweakable* block cipher (TBC) takes an additional (potentially public) input parameter $\mathbf{t} \in \mathcal{T}$ called a *tweak* aiming at providing inherent variability in about the same way an IV or nonce brings variability to an encryption scheme. Some block ciphers such as the Hasty Pudding Cipher [35], Mercy [10], or Threefish (the block cipher underlying the Skein hash function [15]) were designed so as to natively support tweaks. The syntax and security requirements for tweakable block ciphers were formally articulated in a seminal paper by Liskov, Rivest and Wagner [24]. Since then,

© International Association for Cryptologic Research 2015
T. Iwata and J.H. Cheon (Eds.): ASIACRYPT 2015, Part II, LNCS 9453, pp. 134–158, 2015.
DOI: 10.1007/978-3-662-48800-3_6

TBCs have found multiple applications such as (tweakable) length-preserving encryption modes [18,19], online ciphers [1,33], and authenticated encryption modes [24,31,32].

Liskov *et al.* [24] also proposed two generic constructions of a TBC from a standard block cipher, achieving security up to the so-called birthday bound, i.e., when the adversary is allowed at most roughly $2^{n/2}$ queries to the encryption or decryption oracle, where n is the block size (that is, the message space of the TBC is $\mathcal{M} = \{0,1\}^n$). The "black-box" design strategy (i.e., building a TBC on top of an existing standard block cipher, in a black-box way) has since then been the main avenue of research. Earlier proposals, such as XEX [31] and variants [4,26] were related to the second of the two original proposals of Liskov *et al.*, and were limited to birthday-bound security as well. Recently, a number of constructions achieving beyond-birthday-bound security have emerged, such as Minematsu's construction [27], the CLRW construction [22,23,30], and two constructions by Mennink [25]. All those constructions enjoy a security proof in the standard model (i.e., assuming that the underlying block cipher is a pseudo-random permutation), except for Mennink's constructions that were analyzed in the ideal cipher model.

Tweaking Even-Mansour Ciphers. Unfortunately, none of the currently known black-box TBC constructions with beyond-birthday-bound security can be deemed truly practical (even though some of them might come close to it [25]). Hence, it might be beneficial to "open the hood" and to study how to build a TBC from some lower level primitive than a full-fledged conventional block cipher, e.g., a pseudorandom function or a public permutation. For example, Goldenberg *et al.* [16] investigated how to include a tweak in Feistel ciphers. This was extended to generalized Feistel ciphers by Mitsuda and Iwata [28]. Recently, a similar study was undertaken for the second large class of block ciphers besides Feistel ciphers, namely key-alternating ciphers [11], a super-class of Substitution-Permutation Networks (SPNs). An r-round key-alternating cipher based on a tuple of public n-bit permutations $(P_1, \ldots, P_r)$ maps a plaintext $x \in \{0,1\}^n$ to the ciphertext defined as

$$y = k_r \oplus P_r(k_{r-1} \oplus P_{r-1}(\cdots P_2(k_1 \oplus P_1(k_0 \oplus x))\cdots)), \tag{1}$$

where the n-bit round keys $k_0, \ldots, k_r$ are either independent or derived from a master key $\mathbf{k}$. When the P_i's are modeled as public permutation oracles, construction (1) is also referred to as the (iterated) Even-Mansour construction, in reference to Even and Mansour who pioneered the analysis of this construction in the Random Permutation Model [13]. While Even and Mansour limited themselves to proving birthday-bound security in the case $r = 1$, larger numbers of rounds were studied in subsequent works [3,21,36]. The general case has been recently (tightly) settled by Chen and Steinberger [6], who proved that the r-round iterated Even-Mansour cipher with r-wise independent round keys ensures security up to roughly $2^{\frac{rn}{r+1}}$ adversarial queries.

In order to incorporate a tweak $\mathbf{t}$ in the iterated Even-Mansour construction, it is tantalizing to generalize (1) by replacing round keys k_i by some function

$f_i(\mathbf{k}, \mathbf{t})$ of the master key $\mathbf{k}$ *and* the tweak $\mathbf{t}$ (see Fig. 1). We will refer to such a construction as a *Tweakable Even-Mansour* (TEM) construction.[1] This is exactly the spirit of the TWEAKEY framework introduced by Jean *et al.* [20]. In fact, these authors go one step further and propose to unify the key and tweak inputs into what they dub the *tweakey*. The main topic of this paper being provable security (in the traditional model where the key is secret and the tweak is chosen by the adversary), we will not make such a bold move here, since we are not aware of any formal security model adequately capturing what Jean *et al.* had in mind.

The investigation of the theoretical soundness of this design strategy was initiated in three recent papers. First, Cogliati and Seurin [8], and independently Farshim and Procter [14], analyzed the simple case of an n-bit key k and an n-bit tweak t simply xored together at each round, i.e., $f_i(k,t) = k \oplus t$ for each $i = 0, \ldots, r$.[2] They gave attacks up to two rounds, and proved birthday-bound security for three rounds. In fact, the security of this construction caps at $2^{n/2}$ queries independently of the number of rounds. Indeed, it can be written $\widetilde{E}(k,t,x) = E(k \oplus t, x)$, where E is the conventional iterated Even-Mansour cipher with the trivial key-schedule (i.e., the same round key is xored between each round), and by a result of Bellare and Kohno [2, Corollary 5.7], a tweakable block cipher of this form can never offer more than $\kappa/2$ bits of security, where κ is the key-length of E (i.e., $\kappa = n$ in the case at hand). Hence, if we want beyond-birthday-bound security, we have no choice but to consider more complex functions f_i (at the bare minimum, these functions, even if linear, should prevent the TBC construction from being of the form $E(k \oplus t, x)$ for some block cipher E with n-bit keys).

This was undertaken by Cogliati, Lampe, and Seurin [7], who considered nonlinear ways of mixing the key and the tweak. More specifically, they studied the case where $f_i(\mathbf{k}, t) = H_{k_i}(t)$, where the family of functions (H_k) is uniform and almost XOR-universal, and the master key is $\mathbf{k} = (k_0, \ldots, k_r)$. A classical example is multiplication-based hashing, i.e., $f_i(\mathbf{k}, t) = k_i \otimes t$, where $\otimes$ denotes the multiplication in the finite field $\mathbb{F}_{2^n}$, the tweak $t = 0$ being forbidden. Cogliati *et al.* showed that one round is secure up to the birthday bound, and that two rounds are secure up to roughly $2^{2n/3}$ adversarial queries.[3] They also provided a

---

[1] We warn that the naming *Tweakable Even-Mansour* construction was previously used by the designers of Minalpher [34], a candidate to the CAESAR competition, to designate a permutation-based variant of Rogaway's XEX construction [31], i.e., a 1-round Even-Mansour construction where the derivation functions f_0 and f_1 applied to $(\mathbf{k}, \mathbf{t})$ are allowed to depend on the internal permutation P_1 (something we do not consider in this paper).

[2] Actually, the results of [8,14] were stated in terms of xor-induced related-key security of the (conventional) iterated Even-Mansour cipher, but in this case this is equivalent to standard (i.e., single-key) security of the corresponding tweakable construction.

[3] More precisely, the birthday-bound result applies to the variant of the construction were the same key is used before and after permutation P_1, and the $2^{2n/3}$-security bound applies to the cascade of this construction with two independent keys and two independent permutations.

(non-tight) asymptotic security bound improving as the number of rounds grows. However, implementing a xor-universal hash function might be costly, and linear functions f_i's would be highly preferable for obvious efficiency reasons.

Our Results. In this paper, we ask whether it is possible to come with a tweakable Even-Mansour construction achieving both:

1. a linear mixing of the tweak and the key to the state;
2. beyond-birthday-bound security.

We answer positively, by providing a construction with $2n$-bit keys and n-bit tweaks. The starting point is the 4-round iterated Even-Mansour construction with a $2n$-bit master key (k_0, k_1), k_0 and k_1 being both n bits, and what we call the "alternating" key schedule, namely round keys are k_0, k_1, k_0, etc. This is for example how LED-128 is designed [17]. To turn this block cipher into a tweakable Even-Mansour construction, we simply add the n-bit tweak t between each permutation (see Fig. 2). In other words, if we denote $E((k_0, k_1), x)$ the conventional Even-Mansour cipher with alternating round keys, the tweakable construction that we consider can be written

$$\widetilde{E}((k_0, k_1), t, x) = E((k_0 \oplus t, k_1 \oplus t), x).$$

We prove that this construction is secure up to roughly $2^{2n/3}$ adversarial queries. Unsurprisingly, and as in many previous works, our proof uses Patarin's H-coefficients technique [6,29]. In particular, we rely on a key lemma by Cogliati *et al.* [7] to analyze so-called good transcripts.

Application to Related-Key Security. Our result can be rephrased in terms of related-key security [2] of the conventional Even-Mansour cipher: the 4-round conventional Even-Mansour cipher with the alternating key-schedule is secure up to roughly $2^{2n/3}$ adversarial queries against related-key attacks for the set of related-key deriving functions.

$$\Phi^{2-\oplus} \stackrel{\text{def}}{=} \{(k_0, k_1) \mapsto (k_0 \oplus \Delta, k_1 \oplus \Delta) : \Delta \in \{0,1\}^n\}.$$

Note that this set is more restrictive than the set $\Phi^{\oplus}$ that would allow to xor an arbitrary $2n$-bit string to the master key (k_0, k_1). It remains an open problem (already stated in [8]) to find an Even-Mansour construction provably secure beyond the birthday bound against $\Phi^{\oplus}$-related-key attacks.

Open Problems. We propose three challenging open problems, the first two being restricted to the case of n-bit tweaks. First, what would be the analogue of the Chen-Steinberger result [6] in the tweakable setting? In more details, we know how to deliver $n/2$ bits of security with an n-bit master key [8,14] and this paper shows how to reach $2n/3$ bits of security with a $2n$-bit master key. Hence, it is natural to ask whether one can obtain $rn/(r+1)$ bits of security from an rn-bit master key for $r > 2$, and what would be the adequate number of rounds and the corresponding (linear) "tweak-and-key" schedule. Second, Chen *et al.* [5] showed that the 2-round conventional Even-Mansour construction can provably deliver $2n/3$ bits of security even with an n-bit master key (for example, when the two inner permutations are independent, the trivial key-schedule is sufficient). Again, what would be the analogue of this result in the tweakable setting? Can we design a TEM construction with an n-bit master key and an n-bit tweak delivering $2n/3$ bits of security, or even more? Finally, it is natural to ask whether one can extend the construction of this paper to handle larger tweaks, in particular $2n$-bit tweaks. We show in the full version of this paper [9] that the naive way of proceeding, namely adding alternatively t_0 and t_1, is insecure for four rounds. Hence, this seems to require at least five rounds.

We also remark that attacks against the (conventional) iterated Even-Mansour cipher with the alternating key-schedule have been investigated by Dinur et al. [12]. It would be interesting to study whether these attacks can be adapted (and potentially improved) in the tweakable setting.

Organization. In Sect. 2, we introduce the notation, the security definitions, and give some background on the H-coefficients technique. Our main result is proved in Sect. 3.

2 Preliminaries

2.1 Notation and General Definitions

General Notation. In all the following, we fix an integer $n \geq 1$ and denote $N = 2^n$. For integers $1 \leq b \leq a$, we will write $(a)_b = a(a-1)\cdots(a-b+1)$ and $(a)_0 = 1$ by convention. The set of all permutations of $\{0,1\}^n$ will be denoted $\mathsf{P}(n)$.

Tweakable Block Ciphers. A *tweakable block cipher* with key space $\mathcal{K}$, tweak space $\mathcal{T}$, and message space $\mathcal{M}$ is a mapping $\widetilde{E} : \mathcal{K} \times \mathcal{T} \times \mathcal{M} \to \mathcal{M}$ such that for any key $k \in \mathcal{K}$ and any tweak $t \in \mathcal{T}$, $x \mapsto \widetilde{E}(k,t,x)$ is a permutation of $\mathcal{M}$. We denote $\mathsf{TBC}(\mathcal{K}, \mathcal{T}, n)$ the set of all tweakable block ciphers with key space $\mathcal{K}$, tweak space $\mathcal{T}$, and message space $\{0,1\}^n$. A *tweakable permutation* with tweak space $\mathcal{T}$ and message space $\mathcal{M}$ is a mapping $\widetilde{P} : \mathcal{T} \times \mathcal{M} \to \mathcal{M}$ such that for any tweak $t \in \mathcal{T}$, $x \mapsto \widetilde{P}(t,x)$ is a permutation of $\mathcal{M}$. We denote $\mathsf{TP}(\mathcal{T}, n)$ the set of all tweakable permutations with tweak space $\mathcal{T}$ and message space $\{0,1\}^n$.

Tweakable Even-Mansour Constructions. Fix integers $n, r \geq 1$. Let $\mathcal{K}$ and $\mathcal{T}$ be two sets, and let $\mathbf{f} = (f_0, \ldots, f_r)$ be a $(r+1)$-tuple of functions from $\mathcal{K} \times \mathcal{T}$ to $\{0,1\}^n$. The r-round tweakable Even-Mansour construction $\mathsf{TEM}[n, r, \mathbf{f}]$ specifies, from an r-tuple $\mathbf{P} = (P_1, \ldots, P_r)$ of permutations of $\{0,1\}^n$, a tweakable block cipher with key space $\mathcal{K}$, tweak space $\mathcal{T}$, and message space $\{0,1\}^n$, simply denoted $\mathsf{TEM}^{\mathbf{P}}$ in the following (parameters $[n, r, \mathbf{f}]$ will always be clear from the context) which maps a key $\mathbf{k} \in \mathcal{K}$, a tweak $\mathbf{t} \in \mathcal{T}$, and a plaintext $x \in \{0,1\}^n$ to the ciphertext defined as (see Fig. 1):

$$\mathsf{TEM}^{\mathbf{P}}(\mathbf{k}, \mathbf{t}, x) = f_r(\mathbf{k}, \mathbf{t}) \oplus P_r(f_{r-1}(\mathbf{k}, \mathbf{t}) \oplus P_{r-1}(\cdots P_1(f_0(\mathbf{k}, \mathbf{t}) \oplus x)\cdots)).$$

We will denote $\mathsf{TEM}_{\mathbf{k}}^{\mathbf{P}}$ the mapping taking as input $(\mathbf{t}, x) \in \mathcal{T} \times \{0,1\}^n$ and returning $\mathsf{TEM}^{\mathbf{P}}(\mathbf{k}, \mathbf{t}, x)$.

We will mostly be interested in the case where $\mathcal{K} = (\{0,1\}^n)^a$ and $\mathcal{T} = (\{0,1\}^n)^b$ for integers $a, b \geq 1$. In this setting, we will denote $\mathbf{k} = (k_0, \ldots, k_{a-1})$ and $\mathbf{t} = (t_0, \ldots, t_{b-1})$, all k_i's and t_j's being n-bit strings, or simply $\mathbf{k} = k$, resp. $\mathbf{t} = t$ when $a = 1$, resp. $b = 1$. When all f_i's are linear over $(\{0,1\}^n)^{a+b}$, we say that the construction has *linear tweak and key mixing*.

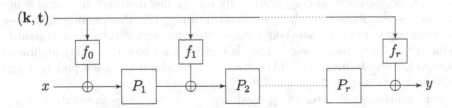

Fig. 1. The r-round tweakable Even-Mansour construction based on a tuple of public permutations $(P_1, \ldots, P_r)$.

Previously Studied Constructions. Two types of TEM constructions have already been studied. In [8], Cogliati and Seurin considered the simplest case where $a = b = 1$ (n-bit keys and n-bit tweaks) and $f_i(k, t) = k \oplus t$ for each $i = 0, \ldots, r$. This construction has linear tweak and key mixing, and is secure up to $2^{n/2}$ adversarial queries starting from $r = 3$. (The results of [8] were formulated in terms of xor-induced related-key attacks against the conventional iterated Even-Mansour construction, but in this simple case the two security notions are in fact equivalent.) In [7], Cogliati, Lampe, and Seurin studied a large class of nonlinear mixing functions, in particular, for n-bit tweaks, finite field multiplication-based ones, i.e., $f(k, t) = k \otimes t$, or more generally, for bn-bit tweaks, polynomial hashing-based functions, i.e., $f(k, (t_0, \ldots, t_{b-1})) = \sum_{i=0}^{b-1} k^{i+1} \otimes t_i$.

2.2 Security Definitions

Fix some family of functions $\mathbf{f} = (f_0, \ldots, f_r)$ from $\mathcal{K} \times \mathcal{T}$ to $\{0,1\}^n$. To study the security of the construction $\mathsf{TEM}[n, r, \mathbf{f}]$ in the Random Permutation Model, we consider a distinguisher $\mathcal{D}$ which interacts with $r + 1$ oracles that we denote generically $(\tilde{P}_0, P_1, \ldots, P_r)$, where syntactically $\tilde{P}_0$ is a tweakable permutation with tweak space $\mathcal{T}$ and message space $\{0,1\}^n$, and $P_1, \ldots, P_r$ are permutations of $\{0,1\}^n$. The goal of $\mathcal{D}$ is to distinguish two "worlds": the so-called *real world*, where $\mathcal{D}$ interacts with $(\mathsf{TEM}_{\mathbf{k}}^{\mathbf{P}}, \mathbf{P})$, where $\mathbf{P} = (P_1, \ldots, P_r)$ is a tuple of public random permutations and the key $\mathbf{k}$ is drawn uniformly at random from $\mathcal{K}$, and the so-called *ideal world* $(\tilde{P}_0, \mathbf{P})$, where $\tilde{P}_0$ is a uniformly random tweakable permutation and $\mathbf{P}$ is a tuple of random permutations of $\{0,1\}^n$ independent from $\tilde{P}_0$. We will refer to $\tilde{P}_0$ as the *construction oracle* and to $P_1, \ldots, P_r$ as the *inner permutation oracles*.

The distinguishing advantage of a distinguisher $\mathcal{D}$ is defined as

$$\mathbf{Adv}(\mathcal{D}) \stackrel{\text{def}}{=} \left| \Pr\left[\mathcal{D}^{\mathsf{TEM}_{\mathbf{k}}^{\mathbf{P}}, \mathbf{P}} = 1 \right] - \Pr\left[\mathcal{D}^{\widetilde{P}_0, \mathbf{P}} = 1 \right] \right|,$$

where the first probability is taken over the random choice of $\mathbf{k}$ and $\mathbf{P}$, and the second probability is taken over the random choice of $\widetilde{P}_0$ and $\mathbf{P}$. In all the following, we consider computationally unbounded distinguishers, and hence we can assume *wlog* that they are deterministic. We also assume that they never make pointless queries (i.e., queries whose answers can be unambiguously deduced from previous answers). The distinguisher is allowed to query all oracles adaptively in both directions; this corresponds to adaptive chosen-plaintext and ciphertext attacks (CCA).

For non-negative integers q_c and q_p, we define the insecurity of the $\mathsf{TEM}[n, r, \mathbf{f}]$ construction against CCA-attacks as

$$\mathbf{Adv}^{\text{cca}}_{\mathsf{TEM}[n,r,\mathbf{f}]}(q_c, q_p) = \max_{\mathcal{D}} \mathbf{Adv}(\mathcal{D}),$$

where the maximum is taken over all distinguishers making exactly q_c queries to the construction oracle and exactly q_p queries to each inner permutation oracle.

2.3   The H-Coefficients Technique

As in many previous works [5–8], our security proof will use the H-coefficients technique [29], which we explain here.

Transcript. Recall that the distinguisher $\mathcal{D}$ interacts with a tuple of $r+1$ oracles denoted $(\widetilde{P}_0, P_1, \ldots, P_r)$. In the real world, the construction oracle $\widetilde{P}_0$ is $\mathsf{TEM}_{\mathbf{k}}^{\mathbf{P}}$ where $\mathbf{P} = (P_1, \ldots, P_r)$ and $\mathbf{k}$ is random, whereas in the ideal world it is a random tweakable permutation independent from $(P_1, \ldots, P_r)$. From the interaction of $\mathcal{D}$ with these oracles, we define the *queries transcript* $(\mathcal{Q}_C, \mathcal{Q}_{P_1}, \ldots, \mathcal{Q}_{P_r})$ of the attack as follows. The list $\mathcal{Q}_C$ records the queries to the construction oracle: if $\mathcal{D}$ made either a direct query $(\mathbf{t}, x)$ to the construction oracle $\widetilde{P}_0$ which was answered by y, or an inverse query $(\mathbf{t}, y)$ which was answered by x, then the triple $(\mathbf{t}, x, y) \in \mathcal{T} \times \{0,1\}^n \times \{0,1\}^n$ is added to $\mathcal{Q}_C$. Similarly, for $1 \leq i \leq r$, $\mathcal{Q}_{P_i}$ contains all pairs $(u, v) \in \{0,1\}^n \times \{0,1\}^n$ such that $\mathcal{D}$ made either a direct query u to permutation P_i which was answered by v, or an inverse query v which was answered by u. Note that queries are recorded in a directionless and unordered way, but by our assumption that the distinguisher is deterministic, the raw interaction of $\mathcal{D}$ with its oracles can unambiguously be reconstructed from the queries transcript (see e.g. [6] for more details). Note also that by our assumption that $\mathcal{D}$ never makes pointless queries, each query to the construction oracle results in a distinct triple in $\mathcal{Q}_C$, and each query to P_i results in a distinct pair in $\mathcal{Q}_{P_i}$. Moreover, since we assume that the distinguisher always makes the maximal number of allowed queries to each oracle, one has $|\mathcal{Q}_C| = q_c$ and $|\mathcal{Q}_{P_i}| = q_p$ for $1 \leq i \leq r$. In all the following, we also denote m the number of distinct tweaks appearing in $\mathcal{Q}_C$, and q_i the number of queries for the i-th tweak, $1 \leq i \leq m$, ordering the tweaks arbitrarily. Note that one always has

$\sum_{i=1}^{m} q_i = q_c$, even though m may depend on the answers received from the oracles.

A queries transcript is said *attainable* (with respect to some fixed distinguisher $\mathcal{D}$) if there exists oracles $(\widetilde{P}_0, \mathbf{P})$ such that the interaction of $\mathcal{D}$ with $(\widetilde{P}_0, \mathbf{P})$ results in this transcript (in other words, the probability to obtain this transcript in the ideal world is non-zero). Moreover, in order to have a simple definition of bad transcripts, the actual key $\mathbf{k}$ is revealed to the adversary at the end of the experiment if we are in the real world, while in the ideal world, a "dummy" key $\mathbf{k} \leftarrow_\$ \mathcal{K}$ is simply drawn uniformly at random independently from the answers of the oracle $\widetilde{P}_0$ (this is obviously without loss of generality since this can only help the distinguisher and increase its advantage). All in all, a transcript τ is a tuple $\tau = (\mathcal{Q}_C, \mathcal{Q}_{P_1}, \ldots, \mathcal{Q}_{P_r}, \mathbf{k})$, and we say that a transcript is attainable if the corresponding queries transcript $(\mathcal{Q}_C, \mathcal{Q}_{P_1}, \ldots, \mathcal{Q}_{P_r})$ is attainable. We denote Θ the set of attainable transcripts. In all the following, we denote T_{re}, resp. T_{id}, the probability distribution of the transcript τ induced by the real world, resp. the ideal world (note that these two probability distributions depend on the distinguisher). By extension, we use the same notation to denote a random variable distributed according to each distribution. The main lemma of the H-coefficients technique is the following one (see e.g. [5,6] for the proof).

Lemma 1. *Fix a distinguisher $\mathcal{D}$. Let $\Theta = \Theta_{\mathrm{good}} \sqcup \Theta_{\mathrm{bad}}$ be a partition of the set of attainable transcripts. Assume that there exists ε_1 such that for any $\tau \in \Theta_{\mathrm{good}}$, one has*[4]

$$\frac{Pr[T_{\mathrm{re}} = \tau]}{Pr[T_{\mathrm{id}} = \tau]} \geq 1 - \varepsilon_1,$$

and that there exists ε_2 such that $Pr[T_{\mathrm{id}} \in \Theta_{\mathrm{bad}}] \leq \varepsilon_2$. Then $\mathbf{Adv}(\mathcal{D}) \leq \varepsilon_1 + \varepsilon_2$.

Useful Observations. We end this section with some useful preliminary observations. First, we introduce some additional notation. Given a permutation queries transcript $\mathcal{Q}$ and a permutation P, we say that P *extends* $\mathcal{Q}$, denoted $P \vdash \mathcal{Q}$, if $P(u) = v$ for all $(u, v) \in \mathcal{Q}$. By extension, given a tuple of permutation queries transcripts $\mathcal{Q}_\mathbf{P} = (\mathcal{Q}_{P_1}, \ldots, \mathcal{Q}_{P_r})$ and a tuple of permutations $\mathbf{P} = (P_1, \ldots, P_r)$, we say that $\mathbf{P}$ extends $\mathcal{Q}_\mathbf{P}$, denoted $\mathbf{P} \vdash \mathcal{Q}_\mathbf{P}$, if $P_i \vdash \mathcal{Q}_{P_i}$ for each $i = 1, \ldots, r$. Note that for a permutation transcript of size q_p, one has

$$Pr[P \leftarrow_\$ \mathsf{P}(n) : P \vdash \mathcal{Q}] = \frac{1}{(N)_{q_p}}. \tag{2}$$

Similarly, given a tweakable permutation transcript $\widetilde{\mathcal{Q}}$ and a tweakable permutation $\widetilde{P}$, we say that $\widetilde{P}$ *extends* $\widetilde{\mathcal{Q}}$, denoted $\widetilde{P} \vdash \widetilde{\mathcal{Q}}$, if $\widetilde{P}(t, x) = y$ for all $(t, x, y) \in \widetilde{\mathcal{Q}}$. For a tweakable permutation transcript $\widetilde{\mathcal{Q}}$ with m distinct tweaks and q_i queries corresponding to the i-th tweak, one has

$$Pr[\widetilde{P} \leftarrow_\$ \mathsf{TP}(\mathcal{T}, n) : \widetilde{P} \vdash \widetilde{\mathcal{Q}}] = \prod_{i=1}^{m} \frac{1}{(N)_{q_i}}. \tag{3}$$

---

[4] Recall that for an attainable transcript, one has $Pr[T_{\mathrm{id}} = \tau] > 0$.

It is easy to see that the interaction of a distinguisher $\mathcal{D}$ with oracles $(\widetilde{P}_0, P_1, \ldots, P_r)$ yields any attainable queries transcript $(\mathcal{Q}_C, \mathcal{Q}_\mathbf{P})$ with $\mathcal{Q}_\mathbf{P} = (\mathcal{Q}_{P_1}, \ldots, \mathcal{Q}_{P_r})$ iff $\widetilde{P}_0 \vdash \mathcal{Q}_C$ and $P_i \vdash \mathcal{Q}_{P_i}$ for $1 \leq i \leq r$. In the ideal world, the key $\mathbf{k}$, the permutations $P_1, \ldots, P_r$, and the tweakable permutation $\widetilde{P}_0$ are all uniformly random and independent, so that, by (2) and (3), the probability of getting any attainable transcript $\tau = (\mathcal{Q}_C, \mathcal{Q}_\mathbf{P}, \mathbf{k})$ in the ideal world is

$$\Pr[T_{\mathrm{id}} = \tau] = \frac{1}{|\mathcal{K}|} \times \left(\frac{1}{(N)_{q_p}} \right)^r \times \prod_{i=1}^{m} \frac{1}{(N)_{q_i}}.$$

In the real world, the probability to obtain τ is

$$\Pr[T_{\mathrm{re}} = \tau] = \frac{1}{|\mathcal{K}|} \times \left(\frac{1}{(N)_{q_p}} \right)^r \times \Pr\left[\mathbf{P} \leftarrow_\$ (P(n))^r : \mathrm{TEM}_\mathbf{k}^\mathbf{P} \vdash \mathcal{Q}_C \,\middle|\, \mathbf{P} \vdash \mathcal{Q}_\mathbf{P} \right].$$

Let

$$\mathsf{p}(\tau) \stackrel{\mathrm{def}}{=} \Pr\left[\mathbf{P} \leftarrow_\$ (P(n))^r : \mathrm{TEM}_\mathbf{k}^\mathbf{P} \vdash \mathcal{Q}_C \,\middle|\, \mathbf{P} \vdash \mathcal{Q}_\mathbf{P} \right].$$

Then we have

$$\frac{\Pr[T_{\mathrm{re}} = \tau]}{\Pr[T_{\mathrm{id}} = \tau]} = \mathsf{p}(\tau) \Big/ \prod_{i=1}^{m} \frac{1}{(N)_{q_i}}. \tag{4}$$

Hence, applying Lemma 1 will require three steps: first, define good and bad transcripts, then upper bound the probability of bad transcripts in the ideal world, and finally lower bound the real world probability $\mathsf{p}(\tau)$ when τ is good in order to use Eq. (4).

2.4 An Extended Sum-Capture Lemma

To upper bound the probability of getting a bad transcript in the ideal world, we will need a generalization of the sum-capture theorem from [5] (that applied to a random permutation) to the case of a *family* of random permutations (in other words, a random tweakable permutation).

We denote $\mathsf{GL}(n)$ the general linear group of degree n over $\mathbb{F}_2$, i.e., the set of all automorphisms (linear bijective mappings) of $\mathbb{F}_2^n$.

Lemma 2. *Fix an automorphism $\Gamma \in \mathsf{GL}(n)$ and a non-empty set $\mathcal{T}$. Let $\widetilde{P}$ be a uniformly random tweakable permutation in $\mathsf{TP}(\mathcal{T}, n)$, and let $\mathcal{A}$ be some probabilistic algorithm making exactly q (two-sided) adaptive queries to $\widetilde{P}$. Let $\widetilde{\mathcal{Q}} = ((t_1, x_1, y_1), \ldots, (t_q, x_q, y_q))$ denote the transcript of the interaction of $\mathcal{A}$ with $\widetilde{P}$. For any two subsets U and V of $\{0,1\}^n$, let*

$$\mu(\widetilde{\mathcal{Q}}, U, V) = |\{((t, x, y), u, v) \in \widetilde{\mathcal{Q}} \times U \times V : x \oplus u = \Gamma(y \oplus v)\}|.$$

Then, assuming $9n \leq q \leq N/2$, one has

$$\Pr_{\widetilde{P}, \omega}\left[\exists U, V \subseteq \{0,1\}^n : \mu(\widetilde{\mathcal{Q}}, U, V) \geq \frac{q|U||V|}{N} + \frac{2q^2 \sqrt{|U||V|}}{N} + 3\sqrt{nq|U||V|} \right]$$

$$\leq \frac{2}{N},$$

where the probability is taken over the random choice of $\widetilde{P}$ and the random coins ω of $\mathcal{A}$.

The proof of this lemma is a simple generalization of the one from [5] and can be found in the full version of this paper [9].

3   Beyond-Birthday-Bound Security

3.1   Statement of the Result and Discussion

In this section, we consider the 4-round tweakable Even-Mansour construction $\mathsf{TEM}[n, 4, \mathbf{f}]$ with $2n$-bit keys and n-bit tweaks depicted on Fig. 2. The main result of this paper is the following one:

Theorem 1. *Let $\mathbf{f} = (f_0, \ldots, f_4)$ where $f_i((k_0, k_1), t) = k_{i \bmod 2} \oplus t$. Let q_c, q_p be two integers such that $9n \leq q_c$ and $q_p + 3q_c + 1 \leq N/2$. Then one has*

$$\mathbf{Adv}^{\mathrm{cca}}_{\mathsf{TEM}[n,4,\mathbf{f}]}(q_c, q_p) \leq \frac{44q_c^{3/2} + 38q_c\sqrt{q_p} + (30 + 3\sqrt{n})q_p\sqrt{q_c} + 4q_p^{3/2} + 2}{N}.$$

Hence, this construction ensures CCA-security as long as q_c and q_p are small compared to $2^{2n/3}$, up to logarithmic terms in $N = 2^n$.

The proof follows the H-coefficients method exposed in Sect. 2.3. In Sect. 3.2, we begin by describing the set of bad transcripts and upper bound the probability to get such a transcript in the ideal world. Then, for any good attainable transcript τ, we prove in Sect. 3.3 that the ratio between the probability to get τ in the real world and in the ideal world is close enough to 1.

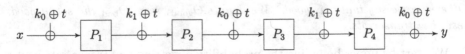

Fig. 2. The 4-round tweakable Even-Mansour construction with a $2n$-bit key (k_0, k_1) and an n-bit tweak t.

3.2   Definition and Probability of Bad Transcripts

The first step is to define the set of bad transcripts. Let $\tau = (\mathcal{Q}_C, \mathcal{Q}_{P_1}, \ldots, \mathcal{Q}_{P_4}, (k_0, k_1))$ be an attainable transcript, with $|\mathcal{Q}_C| = q_c$ and $|\mathcal{Q}_{P_i}| = q_p$ for $i = 1, \ldots, 4$. In all the following, we let, for $i \in \{1, \ldots, 4\}$,

$$U_i = \{u_i \in \{0, 1\}^n : (u_i, v_i) \in \mathcal{Q}_{P_i}\}$$
$$V_i = \{v_i \in \{0, 1\}^n : (u_i, v_i) \in \mathcal{Q}_{P_i}\}$$

denote the domains and ranges of $\mathcal{Q}_{P_i}$ respectively. We also define three quantities characterizing the transcript,

$$\alpha_1 \overset{\text{def}}{=} |\{((t,x,y),u_1) \in \mathcal{Q}_C \times U_1 : x \oplus k_0 \oplus t = u_1\}|$$
$$\alpha_4 \overset{\text{def}}{=} |\{((t,x,y),v_4) \in \mathcal{Q}_C \times V_4 : y \oplus k_0 \oplus t = v_4\}|$$
$$\alpha_{2,3} \overset{\text{def}}{=} |\{((t,x,y),v_2,u_3) \in \mathcal{Q}_C \times V_2 \times U_3 : v_2 \oplus k_0 \oplus t = u_3\}|.$$

We also define two quantities depending respectively on $\mathcal{Q}_{P_2}$ and $\mathcal{Q}_{P_3}$:

$$\nu_2 \overset{\text{def}}{=} |\{((u_2,v_2),(u_2',v_2')) \in (\mathcal{Q}_{P_2})^2 : (u_2,v_2) \neq (u_2',v_2'),\, u_2 \oplus v_2 = u_2' \oplus v_2'\}|$$
$$\nu_3 \overset{\text{def}}{=} |\{((u_3,v_3),(u_3',v_3')) \in (\mathcal{Q}_{P_3})^2 : (u_3,v_3) \neq (u_3',v_3'),\, u_3 \oplus v_3 = u_3' \oplus v_3'\}|.$$

Definition 1. *We say that a transcript τ is bad if at least one of the following conditions is fulfilled:*

(B-1) there exists $(t,x,y) \in \mathcal{Q}_C$, $(u_1,v_1) \in \mathcal{Q}_{P_1}$, and $(u_4,v_4) \in \mathcal{Q}_{P_4}$ such that $k_0 \oplus t = x \oplus u_1 = v_4 \oplus y$;
(B-2) there exists $(t,x,y) \in \mathcal{Q}_C$, $(u_1,v_1) \in \mathcal{Q}_{P_1}$, and $(u_2,v_2) \in \mathcal{Q}_{P_2}$ such that $k_0 \oplus t = x \oplus u_1$ and $k_1 \oplus t = v_1 \oplus u_2$;
(B-3) there exists $(t,x,y) \in \mathcal{Q}_C$, $(u_3,v_3) \in \mathcal{Q}_{P_3}$, and $(u_4,v_4) \in \mathcal{Q}_{P_4}$ such that $k_1 \oplus t = v_3 \oplus u_4$ and $k_0 \oplus t = v_4 \oplus y$;
(B-4) $\alpha_1 \geq \sqrt{q_c}/2$;
(B-5) $\alpha_4 \geq \sqrt{q_c}/2$;
(B-6) $\alpha_{2,3} \geq q_p\sqrt{q_c}$;
(B-7) $\nu_2 \geq \sqrt{q_p}$;
(B-8) $\nu_3 \geq \sqrt{q_p}$.

Otherwise we say that τ is good.[5] *We denote Θ_{good}, resp. Θ_{bad} the set of good, resp. bad transcripts.*

We start by upper bounding the probability of getting bad transcripts in the ideal world.

Lemma 3. *Assume that $9n \leq q_c \leq N/2$ and $q_p \leq N/2$. Then one has*

$$Pr[T_{\text{id}} \in \Theta_{\text{bad}}] \leq \frac{2q_c^2 q_p + 3q_c q_p^2}{N^2} + \frac{(5 + 3\sqrt{n})\sqrt{q_c}q_p + 4q_p^{3/2} + 2}{N}.$$

Proof. We upper bound the probability of each condition in turn. We denote Θ_i the set of attainable transcripts satisfying condition (B-i). Recall that in the ideal world, the key (k_0, k_1) is drawn independently from the queries transcript.

---

[5] We define conditions (B-4) and (B-5) using $\sqrt{q_c}/2$ rather than $\sqrt{q_c}$ in order to be able later to directly apply a previous result by Cogliati *et al.* [7].

Condition (B-1). Let BadK_1 be the set of keys k_0 such that there exists $(t, x, y) \in \mathcal{Q}_C$, $(u_1, v_1) \in \mathcal{Q}_{P_1}$, and $(u_4, v_4) \in \mathcal{Q}_{P_4}$ such that $k_0 \oplus t = x \oplus u_1 = y \oplus v_4$. Note that BadK_1 only depends on the queries transcript, hence for any constant C we have, since k_0 is uniformly random,

$$\Pr\left[T_{\mathrm{id}} \in \Theta_1\right] \leq \Pr\left[\widetilde{P}_0 \leftarrow_\$ \mathsf{TP}(\mathcal{T}, n), \mathbf{P} \leftarrow_\$ (\mathsf{P}(n))^4 \; : \; |\mathsf{BadK}_1| > C\right] + \frac{C}{N}. \quad (5)$$

Moreover, if we let

$$\mu(\mathcal{Q}_C, U_1, V_4) \stackrel{\text{def}}{=} |\{((t, x, y), u_1, v_4) \in \mathcal{Q}_C \times U_1 \times V_4 \; : \; x \oplus u_1 = y \oplus v_4)\}|,$$

then one clearly has

$$|\mathsf{BadK}_1| \leq \mu(\mathcal{Q}_C, U_1, V_4).$$

Hence, we can use Lemma 2 in order to upper-bound $|\mathsf{BadK}_1|$ with overwhelming probability (we consider $\mathcal{D}$ with access to the inner permutations as a probabilistic algorithm $\mathcal{A}$ interacting with the tweakable permutation $\widetilde{P}_0$, resulting in the transcript $\mathcal{Q}_C$, and we let Γ be the identity mapping). For

$$C = \frac{q_c q_p^2}{N} + \frac{2q_c^2 q_p}{N} + 3q_p\sqrt{n q_c},$$

we obtain that

$$\Pr\left[\widetilde{P}_0 \leftarrow_\$ \mathsf{TP}(\mathcal{T}, n), \mathbf{P} \leftarrow_\$ (\mathsf{P}(n))^4 \; : \; |\mathsf{BadK}_1| > C\right] \leq \frac{2}{N}.$$

Using (5) gives

$$\Pr\left[T_{\mathrm{id}} \in \Theta_1\right] \leq \frac{q_c q_p^2}{N^2} + \frac{2q_c^2 q_p}{N^2} + \frac{3q_p\sqrt{n q_c}}{N} + \frac{2}{N}.$$

Conditions (B-2) and (B-3). We consider (B-2). For each $(t, x, y) \in \mathcal{Q}_C$, $(u_1, v_1) \in \mathcal{Q}_{P_1}$, and $(u_2, v_2) \in \mathcal{Q}_{P_2}$, the probability, over the random draw of (k_0, k_1), that $k_0 \oplus t = x \oplus u_1$ and $k_1 \oplus t = v_1 \oplus u_2$ is $1/N^2$ since (k_0, k_1) is uniform and independent from the queries transcript. Summing over the $q_c q_p^2$ possibilities for (t, x, y), (u_1, v_1), and (u_2, v_2) yields

$$\Pr\left[T_{\mathrm{id}} \in \Theta_2\right] \leq \frac{q_c q_p^2}{N^2}.$$

Similarly,

$$\Pr\left[T_{\mathrm{id}} \in \Theta_3\right] \leq \frac{q_c q_p^2}{N^2}.$$

Conditions (B-4) and (B-5). We consider (B-4). Seeing α_1 as a random variable over the random draw of (k_0, k_1), one has

$$\mathbb{E}[\alpha_1] = \sum_{(t,x,y)\in\mathcal{Q}_C} \sum_{u_1\in U_1} \Pr\left[k_0 = x \oplus u_1 \oplus t\right] \leq \frac{q_c q_p}{N}.$$

Then, using Markov's inequality,

$$\Pr\left[T_{\mathrm{id}} \in \Theta_4\right] = \Pr\left[\alpha_1 \geq \frac{\sqrt{q_c}}{2}\right] \leq \frac{2\mathbb{E}[\alpha_1]}{\sqrt{q_c}} \leq \frac{2q_p\sqrt{q_c}}{N}.$$

Similarly,

$$\Pr\left[T_{\mathrm{id}} \in \Theta_5\right] \leq \frac{2q_p\sqrt{q_c}}{N}.$$

Condition (B-6). Again, we see $\alpha_{2,3}$ as a random variable over the random draw of k_0. Then

$$\mathbb{E}[\alpha_{2,3}] = \sum_{(t,x,y)\in\mathcal{Q}_C} \sum_{v_2\in V_2} \sum_{u_3\in U_3} \Pr\left[k_0 = v_2 \oplus u_3 \oplus t\right] \leq \frac{q_c q_p^2}{N}.$$

Then, using Markov's inequality,

$$\Pr\left[T_{\mathrm{id}} \in \Theta_6\right] = \Pr\left[\alpha_{2,3} \geq q_p\sqrt{q_c}\right] \leq \frac{\mathbb{E}[\alpha_{2,3}]}{q_p\sqrt{q_c}} \leq \frac{q_p\sqrt{q_c}}{N}.$$

Conditions (B-7) and (B-8). Consider (B-7). We see the distinguisher combined with $\widetilde{P}_0$ and the inner permutations P_1, P_3, and P_4 as a probabilistic algorithm $\mathcal{A}$ interacting with P_2, and we see ν_2 as a random variable over the random choice of P_2 and the randomness of $\mathcal{A}$. One has

$$\mathbb{E}[\nu_2] = \sum_{\substack{(i,j) \\ 1\leq i\neq j\leq q_c}} \Pr\left[u_{2,i} \oplus v_{2,i} = u_{2,j} \oplus v_{2,j}\right],$$

where the queries to P_2 are ordered as they are issued by $\mathcal{A}$. Consider the i-th and the j-th query, and assume *wlog* that $i < j$. If the j-th is a direct query $u_{2,j}$, then $v_{2,j}$ is uniformly random in a set of size $N - j + 1$. Similarly, if this is a inverse query $v_{2,j}$, then $u_{2,j}$ is uniformly random in a set of size $N - j + 1$. In all cases, the probability that $u_{2,i} \oplus v_{2,i} = u_{2,j} \oplus v_{2,j}$ is at most $1/(N - q_p)$. Hence,

$$\mathbb{E}[\nu_2] \leq \frac{q_p(q_p - 1)}{N - q_p} \leq \frac{2q_p^2}{N}.$$

Using Markov's inequality,

$$\Pr\left[T_{\mathrm{id}} \in \Theta_7\right] = \Pr\left[\nu_2 \geq \sqrt{q_p}\right] \leq \frac{2q_p^{3/2}}{N}.$$

Similarly,

$$\Pr\left[T_{\mathrm{id}} \in \Theta_8\right] \leq \frac{2q_p^{3/2}}{N}.$$

The result follows by a union bound over all cases.   □

3.3 Analysis of Good Transcripts

In this section, we fix a good transcript $\tau = (\mathcal{Q}_C, \mathcal{Q}_{P_1}, \ldots, \mathcal{Q}_{P_4}, (k_0, k_1))$. By (4), we have to lower bound

$$\mathsf{p}(\tau) \stackrel{\text{def}}{=} \Pr\left[\mathbf{P} \leftarrow_\$ (\mathsf{P}(n))^4 : \mathrm{TEM}^{\mathbf{P}}_{k_0, k_1} \vdash \mathcal{Q}_C \middle| P_1 \vdash \mathcal{Q}_{P_1} \wedge \ldots \wedge P_4 \vdash \mathcal{Q}_{P_4}\right].$$

The proof will proceed in two steps: first, we will lower bound the probability that permutations P_1 and P_4 satisfy some conditions given in the definition below, and then, assuming (P_1, P_4) is good, we will lower bound the probability, over the choice of P_2 and P_3, that $\mathrm{TEM}^{\mathbf{P}}_{k_0, k_1} \vdash \mathcal{Q}_C$. For this second step, we will directly appeal to a previous result by Cogliati *et al.* [7].

We start by giving the conditions defining good pairs of permutations (P_1, P_4). We stress that these conditions cannot be accommodated in the definition of bad transcripts since they depend on values of P_1 and P_4 which do *not* appear in the queries transcript, so that they cannot be defined from the transcript τ alone. We also warn the reader upfront that conditions (C-5) and (C-6) are "dummy" conditions that will easily be seen to be impossible to fulfill, yet will allow us to cleanly use the previous result of Cogliati *et al.* [7].

Definition 2. *A pair of permutations* (P_1, P_4) *such that* $P_1 \vdash \mathcal{Q}_{P_1}$ *and* $P_4 \vdash \mathcal{Q}_{P_4}$ *is said bad if at least one of the following conditions is fulfilled (see Fig. 3 for a diagram of the first ten conditions):*

(C-1) there exists $(t, x, y) \in \mathcal{Q}_C$, $u_2 \in U_2$, *and* $v_3 \in V_3$ *such that*

$$\begin{cases} P_1(x \oplus k_0 \oplus t) \oplus k_1 \oplus t = u_2 \\ P_4^{-1}(y \oplus k_0 \oplus t) \oplus k_1 \oplus t = v_3; \end{cases}$$

(C-2) there exists $(t, x, y) \in \mathcal{Q}_C$, $(u_2, v_2) \in \mathcal{Q}_{P_2}$, *and* $u_3 \in U_3$ *such that*

$$\begin{cases} P_1(x \oplus k_0 \oplus t) \oplus k_1 \oplus t = u_2 \\ v_2 \oplus k_0 \oplus t = u_3; \end{cases}$$

(C-3) there exists $(t, x, y) \in \mathcal{Q}_C$, $(u_3, v_3) \in \mathcal{Q}_{P_3}$, *and* $v_2 \in V_2$ *such that*

$$\begin{cases} P_4^{-1}(y \oplus k_0 \oplus t) \oplus k_1 \oplus t = v_3 \\ u_3 \oplus k_0 \oplus t = v_2; \end{cases}$$

(C-4) there exists $(t, x, y), (t', x', y'), (t'', x'', y'') \in \mathcal{Q}_C$ *with* (t, x, y) *distinct from* (t', x', y') *and from* (t'', x'', y'') *such that*

$$\begin{cases} P_1(x \oplus k_0 \oplus t) \oplus t = P_1(x' \oplus k_0 \oplus t') \oplus t' \\ P_4^{-1}(y \oplus k_0 \oplus t) \oplus t = P_4^{-1}(y'' \oplus k_0 \oplus t'') \oplus t''; \end{cases}$$

(C-5) there exists $(t, x, y,) \neq (t', x', y') \in \mathcal{Q}_C$ *such that*

$$\begin{cases} P_1(x \oplus k_0 \oplus t) \oplus t = P_1(x' \oplus k_0 \oplus t') \oplus t' \\ t = t'; \end{cases}$$

(C-6) there exists $(t, x, y,) \neq (t', x', y') \in \mathcal{Q}_C$ *such that*

$$\begin{cases} P_4^{-1}(y \oplus k_0 \oplus t) \oplus t = P_4^{-1}(y' \oplus k_0 \oplus t') \oplus t' \\ t = t'; \end{cases}$$

(C-7) there exists $(t, x, y) \neq (t', x', y') \in \mathcal{Q}_C$ *and* $u_2 \in U_2$ *such that*

$$\begin{cases} P_1(x \oplus k_0 \oplus t) \oplus k_1 \oplus t = u_2 \\ P_4^{-1}(y \oplus k_0 \oplus t) \oplus t = P_4^{-1}(y' \oplus k_0 \oplus t') \oplus t'; \end{cases}$$

(C-8) there exists $(t, x, y) \neq (t', x', y') \in \mathcal{Q}_C$ *and* $v_3 \in V_3$ *such that*

$$\begin{cases} P_4^{-1}(y \oplus k_0 \oplus t) \oplus k_1 \oplus t = v_3 \\ P_1(x \oplus k_0 \oplus t) \oplus t = P_1(x' \oplus k_0 \oplus t') \oplus t'; \end{cases}$$

(C-9) there exists $(t, x, y) \neq (t', x', y') \in \mathcal{Q}_C$ *and* $(u_2, v_2), (u'_2, v'_2) \in \mathcal{Q}_{P_2}$ *such that*

$$\begin{cases} P_1(x \oplus k_0 \oplus t) \oplus k_1 \oplus t = u_2 \\ P_1(x' \oplus k_0 \oplus t') \oplus k_1 \oplus t' = u'_2 \\ v_2 \oplus t = v'_2 \oplus t'; \end{cases}$$

(C-10) there exists $(t, x, y) \neq (t', x', y') \in \mathcal{Q}_C$ *and* $(u_3, v_3), (u'_3, v'_3) \in \mathcal{Q}_{P_3}$ *such that*

$$\begin{cases} P_4^{-1}(y \oplus k_0 \oplus t) \oplus k_1 \oplus t = v_3 \\ P_4^{-1}(y' \oplus k_0 \oplus t') \oplus k_1 \oplus t' = v'_3 \\ u_3 \oplus t = u'_3 \oplus t'; \end{cases}$$

(C-11) $\alpha_2 \geq \sqrt{q_c}$;
(C-12) $\alpha_3 \geq \sqrt{q_c}$;
(C-13) $\beta_2 \geq \sqrt{q_c}$;
(C-14) $\beta_3 \geq \sqrt{q_c}$;

where

$$\alpha_2 \stackrel{\text{def}}{=} |\{(t, x, y) \in \mathcal{Q}_C : P_1(x \oplus k_0 \oplus t) \oplus k_1 \oplus t \in U_2\}|,$$

$$\alpha_3 \stackrel{\text{def}}{=} |\{(t, x, y) \in \mathcal{Q}_C : P_4^{-1}(y \oplus k_0 \oplus t) \oplus k_1 \oplus t \in V_3\}|,$$

$$\beta_2 \stackrel{\text{def}}{=} |\{(t, x, y) \in \mathcal{Q}_C : \exists (t', x', y') \neq (t, x, y),$$
$$P_1(x \oplus k_0 \oplus t) \oplus t = P_1(x' \oplus k_0 \oplus t') \oplus t'\}|,$$

$$\beta_3 \stackrel{\text{def}}{=} |\{(t, x, y) \in \mathcal{Q}_C : \exists (t', x', y') \neq (t, x, y),$$
$$P_4^{-1}(y \oplus k_0 \oplus t) \oplus t = P_4^{-1}(y' \oplus k_0 \oplus t') \oplus t'\}|.$$

Otherwise we say that (P_1, P_4) *is good. We denote* Π_{good}, *resp.* Π_{bad} *the set of good, resp. bad pairs of permutations* (P_1, P_4) *such that* $P_1 \vdash \mathcal{Q}_{P_1}$ *and* $P_4 \vdash \mathcal{Q}_{P_4}$.

In all the following, we denote Π the set of pairs of permutations (P_1, P_4) such that $P_1 \vdash \mathcal{Q}_{P_1}$ and $P_4 \vdash \mathcal{Q}_{P_4}$. The first step towards studying good transcripts will be to upper bound the probability that the pair (P_1, P_4) is bad.

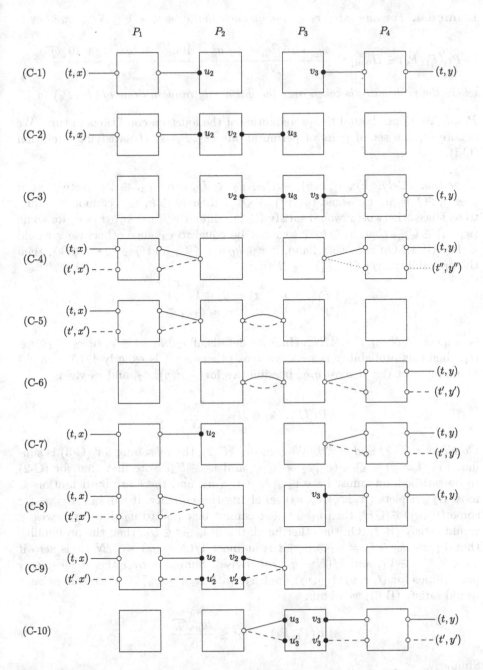

Fig. 3. The ten "collision" conditions characterizing a bad pair of permutations (P_1, P_4). Black dots correspond to pairs $(u_2, v_2) \in \mathcal{Q}_{P_2}$ or $(u_3, v_3) \in \mathcal{Q}_{P_3}$. Note that for (C-4) one might have $(t', x') = (t'', x'')$, and for (C-9) (resp. (C-10)) one might have $x \oplus t = x' \oplus t'$ (resp. $y \oplus t = y' \oplus t'$).

Lemma 4. *For any integers q_c and q_p such that $q_p + q_c + 1 \leq N/2$, one has*

$$Pr[(P_1, P_4) \in \Pi_{\text{bad}}] \leq \frac{4q_c^3 + 16q_c^2 q_p + 4q_c q_p^2}{N^2} + \frac{10q_c^{3/2} + 4q_c\sqrt{q_p} + 10\sqrt{q_c}q_p}{N}$$

where the probability is taken over the uniformly random draw of (P_1, P_4) in Π.

Proof. We upper bound the probabilities of the fourteen conditions in turn. We denote Π_i the set of pairs of permutations $(P_1, P_4) \in \Pi$ satisfying condition (C-i).

Condition (C-1). Fix $(t, x, y) \in \mathcal{Q}_C$, $u_2 \in U_2$, and $v_3 \in V_3$. Note that if $x \oplus k_0 \oplus t = u_1$ for some $(u_1, v_1) \in \mathcal{Q}_{P_1}$, then $v_1 \oplus k_1 \oplus t$ cannot be equal to u_2 since otherwise τ would satisfy (B-2). Similarly, if $y \oplus k_0 \oplus t = v_4$ for some $(u_4, v_4) \in \mathcal{Q}_{P_4}$, then $u_4 \oplus k_1 \oplus t$ cannot be equal to v_3 since otherwise τ would satisfy (B-3). On the other hand, if $x \oplus k_0 \oplus t \notin U_1$ and $y \oplus k_0 \oplus t \notin V_4$, then the probability over $(P_1, P_4) \leftarrow_\$ \Pi$ that

$$\begin{cases} P_1(x \oplus k_0 \oplus t) = u_2 \oplus k_1 \oplus t \\ P_4^{-1}(y \oplus k_0 \oplus t) = v_3 \oplus k_1 \oplus t \end{cases}$$

is at most $1/(N - q_p)^2 \leq 4/N^2$. (In more details, if $u_2 \oplus k_1 \oplus t \in V_1$ or $v_3 \oplus k_1 \oplus t \in U_4$, then this probability is zero, whereas otherwise it is exactly $1/(N - q_p)^2$.) Summing over the at most $q_c q_p^2$ possibilities for (t, x, y), u_2, and v_3 yields

$$Pr[(P_1, P_4) \in \Pi_1] \leq \frac{4q_c q_p^2}{N^2}.$$

Conditions (C-2) and (C-3). We consider (C-2), the reasoning for (C-3) is similar. Fix $(t, x, y) \in \mathcal{Q}_C$, $(u_2, v_2) \in \mathcal{Q}_{P_2}$, and $u_3 \in U_3$. Note first that for (C-2) to be satisfied, one must have $v_2 \oplus k_0 \oplus t = u_3$, and there are by definition at most $\alpha_{2,3}$ triplets $((t, x, y), v_2, u_3)$ satisfying this equality. If $x \oplus k_0 \oplus t = u_1$ for some $(u_1, v_1) \in \mathcal{Q}_{P_1}$, then $v_1 \oplus k_1 \oplus t$ cannot be equal to u_2 since otherwise τ would satisfy (B-2). On the other hand, if $x \oplus k_0 \oplus t \notin U_1$, then the probability that $P_1(x \oplus k_0 \oplus t) = u_2 \oplus k_1 \oplus t$ is at most $1/(N - q_p) \leq 2/N$ (it is zero if $u_2 \oplus k_1 \oplus t \in V_1$, and $1/(N - q_p)$ otherwise). Summing over the at most $\alpha_{2,3}$ possibilities for (t, x, y), (u_2, v_2), and u_3, with $\alpha_{2,3} \leq q_p\sqrt{q_c}$ since otherwise τ would satisfy (B-6), we obtain

$$Pr[(P_1, P_4) \in \Pi_2] \leq \frac{2q_p\sqrt{q_c}}{N}.$$

Similarly,

$$Pr[(P_1, P_4) \in \Pi_3] \leq \frac{2q_p\sqrt{q_c}}{N}.$$

Condition (C-4). Fix $(t, x, y), (t', x', y'), (t'', x'', y'') \in \mathcal{Q}_C$ with (t, x, y) distinct from (t', x', y') and from (t'', x'', y''). First, note that if $x \oplus k_0 \oplus t = x' \oplus k_0 \oplus t'$ or $y \oplus k_0 \oplus t = y'' \oplus k_0 \oplus t''$, then (C-4) cannot be satisfied. Hence, we assume that none of these two equalities holds. We consider three cases. Assume first that $x \oplus k_0 \oplus t = u_1$ for some $(u_1, v_1) \in \mathcal{Q}_{P_1}$. Note that there are at most α_1 possibilities for (t, x, y), and $\alpha_1 \leq \sqrt{q_c}/2$ since otherwise τ would satisfy (B-4). Moreover $y \oplus k_0 \oplus t \notin V_4$ since otherwise τ would satisfy (B-1). Hence, the probability that

$$P_4^{-1}(y \oplus k_0 \oplus t) \oplus t = P_4^{-1}(y'' \oplus k_0 \oplus t'') \oplus t''$$

is at most $1/(N - q_p - 1) \leq 2/N$. (In more details, if $y'' \oplus k_0 \oplus t'' \in V_4$, then this probability is either zero if $P_4^{-1}(y'' \oplus k_0 \oplus t'') \oplus t \oplus t'' \in U_4$, or exactly $1/(N - q_p)$ otherwise, whereas if $y'' \oplus k_0 \oplus t'' \notin V_4$, then this probability is at most $1/(N - q_p - 1)$.) Summing over the at most $\sqrt{q_c}/2 \times q_c$ possibilities for (t, x, y) and (t'', x'', y''), the probability of this first case is at most $q_c^{3/2}/N$. The second case where $y \oplus k_0 \oplus t \in V_4$ is handled similarly. Finally, consider the case where $x \oplus k_0 \oplus t \notin U_1$ and $y \oplus k_0 \oplus t \notin V_4$. Then the probability that

$$\begin{cases} P_1(x \oplus k_0 \oplus t) \oplus t = P_1(x' \oplus k_0 \oplus t') \oplus t' \\ P_4^{-1}(y \oplus k_0 \oplus t) \oplus t = P_4^{-1}(y'' \oplus k_0 \oplus t'') \oplus t''; \end{cases}$$

is at most $1/(N - q_p - 1)^2 \leq 4/N^2$. Summing over the at most q_c^3 possibilities for (t, x, y), (t', x', y'), and (t'', x'', y''), the probability of this third case is at most $4q_c^3/N^2$. Overall, we obtain

$$\Pr[(P_1, P_4) \in \Pi_4] \leq \frac{4q_c^3}{N^2} + \frac{2q_c^{3/2}}{N}.$$

Conditions (C-5) and (C-6). These conditions cannot be satisfied. Indeed, assume that there exits $(t, x, y) \neq (t', x', y') \in \mathcal{Q}_C$ satisfying (C-5). Since $t = t'$, then $x \neq x'$ by the assumption that the distinguisher never makes pointless queries. This obviously implies that $P_1(x \oplus k_0 \oplus t) \oplus t \neq P_1(x' \oplus k_0 \oplus t') \oplus t'$, a contradiction. The reasoning is similar for (C-6). Hence,

$$\Pr[(P_1, P_4) \in \Pi_5] = \Pr[(P_1, P_4) \in \Pi_6] = 0.$$

Conditions (C-7) and (C-8). We consider condition (C-7). Fix queries $(t, x, y) \neq (t', x', y') \in \mathcal{Q}_C$ and $u_2 \in U_2$. We will consider two cases: first, the case where $y \oplus k_0 \oplus t \in V_4$, and then the case where $y \oplus k_0 \oplus t \notin V_4$. For both cases, note that if $x \oplus k_0 \oplus t = u_1$ for some $(u_1, v_1) \in \mathcal{Q}_{P_1}$, then $v_1 \oplus k_1 \oplus t$ cannot be equal to u_2 since otherwise τ would satisfy (B-2). Hence, we can assume that $x \oplus k_0 \oplus t \notin U_1$. It follows that the probability that

$$P_1(x \oplus k_0 \oplus t) \oplus k_1 \oplus t = u_2$$

is at most $1/(N - q_p) \leq 2/N$ (it is zero if $u_2 \oplus k_1 \oplus t \in V_1$, and $1/(N - q_p)$ otherwise). Summing over the at most α_4 queries $(t, x, y) \in \mathcal{Q}_C$ such that

$y \oplus k_0 \oplus t \in V_4$, with $\alpha_4 \leq \sqrt{q_c}/2$ since otherwise τ would satisfy (B-5), and the q_p possibilities for u_2, we see that the first case happens with probability at most $q_p\sqrt{q_c}/N$. Assume now that $y \oplus k_0 \oplus t \notin V_4$. Then the probability that

$$P_4^{-1}(y \oplus k_0 \oplus t) \oplus t = P_4^{-1}(y' \oplus k_0 \oplus t') \oplus t'$$

is at most $1/(N - q_p - 1) \leq 2/N$. (In more details, if $y \oplus k_0 \oplus t = y' \oplus k_0 \oplus t'$, then it can easily be seen that it cannot hold, whereas if $y \oplus k_0 \oplus t \neq y' \oplus k_0 \oplus t'$, the equation holds with probability at most $1/(N - q_p - 1)$.) Summing over the at most $q_c^2 q_p$ possibilities for (t, x, y), (t', x', y'), and u_2, we see that the probability of the second case is at most $4q_c^2 q_p/N^2$. Overall,

$$\Pr\left[(P_1, P_4) \in \Pi_7\right] \leq \frac{q_p\sqrt{q_c}}{N} + \frac{4q_c^2 q_p}{N^2}.$$

Similarly, one has

$$\Pr\left[(P_1, P_4) \in \Pi_8\right] \leq \frac{q_p\sqrt{q_c}}{N} + \frac{4q_c^2 q_p}{N^2}.$$

Conditions (C-9) and (C-10). Consider condition (C-9). First note that, if the condition is satisfied, we have $x \oplus k_0 \oplus t \notin U_1$, $x' \oplus k_0 \oplus t' \notin U_1$, $u_2 \oplus k_1 \oplus t \notin V_1$ and $u_2' \oplus k_1 \oplus t' \notin V_1$, otherwise (B-2) is fulfilled. Moreover, if $(u_2, v_2) = (u_2', v_2')$, then $t = t'$, thus $x = x'$, which is impossible. Hence we must have $(u_2, v_2) \neq (u_2', v_2')$. The condition can be divided into two conditions:

9.1 there exists $(t, x, y) \neq (t', x', y') \in \mathcal{Q}_C$ and $(u_2, v_2) \neq (u_2', v_2') \in \mathcal{Q}_{P_2}$ such that $x \oplus t = x' \oplus t'$, $P_1(x \oplus k_0 \oplus t) = u_2 \oplus k_1 \oplus t$ and $P_1(x' \oplus k_0 \oplus t') = u_2' \oplus k_1 \oplus t'$ and $v_2 \oplus t = v_2' \oplus t'$;

9.2 there exists $(t, x, y) \neq (t', x', y') \in \mathcal{Q}_C$ and $(u_2, v_2) \neq (u_2', v_2') \in \mathcal{Q}_{P_2}$ such that $x \oplus t \neq x' \oplus t'$, $P_1(x \oplus k_0 \oplus t) = u_2 \oplus k_1 \oplus t$ and $P_1(x' \oplus k_0 \oplus t') = u_2' \oplus k_1 \oplus t'$ and $v_2 \oplus t = v_2' \oplus t'$.

In the first case, one has

$$u_2 \oplus k_1 \oplus t = P_1(x \oplus k_0 \oplus t) = P_1(x' \oplus k_0 \oplus t') = u_2' \oplus k_1 \oplus t',$$

thus $u_2 \oplus u_2' = t \oplus t' = v_2 \oplus v_2'$. Hence the first condition implies the following one: there exists $(t, x, y) \in \mathcal{Q}_C$ and $(u_2, v_2) \neq (u_2', v_2') \in \mathcal{Q}_{P_2}$ such that $P_1(x \oplus k_0 \oplus t) = u_2 \oplus k_1 \oplus t$ and $u_2 \oplus u_2' = v_2 \oplus v_2'$, with $x \oplus k_0 \oplus t \notin U_1$ and $u_2 \oplus k_1 \oplus t \notin V_1$. Since $\nu_2 < \sqrt{q_p}$, the number of suitable $u_2 \in U_2$ is lower than $\sqrt{q_p}$, and the probability that this first condition is fulfilled is at most $\frac{q_c\sqrt{q_p}}{N - q_p} \leq \frac{2q_c\sqrt{q_p}}{N}$. For the second condition, fix any queries $(t, x, y) \neq (t', x', y') \in \mathcal{Q}_C$ such that $x \oplus t \neq x' \oplus t'$, $x \oplus k_0 \oplus t \notin U_1$, $x' \oplus k_0 \oplus t' \notin U_1$ and $(u_2, v_2) \in \mathcal{Q}_{P_2}$. If $v_2 \oplus t \oplus t' \notin V_2$, the condition cannot be fulfilled. Otherwise let $(u_2', v_2') \in \mathcal{Q}_{P_2}$ be the unique query such that $v_2 \oplus t = v_2' \oplus t'$. Then the probability that $P_1(x \oplus k_0 \oplus t) = u_2 \oplus k_1 \oplus t$ and $P_1(x' \oplus k_0 \oplus t') = u_2' \oplus k_1 \oplus t'$ is at most $\frac{1}{(N - q_p)(N - q_p - 1)}$. Finally, by summing over every possible tuple of queries, and by taking into account the condition 9.1, one has

$$\Pr\left[(P_1, P_4) \in \Pi_9\right] \leq \frac{2q_c\sqrt{q_p}}{N} + \frac{4q_c^2 q_p}{N^2}.$$

Similarly,

$$\Pr\left[(P_1, P_4) \in \Pi_{10}\right] \leq \frac{2q_c\sqrt{q_p}}{N} + \frac{4q_c^2 q_p}{N^2}.$$

Conditions (C-11) and (C-12). We see α_2 (resp. α_3) as a random variable over the choice of P_1 (resp. P_4). Note that

$$\alpha_2 = |\{(t, x, y) \in \mathcal{Q}_C : P_1(x \oplus k_0 \oplus t) \oplus k_1 \oplus t \in U_2\}|$$
$$= |\{(t, x, y) \in \mathcal{Q}_C : x \oplus k_0 \oplus t \notin U_1, P_1(x \oplus k_0 \oplus t) \oplus k_1 \oplus t \in U_2\}|,$$

because, if $x \oplus k_0 \oplus t \in U_1$ and $P_1(x \oplus k_0 \oplus t) \oplus k_1 \oplus t \in U_2$, then (B-2) is fulfilled. We denote $\mathcal{Q}_{C,1}$ the subset of queries $(t, x, y) \in \mathcal{Q}_C$ such that $x \oplus k_0 \oplus t \notin U_1$. Then

$$\mathbb{E}[\alpha_2] = \sum_{(t,x,y)\in\mathcal{Q}_{C,1}} \sum_{u_2 \in U_2} \Pr\left[P_1(x \oplus k_0 \oplus t) = u_2 \oplus k_1 \oplus t\right]$$
$$\leq \sum_{(t,x,y)\in\mathcal{Q}_{C,1}} \sum_{u_2 \in U_2} \frac{1}{N - q_p}$$
$$\leq \frac{2q_c q_p}{N}.$$

Using Markov's inequality, we get

$$\Pr\left[(P_1, P_4) \in \Pi_{11}\right] \leq \frac{2q_p\sqrt{q_c}}{N}.$$

Similarly,

$$\Pr\left[(P_1, P_4) \in \Pi_{12}\right] \leq \frac{2q_p\sqrt{q_c}}{N}.$$

Conditions (C-13) and (C-14). Consider condition (C-13). Note that

$$\beta_2 = |\{(t, x, y) \in \mathcal{Q}_C : \exists (t', x', y') \neq (t, x, y),$$
$$P_1(x \oplus k_0 \oplus t) \oplus t = P_1(x' \oplus k_0 \oplus t') \oplus t'\}|$$
$$\leq \alpha_1 + |\{(t, x, y) \in \mathcal{Q}_C : x \oplus k_0 \oplus t \notin U_1 \text{ and } \exists (t', x', y') \neq (t, x, y),$$
$$P_1(x \oplus k_0 \oplus t) \oplus t = P_1(x' \oplus k_0 \oplus t') \oplus t'\}|.$$

We denote β_2' the last term of this sum. Thus

$$\mathbb{E}[\beta_2'] = \sum_{(t,x,y)\in\mathcal{Q}_{C,1}} \sum_{(t',x',y')\neq(t,x,y)} \Pr\left[P_1(x \oplus k_0 \oplus t) \oplus t = P_1(x' \oplus k_0 \oplus t') \oplus t'\right]$$
$$\leq \frac{q_c^2}{N - q_p - 1} \leq \frac{2q_c^2}{N}.$$

This inequality holds because, if $x \oplus t = x' \oplus t'$, then $t \neq t'$ since the distinguisher never makes pointless queries, thus $P_1(x \oplus k_0 \oplus t) \oplus t = P_1(x' \oplus k_0 \oplus t') \oplus t'$ cannot be fulfilled. Otherwise,

$$\Pr[P_1(x \oplus k_0 \oplus t) \oplus t = P_1(x' \oplus k_0 \oplus t') \oplus t'] \leq \frac{1}{N - q_p - 1}.$$

Finally, since (B-4) is not fulfilled, $\alpha_1 < \sqrt{q_c}/2$. Thus $\beta_2 \geq \sqrt{q_c}$ implies $\beta'_2 \geq \sqrt{q_c}/2$. Hence, using Markov's inequality,

$$\Pr[(P_1, P_4) \in \Pi_{13}] \leq \Pr[\beta'_2 \geq \sqrt{q_c}/2] \leq \frac{2\mathbb{E}[\beta'_2]}{\sqrt{q_c}} \leq \frac{4q_c^{3/2}}{N}.$$

Similarly,

$$\Pr[(P_1, P_4) \in \Pi_{14}] \leq \frac{4q_c^{3/2}}{N}.$$

The result follows by an union bound over all conditions.   □

We are now ready for the second step of the reasoning.

Definition 3. *Fix any pair of permutations (P_1, P_4) such that $P_1 \vdash \mathcal{Q}_{P_1}$ and $P_4 \vdash \mathcal{Q}_{P_4}$. We define a new query transcript $\mathcal{Q}'_C$ depending on (P_1, P_4) as*

$$\mathcal{Q}'_C = \{(t, P_1(x \oplus k_0 \oplus t), P_4^{-1}(y \oplus k_0 \oplus t)) : (t, x, y) \in \mathcal{Q}_C\}.$$

We also denote

$$\tilde{\mathsf{p}}(\tau, P_1, P_4) = \Pr\left[P_2, P_3 \leftarrow_\$ \mathsf{P}(n) : \mathsf{TEM}_{k_1, k_0}^{P_2, P_3} \vdash \mathcal{Q}'_C \,\middle|\, (P_2 \vdash \mathcal{Q}_{P_2}) \wedge (P_3 \vdash \mathcal{Q}_{P_3})\right].$$

Lemma 5. *One has*

$$\frac{\Pr[T_{\mathrm{re}} = \tau]}{\Pr[T_{\mathrm{id}} = \tau]} \geq \sum_{(P_1, P_4) \in \Pi_{\mathrm{good}}} \frac{\tilde{\mathsf{p}}(\tau, P_1, P_4)}{((N - q_p)!)^2 \prod_{i=1}^m 1/(N)_{q_i}}.$$

Proof. Clearly, once P_1 and P_4 are fixed, $\mathsf{TEM}_{k_0, k_1}^{P_1, P_2, P_3, P_4} \vdash \mathcal{Q}_C$ is equivalent to $\mathsf{TEM}_{k_1, k_0}^{P_2, P_3} \vdash \mathcal{Q}'_C$. Hence,

$$\mathsf{p}(\tau) = \sum_{(\bar{P}_1, \bar{P}_4) \in \Pi} \Pr[(P_1, P_4) \leftarrow_\$ \Pi : (P_1 = \bar{P}_1) \wedge (P_4 = \bar{P}_4)]\, \tilde{\mathsf{p}}(\tau, \bar{P}_1, \bar{P}_4)$$

$$\geq \sum_{(\bar{P}_1, \bar{P}_4) \in \Pi_{\mathrm{good}}} \frac{\tilde{\mathsf{p}}(\tau, \bar{P}_1, \bar{P}_4)}{((N - q_p)!)^2}.$$

The result follows from Eq. (4).   □

We can now directly appeal to a previous result by Cogliati *et al.* [7].

Lemma 6. *Let q_c and q_p be two positive integers such that $q_p + 3q_c \leq N/2$. Fix any pair of permutations $(P_1, P_4) \in \Pi_{\text{good}}$. Then*

$$\frac{\tilde{p}(\tau, P_1, P_4)}{\prod_{i=1}^m 1/(N)_{q_i}} \geq 1 - \left(\frac{4q_c(q_p + 2q_c)^2}{N^2} + \frac{14q_c^{3/2} + 4\sqrt{q_c}q_p}{N}\right).$$

Proof. One can check that the queries transcript $\tau' = (\mathcal{Q}'_C, \mathcal{Q}_{P_2}, \mathcal{Q}_{P_3})$ satisfies exactly the conditions defining a good transcript as per [7, Definition 2]. Moreover, the ratio $\tilde{p}(\tau, P_1, P_4)/\prod_{i=1}^m 1/(N)_{q_i}$ is exactly the ratio of the probabilities to get τ' in the real and in the ideal world once a good pair (P_1, P_4) is fixed. Hence, we can apply [7, Lemma 6] that directly yields the result.[6] $\square$

We are now ready to prove the main lemma of this section.

Lemma 7. *Let q_c and q_p be two positive integers such that $q_p + 3q_c + 1 \leq N/2$. One has*

$$\frac{Pr[T_{\text{re}} = \tau]}{Pr[T_{\text{id}} = \tau]} \geq 1 - \frac{20q_c^3 + 32q_c^2 q_p + 8q_c q_p^2}{N^2} - \frac{24q_c^{3/2} + 4q_c\sqrt{q_p} + 14\sqrt{q_c}q_p}{N}.$$

Proof. From Lemmas 5 and 6, one has

$$\frac{\Pr[T_{\text{re}} = \tau]}{\Pr[T_{\text{id}} = \tau]} \geq \sum_{(P_1,P_4)\in\Pi_{\text{good}}} \frac{\tilde{p}(\tau, P_1, P_4)}{((N - q_p)!)^2 \prod_{i=1}^m 1/(N)_{q_i}}$$

$$\geq \left(1 - \frac{4q_c(q_p + 2q_c)^2}{N^2} - \frac{14q_c^{3/2} + 4\sqrt{q_c}q_p}{N}\right) \sum_{\Pi_{\text{good}}} \frac{1}{((N - q_p)!)^2}$$

$$= \left(1 - \frac{4q_c(q_p + 2q_c)^2}{N^2} - \frac{14q_c^{3/2} + 4\sqrt{q_c}q_p}{N}\right) \frac{|\Pi_{\text{good}}|}{((N - q_p)!)^2}$$

$$= \left(1 - \frac{4q_c(q_p + 2q_c)^2}{N^2} - \frac{14q_c^{3/2} + 4\sqrt{q_c}q_p}{N}\right) \Pr[(P_1, P_4) \in \Pi_{\text{good}}],$$

where the last probability is taken over the random draw of (P_1, P_4) from Π, the set of pairs of permutations satisfying $P_1 \vdash \mathcal{Q}_{P_1}$ and $P_4 \vdash \mathcal{Q}_{P_4}$. Using Lemma 4, one has

$$\frac{\Pr[T_{\text{re}} = \tau]}{\Pr[T_{\text{id}} = \tau]} \geq \left(1 - \frac{4q_c^3 + 16q_c^2 q_p + 4q_c q_p^2}{N^2} - \frac{10q_c^{3/2} + 4q_c\sqrt{q_p} + 10\sqrt{q_c}q_p}{N}\right)$$

$$\times \left(1 - \frac{4q_c(q_p + 2q_c)^2}{N^2} - \frac{14q_c^{3/2} + 4\sqrt{q_c}q_p}{N}\right)$$

$$\geq 1 - \frac{20q_c^3 + 32q_c^2 q_p + 8q_c q_p^2}{N^2} - \frac{24q_c^{3/2} + 4q_c\sqrt{q_p} + 14\sqrt{q_c}q_p}{N}. \qquad \square$$

---

[6] Even though this might not be apparent to the reader unfamiliar with [7], the proof of Lemma 7 in that paper does not rely on the xor-universal hash functions h_1 and h_2 appearing in the definition of good transcripts of [7].

Concluding. We are now ready to prove Theorem 1. Combining Lemmas 1, 3, and 7, one has

$$\mathbf{Adv}^{\mathrm{cca}}_{\mathrm{TEM}[n,4,\mathbf{f}]}(q_c, q_p) \leq \frac{2q_c^2 q_p + 3q_c q_p^2}{N^2} + \frac{(5 + 3\sqrt{n})\sqrt{q_c}q_p + 4q_p^{3/2} + 2}{N}$$

$$+ \frac{20q_c^3 + 32q_c^2 q_p + 8q_c q_p^2}{N^2} + \frac{24q_c^{3/2} + 4q_c\sqrt{q_p} + 14\sqrt{q_c}q_p}{N}$$

$$\leq \frac{20q_c^3 + 34q_c^2 q_p + 11q_c q_p^2}{N^2}$$

$$+ \frac{24q_c^{3/2} + 4q_c\sqrt{q_p} + (19 + 3\sqrt{n})\sqrt{q_c}q_p + 4q_p^{3/2} + 2}{N}.$$

Since the result holds trivially when $q_c^3 > N^2$, $q_c^2 q_p > N^2$, or $q_c q_p^2 > N^2$, we can assume that $q_c^3 \leq N^2$, $q_c^2 q_p \leq N^2$, and $q_c q_p^2 \leq N^2$, so that

$$\frac{q_c^3}{N^2} \leq \frac{q_c^{3/2}}{N}, \quad \frac{q_c^2 q_p}{N^2} \leq \frac{q_c\sqrt{q_p}}{N}, \quad \text{and} \quad \frac{q_c q_p^2}{N^2} \leq \frac{\sqrt{q_c}q_p}{N}.$$

Thus

$$\mathbf{Adv}^{\mathrm{cca}}_{\mathrm{TEM}[n,4,\mathbf{f}]}(q_c, q_p) \leq \frac{44q_c^{3/2} + 38q_c\sqrt{q_p} + (30 + 3\sqrt{n})q_p\sqrt{q_c} + 4q_p^{3/2} + 2}{N},$$

which concludes the proof of Theorem 1.

Acknowledgment. We wish to thank the anonymous reviewers of ASIACRYPT 2015 for their useful suggestions.

References

1. Andreeva, E., Bogdanov, A., Luykx, A., Mennink, B., Tischhauser, E., Yasuda, K.: Parallelizable and authenticated online ciphers. In: Sako, K., Sarkar, P. (eds.) ASIACRYPT 2013, Part I. LNCS, vol. 8269, pp. 424–443. Springer, Heidelberg (2013)
2. Bellare, M., Kohno, T.: A theoretical treatment of related-key attacks: RKA-PRPs, RKA-PRFs, and applications. In: Biham, E. (ed.) EUROCRYPT 2003. LNCS, vol. 2656, pp. 491–506. Springer, Heidelberg (2003)
3. Bogdanov, A., Knudsen, L.R., Leander, G., Standaert, F.-X., Steinberger, J., Tischhauser, E.: Key-alternating ciphers in a provable setting: encryption using a small number of public permutations. In: Pointcheval, D., Johansson, T. (eds.) EUROCRYPT 2012. LNCS, vol. 7237, pp. 45–62. Springer, Heidelberg (2012)
4. Chakraborty, D., Sarkar, P.: A general construction of tweakable block ciphers and different modes of operations. In: Lipmaa, H., Yung, M., Lin, D. (eds.) Inscrypt 2006. LNCS, vol. 4318, pp. 88–102. Springer, Heidelberg (2006)
5. Chen, S., Lampe, R., Lee, J., Seurin, Y., Steinberger, J.: Minimizing the two-round Even-Mansour cipher. In: Garay, J.A., Gennaro, R. (eds.) CRYPTO 2014, Part I. LNCS, vol. 8616, pp. 39–56. Springer, Heidelberg (2014). http://eprint.iacr.org/2014/443

6. Chen, S., Steinberger, J.: Tight security bounds for key-alternating ciphers. In: Nguyen, P.Q., Oswald, E. (eds.) EUROCRYPT 2014. LNCS, vol. 8441, pp. 327–350. Springer, Heidelberg (2014). http://eprint.iacr.org/2013/222

7. Cogliati, B., Lampe, R., Seurin, Y.: Tweaking even-mansour ciphers. In: Gennaro, R., Robshaw, M. (eds.) CRYPTO 2015 - Proceedings, Part I. LNCS, vol. 9215, pp. 189–208. Springer, Heidelberg (2015). http://eprint.iacr.org/2015/539

8. Cogliati, B., Seurin, Y.: On the provable security of the iterated even-mansour cipher against related-key and chosen-key attacks. In: Oswald, E., Fischlin, M. (eds.) EUROCRYPT 2015. LNCS, vol. 9056, pp. 584–613. Springer, Heidelberg (2015). http://eprint.iacr.org/2015/069

9. Cogliati, B., Seurin, Y.: Beyond-Birthday-Bound Security for Tweakable Even-Mansour Ciphers with Linear Tweak and Key Mixing. Full version of this paper. Available at http://eprint.iacr.org/2015/851

10. Crowley, P.: Mercy: a fast large block cipher for disk sector encryption. In: Schneier, B. (ed.) FSE 2000. LNCS, vol. 1978, pp. 49–63. Springer, Heidelberg (2001)

11. Daemen, J., Rijmen, V.: The wide trail design strategy. In: Honary, B. (ed.) Cryptography and Coding 2001. LNCS, vol. 2260, pp. 222–238. Springer, Heidelberg (2001)

12. Dinur, I., Dunkelman, O., Keller, N., Shamir, A.: Cryptanalysis of iterated even-mansour schemes with two keys. In: Sarkar, P., Iwata, T. (eds.) ASIACRYPT 2014. LNCS, vol. 8873, pp. 439–457. Springer, Heidelberg (2014). http://eprint.iacr.org/2013/674

13. Even, S., Mansour, Y.: A construction of a cipher from a single pseudorandom permutation. J. Cryptol. 10(3), 151–162 (1997)

14. Farshim, P., Procter, G.: The related-key security of iterated Even–Mansour ciphers. In: Leander, G. (ed.) FSE 2015. LNCS, vol. 9054, pp. 342–363. Springer, Heidelberg (2015). http://eprint.iacr.org/2014/953

15. Ferguson, N., Lucks, S., Schneier, B., Whiting, D., Bellare, M., Kohno, T., Callas, J., Walker, J.: The Skein Hash Function Family. SHA3 Submission to NIST (Round 3) (2010)

16. Goldenberg, D., Hohenberger, S., Liskov, M., Schwartz, E.C., Seyalioglu, H.: On tweaking luby-rackoff blockciphers. In: Kurosawa, K. (ed.) ASIACRYPT 2007. LNCS, vol. 4833, pp. 342–356. Springer, Heidelberg (2007)

17. Guo, J., Peyrin, T., Poschmann, A., Robshaw, M.: The LED block cipher. In: Preneel, B., Takagi, T. (eds.) CHES 2011. LNCS, vol. 6917, pp. 326–341. Springer, Heidelberg (2011)

18. Halevi, S., Rogaway, P.: A tweakable enciphering mode. In: Boneh, D. (ed.) CRYPTO 2003. LNCS, vol. 2729, pp. 482–499. Springer, Heidelberg (2003)

19. Halevi, S., Rogaway, P.: A parallelizable enciphering mode. In: Okamoto, T. (ed.) CT-RSA 2004. LNCS, vol. 2964, pp. 292–304. Springer, Heidelberg (2004)

20. Jean, J., Nikolic, I., Peyrin, T.: Tweaks and keys for block ciphers: the TWEAKEY framework. In: Sarkar, P., Iwata, T. (eds.) ASIACRYPT 2014 - Proceedings, Part II. LNCS, vol. 8874, pp. 274–288. Springer, Heidelberg (2014)

21. Lampe, R., Patarin, J., Seurin, Y.: An asymptotically tight security analysis of the iterated Even-Mansour cipher. In: Wang, X., Sako, K. (eds.) ASIACRYPT 2012. LNCS, vol. 7658, pp. 278–295. Springer, Heidelberg (2012)

22. Lampe, R., Seurin, Y.: Security analysis of key-alternating feistel ciphers. In: Cid, C., Rechberger, C. (eds.) FSE 2014. LNCS, vol. 8540, pp. 243–264. Springer, Heidelberg (2015)

23. Landecker, W., Shrimpton, T., Terashima, R.S.: Tweakable blockciphers with beyond birthday-bound security. In: Safavi-Naini, R., Canetti, R. (eds.) CRYPTO 2012. LNCS, vol. 7417, pp. 14–30. Springer, Heidelberg (2012). http://eprint.iacr.org/2012/450

24. Liskov, M., Rivest, R.L., Wagner, D.: Tweakable block ciphers. In: Yung, M. (ed.) CRYPTO 2002. LNCS, vol. 2442, pp. 31–46. Springer, Heidelberg (2002)

25. Mennink, B.: Optimally secure tweakable blockciphers. In: Leander, G. (ed.) FSE 2015. LNCS, vol. 9054, pp. 428–448. Springer, Heidelberg (2015). http://eprint.iacr.org/2015/363

26. Minematsu, K.: Improved security analysis of XEX and LRW modes. In: Biham, E., Youssef, A.M. (eds.) SAC 2006. LNCS, vol. 4356, pp. 96–113. Springer, Heidelberg (2007)

27. Minematsu, K.: Beyond-birthday-bound security based on tweakable block cipher. In: Dunkelman, O. (ed.) FSE 2009. LNCS, vol. 5665, pp. 308–326. Springer, Heidelberg (2009)

28. Mitsuda, A., Iwata, T.: Tweakable pseudorandom permutation from generalized feistel structure. In: Baek, J., Bao, F., Chen, K., Lai, X. (eds.) ProvSec 2008. LNCS, vol. 5324, pp. 22–37. Springer, Heidelberg (2008)

29. Patarin, J.: The "Coefficients H" technique. In: Avanzi, R.M., Keliher, L., Sica, F. (eds.) SAC 2008. LNCS, vol. 5381, pp. 328–345. Springer, Heidelberg (2009)

30. Procter, G.: A Note on the CLRW2 Tweakable Block Cipher Construction. IACR Cryptology ePrint Archive, Report 2014/111 (2014). http://eprint.iacr.org/2014/111

31. Rogaway, P.: Efficient instantiations of tweakable blockciphers and refinements to modes OCB and PMAC. In: Lee, P.J. (ed.) ASIACRYPT 2004. LNCS, vol. 3329, pp. 16–31. Springer, Heidelberg (2004)

32. Rogaway, P., Bellare, M., Black, J.: OCB: a block-cipher mode of operation for efficient authenticated encryption. ACM Trans. Inf. Syst. Secur. 6(3), 365–403 (2003)

33. Rogaway, P., Zhang, H.: Online ciphers from tweakable blockciphers. In: Kiayias, A. (ed.) CT-RSA 2011. LNCS, vol. 6558, pp. 237–249. Springer, Heidelberg (2011)

34. Sasaki, Y., Todo, Y., Aoki, K., Naito, Y., Sugawara, T., Murakami, Y., Matsui, M., Hirose, S.: Minalpher v1. Submission to the CAESAR competition (2014)

35. Schroeppel, R.: The Hasty Pudding Cipher. AES submission to NIST (1998)

36. Steinberger, J.: Improved Security Bounds for Key-Alternating Ciphers via Hellinger Distance. IACR Cryptology ePrint Archive, Report 2012/481 (2012). http://eprint.iacr.org/2012/481

An Inverse-Free Single-Keyed Tweakable Enciphering Scheme

Ritam Bhaumik$^{(\boxtimes)}$ and Mridul Nandi

Indian Statistical Institute, Kolkata, India
{bhaumik.ritam,mridul.nandi}@gmail.com

Abstract. In CRYPTO 2003, Halevi and Rogaway proposed CMC, a tweakable enciphering scheme (TES) based on a blockcipher. It requires two blockcipher keys and it is not inverse-free (i.e., the decryption algorithm uses the inverse (decryption) of the underlying blockcipher). We present here a new *inverse-free, single-keyed* TES. Our construction is a tweakable strong pseudorandom permutation (TSPRP), i.e., it is secure against chosen-plaintext-ciphertext adversaries assuming that *the underlying blockcipher is a pseudorandom permutation* (PRP), i.e., secure against chosen-plaintext adversaries. In comparison, SPRP assumption of the blockcipher is required for the TSPRP security of CMC. Our scheme can be viewed as a mixture of type-1 and type-3 Feistel cipher and so we call it FMix or mixed-type Feistel cipher.

Keywords: (Tweakable strong) pseudorandom permutation · Coefficient H Technique · Encipher · CMC · Fiestel cipher

1 Introduction

A **tweakable enciphering scheme** (TES) is a *length-preserving encryption scheme* that takes a *tweak* as an additional input. In other words, for each tweak, TES computes a ciphertext preserving length of the plaintext. Preserving length can be very useful in applications such as disk-sector encryption (as addressed by the IEEE SISWG P1619), where a length-preserving encryption preserves the file size after encryption. When a tweakable enciphering scheme is used, the disk sectors can serve as tweaks. Other applications of enciphering schemes could include bandwidth-efficient network protocols and security-retrofitting of old communication protocols.

Examples based on Paradigms. There are four major paradigms of tweakable enciphering schemes. Almost all enciphering schemes fall in one of the following categories.

– **Feistel Structure:** 2-block Feistel design was used in early block ciphers like Lucifer [4,22] and DES [23]. Luby and Rackoff gave a security proof of Feistel ciphers [12], and later the design was generalised to obtain inverse-free enciphering of longer messages [17]. *Examples:* Naor-Reingold Hash [16], GFN [10], matrix representations [1].

© International Association for Cryptologic Research 2015
T. Iwata and J.H. Cheon (Eds.): ASIACRYPT 2015, Part II, LNCS 9453, pp. 159–180, 2015.
DOI: 10.1007/978-3-662-48800-3_7

- **Hash-Counter-Hash:** Two layers of universal hash with a counter mode of encryption in between. *Examples:* XCB [13], HCTR [25], HCH [2].
- **Hash-Encrypt-Hash:** Two layers of universal hash with an ECB mode of encryption in between. *Examples:* PEP [3], TET [6], HEH [21].
- **Encrypt-Mix-Encrypt:** Two encryption layers with a mixing layer in between. *Examples:* EME [8], EME* [5] (with ECB encryption layer), CMC [7] (with CBC encryption layer).

Among all these constructions, the examples from Feistel cipher and Encrypt-mix-encrypt paradigms are based on blockciphers alone (i.e., no field multiplication or other primitive is used). Now we take a closer look at CMC encryption.

CMC. In CRYPTO 2003, Halevi and Rogaway proposed CMC, a tweakable enciphering scheme (TES) based on a blockcipher (Fig. 1). It accepts only plaintexts of size a multiple of n, the size of the underlying blockcipher. We call each n-bit segment of the plaintext a *block*. The CMC construction has the following problems:

- For an encryption using e_K, the decryption needs e_K^{-1}. In a **combined hardware implementation**, the footprint size (e.g., the number of gates or slices) goes up;
- The security proof of CMC relied on the stronger assumption SPRP (Strong Pseudo-Random Permutation) on the underlying blockcipher;
- Tweak is processed using an **independent key**, and the proposed single-key variant uses an **extra call** to the blockcipher.

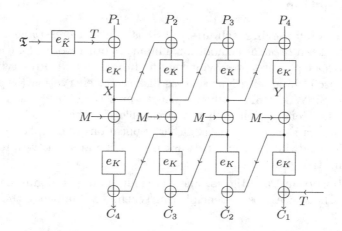

Fig. 1. CMC for four blocks, with tweak $\mathfrak{T}$ and $M = 2(X \oplus Y)$. Here 2 represents a primitive element of a finite field over $\{0, 1\}^n$.

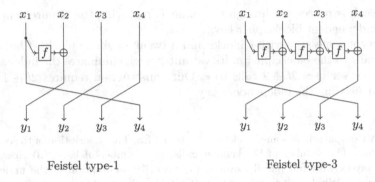

Feistel type-1   Feistel type-3

Fig. 2. The round function of two types of generalised Feistel networks for four block inputs. Similar definition can be applied for any number of blocks.

Feistel Cipher: An Inverse-Free Cipher. To resolve the first issue mentioned above, one can fall back on a Feistel network. For inverse-free constructions, the main approach so far has been to generalise the classical 2-block Feistel network to work for longer messages. Two of the interesting approaches were the type-1 Feistel network and the type-3 Feistel network (Fig. 2). In [10], it is shown that to encrypt ℓ block plaintext, type-1 and type-3 need $4\ell - 2$ and $2\ell + 2$ rounds respectively for achieving birthday security, which translates to $4\ell - 2$ and $2\ell^2 - 2$ invocations of the underlying blockcipher. However, their result is meant for providing a security performance trade-off and there is a provision for having beyond-birthday security.

One recent inverse-free construction based on Feistel networks is the AEZ-core, which forms part of the implementation of AEZ [9]. It belongs to the Encrypt-Mix-Encrypt paradigm, where the encryption uses a Feistel structure. It requires five blockcipher calls for every two plaintext blocks, but is highly parallelizable.

1.1 Our Contribution

In this paper, we address all the issues present in CMC in our construction. We use a mixture of type-1 and type-3 for our construction (hence the name FMix) to have an inverse-free construction which minimizes the number of blockcipher calls. FMix applies a simple balanced regular function b. Except for this, it looks exactly like the composition of $\ell + 1$ rounds of type-1 and one round of type-3 Feistel cipher. The features of FMix can be summarized as follows (see Table 1 for a comparison study):

1. FMix is **inverse-free**, i.e., it needs the same f for both encryption and decryption, having low footprint in the combined hardware implementation.
2. Because it is inverse-free, an important improvement is on the **security requirement** of e_K. CMC relies upon an SPRP-secure e_K, while our construction just needs a PRF-secure e_K. This can have significant practical implications in reducing the cost of implementation.

3. The tweak is processed through the same f, removing the requirement of an extra independent blockcipher key.
4. To encrypt a message with ℓ blocks and a tweak (a single block), CMC needs $2\ell+1$ calls to the blockcipher e. Its variant (which eliminates the independent key), however uses $2\ell + 2$ calls to e. Our construction requires $2\ell + 1$ calls, without needing the independent key.

Table 1. A Comparison of some blockcipher based TES. The description of the columns are as follows: (1) Number of blockcipher calls, (2) Number of keys, (3) How many sequential layers with full parallelization, (4) Security assumption of the underlying blockcipher, (5) Whether it is inverse-free. (CMC' is a "natively tweakable" variant of CMC, as described in [7]).

Schemes	#BC	#Key	#Layers	BC-security	Inverse-free?
CMC	$2\ell + 1$	2	$\ell + 2$	SPRP	NO
CMC'	$2\ell + 2$	2	$\ell + 2$	SPRP	NO
EME	$2\ell + 3$	1	4	SPRP	NO
GFN-1	$4\ell - 2$	$4\ell - 2$	$4\ell - 2$	PRP	YES
GFN-3	$2\ell^2 - 2$	$2\ell^2 - 2$	$2\ell - 2$	PRP	YES
AEZ-core	$\sim \frac{5}{2}\ell$	1	5	PRP	YES
FMix (this paper)	$2\ell + 1$	1	$\ell + 3$	PRP	YES

2   Preliminaries

2.1   Tweakable Encryption Schemes

This paper proposes a new **tweakable encryption scheme**, so we begin by describing what we mean by that. Formally, with a tweakable (deterministic) encryption scheme we associate four finite sets of binary strings: the message space $\mathcal{M}$, the tweak space $\mathcal{T}$, the ciphertext space $\mathcal{C}$, and the key space $\mathcal{K}$. The encryption function $e : \mathcal{K} \times \mathcal{T} \times \mathcal{M} \longrightarrow \mathcal{C}$ and the corresponding decryption function $\mathfrak{d} : \mathcal{K} \times \mathcal{T} \times \mathcal{C} \longrightarrow \mathcal{M}$ are required to satisfy the following (known as the **correctness requirement**):

$$\forall (K, \mathfrak{T}, P) \in \mathcal{K} \times \mathcal{T} \times \mathcal{M}, \quad \mathfrak{d}(K, \mathfrak{T}, e(K, \mathfrak{T}, P)) = P.$$

We also write $e(K, \mathfrak{T}, P)$ by $e_K(\mathfrak{T}, P)$ and $\mathfrak{d}(K, \mathfrak{T}, C)$ by $e_K^{-1}(\mathfrak{T}, C)$. We call a tweakable encryption scheme **tweakable enciphering scheme** (TES) if for all plaintext P, key $K \in \mathcal{K}$ and tweak $\mathfrak{T} \in \mathcal{T}$, $|e(K, \mathfrak{T}, P)| = |P|$ (i.e., it preserves length).

Random Function. In the heart of most encryption schemes lies the notion of a **random function**. Given a domain $\mathcal{D}$ and a range $\mathcal{R}$, a random function

$$f : \mathcal{D} \xrightarrow{*} \mathcal{R}$$

is a function chosen uniformly from the class of all functions from $\mathcal{D}$ to $\mathcal{R}$ (denoted $\mathcal{R}^{\mathcal{D}}$). Some elementary calculations show that for distinct $x_1, ..., x_n \in \mathcal{D}$, $f(x_1), ..., f(x_n)$ are independent and uniformly distributed over $\mathcal{R}$. More generally, we define the following:

Definition 1. *Let $\mathscr{C} \subseteq \mathcal{R}^{\mathcal{D}}$ be a class of functions from $\mathcal{D}$ to $\mathcal{R}$. A random $\mathscr{C}$-function*

$$f : \mathcal{D} \xrightarrow{\mathscr{C}} \mathcal{R}$$

is a function chosen uniformly from $\mathscr{C}$.

Note that choosing a function uniformly from a class $\{f_\alpha\}_{\alpha \in I}$ indexed by some finite set I can be achieved by choosing α_0 uniformly from I and then picking f_{α_0} as the chosen function.

Tweakable Random Permutation. When $\mathcal{R} = \mathcal{D}$, a popular choice of $\mathscr{C}$ is $\Pi_{\mathcal{D}}$, the class of all permutations on $\mathcal{D}$ (i.e., bijections from $\mathcal{D}$ to itself). A random permutation over $\mathcal{D}$ is a $\Pi_{\mathcal{D}}$-random function. It is an ideal choice corresponding to an encryption scheme over $\mathcal{D}$. The ideal choice corresponding to a tweakable enciphering scheme over $\mathcal{D}$ with tweak space $\mathcal{T}$ is called tweakable random permutation $\tilde{\pi}$ which is chosen uniformly from the class $\Pi_{\mathcal{D}}^{\mathcal{T}}$. For each tweak $\mathfrak{T} \in \mathcal{T}$, we choose a random permutation $\pi_{\mathfrak{T}}$ independently, and $\tilde{\pi}$ is a stochastically independent collection of random permutations $\{\pi_{\mathfrak{T}}; \mathfrak{T} \in \mathcal{T}\}$.

2.2 Pseudorandomness and Distinguishing Games

It should be noted that a random function or a random permutation is an ideal concept, since in practice the sizes of $\mathcal{R}^{\mathcal{D}}$ or $\Pi_{\mathcal{D}}$ are so huge that the cost of simulating a uniform random sampling on them is prohibitive. What is used instead of a truly random function is a **pseudorandom function** (PRF), a function whose behaviour is so close to that of a truly random function that no algorithm can effectively distinguish between the two. An adversary for a pseudorandom function f_1 is a deterministic algorithm $\mathcal{A}$ that tries to distinguish f_1 from a truly random f_0.

Security Notions. To test the pseudorandomness of f_1, $\mathcal{A}$ plays the **PRF distinguishing game** with an oracle $\mathcal{O}$ simulating (unknown to $\mathcal{A}$) either f_1 or f_0. For this, $\mathcal{A}$ makes q queries, in a deterministic but possibly adaptive manner. It is well known that there is no loss in assuming a distinguisher deterministic as unbounded time deterministic distinguisher is as powerful as a probabilistic distinguisher. Thus, the first query $x_1 = \mathsf{q}_1()$ is fixed, and given the responses $y_j = \mathcal{O}(x_j), j \in \{1, ..., i-1\}$, the i-th query becomes $x_i = \mathsf{q}_i(y_1, ..., y_{i-1})$, where q_i is a deterministic function for choosing the i-th query for $i \in \{1, ..., q\}$. Finally, a deterministic decision function examines $y_1, ..., y_q$ and chooses the output $b \in \{0, 1\}$ of $\mathcal{A}$. $\mathcal{A}$ wins if $\mathcal{O}$ was simulating f_b. An equivalent way to measure this winning event is called prf-advantage defined as

$$\Delta_{\mathcal{A}}(f_0 \; ; \; f_1) := \mathbf{Adv}_{f_1}^{\mathrm{prf}}(\mathcal{A}) = |\mathrm{Pr}_{f_0}[\mathcal{A}^{f_0} \to 1] - \mathrm{Pr}_{f_1}[\mathcal{A}^{f_1} \to 1]|,$$

where $\mathrm{Pr}_f[.]$ denotes the probability of some event when $\mathcal{O}$ imitates f. The above definition can be extended for more than one oracles. We can analogously define **pseudorandom permutation** (PRP) advantage $\mathbf{Adv}_{f_1}^{\mathrm{prp}}(\mathcal{A})$ of f_1 in which case f_0 is the random permutation. When f_1 is an enciphering scheme and $\mathcal{A}$ is interacting with both f_1 and its inverse f_1^{-1} (or with f_0 and f_0^{-1}) we have **strong pseudorandom permutation** (SPRP) advantage

$$\mathbf{Adv}_{f_1}^{\mathrm{sprp}}(\mathcal{A}) = |\mathrm{Pr}_{f_0}[\mathcal{A}^{f_0, f_0^{-1}} \to 1] - \mathrm{Pr}_{f_1}[\mathcal{A}^{f_1, f_1^{-1}} \to 1]|.$$

Finally, for a tweakable enciphering schemes with the strong pseudorandom property as above, we analogously define the **tweakable strong pseudorandom permutation** (TSPRP) advantage $\mathbf{Adv}_{f_1}^{\mathrm{tsprp}}(\mathcal{A})$.

Pointless Adversaries. In addition to the adversary being deterministic, we also assume that it does not make any pointless queries. An adversary $\mathcal{A}$ making queries to a tweakable encryption scheme f and f^{-1} is called pointless if either it makes a duplicate query or it makes an f-query $(\mathfrak{T}, P)$ and obtains response C and f^{-1}-query $(\mathfrak{T}, C)$ and obtains response P (the order of these two queries can be reversed). We can assume that adversary is not pointless since the responses are uniquely determined for these types of queries.

Theorem 1. *[11] Let f_1 be a TES over a message space $\mathcal{M} \subseteq \{0,1\}^*$ and f_0 and f_0' be two independently chosen random functions. Then for any adversary non-pointless distinguisher $\mathcal{A}$ making at most q queries, we have,*

$$\mathbf{Adv}_{f_1}^{\mathrm{tsprp}}(\mathcal{A}) \le \Delta_{\mathcal{A}}((f_1, f_1^{-1}) \; ; \; (f_0, f_0')) + \frac{q(q-1)}{2^{m+1}}$$

where $m = \min\{\ell : \mathcal{M} \cap \{0,1\}^\ell \ne \emptyset\}$.

The above result says that an uniform length-preserving random permutation is very close to an uniform length-preserving random function.

2.3   Domain Extensions and Coefficient H Technique

The notion of pseudorandomness, while giving us an approximate implementation of random functions, introduces a new problem. In general, it is very hard to decide whether or not there is an adversary that breaks the pseudorandomness of a particular function, since there is no easy way of exhaustively covering all possible adversaries in an analysis, and since there is no true randomness in a practically implemented function, probabilistic arguments cannot be used.

The common get-around is to assume we have PRFs $f_1, ..., f_n$ each with domain $\mathcal{D}$ and use them to obtain an F with domain $\mathcal{D}' \supset \mathcal{D}$, such that a PRF-attack on F leads to a PRF-attack on one of $f_1, ..., f_n$. Now, there are known functions on small domains (like AES, for instance) which have withstood decades of attempted PRF-attacks and are believed to be reasonably secure against PRF-attacks. Choosing $\mathcal{D}$ suitably to begin with and using the known PRFs in our construction, we can find a PRF F with domain $\mathcal{D}'$ that is secure as long as the smaller functions are secure. This technique is known as a **domain extension**.

Here, the central step in a proving the security of F is the **reduction** of an adversary of F to an adversary of one of $f_1, ..., f_n$. This reduction is achieved by assuming $f_1, ..., f_n$ to be truly random, and giving an information-theoretic proof that the distinguishing advantage of any adversary at F is small. Thus, if an adversary thus distinguish F from random with a reasonable advantage, we must conclude that $f_1, ..., f_n$ are not truly random. Thus, all we need to show is that when the underlying functions are truly random, F behaves like a truly random function.

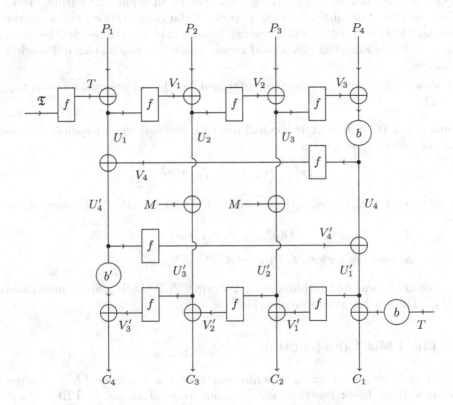

Fig. 3. The FMix construction for four blocks, with $M = V_4 + V_4'$

Patarin's Coefficient H Technique. There are several techniques for showing this. The one we use is based on the **Coefficient H Technique**, due to Jacques Patarin, which we briefly describe here. We look at the queries $x_1, ..., x_q$ and the outputs $y_1, ..., y_q$, and note that the adversary's decision will be based solely on the $2q$-tuple $(x_1, ..., x_q, y_1, ..., y_q)$. Now, if F_0 is the truly random function F is trying to emulate, then $F_0^{(q)}$ is also truly random, so on input $(x_1, ..., x_q)$,

$F_0^{(q)}(x_1, ..., x_q)$ will be uniform over $\mathcal{R}'$, $\mathcal{R}$ being the range of F. Thus when $\mathcal{D}' = \mathcal{R}' = \{0, 1\}^m$,

$$\Pr[F_0^{(q)}(x_1, ..., x_q) = (y_1, ..., y_q)] = \frac{1}{2^{mq}}.$$

If we can now show that $\Pr[F^{(q)}(x_1, ..., x_q) = (y_1, ..., y_q)]$ (which we call its **interpolation probability** after Bernstein) is "very close" to $\frac{1}{2^{mq}}$ for most $2q$-tuples $(x_1, ..., x_q, y_1, ..., y_q)$, we can conclude that no adversary can distinguish F from F_0 with a reasonable advantage. One way to formalize "very close" is that the interpolation probability is at least $(1-\epsilon)2^{-mq}$. Moreover, this may not happen for all possible views. (A view consists of all input and output blocks taken together. Informally, it is the portion of the computations visible to the adversary after completing all the queries.) So we may need to restrict the interpolation calculation on so called good views. This is the central idea of Patarin's technique.

Let $\text{view}(\mathcal{A}^{\mathcal{O}})$ denote the the view obtained by the adversary $\mathcal{A}$ interacting with $\mathcal{O}$.

Theorem 2 (Coefficient H Technique[19]). *Suppose the interpolation probabilities follow the inequality*

$$IP_{FMix}^f(\mathcal{V}) \geq (1 - \epsilon) \cdot 2^{-nL}$$

for all views $\mathcal{V} \in \mathcal{V}_{good}$ *(set of good views). Then for an SPRP-adversary* $\mathcal{A}$, *we have*

$$\boldsymbol{Adv}_{SPRP}^{\mathcal{A}}(F) \leq \epsilon + \epsilon'$$

where ϵ' *denotes the probability* $\Pr[\text{view}(\mathcal{A}^{F_0, F}) \notin \mathcal{V}_{good}]$.

This technique was first introduced by Patarin's PhD thesis [18] (as mentioned in [24]). Later it has been formalized in [19].

3   The FMix Construction

We are now in a position to describe our encryption scheme FMix. We use one underlying block function, chosen from a keyed family of PRFs $\{f_K : \{0, 1\}^n \longrightarrow \{0, 1\}^n\}_{K \in \mathcal{K}}$. The extended domain, which serves as both $\mathcal{M}$ and $\mathcal{C}$, is $\cup_{l \geq 2}\{0, 1\}^{ln}$, all strings consisting of two or more n-bit blocks. In addition to a key and a plaintext, the encryption algorithm also takes a tweak $\mathfrak{T}$ as input, which is also supplied to the decryption algorithm. Encryption is length-preserving: for $m \in \{0, 1\}^{lom}$, $\mathfrak{e}(K, m, \mathfrak{T}) \in \{0, 1\}^{lom}$ as well. The basic structure of the construction is based on that of CMC: a CBC encryption layer, followed by a layer of mixing, followed by a CBC decryption layer. However, using a generalisation of the Feistel scheme, we eliminate the need for f_K^{-1} during decryption, making do with f_K instead, thus making this construction inverse-free (Fig. 3).

input : A tweak $\mathfrak{T}$, an integer $l \geq 2$, l plaintext blocks $P_1, ..., P_l$
output: l ciphertext blocks $C_1, ..., C_l$

begin

$\quad T \leftarrow \mathtt{f}(\mathfrak{T})$

$\quad V_0 \leftarrow T$

$\quad$**for** $i \leftarrow 1$ **to** $l - 1$ **do**

$\quad\quad U_i \leftarrow V_{i-1} \oplus P_i$

$\quad\quad V_i \leftarrow \mathtt{f}(U_i)$

$\quad$**end**

$\quad U_l \leftarrow \mathtt{b}(V_{l-1} \oplus P_l)$

$\quad V_l \leftarrow \mathtt{f}(U_l)$

$\quad U_l' \leftarrow V_l \oplus U_1$

$\quad V_l' \leftarrow \mathtt{f}(U_l')$

$\quad M \leftarrow V_l \oplus V_l'$

$\quad U_{l-1}' \leftarrow U_2 \oplus M$

$\quad V_{l-1}' \leftarrow \mathtt{f}(U_{l-1}')$

$\quad C_l \leftarrow V_{l-1}' \oplus \mathtt{b'}(U_l')$

$\quad$**for** $i \leftarrow 3$ **to** $l - 1$ **do**

$\quad\quad U_{l+1-i}' \leftarrow U_i \oplus M$

$\quad\quad V_{l+1-i}' \leftarrow \mathtt{f}(U_{l+1-i}')$

$\quad\quad C_{l+2-i} \leftarrow V_{l+1-i}' \oplus U_{l+2-i}'$

$\quad$**end**

$\quad U_1' \leftarrow U_l + V_l'$

$\quad V_1' \leftarrow \mathtt{f}(U_1')$

$\quad V_0' \leftarrow T$

$\quad C_2 \leftarrow V_1' \oplus U_2'$

$\quad C_1 \leftarrow \mathtt{b}(V_0') \oplus U_1'$

end

Algorithm 1: FMix Encryption Algorithm. The decryption algorithm is exactly same as the encryption except that the $\mathtt{b}(T)$ is computed in the first layer and only T is used in the second.

The details of the construction are demonstrated in the figure, which shows a four-block FMix construction. The algorithm for general l is described in the box. Here, b is a balanced linear permutation, which we define below, and b' is b^{-1}. Decryption is almost identical, just with T and $b(T)$ switching roles.

Definition 2. *A permutation* $b : \{0,1\}^n \longrightarrow \{0,1\}^n$ *will be called a* **balanced linear permutation** *if both* $t \mapsto b(t)$ *and* $t \mapsto t + b(t)$ *are linear permutations.*

One choice of b could be multiplication by a primitive α, but this is not very software-friendly. A more software-friendly choice is $(t_1, t_2) \mapsto (t_1 \oplus t_2, t_1)$, where t_1 and t_2 are the higher and lower halves of t.

Notation for Our Construction. For our analysis we will assume the underlying PRF to be a truly random function f. We now model the encryption scheme in terms of computations based on f. An encryption is a computation

$$\mathbf{C} \longleftarrow \mathsf{e}^f(\mathfrak{T}, \mathbf{P}),$$

where $\mathfrak{T} \in \{0,1\}^n$, and $\mathbf{P}, \mathbf{C} \in \{0,1\}^{ln}$ for some $l \geq 2$. Similarly, a decryption is a computation

$$\mathbf{P} \longleftarrow \mathfrak{d}^f(\mathfrak{T}, \mathbf{C}),$$

which inverts e^f, for any tweak $\mathfrak{T}$. The plaintext $\mathbf{P}$ is denoted $(P_1, ..., P_l)$, where each P_i is an n-bit block of $\mathbf{P}$. Simlarly, the ciphertext $\mathbf{C}$ is denoted $(C_1, ..., C_l)$.

In the TSPRP game, the adversary makes q queries to the oracle $\mathcal{O}$. Each query is of the form $(\delta, \mathfrak{T}, \mathbf{X})$, where $\delta \in \{e, d\}$ denotes the direction of the query, $\mathfrak{T} \in \{0,1\}^n$ is the tweak, and $\mathbf{X} \in \{0,1\}^{nl}$ for some l is the input. If $\mathcal{O}$ is imitating FMIX, $\mathcal{O}(e, \mathfrak{T}, \mathbf{X})$ returns $\mathcal{E}^f(\mathfrak{T}, \mathbf{X})$, and $\mathcal{O}(d, \mathfrak{T}, \mathbf{X})$ returns $\mathcal{D}^f(\mathfrak{T}, \mathbf{X})$. If $\mathcal{O}$ is imitating a tweaked PRP Π, $\mathcal{O}(e, \mathfrak{T}, \mathbf{X})$ returns $\Pi(\mathfrak{T}, \mathbf{X})$, and $\mathcal{O}(d, \mathfrak{T}, \mathbf{X})$ returns $\Pi^{-1}(\mathfrak{T}, \mathbf{X})$. The output of $\mathcal{O}$ is denoted $\mathbf{Y}$.

All the queries and their outputs taken together form what we call a **view**. We use the following notation in a view. For the i-th query, δ^i denotes the direction of the query, $\mathfrak{T}^i$ denotes the tweak, and l^i denotes the number of blocks in $\mathbf{X}$. When $\delta^i = e$, the blocks of $\mathbf{X}$ are denoted $P_1, ..., P_{l^i}$ and those of $\mathbf{Y}$ are denoted $C_1, ..., C_{l^i}$. When $\delta^i = d$, this notation is reversed, i.e., the blocks of $\mathbf{Y}$ are denoted $P_1, ..., P_{l^i}$ and those of $\mathbf{X}$ are denoted $C_1, ..., C_{l^i}$. In the analysis, the tweak $\mathfrak{T}$ is denoted both P_0^i and C_0^i.

4   TSPRP Security Analysis of FMix

4.1   Good Views and Interpolation

Our first task is to formulate the version of Patarin's Coefficient H Technique we shall use for our proof. We begin by restricting our attention to a particular class of views.

Pointless View. A view is an indexed set of tuples

$$\mathcal{V} = \{(\delta^i, \mathfrak{T}^i, l^i, P_j^i, C_j^i) | 1 \leq i \leq q, 1 \leq j \leq l^i\}.$$

Here δ^i can take values e and d only. The l^i's are positive integers and $\mathfrak{T}^i, P_j^i, C_j^i \in \{0,1\}^n$, called *blocks*. The P_j^i and C_j^i mean the jth block of plaintext and ciphertext respectively on the ith query. We denote $\mathfrak{T}^i$ by both P_0^i and C_0^i. For any $0 \leq a \leq b \leq l_i$, we write $P_{a..b}^i$ to represent the tuple $(P_a^i, ..., P_b^i)$ and P^i to denote $P_{0..l_i}^i$. Similar notation for C^i and $C_{a..b}^i$. A view $\mathcal{V}$ is said to be *pointless* if at least one of the followings holds:

1. $\exists i \neq i'$ such that $\delta^i = \delta^{i'} = e$, $P^i = P^{i'}$.
2. $\exists i \neq i'$ such that $\delta^i = \delta^{i'} = d$, $C^i = C^{i'}$.
3. $\exists i' < i$ such that $\delta^i = e$, $\delta^{i'} = d$, $P^i = P^{i'}$.
4. $\exists i' < i$ such that $\delta^i = d$, $\delta^{i'} = e$, $C^i = C^{i'}$.

The first two cases are for duplicate queries. The third holds when we obtain a response $P^{i'}$ for some decryption query $C^{i'}$ and then make an encryption query $P^i := P^{i'}$. (The fourth case is the third case with the order of the queries reversed.) It is easy to see that when an adversary $\mathcal{A}$ is interacting with a TES, the view obtained is pointless if and only if $\mathcal{A}$ is pointless.

As we do not allow a pointless adversary we can restrict ourselves to non-pointless views only. Now we define good and bad views among this class.

Definition 3. (Good and Bad Views). *A view* $\{(\delta^i, \mathfrak{T}^i, l^i, P_j^i, C_j^i) | 1 \leq i \leq q, 1 \leq j \leq l^i\}$ *is said to be **good** if it is not pointless and*

$$(\forall i \text{ with } \delta^i = e)(\nexists i' < i)(C_1^i = C_1^{i'}), \text{ and } (\forall i \text{ with } \delta^i = d)(\nexists i' < i)(P_1^i = P_1^{i'}).$$

*A view that is not good and not pointless is called **bad**.*

The proof revolves around showing that the good views have a near-random distribution, and the bad views occur with a low probability. For the rest of the analysis, we fix a good view $\mathcal{V}$.

Interpolation Probability. Now we consider the interpolation probability for FMix construction. It is easy to see that

$$\text{IP}_{\text{FMix}}^f(\mathcal{V}) = \Pr_f[\text{FMix}^f(\mathfrak{T}^i, P_{1..l_i}^i) = C_{1..l_i}^i, 1 \leq i \leq q]$$

where the probability is taken under the randomness of f chosen uniformly from the set of all functions from $\{0,1\}^n$ to itself. Similarly, the interpolation probability for an ideal random function $\text{IP}_*(\mathcal{V})$ is 2^{-nL} where $L = \sum_{i=1}^q l_i$. This corresponds to the case where $\mathcal{O}$ imitating a truly random function. Now we state a result the proof of which is deferred to the next section.

Proposition 1. *For any good view* $\mathcal{V}$,

$$\text{IP}_{\text{FMix}}^f(\mathcal{V}) \geq (1 - \epsilon) \times 2^{-nL}, \text{ where } \epsilon = \frac{\binom{2L}{2}}{2^n}.$$

Armed with this result and the Coefficient H Technique, we are now ready to state and prove the main result of this paper.

Theorem 3. *For any SPRP-adversary* $\mathcal{A}$ *making* q *queries with* L *blocks in all,*

$$\text{Adv}_{\text{FMix}}^{\text{tsprp}}(\mathcal{A}) \leq \frac{\binom{2L}{2} + \binom{q}{2}}{2^n}.$$

Proof. When a non-pointless adversary $\mathcal{A}$ is interacting with a pair of independent random functions (f_0, f_0'), it obtains a bad view has probability upper bounded by $\binom{q}{2}$. To see this, let the bad event occurs for the first time at the ith query. If it is an encryption query (similar proof can be carried out for the decryption query) then C_1^i is chosen randomly from $\{0,1\}^n$ and so it matches with one of the previous first ciphertext block is at most $(i-1)/2^n$. So $\Pr_{f_0, f_0'}[\text{view}(\mathcal{A}^{f_0, f_0'})$ is a bad view] $\leq \sum_{i=1}^q \frac{i-1}{2^n} = \frac{q(q-1)}{2^{n+1}}$. By using Coefficient H Technique (see in Sect. 2.3) and the proposition stated above we have proved our theorem. $\square$

Corollary 1. *Let* FMix$_K$ *denote the* FMix *construction based on the keyed blcok-cipher* f_K. *For any TSPRP-adversary* $\mathcal{A}$ *making* q *queries with* L *blocks in all there exists an adversary* $\mathcal{A}'$ *making at most* L *encryption queries (and similar time as* $\mathcal{A}$)

$$Adv_{\mathsf{FMix}_K}^{\mathrm{tsprp}}(\mathcal{A}) \leq Adv_{f_K}^{\mathrm{prp}}(\mathcal{A}') + \frac{\binom{2L}{2} + \binom{q}{2}}{2^n}.$$

This follows from the standard hybrid argument.

4.2   Extension of FMix for Partial Block Input

In Sect. 3, we define our construction only for complete block inputs. In practice, messages-lengths m may not be a multiple of block-length n. For a complete enciphering scheme, our message space needs to be extended to include these partial block inputs. Two known methods for message-space extension of a cipher were XLS [20] and Nandi's scheme [14]. XLS is now known to be insecure [15], so we use Nandi's generic scheme for extending the message-space. The generic construction requires two additional blockcipher keys. We write these blockciphers as f_2 and f_3. The blockcipher f_1 is used in FMix. Given any partial block x, $1 \leq |x| \leq n-1$, we write $\mathsf{pad}(x) = x1\|0^{n-1-|x|}$. Similarly, $\mathsf{chop}_r(x)$ denotes the first r bits of x.

---

input : A tweak $\mathfrak{T}$, an integer $l \geq 2$, $l-1$ complete plaintext blocks
$P_1, ..., P_{l-1}$, partial last plaintext block p_l
output: $l-1$ complete ciphertext blocks $C_1, ..., C_{l-1}$, partial last ciphertext
block c_l
begin
$\quad P'_{l-1} \leftarrow f_2(\mathsf{pad}(p_l)) \oplus P_{l-1}$
$\quad (C_1, ..., C_{l-2}, C'_{l-1}) \leftarrow \mathsf{FMix}^{f_1}(P_1, ..., P_{l-2}, P'_{l-1})$
$\quad c_l \leftarrow \mathsf{chop}_{|p'_l|}(f_3(P'_{l-1} \oplus C'_{l-1})) \oplus p_l$
$\quad C_{l-1} \leftarrow f_2(\mathsf{pad}(c_l)) \oplus C'_{l-1}$
end

Theorem 4. *For any SPRP-adversary* $\mathcal{A}$ *making* q *queries with* L *blocks (including incomplete) in all,*

$$Adv_{\mathsf{FMix}}^{\mathrm{tsprp}}(\mathcal{A}) \leq \frac{\binom{2L}{2} + \binom{q}{2}}{2^n} + \frac{3q(q-1)}{2^{n+1}}.$$

The proof of the statement is immediate from Theorem 1 and the generic conversion as described in [14].

5   Proof of Proposition 1

In this section we provide the proof of Proposition 1.

Proposition. For any good view $\mathcal{V}$,

$$\mathrm{IP}^f_{\mathsf{FMix}}(\mathcal{V}) \geq (1 - \epsilon) \times 2^{-nL} \text{ where } \epsilon = \frac{\binom{2L}{2}}{2^n}.$$

We find a lower bound for the probability on the left by counting the choices of f that give rise to $\mathcal{V}$. For this counting, we find the number of internal states (simulations) σ that can result in $\mathcal{V}$, and for each σ, the number of choices of f compatible with it. As it turns out, slightly undercounting the simulations (counting only what we call admissible simulations) will suffice to prove our security bound.

5.1   Simulations

We shall develop an effective way of calculating the interpolation probability of $\mathcal{V}$. We begin by introducing the notion of variables. Let E be the set of all encryption query indices, i.e., $E = \{i | \delta^i = e\}$. Similarly, let D be the set of all decryption query indices. In identifying and labelling internal blocks, we continue using superscripts to denote query indices. Thus, for a query i, the $2l^i$ inputs of f (other than $\mathfrak{T}^i$) are denoted $U_1^i, ..., U_{l^i}^i, U_1'^i, ..., U_{l^i}'^i$, and the $2l^i + 1$ outputs of f are denoted $V_0^i, V_1^i, ..., V_{l^i}^i, V_1'^i, ..., V_{l^i}'^i$. For ease of notation, we shall write both U_0^i and $U_0'^i$ to denote $\mathfrak{T}^i$.

Variables and Derivables. We pick a set of output blocks

$$\mathcal{S} = \{V_j^i | i \in E, j \in \{1, ..., l^i\}\} \cup \{V_j'^i | i \in D, j \in \{1, ..., l^i\}\}.$$

$\mathcal{S}$ will be our set of **primary variables**, or simply **variables**. Any non-trivial linear combination of variables, optionally including blocks from $\mathcal{V}$ as well, will be called a **derivable**. While the proof will be primarily depend on variables, derivables will serve in the proof mainly to simplify notation and make the proof easier to grasp. Examples of derivables would be U_3^2, $\sum_i V_1'^i$ and $V_2^2 + P_1^1$. Note that a linear combination of view blocks alone, say $C_2^3 + C_3^2$, will not be considered a derivable, since it's value has already been fixed by choosing $\mathcal{V}$.

Let us assume for now that the input block and its corresponding output block are unrelated. We note that all input and output blocks of f are either variables or derivables. Thus, if we assign values to the variables, all the inputs and outputs of f over all queries are linearly determined. Thus, the variables linearly generate the entire set of input and output blocks, while themselves being linearly independent. We now formalise the notion of value assignment to variables.

Definition 4. *A **transcript** τ is a collection of **variable-value** pairs (Z, v) such that no two pairs in the collection contain the same variable. For every $(Z, v) \in \tau$, the variable Z is said to be assigned the value v under τ. We denote this as $Z|_\tau = v$. The **domain** $\mathbb{D}(\tau)$ of a transcript τ is defined as $\{Z | (\exists v)(Z, v) \in \tau\}$. Given a set S of variables, a transcript τ with $\mathbb{D}(\tau) = S$ is said to be an **instantiation of** S.*

For a transcript τ and a derivable Z' whose value only depends on the variables in $\mathbb{D}(\tau)$, τ effectively determines a value for Z'. This value is denoted by $Z'||_\tau$. For ease of notation, for any view block X, $X||_\tau$ will simply denote the value of X fixed in $\mathcal{V}$. An instantiation σ of $\mathcal{S}$ will be called a **simulation**, since it determines all inputs and outputs of f and thus describes a complete simulation of the internal computations that resulted in view $\mathcal{S}$.

Not all simulations make sense, however, when we consider the connection between and input block and its corresponding output block. A dependence now creeps in among the variables, owing to the key observation below, which poses the only non-trivial questions in the entire proof.

Wherever the inputs of f are identical, so are its outputs.

There can be simulations which violate this rule, and thus describe internal computations that can never occur. A simulation which actually describes a possible set of internal computations is called **realisable**. It is immediately clear that our observation holds for all realisable simulations. The problem of calculating the interpolation probability of $\mathcal{V}$ boils down to counting the number of realisable simulations.

5.2   Admissibility

All realisable simulations can be difficult to count, however. We shall focus instead on a smaller class of simulations, called admissible simulations, which are easy to count and yet are abundant enough to give us the desired result. Before that, we let us formulate in specific terms the ramifications of this observation. The immediate consequence is what we call pre-destined collisions. Let $\mathcal{I} = \cup_i \{U_0^i, U_1^i, ..., U_{l^i}^i, U_1'^i, ..., U_{l_i'}'^i\}$ be the set of all input blocks of f.

Definition 5. *A pair of input blocks $Z_1, Z_2 \in \mathcal{I}$ is said to constitute a **pre-destined collision** if for any realisable simulation σ,*

$$Z_1||_\sigma = Z_2||_\sigma.$$

All other collisions between input blocks are called **accidental collisions**. Our next task is to identify all pre-destined collisions. For that we'll need some more definitions.

Definition 6. *Query indices i and i' are called k-**encryption equivalent** for some $k < \min(l^i, l^{i'})$ if either $i = i'$, or*

$$(P_0^i, ..., P_k^i) = (P_0^{i'}, ..., P_k^{i'}).$$

*This is denoted as $i \sim_{e_k} i'$. Similarly, i and i' are called k-**decryption equivalent** for some $k < \min(l^i, l^{i'})$ if either $i = i'$, or*

$$(C_0^i, ..., C_k^i) = (C_0^{i'}, ..., C_k^{i'}).$$

This is denoted as $i \sim_{d_k} i'$.

Note that if $i \sim_{e_k} i'$, then $(\forall k' < k)(i \sim_{e_{k'}} i')$, and similarly for decryption equivalence. Our choice of $\mathcal{V}$ as a good view ensures that $i \in E$ whenever $i \sim_{e_1} i'$ for some $i' < i$, and $i \in D$ whenever $i \sim_{d_1} i'$ for some $i' < i$. We can now make a list of pre-destined collisions:

- $(U_k^i, U_k^{i'}), 0 \leq k < \min(l^i, l^{i'}), i \sim_{e_k} i'$
- $(U_k^{\prime i}, U_k^{\prime i'}), 0 \leq k < \min(l^i, l^{i'}), i \sim_{d_k} i'$

Substituting $V_{k-1}^i + P_k^i$ for U_k^i and $V_{k-1}^{\prime i} + C_k^i$ for $U_k^{\prime i}$, we can re-write the pre-destined collisions as

- $(V_{k-1}^i + P_k^i, V_{k-1}^{i'} + P_k^{i'}), 0 \leq k < \min(l^i, l^{i'}), i \sim_{e_k} i'$
- $(V_{k-1}^{\prime i} + C_k^i, V_{k-1}^{\prime i'} + C_k^{i'}), 0 \leq k < \min(l^i, l^{i'}), i \sim_{d_k} i'$

List of Pre-destined Collision. By our Observation, a pre-destined collision on inputs naturally entails a collision on the corresponding outputs. This leads to a corresponding set of **pre-destined output collisions**, which we write in the form of equations over derivables and view blocks:

(a) $(i \sim_{e_k} i') \rightarrow (V_k^i = V_k^{i'}), 0 \leq k < \min(l^i, l^{i'})$,
(b) $(i \sim_{d_k} i') \rightarrow (V_k^{\prime i} = V_k^{\prime i'}), 0 \leq k < \min(l^i, l^{i'})$.

The pre-destined output collisions linearly follow from the pre-destined collisions, but are formulated separately here, because they'll later be useful as a class of constraints on realisable simulations. Finally, we define the class of admissible simulations.

Definition 7 (Admissible). *A simulation σ is called **admissible** if, for any $Z_1, Z_2 \in \mathcal{I}$ that do not constitute a pre-destined collision, $Z_1\|_\sigma \neq Z_2\|_\sigma$.*

Thus, in an admissible simulation, no two input blocks of f can accidentally collide, and the only collisions are the pre-destined ones.

5.3   Basis and Extension

We now identify a subclass B of the variables which are linearly independent under assumption of admissibility, and such that an instantiation τ_B of B admits a unique extension $\mathbb{E}(\tau_B)$ to a realisable simulation. We shall call B a **basis** of X. First, we'll need one more definition.

Definition 8. *A query index i, $1 \leq i \leq q$, is called k-**fresh**, $k \geq 0$ if $k = l^i$, or $k < l^i$ and $\nexists i' \leq i$ with $k < l^{i'}$ such that $i \sim_{e_k} i'$ or $i \sim_{d_k} i'$.*

The set E_k of k-fresh encryption queries is defined as $\{i | \delta^i = e, i$ k-fresh$\}$. Similarly, the set D_k of k-fresh decryption queries is defined as $\{i | \delta^i = d, i$ k-fresh$\}$. Clearly, $E = \cup_k E_k$, and $D = \cup_k D_k$, since any i is l^i-fresh.

We are now in a position to choose our basis B. Let $l = \max_i l^i$. We define the following:

$$B_j = \{V_j^i | i \in E_j\}, 0 \leq j \leq l,$$

$$B_j' = \{V_j'^i | i \in D_j\}, 0 \leq j \leq l.$$

Finally, we define our basis as

$$B = \bigcup_{j=0}^{l} (B_j \cup B_j').$$

We next show how to obtain $\sigma = \mathbb{E}(\tau_B)$ given instantiation τ_B of B. To simplify the description, we shall use a couple of new definitions.

Definition 9. *The **encryption k-ancestor** of a query index i is defined as*

$$A_k^e(i) = \min_{i \sim_{e_k} i'} i'.$$

*Similarly, the **decryption k-ancestor** of a query index i is defined as*

$$A_k^d(i) = \min_{i \sim_{d_k} i'} i'.$$

Clearly, if i is k-fresh, then i is its own k-ancestor.

Definition 10. *For a query index i and a transcript τ, the **query slice at i** of τ is defined as*

$$Q_i(\tau) = \{(Z^i, v) | (Z^i, v) \in \tau\}.$$

Thus, a query slice is the portion of a transcript that refers to a specific query. The query slices of a transcript form a partition of it.

We are now ready to describe how to uniquely obtain σ. To begin with, for all $Z \in B$, we set

$$Z|_\sigma = Z|_{\tau_B}.$$

This gives us, among other things, the complete $Q_1(\sigma)$. (To see why, assume without loss of generality that $\delta^1 = e$. Then $1 \in E_j$ for every j, so $V_j^1 \in B$ for $0 \leq j \leq l^1$.) We proceed inductively to determine $Q_i(\sigma)$ based on $Q_1(\sigma), ..., Q_{i-1}(\sigma)$.

Suppose we have determined $Q_{i'}(\sigma)$ for all $i' < i$. For $0 \leq j \leq l^i$, let i_j denote $A_j^{\delta^i}(i)$. Clearly, $\{i_j\}_j$ form a non-decreasing sequence, and $i_{l^i} = i$. Let

$$k = \min_{i_j = i} j.$$

Suppose without loss of generality that $\delta^i = d$. Thus, for all $j \geq k$, $i \in D_j$. So $V_j'^i \in B$ for every $k \leq j \leq l^i$. For $0 \leq j < k$, since $i \sim_{d_j} i_j$, and i_j is decryption j-fresh, we use 4.3 (b) to set

$$V_j'^i|_\sigma = V_j'^{i_j}||_\sigma.$$

Finally, we set

$$V_0^i|_\sigma = V_0'^i||_\sigma.$$

This completes our extension of τ_B.

To show that σ indeed is a simulation, we just observe that if $\cup_1^{i-1} Q_i(\sigma)$ is realisable, and $\delta^i = d$, then $Q_i(\sigma)$ cannot violate 4.3 (a) (which concerns encryption queries only), and $Q_i(\sigma)$ is chosen so as to conform to 4.3 (b).

5.4 Extension Equations

We observe that in extending τ_B to $\mathbb{E}(\tau_B)$, once we've set the basis variables in accordance with τ_B, none of the steps we perform thereafter depend on the specific instantiation τ_B. Thus, for each variable we can identify an equation relating it to the basis variables, so that a simulation can be obtained by simply plugging in an appropriate instantiation of B. We call these equations the **extension equations**.

Pick $i \in E, j \in \{0, ..., l^i\}$. Then V_j^i is a variable. Let b_1 be j, and a_1 be $A_j^e(i)$. Having obtained $b_1, ..., b_k$ and $a_1, ..., a_k$, we stop if k is odd and $a_k \in E$, or if k is even and $a_k \in D$. Otherwise, let $b_{k+1} = l^{a_k} - 1 - b_k$, and a_{k+1} be $A_{b_{k+1}}^{\delta^{a_k}}(a_k)$. Since $a_{k+1} > a_k$, this terminates after finitely many steps, say upon obtaining a_{k_0}. Then we call $((b_1, a_1), ..., (b_{k_0}, a_{k_0}))$ the **extension chain** of V_j^i, denoted $\mathfrak{C}(V_j^i)$.

To obtain the extension equation of V_j^i from $\mathfrak{C}(V_j^i)$, note that $V_j^i = V_j^{a_1}$, and for any even $k \le k_0$, $V_j'^{a_k} = V_j'^{a_{k-1}}$, and (if $k < k_0$) $V_j^{a_{k+1}} = V_j'^{a_k}$. To bridge these equations, we just need to recall the equations relating $V_j'^{i'}$ to $V_{l^{i'}-1-j}'^{i'}$ for arbitrary i' with $l^{i'} \ge j$.

From our algorithm, $V_0'^{i'} = V_0^{i'}$, $V_{l^{i'}}'^{i'} = b(V_{l^{i'}-1}^{i'} + V_0^{i'} + P_{l^{i'}}^{i'}) + C_1^{i'}$ and $V_{l^{i'}-1}'^{i'} = b(V_{l^{i'}}^{i'} + V_0^{i'} + P_1^{i'}) + C_{l^{i'}}^{i'}$.

For $1 \le j \le l^{i'} - 2$, recall the masking equation

$$V_j'^{i'} = V_{l_i-j-1}^{i'} + V_{l^{i'}}^{i'} + V_{l^{i'}}'^{i'} + P_{l^{i'}-j}^{i'} + C_{j+1}^{i'}.$$

On replacing $V_{l^{i'}}'^{i'}$ by $b(V_{l^{i'}-1}^{i'} + V_0^{i'} + P_{l^{i'}}^{i'}) + C_1^{i'}$, this becomes

$$V_j'^{i'} = V_{l_i-j-1}^{i'} + V_{l^{i'}}^{i'} + b(V_{l^{i'}-1}^{i'} + V_0^{i'} + P_{l^{i'}}^{i'}) + C_1^{i'} + P_{l^{i'}-j}^{i'} + C_{j+1}^{i'}.$$

The extension equations can be computed inductively using the above. Similarly, for derivables, we can get the extension equations by writing it in terms of variables, and expanding these variables through their corresponding extension equations. We'll mostly be interested in the set of basis variables appearing in the extension equation of an input derivable Z, called the **base** $\mathfrak{B}(Z)$ of Z.

We'll show that whenever for two input derivables Z and Z', $\mathfrak{B}(Z) = \mathfrak{B}(Z')$, (Z, Z') is either a pre-destined collision, or Z and Z' cannot collide. This'll show that every accidental input collision corresponds to a linear equation on the basis variables and view blocks. Note that this linear equation actually corresponds to

n linear equations in terms of the bits, all of which should be dodged. For most of the analysis, this distinction will not matter, and it'll only become important when we deal with two special cases in the very end.

Lemma 1. *Every accidental input collision imposes a non-trivial linear equation on the basis variables.*

The proof of the lemma is postponed to the end of this section. It basically considers all cases for accidental collision and shows that it gives a non-trivial linear equation.

5.5   Bringing It All Together

We are now ready to wrap up our proof of the proposition 1. Let L denote $\sum_i l^i$. Let $\epsilon = \frac{\binom{2L}{2}}{2^n}$. The total number of output bits $\mathcal{V}$ in is nL, so clearly

$$\mathrm{IP}_*(\mathcal{V}) = \frac{1}{2^{nL}}.$$

Now, let $\mathcal{F} \subset (\{0,1\}^n)^{\{0,1\}^n}$ be such that $(f \in \mathcal{F}) \longleftrightarrow$ (choosing f results in $\mathcal{V}$). We see that

$$\mathrm{IP}^f_{\mathsf{FMix}}(\mathcal{V}) = \frac{\# \text{ choices of } f \text{ which result in } \mathcal{V}}{\# \text{ choices of } f \text{ in all}} = \frac{|\mathcal{F}|}{(2^n)^{2^n}}.$$

Let $\mathfrak{A}$ be the set of all admissible simulations. For an admissible simulation σ, let $\mathcal{F}_\sigma$ denote the subset of $\mathcal{F}$ such that $(f \in \mathcal{F}_\sigma) \longleftrightarrow$ (choosing f results in $\mathcal{V}$ and σ). With this notation, we can write

$$|\mathcal{F}| \geq \sum_{\sigma \in \mathfrak{A}} |\mathcal{F}_\sigma|.$$

To calculate $|\mathcal{F}_\sigma|$, we note that σ fixes the values f for $L + |B|$ distinct inputs. Thus,

$$|\mathcal{F}_\sigma| = (2^n)^{2^n - L - |B|}.$$

Since this does not depend on σ, we can write

$$\sum_{\sigma \in \mathfrak{A}} |\mathcal{F}_\sigma| = |\mathfrak{A}| \cdot (2^n)^{2^n - L - |B|}.$$

Now, each admissible simulation is $\mathbb{E}(\tau_B)$ for some instantiation τ_B of B. To ensure $\mathbb{E}(\tau_B) \in \mathfrak{A}$, we just have to choose τ_B such that it dodges all the linear equations corresponding to accidental input collisions. As there can be at most $\binom{2L}{2}$ such equations, we conclude that

$$|\mathfrak{A}| \geq 2^{n|B|} - \binom{2L}{2} \cdot 2^{n(|B|-1)} = 2^{n|B|}(1 - \epsilon).$$

Putting all of this together, we get

$$|\mathcal{F}| \geq (2^n)^{2^n - L} \cdot (1 - \epsilon) = (1 - \epsilon) \cdot \mathrm{IP}_*(\mathcal{V}) \cdot (2^n)^{2^n},$$

from which the Proposition follows.

5.6 Proof of Lemma 1

Proof. We'll divide the possible input pairs into several cases, which we'll further subdivide into groups, and we write out the proof only for the first case in each group, and the rest follow from it. The classifying factors are as follows:

- Whether they both occur in the same layer (**encryption layer** $\{U_j^i\}$ or **decryption layer** $\{U_j^i\}$), or in different layers;
- Whether they occur in the **right layer** (encryption layer of an encryption query, or decryption layer of decryption query) or the **wrong layer**;
- Whether their **first-cross indices** match (this would be the current query index if in the wrong layer, and the index after the first backward jump during extension if in the right layer).

We begin with an easy group of cases, where both occur in the right layer, and their first-cross indices do not match:

Case 1a. $(U_j^i, U_{j'}^{i'}), i, i' \in E, a = A_{j-1}^e(i) < A_{j'-1}^e(i') = a'$

$\mathcal{B}(U_j^i) = \mathcal{B}(V_{j-1}^i)$ can only contain basis variables with query indices $\leq a$. Since $\mathcal{B}(U_{j'}^{i'}) = \mathcal{B}(V_{j'-1}^{i'})$ will contain either $V_{la'}^{\prime a'}$ or $V_{j'}^{a'}$, $\mathcal{B}(U_j^i) \neq \mathcal{B}(U_{j'}^{i'})$.

Case 1b. $(U_j^{\prime i}, U_{j'}^{\prime i'}), i, i' \in D, a = A_{j-1}^d(i) < A_{j'-1}^d(i') = a'$

Case 1c. $(U_j^i, U_{j'}^{\prime i'}), i \in E, i' \in D, a = A_{j-1}^e(i) < A_{j'-1}^d(i') = a'$

Case 1d. $(U_j^{\prime i}, U_{j'}^{i'}), i \in D, i' \in E, a = A_{j-1}^d(i) < A_{j'-1}^e(i') = a'$

We next turn to another easy group, where exactly one of them is in the right layer, and first-cross indices do not match:

Case 2a. $(U_j^i, U_{j'}^{\prime i'}), i, i' \in E, a = A_{j-1}^e(i) \neq i'$

If $a < i'$, $V_{li'}^{i'}$ is in $\mathcal{B}(U_{j'}^{\prime i'})$ but not in $\mathcal{B}(U_j^i)$. If $a > i'$, either V_{j-1}^a is in $\mathcal{B}(U_j^i)$ but not in $\mathcal{B}(U_{j'}^{\prime i'})$, or $V_{la}^{\prime a}$ is in $\mathcal{B}(U_j^i)$ but not in $\mathcal{B}(U_{j'}^{\prime i'})$.

Case 2b. $(U_j^i, U_{j'}^{\prime i'}), i, i' \in D, i \neq A_{j'-1}^d(i') = a'$

Case 2c. $(U_j^i, U_{j'}^{i'}), i \in E, i' \in D, a = A_{j-1}^e(i) \neq i'$

Case 2d. $(U_j^{\prime i}, U_{j'}^{\prime i'}), i \in E, i' \in D, i \neq A_{j'-1}^d(i') = a'$

The next group is even easier: both in the wrong layer, with non-matching first-cross indices. This takes care of all cases with non-matching first-cross indices.

Case 3a. $(U_j^i, U_{j'}^{i'}), i, i' \in D, i < i'$

$V_{li'}^{\prime i'}$ is in $\mathcal{B}(U_{j'}^{i'})$ but not in $\mathcal{B}(U_j^i)$.

Case 3b. $(U_j^{\prime i}, U_{j'}^{\prime i}), i, i' \in E, i < i'$

Case 3c. $(U_j^i, U_{j'}^{\prime i'}), i \in D, i' \in E, i < i'$

Case 3d. $(U_j^{\prime i}, U_{j'}^{i'}), i \in E, i' \in D, i < i'$

Next we turn to a slightly trickier group, where they are in the same layer, both in the right layer, and first-cross indices match.

Case 4. $(U_j^i, U_{j'}^{i'}), i, i' \in E, A_{j-1}^e(i) = A_{j'-1}^e(i')$

Consider $\mathfrak{C}(V_{j-1}^i) = ((b_1, a_1), ..., (b_{k_0}, a_{k_0})), \mathfrak{C}(V_{j'-1}^{i'}) = ((b_1', a_1'), ..., (b_{k_0'}', a_{k_0'}'))$. If the chains follow the same query paths (i.e., if $k_0 = k_0'$ and $(\forall k \leq k_0)(a_k = a_k')$), assuming without loss of generality k_0 is odd and $k_0 \in E$ (from the chain-termination condition), we have $V_{b_{k_0}}^{a_{k_0}} \in \mathfrak{B}(U_j^i)$, and $V_{b_{k_0}'}^{a_{k_0}} \in \mathfrak{B}(U_{j'}^{i'})$, all other basis variables in the two extension equations being the same. Thus, if $b_{k_0} \neq b_{k_0}'$, $\mathfrak{B}(U_j^i) \neq \mathfrak{B}(U_{j'}^{i'})$, and if $b_{k_0} = b_{k_0}'$, $(U_j^i, U_{j'}^{i'})$ is either a pre-destined collision (if $P_j^i = P_{j'}^{i'}$) or it cannot be a collision. If the chains do not follow the same query path, we can find k such that $a_k \neq a_k'$, which reduces to one of the previous cases.

Case 4a. $(U_j'^i, U_{j'}'^{i'}), i, i' \in D, A_{j-1}^d(i) = A_{j'-1}^d(i')$

The next group is much simpler, where they are in different layers, both in the right layer, and first-cross indices match.

Case 5. $(U_j^i, U_{j'}'^{i'}), i \in E, i' \in D, a = A_{j-1}^e(i) = A_{j'-1}^d(i')$

Without loss of generality, $a \in E$. So $V_{l^a}^a$ is in $\mathfrak{B}(U_{j'}^{i'})$ but not in $\mathfrak{B}(U_j^i)$.

Case 5a. $(U_j'^i, U_{j'}^{i'}), i \in D, i' \in E, A_{j-1}^d(i) = A_{j'-1}^e(i')$

We're almost done with the proof at this point. We wrap up with the few remaining cases. In the next group, they come from different layers, exactly one of them in the right layer, and first-cross indices match.

Case 6. $(U_j^i, U_{j'}'^{i'}), i, i' \in E, A_{j-1}^e(i) = i'$

Here, $V_{l^{i'}}^{i'}$ is in $\mathfrak{B}(U_{j'}^{i'})$ but not in $\mathfrak{B}(U_j^i)$.

Case 6a. $(U_j^i, U_{j'}'^{i'}), i, i' \in D, i = A_{j'-1}^d(i')$

The four cases of the final group can be proved using the extension-chain-comparison technique of Case 4. In this group, they are in the same layer, at least one in the wrong layer, and first-cross indices match. (If they are both in the wrong layer, and first-cross indices match, they occur at the same query, so they cannot be in different layers, so this wraps up the case analysis).

Case 7. $(U_j^i, U_{j'}^{i'}), i \in E, i' \in D, A_{j-1}^e(i) = i'$

Case 7a. $(U_j'^i, U_{j'}'^{i'}), i \in E, i' \in D, i = A_{j'-1}^d(i')$

Case 7b. $(U_j^i, U_{j'}^{i'}), i \in D$

Case 7c. $(U_j'^i, U_{j'}'^i), i \in E$

This leaves only a few boundary cases (involving the likes of $U_{l^i}^i$), which can be easily verified. We just point out two special cases which underline the importance of choosing b as a balanced permutation. For the pair $(U_1^i, U_1'^i)$ for some i, if $P_1^i = C_1^i$, the condition for an accidental collision becomes $V_0^i + b(V_0^i) = 0$, which is still n independent linear equations in terms of the bits, by choice of b. Similarly, if $i \sim_{e_{l^i-1}} i'$, and $b(P_{l^i}^i) = P_{l^i}^{i'}$, the pair $(U_{l^i}^i, U_{l^i}^{i'})$ yields the equation

$b(V_{l^i-1}^{i}) + V_{l^i-1}^{i'} = 0$, which again is n independent linear equations in terms of the bits.

Thus we establish our lemma.   □

6   Conclusion and Future Works

In this paper we propose a new Feistel type length preserving tweakable encryption scheme. Our construction, called FMix, has several advantages over CMC and other blockcipher based enciphering scheme. It makes an optimal number of blockcipher calls using single keyed PRP blockcipher. The only drawback compare to EME is that the first layer of encryption, like CMC, is sequential. We can view our construction as a composition of type-1 and type-3 Feistel ciphers.

There are several possible scopes of future work. When we apply a generic method to encrypt last partial block message, we need an independent key. (This is always true for generic construction.) However, one can have a very specific way to handle partial block message keeping only one blockcipher key. The presence of the function b helps us to simplify the security proof. However, we do not know of any attack if we do not use this function (except for handling the tweak in the bottom layer - that use is necessary). So it would be interesting to see whether our proof can be extended for the variant without using the function b.

References

1. Berger, T.P., Minier, M., Thomas, G.: Extended generalized feistel networks using matrix representation. In: Lange, T., Lauter, K., Lisoněk, P. (eds.) SAC 2013. LNCS, vol. 8282, pp. 289–305. Springer, Heidelberg (2014)
2. Chakraborty, D., Sarkar, P.: HCH: a new tweakable enciphering scheme using the hash-encrypt-hash approach. In: Barua, R., Lange, T. (eds.) INDOCRYPT 2006. LNCS, vol. 4329, pp. 287–302. Springer, Heidelberg (2006)
3. Chakraborty, D., Sarkar, P.: A new mode of encryption providing a tweakable strong pseudo-random permutation. In: Robshaw, M. (ed.) FSE 2006. LNCS, vol. 4047, pp. 293–309. Springer, Heidelberg (2006)
4. Feistel, H.: Block cipher cryptographic system, US Patent 3,798,359, 19 March 1974
5. Halevi, S.: EME*: Extending EME to handle arbitrary-length messages with associated data. In: Canteaut, A., Viswanathan, K. (eds.) INDOCRYPT 2004. LNCS, vol. 3348, pp. 315–327. Springer, Heidelberg (2004)
6. Halevi, S.: Invertible universal hashing and the TET encryption mode. In: Menezes, A. (ed.) CRYPTO 2007. LNCS, vol. 4622, pp. 412–429. Springer, Heidelberg (2007)
7. Halevi, S., Rogaway, P.: A tweakable enciphering mode. In: Boneh, D. (ed.) CRYPTO 2003. LNCS, vol. 2729, pp. 482–499. Springer, Heidelberg (2003)
8. Halevi, S., Rogaway, P.: A parallelizable enciphering mode. In: Okamoto, T. (ed.) CT-RSA 2004. LNCS, vol. 2964, pp. 292–304. Springer, Heidelberg (2004)
9. Hoang, V.T., Krovetz, T., Rogaway, P.: Robust authenticated-encryption AEZ and the problem that it solves. In: Oswald, E., Fischlin, M. (eds.) EUROCRYPT 2015. LNCS, vol. 9056, pp. 15–44. Springer, Heidelberg (2015)

10. Hoang, V.T., Rogaway, P.: On generalized feistel networks. In: Rabin, T. (ed.) CRYPTO 2010. LNCS, vol. 6223, pp. 613–630. Springer, Heidelberg (2010)
11. Liskov, M., Rivest, R.L., Wagner, D.: Tweakable block ciphers. In: Yung, M. (ed.) CRYPTO 2002. LNCS, vol. 2442, pp. 31–46. Springer, Heidelberg (2002)
12. Luby, M., Rackoff, C.: How to construct pseudorandom permutations from pseudorandom functions. SIAM J. Comput. 17(2), 373–386 (1988)
13. McGrew, D.A., Fluhrer, S.R.: The security of the extended codebook (XCB) mode of operation. In: Adams, C., Miri, A., Wiener, M. (eds.) SAC 2007. LNCS, vol. 4876, pp. 311–327. Springer, Heidelberg (2007)
14. Nandi, M.: A generic method to extend message space of a strong pseudorandom permutation. Computación y Sistemas 12(3), 285–296 (2009)
15. Nandi, M.: XLS is not a strong pseudorandom permutation. In: Sarkar, P., Iwata, T. (eds.) ASIACRYPT 2014. LNCS, vol. 8873, pp. 478–490. Springer, Heidelberg (2014)
16. Naor, M., Reingold, O.: On the construction of pseudorandom permutations: lubyrackoff revisited. J. Cryptology 12(1), 29–66 (1999)
17. Nyberg, K.: Generalized feistel networks. In: Kim, K., Matsumoto, T. (eds.) Advances in Cryptology ASIACRYPT 1996. LNCS, vol. 1163, pp. 91–104. Springer, Berlin Heidelberg (1996)
18. Patarin, J.: Etude des Générateurs de Permutations Basés sur le Schéma du D.E.S. Ph.D Thèsis de Doctorat de l'Université de Paris 6 1991
19. Patarin, J.: The "Coefficients H" technique. In: Avanzi, R.M., Keliher, L., Sica, F. (eds.) SAC 2008. LNCS, vol. 5381, pp. 328–345. Springer, Heidelberg (2009)
20. Ristenpart, T., Rogaway, P.: How to enrich the message space of a cipher. In: Biryukov, A. (ed.) FSE 2007. LNCS, vol. 4593, pp. 101–118. Springer, Heidelberg (2007)
21. Sarkar, P.: Improving upon the TET mode of operation. In: Nam, K.-H., Rhee, G. (eds.) ICISC 2007. LNCS, vol. 4817, pp. 180–192. Springer, Heidelberg (2007)
22. Sorkin, A.: Lucifer, a cryptographic algorithm. Cryptologia 8(1), 22–42 (1984)
23. Data Encryption Standard: Fips pub 46. Federal Information Processing Standards Publication, Appendix A (1977)
24. Vaudenay, S.: Decorrelation: a theory for block cipher security. In: Journal of Cryptology, Lecture Notes in Computer Science, vol. 16(4), pp. 249–286. Springer-Verlag, New York (2003)
25. Wang, P., Feng, D., Wu, W.: HCTR: a variable-input-length enciphering mode. In: Feng, D., Lin, D., Yung, M. (eds.) CISC 2005. LNCS, vol. 3822, pp. 175–188. Springer, Heidelberg (2005)

Foundations

On Black-Box Complexity of Universally Composable Security in the CRS Model

Carmit Hazay[1]($\boxtimes$) and Muthuramakrishnan Venkitasubramaniam[2]

[1] Faculty of Engineering, Bar-Ilan University, Ramat Gan, Israel
carmit.hazay@biu.ac.il
[2] University of Rochester, Rochester, NY 14611, USA
muthuv@cs.rochester.edu

Abstract. In this work, we study the intrinsic complexity of *black-box Universally Composable (UC) secure computation* based on *general assumptions*. We present a thorough study in various corruption modelings while focusing on achieving security in the common reference string (CRS) model. Our results involve the following:

- **Static UC Secure Computation.** Designing *the first* static UC secure oblivious transfer protocol based on public-key encryption and stand-alone semi-honest oblivious transfer. As a corollary we obtain the first black-box constructions of UC secure computation assuming only two-round semi-honest oblivious transfer.
- **One-sided UC Secure Computation.** Designing adaptive UC secure two-party computation with single corruptions assuming public-key encryption with oblivious ciphertext generation.
- **Adaptive UC Secure Computation.** Designing adaptively secure UC commitment scheme assuming only public-key encryption with oblivious ciphertext generation. As a corollary we obtain the first black-box constructions of adaptive UC secure computation assuming only (trapdoor) simulatable public-key encryption (as well as a variety of concrete assumptions).

We remark that such a result was not known even under non-black-box constructions.

Keywords: UC secure computation · Black-box constructions · Oblivious transfer · UC commitments

1 Introduction

Secure multi-party computation enables a set parties to mutually run a protocol that computes some function f on their private inputs, while preserving a number of security properties. Two of the most important properties are privacy and

C. Hazay—Research partially supported by a grant from the Israel Ministry of Science and Technology (grant No. 3-10883).

M. Venkitasubramaniam—Research supported by Google Faculty Research Grant and NSF Award CNS-1526377.

T. Iwata and J.H. Cheon (Eds.): ASIACRYPT 2015, Part II, LNCS 9453, pp. 183–209, 2015.
DOI: 10.1007/978-3-662-48800-3_8

correctness. The former implies data confidentiality, namely, nothing leaks by the protocol execution but the computed output. The later requirement implies that no corrupted party or parties can cause the output to deviate from the specified function. It is by now well known how to securely compute any efficient functionality [2,4,24,45,50] in various models and under the stringent simulation-based definitions (following the ideal/real paradigm). Security is typically proven with respect to two adversarial models: the semi-honest model (where the adversary follows the instructions of the protocol but tries to learn more than it should from the protocol transcript), and the malicious model (where the adversary follows an arbitrary polynomial-time strategy), and feasibility results are known in the presence of both types of attacks. The initial model considered for secure computation was of a static adversary where the adversary controls a subset of the parties (who are called corrupted) before the protocol begins, and this subset cannot change. In a stronger corruption model the adversary is allowed to choose which parties to corrupt throughout the protocol execution, and as a function of its view; such an adversary is called adaptive.

These feasibility results rely in most cases on stand-alone security, where a *single* set of parties run a *single* execution of the protocol. Moreover, the security of most cryptographic protocols proven in the stand-alone setting does not remain intact if many instances of the protocol are executed concurrently [40]. The strongest (but also the most realistic) setting for concurrent security is known by *Universally Composable* (UC) security [4]. This setting considers the execution of an unbounded number of concurrent protocols in an arbitrary and adversarially controlled network environment. Unfortunately, stand-alone secure protocols typically fail to remain secure in the UC setting. In fact, without assuming some *trusted help*, UC security is impossible to achieve for most tasks [7,8,40]. Consequently, UC secure protocols have been constructed under various *trusted setup* assumptions in a long series of works; see [1,5,10,14,34,38] for few examples.

In this work, we are interested in understanding the intrinsic complexity of *UC secure computation*. Identifying the general assumptions required for a particular cryptographic task provides an abstraction of the functionality and the specific hardness that is exploited to obtain a secure realization of the task. The expressive nature of general assumptions allows the use of a large number of concrete assumptions of our choice, even one that may not have been considered at the time of designing the protocols. Constructions that are based on general assumptions are proven in two flavors:

Black-box Usage: A construction is black-box if it refers only to the input/output behavior of the underlying primitives.

Non-black-box Usage: A construction is non-black box if it uses the code computing the functionality of the underlying primitives.

Typically, non-black-box constructions have been employed to demonstrate feasibility and derive the minimal assumptions required to achieve cryptographic tasks. An important theoretical question is whether or not non-black-box usage of the underlying primitive is necessary in a construction. Besides its theoretical importance, obtaining black-box constructions is related to efficiency as an undesirable effect of non-black-box constructions is that they are typically inefficient and unlikely to be implemented in practice. Fortunately, a recent line of works [25,26,32,47] has narrowed the gap between what is achievable via non-black-box and black-box constructions under minimal assumptions.

More relevant to our context, the work of Ishai, Prabhakaran and Sahai [33] provided the first black-box constructions of UC secure protocols assuming only one-way functions in a model where all parties have access to an ideal oblivious transfer (OT) functionality. Orthogonally, Choi et al. [12] provided a compiler that transforms any semi-honest OT to a protocol that is secure against malicious static adversaries *in the stand-alone* (i.e. not UC) while assuming that all parties have access to the ideal commitment functionality. In the adaptive setting, the work of Choi et al. provides a transformation from adaptively secure semi-honest oblivious transfer to one that is secure *in the stronger UC setting* against malicious adaptive adversaries while assuming that all parties have access to the ideal commitment functionality. In essence, these works provide black-box constructions, however, they fall short of identifying the necessary minimal general computational assumptions in the UC setting.

Loosely speaking, a UC commitment scheme [7] is a fundamental building block in secure computation which is defined in two phases: in the commit phase a committer commits to a value while keeping it hidden, whereas in the decommit phase the committer reveals the value that it previously committed to. In addition to the standard binding and hiding security properties that any commitment must adhere, commitment schemes that are secure in the UC framework must allow straight-line extraction (where a simulator should be able to extract the content of any valid commitment generated by the adversary) and straight-line equivocation (where a simulator should be able to produce many commitments for which it can later decommit to both 0 and 1). We stress that even security in the static setting requires some notion of equivocation. Due to these rigorous requirements, it has been a real challenge to design black-box constructions of UC secure commitment schemes.

In the context of realizing the UC commitments in the CRS model, Damgård and Nielsen introduced the notion of mixed-commitments in [16]. This construction requires a CRS that is linear in the number of parties and can be instantiated under the N-residuosity and p-subgroup hardness assumptions. In the global CRS model (where a single CRS is introduced for any number of executions), the only known constructions are by Damgård and Groth [15] based on the Strong RSA assumption and Lindell [42] based on the DDH assumption, where the former construction guarantees security in the adaptive setting whereas the later construction provides static security.

Another fundamental building block in secure computation which has been widely studied is oblivious transfer [21,49]. Semi-honest two-round oblivious transfer can be constructed based enhanced trapdoor permutations [21] and smooth projective hashing [28], and concretely under Discrete Diffie-Hellman (DDH) [46]. Two-round protocols with malicious UC security are presented in the influential paper by Peikert et al. [48] that presents a black-box framework in the common reference string (CRS) model for oblivious transfer, based on dual-mode public-key encryption (PKE) schemes, which can be concretely instantiated under the DDH, quadratic residuosity and Learning with Errors (LWE) hardness assumptions. In a followup work [13], the authors present UC oblivious transfer constructions in the global CRS model assuming DDH, N-residuosity and the Decision Linear Assumption (DLIN). As pointed out in [13], the [48] constructions require a distinct CRS per party. In the context of adaptive UC oblivious transfer protocols, the works of [12] and [22] give constructions in the UC commitment hybrid model where they additionally rely on an assumption that implies adaptive semi-honest oblivious transfer.

It is worth noting that while the works of [48] and [13] provide abstractions of their assumptions, the assumptions themselves are not general enough to help understand the minimal assumptions required to achieve static UC security. In particular, when restricting attention to black-box constructions based on general assumptions, the state-of-the-art literature seems to indicate that achieving UC security in most trusted setup models reduces to constructing two apparently incomparable primitives: *semi-honest oblivious transfer* and *UC commitment schemes*. This leaves the following important question open:

> *What are the minimal (general) assumptions required to construct UC secure protocols, given only* black-box *access to the underlying primitives?*

We note that this question is already well understood in the static setting when relaxing the black-box requirement. Namely, in [18] Damgård, Nielsen and Orlandi showed how to construct UC commitments assuming only semi-honest oblivious transfer in the global CRS model, while additionally assuming a pre-processing phase where the parties participate in a round-robin manner[1]. More recently, Lin, Pass and Venkitasubramaniam [39] improved this result by removing any restricted pre-processing phase. In the same work the authors showed how to achieve UC security in the global CRS model assuming only the existence of semi-honest oblivious transfer. In particular, this construction shows that static UC security can be achieved without assuming UC commitments when relying on non-black-box techniques.

In the stand-alone (i.e. not UC) setting, assuming only the existence of semi-honest oblivious transfer [26,27,32] show how to construct secure multiparty computation protocols while relying on the underlying primitives in a black-box manner. More recently, [12] provided black-box constructions that are secure against static adversaries, again, in the stand-alone setting, where all parties have access to an ideal commitment functionality (cf. Proposition 1 in [12]). The latter construction achieves a stronger notion of straight-line simulation, however falls short of achieving static UC security (see more details in Sect. 3).

---

[1] In such a pre-processing phase, it is assumed that at most one party is allowed to transmit messages in any round.

In the adaptive setting, the only work that considers a single general assumption that implies adaptive UC security using non-black-box techniques is the result due to Dachman-Soled et al. [14], that shows how to obtain adaptive UC commitments assuming simulatable PKE. Moreover, the best known general assumptions required to achieve black-box UC security are adaptive semi-honest oblivious transfer and UC commitments [12,17]. Known minimal general assumptions that are required to construct these primitives are (trapdoor) simulatable PKE for adaptive semi-honest oblivious transfer [11] and mixed commitments for UC commitments [17].

1.1 Our Results

In this paper we present a thorough study of black-box UC secure computation in the CRS model; details follow.

Static UC Secure Computation. Our first result is given in the static setting, where we demonstrate the feasibility of UC secure computation based on semi-honest oblivious transfer and extractable commitments. More concretely, we prove how to transform any statically semi-honest secure oblivious transfer into one that is secure in the presence of malicious adversaries, giving *only black-box access* to the underlying semi-honest oblivious transfer protocol. Our approach is inspired by the protocols from [27] and [37], where we observe that it is not required to use the full power of static UC commitments. Instead, we employ a weaker primitive that only requires straight-line input extractability. Interestingly, we prove that this weaker notion of security, denoted by extractable commitments [44], can be realized based on any CPA secure PKE. More precisely, we prove the following theorem.

Theorem 11 (*Informally*). *Assuming the existence of PKE and semi-honest oblivious transfer, then any functionality can be realized in the CRS model with static UC security, where the underlying primitives are accessed in a black-box manner.*

We remark here that this theorem makes a significant progress towards reducing the general assumptions required to construct UC secure protocols. Previously, the only general assumptions based on which we knew how to construct UC secure protocols were mixed-commitments [16] and dual-mode PKE [48] both of which were tailor-made for the particular application. Towards understanding the required minimal assumptions, we recall the work Damgård and Groth in [15] who showed that the existence of UC commitments in the CRS model implies a stand-alone key agreement protocol. Moreover, under black-box constructions, the seminal work of Impagliazzo and Rudich [31] implies that key agreement cannot be based on one-way functions. Thus, there is reasonable evidence to believe that some public-key primitive is required for UC commitments. In that sense, our assumption regarding PKE is close to being optimal. Nevertheless, it is unknown whether the semi-honest oblivious transfer assumption is required.

Our result is shown in two phases. At first we compile the semi-honest oblivious transfer protocol into a new protocol with intermediate security properties in the presence of malicious adversaries. This transformation is an extension of the [27] transformation that is only proven for bit oblivious transfer, whereas our proof works for string oblivious transfer. Next, we use the transformed oblivious transfer protocol in order to construct a maliciously fully secure oblivious transfer. By combining our oblivious transfer with the [33] protocol we obtain a statically generic UC secure computation.

An important corollary is deduced from the work by Gertner et al. [23], who provided a black-box construction of PKE based on any two-round semi-honest oblivious transfer protocol. Specifically, the combination of their result with ours implies the following corollary, which demonstrates that two-round semi-honest oblivious transfer is sufficient in the CRS model to achieve black-box constructions of UC secure protocols.

Corollary 12 *(Informally). Assuming the existence of two-round semi-honest oblivious transfer, then any functionality can be UC realized in the CRS model, where the oblivious transfer is accessed in a black-box manner.*

Implications. In what follows, we make a sequence of interesting observations that are implied by our result in the static UC setting.

- The important result by Canetti, Lindell, Ostrovsky and Sahai [9] presents the first *non-black-box* constructions of static UC secure protocols assuming enhanced trapdoor permutations. In fact, their result can be extended assuming only PKE with oblivious ciphertext generation (which is PKE with the special property that a ciphertext can be obliviously sampled without the knowledge of the plaintext, and can be further realized using enhanced trapdoor permutation). In that sense, our result, assuming PKE with oblivious ciphertext generation, can be viewed as an improvement of [9] when relying on this primitive in a *black-box* manner.
- The pair of works by Damgard, Nielsen and Orlandi [18] and Lin, Pass and Venkitasubramaniam [39] demonstrate that *non-black-box* constructions of UC commitments, and more generally static UC secure computation, can be achieved in the CRS model assuming only semi-honest oblivious transfer. In comparison, our result shows that two-round semi-honest oblivious transfer protocols are sufficient for obtaining *black-box* UC secure computation in the CRS model. Note that most semi-honest oblivious transfer protocols anyway require only two-round of communication, e.g., [21].
- In [38,39], Lin, Pass and Venkitasubramaniam provided a unified framework for constructing UC secure protocols in any "trusted-setup" model. Their result is achieved by capturing the minimal requirement that implies UC computations in the setup model. More precisely, they introduced the notion of a UC puzzle and showed that any setup model that admits a UC puzzle can be used to securely realize any functionality in the UC setting, while additionally assuming the existence of semi-honest oblivious transfer. Moreover, they showed how to easily construct such puzzles in most models. We remark

that our approach can be viewed as providing a framework to construct black-box UC secure protocols in other UC models. More precisely, we show that any setup model that admits the extractable commitment functionality can be used to securely realize any functionality assuming the existence of semi-honest oblivious transfer. In fact, our result easily extends to the chosen key registration authority (KRA) model [1], where it is assumed the existence of a trusted authority that samples public key, secret key pairs for each party, and broadcasts the public key to all parties. We leave it for future work to instantiate our framework in other setup models.

- The fact that our construction only requires PKE and semi-honest oblivious transfer allows an easy translation of static UC security to various efficient implementations under a wide range of concrete assumptions. Specifically, both PKE and (two-round) semi-honest oblivious transfer can be realized under RSA, factoring Blum integers, LWE, DDH, N-residuosity, p-subgroup and coding assumptions. This is compared to prior results that could be based on the later five assumptions [13,19,20,48].
- Recently, Maji, Prabhakaran, and Rosulek [44] initiated the study of the cryptographic complexity of secure computation tasks, while characterizing the relative complexity of a task in the UC setting. Specifically, they established a zero-one law that states that any task is either trivial (i.e., it can be reduced to any other task), or complete (i.e., to which any task can be reduced to), where a functionality $\mathcal{F}$ is said to *reduce* to another functionality $\mathcal{G}$, if there is a UC secure protocol for $\mathcal{F}$ using ideal access to $\mathcal{G}$. More precisely, they showed that assuming the existence of semi-honest oblivious transfer, every finite two-party functionality is either trivial or complete. While their main theorem relies on the minimal assumption of semi-honest oblivious transfer, their use of the assumption is non-black-box and they leave it as an open problem to achieve the same while relying on oblivious transfer in a black-box manner. Our result makes progress towards establishing this.

In more details, their high-level approach is to identify complete functionalities using four categories, namely, (1) $\mathcal{F}_{\mathrm{XOR}}$ that abstracts a XOR-type functionality, (2) $\mathcal{F}_{\mathrm{CC}}$ that abstracts a simple cut-and-choose functionality, (3) $\mathcal{F}_{\mathrm{OT}}$ the oblivious transfer functionality, and (4) $\mathcal{F}_{\mathrm{COM}}$ the commitment functionality. They then show that each category can be used to securely realize any computational task[2]. Among these reductions, functionalities $\mathcal{F}_{\mathrm{XOR}}$ and $\mathcal{F}_{\mathrm{CC}}$ rely on oblivious transfer in a non-black-box way. In this work we improve the reduction of functionality $\mathcal{F}_{\mathrm{CC}}$. That is, we obtain this improvement by showing that the extractable commitment functionality $\mathcal{F}_{\mathrm{EXTCOM}}$ and semi-honest oblivious transfer can be used in a black-box way to realize functionality $\mathcal{F}_{\mathrm{OT}}$, and combine this with a reduction presented in [44] that reduces $\mathcal{F}_{\mathrm{CC}}$ to the $\mathcal{F}_{\mathrm{EXTCOM}}$ functionality in a black-box way.

One-Sided UC Secure Computation. In this stronger two-party setting, where at most one of the parties is adaptively corrupted [29,35], we prove that

---

[2] Where it suffices to realize the $\mathcal{F}_{\mathrm{OT}}$ functionality as it is known to be complete [36].

one-sided adaptive UC security is implied by PKE with oblivious ciphertext generation. Here we combine two observations, one where our malicious static oblivious transfer from the previous result requires using the parties' inputs in only one phase, together with the fact that one-sided non-committing encryption (NCE) can be designed based on PKE with oblivious ciphertext generation [6,16]. In particular, NCE allow secure communication in the presence of adaptive attacks, which implies that the communication can be equivocated once the real message is handed to the simulator. Then, by encrypting part of our statically secure protocol using NCE, we obtain a generic protocol for any two-party functionality under the assumption specified above[3]. Namely,

Theorem 13 *(Informally). Assuming the existence of PKE with oblivious ciphertext generation, then any two-party functionality can be realized in the CRS model with one-sided adaptive UC security and black-box access to the PKE.*

Adaptive UC Secure Computation. Our last result is in the strongest corruption setting, where any number of parties can be adaptively corrupted. Here we design a new adaptively secure UC commitment scheme under the assumption of PKE with oblivious ciphertext generation, which is the first construction that achieves the stronger notion of adaptive security based on this hardness assumption. Our construction makes a novel usage of such a PKE together with Reed-Solomon codes, where the polynomial shares are encrypted using the PKE with oblivious ciphertext generation. Plugging-in our UC commitment protocol into the transformation of [12] that generates adaptive malicious oblivious transfer given adaptive semi-honest oblivious transfer and UC commitments, implies an adaptively UC secure oblivious transfer protocol with malicious security based on semi-honest adaptive oblivious transfer and PKE with oblivious ciphertext generation, using only black-box access to the semi-honest oblivious transfer and the PKE. That is,

Theorem 14 *(Informally). Assuming the existence of PKE with oblivious ciphertext generation and adaptive semi-honest oblivious transfer, then any functionality can be realized in the CRS model with adaptive UC security, where the underlying primitives are accessed in a black-box manner.*

We further recall the work of Choi et al. [11] that shows that the weakest general known assumption that is required to construct adaptively secure semi-honest oblivious transfer is trapdoor simulatable PKE. Now, since such an encryption scheme admits PKE with oblivious ciphertext generation, we obtain the following corollary that unifies the two assumptions required to achieve adaptive UC security.

Corollary 15. *Assuming the existence of (trapdoor) simulatable PKE, then any functionality can be realized in the CRS model with adaptive UC security and black-box access to the PKE.*

---

[3] We note that while in the plain model any statically secure protocol can be compiled into one-sided secure protocol by encrypting its entire communication using one-sided NCE, it is not the case in the UC setting due to the additional setup.

An additional interesting observation that is implied by our work is that our UC commitment scheme implies a construction that is secure in the adaptive setting when erasures are allowed, and under the weaker assumption of PKE. Specifically, instead of obliviously sampling ciphertexts in the commitment phase, the committer encrypts arbitrary plaintexts and then erases the plaintexts and randomness used for these computations. Our proof follows easily for this case as well. Combining our UC commitment scheme together with the semi-honest with erasures OT from [41] and the transformation of [12], we obtain the following result

Theorem 16 *(Informally). Assuming the existence of PKE and semi-honest oblivious transfer secure against an adaptive adversary assuming erasures, then any functionality can be realized in the CRS model with adaptive UC security assuming erasures, where the underlying primitives are accessed in a black-box manner.*

Noting that OT secure against adaptive adversaries assuming erasures can be realized under assumptions sufficient for achieving the same with respect to the weaker static adversaries, this theorem shows that achieving UC security against adaptive adversaries in the presence of erasures does not require any additional assumption beyond what is required to secure against static adversaries.

Implications. Next, we specify a sequence of interesting observations that are implied by our result in the adaptive UC setting.

- Previously, Dachman-Soled et al. [14], showed that adaptive UC secure protocols can be constructed in the CRS model assuming the existence of simulatable PKE. Our result improves this result in terms of complexity assumptions by showing that trapdoor simulatable PKE is sufficient, and provides new constructions based on concrete assumptions that were not known before. Nevertheless, we should point out that while the work of Dachman-Soled et al. is constructed in the global CRS model using a non-black-box construction, our result provides a black-box construction in a CRS model where the length of the reference string is linear in the number of parties.
- Analogous to our result on static UC security, it is possible to extend this result to the chosen key-registration authority (KRA) model, where we assume the existence of a trusted-party that samples public keys and secret keys for each party, and broadcasts the public key to all parties.
- Importantly, this result provides the first evidence that adaptively secure UC commitment is theoretically easier to construct than stand-alone adaptively secure semi-honest oblivious transfer. This is due to a separation from [43] (regarding static vs. adaptive oblivious transfer), that proves that adaptive oblivious transfer requires a stronger hardness assumption than enhanced trapdoor permutation.
- Regarding concrete assumptions, previously, adaptive UC commitments without erasures were constructed based on N-residuosity and p-subgroup hardness assumptions [17] and Strong RSA [15]. On the other hand, our result demonstrates the feasibility of this primitive under DDH, LWE, factoring

Blum integers and RSA assumptions. When considering adaptive corruption with erasures, the work of Blazy, et al. [3], extending the work of Lindell [42], shows how to construct highly efficient UC commitments based on the DDH assumption. On the other hand, assuming erasures, we are able to construct an adaptive UC commitment scheme based on any CPA-secure PKE.

2   Preliminaries

We denote the security parameter by n. We use the abbreviation PPT to denote probabilistic polynomial-time. We further denote by $a \leftarrow A$ the random sampling of a from a distribution A, and by $[n]$ the set of elements $\{1, \ldots, n\}$.

Definition 21 (PKE with Oblivious Ciphertext Generation [16]). A PKE Π with oblivious sampling generation is defined by the tuple (Gen, Enc, Dec, $\widetilde{\mathsf{Enc}}, \widetilde{\mathsf{Enc}}^{-1}$) and has the following additional property,

- **Indistinguishability of Oblivious and Real Ciphertexts.** *For any message m in the appropriate domain, consider the experiment* $(\mathrm{PK}, \mathrm{SK}) \leftarrow$ Gen(1^n), $c_1 \leftarrow \widetilde{\mathsf{Enc}}_{\mathrm{PK}}(r_1)$, $c_2 \leftarrow \mathsf{Enc}_{\mathrm{PK}}(m; r_2)$, $r_1' \leftarrow \widetilde{\mathsf{Enc}}_{\mathrm{PK}}^{-1}(c_2)$.
  Then, $(\mathrm{PK}, r_1', c_1, m) \overset{c}{\approx} (\mathrm{PK}, r_2, c_2, m)$.

To this end, we only employ PKE with perfect decryption. This merely simplifies the analysis and can be relaxed by using PKE with a negligible decryption error instead.

2.1   Oblivious Transfer

1-out-of-2 oblivious transfer (OT) is an important functionality in the context of secure computation that is engaged between a sender Sen and a receiver Rec; see Fig. 1 for the description of functionality $\mathcal{F}_{\mathrm{OT}}$. In this paper we are interested in reducing the hardness assumptions for general UC secure computation when using only black-box access to the underlying cryptographic primitives, such as the semi-honest OT. We use semi-honest OT as a building block for designing UC secure protocols in both static and adaptive settings. In the static setting, we refer to the two-round protocol of [21] that is based on PKE with oblivious ciphertext generation (or enhanced trapdoor permutation). In the adaptive setting, we refer to the two-round protocol of [9] that is based on augmented non-committing encryption scheme.

We next recall that any two-round semi-honest OT implies PKE. We demonstrate that in two phases, starting with the claim that semi-honest OT implies a key agreement (KA) protocol, where two parties agree on a secret key over a public channel. This statement has already been proven in [23] in the static setting, and holds for any number of rounds. The idea is simple, the parties execute an OT protocol where the party that plays the sender picks two random inputs s_0, s_1, whereas the party that plays the receiver enters 0. Finally, the parties output s_0 and security follows from the correctness and privacy of the OT.

A simple observation shows that this reduction also holds in the adaptive setting. Namely, starting with an adaptive semi-honest OT, the same reduction implies an adaptively secure KA (where the protocol communication must be consistent with respect to any key). Note that this reduction preserves the number of rounds, thus if the starting point is a two-round OT then the reduction implies a two-round KA. Next, a well established fact shows that in the static setting a two-round key agreement implies PKE (in fact, these primitives are equivalent). Formally,

Theorem 22. *Assume the existence of two-round key agreement protocol with static security, then there exists IND-CPA PKE.*

Functionality $\mathcal{F}_{OT}$

Functionality $\mathcal{F}_{OT}$ communicates with with sender Sen and receiver Rec, and adversary $\mathcal{S}$.

1. Upon receiving input (sender, sid, v_0, v_1) from Sen where $v_0, v_1 \in \{0,1\}^\ell$, record (sid, v_0, v_1).
2. Upon receiving (receiver, sid, u) from Rec, where a tuple (sid, v_0, v_1) is recorded and $u \in \{0,1\}$, send (sid, v_u) to Rec and sid to $\mathcal{S}$. Otherwise, abort.

Fig. 1. The oblivious transfer functionality.

Sender Private Oblivious Transfer. Sender privacy is a weaker notion than malicious security and only requires that the receiver's input be hidden even against a malicious sender. It is weaker than malicious security in that it does not require a simulation of the malicious sender that extracts the sender's inputs. In particular, we will only require that a malicious sender cannot distinguish the cases where the receiver's input is 0 or 1. Formally stated,

Definition 23 (Sender Private OT). *Let π be a two-party protocol that is engaged between a sender Sen and a receiver Rec. We say that π is a sender private oblivious transfer protocol, if for every PPT adversary $\mathcal{A}$ that corrupts Sen, the following ensembles are computationally indistinguishable:*

- *$\{\mathbf{View}_{\mathcal{A},\pi}[\mathcal{A}(1^n), \text{Rec}(1^n, 0)]\}_{n\in\mathbb{N}}$*
- *$\{\mathbf{View}_{\mathcal{A},\pi}[\mathcal{A}(1^n), \text{Rec}(1^n, 1)]\}_{n\in\mathbb{N}}$*

where $\mathbf{View}_{\mathcal{A},\pi}[\mathcal{A}(1^n), \text{Rec}(1^n, b)]$ denotes $\mathcal{A}$'s view within π whenever the receiver Rec inputs the bit b.

We point out that sender privacy protects the receiver against a malicious sender and should be read as privacy against a malicious sender.

Defensibly Private Oblivious Transfer. The notion of *defensible privacy* was introduced by Haitner in [26,27]. A defense in a two-party protocol $\pi = (P_1, P_2)$ execution is an input and random tape provided by the adversary after the execution concludes. A defense for a party controlled by the adversary is said to be *good*, if this party participated honestly in the protocol using this very input and random tape, then it would have resulted in the exact same messages that were sent by the adversary. In essence, this defense serves as a *proof* of honest behavior. It could very well be the case that an adversary deviates from the protocol in the execution but later provides a good defense. The notion of defensible privacy says that a protocol is private in the presence of defensible adversaries if the adversary learns nothing more than its prescribed output when it provides a good defense.

We informally describe the notion of *good defense* for a protocol π; we refer to [27] for the formal definition. Let trans $= (q_1, a_1, \ldots, q_\ell, a_\ell)$ be the transcript of an execution of a protocol π that is engaged between P_1 and P_2 and let $\mathcal{A}$ denote an adversary that controls P_1, where q_i is the ith message from P_1 and a_i is the ith message from P_2 (that is, a_i is the response for q_i). Then we say that (x, r) constitutes a *good defense* of $\mathcal{A}$ relative to trans if the transcript generated by running the honest algorithm for P_1 with input x and random tape r against P_2's messages $a_1, \ldots, a_\ell$ results trans.

The notion of defensible privacy can be defined for any secure computation protocol. Nevertheless, since we are only interested in oblivious transfer protocols, we present a definition below that is restricted to oblivious transfer protocols. The more general definition can be found in [27]. At a high-level, an OT protocol is defensibly private with respect to a corrupted sender if no adversary interacting with an honest receiver with input b should be able to learn b, if at the end of the execution the adversary produces any good defense. Similarly, an OT protocol that is defensibly private with respect to malicious receivers requires that any adversary interacting with an honest sender with input (s_0, s_1) should not be able to learn s_{1-b}, if at the end of the execution the adversary produces a good defense with input b. Below we present a variant of the definition presented in [27]. We stress that while the [27] definition only considers bit OT (i.e. sender's inputs are bits) we consider *string OT*.

Definition 24 (Defensible-Private String OT). *Let π be a two-party protocol that is engaged between a sender Sen and a receiver Rec. We say that π is a defensibly-private string oblivious transfer protocol, if for every PPT adversary $\mathcal{A}$ the following holds,*

1. $\{\Gamma(\mathbf{View}_{\mathcal{A}}[\mathcal{A}(1^n), \mathrm{Rec}(1^n, U)], U)\} \overset{c}{\approx} \{\Gamma(\mathbf{View}_{\mathcal{A}}[\mathcal{A}(1^n), \mathrm{Rec}(1^n, U)], U')\}$, *where $\Gamma(v, *)$ is set to $(v, *)$ if following the execution $\mathcal{A}$ outputs a good defense for π, and $\perp$ otherwise, and U and U' are independent random variables uniformly distributed over $\{0, 1\}$. This property is referred to as defensibly private with respect to a corrupted sender.*

2. $\{\Gamma(\mathbf{View}_{\mathcal{A}}[\mathrm{Sen}(1^n, (U_0^n, U_1^n)), \mathcal{A}(1^n)], U_{1-b}^n)\} \overset{c}{\approx} \{\Gamma(\mathbf{View}_{\mathcal{A}}[\mathrm{Sen}(1^n, (U_0^n, U_1^n)), \mathcal{A}(1^n)], \bar{U}^n)\}$ *where $\Gamma(v, *)$ is set to $(v, *)$ if following the execution*

A outputs a good defense for π, and $\perp$ otherwise, b is the Rec's input in this defense and $U_0^n, U_1^n, \bar{U}^n$ are independent random variables uniformly distributed over $\{0,1\}^n$. This property is referred to as defensibly private with respect to a corrupted receiver.

In our construction from Sect. 3, we will rely on an OT protocol that is sender private and defensibly private with respect to a corrupted receiver. In [27], Haitner et al. showed how to transform any semi-honest bit-OT to one that is defensibly private with respect to a corrupted receiver and malicious secure with respect to a corrupted sender. More formally, the following Lemma is implicit in the work of [27].

Lemma 21 (Implicit in Theorem 4.1 and Corollary 5.3 [27]). Assume the existence of a semi-honest oblivious transfer protocol π. Then there exists an oblivious transfer protocol $\hat{\pi}$ that is defensible-private with respect to the receiver and sender private that relies on the underlying primitive in a black-box manner.

Now, since sender privacy is implied by malicious security with respect to a corrupted sender, this transformation yields a bit OT protocol with the required security guarantees. Nevertheless, our protocol crucially relies on the fact that the underlying OT is a string OT protocol. We therefore show in the full version [30] how to transform any bit OT to a string OT protocol while preserving both defensible private with respect to a maliciously corrupted receiver and sender privacy.

At a high-level, in order to convert any protocol from semi-honest security to defensible privacy, Haitner et al. include a coin-tossing stage at the beginning of the protocol that determines the parties' random tapes. In fact, they let the coin-tossing also determine the parties inputs as they only require OT secure with respect to random inputs for both the sender and receiver. Now, if the receiver has to provide a good defense, then it must reveal the input and randomness used for the semi-honest OT protocol and prove consistency relative to the values generated in the coin-tossing stage. Due to the fact that the commitment schemes that are used in the coin-tossing stage are statistically-binding, the probability that a malicious receiver can deviate from the protocol and provide a good defense is negligible. Using this fact, Haitner et al. argued that the probability that a malicious receiver outputs a good defense and guesses the other sender's input is negligible. Next, to obtain sender private oblivious transfer they first transformed an OT protocol that is defensible-private against malicious receivers to one that is maliciously secure, and then exploited the symmetry of OT in order to obtain a protocol that is sender-private. The first transformation relies on the cut-and-choose approach to ensure that the receiver provides a valid defense, and then using the fact that defensible privacy hides the sender's other input they argued that it is receiver-private.

2.2 UC Commitment Schemes

The notion of UC commitments was introduced by Canetti and Fischlin in [7]. The formal description of functionality $\mathcal{F}_{\text{COM}}$ is depicted in Fig. 2.

Functionality $\mathcal{F}_{\text{COM}}$

Functionality $\mathcal{F}_{\text{COM}}$ communicates with with sender Sen and receiver Rec, and adversary $\mathcal{S}$.

1. Upon receiving input (commit, sid, m) from Sen where $m \in \{0,1\}^t$, internally record (sid, m) and send message $(sid, \text{Sen}, \text{Rec})$ to the adversary. Upon receiving approve from the adversary send sid, to Rec. Ignore subsequent (commit, ., ., .) messages.
2. Upon receiving (reveal, sid) from Sen, where a tuple (sid, m) is recorded, send message m to adversary $\mathcal{S}$ and Rec. Otherwise, ignore.

Fig. 2. The string commitment functionality.

2.3 Extractable Commitments

Our result in the static setting requires the notion of (static) extractable UC commitments, which is a weaker security property than UC commitments in the sense that it does not require equivocality. In what follows, we introduce the definition for the ideal functionality $\mathcal{F}_{\text{EXTCOM}}$ from [44]. Towards introducing this definition, Maji et al. introduced some notions first. More concretely,

Definition 25. *A protocol is a* syntactic commitment protocol *if:*

- *It is a two phase protocol between a sender and a receiver (using only plain communication channels).*
- *At the end of the first phase (commitment phase), the sender and the receiver output a transcript* trans. *Furthermore, the sender receives an output (which will be used for opening the commitment).*
- *In the decommitment phase the sender sends a message γ to the receiver, who extracts an output value* opening(trans, γ) $\in \{0,1\}^n \cup \{\perp\}$.

Definition 26. *Two syntactic commitment protocols (ω_L, ω_R) form a pair of* complementary statistically binding commitment protocols *if the following hold:*

- *ω_R is a statistically binding commitment scheme (with stand-alone security).*
- *In ω_L, at the end of the commitment phase the receiver outputs a string $z \in \{0,1\}^n$. If the receiver is honest, it is only with negligible probability that there exists γ such that* opening(trans, γ) $\neq \perp$ *and* opening(trans, γ) $\neq z$.

As noted in [44], ω_L by itself is not an interesting cryptographic goal, as the sender can simply send the committed string in the clear during the commitment phase. Nevertheless, in defining $\mathcal{F}_{\text{EXTCOM}}$ below, there exists a single protocol that satisfies both the security guarantees. We are now ready to introduce the notion of extractable commitments in Fig. 3 that is parameterized by (ω_L, ω_R). We also include a function pp that will be used as an initialization phase to set up the public-parameters for ω_L and ω_R.

Functionality $\mathcal{F}_{\text{EXTCOM}}$ parameterized by (pp, ω_L, ω_R)

$\mathcal{F}_{\text{EXTCOM}}$ is running with parties $P_1, \ldots, P_n$ and an adversary $\mathcal{S}$: Upon receiving a message $(\text{init} - \text{commit}, sid, ssid, P_i, P_j)$ from P_i, it first checks if there is a tuple $(\text{public} - \text{params}, sid, P_i, (pp, sp))$. If yes, it sends $(\text{init} - \text{commit}, sid, ssid, P_i, P_j)$ to P_j. If not, it runs $(pp, sp) \leftarrow pp(1^n)$ and sends $(\text{init} - \text{commit}, sid, P_i, pp)$ to P_i, P_j and $\mathcal{S}$. It stores $(\text{public} - \text{params}, sid, P_i, (pp, sp))$. We denote P_i by the sender and P_j by the receiver in this interaction. Next, the functionality behaves as follows, depending on which party is corrupted.

- P_i IS HONEST AND P_j IS HONEST.

  Commit Phase: Upon receiving $(\text{commit}, sid, ssid, P_i, P_j, m)$ from P_i, it internally simulates a session of ω_R (simulating both the sender and receiver in ω_R), with the sender's input fixed to m. It gives $(\text{transcript}, sid, ssid, \text{trans}, \gamma)$ to P_i and $(\text{receipt}, sid, ssid, P_i, P_j, \text{trans})$ to P_j and $\mathcal{S}$.

  Reveal Phase: Upon receiving $(\text{decommit}, sid, ssid, \cdot)$ from the sender, it sends $(\text{decommit}, sid, ssid, P_i, P_j, z)$ to P_j and $\mathcal{S}$.

- P_i IS CORRUPTED AND P_j IS HONEST.

  Commit Phase: It runs the commitment ω_L with the sender, playing the part of the receiver in ω_L, to obtain $(sid, ssid, \text{trans}, z)$. It sends $(\text{receipt}, sid, ssid, P_i, P_j, \text{trans})$ to P_j and $\mathcal{S}$.

  Reveal Phase: Upon receiving $(\text{decommit}, sid, ssid, \gamma)$ from the sender, if $\text{opening}(\text{trans}, \gamma) = z$, it sends $(\text{decommit}, sid, ssid, P_i, P_j, z)$ to P_j and $\mathcal{S}$. Otherwise ignore.

- P_i IS HONEST AND P_j IS CORRUPT.

  Commit Phase: Upon receiving $(\text{commit}, sid, ssid, P_i, P_j, m)$ from P_i, it runs the commitment phase of ω_R with P_j, playing the sender's role in ω_R with m as input. It obtains the output (trans, γ) at the end of this phase, and sends $(\text{transcript}, sid, ssid, \text{trans}, \gamma)$ to P_i.

  Reveal Phase: Upon receiving $(\text{decommit}, sid, ssid)$ from the sender it sends $(\text{decommit}, sid, ssid, P_i, P_j, (\gamma, z))$ to P_j and $\mathcal{S}$.

The functionality does not do anything when both the sender and the receiver are corrupted.

Fig. 3. Extractable commitment functionality.

Implementing $\mathcal{F}_{\text{EXTCOM}}$ in the CRS Model. We briefly sketch how to implement the extractable commitment functionality in the $\mathcal{F}_{\text{CRS}}$-hybrid based on the CPA-security of any PKE. Namely, the CRS will be set to a public-key generated using the key-generation function of the PKE scheme. To commit, a sender simply encrypts the message using the public-key in the CRS and sends the ciphertext to the receiver. We can achieve extraction by setting the CRS to a public-key for which the secret-key is available to the extractor (in this case, the extractor is the $\mathcal{F}_{\text{EXTCOM}}$ functionality). Hiding follows from the CPA-security of the encryption scheme. A formal description and proof of this construction can be found in the full version of this paper [30].

3   Static UC Secure Computation

In this section we prove the feasibility of UC secure computation based on semi-honest OT and extractable commitments, where the latter can be constructed based on two-round semi-honest OT (see Sects. 2.1 and 2.3 for more details). More concretely, we prove how to transform any statically semi-honest secure OT into one that is secure in the presence of malicious adversaries, giving *only black-box access* to the underlying semi-honest OT protocol. Our protocol is a variant of the protocol by Lin and Pass from [37] (which in turn is a variant of the protocol of [27]). In particular, in [37], the authors rely on a strong variant of a commitment scheme known as a CCA-secure commitment in order to achieve extraction. We observe that it is not required to use the full power of such commitments, or for that matter UC commitments. Specifically, using a weaker primitive that only implies straight-line input extractability enables to solely rely on semi-honest OT. An important weakening in our commitment scheme compared to CCA-secure commitments from [37] is that we allow invalid commitments to be made by the adversary. We remark here that the work of [37] rely on string OT that are secure against malicious senders and state that the work of [26] provides a black-box construction of such a protocol starting from a semi-honest bit OT. However, the work of [26] only shows how to construct a bit OT secure against malicious senders where the proof crucially relies on the sender's input being only bits. We provide a transformation and complete analysis from bit OT to a string OT for the weaker notion of defensible privacy as this is sufficient for our work. Finally, combining our UC OT protocol with the [33] protocol, we obtain a statically UC secure protocol for any well-formed functionality (see definition in [9]). Namely,

Theorem 31. *Assume the existence of static semi-honest oblivious transfer. Then for any multi-party well-formed functionality $\mathcal{F}$, there exists a protocol that UC realizes $\mathcal{F}$ in the presence of static, malicious adversaries in the $\mathcal{F}_{\mathrm{EXTCOM}}$-hybrid model using black-box access to the oblivious transfer protocol.*

We remark here that the work of [12] shows how starting from a semi-honest oblivious transfer it is possible to obtain a black-box construction of an OT protocol that is secure against stand-alone static adversaries in the $\mathcal{F}_{\mathrm{COM}}$-hybrid model. It is noted in [12] that the (high-level) analysis provided in the work might be extendable to the UC-setting (cf. Footnote 10 in [12]). Furthermore, in the static setting, it is conceivable that $\mathcal{F}_{\mathrm{COM}}$ can be directly realized in the $\mathcal{F}_{\mathrm{EXTCOM}}$-hybrid using the notion of extractable trapdoor commitments [47]. We do not pursue this approach and instead directly realize OT in the $\mathcal{F}_{\mathrm{EXTCOM}}$-hybrid. While the previous works of [12] and [27] require a three step transformation, our transformation is one shot and therefore more direct.

It seems possible to generalize our theorem to multi-session functionalities. Analogous to [7], this will allows us to extend our corollaries to the Global CRS model by additionally assuming CCA encryption scheme and leave it as future work.

3.1 Static UC Oblivious Transfer

In the following, we discuss a secure implementation of the oblivious transfer functionality (see Fig. 1) with static, malicious security in the $\mathcal{F}_{\text{EXTCOM}}$-hybrid model (where $\mathcal{F}_{\text{EXTCOM}}$ is stated formally in Fig. 3). Our goal in this section is to show that the security of malicious UC OT can be based on UC semi-honest OT, denoted by $\pi_{\text{OT}}^{\text{SH}}$, and extractable commitments. Our result is shown in two phases. At first we compile the semi-honest OT protocol $\pi_{\text{OT}}^{\text{SH}}$ into a new protocol with the security properties that are specified in Sect. 2.1, extending the [27] transformation into string OT; denote the compiled OT protocol by $\widehat{\pi}_{\text{OT}}$. Next, we use $\widehat{\pi}_{\text{OT}}$ in order to construct a new protocol $\pi_{\text{OT}}^{\text{ML}}$ that is secure in the presence of malicious adversaries. Details follow,

Protocol 1 (Protocol $\pi_{\text{OT}}^{\text{ML}}$ with Static Security)
Input: *The sender* Sen *has input* (v_0, v_1) *where* $v_0, v_1 \in \{0,1\}^n$ *and the receiver* Rec *has input* $u \in \{0,1\}$.
The protocol:

1. **Coin Tossing:**
   - Receiver's random tape generation: *The parties use a coin tossing protocol in order to generate the inputs and random tapes for the receiver.*
     - *The receiver commits to* $20n$ *strings of appropriate length, denoted by* $a_{\text{Rec}}^1, \ldots, a_{\text{Rec}}^{20n}$, *by sending* $\mathcal{F}_{\text{EXTCOM}}$ *the message* (commit, sid, $\widehat{ssid_i}$, a_{Rec}^i) *for all* $i \in [n]$.
     - *The sender responds with* $20n$ *random strings of appropriate length* $b_{\text{Rec}}^1, \ldots, b_{\text{Rec}}^{20n}$.
     - *The receiver computes* $r_{\text{Rec}}^i = a_{\text{Rec}}^i \oplus b_{\text{Rec}}^i$ *and then interprets* $r_{\text{Rec}}^i = c_i || \tau_{\text{Rec}}^i$ *where* c_i *determines the receiver's input for the* i^{th} *OT protocol, whereas* τ_{Rec}^i *determines the receiver's random tape used for this execution.*
   - Sender's random tape generation: *The parties use a coin tossing protocol in order to generate the inputs and random tapes for the sender.*
     - *The sender commits to* $20n$ *strings of appropriate length, denoted by* $a_{\text{Sen}}^1, \ldots, a_{\text{Sen}}^{20n}$, *by sending* $\mathcal{F}_{\text{EXTCOM}}$ *the message* (commit, sid, $\widehat{ssid_i'}$, a_{Sen}^i) *for all* $i \in [n]$.
     - *The receiver responds with* $20n$ *random strings of appropriate length* $b_{\text{Sen}}^1, \ldots, b_{\text{Sen}}^{20n}$.
     - *The sender computes* $r_{\text{Sen}}^i = a_{\text{Sen}}^i \oplus b_{\text{Sen}}^i$ *and then interprets* $r_{\text{Sen}}^i = s_i^0 || s_i^1 || \tau_{\text{Sen}}^i$ *where* (s_i^0, s_i^1) *determine the sender's input for the* i^{th} *OT protocol, whereas* τ_{Sen}^i *determines the sender's random tape used for this execution.*

2. **Oblivious Transfer:**
   - *The parties participate in* $20n$ *executions of the OT protocol* $\widehat{\pi}_{\text{OT}}$ *with the corresponding inputs and random tapes obtained from Stage 2. Let the output of the receiver in the* i^{th} *execution be* $\tilde{s}_i$.

3. **Cut-and-choose:**
   - Sen *chooses a random subset* $q_{\text{Sen}} = (q_{\text{Sen}}^1, \ldots, q_{\text{Sen}}^n) \in \{1, \ldots, 20\}^n$ *and sends it to* Rec. *The string* q_{Sen} *is used to define a set of indices* $\Gamma_{\text{Sen}} \subset \{1, \ldots, 20n\}$ *of size* n *in the following way:* $\Gamma_{\text{Sen}} = \{20i - q_{\text{Sen}}^i\}_{i \in [n]}$. *The receiver then opens the commitments from Stage 1 that correspond to the indices within* Γ_{Sen}, *namely,*

the receiver decommits a^i_{Rec} for all $i \in \Gamma_{\mathrm{Sen}}$. Sen checks that the decommitted values are consistent with the inputs and randomness used for the OTs in Stage 2 by the receiver, and aborts in case of a mismatch.

- Rec chooses a random subset $q_{\mathrm{Rec}} = (q^1_{\mathrm{Rec}}, \ldots, q^n_{\mathrm{Rec}}) \in \{1, \ldots, 20\}^n$ and sends it to Sen. The string q_{Rec} is used to define a set of indices $\Gamma_{\mathrm{Rec}} \subset \{1, \ldots, 20n\}$ of size n in the following way: $\Gamma_{\mathrm{Rec}} = \{20i - q^i_{\mathrm{Rec}}\}_{i \in [n]}$. The sender then opens the commitments from Stage 1 that correspond to the indices within Γ_{Rec}, namely, the sender decommits a^i_{Sen} for all $i \in \Gamma_{\mathrm{Rec}}$. Rec checks that the decommitted values are consistent with the inputs and randomness used for the OTs in Stage 2 by the sender, and aborts in case of a mismatch.

- Rec commits to another subset $\Gamma \subset [20n]$ denoted by $(\Gamma^1, \ldots, \Gamma^n)$, by sending $\mathcal{F}_{\mathrm{EXTCOM}}$ the message (commit, $sid, ssid'_i, \Gamma^i$) for all $i \in [n]$. (The sender will reveal its inputs and randomness that are used in Stage 2 that correspond to the indices in Γ later in Stage 5.)

4. **Combiner:**

- Let $\Delta = [20n] - \Gamma_{\mathrm{Rec}} - \Gamma_{\mathrm{Sen}}$. Then for every $i \in \Delta$, the receiver computes $\alpha_i = u \oplus c_i$ and sends it to the sender.

- The sender computes a $10n$-out-of-$18n$ secret sharing of v_0, denote the shares by $\{\rho^0_i\}_{i \in \Delta}$. Analogously, it computes a $10n$-out-of-$18n$ secret sharing of v_1, denote the shares by $\{\rho^1_i\}_{i \in \Delta}$. The sender computes $\beta^b_i = \rho^b_i \oplus s^{b \oplus \alpha_i}_i$ for all $b \in \{0, 1\}$ and $i \in \Delta$, and sends the outcome to the receiver.

- The receiver computes $\tilde{\rho}_i = \beta^u_i \oplus \tilde{s}_i$ for all $i \in \Delta$. Denote by ρ these concatenated bits.

5. **Final cut-and-choose:**

- The receiver decommits Γ and the sender sends the inputs and randomness it used in Stage 2 for the coordinates that correspond to $\Delta \cap \Gamma$. (Note that the sender need only reveal the indices that were not decommitted in Stage 3). Rec checks that the sender's values are consistent with the inputs and randomness used for the OTs in Stage 2 by the sender, and aborts in case of a mismatch.

- The receiver checks whether $(\tilde{\rho}_i)_{i \in \Delta}$ agrees with some codeword $w \in \mathcal{W}_{18n,10n}$ on $17n$ locations (where the code $\mathcal{W}_{18n,10n}$ is induced by the secret sharing construction that we use in Stage 4). Recall that the minimum distance of the code $\mathcal{W}_{18n,10n}$ is at least $18n - 10n > 8n$, which implies that there will be at most one such codeword w. Furthermore, since we can correct up to $\frac{18n - 10n}{2} = 4n$ errors, any code that is $17n$ close to a codeword can be efficiently recovered using the Berlekamp-Welch algorithm. The receiver outputs that w as its output in the OT protocol. If no such w exists, the receiver returns a default value.

Theorem 32. *Assume that $\pi^{\mathrm{SH}}_{\mathrm{OT}}$ is static semi-honest secure and that the compiled $\widehat{\pi}_{\mathrm{OT}}$ is secure according to Lemma 21. Then Protocol 1 UC realizes $\mathcal{F}_{\mathrm{OT}}$ in the presence of static malicious adversaries in the $\mathcal{F}_{\mathrm{EXTCOM}}$-hybrid model using black-box access to the oblivious transfer protocol.*

Recalling that our protocol relies on the existence of semi-honest OT and extractable commitments, and that the later can be constructed based on any two-round semi-honest OT, e.g., [21], which implies PKE (see Sects. 2.1 and 2.3 for more details), an immediate corollary from Theorem 32 implies that,

Corollary 33. *Assume the existence of two-round static semi-honest oblivious transfer. Then there exists a protocol that securely realizes $\mathcal{F}_{\mathrm{OT}}$ in the presence of static malicious adversaries in the CRS model using black-box access to the oblivious transfer protocol.*

A High Level Proof. We first provide an overview of the security proof; the complete proof is found in [30]. Loosely speaking, in case the receiver is corrupted the simulator plays the role of the honest sender in Stages 1–4. Next in Stage 5, the simulator extracts the receiver's input u. Specifically, the simulator extracts all the committed values of the receiver within Stage 1 (relying on the fact that the commitment scheme is extractable), and then uses these values in order to obtain the inputs for the OT executions in Stage 2. Upon completing Stage 2, the simulator records the coordinates for which the receiver deviates from the prescribed input and random tape chosen in the coin tossing phase. Denoting these set of coordinates by Φ, we recall that a malicious receiver may obtain both of the sender's inputs with respect to the OT executions that correspond to the coordinates within Φ and Γ. On the other hand, it obtains only one of the two inputs with respect to the rest of the OT executions that correspond to the coordinates within $\Delta - \Phi - \Gamma$. Consequently, the simulator checks how many shares of v_0 and v_1 are obtained by the receiver and proceeds accordingly. In more details,

- If the receiver obtains more than $10n$ shares of both inputs then the simulator halts and outputs fail (we prove in Section [30] that this event only occurs with negligible probability).
- If the receiver obtains less than $10n$ shares of both inputs then the simulator picks two random values for v_0 and v_1 of the appropriate length and completes the interaction, playing the role of the honest sender on these values. Note that in this case the simulator does not need to call the ideal functionality.
- Finally, if the receiver obtains more than $10n$ shares for only one input $u \in \{0, 1\}$, then the simulator sends u to the ideal functionality $\mathcal{F}_{OT}$ and obtains v_u. The simulator then sets v_{1-u} as a random string of the appropriate length and completes the interaction by playing the role of the honest sender on these values.

Recall that the only difference between the simulation and the real execution is in the way the messages in Stage 4 are generated. Specifically, in the simulation a value u is extracted from the malicious receiver and then fed to the $\mathcal{F}_{OT}$ functionality. The simulation is then completed based on the output returned from the functionality. Intuitively, the cut-and-choose mechanism ensures that the receiver cannot deviate from the honest strategy in Stage 2 in more than n OT sessions without getting caught with overwhelming probability. Moreover, the defensible privacy of the OT protocol implies that the receiver can learn at most one of the two inputs of the sender relative to the OT executions in Stage 2 for which the receiver proceeded honestly.

In case the sender is corrupted, the simulator's strategy is to play the role of the honest receiver until Stage 5 where the simulator extracts the sender's inputs. More specifically, the simulator first extracts the sender's input for the OT executions in Stage 1 (relying on the fact that the commitment scheme is extractable). Next, the simulator extracts the shares $\{\rho_i^0\}_{i \in \Delta}$ and $\{\rho_i^1\}_{i \in \Delta}$ that correspond to inputs v_0 and v_1. To obtain the actual values the simulator checks if these shares agree with some codeword relative to $16n$ locations. That is,

– Let w_0 and w_1 denote the corresponding codewords (if there are no such codewords that agree with with v_0 and v_1 on $16n$ locations then the simulator uses a default codeword instead). Next, the simulator checks w_0 and w_1 against the final cut-and-choose. If any of the shares from w_b are inconsistent with the opened shares that are opened by the sender in the final cut-and-choose, then v_b is set to a default value, otherwise v_b is the value corresponding to the shared secret.

Finally, the simulator sends (v_0, v_1) to the ideal functionality for $\mathcal{F}_{\mathrm{OT}}$. Security in this case is reduced to the privacy of the receiver. In addition, the difference between the simulation's strategy and the honest receiver's strategy is that the simulator extracts the sender's both inputs in all $i \in \Delta - \Phi$ and then finds codewords that are $16n$-close to the extracted values, whereas the honest receiver finds a codeword that is $17n$-close based on the inputs it received in the Stages 2 and 5, and returns it. We thus prove that the value u extracted by the simulator is identical the to the reconstructed output of the honest receiver relying on the properties of the secret sharing scheme.

4   One-Sided Adaptive UC Secure Computation

In the two-party one-sided adaptive setting, at most one of the parties is adaptively corrupted [29,35]. In this section we provide a simple transformation of our static UC secure protocol from Sect. 3 to a two-party UC-secure protocol that is secure against one-sided adaptive corruption. Our first observation is that in Protocol 1 the parties use their real inputs to the OT protocol only in Phase 4. Therefore simulation of the first three phases can be easily carried out by simply following the honest strategy. On the other hand, simulating messages in Phase 4 requires some form of equivocation since if corruption takes place after this phase is concluded then the simulator needs to explain this message with respect to the real input of the corrupted party. On a high-level we will transform the protocol so that if no party is corrupted until end of Phase 4, the simulator can equivocate the message in Phase 4. We explain how to achieve equivocation later. First, we describe our simulator: In case either party is statically corrupted the simulation for Protocol 1 follows the strategy of the honest party until Phase 4, where the simulator extracts the corrupted party's input relying on the fact that it knows the adversary's committed input in Phase 1. Therefore, the same proof follows in case the adversary adaptively corrupts one of the parties at any point before Phase 4, as the simulator can pretend that corruption took place statically. On the other hand, if corruption takes place after Phase 4, then the simulator equivocates the communication. It is important to note that while in the plain model any statically secure protocol can be compiled into one-sided secure protocol by encrypting its entire communication, it is not clear that this is the case in the UC setting due to the additional setup, e.g., a CRS that may depend on the identity of the corrupted party. Nevertheless, in Phase 4 the parties only run a combiner for which the computation does not involve any usage of the CRS (which is induced by the extractable commitment). Therefore, the proof follows.

A common approach to achieve equivocation is to rely on non-committing encryption schemes (NCE) [6,11,16], that allow secure communication in the presence of adaptive attacks. This powerful tool has been constructed while relying on (a variant of) simulatable PKE schemes, which, roughly speaking, allows for both the public-key and the ciphertexts to be generated obliviously without the knowledge of the plaintext or the secret key [11,16]. Notably, these constructions achieve a stronger notion of security where both parties may be adaptively corrupted (also referred to as *fully adaptive*). Our second observation is that it is sufficient to rely on a weaker variant of NCE, namely, one that is secure against only one-sided adaptive corruption.

In particular, we take advantage of a construction presented in [6] and later refined in [16], that achieves receiver equivocation under the assumption of semi-honest OT. We will briefly describe it now. Recall that in the fully adaptive case, the high-level idea is for the sender and receiver to mutually agree on a random bit, which is then used by the sender to determine which of two random strings to mask its message. The process of agreeing on a bit requires the ability to both obliviously sample a public-key without the knowledge of the secret key, as well as the ability to obliviously sample a ciphertext without the knowledge of the corresponding plaintext. In the simpler one-sided scenario, Canetti et al. observed that an oblivious transfer protocol can replace the oblivious generation of the public-key. Specifically, the NCE receiver sends two public keys to the sender, and then the parties invoke an OT protocol where the NCE receiver plays the role of the OT sender and enters the corresponding secret keys. To allow equivocation for the NCE sender, the OT must enable equivocation with respect to the OT receiver. The [21] OT protocol is an example for such a protocol. Here the OT receiver can pick the two ciphertexts so that it knows both plaintexts. Then equivocation is carried out by declaring that the corresponding ciphertext is obliviously sampled.

The advantage of this approach is that it removes the requirement of generating the public key obliviously, as now the randomness for its generation is split between the parties, where anyway only one of them is corrupted. This implies that the simulator can equivocate the outcome of the protocol execution without letting the adversary the ability to verify it. To conclude, it is possible to strengthen the security of Protocol 1 into the one-sided setting by simply encrypting the communication within the combiner phase using one-sided NCE which in turn can be constructed based on PKE with oblivious ciphertext generation. This implies the following theorem which further implies black-box one-sided UC secure computation from enhanced trapdoor permutation.

Theorem 41. *Assume the existence of PKE with oblivious ciphertext generation. Then for any two-party well-formed functionality $\mathcal{F}$, there exists a protocol that UC realizes $\mathcal{F}$ in the presence of one-sided adaptive, malicious adversaries in the CRS model using black-box access to the PKE.*

5   Adaptive UC Secure Computation

In this section we demonstrate the feasibility of UC secure commitment schemes based on PKE with oblivious ciphertext generation (namely, where it is possible to obliviously sample the ciphertext without knowing the plaintext). Our construction is secure even in the presence of adaptive corruptions and is the first to achieve the stronger notion of adaptive security based on this hardness assumption. Plugging-in our UC commitment protocol into the transformation of [12] that generates adaptive malicious OT given adaptive semi-honest OT and UC commitments, implies an adaptively UC secure oblivious transfer protocol with malicious security based on semi-honest adaptive OT and PKE with oblivious ciphertext generation using only black-box access to the semi-honest OT and the PKE. Stating formally,

Theorem 51. *Assume the existence of adaptive semi-honest oblivious transfer and PKE with oblivious ciphertext generation. Then for any multi-party well-formed functionality $\mathcal{F}$, there exists a protocol that UC realizes $\mathcal{F}$ in the presence of adaptive, malicious adversaries in the CRS model using black-box access to the oblivious transfer protocol and the PKE.*

Noting that simulatable PKE implies both semi-honest adaptive OT [9,11] and PKE with oblivious ciphertext generation, we derive the following corollary (where simulatable PKE implies oblivious sampling of both public keys and ciphertexts),

Corollary 52. *Assume the existence of simulatable PKE. Then for any multi-party well-formed functionality $\mathcal{F}$, there exists a protocol that UC realizes $\mathcal{F}$ in the presence of adaptive, malicious adversaries in the CRS model using black-box access to the simulatable PKE.*

This in particular improves the result from [14] that relies on simulatable PKE in a non-black-box manner. Note also that our UC commitment can be constructed using a weaker notion than simulatable PKE where the inverting algorithms can require a trapdoor. This notion is denoted by trapdoor simulatable PKE [11] and can be additionally realized based on the hardness assumption of factoring Blum integers. This assumption, however, requires that we modify our commitment scheme so that the CRS includes $3n+1$ public keys of the underlying PKE instead of just one, as otherwise the reduction to the security of the PKE does not follow for multiple ciphertexts. Specifically, at the cost of linear blowup (in the security parameter) of the CRS, we obtain adaptively secure UC commitments under a weaker assumption. Now, since trapdoor simulatable PKE implies adaptive semi-honest OT [11] it holds,

Corollary 53. *Assume the existence of trapdoor simulatable PKE. Then for any multi-party well-formed functionality $\mathcal{F}$, there exists a protocol that UC realizes $\mathcal{F}$ in the presence of adaptive, malicious adversaries in the CRS model using black-box access to the trapdoor simulatable PKE.*

Note that, since the best known general assumptions for realizing adaptive semi-honest OT is trapdoor simulatable PKE, this corollary gives evidence that the assumptions for adaptive semi-honest OT are sufficient for adaptive UC security and makes a step towards identifying the minimal assumptions for achieving UC security in the adaptive setting. To conclude, we note that enhanced trapdoor permutations, which imply PKE with oblivious ciphertext generation, imply the following corollary,

Theorem 54. *Assume the existence of enhanced trapdoor permutation. Then $\mathcal{F}_{\text{COM}}$ (cf. Fig. 2) can be UC realized in the CRS model in the presence of adaptive malicious adversaries.*

5.1 UC Commitments from PKE with Oblivious Ciphertext Generation

In this section we demonstrate the feasibility of adaptively secure UC commitments for the message space $m \in \{0,1\}$ from any public-key encryption scheme $\Pi = (\text{Gen}, \text{Enc}, \text{Dec}, \widetilde{\text{Enc}}, \widetilde{\text{Enc}}^{-1})$ with oblivious ciphertext generation (cf. Definition 21) in the common reference string (CRS) model. In this model [7] the parties have access to a CRS chosen from a specified trusted distribution $\mathcal{D}$. This is captured via the ideal functionality $\mathcal{F}_{\text{CRS}}^{\mathcal{D}}$ (see [30] for the definition). We note that we use Π in two places in our protocol. First, in the encoding phase (where the commitments are computed by the sender) and then in the coin-tossing phase (where the commitments are computed by the receiver). Our complete construction can be found in Fig. 4. Next, we prove

Theorem 55. *Assume that $\Pi = (\text{Gen}, \text{Enc}, \text{Dec}, \widetilde{\text{Enc}}, \widetilde{\text{Enc}}^{-1})$ is a PKE with oblivious ciphertext generation. Then protocol π_{COM} (cf. Fig. 4) UC realizes $\mathcal{F}_{\text{COM}}$ in the CRS model in the presence of adaptive malicious adversaries.*

A High Level Proof. Intuitively, security requires proving both hiding and binding in the presence of static and adaptive corruptions. The hiding property follows from the IND-CPA security of the encryption scheme combined with the fact that the receiver only sees n shares in a n-out-of-$3n+1$ secret-sharing of the message in the commit phase. On the other hand, proving binding is much more challenging and reduces to the facts that a corrupted sender cannot successfully predict exactly the n indices from $\{1, \ldots, 3n+1\}$ that will be chosen in the coin-tossing protocol. In fact, if it can identify these n indices, then it would be possible for the adversary to break binding. An important information-theoretic argument that we prove here is that for a fixed encoding phase, no adversary can equivocate on two continuations from the encoding phase with different outcomes of the coin-tossing phase. Saying differently, for any given encoding phase there is exactly one outcome for the coin-tossing phase that will allow equivocation. Given this claim, binding now follows from the IND-CPA security of the encryption scheme used in the coin-tossing phase. In addition, recall that in the UC setting the scheme must also support a simulation that allows straight-line extraction and equivocation. At a high-level, the simulator sets the CRS to

---

Protocol π_{COM}.

CRS: Two independent keys PK, $\widetilde{\text{PK}}$ that are in the range of $\text{Gen}(1^n)$.

Sender's Input: A message $m \in \{0, 1\}$ and a security parameter 1^n.

[Commitment phase:]

Encoding phase: The sender chooses a random n-degree polynomial $p(\cdot)$ over a field $\mathbb{F}[x]$ such that $p(0) = m$. Namely, it randomly chooses $a_i \leftarrow \mathbb{F}$ for all $i \in [n]$ and sets $a_0 = m$, and defines the polynomial $p(x) = a_0 + a_1 x + \cdots + a_n x^n$. The sender then creates a commitment to m as follows. For every $i = [3n + 1]$, it first pick $b_i \leftarrow \{0, 1\}$ at random and then computes the following pairs:

$$\text{If } b_i = 0 \text{ then } \begin{array}{l} c_i^0 = \text{Enc}_{\text{PK}}(p(i); t_i) \\ c_i^1 = r_i \end{array} \quad \text{else,} \quad \text{if } b_i = 1 \text{ then } \begin{array}{l} c_i^0 = r_i \\ c_i^1 = \text{Enc}_{\text{PK}}(p(i); t_i) \end{array}$$

where $t_i \leftarrow \{0, 1\}^n$ and $r_i \leftarrow \widetilde{\text{Enc}}(\cdot)$ is obliviously sampled. The sender sends $(c_0^0, c_0^1), \ldots, (c_{3n+1}^0, c_{3n+1}^1)$ to the receiver.

Coin-tossing phase: The sender and receiver interact in a coin-tossing protocol that is carried out as follows.

1. The receiver sends $c = \text{Enc}_{\widetilde{\text{PK}}}(\sigma_0; r_{\sigma_0})$ to the sender where $\sigma_0 \leftarrow \{0, 1\}^N$ is chosen uniformly at random.
2. The sender picks $\sigma_1 \leftarrow \{0, 1\}^N$ at random and sends it in the clear to the receiver
3. The receiver decrypts c by revealing σ_0 and r_{σ_0}.

Both the sender and the receiver compute $\sigma = \sigma_0 \oplus \sigma_1$ and use σ as the random string to sample a random subset $S \subset [3n + 1]$ of size n. (Note that such sampling can be done in a simple way by partitioning the set of coordinates into n sets of triples (where the last set includes 4 elements) and picking one element per set. Notably, this technique does not imply that any potential subset of size n will be picked, rather it ensures that a subset is picked with a negligible probability in n, specifically $(1/3)^n$, which suffices for our proof.)

Cut-and-choose phase: The sender decrypts the set $\{c_i^{b_i}\}_{i \in S}$ by sending the sequence $\{b_i, p(i), t_i\}_{i \in S}$. The receiver verifies that all the decryptions are correct and aborts otherwise.

[Decommitment phase:] Let $T = [3n + 1] - S$. The sender reveals its input m and decrypts all the ciphertexts in $\{c_i^{b_i}\}_{i \in T}$. The receiver checks if all the decryptions are correct and aborts otherwise. Using the n polynomial evaluations revealed relative to $i \in S$ and any additional polynomial evaluation that was revealed relative to T, the receiver reconstructs the polynomial $p(\cdot)$ (via polynomial interpolation of $n+1$ points). Next, the receiver verifies whether $p(0) = m$, and that for every $i \in [3n + 1]$ the point $p(i)$ is the decrypted value within $c_i^{m_i}$.

Fig. 4. UC adaptively secure commitment scheme.

public-keys for which it knows the corresponding secret-keys. This will allow the simulator to extract all the values encrypted by the adversary. We observe that the simulator can fix the outcome of the coin-tossing phase to any n-indices of its choice by extracting the random string σ_0 encrypted by the receiver and choosing a random string σ_1 so that $\sigma_0 \oplus \sigma_1$ is a particular string. Next, the simulator generates secret-sharing for both 0 and 1 so that they overlap in the particular n shares. To commit, the simulator encrypts the n common shares within the n indices to be revealed (which it knows in advance), and for the rest of the indices it encrypts two shares, one that corresponds to the sharing of 0 and the other that corresponds to the sharing of 1. Finally, in the decommit phase, the simulator reveals that shares that correspond to the real message m, and exploits the invertible sampling algorithm to prove that the other ciphertexts were obliviously generated.

References

1. Barak, B., Canetti, R., Nielsen, J.B., Pass, R.: Universally composable protocols with relaxed set-up assumptions. In: FOCS, pp. 186–195 (2004)
2. Beaver, D.: Foundations of secure interactive computing. In: Feigenbaum, J. (ed.) CRYPTO 1991. LNCS, vol. 576, pp. 377–391. Springer, Heidelberg (1992)
3. Blazy, O., Chevalier, C., Pointcheval, D., Vergnaud, D.: Analysis and improvement of Lindell's UC-secure commitment schemes. In: Jacobson, M., Locasto, M., Mohassel, P., Safavi-Naini, R. (eds.) ACNS 2013. LNCS, vol. 7954, pp. 534–551. Springer, Heidelberg (2013)
4. Canetti, R.: Universally composable security: a new paradigm for cryptographic protocols. In: FOCS, pp. 136–145 (2001)
5. Canetti, R., Dodis, Y., Pass, R., Walfish, S.: Universally composable security with global setup. IACR Cryptology ePrint Archive 2006, 432 (2006)
6. Canetti, R., Feige, U., Goldreich, O., Naor, M.: Adaptively secure multi-party computation. In: STOC, pp. 639–648 (1996)
7. Canetti, R., Fischlin, M.: Universally composable commitments. In: Kilian, J. (ed.) CRYPTO 2001. LNCS, vol. 2139, pp. 19–40. Springer, Heidelberg (2001)
8. Canetti, R., Kushilevitz, E., Lindell, Y.: On the limitations of universally composable two-party computation without set-up assumptions. J. Cryptol. 19(2), 135–167 (2006)
9. Canetti, R., Lindell, Y., Ostrovsky, R., Sahai, A.:Universally composable two-party and multi-party secure computation. In: STOC, pp. 494–503 (2002)
10. Canetti, R., Pass, R., Shelat, A.:Cryptography from sunspots: how to use an imperfect reference string. In: FOCS, pp. 249–259 (2007)
11. Choi, S.G., Dachman-Soled, D., Malkin, T., Wee, H.: Improved non-committing encryption with applications to adaptively secure protocols. In: Matsui, M. (ed.) ASIACRYPT 2009. LNCS, vol. 5912, pp. 287–302. Springer, Heidelberg (2009)
12. Choi, S.G., Dachman-Soled, D., Malkin, T., Wee, H.: Simple, black-box constructions of adaptively secure protocols. In: Reingold, O. (ed.) TCC 2009. LNCS, vol. 5444, pp. 387–402. Springer, Heidelberg (2009)
13. Choi, S.G., Katz, J., Wee, H., Zhou, H.-S.: Efficient, adaptively secure, and composable oblivious transfer with a single, global CRS. In: PKC, pp. 73–88 (2013)
14. Dachman-Soled, D., Malkin, T., Raykova, M., Venkitasubramaniam, M.: Adaptive and concurrent secure computation from new adaptive, non-malleable commitments. In: Sako, K., Sarkar, P. (eds.) ASIACRYPT 2013, Part I. LNCS, vol. 8269, pp. 316–336. Springer, Heidelberg (2013)
15. Damgård, I., Groth, J.: Non-interactive and reusable non-malleable commitment schemes. In: STOC, pp. 426–437 (2003)
16. Damgård, I.B., Nielsen, J.B.: Improved non-committing encryption schemes based on a general complexity assumption. In: Bellare, M. (ed.) CRYPTO 2000. LNCS, vol. 1880, pp. 432–450. Springer, Heidelberg (2000)

17. Damgård, I.B., Nielsen, J.B.: Perfect hiding and perfect binding universally composable commitment schemes with constant expansion factor. In: Yung, M. (ed.) CRYPTO 2002. LNCS, vol. 2442, pp. 581–596. Springer, Heidelberg (2002)
18. Damgård, I., Nielsen, J.B., Orlandi, C.: On the necessary and sufficient assumptions for UC computation. In: Micciancio, D. (ed.) TCC 2010. LNCS, vol. 5978, pp. 109–127. Springer, Heidelberg (2010)
19. David, B., Dowsley, R., Nascimento, A.C.A.: Universally composable oblivious transfer based on a variant of LPN. In: Gritzalis, D., Kiayias, A., Askoxylakis, I. (eds.) CANS 2014. LNCS, vol. 8813, pp. 143–158. Springer, Heidelberg (2014)
20. David, B.M., Nascimento, A.C.A., Müller-Quade, J.: Universally composable oblivious transfer from lossy encryption and the McEliece assumptions. In: Smith, A. (ed.) ICITS 2012. LNCS, vol. 7412, pp. 80–99. Springer, Heidelberg (2012)
21. Even, S., Goldreich, O., Lempel, A.: A randomized protocol for signing contracts. Commun. ACM 28(6), 637–647 (1985)
22. Garay, J.A., Wichs, D., Zhou, H.-S.: Somewhat non-committing encryption and efficient adaptively secure oblivious transfer. In: Halevi, S. (ed.) CRYPTO 2009. LNCS, vol. 5677, pp. 505–523. Springer, Heidelberg (2009)
23. Gertner, Y., Kannan, S., Malkin, T., Reingold, O., Viswanathan, M.: The relationship between public key encryption and oblivious transfer. In: FOCS, pp. 325–335 (2000)
24. Goldreich, O., Micali, S., Wigderson, A.: How to play any mental game or A completeness theorem for protocols with honest majority. In: STOC, pp. 218–229 (1987)
25. Goyal, V., Lee, C.-K., Ostrovsky, R., Visconti, I.: Constructing non-malleable commitments: a black-box approach. In: FOCS, pp. 51–60 (2012)
26. Haitner, I.: Semi-honest to malicious oblivious transfer—the black-box way. In: Canetti, R. (ed.) TCC 2008. LNCS, vol. 4948, pp. 412–426. Springer, Heidelberg (2008)
27. Haitner, I., Ishai, Y., Kushilevitz, E., Lindell, Y., Petrank, E.: Black-box constructions of protocols for secure computation. SIAM J. Comput. 40(2), 225–266 (2011)
28. J. Cryptol. Smooth projective hashing and two-message oblivious transfer. 25(1), 158–193 (2012)
29. Hazay, C., Patra, A.: One-sided adaptively secure two-party computation. In: Lindell, Y. (ed.) TCC 2014. LNCS, vol. 8349, pp. 368–393. Springer, Heidelberg (2014)
30. Hazay, C., Venkitasubramaniam, M.: On black-box complexity of universally composable security in the CRS model. IACR Cryptology ePrint Archive 2015, 488 (2015)
31. Impagliazzo, R., Rudich, S.: Limits on the provable consequences of one-way permutations. In: Goldwasser, S. (ed.) CRYPTO 1988. LNCS, vol. 403, pp. 8–26. Springer, Heidelberg (1990)
32. Ishai, Y., Kushilevitz, E., Lindell, Y., Petrank, E.: Black-box constructions for secure computation. In: STOC, pp. 99–108 (2006)
33. Ishai, Y., Prabhakaran, M., Sahai, A.: Founding cryptography on oblivious transfer – efficiently. In: Wagner, D. (ed.) CRYPTO 2008. LNCS, vol. 5157, pp. 572–591. Springer, Heidelberg (2008)
34. Kalai, Y.T., Lindell, Y., Prabhakaran, M.: Concurrent composition of secure protocols in the timing model. J. Cryptol. 20(4), 431–492 (2007)
35. Katz, J., Ostrovsky, R.: Round-optimal secure two-party computation. In: Franklin, M. (ed.) CRYPTO 2004. LNCS, vol. 3152, pp. 335–354. Springer, Heidelberg (2004)

36. Kilian, J.: Founding cryptography on oblivious transfer. In: STOC, pp. 20–31 (1988)
37. Lin, H., Pass, R.: Black-box constructions of composable protocols without setup. In: Safavi-Naini, R., Canetti, R. (eds.) CRYPTO 2012. LNCS, vol. 7417, pp. 461–478. Springer, Heidelberg (2012)
38. Lin, H., Pass, R., Venkitasubramaniam, M.: A unified framework for concurrent security: universal composability from stand-alone non-malleability. In: STOC, pp. 179–188 (2009)
39. Pass, R., Lin, H., Venkitasubramaniam, M.: A unified framework for UC from only OT. In: Wang, X., Sako, K. (eds.) ASIACRYPT 2012. LNCS, vol. 7658, pp. 699–717. Springer, Heidelberg (2012)
40. Lindell, Y.: General composition and universal composability in secure multi-party computation. In: FOCS, pp. 394–403 (2003)
41. Lindell, A.Y.: Adaptively secure two-party computation with erasures. In: Fischlin, M. (ed.) CT-RSA 2009. LNCS, vol. 5473, pp. 117–132. Springer, Heidelberg (2009)
42. Lindell, Y.: Highly-efficient universally-composable commitments based on the DDH assumption. In: Paterson, K.G. (ed.) EUROCRYPT 2011. LNCS, vol. 6632, pp. 446–466. Springer, Heidelberg (2011)
43. Lindell, Y., Zarosim, H.: Adaptive zero-knowledge proofs and adaptively secure oblivious transfer. In: Reingold, O. (ed.) TCC 2009. LNCS, vol. 5444, pp. 183–201. Springer, Heidelberg (2009)
44. Maji, H.K., Prabhakaran, M., Rosulek, M.: A zero-one law for cryptographic complexity with respect to computational UC security. In: Rabin, T. (ed.) CRYPTO 2010. LNCS, vol. 6223, pp. 595–612. Springer, Heidelberg (2010)
45. Micali, S., Rogaway, P.: Secure computation. In: Feigenbaum, J. (ed.) CRYPTO 1991. LNCS, vol. 576, pp. 392–404. Springer, Heidelberg (1992)
46. Naor, M., Pinkas, B.: Efficient oblivious transfer protocols. In: Proceedings of the Twelfth Annual Symposium on Discrete Algorithms, Washington, DC, USA, pp. 448–457, 7–9 Jan 2001
47. Pass, R., Wee, H.: Black-box constructions of two-party protocols from one-way functions. In: Reingold, O. (ed.) TCC 2009. LNCS, vol. 5444, pp. 403–418. Springer, Heidelberg (2009)
48. Peikert, C., Vaikuntanathan, V., Waters, B.: A framework for efficient and composable oblivious transfer. In: Wagner, D. (ed.) CRYPTO 2008. LNCS, vol. 5157, pp. 554–571. Springer, Heidelberg (2008)
49. Rabin, M.: How to exchange secrets by oblivious transfer. Technical memo TR-81, Aiken Computation Laboratory, Harvard University (1981)
50. Yao, A.C.-C.: How to generate and exchange secrets (extended abstract). In: FCOS, pp. 162–167 (1986)

Public Verifiability in the Covert Model (Almost) for Free

Vladimir Kolesnikov[1]([⊠]) and Alex J. Malozemoff[2]

[1] Bell Labs, Murray Hill, USA
kolesnikov@research.bell-labs.com
[2] University of Maryland, College Park, USA
amaloz@cs.umd.edu

Abstract. The covert security model (Aumann and Lindell, TCC 2007) offers an important security/efficiency trade-off: a covert player may arbitrarily cheat, but is caught with a certain fixed probability. This permits more efficient protocols than the malicious setting while still giving meaningful security guarantees. However, one drawback is that cheating cannot be proven to a third party, which prevents the use of covert protocols in many practical settings. Recently, Asharov and Orlandi (ASIACRYPT 2012) enhanced the covert model by allowing the honest player to generate a *proof of cheating*, checkable by any third party. Their model, which we call the PVC (*publicly verifiable covert*) model, offers a very compelling trade-off.

Asharov and Orlandi (AO) propose a practical protocol in the PVC model, which, however, relies on a specific expensive oblivious transfer (OT) protocol incompatible with OT extension. In this work, we improve the performance of the PVC model by constructing a PVC-compatible OT extension as well as making several practical improvements to the AO protocol. As compared to the state-of-the-art OT extension-based two-party covert protocol, our PVC protocol adds relatively little: four signatures and an $\approx 67\,\%$ wider OT extension matrix. This is a significant improvement over the AO protocol, which requires public-key-based OTs per input bit. We present detailed estimates showing (up to orders of magnitude) concrete performance improvements over the AO protocol and a recent malicious protocol.

Keywords: Secure computation · Publicly verifiable covert security

1 Introduction

Two-party secure computation addresses the problem where two parties need to evaluate a common function f on their inputs while keeping the inputs private. Several security models for secure computation have been proposed. The most

---

A.J. Malozemoff—Work partially done while the author was at Bell Labs.

T. Iwata and J.H. Cheon (Eds.): ASIACRYPT 2015, Part II, LNCS 9453, pp. 210–235, 2015.
DOI: 10.1007/978-3-662-48800-3_9

basic is the *semi-honest* model, where the parties are expected to follow the protocol description but must not be able to learn anything about the other party's input from the protocol transcript. A much stronger guarantee is provided by the *malicious* model, where parties may deviate arbitrarily from the protocol description. This additional security comes at a cost. Recent garbled circuit-based protocols [3,17] have an overhead of at least 40× that of their semi-honest counterparts, and are considerably more complex.

Aumann and Lindell [8] introduced a very practical compromise between these two models, that of *covert* security. In the covert security model, a party can deviate arbitrarily from the protocol description but is caught with a fixed probability ϵ, called the *deterrence factor*. In many practical scenarios, this guaranteed risk of being caught (likely resulting in loss of business and/or embarrassment) is sufficient to deter would-be cheaters. Importantly, covert protocols are much more efficient and simpler than their malicious counterparts.

Motivating the Publicly Verifiable Covert (PVC) Model. At the same time, the cheating deterrent introduced by the covert model is relatively weak. Indeed, a party catching a cheater certainly knows what happened and can respond accordingly, e.g., by taking their business elsewhere. However, the impact is largely limited to this, since the honest player cannot credibly *accuse* the cheater publicly. If, however, credible public accusation were possible, the deterrent for the cheater would be immeasurably greater: suddenly, *all* the cheater's customers would be aware of the cheating and thus any cheating may affect the cheater's global customer base.

The addition of credible accusation greatly improves the covert model even in scenarios with a small number of players, such as those involving the government. Consider, for example, the setting where two agencies are engaged in secure computation on their respective classified data. The covert model may often be insufficient here. Indeed, consider the case where one of the two players deviates from the protocol, perhaps due to an insider attack. The honest player detects this, but we are now faced with the problem of identifying the culprit across two domains, where the communication is greatly restricted due to trust, policy, data privacy legislation, or all of the above. On the other hand, credible accusation immediately provides the ability to exclude the honest player from the suspect list, and focus on tracking the problem *within one organization/trust domain*, which is dramatically simpler.

PVC Definition and Protocol. Asharov and Orlandi [7] proposed a security model, *covert with public verifiability*, and an associated protocol, addressing these concerns. At a high level, they proposed that when cheating is detected, the honest player is able to publish a "certificate of cheating" which can be checked by any third party. In this work, we abbreviate their model as *PVC: publicly verifiable covert*. Their proposed protocol (which we call the "AO protocol") has performance similar to the original covert protocol of Aumann and Lindell [8], with the exception of requiring signed-OT, a special form of oblivious transfer (OT). Their signed-OT construction is based on the OT of Peikert et al. [18], and thus requires several expensive public-key operations.

In this work, we propose several critical performance improvements to the AO protocol. Our most technically involved contribution is a novel signed-OT *extension* protocol which eliminates per-instance public-key operations. Before discussing our contributions and technical approach in Sect. 1.1, we review the AO protocol.

The Asharov-Orlandi (AO) PVC Protocol [7]. The AO protocol is based on the covert construction of Aumann and Lindell [8]. Let P_1 be the circuit generator, P_2 be the evaluator, and $f(\cdot, \cdot)$ be the function to be computed. Recall the standard garbled circuit (GC) construction in the semi-honest model: P_1 constructs a garbling of f and sends it to P_2 along with the wire labels associated with its input. The parties then run OT, with P_1 acting as the sender and inputting the wire labels associated with P_2's input, and P_2 acting as the receiver and inputting as its choice bits the associated bits of its input.

We now adapt this protocol to the PVC setting. Recall the "selective failure" attack on P_2's input wires, where P_1 can send P_2 via OT an invalid wire label for one P_2's two inputs and learn one of P_2's input bits based on whether P_2 aborts. To protect against this attack, the parties construct $f'(\mathbf{x}_1, \mathbf{x}_2^1, \ldots, \mathbf{x}_2^\nu) = f(\mathbf{x}_1, \bigoplus_{i \in [\nu]} \mathbf{x}_2^i)$, where ν is the *XOR-tree replication factor*, and compute f' instead of f. Party P_1 then constructs λ (the *GC replication factor*) garblings of f' and P_2 checks that $\lambda - 1$ of the GCs are correctly constructed, evaluating the remaining GC to derive the output. The main difficulty of satisfying the PVC model is ensuring that neither party can improve its odds by aborting (e.g., based on the other party's challenge). For example, if P_1 could abort whenever P_2's challenge would reveal P_1's cheating, this would enable P_1 to cheat without the risk of generating a proof of cheating. Thus, P_1 sends the GCs to P_2 through a 1-out-of-λ OT; namely, in the ith input to the OT P_1 provides openings for all the GCs but the ith, as well as the input wire labels needed to evaluate GC_i. Party P_2 inputs a random γ, checks that all GCs besides GC_γ are constructed correctly, and if so, evaluates GC_γ.

Finally, it is necessary for P_1 to operate in a *verifiable* manner, so that an honest P_2 has proof if P_1 tries to cheat and gets caught. (Note that GCs guarantee that P_2 cannot cheat in the GC evaluation at all, so we only worry about catching P_1.) The AO protocol addresses this by having P_1 sign all its messages and the parties using *signed*-OT in place of all standard OTs (including wire label transfers and GC openings). Informally, the signed-OT functionality proceeds as follows: rather than the receiver R getting message $\mathbf{m}_b$ from the sender S for choice bit b, R receives $((b, \mathbf{m}_b), \sigma)$, where σ is S's signature of $(b, \mathbf{m}_b)$. This guarantees that if R detects any cheating by S, it has S's signature on an inconsistent set of messages, which can be used as proof of this cheating. Asharov and Orlandi show that this construction is ϵ-PVC-secure for $\epsilon = (1 - 1/\lambda)(1 - 2^{-\nu+1})$.

1.1   Our Contribution

Our main contribution is a signed-OT extension protocol built on the recent malicious OT extension of Asharov et al. [6]. Informally, signed-OT extension ensures that (1) a cheating sender S is held accountable in the form of a "certificate of cheating" that the honest receiver R can generate, and (2) a malicious R cannot *defame* an honest S by presenting a false "certificate of cheating". Achieving the first goal is fairly straightforward by having S simply sign all its messages. The challenge is in simultaneously protecting against a malicious R. In particular, we need to commit R to its particular choices throughout the OT extension protocol to prevent it from defaming an honest S, while maintaining that those commitments do not leak any information about R's choices.

Recall that in the standard OT extension protocol of Ishai et al. [12] (cf. Fig. 3), R constructs a random matrix M, and S obtains a matrix M' derived from M, S's random string $\mathbf{s}$ and R's vector of OT inputs $\mathbf{r}$. The key challenge of adapting this protocol to the signed variant is to efficiently prevent R from submitting a malleated M as part of the proof without it ever explicitly revealing M to S (as this would leak R's choice bits). We achieve this by observing that S does in fact learn some of M, as in the OT extension construction some of the columns of M and M' are the same (i.e., those corresponding to zero bits of S's string $\mathbf{s}$). We prevent R from cheating by having S include in its signature carefully selected information from the columns in M which S sees. Finally, we require that R generates each row of M from a seed, and that R's proof of cheating includes this seed such that the row rebuilt from the seed is consistent with the columns included in S's signature. We show that this makes it infeasible for R to successfully present an invalid row in the proof of cheating. We describe this approach in greater detail in Sect. 3[1].

As another contribution, we present a new more communication efficient PVC protocol, building off of the AO protocol; see Sect. 4. Our main (simple) trick there is a careful amendment allowing us to send GC hashes instead of GCs; this is based on an idea from Goyal et al. [11].

We work in the random oracle model, a slight strengthening of the assumptions needed for standard OT extension and free-XOR, two standard secure computation tools.

Comparison with Existing Approaches. The cost of our protocol is almost the same as that of the covert protocol of Goyal et al. [11]; the only extra cost is essentially a $\approx 67\%$ wider OT extension matrix and four signatures. This often negligible additional overhead (versus covert protocols) provides us with dramatically stronger (than covert) deterrent. We believe that our PVC protocol could be used in many applications where covert security is insufficient at the order-of-magnitude cost advantage over previously-needed malicious protocols or the PVC protocol of Asharov and Orlandi [7]. See Sect. 5 for more details.

Related Work. The only directly related work is that of Asharov and Orlandi [7], already discussed at length. We also note a recent line of work on secure

---

[1] Our construction is also interesting from a theoretical perspective in that we construct signed-OT from *any* maliciously secure OT protocol, whereas Asharov and Orlandi [7] build a specific construction based on the Decisional Diffie-Hellman problem.

computation with cheaters (including fairness violators) punished by an external entity, such as the Bitcoin network [4, 10, 16]. Similarly to the PVC model and our protocols, this line of work relies on generating proofs of misbehavior which could be accepted by a third-party authority. However, these works address a different setting and use different techniques; in particular, they build on maliciously-secure computation and require the Bitcoin framework.

2 Preliminaries

Let κ denote the (computational) security parameter, let ρ denote the statistical security parameter, and let τ denote the field size. When considering concrete costs, we utilize the security parameter and field size settings for key lengths recommended by NIST [9]; see Fig. 1. We use PPT to denote "probabilistic polynomial time" and let $\mathsf{negl}(\cdot)$ denote a negligible function in its input. We consider two-party protocols between parties P_1 and P_2, and when we use subscript $i \in \{1, 2\}$ to denote a party we let subscript $-i = 3 - i$ denote the other party. We use $i^* \in \{1, 2\}$ to denote a malicious party and $-i^* = 3 - i^*$ to denote the associated honest party.

Security	κ	FCC	ECC
Short	80	1024	160
Long	128	3072	256

Fig. 1. Settings for (computational) security parameter κ and field size τ for various security settings as recommended by NIST [9]. FCC denotes the setting of τ when using finite field cryptography and ECC denotes the setting of τ when using elliptic curve cryptography.

We use bold lowercase letters (e.g., $\mathbf{x}$) to denote bitstrings and use the notation $\mathbf{x}[i]$ to denote the ith bit in bitstring $\mathbf{x}$. Likewise, we use bold uppercase letters (e.g., $\mathbf{T}$) to denote matrices over bits. We use $[n]$ to denote $\{1, \ldots, n\}$. Let "$a \leftarrow f(x_1, x_2, \ldots)$" denote setting a to be the deterministic output of f on inputs $x_1, x_2, \ldots$; the notation "$a \leftarrow_\$ f(x_1, x_2, \ldots)$" is the same except that f here is randomized. We abuse notation and let $a \leftarrow_\$ S$ denote selecting a uniformly at random from set S.

Our constructions are in the $\mathcal{F}_{\mathsf{PKI}}$ model, where each party P_i can register a verification key, and other parties can retrieve P_i's verification key by querying $\mathcal{F}_{\mathsf{PKI}}$ on id_i. We use the notation $\mathsf{Sign}_{P_i}(\cdot)$ to denote a signature signed by P_i's secret key, and we assume that this signature can be verified by any third party. We often leave off the subscript if the identity of the signing party is clear.

2.1 Publicly Verifiable Covert Security

We assume the reader is familiar with the covert security model; however, we review the less familiar *publicly verifiable covert* (PVC) security model of

Asharov and Orlandi [7] below. When we say a protocol is "secure in the covert model," we assume it is secure under the strong explicit cheat formulation with ϵ-deterrent [8, §3.4], for some value of ϵ.

Let π be a two-party protocol between parties P_1 and P_2 implementing function f. Following Aumann and Lindell [8], we call π *non-halting* if for honest P_i and fail-stop adversary[2] P_{-i}, the probability that P_i outputs corrupted$_{-i}$ is negligible. Consider the triple of algorithms $(\pi', \mathsf{Blame}, \mathsf{Judgment})$ defined as follows:

- Protocol π' is the same as π except that if an honest party P_{-i^*} outputs corrupted$_{i^*}$ when executing π, it computes $\mathsf{Cert} \leftarrow \mathsf{Blame}(\mathsf{id}_{i^*}, \mathsf{key}, \mathsf{View}_{-i^*})$, where key denotes the type of cheating detected, and sends Cert to P_{i^*}.
- Algorithm Blame is a deterministic algorithm which takes as input a cheating identity id, a cheating type key, and a view View of a protocol execution, and outputs a certificate Cert.
- Algorithm $\mathsf{Judgment}$ is a deterministic algorithm which takes as input a certificate Cert and outputs either an identity id or $\perp$.

Before proceeding to the definition, we first introduce some notation. Let $\mathsf{Exec}_{\pi,\mathcal{A}(z)}(x_1, x_2; 1^\kappa)$ denote the transcript (i.e., messages and output) produced by P_1 with input x_1 and P_2 with input x_2 running protocol π, where adversary $\mathcal{A}$ with auxiliary input z can corrupt parties before execution begins. Let $\mathsf{Output}_{P_i}(\mathsf{Exec}_{\pi,\mathcal{A}(z)}(x_1, x_2; 1^\kappa))$ denote the output of P_i on the input transcript.

Definition 1. *We say that* $(\pi', \mathsf{Blame}, \mathsf{Judgment})$ *securely computes* f *in the presence of a publicly verifiable covert adversary with* ϵ-*deterrent (or, is* ϵ-*PVC-secure) if the following conditions hold:*

1. *The protocol* π' *is a non-halting and secure realization of* f *in the covert model with* ϵ-*deterrent.*
2. *(Accountability) For every* PPT *adversary* $\mathcal{A}$ *corrupting party* P_{i^*}, *there exists a negligible function* $\mathsf{negl}(\cdot)$ *such that if* $\mathsf{Output}_{P_{-i^*}}(\mathsf{Exec}_{\pi,\mathcal{A}(z)}(x_1, x_2; 1^\kappa)) = $ corrupted$_{i^*}$ *then* $\Pr[\mathsf{Judgment}(\mathsf{Cert}) = \mathsf{id}_{i^*}] > 1 - \mathsf{negl}(\kappa)$, *where* $\mathsf{Cert} \leftarrow \mathsf{Blame}(\mathsf{id}_{i^*}, \mathsf{key}, \mathsf{View}_{-i^*})$ *and the probability is over the randomness used in the protocol execution.*
3. *(Defamation-free) For every* PPT *adversary* $\mathcal{A}$ *corrupting party* P_{i^*} *and outputting a certificate* Cert, *there exists a negligible function* $\mathsf{negl}(\cdot)$ *such that* $\Pr[\mathsf{Judgment}(\mathsf{Cert}) = \mathsf{id}_{-i^*}] < \mathsf{negl}(\kappa)$, *where the probability is over the randomness used by* $\mathcal{A}$.

Note that, in particular, the PVC definition implicitly disallows Blame to reveal P_{-i^*}'s input. This is because π' specifies that Cert is sent to P_{i^*}.

2.2 Signed Oblivious Transfer

A central functionality for constructing PVC protocols is *signed oblivious transfer* (signed-OT). Introduced by Asharov and Orlandi [7], we can define the basic

---

[2] A *fail-stop adversary* is one which acts semi-honestly but may halt at any time.

signed-OT functionality $\mathcal{F}$ as

$$(\bot, (m_b, \mathsf{Sign}_{\mathsf{sk}}(b, m_b))) \leftarrow_{\$} \mathcal{F}((m_0, m_1, \mathsf{sk}), (b, \mathsf{vk})),$$

where the signature scheme is assumed to be existentially unforgeable under adaptive chosen message attack (EU-CMA). Namely, the sender S inputs two messages m_0 and m_1 along with a signing key sk; the receiver R inputs a choice bit b and a verification key vk; S receives no output whereas R receives m_b alongside a signature on (b, m_b).

However, as in prior work [7], this definition is too strong for our signed-OT extension construction to satisfy. We introduce a relaxed signed-OT variant (slightly different from Asharov and Orlandi's variant [7]) which is tailored for OT extension and is sufficient for obtaining PVC-security. Essentially, we need a signature scheme that satisfies a weaker notion than EU-CMA in which the signing algorithm takes randomness, a portion of which can be controlled by the adversary[3]. This is because in our signed-OT extension construction, a malicious party can influence the randomness used in the signing algorithm. In addition, we introduce an *associated data* parameter to the signing algorithm which allows the signer to specify some additional information unrelated to the message being signed but used in the signature. In our construction, we use the associated data to tie the signature to a specific counter (such as a session ID or message ID), preventing a malicious receiver from "mixing" properly signed values to defame an honest sender.

Let $\Pi = (\mathsf{Gen}, \mathsf{Sign}, \mathsf{Verify})$ be a tuple of PPT algorithms over message space $\mathcal{M}$, associated data space $\mathcal{D}$, and randomness spaces $\mathcal{R}_1$ and $\mathcal{R}_2$, defined as follows:

1. $\mathsf{Gen}(1^\kappa)$: On input security parameter 1^κ, output key pair $(\mathsf{vk}, \mathsf{sk})$.
2. $\mathsf{Sign}_{\mathsf{sk}}(m, a; (r_1, r_2))$: On input secret key sk, message $m \in \mathcal{M}$, associated data $a \in \mathcal{D}$, and randomness $r_1 \in \mathcal{R}_1$ and $r_2 \in \mathcal{R}_2$, output signature $\sigma = (a, \sigma')$.
3. $\mathsf{Verify}_{\mathsf{vk}}(m, \sigma)$: On input verification key vk, message $m \in \mathcal{M}$, and signature σ, output 1 if σ is a valid signature for m and 0 otherwise.

For security, we need the condition that unforgeability remains even if the adversary inputs some arbitrary r_1 or r_2. However, the adversary is prevented from inputting values for *both* r_1 and r_2. This reflects the fact that in our signed-OT extension construction, a malicious sender can control only r_1 and a malicious receiver can control only r_2. We place a further restriction that the choice of r_1 must be *consistent*; namely, all queries to Sign must use the same value for r_1. Looking ahead, this property exactly captures the condition we need (r_1 corresponds to the zero bits in the sender's column selection string in the OT

---

[3] Our notion is similar to the ρ-EU-CMRA notion introduced by Asharov and Orlandi [7]. It differs in that we allow different portions of the randomness to be corrupted, but not both portions at once. Looking forward, this is needed because the sender in our signed-OT functionality is only allowed to control some of the randomness.

extension), where the choice of r_1 is made once and then fixed throughout the protocol execution.

Towards our definition, we define an oracle $\mathcal{O}_{sk}(\cdot, \cdot, \cdot, \cdot)$ as follows. Let $\bot$ be a special symbol. On input (m, a, r_1, r_2), proceed as follows. If neither r_1 nor r_2 equal $\bot$, output $\bot$. Otherwise, proceed as follows. If $r_1 = \bot$ and r_1' has not been set, set r_1' uniformly at random; if $r_1 \neq \bot$ and r_1' has not been set, set $r_1' = r_1$; if $r_2 = \bot$, set r_2' uniformly at random; otherwise, set $r_2' = r_2$. Finally, output $\mathsf{Sign}_{sk}(m, a; (r_1', r_2'))$.

Now, consider the following game $\mathsf{Sig\text{-}forge}_{\mathcal{A}, \Pi}^{\mathsf{CMPRA}}(\kappa)$ for signature scheme Π between PPT adversary $\mathcal{A}$ and PPT challenger $\mathcal{C}$.

1. $\mathcal{C}$ runs $(\mathsf{vk}, \mathsf{sk}) \leftarrow_\$ \mathsf{Gen}(1^\kappa)$ and sends vk to $\mathcal{A}$.
2. $\mathcal{A}$, who has oracle access to $\mathcal{O}_{sk}(\cdot, \cdot, \cdot, \cdot)$, outputs a tuple $(m, (a, \sigma'))$. Let $\mathcal{Q}$ be the set of messages and associated data pairs input to $\mathcal{O}_{sk}(\cdot, \cdot, \cdot, \cdot)$.
3. $\mathcal{A}$ succeeds if and only if (1) $\mathsf{Verify}_{vk}(m, (a, \sigma')) = 1$ and (2) $(m, a) \notin \mathcal{Q}$.

Definition 2. *Signature scheme $\Pi = (\mathsf{Gen}, \mathsf{Sign}, \mathsf{Verify})$ is existentially unforgeable under adaptive chosen message and partial randomness attack (EU-CMPRA) if for all PPT adversaries $\mathcal{A}$ there exists a negligible function $\mathsf{negl}(\cdot)$ such that* $\Pr[\mathsf{Sig\text{-}forge}_{\mathcal{A}, \Pi}^{\mathsf{CMPRA}}(\kappa)] < \mathsf{negl}(\kappa)$.

Functionality $\mathcal{F}_{\mathsf{signedOT}}^{\Pi}$

The functionality is parameterized by an EU-CMPRA signature scheme $\Pi = (\mathsf{Gen}, \mathsf{Sign}, \mathsf{Verify})$.

Input: The sender inputs messages m_0 and m_1 such that $|m_0| = |m_1|$, secret key sk, associated data a, randomness r_1^*, and signatures σ_0^* and σ_1^*. The receiver inputs choice bit b, verification key vk, and randomness r_2^*. If the sender (resp., the receiver) is honest, then $r_1^* = \sigma_0^* = \sigma_1^* = \bot$ (resp., $r_2^* = \bot$).

Output: The functionality computes $\sigma_b = \mathsf{Sign}_{sk}((b, m_b), a; (r_1^*, r_2^*))$ for $b \in \{0, 1\}$. The sender receives no output. The receiver receives the following output based on if the sender is corrupt or not:

- If either $\sigma_0^* \neq \bot$ or $\sigma_1^* \neq \bot$, the functionality outputs $((b, m_b), \sigma_b^*)$ if and only if $\mathsf{Verify}_{vk}((0, m_0), \sigma_0^*) = \mathsf{Verify}_{vk}((1, m_1), \sigma_1^*) = 1$, where $\sigma_b^* \leftarrow \sigma_b$ if $\sigma_b^* = \bot$; otherwise it outputs abort.
- If $\sigma_0^* = \sigma_1^* = \bot$, the functionality outputs $((b, m_b), \sigma_b)$.

Fig. 2. Signed oblivious transfer functionality.

Signed-OT Functionality. We are now ready to introduce our relaxed signed-OT functionality. As is our EU-CMPRA signature, it is tailored for OT extension, and is sufficient for building PVC protocols. This functionality, denoted by $\mathcal{F}^{\Pi}_{\text{signedOT}}$, is parameterized by an EU-CMPRA signature scheme Π and is defined in Fig. 2. As in standard OT, the sender inputs two messages (of equal length) and the receiver inputs a choice bit. However, in this formulation we allow a malicious sender to specify some random value r_1^* as well as signatures σ_0^* and σ_1^*. Likewise, a malicious receiver can specify some random value r_2^*. (Honest players input $\perp$ for these values.) If both players are honest, the functionality computes $\sigma \leftarrow \text{Sign}((b, m_b); (r_1, r_2))$ with uniformly random values r_1 and r_2 and outputs $((b, m_b), \sigma)$ to the receiver. However, if either party is malicious and specifies·some random value, this is fed into the Sign algorithm. Likewise, if the sender is malicious and specifies some signature $\sigma_b^* \neq \perp$, this value is used as the signature sent to the receiver.

Note that $\mathcal{F}^{\Pi}_{\text{signedOT}}$ is nearly identical to the signed-OT functionality presented by Asharov and Orlandi [7, Functionality 2]; it differs in the use of EU-CMPRA signature schemes instead of ρ-EU-CMRA schemes. We also note that it is straightforward to adapt $\mathcal{F}^{\Pi}_{\text{signedOT}}$ to realize OTs with more than two inputs from the sender. We let $\binom{\lambda}{1}$-$\mathcal{F}^{\Pi}_{\text{signedOT}}$ denote a 1-out-of-λ variant of $\mathcal{F}^{\Pi}_{\text{signedOT}}$.

A Compatible Commitment Scheme. Our construction of an EU-CMPRA signature scheme (cf. Sect. 3.3) uses a non-interactive commitment scheme, which we define here. Our definition follows the standard commitment definition, except we tweak the Com algorithm to take an additional associated data value.

Let $\Pi_{\text{Com}} = (\text{ComGen}, \text{Com})$ be a tuple of PPT algorithms over message space $\mathcal{M}$ and associated data space $\mathcal{D}$, defined as follows:

1. $\text{ComGen}(1^\kappa)$: On input security parameter 1^κ, compute parameters params.
2. $\text{Com}(m, a; r)$: On input message $m \in \mathcal{M}$, associated data $a \in \mathcal{D}$, and randomness r, output commitment com.

A commitment can be opened by revealing the randomness r used to construct that commitment.

We now define security for our commitment scheme. We only consider the *binding* property; namely, the inability for a PPT adversary to open a commitment to some other value than that committed to. Security is the same as for standard commitment schemes, except we allow the adversary to control the randomness used in ComGen.

Consider the game $\text{Com-bind}_{\mathcal{A}, \Pi_{\text{Com}}}(\kappa)$ for commitment scheme Π_{Com} between a PPT adversary $\mathcal{A}$ and a PPT challenger $\mathcal{C}$, defined as follows.

1. $\mathcal{A}$ sends randomness r to $\mathcal{C}$.
2. $\mathcal{C}$ computes params $\leftarrow \text{ComGen}(1^\kappa; r)$ and sends params to $\mathcal{A}$.
3. $\mathcal{A}$ outputs $(\text{com}, m_1, a_1, r_1, m_2, a_2, r_2)$ and wins if and only if (1) $m_1 \neq m_2$, and (2) com $= \text{Com}(\text{params}, m_1, a_1; r_1) = \text{Com}(\text{params}, m_2, a_2; r_2)$.

Definition 3. *A commitment scheme* $\Pi_{\text{Com}} = (\text{ComGen}, \text{Com})$ *is (computationally) binding if for all PPT adversaries* $\mathcal{A}$, *there exists a negligible function* $\text{negl}(\cdot)$ *such that* $\Pr[\text{Com-bind}_{\mathcal{A}, \Pi_{\text{Com}}}(\kappa)] < \text{negl}(\kappa)$.

3 Signed Oblivious Transfer Extension

We now present our main contribution: an efficient instantiation of signed oblivious transfer (signed-OT) extension. We begin in Sect. 3.1 by describing in detail the logic of the construction, iteratively building it up from the passively secure protocol of Ishai et al. [12]. We motivate the need for EU-CMPRA signature schemes in Sect. 3.2 and present a compatible such scheme in Sect. 3.3. In Sect. 3.4 we present the proof of security.

3.1 Intuition for the Construction

Consider the OT extension protocol of Ishai et al. [12] in Fig. 3, run between sender S and receiver R. This protocol is secure against a semi-honest R and malicious S. We show how to convert this protocol into one which satisfies the $\mathcal{F}_{\mathsf{signedOT}}^{\Pi}$ functionality defined in Fig. 2. For clarity of presentation, we build on the protocol of Fig. 3 and later discuss how to support a malicious R as well, based on the malicious OT extension protocol of Asharov et al. [6].

S's inputs: Message pairs $\{(\mathbf{x}_j^0, \mathbf{x}_j^1)\}_{j \in [m]}$, where each $\mathbf{x}_j^0, \mathbf{x}_j^1 \in \{0,1\}^n$.
R's inputs: Selection bits $\mathbf{r} = (r_1, \ldots, r_m)$.
Common inputs: Security parameter κ; number of base OTs $\ell\ (= \kappa)$; hash function $H : \mathbb{N} \times \{0,1\}^\ell \to \{0,1\}^n$; ideal functionality $\mathcal{F}_{\mathsf{OT}}$.

1. Initial OT Phase:
   - S computes $\mathbf{s} \leftarrow_{\$} \{0,1\}^\ell$.
   - R generates a random $m \times \ell$ matrix $\mathbf{T}$, where the jth row is $\mathbf{t}_j$ and the ith column is $\mathbf{t}^i$. Likewise, R generates a random $m \times \ell$ matrix $\mathbf{V}$, where the jth row is $\mathbf{v}_j$ and the ith column is $\mathbf{v}^i$.
   - S and R run $\mathcal{F}_{\mathsf{OT}}$ ℓ times in parallel, where S acts as the *receiver* with input s_i in the ith OT and R acts as the *sender* with input $(\mathbf{t}^i, \mathbf{v}^i)$ in the ith OT.
2. OT Extension Phase (Part I):
   - For $i \in [m]$, R sets $\mathbf{u}^i \leftarrow \mathbf{t}^i \oplus \mathbf{v}^i \oplus \mathbf{r}$, and sends $\mathbf{u}^i$ to S.
3. OT Extension Phase (Part II):
   - Let $\mathbf{Q}$ be the $m \times \ell$ matrix where each column $\mathbf{q}^i = (s_i \cdot (\mathbf{u}^i \oplus \mathbf{v}^i)) \oplus ((1 - s_i) \cdot \mathbf{t}^i)$. Note that $\mathbf{q}^i = (s_i \cdot \mathbf{r}) \oplus \mathbf{t}^i$ and $\mathbf{q}_j = (r_j \cdot \mathbf{s}) \oplus \mathbf{t}_j$.
   - For $j \in [m]$, S computes $\mathbf{y}_j^0 \leftarrow \mathbf{x}_j^0 \oplus H(j, \mathbf{q}_j)$ and $\mathbf{y}_j^1 \leftarrow \mathbf{x}_j^1 \oplus H(j, \mathbf{q}_j \oplus \mathbf{s})$, and sends $\mathbf{y}_j^0$ and $\mathbf{y}_j^1$ to R.
   - For $j \in [m]$, R computes $\mathbf{x}_j \leftarrow \mathbf{y}_j^{r_j} \oplus H(j, \mathbf{t}_j)$.
4. Output:
   - S outputs $\perp$ and R outputs $\{\mathbf{x}_j\}_{j \in [m]}$.

Fig. 3. Protocol implementing passively secure OT extension [5,12].

As a first attempt, suppose S simply signs all its messages in Step 3. Recall that we will use this construction to have P_1 send the appropriate input wire labels to P_2; namely, P_1 acts as S in the OT extension and inputs the wire labels for P_2's input wires whereas P_2 acts as R and inputs its input bits. Thus, our first step is to enhance the protocol in Fig. 3 to have S send $\sigma' \leftarrow_\$ \mathsf{Sign}((j, \mathbf{y}_j^0))$ and $\sigma'' \leftarrow_\$ \mathsf{Sign}((j, \mathbf{y}_j^1))$ in Step 3.

Now, if P_2 gets an invalid (with respect to a signed GC sent in the PVC protocol of Sect. 4) wire label $\mathbf{x}_j$, it can easily construct a certificate Cert which demonstrates P_1's cheating. Namely, it outputs as its certificate the tuple $(b, j, \mathbf{y}_j^0, \mathbf{y}_j^1, \sigma', \sigma'', \mathbf{t}_j)$ along with the (signed by P_1 and opened) GC containing the invalid wire label. A third party can (1) check that σ' and σ'' are valid signatures and (2) compute $\mathbf{x}_j^b \leftarrow H(j, \mathbf{t}_j) \oplus \mathbf{y}_j^b$ and check that $\mathbf{x}_j^b$ is indeed an invalid wire label for the given garbled circuit.

This works for protecting against a malicious P_1; however, note that P_2 can easily *defame* an honest P_1 by outputting $\mathbf{t}_j^* \neq \mathbf{t}_j$ as part of its certificate (in which case $\mathbf{x}_j^b \leftarrow H(j, \mathbf{t}_j^*) \oplus \mathbf{y}_j^b$ will very likely be an invalid wire label). Thus, the main difficulty in constructing signed-OT extension is tying P_2 to its choice of the matrix $\mathbf{T}$ generated in Step 1 of the protocol so it cannot blame an honest P_1 by using invalid rows $\mathbf{t}_j^*$ in its certificate.

Towards this end, consider the following modification. In Step 1, R now additionally sends commitments to each $\mathbf{t}_j$ to S, and S signs these and sends them as part of its messages in Step 3. This prevents R from later changing $\mathbf{t}_j$ to blame S. This does not quite work, however, as R could simply commit to an incorrect $\mathbf{t}_j^*$ in the first place! Clearly, R cannot send $\mathbf{T}$ to S, as this would leak R's selection bits, yet we still need R to somehow be committed to its choice of the matrix $\mathbf{T}$. The key insight is noting that S does in fact know *some* of the bits of $\mathbf{T}$; namely, it knows those columns at which $s_i = 0$ (as it learns $\mathbf{t}^i$ in the base OT). We can use this information to tie R to its choice of $\mathbf{T}$ such that it cannot later construct some matrix $\mathbf{T}^* \neq \mathbf{T}$ to defame S.

We do this by enhancing Step 3 as follows. Let I^0 be the set of indices i such that $s_i = 0$ (recall that $\mathbf{s}$ is the random selection bits of S input to the base OTs in Step 1). Let $t_{j,i}$ denote the ith bit in row $\mathbf{t}_j$. Note that S knows the values of $t_{j,i}$ for $i \in I^0$, and could thus compute $\{(i, t_{j,i})\}_{i \in I^0}$ as a "binding" of R's choice of $\mathbf{t}_j$. By including this information in its signature, S enforces that any $\mathbf{t}_j^*$ that R tries to use to blame S must match in the given positions. This brings us closer to our goal; however, there are still two issues that we need to resolve:

1. Sending $\{(i, t_{j,i})\}_{i \in I}$ to R leaks $\mathbf{s}$, which allows R to learn both of S's inputs. We address this by increasing the number of base OTs in Step 1 and having S only send some subset $I \subseteq I^0$ such that $|I| = \kappa$. Thus, while R learns that $s_i = 0$ for $i \in I$, by increasing the number of base OTs enough, R does not have enough information to recover $\mathbf{s}$.
2. R can still flip one bit in $\mathbf{t}_j$ and pass the check with high probability. We fix this by having each $\mathbf{t}_j$ be generated by a seed $\mathbf{k}_j$. Namely, R computes

$t_j \leftarrow G(k_j)$ in Step 1, where G is a random oracle[4]. Then, when blaming S, R must reveal k_j instead of t_j. Thus, with high probability a malicious polytime R cannot find some $k_j^* \neq k_j$ such that the Hamming distance between $G(k_j^*)$ and $G(k_j)$ is small enough that the above check succeeds.

Finally, note that we have thus far considered the passively secure OT extension protocol, which is insecure against a malicious R. We thus utilize the maliciously secure OT extension protocol of Asharov et al. [6]. The only way R can cheat in passively secure OT extension is by using different r values in Step 2. Asharov et al. add a "consistency check" phase between Steps 1 and 2 to enforce that r is consistent. This does not affect our construction, and thus we can include this step to complete the protocol[5]. We refer the reader to Asharov et al. [6] for the justification and intuition of this step; as far as this work is concerned we can treat this consistency check as a "black box".

Observation 1 (OT Extension Matrix Size). *We set ℓ, the number of base OTs, so that leaking κ bits to R does not allow it to recover s and thus both messages. We do this as follows. Let ℓ' be the number of base OTs required in malicious OT extension [6]. We set $\ell = \ell' + \kappa$ and require that when S chooses s, it first fixes κ randomly selected bits to zero before randomly setting the rest of the bits. Now, when S reveals I to R, the number of unknown bits in s is equal to ℓ' and thus the security of the Asharov et al. scheme carries over to our setting. Asharov et al. set $\ell' \approx 1.6\kappa$, and thus us using κ extra columns results in an $\approx 67\%$ matrix size increase.*

Observation 2 (Batching Signatures). *The main computational cost of our protocol is the signatures sent by S in Step 4. This cost can easily be brought to negligible, as follows. Recall that when using our protocol for transferring the input wire labels of a GC using free-XOR we can optimize the communication slightly by setting $x_j^0 \leftarrow H(j, q_j)$ and $y_j^1 \leftarrow x_j^0 \oplus \Delta \oplus H(j, q_j \oplus s)$, where Δ is the free-XOR global offset. Thus, S only needs to send (and sign) y_j^1.*

The most important idea, however, is to batch messages across OT executions and have S sign (and send) only one signature which includes all the necessary information across many OTs. Namely, using the free-XOR optimization above, S signs and sends the tuple $(I, \{y_j^1, \{t_{j,i}\}_{i \in I}\}_{j \in [m]})$ to R. We note that the j values need not be sent as they are implied by the protocol execution.

Figure 4 gives the full protocol for signed-OT extension. For clarity of presentation, this description, and the following proof of security, does not take into account the optimizations described in Observation 2.

---

[4] Note that G *cannot* be a pseudorandom generator because the input to G is not necessarily uniform as the inputs may be adversarially chosen by R.

[5] The reason this does not affect our construction is because the consistency check phase only involves R sending messages to S. A malicious R cannot defame S because we are only enforcing that R's value r is consistent.

S's inputs: Messages $\{(\mathbf{x}_j^0, \mathbf{x}_j^1)\}_{j \in [m]}$ where $\mathbf{x}_j^0, \mathbf{x}_j^1 \in \{0,1\}^n$; signing key sk.

R's inputs: Selection bits $\mathbf{r} = (r_1, \ldots, r_m)$; verification key vk.

Common inputs: Security parameter κ; statistical security parameter ρ; number of base OTs ℓ; number of check functions μ; random oracle $G : \{0,1\}^\kappa \to \{0,1\}^\ell$; random oracle $H : \mathbb{N} \times \{0,1\}^\ell \to \{0,1\}^n$; random oracle $H' : \{0,1\}^m \to \{0,1\}^\kappa$; EU-CMA signature scheme $\Pi = (\mathsf{KeyGen}', \mathsf{Sign}', \mathsf{Verify}')$; ideal functionality $\mathcal{F}_{\mathsf{OT}}$.

1. Initial OT Phase:
   - S computes $\mathbf{s} \in \{0,1\}^\ell$ as follows. Let I be a set of indices, where $|I| = \kappa$. For $i \in I$, S sets $s_i = 0$. Then, S fills the remaining bits at random.
   - For $j \in [m]$, R computes $\mathbf{k}_j \leftarrow_\$ \{0,1\}^\kappa$ and sets $\mathbf{t}_j \leftarrow G(\mathbf{k}_j)$.
   - Let $\mathbf{T}$ be an $m \times \ell$ matrix, where the jth row is $\mathbf{t}_j$ and the ith column is $\mathbf{t}^i$. Let $\mathbf{V}$ be an $m \times \ell$ matrix, where the jth row is $\mathbf{v}_j$ and the ith column is $\mathbf{v}^i$. S and R run $\mathcal{F}_{\mathsf{OT}}$ ℓ times in parallel, where S acts as the *receiver* with input s_i and R acts as the *sender* with input $(\mathbf{t}^i, \mathbf{v}^i)$.

2. OT Extension Phase (Part I):
   - For $i \in [\ell]$, R sets $\mathbf{u}^i \leftarrow \mathbf{t}^i \oplus \mathbf{v}^i \oplus \mathbf{r}$, and sends $\mathbf{u}^i$ to S.

3. Consistency check of $\mathbf{r}$:
   - Same as in maliciously-secure OT extension protocol of Asharov et al. [6].

4. OT Extension Phase (Part II):
   - Let $\mathbf{Q}$ be the $m \times \ell$ matrix where each column $\mathbf{q}^i = (s_i \cdot (\mathbf{u}^i \oplus \mathbf{v}^i)) \oplus ((1 - s_i) \cdot \mathbf{t}^i)$. Note that $\mathbf{q}^i = (s_i \cdot \mathbf{r}) \oplus \mathbf{t}^i$ and $\mathbf{q}_j = (r_j \cdot \mathbf{s}) \oplus \mathbf{t}_j$.
   - Let I be the set defined in Step 1, and let $t_{j,i}$ denote the ith bit in row $\mathbf{t}_j$. S sends I to R, who checks that $|I| = \kappa$ and otherwise aborts.
   - For $j \in [m]$, S computes $\mathbf{y}_j^0 \leftarrow \mathbf{x}_j^0 \oplus H(j, \mathbf{q}_j)$ and $\mathbf{y}_j^1 \leftarrow \mathbf{x}_j^1 \oplus H(j, \mathbf{q}_j \oplus \mathbf{s})$ and signatures $\sigma_j' \leftarrow \mathsf{Sign}'_{\mathsf{sk}}((I, j, \mathbf{y}_j^0, \{t_{j,i}\}_{i \in I}))$, and $\sigma_j'' \leftarrow \mathsf{Sign}'_{\mathsf{sk}}((I, j, \mathbf{y}_j^1, \{t_{j,i}\}_{i \in I}))$, and sends $(j, \mathbf{y}_j^0, \mathbf{y}_j^1, \{t_{j,i}\}_{i \in I}, \sigma_j', \sigma_j'')$ to R.
   - For $j \in [m]$, R computes $\mathbf{x}_j \leftarrow \mathbf{y}_j^{r_j} \oplus H(j, \mathbf{t}_j)$.

5. Output:
   - S outputs $\perp$; R outputs $\{\mathbf{x}_j, (j, r_j, \mathbf{k}_j, I, \mathbf{y}_j^0, \mathbf{y}_j^1, \{t_{j,i}\}_{i \in I}, \sigma_j', \sigma_j'')\}_{j \in [m]}$.

Fig. 4. Signed-OT extension, based on the OT extension protocol of Asharov et al. [6].

3.2 Towards a Proof of Security

Before presenting the security proof, we first motivate the need for EU-CMPRA signature schemes. As mentioned in Sect. 3.1, ideally we could just have S sign everything using an EU-CMA signature scheme; however, this presents opportunities for R to defame S. Thus, we need to enforce that R cannot output an $\mathbf{x}_j^b$ value different from the one sent by S. We do so by using a binding commitment scheme $\Pi_{\mathsf{Com}} = (\mathsf{ComGen}, \mathsf{Com})$, and show that the messages sent by S in Step 4 are essentially binding commitments to the underlying $\mathbf{x}_j^b$ values.

We define Π_{Com} as follows, where $G : \{0,1\}^\kappa \to \{0,1\}^\ell$ and $H : \mathbb{N} \times \{0,1\}^\ell \to \{0,1\}^\kappa$ are random oracles, and $\ell \geq \kappa$.

1. ComGen(1^κ): choose set $I \subseteq [\ell]$ uniformly at random subject to $|I| = \kappa$; output params $\leftarrow I$.
2. Com(params, $\mathbf{m}, j; \mathbf{r}$): On input parameters $I \leftarrow$ params, message $\mathbf{m}$, counter j, and randomness $\mathbf{r} \in \{0,1\}^\kappa$, proceed as follows. Compute $\mathbf{t} \leftarrow G(\mathbf{r})$, set com $\leftarrow (j, \mathbf{m} \oplus H(j, \mathbf{t}), I, \{t_i\}_{i \in I})$, and output com.

We make the assumption that given I, one can derive the randomness input to ComGen. (We use this when defining our EU-CMPRA signature scheme below, which uses a generic binding commitment scheme). We can satisfy this by simply letting the randomness input to ComGen be the set I.

In our signed-OT extension protocol, the set I chosen by S is used as params and the $\mathbf{k}_j$ values chosen by R are used as the randomness to Com. The commitment value com is exactly the message signed and sent by S in Step 4. Thus, ignoring the signatures for now, we have an OT extension protocol that binds S to its $\mathbf{x}_j^b$ values, and thus prevents a malicious R from defaming an honest S. Adding in the signatures (cf. Sect. 3.3) gives us an EU-CMPRA signature scheme. Namely, S is tied to its messages due to the signatures and R is prevented from "changing" the messages to defame S due to the binding property of the commitment scheme.

We now prove that the commitment scheme described above is binding. We actually prove something stronger than what is required in our protocol. Namely, we prove that an adversary who can control *both* random values still cannot win, whereas when we use this commitment scheme in our signed-OT extension protocol, only one of the two random values can be controlled by any one party.

Theorem 1. *Protocol Π_{Com} is binding according to Definition 3.*

Proof. Adversary $\mathcal{A}$ needs to come up with choices of I, $\mathbf{m}$, $\mathbf{m}'$, j, j', $\mathbf{r}$, and $\mathbf{r}'$ such that $(j, \mathbf{m} \oplus H(j, \mathbf{t}), I, \{t_i\}_{i \in I}) = (j', \mathbf{m}' \oplus H(j', \mathbf{t}'), I, \{t_i'\}_{i \in I'})$, where $\mathbf{t} \leftarrow G(\mathbf{r})$ and $\mathbf{t}' \leftarrow G(\mathbf{r}')$. Clearly, $j = j'$. Thus, $\mathcal{A}$ must find $\mathbf{t}$ and $\mathbf{t}'$ such that $t_i = t_i'$ for all $i \in I$. However, by the property that G is a random oracle, the values $\mathbf{t}$ and $\mathbf{t}'$ are distributed uniformly at random in $\{0,1\}^\ell$. Thus, the probability that $\mathcal{A}$ finds two bitstrings $\mathbf{t}$ and $\mathbf{t}'$ that match in κ bits is negligible, regardless of the choice of I. ∎

3.3   An EU-CMPRA Signature Scheme

We now show that the messages sent by S in Step 4 form an EU-CMPRA signature scheme. Let $\Pi' = (\mathsf{Gen}', \mathsf{Sign}', \mathsf{Verify}')$ be an EU-CMA signature scheme and $\Pi_{\mathsf{Com}} = (\mathsf{ComGen}, \mathsf{Com})$ be a commitment scheme satisfying Definition 3 (e.g., the scheme presented in Sect. 3.2). Consider the scheme $\Pi = (\mathsf{Gen}, \mathsf{Sign}, \mathsf{Verify})$ defined as follows.

1. Gen(1^κ): On input 1^κ, run $(\mathsf{vk}, \mathsf{sk}) \leftarrow_\$ \mathsf{Gen}'(1^\kappa)$ and output $(\mathsf{vk}, \mathsf{sk})$.
2. Sign$_{\mathsf{sk}}(\mathbf{m}, j; (\mathbf{r}_1^*, \mathbf{r}_2^*))$: On input message $\mathbf{m} \in \{0,1\}^\kappa$, counter $j \in \mathbb{N}$, and randomness $\mathbf{r}_1^*$ and $\mathbf{r}_2^*$, proceed as follows. Compute params $\leftarrow \mathsf{ComGen}(1^\kappa; \mathbf{r}_1^*)$ and com $\leftarrow \mathsf{Com}(\mathsf{params}, \mathbf{m}, j; \mathbf{r}_2^*)$. Next, choose $\mathbf{m}' \leftarrow_\$ \{0,1\}^\kappa$ and compute

com$'$ $\leftarrow$ Com(params, $\mathbf{m}'$, j; $\mathbf{r}_2^*$)^6. Output σ $\leftarrow$ (j, params, $\mathbf{r}_2^*$, com, com$'$, Sign$'_{sk}$ ((params, com)), Sign$'_{sk}$((params, com$'$))).

3. Verify$_{pk}$($\mathbf{m}$, σ): On input message $\mathbf{m}$ and signature σ, parse σ as (j, params, $\mathbf{r}$, com$'$, com$''$, σ', σ''), and output 1 if and only if (1) Com(params, $\mathbf{m}$; $\mathbf{r}$) = com$'$, (2) Verify$'_{vk}$((params, com$'$), σ') = 1, and (3) Verify$'_{vk}$((params, com$''$), σ'') = 1; otherwise output 0.

As explained in Sect. 3.2, this signature scheme exactly captures the behavior of S in our signed-OT extension protocol. We now prove that this is indeed an EU-CMPRA signature scheme.

Theorem 2. *Given an EU-CMA signature scheme $\Pi' = (\mathsf{Gen}', \mathsf{Sign}', \mathsf{Verify}')$ and a commitment scheme $\Pi_{\mathsf{Com}} = (\mathsf{ComGen}, \mathsf{Com})$ secure according to Definition 3, then $\Pi = (\mathsf{Gen}, \mathsf{Sign}, \mathsf{Verify})$ described above is an EU-CMPRA signature scheme.*

Proof. Let $\mathcal{A}$ be a PPT adversary attacking Π. We construct an adversary $\mathcal{B}$ attacking Π'. Adversary $\mathcal{B}$ receives vk from the challenger and initializes $\mathcal{A}$ with vk as input. Let $(\mathbf{m}, j, \mathbf{r}_1^*, \mathbf{r}_2^*)$ be the input of $\mathcal{A}$ to its signing oracle. Adversary $\mathcal{B}$ emulates the execution of $\mathcal{A}$'s signing oracle as follows: it computes params $\leftarrow$ ComGen(1^κ; $\mathbf{r}_1^*$) and com $\leftarrow$ Com(params, $\mathbf{m}$, j; $\mathbf{r}_2^*$), chooses $\mathbf{m}'$ uniformly at random and computes com$'$ $\leftarrow$ Com(params, $\mathbf{m}'$, j; $\mathbf{r}_2^*$), constructs σ $\leftarrow$ (j, params, $\mathbf{r}_2^*$, com, com$'$, Sign$'_{sk}$((params, com)), Sign$'_{sk}$((params, com$'$))), and sends σ to $\mathcal{A}$. After each of $\mathcal{A}$'s queries, $\mathcal{B}$ stores $(\mathbf{m}, j)$ in set $\mathcal{Q}_{\mathcal{A}}$ and stores all the messages it sent to its signing oracle in set $\mathcal{Q}_{\mathcal{B}}$.

Eventually, $\mathcal{A}$ outputs $(\mathbf{m}, (j, \sigma'))$ as its forgery. Adversary $\mathcal{B}$ checks that Verify$_{vk}$($\mathbf{m}$, (j, σ')) = 1 and that $(\mathbf{m}, j) \notin \mathcal{Q}_{\mathcal{A}}$. If not, $\mathcal{B}$ outputs 0. Otherwise, $\mathcal{B}$ parses σ' as (params, $\mathbf{r}$, com$'$, com$''$, σ', σ'') and checks that com$'$ $\notin \mathcal{Q}_{\mathcal{B}}$. If so, it outputs (com$'$, σ'); otherwise it outputs 0.

Note that Sig-forge$_{\mathcal{A}, \Pi}^{\mathsf{CMPRA}}(\kappa)$ = 1 and Sig-forge$_{\mathcal{B}, \Pi'}^{\mathsf{CMA}}(\kappa)$ = 0 if and only if Verify$_{vk}$($\mathbf{m}$, $(j, \text{params}, \mathbf{r}, \text{com}', \text{com}'', \sigma', \sigma'')$) = 1 and $(\mathbf{m}, j) \notin \mathcal{Q}_{\mathcal{A}}$ but com$'$ $\in \mathcal{Q}_{\mathcal{B}}$. Fix some $(\mathbf{m}, (j, \text{params}, \mathbf{r}, \text{com}_1, \text{com}_{1'}, \sigma_1, \sigma_{1'}))$ such that this is the case. Thus it holds that com$_1$ $\in \mathcal{Q}_{\mathcal{B}}$. This implies that $\mathcal{B}$ queried Sign$'$ on com$_1$, which means that $\mathcal{A}$ queried *its* signing oracle on some $(\mathbf{m}', j', \mathbf{r}_1^*, \mathbf{r}_2^*)$, where $\mathbf{m}' \neq \mathbf{m}$, and received back $(j', \text{params}, \mathbf{r}', \text{com}_1, \text{com}_{2'}, \sigma_{1''}, \sigma_{2'})$. However, this implies that Com(params, com$_1$; $\mathbf{r}$) = $\mathbf{m}$ and Com(params, com$_1$; $\mathbf{r}'$) = $\mathbf{m}'$. Thus, Pr[Sig-forge$_{\mathcal{A}, \Pi}^{\mathsf{CMPRA}}(\kappa)$] = Pr[Sig-forge$_{\mathcal{B}, \Pi}^{\mathsf{CMA}}(\kappa)$] + Pr[Com-bind$_{\mathcal{B}', \Pi_{\mathsf{Com}}}(\kappa)$] for some PPT adversary $\mathcal{B}'$. We now bound Pr[Com-bind$_{\mathcal{B}', \Pi_{\mathsf{Com}}}(\kappa)$].

Adversary $\mathcal{B}'$ runs almost exactly like $\mathcal{B}$. On the first query $(\mathbf{m}, j, \mathbf{r}_1^*, \mathbf{r}_2)$ by $\mathcal{A}$, it sets $\mathbf{r} = \mathbf{r}_1^*$ if $\mathbf{r}_1^* \neq \perp$ and otherwise it sets $\mathbf{r}$ uniformly at random; $\mathcal{B}'$ then sends $\mathbf{r}$ to $\mathcal{C}$, receiving back params.

Let $(\mathbf{m}_1, j_1, \mathbf{r}_1^*, \mathbf{r}_2^*)$ and $(\mathbf{m}_2, j_2, \mathbf{r}_1^*, \mathbf{r}_2^{*'})$ be the two queries made by $\mathcal{A}$ resulting in a common commitment value. Let $(j_1, \text{params}, \mathbf{r}_1, \text{com}_1, \text{com}_1', \sigma_1, \sigma_{1'})$ and $(j_2, \text{params}, \mathbf{r}_2, \text{com}_1, \text{com}_2', \sigma_{1''}, \sigma_{2'})$ be the corresponding signatures resulting

---

^{6} This extra commitment on a random message is needed for our signed-OT extension proof.

from $\mathcal{A}$'s queries. Adversary $\mathcal{B}'$ sends $(\mathrm{com}_1, \mathbf{m}_1, j_1, \mathbf{r}_2^*, \mathbf{m}_2, j_2, \mathbf{r}_2^{*'})$ to its challenger and wins with probability one, contradicting the security of the commitment scheme. Thus, we have that $\Pr[\mathsf{Com\text{-}bind}_{\mathcal{B}', \Pi_{\mathsf{Com}}}(\kappa)] < \mathsf{negl}(\kappa)$, completing the proof. $\blacksquare$

3.4 Proof of Security

We are now ready to prove the security of our signed-OT extension protocol. Most of the proof complexity is hidden in the proofs of the associated EU-CMPRA signature scheme and commitment scheme. Thus, the signed-OT extension simulator is relatively straightforward, and mostly involves parsing the output of $\mathcal{F}_{\mathsf{signedOT}}^{\Pi}$ and passing the correct values to the adversary. The analysis follows almost exactly that of Asharov et al. [6] and thus we elide most of the details.

Theorem 3. *Let* $\Pi = (\mathsf{Gen}, \mathsf{Sign}, \mathsf{Verify})$ *be the EU-CMPRA signature scheme in Sect. 3.3. Then the protocol in Fig. 4 is a secure realization of* $\mathcal{F}_{\mathsf{signedOT}}^{\Pi}$ *in the* $\mathcal{F}_{\mathsf{OT}}$*-hybrid model.*

Proof. We separately consider the case where S is malicious and R is malicious. The case where the parties are either both honest or both malicious is straightforward.

Malicious S. Let $\mathcal{A}$ be a PPT adversary corrupting S. We construct a simulator $\mathcal{S}$ as follows.

1. The simulator $\mathcal{S}$ acts as an honest R would in Step 1, extracting $\mathbf{s}$ from $\mathcal{A}$'s input to $\mathcal{F}_{\mathsf{OT}}$.
2. The simulator $\mathcal{S}$ acts as an honest R would in Steps 2 and 3.
3. Let I and $(j, \mathbf{y}_j^0, \mathbf{y}_j^1, \{t_{j,i}\}_{i \in I}, \sigma_{j,0}', \sigma_{j,1}')$, for $j \in [m]$, be the messages sent by $\mathcal{A}$ in Step 4. If any of these are invalid, $\mathcal{S}$ sends abort to $\mathcal{F}_{\mathsf{signedOT}}^{\Pi}$ and simulates R aborting, outputting whatever $\mathcal{A}$ outputs.
4. For $j \in [m]$, proceed as follows. The simulator $\mathcal{S}$ extracts $\mathbf{x}_j^0 \leftarrow \mathbf{y}_j^0 \oplus H(j, \mathbf{q}_j)$ and $\mathbf{x}_j^1 \leftarrow \mathbf{y}_j^1 \oplus H(j, \mathbf{q}_j \oplus \mathbf{s})$, constructs $\sigma_{j,b}^* \leftarrow (j, I, \mathbf{k}_j, (I, (j, \mathbf{y}_j^b, I, \{t_{j,i}\}_{i \in I})),$ $(I, (j, \mathbf{y}_j^{1-b}, I, \{t_{j,i}\}_{i \in I})), \sigma_{j,b}', \sigma_{j,1-b}')$ for $b \in \{0, 1\}$, and sends $\mathbf{x}_j^0$, $\mathbf{x}_j^1$, $\sigma_{j,0}^*$, and $\sigma_{j,1}^*$ to $\mathcal{F}_{\mathsf{signedOT}}^{\Pi}$, receiving back either $((b, m_b), \sigma_{j,b})$ or abort.
5. If $\mathcal{S}$ received abort in any of the above iterations, it simulates R aborting, outputting whatever $\mathcal{A}$ outputs. Otherwise, for $j \in [m]$, $\mathcal{S}$ parses $\sigma_{j,b}$ as $(j, I, \mathbf{k}_j,$ $(I, (j, \mathbf{y}_j^b, I, \{t_{j,i}\}_{i \in I})), (I, (j, \mathbf{y}_j^{1-b}, I, \{t_{j,i}\}_{i \in I})), \sigma_{j,b}', \sigma_{j,1-b}')$, constructs message $\sigma_j \leftarrow (j, \mathbf{y}_j^0, \mathbf{y}_j^1, \{t_{j,i}\}_{i \in I}, \sigma_{j,0}', \sigma_{j,1}')$, and acts as an honest R would when receiving messages I and $\{\sigma_j\}_{j \in [m]}$.
6. The simulator $\mathcal{S}$ outputs whatever $\mathcal{A}$ outputs.

It is easy to see that this protocol perfectly simulates a malicious sender since $\mathcal{S}$ acts exactly as an honest R would (beyond feeding the appropriate messages to $\mathcal{F}_{\mathsf{signedOT}}^{\Pi}$).

Malicious R. Let $\mathcal{A}$ be a PPT adversary corrupting R. We construct a simulator $\mathcal{S}$ as follows.

1. The simulator S acts as an honest S would in Step 1, extracting matrices $\mathbf{T}$ and $\mathbf{V}$ through S's $\mathcal{F}_{\mathsf{OT}}$ inputs, and thus the values $\{\mathbf{k}_j\}_{j\in[m]}$.
2. The simulator S uses the values extracted above to extract selection bits $\mathbf{r}$ after receiving the $\mathbf{u}^i$ values from $\mathcal{A}$ in Step 2.
3. The simulator S acts as an honest S would in Step 3.
4. Let I^0 be the indices at which $\mathbf{s}$ (generated in Step 1) is zero, and let $I \subseteq I^0$ be a set of size κ. For $j \in [m]$, S sends r_j, vk, and I to $\mathcal{F}^{\Pi}_{\mathsf{signedOT}}$, receiving back $((r_j, \mathbf{x}_j^{r_j}), \sigma_{j,r_j})$; S parses σ_{j,r_j} as $(j, I, \mathbf{r}, (I, (j, \mathbf{c}_{r_j}, I, \{t_{j,i}\}_{i\in I})), (I, (j, \mathbf{c}_{1-r_j}, I, \{t_{j,i}\}_{i\in I})), \sigma'_{j,r_j}, \sigma'_{j,1-r_j})$.
5. In Step 4, S sends I and $(j, \mathbf{c}_0, \mathbf{c}_1, \{t_{j,i'}\}_{i'\in I'}, \sigma'_{j,0}, \sigma'_{j,1})$, for $j \in [m]$, to $\mathcal{A}$.
6. The simulator S outputs whatever $\mathcal{A}$ outputs.

The analysis is almost exactly that of the malicious receiver proof in the construction of Asharov et al. [6]; we thus give an informal security argument here and refer the reader to the aforementioned work for the full details.

A malicious R has two main attacks: using inconsistent choices of its selection bits $\mathbf{r}$ and trying to cheat in the signature creation in Step 4. This latter attack is prevented by the security of our EU-CMPRA signature scheme. The former is prevented by the consistency check in Step 3. Namely, Asharov et al. show that the consistency check guarantees that: (1) most inputs are consistent with some string $\mathbf{r}$, and (2) the number of inconsistent inputs is small and thus allow R to only learn a small number of bits of $\mathbf{s}$. Thus, for specific choices of ℓ and μ, the probability of a malicious R cheating is negligible. Asharov et al. provide concrete parameters for various settings of the security parameter [6, §3.2]; let ℓ' denote the number of base OTs used in their protocol. Now, in our protocol we set $\ell = \ell' + \kappa$; S leaks κ bits of $\mathbf{s}$ when revealing the set I in Step 4, and so is left with ℓ' unknown bits of $\mathbf{s}$. Thus, the security argument presented by Asharov et al. carries over into our setting. ∎

4   Our Complete PVC Protocol

As noted above, the main technical challenge of the PVC model is in the signed-OT construction and model definitions. The AO protocol in the $\mathcal{F}^{\Pi}_{\mathsf{signedOT}}$-hybrid model is relatively straightforward: the natural (but careful) combination of taking a non-halting covert protocol, having the GC generator P_1 sign appropriate messages, and replacing OTs with signed-OTs works. In particular, our signed-OT extension can be naturally modified and used in place of the signed-OT primitive in the AO protocol.

In this section we present a new PVC protocol based on signed-OT extension. Our protocol is similar to the AO protocol in the $\mathcal{F}^{\Pi}_{\mathsf{signedOT}}$-hybrid model, but with applying several simple yet very effective optimizations, resulting in a much lower communication cost.

We present our protocol by starting off with the AO protocol and pointing out the differences. We presented the AO protocol intuition in the Introduction; see Fig. 5 for its formal description; due to lack of space, we omit the (straightforward) Blame and Judgment algorithms. In presenting our changes, we sketch

Private inputs: P_1 has input $\mathbf{x}_1 \in \{0,1\}^n$ and P_2 has input $\mathbf{x}_2 \in \{0,1\}^n$.

Common inputs: Security parameter κ; XOR-tree replication factor ν; garbled circuit replication factor λ; circuit $C(\cdot,\cdot)$; commitment scheme $\Pi_{\mathsf{Com}} = (\mathsf{Com}, \mathsf{Open})$; ideal functionalities $\mathcal{F}^{\Pi}_{\mathsf{signedOT}}$ and $\binom{\lambda}{1}\text{-}\mathcal{F}^{\Pi}_{\mathsf{signedOT}}$ for EU-CMPRA signature scheme Π.

1. P_1 and P_2 define a new circuit $C'(\mathbf{x}_1, \mathbf{x}_2^1, \ldots, \mathbf{x}_2^{\nu}) = C(\mathbf{x}_1, \bigoplus_{i \in [\nu]} \mathbf{x}_2^i)$. Let $w_1, \ldots, w_n$ denote the input wires of $\mathbf{x}_1$ and let $w_{n+(i-1)\nu}, \ldots, w_{n+i\nu}$ denote the input wires of $\mathbf{x}_2^i$.

2. For $i \in [\nu - 1]$, P_2 chooses $\mathbf{x}_2^i \leftarrow\!\!\$\ \{0,1\}^n$. P_2 sets $\mathbf{x}_2^{\nu} \leftarrow (\bigoplus_{i \in [\nu-1]} \mathbf{x}_2^i) \oplus \mathbf{x}_2$.

3. For $j \in [\lambda]$, $i \in [n + \nu n]$, and $b \in \{0,1\}$, P_1 chooses $\mathbf{k}^j_{w_{n+i},b} \leftarrow\!\!\$\ \{0,1\}^{\kappa}$.

4. P_1 and P_2 run νn instantiations of $\mathcal{F}^{\Pi}_{\mathsf{signedOT}}$, where in the ith execution P_1 acts as the sender with input $(\mathbf{k}^1_{w_{n+i},0}\| \cdots \|\mathbf{k}^{\lambda}_{w_{n+i},0}, \mathbf{k}^1_{w_{n+i},1}\| \cdots \|\mathbf{k}^{\lambda}_{w_{n+i},1})$ and P_2 acts as the receiver with input $\mathbf{x}_2^{\lceil i/n \rceil}[i \bmod \nu]$. If P_2's output is abort_1, it outputs abort_1.

5. For $j \in [\lambda]$, P_1 constructs garbled circuit GC_j of circuit C', where for $i \in [n + \nu n]$ the keys for input wire w_i are $\mathbf{k}^j_{w_i,0}$ and $\mathbf{k}^j_{w_i,1}$. P_1 sends $(\mathsf{GC}_j, \mathsf{Sign}(\mathsf{GC}_j))$ to P_2, who checks that the signature is valid; if not, P_2 outputs abort_1.

6. For $i \in [n]$ and $j \in [\lambda]$, P_1 chooses $b \leftarrow\!\!\$\ \{0,1\}$, computes commitments $(\mathbf{c}^j_{w_i,0}, \mathbf{o}^j_{w_i,0}) \leftarrow\!\!\$\ \mathsf{Com}(\mathbf{k}^j_{w_i,0})$ and $(\mathbf{c}^j_{w_i,1}, \mathbf{o}^j_{w_i,0}) \leftarrow\!\!\$\ \mathsf{Com}(\mathbf{k}^j_{w_i,1})$, and sends $(\mathbf{c}_{w_i,b}, \mathsf{Sign}(\mathbf{c}_{w_i,b}))$ and $(\mathbf{c}_{w_i,\bar{b}}, \mathsf{Sign}(\mathbf{c}_{w_i,\bar{b}}))$ to P_2, who checks that the signatures are valid; if not, P_2 outputs abort_1.

7. P_1 and P_2 run $\binom{\lambda}{1}\text{-}\mathcal{F}^{\Pi}_{\mathsf{signedOT}}$ with P_1 as the sender inputting $(\{\mathbf{k}^j_{w_p,b}\}_{i \in [\lambda]\backslash\{j\}, p \in [n+\nu n], b \in \{0,1\}}, \{\mathbf{o}^j_{w_p,b}\}_{i \in [\lambda]\backslash\{j\}, p \in [n], b \in \{0,1\}}, \{\mathbf{k}^j_{w_i, \mathbf{x}_1[i]}\}_{i \in [n]})$ as its jth input and P_2 as the receiver inputting $\gamma \leftarrow\!\!\$\ [\lambda]$ as its input; if P_2's output is abort_1, it outputs abort_1.

8. P_2 does the following:
   - For $j \in [\lambda]\backslash\{\gamma\}$, $i \in [n]$, and $b \in \{0,1\}$, P_2 checks that $\mathsf{Open}(\mathbf{c}^j_{w_i,b}, \mathbf{o}^j_{w_i,b}) = \mathbf{k}^j_{w_i,b}$. If not, P_2 sets $\mathsf{key} \leftarrow \mathsf{InvalidDecommitment}$ and moves to Step 9.
   - For $j \in [\lambda]\backslash\{\gamma\}$, P_2 uses the input wire keys received from the signed-OT in Step 7 to check that GC_j is a correctly garbled circuit. If not, P_2 sets $\mathsf{key} \leftarrow \mathsf{InvalidCircuit}$ and moves to Step 9.
   - For $j \in [\lambda]\backslash\{\gamma\}$, P_2 checks that the keys received in the signed-OT in Step 4 match the keys sent by P_1 in Step 7. If not, P_2 sets $\mathsf{key} \leftarrow \mathsf{SelectiveOTAttack}$ and moves to Step 9.

9. If any of the above checks fail, P_2 computes $\mathsf{Cert} \leftarrow \mathsf{Blame}(\mathsf{id}_1, \mathsf{key}, \mathsf{View}_2)$, publishes Cert, and outputs $\mathsf{corrupted}_1$. Otherwise, P_2 uses the keys to compute $C'(\mathbf{x}_1, \mathbf{x}_2^1, \ldots, \mathbf{x}_2^{\nu})$ and outputs the result.

Fig. 5. The AO PVC protocol [7, Protocol 3].

the improvement each of them brings. Thus, we start by reviewing the communication cost of the AO protocol.

Communication Cost of the AO Protocol. Using state-of-the-art optimizations [13,19,20], the size of each GC sent in Step 5 is $2\kappa|G_C|$, where $|G_C|$ is the number of non-XOR gates in circuit C (note that $|G_C| = |G_{C'}|$ for circuit C' generated in Step 1 since the XOR-tree only adds XOR gates to the circuit, which are "free" [13]). Let τ be the field size (in bits), ν the XOR-tree replication factor, λ the GC replication factor, and n the length of the inputs, and assume that each signature is of length τ and the commitment and decommitment values are of length κ. Using the signed-OT instantiations of Asharov and Orlandi [7, Protocols 1 and 2], we get a total communication cost of $\tau(7\nu n + 11) + 2\lambda\kappa\nu n + \ell(2\kappa|G_C| + \tau) + 2n\lambda(\kappa + \tau) + \tau(3 + 2\lambda + 11(\lambda - 1)) + \lambda\kappa(2(n + \nu n)(\lambda - 1) + 2n(\lambda - 1) + n)$.

As an example, consider the secure computation of $\mathrm{AES}(\mathbf{m}, \mathbf{k})$, where P_1 inputs message $\mathbf{m} \in \{0, 1\}^{128}$ and P_2 inputs key $\mathbf{k} \in \{0, 1\}^{128}$, and suppose we set both the GC replication factor λ and the XOR-tree replication factor ν to 3, giving a cheating probability of $\epsilon = 1/2$. Letting $\kappa = 128$ and $\tau = 256$, we have a total communication cost of 9.3 Mbit (where we assume that the AES circuit has 9,100 non-XOR gates [15]).

Our Modifications. We make the following modifications to the AO protocol:

- In Step 6, instead of using a commitment scheme we can use a hash function. This saves on communication in Step 7 as P_1 no longer needs to send the openings $\{o^i_{w_p, b}\}$ to the commitments in the signed-OT, and is secure when treating H as a random oracle since the keys are generated uniformly at random and thus it is infeasible for P_2 to guess the committed values. The total savings are $2n(\lambda - 1)\kappa\lambda$ bits; in our example, this saves us 196 kbit.
- In Step 3, we use a random seed to generate the input wire keys. Namely, for all $j \in [\lambda]$ we compute $s_j \leftarrow_\$ \{0, 1\}^\kappa$, and compute the input wire keys for circuit j as $k^j_{w_1, 0} \| k^j_{w_1, 1} \| \cdots \| k^j_{w_{n + \nu n}, 0} \| k^j_{w_{n + \nu n}, 1} \leftarrow G(s_j)$, where G is a pseudorandom generator. Now, in the 1-out-of-λ signed-OT in Step 7 we can just send the seeds to the input wire keys rather than the input wire keys themselves. The total savings are $2(n + \nu n)(\lambda - 1)\lambda\kappa - n(\lambda - 1)\lambda\kappa$ bits; in our example, this saves us 688 kbit.
- In Step 5, P_1 generates each GC_j from a seed s^j_{GC}. (This idea was first put forward by Goyal et al. [11].) That is, s^j_{GC} specifies the randomness used to construct all wire keys *except* for the input wire keys which were set in Step 3. Instead of P_1 sending each GC to P_2 in Step 5, P_1 instead sends a commitment $c^j_{\mathrm{GC}} \leftarrow H(\mathrm{GC}_j)$. Now, in Step 7, P_1 can send the appropriate seeds $\{s^j_{\mathrm{GC}}\}_{j \in [\lambda] \setminus \{j\}}$ in the jth input of the 1-out-of-λ signed-OT to allow P_2 to check the correctness of the check GCs. We then add an additional step where, if the checks pass, P_1 sends GC_γ (along with a signature on GC_γ) to P_2, who can check whether $H(\mathrm{GC}_\gamma) = c^\gamma_{\mathrm{GC}}$. Note that this does not violate the security conditions required by the PVC model because P_2 catches any cheating of P_1 before the evaluation circuit is sent. If P_1 tries to cheat here, P_2 already has a commitment to the circuit so can detect any cheating. The total savings are $(\lambda - 1)2\kappa|G_C| - \lambda\tau - \lambda\kappa(\lambda - 1)$ bits; in our example, this saves us 4.6 Mbit.

Private inputs: P_1 has input $\mathbf{x}_1 \in \{0,1\}^n$; P_2 has input $\mathbf{x}_2 \in \{0,1\}^n$.
Common inputs: Security parameter κ; XOR-tree replication factor ν; garbled circuit replication factor λ; circuit $C(\cdot, \cdot)$; hash function $H : \{0,1\}^* \to \{0,1\}^\kappa$; pseudorandom generator $G : \{0,1\}^\kappa \to \{0,1\}^{2(n+\nu n)\kappa}$; ideal functionalities $\mathcal{F}_{\mathsf{signedOT}}^\Pi$ and $\binom{\lambda}{1}$-$\mathcal{F}_{\mathsf{signedOT}}^\Pi$ for EU-CMPRA signature scheme Π.

1. P_1 and P_2 define a new circuit $C'(\mathbf{x}_1, \mathbf{x}_2^1, \ldots, \mathbf{x}_2^\nu) = C(\mathbf{x}_1, \bigoplus_{i \in [\nu]} \mathbf{x}_2^i)$. Let $w_1, \ldots, w_n$ denote the input wires of $\mathbf{x}_1$ and let $w_{n+(i-1)\nu}, \ldots, w_{n+i\nu}$ denote the input wires of $\mathbf{x}_2^i$.

2. For $i \in [\nu - 1]$, P_2 chooses $\mathbf{x}_2^i \leftarrow_\$ \{0,1\}^n$ and sets $\mathbf{x}_2^\nu \leftarrow (\bigoplus_{i \in [\nu-1]} \mathbf{x}_2^i) \oplus \mathbf{x}_2$.

3. For $j \in [\lambda]$, P_1 chooses $\mathbf{s}_j \leftarrow_\$ \{0,1\}^\kappa$ and computes $\mathbf{k}_{w_1,0}^j \| \mathbf{k}_{w_1,1}^j \| \cdots \| \mathbf{k}_{w_{n+\nu n},0}^j \| \mathbf{k}_{w_{n+\nu n},1}^j \leftarrow G(\mathbf{s}_j)$.

4. P_1 and P_2 run νn instantiations of $\mathcal{F}_{\mathsf{signedOT}}^\Pi$, where in the ith execution P_1 acts as the sender with input $(\mathbf{k}_{w_{n+i},0}^1 \| \cdots \| \mathbf{k}_{w_{n+i},0}^\lambda, \mathbf{k}_{w_{n+i},1}^1 \| \cdots \| \mathbf{k}_{w_{n+i},1}^\lambda)$ and P_2 acts as the receiver with input $\mathbf{x}_2^{\lceil i/n \rceil}[i \bmod \nu]$. If P_i's output is abort_i, it outputs abort_i.

5. For $j \in [\lambda]$, P_1 computes $\mathbf{s}_{\mathsf{GC}}^j \leftarrow_\$ \{0,1\}^\kappa$ and uses $\mathbf{s}_{\mathsf{GC}}^j$ as the randomness used to generate garbled circuit GC_j, where for $i \in [n + \nu n]$ the keys for input wire w_i are $\mathbf{k}_{w_i,0}^j$ and $\mathbf{k}_{w_i,1}^j$. P_1 computes $\mathbf{c}_{\mathsf{GC}}^j \leftarrow H(\mathsf{GC}_j)$ and sends $(\mathbf{c}_{\mathsf{GC}}^j, \mathsf{Sign}(\mathbf{c}_{\mathsf{GC}}^j))$ to P_2, who checks that the signature is valid; if not, P_2 outputs abort_1.

6. For $i \in [n]$ and $j \in [\lambda]$, P_1 computes $\mathbf{c}_{w_i,0}^j \leftarrow H(\mathbf{k}_{w_i,0}^j)$ and $\mathbf{c}_{w_i,1}^j \leftarrow H(\mathbf{k}_{w_i,1}^j)$, and sends $(\mathbf{c}_{w_i,b}, \mathsf{Sign}(\mathbf{c}_{w_i,b})), (\mathbf{c}_{w_i,\bar{b}}, \mathsf{Sign}(\mathbf{c}_{w_i,\bar{b}}))$ to P_2, where $b \leftarrow_\$ \{0,1\}$. P_2 checks that the signatures are valid; if not, P_2 outputs abort_1.

7. P_1 and P_2 run $\binom{\lambda}{1}$-$\mathcal{F}_{\mathsf{signedOT}}^\Pi$ with P_1 as the sender and P_2 as the receiver. P_2 uses $\gamma \leftarrow_\$ [\lambda]$ as its input and P_1 uses $(\{\mathbf{s}_i, \mathbf{s}_{\mathsf{GC}}^i\}_{i \in [\lambda] \setminus \{j\}}, \{\mathbf{k}_{w_i, \mathbf{x}_1[i]}^j\}_{i \in [n]})$ as its jth input. If P_i's output is abort_i, it outputs abort_i.

8. P_2 does the following:
   - For $j \in [\lambda] \setminus \{\gamma\}$, $i \in [n]$, and $b \in \{0,1\}$, P_2 checks that $H(\mathbf{k}_{w_i,b}^j) = \mathbf{c}_{w_i,b}^j$. If not, P_2 sets $\mathsf{key} \leftarrow \mathsf{InvalidDecommitment}$ and moves to Step 12.
   - For $j \in [\lambda] \setminus \{\gamma\}$, P_2 uses $\mathbf{s}_j$ and $\mathbf{s}_{\mathsf{GC}}^j$ received from $\binom{\lambda}{1}$-$\mathcal{F}_{\mathsf{signedOT}}^\Pi$ to check that GC_j is a correctly garbled circuit and that $H(\mathsf{GC}_j) = \mathbf{c}_{\mathsf{GC}}^j$. If not, P_2 sets $\mathsf{key} \leftarrow \mathsf{InvalidCircuit}$ and moves to Step 12.
   - For $j \in [\lambda] \setminus \{\gamma\}$, P_2 checks that the keys received in $\mathcal{F}_{\mathsf{signedOT}}^\Pi$ match the keys generated by $\mathbf{s}_j$ received in Step 7. If not, P_2 sets $\mathsf{key} \leftarrow \mathsf{SelectiveOTAttack}$ and moves to Step 12.

9. Let $((\gamma, m_\gamma), \sigma)$ be P_2's output of $\binom{\lambda}{1}$-$\mathcal{F}_{\mathsf{signedOT}}^\Pi$. P_2 sends (γ, σ) to P_1, who checks that the signature is valid and otherwise outputs abort_2.

10. P_1 sends $(\mathsf{GC}_\gamma, \mathsf{Sign}(\mathsf{GC}_\gamma))$ to P_2, who checks that the signature is valid; if not, P_2 outputs abort_1.

11. P_2 checks that $H(\mathsf{GC}_\gamma) = \mathbf{c}_{\mathsf{GC}}^\gamma$. If not, P_2 sets $\mathsf{key} \leftarrow \mathsf{InvalidCircuitHash}$ and moves to Step 12.

12. If any of the above checks fail, P_2 computes $\mathsf{Cert} \leftarrow \mathsf{Blame}(\mathsf{id}_1, \mathsf{key}, \mathsf{View}_2)$, publishes Cert, and outputs $\mathsf{corrupted}_1$. Otherwise, P_2 uses the keys to compute $C'(\mathbf{x}_1, \mathbf{x}_2^1, \ldots, \mathbf{x}_2^\nu)$ and outputs the result.

Fig. 6. Our PVC protocol.

Our PVC Protocol and Its Cost. Fig. 6 presents our optimized protocol. For simplicity, we sign each message in Steps 5 and 6 separately; however, we note that we can group all the messages in a given step into a single signature (cf. Observation 2). The Blame and Judgment algorithms are straightforward and similar to the AO protocol (Blame outputs the relevant parts of the view, including the cheater's signatures, and Judgment checks the signatures). We prove the following theorem in the full version.

Theorem 4. *Let $\lambda < p(\kappa)$ and $\nu < p(\kappa)$, for some polynomial $p(\cdot)$, be parameters to the protocol, and set $\epsilon = (1 - 1/\lambda)(1 - 2^{-\nu+1})$. Let f be a PPT function, let H be a random oracle, let $\mathcal{F}_{\mathbf{signedOT}}^{\Pi}$ and $\binom{\lambda}{1}$-$\mathcal{F}_{\mathbf{signedOT}}^{\Pi}$ be the $\binom{2}{1}$-signed-OT and $\binom{\lambda}{1}$-signed-OT ideal functionalities, respectively, where Π is an EU-CMPRA signature scheme. Then the protocol in Fig. 6 securely computes f in the presence of (1) an ϵ-PVC adversary corrupting P_1 and (2) a malicious adversary corrupting P_2.*

Using our AES circuit example, we find that the total communication cost is now 2.5 Mbit, plus the cost of signed-OT/signed-OT extension. In this particular example, signed-OT requires around 1 Mbit and signed-OT extension requires around 1.4 Mbit. However, as we show below, as the number of OTs required grows, signed-OT extension quickly outperforms signed-OT, both in communication and computation.

5   Comparison with Prior Work

We now compare our signed-OT extension construction (including optimizations, and in particular, the signature batching of Observation 2) with the signed-OT protocol of Asharov and Orlandi [7], along with a comparison of existing covert and malicious protocols and our PVC protocol using both signed-OT and signed-OT extension. All comparisons are done through calculating the number of bits transferred and estimated running times based on the relative cost of public-key versus symmetric-key operations. We use a very conservative (low-end) estimate on the public/symmetric speed ratio. We note that this ratio does vary greatly across platforms, being much higher on low power mobile devices, which often employ a weak CPU but have hardware AES support. For such platforms our numbers would be even better.

Recall that τ is the field size (in bits), ν is the XOR-tree replication factor, λ is the GC replication factor, n is the input length, and we assume that each signature is of length τ.

Communication Cost. We first focus on the *communication cost* of the two protocols. The signed-OT protocol of Asharov and Orlandi [7] is based on the maliciously secure OT protocol of Peikert et al. [18], and inherits similar costs. Namely, the communication cost of executing ℓ OTs each of length n is $(6\ell+11)\tau$ if $n \leq \tau$, and $(6\ell + 11)\tau + 2n\ell$ if $n > \tau$. Signed-OT requires the additional communication of a signature per OT, adding an additional $\tau\ell$ bits. In the underlying secure computation protocol we have that $n = \lambda\kappa$, where λ is the garbled circuit replication factor. For simplicity, we set $\lambda = 3$ (which along with an XOR-tree replication factor of three equates to a deterrence factor of $\epsilon = 1/2$) and thus $n = 3\kappa$. Thus, the total communication cost of executing t signed-OTs is $\tau(7t + 11)$ bits if $3\kappa \leq \tau$ and $\tau(7t + 11) + 6\kappa t$ bits otherwise.

On the other hand, the cost of signed-OT extension for t OTs is $(6\ell + 11)\tau + 2\ell t + \ell t + \mu\ell \log\ell + 4\mu\ell\kappa + \kappa\log\ell + (n + \kappa)t + \tau$. Asharov et al. [6, §3.2] present concrete choices of μ and ℓ for various security parameters. However, in our setting we need to increase ℓ by κ bits. Thus, let ℓ' be the particular choice of ℓ specified by Asharov et al. We then set $\ell = \ell' + \kappa$. Thus, for short security parameter we set $\ell = 133 + 80 = 213$ and $\mu = 3$, and for long security parameter we set $\ell = 190 + 128 = 318$ and $\mu = 2$. Thus, the total communication cost of executing t signed-OTs when using signed-OT extension is $(6\ell + 12)\tau + (3\ell + n + \kappa)t + \mu\ell\log\ell + 4\mu\ell\kappa + \kappa\log\ell$ bits.

	1,000 OTs			**10,000 OTs**		
Security	sOT	sOT-ext	Improvement	sOT	sOT-ext	Improvement
Short (FFC)	7,179	2,539	2.8×	71,691	11,305	6.3×
Short (ECC)	1,602	1,398	1.1×	16,002	10,164	1.6×
Long (FFC)	21,538	7,694	2.8×	215,074	20,888	10.3×
Long (ECC)	2,563	2,288	1.1×	25,603	15,482	1.7×

Fig. 7. Communication cost (in kbits) of transferring the input wire labels for P_2 when using signed-OT (sOT) versus signed-OT extension (sOT-ext) for 1,000 and 10,000 OTs.

Figure 7 presents a comparison of the communication cost of both approaches when executing 1,000 and 10,000 OTs, for various keylength settings and underlying public-key cryptosystems. We see improvements from 1.1–10.3×, depending on the number of OTs, the underlying public-key cryptosystem, and the size of the security parameter. Note that for a smaller number of OTs (such as 100), signed-OT is more efficient, which makes sense due to the overhead of OT extension and the need to compute the base OTs. However, as the number of OTs grows, we see that signed-OT extension is superior across the board.

Computational Cost. We now look at the *computational cost* of the two protocols. Let ξ denote the cost of a public-key operation (we assume exponentiations and signing take the same amount of time), and let ζ denote the cost of a symmetric-key operation (where we let ζ denote the cost of operating over κ bits; e.g., hashing a 2κ-bit value costs 2ζ). We assume all other operations are "free". This is obviously a very coarse analysis; however, it gives a general idea of the performance characteristics of the two approaches.

The cost of executing ℓ OTs on n-bit messages is $(14\ell + 12)\xi$ if $n \leq \tau$ and $(14\ell + 12)\xi + 2\ell\frac{n}{\kappa}\zeta$ if $n > \tau$. Signed-OT requires an additional $2\ell\xi$ operations (for signing and verifying). We again set $n = 3\kappa$, and thus the cost of executing t signed-OTs is $(16t + 12)\xi$ if $3\kappa \leq \tau$ and $(16t + 12)\xi + 6t\zeta$ otherwise.

The cost of our signed-OT extension protocol for t OTs (where we assume $t > \kappa$ and we hash the input prior to signing in Step 4) is $\frac{\ell}{\kappa}t\zeta + (14\ell + 12)\xi + 2\ell\frac{t}{\kappa}\zeta + 6\ell\mu\frac{t}{\kappa}\zeta + 2\log\ell + 2t\frac{\ell+n+\kappa}{\kappa}\zeta + 2\xi$. As above, we set $\ell = 213$ and $\mu = 3$ for short security parameter, $\ell = 318$ and $\mu = 2$ for long security parameter, and $n = 3\kappa$. Thus, the cost of executing t signed-OTs is $(14\ell + 14)\xi + ((5 + 6\mu)\frac{\ell}{\kappa} + 8)t\zeta + 2\log\ell\zeta$.

Security	1,000 OTs			10,000 OTs		
	sOT	sOT-ext	Improvement	sOT	sOT-ext	Improvement
Short (FFC)	16.0	3.1	5.1×	160.0	3.8	42.4×
Short (ECC)	5.3	1.1	4.9×	53.3	1.7	30.9×
Long (FFC)	144.1	40.2	3.6×	1440.1	40.7	35.4×
Long (ECC)	14.4	4.1	3.5×	144.1	4.5	31.9×

Fig. 8. Computation cost (in millions of "time units") of transferring the input wire labels for P_2 when using signed-OT (sOT) versus signed-OT extension (sOT-ext) for 1,000 and 10,000 OTs. We assume symmetric-key operations take 1 "time unit", FFC (resp., ECC) operations take 1000 (resp., 333) "time units" for the short security parameter, and FFC (resp., ECC) operations take 9000 (resp., 900) "time units" for the long security parameter [1].

Figure 8 presents a comparison of the computational cost of both approaches when executing 1,000 and 10,000 OTs, for various keylength settings and underlying public-key cryptosystems. Here we see that regardless of the number of OTs and public-key cryptosystem used, signed-OT extension is (often much) more efficient, and as the number of OTs increases so does this improvement. For as few as 1,000 OTs we already see a 3.5–5.1× improvement, and for 10,000 OTs we see a 30.9–42.4× improvement.

Comparing Covert, PVC, and Malicious Protocols. We now compare the *computation* cost of our PVC protocol in Fig. 6, using both signed-OT and signed-OT extension, with the covert protocol of Goyal et al. [11] and the malicious protocol of Lindell [17][7].

Figure 9 presents a comparison of the computation cost of our protocol using both signed-OT (Ours^{sOT}) and signed-OT extension (Ours$^{sOT\text{-}ext}$), as well as comparisons to the Goyal et al. protocol (GMS) and Lindell protocol (Lin). Due to lack of space, the detailed cost formulas appear in the full version. We fix $\kappa = 128$, $\lambda = \nu = 3$ (giving a deterrence factor of $\epsilon = 1/2$), and assume the

---

[7] Lindell's malicious protocol can also be adapted into a covert protocol; however, we found that the computation cost is much more than that of Goyal et al., at least for deterrence factor 1/2.

f	# inputs	# gates	$\frac{GMS}{Ours^{sOT\text{-}ext}}$	$\frac{Ours^{sOT}}{Ours^{sOT\text{-}ext}}$	$\frac{Lin}{Ours^{sOT\text{-}ext}}$
16384-bit Comp.	16,384	32,229	0.85–0.73	17.1–86.7	357.0–1887.2
Hamming 16000	16,000	97,175	0.90–0.79	11.0–67.0	224.7–1408.4
16×16 Matrix Mult.	8192	4,186,368	1.00–0.98	1.2–3.1	14.2–54.3
1024-bit Sum	1,024	2,977	0.71–0.61	6.7–10.2	166.6–258.2
1024-bit Mult.	1,024	6,371,746	1.00–0.99	1.0–1.2	10.1–13.9
1024-bit RSA	1,024	15,149,856,895	1.00–1.00	1.0–1.0	9.6–9.6

Fig. 9. Ratio of computation cost of various secure computation protocols with our signed-OT extension construction, using a deterrence factor of $1/2$ for the covert and PVC protocols. *GMS* denotes the covert protocol of Goyal et al. [11], $Ours^{sOT}$ denotes the optimized Asharov-Orlandi protocol run using signed-OT, $Ours^{sOT\text{-}ext}$ denotes the same protocol using signed-OT extension, and *Lin* denotes Lindell's malicious protocol [17]. We let f denote the function being computed, # *inputs* denote the number of input bits required as input by P_2, and # *gates* denote the number of non-XOR gates in the resulting circuit. All circuit information is taken from the PCF compiler [14, Table5]. We report each ratio as a range; the first number uses $\xi = 125$ as the cost of public-key operations and the second number uses $\xi = 1250$, where we assume a symmetric-key operation costs $\zeta = 1$.

use of elliptic curve cryptography (and thus $\tau = 256$). We expect public-key operations to take between 125–1250× more than symmetric-key operations, depending on implementation details, whether one uses AES-NI, etc. This range is a very conservative estimate using the Crypto++ benchmark [2], experiments using OpenSSL, and estimated ratios of running times between finite field and elliptic curve cryptography [1].

When comparing against GMS, we find that $Ours^{sOT\text{-}ext}$ is slightly more expensive, due almost entirely to the larger number of base OTs in the signed-OT extension. We note that in practice, however, a deterrence factor of $1/2$ may not be sufficient for a covert protocol but may be sufficient for a PVC protocol, due to the latter's ability to "name-and-shame" the perpetrator. When increasing the deterrence factor for the covert protocol to $\epsilon \approx .9$, the cost ratios favor $Ours^{sOT\text{-}ext}$. For example, for 16×16 matrix multiplication, the ratio becomes 3.60–3.53×, depending on the cost of public-key operations (versus 1.00–0.98×).

Comparing $Ours^{sOT\text{-}ext}$ with $Ours^{sOT}$, we find that the former is 1.0–86.7× more efficient, depending largely on the characteristics of the underlying circuit. For circuits with a large number of inputs but a relatively small number of gates (e.g., 16384-bit Comp., Hamming 16000, and 1024-bit Sum) this difference is greatest, which makes sense, as the cost of the OT operations dominates. The circuits for which the ratio is around 1.0 (e.g., 1024-bit RSA) are those that have a huge number of gates compared to the number of inputs, and thus the cost of processing the GC far outweighs the cost of signed-OT/signed-OT extension.

Finally, comparing $Ours^{sOT\text{-}ext}$ with Lin, the former is 9.6–1887.2× more efficient, again depending in a large part on the characteristics of the circuit. We see that for circuits with a large number of inputs this difference is starkest;

e.g., for the Hamming 16000 circuit, we get an improvement of 224.7–1408.4×. The reason we see such large improvements for these circuits is that Lin requires cut-and-choose oblivious transfer, which cannot take advantage of OT extension. Thus, the number of public-key operations is huge compared to the circuit size, and this cost has a large impact on the overall running time. Note, however, that even for circuits where the number of gates dominates, we still see a relatively significant improvement (e.g., 14.2–54.3× for 16×16 Matrix Mult.). These results demonstrate that for settings where public shaming is enough of a deterrent from cheating, Ours$^{\text{sOT-ext}}$ presents a better choice than malicious protocols.

Acknowledgments. The authors thank Michael Zohner for a brief discussion on the relative performance of public- and symmetric-key primitives, and the anonymous reviewers for helpful suggestions.

The authors acknowledge the Office of Naval Research and its support of this work under contract N00014-14-C-0113. Work of Alex J. Malozemoff was also supported by the Department of Defense (DoD) through the National Defense Science & Engineering Graduate (NDSEG) Fellowship.

References

1. The case for elliptic curve cryptography. https://www.nsa.gov/business/programs/elliptic_curve.shtml
2. Crypto++ 5.6.0 benchmarks. http://www.cryptopp.com/benchmarks.html
3. Afshar, A., Mohassel, P., Pinkas, B., Riva, B.: Non-interactive secure computation based on cut-and-choose. In: Nguyen, P.Q., Oswald, E. (eds.) EUROCRYPT 2014. LNCS, vol. 8441, pp. 387–404. Springer, Heidelberg (2014)
4. Andrychowicz, M., Dziembowski, S., Malinowski, D., Mazurek, L.: Secure multi-party computations on bitcoin. In: 2014 IEEE Symposium on Security and Privacy, pp. 443–458. IEEE Computer Society Press, May 2014
5. Asharov, G., Lindell, Y., Schneider, T., Zohner, M.: More efficient oblivious transfer and extensions for faster secure computation. In: ACM CCS 13, pp. 535–548 (2013)
6. Asharov, G., Lindell, Y., Schneider, T., Zohner, M.: More efficient oblivious transfer extensions with security for malicious adversaries. In: Oswald, E., Fischlin, M. (eds.) EUROCRYPT 2015. LNCS, vol. 9056, pp. 673–701. Springer, Heidelberg (2015)
7. Asharov, G., Orlandi, C.: Calling out cheaters: covert security with public verifiability. In: Wang, X., Sako, K. (eds.) ASIACRYPT 2012. LNCS, vol. 7658, pp. 681–698. Springer, Heidelberg (2012)
8. Aumann, Y., Lindell, Y.: Security against covert adversaries: efficient protocols for realistic adversaries. J. Cryptol. **23**(2), 281–343 (2010)
9. Barker, E., Barker, W., Burr, W., Polk, W., Smid, M.: Recommendation for key management – Part 1: General (Revision 3). NIST Special Publication 800–57, July 2012
10. Bentov, I., Kumaresan, R.: How to use bitcoin to design fair protocols. In: Garay, J.A., Gennaro, R. (eds.) CRYPTO 2014, Part II. LNCS, vol. 8617, pp. 421–439. Springer, Heidelberg (2014)
11. Goyal, V., Mohassel, P., Smith, A.: Efficient two party and multi party computation against covert adversaries. In: Smart, N.P. (ed.) EUROCRYPT 2008. LNCS, vol. 4965, pp. 289–306. Springer, Heidelberg (2008)

12. Ishai, Y., Kilian, J., Nissim, K., Petrank, E.: Extending oblivious transfers efficiently. In: Boneh, D. (ed.) CRYPTO 2003. LNCS, vol. 2729, pp. 145–161. Springer, Heidelberg (2003)
13. Kolesnikov, V., Schneider, T.: Improved garbled circuit: free XOR gates and applications. In: Aceto, L., Damgård, I., Goldberg, L.A., Halldórsson, M.M., Ingólfsdóttir, A., Walukiewicz, I. (eds.) ICALP 2008, Part II. LNCS, vol. 5126, pp. 486–498. Springer, Heidelberg (2008)
14. Kreuter, B., Mood, B., Shelat, A., Butler, K.: PCF: a portable circuit format for scalable two-party secure computation. In: 22nd USENIX Security Symposium, August 2013
15. Kreuter, B., Shelat, A., Shen, C.H.: Towards billion-gate secure computation with malicious adversaries. In: 21st USENIX Security Symposium, August 2012
16. Kumaresan, R., Bentov, I.: How to use bitcoin to incentivize correct computations. In: Ahn, G.J., Yung, M., Li, N. (eds.) ACM CCS 14, pp. 30–41, November 2014
17. Lindell, Y.: Fast cut-and-choose based protocols for malicious and covert adversaries. In: Canetti, R., Garay, J.A. (eds.) CRYPTO 2013, Part II. LNCS, vol. 8043, pp. 1–17. Springer, Heidelberg (2013)
18. Peikert, C., Vaikuntanathan, V., Waters, B.: A framework for efficient and composable oblivious transfer. In: Wagner, D. (ed.) CRYPTO 2008. LNCS, vol. 5157, pp. 554–571. Springer, Heidelberg (2008)
19. Pinkas, B., Schneider, T., Smart, N.P., Williams, S.C.: Secure two-party computation is practical. In: Matsui, M. (ed.) ASIACRYPT 2009. LNCS, vol. 5912, pp. 250–267. Springer, Heidelberg (2009)
20. Zahur, S., Rosulek, M., Evans, D.: Two halves make a whole. In: Oswald, E., Fischlin, M. (eds.) EUROCRYPT 2015. LNCS, vol. 9057, pp. 220–250. Springer, Heidelberg (2015)

Limits of Extractability Assumptions
with Distributional Auxiliary Input

Elette Boyle[1]([⊠]) and Rafael Pass[2]

[1] Technion Israel, Haifa, Israel
eboyle@alum.mit.edu
[2] Cornell University, Ithaca, USA
rafael@cs.cornell.edu

Abstract. Extractability, or "knowledge," assumptions have recently gained popularity in the cryptographic community, leading to the study of primitives such as extractable one-way functions, extractable hash functions, succinct non-interactive arguments of knowledge (SNARKs), and (public-coin) differing-inputs obfuscation ((PC-)$di\mathcal{O}$), and spurring the development of a wide spectrum of new applications relying on these primitives. For most of these applications, it is required that the extractability assumption holds even in the presence of attackers receiving some *auxiliary information* that is sampled from some *fixed* efficiently computable distribution $\mathcal{Z}$.

We show that, assuming the existence of public-coin collision-resistant hash functions, there exists an efficient distributions $\mathcal{Z}$ such that either

- PC-$di\mathcal{O}$ for Turing machines does not exist, or
- extractable one-way functions w.r.t. auxiliary input $\mathcal{Z}$ do not exist.

A corollary of this result shows that additionally assuming existence of fully homomorphic encryption with decryption in NC^1, there exists an efficient distribution $\mathcal{Z}$ such that either

- SNARKs for NP w.r.t. auxiliary input $\mathcal{Z}$ do not exist, or
- PC-$di\mathcal{O}$ for NC^1 circuits does not exist.

E. Boyle—This work was primarily completed while at Cornell University, supported in part by AFOSR YIP Award FA9550-10-1-0093. The research of the first author has received funding from the European Union's Tenth Framework Programme (FP10/2010-2016) under grant agreement no. 259426 ERC-CaC, and ISF grant 1709/14. Part of this work was done while visiting the Simons Institute for the Theory of Computing, supported by the Simons Foundation and by the DIMACS/Simons Collaboration in Cryptography through NSF grant #CNS-1523467. Supported in part by the ERC under the EU's Seventh Framework Programme (FP/2007-2013) ERC Grant Agreement no. 307952.

R. Pass—Pass is supported in part by a Alfred P. Sloan Fellowship, Microsoft New Faculty Fellowship, NSF Award CNS-1217821, NSF CAREER Award CCF-0746990, NSF Award CCF-1214844, AFOSR YIP Award FA9550-10-1-0093, and DARPA and AFRL under contract FA8750-11-2-0211. The views and conclusions contained in this document are those of the authors and should not be interpreted as representing the official policies, either expressed or implied, of the Defense Advanced Research Projects Agency or the US Government.

© International Association for Cryptologic Research 2015
T. Iwata and J.H. Cheon (Eds.): ASIACRYPT 2015, Part II, LNCS 9453, pp. 236–261, 2015.
DOI: 10.1007/978-3-662-48800-3_10

To achieve our results, we develop a "succinct punctured program" technique, mirroring the powerful punctured program technique of Sahai and Waters (STOC'14), and present several other applications of this new technique. In particular, we construct succinct perfect zero knowledge SNARGs and give a universal instantiation of random oracles in full-domain hash applications, based on PC-diO.

As a final contribution, we demonstrate that *even in the absence of auxiliary input*, care must be taken when making use of extractability assumptions. We show that (standard) diO w.r.t. any distribution $\mathcal{D}$ over programs and bounded-length auxiliary input is directly implied by any obfuscator that satisfies the weaker indistinguishability obfuscation (iO) security notion and diO for a slightly modified distribution $\mathcal{D}'$ of programs (of slightly greater size) and no auxiliary input. As a consequence, we directly obtain negative results for (standard) diO in the absence of auxiliary input.

1 Introduction

Extractability Assumptions. Extractability, or "knowledge," assumptions (such as the "knowledge-of-exponent" assumption), have recently gained in popularity, leading to the study of primitives such as extractable one-way functions, extractable hash-functions, SNARKs (succinct non-interactive arguments of knowledge), and differing-inputs obfuscation:

- **Extractable OWF:** An extractable family of one-way (resp. collision-resistant) functions [14,15,27], is a family of one-way (resp. collision-resistant) functions $\{f_i\}$ such that any attacker who outputs an element y in the range of a randomly chosen function f_i given the index i must "know" a pre-image x of y (i.e., $f_i(x) = y$). This is formalized by requiring for every adversary $\mathcal{A}$, the existence of an "extractor" $\mathcal{E}$ that (with overwhelming probability) given the view of $\mathcal{A}$ outputs a pre-image x whenever $\mathcal{A}$ outputs an element y in the range of the function.
  For example, the "knowledge-of-exponent" assumption of Damgard [15] stipulates the existence of a particular such extractable one-way function.
- **SNARKs:** Succinct non-interactive arguments of knowledge (SNARKs) [5,32,35] are communication-efficient (i.e., "short" or "succinct") arguments for NP with the property that if a prover generates an accepting (short) proof, it must "know" a corresponding (potentially long) witness for the statement proved, and this witness can be efficiently "extracted" out from the prover.
- **Differing-Inputs Obfuscation:** [1,2,10] A differing-inputs obfuscator $\mathcal{O}$ for program-pair distribution $\mathcal{D}$ is an efficient procedure which ensures if any efficient attacker $\mathcal{A}$ can distinguish obfuscations $\mathcal{O}(C_1)$ and $\mathcal{O}(C_2)$ of programs C_1, C_2 generated via $\mathcal{D}$ given the randomness r used in sampling, then it must "know" an input x such that $C_1(x) \neq C_2(x)$, and this input can be efficiently "extracted" from $\mathcal{A}$.

A recently proposed (weaker) variant known as *public-coin differing-inputs obfuscation* [30] additionally provides the randomness used to sample the programs $(C_0, C_1) \leftarrow \mathcal{D}$ to the extraction algorithm (and to the attacker $\mathcal{A}$).

The above primitives have proven extremely useful in constructing cryptographic tools for which instantiations under complexity-theoretic hardness assumptions are not known (e.g., [1,5,10,16,24,27,30]).

Extraction with (Distribution-Specific) Auxiliary Input. In all of these applications, we require a notion of an *auxiliary-input* extractable one-way function [14,27], where both the attacker and the extractor may receive an auxiliary input. The strongest formulation requires extractability in the presence of an *arbitrary* auxiliary input. Yet, as informally discussed already in the original work by Hada and Tanaka [27], extractability w.r.t. an arbitrary auxiliary input is an "overly strong" (or in the language of [27], "unreasonable") assumption. Indeed, a recent result of Bitansky, Canetti, Rosen and Paneth [7] (formalizing earlier intuitions from [5,27]) demonstrates that assuming the existence of indistinguishability obfuscators for the class of polynomial-size circuits[1] there cannot exist auxiliary-input extractable one-way functions that remain secure for an arbitrary auxiliary input.

However, for most of the above applications, we actually do not require extractability to hold w.r.t. an arbitrary auxiliary input. Rather, as proposed by Bitansky et al. [5,6], it often suffices to consider extractability with respect to specific distributions $\mathcal{Z}$ of auxiliary input.[2] More precisely, it would suffice to show that for every desired output length $\ell(\cdot)$ and distribution $\mathcal{Z}$ there exists a function family $\mathcal{F}_{\mathcal{Z}}$ (which, in particular, may be tailored for $\mathcal{Z}$) such that $\mathcal{F}_{\mathcal{Z}}$ is a family of extractable one-way (or collision-resistant) functions $\{0,1\}^k \rightarrow \{0,1\}^{\ell(k)}$ with respect to $\mathcal{Z}$. In fact, for some of these results (e.g., [5,6]), it suffices to just assume that extraction works for just for the *uniform* distribution.

In contrast, the result of [7] can be interpreted as saying that (assuming $i\mathcal{O}$), there do not exist extractable one-way functions with respect to *every* distribution of auxiliary input: That is, for every candidate extractable one-way function family $\mathcal{F}$, there exists *some* distribution $\mathcal{Z}_{\mathcal{F}}$ of auxiliary input that breaks it.

---

[1] The notion of indistinguishability obfuscation [2] requires that obfuscations $\mathcal{O}(C_1)$ and $\mathcal{O}(C_2)$ of any two *equivalent* circuits C_1 and C_2 (i.e., whose outputs agree on all inputs) from some class $\mathcal{C}$ are computationally indistinguishable. A candidate construction for general-purpose indistinguishability obfuscation was recently given by Garg et al. [18].

[2] As far as we know, the only exceptions are in the context of zero-knowledge simulation, where the extractor is used in the simulation (as opposed to being used as part of a reduction), and we require simulation w.r.t. arbitrary auxiliary inputs. Nevertheless, as pointed out in the works on zero-knowledge [26,27], to acheive "plain" zero-knowledge [3,25] (where the verifier does not receive any auxiliary input), weaker "bounded" auxiliary input assumptions suffice.

Our Results. In this paper, we show limitations of extractability primitives with respect to *distribution-specific* auxiliary input (assuming the existence of public-coin collision-resistant hash functions (CRHF) [29]). Our main result shows a conflict between public-coin differing-inputs obfuscation for Turing machines [30] and extractable one-way functions.

Theorem 1 (Main Theorem − Informal). *Assume the existence of public-coin collision-resistant hash functions. Then for every polynomial ℓ, there exists an efficiently computable distribution $\mathcal{Z}$ such that one of the following two primitives does not exist:*

- *extractable one-way functions $\{0,1\}^k \to \{0,1\}^{\ell(k)}$ w.r.t. auxiliary input from $\mathcal{Z}$.*
- *public-coin differing-inputs obfuscation for Turing machines.*

By combining our main theorem with results from [5,30], we obtain the following corollary:

Theorem 2 (Informal). *Assume the existence of public-coin CRHF and fully homomorphic encryption with decryption in NC^1.[3] Then there exists an efficiently computable distribution $\mathcal{Z}$ such that one of the following two primitives does not exist:*

- *SNARKs w.r.t. auxiliary input from $\mathcal{Z}$.*
- *public-coin differing-inputs obfuscation for NC^1 circuits.*

To prove our results, we develop a new proof technique, which we refer to as the "succinct punctured program" technique, extending the "punctured program" paradigm of Sahai and Waters [34]; see Sect. 1.1 for more details. This technique has several other interesting applications, as we discuss in Sect. 1.3.

As a final contribution, we demonstrate that *even in the absence of auxiliary input*, care must be taken when making use of extractability assumptions. Specifically, we show that differing-inputs obfuscation ($di\mathcal{O}$) for any distribution $\mathcal{D}$ of programs and bounded-length auxiliary inputs, is directly implied by any obfuscator that satisfies a weaker indistinguishability obfuscation ($i\mathcal{O}$) security notion (which is not an extractability assumption) and $di\mathcal{O}$ security for a related distribution $\mathcal{D}'$ of programs (of slightly greater size) which does *not contain auxiliary input*. Thus, negative results ruling out existence of $di\mathcal{O}$ with bounded-length auxiliary input directly imply negative results for $di\mathcal{O}$ in a setting without auxiliary input.

Theorem 3 (Informal). *Let $\mathcal{D}$ be a distribution over pairs of programs and ℓ-bounded auxiliary input information $\mathcal{P} \times \mathcal{P} \times \{0,1\}^\ell$. There exists $di\mathcal{O}$ with respect to $\mathcal{D}$ if there exists an obfuscator satisfying $i\mathcal{O}$ in addition to $di\mathcal{O}$ with respect to a modified distribution $\mathcal{D}'$ over $\mathcal{P}' \times \mathcal{P}'$ for slightly enriched program class $\mathcal{P}'$, and no auxiliary input.*

---

[3] As is the case for nearly all existing FHE constructions (e.g., [13,21]).

Our transformation applies to a recent result of Garg *et al.* [20], which shows that based on a new assumption (pertaining to special-purpose obfuscation of Turing machines) general-purpose diO w.r.t. auxiliary input cannot exist, by constructing a distribution over circuits and bounded-length auxiliary inputs for which no obfuscator can be diO-secure. Our resulting conclusion is that, assuming such special-purpose obfuscation exists, then general-purpose diO cannot exist, *even in the absence of auxiliary input.*

We view this as evidence that public-coin differing inputs may be the "right" approach definitionally, as restrictions on auxiliary input without regard to the programs themselves will not suffice.

Interpretation of Our Results. Our results suggest that one must take care when making extractability assumptions, even in the presence of specific distributions of auxiliary inputs, and in certain cases even in the absence of auxiliary input. In particular, we must develop a way to distinguish "good" distributions of instances and auxiliary inputs (for which extractability assumptions may make sense) and "bad" ones (for which extractability assumptions are unlikely to hold). As mentioned above, for some applications of extractability assumptions, it in fact suffices to consider a particularly simple distribution of auxiliary inputs— namely the *uniform* distribution.[4] We emphasize that our results do not present any limitations of extractable one-way functions in the presence of uniform auxiliary input, and as such, this still seems like a plausible assumption.

Comparison to [20]. An interesting subsequent[5] work of Garg *et al.* [19,20] contains a related study of differing-inputs obfuscation. In [20], the authors propose a new "special-purpose" circuit obfuscation assumption, and demonstrate based on this assumption an auxiliary input distribution (whose size grows with the desired circuit size of circuits to be obfuscated) for which general-purpose diO cannot exist. Using similar techniques of hashing and obfuscating Turing machines as in the current work, they further conclude that if the new obfuscation assumption holds also for *Turing machines*, then the "bad" auxiliary input distribution can have bounded length (irrespective of the circuit size).

Garg *et al.* [20] show the "special-purpose" obfuscation assumption is a falsifiable assumption (in the sense of [33]) and is implied by virtual black-box obfuscation for the relevant restricted class of programs, but plausibility of the notion in relation to other primitives is otherwise unknown. In contrast, our results provide a direct relation between existing, studied topics (namely, diO, EOWFs, and SNARKs). Even in the case that the special-purpose obfuscation assumption *does* hold, our primary results provide conclusions for *public-coin* diO, whereas Garg *et al.* [20] consider (stronger) standard diO, with respect to auxiliary input.

---

[4] Note that this is not the case for all applications; e.g. [11,23,26,27] require considering more complicated distributions.

[5] A version of our paper with Theorems 1 and 2 for (standard) differing-inputs obfuscation in the place of public-coin diO has been on ePrint since October 2013 [12].

And, utilizing our final observation (which occurred subsequent to [20]), we show that based on their same special-purpose obfuscation assumption for Turing machines, we can in fact rule out general-purpose $di\mathcal{O}$ for circuits even *in the absence of auxiliary input*.

1.1   Proof Techniques

To explain our techniques, let us first explain earlier arguments against the plausibility of extractable one-way functions with auxiliary input. For simplicity of notation, we focus on extractable one-way function over $\{0,1\}^k \rightarrow \{0,1\}^k$ (as opposed to over $\{0,1\}^k \rightarrow \{0,1\}^{\ell(k)}$ for some polynomial ℓ), but emphasize that the approach described directly extends to the more general setting.

Early Intuitions. As mentioned above, already the original work of Hada and Tanaka [27], which introduced auxiliary input extractable one-way functions (EOWFs) (for the specific case of exponentiation), argued the "unreasonable-ness" of such functions, reasoning informally that the auxiliary input could con-tain a program that evaluates the function, and thus a corresponding extractor must be able to "reverse-engineer" *any* such program. Bitansky et al. [5] made this idea more explicit: Given some candidate EOWF family $\mathcal{F}$, consider the distribution $\mathcal{Z}_{\mathcal{F}}$ over auxiliary input formed by "obfuscating" a program $\Pi^s(\cdot)$ for uniformly chosen s, where $\Pi^s(\cdot)$ takes as input a function index e from the alleged EOWF family $\mathcal{F} = \{f_i\}$, applies a pseudorandom function (PRF) with hardcoded seed s to the index i, and then outputs the evaluation $f_i(\mathsf{PRF}_s(i))$. Now, consider an attacker $\mathcal{A}$ who, given an index i, simply runs the obfuscated program to obtain a "random" point in the range of f_i. If it were possible to obfuscate Π^s in a "virtual black-box (VBB)" way (as in [2]), then it easily fol-lows that any extractor $\mathcal{E}$ for this particular attacker $\mathcal{A}$ can invert f_i. Intuitively, the VBB-obfuscated program hides the PRF seed s (revealing, in essence, only black-box access to Π^s), and so if $\mathcal{E}$ can successfully invert f_i on $\mathcal{A}$'s output $f_i(\mathsf{PRF}_s(i))$ on a pseudorandom input $\mathsf{PRF}_s(i)$, he must also be able to invert for a *truly* random input. Formally, given an index i and a random point y in the image of f_i, we can "program" the output of $\Pi^s(i)$ to simply be y, and thus E will be forced to invert y.

The problem with this argument is that (as shown by Barak et al. [2]), for large classes of functions VBB program obfuscation simply does not exist.

The Work of [7] and the "Punctured Program" Paradigm of [34]. Intriguingly, Bitansky, Canetti, Rosen and Paneth [7] show that by using a particular PRF and instead relying on indistinguishability obfuscation, the above argument still applies! To do so, they rely on the powerful "punctured-program" paradigm of Sahai and Waters [34] (and the closely related work of Hohenberger, Sahai and Waters [28] on "instantiating random oracles"). Roughly speaking, the punc-tured program paradigm shows that if we use indistinguishability obfuscation to obfuscate a (function of) a special kind of "puncturable" PRF[6] [8,11,31], we can still "program" the output of the program on *one* input (which was used in [28,34] to show various applications of indistinguishability obfuscation). Bitansky et al. [7] show that by using this approach, then from any alleged extractor $\mathcal{E}$ we can construct a one-way function inverter Inv by "programming" the output of the program Π^s at the input i with the challenge value y. More explicitly, mirroring [28,34], they consider a hybrid experiment where $\mathcal{E}$ is executed with fake (but indistinguishable) auxiliary input, formed by obfuscating a *"punctured"* variant $\Pi^s_{i,y}$ of the program Π^s that contains an i-punctured PRF seed s^* (enabling evaluation of $\mathsf{PRF}_s(j)$ for any $j \neq i$) and directly outputs the hardcoded value $y := f_i(\mathsf{PRF}_s(i))$ on input i: indistinguishability of this auxiliary input follows by the security of indistinguishability obfuscation since the programs $\Pi^s_{i,y}$ and Π^s are equivalent when $y = f_i(\mathsf{PRF}_s(i)) = \Pi^s(i)$. In a second hybrid experiment, the "correct" hardcoded value y is replaced by a *random* evaluation $f_i(u)$ for uniform u; here, indistinguishability of the auxiliary inputs follows directly by the security of the punctured PRF. Finally, by indistinguishability of the three distributions of auxiliary input in the three experiments, it must be that $\mathcal{E}$ can extract an inverse to y with non-negligible probability in each hybrid; but, in the final experiment this implies the ability to invert a random evaluation, breaking one-wayness of the EOWF.

The Problem: Dependence on $\mathcal{F}$. Note that in the above approach, the auxiliary input distribution is selected *as a function* of the family $\mathcal{F} = \{f_j\}$ of (alleged) extractable one-way functions. Indeed, the obfuscated program Π^s must be able to evaluate f_j given j. One may attempt to mitigate this situation by instead obfuscating a universal circuit that takes as input both $\mathcal{F}$ and the index j, and appropriately evaluates f_j. But here still the *size* of the universal circuit must be greater than the running time of f_j, and thus such an auxiliary input distribution would only rule out EOWFs with a-priori bounded running time. This does not suffice for what we aim to achieve: in particular, it still leaves open the possibility that for every distribution of auxiliary inputs, there may exist a family of extractable one-way functions that remains secure for that particular auxiliary input distribution (although the running time of the extractable one-way function needs to be greater than the length of the auxiliary input).

A First Idea: Using Turing Machine Obfuscators. At first sight, it would appear this problem could be solved if we could obfuscate *Turing machines*. Namely, by obfuscating a universal Turing machine in the place of a universal circuit in the construction above, the resulting program Π^s would depend only on the size of the PRF seed s, and not on the runtime of $f_j \in F$.

But there is a catch. To rely on the punctured program paradigm, we must be able to obfuscate the program Π^s in such a way that the result is indistinguishable

---

[6] That is, a PRF where we can surgically remove one point in the domain of the PRF, keeping the rest of the PRF intact, and yet, even if we are given the seed of the punctured PRF, the value of the original PRF on the surgically removed point remains computationally indistinguishable from random.

from an obfuscation of a related "punctured" program $\Pi_{i,y}^s$; in particular, the *size* of the obfuscation must be at least as large as $|\Pi_{i,y}^s|$. Whereas the size of Π^s is now bounded by a polynomial in the size of the PRF seed s, the description of this punctured program must specify a punctured input i (corresponding to an index of the candidate EOWF $\mathcal{F}$) and hardcoded output value y, and hence must grow with the size of $\mathcal{F}$. We thus run into a similar wall: even with obfuscation of Turing machines, the resulting auxiliary input distribution $\mathcal{Z}$ would only rule out EOWF with a-priori bounded index length.

Our "Succinct Punctured Program" Technique. To deal with this issue, we develop a "succinct punctured program" technique. That is, we show how to make the size of the obfuscation be independent of the length of the input, while still retaining its usability as an obfuscator. The idea is two-fold: First, we modify the program Π^s to *hash* the input to the PRF, using a collision-resistant hash function h. That is, we now consider a program $\Pi^{h,s}(j) = f_j(PRF_s(h(j)))$. Second, we make use of *differing-inputs obfuscation*, as opposed to just indistinguishability obfuscation. Specifically, our constructed auxiliary input distribution $\mathcal{Z}$ will sample a uniform s and a random hash function h (from some appropriate collection of collision-resistant hash functions) and then output a differing-inputs obfuscation of $\Pi^{h,s}$.

To prove that this "universal" distribution $\mathcal{Z}$ over auxiliary input breaks *all* alleged extractable one-way functions over $\{0,1\}^k \to \{0,1\}^k$, we define a one-way function inverter Inv just as before, except that we now feed the EOWF extractor $\mathcal{E}$ the obfuscation of the "punctured" variant $\Pi_{i,y}^{h,s}$ which contains a PRF seed punctured at point $h(i)$. The program $\Pi_{i,y}^{h,s}$ proceeds just as $\Pi^{h,s}$ except on all inputs j such that $h(j)$ is equal to this special value $h(i)$; for those inputs it simply outputs the hardcoded value y. (Note that the index i is no longer needed to specify the function $\Pi_{i,y}^{h,s}$—rather, just its hash $h(i)$—but is included for notational convenience). As before, consider a hybrid experiment where y is selected as $y := \Pi^{h,s}(i)$.

Whereas before the punctured program was equivalent to the original, and thus indistinguishability of auxiliary inputs in the different experiments followed by the definition of indistinguishability obfuscation, here it is *no longer* the case that if $y = \Pi^{h,s}(i)$, then $\Pi_{i,y}^{h,s}$ is equivalent to $\Pi^{h,s}$—in fact, they may differ on many points. More precisely, the programs may differ in all points j such that $h(j) = h(i)$, but $j \neq i$ (since f_j and f_i may differ on the input $PRF_s(h(i))$). Thus, we can no longer rely on indistinguishability obfuscation to provide indistinguishability of these two hybrids.

We resolve this issue by relying *differing-inputs obfuscation* instead of just indistinguishability obfuscation. Intuitively, if obfuscations of $\Pi^{h,s}$ and $\Pi_{i,y}^{h,s}$ can be distinguished when y is set to $\Pi^{h,s}(i)$, then we can efficiently recover some input j where the two programs differ. But, by construction, this must be some point j for which $h(j) = h(i)$ (or else the two program are the same), *and $j \neq i$* (since we chose the hardcoded value $y = \Pi^{h,s}(i)$ to be consistent with $\Pi^{h,s}$ on input i. Thus, if the obfuscations can be distinguished, we can find a collision in h, contradicting its collision resistance.

To formalize this argument using just public-coin diO, we require that h is a *public-coin* collision-resistant hash function [29].

1.2   Removing Auxiliary Input in diO

The notion of public-coin diO is weaker than "general" (not necessarily public-coin) diO in two aspects: (1) the programs M_0, M_1 are sampled using only public randomness, and (2) we consider only a very specific auxiliary input that is given to the attacker—namely the randomness of the sampling procedure.

In this section, we explore another natural restriction of diO where we simply disallow auxiliary input, but allow for "private" sampling of M_0, M_1. We show that "bad side information" cannot be circumvented simply by simply disallowing auxiliary input, but rather such information can appear in the *input-output behavior* of the programs to be obfuscated.

More precisely, we show that for any distribution $\mathcal{D}$ over $\mathcal{P} \times \mathcal{P} \times \{0,1\}^\ell$ of programs $\mathcal{P}$ and bounded-length auxiliary input, the existence of diO w.r.t. $\mathcal{D}$ is directly implied by the existence of any *indistinguishability* obfuscator (iO) that is diO-secure for a slightly enriched distribution of programs $\mathcal{D}'$ over $\mathcal{P}' \times \mathcal{P}'$, *without* auxiliary input.

Intuitively, this transformation works by embedding the "bad auxiliary input" into the *input-output behavior* of the circuits to be obfuscated themselves. That is, the new distribution $\mathcal{D}'$ is formed by sampling first a triple (P_0, P_1, z) of programs and auxiliary input from the original distribution $\mathcal{D}$, and then instead considering the tweaked programs P_0^z, P_1^z that have a special additional input x^* (denoted later as "mode $= *$") for which $P_0^z(x^*) = P_1^z(x^*)$ is defined to be z. This introduces no new differing inputs to the original program pair P_0, P_1, but now there is no hope of preventing the adversary from learning z without sacrificing correctness of the obfuscation scheme.

A technical challenge arises in the security reduction, however, in which we must modify the obfuscation of the z-embedded program P_b^z to "look like" an obfuscation of the original program P_b. Interestingly, this issue is solved by making use of a *second layer* of obfuscation, and is where the iO security of the obfuscator is required. We refer the reader to the full version of this work for details.

1.3   Other Applications of the "Succinct Punctured Program" Technique

As mentioned above, the "punctured program" paradigm of [34] has been used in multiple applications (e.g., [9,17,28,34]). Many of them rely on punctured programs in an essentially identical way to the approach described above, and in particular follow the same hybrids within the security proof. Furthermore, for some of these applications, there are significant gains in making the obfuscation succinct (i.e., independent of the input size of the obfuscated program). Thus, for these applications, if we instead rely on public-coin differing-inputs obfuscation (and the existence of public-coin collision-resistant hash functions), by using our succinct punctured program technique, we can obtain significant improvements. For instance, relying on the same approach as above, we can show based on these assumptions:

- "Succinct" Perfect Zero-Knowledge Non-Interactive *Universal* Argument System (with communication complexity k^ϵ for every ϵ), by relying on the nonsuccinct Perfect NIZK construction of [34].
- A *universal* instantiation of Random Oracles, for which the Full Domain Hash (FDH) signature paradigm [4] is (selectively) secure for *every* trapdoor (one-to-one) function (if hashing not only the message but also the index of the trapdoor function), by relying on the results of [28] showing how to provide a trapdoor-function specific instantiation of the random oracle in the FDH.[7]

1.4   Overview of Paper

We focus in this extended abstract on the primary result: the conflict between public-coin differing inputs obfuscation and extractable OWFs (and SNARKs). Further preliminaries, applications of our succinct punctured programs technique, and our transformation removing auxiliary input in differing-inputs obfuscation are deferred to the full version [12].

2   Preliminaries

2.1   Public-Coin Differing-Inputs Obfuscation

The notion of public-coin differing-inputs obfuscation (PC-$di\mathcal{O}$) was introduced by Ishai *et al.* [30] as a refinement of (standard) differing-inputs obfuscation [2] to exclude certain cases whose feasibility has been called into question. (Note that we also consider "standard" differing-inputs obfuscation as described in Sect. 1.2. For a full treatment of the notion and our result, we refer the reader to the full version of this work [12]).

We now present the PC-$di\mathcal{O}$ definition of [30], focusing only on Turing machine obfuscation; the definition easily extends also to circuits.

Definition 1 (Public-Coin Differing-Inputs Sampler for TMs). *An efficient non-uniform sampling algorithm* Samp $= \{$Samp$_k\}$ *is called a public-coin differing inputs sampler for the parameterized collection of TMs* $\mathcal{M} = \{\mathcal{M}_k\}$ *if the output of* Samp$_k$ *is always a pair of Turing machines* $(M_0, M_1) \in \mathcal{M}_k \times \mathcal{M}_k$

---

[7] That is, [28] shows that for every trapdoor one-to-one function, there exists some way to instantiate the random oracle so that the resulting scheme is secure. In contrast, our results shows that there exists a single instantiation that works no matter what the trapdoor function is.

such that $|M_0| = |M_1|$ and for every efficient non-uniform algorithm $\mathcal{A} = \{\mathcal{A}_k\}$ there exists a negligible function ϵ such that for all $k \in \mathbb{N}$,

$$\Pr\Big[r \leftarrow \{0,1\}^*; (M_0, M_1) \leftarrow \mathsf{Samp}_k(r); (x,,1^t) \leftarrow \mathcal{A}_k(r)$$
$$: \big(M_0(x) \neq M_{(}x)\big) \wedge \big(\mathsf{steps}(M_0, x) = \mathsf{steps}(M_1, x)\big)\Big] \leq \epsilon(k).$$

Definition 2 (Public-Coin Differing-Inputs Obfuscator for TMs).
A uniform PPT algorithm $\mathcal{O}$ is a public-coin differing-inputs obfuscator for the collection $\mathcal{M} = \{\mathcal{M}_k\}$ *if the following requirements hold:*

- **Correctness:** *For every $k \in \mathbb{N}$, every $M \in \mathcal{M}_k$, and every x, we have that* $\Pr[\tilde{M} \leftarrow \mathcal{O}(1^k, M) : \tilde{M}(x) = M(x)] = 1.$
- **Security:** *For every public-coin differing-inputs sampler $\mathsf{Samp} = \{\mathsf{Samp}_k\}$ for the ensemble $\mathcal{M}$, every efficient non-uniform distinguishing algorithm $\mathcal{D} = \{\mathcal{D}_k\}$, there exists a negligible function ϵ such that for all k,*

$$\big| \Pr[r \leftarrow \{0,1\}^*; (M_0, M_1) \leftarrow \mathsf{Samp}_k(r); \tilde{M} \leftarrow \mathcal{O}(1^k, M_0) : \mathcal{D}_k(r, \tilde{M}) = 1] -$$
$$\Pr[r \leftarrow \{0,1\}^*; (M_0, M_1) \leftarrow \mathsf{Samp}_k(r); \tilde{M} \leftarrow \mathcal{O}(1^k, M_1) : \mathcal{D}_k(r, \tilde{M}) = 1]\big| \leq \epsilon(k).$$

2.2 Extractable One-Way Functions

We present a non-uniform version of the definition, in which both one-wayness and extractability are with respect to *non-uniform* polynomial-time adversaries.

Definition 3 ($\mathcal{Z}$-Auxiliary-Input EOWF). *Let ℓ, m be polynomially bounded length functions. An efficiently computable family of functions*

$$\mathcal{F} = \Big\{f_i : \{0,1\}^k \to \{0,1\}^{\ell(k)} \ \Big| \ i \in \{0,1\}^{m(k)}, k \in \mathbb{N}\Big\},$$

associated with an efficient probabilistic key sampler $\mathcal{K}_\mathcal{F}$, is a $\mathcal{Z}$-auxiliary-input extractable one-way function if it satisfies:

- **One-wayness:** *For non-uniform poly-time $\mathcal{A}$ and sufficiently large $k \in \mathbb{N}$,*

$$\Pr\Big[z \leftarrow \mathcal{Z}_k; i \leftarrow \mathcal{K}_\mathcal{F}(1^k); x \leftarrow \{0,1\}^k; x' \leftarrow \mathcal{A}(i, f_i(x); z)$$
$$: f_i(x') = f_i(x)\Big] \leq \mathsf{negl}(k).$$

- **Extractability:** *For any non-uniform polynomial-time adversary $\mathcal{A}$, there exists a non-uniform polynomial-time extractor $\mathcal{E}$ such that, for sufficiently large security parameter $k \in \mathbb{N}$:*

$$\Pr\Big[z \leftarrow \mathcal{Z}_k; i \leftarrow \mathcal{K}_\mathcal{F}(1^k); y \leftarrow \mathcal{A}(i; z); x' \leftarrow \mathcal{E}(i; z)$$
$$: \exists x \ s.t. \ f_i(x) = y \wedge f_i(x') \neq y\Big] \leq \mathsf{negl}(k).$$

2.3 Succinct Non-Interactive Arguments of Knowledge (SNARKs)

We focus attention to *publicly verifiable* succinct arguments. We consider succinct non-interactive arguments of knowledge (SNARKs) with adaptive soundness in Sect. 3.2, and consider the case of specific distributional auxiliary input.

Definition 4 ($\mathcal{Z}$-Auxiliary Input Adaptive SNARK). *A triple of algorithms* (CRSGen, Prove, Verify) *is a* publicly verifiable, adaptively sound succinct non-interactive argument of knowledge (SNARK) *for the relation $\mathcal{R}$ if the following conditions are satisfied for security parameter k:*

- **Completeness:** *For any $(x, w) \in \mathcal{R}$,*

$$\Pr[\mathsf{crs} \leftarrow \mathsf{CRSGen}(1^k); \pi \leftarrow \mathsf{Prove}(x, w, \mathsf{crs}) : \mathsf{Verify}(x, \pi, \mathsf{crs}) = 1] = 1.$$

  In addition, Prove(x, w, crs) *runs in time* poly$(k, |y|, t)$.
- **Succinctness:** *The length of the proof π output by* Prove(x, w, crs), *as well as the running time of* Verify(x, π, crs), *is bounded by $p(k + |X|)$, where p is a universal polynomial that does not depend on $\mathcal{R}$. In addition,* CRSGen(1^k) *runs in time* poly(k): *in particular,* crs *is of length* poly(k).
- **Adaptive Proof of Knowledge:** *For any non-uniform polynomial-size prover P^* there exists a non-uniform polynomial-size extractor $\mathcal{E}_{P^*}$, such that for all sufficiently large $k \in \mathbb{N}$ and auxiliary input $z \leftarrow \mathcal{Z}$, it holds that*

$$\Pr[z \leftarrow \mathcal{Z}; \mathsf{crs} \leftarrow \mathsf{CRSGen}(1^k); (x, \pi) \leftarrow P^*(z, \mathsf{crs});$$
$$(x, w) \leftarrow \mathcal{E}_{P^*}(z, \mathsf{crs}) : \mathsf{Verify}(\mathsf{crs}, x, \pi) = 1 \wedge w \notin R(x)] \leq \mathsf{negl}(k).$$

In the full version of this work, we obtain as an application of our succinct programs technique zero-knowledge (ZK) succinct non-interactive arguments (SNARGs), without the extraction property. We refer the reader to [12] for a full treatment.

2.4 Puncturable PRFs

Our result makes use of puncturable PRFs, which are PRFs with an extra capability to generate keys that allow one to evaluate the function on all bit strings of a certain length, except for any polynomial-size set of inputs. We focus on the simple case of puncturing PRFs at a single point: that is, given a punctured key k^* with respect to input x, one can efficiently evaluate the PRF at all points *except* x, whose evaluation remains pseudorandom. We refer the reader to [34] for a formal definition.

As observed in [8,11,31], the GGM tree-based PRF construction [22] yields puncturable PRFs, based on any one-way function.

Theorem 4 ([8,11,31]). *If one-way functions exist, then for all efficiently computable $m'(k)$ and $\ell(k)$, there exists a puncturable PRF family that maps $m'(k)$ bits to $\ell(k)$ bits, such that the size of a punctured key is $O(m'(k) \cdot \ell(k))$.*

3   Public-Coin Differing-Inputs Obfuscation or Extractable One-Way Functions

In this section, we present our main result: a conflict between extractable one-way functions (EOWF) w.r.t. a particular distribution of auxiliary information and public-coin differing-inputs obfuscation ("PC$-di\mathcal{O}$") (for Turing Machines).

3.1   From PC-$di\mathcal{O}$ to Impossibility of $\mathcal{Z}$-Auxiliary-Input EOWF

We demonstrate a bounded polynomial-time uniformly samplable distribution $\mathcal{Z}$ (with bounded poly-size output length) and a public-coin differing-inputs sampler for Turing Machines $\mathcal{D}$ (over TM $\times$ TM) such that if there exists public-coin differing-inputs obfuscation for Turing machines (and, in particular, for the program sampler $\mathcal{D}$), and there exist public-coin collision-resistant hash functions (CRHF), then there do *not* exist extractable one-way functions (EOWF) w.r.t. auxiliary information sampled from distribution $\mathcal{Z}$. In our construction, $\mathcal{Z}$ consists of an obfuscated Turing machine.

We emphasize that we provide a single distribution $\mathcal{Z}$ of auxiliary inputs for which *all* candidate EOWF families $\mathcal{F}$ with given output length will fail. This is in contrast to the result of [7], which show for each candidate family $\mathcal{F}$ that there exists a tailored distribution $\mathcal{Z}_{\mathcal{F}}$ (whose size grows with $|\mathcal{F}|$) for which $\mathcal{F}$ will fail.

Theorem 5. *For every polynomial ℓ, there exists an efficient, uniformly samplable distribution $\mathcal{Z}$ such that, assuming the existence of public-coin collision-resistant hash functions and public-coin differing-inputs obfuscation for Turing machines, then there cannot exist $\mathcal{Z}$-auxiliary-input extractable one-way functions $\{f_i : \{0,1\}^k \rightarrow \{0,1\}^{\ell(k)}\}$.*

Proof. We construct an adversary $\mathcal{A}$ and desired distribution $\mathcal{Z}$ on auxiliary inputs, such that for any alleged EOWF family $\mathcal{F}$, there cannot exist an efficient extractor corresponding to $\mathcal{A}$ given auxiliary input from $\mathcal{Z}$ (assuming public-coin CRHFs and PC $- di\mathcal{O}$).

The Universal Adversary $\mathcal{A}$. We consider a universal PPT adversary $\mathcal{A}$ that, given $(i, z) \in \{0,1\}^{\mathsf{poly}(k)} \times \{0,1\}^{n(k)}$, parses z as a Turing machine and returns $z(i)$. Note that in our setting, i corresponds to the index of the selected function $f_i \in \mathcal{F}$, and (looking ahead) the auxiliary input z will contain an obfuscated program.

The Auxiliary Input Distribution $\mathcal{Z}$. Let $\mathcal{PRF} = \{\mathrm{PRF}_s : \{0,1\}^{m(k)} \rightarrow \{0,1\}^k\}_{s \in \{0,1\}^k}$ be a puncturable pseudorandom function family, and $\mathcal{H} = \{\mathcal{H}_k\}$ a public-coin collision-resistant hash function family with $h : \{0,1\}^* \rightarrow \{0,1\}^{m(k)}$ for each $h \in \mathcal{H}_k$. (Note that by Theorem 4, punctured PRFs for these parameters exist based on OWFs, which are implied by CRHF). We begin by defining two classes of *Turing machines*:

$$\mathcal{M} = \left\{ \Pi^{h,s} \ \middle| \ s \in \{0,1\}^k, \ h \in \mathcal{H}_k, \ k \in \mathbb{N} \right\},$$

$$\mathcal{M}^* = \left\{ \Pi^{h,s}_{i,y} \ \middle| \ s \in \{0,1\}^k, \ y \in \{0,1\}^{\ell(k)}, \ h \in \mathcal{H}_k, \ k \in \mathbb{N} \right\},$$

which we now describe. We assume without loss of generality for each k that the corresponding collection of Turing machines $\Pi^{h,s} \in \mathcal{M}_k$, $\Pi^{h,s}_{i,y} \in \mathcal{M}^*_k$ are of the *same size*; this can be achieved by padding. (We address the size bound of each class of machines below). In a similar fashion, we may further assume that for each k the runtime of each $\Pi^{h,s}$ and $\Pi^{h,s}_{i,y}$ on any given input f_i is equal.

At a high level, each machine $\Pi^{h,s}$ accepts as input a poly-size circuit description of a function f_i (with canonical description, including a function index i), computes the hash of the corresponding index i w.r.t. the hardcoded hash function h, applies a PRF with hardcoded seed s to the hash, and then evaluates the circuit f_i on the resulting PRF output value x: that is, $\Pi^{h,s}_{i,y}(f_i)$ outputs $U_k(f_i, \mathsf{PRF}_s(h(i)))$, where U_k is the universal Turing machine. See Fig. 1. Note that each $\Pi^{h,s}$ can be described by a Turing machine of size $O(|s| + |h| + |U_k|)$, which is bounded by $p(k)$ for some fixed polynomial p.

Turing Machine $\Pi^{h,s}$:

Hardwired: Hash function $h : \{0,1\}^* \to \{0,1\}^{m(k)}$, PRF seed $s \in \{0,1\}^k$.
Inputs: Circuit description f_i
  1. Hash the index: $v = h(i)$.
  2. Compute the PRF on this hash: $x = \mathsf{PRF}_s(v)$.
  3. Output the evaluation of the universal Turing machine on inputs f_i, x: i.e., $y = U_k(f_i, x)$.

Fig. 1. Turing machines $\Pi^{h,s} \in \mathcal{M}$.

Auxiliary Input Distribution $\mathcal{Z}_k$:

  1. Sample a hash function $h \leftarrow \mathcal{H}_k$ and PRF seed $s \leftarrow \mathcal{K}_{\mathcal{PRF}}(1^k)$.
  2. Output an obfuscation $\tilde{\Pi} \leftarrow \textsf{PC-}di\mathcal{O}(\Pi^{h,s})$.

Fig. 2. The auxiliary input distribution $\mathcal{Z}_k$.

The machines $\Pi^{h,s}_{i,y}$ perform a similar task, except that instead of having the entire PRF seed s hardcoded, they instead only have a *punctured* seed s^* derived from s by puncturing it at the point $h(i)$ (i.e., enabling evaluation of the PRF on all points except $h(i)$). In addition, it has hardwired an output y to replace the punctured result. More specifically, on input a circuit description f_j (with explicitly specified index j), the program $\Pi_{i,y}^{h,s}$ first computes the hash $h = h(j)$, continues computation as usual for any $h \neq h(i)$ using the punctured PRF key, and for $h = h(i)$, it skips the PRF and U_k evaluation steps and directly outputs y. Note that because h is not injective, this puncturing may change the value of the program on multiple inputs f_j (corresponding to functions $f_j \in \mathcal{F}$ with $h(j) = h(i)$). When the hardcoded value y is set to $y = f_i(\mathsf{PRF}_s(h(i)))$, then $\Pi_{i,y}^{h,s}$ agrees with $\Pi^{h,s}$ additionally on the input f_i, but not necessarily on the other inputs f_j for which $h(j) = h(i)$. (Indeed, whereas the hash of their indices collide, and thus their corresponding PRF outputs, $\mathsf{PRF}(h(j))$, will agree, the final step will apply *different* functions f_j to this value).

We first remark that indistinguishability obfuscation arguments will thus not apply to this scenario, since we are modifying the computed functionality. In contrast, differing-inputs obfuscation would guarantee that the two obfuscated programs are indistinguishable, since otherwise we could efficiently *find* one of the disagreeing inputs, which would correspond to a collision in the CRHF. But, most importantly, this argument holds even if *the randomness used to sample the program pair* $(\Pi^{h,s}, \Pi_{i,y}^{h,s})$ *is revealed*. Namely, we consider a program sampler that generates pairs $(\Pi^{h,s}, \Pi_{i,y}^{h,s})$ of the corresponding distribution; this amounts to sampling a hash function h, an EOWF challenge index i, and a PRF seed s, and a $h(i)$-puncturing of the seed, s^*. All remaining values specifying the programs, such as $y = f_i(\mathsf{PRF}_s(h(i)))$, are deterministically computed given (h, i, s, s^*). Now, since $\mathcal{H}$ is a public-coin CRHF family, revealing the randomness used to sample $h \leftarrow \mathcal{H}$ is not detrimental to its collision resistance. And, the values i, s, and s^* are completely *independent* of the CRHF security (i.e., a CRHF adversary reduction could simply generate them on its own in order to break h). Therefore, we ultimately need only rely on *public-coin diO*.

We finally consider the size of the program(s) to be obfuscated. Note that each $\Pi_{i,y}^{h,s}$ can be described by a Turing machine of size $O(|s^*| + |h| + |y| + |U_k|)$. Recall by Theorem 4 the size of the punctured PRF key $|s^*| \in O(m'(k)\ell(k))$, where the PRF has input and output lengths $m'(k)$ and $\ell(k)$. In our application, note that the input to the PRF is not the function index i itself (in which case the machine $\Pi_{i,y}^{h,s}$ would need to grow with the size of the alleged EOWF family), but rather the *hashed* index $h(i)$, which is of fixed polynomial length. Thus, collectively, we have $|\Pi_{i,y}^{h,s}|$ is bounded by a fixed polynomial $p'(k)$, and finally that there exists a single fixed polynomial bound on the size of *all* programs $\Pi^{h,s} \in \mathcal{M}, \Pi_{i,y}^{h,s} \in \mathcal{M}^*$. This completely determines the auxiliary input distribution $\mathcal{Z} = \{\mathcal{Z}_k\}$, described in full in Fig. 2. (Note that the size of the auxiliary output generated by $\mathcal{Z}$, which corresponds to an obfuscation of an appropriately padded program $\Pi^{h,s}$ is thus also bounded by a fixed polynomial in k).

A Has No Extractor. We show that, based on the assumed security of the underlying tools, the constructed adversary $\mathcal{A}$ given auxiliary input from the constructed distribution $\mathcal{Z} = \{Z_k\}$, cannot have an extractor $\mathcal{E}$ satisfying Definition 3:

Turing Machine $\Pi_{i,y}^{h,s}$:

Hardwired: Hash function $h : \{0,1\}^* \to \{0,1\}^{m(k)}$, punctured PRF seed $s^* \in \{0,1\}^k$, punctured point $h(i)$, bit string $y \in \{0,1\}^{\ell(k)}$.
Input: Circuit description f_j (containing index j)
1. Hash the index: $v = h(j)$.
2. If $v \neq h(i)$, compute $x = \mathsf{PRF}_{s^*}(v)$, and output $U_k(f_j, x)$.
3. If $v = h(i)$, output y.

Fig. 3. "Punctured" Turing machines $\Pi_{i,y}^{h,s} \in \mathcal{M}^*$.

Auxiliary Input Distribution $\mathcal{Z}_k(i,y)$:

1. Sample a hash function $h \leftarrow \mathcal{H}_k$ and PRF seed $s \leftarrow \mathcal{K}_{\mathcal{PRF}}(1^k)$.
2. Sample a punctured PRF seed $s^* \leftarrow \mathsf{Punct}(s, h(i))$, punctured at point $h(i)$.
3. Compute the "correct" punctured evaluation: $y = f_i(\mathsf{PRF}_s(h(i)))$.
4. Output an obfuscation $\tilde{M} \leftarrow \mathsf{PC}\text{-}di\mathcal{O}(\Pi_{i,y}^{h,s})$, where $\Pi_{i,y}^{h,s}$ is defined from (h, s^*, y), as in Figure 3.

Fig. 4. The "punctured" distribution $\mathcal{Z}_k(i,y)$.

Proposition 1. *For any non-uniform polynomial-time candidate extractor $\mathcal{E}$ for $\mathcal{A}$, it holds that $\mathcal{E}$ fails with overwhelming probability: i.e.,*

$$\Pr\left[z \leftarrow \mathcal{Z}_k; \ i \leftarrow \mathcal{K}_{\mathcal{F}}(1^k); \ y \leftarrow \mathcal{A}(i;z); \ x' \leftarrow \mathcal{E}(i;z)\right.$$
$$\left. : \exists x \ s.t. \ f_i(x) = y \wedge f_i(x') \neq y\right] \geq 1 - \mathsf{negl}(k).$$

Proof. First note that given auxiliary input $z \leftarrow \mathcal{Z}_k$, $\mathcal{A}$ produces an element in the image of the selected f_i with high probability. That is,

$$\Pr\left[z \leftarrow \mathcal{Z}_k; i \leftarrow \mathcal{K}_{\mathcal{F}}(1^k); y \leftarrow \mathcal{A}(i;z) : \exists x \text{ s.t. } f_i(x) = y\right] \geq 1 - \mathsf{negl}(k).$$

Indeed, by the definition of $\mathcal{A}$ and $\mathcal{Z}_k$, and the correctness of the obfuscator $\mathsf{PC} - di\mathcal{O}$, then we have with overwhelming probability

$$\mathcal{A}(i;z) = \tilde{M}(f_i) = \Pi^{h,s}(f_i) = f_i(\mathsf{PRF}_s(h(i))),$$

where $z = \tilde{M}$ is an obfuscation of $\Pi^{h,s} \in \mathcal{M}$; i.e., $z = \tilde{M} \leftarrow \mathsf{PC} - di\mathcal{O}(\Pi^{h,s})$.

Now, suppose for contradiction that there exists a non-negligible function $\epsilon(k)$ such that for all $k \in \mathbb{N}$ the extractor $\mathcal{E}$ successfully outputs a preimage corresponding to the output $\mathcal{A}(i;z) \in Range(f_i)$ with probability $\epsilon(k)$: i.e.,

$$\Pr\left[z \leftarrow \mathcal{Z}_k; \ i \leftarrow \mathcal{K}_{\mathcal{F}}(1^k); \ x' \leftarrow \mathcal{E}(i;z)\right.$$
$$\left. : f_i(x') = \mathcal{A}(i;z) = f_i(\mathsf{PRF}_s(h(i)))\right] \geq \epsilon(k).$$

where as before, s, h are such that $z = \mathsf{PC} - di\mathcal{O}(\Pi^{h,s})$. We show that this cannot be the case, via three steps.

Step 1: Replace $\mathcal{Z}$ with "punctured" distribution $\mathcal{Z}(i, y)$. For every index i of the EOWF family $\mathcal{F}$ and $k \in \mathbb{N}$, consider an alternative distribution $\mathcal{Z}_k(i, y)$ that, instead of sampling and obfuscating a Turing machine $\Pi^{h,s}$ from the class $\mathcal{M}$, as is done for $\mathcal{Z}$, it does so with a Turing machine $\Pi_{i,y}^{h,s} \in \mathcal{M}^*$ as follows. First, it samples a hash function $h \leftarrow \mathcal{H}_k$ and PRF seed s as usual. It then generates a *punctured* PRF key $s^* \leftarrow \mathsf{Punct}(s, h(i))$ that enables evaluation of the PRF on all points except the value $h(i)$. For the specific index i, it computes the correct full evaluation $y := f_i(\mathsf{PRF}_s(h(i)))$. Finally, $\mathcal{Z}_k(i, y)$ outputs an obfuscation of the constructed program $\Pi_{i,y}^{h,s}$ as specified in Fig. 3 from the values (h, s^*, y): i.e., $\tilde{M} \leftarrow \mathsf{PC} - di\mathcal{O}(\Pi_{i,y}^{h,s})$. See Fig. 4 for a full description of $\mathcal{Z}(i, y)$.

We now argue that the extractor $\mathcal{E}$ must also succeed in extracting a preimage when given a value $z^* \leftarrow \mathcal{Z}_k(i, y)$ from this modified distribution instead of $\mathcal{Z}_k$.

Consider the Turing Machine program sampler algorithm Samp as in Fig. 5.

Program Pair Sampler $\mathsf{Samp}(1^k, r)$:

1. Sample a hash function $h = \mathcal{H}_k(r_h)$.
2. Sample an EOWF index $i = K_{\mathcal{F}}(1^k; r_i)$.
3. Sample a PRF seed $s = K_{\mathsf{PRF}}(1^k; r_s)$.
4. Sample a punctured PRF seed $s^* = \mathsf{Punct}(s, h(i); r_*)$.
5. Let $y = f_i(\mathsf{PRF}_s(h(i)))$.
6. Denote $r := (r_h, r_i, r_s, r_*)$.
7. Output program pair $(\Pi^{h,s}, \Pi_{i,y}^{h,s})$, defined by h, i, s, s^*, y as above (and padded to equal length).

Fig. 5. Program pair sampler algorithm, to be used in public-coin differing inputs security step.

We first argue that, by the (public-coin) collision resistance of the hash family $\mathcal{H}$, the sampler algorithm Samp is a *public-coin differing-inputs sampler*, as per Definition 1.

Claim. Samp is a public-coin differing-inputs sampler. That is, for all efficient non-uniform $\mathcal{A}_{\mathsf{PC}}$, there exists a negligible function ϵ such that for all $k \in \mathbb{N}$,

$$\Pr\left[r \leftarrow \{0,1\}^*; (M_0, M_1) \leftarrow \mathsf{Samp}(1^k, r); (x, 1^t) \leftarrow \mathcal{A}_{\mathsf{PC}}(1^k, r) :\right.$$
$$\left. M_0(x) \neq M_1(x) \land \mathsf{steps}(M_0, x) = \mathsf{steps}(M_1, x) = t\right] \leq \epsilon(k). \quad (1)$$

Proof. Suppose, to the contrary, there exists an efficient (non-uniform) adversary $\mathcal{A}_{\mathsf{PC}}$ and non-negligible function $\alpha(k)$ for which the probability in Eq. 1 is greater than $\alpha(k)$. We show such an adversary contradicts the security of the (public-coin) CRHF. Consider an adversary $\mathcal{A}_{\mathsf{CR}}$ in the CRHF security challenge. Namely, for a challenge hash function $h \leftarrow \mathcal{H}_k(r_h)$, the adversary $\mathcal{A}_{\mathsf{CR}}$ receives h, r_h, and performs the following steps:

CRHF adversary $\mathcal{A}_{\mathsf{CR}}(1^k, h, r_h)$:

1. Imitate the remaining steps of Samp. That is, sample an EOWF index $i = K_{\mathcal{F}}(1^k; r_i)$; a PRF seed $s = K_{\mathsf{PRF}}(1^k; r_s)$; and a punctured PRF seed $s^* = \mathsf{Punct}(s, h(i); r_*)$. Define $y = f_i(\mathsf{PRF}_s(h(i)))$ and $r = (r_h, r_i, r_s, r_*)$, and let $M_0 = \Pi^{h,s}$ and $M_1 = \Pi^{h,s}_{i,y}$.

2. Run $\mathcal{A}_{\mathsf{PC}}(1^k, r)$ on the collection of randomness r used above. In response, $\mathcal{A}_{\mathsf{PC}}$ returns a pair $(x, 1^t)$.

3. $\mathcal{A}_{\mathsf{CR}}$ outputs the pair (i, x) as an alleged collision in the challenge hash function h.

Now, by assumption, the value x generated by $\mathcal{A}_{\mathsf{PC}}$ satisfies (in particular) that $M_0(x) \neq M_1(x)$. From the definition of M_0, M_1 (i.e., $\Pi^{h,s}, \Pi^{h,s}_{i,y}$), this must mean that $h(i) = h(x)$ (since all values with $h(x) \neq h(i)$ were not changed from $\Pi^{h,s}$ to $\Pi^{h,s}_{i,y}$), and that $i \neq x$ (since $\Pi^{h,s}_{i,y}(i)$ was specifically "patched" to the correct output value $\Pi^{h,s}(i)$). That is, $\mathcal{A}_{\mathsf{CR}}$ successfully identifies a collision with the same probability $\alpha(k)$, which must thus be negligible.

We now show that this implies, by the security of the public-coin $di\mathcal{O}$, that our original EOWF extractor $\mathcal{E}$ must succeed with nearly equivalent probability in the EOWF challenge when instead of receiving (real) auxiliary input from $\mathcal{Z}_k$, both $\mathcal{E}$ and $\mathcal{A}$ are given auxiliary input from the fake distribution $\mathcal{Z}_k(i, y)$. (Recall that ϵ is assumed to be $\mathcal{E}$'s success in the same experiment as below but with $z \leftarrow \mathcal{Z}_k$ instead of $z^* \leftarrow \mathcal{Z}_k(i, y)$).

Lemma 1. *It holds that*

$$\Pr\left[i \leftarrow K_{\mathcal{F}}(1^k);\ z^* \leftarrow \mathcal{Z}_k(i, y);\ x' \leftarrow \mathcal{E}(i; z^*):\right.$$
$$\left. f_i(x') = \mathcal{A}(i; z^*) = f_i(\mathsf{PRF}_s\,(h(i)))\right] \geq \epsilon(k) - \mathsf{negl}(k). \quad (2)$$

Proof. Note that given $z^* \leftarrow \mathcal{Z}_k(i, y)$ (which corresponds to an obfuscated program of the form $\Pi^{h,s}_{i,y}$) our EOWF adversary $\mathcal{A}$ indeed will still output $\Pi^{h,s}_{i,y}(i) = y := f_i(\mathsf{PRF}_s(h(i)))$ (see Figs. 3,4).

Now, suppose there exists a non-negligible function $\alpha(k)$ for which the probability in Eq. (2) is less than $\epsilon(k) - \alpha(k)$. We directly use such $\mathcal{E}$ to design another adversary $\mathcal{A}_{di\mathcal{O}}$ to contradict the security of the public-coin $di\mathcal{O}$ with respect to the program pair sampler Samp (which we showed in Claim 3.1 to be a void public-coin differing inputs sampler). Recall the $di\mathcal{O}$ challenge samples a program pair $(\Pi^{h,s}, \Pi^{h,s}_{i,y}) \leftarrow \mathsf{Samp}(1^k, r)$, selects a random $M \leftarrow \{\Pi^{h,s}, \Pi^{h,s}_{i,y}\}$ to obfuscate as $\tilde{M} \leftarrow \mathsf{PC} - di\mathcal{O}(1^k, M)$, and gives as a challenge the pair $(r, \tilde{M})$ of the randomness used by Samp and obfuscated program. Define $\mathcal{A}_{di\mathcal{O}}$ (who wishes to distinguish which program was selected) as follows.

PC-$di\mathcal{O}$ adversary $\mathcal{A}_{di\mathcal{O}}(1^k, r, \tilde{M})$:

1. Parse the given randomness r used in Samp as $r = (r_h, r_i, r_s, r_*)$ (see Fig. 5).
2. Recompute the "challenge index" $i = K_{\mathcal{F}}(1^k; r_i)$. Let $z^* = \tilde{M}$.
3. Run the extractor algorithm $\mathcal{E}(i; z^*)$, and receive an alleged preimage x'.
4. Recompute $h = \mathcal{H}_k(r_h)$, $s = K_{\mathsf{PRF}}(1^i r_s)$, again using the randomness from r.
5. If $f_i(x') = f_i(\mathsf{PRF}_s(h(i)))$ — i.e., if $\mathcal{E}$ succeeded in extracting a preimage — then $\mathcal{A}_{di\mathcal{O}}$ outputs 1. Otherwise, $\mathcal{A}_{di\mathcal{O}}$ outputs 0.

Now, if $\tilde{M}$ is an obfuscation of $\Pi^{h,s}$, then this experiment corresponds directly to the EOWF challenge where $\mathcal{E}$ (and $\mathcal{A}$) is given auxiliary input $z \leftarrow \mathcal{Z}_k$. On the other hand, if $\tilde{M}$ is an obfuscation of $\Pi_{i,y}^{h,s}$, then the experiment corresponds directly to the same challenge where $\mathcal{E}$ (and $\mathcal{A}$) is given auxiliary input $z^* \leftarrow \mathcal{Z}_k(i, y)$. Thus, $\mathcal{A}_{di\mathcal{O}}$ will succeed in distinguishing these two cases with probability at least $[\epsilon(k)] - [\epsilon(k) - \alpha(k)] = \alpha(k)$. By the security of $\mathsf{PC} - di\mathcal{O}$, it hence follows that $\alpha(k)$ must be negligible.

Step 2: Replace "correct" hardcoded y in $\mathcal{Z}(i, y)$ with random f_i evaluation. Next, we consider another experiment where $\mathcal{Z}_k(i, y)$ is altered to a nearly identical distribution $\mathcal{Z}_k(i, u)$ where, instead of hardcoding the "correct" i-evaluation value $y = f_i(\mathsf{PRF}_s(h(i)))$ in the generated "punctured" program $\Pi_{i,y}^{h,s}$, the distribution $\mathcal{Z}_k(i, u)$ now simply samples a random f_i output $y = f_i(u)$ for an independent random $u \leftarrow \{0,1\}^k$. We claim that the original EOWF extractor $\mathcal{E}$ still succeeds in finding a preimage when given this new auxiliary input distribution:

Lemma 2. *It holds that*

$$\Pr\left[i \leftarrow K_{\mathcal{F}}(1^k); \ z^{**} \leftarrow \mathcal{Z}_k(i, u); \ x' \leftarrow \mathcal{E}(i; z^{**}) : \right.$$
$$\left. f_i(x') = \mathcal{A}(i; z^{**}) = f_i(u)\right] \geq \epsilon(k) - \mathsf{negl}(k). \quad (3)$$

Proof. This follows from the fact that $\mathsf{PRF}_s(h(i))$ is pseudorandom, even given the $h(i)$-punctured key s^*.

Formally, consider an algorithm $\mathcal{A}_{\mathsf{PRF}}^0$ which, on input the security parameter 1^k, a pair of values i, h, and a pair s^*, x (that will eventually correspond to a challenge punctured PRF key, and either $\mathsf{PRF}_s(h(i))$ or random u), performs the following steps.

Algorithm $\mathcal{A}_{\mathsf{PRF}}^0(1^k, i, h, s^*, x)$:

1. Take $y = f_i(x)$, and obfuscate the associated program $\Pi_{i,y}^{h,s}$: i.e., $z^{**} \leftarrow \mathsf{PC} - di\mathcal{O}(1^k, \Pi_{i,y}^{h,s})$.
2. Run the EOWF extractor given index i and auxiliary input z^{**}: $x' \leftarrow \mathcal{E}(i; z^{**})$.

3. Output 0 if $\mathcal{E}$ succeeds in extracting a valid preimage: i.e., if $f_i(x') = y^* = f_i(x)$. Otherwise, output a random bit $b \leftarrow \{0,1\}$.

Now, suppose Lemma 2 does not hold: i.e., the probability in Eq. (3) differs by some non-negligible amount from $\epsilon(k)$. Then, expanding out the sampling procedure of $\mathcal{Z}_k(i,y)$ and $\mathcal{Z}_k(i,u)$, we have for some non-negligible function $\alpha(k)$ that

$$\Pr\left[i \leftarrow \mathcal{K}_{\mathcal{F}}(1^k); \ h \leftarrow \mathcal{H}_k; \ s \leftarrow \mathcal{K}_{\mathcal{PRF}}(1^k); \ s^* \leftarrow \mathsf{Punct}(s, h(i));\right.$$
$$\left. u \leftarrow \{0,1\}^k; b \leftarrow \{0,1\} : \mathcal{A}^0_{\mathsf{PRF}}(1^k, i, h, x_b) = b\right] \geq \frac{1}{2} + \alpha(k), \quad (4)$$

where $x_0 := \mathsf{PRF}_s(h(i))$ and $x_1 := u$. Indeed, in the case $b = 0$, the auxiliary input z^{**} generated by $\mathcal{A}_{\mathsf{PRF}}$ and given to $\mathcal{E}$ has distribution exactly $\mathcal{Z}(i,y)$, whereas in the case $b = 1$, the generated z^{**} has distribution exactly $\mathcal{Z}(i,u)$.

In particular, there exists a polynomial $p(k)$ such that for infinitely many k, there exists an index i_k and hash function $h_k \in \mathcal{H}_k$ with

$$\Pr\left[s \leftarrow \mathcal{K}_{\mathcal{PRF}}(1^k); \ s^* \leftarrow \mathsf{Punct}(s, h(i_k)); \ u \leftarrow \{0,1\}^k;\right.$$
$$\left. b \leftarrow \{0,1\} : \mathcal{A}^0_{\mathsf{PRF}}(1^k, i_k, h, x_b) = b\right] \geq \frac{1}{2} + \frac{1}{p(k)}, \quad (5)$$

where x_0, x_1 are as before.

Consider a non-uniform punctured-PRF adversary $\mathcal{A}^I_{\mathsf{PRF}}$ (with the ensemble $I = \{i_k, h_k\}$ hardcoded) that first selects the challenge point $h_k(i_k)$; receives the PRF challenge information (s^*, x) for this point; executes $\mathcal{A}^0_{\mathsf{PRF}}$ on input $(1^k, i_k, h_k, s^*, x)$, and outputs the corresponding bit b output by $\mathcal{A}^0_{\mathsf{PRF}}$. Then by (5), it follows that $\mathcal{A}^I_{\mathsf{PRF}}$ breaks the security of the punctured PRF.

Step 3: Such an extractor breaks one-wayness of EOWF. Finally, we observe that this means that $\mathcal{E}$ can be used to break the one-wayness of the original function family $\mathcal{F}$. Indeed, given a random key i and a challenge output $y = f_i(u)$, an inverter can simply sample a hash function h and $h(i)$-punctured PRF seed s^* on its own, construct the program $\Pi^{h,s}_{i,y}$ with its challenge y hardcoded in, and sample an obfuscation $z^{**} \leftarrow \mathsf{PC} - di\mathcal{O}(\Pi^{h,s}_{i,y})$. Finally, it runs $\mathcal{E}(i, z^{**})$ to invert y^*, with the same probability $\epsilon(k) - \mathsf{negl}(k)$.

This concludes the proof of Theorem 5.

3.2   PC-$di\mathcal{O}$ or SNARKs

We link the existence of public-coin differing-inputs obfuscation for NC^1 and the existence of succinct non-interactive arguments of knowledge (SNARKs), via an intermediate step of *proximity* extractable one-way functions (PEOWFs), a notion related to EOWFs, introduced in [5]. Namely, *assume the existence of* *fully homomorphic encryption (FHE) with decryption in NC^1 and public-coin collision-resistant hash functions.* Then, building upon the results of the previous subsection, and the results of [5,30], we show:

1. Assuming SNARKs for NP, there exists an efficient distribution $\mathcal{Z}$ such that public-coin differing-inputs obfuscation for NC^1 implies that there *cannot* exist PEOWFs $\{f : \{0,1\}^k \to \{0,1\}^k\}$ w.r.t. $\mathcal{Z}$.
2. PEOWFs $\{f : \{0,1\}^k \to \{0,1\}^k\}$ w.r.t. this auxiliary input distribution $\mathcal{Z}$ are *implied by* the existence of SNARKs for NP secure w.r.t. a second efficient auxiliary input distribution $\mathcal{Z}'$, as shown in [5].
3. Thus, one of these conflicting hypotheses must be false. That is, there exists an efficient distribution $\mathcal{Z}'$ such that assuming existence of FHE with decryption in NC^1 and collision-resistant hash functions, then either: (1) public-coin differing-inputs obfuscation for NC^1 does not exist, or (2) SNARKS for NP w.r.t. $\mathcal{Z}'$ do not exist.

Note that we focus on the specific case of PEOWFs with k-bit inputs and k-bit outputs, as this suffices to derive the desired contradiction; however, the theorems following extend also to the more general case of PEOWF output length (demonstrating an efficient distribution $\mathcal{Z}$ to rule out each potential output length $\ell(k)$).

Proximity EOWFs. We begin by defining Proximity EOWFs.

Proximity Extractable One-Way Functions (PEOWFs). In a Proximity EOWF (PEOWF), the extractable function family $\{f_i\}$ is associated with a "proximity" equivalence relation $\sim$ on the range of f_i, and the one-wayness and extractability properties are modified with respect to this relation. The one-wayness is strengthened: not only must it be hard to find an exact preimage of v, but it is also hard to find a preimage of any equivalent $v \sim v'$. The extractability requirement is weakened accordingly: the extractor does not have to output an exact preimage of v, but only a preimage of of some equivalent value $v' \sim v$.

As an example, consider functions of the form $f : x \mapsto (f_1(x), f_2(x))$ and equivalence relation on range elements $(a, b) \sim (a, b')$ whose first components agree. Then the proximity extraction property requires for any adversary $\mathcal{A}$ who outputs an image element $(a, b) \in Range(f)$ that there exists an extractor $\mathcal{E}$ finding an input x s.t. $f(x) = (a, b')$ for some b' not necessarily equal to b.

In this work, we allow the relation $\sim$ to depend on the function index i, but require that the relation $\sim$ is *publicly* (and efficiently) testable. We further consider non-uniform adversaries and extraction algorithms, and (in line with this work) auxiliary inputs coming from a specified distribution $\mathcal{Z}$.

Definition 5 ($\mathcal{Z}$-Auxiliary-Input Proximity EOWFs). *Let ℓ, m be polynomially bounded length functions. An efficiently computable family of functions*

$$\mathcal{F} = \left\{ f_i : \{0,1\}^k \to \{0,1\}^{\ell(k)} \;\middle|\; i \in \{0,1\}^{m(k)}, k \in \mathbb{N} \right\},$$

associated with an efficient probabilistic key sampler $\mathcal{K}_{\mathcal{F}}$, is a $\mathcal{Z}$-auxiliary-input proximity extractable one-way function if it satisfies the following (strong) one-wayness, (weak) extraction, and public testability properties:

- **(Strengthened) One-wayness:** *For non-uniform polynomial-time $\mathcal{A}$ and sufficiently large security parameter $k \in \mathbb{N}$,*

$$\Pr\left[z \leftarrow \mathcal{Z}_k;\ i \leftarrow \mathcal{K}_{\mathcal{F}}(1^k);\ x \leftarrow \{0,1\}^k;\ x' \leftarrow \mathcal{A}(i, f_i(x); z) \right.$$
$$\left. : f_i(x') \sim f_i(x) \right] \leq \mathsf{negl}(k).$$

- **(Weakened) Extractability:** *For any non-uniform polynomial-time adversary $\mathcal{A}$, there exists a non-uniform polynomial-time extractor $\mathcal{E}$ such that, for sufficiently large security parameter $k \in \mathbb{N}$,*

$$\Pr\left[z \leftarrow \mathcal{Z}_k;\ i \leftarrow \mathcal{K}_{\mathcal{F}}(1^k);\ y \leftarrow \mathcal{A}(i; z);\ x' \leftarrow \mathcal{E}(i; z) \right.$$
$$\left. : \exists x \ s.t.\ f_i(x) = y \wedge f_i\left(x'\right) \not\sim y \right] \leq \mathsf{negl}(k).$$

- **Publicly Testable Relation:** *There exists a deterministic polytime machine $\mathcal{T}$ such that, given the function index i, $\mathcal{T}$ accepts $y, y' \in \{0,1\}^{\ell(k)}$ if and only if $y \sim_k y'$.*

(PC − $di\mathcal{O}$ for NC^1 + PC-CRHF + FHE + SNARK) $\Rightarrow$ No $\mathcal{Z}$-PEOWF. We now show that, assuming the existence of public-coin collision-resistant hash functions (CRHF) and fully homomorphic encryption (FHE) with decryption in NC^1,[8] then for some efficiently computable distributions $\mathcal{Z}_{\mathsf{SNARK}}$, $\mathcal{Z}_{\mathsf{PEOWF}}$, if there exist public-coin differing-inputs obfuscators for NC^1 circuits, and SNARKs w.r.t. auxiliary input $\mathcal{Z}_{\mathsf{SNARK}}$, then there *cannot* exist PEOWFs w.r.t. auxiliary input $\mathcal{Z}_{\mathsf{PEOWF}}$. This takes place in two steps.

First, we remark that an identical proof to that of Theorem 5 rules out the existence of $\mathcal{Z}$-auxiliary-input *proximity EOWFs* in addition to standard EOWFs, based on the same assumptions: namely, assuming public-coin differing-inputs obfuscation for Turing machines, and public-coin collision-resistant hash functions. Indeed, assuming the existence of a PEOWF extractor $\mathcal{E}$ for the adversary $\mathcal{A}$ and auxiliary input distribution $\mathcal{Z}$ (who extracts a "related" preimage to the target value), the same procedure yields a PEOWF inverter who similarly extracts a "related" preimage to any challenge output. In the reduction, it is merely required that the success of $\mathcal{E}$ is efficiently and publicly testable (this is used to construct a distinguishing adversary for the differing-inputs obfuscation scheme, in Step 1). However, this is directly implied by the public testability of the PEOWF relation $\sim$, as specified in Definition 5.

---

[8] As is the case for nearly all existing FHE constructions (e.g., [13,21]).

Theorem 6. *There exist an efficient, uniformly samplable distribution $\mathcal{Z}$ such that, assuming the existence of public-coin collision-resistant hash functions and public-coin differing-inputs obfuscation for polynomial-size Turing machines, there cannot exist (publicly testable) $\mathcal{Z}$-auxiliary-input PEOWFs $\{f_i : \{0,1\}^k \to \{0,1\}^k\}$.*

Now, in [30], it was shown that public-coin differing-inputs obfuscation for the class of all polynomial-time Turing machines can be achieved by bootstrapping up from public-coin differing-inputs obfuscation for circuits in the class NC^1, assuming the existence of FHE with decryption in NC^1, public-coin CRHF, and public-coin SNARKs for NP.

Putting this together with Theorem 6, we thus have the following corollary.

Corollary 1. *There exists an efficient, uniformly samplable distribution $\mathcal{Z}$ s.t., assuming existence of public-coin SNARKs and FHE with decryption in NC^1, then assuming the existence of public-coin differing-inputs obfuscation for NC^1, there cannot exist PEOWFs $\{f_i : \{0,1\}^k \to \{0,1\}^k\}$ w.r.t. auxiliary input $\mathcal{Z}$.*

(SNARK + CRHF) $\implies$ $\mathcal{Z}$-PEOWF. As shown in [5], Proximity EOWFs (PEOWFs) with respect to an auxiliary input distribution $\mathcal{Z}$ are *implied by* collision-resistant hash functions (CRHF) and SNARKs secure with respect to a related auxiliary input distribution $\mathcal{Z}'$.[9]

Loosely, the transformation converts any CRHF family $\mathcal{F}$ into a PEOWF by appending to the output of each $f \in \mathcal{F}$ a succinct SNARK argument π_x that there exists a preimage x yielding output $f(x)$. (If the Prover algorithm of the SNARK system is randomized, then the function is also modified to take an additional input, which is used as the random coins for the SNARK generation). The equivalence relation on outputs is defined by $(y, \pi) \sim (y', \pi')$ if $y = y'$ (note that this relation is publicly testable). More explicitly, consider the new function family $\mathcal{F}'$ composed of functions

$$f'_{\mathsf{crs}}(x, r) = \left(f(x), \mathsf{Prove}(1^k, \mathsf{crs}, f(x), x; r) \right),$$

where a function $f'_{\mathsf{crs}} \in \mathcal{F}'$ is sampled by first sampling a function $f \leftarrow \mathcal{F}$ from the original CRHF family, and then sampling a CRS for the SNARK scheme, $\mathsf{crs} \leftarrow \mathsf{CRSGen}(1^k)$.

Now (as proved in [5]), the resulting function family will be a PEOWF with respect to auxiliary input $\mathcal{Z}$ if the underlying SNARK system is secure with respect to an augmented auxiliary input distribution $\mathcal{Z}_{\mathsf{SNARK}} := (\mathcal{Z}, h)$, formed by concatenating a sample from $\mathcal{Z}$ with a function index h sampled from the collision-resistant hash function family $\mathcal{F}$. (Note that we will be considering public-coin CRHF, in which case h is uniform).

Theorem 7 ([5]). *There exist efficient, uniformly samplable distributions $\mathcal{Z}, \mathcal{Z}_{\mathsf{SNARK}}$ such that, assuming the existence of collision-resistant hash functions and SNARKs for NP secure w.r.t. auxiliary input distribution $\mathcal{Z}_{\mathsf{SNARK}}$, then there exist PEOWFs $\{f_i : \{0,1\}^k \to \{0,1\}^k\}$ w.r.t. $\mathcal{Z}$.*

---

[9] [5] consider the setting of arbitrary auxiliary input; however, their construction directly implies similar results for specific auxiliary input distributions.

Reaching a Standoff. Observe that the conclusions of Corollary 1 and Theorem 7 are in direct contradiction. Thus, it must be that one of the two sets of assumptions is false. Namely,

Corollary 2. *Assuming the existence of public-coin collision-resistant hash functions and fully homomorphic encryption with decryption in NC^1, there exists an efficiently samplable distribution $\mathcal{Z}_{\mathsf{SNARK}}$ such that one of the following two objects cannot exist:*

- *SNARKs w.r.t. auxiliary input distribution $\mathcal{Z}_{\mathsf{SNARK}}$.*
- *Public-coin differing-inputs obfuscation for NC^1.*

More explicitly, we have that $\mathcal{Z}_{\mathsf{SNARK}} = (\mathcal{Z}, U)$, where $\mathcal{Z}$ is composed of an obfuscated program, and U is a uniform string (corresponding to a randomly sampled index from a public-coin CRHF family).

Acknowledgements. The authors would like to thank Kai-Min Chung for several insightful discussions.

References

1. Ananth, P., Boneh, D., Garg, S., Sahai, A., Zhandry, M.: Differing-inputs obfuscation and applications. Cryptology ePrint Archive, Report 2013/689 (2013)
2. Barak, B., Goldreich, O., Impagliazzo, R., Rudich, S., Sahai, A., Vadhan, S.P., Yang, K.: On the (im)possibility of obfuscating programs. J. ACM **59**(2), Article No. 6 (2012)
3. Barak, B., Lindell, Y., Vadhan, S.P.: Lower bounds for non-black-box zero knowledge. J. Comput. Syst. Sci. **72**(2), 321–391 (2006)
4. Bellare, M., Rogaway, P.: Random oracles are practical: A paradigm for designing efficient protocols. In: ACM Conference on Computer and Communications Security, pp. 62–73 (1993)
5. Bitansky, N., Canetti, R., Chiesa, A., Tromer, E.: From extractable collision resistance to succinct non-interactive arguments of knowledge, and back again. In: ITCS, pp. 326–349 (2012)
6. Bitansky, N., Canetti, R., Chiesa, A., Tromer, E.: Recursive composition and bootstrapping for snarks and proof-carrying data. In: STOC, pp. 111–120 (2013)
7. Bitansky, Nir, Canetti, Ran, Paneth, Omer, Rosen, Alon: On the existence of extractable one-way functions. In: STOC 2014, pp. 505–514 (2014)
8. Boneh, D., Waters, B.: Constrained pseudorandom functions and their applications. In: Sako, K., Sarkar, P. (eds.) ASIACRYPT 2013, Part II. LNCS, vol. 8270, pp. 280–300. Springer, Heidelberg (2013)
9. Boneh, D., Zhandry, M.: Multiparty key exchange, efficient traitor tracing, and more from indistinguishability obfuscation. In: Garay, J.A., Gennaro, R. (eds.) CRYPTO 2014, Part I. LNCS, vol. 8616, pp. 480–499. Springer, Heidelberg (2014)
10. Boyle, E., Chung, K.-M., Pass, R.: On extractability obfuscation. In: Lindell, Y. (ed.) TCC 2014. LNCS, vol. 8349, pp. 52–73. Springer, Heidelberg (2014)
11. Boyle, E., Goldwasser, S., Ivan, I.: Functional signatures and pseudorandom functions. In: Krawczyk, H. (ed.) PKC 2014. LNCS, vol. 8383, pp. 501–519. Springer, Heidelberg (2014)

12. Boyle, E., Pass, R.: Limits of extractability assumptions with distributional auxiliary input. Cryptology ePrint Archive, Report 2013/703 (2013)
13. Brakerski, Z., Vaikuntanathan, V.: Lattice-based FHE as secure as PKE. In: Innovations in Theoretical Computer Science, ITCS 2014, pp. 1–12 (2014)
14. Canetti, R., Dakdouk, R.R.: Towards a theory of extractable functions. In: Reingold, O. (ed.) TCC 2009. LNCS, vol. 5444, pp. 595–613. Springer, Heidelberg (2009)
15. Damgård, I.B.: Towards practical public key systems secure against chosen ciphertext attacks. In: Feigenbaum, J. (ed.) CRYPTO 1991. LNCS, vol. 576, pp. 445–456. Springer, Heidelberg (1992)
16. Damgård, I., Faust, S., Hazay, C.: Secure two-party computation with low communication. In: Cramer, R. (ed.) TCC 2012. LNCS, vol. 7194, pp. 54–74. Springer, Heidelberg (2012)
17. Garg, S., Gentry, C., Halevi, S., Raykova, M.: Two-round secure MPC from indistinguishability obfuscation. In: Lindell, Y. (ed.) TCC 2014. LNCS, vol. 8349, pp. 74–94. Springer, Heidelberg (2014)
18. Garg, S., Gentry, C., Halevi, S., Raykova, M., Sahai, A., Waters, B.: Candidate indistinguishability obfuscation and functional encryption for all circuits. In: FOCS (2013)
19. Garg, S., Gentry, C., Halevi, S., Raykova, M., Sahai, A., Wichs, D.: On the implausibility of differing-inputs obfuscation and extractable witness encryption with auxiliary input. Cryptology ePrint Archive, Report 2013/860 (2013)
20. Garg, S., Gentry, C., Halevi, S., Wichs, D.: On the implausibility of differing-inputs obfuscation and extractable witness encryption with auxiliary input. In: Garay, J.A., Gennaro, R. (eds.) CRYPTO 2014, Part I. LNCS, vol. 8616, pp. 518–535. Springer, Heidelberg (2014)
21. Gentry, C., Sahai, A., Waters, B.: Homomorphic encryption from learning with errors: conceptually-simpler, asymptotically-faster, attribute-based. In: Canetti, R., Garay, J.A. (eds.) CRYPTO 2013, Part I. LNCS, vol. 8042, pp. 75–92. Springer, Heidelberg (2013)
22. Goldreich, O., Goldwasser, S., Micali, S.: How to construct random functions. J. ACM 33(4), 792–807 (1986)
23. Goldwasser, S., Kalai, Y.T., Popa, R.A., Vaikuntanathan, V., Zeldovich, N.: How to run turing machines on encrypted data. In: Canetti, R., Garay, J.A. (eds.) CRYPTO 2013, Part II. LNCS, vol. 8043, pp. 536–553. Springer, Heidelberg (2013)
24. Goldwasser, S., Lin, H., Rubinstein, A.: Delegation of computation without rejection problem from designated verifier cs-proofs. IACR Cryptology ePrint Archive 2011, 456 (2011)
25. Goldwasser, S., Micali, S., Rackoff, C.: The knowledge complexity of interactive proof systems. SIAM J. Comput. 18(1), 186–208 (1989)
26. Gupta, D., Sahai, A.: On constant-round concurrent zero-knowledge from a knowledge assumption. Progress in Cryptology - INDOCRYPT 2014, pp. 71–88 (2014)
27. Hada, S., Tanaka, T.: On the existence of 3-round zero-knowledge protocols. In: Krawczyk, H. (ed.) CRYPTO 1998. LNCS, vol. 1462, pp. 408–423. Springer, Heidelberg (1998)
28. Hohenberger, S., Sahai, A., Waters, B.: Replacing a random oracle: full domain hash from indistinguishability obfuscation. In: Nguyen, P.Q., Oswald, E. (eds.) EUROCRYPT 2014. LNCS, vol. 8441, pp. 201–220. Springer, Heidelberg (2014)
29. Hsiao, C.-Y., Reyzin, L.: Finding collisions on a public road, or do secure hash functions need secret coins? In: Franklin, M. (ed.) CRYPTO 2004. LNCS, vol. 3152, pp. 92–105. Springer, Heidelberg (2004)

30. Ishai, Y., Pandey, O., Sahai, A.: Public-coin differing-inputs obfuscation and its applications. In: Dodis, Y., Nielsen, J.B. (eds.) TCC 2015, Part II. LNCS, vol. 9015, pp. 668–697. Springer, Heidelberg (2015)
31. Kiayias, A., Papadopoulos, S., Triandopoulos, N., Zacharias, T.: Delegatable pseudorandom functions and applications. In: CCS2013, pp. 669–684 (2013)
32. Micali, S.: CS proofs (extended abstracts). In: FOCS, pp. 436–453 (1994)
33. Naor, M.: On cryptographic assumptions and challenges. In: Boneh, D. (ed.) CRYPTO 2003. LNCS, vol. 2729, pp. 96–109. Springer, Heidelberg (2003)
34. Sahai, Amit, Waters, Brent: How to use indistinguishability obfuscation: deniable encryption, and more. In: STOC 2014, pp. 475–484 (2014)
35. Valiant, P.: Incrementally verifiable computation or proofs of knowledge imply time/space efficiency. In: Canetti, R. (ed.) TCC 2008. LNCS, vol. 4948, pp. 1–18. Springer, Heidelberg (2008)

Composable and Modular Anonymous Credentials: Definitions and Practical Constructions

Jan Camenisch[1](✉), Maria Dubovitskaya[1](✉), Kristiyan Haralambiev[2](✉), and Markulf Kohlweiss[3](✉)

[1] IBM Research – Zurich, Rüschlikon, Switzerland
jan@camenisch.org, maria.dubovitskaya@gmail.com
[2] Google Inc., Zurich, Switzerland
haralambiev@gmail.com
[3] Microsoft Research, Cambridge, UK
markulf@microsoft.com

Abstract. It takes time for theoretical advances to get used in practical schemes. Anonymous credential schemes are no exception. For instance, existing schemes suited for real-world use lack formal, composable definitions, partly because they do not support straight-line extraction and rely on random oracles for their security arguments. To address this gap, we propose *unlinkable redactable signatures* (URS), a new building block for privacy-enhancing protocols, which we use to construct the first efficient UC-secure anonymous credential system that supports multiple issuers, selective disclosure of attributes, and pseudonyms. Our scheme is one of the first such systems for which both the size of a credential and its presentation proof are independent of the number of attributes issued in a credential. Moreover, our new credential scheme does not rely on random oracles. As an important intermediary step, we address the problem of building a functionality for a complex credential system that can cover many different features. Namely, we design a core building block for a single issuer that supports credential issuance and presentation with respect to pseudonyms and then show how to construct a full-fledged credential system with multiple issuers in a modular way. We expect this definitional approach to be of independent interest.

Keywords: Structure preserving signatures · Vector commitments · Anonymous credentials · Universal composability · Groth-Sahai proofs

1 Introduction

Digital signature schemes are a fundamental cryptographic primitive. Besides their use for signing digital items, they are used as building blocks in more complex cryptographic schemes such as blind signatures [6,42], group signatures [15,52], direct anonymous attestation [20], electronic cash [40], voting schemes [48], adaptive oblivious transfer [23,32], and anonymous credentials [12].

© International Association for Cryptologic Research 2015
T. Iwata and J.H. Cheon (Eds.): ASIACRYPT 2015, Part II, LNCS 9453, pp. 262–288, 2015.
DOI: 10.1007/978-3-662-48800-3_11

For protocols constructed like this to be efficient, special properties are demanded from a signature scheme, in particular when the protocol needs to achieve strong privacy properties. The most important such properties seem to be that the issuance of a signature and its later use in a protocol is *unlinkable* as well as that the scheme is able to sign *multiple messages* (without employing a hash function). The first signature scheme that met these requirements is a blind signature scheme by Brands [18]. The drawback of blind signatures, however, is that when using the signature later in a higher-level protocol it must typically be revealed so that a third party can be convinced of its validity. Thus, a signature can be used only once, which turns out to be quite limiting for applications such as group signatures, multi-show anonymous credentials, and compact e-cash [25].

Camenisch and Lysyanskaya [30, 31] were the first to design signature schemes (CL-signatures) overcoming these drawbacks. Their schemes are secure under the Strong RSA, the q-SDH, or the LRSW assumption, respectively, and allow for an alternative approach when using a signature in a protocol: instead of revealing the signature to a party, the user employs zero-knowledge proofs to convince the party that she possesses a valid signature. While in theory this is possible for any signature scheme, CL-signatures were the first that enabled efficient proofs using generalized Schnorr proofs of knowledge. This is due to the algebraic properties of CL-signatures, i.e., no hash function is applied to the message and the signature and message values are either exponents or group elements.

Since the introduction of CL-signatures, the area of privacy preserving protocols flourished considerably and numerous new protocols based on them have been proposed. This has also made it apparent, however, that CL-signatures still have a number of drawbacks:

1. *Random oracles.* To make generalized Schnorr proofs of knowledge non-interactive (which is often required), one needs to resort to the Fiat-Shamir heuristic, i.e., to the random oracle model, and thus looses all provable security guarantees when the oracle is instantiated by a hash function [36].
2. *Straight-line extraction.* When designing a protocol to be secure in the UC model [35], rewinding can not be used to prove security. As a result, witnesses in generalized Schnorr proofs of knowledge need to be encrypted under a public key encoded in the common reference string (CRS). As the witnesses (messages signed with CL-signatures) are discrete logarithms, this is rather expensive [26] and may render the overall protocol impractical.
3. *Linear size.* When proving ownership of a CL-signature on many messages, all of them are needed for the verification of the signature and therefore a proof of possession of a signature will be linear in the number of messages.

A promising ingredient to overcome these drawbacks is the work by Groth and Sahai [45], who for the first time constructed efficient non-interactive zero-knowledge proofs (NIZK) without using random oracles, albeit for a limited set of languages. Indeed, the set of languages covered by these so-called GS-proofs does not include the ones covered by generalized Schnorr protocols and therefore many authors started to look for a compatible CL-signatures replacement, i.e., structure-preserving signature schemes [1–3]. Together with GS-proofs, these new schemes can also be used as signatures of knowledge [39] and thus are applicable in scenarios previously addressed with CL-signatures.

However, structure-preserving signatures still suffer in terms of performance when signing multiple messages (cf. drawback (3)), which is a typical requirement in applications such as anonymous credentials. Indeed, as for CL-signatures, the size of proofs with structure-preserving signatures grows linearly with the number of messages. As the constant factor for GS-proofs is larger than for generalized Schnorr proofs, structure-preserving signatures loose their attractiveness as a building block for such applications.

Our Contribution. In this paper, our goal is to address the three drawbacks of CL-signatures discussed above. To this end, we propose a new type of signature scheme, *unlinkable redactable signatures* (URS), in which one can redact message-signature pairs and reveal only their relevant parts each time they are used. Moreover, signatures in URS are unlinkable and the same message-signature pair can be redacted and revealed multiple times without being linked back to its origin. The real-world efficiency of URS is comparable to that of CL-signatures when a single message is signed and becomes superior when the number of messages signed grows. We view our contribution as threefold: First, in Sect. 2, we formally define URS. We present property-based security definitions for *unlinkability* and *unforgeability* and also a UC functionality for URS. Comparing the two definitions we find the seemingly common phenomenon that the functionality-based definition requires a key-registration process (allowing for the extraction of keys in the proof) while the property based definition per se does not require that. We validate our definitions by showing that an URS scheme satisfying strengthened property-based security definitions with key extraction securely implements our UC functionality.

Second, in Sect. 3, we construct an efficient URS scheme from vector commitments [37,51,56], structure-preserving signatures [2,3], and (a minimal dose of) non-interactive proofs of knowledge (NIPoK), which in practice can be instantiated by *witness-indistinguishable* Groth-Sahai proofs [45]. As we are interested in practical efficiency, we instantiate our scheme with concrete building block that deliberately rely on stronger assumptions (see Sect. 4.3). However, if one is willing to accept a less efficient scheme, a CDH-based vector commitment scheme [37] secure under less strong assumptions. We show how to make use of algebraic properties in our building blocks to minimize the witness size of the NIPoK.

Third, in Sect. 4, to demonstrate the versatility of our URS scheme as a CL-signature scheme 'replacement', we employ it to design the first efficient universally composable anonymous credential system that supports multiple issuers, pseudonyms, and selective disclosure of attributes.

Anonymous credential systems usually need to support an ecosystem of different features. Therefore, a single ideal functionality providing all the features such as pseudonyms, selective attribute disclosure, predicates over attributes, revocation, inspection, etc. would be very complex and hard to both create and use in a modular way—not to mention credible security proofs.

Nevertheless, ideal functionalities are very attractive for modeling the complexity of anonymous credential schemes. Indeed an early seminal paper [29] attempted exactly this, but was foiled by drawback (2)—as well as by the immaturity of the UC framework at the time. To overcome this complexity, we present a flexible and modular approach for constructing UC-secure anonymous credentials. Namely, we design a core building block for a single issuer that supports credential issuance and presentation with respect to pseudonyms. We then show how to compose multiple such blocks to construct in a modular way a full-fledged credential system with multiple issuers.

Besides being composable, our system is also arguably one of the first schemes to support efficient non-interactive attribute disclosure with cost independent of the number of attributes issued without having to rely on random oracles. Even in the random oracle model this has been an elusive goal. Therefore, because of the composability and efficient selective disclose, our scheme is very attractive and quickly surpasses schemes based on blind signatures and CL-signatures [19,31] when the number of attributes grows.

Related Work. We compare our signatures and credential schemes with other related work, a full comparison is deferred to the full paper [9]. As there are a multitude of papers on redactable, quotable, and sanitizable signatures [7,21,46, 58], we focus on the most influential definitional work and the most promising approaches in terms of efficiency.

A variety of signature schemes with flexible signing capabilities and strong privacy properties have been proposed [7,8,10,14,17,34,38]. While these works provide a fresh definitional approach, their schemes are very inefficient, especially when redacting a message vector with a large number of attributes. Some schemes built on vector commitments [51,55] achieve better efficiency but only consider one-time-show credentials, and while the scheme in [51] is not defined formally, the scheme in [55] involves interaction.

The first efficient multi-show anonymous credential scheme is [29]. It was extended with efficient attribute disclosure [24] and had real-world exposure [20,33]. It can, however, only be non-interactive in the random oracle model. Non-interactive credentials in the standard model have been built from P-signatures [12,13]. An instantiation of our URS scheme, however, is almost twice more efficient than [12] despite the fact that the latter does not support signing multiple messages. Belenkiy et al. [11] show how to use the randomizability of P-signatures for delegation and Chow et al. [41] extend the randomizable group signatures scheme underlying [11] with a flexible attribute mechanism. Izabachène et al. [50] extends the work of [12] with vector commitments; their scheme is, however, not secure under our definitions. In independent work, Hanser and Slamanig [47] present a credential system with efficient (independent of the number of attributes) attribute disclosure. However, their system is only secure in the generic group model [43]. Furthermore, it uses hash function to encode attributes and thus does not enable efficient protocol design. None of these schemes is (universally) composable. Camenisch et al. [27] have recently proposed property-based definitions of anonymous credentials and of the necessary building blocks, given a construction and proved it secure. Their definitions turn out to be rather complex, indicating that for complex systems functionality-based definitions might be easier to handle. Indeed, for their definition of privacy, Camenisch et al. make use of what they call 'filter' which is very reminiscent of an ideal functionality. Finally, the construction they provide is based on CL-signatures and thus suffers from the drawbacks of that approach.

An important factor that is often neglected is the compatibility of schemes with zero-knowledge proofs to enable efficient protocol design. Because of its compatibility with Groth-Sahai proofs, efficiency and composability, immediate further applications of our URS scheme include efficient e-cash, credential-based key exchange, e-voting, auditing, and others.

2   Definitions of Unlinkable Redactable Signatures

Redactable signatures are an instance of homomorphic [7] or controlled-malleable signatures [38]. For our credentials application the most useful redaction operation is to selectively white-list or quote a subset of messages and their positions from a message vector of length n ([7] consider the quoting of sub-sentences). We denote the message space of all valid message vectors as $\mathcal{M}$. We also refer to the redacted message as a quote of the original message. To distinguish the original vector from the quote of all messages we denote the original vector as $\boldsymbol{m} = (1, m_1, \ldots, m_n)$ and a quote as $\boldsymbol{m}_I = (2, m'_1, \ldots, m'_n)$. We represent each valid quoting transformation by a set $I \subseteq [1, n]$ of message positions and denote quoting either by $I(\boldsymbol{m})$ or $\boldsymbol{m}_I$. We denote the ith message element either by $\boldsymbol{m}[i]$ or m_i. A quote $\boldsymbol{m}_I$ from $\boldsymbol{m}$ is of the form

$$\boldsymbol{m}_I[i] = m'_i = \begin{cases} m_i & i \in I \\ \bot & \text{otherwise} \end{cases}.$$

Note that the message itself already reveals whether it is a quote. Chase et al. [38] call such a scheme tiered and we refer to the vectors $\boldsymbol{m}$ and $\boldsymbol{m}_I$ as Tier 1 and Tier 2 vectors respectively. The vector $\boldsymbol{m}_I$ can be sparse and can have a much shorter encoding than $\boldsymbol{m}$. Finally, we define $\mathsf{Zero}(\boldsymbol{m}, I) = (1, \tilde{m}_1, \ldots, \tilde{m}_n)$, with $\tilde{m}_j = m_j$ for $j \in I$ and $\tilde{m}_j = 0$ for $j \notin I$. This should not be confused with the operator I that outputs a Tier 2 message.

2.1   Property-Based Definitions for Unlinkable Redactable Signatures

One can define the security of redactable signatures by instantiating controlled-malleable signature definitions for simulatability, simulation unforgeability, and simulation context-hiding of Chase et al. [38] with the quoting transformation class $\mathcal{T} = \{I(\cdot) | I \subseteq [1..n]\}$ above. We prefer, however, to give our own unforgeability and unlinkability definitions that are more specific and do not rely on simulation and extraction. This makes them simpler and easier to prove, and thus more efficiently realizable. Together with key extractability they are nevertheless sufficient to realize the strong guarantees of our UC functionality.

Definition 1 (Unlinkable Redactable Signatures). *An unlinkable redactable signature scheme* URS *consists of the following algorithms:*

URS.SGen(1^κ) → *SP*. SGen *takes the security parameter* 1^κ *as input and outputs the system parameters SP.*

URS.Kg(SP) → (pk, sk). Kg *takes the system parameters SP as an input and outputs public verification and private signing keys* (pk, sk). *The verification key pk defines the message space* $\mathcal{M}$.

URS.Sign($sk, \boldsymbol{m}$) → σ. Sign *takes the signing key sk and a message* $\boldsymbol{m} \in \mathcal{M}$ *as input and produces a signature* σ.

URS.Derive($pk, I, \boldsymbol{m}, \sigma$) → σ_I. Derive *takes the public key pk, a selection vector I, a message* $\boldsymbol{m}$ *and a signature* σ *(both of Tier 1) as input. It produces a Tier 2 signature* σ_I *for* $\boldsymbol{m}_I$.

URS.Verify($pk, \sigma, \boldsymbol{m}$) → 0/1. Verify *takes the verification key pk, a signature* σ, *and a message* $\boldsymbol{m}$ *of Tier 1 or Tier 2 as input and checks the signature.*

We omit the URS qualifier when it is clear from the context.

Correctness. Informally, correctness requires that for honestly generated keys, both honestly generated and honestly derived signatures must always verify.

Unforgeability. Unforgeability captures the requirement that an attacker, who is given Tier 1 and Tier 2 signatures on messages of his choice, should not be able to produce a signature on a message that is not derivable from the set of signed messages in his possession. More formally:

Definition 2 (Unforgeability). *Let* H *output unique handles, for instance implemented using a counter. For a redactable signature scheme* URS.{SGen, Kg, Sign, Derive, Verify}, *tables* Q_1, Q_2, Q_3, *and an adversary* $\mathcal{A}$, *consider the following game:*

- *Step 1. SP* ← SGen(1^k); (pk, sk) $\xleftarrow{\$}$ Kg(SP); Q_1, Q_2, Q_3 ← $\emptyset$.
- *Step 2.* ($\boldsymbol{m}_I^*, \sigma^*$) $\xleftarrow{\$}$ $\mathcal{A}^{\mathcal{O}_\text{Sign}(\cdot), \mathcal{O}_\text{Derive}(\cdot, \cdot), \mathcal{O}_\text{Reveal}(\cdot)}(pk)$, *where* $\mathcal{O}_\text{Sign}$, $\mathcal{O}_\text{Derive}$, *and* $\mathcal{O}_\text{Reveal}$ *behave as follows:*

$\mathcal{O}_\text{Sign}(\boldsymbol{m})$	$\mathcal{O}_\text{Derive}(h, I)$	$\mathcal{O}_\text{Reveal}(h)$
$h \leftarrow$ H; $\sigma \leftarrow$ Sign($sk, \boldsymbol{m}$)	*if* ($h, \boldsymbol{m}, \sigma$) ∈ Q_1	*if* ($h, \boldsymbol{m}, \sigma$) ∈ Q_1
add ($h, \boldsymbol{m}, \sigma$) *to* Q_1	$\sigma' \leftarrow$ Derive($pk, I, \boldsymbol{m}, \sigma$)	*add* $\boldsymbol{m}$ *to* Q_3
return h	*add* $\boldsymbol{m}_I$ *to* Q_2; *return* σ'	*return* σ

A signature scheme URS *satisfies* unforgeability *if for all such PPT algorithms* $\mathcal{A}$ *there exists a negligible function* $\nu(\cdot)$ *such that in the above game the probability (over the random choices of* Kg, Sign, Derive *and* $\mathcal{A}$*) that* Verify($pk, \sigma^*, \boldsymbol{m}_I^*$) = 1 *and* $\forall \boldsymbol{m} \in Q_3$, $\boldsymbol{m}_I^* \neq \boldsymbol{m}_I$, *and* $\boldsymbol{m}_I^* \notin Q_2$ *is less than* $\nu(\kappa)$.

Note that we do not consider a Tier 1 signature itself as a forgery. However, if the adversary manages to produce a valid Tier 1 signature on a message $\boldsymbol{m}$ without calling Sign($\boldsymbol{m}$) and either Reveal(h) or Derive(h, I) on all subsets $I \subseteq [1..n]$ for the corresponding handle h, he can use this Tier 1 signature to break unforgeability by deriving a Tier 2 signature from it.

Unlinkability. Informally, unlinkability ensures that an adversarial signer cannot distinguish which of two Tier 1 signatures of his choosing was used to derive a Tier 2 signature. More formally:

Definition 3 (Unlinkability). *For the signature scheme* URS.{SGen, Kg, Sign, Derive, Verify} *and a stateful adversary* $\mathcal{A}$, *consider the following game:*

- *Step 1.* $SP \leftarrow \mathsf{SGen}(1^k)$.
- *Step 2.* $(pk, I, \boldsymbol{m}^{(0)}, \boldsymbol{m}^{(1)}, \sigma^{(0)}, \sigma^{(1)}) \xleftarrow{\$} \mathcal{A}(SP)$, *where* $\boldsymbol{m}_I^{(0)} = \boldsymbol{m}_I^{(1)}$, $\mathsf{Verify}(pk, \sigma^{(0)}, \boldsymbol{m}^{(0)}) = 1$, *and* $\mathsf{Verify}(pk, \sigma^{(1)}, \boldsymbol{m}^{(1)}) = 1$.
- *Step 3.* *Pick* $b \leftarrow \{0, 1\}$ *and form* $\sigma_I^{(b)} \xleftarrow{\$} \mathsf{Derive}(pk, I, \boldsymbol{m}^{(b)}, \sigma^{(b)})$.
- *Step 4.* $b' \xleftarrow{\$} \mathcal{A}(\sigma_I^{(b)})$.

The signature scheme URS *is unlinkable if for any polynomial time adversary* $\mathcal{A}$ *there exists a negligible function* $\nu(\cdot)$ *such that* $\Pr[b = b'] < \frac{1 + \nu(\kappa)}{2}$.

Note that this definition is very strong, as the adversary can even pick pk.

2.2   Ideal Functionality for Unlinkable Redactable Signatures

We now give an alternative characterization of unlinkable redactable signatures using an ideal functionality $\mathcal{F}_{\mathsf{URS}}$ defined as follows:

Functionality $\mathcal{F}_{\mathsf{URS}}$

The functionality maintains tables $\mathcal{K}$ and $\mathcal{Q}$ initialized to $\emptyset$ and flags kg and $keyleak$ which are initially unset.

- On input (**keygen**, sid) from S, verify that $sid = (S, sid')$ for some sid' and that flag kg is unset. If not, then return $\bot$. Else, send (**initF**, sid) to $\mathcal{SIM}$ and wait for a message (**initF**, sid, SP, Kg, Sign, Derive, Verify) from $\mathcal{SIM}$, where Kg, Sign, and Derive are PPT algorithms and Verify is a deterministic algorithm. Then, store SP, Kg, Sign, Derive, and Verify, run $(pk, sk) \leftarrow \mathsf{Kg}(SP)$, set flag kg, store (pk, sk) in $\mathcal{K}$, and return (**verificationKey**, sid, pk) to S.
- On input (**checkPK**, sid, pk') from some party P, verify that the flag kg is set. Check whether $pk' = pk$ or whether (pk', sk') for some sk' was stored in $\mathcal{K}$. In this case, return (**checkedPK**, sid, $true$). Else, if $(pk', \bot)$ was stored in $\mathcal{K}$ return (**checkedPK**, sid, $false$). Else, send (**checkPK**, sid, pk') to $\mathcal{SIM}$, wait for (**checkedPK**, sid, sk') from $\mathcal{SIM}$, add (pk', sk') to $\mathcal{K}$. If $sk' \neq \bot$, return (**checkedPK**, sid, $true$) to P. Otherwise, return (**checkedPK**, sid, $false$) to P.
- On input (**leakSK**, sid) from S verify that $sid = (S, sid')$ for some sid'. If not, return $\bot$. Else, if flag kg is set, set flag $keyleak$ and return (**leakSK**, sid, sk), otherwise - abort.
- On input (**sign**, sid, m) from S, verify that $sid = (S, sid')$ for some sid' and that the flag kg is set. If not, return $\bot$. Else, run $\sigma \leftarrow \mathsf{Sign}(sk, m)$ and $\mathsf{Verify}(pk, \sigma, m)$. If Verify is successful, return (**signature**, sid, m, σ) to S and add m to $\mathcal{Q}$, otherwise return $\bot$.   (Continue on the next page.)

- On input (derive, $sid, pk', I, \boldsymbol{m}, \sigma$) from some party P, run Derive($pk', I, \boldsymbol{m}, \sigma$) and if it fails, return $\perp$. Otherwise, if the flag kg is set and $pk = pk'$ then set $sk_{tmp} = sk$. If there is an entry $(pk', sk') \in \mathcal{K}$ recorded, set $sk_{tmp} = sk'$. If sk_{tmp} was set run $\sigma' \leftarrow$ Sign(sk_{tmp}, Zero($\boldsymbol{m}, I$)) and return Derive($pk', I,$ Zero($\boldsymbol{m}, I$), σ'). Otherwise, return the output of Derive($pk', I, \boldsymbol{m}, \sigma$).
- On input (verify, $sid, pk', \sigma, \boldsymbol{m}_I$) from some party P, compute $result \leftarrow$ Verify($pk', \sigma, \boldsymbol{m}_I$). If the flag kg is set, $pk' = pk$, flag $keyleak$ is not set, and $\not\exists \boldsymbol{m} \in \mathcal{Q}$ such that $\boldsymbol{m}_I = I(\boldsymbol{m})$, then output (verified, $sid, \boldsymbol{m}_I, 0$). Otherwise, output (verified, $sid, \boldsymbol{m}_I, result$).

We point out some aspects of the ideal functionality. The functionality needs to output concrete values as signatures of messages and redacted signatures, as well as key material. To generate and verify these values, $\mathcal{F}_{\mathsf{URS}}$ requires the adversary/simulator $\mathcal{SIM}$ to provide it with a number of polynomial-time algorithms. This is in line with how ideal functionalities for signatures, and in particular blind signatures, have been defined before [6,35,42,49,53]. We consider static corruptions of protocol machines, but allow the simulator to request the signing key at any time by sending the leakSK message. This allows us to ensure that the privacy properties for users are still enforced even if the signer leaks its secret key. The functional and security properties are enforced by the functionality no matter how these (adversarial) algorithms compute the values. Unforgeability is enforced by the fact that $\mathcal{F}_{\mathsf{URS}}$ will output false (0) for verification queries for which the message (or a corresponding original message) has not been signed, provided that the signer is not corrupted and the signing key not leaked. (If the signer is corrupted statically, (keygen, sid) will not be sent and hence kg not set and unforgeability not enforced.) Unlinkability of redacted signature is enforced by $\mathcal{F}_{\mathsf{URS}}$ as follows. It generates a fresh redacted signature only from those parts of the original message that are quoted, i.e., the hidden message parts are set to zero, and thus redacted signatures from $\mathcal{F}_{\mathsf{URS}}$ do not contain any information about the hidden parts of messages. More precisely, this is enforced for the keys generated by $\mathcal{F}_{\mathsf{URS}}$ and for any keys that an honest party successfully checked before generating a redacted signature. Unless mentioned otherwise, the reply of the functionality upon a failed check or verification is $\perp$.

2.3 Key Registration and UC Realizability

We now want to construct a protocol $\mathcal{R}_{\mathsf{URS}}$ that realizes $\mathcal{F}_{\mathsf{URS}}$ using a URS scheme in the $\mathcal{F}_{\mathsf{CRS}}$-hybrid model where SP is the reference string and each call to $\mathcal{F}_{\mathsf{URS}}$ is essentially replaced by running one of the algorithms of URS. While this can be done (the detailed description of $\mathcal{R}_{\mathsf{URS}}$ is given in the full version [9]), there are a number of hurdles that need to be overcome. These hurdles are quite typical and include, e.g., that we need to be able to extract the secret keys from the adversary to be able to simulate properly. They are, however, often treated only informally in security proofs. Here we want to make them explicit and treat them formally correct. So our goal is to prove a statement (Theorem) of the form:

> If URS *is correct, unforgeable, unlinkable, and* X *then* $\mathcal{R}_{\mathsf{URS}}$ *securely realizes* $\mathcal{F}_{\mathsf{URS}}$ *in the* $\mathcal{F}_{\mathsf{CRS}}$-*hybrid model.*

What do we have to require from X to make this theorem true? To prove the theorem we have to show indistinguishability between the ideal world and the real world. In the ideal world, an environment $\mathcal{Z}$ interacts with the simulator $\mathcal{SIM}$ and functionality $\mathcal{F}_{\mathsf{URS}}$. In the real world, the environment $\mathcal{Z}$ interacts with the real adversary $\mathcal{A}$ and the protocol $\mathcal{R}_{\mathsf{URS}}$.

We provide a tentative description of $\mathcal{SIM}$ in the ideal world: when receiving the $(\texttt{initF}, sid)$ message from $\mathcal{F}_{\mathsf{URS}}$, it generates a trapdoor td (in addition to SP) and returns $(\texttt{initF}, sid, SP, \mathsf{Kg}, \mathsf{Sign}, \mathsf{Derive}, \mathsf{Verify})$. On receiving the $(\texttt{checkPK}, sid, pk)$ message, is uses the trapdoor to extract the secret key sk and returns sk to $\mathcal{F}_{\mathsf{URS}}$.

To make this work, we extend URS with several algorithms: CheckPK is run by $\mathcal{R}_{\mathsf{URS}}$ on receiving a message $(\texttt{checkPK}, sid, pk)$. SGenT and ExtractKey are the trapdoored parameter generation and key extraction algorithm for $\mathcal{SIM}$. CheckKeys is used to define what it means to extract a valid key.

URS.CheckPK(pk) $\rightarrow$ 0/1. CheckPK *is a deterministic algorithm that takes a public key pk as an input and checks that it is correctly formed. It outputs 1 if pk is correct, and 0 otherwise.*

URS.SGenT(1^κ) $\rightarrow$ (SP, td). SGenT *is a system parameters generation algorithm that takes the security parameter 1^κ as input and outputs the system parameters SP and a trapdoor td for the key extraction algorithm.*

URS.ExtractKey(pk, td) $\rightarrow$ sk. ExtractKey *is an algorithm that takes a public key pk and a trapdoor td as input. It extracts the corresponding secret key sk.*

URS.CheckKeys(pk, sk) $\rightarrow$ 0/1. CheckKeys *is an algorithm that takes a public pk and a private sk keys and checks if they constitute a valid signing key pair. It outputs 1 if they do, and 0 otherwise.*

Strengthened Correctness requires that honestly generated keys, but also keys for which predicate CheckKeys(pk, sk) holds can be used to create signatures that will verify. It moreover guarantees that CheckPK(pk) holds for honestly generated public keys.

Parameter Indistinguishability. Informally, parameter indistinguishability ensures that the SP produced by SGenT and SGen are computationally indistinguishable. It is formally defined as follows:

Definition 4 (Parameter Indistinguishability). *A redactable signature scheme* URS.$\{\mathsf{SGen}, \mathsf{Kg}, \mathsf{Sign}, \mathsf{Derive}, \mathsf{Verify}\}$ *with alternative parameter generation* SGenT *is parameter indistinguishable if for any polynomial time adversary $\mathcal{A}$ there exists a negligible function $\nu(\cdot)$ such that* $\Pr[(SP_0, td) \leftarrow \mathsf{SGenT}(1^k); SP_1 \leftarrow \mathsf{SGen}(1^k); b \leftarrow \{0,1\}; b' \leftarrow \mathcal{A}(SP_b) : b = b'] < \frac{1+\nu(\kappa)}{2}$.

Key Extractability. Informally, the key extractability ensures that if SGenT was run and if CheckPK outputs 1, then the extraction algorithm ExtractKey(pk, td) will output a valid secret key sk, i.e. CheckKeys(pk, sk) = 1.

Definition 5 (Key Extractability). *A redactable signature scheme* URS. {SGen, Kg, Sign, Derive, Verify} *with additional algorithms* (CheckPK, SGenT, CheckKeys, ExtractKey) *is key extractable if* CheckKeys *is correct and for any polynomial time adversary* $\mathcal{A}$ *there exists a negligible function* $\nu(\cdot)$ *such that* $\Pr[(SP, td) \leftarrow \mathsf{SGenT}(1^k); pk \leftarrow \mathcal{A}(SP, td); sk \leftarrow \mathsf{ExtractKey}(pk, td) : (\mathsf{CheckPK}(pk) = 1 \wedge \mathsf{CheckKeys}(pk, sk) = 0))] < \nu(\kappa)$.

Composable Unlinkability holds even when parameters in the unlinkability game are generated using $(SP, td) \leftarrow \mathsf{SGenT}(1^k)$ and $\mathcal{A}$ is handed td. This allows for the use of the game in a hybrid argument when proving the security of the simulator. We note that in such an adapted unlinkability game the trapdoor td must only enable key-extraction, and crucially does not allow the adversary to extract a Tier 1 signature from a Tier 2 signature (this would break unlinkability). In our instantiation this is achieved by splitting SP into several parts. The trapdoor is only generated for the part used for key extraction.

UC Realization. We prove that if an unlinkable redactable signature URS is correct, parameter indistinguishable, key extractable, unforgeable, and unlinkable, then $\mathcal{R}_{\mathsf{URS}}$ securely realizes $\mathcal{F}_{\mathsf{URS}}$. More formally, we have the following theorem (which is proven in the full version of this paper [9]).

Theorem 1. *Let* URS *be an unlinkable redactable signature scheme. If* URS *is correct, parameter indistinguishable, key-extractable, unforgeable, and composable unlinkable then* $\mathcal{R}_{\mathsf{URS}}$ *securely realizes* $\mathcal{F}_{\mathsf{URS}}$ *in the* $\mathcal{F}_{\mathsf{CRS}}$*-hybrid model.*

3 The Construction of Our Redactable Signature Scheme

As a first step toward our full solution, we will construct an unforgeable and unlinkable URS scheme without key extraction. The scheme should be of independent interest, in case universal composability is not a design requirement. This isolation of key extraction, seemingly only needed for universal composition, is a nice feature of our definitions.

Let $\mathcal{G}$ be a bilinear group generator that takes as an input security parameter 1^κ and outputs the descriptions of multiplicative groups $grp = (p, \mathbb{G}, \tilde{\mathbb{G}}, \mathbb{G}_t, e, G, \tilde{G})$ where $\mathbb{G}$, $\tilde{\mathbb{G}}$, and $\mathbb{G}_t$ are groups of prime order p, e is an efficient, non-degenerating bilinear map $e : \mathbb{G} \times \tilde{\mathbb{G}} \rightarrow \mathbb{G}_t$, and G and $\tilde{G}$ are generators of the groups $\mathbb{G}$ and $\tilde{\mathbb{G}}$, respectively.

Our construction makes use of a structure preserving signature (SPS) scheme SPS.{Kg, Sign, Verify} and a vector commitment scheme VC.{Setup, Commit, Open, Check}. We recall that the structure-preserving property of the signature scheme requires that verification keys, messages, and signatures are group elements and the verification predicate is a conjunction of pairing product equations. The intuition behind our construction is susceptibly simple: Use SPS.Kg to generate a signing key pair and VC.Setup to add commitment parameters to the public key. To sign a vector m, first, commit to m and then sign the resulting commitment C. To derive a quote for a subset I of the messages, simply open the commitment to the messages in m_I. We verify a signature on a quote by verifying both the structure-preserving signature (SPS.Verify) and checking the opening of the commitment (VC.Check).

Such a scheme has, however, several shortcomings. First, it is linkable, as the same commitment is reused across multiple quotes of the same message. Even if both the underlying SPS scheme and the commitment scheme are individually re-randomizable, this seems hard to avoid as the unforgeability of the SPS scheme prevents randomization of the message. Second, such a construction is only heuristically secure. Existing vector commitments do not guarantee that multiple openings cannot be combined and mauled into an opening for a different sub-vector. We call vector commitment schemes that prevent this *opening nonmalleable*. (Recently, [47] constructed an SPS scheme allowing for randomization within an equivalence class. However, their commitments cannot be opened to arbitrary vectors of $\mathbb{Z}_p$ and are not provably opening non-malleable.)

Our main design goal is to address both of these weaknesses while avoiding a large performance overhead. Our main tool for this is an efficient non-interactive proof-of-knowledge. Intuitively, we hide the commitments and their openings, as well as a small part of the signature to achieve unlinkability. Hiding the commitment opening also helps solve the malleability problems for commitments. To achieve real-world efficiency we show how to exploit the re-randomizability of the SPS (and optionally the commitment scheme as described in the full version [9]).

Before describing our redactable signature scheme in more detail, we present a vector commitment scheme that uses a variant of polynomial commitments from [51]. While our changes are partly cosmetic, they simplify the assumption needed for opening non-malleability.

3.1   Vector Commitments Simplified

A vector of messages $m \in \mathbb{Z}_p^n$ is committed using a polynomial $f(x)$ that has a value $f(i) = m_i$ at the position i. In Lagrange form such a polynomial is a linear combination $f(x) = \sum_{i=1}^n m_i f_i(x)$ of Lagrange basis polynomials $f_i(x) = \prod_{j=0, j \neq i}^n \frac{x-j}{i-j}$. To batch-open a vector commitment for a position set $I \subseteq \{1, \ldots, n\}$, one uses a polynomial $f_I(x) = \sum_{i \in I} m_i f_i(x)$. For such a polynomial, it holds that $f_I(i) = m_i$ for $i \in I$; and $f_I(0) = 0$. (The additional root at 0 is necessary to achieve opening non-malleability). The reuse of the same Lagrange basis polynomials, which yields polynomials of not the lowest possible degree, reduces the number of variable bases in the equation of Check below and increases efficiency when used for the construction of bigger protocols such as anonymous credential. Also, note that $f(x) - f_I(x)$ is divisible by the polynomial $p_I(x) = x \cdot \prod_{i \in I}(x - i)$. We use the polynomial $p(x) = x \cdot \prod_{i=1}^n (x - i)$ which is divisible by $p_I(x)$ for any $I \subseteq \{1, \ldots, n\}$ to randomize commitments to make them perfectly hiding.

Construction. We reuse the notation of Sect. 2 and use Tier 1 vectors $\boldsymbol{m}$ for the vectors being committed and Tier 2 vectors $\boldsymbol{m}_I$ for batch openings at positions I. We also let $grp = (p, \mathbb{G}, \tilde{\mathbb{G}}, \mathbb{G}_t, e, G, \tilde{G})$ be bilinear map parameters generated by a bilinear group generator $\mathcal{G}(1^\kappa)$.

VC.Setup(grp). Pick $\alpha \leftarrow \mathbb{Z}_p$ and compute $(G_1, \tilde{G}_1, \ldots, G_{n+1}, \tilde{G}_{n+1})$, where $G_i = G^{(\alpha^i)}$ and $\tilde{G}_i = \tilde{G}^{(\alpha^i)}$. Output parameters $pp = (grp, G_1, \tilde{G}_1, \ldots, G_{n+1}, \tilde{G}_{n+1})$. Values $G_1, \ldots, G_{n+1}$ suffice to compute $G^{\phi(\alpha)}$ for any polynomial $\phi(x)$ of maximum degree $n + 1$ (and similarly for $\tilde{G}^{\phi(\alpha)}$).

Furthermore, for the above defined $f_i(x)$, $p(x)$, and $p_I(x)$, we implicitly define $F_i = G^{f_i(\alpha)}$, $P = G^{p(\alpha)}$, $P_I = G^{p_I(\alpha)}$, and $\tilde{P}_I = \tilde{G}^{p_I(\alpha)}$. These group elements can be computed from the parameters pp.

VC.Commit($pp, \boldsymbol{m}, r$). Output $C = \prod_{i=1}^{n} F_i^{m_i} P^r$.

VC.Open($pp, I, \boldsymbol{m}, r$). Let $w(x) = \frac{f(x) - f_I(x) + r \cdot p(x)}{p_I(x)}$ and compute the witness $W = G^{w(\alpha)}$ using parameters pp.

VC.Check($pp, C, \boldsymbol{m}_I, W$). Accept if $e(C, \tilde{G}) = e(W, \tilde{P}_I)e(\prod_{i \in I} F_i^{m_i}, \tilde{G})$.

Note that $p_I(x)$ always has the factor x. This is essential for achieving opening non-malleability. If $p_I(x)$ would be 1 for $I = \emptyset$, as in the original polynomial commitment scheme of [51], then C would be a valid batch opening witness for the empty set of messages.

Security Analysis. We require the commitment scheme to be *complete, batch binding*, and *opening non-malleable*. Completeness is standard for a commitment scheme follows easily from the following equation: $e(C, \tilde{G}) = e(G, \tilde{G})^{f(\alpha) + r \cdot p(\alpha)} = e(G, \tilde{P}_I)^{\frac{f(\alpha) - f_I(\alpha) + r \cdot p(\alpha)}{p_I(\alpha)}} e(G, \tilde{G})^{f_I(\alpha)} = e(W, \tilde{P}_I)e(\prod_{i \in I} F_i^{m_i}, \tilde{G})$.

Next, we define the batch binding and opening non-malleability properties:

Definition 6 (Batch Binding). *For a vector commitment scheme* VC.{Setup, Commit, Open, Check} *and an adversary* $\mathcal{A}$ *consider the following game:*

- *Step 1.* $grp \leftarrow \mathcal{G}(1^\kappa)$ *and* $pp \leftarrow$ VC.Setup(grp)
- *Step 2.* $C, \boldsymbol{m}_I, W, \boldsymbol{m}'_{I'}, W' \leftarrow \mathcal{A}(pp)$

Then, the commitment scheme satisfies batch binding *if for all such PPT algorithms* $\mathcal{A}$ *there exists a negligible function* $\nu(\cdot)$ *such that the probability (over the choices of* $\mathcal{G}$, Setup, *and* $\mathcal{A}$*) that* $1 = $ VC.Check($pp, C, \boldsymbol{m}_I, W$) $ = $ VC.Check ($pp, C, \boldsymbol{m}'_{I'}, W'$) *and that there exist* $i \in I \cap I'$ *such that* $m_i \neq m'_i$ *is at most* $\nu(\kappa)$. *(Note that* $\boldsymbol{m}_I$ *and* $\boldsymbol{m}'_{I'}$ *are Tier 2 vectors, and thus encode the sets* I *and* I' *respectively.)*

Definition 7 (Opening Non-malleability). *For a vector commitment scheme* VC.{Setup, Commit, Open, Check} *and an adversary* $\mathcal{A}$ *consider the following game:*

- *Step 1.* $grp \leftarrow \mathcal{G}(1^\kappa)$ *and* $pp \leftarrow$ VC.Setup(grp)
- *Step 2.* $\boldsymbol{m}, I \leftarrow \mathcal{A}(pp)$

- *Step 3. Pick random r, compute $C \leftarrow$ VC.Commit$(pp, \boldsymbol{m}, r)$, and $W \leftarrow$ VC.Open$(pp, I, \boldsymbol{m}, r)$.*
- *Step 4. $W', I' \leftarrow \mathcal{A}(C, W)$*

Then the commitment scheme satisfies opening non-malleability if for all such PPT algorithms $\mathcal{A}$ there exists a negligible function $\nu(\cdot)$ such that the probability (over the choices of $\mathcal{G}$, Setup, Commit, and $\mathcal{A}$ that $1 = $ VC.Check$(pp, C, \boldsymbol{m}_{I'}, W')$, and $I \neq I'$ is at most $\nu(\kappa)$.

In the following theorems we make use of the n-BSDH assumption [44] and the n-RootDH assumption that are defined next. See the full version of this paper [9] for its generic group model proof. (We note that this assumption is only required for opening non-malleability, which is ignored by most existing constructions of anonymous credentials from vector commitments.)

Definition 8 (n-SDH Assumption). *The n-strong Diffie-Hellman (n-SDH) assumption [16] states that there exists a $\mathcal{G}$ that for all algorithms $\mathcal{A}$, the following advantage*

$$\mathbf{Adv}_{\mathcal{G}}^{\mathrm{nSDH}}(\lambda) = \Pr\big[(p, e, \mathbb{G}, G) \xleftarrow{\$} \mathcal{G} ; \ x, c \xleftarrow{\$} \mathcal{Z}_p ;$$
$$\mathcal{A}(1^{\lambda}, p, \mathbb{G}, G, G^x, \dots, G^{x^n}) = (c, G^{1/(x+c)})\big] \leq \mathsf{negl}(\lambda).$$

The n-BSDH assumption is defined identically to n-SDH except that now $\mathcal{A}$ is challenged to compute $(c, e(G, \tilde{G}^{1/(x+c)})$. Note that the n-BSDH assumption is already implied by the n-SDH assumption.

Definition 9 (n-RootDH Assumption).
For an adversary $\mathcal{A}$ consider the following game:

- *Step 1. $grp \leftarrow \mathcal{G}(1^{\kappa})$*
- *Step 2. Pick random $\alpha, r \leftarrow \mathbb{Z}_p^*$, compute $X = (G^{\alpha \cdot \prod_{i=1}^{n}(\alpha - i)})^r$.*
- *Step 3. $(J, state) \leftarrow \mathcal{A}(G, \tilde{G}, \{G^{\alpha^i}, \tilde{G}^{\alpha^i}\}_{i=1}^{n+1}, X)$*
- *Step 4. Compute $Y = (G^{\prod_{i \in J}(\alpha - i)})^r$.*
- *Step 5. $J', Z \leftarrow \mathcal{A}(state, Y)$*

Then the group generator $\mathcal{G}$ satisfies the n-RootDH assumption if for all such PPT algorithms $\mathcal{A}$ there exists a negligible function $\nu(\cdot)$ such that the probability (over the choices of $\mathcal{G}$, α, r, and $\mathcal{A}$ that J and J' are subsets of $[1..n]$, $J' \neq J$, and $Z = (G^{\prod_{i \in J'}(\alpha - i)})^r$ is at most a negligible function $\nu(\kappa)$.

Theorem 2. *The commitment scheme* VC *defined above is batch binding under the $(n + 1)$-BSDH assumption.* The proof is similar to that of [51] and can be found in the full version [9].

Theorem 3. *The commitment scheme* VC *defined above is opening non-malleable under the n-RootDH assumption.* The proofs can be found in the full version [9].

3.2 Non-interactive Zero-Knowledge and Witness Indistinguishable Proof Systems

Let R be an efficiently computable binary relation. For pairs $(W, Stmt) \in R$ we call $Stmt$ the statement and W the witness. Let $\mathcal{L}$ be the language consisting of statements in R. A non-interactive zero-knowledge (NIZK) proof-of-knowledge system for a language $\mathcal{L}$ consists of the following algorithms and protocols:

Π.Setup$(grp) \to CRS$. On input $grp \leftarrow (1^\kappa)$, it outputs common parameters (a common reference string) CRS for the proof system.

Π.Prove$(CRS, W, Stmt) \to \pi$. On input a statement $Stmt$ and a witness W, it generates a zero-knowledge proof π that the witness satisfies the statement.

Π.Verify$(CRS, \pi, Stmt) \to 0/1$. On input $Stmt$ and π, it outputs 1 if π is valid, and 0 otherwise.

We explain the notation for the statements $Stmt$. We call extractable (f-extractable [12]) witnesses that can be (only partially) extracted from the corresponding proof, respectively. To express the "extractability" property of the witnesses we use notation introduced by Camenisch et al. [28]. For the extractable witnesses we use the "knowledge" notation ($\maltese$), and for the f-extractable witnesses we use "existence" ($\exists$) notation. (If function f is constant, nothing can be extracted.) We define $\mathfrak{K}$ as a set of extractable witnesses and $\mathfrak{E}$ as a set of the witnesses that we can only prove existence about. We only consider proofs for multi-exponentiations (for existence) and pairing products (for existence and knowledge) equations. The following is an examplary statement:

$$Stmt = \maltese \left\{ Y_i, \tilde{Y}_i \in \mathfrak{K} \right\}_{i=1}^n \; ; \exists \left\{ x_j \in \mathfrak{E} \right\}_{j=1}^m : z = \prod_{j=1}^m G^{x_j}$$

$$\wedge \, e(G, \tilde{G}) = \prod_{i=1}^n \left(e(Y_i, \tilde{B}_i) \cdot e(A_i, \tilde{Y}_i) \right).$$

For simplicity of presentation, we do not explicitly specify public values of a statement as additional input to the algorithms, since they are clear from the description of the statement and the list of witnesses.

We employ different proof systems that are either witness indistinguishable or zero-knowledge in terms of privacy, and either extractable or simulation-extractable in term of soundness. For the security proofs we introduce the following algorithms:

Π.ExtSetup$(grp) \to (CRS, td_{ext})$. On input grp, it outputs a common reference string CRS and a trapdoor td_{ext} for extraction of valid witnesses from valid proofs. This is for witness-indistinguishable extractable proofs.

Π.SimSetup$(grp) \to (CRS, td_{ext}, td_{sim})$ It outputs a CRS and the extraction and simulation trapdoors. This is for proofs that are also zero-knowledge.

Π.SimProve$(CRS, td_{sim}, Stmt) \to \pi$. On input CRS and a trapdoor td_{sim}, it outputs a simulated proof π such that Π.Verify$(CRS, \pi, Stmt) = 1$.

Π.Extract(CRS, td_{ext}, π, $Stmt$) → W. On input a proof π and a trapdoor td_{ext}, it extracts a witness W that satisfies the statement $Stmt$ of the proof π.

For simulation-extractable NIZK proofs (that are non-malleable) we also allow an additional public input to the Prove, Extract, SimProve, and Verify algorithms – a message (label) L, which is non-malleably attached to the proof (i.e. the signature of knowledge is computed on this message). We provide a formal definition below.

Definition 10 (Simulation Extractability). *A proof system Π is called simulation extractable with labels if for any PPT adversary $\mathcal{A}$ and security parameter λ there exists a negligible function* $\mathsf{negl}(\cdot)$ *such that:*

$$\Pr[(CRS, td_{\text{sim}}, td_{\text{ext}}) \xleftarrow{\$} \mathsf{SimSetup}(1^{\lambda}); (Stmt^*, L^*, \pi) \leftarrow \mathcal{A}^{\mathcal{O}_{Sim}(td_{\text{sim}}, \cdot, \cdot)}(CRS);$$
$$W \leftarrow \mathsf{Extract}(CRS, td_{\text{ext}}, \pi, Stmt^*, L^*) : \mathsf{Verify}(CRS, \pi, Stmt^*, L^*) = 1 \wedge$$
$$(W, Stmt^*) \notin R \wedge \mathcal{O}_{Sim} \text{ was never queried with } (Stmt^*, L^*)] \leq \mathsf{negl}(\lambda).$$

3.3 Our Redactable Signature Scheme

We construct our redactable signature scheme URS from a structure-preserving signature scheme SPS, a vector commitment scheme VC, and an extractable and witness-indistinguishable non-interactive proof-of-knowledge system Π described in the previous section. Some SPS and vector commitment schemes might also support randomization; we already discussed such a property for vector commitments in the last sub-section; for signatures we refer the reader to [2,3]. We denote the randomization algorithm of signatures by SPS.Rand. We denote the randomizable elements of a SPS signature Σ by $\psi_{\text{rnd}}(\Sigma)$ and the other elements by $\psi_{\text{wit}}(\Sigma)$. (For a non-randomizable SPS signature $\psi_{\text{wit}}(\Sigma) = \Sigma$.)

Our construction does not utilize any randomizability in the vector commitment scheme itself. In the full version [9] we analyze batch-binding and opening non-malleability in presence of such a randomization algorithm.

Construction.

URS.SGen(1^{κ}). Compute $grp \leftarrow \mathcal{G}(1^{\kappa})$, $pp \leftarrow$ VC.Setup(grp), $CRS \leftarrow \Pi$.Setup (grp), output $SP = (grp, pp, CRS)$.

URS.Kg(SP). Obtain grp from SP, generate $(pk_{sps}, sk_{sps}) \leftarrow$ SPS.Kg(grp), output $pk = (pk_{sps}, SP)$ and $sk = (sk_{sps}, pk)$.

URS.Sign($sk, \boldsymbol{m}$). Pick $r \leftarrow \mathbb{Z}_p$, compute $C =$ VC.Commit($pp, \boldsymbol{m}, r$) and $\Sigma \leftarrow$ SPS.Sign(sk_{sps}, C), and return $\sigma = (\Sigma, C, r)$.

URS.Derive($pk, I, \boldsymbol{m}, \sigma$). First, compute $W =$ VC.Open($pp, I, \boldsymbol{m}, r$). Then, if a SPS.Rand algorithm is present, randomize the signature as $\Sigma' \leftarrow$ SPS.Rand (pk_{sps}, Σ); otherwise, set $\Sigma' \leftarrow \Sigma$. And compute the proof $\pi \leftarrow \Pi$.Prove ($CRS; C, W, \psi_{\text{wit}}(\Sigma'); \nexists C, W, \psi_{\text{wit}}(\Sigma') :$ SPS.Verify(pk_{sps}, Σ', C) $\wedge$ VC.Check ($pp, C, \boldsymbol{m}_I, W$)). Return $\sigma = (\psi_{\text{rnd}}(\Sigma'), \pi)$ as the signature on $\boldsymbol{m}_I$.

URS.Verify($pk, \sigma, \boldsymbol{m}_I$). Check that Π.Verify($CRS; \pi; \nexists C, W, \psi_{\text{wit}}(\Sigma') :$ SPS. Verify(pk_{sps}, Σ', C)) = VC.Check($pp, C, \boldsymbol{m}_I, W$) = 1.

Theorem 4. URS *is an unforgeable redactable signature scheme, if the SPS scheme is unforgeable, the vector commitment scheme satisfies the batch binding and opening non-malleability property, and the proof-of-knowledge system is extractable and witness indistinguishable.* The proofs of Theorems 4 is provided in the full version [9].

Theorem 5. URS *is an unlinkable redactable signature scheme if the proof-of-knowledge system is witness indistinguishable.* The proofs are given in the full version of this paper [9].

Strengthened Scheme for an Universally Composable Construction. To be able to satisfy the UC functionality, we require an additional key-extraction property. We thus build an augmented redactable signature scheme URS from a redactable signature scheme URS* (without key extraction) and a zero-knowledge non-interactive proof-of-knowledge system Π^*.

URS.SGen(1^κ). Run $SP^* \leftarrow$ URS*.SGen(1^κ), get grp from SP^*, run $CRS_{sk} \leftarrow$
$\quad \Pi^*$.Setup(grp), and output $SP = (SP^*, CRS_{sk})$.
URS.Kg(SP). Obtain SP^* and CRS_{sk} from SP, sample randomness r, and run
$\quad (pk^*, sk^*) \leftarrow$ URS*.Kg($SP^*; r$). Compute the proof
$\quad \pi_{sk} \leftarrow \Pi^*$.Prove $\left(CRS_{sk}; (sk^*, r); \lambda\, sk^* \, \exists\, r : (pk^*, sk^*) = \text{URS*.Kg}(SP^*; r)\right)$.
$\quad$ Output $pk = (SP, pk^*, \pi_{sk})$ and $sk = (sk^*, pk)$. We note that URS*.Kg
$\quad (SP^*; r)$ fixes the randomness of the a key generation algorithm.
URS.CheckPK(SP, pk). Check Π^*.Verify($CRS_{sk}; \pi_{sk}; \lambda\, sk \, \exists\, r : (pk, sk) =$
$\quad$ URS*.Kg($SP^*; r$)) = 1.

Sign, Derive, Verify are almost unchanged and use pk^* internally. SGenT and ExtractKey use the extraction setup and extractor of the proof system respectively, while CheckKeys checks that the relation R holds for pk and sk.

Note that Groth-Sahai proofs can be used to implement key-extraction by proving a binary, or n-ary decomposition of the secret key [57]. But this comes at a huge cost of more than 61,000 group elements at 128-bit security, even if this cost is only incurred once by every user per public key. We propose instead to use *fully* structure-preserving signatures (FSPS) [5] such that sk consists of group elements and can be easily extracted. FSPS for signing single group elements can be as cheap as 15 elements per signature and proofs of key possession consist of just 18 elements.

Theorem 6. *The strengthened scheme* URS *is an unforgeable, unlinkable, and key extractable redactable signature scheme, if the underlying redactable signature scheme* URS* *is unforgeable and unlinkable, and the proof-of-knowledge system Π^* is zero-knowledge and extractable.*

Unforgeability and unlinkability are corollaries of Theorem 4 and Theorem 5. Key-extractability follows directly from the extractability of the proof system.

Signing Group Elements as Additional Parts of the Message. While the presented redactable signature scheme can sign and quote a large number of values in $\mathbb{Z}_p$ very efficiently, in certain applications, like the one presented in the next section, one might also need to sign a small number of additional group elements. In the Derive algorithm these elements will either be part of the derived message, and given in the clear after derivation, or be treated as part of the witness, i.e., hidden from the verifier. The detailed construction and the security proofs are given in the full version [9].

4 From Unlinkable Redactable Signatures to Anonymous Credentials

As we designed our UC-secure URS scheme as a building block for privacy-preserving protocols, anonymous credentials are a natural application. Indeed, an (unlinkable) redactable signature scheme is already a simple selective-disclosure credential system where the attributes issued to users are the messages signed in Tier 1 signatures and a user can later reveal a subset of her attributes by deriving a Tier 2 signature. However, in an anonymous credential system, users also require secret keys and pseudonyms (pseudonymous public keys), on which credentials can be issued and with respect to which credentials can be presented. This allows users to prove that they possess several credentials issued from different parties on the same secret key [19,31].

In this section, we extend the functionality of URS in two ways: (1) we bind Tier 1 signatures to user secret keys in a way that prevents the derivation of signatures without knowledge of the secret and (2) we bind Tier 2 derived signatures to the unique context, *cxt* (nonce), to prevent replay attacks in which an attacker shows the same signature derived twice.

We first recall the algorithms of a multi-issuer anonymous credential system and then provide an instantiation using URS. To be modular and to simplify the analysis, we then provide an ideal functionality for a single issuer. The functionality is carefully designed to self-compose naturally into a full-fledged credential system with multiple issuers. Finally, we provide a concrete instantiation of our generic construction and analyze its efficiency.

4.1 Algorithms of Our Anonymous Credential System

Let us first introduce the parties and the algorithms of a multi-issuer anonymous credential system supporting user attributes (cf. [19,31]). Its protagonists are *users* ($\mathcal{U}$), *issuers* ($\mathcal{I}$), and *verifiers* ($\mathcal{V}$). Each user has a secret key X, from which she can derive (cryptographic) pseudonyms P. To get a credential issued, a user sends to the issuer a pseudonym P together with a (non-interactive) proof $\pi_{X,P}$ that she is privy to the underlying secret key. The issuer will then issue her a credential *Cred* on P containing the attributes $\boldsymbol{a}$ the issuer vouches for. The user can then present the credential to a verifier under a potentially different pseudonym P' by sending, together with P', a (non-interactive) proof $\pi_{X,Cred}$

that she possesses a credential on the attributes a_I. Recall that I defines which attributes shall be revealed.

A credential system Cred defines a set of algorithms: a system parameters generation algorithm SGen; an issuer setup algorithm Kg; a user secret generation algorithm SecGen; algorithms for pseudonym generation and verification NymGen and NymVerify, respectively; an algorithm to request a credential RequestCred; an algorithm for issuing a credential IssueCred; an algorithm to check a newly issued credential for correctness CheckCred; an algorithm to show a credential with respect to a pseudonym (to create a credential proof) Prove; and an algorithm to verify a credential proof Verify.

A more detailed discussion of these algorithms is given in the full version [9]. We instantiate these algorithms by adding support for user secrets, pseudonyms and contexts to our redactable signature scheme. Besides the URS algorithms, we use pseudonym generation and verification algorithms based on a structure preserving commitment scheme SPC and a hard relation to generate credential specific secrets. A hard relation has a generator K_{Rgap} that generates a witness $(X_{Cred}$ and a public value $Y_{Cred})$, and a verification algorithm V_{Rgap}, such that it is easy to verify (X_{Cred}, Y_{Cred}) but hard to compute X_{Cred} from Y_{Cred}. This hardness is used to prevent a network adversary that observes the issuing protocol from impersonating the user.

Table 1 gives the construction of our credential scheme. We group the core credential algorithms into those used for setup, issuing and presentation. In our security definition and the proof we will make use of additional algorithms for simulation and extraction.

4.2   Ideal Functionality for Credentials

To tame the complexity of definitions for credential systems with many different issuers, we chose to give a definition $\mathcal{F}_{Cred}$ of a scheme for a single issuer only, that then can be used as building block to a a full-fledged credential system with multiple issuers. The single issuer functionally $\mathcal{F}_{Cred}$ will just allow users to get a credential on a pseudonym from the issuer and to prove ownership of a credential by the issuer w.r.t. a given pseudonym.

To serve as a secure building block, $\mathcal{F}_{Cred}$ must be carefully designed. On the one hand it must deal with the unforgeability of credentials and on the other hand it must provide the hooks such that colluding users cannot mix and match credentials from different issuers. To address the former $\mathcal{F}_{Cred}$ binds issued credentials to the respective users' secret key X and for the latter $\mathcal{F}_{Cred}$ will enforce that credential proofs will not verify w.r.t. a pseudonym P unless a corresponding credential got issued to the X underlying that pseudonym. Then, provided adversarial users are unable to provide different X's for the same pseudonym, credentials from different issuers issued to different users (i.e., different X's) cannot be matched. As a consequence of this, the generation of user secret keys and pseudonyms is not done inside $\mathcal{F}_{Cred}$ but users are require to input their secret key X the pseudonym P (as cryptographic values) to $\mathcal{F}_{Cred}$ on the calls to get

Table 1. Algorithms of our credential system

Setup algorithms

Cred.SGen(1^κ): Compute $SP_{\mathsf{URS}} \leftarrow \mathsf{URS.SGen}(1^\kappa)$; $CRS_X \leftarrow \Pi.\mathsf{Setup}(1^\kappa)$; pp_{SPC}
  $\leftarrow \mathsf{SPC.Setup}(SP_{\mathsf{URS}})$; and output $SP = (SP_{\mathsf{URS}}, CRS_X, pp_{\mathsf{SPC}})$.
Cred.Kg(SP): Compute $(pk_{\mathsf{URS}}, sk_{\mathsf{URS}}) \leftarrow \mathsf{URS.KeyGen}(SP_{\mathsf{URS}})$, and output
  $(sk, pk) = (sk_{\mathsf{URS}}, pk_{\mathsf{URS}})$.
Cred.SecGen(SP) : Take G from SP, pick random $x \leftarrow \mathbb{Z}_p, X = G^x$. Output X.
Cred.NymGen(SP, X) : $(P, O) \leftarrow \mathsf{SPC.Commit}(pp_{\mathsf{SPC}}, X)$. Output $(P, aux(P) = O)$.
Cred.NymVerify($SP, X, P, aux(P)$) : Parse $aux(P)$ as O. Output the result of
  $\mathsf{SPC.Check}(pp_{\mathsf{SPC}}, P, O)$.

Issuing algorithms

Cred.RequestCred($SP, pk, X, P, aux(P)$) :
  $(X_{Cred}, Y_{Cred}) \leftarrow \mathsf{K_{Rgap}}$; $\pi_{X,P} \leftarrow \Pi.\mathsf{Prove}(CRS_X; (X, X_{Cred}, aux(P)); Stmt_P)$,
  where $Stmt_P = (\nexists X, aux(P) : \mathsf{NymVerify}(SP, X, P, aux(P)) = 1)$.
  Add $X_{Cred}, Y_{Cred}, P, aux(P)$ to $aux(Cred)$ and Y_{Cred} to $\pi_{X,P}$.
Cred.IssueCred($SP, sk, P, \boldsymbol{a}, \pi_{X,P}$):
  1. Verify the request for issuance $\pi_{X,P}$:
     If $\Pi.\mathsf{Verify}(CRS_X; \pi_{X,P}; Stmt_P) = 0$, return $\perp$.
  2. Else, generate a credential by creating a Tier 1 signature on the vector of
     messages, providing the pseudonym and a gap problem challenge, and
     calling $\sigma \leftarrow \mathsf{URS.Sign}(sk, (P, Y_{Cred}, \boldsymbol{a}))$ and output $Cred = \sigma$.
Cred.CheckCred($SP, pk, X, P, aux(P), Cred, aux(Cred), \boldsymbol{a}$) : Output the result of
  $\mathsf{URS.Verify}(pk, Cred, (P, Y_{Cred}, \boldsymbol{a}))$.

Presentation algorithms

Cred.Prove($SP, pk, X, P', aux(P)', Cred, aux(Cred), \boldsymbol{a}, I, cxt$) $\rightarrow \pi_{X,Cred}$:
  1. Obtain $X_{Cred}, Y_{Cred}, P, aux(P)$ from $aux(Cred)$ and σ from $Cred$.
  2. Run $\sigma_I \leftarrow \mathsf{URS.Derive}(pk, I, (P, Y_{Cred}, \boldsymbol{a}), \sigma))$.
  3. Compute a proof of knowledge of the secret, pseudonym, where the context
     is non-malleably attached as a label to the proof:
     $\pi_{X,Cred} = \Pi.\mathsf{Prove}(CRS_X; (X, P,$
     $aux(P), aux(P)', Y_{Cred}, X_{Cred}); Stmt, cxt); Stmt =$
     $(\nexists X, P, aux(P), aux(P)', Y_{Cred}, X_{Cred} : \mathsf{NymVerify}(SP, X, P', aux(P)') =$
     $1 \wedge \mathsf{NymVerify}(SP, X, P, aux(P)) = 1 \wedge \mathsf{URS.Verify}(pk, \sigma_I, (P, Y_{Cred}, \boldsymbol{a})_I)) =$
     $1 \wedge \mathsf{V_{Rgap}}(X_{Cred}, Y_{Cred}) = 1)$.
     Add σ_I to $\pi_{X,Cred}$ as a part of the public input.
Cred.Verify($SP, pk, P', \pi_{X,Cred}, \boldsymbol{a}_I, cxt$) : Output the result of $\Pi.\mathsf{Verify}(CRS_X;$
  $\pi_{X,Cred}; Stmt(SP, P', \sigma_I, \boldsymbol{a}_I), cxt)$.

Simulation and extraction algorithms

Cred.SGenT(1^κ) : $(SP_{\mathsf{URS}}, td) \leftarrow \mathsf{URS.SGenT}(1^\kappa)$;
  $(CRS_X, td_{ext}, td_{sim}) \leftarrow \Pi.\mathsf{SimSetup}(1^\kappa)$; $pp_{\mathsf{SPC}} \leftarrow \mathsf{SPC.Setup}(SP_{\mathsf{URS}})$.
  Output $(SP = (SP_{\mathsf{URS}}, CRS_X, pp_{\mathsf{SPC}}), td_{ext} = (td, td_{ext}), td_{sim})$.
Cred.Extract($SP, td_{ext}, pk, P', \pi_{X,Cred}, \boldsymbol{a}_I, cxt$):
  Take $(X, aux(P))$ from $\Pi.\mathsf{Extract}(SP; td_{ext}; \pi_{X,Cred}; Stmt))$.
Cred.SimProve($SP, sk, td_{sim}, P', \boldsymbol{a}_I, cxt$) $\rightarrow \pi_{X,Cred}$:
  1. $X \leftarrow \mathsf{SecGen}(SP)$; $(P, aux(P)) \leftarrow \mathsf{NymGen}(SP, X)$.
  2. Let $\boldsymbol{a}_0$ be a Tier 1 message restored from $\boldsymbol{a}_I$ by replacing $\perp$-s with 0-s as if
     it was derived from the original message $\boldsymbol{a}$ by applying $\mathsf{Zero}(\boldsymbol{a}, I)$.
  3. $\sigma \leftarrow \mathsf{URS.Sign}(sk, (P, Y_{Cred}, \boldsymbol{a}_0))$
  4. $\sigma_I \leftarrow \mathsf{URS.Derive}(pk, I, (P, Y_{Cred}, \boldsymbol{a}_0), \sigma))$.
  5. Compute a proof of knowledge of the secret, pseudonym, and the
     correctness of the signature on a context:
     $\pi_{X,Cred} \leftarrow \Pi.\mathsf{SimProve}(CRS_X; td_{sim}; P'; Stmt, cxt)$.

credentials issued or to generate a credential proof. Thus we assume that algorithms (SecGen, NymGen, NymVerify) to generate user secret keys, to generate pseudonyms, and to verify pseudonyms are available. $\mathcal{F}_{\mathsf{Cred}}$ is given NymVerify as an input parameter and will use this algorithm, to check the relation between P and X. For the security properties guaranteed by $\mathcal{F}_{\mathsf{Cred}}$, we do not make any assumptions about the security properties of these algorithms. However, for the security of the overall credential scheme, pseudonym need to be commitments to X, i.e., to be binding and hiding w.r.t. X.

In the following we provide the definition of $\mathcal{F}_{\mathsf{Cred}}$ and a protocol $\mathcal{R}_{\mathsf{Cred}}$ that realizes $\mathcal{F}_{\mathsf{Cred}}$ using $\mathcal{F}_{\mathsf{CA}}$ and $\mathcal{F}_{\mathsf{CRS}}$, assuming static corruptions.

Single Issuer Ideal Functionality. The starting point for our credential functionality is the ideal functionality of unlinkable redactable signatures, extended in a number of ways. Similar to $\mathcal{F}_{\mathsf{URS}}$ (and in line with other UC-functionalities such as $\mathcal{F}_{\mathsf{sig}}$ that need to output cryptographic values), $\mathcal{F}_{\mathsf{Cred}}$ is handed a number of cryptographic algorithms by the simulator. These algorithms allow $\mathcal{F}_{\mathsf{Cred}}$ to produce cryptographic artifacts for proofs of credential ownership and attribute disclosure, to verify such proofs (when they are coming from the adversary), and to extract values from proofs. (We note that there are no artifacts for the credentials themselves.) While these algorithms can be completely adversarial, $\mathcal{F}_{\mathsf{Cred}}$ will enforce that algorithms and the artifacts produced by them) satisfy the required unforgeability and privacy properties. In fact, because of the privacy properties, $\mathcal{F}_{\mathsf{Cred}}$ needs to run these algorithms itself and cannot ask the simulator for the artifacts as is sometimes done (cf. $\mathcal{F}_{\mathsf{URS}}$ and the UC-functionalities for blind signatures [6,42]).

We now describe the steps of our ideal functionality $\mathcal{F}_{\mathsf{Cred}}$ (cf. Fig. 1) and explain the security properties it ensures and how it does so. Note that because we consider static corruption, $\mathcal{F}_{\mathsf{Cred}}$ and $\mathcal{SIM}$ are aware of which parties are corrupted.

$\mathcal{F}_{\mathsf{Cred}}$ maintains two tables for bookkeeping: $\mathcal{M}_{ISS}$ stores information about issued credentials and $\mathcal{M}_{PRES}$ stores information about credentials that produced presentation proofs. It then handles requests as follows. Upon receiving a (keygen, sid) message, $\mathcal{F}_{\mathsf{Cred}}$ performs a setup by asking the simulator for the system parameters, the keys of the issuer, trapdoors, a set of algorithms and a list of corrupted parties. The message (leakSK, sid) is handled in exactly the same way as for redactable signatures.

Next, upon receiving a (issueCred, sid, qid, $\mathcal{U}$, X, P, $aux(P)$) message from a user $\mathcal{U}$, $\mathcal{F}_{\mathsf{Cred}}$ initiates credential issuing by sending a corresponding message to the issuer specified in $sid = (\mathcal{I}, sid')$. If $\mathcal{I}$ responds to the request with a list of attributes $\boldsymbol{a}$, $\mathcal{F}_{\mathsf{Cred}}$ verifies that X, P, and $aux(P)$ form a valid pseudonym (i.e., NymVerify outputs 1), and, if so, records in $\mathcal{M}_{ISS}$ that a credential with attributes $\boldsymbol{a}$ to user $\mathcal{U}$ w.r.t. secret X has been issued.

Upon receiving a credential proof request in the form of a (proveCred, ...) message, $\mathcal{F}_{\mathsf{Cred}}$ verifies whether the provided X, P, and $aux(P)$ form a valid pseudonym and whether a credential with attributes $\boldsymbol{a}$ to user $\mathcal{U}$ w.r.t. secret X has been issued. Then, $\mathcal{F}_{\mathsf{Cred}}$ creates a cryptographic artifact for the proof using the Cred.SimProve algorithm where no information that must not be revealed is input to the algorithm. This will guarantee the privacy properties of the credential proof for honest users. Furthermore, before outputting the proof to the user, $\mathcal{F}_{\mathsf{Cred}}$ will verify it using Cred.Verify as to ensure correctness.

Finally, upon receiving the (verifyCredProof,...) message, $\mathcal{F}_{\mathsf{Cred}}$ has to determine whether or not the proof should be accepted. Here we need to deal with unforgeability of credential proofs (and thus of credentials) if the issuer is honest and its secret key has not been leaked. Naturally, $\mathcal{F}_{\mathsf{Cred}}$ should accept proofs that it has generated itself. Apart from that, $\mathcal{F}_{\mathsf{Cred}}$ could in principle just accept all proofs for which the revealed attributes correspond to a credential that was issued. This would allow the adversary to also produce proofs that match credentials that were not issued to dishonest users but only to an honest user. To prevent this, we require an extraction algorithm Cred.Extract which, on input a credential proof, will generate a user secret. Then, $\mathcal{F}_{\mathsf{Cred}}$ will accept a credential proof only if the revealed attributes correspond to a credential that was issued to a corrupted users w.r.t. the X' extracted from the proof. That, however, would still allow (dishonest) users to mix and match their credentials. Therefore, $\mathcal{F}_{\mathsf{Cred}}$ will accept the proofs only if the extracted X' underlies the pseudonym P' w.r.t. which the proof verifies. Therefore, $\mathcal{F}_{\mathsf{Cred}}$ checks the latter using NymVerify.

Realization of $\mathcal{F}_{\mathsf{Cred}}$. A protocol $\mathcal{R}_{\mathsf{Cred}}$ that realizes $\mathcal{F}_{\mathsf{Cred}}$ can be obtained from the algorithms described in Sect. 4.1 in the $(\mathcal{F}_{\mathsf{CRS}}, \mathcal{F}_{\mathsf{CA}})$-hybrid model where SP is the reference string and each call to $\mathcal{F}_{\mathsf{Cred}}$ is replaced by running the corresponding algorithms. The detailed description of $\mathcal{R}_{\mathsf{Cred}}$ is given in the full version [9].

For efficiency reasons related to the integration of pseudonyms (which requires zero-knowledge proofs and thus whitebox techniques), $\mathcal{R}_{\mathsf{Cred}}$ does not use $\mathcal{R}_{\mathsf{URS}}$ as a (blackbox) subroutine. We will, however, carefully align the internals of $\mathcal{F}_{\mathsf{Cred}}$ and $\mathcal{R}_{\mathsf{Cred}}$ with those of $\mathcal{F}_{\mathsf{URS}}$ and $\mathcal{R}_{\mathsf{URS}}$ respectively, such that we can use the UC emulation theorem in one of the hybrid steps of our security proof.

Theorem 7. *Let* URS *be an unlinkable redactable signature scheme according to Definition 1,* SPC *be a structure-preserving commitment scheme,* Rgap *be a verifiable relation,* Π *be a non-interactive proof of knowledge system. Then* $\mathcal{R}_{\mathsf{Cred}}$ *securely realizes* $\mathcal{F}_{\mathsf{Cred}}$ *in the* $(\mathcal{F}_{\mathsf{CRS}}, \mathcal{F}_{\mathsf{CA}})$-*hybrid model if* URS *is correct, unlinkable, unforgeable, and key extractable,* SPC *is binding, the non-interactive proof-of-knowledge system is zero-knowledge and simulation extractable, and the* Rgap *relation is hard. The proof is provided in the full version [9].*

Building a Full-Fledged Credential System with Multiple Issuers. We now explain how to use our credential functionality to support multiple issuers using multiple sessions of $\mathcal{F}_{\mathsf{Cred}}$, one for each issuer, together with algorithms (SecGen, NymGen, NymVerify) to generate user secret keys, to generate pseudonyms, and to verify pseudonyms. The pseudonyms are required to be both hiding and binding w.r.t. the user secret to provide privacy to the honest users and to prevent colluding users from sharing credentials unless they all user the same user secret. A user

Functionality $\mathcal{F}_{\mathsf{Cred}}(\mathsf{NymVerify})$

The functionality maintains tables $\mathcal{M}_{ISS}$ and $\mathcal{M}_{PRES}$ initialized to $\emptyset$ and flags kg and $keyleak$ which are initially unset.

- On input (keygen, sid) from $\mathcal{I}$, verify that $sid = (\mathcal{I}, sid')$ for some sid' and that flag kg is unset. If not, then return $\bot$. Else, do the following:
  1. Send (initF, sid) to $\mathcal{SIM}$ and wait for a message $(\mathsf{initF}, sid, SP, sk, pk, td_{sim}, td_{ext}, \mathsf{Cred.SimProve}, \mathsf{Cred.Verify}, \mathsf{Cred.Extract})$ from $\mathcal{SIM}$, where SP are the system parameters, td_{sim} and td_{ext} are the simulation and extraction trapdoors respectively, and the rest are polynomial-time algorithms. Store all of these values and set flag kg.
  2. Return $(\mathsf{verificationKey}, sid, pk)$ to $\mathcal{I}$.
- On input (leakSK, sid) from $\mathcal{I}$ verify that $sid = (\mathcal{I}, sid')$ for some sid'. If not, return $\bot$. Else, if flag kg is set, set flag $keyleak$ and return $(\mathsf{leakSK}, sid, sk)$, otherwise - abort.
- On input $(\mathsf{issueCred}, sid, qid, X, P, aux(P))$ from $\mathcal{U}$, check $sid = (\mathcal{I}, sid')$ for some sid', and that flag kg is set. If not, return $\bot$. Else send a public delayed output $(\mathsf{issueCred}, sid, qid, P)$ to $\mathcal{I}$.
- On input $(\mathsf{issueCred}, sid, qid, \boldsymbol{a})$ from $\mathcal{I}$, check for $(\mathsf{issueCred}, sid, qid, X, P, aux(P))$ from $\mathcal{U}$, and verify that $sid = (\mathcal{I}, sid')$ for some sid' and that the flag kg is set. If not, return $\bot$. Else, do the following:
  1. Run $b \leftarrow \mathsf{NymVerify}(SP, P, X, aux(P))$. If $b = 0$, return $\bot$.
  2. Add $(ISS, \bot, X, \boldsymbol{a})$ to $\mathcal{M}_{ISS}$.
  3. Send a public delayed output $(\mathsf{credIssued}, sid, qid, \boldsymbol{a})$ to $\mathcal{U}$.
  4. When $(\mathsf{credIssued}, sid, qid, \boldsymbol{a})$ is delivered to $\mathcal{U}$, update the issuance record by adding the user to $(ISS, \mathcal{U}, X, \boldsymbol{a})$ of $\mathcal{M}_{ISS}$.
- On input $(\mathsf{proveCred}, sid, X, P', aux(P)', I, \boldsymbol{a}, cxt)$ from $\mathcal{U}$, do the following:
  1. Check if kg is set. If not, return $\bot$.
  2. Check if $\mathsf{NymVerify}(SP, P', X, aux(P)') = 1$. If not, return $\bot$.
  3. Check if $(ISS, \mathcal{U}, X, \boldsymbol{a})$ exists. If not, return $\bot$.
  4. $\pi_{X, Cred} \leftarrow \mathsf{Cred.SimProve}(SP, sk, td_{sim}, P', \boldsymbol{a}_I, cxt)$.
  5. Check if $\mathsf{Cred.Verify}(SP, pk, P', \pi_{X, Cred}, \boldsymbol{a}_I, cxt) = 0$, then output $\bot$.
  6. Add $(PRES, \mathcal{U}, cxt, X, P', aux(P)', \boldsymbol{a}_I, \pi_{X, Cred})$ to $\mathcal{M}_{PRES}$.
  7. Return $(\mathsf{credProved}, sid, \boldsymbol{a}_I, \pi_{X, Cred})$ to $\mathcal{U}$.
- On input $(\mathsf{verifyCredProof}, sid, pk', P', \pi'_{X, Cred}, \boldsymbol{a}'_I, cxt')$ from some party $\mathcal{P}$, do the following:
  1. Verify the proof $result = \mathsf{Cred.Verify}(SP, pk', P', \pi'_{X, Cred}, \boldsymbol{a}'_I, cxt')$.
  2. If $pk \neq pk'$, or $keyleak$ is set, or $\mathcal{I}$ is dishonest, or $result = 0$, send $(\mathsf{verified}, sid, \boldsymbol{a}'_I, result)$ to $\mathcal{P}$.
  3. Else, if there is a record $(PRES, *, cxt', *, P', *, \boldsymbol{a}'_I, \pi'_{X, Cred})$ return $(\mathsf{verified}, sid, \boldsymbol{a}'_I, 1)$ to $\mathcal{P}$.
  4. Else, run $(X', aux(P)') \leftarrow \mathsf{Cred.Extract}(SP, td_{ext}, pk, P', \pi'_{X, Cred}, \boldsymbol{a}'_I, cxt')$.
  5. If $\mathsf{NymVerify}(SP, P', X', aux(P)') = 0$, return $(\mathsf{verified}, sid, \boldsymbol{a}'_I, 0)$ to $\mathcal{P}$.
  6. Else, if there is a record $(ISS, \mathcal{U}, X', \boldsymbol{a})$ in $\mathcal{M}_{ISS}$ for a corrupted user $\mathcal{U}$ such that $\boldsymbol{a}_I = \boldsymbol{a}'_I$, return $(\mathsf{verified}, sid, \boldsymbol{a}'_I, 1)$ to $\mathcal{P}$.
  7. Otherwise return $(\mathsf{verified}, sid, \boldsymbol{a}'_I, 0)$ to $\mathcal{P}$.

Fig. 1. The ideal functionality for single issuer anonymous credentials

now can generate a user secret and different pseudonyms on them and then use multiple calls to the $\mathcal{F}_{\mathsf{Cred}}$ instances for different issuers to get credentials on her pseudonyms. To compose a presentation proof that reveals attributes from different credentials, the user creates a pseudonym P' and uses the corresponding $\mathcal{F}_{\mathsf{Cred}}$ instances to generate the required proofs with respect to this pseudonym. Because the pseudonym is the same in different proofs and each proof guarantees the same underlying secret in the credential and the pseudonym, the collection of these proofs together results in a single proof for multiple credentials. Each proof block guarantees unlinkability and unforgeability, and because the pseudonym is both binding and hiding this composed proof is also unforgeable and unlinkable with respect to other proof collections. The verification is done by querying the corresponding $\mathcal{F}_{\mathsf{Cred}}$ instances for verification of each particular proof part and by checking that the pseudonym is the same in each proof part. A formal definition of a full-fledged credential scheme and proof that the scheme just sketched meets it is left as future work.

4.3   Instantiation and Efficiency Analysis

To analyze the efficiency of our scheme we consider a concrete instantiation scenario. We instantiate our non-interactive construction with Groth-Sahai proofs [45], the structure-preserving commitment scheme of [4], and our unlinkable redactable signature scheme presented in Sect. 3.3. We use disjunctive proofs to make the proof system simulation-extractable [22], see [54] for the efficient instantiation with 48 group elements overhead in the XDH setting that forms the basis of our efficiency analysis. As a hard relation we pick the Computational Diffie-Hellman problem. The URS scheme is instantiate with the fully structure-preserving signature scheme by Abe et al. [5], Groth-Sahai proofs, and the vector commitment scheme from Sect. 3.1. The proof of Theorem 8 follows from Theorems 6-7.

Theorem 8. *The credential system described above securely realizes* $\mathcal{F}_{\mathsf{Cred}}$ *defined in Sect. 4.2 if the SXDH, n-RootDH, n-BSDH, q-SDH, XDLIN, co-CDH, and DBP assumptions hold.* Consult building blocks for definitions of assumptions.

We refer to the full version [9] for the comparison with prior work. We stress that the complexity of the Prove and Verify algorithms is independent of the number of all attributes contained in a credential.

The size of the credential proof is roughly 178 group elements (148 when using the SPS of [2] instead of FSPS). This means that the communication efficiency for showing a credential with respect to a pseudonym is around 11 KB (9 KB for SPS) at 128-bit security level, which is close to Idemix credentials [31] as the size of pairing groups is much smaller than the size of RSA groups and because the size of Idemix credential proofs is linear in the number of attributes. Besides, Idemix credentials do not provide such strong formal security guarantees, i.e. they require random oracles for non-interactive proofs and are not universally composable. Our non-UC scheme is comparable in efficiency with the credential system of Izabachène et al. [50] that has credential proofs of around 8 KB, while our UC scheme has larger proof sizes. Our scheme is much less efficient than the scheme of [47] but their scheme relies on hash functions in their construction and thus does not enable efficient protocol design.

Open Questions. We leave the construction of a scheme that achieves the same functionality as ours with the efficiency of [47]—perhaps using fully structure preserving signatures of equivalence classes—as an interesting open problem. Other interesting questions are exploiting the lack of opening non-malleability for attacks on existing constructions and efficiently basing the opening non-malleability property of vector commitments on a more standard cryptographic assumption than the n-RootDH assumption of Definition 9.

Acknowledgments. The authors would like to thank Sherman Chow and the anonymous reviewers for their helpful comments and suggestions. The research leading to these results was supported in part by the European Community's Seventh Framework Programme for the project FutureID (grant agreement no. 318424).

References

1. Abe, M., Chase, M., David, B., Kohlweiss, M., Nishimaki, R., Ohkubo, M.: Constant-size structure-preserving signatures: generic constructions and simple assumptions. In: Wang, X., Sako, K. (eds.) ASIACRYPT 2012. LNCS, vol. 7658, pp. 4–24. Springer, Heidelberg (2012)
2. Abe, M., Fuchsbauer, G., Groth, J., Haralambiev, K., Ohkubo, M.: Structure-preserving signatures and commitments to group elements. In: Rabin, T. (ed.) CRYPTO 2010. LNCS, vol. 6223, pp. 209–236. Springer, Heidelberg (2010)
3. Abe, M., Groth, J., Haralambiev, K., Ohkubo, M.: Optimal structure-preserving signatures in asymmetric bilinear groups. In: Rogaway, P. (ed.) CRYPTO 2011. LNCS, vol. 6841, pp. 649–666. Springer, Heidelberg (2011)
4. Abe, M., Haralambiev, K., Ohkubo, M.: Group to group commitments do not shrink. In: Pointcheval, D., Johansson, T. (eds.) EUROCRYPT 2012. LNCS, vol. 7237, pp. 301–317. Springer, Heidelberg (2012)
5. Abe, M., Kohlweiss, M., Ohkubo, M., Tibouchi, M.: Fully structure-preserving signatures and shrinking commitments. In: Oswald, E., Fischlin, M. (eds.) EUROCRYPT 2015. LNCS, vol. 9057, pp. 35–65. Springer, Heidelberg (2015)
6. Abe, M., Ohkubo, M.: A framework for universally composable non-committing blind signatures. In: Matsui, M. (ed.) ASIACRYPT 2009. LNCS, vol. 5912, pp. 435–450. Springer, Heidelberg (2009)
7. Ahn, J.H., Boneh, D., Camenisch, J., Hohenberger, S., Shelat, a, Waters, B.: Computing on authenticated data. In: Cramer, R. (ed.) TCC 2012. LNCS, vol. 7194, pp. 1–20. Springer, Heidelberg (2012)
8. Attrapadung, N., Libert, B., Peters, T.: Computing on authenticated data: new privacy definitions and constructions. In: Wang, X., Sako, K. (eds.) ASIACRYPT 2012. LNCS, vol. 7658, pp. 367–385. Springer, Heidelberg (2012)
9. Camenisch, J., Dubovitskaya, M., Haralambiev, K., Kohlweiss, K.: Composable & Modular Anonymous Credentials: Definitions and Practical Constructions. IACR Cryptology ePrint Archive, Report 2015/580

10. Backes, M., Meiser, S., Schröder, D.: Delegatable functional signatures. Cryptology ePrint Archive, Report 2013/408
11. Belenkiy, M., Camenisch, J., Chase, M., Kohlweiss, M., Lysyanskaya, A., Shacham, H.: Randomizable proofs and delegatable anonymous credentials. In: Halevi, S. (ed.) CRYPTO 2009. LNCS, vol. 5677, pp. 108–125. Springer, Heidelberg (2009)
12. Belenkiy, M., Chase, M., Kohlweiss, M., Lysyanskaya, A.: P-signatures and noninteractive anonymous credentials. In: Canetti, R. (ed.) TCC 2008. LNCS, vol. 4948, pp. 356–374. Springer, Heidelberg (2008)
13. Belenkiy, M., Chase, M., Kohlweiss, M., Lysyanskaya, A.: Compact e-Cash and simulatable VRFs revisited. In: Shacham, H., Waters, B. (eds.) Pairing 2009. LNCS, vol. 5671, pp. 114–131. Springer, Heidelberg (2009)
14. Bellare, M., Fuchsbauer, G.: Policy-based signatures. In: Krawczyk, H. (ed.) PKC 2014. LNCS, vol. 8383, pp. 520–537. Springer, Heidelberg (2014)
15. Bellare, M., Micciancio, D., Warinschi, B.: Foundations of group signatures: formal definitions, simplified requirements, and a construction based on general assumptions. In: Biham, E. (ed.) EUROCRYPT 2003. LNCS, vol. 2656, pp. 614–629. Springer, Heidelberg (2003)
16. Boneh, D., Boyen, X.: Short signatures without random oracles. In: Cachin, C., Camenisch, J.L. (eds.) EUROCRYPT 2004. LNCS, vol. 3027, pp. 56–73. Springer, Heidelberg (2004)
17. Boyle, E., Goldwasser, S., Ivan, I.: Functional signatures and pseudorandom functions. In: Krawczyk, H. (ed.) PKC 2014. LNCS, vol. 8383, pp. 501–519. Springer, Heidelberg (2014)
18. Brands, S.: Untraceable off-line cash in wallets with observers: extended abstract. In: Stinson, D.R. (ed.) CRYPTO 1993. LNCS, vol. 773, pp. 302–318. Springer, Heidelberg (1994)
19. Brands, S.: Restrictive blinding of secret-key certificates. In: Guillou, L.C., Quisquater, J.-J. (eds.) EUROCRYPT 1995. LNCS, vol. 921, pp. 231–247. Springer, Heidelberg (1995)
20. Brickell, E.F., Camenisch, J., Chen, L.: Direct anonymous attestation. In: Atluri, V., Pfitzmann, B., McDaniel, P. (eds.) Proceedings of the 11th Conference on Computer and Communications Security, ACM CCS 2004, pp. 132–145. ACM Press, October 2004
21. Brzuska, C., Busch, H., Dagdelen, O., Fischlin, M., Franz, M., Katzenbeisser, S., Manulis, M., Onete, C., Peter, A., Poettering, B., Schröder, D.: Redactable signatures for tree-structured data: definitions and constructions. In: Zhou, J., Yung, M. (eds.) ACNS 2010. LNCS, vol. 6123, pp. 87–104. Springer, Heidelberg (2010)
22. Camenisch, J., Chandran, N., Shoup, V.: A public key encryption scheme secure against key dependent chosen plaintext and adaptive chosen ciphertext attacks. In: Joux, A. (ed.) EUROCRYPT 2009. LNCS, vol. 5479, pp. 351–368. Springer, Heidelberg (2009)
23. Camenisch, J., Dubovitskaya, M., Neven, G., Zaverucha, G.M.: Oblivious transfer with hidden access control policies. In: Catalano, D., Fazio, N., Gennaro, R., Nicolosi, A. (eds.) PKC 2011. LNCS, vol. 6571, pp. 192–209. Springer, Heidelberg (2011)
24. Camenisch, J., Groß, T.: Efficient attributes for anonymous credentials. In: Ning, P., Syverson, P.F., Jha, S. (eds.) Proceedings of the 15th Conference on Computer and Communications Security, ACM CCS 2008, pp. 345–356. ACM Press, October 2008

25. Camenisch, J.L., Hohenberger, S., Lysyanskaya, A.: Compact e-Cash. In: Cramer, R. (ed.) EUROCRYPT 2005. LNCS, vol. 3494, pp. 302–321. Springer, Heidelberg (2005)
26. Camenisch, J., Kiayias, A., Yung, M.: On the portability of generalized Schnorr proofs. In: Joux, A. (ed.) EUROCRYPT 2009. LNCS, vol. 5479, pp. 425–442. Springer, Heidelberg (2009)
27. Camenisch, J., Krenn, S., Lehmann, A., Mikkelsen, G.L., Neven, G., Pedersen, M.: Formal treatment of privacy-enhancing credential systems. In: SAC 2015. Cryptology ePrint Archive, Report 2014/708
28. Camenisch, J., Krenn, S., Shoup, V.: A framework for practical universally composable zero-knowledge protocols. In: Lee, D.H., Wang, X. (eds.) ASIACRYPT 2011. LNCS, vol. 7073, pp. 449–467. Springer, Heidelberg (2011)
29. Camenisch, J.L., Lysyanskaya, A.: An efficient system for non-transferable anonymous credentials with optional anonymity revocation. In: Pfitzmann, B. (ed.) EUROCRYPT 2001. LNCS, vol. 2045, p. 93. Springer, Heidelberg (2001)
30. Camenisch, J.L., Lysyanskaya, A.: A signature scheme with efficient protocols. In: Cimato, S., Galdi, C., Persiano, G. (eds.) SCN 2002. LNCS, vol. 2576, pp. 268–289. Springer, Heidelberg (2003)
31. Camenisch, J.L., Lysyanskaya, A.: Signature schemes and anonymous credentials from bilinear maps. In: Franklin, M. (ed.) CRYPTO 2004. LNCS, vol. 3152, pp. 56–72. Springer, Heidelberg (2004)
32. Camenisch, J.L., Neven, G., Shelat, A.: Simulatable adaptive oblivious transfer. In: Naor, M. (ed.) EUROCRYPT 2007. LNCS, vol. 4515, pp. 573–590. Springer, Heidelberg (2007)
33. Camenisch, J., Van Herreweghen, E.: Design and implementation of the idemix anonymous credential system. In: Atluri, V. (ed.) Proceedings of the 9th Conference on Computer and Communications Security, ACM CCS 2002, pp. 21–30. ACM Press, November 2002
34. Canard, S., Lescuyer, R.: Protecting privacy by sanitizing personal data: a new approach to anonymous credentials. In: Proceedings of the 8th ACM SIGSAC Symposium on Information, Computer and Communications Security, ASIACCS 2013, pp. 381–392. ACM, New York, NY, USA (2013)
35. Canetti, R.: Universally composable security: a new paradigm for cryptographic protocols. In: 42nd Annual Symposium on Foundations of Computer Science, pp. 136–145. IEEE Computer Society Press, October 2001
36. Canetti, R., Goldreich, O., Halevi, S.: The random oracle methodology, revisited (preliminary version). In: Proceedings of the 30th Annual ACM Symposium on Theory of Computing, pp. 209–218. ACM Press, May 1998
37. Catalano, D., Fiore, D.: Vector commitments and their applications. In: Kurosawa, K., Hanaoka, G. (eds.) PKC 2013. LNCS, vol. 7778, pp. 55–72. Springer, Heidelberg (2013)
38. Chase, M., Kohlweiss, M., Lysyanskaya, A., Meiklejohn, S.: Malleable signatures: new definitions and delegatable anonymous credentials. In: 2013 IEEE 27th Computer Security Foundations Symposium, IEEE (2014)
39. Chase, M., Lysyanskaya, A.: On signatures of knowledge. In: Dwork, C. (ed.) CRYPTO 2006. LNCS, vol. 4117, pp. 78–96. Springer, Heidelberg (2006)
40. Chaum, D., Fiat, A., Naor, M.: Untraceable electronic cash. In: Goldwasser, S. (ed.) CRYPTO 1988. LNCS, vol. 403, pp. 319–327. Springer, Heidelberg (1990)
41. Chow, S.S.M., He, Y.J., Hui, L.C.K., Yiu, S.M.: SPICE - simple privacy-preserving identity-management for cloud environment. In: Bao, F., Samarati, P., Zhou, J. (eds.) ACNS 2012. LNCS, vol. 4731, pp. 516–543. Springer, Heidelberg (2012)

42. Fischlin, M.: Round-optimal composable blind signatures in the common reference string model. In: Dwork, C. (ed.) CRYPTO 2006. LNCS, vol. 4117, pp. 60–77. Springer, Heidelberg (2006)

43. Fuchsbauer, G., Hanser, C., Slamanig, D.: EUF-CMA-secure structure-preserving signatures on equivalence classes. Cryptology ePrint Archive, Report 2014/944

44. Goyal, V.: Reducing trust in the PKG in identity based cryptosystems. In: Menezes, A. (ed.) CRYPTO 2007. LNCS, vol. 4622, pp. 430–447. Springer, Heidelberg (2007)

45. Groth, J., Sahai, A.: Efficient non-interactive proof systems for bilinear groups. In: Smart, N.P. (ed.) EUROCRYPT 2008. LNCS, vol. 4965, pp. 415–432. Springer, Heidelberg (2008)

46. Haber, S., Hatano, Y., Honda, Y., Horne, W., Miyazaki, K., Sander, T., Tezoku, S., Yao, D.: Efficient signature schemes supporting redaction, pseudonymization, and data deidentification. In: Abe, M., Gligor, V. (eds.) 3rd Conference on Computer and Communications Security, ASIACCS 2008, pp. 353–362. ACM Press, March 2008

47. Hanser, C., Slamanig, D.: Structure-preserving signatures on equivalence classes and their application to anonymous credentials. In: Sarkar, P., Iwata, T. (eds.) ASIACRYPT 2014. LNCS, vol. 8873, pp. 491–511. Springer, Heidelberg (2014)

48. Hirt, M., Sako, K.: Efficient receipt-free voting based on homomorphic encryption. In: Preneel, B. (ed.) EUROCRYPT 2000. LNCS, vol. 1807, pp. 539–556. Springer, Heidelberg (2000)

49. Hofheinz, D., Shoup, V.: GNUC: a new universal composability framework. J. Cryptol. **28**, 423–508 (2015). IACR Cryptology ePrint Archive, Report 2011/303

50. Izabachène, M., Libert, B., Vergnaud, D.: Block-wise P-signatures and non-interactive anonymous credentials with efficient attributes. In: Chen, L. (ed.) IMACC 2011. LNCS, vol. 7089, pp. 431–450. Springer, Heidelberg (2011)

51. Kate, A., Zaverucha, G.M., Goldberg, I.: Constant-size commitments to polynomials and their applications. In: Abe, M. (ed.) ASIACRYPT 2010. LNCS, vol. 6477, pp. 177–194. Springer, Heidelberg (2010)

52. Kiayias, A., Yung, M.: Group signatures with efficient concurrent join. In: Cramer, R. (ed.) EUROCRYPT 2005. LNCS, vol. 3494, pp. 198–214. Springer, Heidelberg (2005)

53. Kiayias, A., Zhou, H.-S.: Equivocal blind signatures and adaptive UC-security. In: Canetti, R. (ed.) TCC 2008. LNCS, vol. 4948, pp. 340–355. Springer, Heidelberg (2008)

54. Kohlweiss, M., Miers, I.: Accountable tracing signatures. Cryptology ePrint Archive, Report 2014/824

55. Kohlweiss, M., Rial, A.: Optimally private access control. In: Proceedings of the 12th ACM Workshop on Privacy in the Electronic Society, WPES 2013, pp. 37–48, ACM, New York, NY, USA (2013)

56. Libert, B., Yung, M.: Concise mercurial vector commitments and independent zero-knowledge sets with short proofs. In: Micciancio, D. (ed.) TCC 2010. LNCS, vol. 5978, pp. 499–517. Springer, Heidelberg (2010)

57. Meiklejohn, S.: An extension of the groth-sahai proof system. Master's thesis, Brown University (2009)

58. Nojima, R., Tamura, J., Kadobayashi, Y., Kikuchi, H.: A storage efficient redactable signature in the standard model. In: Samarati, P., Yung, M., Martinelli, F., Ardagna, C.A. (eds.) ISC 2009. LNCS, vol. 5735, pp. 326–337. Springer, Heidelberg (2009)

Side-Channel Attacks

ASCA, SASCA and DPA with Enumeration: Which One Beats the Other and When?

Vincent Grosso[✉] and François-Xavier Standaert

ICTEAM/ELEN/Crypto Group, Université catholique de Louvain,
Louvain-la-Neuve, Belgium
vincent.grosso@uclouvain.be

Abstract. We describe three contributions regarding the Soft Analytical Side-Channel Attacks (SASCA) introduced at Asiacrypt 2014. First, we compare them with Algebraic Side-Channel Attacks (ASCA) in a noise-free simulated setting. We observe that SASCA allow more efficient key recoveries than ASCA, even in this context (favorable to the latter). Second, we describe the first working experiments of SASCA against an actual AES implementation. Doing so, we analyse their profiling requirements, put forward the significant gains they provide over profiled Differential Power Analysis (DPA) in terms of number of traces needed for key recoveries, and discuss the specificities of such concrete attacks compared to simulated ones. Third, we evaluate the distance between SASCA and DPA enhanced with computational power to perform enumeration, and show that the gap between both attacks can be quite reduced in this case. Therefore, our results bring interesting feedback for evaluation laboratories. They suggest that in several relevant scenarios (e.g. attacks exploiting many known plaintexts), taking a small margin over the security level indicated by standard DPA with enumeration should be sufficient to prevent more elaborate attacks such as SASCA. By contrast, SASCA may remain the only option in more extreme scenarios (e.g. attacks with unknown plaintexts/ciphertexts or against leakage-resilient primitives). We conclude by recalling the algorithmic dependency of the latter attacks, and therefore that our conclusions are specific to the AES.

1 Introduction

State-of-the-art. Strategies to exploit side-channel leakages can be classified as Divide and Conquer (DC) and analytical. In the first case, the adversary recovers information about different bytes of (e.g.) a block cipher key independently, and then combines this information, e.g. via enumeration [36]. In the second case, she rather tries to recover the full key at once, exploiting more algorithmic approaches to cryptanalysis with leakage. Rephrasing Banciu et al., one can see these different strategies as a tradeoff between pragmatism and elegance [2].

In brief, the "DC+enumeration" approach is pragmatic, i.e. it is easy to implement, requires little knowledge about the target implementation, and can take advantage of a variety of popular (profiled and non-profiled) distinguishers,

© International Association for Cryptologic Research 2015
T. Iwata and J.H. Cheon (Eds.): ASIACRYPT 2015, Part II, LNCS 9453, pp. 291–312, 2015.
DOI: 10.1007/978-3-662-48800-3_12

such as Correlation Power Analysis (CPA) [6], Mutual Information Analysis (MIA) [14], Linear Regression (LR) [34] or Template Attacks (TA) [8]. We will use the term Differential Power Analysis (DPA) to denote them all [22].

By contrast, analytical approaches are (more) elegant, since they theoretically exploit all the information leaked by an implementation (vs. the leakages of the first and/or last rounds independently for DC attacks). As a result, these attacks can (theoretically) succeed in conditions where the number of measurements available to the adversary is very limited. But this elegance (and the power that comes with it) usually implies stronger assumptions on the target implementation (e.g. most of them require some type of profiling). The Algebraic Side-Channel Attacks (ASCA) described in [30] and further analyzed in [7,32] are an extreme solution in this direction. In this case, the target block cipher and its leakages are represented as a set of equations that are then solved (e.g. with a SAT solver, or Groebner bases). This typically implies a weak resistance to the noise that is usually observed in side-channel measurements. As a result, various heuristics have been suggested to better deal with errors in the information leakages, such as [24,39]. The Tolerant Algebraic Side-Channel Attacks (TASCA) proposed in [25,26] made one additional step in this direction, by replacing the solvers used in ASCA by an optimizer. But they were limited by their high memory complexity (since they essentially deal with noise by exhaustively encoding the errors they may cause). More recently, two independent proposals suggested to design a dedicated solver specialized to byte-oriented ciphers such as the AES [16,27]. The latter ones were more efficient and based on smart heuristics exploiting enumeration. Eventually, Soft Analytical Side-Channel Attacks (SASCA) were introduced at Asiacrypt 2014 as a conceptually different way to exploit side-channel leakages analytically [38]. Namely, rather than encoding them as equations, SASCA describe an implementation and its leakages as a code, that one can efficiently decode using the Belief Propagation (BP) algorithm. As a result, they can directly exploit the (soft) information provided by profiled side-channel attacks (such as LR or TA), in an efficient manner, with limited memory complexity, and for multiple plaintexts. Concretely, this implies that they provide a natural bridge between DC attacks and analytical ones.

Our Contribution. In view of this state-of-the-art, we consider three open problems regarding DC and analytical strategies in side-channel analysis.

First, we observe that the recent work in [38] experimented SASCA in the context of noisy AES leakages. While this context allowed showing that SASCA are indeed applicable in environments where ASCA would fail, it leaves the question whether this comes at the cost of a lower efficiency in a noise-free context open. Therefore, we launched various experiments with noise-free AES leakages to compare ASCA and SASCA. These experiments allowed us to confirm that also in this context, SASCA are equally (even slightly more) efficient.

Second, the experiments in [38] exploited simulations in order to exhibit the strong noise-resilience of SASCA (since the amount of noise can then be used as a parameter of such simulations). But this naturally eludes the question of the profiling of a concrete device, which can be a challenging task, and for which the leakage functions of different target intermediate values may turn out to be quite different [13]. Therefore, we describe the first working experiments of SASCA against an actual AES implementation, for which a bivariate TA exploiting the S-box input/output leakages would typically be successful after more than 50 measurements. We further consider two cases for the adversary's knowledge about the implementation. In the first one, she has a precise description in hand (i.e. the assembly code, typically). In the second one, she only knows AES is running, and therefore only exploits the generic operations that one can assume from the algorithm specification.[1] Our experiments confirm that SASCA are applicable in a simple profiled scenario, and lead to successful key recoveries with less traces than a DC attack (by an approximate factor up to 5). They also allow us to discuss the profiling cost, and the consequences of the different leakage functions in our target implementation. A relevant observation regarding them is that weak leakages in the MixColumns operations are especially damaging for the adversary, which can be explained by the (factor) graph describing an AES implementation: indeed, XORing two values with limited information significantly reduces the information propagation of the BP algorithm execution. This suggest interesting research directions for preventing such attacks, since protecting the linear parts of a block cipher is usually easier/cheaper.

Third, we note that SASCA are in general more computationally intensive than DC attacks. Therefore, a fair comparison should allow some enumeration power to the DC attacks as well. We complement our previous experimental attacks by considering this last scenario. That is, we compare the success rate of SASCA with the ones of DC attacks exploiting a computational power corresponding to up to 2^{30} encryptions (which corresponds to more than the execution time of SASCA on our computing platform). Our results put forward that SASCA remain the most powerful attack in this case, but with a lower gain.

Summary. These contributions allow answering the question of our title. First, SASCA are in general preferable to ASCA, with both noise-free and noisy AES leakages. Second, the tradeoff between SASCA and DC attacks is more balanced. As previously mentioned, DC attacks are more pragmatic. So the interest of SASCA essentially depends on the success rate gains it provides, which itself depends on the scenarios. If multiple plaintexts/ciphertext pairs are available, our experiments suggest that the gain of SASCA over DPA with enumeration is somewhat limited, and may not justify such an elegant approach. This conclusion backs up the results in [2], but in a more general scenario, since we consider multiple-queries attacks rather than single-query ones, together with more a powerful analytical strategy. By contrast, if plaintexts/ciphertexts are unknown (which renders DPA [17] and enumeration more challenging to apply), or if the number of plaintexts one can observe is very limited (e.g. by design, due to a leakage-resilient primitive [10]), SASCA may be the best/only option.

---

[1] Admittedly, such a generic scenario still assumes that the target implementation closely follows the specifications given in [11] which may not always be the case, e.g. for bitslice implementations [29], or T-table based implementations [9].

Preliminary Remark. Our focus in this paper is on a couple of extreme approaches to side-channel analysis, i.e. the most pragmatic DC attacks against 8-bit targets of the first AES round, and the most elegant ASCA/SASCA exploiting most/all such targets in the implementation. Quite naturally, the other analytical attacks mentioned in this introduction would provide various tradeoffs between these extremes. Besides, more computationally-intensive DPA attacks (based on larger key hypotheses) are also possible, as recently discussed by Mather et al. [23]. Such attacks are complementary and may further reduce the gain of SASCA over DPA, possibly at the cost of increased computational requirements (e.g. the latter work exploited high-performance computing whereas all our experiments were carried out on a single desktop computer).

2 Background

In this section we first describe the measurement setup used in our experiments. Then, we describe two tools we used to identify and evaluate information leakages in the traces. Finally, we recall the basics of the different attacks we compare.

2.1 Measurement Setup

Our measurements are based on the open source AES FURIOUS implementation (http://point-at-infinity.org/avraes) run by an 8-bit Atmel ATMEGA644p microcontroller at a 20 MHz clock frequency. We monitored the power consumption across a 22Ω resistor. Acquisitions were performed using a Lecroy WaveRunner HRO 66 ZI providing 8-bit samples, running at 400 Msamples/second. For SASCA, we can exploit any intermediate values that appear during the AES computation. Hence, we measured the full encryption. Our traces are composed of 94 000 points, containing the key scheduling and encryption rounds. Our profiling is based on 256 000 traces corresponding to random plaintexts and keys. As a result, we expect around 1 000 traces for each value of each intermediate computation. We use $l^i_{n,x}$ for the value x of the nth intermediate value in the ith leakage trace, and $l^i_{n,x}(t)$ when we access at the tth point (sample) of this trace.

2.2 Information Detection Tools

Since SASCA can exploit many target intermediate values, we need to identify the time samples that contain information about them in our traces, next referred to as Points Of Interest (POI). We recall two simple methods for this purpose, and denote the POI of the nth intermediate value in our traces with t_n.

(a) Correlation Power Analysis (CPA) [6]. is a standard side-channel distinguisher that estimates the correlation between the measured leakages and some key-dependent model for a target intermediate value. In its standard version, an a-priori (here, Hamming weight) model is used for this purpose.

In practice, this estimation is performed by sampling (i.e. measuring) traces from a leakage variable L and a model variable M_k, using Pearson's correlation coefficient:

$$\rho_k(L, M_k) = \frac{\hat{\mathsf{E}}[(L - \hat{\mu}_L)(M_k - \hat{\mu}_{M_k})]}{\sqrt{\hat{\mathsf{var}}(L)\hat{\mathsf{var}}(M_k)}}.$$

In this equation, $\hat{\mathsf{E}}$ and $\hat{\mathsf{var}}$ respectively denote the sample mean and variance operators, and $\hat{\mu}_L$ is the sample mean of the leakage distribution L. CPA is a univariate distinguisher and therefore launched sample by sample.

(b) The Signal-to-Noise Ratio (SNR) [21]. of the nth intermediate value at the time sample t can be defined according to Mangard's formula [21]:

$$\mathsf{SNR}_n(t) = \frac{\hat{\mathsf{var}}_x\left(\hat{\mathsf{E}}_i\left(l_{n,x}^i(t)\right)\right)}{\hat{\mathsf{E}}_x\left(\hat{\mathsf{var}}_i\left(l_{n,x}^i(t)\right)\right)}.$$

Despite connected (high SNRs imply efficient CPA if the right model is used), these metrics allow slightly different intuitions. In particular, the SNR cannot tell apart the input and output leakages of a bijective operation (such as an S-box), since both intermediate values will generate useful signal. This separation can be achieved by CPA thanks to its a-priori leakage predictions.

2.3   Gaussian Templates Attacks

Gaussian TA [8] are the most popular profiled distinguisher. They assume that the leakages can be interpreted as the realizations of a random variable which generates samples according a Gaussian distribution and work in two steps. In a profiling phase, the adversary estimates a mean $\hat{\mu}_{n,x}$ and variance $\hat{\sigma}_{n,x}^2$ for each value x of the nth intermediate computation. In practice, this is done for the time sample t_n obtained thanks to the previously mentioned POI detection tools. Next, in the attack phase and for each trace l, she can calculate the likelihood to observe this leakage at the time t_n for each x as:

$$\hat{\mathsf{Pr}}[l(t_n)|x] \sim \mathcal{N}(\hat{\mu}_{n,x}, \hat{\sigma}_{n,x}^2).$$

In the context of standard DPA, we typically have $x = p \oplus k$, with p a known plaintext and k the target subkey. Therefore, the adversary can easily calculate $\hat{\mathsf{Pr}}[k^*|p, l(t_n)]$ using Bayes theorem, for each subkey candidate k^*:

$$\hat{\mathsf{Pr}}[k^*] = \prod_i \hat{\mathsf{Pr}}[k^*|p, l^i(t_n)].$$

To recover the full key, she can run a TA on each subkey independently.

By contrast, in the context of SASCA, we will directly insert the knowledge (i.e. probabilities) about any intermediate value x in the (factor) graph describing the implementation, and try to recover the full key at once.

Note that our SASCA experiments consider univariate Gaussian TA whereas our comparisons with DPA also consider bivariate TA exploiting the S-box input and output leakages (i.e. the typical operations that a divide-and-conquer adversary would exploit). In the latter case, the previous means and variances just have to be replaced by mean vectors and covariance matrices. This choice is motivated by our focus on the exploitation of multiple intermediate AES computations. It could be further combined with the exploitation of more samples per intermerdiate computation, e.g. thanks to dimensionality reduction [1].

2.4   Key Enumeration and Rank Estimation

At the end of a DC side-channel attack (as the previous TA), the attacker has probabilities on each subkey. If the master key is not the most probable one, she can perform enumeration up to some threshold thanks to enumeration algorithms, e.g. [36]. This threshold depends on the computational power of the adversary, since enumerating all keys is computationally impossible. If the key is beyond the threshold of computationally feasible enumeration, and in order to gain intuition about the computational security remaining after an attack, key rank estimation algorithms can be used [15,37]. A key rank estimation takes in input the list of probabilities of all subkeys and the probability of the correct key (which is only available in an evaluation context), and returns an estimation on the number of keys that are more likely than the actual key. Rank estimation allows to approximate d^{th}-order success rates (i.e. the probability that the correct key lies among the d first ones rated by the attack) efficiently and quite accurately. The security graphs introduced in [37] provide a visual representation of higher-order success rates in function of the number attack traces.

2.5   Algebraic Side-Channel Attacks

ASCA were introduced in [30] as one of the (if not the) first method to efficiently exploit all the informative samples in a leakage trace. We briefly recall their three main steps and refer to previous publications for the details.

1. *Construction* consists in representing the cipher as an instance of an algebraic problem (e.g. Boolean satisfiability, Groebner bases). Because of their large memory (RAM) requirements, ASCA generally build a system corresponding to one (or a few) traces only. For example, the SAT representation of a single AES trace in [32] has approximatively $18,000$ equations in $10,000$ variables.

2. *Information extraction* consists in getting exploitable leakages from the measurements. For ASCA, the main constraint is that actual solvers require hard information. Therefore, this phase usually translates the result of a TA into deterministic leakages such as the Hamming weight of the target intermediate

values. Note that the attack is (in principle) applicable with any type of lekages given that they are sufficiently informative and error-free.

3. *Solving.* Eventually, the side-channel information extracted in the second phase is added to the system of equations constructed in the first phase, and generic solvers are launched to solve the system and recover the key. In practice, this last phase generally has large RAM requirements causing ASCA to be limited to the exploitation of one (or two) measurement traces.

Summarizing, ASCA are powerful attacks since they can theoretically recover a key from very few leakage traces, but this comes at the cost of low noise-resilience, which motivated various heuristic improvements listed in introduction. The next SASCA are a more founded solution to get rid of this limitation.

2.6   Soft Analytical Side-Channel Attacks

SASCA [38] describe the target block cipher implementation and its leakages in a way similar to a Low-Density Parity Check code (LDPC) [12]. Since the latter can be decoded using soft decoding algorithms, it implies that SASCA can directly use the posterior probabilities obtained during a TA. Similar to ASCA, they can also be described in three main steps.

1. *Construction.* The cipher is represented as a so-called "factor graph" with two types of nodes and bidirectional edges. First, variable nodes represent the intermediate values. Second, function nodes represent the a-priori knowledge about the variables (e.g. the known plaintexts and leakages) and the operations connecting the different variables. Those nodes are connected with bidirectional edges that carry two types of messages (i.e. propagate the information) through the graph: the type q message are from variables to functions and the type r messages are from functions to variables (see [20] for more details).

2. *Information extraction.* The description of this phase is trivial. The probabilities provided by TA on any intermediate variable of the encryption process can be directly exploited, and added as a function node to the factor graph.

3. *Decoding.* Similar to LDPC codes, the factor graph is then decoded using the BP algorithm [28]. Intuitively, it essentially iterates the local propagation of the information about the variable nodes of the target implementation.

Since our work is mostly focused on concrete investigations of SASCA, we now describe the BP algorithm in more details. Our description is largely inspired by the description of [20, Chapter 26]. For this purpose, we denote by x_i the ith intermediate value and by f_i the ith function node. As just mentioned, the nodes will be connected by edges that carry two types of messages. The first ones go from a variable node to a function node, and are denoted as $q_{v_n \to f_m}$. The second ones go from a function node to a variable node, and are denoted as $r_{f_n \to v_m}$. In both cases, n is the index of the sending node and m the index of the recipient node. The messages carried correspond to the scores for the different values of the variable nodes. At the beginning of the algorithm execution, the messages from variable nodes to function nodes are initialized with no information on the variable. That is, for all n, m and for all x_n we have:

$$q_{v_n \to f_m}(x_n) = 1.$$

The scores are then updated according to two rules (one per type of messages):

$$r_{f_m \to v_n}(x_n) = \sum_{x_{n'}, n' \neq n} \left(f_m(x_{n'}, x_n) \prod_{n'} q_{v_{n'} \to f_m}(x_{n'}) \right). \qquad (1)$$

$$q_{v_n \to f_m}(x_n) = \prod_{m' \neq m} r_{f_{m'} \to v_n}(x_n). \qquad (2)$$

In Eq. 2, the variable node v_n sends the product of the messages about x_n received from the others function nodes ($m' \neq m$) to the function node f_m, for each value of x_n. And in Eq. 1, the function node f_m sends a sum over all the possible input values of f_m of the value of f_m evaluated on the vector of $(x_{n'}, n' \neq n)$'s, multiplied by the product of the messages received by f_m for the considered values of $x_{n'}$. The BP algorithm essentially works by iteratively applying these rules on all nodes. If the factor graph is a tree (i.e. if it has no loop), a convergence should occur after a number of iterations at most equal to the diameter of the graph. In case the graph includes loops (e.g. as in our AES implementation case), convergence is not guaranteed, but usually occurs after a number of iterations slightly larger than the graph diameter. The main parameters influencing the time and memory complexity of the BP algorithm are the number of possible values for each variable (i.e. 2^8 in our 8-bit example) and the number of edges. The time complexity additionally depends on the number of inputs of the function nodes representing the block cipher operations (since the first rule sums over all the input combinations of these operations).

3   Comparison with ASCA

ASCA and SASCA are both analytical attacks with very similar descriptions. As previously shown in [38], SASCA have a clear advantage when only noisy information is available. But when the information is noise-free, the advantage of one over the other has not been studied yet. In this section, we therefore tackle the question "which analytical attack is most efficient in noise-free scenario?". To this end, we compare the results of SASCA and ASCA against a simulated AES implementation with noise-free (Hamming weight) leakages. We first describe the AES representation we used in our SASCA (which will also be used in the following sections), then describe the different settings we considered for our simulated attacks, and finally provide the results of our experiments.

3.1   Our Representation for SASCA

As usual in analytical attacks, our description of the AES is based on its target implementation. This allows us to easily integrate the information obtained

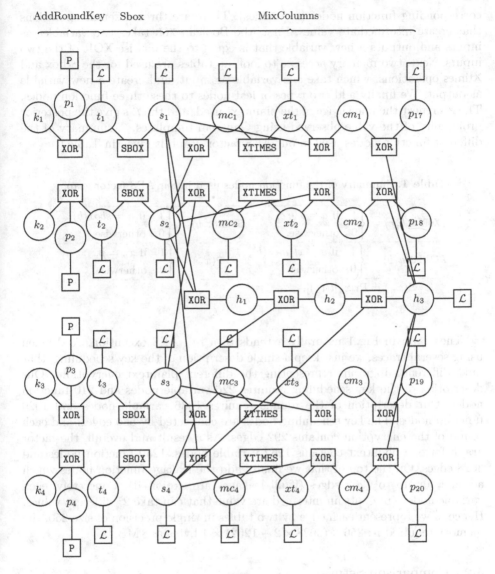

Fig. 1. Graph representation of one column of the first AES round.

during its execution. For readability purposes, we start by illustrating the graph representation for the first round of one column of the AES in Fig. 1. To build this graph for one plaintext, we start with 32 variable nodes (circles), 16 for the 8-bit subplaintexts (p_i), and 16 for the 8-bit subkeys (k_i). We first add a new variable node in the graph representation each time a new intermediate value is computed in the AES FURIOUS implementation,[2] together with the

---

[2] Excluding memory copies which only increase the graph diameter.

corresponding function nodes (rectangles). There are three different operations that create intermediate values. First, the Boolean XOR takes two variables as inputs and outputs a new variable that is equal to the bitwise XOR of the two inputs. Next, two memory accesses to look-up tables are used for the S-box and Xtimes operations, which take one variable as input, and create a new variable as output. We finally add two types of leaf nodes to these three function nodes. The P's reflect the knowledge of the plaintext used, and the $\mathcal{L}$'s give the posterior probability of the value observed using Gaussian templates. A summary of the different function nodes used in our AES factor graph is given in Table 1.

Table 1. Summary of the function nodes used in our AES factor graph.

$$\text{XOR}(a,b,c) = \begin{cases} 1 & \text{if } a = b \oplus c, \\ 0 & \text{otherwise.} \end{cases} \qquad \text{SBOX}(a,b) = \begin{cases} 1 & \text{if } a = sbox(b), \\ 0 & \text{otherwise.} \end{cases}$$

$$\text{XTIMES}(a,b) = \begin{cases} 1 & \text{if } a = xtimes(b), \\ 0 & \text{otherwise.} \end{cases} \qquad \text{P}(x_n) = \begin{cases} 1 & \text{if } x_n = p, \\ 0 & \text{otherwise.} \end{cases}$$

$$\mathcal{L}(x_n) = \Pr[x_n | l(t_n)].$$

The graph in Fig. 1 naturally extends to a full AES execution. And when using several traces, we just keep a single description of the key scheduling, that links different subgraphs representing the different plaintext encryptions. Our description of the key scheduling requires 226 variable nodes and 210 function nodes. Our description of the rounds requires 1036 variable nodes and 1020 function nodes. The key scheduling nodes are connected by 580 edges, and each round of the encryption contains 292 edges. As a result and overall, the factor graph for one plaintext contains 1262 variable nodes, 1230 function nodes and 3628 edges. On the top of that we finally add the leakage function nodes which account for up to 1262 edges (if all leakages are exploited). Concretely, each variable node represents an intermediate value that can take 2^8 different values. Hence, if we represent each edge by two tables in single precision of size 256, the memory required is: $256 \times (3628 \times 2 + 1262) \times 4$ bytes ≈ 8 MB.[3]

3.2   Comparison Setup

Our noise-free evaluations of ASCA and SASCA are based on single-plaintext attacks, which is due to the high memory requirements of ASCA (that hardly extend to more plaintexts). In order to stay comparable with the previous work in [32], we consider a Hamming weight (W_H) leakage function and specify the location of the leakages as follows:

- 16 W_H's for AddRoundKey,
- 16 W_H's for the output of SubBytes and ShiftRows,
- 36 W_H's for the XORs and 16 W_H for the look-up tables in MixColumns.

---

[3] For the leakage nodes, messages from variable to function ($q_{v_n \to f_m}$) are not necessary.

As previously mentioned, these leakages are represented by $\mathcal{L}$ boxes in Fig. 1. We also consider two different contexts for the information extraction:

- Consecutive weights (cw), i.e. the W_H's are obtained for consecutive rounds.
- Random weights (rw), i.e. we assume the knowledge of W_H's for randomly distributed intermediate values among the 804 possible ones.

Eventually, we analyzed attacks in a Known Plaintext (KP) and Unknown Plaintext (UP) scenario. And in all cases, we excluded the key scheduling leakages, as in [32]. Based on these settings, we evaluated the success rate in function of the quantity of information collected, counted in terms of "rounds of information", where one round corresponds to 84 W_H's of 8-bit values.

3.3 Experimental Results

The results of our SASCA with noise-free leakages are reported in Fig. 2, and compared to the similar ASCA experiments provided in Reference [32].

We first observe that 2 consecutive rounds of W_H's are enough to recover the key for SASCA with the knowledge of plaintext and when the leakages are located in the first rounds.[4] Next, if we do not have access to the plaintext, SASCA requires 3 consecutive rounds of leakage, as for ASCA. By contrast, and as previously underlined, the solving/decoding phase is significantly more challenging in case the leakage information is randomly distributed among the intermediate variables. This is intuitively connected to the fact that the solver and decoder both require to propagate information through the rounds, and that this information can rapidly vanish in case some intermediate variables are unknown. The simplest example is a XOR operation within MixColumns, as mentioned in introduction. So accumulating information on closely connected intermediate computations is always the best approach in such analytical attacks. This effect is of course amplified if the leakages are located in the middle rounds and the plaintext/ciphertext are unknown, as clear from Fig. 2.

Overall, and since both SAT-solvers and the BP algorithm with loops in the factor graph are highly heuristic tools, it is of course difficult to make strong statements about their respective leakage requirements. However, these experiments confirm that at least in the relevant case-study of Hamming weight AES leakages, the better noise-resilience of SASCA does not imply weaker performances in a noise-free setting. Besides, and in terms of time complexity, the attacks also differ. Namely, the resolution time for ASCA depends of the quantity of information, whereas it is independent of this quantity in SASCA, and approximately 20 times lower than the fastest resolution times for ASCA.

Note finally that moving to a noisy scenario can only be detrimental to ASCA. Indeed, and as discussed in [26], ASCA requires correct hard information for the

---

[4] We considered leakages for the two first rounds in this case, which seems more natural, and is the only minor differences with the experiments in [32], which considered middle rounds. However, we note that by considering middle round leakages with known plaintext, we then require three rounds of W_H's, as for ASCA.

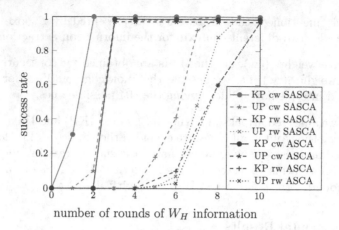

number of rounds of W_H information

Fig. 2. Experimental results of comparison of ASCA and SASCA.

key recovery to succeed. In case of noisy measurements, this can only be guaranteed by considering less informative classes of leakages or similar heuristics. For example, previous works in this direction considered Hamming weights h's between $h - d$ and $h + d$ for increasing distances d's, which rapidly makes the attack computationally hard (and cannot be mitigated with multiple plaintext leakages because of the high RAM requirements of ASCA). So the efficiency gain of SASCA over ASCA generally increases with the measurement noise.

4   SASCA Against a Concrete AES Implementation

In this section, we complete the previous simulated experiments and explore whether SASCA can be transposed in the more realistic context of measured leakages. To the best of our knowledge, we describe the first uses of SASCA against a concrete AES implementation, and take advantage of this case-study to answer several questions such as (i) how to perform the profiling of the many target intermediate values in SASCA?, (ii) what happens when the implementation details (such as the source code) are unknown?, and (iii) are there significant differences (or even gaps) between concrete and simulated experiments?

4.1   Profiling Step

We first describe how to exploit the tools from Sect. 2.2 in order to detect POIs for our 1230 target intermediate values (which correspond to 1262 variable nodes minus 32 corresponding to the 16 bytes of plaintext and ciphertext). In this context, directly computing the SNRs or CPAs in parallel for all our samples turns out to be difficult. Indeed, the memory requirements to compute the mean trace of an intermediate value with simple precision requires 94,000 (samples) $\times$ 256 (values) $\times$ 4 (bytes) $\approx$ 91MB, which means approximately 100 GB for the 1,230 values. For similar reasons, computing all these SNRs or CPAs sequentially is not possible (i.e. would require too much time). So the natural option is to trade time and memory by cutting the traces in a number of pieces that fit in RAM. This is easily done if we can assume some knowledge about the implementation (which we did), resulting in a relatively easy profiling step carried out in a dozen of hours on a single desktop computer. A similar profiling could be performed without implementation knowledge, by iteratively testing the intermediate values that appear sequentially in an AES implementation.

A typical outcome of this profiling is given in Fig. 3, where we show the SNR we observed for the intermediate value t_1 from the factor graph in Fig. 1 (i.e. the value of the bitwise XOR of the first subkey and the first subplaintext). As intuitively expected, we can identify significant leakages at three different times. The first one, at $t = 20,779$, corresponds to the computation of the value t_1, i.e. the XOR between p_1 and k_1. The second one, at $t = 22,077$, corresponds to the computation of the value s_1, i.e. a memory access to the look-up table of the S-box. The third one, at $t = 24,004$, corresponds to memory copies of s_1 during the computation of MixColumns. Indeed, the SNR cannot tell apart intermediate values that are bijectively related. So we used the CPA distinguisher to get rid of this limitation (taking advantage of the fact that a simple Hamming weight leakage model was applicable against our target implementation).

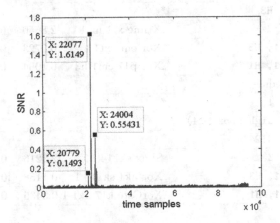

Fig. 3. SNR-based profiling of a single intermediate value.

A summary of the results obtained after our profiling step is given in Table 2, where the most interesting observation is that the informativeness of the leakage samples strongly depends on the target intermediate values. In particular, we see that memory accesses allow SNRs over 2, while XOR operations lead to SNRs below 0.4 (and this SNR is further reduced in case of consecutive XOR operations). This is in strong contrast, with the simulated cases (in the previous section and in [38]), where all the variables were assumed to leak with the same SNR. Note that the table mentions both SNR and CPA values, though our

Table 2. Summary of profiling step results.

Assembly code	Graph description	SNR	$\rho(W_H)$
Add round key			
ld H1, Y+	*	*	*
eor ST11, H1	_Xor t1 p1 k1	0.1493	0.5186
Sbox			
ldi ZH, high(sbox<<1)	*	*	*
mov ZL, ST11	*	*	*
lpm ST11, Z	_Sbox s1 t1	1.6301	0.4766
MixColumns			
ldi ZH, high(xtime<<1)	*	*	*
mov H1, ST11	*	*	*
eor H1, ST21	_Xor h1 s1 s2	0.1261	0.6158
eor H1, ST31	_Xor h2 h1 s3	0.0391	0.1449
eor H1, ST41	_Xor h3 h2 s4	0.3293	0.5261
mov H2, ST11	*	*	*
mov H3, ST11	*	*	*
eor H3, ST21	_Xor mc1 s1 s2	0.2802	0.6163
mov ZL, H3	*	*	*
lpm H3, Z	_Xtime xt1 mc1	2.8650	0.6199
eor ST11, H3	_Xor cm1 xt1 s1	0.0723	0.2508
eor ST11, H1	_Xor p17 cm1 h3	0.1064	0.3492
Key schedule			
ldi H1, 1	*	*	*
ldi ZH, high(sbox<<1)	*	*	*
mov ZL, ST24	*	*	*
lpm H3, Z	_Sbox sk14 k14	2.2216	0.5553
eor ST11, H3	_Xor ak1 sk14 k1	0.1158	0.5291
eor ST11, H1	_XorCste k17 ak1 1	0.3435	0.5140

selection of POIs was based on the (more generic) first criteria, and CPA was only used to separate the POIs of bijectively related intermediate values.[5]

4.2   Experimental Results

Taking advantage of the previous POI detection, we now want to discuss the consequences of different assumptions about the implementation knowledge. These investigations are motivated by the usual gap between Kerckhoff's laws [18], which

---

[5] We used a relatively noisy setup on purpose (e.g. we did not filter our measurements), in order to magnify the effectiveness of SASCA in such challenging contexts.

advises to keep the key as only secret in cryptography, and the practice in embedded security, that usually takes advantage of some obscurity regarding the implementations. For this purpose, we considered three adversaries:

1. *Informed.* The adversary has access to the implementation details (i.e. source code), and can exploit the leakages of all the target intermediate values.
2. *Informed, but excluding the key scheduling.* This is the same case as the previous one, but we exclude the key scheduling leakages as in the simulations of the previous section (e.g. because round keys are precomputed).
3. *Uninformed.* Here the adversary only knows the AES is running, assumes it is implemented following the specifications in [11], and only exploits generic operations (i.e. the inputs and outputs of AddRoundKey, SubByte, ShiftRows and MixColumns, together with the key rounds' inputs and outputs).

In order to have fair comparisons, we used the same profiling for all three cases (i.e. we just excluded some POIs for cases 2 and 3), and we used 100 sets of 30 traces with different keys and plaintexts to calculate the success rate of SASCA in these different conditions. The results of our experiments are in Fig. 4. Our first and main observation is that SASCA are applicable to actual implementations, for which the leakages observed provide more or less information (and SNR) depending on the intermediate values. As expected, the informed adversary is the most powerful. But we also see that excluding the key scheduling leakages, or considering an uninformed adversary, only marginally reduces the attack success rates. Interestingly, there is a strong correlation between this success rate and the number of leakage samples exploited, since excluding the key scheduling implies the removal of 226 leakage function nodes, and the uninformed adversary has 540 leakage function nodes less than the informed one (mostly corresponding to the MixColumns operation). So we can conclude that SASCA are not only a threat for highly informed adversaries, and in fact quite generically apply to unprotected software implementations with many leaking points.

Simulation Vs. Measurement. In view of the previous results, with information leakages depending on the target intermediate values, a natural question is whether security against SASCA was reasonably predicted with a simulated analysis. Of course, we know that in general, analytical attacks are much harder to predict than DPA [31], and do not enjoy simple formulas for the prediction of their success rates [22]. Yet, we would like to study informally the possible connection between simple simulated analyses and concrete ones. For this purpose, we compare the results obtained in these two cases in Fig. 5. For readability, we only report results for the informed and uninformed cases, and consider different SNRs for the simulated attacks. In this context, we first recall Table 2 where the SNRs observed for our AES implementation vary between 2^1 and 2^{-2}. Interestingly, we see from Fig. 5 that the experimental success rate is indeed bounded by these extremes. (Tighter and more rigorous bounds are probably hard to obtain for such heuristic attacks). Besides, we also observe that the success rates of the

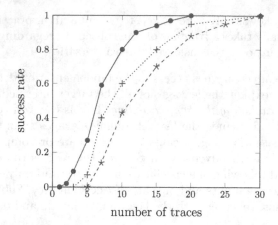

Fig. 4. Success rate in function of the # of traces for different adversaries: informed one (─●─), informed one without key scheduling leakages (··+··) and uninformed one (-✳-).

measurements and simulations are closer in the case of the uninformed adversary, which can be explained by the fact that we essentially ignore MixColumns leakages in this case, for which the SNRs are lower.

5 Comparison with DPA and Enumeration

In this section, we start from the observation that elegant approaches to side-channel analysis generally require more computational power than standard DPA. Thus, a fair comparison between both approaches should not only look at the success rate in function of the number of traces, but also take into account the resolution time as a parameter. As a result, and in order to compare SASCA and the pragmatic DPA on a sound basis, this section investigates the result of DC attacks combined with computational power for key enumeration.

5.1 Evaluation of Profiled Template Attacks

In order to be as comparable as possible with the previous SASCA, our comparison will be based on the profiled TA described in Sect. 2.3.[6] More precisely, we considered a quite pragmatic DC attack exploiting the bivariate leakages corresponding to the AddRoundKey and SubByte operations (i.e. $\{s_i\}_{i=1}^{16}$ and $\{t_i\}_{i=1}^{16}$ in Fig. 1). We can take advantage of the same detection of POIs as described in the previous section for this purpose. This choice allows us to keep the computational complexity of the TA itself very minimal (since relying only on 8-bit hypotheses). As previously mentioned, it also aims to make comparison

---

[6] We considered TA for our DPA comparison because they share the same profiled setting as SASCA. Comparisons with a non-profiled CPA can only be beneficial to SASCA. More precisely, we expect a typical loss factor of 2 to 5 between (W_H-based) CPA and TA, according to the results in [35] obtained on the same device.

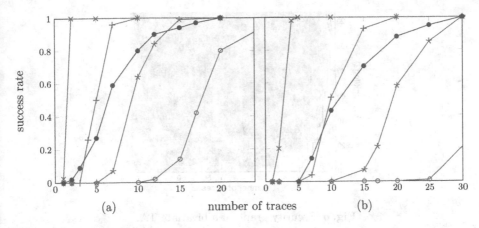

Fig. 5. Experimental results for SASCA for an informed adversary (a) and uninformed adversary (b). Red curves are for simulated cases ($-\times-,-+-,-*-,-\ominus-$) for SNR ($2^1, 2^{-1}, 2^{-2}, 2^{-3}$). Blue curves ($-\bullet-$) are for experiments on real traces (Color figure online).

as meaningful as possible (since we compare two attacks with one sample per target operation that only differ by their number of target operations). Following, we built the security graph of our bivariate TA, as represented in Fig. 6, where the white (resp. black) curve corresponds to the maximum (resp. minimum) rank observed, and the red curve is for the average rank. It indicates that approximately 60 plaintexts are required to recover the key without any enumeration (which is in line with Footnote 5). But more interestingly, the graph also highlights that allowing enumeration up to ranks (e.g.) 2^{30} allows to reduce the required number of measured traces down to approximately 10.

5.2 Comparing SASCA and DPA with Enumeration

In our prototype implementation running on a desktop computer, SASCA requires roughly one second per plaintext, and reaches a success rate of one after 20 plaintexts (for the informed adversary). In order to allow reasonably fair comparisons, we first measured that the same desktop computer can perform a bit more than 2^{20} AES encryptions in 20 seconds. So this is typically the amount of enumeration that we should grant the bivariate TA for comparisons with SASCA.[7] For completeness, we also considered the success rates of bivariate TA without enumeration and with 2^{30} enumeration power.[8] The results of these last experiments

---

[7] We omit to take the (time and memory) resources required for the generation of the list of the most probable keys to enumerate into account in our comparisons, since these resources remain small in the total enumeration cost. Using the state-of-the-art enumeration algorithm [36], we required 2.7MB + 0.55 seconds to generate a list of 2^{20} keys, and 1.8GB + 3130 seconds to generate a list of 2^{32} keys.

[8] Which is also more than allowed by the new suboptimal key enumeration in [3].

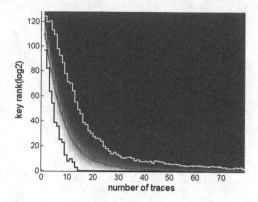

Fig. 6. Security graph of a bivariate TA.

are in Fig. 7. Overall, they bring an interesting counterpart to our previous investigations. On the one hand, we see that SASCA remains the most powerful attack when the adversary has enough knowledge of the implementation. By contrast in the uninformed case, the gain over the pragmatic TA with enumeration is lower. So as expected, it is really the amount and type of leakage samples exploitable by the adversary that make SASCA more or less powerful, and determine their interest (or lack thereof) compared to DC attacks. In this respect, a meaningful observation is that the gap between SASCA and DPA without enumeration (here approximately 5) is lower than the approximate factor 10 that was observed in the previous simulations of [38]. This difference is mainly due to the lower SNRs observed in the MixColumns transform.

Eventually, we note that in view of these results, another natural approach would be to use enumeration for SASCA. Unfortunately, our experiments have shown that enumeration is much less effective in the context of analytical attacks. This is essentially caused by the fact that DC attacks consider key bytes independently, whereas SASCA decode the full key at once, which implies that the subkey probabilities are not independent in this case, and can be degraded when running the loopy BP too long. Possible tracks to improve this issue include the use of list decoding algorithms for LDPC codes (as already mentioned in [13]), or enumeration algorithms that can better take subkey dependencies into account (as suggested in [19] for elliptic curve implementations).

6   Conclusion and Open Problems

This paper puts forward that the technicalities involved in elaborate analytical side-channel attacks, such as the recent SASCA, are possible to solve in practice. In particular, our results show that the intensive profiling of many target intermediate values within an implementation is achievable with the same (SNR &CPA) tools as any profiled attack (such as the bivariate TA we considered). This profiling only requires a dozen of hours to complete, and then enables very efficient SASCA that recover the key of our AES implementation in a couple

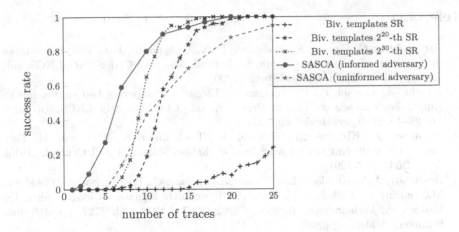

Fig. 7. Comparison between elegant and pragmatic approaches.

of seconds and traces, using a single desktop computer. Furthermore, these successful attacks are even possible in a context where limited knowledge about the target implementation is available, hence mitigating previous intuitions regarding analytical attacks being "only theoretical". Besides this positive conclusion, a fair comparison with DC attacks also highlights that the gap between a bivariate TA and a SASCA can be quite reduced in case enumeration power is granted to the DC adversary, and several known plaintexts are available. Intuitively, the important observation in this respect is that the advantage of SASCA really depends on the amount and type of intermediate values leaking information, which highly depends on the algorithms and implementations analyzed.

The latter observation suggests two interesting directions for further research. On the one hand, the AES Rijndael is probably among the most challenging targets for SASCA. Indeed, it includes a strong linear diffusion layer, with many XOR operations through which the information propagation is rapidly amortized. Besides, it also relies on a non-trivial key scheduling, which prevents the direct combination of information leaked from multiple rounds. So it is not impossible that the gap between SASCA and standard DPA could be larger for other ciphers (e.g. with permutation based diffusion layers [4], and very minimum key scheduling algorithms [5]). On the other hand, since the propagation of the leakage information through the MixColumns operation is hard(er), one natural solution to protect the AES against such attacks would be to enforce good countermeasures for this part of the cipher, which would guarantee that SASCA do not exploit more information than the one of a single round. Ideally, and if one can prevent any information propagation beyond the cipher rounds, we would then have a formal guarantee that SASCA is equivalent to DPA.

Acknowledgements. F.-X. Standaert is a research associate of the Belgian Fund for Scientific Research (FNRS-F.R.S.). This work has been funded in parts by the European Commission through the ERC project 280141 (CRASH).

References

1. Archambeau, C., Peeters, E., Standaert, F.-X., Quisquater, J.-J.: Template attacks in principal subspaces. In: Goubin, L., Matsui, M. (eds.) CHES 2006. LNCS, vol. 4249, pp. 1–14. Springer, Heidelberg (2006)
2. Banciu, V., Oswald, E.: Pragmatism vs. Elegance: comparing two approaches to simple power attacks on AES. In: Prouff, E. (ed.) COSADE 2014. LNCS, vol. 8622, pp. 29–40. Springer, Heidelberg (2014)
3. Bogdanov, A., Kizhvatov, I., Manzoor, K., Tischhauser, E., Witteman, M.: Fast and memory-efficient key recovery in side-channel attacks. IACR Cryptol. ePrint Arch. **2015**, 795 (2015)
4. Bogdanov, A.A., Knudsen, L.R., Leander, G., Paar, C., Poschmann, A., Robshaw, M., Seurin, Y., Vikkelsoe, C.: PRESENT: an ultra-lightweight block cipher. In: Paillier, P., Verbauwhede, I. (eds.) CHES 2007. LNCS, vol. 4727, pp. 450–466. Springer, Heidelberg (2007)
5. Bogdanov, A., Knudsen, L.R., Leander, G., Standaert, F.-X., Steinberger, J., Tischhauser, E.: Key-alternating ciphers in a provable setting: encryption using a small number of public permutations. In: Pointcheval, D., Johansson, T. (eds.) EUROCRYPT 2012. LNCS, vol. 7237, pp. 45–62. Springer, Heidelberg (2012)
6. Brier, E., Clavier, C., Olivier, F.: Correlation power analysis with a leakage model. In: Joye, M., Quisquater, J.-J. (eds.) CHES 2004. LNCS, vol. 3156, pp. 16–29. Springer, Heidelberg (2004)
7. Carlet, C., Faugère, J.-C., Goyet, C., Renault, G.: Analysis of the algebraic side channel attack. J. Crypt. Eng. **2**(1), 45–62 (2012)
8. Chari, S., Rao, J.R., Rohatgi, P.: Template attacks. In: Kaliski Jr, B.S., Koç, Ç.K., Paar, C. (eds.) Cryptographic Hardware and Embedded Systems - CHES 2002. LNCS, vol. 2523. Springer, Heidelberg (2002)
9. Daemen, J., Rijmen, V.: The Design of Rijndael: AES - The Advanced Encryption Standard. Information Security and Cryptography. Springer, Heidelberg (2002)
10. Dziembowski, S., Pietrzak, K.: Leakage-resilient cryptography. In: 49th Annual IEEE Symposium on Foundations of Computer Science, FOCS 2008, October 25–28, 2008, Philadelphia, PA, USA, pp. 293–302. IEEE Computer Society (2008)
11. Pub, FIPS 197. Advanced encryption standard (AES). http://csrc.nist.gov/publications/fips/fips197/fips-197.pdf
12. Gallager, R.G.: Low-density parity-check codes. IRE Trans. Inf. Theor. **8**(1), 21–28 (1962)
13. Gérard, B., Standaert, F.-X.: Unified and optimized linear collision attacks and their application in a non-profiled setting: extended version. J. Crypt. Eng. **3**(1), 45–58 (2013)
14. Gierlichs, B., Batina, L., Tuyls, P., Preneel, B.: Mutual information analysis. In: Oswald, E., Rohatgi, P. (eds.) CHES 2008. LNCS, vol. 5154, pp. 426–442. Springer, Heidelberg (2008)
15. Glowacz, C., Grosso, V., Poussier, R., Schueth, J., Standaert, F.-X.: Simpler and more efficient rank estimation for side-channel security assessment. IACR Cryptol. ePrint Arch. **2014**, 920 (2014)
16. Guo, S., Zhao, X., Zhang, F., Wang, T., Shi, Z.J., Standaert, F.X., Ma, C.: Exploiting the incomplete diffusion feature: A specialized analytical side-channel attack against the AES and its application to microcontroller implementations. IEEE Trans. Inf. Forensics Secur. **9**(6), 999–1014 (2014)

17. Hanley, N., Tunstall, M., Marnane, W.P.: Unknown plaintext template attacks. In: Youm, H.Y., Yung, M. (eds.) WISA 2009. LNCS, vol. 5932, pp. 148–162. Springer, Heidelberg (2009)

18. Kerckhoffs, A.: La cryptographie militaire, ou, Des chiffres usités en temps de guerre: avec un nouveau procédé de déchiffrement applicable aux systèmes à double clef. Librairie militaire de L, Baudoin (1883)

19. Lange, T., van Vredendaal, C., Wakker, M.: Kangaroos in side-channel attacks. In: Joye, M., Moradi, A. (eds.) CARDIS 2014. LNCS, vol. 8968, pp. 104–121. Springer, Heidelberg (2015)

20. MacKay, D.J.C.: Information Theory, Inference, and Learning Algorithms, vol. 7. Cambridge University Press, Cambridge (2003)

21. Mangard, S.: Hardware countermeasures against DPA – a statistical analysis of their effectiveness. In: Okamoto, T. (ed.) CT-RSA 2004. LNCS, vol. 2964, pp. 222–235. Springer, Heidelberg (2004)

22. Mangard, S., Oswald, E., Standaert, F.-X.: One for all - all for one: unifying standard differential power analysis attacks. IET Inf. Secur. 5(2), 100–110 (2011)

23. Mather, L., Oswald, E., Whitnall, C: Multi-target DPA attacks: pushing DPA beyond the limits of a desktop computer. In: Sarkar and Iwata [33], pp. 243–261

24. Mohamed, M.S.E., Bulygin, S., Zohner, M., Heuser, A., Walter, M., Buchmann, J.: Improved algebraic side-channel attack on AES. J. Crypt. Eng. 3(3), 139–156 (2013)

25. Oren, Y., Kirschbaum, M., Popp, T., Wool, A.: Algebraic side-channel analysis in the presence of errors. In: Mangard, S., Standaert, F.-X. (eds.) CHES 2010. LNCS, vol. 6225, pp. 428–442. Springer, Heidelberg (2010)

26. Oren, Y., Renauld, M., Standaert, F.-X., Wool, A.: Algebraic side-channel attacks beyond the hamming weight leakage model. In: Prouff, E., Schaumont, P. (eds.) CHES 2012. LNCS, vol. 7428, pp. 140–154. Springer, Heidelberg (2012)

27. Oren, Y., Weisse, O., Wool, A.: A new framework for constraint-based probabilistic template side channel attacks. In: Batina, L., Robshaw, M. (eds.) CHES 2014. LNCS, vol. 8731, pp. 17–34. Springer, Heidelberg (2014)

28. Pearl, J.: Reverend Bayes on inference engines: a distributed hierarchical approach. In: Waltz, D.L. (ed) Proceedings of the National Conference on Artificial Intelligence, Pittsburgh, PA, August 18–20, 1982, pp. 133–136. AAAI Press (1982)

29. Rebeiro, C., Selvakumar, D., Devi, A.S.L.: Bitslice implementation of AES. In: Pointcheval, D., Mu, Y., Chen, K. (eds.) CANS 2006. LNCS, vol. 4301, pp. 203–212. Springer, Heidelberg (2006)

30. Renauld, M., Standaert, F.-X.: Algebraic side-channel attacks. In: Bao, F., Yung, M., Lin, D., Jing, J. (eds.) Inscrypt 2009. LNCS, vol. 6151, pp. 393–410. Springer, Heidelberg (2010)

31. Bauer, A., Coron, J.-S., Naccache, D., Tibouchi, M., Vergnaud, D.: On the broadcast and validity-checking security of pkcs#1 v1.5 encryption. In: Zhou, J., Yung, M. (eds.) ACNS 2010. LNCS, vol. 6123, pp. 1–18. Springer, Heidelberg (2010)

32. Renauld, M., Standaert, F.-X., Veyrat-Charvillon, N.: Algebraic side-channel attacks on the AES: why time also matters in DPA. In: Clavier, C., Gaj, K. (eds.) CHES 2009. LNCS, vol. 5747, pp. 97–111. Springer, Heidelberg (2009)

33. Sarkar, P., Iwata, T. (eds.): Advances in Cryptology - ASIACRYPT 2014. LNCS, vol. 8873. Springer, Berlin Heidelberg (2014)

34. Schindler, W., Lemke, K., Paar, C.: A stochastic model for differential side channel cryptanalysis. In: Rao, J.R., Sunar, B. (eds.) CHES 2005. LNCS, vol. 3659, pp. 30–46. Springer, Heidelberg (2005)

35. Standaert, F.-X., Gierlichs, B., Verbauwhede, I.: Partition *vs.* Comparison side-channel distinguishers: an empirical evaluation of statistical tests for univariate side-channel attacks against two unprotected CMOS devices. In: Lee, P.J., Cheon, J.H. (eds.) ICISC 2008. LNCS, vol. 5461, pp. 253–267. Springer, Heidelberg (2009)
36. Veyrat-Charvillon, N., Gérard, B., Renauld, M., Standaert, F.-X.: An optimal key enumeration algorithm and its application to side-channel attacks. In: Knudsen, L.R., Wu, H. (eds.) SAC 2012. LNCS, vol. 7707, pp. 390–406. Springer, Heidelberg (2013)
37. Veyrat-Charvillon, N., Gérard, B., Standaert, F.-X.: Security evaluations beyond computing power. In: Johansson, T., Nguyen, P.Q. (eds.) EUROCRYPT 2013. LNCS, vol. 7881, pp. 126–141. Springer, Heidelberg (2013)
38. Veyrat-Charvillon, N., Gérard, B., Standaert, F.-X.: Soft analytical side-channel attacks. In: Sarkar and Iwata [33], pp. 282–296
39. Zhao, X., Zhang, F., Guo, S., Wang, T., Shi, Z., Liu, H., Ji, K.: MDASCA: an enhanced algebraic side-channel attack for error tolerance and new leakage model exploitation. In: Schindler, W., Huss, S.A. (eds.) COSADE 2012. LNCS, vol. 7275, pp. 231–248. Springer, Heidelberg (2012)

Counting Keys in Parallel After a Side Channel Attack

Daniel P. Martin[✉], Jonathan F. O'Connell, Elisabeth Oswald,
and Martijn Stam

Department of Computer Science, University of Bristol,
Merchant Venturers Building, Woodland Road, Bristol BS8 1UB, UK
{dan.martin,j.oconnell,elisabeth.oswald,martijn.stam}@bris.ac.uk

Abstract. Side channels provide additional information to skilled adversaries that reduce the effort to determine an unknown key. If sufficient side channel information is available, identification of the secret key can even become trivial. However, if not enough side information is available, some effort is still required to find the key in the key space (which now has reduced entropy). To understand the security implications of side channel attacks it is then crucial to evaluate this remaining effort in a meaningful manner. Quantifying this effort can be done by looking at two key questions: first, how 'deep' (at most) is the unknown key in the remaining key space, and second, how 'expensive' is it to enumerate keys up to a certain depth?

We provide results for these two challenges. Firstly, we show how to construct an extremely efficient algorithm that accurately computes the rank of a (known) key in the list of all keys, when ordered according to some side channel attack scores. Secondly, we show how our approach can be tweaked such that it can be also utilised to enumerate the most likely keys in a parallel fashion. We are hence the first to demonstrate that a smart and parallel key enumeration algorithm exists.

Keywords: Key enumeration · Key rank · Side channels

1 Introduction

Side channel attacks have proven to be a hugely popular research topic, as the proliferation of new venues such as CHES, COSADE and HOST shows. Much of the published research is about key recovery attacks utilising side channel information. Key recovery attacks essentially take a number of side channel observations, colloquially referred to as 'traces', and apply a so-called distinguisher to traces that assigns scores to keys. An attack is considered (first-order) successful given a set of traces, if the actual secret key receives the highest score. Besides describing methods (*i.e.* the distinguishers) that recover the secret key from the available data, papers focus on the question of how many traces are required for successful first-order attacks.

© International Association for Cryptologic Research 2015
T. Iwata and J.H. Cheon (Eds.): ASIACRYPT 2015, Part II, LNCS 9453, pp. 313–337, 2015.
DOI: 10.1007/978-3-662-48800-3_13

The trade-off chosen in most work is, hence, to increase the number of traces to ensure that the secret key is recovered successfully with almost certainty. As observed by Veyrat-Charvillon et al. [13] in their seminal paper on optimal key enumeration, this might not be the trade-off that a well resourced adversary would choose. Suppose that access to the side channel is scarce or difficult. In such a case the actual secret key might not have the highest score after utilising the leakage traces, but it might still have a higher score than many other keys. Now imagine that the adversary can utilise substantial computational resources. This implies that by searching through the key space (in order of decreasing scores; we call this smart key enumeration) the adversary would find the secret key much faster than by a naïve brute-force search (i.e. one that treats all keys as equally likely). Consequently, the true security level of an implementation cannot be judged solely by its security against first-order side channel attacks. Instead it is important to understand how the number of traces impacts on the effort required for a smart key enumeration.

We now illustrate this motivation by linking it to evaluating the impact of the most influential type of side channel attack: Differential Power Analysis.

1.1   Evaluating Resistance Against Differential Power Analysis

Differential Power Analysis (DPA) [9] consists of predicting a so-called target function, e.g. the output of the Substitution Boxes, and mapping the output of this function to 'predicted side channel values' using a power model. For this process it is not necessary to know or guess the whole secret key, SK. One only needs to make guesses about 'enough bits'. The predicted values for a key chunk are then 'compared' to the real traces (point-wise) using a distinguisher. Assuming enough traces are available, the value that represents the correct key guess will lead to a 'higher' distinguishing value. In Kocher et al.'s original paper [9] this was illustrated for the DES cipher, but most contemporary research uses AES as running example.

With respect to AES: Kocher's attack consists of using a t-test statistic as a distinguisher to compute scores for the values of each 8-bit chunk of the 128-bit key; see Fig. 1 for a visual example. Here we have $m = 16$ chunks, each containing $n = 256$ values, with associated distinguishing scores as derived via a t-test statistic. In the graphical illustration, the secret key values are marked out in grey.

If sufficient side information is available, the values of the chunks that correspond to the secret key will have by far the highest distinguishing scores, such as the majority of key chunks in our graphical illustration. In this case the secret key can be trivially found (it is the concatenation of the chunks that lead to the uniquely highest score). However, if less side information is available, the scores may not necessarily favour a single key. Nevertheless, an adversary is still able to utilise these scores to smartly enumerate and then test keys (by using a known plaintext-ciphertext pair).

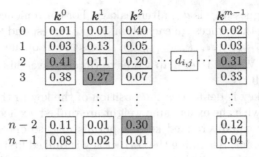

Fig. 1. Score vectors for m key chunks. Each chunk can take values from 0 to $n-1$, and scores $d_{i,j}$ are on a scale that depends on the side channel distinguisher. The values that correspond to the (hypothetical) secret key are highlighted in grey.

Security Evaluations. Considering the perspective of a security evaluator, it is obviously important to characterise the remaining security of an implementation *after* leakage. The evaluator (who can run experiments with a known key, and a varying number of traces) wants to compute its position in a ranked list of all keys. Knowing this position allows the evaluator to assess the amount of effort required by an adversary performing a smart search (given some distinguishing vectors). Ideally, the evaluator is able to compute the ranks of arbitrarily deep keys.

Accuracy and Efficiency are Key Requirements: Naturally, because the evaluator performs concrete statistical experiments, a single run of a single attack is not sufficient to gather sound evidence. In practice, any side channel experiment needs to be repeated multiple times by the evaluation lab, and many different attacks need to be carried out, utilising different amounts of side channel traces. Having the capability to determine the position of the key in a ranked list accurately (rather than just giving an estimation), and efficiently, is crucial to correctly assess the effort of a real world adversary. Previous works' algorithms [6,14] were capable of estimating the key rank within some bound. We demonstrate that we are accurate when enough precision is used, and importantly, we put forward the first approach for parallel and smart key enumeration.

1.2 Problem Statement and Notation

We use a bold type face to denote multi-dimensional entities. Indices in superscript refer to column vectors (we use the variable j for this purpose), and indices in subscript refer to row vectors (we use i for this purpose). Two indices i, j refer to an element in row i and column j. To maintain an elegant layout, we sometimes typeset column vectors 'in line', and then indicate transposition via a superscript $k = (\dots)^T$.

We partition a key guess k into m chunks, each able to take one of n possible values, *i.e.* $k = (k^0, \dots, k^{m-1})$, and $k^j = (d_{0,j}, d_{1,j}, \dots, d_{n-1,j})^T$. After exploiting some leakage L all chunks k^j have some corresponding score vectors, *i.e.* we know the score for each guess $k_{i,j}$ is $d_{i,j}$ after leakage. For convenience we use the variable sk_j to refer to the indices (in each chunk) that correspond to the correct key, i.e. $SK = (k_{sk_1,1}, k_{sk_2,2}, \ldots, k_{sk_{m-1},m-1})$. The score D of the secret key is then $D = \sum_{j=0}^{m-1} d_{sk_j,j}$. We will later map scores to (integer) weights and the weight of the secret key will be W.

The rank of a key is defined as the position of the key in the ordering (of all keys), where keys with the exact same weight are ranked 'ex aequo'. In principle, any order of these equally ranked keys is permissible, so one is free to make a choice about this order. Assuming the correct key is ranked first among all keys of the same weight requires us to count all keys with weight less than W. It implies that the rank we return is conservative in the following sense: key ranking is used to evaluate the security of side-channel attacks; our assumption on the ordering implies we give a side-channel adversary the benefit of the doubt (and so we deem it slightly more successful than it in reality can be). As an alternative, one could assume the correct key is ranked last among all keys of the same weight (since we use integer weights, this can be done by increasing the weight by one, counting all keys according to the ranked-first method, and subtracting one from the returned rank); ranking the candidate key both as first and last of its weight will lead to an interval of ranks containing all keys of that rank. Thus our choice (rank first) is effectively without loss of generality: run once it gives a conservative estimate, run twice it gives the exact interval of possible ranks for the candidate key.

Definition 1 (Key Rank Computation). *Given m vectors of n distinguishing scores, and the score D of the secret key SK, count the number of keys with score strictly larger than D.*

Definition 2 (Smart Key Enumeration). *Given m vectors of n distinguishing scores, list the B keys with the highest score.*

1.3   Outline and Our Contributions

In a nutshell, we utilise an elegant mapping of the key rank computation problem to a knapsack problem, which can be simplified and expressed as (efficient) path counting. As a result, we can compute *accurate* key ranks, and importantly, this enables us to put forward the first algorithm that can perform smart key enumeration in a parallel manner.

Our contribution is structured in four main sections as follows:

Casting the Key Enumeration as an Integer Knapsack. In Sect. 2 we show how to cast the key enumeration problem as a solution to counting knapsack solutions. In particular, we develop the representation of key rank as a multi-dimensional knapsack, and discuss its resulting graph representation. Whilst the final definition can be represented as an integer programming problem, we choose to frame each step as an extension of the knapsack problem, for intuition.

A Key Rank Algorithm. In Sect. 3 we map the multi-dimensional knapsack to a directed acyclic graph. We can therefore count solutions to the multi-dimensional knapsack problem by counting paths in the graph. The restriction of picking one item per chunk keeps the number of vertices in the directed acyclic graph small. As the graph is compact, and each node has at most two outgoing edges, the path counting problem can be solved efficiently in $O\left(m^2 \cdot n \cdot W \cdot \log n\right)$.

Smart Key Enumeration. In Sect. 4, with the additional book-keeping of storing the vertices we visit, we can enumerate the B most likely keys with complexity $O\left(m^2 \cdot n \cdot W \cdot B \cdot \log n\right)$. We then show several techniques to make this process as efficient as possible.

Practical Evaluation and Comparison with Previous Work. In Sect. 5 we discuss requirements around precision. The main factor that influences performance is the size of the key rank graph, which is determined by the precision of the initial mapping and the weight of the target key. We compare our work with previous works in terms of precision and speed with regards to the key rank algorithm, and in terms of speed with regards to smart key enumeration.

A full version of this paper can be found on ePrint[1], where we consider additional alternative topological sorting methods and provide pseduo-code for each. Also implementation details, and testing methodologies are considered in greater depth.

1.4   Previous Work

Key Rank. An naïve approach is that by simply removing a number of the least likely key values from each key chunk, the size of the search space is then restricted as n is reduced. However there are inherit problems with the approach; firstly this may be removing valid high ranking keys, as it is possible that a key may be constructed from one very low ranked value in one key chunk, and very high in others. Secondly, it is still reliant on a simple brute force approach, and even with a reduced n value this approach is thus too expensive to be practical. Finally, if the target key is deep, this approach won't work at all as it is possible that the correct key values have been removed.

Veyrat-Charvillon et al. [14] were the first to demonstrate an algorithm to estimate the rank of the key without using full key enumeration. The search space can be represented as a multidimensional space, with each dimension corresponding to a key chunk, sorted by decreasing likelihoods. The space can be divided into two, those keys ranked above the target and those ranked below. Using the property that the 'frontier' between these two spaces is convex, they are able to 'trim' each space down until the key rank has been estimated to within 10 bits of accuracy.

Bernstein et al. [1] propose two key ranking algorithms. The first is based on [14] and adds a post processing phase which has been shown to tighten the

---

bounds from 10 bits to 5 bits. The second algorithm ranks keys using techniques similar to those used to count all y-smooth numbers less than x. By having an accuracy parameter they are able to get their bounds arbitrarily tight (at the expense of run time).

Glowacz et al. [6] construct a novel rank estimation algorithm using convolution of histograms. Using the property that $S_1 + S_2 := \{x_1 + x_2 | x_1 \in S_1, x_2 \in S_2\}$ can be approximated by histogram convolution, by creating a histogram per key chunk and convoluting them all together, they are able to efficiently estimate the key rank to within 1 bit of precision.

Duc et al. [4] perform key rank using a method inspired by Glowacz et al. [6]. They repeatedly 'merge' the data in one column at a time (as the histograms were convoluted in one at a time). Each piece of information is downsampled to one of a series of discrete values (similar to putting into a histogram bin). The major difference is that instead of just downsampling the orginial data, they also downsample after each key chunk is merged in.

Key Enumeration. Veyrat-Charvillon et al. [13] propose a deterministic algorithm to enumerate keys based on a divide-and-conquer approach. Using a tree-like recursion (starting with two subkeys, then four, all the way to sixteen) and keeping track of what they call the frontier set (similarities can be drawn to the frontier of Veyrat-Charvillon et al. [14]), they are able to efficiently enumerate keys.

Ye et al. [15] present what they describe as a Key Space Finding algorithm. A Key Space Finding algorithm takes in the distinguishing score vector and returns two outputs: the minimum verification complexity to ensure a desired success probability, along with the optimal effort distributor which achieves this lower bound. Given this it is straightforward to run a (probabilistic) key enumeration algorithm. The distinguisher intuitively moves the boundary of which subkeys to enumerate until the desired probability is achieved.

Bogdanov et al. [2] create a score based key enumeration algorithm which can be seen as a variation of depth first search. Potential keys are generated via *score paths*, each of which has a score associated with them, which in conjunction with a precomputed score table allows for efficient pruning of impossible paths. From here it is possible to efficiently enumerate possible values.

Reflecting on the Approaches Taken by Previous Work. All of the previous work has treated key rank and key enumeration as two disjoint problems and hence approached them using different techniques. For instance, it is unclear how to extend the existing key rank algorithms to enumerate keys, and conversely, it is not apparent how to simplify the enumeration algorithms to compute key ranks efficiently (*i.e.* without just counting as you enumerate). We however believe that both of these problems are highly similar in nature and by maintaining some structure within the key rank it should be possible to enumerate without making the ranking inefficient. In the remainder of the paper we explain how to do just that: efficient key ranking with enough structure to make (fully parallel) enumeration possible.

2   Casting the Key Enumeration as a Knapsack

We now explain how the key enumeration problem can be formulated as a variant of a knapsack problem. In its most basic form a knapsack problem takes a set of n items that have a profit p_i and a weight w_i. A binary variable x_i is used to select items from the set. The objective is then to select items such that the profit is maximised whilst the total weight of the items does not exceed a set maximum W:

$$\text{maximize: } \sum_{i=0}^{n-1} p_i \cdot x_i$$

$$\text{subject to: } \sum_{i=0}^{n-1} w_i \cdot x_i \leq W$$

$$x_i \in \{0, 1\}, \forall i$$

The counting knapsack (#knapsack) problem is then understood to be the associated counting problem: given a knapsack definition, count how many solutions there are to the knapsack problem.

Intuitively, we should be able to frame the key rank computation problem as a knapsack variant. In contrast to a basic knapsack, however, we have classes of items (these are the distinguishing vectors k^j), profits can be dropped since we are counting the number of solutions, and we must take exactly one item from each class. The weight $w_{i,j}$ for each item can be derived[2] from the distinguishing score $w_{i,j} = MapToWeight(d_{i,j})$ in such a way that higher distinguishing scores lead to lower weights[3]. We define the maximum weight W as the sum of the weights associated with the secret key chunks, i.e. $W = \sum_{j=0}^{m-1} w_{sk_j,j}$. Recall that we assume if several keys have weight W the secret key (which must be among those) is listed first. This enforces W as a strict upper bound in the knapsack definition.

The multiple-choice knapsack problem that identifies keys with weight lower than W is then defined as follows:

$$\sum_{j=0}^{m-1} \sum_{i=0}^{n-1} w_{i,j} \cdot x_{i,j} < W$$

$$\sum_{i=0}^{n-1} x_{i,j} = 1, \forall j$$

$$x_{i,j} \in \{0, 1\}, \forall i, j$$

---

[2] For the sake of readability, we do not discuss the implications of needing to map distinguishing scores (which are floating point values) to weights at this point, but refer the reader to Sect. 5.1 for a discussion.

[3] This ensures compatibility with knapsack notation.

The first constraint ensures that all keys (*i.e.* selections of items) have a weight lower than the secret key. The second constraint ensures that only one item per distinguishing vector is selected. The counting version of this multiple-choice knapsack equals to computing the key rank.

Counting solutions to knapsack problems in general is known to be a computationally hard problem, and known classical solutions [5] rely on combinations of dynamic programming and rejection sampling to construct an FPRAS. Gopalan *et al.* [7] more recently utilise branching programs for efficient counting, and we took inspiration from this paper to approach the solution to our counting problem.

To illustrate our solution, we have to slightly modify the knapsack representation. It will be convenient to express the multiple-choice knapsack as a multi-dimensional knapsack variation as follows. Each key chunk corresponds to 'one dimension'. Each item $k_{i,j}$ has an associated weight vector $w_{i,j}$ of length $m + 1$ of the form $(w_{i,j}, 0, \ldots, 1, 0, \ldots, 0)$, where the 1 is in position j. The global weight is also expressed as a vector $W = (W, 2, \ldots, 2)$ of length $m + 1$. The key rank problem is then to count the number of solutions (that satisfy all constraints simultaneously) to

$$\sum_{j=0}^{m-1} \sum_{i=0}^{n-1} w_{i,j} \cdot x_{i,j} < W$$

$$x_{i,j} \in \{0, 1\}, \forall i, j$$

The constraint W ensures that all keys that are counted have a strictly lower weight than the secret key. If the weight vector has a 1 in position j, it means that this is a value for the jth key chunk. Since the weight limit is 2 in the constraint vector W, it means that only a single value for any key chunk can be chosen. We now illustrate this by a simple example.

Example 1. Our illustrative example, which will run throughout the paper, consists of two distinguishing vectors with three elements each: $k^0 = (0, 1, 3)^T$, and $k^1 = (0, 2, 3)^T$. We assume that the secret key, SK, is $(2, 1)$. First we derive the global weight constraint vector. In this case it has length $m + 1 = 3$ and contains the maximum weight $W = w_{0,2} + w_{1,1} = 3 + 2 = 5$, which results in $W = (5, 2, 2)$. The weight vectors of the $k_{i,j}$ are:

$$w_{0,0} = (0, 1, 0), \ w_{0,1} = (1, 1, 0), \ w_{0,2} = (3, 1, 0)$$
$$w_{1,0} = (0, 0, 1), \ w_{1,1} = (2, 0, 1), \ w_{1,2} = (3, 0, 1)$$

Given that $W = 5$, all except two of the combinations are below this threshold. Hence the solutions to the knapsack are:

$$(k_{0,0}, k_{0,1}), \ (k_{0,0}, k_{1,1}), \ (k_{0,0}, k_{2,1}),$$
$$(k_{1,0}, k_{0,1}), \ (k_{1,0}, k_{1,1}), \ (k_{1,0}, k_{2,1}),$$
$$(k_{2,0}, k_{0,1})$$

Notice that the knapsack solution will never contain the secret key itself, as it returns all keys with weight *strictly* less than the weight of the secret key. For the ranking problem this would give us that our secret key has rank 8.

For standard knapsack problems it is well known [3] that solutions can be found via finding longest paths on a directed acyclic graph. In the following section we will show that such a graph exists also for our knapsack, and importantly, that the resulting graph allows for a particularly efficient path counting solution, which gives us a solution to the Key Rank problem.

3   An Accurate Key Rank Algorithm

In this section we first define the graph and illustrate how it relates to the multidimensional knapsack via intuition and a working example. We then explain our fast path counting algorithm for a compact representation of the graph.

3.1   Key Rank Graph

Recall that our multi-dimensional knapsack has $m \cdot n$ elements, and for each element we have a weight vector. Also, a correct solution to the multi-dimensional knapsack must have a weight that is strictly smaller than W. Since we need to be able to represent all permissible solutions we need W extra vertices (per element). This means that we 'encode' all solutions to the knapsack in a graph with $m \cdot n \cdot W$ vertices (plus an extra two for accept and reject nodes). The vertices corresponding to item $k_{i,j}$ are labelled $V_{i,j}^w$, where the variable w denotes the 'current weight'. The key rank graph contains a start node S, an accept node A and a reject node R. The edges are constructed as follows:

- $\left(V_{i,j}^w, V_{i+1,j}^w\right)$ which corresponds to the item not being chosen in this set
- $\left(V_{i,j}^w, V_{0,j+1}^{w+w_{i,j}}\right)$ if the item is chosen for this set and $w + w_{i,j} < W$
- $\left(V_{n-1,j}^w, R\right)$ if no elements are chosen from the set
- $\left(V_{i,m-1}^w, A\right)$ if the item is chosen for the last set and $w + w_{i,m-1} < W$
- $\left(V_{i,j}^w, R\right)$ if the item is chosen for this set and $w + w_{i,j} \geq W$
- $S = V_{0,0}^0$ to set up the start node

When visualising the key rank graph it will be convenient to think of the indices i, j as though they are 'flattened' (*i.e.* they are topologically sorted and occur in a linear order). In this representation the graph is $n \cdot m$ deep, W wide, where the width of the key rank graph essentially tracks the current weight (of the partial keys). Each vertex has exactly two edges coming out of it (with the exception of A and R): either the vertex was 'included' (this corresponds to the choice of selecting the corresponding value of the key chunk to become part of the key) or not. If the answer is yes then the edge must point to the first item in the next key chunk, as we can only choose one item per key chunk, and the weight must be incremented by the weight of this key item. If the item/vertex is not chosen then the edge must go to the next item in the chunk (or reject if this was the last item) and the weight must not be incremented as the item was not selected. For any partial key, if a new key chunk is added and the new weight exceeds W then this is not a valid key and thus the path will go to the reject node R.

Graph Example. To illustrate the working principle we provide in Fig. 2 the process of constructing the graph for the example provided above.

Initially the graph is constructed (top right) and the start node is initialised based on the rule $S = V_{0,0}^0$. The width of the graph is set to be 5 (0–4) as this matches our maximum weight ($W = 5$). The depth of the graph becomes 6 as each chunk contains 3 items.

Next (middle left) the first two children from the start node are created. The edge that denotes the chunk is, in fact, selected (the right child) is built following the rule $\left(V_{0,0}^0, V_{0,1}^{0+0}\right)$, which creates the edge from the start node, to an element in the next chunk. The edge that denotes the chunk is not selected (the left child) is built following the rule $\left(V_{0,0}^0, V_{1,0}^0\right)$, which creates the edge from the start node to the next item within the same chunk.

Moving onto the following step (middle right), children continue to be created through the same set of rules. However, note that at the point a link is created to the accept node based on the rule $\left(V_{0,1}^0, A\right)$, this demonstrates that the selection of key chunk 0 followed by 0 is a valid solution to the problem.

In the following steps links continue to be created based on the rules until all paths in the graph are created. Please note that throughout the construction of the graph, the last item in each chunk will have a left child that points to reject (as obviously there are no further chunks to select) but these have been omitted from the example diagram for the sake of clarity.

All the greyed out nodes also have their children calculated. However, as they do not alter the path count, we have excluded them from the example figures to aid clarity.

Each path from S to A if, corresponds to a key with lower weight than our secret key. Thus, counting these paths will yield the rank of the secret key. While in general path counting is hard [12], we explain how our graph structure, having at most two outgoing edges per node, lends itself to efficient counting.

3.2   Counting Valid Paths

Clearly our key rank graph is a directed acyclic graph. We have already mentioned that it is convenient to 'flatten' the graph (as it has been presented in the example). This 'flat' graph is also more suited for an efficient counting algorithm.

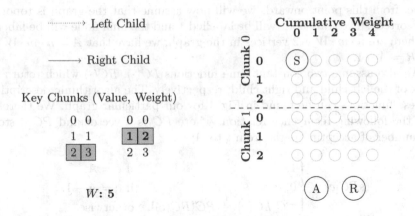

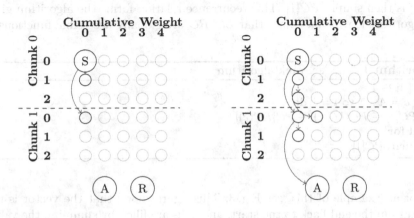

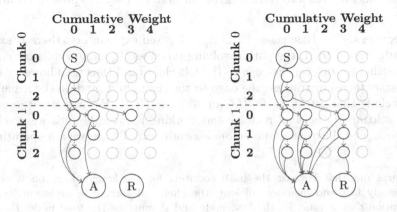

Fig. 2. An example showing the construction of the graph for the small example instance provided

Hence, from this point onwards we will now assume that the graph is topologically sorted[4]. The start node will be labelled 1 and the final node will be labelled A. There are $n \cdot m \cdot W + 2$ vertices in the graph, we have that $A = n \cdot m \cdot W + 2$ and $R = A - 1$.

We also assume two constant time functions $LC(\cdot)$, $RC(\cdot)$ which return the index of the left child and right child, respectively. The algorithmic descriptions of these functions can be found in Fig. 4 for our particular graph. We therefore have the following recurrence relation, where PC is a vector and $PC[i]$ stores the number of accepting paths from i to A:

$$sPC[c] = \begin{cases} 1, & \text{if } c = A. \\ 0, & \text{if } c = A - 1. \\ PC[LC(c)] + PC[RC(c)], & \text{otherwise.} \end{cases}$$

The total number of paths between 1 (our start node) and A (our accept node) is then simply $PC[1]$. This recurrence relation forms the algorithm given in Algorithm 1, which assumes that LC, RC are globally accessible functions.

Algorithm 1. The key rank algorithm

$PC[A] \leftarrow 1$
for $c = A - 1$ **to** $i = 1$ **do**
    $PC[c] \leftarrow PC[LC(c)] + PC[RC(c)]$
end for
return $PC[1]$

For an example of this, see Fig. 3. This figure shows that the vector is traversed from the end back to the start, and cells are filled by summing the values in their left and right children cells. For clarity in the figure, we only show example links betwen two cells, whereas in practise they are present on all.

Correctness. The base case of $PC[A] = 1$ is self explanatory; there is exactly one path from A to A, the path involving no edges. From an arbitrary node c, it is possible to traverse the edge to the left child (and thus take however many paths start there) or traverse the edge to the right child (and do the same) and we conclude that $PC[c] = PC[LC(c)] + PC[RC(c)]$. We can iterate over all nodes starting at the final node A and working backwards until we reach the start node, and since our graph is topologically sorted when we are operating on

---

[4] It turns out that because the path counting for the key rank graph is already extremely fast and memory efficient, the choice of sorting is irrelevant, bar the exception that S must be the first node and A must be the final node. However, when it comes to key enumeration this will be an important consideration and thus will be discussed in further detail in the corresponding section.

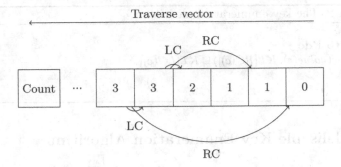

Fig. 3. Example demonstrating how the path count is calculated. Two nodes children links have been included to demonstrate the process.

a node, the values for the nodes' children will already have been calculated as they come later in the topological sorting.

We note at this point that this counting algorithm is *exact*. However, as we pointed out before, we need to convert floating point distinguishing scores to integer weights, and this conversion may incur a loss of precision and hence could cause a loss of accuracy. We discuss this in Sect. 5.

Time Complexity. The time complexity of the key rank algorithm depends on the number of vertices in the key rank graph and their size. Our graph contains $A = m \cdot n \cdot W$ vertices. The integers stored in the vertices could be up to $O(2^A)$ (and thus be of size $O(A)$) because in the worst case each value can be double the previous value (if $PC[LC(i)] = PC[LC(i)]$). Hence, given that we have A vertices and perform an integer addition with an A-bit variable for each vertex, we have worst-case time complexity of $O(A^2) = O(m^2 \cdot n^2 \cdot W^2)$.

However, whilst we touch each vertex once, we know that there are at most $O(n^m)$ keys. Consequently, we need no more than $O(m \log n)$ bits to store the path count (in contrast to the $O(A) = O(m \cdot n \cdot W)$ bits for the worst case). Hence the time complexity for computing the key rank via the key rank graph is $O(m^2 \cdot n \cdot W \cdot \log n)$.

It is worth noting that the key depth does not factor into the time complexity and the following example will help to clarify this. Consider the target key which has weight 1 in every column (which gives $W = 16$ and a grid with 65536 nodes). If all other key chunks have weight 0 then the target key will have rank 2^{128} since all other keys have a lower weight. However, if all other key chunks have weight 2, then our target key will have rank 0 because all other keys have a higher weight. None of the other values affect the size of the graph and thus it is clear that the runtime is not changed by the key depth.

In fact for AES-128 the values of m, n are also fixed and thus we get that the algorithm runs in $O(W)$, that is to say it is linear in the weight of the secret key. See Sect. 5.1 for experiments supporting this.

Algorithm 2. The key enumeration algorithm

$KL[A] \leftarrow \emptyset$
for $c = A$ **to** 1 **do**
  $KL[c] \leftarrow (value(c), KL[RC(c)]) \cup KL[LC(c)]$
end for
return $KL[1]$

4   Parallelisable Key Enumeration Algorithm

We are able to further modify our algorithm such that with minor (standard book-keeping) adjustments, we are able to list all valid paths, as opposed to just counting them, with reasonable efficiency. The algorithm is given in Algorithm 2 and requires an additional (constant time) function call *value* which, given an index c, returns the value of a vertex. We write $(a, \{x_c\}_c)$ to mean $\{(a, x_c)\}_c$. That is to say if we concatenate an item a with a set, we are really concatenating the item a with every item in the set to form a new set.

It is now easy to use this for key enumeration. Assume we have the resources to enumerate/test up to B keys. Then, we choose some weights (which correspond to key guesses) and use the key rank algorithm to determine their ranks and compare them to B. This allows us to quickly select the appropriate W for the given B. Then, Algorithm 2 proceeds as follows: for any valid path (in the key rank graph), every time a right child is taken (this can be determined by the node indices) the corresponding value for the respective key chunk is chosen. A left child means that we are not taking a particular value for key chunk. In this manner the keys are effectively reconstructed from the key rank graph.

If one wanted to enumerate the keys in a smart order, this would simply be a case of altering the construction of the tree which stores the valid key chunks for enumeration. Currently the valid key chunks are stored in *numerical* order within the tree, however if this was changed such that they were stored in order of scores, the keys would be rebuilt in a near optimal order.

In the rest of this section we discuss run time and memory requirements. Whilst the run time is bounded by the number of keys that we want to enumerate, we show there are different strategies to improve the memory performance. Finally, we show that with a further simple observation, we can parallelise the key enumeration algorithm.

4.1   Time Complexity

We begin with a worst case analysis considering a general graph. In this case, the enumeration algorithm would be exponential in the length of the number of vertices, because to generate all paths (each vertex has two children) the algorithm clearly must take $O(2^A)$ time.

However, in our key rank graph, each path corresponds to a valid key with weight lower than W. Considering this, the run time of this algorithm is relative to the rank (which is determined by W) and *not* to the total number of keys;

hence this algorithm can be used to enumerate keys for a given workload in time $O(m^2 \cdot n \cdot W \cdot B \cdot \log n)$. This is because all $O(n \cdot m \cdot W)$ nodes are touched once, and B keys are reconstructed which are of length $m \cdot \log n$.

4.2   Memory Efficiency

How we topologically sort the key rank graph has a major impact on the memory efficiency of the key enumeration. While there are a variety of explicit topological sorting algorithms in the literature [8,11], we are able to avoid explicit sorting because we know our graph structure in advance. Hence, we show that our graph can be sorted implicitly by how the nodes are numbered within the calculation of the left and right child functions. The remaining question is what method of sorting is the most desirable.

In Fig. 4 we demonstrate topologically sorting the example graph previously considered in Fig. 2, as well as present the associated pseudo code. There are alternative sorting methods available which were considered, and we discuss the pros and cons of these in the extended version of this paper available on ePrint[5]. We also discuss how to improve memory efficiency further by appropriately storing the generated keys.

Wide Sort. In this sorting the graph is numbered one chunk at a time, one item at a time, along the weight in increasing order (see Fig. 4). Formally given a chunk, item and weight (x, y, z) the index is $i = x \cdot W \cdot n + y \cdot W + z$. This is a valid topological sorting of the graph, since a nodes' children will be either one item lower in the same chunk (for the left child) or the first item in the next chunk (right child) both of which have a higher number.

This is the topological sort we described for key rank. Note that, since key rank is extremely fast we describe the most intuitive sort since it did not have an impact on performance, while with enumeration this is no longer the case and must now be taken into consideration.

The advantage of this sorting is that it is due to the fact an element will only need to look at the item below and the item at the top of the next chunk; these are the only things needing to be stored in memory. This makes it very memory efficient requiring $O(W)$ memory.

The disadvantage of this method is that it is highly serial and it does not seem possible to (easily) parallelise.

Key Storage. The topological sorting of the graph is clearly a crucial factor for memory efficiency. The other factor is how keys are represented/stored within our graph.

In the algorithm as described all (partial) keys are stored at each point in the algorithm. This will become very inefficient. Consider, for example, the case where you want to enumerate all keys. There are 2^{120} keys which have the first

---

[5] https://eprint.iacr.org/2015/689.

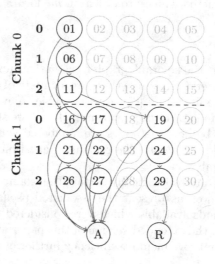

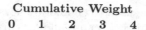

Left Child	Right Child
if $(n \cdot W) - (c \mod (n \cdot W)) \leq W$ **then**	$w' \leftarrow c \mod W$
return R	$i \leftarrow \frac{(c - w') \mod (n \cdot W)}{W}$
else	$j \leftarrow \frac{c - w' - i \cdot W}{n \cdot W}$
return $c + W$	**if** $w + w_{i,j} \geq W$ **then**
end if	**return** R
	else if $i \neq m - 1$ **then**
	return $(i + 1) \cdot n \cdot w' + W + w_{i,j}$
	else
	return A
	end if

Fig. 4. Topological sorting of our previous example. Note that the deepest node in each chunk will be guaranteed to have a left child leading to R; for clarity these paths are omitted (top). Pseudo code of how the child indices are calculated for each node in the tree (bottom).

key chunk set to zero (hence this chunk would be duplicated 2^{120} times). Clearly, one needs to choose an appropriate data structure, and we use a tree, see Fig. 5. This key tree is passed to a separate algorithm that converts it into a series of keys for testing. The advantage of this is threefold. First, it greatly speeds up the enumeration. Second, the conversion of the key tree into a list of keys is trivially parallelisable, and third, the actual testing (in our case checking the

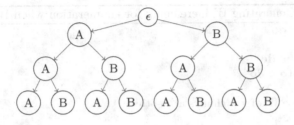

Fig. 5. The key tree for all possible three character keys containing 'A' or 'B'

AES encryption using a given plaintext/ciphertext pair) can be amortised into this cost.

4.3 Parallelisation

We can achieve parallelisation with a simple observation: by adjusting the graph such that instead of vertices with a weight lower than W going to the accept state, we only allow vertices with weight in the range between the two weights W_1 and W_2 to reach the accept state. The width of the graph is defined by W_2; W_1 has no impact on the graph size. This results in an algorithm that enumerates 'batches' of likely keys. Hence, one can run multiple instances of the key enumeration algorithm in parallel, where each instance enumerates a unique batch of keys.

All ranges of keys can be computed in parallel and require no communication between threads except for the initial passing of the distinguishing scores and a notification when the key has been found. It is hence trivial to utilise multiple machines (or cores).

Setting W In an enumeration setting, the correct key, and therefore W is unknown. We create a series of 'steps' in W (to bound using W_1, W_2 as introduced previously), which are enumerated in order until the correct key is found.

Iterating across these W increments, we select the weights by first taking the most probable across all distinguishing vectors, *i.e.* the weight at the top of each column. If the correct key is not located, the weight limit is increased by an amount equal to moving down by one key chunk in a column. The generation of each W step is done according to the following:

More complex methods of bounding the weights could be used, such as binary searches or similar, but this would increase the cost of calculating the capacities before the enumeration begins, with little tangible benefit.

Also, it should be noted that if we simply incremented the capacity in the smallest possible steps, then the algorithm would then be guaranteed to be accurate, enumerating keys in the correct order. However, this would make parallelism nearly impossible as each unit of work would be too small causing the overhead from the parallel computation to dominate the runtime.

Algorithm 3. Generating W increments for enumeration when W is unknown

---

for $k = 0$ to m do
    $c \leftarrow 0_{0,...,m}$
    for $i = k$ to m do
        for $j = 0$ to n do
            $c_i + 1$
            Calculate W of key chunks at depths c
        end for
    end for
end for

Further Speed Optimisations. Currently the algorithm operates on every node of the graph. However, some of the nodes are not even reachable from the start node (for example the greyed out nodes in Fig. 2). Hence any computation done on these nodes is wasted because it will never be combined into a solution. By precalculating the number of valid paths from S to all other nodes in the graph (a reasonably cheap operation compared to a large key enumeration – this is done using the key rank algorithm), we can skip over a node if the number of paths from the start node to here is 0 because any work here will not be combined with the final solution.

5   Practical Evaluation and Comparison with Previous Work

Our key enumeration and key rank algorithms are both based on a graph representation of a multi-dimensional knapsack. To define this multi-dimensional knapsack it is necessary to map distinguishing scores, which typically are floating point values, to integer weights. This is a very simple process of multiplying the raw score $d_{i,j}$, of value most 2^α, in the distinguishing vector by $2^{p-\alpha}$ where p is the bit value of precision we wish to maintain. Then performing an *abs* has the double effect of removing the negative sign, and making the most probable (the most negative numbers) the smallest, meaning they have the lightest weight which maps to our knapsack representation perfectly. Formally $w_{i,j} = MapToWeight(d_{i,j})$ where $MapToWeight(d_{i,j}) = \lfloor abs(d_{i,j} \cdot 2^{p-\alpha}) \rfloor$ for p bits of precision.

This requirement has implications for the performance of our algorithms, as the time complexities for both algorithms strongly depend on the parameters m (the number of key chunks), n (the number of items per chunk), and W (the maximum weight). In particular, for any fixed key size (and number of chunks) the size of the graph (*i.e.* the width) grows with W, and W grows with the precision that we allow for in the conversion from floating point values to integers.

We hence focus our practical evaluation on the impact of the precision[6], on accuracy[7] and on performance. First, we discuss the precision requirements for practical DPA outcomes. Second, we explore the practical impact on the performance of key rank when we increase the precision. Third, assuming we allow for sufficient precision, we ask what are the best performance results that we can achieve on single but also many core platforms for the key enumeration.

It is clear that to answer these questions we need to be able to generate many practically relevant distinguishing vectors in a manner that is comparable to previous work. We hence decided to adopt the simulator used by Veyrat-Charvillon *et al.* [13]. Veyrat-Charvillon *et al.* create distinguishing vectors based on attacking the AES SubBytes output, assuming noisy Hamming weight leaks, and using the Hamming weight as power model. Their DPA simulator allows us to manipulate the level of noise, and the number of measurements. The simulator then performs a standard DPA by utilising template matching as a distinguisher (this has been shown by Mangard *et al.* [10] to be equivalent to performing a correlation based DPA with a perfect model). They output 'additive' scores (by taking the logarithm of the raw matching scores), which we pass directly to our *MapToWeight* function. In all experiments we keep the number of traces constant at 30 (which matches [13]) and changed the variance of the noise to create 'deeper' keys.

5.1   Evaluating and Comparing Precision

In practical DPA attacks the combination of measured power traces, model values, number of traces and distinguisher will influence the *effective* precision of the

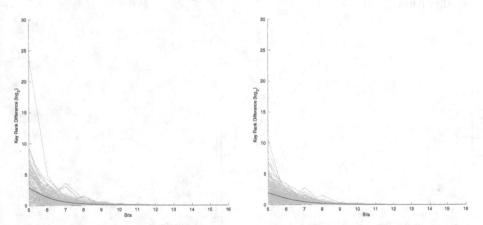

Fig. 6. Impact of the distinguisher (left: correlation, right: Gaussian templates without log2) on the precision requirements when considering up to 16 bits of precision.

---

[6] Precision is the ability to reproduce a measurement result, *i.e.* if several measurements of a variable give very close values then the measurement is precise.

[7] Accuracy is the closeness of a measurement to a true value, *i.e.* this relates to the 'trueness' of a measurement.

distinguisher scores[8]. We discuss the mentioned factors briefly. Then we experimentally determine the necessary level of precision for our key rank algorithm, and compare this to the number of bins for the method by Glowacz et al. [6].

Precision in Factors Influencing DPA Outcomes. Various factors can influence the outcome of a DPA attack, and also have an affect on the amount of precision required to accurately represent distinguishing scores. These can include the resolution of the leakage traces, the power model used, and the distinguisher applied. In our experiments we vary the precision from four to sixteen bits.

Experimentally Measuring Precision for Key Rank and Glowacz *et al.*
We ran precision tests using Veyrat-Charvillon's simulator, using $N = 30$ and variance two) to determine the appropriate level of precision for further experiments. We plot the difference in ranking outcomes for increasing precision in Fig. 7 (left). In this figure, and in all figures that will follow, we plot outcomes of individual experiments in gray, and average outcomes in black. The x-axis show the precision in w_i, j. The y-axis refers to the change in ranking outcomes when increasing the precision by 0.1 bits from the previous step. From 11 bits onwards the outcomes do not change anymore. Because our ranking method is exact with enough precision, we can infer that with 11 bits of precision in $w_{i,j}$ we produce *exact* ranks. Already from 4 bits of precision (on average, as plotted in black) we are within five bits of accuracy from the real result. From about 8 bits onwards, increasing the precision changes the ranking outcomes by just under a bit for our algorithm.

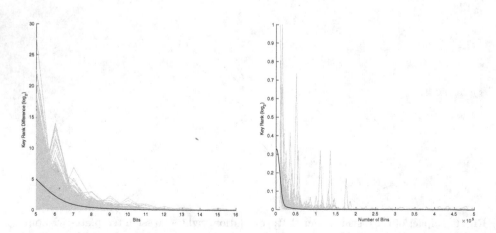

Fig. 7. Bits of precision for key rank (left) and number of bins for Glowacz et al.(right).

---

[8] Veyrat's simulator stores values in variables with double precision (*i.e.* one has 53 bits of precision). But effectively, only a few of them are necessary to contain the effective precision.

We implemented the convolution based method by Glowacz et al. [6]. Their method is essentially based on building m histograms (one from each of the distinguishing vectors) and counting the keys by counting items in the 'amalgamated' histogram efficiently via convolution. Figure 7 (right) shows that they achieve very high average precision (plotted in black) from about 50,000 bins onward. We can therefore conclude that using 50,000 bins roughly corresponds to 11 bits of precision in $w_{i,j}$. Glowacz et al. [6] actually recommend to use 500,000 bins in their paper.

Recall that we hypothesised that different distinguishers would lead to different precision requirements. To test this hypothesis we implemented two further distinguishers for the simulator: one distinguisher was based on correlation and one was based on Veyrat-Charvillon's method but without applying the logarithm. Figure 6 shows the results for them, this time we allowed up the 16 bit precision. The plots show that indeed, different distinguishers require different levels of precision, and that correlation has the least requirements.

To provide further evidence for the exactness of our ranking algorithm (provided enough precision), we considered the difference between the key rank output by our algorithm, and the key rank output by Glowacz et al. In this experiment, we used 16 bits for our algorithm and 500,000 bins for Glowacz et al. Fig. 8 shows the identical trend as Fig. 1 (right panel) of Glowacz et al. Hence the difference between our ranking outcomes and their ranking outcomes are identical to the rank estimation tightness that they measure, reinforcing the exactness of our ranking outcomes.

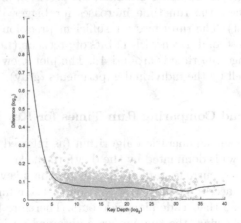

Fig. 8. Observed difference in calculated key rank between our algorithm and Glowacz et al.

5.2   Evaluating and Comparing Run Times for Key Rank

We explained in Sect. 3 that the run time of the Key Rank algorithm is independent of the actual depth of the key. The run time depends on the size of the graph, which is fixed for a certain choice of m and n, and hence depends on the

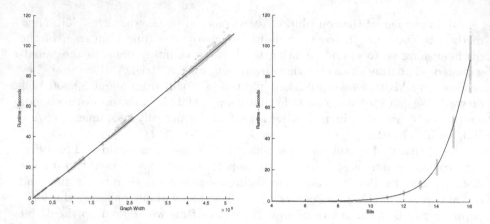

Fig. 9. Impact of the size of W (left) and precision in $w_{i,j}$ (right) on the run time of Key Rank

size of W. Since W is derived from summing the weights of SK, its precision will be determined by the precision that we allow during the conversion of the distinguishing scores.

We hence experimented with the relationship between run time and size of W and also precision. We did this by fixing all parameters for Veyrat-Charvillon's simulator and only varying the precision allowed in the function $MapToWeight$. As in the previous graph, we upper bounded the precision in W at 16 bits.

Figure 9 shows how the run time increases for bigger W (left) and more precision in W (right). The run times for sufficient precision (*i.e.* 8 bits for W) are well below half a second. Even with 11 bits of precision (*i.e.* accurate ranking outcomes) our average run time is around 4 s. The plot shows that this average (black) is tracked well by the individual experiments (gray).

5.3   Evaluating and Comparing Run Times for Key Enumeration

The run time of the key enumeration algorithm (as referred to by KEA in the graphs that will follow) is dominated by the depth to the key. Veyrat-Charvillon *et al.* [13] presented the current state of the art for smart key enumeration, and they kindly gave us access to the latest version of their implementation. We were hence able to run their code alongside ours. Therefore for all graphs that we provide in the following, the timings were obtained on identical platforms. Note that for all experiements, as the toolbox provided a known secret key on which the simulated attack was based, we knew at which point the enumeration had found the correct key without needing to performm an AES operation on a known plaintext/ciphertext pair (this is common within the literature).

Single Core vs Multi Core Comparisons. Figure 10 gives a comparison of run times of Veyrat-Charvillon *et al.*'s algorithm and our algorithm on a single core (left). We sampled multiple distinguishing scores for each key depth and ran our

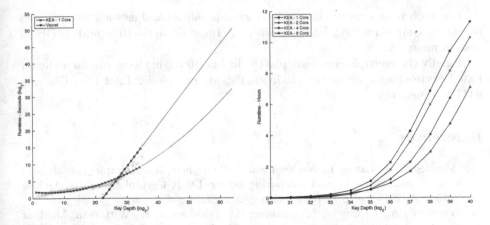

Fig. 10. Comparison between Veyrat-Charvillon *et al.*'s enumeration algorithm and our algorithm for increasing key depths on a single core (left), and run times for parallel instances of the key enumeration (right).

respective key rank algorithms. The graphs show that from key depths just under 30 bits onwards we clearly outperform Veyrat-Charvillon *et al.*'s algorithm, even on a single core. On the right, we provide some performance graphs when running our key enumeration algorithm on multiple cores. The graph shows that eight cores can enumerate 2^{40} keys in the same time as one core enumerates 2^{37}, which is a vast difference. Also another result of note is a single core run enumerates 2^{38} keys in 13.9 h and four cores performs the same enumeration in 6.4 h.

Acknowledgements. We would like to thank Benjamin Sach and Raphael Clifford for there valuable insight and advice during the developement of the algorithm. This work was carried out using the computational facilities of the Advanced Computing Research Centre, University of Bristol - http://www.bris.ac.uk/acrc/. Daniel, Jonathan and Elisabeth have been supported by an EPSRC Leadership Fellowship EP/I005226/1.

A   Computing Environment

All code was implemented using Java 1.7, with the exception of the Glowacz *et al.*'s algorithm [6] which was implemented in Matlab to enable very fast convolution of the histograms. The language difference here was not an issue because key rank is so that fast we only ran accuracy comparisons and not timing comparisons. The implementation of Veyrat-Charvillon *et al.*'s key enumeration algorithm [13] was provided by the author, and translated into Java allowing for direct speed comparisons.

Running the single core enumeration tests, compared to Veyrat-Charvillon which are plotted in Fig. 10 (left), took place on a system running Arch Linux, with an Intel i7-4790S and 8 GB of system memory.

Precision tests required larger memory capabilities and as such were carried out on a system running Ubuntu, with an Intel Xeon E5-1650 and 32 GB of system memory.

Finally the multiple core tests plotted in Fig. 10 (right) were run on a cluster based environment, where each individual node provided 2 Intel E5-2670s and 64 GB of memory.

References

1. Bernstein, D.J., Lange, T., van Vredendaal, C.: Tighter, faster, simpler side-channel security evaluations beyond computing power. IACR Cryptology ePrint Archive 2015, 221 (2015). http://eprint.iacr.org/2015/221
2. Bogdanov, A., Kizhvatov, I., Manzoor, K., Tischhauser, E., Witteman, M.: Fast and memory-efficient key recovery in side-channel attacks. Cryptology ePrint Archive, Report 2015/795 (2015)
3. Dasgupta, S., Papadimitriou, C.H., Vazirani, U.V.: Algorithms. McGraw-Hill, New York (2008)
4. Duc, A., Faust, S., Standaert, F.-X.: Making masking security proofs concrete. In: Oswald, E., Fischlin, M. (eds.) EUROCRYPT 2015. LNCS, vol. 9056, pp. 401–429. Springer, Heidelberg (2015)
5. Dyer, M.E.: Approximate counting by dynamic programming. In: Larmore, L.L., Goemans, M.X. (eds.) Proceedings of the 35th Annual ACM Symposium on Theory of Computing, San Diego, CA, USA, 9–11 June 2003, pp. 693–699. ACM (2003). http://doi.acm.org/10.1145/780542.780643
6. Glowacz, C., Grosso, V., Poussier, R., Schueth, J., Standaert, F.: Simpler and more efficient rank estimation for side-channel security assessment. IACR Cryptology ePrint Archive 2014, 920 (2014). Accepted for publication at FSE 2015
7. Gopalan, P., Klivans, A., Meka, R., Stefankovic, D., Vempala, S., Vigoda, E.: An FPTAS for #Knapsack and related counting problems. In: 2011 IEEE 52nd Annual Symposium on Foundations of Computer Science (FOCS), pp. 817–826, October 2011
8. Kahn, A.B.: Topological sorting of large networks. Commun. ACM 5(11), 558–562 (1962). http://doi.acm.org/10.1145/368996.369025
9. Kocher, P.C., Jaffe, J., Jun, B.: Differential power analysis. In: Wiener, M. (ed.) CRYPTO 1999. LNCS, vol. 1666, pp. 388–397. Springer, Heidelberg (1999)
10. Mangard, S., Oswald, E., Standaert, F.X.: One for all - all for one: unifying standard differential power analysis attacks. IET Inf. Secur. 5(2), 100–110 (2011)
11. Tarjan, R.: Edge-disjoint spanning trees and depth-first search. Acta Informatica 6(2), 171–185 (1976). http://dx.doi.org/10.1007/BF00268499
12. Valiant, L.G.: The complexity of enumeration and reliability problems. SIAM J. Comput. 8(3), 410–421 (1979). http://dx.doi.org/10.1137/0208032
13. Veyrat-Charvillon, N., Gérard, B., Renauld, M., Standaert, F.-X.: An optimal key enumeration algorithm and its application to side-channel attacks. In: Knudsen, L.R., Wu, H. (eds.) SAC 2012. LNCS, vol. 7707, pp. 390–406. Springer, Heidelberg (2013)

14. Veyrat-Charvillon, N., Gérard, B., Standaert, F.-X.: Security evaluations beyond computing power. In: Johansson, T., Nguyen, P.Q. (eds.) EUROCRYPT 2013. LNCS, vol. 7881, pp. 126–141. Springer, Heidelberg (2013)
15. Ye, X., Eisenbarth, T., Martin, W.: Bounded, yet sufficient? how to determine whether limited side channel information enables key recovery. In: Joye, M., Moradi, A. (eds.) CARDIS 2014. LNCS, vol. 8968, pp. 215–232. Springer, Heidelberg (2015)

A Unified Metric for Quantifying Information Leakage of Cryptographic Devices Under Power Analysis Attacks

Liwei Zhang[1], A. Adam Ding[1]($\boxtimes$), Yunsi Fei[2], and Pei Luo[2]

[1] Department of Mathematics, Northeastern University, Boston, MA 02115, USA
zhang.liw@husky.neu.edu, a.ding@neu.edu
[2] Department of Electrical and Computer Engineering, Northeastern University, Boston, MA 02115, USA
yfei@ece.neu.edu, silenceluo@coe.neu.edu

Abstract. To design effective countermeasures for cryptosystems against side-channel power analysis attacks, the evaluation of the system leakage has to be lightweight and often times at the early stage like on cryptographic algorithm or source code. When real implementations and power leakage measurements are not available, security evaluation has to be through metrics for the information leakage of algorithms. In this work, we propose such a general and unified metric, information leakage amount - ILA. ILA has several distinct advantages over existing metrics. It unifies the measure of information leakage to various attacks: first-order and higher-order DPA and CPA attacks. It works on algorithms with no mask protection or perfect/imperfect masking countermeasure. It is explicitly connected to the success rates of attacks, the ultimate security metric on physical implementations. Therefore, we believe ILA is an accurate indicator of the side-channel security level of the physical system, and can be used during the countermeasure design stage effectively and efficiently for choosing the best countermeasure.

Keywords: Information leakage amount · Side-channel security · Power analysis attack

1 Introduction

In the past decade, various side channel attacks (SCAs) utilizing the system power consumption information, such as differential power analysis (DPA) [16], correlation power analysis (CPA) [5], mutual information (MI) attacks [14] and template attacks [6], have been presented to exploit the weakness in cryptographic implementations to recover the secret key. Masking is one of the most popular SCA countermeasures used to randomize sensitive variables [7]. When applying masking at a higher level, e.g., algorithmic or source code level, every key-sensitive intermediate variable is masked with at least one random value M by a carefully designed masking function f, e.g., normally exclusive OR or

© International Association for Cryptologic Research 2015
T. Iwata and J.H. Cheon (Eds.): ASIACRYPT 2015, Part II, LNCS 9453, pp. 338–360, 2015.
DOI: 10.1007/978-3-662-48800-3_14

multiplication. Therefore, during the cryptographic execution, any intermediate variable Z is substituted by its masked counterpart, $f(Z, M)$, to prevent side-channel leakage. Perfectly masked devices with appropriate masking functions and unbiased random masks can eliminate first-order leakage, e.g., it is not feasible to break the system by exploiting only one time point of the power leakage traces which corresponds to one intermediate variable. However, they are still susceptible to second-order and higher-order attacks which combine two or more time points of power leakage to retrieve the secret key. Some practical masking schemes with limited implementation resources are not perfect and may still have some first-order leakage.

How to evaluate a system's SCA vulnerability/resilience comprehensively and accurately under different attacks is an important research issue. Sound quantitative metrics will be used to guide the implementation of countermeasures and fairly compare the overall strength of countermeasures. One widely used metric is success rate, the probability that an attack succeeds given a number of side-channel leakage measurements [21]. This is indeed the ultimate practical measure of a system's SCA vulnerability/resilience, which depends on the cryptographic algorithm, the specific implementation (with power measurement data available), and the attack model (whether it is DPA, CPA, MIA, etc.) as illustrated in [12,18]. We classify this metric as one for measuring the system *physical leakage*. In recent years, there are research interests in using other physical leakage metrics on instructions of cryptographic software and therefore pinpoint the location of leakage to guide automatic implementation of countermeasures. Bayrak et al. [2] introduced a methodology for detecting power leakage, using an information theoretic metric - mutual information, MI_L, between the key and leakage measurements. Although not explicitly related to the success rate, the metric MI_L can be used to bound the success rate [10,21] in some models. However, it requires power consumption data. There are also other efforts in evaluating the cryptosystem *information leakage* at an early stage, i.e., on source code of cipher software or even algorithms and with no need of power measurement data. The automatic software verification tools for SCA vulnerabilities [3,8] employ mutual information between the secret key and intermediate variables, denoted as MI_A. The metric of quantitative masking strength, QMS, is defined by [11] to quantify the software leakage amount under imperfect masking, and a verification process is formulated to find the QMS value of cryptographic software source code. However, none of the prior work has shown the relationship between these system information leakage metrics and the success rate. It is not easy to translate the bound on these information leakage (MI_A and QMS) to the final security measure of the implemented physical system, the success rate.

In this work, we propose a new unified metric, information leakage amount (ILA), to quantify the *system information leakage* under various power analysis attacks at the early cryptographic algorithm or software code level, whether the cipher is unprotected or protected with masking. What is more, we also relate this metric to the success rate of DPA/CPA attacks in analytic models. Note that in this work we choose DPA/CPA because it has been shown both theoretically and empirically that the first-order and second-order CPA attacks are equivalent to the strongest maximum-likelihood attacks under Gaussian noise models [9,13,15]. Our metric is unified, in the sense that it works on original algorithms with no masking, perfect masking, or imperfect masking under first-order DPA/CPA or second-order CPA. The success rate formulas are more general and simpler than the formulas in [9,12,13], which are only for first-order DPA/CPA on unmasked devices and for higher-order attacks on perfectly masked devices. Our explicit success rate formulas in terms of ILA bridge the gap between the system information leakage measure and the physical leakage measure. The metric ILA, as a great indicator of the ultimate side-channel security level of the physical system, can therefore be used during the countermeasure design stage (without real implementations and power measurements) effectively and efficiently for choosing the best countermeasure.

Table 1 summarizes the properties of our metric and compares it with other three metrics, QMS, MI_A, and MI_L. A question mark means that the metric on the column may be able to achieve the objective on the row, but it has not been demonstrated in literature. For example, work in [21] shows that the mutual information MI_L has a monotonic relationship with the success rate of an attack with only two candidate keys, but generally the MI_L may not be converted to the success rate explicitly.

Table 1. Comparison among ILA, QMS and MI as leakage evaluation metrics

		ILA	QMS	MI_A	MI_L
1	First-order DPA/CPA Metric on Software Code/Algorithm	✓	✓	✓	×
2	Relate to First-order DPA Success Rate	✓	✓	✓	✓
3	Relate to First-order CPA Success Rate	✓	×	?	?
4	Second-order DPA/CPA Metric on Software Code/Algorithm	✓	×	?	×
5	Relate to Second-order DPA/CPA Success Rate	✓	×	?	?

The rest of the paper is organized as follows. Section 2 gives an overview of the existing leakage metrics and defines our proposed metric. Section 3 establishes the success rate formula for CPAs in terms of our metrics. Section 4 presents experimental results to evaluate the metrics and compare them with others. Section 5 concludes the paper.

2 Leakage Metrics for Cryptosystems with Masking Countermeasure

In this section, we first introduce the notations used and existing metrics, and propose our unified metric ILA for first-order and second-order attacks on cryptographic algorithm with imperfect/perfect masking. We then analyze these metrics in the case of Boolean masking.

2.1 Notations and Existing First-Order Metrics

We denote sets by calligraphic letters (e.g., $\mathcal{X}$), denote random variables by capital letters (e.g., X) which take values on the set ($\mathcal{X}$), and denote observations of the random variables by lowercase letters (e.g., x). We let $X_{(i)}$ denote the ith bit of X. $\mathbb{P}_X$ and $\mathbb{E}_X$ are the notations for the probability and the expectation with respect to X, respectively. For a cryptographic system with masking protection, K, X, M denote the random variables for the key, the plaintext, and the mask, respectively, and each takes values in sets $\mathcal{X}$, $\mathcal{K}$, $\mathcal{M}$. Let $F = f(X, K, M)$ denote the algorithmic intermediate variable that possibly leaks, which is an algorithmic function of the known input X, unknown key K and the random mask M. For a second-order attack on masked devices, there are two select functions. One is $V_0(X, K, M) = g(F)$, which works on a key-sensitive intermediate variable and therefore is also a function of the input X, the key K and the mask M. Note the select function for an attack is determined by the system's power model, and $g(\cdot)$ is usually Hamming weight or Hamming distance. Without loss of generality, the other select function is $V_1 = g(M)$ which depends on the mask M only. The mask may be biased, i.e., not following the uniform distribution. If the mask is unbiased and the masking operation is appropriate, we call it perfect masking. Let k_c be the secret key, $k_g \in \mathcal{K} \backslash \{k_c\}$ be any other possible key hypothesis, and $N_k = |\mathcal{K}|$ be the dimension of the key set.

A first-order attack uses only one select function V_0 that corresponds to one time point on power traces. Therefore a *first-order leakage metric* measures the leakage of one select function that can be sensitive to both key and mask. Given a plaintext x, the secret key k_c and a random number m, the target select function is $v_0^c = V_0(x, k_c, m)$. The information leakage is measured by the dependency of v_0^c on k_c. Under perfect masking, the distribution of v_0^c is independent of k_c, and hence the secret key could not be recovered from the leakage measurements of v_0^c. Otherwise, v_0^c is still vulnerable to first-order power analysis attacks. There are mainly two existing first-order information leakage metrics.

Eldib et al. [11] proposed to quantify the masking strength under DPA by

$$\text{QMS} = (1 - \Delta_{qms}), \quad \text{with } \Delta_{qms} = \max_{x,x' \in \mathcal{X}, k, k' \in \mathcal{K}} |D_{x,k}(F) - D_{x',k'}(F)|, \quad (1)$$

where $D_{x,k}$ denotes the distribution of F given (x, k), and Δ_{qms} is the maximum distribution difference. For perfect masking, QMS is maximum and reaches one, which indicates that the key K and the intermediate variable F are statistically independent. Without masking, QMS=0. For imperfect masking schemes, QMS is in the range of (0,1).

The other metric uses the mutual information, an information theoretic quantity commonly used for leakage evaluation. The mutual information between two discrete random variables X and Y is defined as:

$$\text{MI}(X, Y) = \sum_{x \in \mathcal{X}, y \in \mathcal{Y}} p(x, y) \log \left(\frac{p(x, y)}{p(x)p(y)} \right), \quad (2)$$

where $p(x, y)$ is the joint probability distribution function of X and Y, with $p(x)$ and $p(y)$ as the corresponding marginal functions. For continuous random variables, the summation in definition (2) is replaced by integrations. Work in [3, 8] uses the mutual information between K and F to measure the information leakage at the software code level. This mutual information only depends on the algorithm and we denote it by $\mathrm{MI_A} = \mathrm{MI}(K, F)$. In contrast, work in [2] uses the mutual information between K and the leakage measurements L. We denote it by $\mathrm{MI_L} = \mathrm{MI}(K, L)$, which is a physical leakage measure.

Note that there is no second-order system information leakage metric based on QMS or MI shown in literature. In this work, we propose a general and unified metric on the selection functions (V_0 for first-order attacks, V_0 and V_1 for second-order attacks), which reflects the system susceptibility to attacks.

2.2   Our Proposed Information Leakage Metric

Eldib et al. empirically [11] showed that there is a relationship between QMS and the number of traces needed in DPA. However, there is no theoretical proof for such relation, and how QMS relates with multi-bit CPA or higher-order attacks is unknown. We are seeking a new unified metric to reflect the information leakage at the algorithm level, similar to QMS and $\mathrm{MI_A}$, and meanwhile can explicitly relate to the success rate of different attacks, including DPA, CPA, and high-order attacks.

Fei et al. [13] defined the confusion coefficient, for unmasked algorithm, as $\kappa(k_c, k_g) = \mathbb{E}_X\{[V(X, k_c) - V(X, k_g)]^2\}$ for the selection function V and the expectation being taken over X. Each confusion coefficient is defined between two key values. They showed that the confusion coefficients and the implementation signal-noise-ratio (SNR) together explicitly determine the success rates for DPA and CPA. However, these confusion coefficients do not reflect the masking strength as they are defined for unmasked algorithms only. The confusion coefficients are also used to model the success rates for higher-order attacks with perfect masking in [9].

We propose to generalize the confusion coefficient definition for masked algorithms (possibly imperfect). We then propose the new metric ILA based on the generalized confusion coefficients. The ILA measures the information leakage of V_0 (and V_1) under the protection of any masking countermeasure.

Definition 1. *We define the new first-order confusion coefficient $\kappa_{1O}(k_c, k_g)$ of masked algorithm as*

$$\kappa_{1O}(k_c, k_g) = \mathbb{E}_X\{[\mathbb{E}_M(V_0|(X, k_c)) - \mathbb{E}_M(V_0|(X, k_g))]^2\}, \tag{3}$$

where $\mathbb{E}_M(V_0|(X, k))$ is the conditional expectation of V_0 given (X, k) over $\mathcal{M}$, and $\mathbb{E}_X$ is the expectation over $\mathcal{X}$.

Definition 2. *The first-order information leakage amount $\mathrm{ILA_{1O}}$ is defined as*

$$\mathrm{ILA_{1O}} = \mathbb{E}_{\mathcal{K}\backslash\{k_c\}}[\kappa_{1O}(k_c, k_g)], \tag{4}$$

where $\mathbb{E}_{\mathcal{K}\backslash\{k_c\}}$ is the expectation over all possible key hypothesis k_g in $\mathcal{K}\backslash\{k_c\}$.

ILA_{1O}, MI_A and QMS are all metrics for sensitivity evaluation at the algorithm level that do not require leakage measurements. QMS focuses on the extreme value among the differences of distributions of any pair $(x, k), (x', k') \in (\mathcal{X}, \mathcal{K})$, but ignores the other differences. The extreme value indicates the probability distance between the secret key to the one guessed key which is easiest to distinguish. However, the SCA succeeds only if the secret key is distinguished from all other guessed keys, not just one. Hence the expectation would be a better measure for information leakage than the extreme value. We can see that ILA_{1O} is an expectation of squared distances:

$$
\begin{aligned}
ILA_{1O} &= \sum_{k_g \in \mathcal{K} \setminus \{k_c\}} p(k_g) \kappa_{1O}(k_c, k_g) \\
&= \sum_{k_g \in \mathcal{K} \setminus \{k_c\}} p(k_g) \sum_{x \in \mathcal{X}} p(x) \cdot \{\mathbb{E}_M[V_0|(x, k_c)] - \mathbb{E}_M[V_0|(x, k_g)]\}^2.
\end{aligned} \tag{5}
$$

The calculation of ILA_{1O} through Eq. (5) involves iterations over $k_g \in \mathcal{K} \setminus k_c$ and $x \in \mathcal{X}$, which can be time-consuming for large sets of $\mathcal{K}$ and $\mathcal{X}$. These same iterations appear in MI_A and QMS definitions too. As recommended for MI calculations by [2,8], the exhaustive iterations in calculating ILA_{1O} can be replaced by averaging over a random subset of sufficiently large size. Thus the computational complexity is similar for the three metrics ILA_{1O}, MI_A and QMS.

Different from MI_A and QMS, we find that ILA_{1O} can be related to the success rates of DPA and CPA in explicit formulas, similar to the work in [13]. In addition, ILA_{1O} can be extended to a second-order metric ILA_{2O} as well, while there is no such work on MI_A and QMS yet.

A second-order attack retrieves the secret key by combining the information leakage at two leakage points, $V_0(X, K, M)$ and $V_1(M)$. A second-order metric measures the leakage under second-order CPA attacks.

Definition 3. *For a key hypothesis $k_g \in \mathcal{K} \setminus \{k_c\}$, we define the second-order confusion coefficient of masked algorithm as*

$$
\kappa_{2O}(k_c, k_g) = \mathbb{E}_X \{[\mathbb{E}_M(\widetilde{V_0}\widetilde{V_1}|(X, k_c)) - \mathbb{E}_M(\widetilde{V_0}\widetilde{V_1}|(X, k_g))]^2\}, \tag{6}
$$

where $\widetilde{V_i} = V_i - \mathbb{E}_{X,M}[V_i], i = 0, 1$, are the centered select function values.

Definition 4. *The second-order information leakage amount ILA_{2O} is defined as*

$$
ILA_{2O} = \mathbb{E}_{\mathcal{K} \setminus \{k_c\}}[\kappa_{2O}(k_c, k_g)]. \tag{7}
$$

Comment: Although the definitions (4) and (7) of ILA depend on the correct key k_c, in many practical situations ILA is key-independent. The leaked intermediate values often depend on key k_c only through $X \oplus k_c$. In that case, for uniformly distributed plaintext X, the ILA is in fact independent of k_c since $k_g \oplus k_c$ iterates over the same values for all k_c.

2.3   Analysis of the Metrics Under Boolean Masking

To better understand the metrics ILA, MI_A and QMS, we compare them in detail for a specific setting of biased Boolean masking $F = Z(X,k) \oplus M$ as in [11], under several commonly used assumptions on the distribution of unmasked $Z(X,k)$ and keys. Here $Z(X,k)$ denotes an unmasked intermediate variable with X being the random plaintext. Hence $V_0 = g(Z(X,k) \oplus M)$.

Assumption 1 *(Uniform Intermediate Variable). Given a key $k \in \mathcal{K}$, for random plaintext X, the unmasked intermediate variable $Z(X,k)$ is uniformly distributed. That is, $Z(X,k) \sim U(0, 2^b - 1)$, for all $k \in \mathcal{K}$, where $U(0, 2^b - 1)$ denotes the discrete uniform distribution on $\{0, 1, ..., 2^b - 1\}$ with b being the number of bits for $Z(X,k)$.*

Let $V_0^*(X,k) = g(Z(X,k))$ denote the unmasked select function. Under Assumption 1, $\mathbb{E}_X[V_0^*(X,k)]$ is a constant independent of keys k. In general, we would like the unmasked select function values under two different keys to be uncorrelated.

Assumption 2 *(Uncorrelated Keys). For any pair of keys $k_1, k_2 \in \mathcal{K}$, and random plaintext X, the select functions $V_0^*(X,k_1)$ and $V_0^*(X,k_2)$ are uncorrelated so that $\mathbb{E}_X[V_0^*(X,k_1)V_0^*(X,k_2)] = \mathbb{E}_X[V_0^*(X,k_1)]\mathbb{E}_X[V_0^*(X,k_2)]$.*

Under Assumptions 1 and 2, $\mathbb{E}_X[V_0^*(X,k_1)V_0^*(X,k_2)] = \{\mathbb{E}_X[V_0^*(X,k_1)]\}^2$ will also be a constant independent of keys k_1 and k_2. Unfortunately, many select functions (e.g., the Hamming weights of an AES S-Box output) do not satisfy Assumption 2. However, for a random key k_2, a weaker assumption often holds.

Assumption 3 *(Weak Uncorrelated Keys). For any fixed key k_1, let k_2 be a random key $\in \mathcal{K}\backslash\{k_1\}$. For a random plaintext X, the intermediate variables $Z(X,k_1)$ and $Z(X,k_2)$ are uncorrelated so that $\mathbb{E}_{X,k_2}[V_0^*(X,k_1)V_0^*(X,k_2)] = \{\mathbb{E}_X[V_0^*(X,k_1)]\}^2$.*

Under Assumptions 1 and 3, $\mathbb{E}_{X,k_2}[V_0^*(X,k_1)V_0^*(X,k_2)]$ is a constant, which helps us to derive simple explicit formulas of ILA in this section. Assumption 3 makes the calculation of the metrics easier here, as it removes ILA's dependence on many aspects of the algorithm including k_c value. The leakage metrics ILA under these assumptions reflect the masking strength only. In the next section, Assumption 3 will not be assumed for DPA/CPA success rates derivations though.

We first consider the DPA attack, where V_0 is on a single bit. Since $\oplus$ is taken bit by bit, we can take both $Z(X, k_c)$ and M as variables with one single bit, and $V_0 = Z \oplus M$. Let the distribution of the mask bit be $\mathbb{P}(M = 1) = p$ and $\mathbb{P}(M = 0) = 1 - p$, we have the following property.

Property 1. *For the DPA model under Assumptions 1 and 3, if $\mathbb{P}(M = 1) = p$, then*

- $ILA_{1O} = (1 - 2p)^2/2$,
- $ILA_{2O} = 2p^2(1 - p)^2$,
- $MI_A = 1 + (1 - p)\log_2(1 - p) + p\log_2(p)$,
- $QMS = 1 - |1 - 2p|$.

The detailed calculations are given in Appendix A. Note that although the generalized confusion coefficients $\kappa_{1O}(k_c, k_g)$ (Eq. 3) and $\kappa_{2O}(k_c, k_g)$ (Eq. 6) are determined by the algorithm, their average terms ILA_{1O} and ILA_{2O} become algorithm-independent and are only determined by the bias of the mask distribution, p, according to Assumption 3. For perfect masking, $p = 1/2$; unmasked, $p = 0$ or $p = 1$; imperfect masking, p takes other values. All metrics change with p and have one-to-one correspondence between each other. Particularly, $ILA_{1O} = (1 - QMS)^2/2$. Work in [11] empirically finds that the number of traces needed for DPA is approximately $N_{trace} = 1/(1 - QMS)^{2.2}$. In Sect. 4.1, we will show that number of traces $N_{trace} \propto 1/ILA_{1O} \propto 1/(1 - QMS)^2$ instead.

Figure 1 shows the relationship between these metrics and the probability p. It is symmetric about the x-axis which implies the same effect of the mask bit being 0 and 1. From Fig. 1, we see that ILA_{1O} and MI_A have the same pattern, but ILA_{1O} increases from 0 to 1/2 and MI_A increases from 0 to 1 as p goes from 1/2 to 0 (or 1). When $p = 0$ or $p = 1$, the device is without any masking protection, $QMS = 0$ while ILA_{1O} and MI_A both reach their maximum. When $p = 1/2$, the devices is protected by perfect masking, $ILA_{1O} = MI_A = 0$ and $QMS = 1$ which are consistent with no first-order information leakage. However, the second-order leakage still exists under perfect masking, and actually reaches its maximum (biggest leakage) 1/8. As the mask gets more biased, the first-order leakage increases while the second-order leakage decreases.

Next we consider CPA in this setting. For CPA, $V_0 = HW(Z \oplus M)$ is the Hamming weight function of a b-bit variable. We assume that the bits in the mask

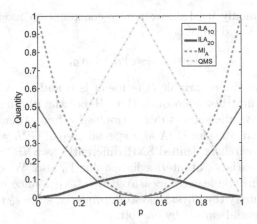

Fig. 1. The quantities of several metrics under the biased masking for DPA.

are independent from the same distribution with $\mathbb{P}(M_{(i)} = 1) = p$, $i = 1, ..., b$. Here $M_{(i)}$ denotes the ith bit of the b-bit mask variable M.

Property 2. *For the CPA model under Assumptions 1 and 3,*

$$\text{ILA}_{1O} = b(1 - 2p)^2/2, \qquad \text{ILA}_{2O} = 2bp^2(1 - p)^2. \qquad (8)$$

The proof is provided in Appendix B.

For the CPA model, the ILA_{1O} and ILA_{2O} follow the similar pattern as in the DPA model, just differing by a factor of b, the number of bits. In fact, the DPA model is a special case of the CPA model with $b = 1$. The other two metrics MI_A and QMS are harder to derive for CPA. It is hard, if not impossible, to relate MI_A and QMS to the success rate of CPA.

3 Relating ILA to DPA and CPA Success Rates

As shown in [9,12,13], the success rates of first-order DPA and CPA on unmasked devices and second-order CPA on perfectly masked devices can all be expressed in terms of the confusion coefficients and the implementation signal-to-noise-ratio (SNR). Our metrics ILA_{1O} and ILA_{2O} are algorithmic properties like the confusion coefficients. We generalize the results of [9,12,13] to masked implementations (possibly with imperfect masking), and show that the success rates of CPA/DPA should be determined by the SNRs and our generalized confusion coefficients. The formulas are further simplified to consist of ILA_{1O} and ILA_{2O}. We show derivations for the success rates of first-order and second-order DPA and CPA on masked devices in this section. We then use these metrics to compare the first-order leakage and second-order leakage.

3.1 First-Order Power Analysis Attack Model

We assume a commonly used linear power consumption model with additive noises for both DPA and CPA,

$$L_0 = c_0 + \epsilon_0 V_0 + \sigma_0 r_0, \qquad (9)$$

where r_0 is the unit noise variable (the mean is 0 and the variance is 1) and ϵ_0 is the single-bit unit power consumption. Hence the physical system SNR is $\delta_0 = \epsilon_0/\sigma_0$. We derive the success rate formulas for first-order CPA in terms of SNR and ILA_{1O}, and consider DPA as a special case of CPA with $b = 1$. Notice that some other researchers defined SNR differently as $\text{SNR}^* = \epsilon_0^2 Var(V_0)/\sigma_0^2$, which includes the variance of intermediate value V_0 also. We consider $Var(V_0)$ to be part of algorithmic leakage measured by ILA_{1O}, since it depends on V_0. Our SNR reflects purely the physical system property, since ϵ_0 reflects the power consumption differential caused by one-bit.

The leakage measurements of L_0 are denoted as $\mathcal{L} = \{l_{1,0}, l_{2,0}, ..., l_{n,0}\}$, where n is the number of traces. For unmasked devices, the CPA exploits the correlation between the leakage L and unmasked select function $V_0^* = g(Z(X, k))$

to discover the secret key. For masked devices, the attacker does not know M value, and therefore does not know the value of $V_0 = g(F(X, k, M))$. To conduct CPA, the attacker has to correlate L with $\mathbb{E}_M[V_0|(X, k, M)]$, the expectation of V_0 over all possible mask values. This value is $V_0^*(X, k)$ for unmasked devices, and is a constant (thus no leakage) for perfectly masked devices. Let $v_{m,i,0}^g$ denote $V_0(x_i, k_g, m_i)$ for the i-th power trace, the selection function value under plaintext x_i, guess key k_g and the mask m_i, $\mathbb{E}_M[v_{m,i,0}^g]$ denote the targeted expectation of $V_0(x_i, k_g, m)$ over all $m \in \mathcal{M}$, and $\mathbb{E}[V_0^g]$ denote the expectation of $V_0(x, k_g, m)$ over all $x \in \mathcal{X}$ and $m \in \mathcal{M}$. Under the power model (9) with imperfect masking, the first-order CPA distinguishes the key k_g by the Pearson's correlation:

$$\hat{\rho}^g = \frac{\sum_{i=1}^n (l_{i,0} - \bar{l}_{.,0})[\mathbb{E}_M(v_{m,i,0}^g) - \mathbb{E}(V_0^g)]}{\sqrt{\sum_{i=1}^n (l_{i,0} - \bar{l}_{.,0})^2 \sum_{i=1}^n [\mathbb{E}_M(v_{m,i,0}^g) - \mathbb{E}(V_0^g)]^2}}, \tag{10}$$

where $\bar{l}_{.,0} = \sum_{i=1}^n l_{i,0}/n$ is the mean of power leakage.

The CPA succeeds when $\hat{\rho}^c - \hat{\rho}^g > 0$ for all $k_g \in \mathcal{K}\backslash\{k_c\}$. For a random plaintext attack with a large number of traces, under Assumption 1, the denominator of (10) converges to the same limit for all k_g, since $\mathbb{E}[\mathbb{E}_M(v_{m,i,0}^g)] = \mathbb{E}(V_0^g) = \mathbb{E}(V_0^c)$ and $\mathbb{E}\{\mathbb{E}_M[(v_{m,i,0}^g)^2]\} = \mathbb{E}\{\mathbb{E}_M[(v_{m,i,0}^c)^2]\}$. Hence $\hat{\rho}^c - \hat{\rho}^g > 0$ is equivalent to that the difference in the numerators of (10) is positive. That is, $\hat{\rho}^c - \hat{\rho}^g > 0$ when $\Delta_n^{1O}(k_c, k_g) > 0$, where

$$\Delta_n^{1O}(k_c, k_g) = \sum_{i=1}^n \frac{(l_{i,0} - \bar{l}_{.,0})}{\sigma_0}[\mathbb{E}_M(v_{m,i,0}^c) - \mathbb{E}_M(v_{m,i,0}^g)]. \tag{11}$$

Let Δ_n^{1O} denote the (N_k-1)-dimension vector consisting of these $\Delta_n^{1O}(k_c, k_g)$ for all $k_g \in \mathcal{K}\backslash\{k_c\}$. Let $\boldsymbol{\mu}$ and $\boldsymbol{\Sigma}$ denote the mean and variance of $\Delta_1^{1O}(k_c, k_g)$. Then following the work in [13,20], the success rate can be described with a multivariate Gaussian distribution $N(\boldsymbol{\mu}, \boldsymbol{\Sigma}/n)$ using the Central Limit Theorem. That is,

$$SR = \Phi_{\boldsymbol{\Sigma}}(\sqrt{n}\boldsymbol{\mu}). \tag{12}$$

where $\Phi_{\boldsymbol{\Sigma}}$ is the cumulative distribution function (CDF) of the $N_k - 1$ dimensional Gaussian distribution with mean $\mathbf{0}$ and variance $\boldsymbol{\Sigma}$.

For unmasked devices, the mean vector $\boldsymbol{\mu}$ and the variance matrix $\boldsymbol{\Sigma}$ are expressed by Fei et al. [13] in terms of their confusion coefficients κ. With imperfect masking, we show (in Appendix C) that similar expressions hold with our generalized confusion coefficients κ_{1O}.

Theorem 1. *Under CPA leakage model (9), the success rate of the CPA is given by Eq. (12). Under Assumption 1, the element in the mean vector $\boldsymbol{\mu}$ corresponding to key k_{gi} is*

$$\mu_{gi} = \frac{\delta_0}{2}\kappa_{1O}(k_c, k_{gi}); \tag{13}$$

And the elements of covariance matrix Σ are

$$\sigma_{k_{gi},k_{gi}} = \kappa_{1O}(k_c, k_{gi}), \qquad \sigma_{k_{gi},k_{gj}} = \kappa_{1O}(k_c, k_{gi}, k_{gj}) \ for \ k_{gi} \neq k_{gj}, \qquad (14)$$

where $\kappa_{1O}(k_c, k_{gi}, k_{gj}) = \mathbb{E}_X\{[\mathbb{E}_M(v_{m,1,0}^c) - \mathbb{E}_M(v_{m,1,0}^{gi})][\mathbb{E}_M(v_{m,1,0}^c) - \mathbb{E}_M(v_{m,1,0}^{gj})]\}.$

Similar to [13], we can get the above three-way generalized confusion coefficients $\kappa_{1O}(k_1, k_2, k_3)$ from two-way generalized confusion coefficients $\kappa_{1O}(k_1, k_2)$ (see more details in Appendix D).

Lemma 1. *Given* $k_c, k_{gi}, k_{gj} \in \mathcal{K}$,

$$\kappa_{1O}(k_c, k_{gi}, k_{gj}) = \frac{1}{2}[\kappa_{1O}(k_c, k_{gi}) + \kappa_{1O}(k_c, k_{gj}) - \kappa_{1O}(k_{gi}, k_{gj})]. \qquad (15)$$

The average of $\kappa_{1O}(k_c, k_{gi})$ over all k_{gi} is ILA_{1O}. By Lemma 1, the average of $\kappa_{1O}(k_c, k_{gi}, k_{gj})$ over all $k_{gi} \neq k_{gj}$ is $\mathrm{ILA}_{1O}/2$. Replacing all the confusion coefficient terms in Eqs. (13) and (14) by their averages, we get an approximate asymptotic success rate for first-order CPA on masked devices:

$$SR = \Phi_{\frac{1}{2}[I_{N_k-1}+J_{N_k-1}]}\left(\frac{\delta_0\sqrt{n}\sqrt{\mathrm{ILA}_{1O}}}{2}1_{N_k-1}\right), \qquad (16)$$

where I_{N_k-1} is the $(N_k - 1) \times (N_k - 1)$ identity matrix with diagonal entries of ones and off-diagonal entries of zeros, J_{N_k-1} is the $(N_k - 1) \times (N_k - 1)$ matrix with all entries of ones, and 1_{N_k-1} is the $(N_k - 1)$ dimensional vector of ones.

The approximation SR formula (16) is very close to the SR formula (12) for small SNR δ_0. We will examine the approximation in Sect. 3.3.

3.2   Second-Order Power Analysis Attack Model

Second-order power analysis attack combines the two leakage measurements of V_0 and V_1 at two different positions involving the same mask M to break the masking protection. Similar to (9), we assume linear leakage for V_1

$$L_1 = c_1 + \epsilon_1 V_1 + \sigma_1 r_1, \qquad (17)$$

where r_1 is the unit noise.

Second-order CPA uses n pairs of independent realizations of noisy physical leakage $(l_{1,0}, l_{1,1}), (l_{2,0}, l_{2,1}), ..., (l_{n,0}, l_{n,1})$ for (L_0, L_1). Here $l_{i,j} = c_j + \epsilon_j v_{i,j} + \sigma_j r_{i,j}, i = 1, ..., n, j = 0, 1$. Denote the centered version of L_j and V_j by $\tilde{L}_j = L_j - \mathbb{E}(L_j)$ and $\tilde{V}_j = V_j - \mathbb{E}(V_j)$, for $j = 0, 1$. While the first-order CPA exploits the correlation between $\tilde{L}_0$ and $\tilde{V}_0$, the second-order CPA exploits the correlation between $\tilde{L}_0\tilde{L}_1$ and $\tilde{V}_0\tilde{V}_1$. That is, it uses the centered product statistic:

$$\frac{1}{n}\sum_{i=1}^{n}\tilde{l}_{i0}\tilde{l}_{i1}\mathbb{E}_M[\tilde{v}_{m,i,0}^g\tilde{v}_{m,1}], \qquad (18)$$

where $\widetilde{l}_{ij} = (l_{i,j} - \overline{l}_{.,j})/\sigma_j, j = 0, 1$, is the centered leakage, $\widetilde{v}^g_{m,i,0} = v^g_{m,i,0} - \mathbb{E}[V^g_0]$ and $\widetilde{v}_{m,1} = v_{m,1} - \mathbb{E}[V_1]$ are the centered select functions values under guessed key k_g given mask m, and $v_{m,1} = V_1(m)$.

We denote the difference between the centered product statistics under secret key k_c and guessed key k_g as

$$\Delta_n^{2O}(k_c, k_g) = \frac{1}{n} \sum_{i=1}^{n} \widetilde{l}_{i0}\widetilde{l}_{i1}[\mathbb{E}_M(\widetilde{v}^c_{m,i,0}\widetilde{v}_{m,1}) - \mathbb{E}_M(\widetilde{v}^g_{m,i,0}\widetilde{v}_{m,1})]. \tag{19}$$

The second-order CPA succeeds when $\Delta_n^{2O}(k_c, k_g) > 0$ for all $k_g \in \mathcal{K}\backslash\{k_c\}$. Using derivations in [9,17,19], the success rate of second-order CPA also follows Eq. (12): $SR = \Phi_{\Sigma}(\sqrt{n}\mu)$.

Ding et al. [9] expressed μ and Σ in terms of confusion coefficients κ under perfect masking. With possibly imperfect masking, we generalize the formula in terms of our generalized confusion coefficients κ_{2O} (see details in Appendix E).

Theorem 2. *Under CPA leakage model (9) and (24), the success rate of the second-order CPA is given by Eq. (12). Under Assumption 1, the element in μ corresponding to key k_{gi} is*

$$\mu_{gi} = \frac{\delta_0 \delta_1}{2} \kappa_{2O}(k_c, k_{gi}); \tag{20}$$

And the elements of covariance Σ are

$$\sigma_{k_{gi},k_{gi}} = \kappa_{2O}(k_c, k_{gi}), \qquad \sigma_{k_{gi},k_{gj}} = \kappa_{2O}(k_c, k_{gi}, k_{gj}) \, for \, k_{gi} \neq k_{gj}, \tag{21}$$

where $\kappa_{2O}(k_c, k_{gi}, k_{gj}) = \mathbb{E}_X\{[\mathbb{E}_M(\widetilde{v}^c_{m,1,0}\widetilde{v}_{m,1}) - \mathbb{E}_M(\widetilde{v}^{gi}_{m,1,0}\widetilde{v}_{m,1})][\mathbb{E}_M(\widetilde{v}^c_{m,1,0}\widetilde{v}_{m,1}) - \mathbb{E}_M(\widetilde{v}^{gj}_{m,1,0}\widetilde{v}_{m,1})]\}$.

Similar with Lemma 1, for $k_c, k_{gi}, k_{gj} \in \mathcal{K}$,

$$\kappa_{2O}(k_c, k_{gi}, k_{gj}) = \frac{1}{2}[\kappa_{2O}(k_c, k_{gi}) + \kappa_{2O}(k_c, k_{gj}) - \kappa_{2O}(k_{gi}, k_{gj})]. \tag{22}$$

As in the first-order analysis, replacing the generalized confusion coefficients κ_{2O} by ILA_{2O}, we get the approximate asymptotic success rate:

$$SR = \Phi_{\frac{1}{2}[I_{N_k-1}+J_{N_k-1}]}\left(\frac{\delta_0 \delta_1 \sqrt{n}\sqrt{ILA_{2O}}}{2}\mathbf{1}_{N_k-1}\right). \tag{23}$$

Next we evaluate the above approximations.

3.3 Approximation Errors in the Simple Success Rate Formulas

Work in [9,13] gives the explicit theoretical success rate formulas for two cases: the first-order CPA on unmasked devices and the second-order CPA on perfectly masked devices, respectively. By plugging ILA_{1O} when $p = 0$ in (16) and ILA_{2O} when $p = 1/2$ in (23), we get the two corresponding simple success rate formulas.

Compared to the formulas in [9, 13], our simple formulas ignore some higher order terms and replace the confusion coefficients by ILA. Here we study the effect of the simplification for CPA on unmasked and perfect masked AES.

We show the difference between our simplified success rate formulas and the explicit success rate formulas of [9, 13] in Fig. 2. The average error-ratio is defined as: $\mathbb{E}_{SR}[|N_{\text{Explicit,SR}} - N_{\text{Simple,SR}}|/N_{\text{Explicit,SR}}]$, where $N_{\text{Explicit,SR}}$ and $N_{\text{Simple,SR}}$ are numbers of traces needed to achieve a fixed SR value by the explicit and simplified theoretical success rate formulas respectively, and $\mathbb{E}_{SR}$ is the expectation over all success rate values SR ranging from 0 to 1. Here, we take the expectation over discrete success rate values SR = $[0.1, 0.2, 0.3, ..., 0.9]$.

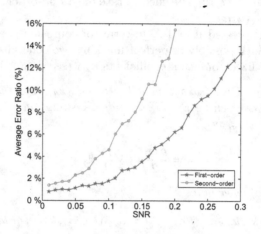

Fig. 2. The average error-ratio of number of measurements between explicit and simplified success rate formulas on AES S-Box

Figure 2 shows that as the SNR grows, both error ratios increase. The error-ratio ≤ 10 % when SNR ≤ 0.26 for the first-order attack, and when SNR ≤ 0.16 for the second-order attack. Hence the simplified success rate formulas in Eqs. (16) and (23) work well for small SNR values. For practical physical implementations, devices with large SNR values are very leaky and not considered secure. The success rate analysis is only meaningful when the SNR is small.

3.4 Comparing Effectiveness of the First-Order Attack and the Second-Order Attack

For unmasked devices, first-order leakage is sufficient to discover the secret key. With perfect masking, only second-order leakage can be used to discover the secret key. However, for imperfect masking implementations, both first-order and second-order leakage exist. Which leakage is more effective to exploit? We can compare them using the proposed metrics through formulas (16) and (23).

Property 3. *For a masked implementation*

- *The first-order attack is more effective when $\delta_1 < \sqrt{\frac{\text{ILA}_{1O}}{\text{ILA}_{2O}}}$;*
- *The second-order attack is better when $\delta_1 > \sqrt{\frac{\text{ILA}_{1O}}{\text{ILA}_{2O}}}$.*

For very small SNR, the first order leakage will dominate. The threshold SNR value to determine dominance by the first-order or the second-order leakage is given by the square root of the ratio between the two information leakages: $\sqrt{\frac{\text{ILA}_{1O}}{\text{ILA}_{2O}}}$. If the typical SNR value is known for certain physical devices, we can predict which type of leakage dominates and therefore guide the software designer in effective leakage reduction.

3.5 Extension to Higher-Order Power Analysis Attack Model

We now consider a cryptography algorithm protected by J-th order masking, with mask shares $M_1, M_2, ..., M_J$. A J-th order attack combines the information leakage of $V_0(X, K, M_1, ..., M_J)$ and the leakage of $V_1(M_1), ..., V_J(M_J)$ to retrieve the secret key. J-th order power analysis attack combines the $J + 1$ leakage measurements of $V_0, V_1, ..., V_J$ at $J + 1$ different positions to break the masking protection. The leakage vector is $l_i = (l_{i,0}, ..., l_{i,J})$. Similar to (9) and (24), the leakage model is now:

$$L_j = c_j + \epsilon_1 V_j + \sigma_j r_j, \qquad j = 0, ..., J. \tag{24}$$

where r_j is the unit noise.

For a key hypothesis $k_g \in \mathcal{K} \backslash \{k_c\}$, we define the J-th order confusion coefficient of masked algorithm as

$$\kappa_{JO}(k_c, k_g) = \mathbb{E}_X\{[\mathbb{E}_M(\widetilde{V_0}\widetilde{V_1}...\widetilde{V_J}|(X, k_c)) - \mathbb{E}_M(\widetilde{V_0}\widetilde{V_1}...\widetilde{V_J}|(X, k_{gi}))]^2\}, \tag{25}$$

where $\widetilde{V_i} = V_i - \mathbb{E}_{X,M}[V_i], i = 0, 1, ..., J$, are the centered select function values.

Definition 5. *The J-th order information leakage amount ILA_{JO} is defined as*

$$\text{ILA}_{JO} = \mathbb{E}_{\mathcal{K} \backslash \{k_c\}}[\kappa_{JO}(k_c, k_g)]. \tag{26}$$

As in [9] and in Sect. 3.2, we can derive the approximate asymptotic success rate as:

$$SR = \Phi_{\frac{1}{2}[I_{N_k-1} + J_{N_k-1}]}\left(\frac{\sqrt{n}\sqrt{\text{ILA}_{JO}}\prod\limits_{j=0}^{J}\delta_j}{2} 1_{N_k-1}\right). \tag{27}$$

4   Numerical Results

In this section, we first numerically investigate the relationship between success rates of DPA/CPA and the metrics ILA, MI_A and QMS on synthetic data examples. We also evaluate our metrics and the simplified success rates of DPA/CPA on realistic measurement data.

4.1   Numerical Comparison of Metrics Versus Success Rates

We first show, by numerical examples, that ILA_{1O} measures the leakage information amount under CPA, but MI_A and QMS do not. We consider synthetic data examples with biased masking on the outputs of an AES S-Box, where the masking bits are independent with $p_i = \mathbb{P}(M_{(i)} = 1), i = 1, 2, ..., 8$.

In the first example, the last 4-bits are perfectly masked with $p_5 = p_6 = p_7 = p_8 = 0.5$, and the information leakage is through the Hamming weights of the first 4-bits according to model (9). We consider two cases where $\overrightarrow{p}_4 = [p_1, p_2, p_3, p_4] = [0.5, 0.2, 0.2, 0.1]$ and $\overrightarrow{p}_4 = [0, 0.4, 0.4, 0.4]$ respectively. We calculate the values of the different metrics through definitions in Eqs. (1), (2) and (5), rather than using specialized formulas in Properties 1 and 2 (which only apply to Boolean masking with equal p_i's for each bit). Detailed algorithms are provided in Appendix F. In both cases $MI_A = 1.09$, but the information leakage amount differs with $ILA_{1O} = 0.68$ and $ILA_{1O} = 0.56$, respectively. Figure 3 (a) shows the success rates of CPA in both cases on synthetic data generated from the power model (9) with SNR $= 0.1$. The empirical success rate for a fixed number of measurements N_{trace} is found by repeatedly randomly sampling N_{trace} traces for an attack, and calculating the proportion of attacks that retrieves the correct secret key. We see that the ILA_{1O} correctly predicts the two different CPA success rates curves (with difference about 10%), while by MI_A the information leakage should be the same in these two cases. Note that from Fig. 2, the error ratio of our simplified SR formula under first-order CPAs is only 1.5% when SNR $= 0.1$.

In the second example, the last 6-bits are perfectly masked. For two cases of $\overrightarrow{p}_2 = [p_1, p_2] = [0.3, 0.3]$ and $\overrightarrow{p}_2 = [0.1, 0.5]$, QMS $= 0.4$, but ILA $= 0.16$ and 0.32 respectively. Figure 3 (b) shows that ILA_{1O} correctly predicts the different empirical CPA success rate curves, while QMS incorrectly labels the two cases as equally leaky. Therefore, only ILA_{1O} correctly measures the CPA leakage in these examples.

The formulas (16) and (23) give the CPA success rates using ILA and SNR. Figure 4 plots the number of traces N_{trace} needed to achieve success rate of SR $= 80\%$, when ILA and SNR vary. Figure 4 (a) is for the first-order CPA attack (16) and (b) is for the second-order CPA attack (23). As ILA increases or SNR increases, less traces are needed to get SR $= 80\%$. For a fixed SNR value, the number of traces N_{trace} is inverse proportional to ILA.

For the special case of single-bit DPA, all three metrics are monotonic functions of each other (Property 1). Thus, MI_A and QMS can predict the DPA success rate through their relationship with ILA. Particularly, for DPA, $ILA_{1O} = (1 - QMS)^2/2$ and the N_{trace} traces needed for DPA is inverse proportional to $(1 - QMS)^2$.

4.2   Experimental Results on Physical Implementations

We next verify the prediction of success rates by ILA, and show that it also correctly predicts the dominance by first-order or second-order CPA leakage on

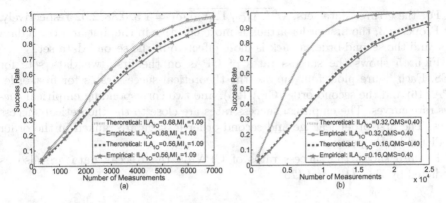

Fig. 3. First-order CPA attacks under two different biased masking schemes with SNR = 0.1 (a) with the same MI_A value but different ILA_{1O} values; (b) with the same QMS value but different ILA_{1O} values

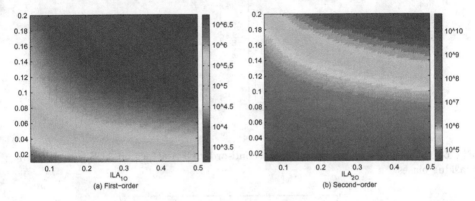

Fig. 4. The theoretical number of traces needed for SR = 80 % under first-order and second-order CPA.

real physical systems. Two physical implementations of masked Keccak and AES algorithms are considered. The masked AES [1] is implemented on an SASEBO-GII board [22]. The protected Keccak implementation with secret sharing [4] is on the 32-bit Microblaze processor of the SASEBO-GII board. All the power traces are collected using a LeCroy WaveRunner 640Zi oscilloscope.

We get several power data sets with biased masking through choosing parts of the fully masking data set according to biased masks distributions. The first two data sets are on the same AES implementation with $\delta_0 = 0.10, \delta_1 = 0.12$ but with different biased masks. The leakage amount on the first data set is $ILA_{1O} = 0.338, ILA_{2O} = 13.8$, while the leakage amount on the second data set is $ILA_{1O} = 0.174, ILA_{2O} = 15.7$ for CPA attacks. For the third data set on Keccak, $\delta_0 = 0.10, \delta_1 = 0.10, ILA_{1O} = 0.010, ILA_{2O} = 0.006$ for DPA attacks.

For these three data sets, $\sqrt{\mathrm{ILA_{1O}}\,/\,\mathrm{ILA_{2O}}}/\delta_1 = 1.3, 0.88, 2.02$ respectively. By Property 3, the first-order attack is more effective in the first and third data sets, and the second-order attack is more effective in the second data set.

Figure 5 shows the success rates of CPAs on the first two data sets for AES. Each figure plots four curves, the theoretical success rates for first-order CPA (16) and the second-order CPA (23), and two corresponding empirical success rate curves. The empirical success rates are close to the theoretical success rates. The first-order leakage and second-order leakage are ranked in the order predicted by Property 3.

Figure 6 shows the success rates of CPA on the Keccak data are also as predicted by Eqs. (16) and (23).

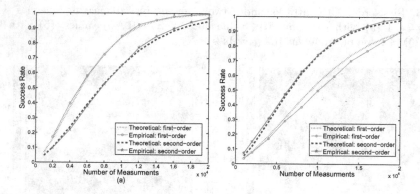

Fig. 5. The first-order CPA attack and second-order CPA attack on AES with different masking biases.

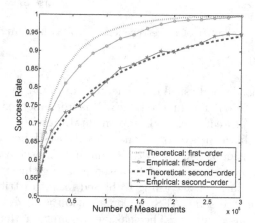

Fig. 6. The first-order CPA attack and the second-order CPA attacks on Keccak data subset.

5   Conclusion

In this work, we propose a new unified metric, ILA, to measure the information leakage at the early stage of cryptographic software under different power analysis attacks. It quantifies the leakage amount of algorithms with various masking strength to first-order or second-order power analysis attacks. Unlike existing metrics, ILA relates to the attack success rate on the physical implementations through a simple explicit formula. We demonstrate that it accurately quantifies the leakage amount comparing to existing metrics on both synthetic data and real physical implementation data. Therefore, it would be a reliable metric for system designers to predict the system leakage and develop better protections.

Acknowledgments. This work is supported in part by the National Science Foundation under grants CNS-1314655 and CNS-1337854.

Appendices

A   Derivation of ILA, QMS and MI$_A$ for DPA Model

For the DPA model, Z is one single bit, as well as M. Under Assumption 1, $\mathbb{P}(Z = 0) = \mathbb{P}(Z = 1) = 1/2$. For the Boolean masking, $V_0 = F = Z \oplus M$. Hence $\mathbb{P}(Z \oplus M = 0) = \mathbb{P}(Z \oplus M = 1) = 1/2$,

$$\mathbb{P}(Z \oplus M = 1|Z) = (1 - 2p)Z + p = p \text{ or } 1 - p. \tag{28}$$

Using Eq. (28), $D_{x,k}(F) = p$ or $1 - p$, which implies $\max\{|D_{x,k}(F) - D_{x',k'}(F)|\} = |1 - 2p|$. Hence QMS $= 1 - |1 - 2p|$.

For MI$_A$, we calculate the entropies first.

$$H(K) = -\sum_{k \in \mathcal{K}} p(k) \log_2 p(k) = -\sum_{k \in \mathcal{K}} \frac{1}{N_k} \log_2 \frac{1}{N_k} = \log_2 N_k.$$

$$H(K|V_0) = -\sum_{k \in \mathcal{K}} p(k). \sum_{x \in \{0,1\}} p(x). \sum_{v_0 \in \{0,1\}} p(v_0|k,x). \log_2 p(k|v_0, x)$$

$$= -\sum_{k \in \mathcal{K}} \frac{1}{N_k}. \sum_{x \in \{0,1\}} \frac{1}{2}. [p \log_2 \frac{\frac{p}{2N_k}}{\frac{1}{4}} + (1-p) \log_2 \frac{\frac{1-p}{2N_k}}{\frac{1}{4}}]$$

$$= \log_2 N_k - [1 + (1-p) \log_2(1-p) + p \log_2 p].$$

Therefore,

$$\text{MI}_A = H(K) - H(V_0|K) = 1 + (1-p) \log_2(1-p) + p \log_2 p. \tag{29}$$

We will derive the ILA_{1O} and ILA_{2O} expressions in the CPA model in Appendix B. Plugging-in $b = 1$, we get their DPA model expressions.

B   Derivation of ILA_{1O} and ILA_{2O} for CPA Model

For the CPA model, the selection is Hamming weights $V_0 = H(Z \oplus M)$, $V_1 = H(M)$, and both M and Z are b-bit variables. Since $\mathbb{P}(M_{(i)} = 1) = p, i = 1, 2, ..., b$, we have:

$$\mathbb{E}_M[H(M)] = bp, \qquad \mathbb{E}_M[H(M)^2] = bp + b(b-1)p^2. \tag{30}$$

Under Assumption 1, Z has uniform distribution for any key k_g so that always

$$\mathbb{E}_X[H(Z)] = b/2, \qquad \mathbb{E}_X[H(Z)^2] = (b^2 + b)/4. \tag{31}$$

Here $V_0^*(X, k) = H[Z(X, k)]$. Under Assumptions 1 and 3,

$$\begin{aligned}
&\mathbb{E}_{k_g}\kappa(k_c, k_g) = \mathbb{E}_{k_g}\mathbb{E}_X\{[V_0^*(X, k_c) - V_0^*(X, k_g)]^2\}\\
&= \mathbb{E}_{k_g}\mathbb{E}_X[V_0^*(X, k_c)^2] + \mathbb{E}_{k_g}\mathbb{E}_X[V_0^*(X, k_g)^2] - 2\mathbb{E}_{k_g}\mathbb{E}_X[V_0^*(X, k_c)V_0^*(X, k_g)]\\
&= 2\mathbb{E}_X[V_0^*(X, k_c)^2] - 2\{\mathbb{E}_X[V_0^*(X, k_c)]\}^2.
\end{aligned} \tag{32}$$

Using (31), this becomes

$$\mathbb{E}_{k_g}\kappa(k_c, k_g) = 2(\tfrac{b^2+b}{4}) - 2(\tfrac{b^2}{4}) = \tfrac{b}{2}. \tag{33}$$

By the property 2 in [19], with $\wedge$ denoting the bit-wise multiplication,

$$\begin{aligned}
\mathbb{E}_M[H(Z \oplus M)|(X, k_c)] &= \mathbb{E}_M[H(Z) + H(M) - 2H(Z \wedge M)|(X, k_c)]\\
&= (1 - 2p)H(Z) + bp.
\end{aligned} \tag{34}$$

Then for the first-order CPA, using Eqs. (34) and (33)

$$\begin{aligned}
\text{ILA}_{1O} &= \mathbb{E}_{k_g}[\kappa_{1O}(k_c, k_g)]\\
&= \mathbb{E}_{k_g}[\mathbb{E}_X\{[\mathbb{E}_M(H(Z \oplus M)|(X, k_c)) - \mathbb{E}_M(H(Z \oplus M)|(X, k_g))]^2\}]\\
&= \mathbb{E}_{k_g}[(1 - 2p)^2\kappa(k_c, k_g)] = \tfrac{b(1-2p)^2}{2}.
\end{aligned} \tag{35}$$

Similar to (34), using (30),

$$\begin{aligned}
&\mathbb{E}_M\{[H(Z \oplus M) - \tfrac{b}{2}][H(M) - bp]|(X, k_c)\}\\
&= \mathbb{E}_M\{[H(Z \oplus M)H(M) - bpH(Z \oplus M)]|(X, k_c)\}\\
&= \mathbb{E}_M\{[H(Z)H(M) + H(M)^2 - 2H(Z \wedge M)H(M) - bpH(Z \oplus M)]|Z\}\\
&= H(Z)bp + [bp + b(b-1)p^2] - 2[p + (b-1)p^2]H(Z)\\
&\quad - bp[(1 - 2p)H(Z) + bp]\\
&= -2p(1 - p)[H(Z) - \tfrac{b}{2}].
\end{aligned} \tag{36}$$

Hence for the second-order CPA, using Eqs. (36) and (33)

$$\begin{aligned}
\text{ILA}_{2O} &= \mathbb{E}_{k_g}[\kappa_{2O}(k_c, k_g)]\\
&= \mathbb{E}_{k_g}[\mathbb{E}_X\{[\mathbb{E}_M(\widetilde{V_0}\widetilde{V_1}|(X, k_c)) - \mathbb{E}_M(\widetilde{V_0}\widetilde{V_1}|(X, k_g))]^2\}]\\
&= \mathbb{E}_{k_g}[4p^2(1 - p)^2\kappa(k_c, k_g)] = 2bp^2(1 - p)^2.
\end{aligned} \tag{37}$$

C   Theorem 1: μ and Σ in the first-order CPA (12)

Denote $v_{m,1,0}^g = V_0(x_1, k_g, m)$ and $v_{1,0} = V_0(x_1, k_c, m_1)$. Recall that, under Assumption 1, $\mathbb{E}[v_{m,1,0}^g] = \mathbb{E}[V_0^g] = \mathbb{E}[V_0^c]$ and $\mathbb{E}_X\{[\mathbb{E}_M(v_{m,1,0}^g)]^2\} = \mathbb{E}_X\{[\mathbb{E}_M(v_{m,1,0}^c)]^2\}$ for any k_g. Hence we have an useful expression that will be used later,

$$
\begin{aligned}
&\mathbb{E}_X\{\mathbb{E}_M(v_{m,1,0}^c)[\mathbb{E}_M(v_{m,1,0}^c) - \mathbb{E}_M(v_{m,1,0}^g)]\} \\
&= \tfrac{1}{2}\mathbb{E}_X\{[\mathbb{E}_M(v_{m,1,0}^c)]^2 + [\mathbb{E}_M(v_{m,1,0}^g)]^2 - 2\mathbb{E}_M(v_{m,1,0}^g)\mathbb{E}_M(v_{m,1,0}^g)]\} \\
&= \tfrac{1}{2}\mathbb{E}_X\{[\mathbb{E}_M(v_{m,1,0}^c) - \mathbb{E}_M(v_{m,1,0}^g)]^2\} = \tfrac{1}{2}\kappa_{10}(k_c, k_g).
\end{aligned}
\tag{38}
$$

For large n, $\bar{l}_{.,0} = c_0 + \epsilon_0\mathbb{E}(v_{1,0})$ and $l_{1,0} = c_0 + \epsilon_0 v_{1,0} + \sigma_0 r_{1,0}$, then Eq. (11) becomes

$$
\Delta_1^{1O}(k_c, k_g) = \{\delta_0[v_{1,0} - \mathbb{E}(v_{1,0})] + r_{1,0}\}[\mathbb{E}_M(v_{m,1,0}^c) - \mathbb{E}_M(v_{m,1,0}^g)].
\tag{39}
$$

Since $\mathbb{E}[r_{1,0}] = 0$, we have:

$$
\begin{aligned}
\mu_{k_g} &= \delta_0\mathbb{E}\{(v_{1,0} - \mathbb{E}[v_{1,0}])(\mathbb{E}_M[v_{m,1,0}^c] - \mathbb{E}_M[v_{m,1,0}^g])\} \\
&= \delta_0\mathbb{E}\{v_{1,0}(\mathbb{E}_M[v_{m,1,0}^c] - \mathbb{E}_M[v_{m,1,0}^g])\} = \tfrac{\delta_0}{2}\kappa_{10}(k_c, k_g).
\end{aligned}
\tag{40}
$$

The last equality uses the fact that $\mathbb{E}_M[v_{1,0}] = \mathbb{E}_M[v_{m,1,0}^c]$ and Eq. (38).

The element in covariance Σ corresponding to k_{gi} and k_{gj} is:

$$
\sigma_{k_{gi},k_{gj}} = COV(\Delta_1^{1O}(k_c, k_{gi}), \Delta_1^{1O}(k_c, k_{gj})) = E[\Delta_1^{1O}(k_c, k_{gi})\Delta_1^{1O}(k_c, k_{gj})] - \mu_{k_{gi}}\mu_{k_{gj}}.
\tag{41}
$$

Since $\mathbb{E}[r_{1,0}^2] = 1$, keep the leading term (dropping the terms with δ_0), we have

$$
\sigma_{k_{gi},k_{gj}} = \mathbb{E}_X\{(\mathbb{E}_M[v_{m,1,0}^c] - \mathbb{E}_M[v_{m,1,0}^{gi}])(\mathbb{E}_M[v_{m,1,0}^c] - \mathbb{E}_M[v_{m,1,0}^{gj}])\} = \kappa_{10}(k_c, k_{gi}, k_{gj}).
\tag{42}
$$

D   Proof of Lemma 1

Similar to the derivation of (38),

$$
\begin{aligned}
&\kappa_{10}(k_c, k_{gi}, k_{gj}) \\
&= \mathbb{E}_X\{(\mathbb{E}_M[v_{m,1,0}^c] - \mathbb{E}_M[v_{m,1,0}^{gi}])(\mathbb{E}_M[v_{m,1,0}^c] - \mathbb{E}_M[v_{m,1,0}^{gj}])\} \\
&= \mathbb{E}_X\{(\mathbb{E}_M[v_{m,1,0}^c])^2 - \mathbb{E}_M[v_{m,1,0}^c]\mathbb{E}_M[v_{m,1,0}^{gi}] \\
&\quad - \mathbb{E}_M[v_{m,1,0}^c]\mathbb{E}_M[v_{m,1,0}^{gj}] + \mathbb{E}_M[v_{m,1,0}^{gi}]\mathbb{E}_M[v_{m,1,0}^{gj}]\} \\
&= \tfrac{1}{2}\mathbb{E}_X\{(\mathbb{E}_M[v_{m,1,0}^c] - \mathbb{E}_M[v_{m,1,0}^{gi}])^2 + (\mathbb{E}_M[v_{m,1,0}^c] - \mathbb{E}_M[v_{m,1,0}^{gj}])^2 \\
&\quad - (\mathbb{E}_M[v_{m,1,0}^{gi}] - \mathbb{E}_M[v_{m,1,0}^{gj}])^2\} \\
&= \tfrac{1}{2}[\kappa_{10}(k_c, k_{gi}) + \kappa_{10}(k_c, k_{gj}) - \kappa_{10}(k_{gi}, k_{gj})].
\end{aligned}
\tag{43}
$$

E   Theorem 2: μ and Σ in the Second-order CPA (12)

For large sample n, $\bar{l}_{.,j} = c_j + \epsilon_j E[v_{1,j}]$, then $l_{1,j} = c_j + \epsilon_j \tilde{v}_{1,j} + \sigma_j r_{1,j}, j = 0, 1$, where $\tilde{v}_{1,j} = v_{1,j} - \mathbb{E}(v_{1,j})$ are the centered version of $v_{1,0} = V_0(x_1, k_c, m_1)$ and $v_{1,1} = V_1(m1)$. Similarly, let $\tilde{v}_{m,1,0}$, $\tilde{v}_{m,1,0}^g$, and $\tilde{v}_{m,1}$ denote the centered versions of corresponding quantities $v_{m,1,0}$, $v_{m,1,0}^g$, and $v_{m,1}$. We have

$$\Delta_1^{2O}(k_c, k_g) = (\delta_0 \tilde{v}_{1,0} + r_{1,0})(\delta_1 \tilde{v}_{1,1} + r_{1,1})(\mathbb{E}_M[\tilde{v}_{m,1,0}^c \tilde{v}_{m,1}] - \mathbb{E}_M[\tilde{v}_{m,1,0}^g \tilde{v}_{m,1}]).$$
(44)

Since $\mathbb{E}[r_{1,0}] = \mathbb{E}[r_{1,1}] = 0$,

$$\begin{aligned}
\mu_{k_g} &= \delta_0 \delta_1 \mathbb{E}\{\tilde{v}_{1,0} \tilde{v}_{1,1}(\mathbb{E}_M[\tilde{v}_{m,1,0}^c \tilde{v}_{m,1}] - \mathbb{E}_M[\tilde{v}_{m,1,0}^g \tilde{v}_{m,1}])\} \\
&= \delta_0 \delta_1 \mathbb{E}_X\{\mathbb{E}_M\{\tilde{v}_{1,0} \tilde{v}_{1,1}(\mathbb{E}_M[\tilde{v}_{m,1,0}^c \tilde{v}_{m,1}] - \mathbb{E}_M[\tilde{v}_{m,1,0}^g \tilde{v}_{m,1}])\}\}.
\end{aligned}$$
(45)

By assumption 1, $\mathbb{E}[\tilde{v}_{1,0} \tilde{v}_{1,1}] = \mathbb{E}[\tilde{v}_{m,1,0}^c \tilde{v}_{m,1}] = \mathbb{E}[\tilde{v}_{m,1,0}^g \tilde{v}_{m,1}]$. Similar to the derivation of (38),

$$\begin{aligned}
\mu_{k_g} &= \delta_0 \delta_1 \mathbb{E}_X\{\mathbb{E}_M[\tilde{v}_{1,0} \tilde{v}_{1,1}](\mathbb{E}_M[\tilde{v}_{m,1,0}^c \tilde{v}_{m,1}] - \mathbb{E}_M[\tilde{v}_{m,1,0}^g \tilde{v}_{m,1}])\} \\
&= \frac{\delta_0 \delta_1}{2} \mathbb{E}_X\{\{\mathbb{E}_M[\tilde{v}_{m,1,0}^c \tilde{v}_{m,1}] - \mathbb{E}_M[\tilde{v}_{m,1,0}^g \tilde{v}_{m,1}]\}^2\} \\
&= \frac{\delta_0 \delta_1}{2} \kappa_{2O}(k_c, k_g).
\end{aligned}$$
(46)

The element in covariance Σ corresponding to k_{gi} and k_{gj} is:

$$\sigma_{k_{gi}, k_{gj}} = COV(\Delta_1(k_c, k_{gi}), \Delta_1(k_c, k_{gj})) = E[\Delta_1(k_c, k_{gi})\Delta_1(k_c, k_{gj})] - \mu_{k_{gi}}\mu_{k_{gj}}.$$
(47)

Since $\mathbb{E}[r_{1,0}^2] = \mathbb{E}[r_{1,1}^2] = 1$, the leading term (dropping terms with δ_0 or δ_1) is ,

$$\begin{aligned}
&\sigma_{k_{gi}, k_{gj}} \\
&= \mathbb{E}_X\{(\mathbb{E}_M[\tilde{v}_{m,1,0}^c \tilde{v}_{m,1}] - \mathbb{E}_M[\tilde{v}_{m,1,0}^{gi} \tilde{v}_{m,1}])(\mathbb{E}_M[\tilde{v}_{m,1,0}^c \tilde{v}_{m,1}] - \mathbb{E}_M[\tilde{v}_{m,1,0}^{gj} \tilde{v}_{m,1}])\} \\
&= \kappa_{2O}(k_c, k_{gi}, k_{gj}).
\end{aligned}$$
(48)

F   Algorithms for Calculating ILA$_{1O}$

Here, we describe the algorithm of computing ILA$_{1O}$ knowing the mask distribution. Algorithm 1 assigns the probability distribution of mask with the known probability for each masking bit. Algorithm 2 calculates the first-order information leakage amount based on this probability distribution. These algorithms are used to calculate the ILA$_{1O}$ values in Sect. 4.1.

Algorithm 1. Probability Distribution of Mask

Input: Probability distribution of masking bits $\overrightarrow{p}$
Output: Probability distribution of mask f_M
1: $N_m \leftarrow$ size of key space $|\mathcal{M}|$
2: $N_{bit} \leftarrow$ size of byte $|\overrightarrow{p}|$
3: **for** $m = 0 \rightarrow N_m - 1$ **do**
4:      $f_M[m] = 1$
5:      **for** $i = 0 \rightarrow N_{bit} - 1$ **do**
6:          **if** $m_{(i)} = 1$ **then** $\triangleright\ m_{(i)}$ the $(i+1)$th bit of m
7:              $f_M[m] \leftarrow f_M[m] * p_i$ $\triangleright\ p_i$ the $(i+1)$th value of $\overrightarrow{p}$
8:          **end if**
9:          **if** $m_{(i)} = 0$ **then**
10:             $f_M[m] \leftarrow f_M[m] * (1 - p_i)$
11:         **end if**
12:     **end for**
13: **end for**

Algorithm 2. Calculation of ILA_{1O}

Input: Correct Key k_c, probability distribution of mask f_M, intermediate value V (a $N_k \times N_x \times N_m$ dimension matrix)
Output: ILA_{1O}
1: $N_k \leftarrow$ size of key space $|\mathcal{K}|$
2: $N_x \leftarrow$ size of plaintext (ciphertext) $|\mathcal{X}|$
3: $N_m \leftarrow$ size of mask $|\mathcal{M}|$
4: $\text{ILA}_{1O} \leftarrow 0$
5: **for** $k_g = 0 \rightarrow N_k - 1$ **do**
6:      $E_2[k_g] \leftarrow 0$
7:      **for** $x = 0 \rightarrow N_x - 1$ **do**
8:          $E_1[k_g][x] \leftarrow 0$
9:          **for** $m = 0 \rightarrow N_m - 1$ **do**
10:             $E_1[k_g][x] \leftarrow E_1[k_g][x] + (V[k_c][x][m] * f_M[m] - V[k_g][x][m] * f_M[m])$
11:         **end for**
12:         $E_2[k_g] \leftarrow E_2[k_g] + E_1[k_g][x] * E_1[k_g][x] * \frac{1}{N_x}$ $\triangleright\ E_2[k_g] = \kappa_{1O}(k_c, k_g)$
13:     **end for**
14:     $\text{ILA}_{1O} \leftarrow \text{ILA}_{1O} + E_2[k_g] * \frac{1}{N_k - 1}$
15: **end for**

References

1. Akkar, M.L., Giraud, C.: An implementation of DES and AES, secure against some attacks. In: Koç, Ç.K., Naccache, D., Paar, C. (eds.) CHES 2001. LNCS, vol. 2162, pp. 309–318. Springer, Heidelberg (2001)

2. Bayrak, A.G., Regazzoni, F., Brisk, P., Standaert, F.X., Ienne, P.: A first step towards automatic application of power analysis countermeasures. In: Proceedings of the 48th Design Automation Conference, pp. 230–235 (2011)

3. Bayrak, A., Regazzoni, F., Novo, D., Ienne, P.: Sleuth: automated verification of software power analysis countermeasures. In: Bertoni, G., Coron, J.-S. (eds.) CHES 2003. LNCS, vol. 8086, pp. 293–310. Springer, Heidelberg (2013)

4. Bertoni, G., Daemen, J., Peeters, M., Van Assche, G.: Building power analysis resistant implementations of Keccak. In: Second SHA-3 Candidate Conference (2010)
5. Brier, E., Clavier, C., Olivier, F.: Correlation power analysis with a leakage model. In: Joye, M., Quisquater, J.J. (eds.) CHES 2004. LNCS, vol. 3156, pp. 16–29. Springer, Heidelberg (2004)
6. Chari, S., Rao, J., Rohatgi, P.: Template attacks. In: Kaliski, B.S., Koç, Ç.K., Paar, C. (eds.) CHES 2002. LNCS, vol. 2523, pp. 13–28. Springer, Heidelberg (2003)
7. Chari, S., Jutla, C.S., Rao, J.R., Rohatgi, P.: Towards sound approaches to counteract power-analysis attacks. In: Wiener, M. (ed.) CRYPTO 1999. LNCS, vol. 1666, pp. 398–412. Springer, Heidelberg (1999)
8. Dan Walters, A.H., Kedaigle, E.: Sleak: a side -channel leakage evaluator and analysis kit. In: International Cryptographic Module Conference (2014)
9. Ding, A.A., Zhang, L., Fei, Y., Luo, P.: A statistical model for multivariate DPA on masked devices. In: Batina, L., Robshaw, M. (eds.) CHES 2014. LNCS, vol. 8731, pp. 147–169. Springer, Heidelberg (2014)
10. Duc, A., Faust, S., Standaert, F.-X.: Making masking security proofs concrete. In: Oswald, E., Fischlin, M. (eds.) EUROCRYPT 2015. LNCS, vol. 9056, pp. 401–429. Springer, Heidelberg (2015)
11. Eldib, H., Wang, C., Taha, M., Schaumont, P.: QMS: evaluating the side-channel resistance of masked software from source code. In: Proceedings of the 51st Annual Design Automation Conference, pp. 209:1–209:6 (2014)
12. Fei, Y., Luo, Q., Ding, A.A.: A statistical model for DPA with novel algorithmic confusion analysis. In: International Workshop on Cryptographic Hardware & Embedded Systems, pp. 233–250 (2012)
13. Fei, Y., Ding, A.A., Lao, J., Zhang, L.: A statistics-based fundamental model for side-channel attack analysis. Cryptology ePrint Archive (2014). http://eprint.iacr.org/
14. Gierlichs, B., Batina, L., Tuyls, P., Preneel, B.: Mutual information analysis. In: International Workshop on Cryptographic Hardware & Embedded Systems, pp. 426–442 (2008)
15. Heuser, A., Rioul, O., Guilley, S.: Good is not good enough: deriving optimal distinguishers from communication theory. In: International Workshop on Cryptographic Hardware & Embedded Systems, pp. 55–74 (2014)
16. Kocher, P.C., Jaffe, J., Jun, B.: Differential power analysis. In: Proceedings International Cryptology Conference on Advances in Cryptology, pp. 388–397 (1999)
17. Lomné, V., Prouff, E., Rivain, M., Roche, T., Thillard, A.: How to estimate the success rate of higher-order side-channel attacks. In: International Workshop on Cryptographic Hardware & Embedded Systems, pp. 35–54 (2014)
18. Luo, Q., Fei, Y.: Algorithmic collision analysis for evaluating cryptographic systems and side-channel attacks. In: IEEE International Symposium Hardware Oriented Security & Trust, pp. 75–80 (2011)
19. Prouff, E., Rivain, M., Bevan, R.: Statistical analysis of second order differential power analysis, pp. 799–811 (2009)
20. Rivain, M.: On the exact success rate of side channel analysis in the gaussian model. In: Avanzi, R.M., Keliher, L., Sica, F. (eds.) SAC 2008. LNCS, vol. 5381, pp. 165–183. Springer, Heidelberg (2009)
21. Standaert, F.-X., Malkin, T.G., Yung, M.: A unified framework for the analysis of side-channel key recovery attacks. In: Joux, A. (ed.) EUROCRYPT 2009. LNCS, vol. 5479, pp. 443–461. Springer, Heidelberg (2009)
22. Evaluation environment for side-channel attacks. http://www.risec.aist.go.jp/project/sasebo/

How Secure is AES Under Leakage

Andrey Bogdanov[1](✉) and Takanori Isobe[2]

[1] Technical University of Denmark, Kongens Lyngby, Denmark
anbog@dtu.dk
[2] Sony Corporation, Tokyo, Japan
Takanori.Isobe@jp.sony.com

Abstract. While traditionally cryptographic algorithms have been designed with the black-box security in mind, they often have to deal with a much stronger adversary – namely, an attacker that has some access to the execution environment of a cryptographic algorithm. This can happen in such grey-box settings as physical side-channel attacks or digital forensics as well as due to Trojans.

In this paper, we aim to address this challenge for symmetric-key cryptography. We study the security of the Advanced Encryption Standard (AES) in the presence of explicit leakage: We let a part of the internal secret state leak in each operation. We consider a wide spectrum of settings – from adversaries with limited control all the way to the more powerful attacks with more knowledge of the computational platform. To mount key recoveries under leakage, we develop several novel cryptanalytic techniques such as *differential bias attacks*. Moreover, we demonstrate and quantify the effect of uncertainty and implementation countermeasures under such attacks: *black-boxed rounds, space randomization, time randomization, and dummy operations*. We observe that the residual security of AES can be considerable, especially with uncertainty and basic countermeasures in place.

Keywords: Grey-box · Side-channel attacks · Leakage · AES · Bitwise multiset attacks · Differential bias attacks · Malware · Mass surveillance

1 Introduction

1.1 Background: Black Box, Grey Box and White Box

It is symmetric-key algorithms that are in charge of bulk data encryption and authentication in the field. Plenty of multiple wide-spread applications such as mobile networks, access control, banking, content protection, and storage encryption often feature only symmetric-key algorithms, with no public-key cryptography involved.

Traditionally, the security of symmetric-key cryptographic primitives has been analyzed in the *black-box* model, where the adversary is mainly limited to observing and manipulating the inputs and outputs of the algorithm, the related-key model [2] being a notable extension. Multiple techniques have been

© International Association for Cryptologic Research 2015
T. Iwata and J.H. Cheon (Eds.): ASIACRYPT 2015, Part II, LNCS 9453, pp. 361–385, 2015.
DOI: 10.1007/978-3-662-48800-3_15

extensively elaborated upon, such as differential and linear cryptanalysis, integral and algebraic attacks, to call a small subset of the cryptanalytic tools available today. Cryptographers have excelled at preventing those by design [8].

In late 1990s, with the introduction of timing attacks [13] by Kocher, differential fault analysis [1] by Boneh, DeMillo and Lipton, simple power analysis as well as differential power analysis [14] by Kocher, Jaffe and Jun, the research community has become aware of side-channel attacks that operate in the *grey-box* model: Now the attacker has access to the physical parameters of cryptographic implementations or can even inject faults into their execution. Numerous countermeasures have been proposed to hamper those attacks, providing a practical level of security in many cases.

Since mid 2000s, a trend of side-channel analysis has been towards analytical side-channel attacks that assume leakage of fixed values of variables instead of stochastic variables and whose techniques border the black-box cryptanalysis. So, collision attacks [22] by Shramm et al. observe an equation within one or several executions of an algorithm. Algebraic side-channel attacks [21] by Renauld and Standaert work under the assumption that the attacker can see the Hamming weight of the internal variables of an algorithm. The attacker uses the techniques of algebraic cryptanalysis to solve the systems of nonlinear equations arising from collisions and algebraic side-channel attacks [5,19,20]. Dinur and Shamir [9] apply integral and cube attacks to block ciphers in a setting where a fixed bit after a round is leaked due to physical probing, power analysis or similar. Also differential fault analysis uses elements of differential cryptanalysis.

As an extreme development of the grey-box setting, the *white-box* model [7] by Chow et al. poses the assumption that the adversary has full control over the implementation of the cryptographic algorithm. The major goal of white-box cryptography is to protect the *confidentiality of secret keys* in such a white-box environment. However, all published white-box implementations of standard symmetric-key algorithms such as AES to date have been practically broken in this model [18]. The white-box setting may be too strong for standard symmetric-key algorithms such as AES, because such a cipher was designed with the black-box security in mind.

1.2   Leakage and AES

In this paper, we enhance the Dinur-Shamir setting [9] and aim to bridge the gap between the physical side-channel attacks, the techniques of provable leakage resilience [17] and white-box setting (dealing with attackers too hard to protect against). Namely, we let the AES implementation leak some information during its execution which is defined as follows.

Definition 1 (Leakage Model). *A malicious agent leaks a part of the intermediate internal secret state (including the key state) of a cryptographic algorithm in each algorithm execution.*

To apply this setting to AES (we will talk about AES-128 most of the time), we make it more concrete and fix several important parameters of the leak:

Frequency: There is a single leak per encryption/decryption. This simplifies complexity estimations in our analysis. If more leaks are available in each execution, the complexities can be adjusted accordingly.

Granularity: A leak can only happen after a full round. This situation corresponds e.g., to a 32-bit serial or round-based hardware implementation of AES or a software implementation using an instruction set extension such as AES-NI available on most Intel/AMD CPUs or the Cryptography Extension on ARMv8.

Knowledge: The attacker does not have any knowledge of the location of leaked bits, i.e., it does not know the bit position and the number of round of leaked bits. He also does not know whether the leak is from the key schedule or data processing part. This circumstance models the limited control of the adversary over the platform.

We let several parameters vary in our analysis[1]:

Time and space: The location of the leak in terms of the round number (time) and bit position within the round (space) can either be fixed or vary.

Known/chosen plaintext/ciphertext: We consider both known and chosen text models. In case of a passive attacker, we talk about the known text setting. Otherwise, the attacker is allowed to choose text.

Alignment: We consider single-bit leaks, byte leaks and multiple-bit leaks. While single-bit leaks are more likely to happen due to physical probing, byte leaks correspond more to software settings.

1.3 Our Contributions

The contributions of this paper are as follows. The cryptanalytic results are also summarized in Table 1.

AES Under Basic Leakage and Bitwise Multiset Attacks. We develop a bitwise multiset attack, which exploits relations of sets of plaintexts and internal states, to evaluate the security of AES if the time and space of the leakage is fixed. Our attack utilizes a *bitwise multiset characteristic* which is an extension of Dinur-Shamir integral attacks [9]. Unlike their attacks, our attack is feasible even if an attacker does not have any knowledge of the location of leaked bits. See Sect. 2 and Table 1 for the details.

---

[1] Further models are worth consideration as well. For instance, the *Dinur-Shamir model* of the side-channel cube attacks [9] can be seen as a special case of our leakage model, with the following differences: First, in the Dinur-Shamir model, the adversary knows the location of the leak. Second, the Dinur-Shamir model does not consider leaks of more that a single bit. Third, Dinur-Shamir do not allow for leaks from the key schedule. Finally, the time and location of a leak are fixed, while we allow for time and space uncertainty in our consideration.

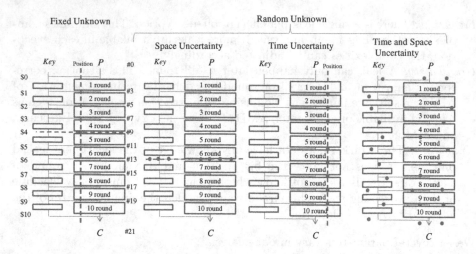

Fig. 1. AES with space and/or time uncertainty

AES Under Leakage with Space/Time Uncertainty and Differential Bias Attacks. We let time, space or both be randomized. The *space randomization* makes the position of leaked bits random in each execution. The *time randomization* makes the round number of leaked bits random in each execution. A combination of time and space randomization is also an advanced model we consider. See Fig. 1 for an illustration.

This setting takes account of a more realistic environment, such as the lack of knowledge of the implementation, and the presence of countermeasures. Here, our multiset attacks are infeasible, as no clean multiset is available. To cope with that, we develop a *differential bias attack* and a *biased state attack* inspired by techniques for distinguishing attacks against stream ciphers [15, 16, 23]. More specifically, by properly choosing differences and values of plaintexts, we create *biased (differential) states*, where the distribution of bitwise differences or value is strongly biased only if the key is correctly guessed. Thus, we are able to distinguish the leak corresponding to the correct key. See Sect. 3 for the techniques as well as Sect. 4 and Table 1 for the results.

AES Under Noisy Leakage. We consider leakage with noise, where the attacker does not know exactly if the variable it accesses corresponds to the execution of the algorithm under attack. For example, it can be the case if multiple instances of encryption (with different keys) are run simultaneously or if the implementation uses dummy operations to hide the AES execution. The differential bias attack remains applicable in this setting, with adjusted complexities. See Sect. 6 and Table 1 for details. To characterize noise, we define π to be the probability that the leak is correctly read. The complexities of our attacks grow quadratically with the increase of $1/\pi$.

Further Results. We discuss the applicability of our attacks to AES-192 and AES-256, multiple-bit leakage, and other granularities of the leaks in Sect. 8.

Table 1. Security of AES-128 under leakage in various settings

Time and space of leaked bits	BB round(s)	Best attack	Bit alignment			Byte alignment		
			Time	Data	Section	Time	Data	Section
Fixed time/space	none	MA*2	2^{18}	2^8 CC	Sect. 3	2^{12}	2^8 CC	Sect. 7.1
	round 9	MA*1	2^{44}	2^{34} CP	Sect. 3	2^{42}	2^{34} CP	Sect. 7.1
	rounds 1, 9	MA	2^{47}	2^8 CP	Sect. 3	2^{44}	2^{34} CP	Sect. 7.1
Uncertain space	none	BSA	2^{26}	2^{26} CC	Sect. 5.1	2^{23}	2^{23} CC	Sect. 7.2
	round 9	DBA	2^{48}	2^{42} CP	Sect. 5.1	2^{41}	2^{39} CP	Sect. 7.2
	rounds 1, 2, 8, 9	DBA	2^{63}	2^{42} CP	Sect. 5.1	2^{56}	2^{42} CP	Sect. 7.2
Uncertain time	none	BSA	2^{23}	2^{23} CC	Sect. 5.2	2^{23}	2^{23} CC	Sect. 7.2
	round 9	DBA	2^{48}	2^{38} CP	Sect. 5.2	2^{44}	2^{38} CP	Sect. 7.2
	rounds 1, 2, 8, 9	DBA	2^{61}	2^{37} CP	Sect. 5.2	2^{53}	2^{37} CP	Sect. 7.2
Uncertain space and time	none	BSA	2^{33}	2^{33} CC	Sect. 5.3	2^{24}	2^{24} CC	Sect. 7.2
	round 9	DBA	2^{58}	2^{46} CP	Sect. 5.3	2^{47}	2^{43} CP	Sect. 7.2
	rounds 1, 2, 8, 9	DBA	2^{72}	2^{45} CP	Sect. 5.3	2^{62}	2^{42} CP	Sect. 7.2
Random space and time w/ $\pi = 2^{-10}$	none	BSA	2^{53}	2^{53} CC	Sect. 6	2^{44}	2^{44} CC	Sect. 7.2
	round 9	DBA	2^{68}	2^{56} CP	Sect. 6	2^{57}	2^{53} CP	Sect. 7.2
	rounds 1, 2, 8, 9	DBA	2^{82}	2^{55} CP	Sect. 6	2^{72}	2^{52} CP	Sect. 7.2
Random space and time w/ $\pi = 2^{-20}$	none	BSA	2^{73}	2^{73} CC	Sect. 6	2^{64}	2^{64} CC	Sect. 7.2
	round 9	DBA	2^{78}	2^{66} CP	Sect. 6	2^{67}	2^{63} CP	Sect. 7.2
	rounds 1, 2, 8, 9	DBA	2^{92}	2^{65} CP	Sect. 6	2^{82}	2^{62} CP	Sect. 7.2

*1 : 32-bit partial key recovery attack, *2 : 8-bit partial key recovery attack
BB round(s): Black-boxed round(s), KP: Known Plaintext, CP: Chosen Plaintext
CC: Chosen Ciphertext, MA: Multiset Attack, DBA: Differential Bias Attack
BSA: Biased State Attack, π is the probability to read a correct leak

Our Observations and Recommendations. To summarize the residual security of AES under leakage in the various settings, we observe the following. First, if no rounds are black-boxed and all intermediate internal states can be visible to the attacker, there are practical attacks, even with uncertain time and space. Second, to approach practical infeasibility of attacks in our leakage model without black-boxing, a substantial level of noise are be needed, $\pi = 2^{-10}$ and lower when combined with randomized time and space.

On the other hand, the black-boxing of round 9 is very effective. Indeed, if round 9 is black-boxed[2] (i.e., when the state between round 9 and round 10 is invisible to the attacker), the complexities of our attacks grow beyond 2^{44} even with fixed time and space. Third, if uncertainty in time and space is combined with the black-boxed 9th round, our attacks require more than 2^{58} operations, even with clean leaks. Then, if more rounds (1,2,8, and 9) are black-boxed, the complexities increase to 2^{72}. If noise is applied as countermeasure on top of that, it is possible to attain security levels of 2^{80} and beyond against our attacks.

Thus, black-boxed round 9, noise or both are needed to hamper our attacks at a practical security level under leakage. Note that a high-budget organization

---

[2] E.g., partly unrolled hardware implementations aimed to reduce latency [6] may have this property.

can practically afford an attack of complexity 2^{80} and higher [12]. However, the countermeasures considered here may still be effective against a mass surveillance attacker.

2 Preliminaries

This section fixes AES notations that we will use throughout the paper and describes the leakage attack by Dinur-Shamir on AES as a starting point.

2.1 Notations of AES

AES is a block cipher with a 128-bit internal state and a 128/192/256-bit key K, referred to as AES-128, AES-192 and AES-256, respectively. In most parts of this paper, we refer to AES-128 whenever speaking of AES. The internal state is represented by a 4×4 byte matrix, and the key is represented by a $4\times4/4\times6/4\times8$ matrix. For example, a 4×4 internal state consisting of 16 byte cells is expressed as follows.

$$S = \begin{bmatrix} s_0 & s_4 & s_8 & s_{12} \\ s_1 & s_5 & s_9 & s_{13} \\ s_2 & s_6 & s_{10} & s_{14} \\ s_3 & s_7 & s_{11} & s_{15} \end{bmatrix}$$

AES consists of a data processing part and a key schedule. The data processing part adopts a substitution-permutation network whose round function consists of four layers: SubBytes, ShiftRow, MixColumns and AddRoundKey. SubBytes is a nonlinear transformation applying a 8-bit S-box to each cell. ShiftRow rotates bytes in row r by r positions to the left. MixColumns is a linear transformation applying a 4×4 diffusion matrix with branch number 5 to each column. AddRoundKey adds a 128-bit subkey to a 128-bit state by an XOR operation. Note that AddRoundKey is also performed before the first round as whitening and that MixColumns is omitted in the last round. Subkeys are generated by a key schedule. For the details of the key schedule of AES, we refer to [11].

Two types of internal states in each round of AES-128 are defined as follows: #1 is the state before SubBytes in round 1, #2 is the state after MixColumns in round 1, #3 is the state before SubBytes in round 2,..., #19 is the state before SubBytes in round 10, and #20 is the state after ShiftRow in round 10 (MixColumns is omitted in the last round). The states in the last round of AES-192 are addressed as #23 and #24, and of AES-256 as #27 and #28. We let #0 be a plaintext and #21, #25 and #29 be a ciphertext in AES-128/192/256, respectively. 128-bit subkeys are denoted as \$0 , \$1, ..., and so on. The i-th byte in the state x is denoted as x_i and the j-th bit in x_i is represented as $x_i[j]$.

2.2 Dinur-Shamir Chosen-Plaintext Attack on AES-128 with Leakage

As a starting point of our analysis, we outline the leakage attack proposed by Dinur and Shamir in [9]. As explained above, the Dinur-Shamir model is different from our leakage models as the adversary knows the time (round number) and space (bit position inside the round) of the leak, only single-bit leaks are considered there, and no leaks from the key schedule are allowed.

In the attack of [9], one uses the following multiset properties of a byte: In set A, all 2^8 values appear exactly once; In set C, all 2^8 values are fixed to a constant; In set B, the XOR sum of all 2^8 values is zero; In set U, all 2^8 values is not A, C or B. Let an N-round attack be an attack based on leaked bits after the N-th round function, e.g., a 2-round attack is based on only leaked bits of #5.

In the first step, the attacker guesses 4 bytes of the key $0, and chooses a set of 2^8 plaintexts, so that #2 consists of Λ-set in which only one byte is A and the other 15 bytes are C. If 4 bytes of $0 are correctly guessed, #5 consists of 4 bytes of A and 12 bytes of C, while in a wrong key, all bytes in #5 become U. Thus, by checking whether the all 2^8 values of #5 are fixed, an attacker is able to sieve wrong keys after 2^{32} operations. The procedure can be repeated for three times with the three 4-byte sets of the key $0 depending on the position of the leaked bit. The remaining 4 bytes of $0 are exhaustively searched. Time complexity is estimated as 2^{42} ($\approx (2^{32} \times 2^8 \times 3)$) encryptions and the required data is 2^{34} ($= 2^{32} \times 4$) chosen plaintexts. The work [9] also proposes other types of 2-round attack using cubes, with a time complexity of 2^{35}. However, the details are not given.

The paper [9] mentioned that 3- and 4-round attacks were possible by using similar techniques but omitted the details. As A expands into all state after 3 rounds even if the key is correctly guessed, at least the 2-round attack has a limited application to 3 and 4 rounds.

3   AES Under Leakage with Fixed Time and Space

In this section, we present new key recovery attacks on AES under leakage with fixed time and space. That is, a bit of the internal state is leaked whose location (round and bit position) is unknown but fixed for the entire attack. Our attack is an extension of the Dinur-Shamir integral attacks [9]. While their attack requires the location of leaked bits in advance, our attack is feasible even if an attacker does not have any knowledge of it. First, we describe a technique to detect whether leaked bits come from the key schedule or the data transformation, and show that leaked bits from the key schedule are of very limited use for a key recovery attack in this setting. Then we introduce key recovery attacks based on leaked bits from the data transformation. Our attacks utilize a bitwise multiset characteristic.

Formalization of Fixed Time and Space. The fixed (unknown) location setting assumes that each execution of encryption leaks only *one bit* of the internal state at the fixed location. Specifically, leaked bits are assumed to come from internal states after each round function of the data processing part: #3, #5, ..., #19 or each state of the key schedule (i.e., subkeys): $0 , $1, ..., $10 at the *fixed* position of the *fixed* rounds in each encryption, e.g., $#9_{11}[2]$ or $\$5_8[5]$. The adversary is able to access the encryption function with known/chosen plaintexts/ciphertexts and obtain corresponding leaked bits.

Leakage From Key Schedule. The states in the key schedule, $0, $1, ..., $10, are deterministic with respect to the value of the key, i.e., if a key is fixed, all states of the key schedule are fixed independently of the values of plaintexts. On the other hand, the states in the data processing part depend on the values of plaintexts. This difference allows us to detect whether leaked bits come from the key schedule. More specifically, we encrypt N different plaintexts and obtain N leaked bits. If all N bits are the same, they come from the key schedule with probability $(1 - 2^{-N})$.

If the leaked bits come from the key schedule, information theoretically, the attacker is able to get at most one bit of the subkey information, as each encryption leaks the same state information at the fixed location. In addition, since an attacker does not know where leaked bits come from, leaked information from the leaked bits is negligible. Therefore, we will focus on the case where leaked bits come from the data processing part in the following.

3.1 Bitwise Multiset Characteristic

Our attacks utilize the following bitwise multiset property in the data transform.

Proposition 1 (Bitwise Zero-Sum Property). *If only one byte of #2 is* A *and the other 15 bytes are* C *(Λ set), the bitwise XOR-sum of 2^8 multiset of any bits in #3 to #10 is zero.*

Proof. As shown in Fig. 2, if #2 consists of a Λ set, #3 is also a Λ set, and #5 consists of 4 bytes of A and 12 bytes of C. Then, #7 and #9 consist of 16 bytes of A and B, respectively. In the 2^8 multiset of each bit of A, C and B, the XOR sum becomes zero [4]. □

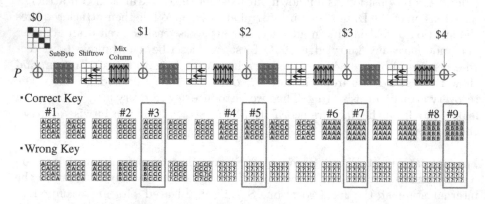

Fig. 2. Bitwise multiset characteristics over 4-round AES-128

3.2   Chosen-Plaintext Bitwise Multiset Attack

The bitwise zero-sum property allows us to develop chosen-plaintext key recovery attacks using leaked bits at a fixed position in #3, #5, #7 or #9. Our attack firstly guesses 4 bytes of the key $0, and chooses a set of 2^8 plaintexts resulting in Λ set in #2. If 4 bytes of $0 are correctly guessed, the bitwise XOR sum of 2^8 leaked bits in any bit position of #3 to #10 is zero (Proposition 1). Otherwise, the probability that the bitwise XOR sum of leaked bits of #5, #7 and #9 is zero is 2^{-1}. If this procedure repeats with N different sets of 2^8 plaintexts, wrong keys can be detected with a probability of $(1 - 2^{-N})$.

First, we prepare a table of 2^{32} plaintexts in which all values of $\#0_0$, $\#0_5$, $\#0_{10}$, $\#0_{15}$ appear once and the other 12 bytes are fixed, and corresponding leaked bits. Assuming that the leaked bits can come from any position of #5, #7 or #9, our attack is performed as follows:

1. Guess $\$0_0$, $\$0_5$, $\$0_{10}$, $\$0_{15}$ (4 bytes) and choose $\#2_1$, $\#2_2$, $\#2_3$ (3 bytes).
2. Compute 2^8 the 4 bytes of $\#0_0$, $\#0_5$, $\#0_{10}$, $\#0_{15}$ backwards with all 2^8 values of $\#2_0$.
3. Get 2^8 leaked bits by accessing the prepared table, and compute the XOR sum of 2^8 leaked bits.
4. Repeat steps 1 to 3 N times with different values of $\#2_1$, $\#2_2$, $\#2_3$. If all N sets of XOR-sums are zero, regard it as a correct key.
5. Repeat steps 1 to 4 with all 2^{32} key candidates for $\$0_0$, $\$0_5$, $\$0_{10}$, $\$0_{15}$.
6. Repeat steps 1 to 5 for three times with the other three 4-byte sets of the key $0 and corresponding bitwise multiset characteristics and tables.

The number of surviving keys after the above procedure is estimated as $(1+2^{-N} \times (2^{32} - 1))^4$. If the remaining key candidates are exhaustively searched, time complexity is estimated as $\{(2^{32} \times 2^8 \times N) \times 4\} + (1 + 2^{-N} \times (2^{32} - 1))^4$ encryptions. When $N = 22$, the time complexity is estimated as $2^{46.46}$ encryptions, the required data is $2^{34} (= 2^{32} \times 4)$ chosen plaintexts and the required memory is 2^{34} bits. This attack is successful if leaked bits come from any bits of #5, #7 and #9 without any knowledge of the location of leaked bits.

3.3   Partial Key Recovery Attack Using Leaked Bits from #3

If leaked bits come from #3, a 32-bit partial key-recovery attack is feasible as AES takes 2 rounds to achieve the full diffusion. If 4 bytes of keys $0 are guessed correctly, 2^8 multiset in only one byte of #3 is not C as shown in Fig. 2, while for a wrong key, 2^8 multisets in 4 bytes of one column are not C. We exploit the gap of the number of C in #3 between a correct key and a wrong key.

We guess the column in #3 where leaked bits come from and then guess corresponding 4 bytes of $0. We check whether the 2^8 multiset of leaked bits is fixed with N different sets of 2^8 plaintexts. A correct key can be detected with probability of $(1 - 2^{-8N})$ if leaked bits come from the byte which is C for a correct key and B for a wrong key. We repeat this $4 \times 4/3$ times by guessing different columns and the byte position of leaked bits in #3 and corresponding

4 bytes of $0. The corresponding 32 bits of the key $0 can be recovered with about 2^{44} ($\approx 2^{32} \times 2^8 \times 4 \times 4 \times 4/3$) encryptions when $N = 4$, 2^{34} chosen ciphertexts and 2^{34} memory.

3.4   Chosen-Ciphertext Bitwise Multiset Attack

In the chosen-ciphertext setting, backward direction attacks are feasible by using leaked bits from #13, #15, #17 or #19. As shown in Fig. 3, if 4 bytes of $10 are correctly guessed and a set of ciphertexts is properly chosen, the XOR-sum of 2^8 multiset of any bit in #12 to #17 is zero (Proposition 1). Since states #13, #15 and #17 correspond to #7, #5 and #3, respectively, chosen-ciphertext attacks using these bits are feasible in the same manner as for chosen-plaintext attacks.

Also, #19 is affected by only one byte of $10. Thus, one byte of $10 can be recovered by the exhaustive search with 8 leaked bits from different ciphertexts after guessing 128 positions of the leaked bit. Time complexity is estimated as 2^{18} ($= 2^8 \times 128 \times 8$) encryptions, the required data is about 2^8 known ciphertexts, and the memory consumption is negligible.

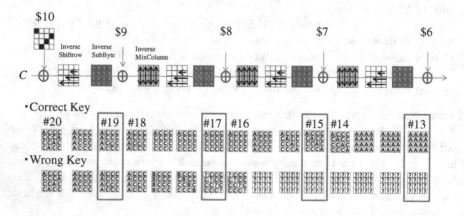

Fig. 3. Bitwise multiset characteristics in 4-round AES-128 in backward direction

3.5   Combined Key Recovery Attacks on AES

Finally, we introduce a key recovery attack on the full AES-128 by combining the forward and the backward direction attacks. Since we do not know in which round the bits leak, we guess it and then mount each round attack in the following order: #19 → #17 → #3 → #5 → #7 → #9 → #13 → #15, i.e., if a correct key is not found by the guessed-round attack, the next round attack is applied in that given order. Our attacks find a correct key successfully except the case where the leaked bits come from #11. Thus the success probability without any knowledge of locations of leaked bits is 0.899 (= 8/9).

Time complexity is estimated as 2^{48} ($\approx 2^{18} + 2^{44} + 2^{44} + 2^{46.46} + 2^{46.46}$) encryptions. The required data is about 2^{35} ($= 2^{34} + 2^{34}$) chosen plaintexts and 2^{34} chosen ciphertexts and the required memory is 2^{34} bits. Note that if the leaked bits come from #3, #17, #19, partial key recovery attacks are possible.

4   Uncertainty and Differential Bias Attacks

The attacker can also have limited control over the execution environment. In particular, the time and space can be uncertain. We assume now that the attacker does not know bit positions and/or the number of rounds of leaked bits. Moreover, the values leaked can be incorrect due to noise or other operations executed in parallel to encryption/decryption. This can happen both for purely technical reasons on a complex multi-process platform and due to countermeasures. This section deals with these uncertainties and develops a cryptanalytic technique that is coined *differential bias attack*.

In a nutshell, the technique works as follows. Let Z_i be a leaked bit from an i-th execution of the encryption function. Our attacks observe a stream of leaked bits $Z_0, Z_1, Z_2, Z_3, \ldots$ and recover the correct key by applying techniques of distinguishing attacks from the domain of stream ciphers [15,16,23]. More specifically, we guess a part of the key $0, and set well-chosen differences for a pair of plaintexts resulting in *biased differential states*, where the distribution of bitwise differences is biased, if the part of key $0 is correctly guessed. As a leaked bit stream from *biased differential states* is also biased, we are able to detect the bit stream corresponding to the correct key by checking bias on the differences of bits. Also, if leakage after round 9 is available, a more powerful attack, called *biased state attack*, is feasible by using similar techniques.

Formalization of Uncertain Time and Space. We assume a random unknown round (time) and/or bit position (space) within the round of the leak. Again, each execution of encryptions leaks only *one bit* of internal states at the random location. More formally, leaked bits are assumed to randomly come from the target space of internal states. For example, if the target space consists of all states after each round function of the data transform and key schedule, it is the leakage from states #0, #3, ..., #19, #21 and states $0, $1, ..., $10. A target space can be a subset of those states if some rounds are black-boxed (and, thus, not visible to the attacker).

4.1   Truncated Differential Characteristic

Our attacks utilize a bytewise truncated differential characteristic of Fig. 4, where a colored-cell is a probability-one non-zero truncated difference, a blank cell is a probability-one zero truncated difference, and ? is an unknown truncated difference. Define 4 bytes of differences $\{\Delta\#0_0, \Delta\#0_5, \Delta\#0_{10}, \Delta\#0_{15}\}$ in a pair of plaintexts as $(\Delta\#0_0, \Delta\#0_5, \Delta\#0_{10}, \Delta\#0_{15}) = S^{-1}(MC^{-1}(\Delta\#2_0, 0, 0, 0))$, where S^{-1} and MC^{-1} are the inverses of SubBytes and MixColumns in a column,

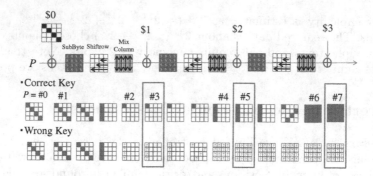

Fig. 4. Truncated differential characteristic over 3-round AES-128

respectively, and $\Delta\#2_0$ is an arbitrary byte difference in $\#2_0$. Given $\{\Delta\#2_0, \#2_0, \ldots, \#2_3\}$ and $\{\$0_0, \$0_5, \$0_{10}, \$0_{15}\}$, $\{\Delta\#0_0, \Delta\#0_5, \Delta\#0_{10}, \Delta\#0_{15}\}$ and $\{\#0_0, \#0_5, \#0_{10}, \#0_{15}\}$ are determined. Let $\#0'$ be a plaintext having differences $\{\Delta\#0_0, \Delta\#0_5, \Delta\#0_{10}, \Delta\#0_{15}\}$, i.e., $\#0'_0 = \#0_0 \oplus \Delta\#0_0$, $\#0'_5 = \#0_5 \oplus \Delta\#0_5$, $\#0'_{10} = \#0_{10} \oplus \Delta\#0_{10}$, $\#0'_{15} = \#0_{15} \oplus \Delta\#0_{15}$. Also, let $\#'1, \ldots, \#'21$ be the corresponding states of $\#0'$, and $Z'_0, Z'_1, Z'_2, Z'_3, \ldots$ be leaked bits of each execution of $\#0'$.

4.2 Biased Differential State

Choosing 4-byte differences $\{\Delta\#0_0, \Delta\#0_5, \Delta\#0_{10}, \Delta\#0_{15}\}$ properly and guessing the 4 bytes of $\{\$0_0, \$0_5, \$0_{10}, \$0_{15}\}$ correctly, we are able to create *biased differential states* in $\#3$: consisting of 15 bytes of probability-one zero differences and 1 byte of a probability-one non-zero difference, $\#5$: consisting of 12 bytes of probability-one zero differences and 4 bytes of probability-one non-zero differences, and $\#7$: consisting of 16 bytes of probability-one non-zero differences. As shown in Fig. 4, a correct key has 27 bytes of probability-one zero differences $\#3_1, \ldots, \#3_{15}$ and $\#5_4, \ldots, \#5_{15}$ and 21 bytes of probability-one non-zero differences $\#3_1, \#5_0, \ldots, \#5_3$, and $\#7_0, \ldots, \#7_{15}$, while a wrong key has only 12 bytes of probability-one zero differences $\#3_4, \ldots, \#3_{15}$ and does not have any probability-one non-zero difference in the state of the data processing part.

In addition, a pair of plaintexts has 12 bytes of probability-one zero differences and 4 bytes of probability-one non-zero differences for both a correct key and a wrong key. Also, the key schedule has 176 $(= 16 \times 11)$ bytes of probability-one zero differences, as the subkeys are always fixed under the same key.

4.3 Bitwise Differential Bias in Biased Differential State

For a probability-one zero/non-zero truncated difference, we derive positive and negative bitwise differential biases. Our attack exploits the gap of the number of positive and negative biases between a correct key and a wrong key when a pair of $\#0$ and $\#'0$ is encrypted.

Positive Bitwise Bias for Probability-One Zero Truncated Difference.
If a bytewise pair $\#x_y$ and $\#x'_y$ has a probability-one zero truncated difference, a bitwise difference at the same position is also zero with probability one:
$Pr(\Delta[\#x_y[j], \#'x_y[j]] = 0) = 1, 0 \le j \le 7$, where $\Delta[a, b] = a \oplus b$. A correct key
has 1720 ($= 27 \times 8 + 176 \times 8 + 12 \times 8$) *positive* bitwise differential biases, while
a wrong key has only 1600 ($= 12 \times 8 + 176 \times 8 + 12 \times 8$) such biases.

**Negative Bitwise Bias for Probability-One Non-zero Truncated
Difference.** If a pair $\#x_y$ and $\#x'_y$ has a probability-one non-zero truncated difference, the probability that a bitwise difference at the same bit position is zero is
estimated as follows: $Pr(\Delta[\#x_y[j], \#'x_y[j]] = 0) = 127/255 = 1/2 \cdot (1 - 2^{-7.99})$
In experiments with 2^{40} randomly-chosen plaintexts and keys, we confirmed that
these negative biases toward zero exist in each bit of the probability-one non-zero
truncated difference, where the experimental value is $Pr(\Delta[\#7_i[j], \#'7_i[j]] =
0) = 1/2 \cdot (1 - 2^{-7.92})$.

A correct key has 200 ($= 21 \times 8 + 4 \times 8$) *negative* bitwise differential biases,
while a wrong key has 32 ($= 0 + 4 \times 8$) ones. The summary of bitwise posi-
tive/negative differential biases for the truncated differential of Fig. 4 is shown
in Table 2.

Table 2. Bitwise differential biases for truncated differential of Fig. 4

	Positive biases toward zero	Negative biases toward zero
Correct key	$\#3_i[j]$ ($1 \le i \le 15, 0 \le j \le 7$) $\#5_i[j]$ ($4 \le i \le 15, 0 \le j \le 7$)	$\#3_0[j]$ ($0 \le j \le 7$) $\#5_i[j]$ ($0 \le i \le 3, 0 \le j \le 7$) $\#7_i[j]$ ($0 \le i \le 15, 0 \le j \le 7$)
Wrong key	$\#3_i[j]$ ($4 \le i \le 15, 0 \le j \le 7$)	-
Both keys	$\#0_i[j]$ ($i \ne 0, 5, 9, 15 \le j \le 7$) $\$x_i[j]$($0 \le x \le 10, 1 \le i \le 15, 0 \le j \le 7$)	$\#0_i[j]$ ($i = 0, 5, 9, 15, 0 \le j \le 7$)

4.4   Bitwise Differential Biases in the Stream of Leaked Bits

Suppose that values of the other bits of the states in the data processing part
and the key schedule are randomly distributed, i.e., the probability that differences of other bitwise pairs become zero is 2^{-1}. Let N_{all}, N_{bias_p}, N_{bias_n}, and
N_{random} be the number of bitwise pairs in entire space, positive biased space
(toward zero), negative biased space (toward zero) and randomly-distributed
space, respectively, and x^c and x^w be those of a correct key and a wrong key,
respectively (see Fig. 5). The probabilities that a difference of a bitwise pair of
randomly-chosen leaked bits is zero ($\Delta[Z_i, Z'_j] = 0$) for a correct key and a wrong
key are estimated as follows:

$$Pr^c(\Delta[Z_i, Z'_j] = 0) = 1/2 \cdot (N^c_{random}/N_{all}) + N^c_{bias_n}/N_{all} \cdot (127/255) + N^c_{bias_p}/N_{all},$$

$$Pr^w(\Delta[Z_i, Z_j'] = 0) = 1/2 \cdot (N_{random}^w/N_{all}) + N_{bias_n}^w/N_{all} \cdot (127/255) + N_{bias_p}^w/N_{all}.$$

Our attack observes leaked bits $Z_0, Z_1, Z_2, Z_3, \ldots$ and $Z_0', Z_1', Z_2', Z_3', \ldots$, and then computes the probability of $\Delta[Z_i, Z_j'] = 0$ in order to distinguish a stream coming from the distribution for a correct key from streams coming from the distribution for a wrong key.

The number of required samples for distinguishing the two distributions with probability of $1 - \alpha$ is given by the following lemmata.

Lemma 1. [15,16] *Let X and Y be two distributions and suppose that the independent events E occur with probabilities $Pr_X(E) = p$ in X and $Pr_Y(E) = (1 + q) \cdot p$ in Y. Then the discrimination D of the distributions is $p \cdot q^2$.*

Lemma 2. [15] *The number of samples N_{sample} that is required for distinguishing two distributions that have discrimination D with success probability $1 - \alpha$ is $(1/D) \cdot (1 - 2\alpha) \cdot log_2 \frac{1-\alpha}{\alpha}$.*

Assuming that the target event E is $\Delta[Z_i, Z_j'] = 0$, X is the distribution for a wrong key, and Y is the distribution for a correct key, p and q are estimated as $p = Pr^w(\Delta[Z_i, Z_j'] = 0)$ and

$$q = \frac{-N_{bias_n}^c + N_{bias_n}^w + 255(N_{bias_p}^c - N_{bias_p}^w)}{255N_{all} - N_{bias_n}^w + 255N_{bias_p}^w}.$$

For success probability $1 - 2^{-32}$, the estimated number of required samples is:

$$N_{sample} = (pq^2)^{-1} \cdot (1 - 2 \cdot 2^{-32}) \cdot log_2 \frac{1 - 2^{-32}}{2^{-32}} \approx 2 \cdot 32 \cdot q^{-2} = 2^6 \cdot q^{-2}.$$

4.5   Chosen-Plaintext Differential Bias Attack

First, this attack prepares 2^{32} chosen plaintexts in which all 2^{32} values of $\#0_0$, $\#0_5$, $\#0_{10}$, $\#0_{15}$ appear once and the other 12 bytes are fixed, and obtains N_s leaked bits in each plaintext, i.e., each plaintext is encrypted N_s times. Given a pair of P and P', $N_s^2 (= N_s \times N_s)$ pairs of leaked bits are obtained as shown in Fig. 6. After we make a table of the values of $\{\#0_0, \#0_5, \#0_{10}, \#0_{15}\}$ and corresponding N_s leaked bits, our attack is performed as follows:

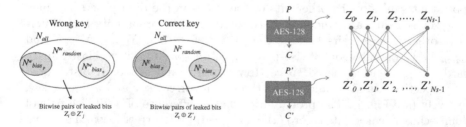

Fig. 5. Bias in leaked stream   **Fig. 6.** Bitwise pairs of leaked bits

1. Guess the 4 bytes of key $\$0_0$, $\$0_5$, $\$0_{10}$, $\$0_{15}$, and choose $\Delta\#2_0$, $\#2_0, \ldots, \#2_3$.
2. Compute a pair of 4 bytes of plaintexts, $\#0_0$, $\#0_5$, $\#0_{10}$, $\#0_{15}$ and $\#0'_0$, $\#0'_5$, $\#0'_{10}$, $\#0'_{15}$, resulting in biased $\#3$, $\#5$ and $\#7$ states if a key is correctly guessed.
3. Get N_s^2 pairs of leaked bits $\Delta[Z_i, Z'_j]$, $0 \le i, j < N_s$ by accessing the prepared table.
4. Repeat steps 2-3 N_{sample}/N_s^2 times with different values of $\#2$.
5. Check whether a distribution of N_{sample} pairs is the one for a correct key. If so, regard it as a candidate for the correct key.
6. Repeat steps 1 to 5 with all 2^{32} candidates of keys $\$0_0$, $\$0_5$, $\$0_{10}$, $\$0_{15}$.
7. Repeat steps 1 to 6 for three times with the other three 4-byte sets of the key $\$0$, corresponding truncated differential characteristics, and the tables of plaintexts and leaked bits.

In steps 3 to 5, we check N_{sample} pairs to detect a stream coming from the biased distribution for a correct key. In the step 3, we count the number of the events $\Delta[Z_i, Z'_j] = 0$, and estimate the probability $Pr(\Delta[Z_i, Z'_j] = 0)$. The straight forward method requires N_s^2 operations to check all N_s^2 pairs. To improve it, we first calculate the number of $Z_i = 0$, $0 \le i < N_s$, defined as N_{zero}. Then the number of $\Delta[Z_i, Z'_j] = 0$ is estimated as

$$\left(N_{zero} \times (\overline{Z'_0} + \ldots, +\overline{Z'}_{N_s-1}) + ((N_s - N_{zero}) \times (Z'_0 + \ldots, +Z'_{N_s-1})\right)/N_{all},$$

where $\overline{a}$ is the complement of a. These costs are estimated as $N_s + (N_s + N_s)$ additions and multiplications. It is assumed to be less than N_s one-round encryptions. The number of surviving keys after the above procedure is estimated as $(1+2^{-\alpha} \times (2^{32}-1))^4$. If the remaining key candidates are exhaustively searched, the entire time complexity is estimated as $(2^{32} \times 4 \times N_{sample}/N_s \times 1/10) + (1 + 2^{-32} \times (2^{32} - 1))^4 \approx 2^{31} \times N_{sample}/N_s$ encryptions and the required data is $2^{34} \times N_s$ $(= 4 \times 2^{32} \times N_s)$ chosen plaintexts with leaked bits. The memory requirement is $2^{34} \times N_s$ bits.

4.6 Chosen-Ciphertext Differential Bias Attack

If the decryption function is accessible, chosen-ciphertext attacks are applicable. Similarly to the setting of bitwise mutiset attacks before, the chosen-ciphertext attacks are more efficient and it makes sense to black-box the output of round 9 also in the cases with time and space uncertainty.

As shown in Fig. 7, the states $\#13$, $\#15$ and $\#17$ correspond to $\#7$, $\#5$ and $\#3$, respectively. Since the state $\#19$ consists of 12 probability-one zero truncated differences and 4 probability-one non-zero truncated differences, both for a correct key and a wrong key, one additionally has 96 positive and 32 negative bitwise differential biases in the chosen-ciphertext attack.

Biased State Attack of $\#19$: Leakage After Round 9. If leaked bits from $\#19$ are obtained, a more powerful attack is feasible. Each byte in $\#19$ can be controlled by one byte of $\$10$ and one byte of a ciphertext. Thus, we are

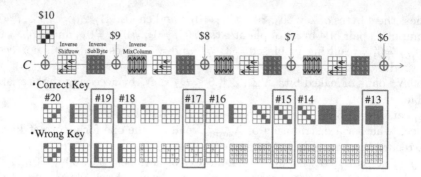

Fig. 7. Differential characteristic over 4-round AES-128 in backward direction

able to create a *biased state* in #19 whose one byte (8 bits) is fixed to 0, if the corresponding byte of $10 is correctly guessed and the respective byte of the ciphertext is property chosen. Suppose that the values of the other bits of the states are randomly distributed. The probabilities that each leaked bit is zero ($Z_i = 0$) for a correct key is estimated as $Pr^c(Z_i = 0) = 1/2 \cdot (N'^c_{random}/N'_{all}) + N'^c_{bias_p}/N'_{all}$, where N'_{all}, N'_{bias_p}, and N'_{random} are the numbers of bits in entire space, positive biased space and randomly-distributed space, respectively. Also, $Pr^w(Z_i = 0)$ is assumed to be $1/2$.

Assuming that the target event E is $Z_i = 0$, p and q are estimated as $p = 1/2$ and $q = N'_{bias_p}/N'_{all}$. For the success rate of $1 - 2^{-8}$ ($\alpha = 2^{-8}$), the sample requirement is estimated as $N'_{sample} \approx 2 \cdot 8 \cdot (q)^{-2} = 2^4 \cdot (q)^{-2}$. We repeat the procedure for all 16 bytes of $10. Therefore, time complexity is estimated as $2^{12} \times N'_{sample}$ ($= 16 \times 2^8 \times N'_{sample}$) encryptions and the required data is $2^{12} \times N'_{sample}$ ($= 16 \times 2^8 \times N'_{sample}$) chosen ciphertexts. The memory requirement is negligible.

4.7   Known-Plaintext Differential Bias Attack

Finally, we introduce a known-plaintext differential bias attack using a truncated differential characteristic of Fig. 8. For a correct key, one has 24 ($= 3 \times 8$) positive bitwise differential biases toward zero and 8 negative bitwise differential biases in #3, while for a wrong key, there are not such biases. The key schedule has the same number of positive biases of chosen-plaintext attacks and the plaintext has 32 ($= 4 \times 8$) negative biases in both of a correct and a wrong key.

This attack prepares 2^{33} known plaintexts and makes a table of #0_0, #0_5, #0_{10}, #0_{15} and the corresponding N_s leaked bits. The expected number of the entries of each value of #0_0, #0_5, #0_{10}, #0_{15} is more than 1. We mount key recovery attacks for 0_0, 0_5, 0_{10}, 0_{15} in the same manner as in the chosen-plaintext attack. In step 3, the prepared table contains the corresponding values of #0_0, #0_5, #0_{10}, #0_{15} with high probability. Thus, time complexity is estimated as $2^{31} \times N_{sample}/N_s$ encryptions and the required data is $2^{35} \times N_s$ ($= 4 \times 2^{33} \times N_s$) known plaintexts with leaked bits and the required memory is about $2^{35} \times N_s$ bits.

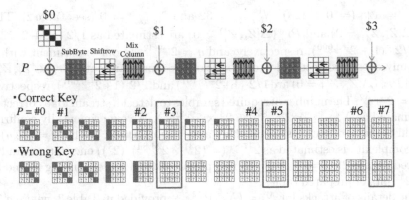

Fig. 8. Differential characteristic over 3-round AES-128 for known plaintext attack

5   AES Under Leakage with Uncertainty in Time/Space

This section evaluates the security of AES if the attacker is uncertain about time and space, that is, if the round of leak and/or the bit position of leak within the round are randomized. Since the multiset of leaked bits at the fixed location is not available in the random unknown setting, our bitwise multiset attacks are not applicable to these variants. Thus, we estimate the costs of differential (state) bias attacks on each variant of AES with countermeasures as shown in Fig. 1.

Formalization of Time/Space Uncertainty for AES. We speak of *randomized time*, when one bit of the state information is leaked at a *fixed* bit position after a *random* number of rounds, e.g., $\#(2x+1)_{10}[7]$ ($0 \leq x \leq 10$) or $\$x_3[4]$ ($0 \leq x \leq 10$). We speak of *randomized space*, when one bit of the state information is leaked at a *random* bit position after a *fixed* number of rounds, e.g., $\{\#17_i[j], \$8_i[j]\}$ ($0 \leq i \leq 15, 0 \leq j \leq 7$). *Randomized time and space* occur, when one bit of state information is leaked at a *random* bit position after a *random* number of rounds, e.g., $\#(2x+1)_i[j]$ ($0 \leq x \leq 10, 0 \leq i \leq 15$ and $0 \leq j \leq 7$) or $\$x_i[j]$ ($0 \leq x \leq 10, 0 \leq i \leq 15$ and $0 \leq j \leq 7$).

5.1   Uncertainty in Space

The space randomization makes the bit position of leaked bits random in each execution, i.e., Z_i and Z_i' randomly come from two 256-bit spaces consisting of a 128-bit state in the data processing part and a 128-bit state in the key schedule at the unknown fixed round, assuming encryptions are executed with a 256-bit working memory for a internal state and a subkey.

Assuming that leaked bits come from the states after round 2, i.e., $\{\#5_i[j]$ and $\$2_i[j]\}$ and $\{\#'5_i[j]$ and $\$'2_i[j]\}$ ($0 \leq i \leq 15, 0 \leq j \leq 7$), the parameters of our differential bias attacks are chosen as $N_{all} = (256)^2$, $N_{bias_p}^{(c)} = 224 (= 96 + 128)$,

$N_{bias_p}^{(w)} = 128 (= 0 + 128)$, $N_{bias_n}^{(c)} = 32$ and $N_{bias_n}^{(w)} = 0$ (see Table 2). Then, $Pr^c(\Delta[Z_i, Z_j'] = 0)$ and $Pr^w(\Delta[Z_i, Z_j'] = 0)$ are estimated as $1/2 \cdot (1 + 2^{-8.192})$, and $1/2 \cdot (1 + 2^{-9.000})$, respectively, and $q = 2^{-9.42}$. In our experiment with 2^{40} randomly-chosen correct and wrong pairs of keys and plaintexts, $Pr_c(\Delta[Z_i, Z_j'] = 0)$ and $Pr_w(\Delta[Z_i, Z_j'] = 0)$ are $1/2 \cdot (1 + 2^{-8.191})$ and $1/2 \cdot (1 + 2^{-9.001})$, respectively, and $q = 2^{-9.42}$. The number of required samples to detect a stream for a correct key is estimated as $N_{sample} = 2^{24.84} (= 2^6 \times 2^{9.42 \times 2})$. We experimentally confirmed that this number of samples is enough for a successful attack. With $N_s = (N_{all})^{1/2}$, time complexity is estimated as $2^{47.84} (= (2^{31} \times 2^{24.84})/(2^8))$ encryptions and the required data is $2^{42} (= 2^{34} \times 2^8)$ chosen plaintexts and corresponding leaked bits with 2^{42} bits of prepared tables.

The details of attacks for $N_s = (N_{all})^{1/2}$ are provided in Table 3, where $q^{(e)}$ is our experimental value with 2^{40} randomly-chosen correct and wrong pairs of keys and plaintexts/ciphertexts, and T and D are time complexity and the amount of the required data, respectively. Our theoretical values closely approximate the experimental data in all cases. Since an attacker does not know the round number of leaked bits, he firstly guesses the round of leaked bits and then mounts an attack similar to the combined attack of the bitwise multiset attacks. If the decryption is accessible, our attacks are successful except the case where leaked bits come from states after 4 or 5 round only. Also, a known plaintext attack is possible if leaked bits from #3 are available.

5.2 Uncertainty in Time

The time randomization makes the round number of leaked bits random in each execution, i.e., Z_i and Z_i' come from the fixed bit position at a random round of the data processing part. Additionally, we take into account the leaked bits from plaintexts #0 or ciphertexts #21 in the data processing part. For instance, assuming that leaked bits come from 33-th bits of the data processing part, i.e., $\#0_4[1]$, $\#3_4[1]$, ..., $\#19_4[1]$ or $\#21_4[1]$, the attack parameters are given as $N_{all} = 11^2$, $N_{bias_p}^{(c)} = 3$, $N_{bias_p}^{(w)} = 2$, $N_{bias_n}^{(c)} = 1$, $N_{bias_n}^{(w)} = 0$. Then $q = 2^{-6.45}$, $N_{sample} = 2^{19.9}$, $T = 2^{47.44}$ and $D = 2^{37.46}$.

The details of our attacks using leaked bits from the data processing part are provided in Table 4, where the attack parameters of chosen-plaintext differential bias attacks depend on the positions of leaked bits, but time and data complexities are almost same for each position. We also evaluate a chosen-plaintext attack when round 9 and round 1, 2, 8 and 9 rounds are black-boxed. i.e., {#19, $9} and {#3,#5,#17, #19, $1, $2, $8, $9} are not available, respectively. Other black-boxed variants are also evaluated by properly choosing attack parameters. Since an attacker does not know the bit position of leaked bits, he firstly guesses it and then mounts an attack. If the decryption is accessible, our attacks are feasible as long as leaked bits after round 1, 2, 3, 6, 7, 8 or 9 in the data processing part are available. A known plaintext attack is applicable if leaked bits from #3 are obtained. However, it is a 32-bit key recovery attack, because a bit of #3 is affected by 32 bits of $0.

Table 3. Differential bias and biased state attacks for space randomization

Chosen-plaintext(ciphertext) differential bias attack

Round	N_{all}	$N^c_{bias_p}$	$N^w_{bias_p}$	$N^c_{bias_n}$	$N^w_{bias_n}$	q	$q^{(e)}$	N_{sample}	T	D	
1 (8)	256^2	248	224	8	0	$2^{-11.42}$	$2^{-11.38}$	$2^{28.84}$	$2^{51.84}$	$2^{42.00}$	CP(CC)
2 (7)	256^2	224	128	32	0	$2^{-9.42}$	$2^{-9.42}$	$2^{24.84}$	$2^{47.84}$	$2^{42.00}$	CP(CC)
3 (6)	256^2	128	128	128	0	$2^{-16.99}$	$2^{-16.84}$	$2^{39.98}$	$2^{62.98}$	$2^{42.00}$	CP(CC)

Known-plaintext differential bias attack

| 1 | 256^2 | 152 | 128 | 8 | 0 | $2^{-11.42}$ | $2^{-11.10}$ | $2^{28.84}$ | $2^{51.84}$ | $2^{43.00}$ | KP |

Chosen-ciphertext biased state attack

Round	N'_{all}	$N'^c_{bias_p}$	$N'^w_{bias_p}$	-	-	q	$q^{(e)}$	N'_{sample}	T	D	
9	256	8	0	-	-	$2^{-5.00}$	$2^{-5.00}$	2^{14}	$2^{26.00}$	$2^{26.00}$	CC

Table 4. Differential bias and biased state attacks for time randomization

Chosen-plaintext differential bias attack

BB round	N_{all}	$N^c_{bias_p}$	$N^w_{bias_p}$	$N^c_{bias_n}$	$N^w_{bias_n}$	q	$q^{(e)}$	N_{sample}	T	D	
None	11^2	3	2	1	0	$2^{-6.95}$	$2^{-6.94}$	$2^{19.90}$	$2^{47.44}$	$2^{37.46}$	CP
9	10^2	3	2	1	0	$2^{-6.68}$	$2^{-6.68}$	$2^{19.36}$	$2^{47.04}$	$2^{37.32}$	CP
1, 2, 8, 9	7^2	0	0	1	0	$2^{-13.61}$	$2^{-13.23}$	$2^{33.22}$	$2^{60.41}$	$2^{36.81}$	CP

Known-plaintext differential bias attack

| None | 11^2 | 1 | 0 | 0 | 0 | $2^{-6.92}$ | $2^{-7.30}$ | $2^{19.84}$ | $2^{47.38}$ | $2^{38.46}$ | KP |

Chosen-ciphertext biased state attack

BB round	N'_{all}	$N'^c_{bias_p}$	$N'^w_{bias_p}$	-	-	q	$q^{(e)}$	N'_{sample}	T	D	
None	11	1	0	-	-	$2^{-3.46}$	$2^{-3.45}$	$2^{10.92}$	$2^{22.92}$	$2^{22.92}$	CC

5.3 Uncertainty in Both Space and Time

The space and time randomization makes the both the bit position and the round number of leaked bits random in each execution, i.e., Z_i and Z'_i randomly come from any bit of any states in the data processing part $\{\#0, \#3, \#5, \ldots, \#19, \#21\}$ and in the key schedule $\{\$0, \ldots, \$10\}$. The parameters of the chosen-plaintext differential bias attacks are estimated as $N_{all} = (256 \times 11)^2$, $N^{(c)}_{bias_p} = 1720 \ (= 27 \times 8 + 176 \times 8 + 12 \times 8)$, $N^{(w)}_{bias_p} = 1600 \ (= 12 \times 8 + 176 \times 8 + 12 \times 8)$, $N^{(c)}_{bias_n} = 200 \ (= 21 \times 8 + 0 + 4 \times 8)$ and $N^{(w)}_{bias_n} = 32 \ (= 0 + 0 + 4 \times 8)$.

The details of our attacks are given in Table 5. We also provide a chosen-plaintext attack when round 9 and round 1, 2, 8 and 9 are black-boxed. If the decryption is accessible, our attacks work as long as leaked bits after round 1, 2, 3, 6, 7, 8 or 9 of the data processing part are available. Also, a known-plaintext attack is applicable if leaked bits from $\#3$ are observable.

Table 5. Differential bias and biased state attacks for space and time randomization

Chosen-plaintext differential bias attack									
BB round	N_{all}	$N_{bias_p}^c$	$N_{bias_p}^w$	$N_{bias_n}^c$	$N_{bias_n}^w$	q	$q^{(e)}$	N_{sample}	T $\qquad$ D
None	$(256 \cdot 11)^2$	1720	1600	200	32	$2^{-16.02}$	$2^{-15.92}$	$2^{38.04}$	$2^{57.58}$ $2^{45.46}$ CP
9	$(256 \cdot 10)^2$	1592	1472	200	32	$2^{-15.75}$	$2^{-15.70}$	$2^{37.49}$	$2^{57.17}$ $2^{45.32}$ CP
1, 2 8, 9	$(256 \cdot 7)^2$	896	896	128	0	$2^{-22.61}$	$2^{-23.07}$	$2^{51.22}$	$2^{71.41}$ $2^{44.81}$ CP
Known-plaintext differential bias attack									
None	$(256 \cdot 11)^2$	1440	1408	40	32	$2^{-17.92}$	$2^{-17.69}$	$2^{41.84}$	$2^{61.38}$ $2^{46.46}$ KP
Chosen-ciphertext biased state attack									
BB round	N_{all}'	$N_{bias_p}'^c$	$N_{bias_p}'^w$	-	-	q	$q^{(e)}$	N_{sample}'	T $\qquad$ D
None	$(256 \cdot 11)$	8	0	-	-	$2^{-8.46}$	$2^{-8.44}$	$2^{20.92}$	$2^{32.92}$ $2^{32.92}$ CC

Table 6. Differential bias and biased state attacks for leakage with noise

BB round	Time	Data	Time	Data	Time	Data	Time	Data
	$\pi = 1$		$\pi = 2^{-10}$		$\pi = 2^{-20}$		$\pi = 2^{-30}$	
Chosen-plaintext differential bias attack								
None	$2^{57.58}$	$2^{45.46}$ CP	$2^{67.58}$	$2^{55.46}$ CP	$2^{77.58}$	$2^{65.46}$ CP	$2^{87.58}$	$2^{75.46}$ CP
1, 2, 8, 9	$2^{71.41}$	$2^{44.81}$ CP	$2^{81.41}$	$2^{54.81}$ CP	$2^{91.41}$	$2^{64.81}$ CP	$2^{101.41}$	$2^{74.81}$ CP
Known-plaintext differential bias attack								
None	$2^{61.38}$	$2^{46.46}$ KP	$2^{71.38}$	$2^{56.46}$ KP	$2^{81.38}$	$2^{66.46}$ KP	$2^{91.38}$	$2^{76.46}$ KP
Chosen-ciphertext biased state attack								
None	$2^{32.92}$	$2^{32.92}$ CC	$2^{52.92}$	$2^{52.92}$ CC	$2^{72.92}$	$2^{72.92}$ CC	$2^{92.92}$	$2^{92.92}$ CC

6 AES Under Noisy Leakage

This section studies the effect of additional noise on top of the time and space randomization. The noise can be due to the limited knowledge of the platform by the adversary or due to the implemented countermeasures such as insertion of dummy operations. In the differential bias attack, this reduces the rate of positive/negative biased bits by adding noise bits into the space of the actually leaked bits. To quantify the amount of noise present in the attack, we define π as the probability that an observed bit is not a noise bit. Suppose that the values of the noise bits are randomly distributed, the bias of a leaked bit stream of the correct key with noise bits is estimated as $q' = q \times \pi$, and the required number of sample bits to distinguish a stream for a correct key increases by the multiple of $(\pi^2)^{-1}$ to $N_{sample}' = N_{sample} \times (\pi^2)^{-1}$. With $N_s = (N_{all})^{1/2} \times \pi^{-1}$, the time and data complexities of our known/chosen plaintext differential bias attacks increase by the multiple of $(\pi)^{-1}$ as $T \approx 2^{31} \times (N_{sample} \times \pi^{-2})/N_s \times \pi^{-1} = 2^{31} \times (N_{sample} \times \pi^{-1})/N_s$ encryptions and $D \approx 2^{34}(2^{35}) \times (N_s \times \pi^{-1})$ chosen/known-plaintexts with leaked bits. Also, the time and data complexities of chosen-ciphertext biased state attacks increase by the multiple of $(\pi)^{-2}$. The detailed evaluations for each values of π are shown in Table 6.

7 Towards More Alignment: Bytewise Leakage

Here we deal with the case where each execution leaks *one byte of a byte-aligned* state. In other words, now we let aligned bytes of internal states leak. Such leaks reflect the realities of a byte-oriented software implementation better.[3] In both settings – leakage with fixed and uncertain time/space – our techniques still apply. However, some adjustments are needed, see below.

7.1 Fixed Time/space: Bytewise Multiset Attack

Our bitwise multiset attacks naturally extend to *bytewise* multiset attacks, because the multiset characteristics are based on the bytewise XOR-sum property. The success probability for detecting wrong keys increases from $(1-2^{-1})$ to $(1-2^{-8})$ by using the bytewise zero-sum property. Then the time complexities of 2, 3, 4, 6 and 7-round attacks are estimated as $\{(2^{32} \times 2^8 \times N) \times 4\} + (1 + 2^{-8N} \times (2^{32} - 1))^4$ encryptions. With $N = 4$, it is about 2^{44}. The time complexities of 1 and 8-round attacks and the 9-round attack also improve to 2^{42} ($\approx 2^{32} \times 2^8 \times 4 \times 4/3$) and 2^{12} ($= 2^8 \times 16$) encryptions, respectively. The time complexity of the combined attack is 2^{45} ($\approx 2^{12}+2^{42}+2^{42}+2^{44}+2^{44}$) encryptions and the required data is 2^{35} chosen plaintexts and 2^{34} chosen ciphertexts.

7.2 Uncertain Time/Space: Differential Bias Attack

Our differential bias attacks also extend to *bytewise* attacks using bytewise differential biases of truncated differential characteristics of Fig. 4, 7 and 8.

Chosen/Known-Plaintext Differential Bias Attack. Let a leaked byte from the i-th execution be Z_i^*, and N_{all}^*, $N_{bias_p}^*$, $N_{bias_n}^*$, N_{random}^* be the number of bytewise pairs in the entire space, positive biased space, negative biased space and randomly-distributed space, respectively. The probabilities that a difference of a bytewise pair of randomly chosen leaked bytes is zero ($\Delta[Z_i^*, Z_j'^*] = 0$) for a correct key and a wrong key are estimated as follows.

$$Pr^c(\Delta[Z_i^*, Z_j'^*] = 0) = 1/2^8 \cdot (N_{random}^{*c}/N_{all}^*) + N_{bias_p}^{*c}/N_{all}^*,$$

$$Pr^w(\Delta[Z_i^*, Z_j'^*] = 0) = 1/2^8 \cdot (N_{random}^{*w}/N_{all}^*) + N_{bias_p}^{*w}/N_{all}^*.$$

Assuming that the target event E is $\Delta[Z_i^*, Z_j^*] = 0$, X is a distribution for a wrong key, and Y is a distribution for a correct key, p and q are estimated as $p = Pr^w(\Delta[Z_i^*, Z_j^*] = 0)$ and $q = \frac{-N_{bias_n}^c + N_{bias_n}^w + 255(N_{bias_p}^c - N_{bias_p}^w)}{N_{all} - N_{bias_n}^w + 255 N_{bias_p}^w}$. For the success probability of $1 - 2^{-32}$, the required sample size is estimated as

---

[3] The stream cipher LEX can be regarded as a bytewise leakage model at the fixed space [3] but the locations of leaked bytes are known for the attacker. Thus, the attack against LEX [10] is not directly applicable to our unknown location model.

Table 7. Evaluation for byte-aligned space randomization ($N_s = (N_{all})^{1/2}$)

Chosen-plaintext(ciphertext) differential bias attack

Round	N_{all}	$N^c_{bias_p}$	$N^w_{bias_p}$	$N^c_{bias_n}$	$N^w_{bias_n}$	q	$q^{(e)}$	N_{sample}	T	D	
1 (8)	32^2	31	28	1	1	$2^{-3.41}$	$2^{-3.42}$	$2^{19.84}$	$2^{45.84}$	$2^{39.00}$	CP(CC)
2 (7)	32^2	28	16	4	4	$2^{-0.74}$	$2^{-0.74}$	$2^{14.48}$	$2^{40.48}$	$2^{39.00}$	CP(CC)
3 (6)	32^2	16	16	16	0	$2^{-8.32}$	$2^{-8.38}$	$2^{29.64}$	$2^{55.64}$	$2^{42.00}$	CP

Known-plaintext differential bias attack

1	32^2	19	16	1	0	$2^{-2.74}$	$2^{-2.74}$	$2^{18.48}$	$2^{44.48}$	$2^{40.00}$	KP

Chosen-ciphertext biased state attack

Round	N'_{all}	$N'^c_{bias_p}$	$N'^w_{bias_p}$	-	-	q	$q^{(e)}$	N_{sample}	T	D	
9	32	1	0	-	-	$2^{2.99}$	$2^{2.99}$	$2^{11.00}$	$2^{23.00}$	$2^{23.00}$	CC

Table 8. Evaluation for byte-aligned time randomization ($N_s = (N_{all})^{1/2}$)

Chosen-plaintext differential bias attack

BB round	N_{all}	$N^c_{bias_p}$	$N^w_{bias_p}$	$N^c_{bias_n}$	$N^w_{bias_n}$	q	$q^{(e)}$	N_{sample}	T	D	
None	11^2	3	2	1	0	$2^{-1.31}$	$2^{-1.31}$	$2^{15.62}$	$2^{43.16}$	$2^{37.46}$	CP
9	10^2	3	2	1	0	$2^{-1.26}$	$2^{-1.26}$	$2^{15.52}$	$2^{43.19}$	$2^{37.32}$	CP
1, 2, 8, 9	7^2	0	0	1	0	$2^{-5.61}$	$2^{-5.50}$	$2^{24.22}$	$2^{52.41}$	$2^{36.81}$	CP

Known-plaintext differential bias attack

None	$(11)^2$	1	0	0	0	$2^{1.08}$	$2^{1.08}$	$2^{13.00}$	$2^{40.54}$	$2^{38.46}$	KP

Chosen-ciphertext biased state attack

BB round	N'_{all}	$N'^c_{bias_p}$	$N'^w_{bias_p}$	-	-	q	$q^{(e)}$	N_{sample}	T	D	
None	11	1	0	-	-	$2^{4.50}$	$2^{4.50}$	$2^{11.00}$	$2^{23.00}$	$2^{23.00}$	CP

$N^*_{sample} \approx 32 \cdot 256 \cdot q^2 = 2^{13} \cdot q^2$. Time complexity is estimated as $2^{31} \times N_{sample}/N_s$ encryptions and the required data is $2^{34}(2^{35}) \times N_s$ chosen/known plaintexts with leaked bits.

Chosen-Ciphertext Biased-State Attack. Assuming that the target event E is $Z_i = 0$, p and q are estimated as $p = 1/2^8$ and $q = (255 \times N^c_{bias_p})/N_{all}$. The number of required samples is estimated as $N_{sample} \approx 8 \cdot 2^8 \cdot (q)^{-2}$. We repeat the procedure for all 16 byte of \$10. Therefore, time complexity is estimated as $2^{12} \times N_{sample}$ ($= 16 \times 2^8 \times N_{sample}$) encryptions and the number of required data is $2^{12} \times N_{sample}$ ($= 16 \times 2^8 \times N_{sample}$) chosen ciphertexts.

Security Under Time and Space Randomization and with Leakage Noise. The results of security evaluations under time and space randomization

Table 9. Evaluation for byte-aligned space and time randomization ($N_s = (N_{all})^{1/2}$)

Chosen-plaintext differential bias attack									
BB round	N_{all}	$N_{bias_p}^c$	$N_{bias_p}^w$	$N_{bias_n}^c$	$N_{bias_n}^w$	q	$q^{(e)}$	N_{sample}	T	D
None	$(32 \cdot 11)^2$	215	200	25	4	$2^{-5.52}$	$2^{-5.53}$	$2^{24.04}$	$2^{46.58}$	$2^{42.45}$ CP
9	$(32 \cdot 10)^2$	199	184	25	4	$2^{-5.29}$	$2^{-5.29}$	$2^{23.58}$	$2^{46.26}$	$2^{42.32}$ CP
1, 2, 8, 9	$(32 \cdot 8)^2$	112	112	16	0	$2^{-12.52}$	$2^{-12.58}$	$2^{38.04}$	$2^{61.23}$	$2^{41.80}$ CP
Known-plaintext differential bias attack									
None	$(32 \cdot 11)^2$	179	176	5	4	$2^{-7.79}$	$2^{-7.74}$	$2^{28.58}$	$2^{52.12}$	$2^{43.45}$ KP
Chosen-ciphertext biased state attack									
BB round	N'_{all}	$N_{bias_p}^{lc}$	$N_{bias_p}^{lw}$	-	-	q	$q^{(e)}$	N_{sample}	T	D
None	$(32 \cdot 11)$	1	0	-	-	$2^{-0.46}$	$2^{-0.46}$	$2^{11.92}$	$2^{23.92}$	$2^{23.92}$ CC

Table 10. Evaluation for byte-aligned leakage with noise ($N_s = (N_{all})^{1/2} \times \pi^{-1}$)

BB round	Time	Data	Time	Data	Time	Data	Time	Data
	$\pi = 1$		$\pi = 2^{-10}$		$\pi = 2^{-20}$		$\pi = 2^{-30}$	
Chosen-plaintext differential bias attack								
None	$2^{46.58}$	$2^{42.45}$ CP	$2^{56.58}$	$2^{52.45}$ CP	$2^{66.58}$	$2^{62.45}$ CP	$2^{76.58}$	$2^{72.45}$ CP
1, 2, 8, 9	$2^{61.23}$	$2^{41.80}$ CP	$2^{71.23}$	$2^{51.80}$ CP	$2^{81.23}$	$2^{61.80}$ CP	$2^{91.23}$	$2^{71.80}$ CP
Known-plaintext differential bias attack								
None	$2^{50.28}$	$2^{43.45}$ KP	$2^{60.28}$	$2^{53.45}$ KP	$2^{70.28}$	$2^{63.45}$ KP	$2^{80.28}$	$2^{73.45}$ KP
Chosen-ciphertext biased state attack								
None	$2^{23.92}$	$2^{23.92}$ CC	$2^{43.92}$	$2^{43.92}$ CC	$2^{63.92}$	$2^{63.92}$ CC	$2^{83.92}$	$2^{83.92}$ CC

with noisy leakage are provided in Tables 7, 8, 9 and 10.[4] In all cases, time complexity and data requirements are improved compared to the bit-aligned attacks.

8 Some Extensions

8.1 AES-192 and 256

Bitwise multiset attacks and differential bias attacks on AES-128 are directly applicable to AES-192 and AES-256 in both fixed and random settings. In the backward direction, 6- to 9- round attacks on AES-128 are corresponded to 8- to 11-round ones on AES-192 and 10- to 13- round ones on AES-256, respectively.

8.2 Multiple-Bit Leakage

Here we consider the case where M bits of the bit-aligned state information leak in each execution for a small M. Let $Z_1^i, Z_2^i, \dots, Z_M^i$ be M leaked bits of the i-th execution.

---

[4] If q is not small, Lemmata 1 and 2 are not applicable [16]. In this case we estimate $N_{sample} = 2^{11}$ and 2^{13} for known-plaintext differential bias attacks and chosen-ciphertext biased state attacks, respectively. We confirmed experimentally that these numbers of samples were enough for successful attacks.

Bitwise Multiset Attack: Assume that $Z_0^i, Z_1^i, \ldots, Z_{M-1}^i$ come from different but fixed locations of the state. If the XOR sum of 2^8 multiset of each location is zero, the XOR-sum of all set of $2^8 \times M$ bits is also zero. Thus, bitwise multiset attacks are feasible as long as leaked bits come from space where each XOR sum is zero only in a correct key. Time and date complexities are almost the same.

Differential Bias Attack: Assume that $Z_1^i, Z_2^i, \ldots, Z_M^i$ come from randomly-chosen different locations of the state. Since the attacker is able to obtain M bits in each execution, the required data reduces by a factor of M.

8.3   Other Granularities

So far, we have assumed that a leak can only occur after a full round. However, in other granularities such as leaks after SubBytes or MixColumns, our bitwise multiset attacks and differential bias attack still work.

Bitwise Multiset Attack: According to Proposition 1, any bit of the states between #3 and #10 has the zero-sum property if the key is correctly guessed. Using the difference of zero-sum properties between correct and wrong key cases, bitwise multiset attacks are applicable to other states in the same manner.

Differential Bias Attack: By properly choosing attack parameters, our differential bias attacks are also made feasible. For instance, if bits of the states after SubBytes are additionally leaked, the parameters of chosen-plaintext differential attacks on AES-128 with the space and time randomization are estimated as $N_{all} = (256 \times 11 + 128 \times 10)^2$, $N_{bias_p}^{(c)} = 2032 \ (= 216 + 216 + 1408 + 96 + 96)$, $N_{bias_p}^{(w)} = 1792 \ (= 96+96+1408+96+96)$, $N_{bias_n}^{(c)} = 400 \ (= 168+168+0+32+32)$, $N_{bias_n}^{(w)} = 64 \ (= 0+0+32+32)$, and $q = 2^{-16.10}$. The number of required samples is estimated as $N_{sample} = 2^{38.02} (= 2^6 \times 2^{16.01 \cdot 2})$. With $N_s = (N_{all})^{1/2}$, time complexity is $2^{57.02} \ (= (2^{31} \times 2^{38.02})/(256 \times 11 + 128 \times 10))$ encryptions and the required data is $2^{46} \ (= 2^{34} \times (256 \times 11 + 128 \times 10))$ chosen plaintexts.

References

1. Boneh, D., DeMillo, R.A., Lipton, R.J.: On the importance of eliminating errors in cryptographic computations. J. Cryptol. **14**(2), 101–119 (2001)
2. Biham, E.: New types of cryptanalytic attacks using related keys. J. Cryptol. **7**(4), 229–246 (1994)
3. Biryukov, A.: The design of a stream cipher LEX. In: Biham, E., Youssef, A.M. (eds.) SAC 2006. LNCS, vol. 4356, pp. 67–75. Springer, Heidelberg (2007)
4. Biryukov, A., Shamir, A.: Structural cryptanalysis of SASAS. J. Cryptol. **23**(4), 505–518 (2010)
5. Bogdanov, A., Kizhvatov, I., Pyshkin, A.: Algebraic methods in side-channel collision attacks and practical collision detection. In: Chowdhury, D.R., Rijmen, V., Das, A. (eds.) INDOCRYPT 2008. LNCS, vol. 5365, pp. 251–265. Springer, Heidelberg (2008)

6. Borghoff, J., Canteaut, A., Güneysu, T., Kavun, E.B., Knezevic, M., Knudsen, L.R., Leander, G., Nikov, V., Paar, C., Rechberger, C., Rombouts, P., Thomsen, S.S., Yalçın, T.: PRINCE – a low-latency block cipher for pervasive computing applications. In: Wang, X., Sako, K. (eds.) ASIACRYPT 2012. LNCS, vol. 7658, pp. 208–225. Springer, Heidelberg (2012)

7. Chow, S., Eisen, P.A., Johnson, H., van Oorschot, P.C.: White-box cryptography and an AES implementation. In: Nyberg, K., Heys, H. (eds.) SAC 2002. LNCS, vol. 2595, pp. 250–270. Springer, Heidelberg (2002)

8. Daemen, J., Rijmen, V.: The Design of Rijndael: AES - The Advanced Encryption Standard. Information Security and Cryptography. Springer, Heidelberg (2002)

9. Dinur, I., Shamir, A.: Side channel cube attacks on block ciphers. In: Cryptology ePrint Archive, Report 2009/127 (2009). http://eprint.iacr.org/

10. Dunkelman, O., Keller, N.: A new attack on the LEX stream cipher. In: Pieprzyk, J. (ed.) ASIACRYPT 2008. LNCS, vol. 5350, pp. 539–556. Springer, Heidelberg (2008)

11. FIPS PUB 197, Advanced Encryption Standard (AES), U.S.Department of Commerce/National Institute of Standards and Technology (2001)

12. Kleinjung, T., Lenstra, A.K., Page, D., Smart, N.P.: Using the cloud to determine key strengths. In: Galbraith, S., Nandi, M. (eds.) INDOCRYPT 2012. LNCS, vol. 7668, pp. 17–39. Springer, Heidelberg (2012)

13. Kocher, P.C.: Timing attacks on implementations of diffie-hellman, RSA, DSS, and other systems. In: Koblitz, N. (ed.) CRYPTO 1996. LNCS, vol. 1109, pp. 104–113. Springer, Heidelberg (1996)

14. Kocher, P.C., Jaffe, J., Jun, B.: Differential power analysis. In: Wiener, M. (ed.) CRYPTO 1999. LNCS, vol. 1666, pp. 388–397. Springer, Heidelberg (1999)

15. Mantin, I.: Predicting and distinguishing attacks on RC4 keystream generator. In: Cramer, R. (ed.) EUROCRYPT 2005. LNCS, vol. 3494, pp. 491–506. Springer, Heidelberg (2005)

16. Mantin, I., Shamir, A.: A practical attack on broadcast RC4. In: Matsui, M. (ed.) FSE 2001. LNCS, vol. 2355, pp. 152–164. Springer, Heidelberg (2002)

17. Micali, S., Reyzin, L.: Physically observable cryptography. In: Naor, M. (ed.) TCC 2004. LNCS, vol. 2951, pp. 278–296. Springer, Heidelberg (2004)

18. De Mulder, Y.: White-box cryptography: analysis of white-box AES implementations. Ph. D. thesis, KU Leuven (2014)

19. Oren, Y., Renauld, M., Standaert, F.-X., Wool, A.: Algebraic side-channel attacks beyond the hamming weight leakage model. In: Prouff, E., Schaumont, P. (eds.) CHES 2012. LNCS, vol. 7428, pp. 140–154. Springer, Heidelberg (2012)

20. Renauld, M., Standaert, F.-X.: Representation-, leakage- and cipher-dependencies in algebraic side-channel attacks. In: The Proceedings of the ACNS 2010 Industrial Track (2010)

21. Renauld, M., Standaert, F.-X., Veyrat-Charvillon, N.: Algebraic side-channel attacks on the AES: why time also matters in DPA. In: Clavier, C., Gaj, K. (eds.) CHES 2009. LNCS, vol. 5747, pp. 97–111. Springer, Heidelberg (2009)

22. Schramm, K., Wollinger, T., Paar, C.: A new class of collision attacks and its application to DES. In: Johansson, T. (ed.) FSE 2003. LNCS, vol. 2887, pp. 206–222. Springer, Heidelberg (2003)

23. Sepehrdad, P., Vaudenay, S., Vuagnoux, M.: Statistical attack on RC4. In: Paterson, K.G. (ed.) EUROCRYPT 2011. LNCS, vol. 6632, pp. 343–363. Springer, Heidelberg (2011)

Design of Block Ciphers

A Synthetic Indifferentiability Analysis of Interleaved Double-Key Even-Mansour Ciphers

Chun Guo[1,2] and Dongdai Lin[1(✉)]

[1] State Key Laboratory of Information Security,
Institute of Information Engineering,
Chinese Academy of Sciences, Beijing 100093, China
[2] University of Chinese Academy of Sciences, Beijing, China
{guochun,ddlin}@iie.ac.cn

Abstract. Iterated Even-Mansour scheme (IEM) is a generalization of the basic 1-round proposal (ASIACRYPT '91). The scheme can use one key, two keys, or completely independent keys.

Most of the published security proofs for IEM against relate-key and chosen-key attacks focus on the case where all the round-keys are derived from a single master key. Whereas results beyond this barrier are relevant to the cryptographic problem whether a secure blockcipher with key-size twice the block-size can be built by mixing two *relatively independent* keys into IEM and iterating sufficiently many rounds, and this strategy actually has been used in designing blockciphers for a long-time.

This work makes the first step towards breaking this barrier and considers IEM with Interleaved Double *independent* round-keys:

$$\text{IDEM}_r((k_1, k_2), m) = k_i \oplus (P_r(\ldots k_1 \oplus P_2(k_2 \oplus P_1(k_1 \oplus m))\ldots)),$$

where $i = 2$ when r is odd, and $i = 1$ when r is even. As results, this work proves that 15 rounds can achieve (full) indifferentiability from an ideal cipher with $O(q^8/2^n)$ security bound. This work also proves that 7 rounds is sufficient and necessary to achieve sequential-indifferentiability (a notion introduced at TCC 2012) with $O(q^6/2^n)$ security bound, so that IDEM_7 is already correlation intractable and secure against any attack that exploits evasive relations between its input-output pairs.

Keywords: Blockcipher · Ideal cipher · Indifferentiability · Key-alternating cipher · Even-mansour cipher · Correlation intractability

1 Introduction

Blockciphers are arguably the most important primitives in cryptography. A blockcipher $\text{BC}[\kappa, n] : \{0,1\}^\kappa \times \{0,1\}^n \to \{0,1\}^n$ maps a κ-bit key K and

---

D. Lin—A full version is available [GL15b].

© International Association for Cryptologic Research 2015
T. Iwata and J.H. Cheon (Eds.): ASIACRYPT 2015, Part II, LNCS 9453, pp. 389–410, 2015.
DOI: 10.1007/978-3-662-48800-3_16

an n-bit input x to an n-bit output y. For each key K, the map $\mathbf{BC}[\kappa, n](K, \cdot)$ is a permutation, and is efficiently invertible.

Most of the existing blockcipher designs can be roughly split into two families, namely Feistel ciphers and substitution-permutation networks. The latter are known as the structure of AES, and can be generalized as *key-alternating ciphers* [DR02]/*iterated Even-Mansour ciphers* (IEM for short). An r-round IEM cipher IEM_r consists of r fixed n-bit permutations P_i separated by key addition

$$\mathrm{IEM}_r(K, m) = k_r \oplus P_r(\ldots k_2 \oplus P_2(k_1 \oplus P_1(k_0 \oplus m))\ldots).$$

The single round Even-Mansour (the case $r = 1$) was developed in 1991 [EM93] in an attempt to turn a single permutation into a family of permutations (blockcipher). IEM_1 has been proved pseudorandom when the underlying permutation is random and public while the keys are secret. Since then, a soar of studies on IEM has been witnessed (especially in the recent half decade), for instance, on minimization [DKS13, CLL+14], on pseudorandomness [BKL+12, Ste12, LPS12, CS14], on related-key (RK) security [FP15, CS15], and on attacks (notable examples include [DKS13, DDKS15, DDKS14]). The pseudorandomness results showed that IEM is provably secure in traditional single secret key settings.

Indifferentiability of IEM. The studies on *indifferentiability* and *sequential-indifferentiability* (*seq-indifferentiability*) of IEM are mainly motivated by further validating the SPN-based blockcipher design methodology by proving IEM secure against *known-key* and *chosen-key (CK) attacks*, in which the adversary knows and chooses keys and tries to exhibit non-randomness. Roughly speaking, indifferentiability of IEM means that IEM can be as secure as an *ideal cipher* [MRH04], whereas seq-indifferentiability of IEM implies that IEM is *correlation intractable* [CGH04], and there is no relation between the inputs and outputs of IEM that can be exploited by an attack (even a chosen-key one) [MPS12]. Here the ideal cipher $\mathbf{IC}[\kappa, n] : \{0,1\}^\kappa \times \{0,1\}^n \rightarrow \{0,1\}^n$ is taken randomly from the set of $(2^n!)^{2^\kappa}$ possible choices of $\mathbf{BC}[\kappa, n]$. In this work, $\mathbf{IC}[2n, n]$ will be referred by $\mathbf{E}$.

As to (seq-)indifferentiability, we have been aware of four works: [ABD+13], [LS13], [CS15], and [Ste15]. [ABD+13] showed that IEM_5 is indifferentiable from $\mathbf{IC}[\kappa, n]$, if the round-key is derived from a preimage-aware key derivation function (KDF). On the other hand, [LS13] and [CS15] concentrated on *single-key EM* (SEM) in which the user-provided n-bit master key is directly used at each round: [LS13] proved that SEM_{12} (12-round SEM; similarly for SEM_4 and SEM_9) is indifferentiable, while [CS15] proved that SEM_4 is seq-indifferentiable. In [Ste15], Steinberger proved the indifferentiability of SEM_9. Results on SEM are closer to concrete designs, since they can be easily generalized to the case where each round-key is derived by an *efficiently invertible* permutation.

Problem: Even-Mansour with Two Keys. Existing works on provable security of IEM in RK and CK settings almost all focus on the SEM context: [LS13] (ASIACRYPT 2013), [FP15] (FSE 2015), [CS15] (EUROCRYPT 2015) (except for those considered random oracle modeled KDF, e.g. [ABD+13]). This work makes the first step towards breaking this barrier and considers the following problem: *can we obtain an ideal cipher by mixing two independent keys into IEM and iterating enough rounds?* (a problem left open by Lampe and Seurin (LS) [LS13])[1]. This problem is far from being trivial because all the works on SEM (in RK and CK settings) crucially rely on the correlation between *all* round-keys, so that they cannot be directly generalized to double-key case. Also, the independence between round-keys may bring in weakness – the most extreme case is IEM with completely independent round-keys, which is vulnerable to trivial related-key attacks. This problem is also practical since the idea is really used in existing designs such as AES-256 [DR02], Serpent [ABK98], and LED-128 [GPPR11] – note that they (certainly) mix the keys into the state by lightweight and efficient operations and iterate, rather than use some very complex hash function to seal the $2n$ key bits first. The intuition is that by iterating enough rounds, such designs will be "secure"; but the fact that the diffusion of the $2n$ key bits is relatively slow brings in doubts (e.g. doubts on AES-256 [KHP07,BDK+10]). The fact that among the three AES variants, AES-256 was the first that is theoretically broken [BK09] seems to support such doubts, and this attack raises a problem whether there exists a $BC[2n, n]$ design behaving like $IC[2n, n]$;[2] due to this, it is necessary to either validate (using a security proof) or negate (using a generic attack) this intuitive methodology.

To dig out a solution, note that using one key in the first $n/2$ rounds while using the other in the last $n/2$ rounds is trivially insecure [LS13]. Instead, a (seemingly) more promising approach to mixing two keys into IEM is the idea behind LED-128 [GPPR11], that is, interleaving the xoring of them: we name it *interleaved double-key Even-Mansour cipher* (IDEM for short; see Fig. 1 for an illustration). More formally, the r-round variant is written as follows:

$$IDEM_r((k_1, k_2), m) = k_t \oplus P_r(\ldots k_2 \oplus P_3(k_1 \oplus P_2(k_2 \oplus P_1(k_1 \oplus m))))),$$

where $t = 2$ when r is odd, and $t = 1$ when r is even. LS viewed IDEM as a promising solution to the problem mentioned before, and gave an extremely preliminary analysis, which led to the *conjecture* that 15 rounds is sufficient to achieve indifferentiability; but no concrete proof exists. Moreover, the provable security of IDEM with shorter rounds has not been considered yet.

Contributions. We give the first indifferentiability proof for 15-round IDEM. This is the first main result of this work. Interestingly, this matches LS's conjecture, but the proof is obtained by an approach quite different from they expected.

To obtain security guarantees for shorter round cases, we prove that $IDEM_7$ is seq-indifferentiable from $IC[2n, n]$; therefore, $IDEM_7$ is also correlation intractable in the random permutation model [MPS12], and resists all attacks

---

[1] A trivial solution to building $IC[2n, n]$ by IEM is hash-than-encrypt, which has been included in [ABD+13]. It was also discussed in [CDMS10]. But this solution imposes strong burden on the key derivation and is far from practical designs.

[2] Please see [CDMS10], page 275: *as of 2009 it is unclear if we have a candidate block-cipher with key-size larger than block-size that behaves like an ideal cipher.*

that exploit evasive relation between its inputs and outputs. We also find a sequential distinguisher against $IDEM_6$ (which is actually an easy extension of LS's attack against SEM_3 [LS13]), so that 7 rounds is also necessary. All the results are summarized by the following informal theorem.

Theorem. *For the construction IDEM based on completely independent random permutations, 6 rounds is not (seq-)indifferentiable; 7 rounds is seq-indifferentiable from* $\mathbf{IC}[2n, n]$ *with* $O(q^6/2^n)$ *security bound, and is also correlation intractable in the random permutation model; 15 rounds is indifferentiable from* $\mathbf{IC}[2n, n]$ *with* $O(q^8/2^n)$ *security bound.*

Due to the independence between the two n-bit round keys, at current time, we are not sure whether the results can be generalized to IEM with "very general" key schedules; however, for the first time, these results indeed validate the (seemingly long standing) *design principle* to some extent in the open-key model, i.e. a secure blockcipher $BC[2n, n]$ can be built from key-alternating ciphers without using very complex KDFs, or even without any KDF. Especially, they show that the intuition behind the key schedule of LED-128 is sound. However, they certainly cannot provide direct security guarantee for LED-128 – in fact, as theoretical results, they do not guarantee the security of *ANY concrete blockcipher*. As already mentioned, whether there exist some designs that "behave like" $\mathbf{IC}[2n, n]$ have to be supported by more (cryptanalysis) works.

Techniques. To prove indifferentiability and seq-indifferentiability, one first builds a simulator to mimic the behaviors of all the underlying permutations. Taking $IDEM_{15}$ as an example, consider a sequence of pairs of input and output (IO for short) $(x_1, y_1), \ldots, (x_{15}, y_{15})$ (called a *computation path/chain*) of the 15 permutations simulated by the simulator, which satisfies $y_i \oplus x_{i+1} = k_2$ when i is odd, and $y_i \oplus x_{i+1} = k_1$ when i is even. The simulator should ensure that each such chain simulated by it matches the ideal cipher $\mathbf{E}$, i.e. $\mathbf{E}((k_1, k_2), x_1 \oplus k_1) = y_{15} \oplus k_2$. The basic idea to reach this goal is Coron et al.'s *simulation via chain completion* technique [CHK+14], which has achieved success in (weaker) indifferentiability proofs for a variety of idealized blockciphers. It requires the simulator $\mathbf{S}$ to *detect* partial computation chains formed by the queries of the distinguisher, and *completes* the chains in advance by querying the ideal cipher $\mathbf{E}$, such that $\mathbf{S}$ is ready for answering queries in the future. To simulate answers that are consistent with $\mathbf{E}$, $\mathbf{S}$ has to use the answer from $\mathbf{E}$ to define some simulated answers; this action is called *adaptation*.

Detect Chains. To detect the so-called partial chains, note that the construction IDEM has the following property: given 4 values of 3 permutations y_i, x_{i+1}, y_{i+1}, and x_{i+2} (namely, an output of P_i, a pair of IO of P_{i+1}, and an input to P_{i+2}), the two associated keys can be derived as $k = y_i \oplus x_{i+1}$ and $k' = y_{i+1} \oplus x_{i+2}$, and it is possible to move forward and backward along the path. By this, at some place, using three rounds for chain detection is necessary – this idea has already appeared in [LS13].

Overall Strategy of the Simulators. As to the overall strategy, the simulator used to prove seq-indifferentiability of $IDEM_7$ is quite close to those for 6-round Feistel [MPS12] and SEM_4 [CS15]: the simulator *detects* partial chains at the *three middle round* of $IDEM_7$, completes them forward or backward, and finally adapts them at the first or last round – depending on concrete contexts.

On the other hand, the simulator used to prove the indifferentiability of $IDEM_{15}$ is motivated by Steinberger's illustration of indifferentiability proof for SEM_9 [Ste15]. The overall strategy requires detecting chains both at the two sides and at the middle – which is similar to several previous works (e.g. [CHK+14]). The core idea in this part is a so-called "pureness" property which is different from [CHK+14]: the simulator may fall into a recursive chain completion process; however, *during each such recursive completion process, all the partial chains are to be adapted at the same round*; as a consequence, *when a partial chain is to be completed, its extending is necessarily due to simulator defining new simulated answers to random ones rather than the adaptation of some other chains, so that the "endpoints" of this chain are always random.* Whereas in the context of IDEM, to uniquely specify a chain requires at least 3 values of 3 consecutive permutations, so that the adversary has more freedom to choose values and make different chains collide. With this in mind, we arrange two rounds to surround each adaptation zone to ensure different chains diverge in the adaptation zone; following an old convention [CHK+14], we call them *buffer rounds.*[3] For a more detailed overview of the simulator, we refer to Sect. 3.1.

In the indifferentiability proof for $IDEM_{15}$, we used an *active-chain-oriented bad events define strategy*, which is motivated by the analysis of $IDEM_7$: we directly define some bad events to be with respect to the chains that are active during the completion process. This helps us achieving the $O(q^8/2^n)$ indifferentiability security bound in spite of the complex character of IDEM. Albeit loose, this bound has been quite well-looking compared to similar (full) indifferentiability proofs for idealized blockciphers (the best non-flawed one(s) among them reached the level of $O(q^{10}/2^n)$ [ABD+13]).

Summary: What are Inherited and What are Novel? Technically speaking, we inherit the simulation via chain completion technique, the randomness mapping argument, and the basic idea for simulator termination argument from [CHK+14]; we also inherit (and adapt) the overall frameworks of Steinberger (which dates back to Seurin [Seu09]) and Cogliati et al. [CS15] (which dates back to Mandal et al. [MPS12]). Our novelties mainly lie in the proof for $IDEM_{15}$: first, we use a bad event to establish a slightly tighter bound on the size of the history $(O(q^2/2^n))$ and the simulator's complexity; second, we define the bad events to be so-called active-chain-oriented, so that the probability can be very low $(O(q^6/2^n))$. They two together enable to establish the $O(q^8/2^n)$ security bound.

---

[3] But our "buffer" rounds deviate from those in [CHK+14], in the sense that the values in them **can be defined** when the simulator is completing other chains.

Organization. Section 2 presents preliminaries. Section 3 contains the first main result – the indifferentiability of IDEM$_{15}$, and the proof sketch. Section 4 contains the second main result – the seq-indifferentiability of IDEM$_7$. Section 5 concludes. Due to page constraints, the full proofs of the two main results have to be deferred to the full version of this paper [GL15b].

2   Preliminaries and Notations

This work focuses on BC$[2n, n]$, say, blockciphers with n-bit blocks and $2n$-bit keys. Throughout the remaining, the n-bit round-keys are denoted by lower-case letters, i.e. k_1 and k_2, while the $2n$-bit master key is interchangeably denoted by the capital letter K or the concatenation (k_1, k_2) (as the reader has seen).

An n-bit random permutation is a permutation that is uniformly selected from all $(2^n)!$ possible choices. In this work, the notation $\mathbf{P} = (\mathbf{P}_1, \ldots, \mathbf{P}_j)$ is used to denote *a tuple of random permutations* ($j = 15$ in the context of IDEM$_{15}$, and $j = 7$ in the context of IDEM$_7$), and we let $\mathbf{P}$ provide a unified interface, i.e. $\mathbf{P}.\mathrm{P}(i, \delta, z) := \{1, \ldots, j\} \times \{+, -\} \times \{0,1\}^n \to \{0,1\}^n$, i indicates the index, $\delta \in \{+, -\}$ indicates direct query or inverse query, and $z \in \{0,1\}^n$ is the queries value). On the other hand, the interface of the ideal cipher $\mathbf{E}$ is $\mathbf{E}.\mathrm{E}(\delta, K, z) := \{+, -\} \times \{0,1\}^{2n} \times \{0,1\}^n \to \{0,1\}^n$.

Indifferentiability. The indifferentiability framework [MRH04] addresses the idealized construction in settings where the underlying parameters are exposed to the adversary. For concreteness, consider IDEM$_{15}^{\mathbf{P}}$: a distinguisher $\mathbf{D}^{\mathrm{IDEM}_{15}^{\mathbf{P}}, \mathbf{P}}$ with oracle access to the cipher and the underlying primitives is trying to distinguish IDEM$_{15}^{\mathbf{P}}$ from $\mathbf{E}$. Then the formal definition is as follows.

Definition 1 (Indifferentiability). *The idealized blockcipher* IDEM$_{15}^{\mathbf{P}}$ *with oracle access to ideal primitives* $\mathbf{P}$ *is said to be statistically and strongly* $(q, \sigma, t, \varepsilon)$-*indifferentiable from an ideal cipher* $\mathbf{E}$ *if there exists a simulator* $\mathbf{S}^{\mathbf{E}}$ *s.t.* $\mathbf{S}$ *makes at most* σ *queries to* $\mathbf{E}$, *runs in time at most* t, *and for any distinguisher* $\mathbf{D}$ *which issues at most* q *queries, it holds*

$$\left| Pr[\mathbf{D}^{\mathrm{IDEM}_{15}^{\mathbf{P}}, \mathbf{P}} = 1] - Pr[\mathbf{D}^{\mathbf{E}, \mathbf{S}^{\mathbf{E}}} = 1] \right| \leq \varepsilon$$

Such a result means that IDEM$_{15}^{\mathbf{P}}$ can safely replace $\mathbf{E}$ in most "natural" settings – although this belief does not necessarily hold when the resource of the adversary is limited [RSS11, DGHM13]. Since introduced, indifferentiability framework has been applied to various constructions, including variants of Merkle-Damgård, Feistel [CHK+14], Sponge [BDPVA08], and IEM [ABD+13, LS13].

To formally define seq-indifferentiability, we first specify a restricted distinguisher class, namely the *sequential distinguisher* (*seq-distinguisher*) [MPS12]. Consider IDEM$_7^{\mathbf{P}}$ and $D^{\mathrm{IDEM}_7^{\mathbf{P}}, \mathbf{P}}$. D is *sequential* if it issues queries in a specific order: (1) queries the underlying primitives $\mathbf{P}$ as it wishes; (2) queries IDEM$_7^{\mathbf{P}}$ as it wishes; (3) outputs, and cannot query $\mathbf{P}$ again. This order is illustrated in the italic numbers in Fig. 3. We then define the notion *total oracle query cost* of D, which equals the total number of queries received by $\mathbf{P}$ (from D or $\mathrm{IDEM}_7^{\mathbf{P}}$) when D interacts with $(\mathrm{IDEM}_7^{\mathbf{P}}, \mathbf{P})$ [MPS12]. Then, the definition of seq-indifferentiability can be obtained by tweaking the definition of (full) indifferentiability by restricting the distinguisher to the range of sequential ones, and replacing the query cost of the distinguisher by the *total oracle query cost*.

Definition 2 (Seq-indifferentiability). *The idealized blockcipher* $\mathrm{IDEM}_7^{\mathbf{P}}$ *with oracle access to ideal primitives* $\mathbf{P}$ *is said to be statistically and strongly* $(q, \sigma, t, \varepsilon)$-*seq-indifferentiable from an ideal cipher* $\mathbf{E}$ *if there exists a simulator* $\mathcal{S}^{\mathbf{E}}$ *s.t.* $\mathcal{S}$ *makes at most* σ *queries to* $\mathbf{E}$, *runs in time at most* t, *and for any sequential distinguisher* D *of total oracle query cost at most* q, *it holds*

$$\left| Pr[D^{\mathrm{IDEM}_7^{\mathbf{P}}, \mathbf{P}} = 1] - Pr[D^{\mathbf{E}, \mathcal{S}^{\mathbf{E}}} = 1] \right| \le \varepsilon$$

Sequential-indifferentiability implies *correlation intractability* [MPS12, CS15].

3   Indifferentiability for 15-Round IDEM

We prove the first main theorem of this paper in this section, which is:

Theorem 1. *The 15-round Even-Mansour cipher* IDEM_{15} *from fifteen independent random permutations* $\mathbf{P} = (\mathbf{P}_1, \dots, \mathbf{P}_{15})$ *and two n-bit keys* (k_1, k_2) *alternatively xored is strongly and statistically* $(q, \sigma, t, \varepsilon)$-*indifferentiable from an ideal cipher* $\mathbf{IC}[2n, n]$, *where* $\sigma = 2^{10} \cdot q^8$, $t = O(q^8)$, *and* $\varepsilon \le \frac{2^{11} \cdot q^8}{2^n} + \frac{2^{14} \cdot q^6}{2^n} = O(\frac{q^8}{2^n})$.

As usual, we first present the simulator, then sketch the proof.

3.1   The Simulator

To build the simulator, we borrow a variant of the *explicit randomness* technique [CHK+14] from [CS15], that is, letting the simulator $\mathbf{S}$ query $\mathbf{P}$ as explicit randomness. We denote by $\mathbf{S}^{\mathbf{E}, \mathbf{P}}$ the simulator for IDEM_{15} which takes $\mathbf{P}$ as randomness source and interacts with $\mathbf{E}$. $\mathbf{S}^{\mathbf{E}, \mathbf{P}}$ provides an interface $\mathbf{S}.\mathbf{P}(i, \delta, z)$ ($i \in \{1, \dots, 15\}$) which is exactly the same as $\mathbf{P}$. As argued [ABD+13, CS15], using such explicit randomness is actually equivalent to lazily sampling in advance before the experiment.

· We now give a high-level overview of the simulator $\mathbf{S}^{\mathbf{E}, \mathbf{P}}$ (depicted in Fig. 1 (left)). $\mathbf{S}$ maintains a history for each simulated permutation under the form of fifteen sets $P_1, \dots, P_{15}$. Each of the sets has entries in the form of (x, y) for $x, y \in \{0, 1\}^n$. $\mathbf{S}$ will ensure that for any $z \in \{0, 1\}^n$ and $i \in \{1, \dots, 15\}$, there is *at most one* $z' \in \{0, 1\}^n$ such that $(z, z') \in P_i$, and vice versa; *once such consistency cannot be kept,* $\mathbf{S}$ *aborts* (will be discussed later). By this, the sets $\{P\} = \{P_1, \dots, P_{15}\}$ are expected to define fifteen *partial permutations*, and we denote by P_i^+ (P_i^-, resp.) the (time-dependent) set of all n-bit values x (y, resp.) satisfying that $\exists z \in \{0, 1\}^n$ s.t. $(x, z) \in P_i$ ($(z, y) \in P_i$, resp.); denote by $P_i^+(x)$ ($P_i^-(y)$, resp.) the corresponding value of z (as mentioned *the uniqueness of* z *is ensured by* $\mathbf{S}$).

Queries that have already appeared in the history will be instantly answered with the contents in $\{P\}$. Upon a new query $\mathbf{S}^{\mathbf{E},\mathbf{P}}.\mathrm{P}(i,\delta,z)$, $\mathbf{S}^{\mathbf{E},\mathbf{P}}$ queries $\mathbf{P}$ to draw $z' = \mathbf{P}.\mathrm{P}(i,\delta,z)$ as the answer and calls a procedure $\textsc{ForceVal}(z,z',\delta,i)$ to add z and z' to P_i – inside this procedure, if z' is found already in $P_i^{\overline{\delta}}$, $\mathbf{S}^{\mathbf{E},\mathbf{P}}$ aborts due to the broken consistency (as mentioned). Then, if $(i,\delta) \in \{(2,+),(6,-),(10,+),(14,-)\}$ it satisfies the *chain detection conditions*, so that $\mathbf{S}^{\mathbf{E},\mathbf{P}}$ enqueues and completes chains formed by previous queries to ensure that it is ready to simulate answers consistent with those of $\mathbf{E}$ in the future.

The cases (i,δ) equals $(2,+)$ and $(14,-)$ are similar: taking the former $\mathrm{P}(2,+,x_2)$ as an example, $\mathbf{S}^{\mathbf{E},\mathbf{P}}$ considers all tuples $(x_1,y_1,x_{14},y_{14},x_{15},y_{15})$ such that $(x_j,y_j) \in P_j$ for $j \in \{1,14,15\}$, recovers two keys $k_2 := y_1 \oplus x_2$ and $k_1 := y_{14} \oplus x_{15}$, computes $y_0 := x_1 \oplus k_1$ and $x_{16} := y_{15} \oplus k_2$, checks whether $\mathbf{E}.\mathrm{E}(+,(k_1,k_2),y_0) = x_{16}$ via an inner procedure $\mathbf{S}.\textsc{Check}$ and enqueues a 5-tuple $(y_0,k_1,k_2,0,4)$ into a queue *ChainQueue* when this is the case. In this tuple, the 4-th value 0 informs $\mathbf{S}$ that the first value of the tuple is y_0, and the last value 4 describes that when completing the chain characterized by the tuple $(y_0,k_1,k_2,0)$, $\mathbf{S}$ should add the adapted pair to P_4 to ensure consistency with $\mathbf{E}$. The action towards answering new query $\mathrm{P}(14,-,y_{14})$ is symmetric: $\mathbf{S}$ considers all tuples $(x_1,y_1,x_2,y_2,x_{15},y_{15})$ such that $(x_j,y_j) \in P_j$ for $j \in \{1,2,15\}$, recovers the two keys, calls $\mathbf{S}.\textsc{Check}$ and enqueues $(y_0,k_1,k_2,0,12)$ into *ChainQueue* when $\textsc{Check}$ returns true. The chain represented by this 5-tuple will be adapted at P_{12}, which is different from the case $(i,\delta)=(2,+)$.

The other two cases $\mathrm{P}(6,-,y_6)$ and $\mathrm{P}(10,+,x_{10})$ are similar by symmetry: in each case, $\mathbf{S}$ considers all tuples $(x_7,y_7,x_8,y_8,x_9,y_9)$ such that $(x_j,y_j) \in P_j$ for $j \in \{7,8,9\}$, computes $k_1 := y_8 \oplus x_9$ and $k_2 := y_7 \oplus x_8$, checks whether $x_7 \oplus k_1 = y_6 \wedge y_9 \oplus k_2 \in P_{10}^+$ (in case $\mathrm{P}(6,-,y_6)$) or $x_7 \oplus k_1 \in P_6^- \wedge y_9 \oplus k_2 = x_{10}$ (in case $\mathrm{P}(10,+,x_{10})$), and enqueues $(y_7,k_1,k_2,7,l)$ into *ChainQueue* when this is the case, where $l = 4$ in case $\mathrm{P}(6,-,y_6)$ and $l = 12$ in case $\mathrm{P}(10,+,x_{10})$.

After enqueuing, $\mathbf{S}$ starts an execution of $\textsc{RecursiveCompletion}$, during which it continues taking the tuples out of *ChainQueue* and completing the associated partial chains till *ChainQueue* is empty again. More clearly, for each chain C dequeued from the queue, $\mathbf{S}$ evaluates in the IDEM_{15} computation path both forward and backward and queries $\mathbf{E}$ once to "wrap" around, until obtaining x_l (when moving forward) or y_l (when moving backward). $\mathbf{S}$ then calls $\textsc{ForceVal}(x_l,y_l,\perp,l)$ to add (x_l,y_l) to P_l as a newly defined pair of IO, so that the entire computation path is consistent with the answers of $\mathbf{E}$. Inside this call to $\textsc{ForceVal}$, if $x_l \in P_l^+$ or $y_l \in P_l^-$ before they are to be added, $\mathbf{S}$ aborts (also as mentioned).

During the completion of a chain, $\mathbf{S}$ adds new entries to P_i which are necessary for its evaluation. Such new values also trigger new chains to be enqueued when they satisfy the chain detection conditions mentioned before. For this, note that $\mathbf{S}$ *continues* dequeuing and completing chains till *ChainQueue* is empty again. To avoid re-completing the same chain, $\mathbf{S}$ maintains a set *CompSet* to keep a record of what it has completed, and a chain C dequeued from the queue will be completed only if $C \notin CompSet$. After all the works above are finished, $\mathbf{S}$ answers the original query with $P_i^\delta(z)$.

Note that throughout the process, the entries in $\mathbf{S}.\{P\}$ are never overwritten; once $\mathbf{S}$ finds itself unable to maintain consistency any more, $\mathbf{S}$ just aborts.

The pseudocode of $\mathbf{S}^{\mathbf{E},\mathbf{P}}$ and a modified simulator $\widetilde{S}^{\widetilde{E}^{\mathbf{E}},\mathbf{P}}$ (please see Sect. 3.2) is presented as follows. When a line has a boxed variant next to it, $\mathbf{S}^{\mathbf{E},\mathbf{P}}$ uses the original code, whereas $\widetilde{S}^{\widetilde{E}^{\mathbf{E}},\mathbf{P}}$ uses the boxed one.

3.2   The Key Points of the Proof

As in previous works, for any fixed, deterministic,[4] and computationally unbounded distinguisher $\mathbf{D}$, we first show that the complexity of $\mathbf{S}^{\mathbf{E},\mathbf{P}}$ is polynomial except with negligible probability, then show that the simulated system $\Sigma_1(\mathbf{E}, \mathbf{S}^{\mathbf{E},\mathbf{P}})$ and the real system $\Sigma_3(\mathrm{IDEM}_{15}^{\mathbf{P}}, \mathbf{P})$ are indistinguishable.

Intermediate System. The proof uses an intermediate system $\Sigma_2(\widetilde{E}^{\mathbf{E}}, \widetilde{S}^{\widetilde{E}^{\mathbf{E}},\mathbf{P}})$ which consists of two modified constructions $\widetilde{E}^{\mathbf{E}}$ and $\widetilde{S}^{\widetilde{E}^{\mathbf{E}},\mathbf{P}}$. $\widetilde{E}^{\mathbf{E}}$ can be seen as an ideal cipher $\mathbf{E}$ enhanced with a history maintaining mechanism and a CHECK procedure. More clearly, $\widetilde{E}^{\mathbf{E}}$ offers the same interface as $\mathbf{E}$, relays the answers of $\mathbf{E}$ except that it uses a set ES to keep the history of these queries. The entries in ES are of the form $(x, y, (k_1, k_2))$. $\widetilde{E}^{\mathbf{E}}$ provides an additional interface CHECK to $\widetilde{S}$; upon a call to CHECK(x, y, K), $\widetilde{E}^{\mathbf{E}}$ checks whether $(x, y, K) \in ES$ and answers accordingly. On the other hand, the modified simulator $\widetilde{S}^{\widetilde{E}^{\mathbf{E}},\mathbf{P}}$ shares the same main strategy with $\mathbf{S}^{\mathbf{E},\mathbf{P}}$ except that it queries $\widetilde{E}^{\mathbf{E}}$ – particularly, it calls $\widetilde{E}^{\mathbf{E}}$.CHECK whenever necessary. The code of $\widetilde{S}$ is presented along with $\mathbf{S}$ in Sect. 3.1. The three systems are depicted in Fig. 2.

Since all the entries of ES actually come from (an ideal cipher) $\mathbf{E}$, ES always defines a partial cipher, and we use a notation system similar to that for $\{P\}$. More clearly, we denote by ES^+ the set of tuples (K, x) s.t. $\exists y : (x, y, K) \in ES$, and denote by $ES^+(K, x)$ the corresponding y. Similarly for ES^- and $ES^-(K, y)$. Finally, denote by $|\widetilde{E}.ES|$ the size of $\widetilde{E}.ES$.

Complexity of $\widetilde{S}$ in Σ_2, and Transition from Σ_1 to Σ_2. The starting point is the same as [CHK+14]: the number of "external" chains $(y_0, k_1, k_2, 0)$ completed by $\widetilde{S}$ is bounded by the number of queries of $\mathbf{D}$ to $\widetilde{E}$, which is at most q; by this, for $i \in \{6, 7, 8, 9, 10\}$, P_i consists of entries due to $\mathbf{D}$'s queries and $\widetilde{S}$ completing chains $(y_0, k_1, k_2, 0)$, so that $|P_i| \leq 2q$.

Then the argument deviates from [CHK+14]: by construction, $\widetilde{S}$ enqueues a "middle" chain $(y_7, k_1, k_2, 7, l)$ only if there are 5 entries $(x_i, y_i) \in P_i$ for $i = 6, 7, 8, 9, 10$ s.t. $y_6 = x_7 \oplus y_8 \oplus x_9$ and $y_7 \oplus x_8 \oplus y_9 = x_{10}$. Consider a bad event BadLockMid, which happens in $\mathbf{D}^{\Sigma_2}$ if any call to INNERP creates a new pair of

---

[4] This is *wlog* since the advantage of any probabilistic distinguisher cannot exceed the advantage of the corresponding optimal deterministic version.

1: **Simulator $\mathbf{S}^{\mathbf{E},\mathbf{P}}$:** $\boxed{\textbf{Simulator } \widetilde{S}^{\widetilde{E}^{\mathbf{E}},\mathbf{P}}:}$

2: **Variables**

3:   Sets $\{P\} = \{P_1, \ldots, P_{15}\}$ and *CompSet*, initially empty

4:   Queue *ChainQueue*, initially empty

5: **public procedure** $\mathrm{P}(i, \delta, z)$

6:   $z' := \mathrm{INNERP}(i, \delta, z)$ // Chains are enqueued in this step

7:   RECURSIVECOMPLETION()

8:   **return** z'

9: // The recursive completion process is extracted as an individual procedure.

10: **private procedure** RECURSIVECOMPLETION()

11:   **while** *ChainQueue* $\neq \emptyset$ **do**

12:     $(y_j, k_1, k_2, j, l) := ChainQueue.\mathrm{DEQUEUE}()$

13:     **if** $(y_j, k_1, k_2, j) \notin CompSet$ **then**

14:       COMPLETE(y_j, k_1, k_2, j, l)

15: // The "inner" permutation interface used by **S** itself.

16: **private procedure** INNERP(i, δ, z)

17:   **if** $z \notin P_i^\delta$ **then**

18:     $z' := \mathbf{P}.\mathrm{P}(i, \delta, z)$

19:     FORCEVAL(z, z', δ, i)

20:     ENQUEUECHAINS(i, δ, z)

21:   **return** $P_i^\delta(z)$

22: // Procedure that enqueues chains.

23: **private procedure** ENQUEUECHAINS(i, δ, z)

24:   **if** $(i, \delta) = (2, +)$ **then**

25:     **forall** $((x_1, y_1), x_2, y_{14}, (x_{15}, y_{15})) \in P_1 \times \{z\} \times P_{14}^- \times P_{15}$ **do**

26:       $k_2 := y_1 \oplus x_2$

27:       $k_1 := y_{14} \oplus x_{15}$

28:       $y_0 := x_1 \oplus k_1$

29:       $x_{16} := y_{15} \oplus k_2$

30:       $flag := \mathrm{CHECK}(y_0, x_{16}, (k_1, k_2))$ $\boxed{flag := \widetilde{E}^{\mathbf{E}}.\mathrm{CHECK}(y_0, x_{16}, (k_1, k_2))}$

31:       **if** $flag = true$ **then**

32:         $ChainQueue.\mathrm{ENQUEUE}(y_0, k_1, k_2, 0, 4)$

33:   **else if** $(i, \delta) = (14, -)$ **then**

34:     **forall** $((x_1, y_1), x_2, y_{14}, (x_{15}, y_{15})) \in P_1 \times P_2^+ \times \{z\} \times P_{15}$ **do**

35:       $k_2 := y_1 \oplus x_2$

36:       $k_1 := y_{14} \oplus x_{15}$

37:       $y_0 := x_1 \oplus k_1$

38:       $x_{16} := y_{15} \oplus k_2$

39:       $flag := \mathrm{CHECK}(y_0, x_{16}, (k_1, k_2))$ $\boxed{flag := \widetilde{E}^{\mathbf{E}}.\mathrm{CHECK}(y_0, x_{16}, (k_1, k_2))}$

40:       **if** $flag = true$ **then**

41:         $ChainQueue.\mathrm{ENQUEUE}(y_0, k_1, k_2, 0, 12)$

42:   **else if** $(i, \delta) = (6, -) \vee (i, \delta) = (10, +)$ **then**

43:     **forall** $((x_7, y_7), (x_8, y_8), (x_9, y_9)) \in P_7 \times P_8 \times P_9$ **do**

44:       $k_1 := y_8 \oplus x_9$

45:       $k_2 := y_7 \oplus x_8$

46:       **if** $(i, \delta) = (6, -) \wedge z = x_7 \oplus k_1 \wedge y_9 \oplus k_2 \in P_{10}^+$ **then**

47:         $ChainQueue.\mathrm{ENQUEUE}(y_7, k_1, k_2, 7, 4)$

48:    **else if** $z = y_9 \oplus k_2 \wedge x_7 \oplus k_1 \in P_6^-$ **then** $// (i, \delta) = (10, +)$
49:       $ChainQueue.\text{ENQUEUE}(y_7, k_1, k_2, 7, 12)$
50: **private procedure** $\text{COMPLETE}(y_j, k_1, k_2, j, l)$
51:    $(y_{l-1}, k_1, k_2, l-1) := \text{EVALFWD}(y_j, k_1, k_2, j, l-1)$
52:    $(y_l, k_1, k_2, l) := \text{EVALBWD}(y_j, k_1, k_2, j, l)$
53:    $\text{FORCEVAL}(y_{l-1} \oplus k_2, y_l, \bot, l)$ // Always k_2, since $l \in \{4, 12\}$.
54:    $(y_0, k_1, k_2, 0) := \text{EVALFWD}(y_j, k_1, k_2, j, 0)$
55:    $(y_7, k_1, k_2, 7) := \text{EVALFWD}(y_0, k_1, k_2, 0, 7)$
56:    $CompSet := CompSet \cup \{(y_0, k_1, k_2, 0), (y_7, k_1, k_2, 7)\}$
57: // Procedure that adds entries to $\{P\}$.
58: **private procedure** $\text{FORCEVAL}(z, z', \delta, l)$
59:    // When $\delta = \bot$ then it's an adaptation
60:    **if** $z \in P_l^\delta \vee z' \in P_l^{\bar{\delta}}$ **then abort**
61:    **else if** $\delta = -$ **then** $P_l := P_l \cup \{(z', z)\}$
62:    **else** $P_l := P_l \cup \{(z, z')\}$ // $\delta = +$ or $\bot$
63: **private procedure** $\text{CHECK}(x, y, K)$ // $\widetilde{S}$ does not own such a procedure
64:    **return** $\mathbf{E}.E(+, K, x) = y$
65: // Two procedures that help evaluate forward and backward respectively in IDEM.
66: **private procedure** $\text{EVALFWD}(y_j, k_1, k_2, j, l)$
67:    **while** $j \neq l$ **do**
68:    **if** $j = 15$ **then**
69:       $x_{16} := y_{15} \oplus k_2$
70:       $y_0 := \mathbf{E}.E(-, (k_1, k_2), x_{16})$ $\boxed{y_0 := \widetilde{E}^{\mathbf{E}}.E(-, (k_1, k_2), x_{16})}$
71:       $j := 0$
72:    **else**
73:       **if** j is odd **then**
74:          $y_{j+1} := \text{INNERP}(j+1, +, y_j \oplus k_2)$
75:       **else**
76:          $y_{j+1} := \text{INNERP}(j+1, +, y_j \oplus k_1)$
77:       $j := j + 1$
78:    **return** (y_l, k_1, k_2, l)
79: **private procedure** $\text{EVALBWD}(y_j, k_1, k_2, j, l)$
80:    **while** $j \neq l$ **do**
81:    **if** $j = 0$ **then**
82:       $x_{16} := \mathbf{E}.E(+, (k_1, k_2), y_0)$ $\boxed{x_{16} := \widetilde{E}^{\mathbf{E}}.E(+, (k_1, k_2), y_0)}$
83:       $y_{15} := x_{16} \oplus k_2$
84:       $j := 15$
85:    **else**
86:       **if** j is odd **then**
87:          $y_{j-1} := \text{INNERP}(j, -, y_j) \oplus k_1$
88:       **else**
89:          $y_{j-1} := \text{INNERP}(j, -, y_j) \oplus k_2$
90:       $j := j - 1$
91:    **return** (y_l, k_1, k_2, l)

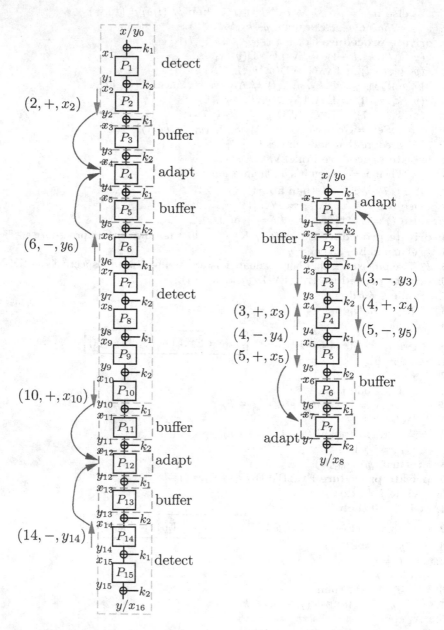

Fig. 1. (left) $IDEM_{15}$ with the zones where the simulator detects chains and adapts them; (right) $IDEM_7$ and how the simulator for sequential indifferentiability works.

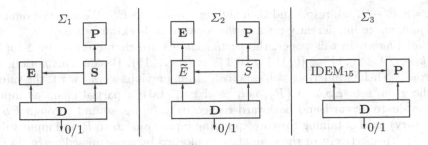

Fig. 2. Systems used in the indifferentiability proof for IDEM$_{15}$.

3-tuples $((x_7, y_7), (x_8, y_8), (x_9, y_9)) \in P_7 \times P_8 \times P_9$ and $((x_7', y_7'), (x_8', y_8'), (x_9', y_9')) \in P_7 \times P_8 \times P_9$ such that $x_7 \oplus y_8 \oplus x_9 = x_7' \oplus y_8' \oplus x_9'$ and $y_7 \oplus x_8 \oplus y_9 = y_7' \oplus x_8' \oplus y_9'$. Taking all possibilities into consideration, its overall probability is at most $\frac{2^7 \cdot q^6}{2^n} + \frac{2^6 \cdot q^4}{2^n} + \frac{2^5 \cdot q^4}{2^n} \leq \frac{2^8 \cdot q^6}{2^n}$; and conditioned on $\neg$BadLockMid, each pair $(y_6, x_{10}) \in P_6^- \times P_{10}^+$ corresponds to at most one tuple $((x_7, y_7), (x_8, y_8), (x_9, y_9)) \in P_7 \times P_8 \times P_9$ s.t. $y_6 = x_7 \oplus y_8 \oplus x_9$ and $y_7 \oplus x_8 \oplus y_9 = x_{10}$, hence $\widetilde{S}$ enqueues at most $|P_6| \cdot |P_{10}| \leq 4q^2$ "middle" chains $(y_7, k_1, k_2, 7, l)$. By this, each $|P_i|$ is bounded to $O(q^2)$, $|\widetilde{E}.ES|$ to $5q^2$, and $\widetilde{S}$ issues at most $(5q^2)^4$ queries to $\widetilde{E}$.CHECK.

The rest part of the first transition is very close to [CHK+14] (and is almost the same as [GL15a]): for two executions $\mathbf{D}^{\Sigma_1}$ and $\mathbf{D}^{\Sigma_2}$ which share the same random primitives $(\mathbf{E}, \mathbf{P})$, conditioned on $\neg$BadLockMid, if the first $(5q^2)^4$ calls to $\mathbf{S}$.CHECK in $\mathbf{D}^{\Sigma_1}$ obtain the same answers as the first $(5q^2)^4$ calls to $\widetilde{E}$.CHECK in $\mathbf{D}^{\Sigma_2}$ ($Pr \geq 1 - 1250q^8/2^n$), then $\mathbf{D}$ outputs the same in $\mathbf{D}^{\Sigma_1}$ and $\mathbf{D}^{\Sigma_2}$.

Lemma 1. *For any distinguisher* $\mathbf{D}$ *which issues at most q queries, it holds:*

(i) $|Pr[\mathbf{D}^{\Sigma_1(\mathbf{E}, \mathbf{S})} = 1] - Pr[\mathbf{D}^{\Sigma_2(\widetilde{E}, \widetilde{S})} = 1]| \leq \frac{2^{11} \cdot q^8}{2^n}$;

(ii) *during the execution* $\mathbf{D}^{\Sigma_1(\mathbf{E}, \mathbf{S})}$, *with probability at least* $1 - \frac{2^{11} \cdot q^8}{2^n}$, $\mathbf{S}$ *issues at most* $2^{10} \cdot q^8$ *queries to* $\mathbf{E}$, *and runs in time at most* $O(q^8)$.

The most difficult part of the proof is the transition from Σ_2 to Σ_3, which will be presented in the next paragraph.

Transition from Σ_2 to Σ_3: Non-abortion Argument for $\widetilde{S}$. $\widetilde{S}$ is built with the expectation that if it does not abort, then the outputs of Σ_2 and Σ_3 are indistinguishable; we will see that this intuition is true, so that the first (and actually core) step of the transition is to show that the abort probability of $\widetilde{S}$ is negligible. For this, we introduce several notions and (improbable) bad events, then show that if neither of them happens, then $\widetilde{S}$ does not abort.

Random Assignments. Similarly to [LS13], we use the notion *random forward assignment in set P_i* (*random backward assignment in set P_i*, resp.) to refer to line 62 inside any execution of FORCEVAL$(z, z', +, i)$ (line 61 inside any execution of FORCEVAL$(z, z', -, i)$, resp.), and use the notion *random forward (backward, resp.) assignment in set ES* to refer to any operation sequence $z' := \mathbf{E}.\mathbf{E}(+, K, z)$

($z' := \mathbf{E}.\mathrm{E}(-, K, z)$, resp.) and then adding z and z' to ES. We also use *random assignments* to indifferently refer to the forward or backward case.

Partial Chains. In this paper, the *partial chains* are characterized by 4-tuples $(y_i, k_1, k_2, i) \in \{0,1\}^n \times \{0,1\}^n \times \{0,1\}^n \times \{0, \ldots, 15\}$. Besides this notion, we borrow two helper functions val_l^+ and val_l^- from previous works: w.r.t. the values in the given sets ES and $\{P\}$, val_l^+ and val_l^- take a partial chain as input, and evaluate forward and backward respectively (wrap around through ES if necessary) until obtaining the corresponding input value x_l to P_l and input value y_l to P_l^{-1} respectively, or the evaluation is blocked by some missed entry (in the sets), and return x_l and y_l respectively in the former case while $\bot$ in the latter case. Based on val_l^+ and val_l^-, we borrow (and redefine) the notion of *equivalent* partial chains: w.r.t. $\{P\}$ and $\widetilde{E}.ES$, two partial chains $C = (y_i, k_1, k_2, i)$ and $D = (y_j, k_1, k_2, j)$ (with the same keys) are *equivalent* (denoted $C \equiv D$) if $y_i = val_i^-(D)$ or $y_j = val_j^-(C)$.[5]

Bad Events, and Non-abortion. A random answer from $\mathbf{P}$ or $\mathbf{E}$ is bad if it collides with some value relevant to the "active" chains. To specify such "active" chains, we define a notion *history for partial chains* $\mathcal{CH}$ from the sets ES and $\{P_7, P_8, P_9\}$ at the moment where the random answer is drawn: $\mathcal{CH}$ is the union of two sets $\mathcal{ECH}$ and $\mathcal{MCH}$, where $\mathcal{ECH}$ includes all the tuples $(y_0, k_1, k_2, 0)$ with $((k_1, k_2), y_0) \in ES^+$, and $\mathcal{MCH}$ includes all $(y_7, k_1, k_2, 7)$ with $y_7 \in P_7^-$, $x_8 = y_7 \oplus k_2 \in P_8^+$, and $x_9 = P_8^+(x_8) \oplus k_1 \in P_9^+$. By the complexity analysis, conditioned on $\neg\mathsf{BadLockMid}$, $|\mathcal{CH}| \le 5q^2 + (2q)^3 \le 13q^3$.

We then list the bad events (more precisely, their ideas). Due to space constraints, we have to defer their formal definitions to the full version [GL15b].

- $\mathsf{BadHitAdapt}$: an answer from $\mathbf{P}$ collides with a previous adapted value;
- BadE: an answer from $\mathbf{E}$ collides with a value in P_1 or P_{15} xored the key, i.e. $\mathbf{E}.\mathrm{E}(-, (k_1, k_2), x_{16}) \oplus k_1 \in P_1^+$, or $\mathbf{E}.\mathrm{E}(+, (k_1, k_2), y_0) \oplus k_2 \in P_{15}^-$;
- BadP: extension of some chain C meets an old P-tuple after a random assignment in $\{P\}$ with the same direction, i.e. $\exists C \in \mathcal{CH}$ s.t. for more than one value i, $val_i^+(C)$ ($val_i^-(C)$, resp.) differs after a random forward (backward, resp.) assignment in $\{P\}$ from before the assignment;
- $\mathsf{BadInvP}$: some chain C extends after a random assignment in $\{P\}$ with the opposite direction, i.e. $\exists C \in \mathcal{CH}$ s.t. for some value i, $val_i^+(C)$ ($val_i^-(C)$, resp.) differs after a random backward (forward, resp.) assignment in $\{P\}$ from before the assignment;
- $\mathsf{BadMidP}$: a random assignment in P_7, P_8, or P_9 creates a new 5-tuple $(y_6, (x_7, y_7), (x_8, y_8), (x_9, y_9), x_{10}) \in P_6^- \times P_7 \times P_8 \times P_9 \times P_{10}^+$ such that $y_6 \oplus x_7 = y_8 \oplus x_9$ and $y_7 \oplus x_8 = y_9 \oplus x_{10}$;
- $\mathsf{BadlyCollide}$ (a term from [CHK+14]): two chains $C, D \in \mathcal{CH}$ that are not equivalent suddenly satisfies $val_i^\delta(C) = val_i^\delta(D)$ after a random assignment.

The overall probability (the event $\mathsf{BadLockMid}$ included) cumulates to $\frac{2^{13.4} \cdot q^6}{2^n}$.

---

[5] Note that if $C = D$ then both $y_i = val_i^-(D)$ and $y_j = val_j^-(C)$.

We call a pair of primitive $(\mathbf{E}, \mathbf{P})$ a *good Σ_2-tuple* if none of the bad events above (BadLockMid included) happens during the execution $D^{\Sigma_2}(\widetilde{E}^{\mathbf{E}}, \widetilde{S}^{\widetilde{E}^{\mathbf{E}}, \mathbf{P}})$, and call D^{Σ_2} with good Σ_2-tuples *good Σ_2 executions*. During good Σ_2 executions, $\widetilde{S}$ never aborts due to calls to FORCEVAL$(x_i, y_i, +, i)$/FORCEVAL$(x_i, y_i, -, i)$, as otherwise BadHitAdapt happens. We then proceed to argue that $\widetilde{S}$ never aborts due to calls to FORCEVAL$(x_l, y_l, \perp, l)$ (i.e. adaptations: $l \in \{4, 12\}$), to complete the non-abortion argument.

Lemma 2. *In a good execution $\overline{D}^{\Sigma_2}$, before any call to FORCEVAL$(x_l, y_l, \perp, l)$ ($l \in \{4, 12\}$), $x_l \notin P_l^+ \wedge y_l \notin P_l^-$ must hold.*

Proof. The proof flow is very similar to [CHK+14], while some ideas slightly deviate. Let's sketch the flow: *wlog* consider a call FORCEVAL$(x_4, y_4, \perp, 4)$, and suppose that it is made during an execution of COMPLETE$(C, 4)$. We argue that $val_4^+(C) \notin P_4^+$ right before the call to FORCEVAL$(x_4, y_4, \perp, 4)$, and the argument for $val_4^-(C) \notin P_4^-$ is similar by symmetry. The ideas are as follows:

First, before C is enqueued, $val_3^+(C) = \perp$ (this implies $val_4^+(C) = \perp \notin P_4^+$): if $C = (y_0, k_1, k_2, 0)$ is enqueued by INNERP$(2, +, x_2)$, then $val_3^+(C) = \perp$ is clear; if $C = (y_7, k_1, k_2, 7)$ is enqueued by INNERP$(6, -, y_6)$, then if $val_3^+(C) \neq \perp$, a chain $(y_0, k_1, k_2, 0)$ equivalent to C must have been previously enqueued and completed due to the call to INNERP$(2, +, val_2^+(C))$, and C would have been in *CompletedSet* when C is dequeued, as a consequence the purported call to FORCEVAL$(x_4, y_4, \perp, 4)$ would not happen.

Second, if $val_4^+(C) \in P_4^+$ when C is dequeued, it can only be that for another chain D enqueued before C is enqueued and dequeued after C is enqueued, it holds $val_4^+(D) = val_4^+(C) \neq \perp$ so that $val_4^+(C)$ was added to P_4^+ during D's completion.

Then, we argue that $val_4^+(D) = val_4^+(C) \neq \perp$ is not possible for any such chain D, so that when C is dequeued, either $val_4^+(C) = \perp$, or $val_4^+(C) \neq \perp \wedge val_4^+(C) \notin P_4^+$. To argue about this, we exclude the possibility for each of the following cases:

(i) if $val_2^+(C) \neq val_2^+(D)$ at some point, then $val_3^+(C) \neq val_3^+(D)$. Otherwise, consider the last assignment before $val_3^+(C) = val_3^+(D) \neq \perp$ holds. This assignment happens earliest right before C is enqueued, at which point both C and D have been in $\mathcal{CH}$ (by construction and definition). Then:
  - it cannot have been in ES, otherwise BadE happens;
  - it cannot have been a random backward assignment in $\{P\}$, otherwise BadInvP happens;
  - it cannot have been a random forward assignment in $\{P\}$, otherwise BadlyCollide happens;
  - it cannot have been due to a previous adaptation, since by construction, when C is enqueued, all the chains in *ChainQueue* are to be adapted at P_4 which is the same as C, so that it cannot be that $val_3^+(C) = \perp$ or $val_3^+(D) = \perp$ due to a missed entry in P_{12} which is later added by an adaptation in this period.

Then a similar discussion further establishes $val_4^+(C) \neq val_4^+(D)$;

(ii) if $val_2^+(C) = val_2^+(D) \neq \perp$ while $val_3^+(C) \neq val_3^+(D)$ at some point, then similarly to Case (i), $val_4^+(C) \neq val_4^+(D)$ will hold;

(iii) if $val_2^+(C) = val_2^+(D) \neq \perp$ and $val_3^+(C) = val_3^+(D) \neq \perp$, then $val_4^+(C) \neq val_4^+(D)$ otherwise $C \equiv D$ and $C \in CompletedSet$ when C is dequeued.

Finally, after C is dequeued, if $val_4^+(C) = \perp$, then since $\neg$BadP, it can only be changed non-empty by a random forward assignment in P_3 which occurs during the completion of C, after which $val_4^+(C) \notin P_4^+$. These complete the proof. $\quad\square$

The Rest Part. During $\mathbf{D}^{\Sigma_2}$, if $\widetilde{S}$ does not abort, then the answers are consistent with some Σ_3 executions. By a randomness mapping argument [CHK+14], the advantage of $\mathbf{D}$ in distinguishing Σ_2 and Σ_3 is bounded.

Lemma 3. *For any distinguisher $\mathbf{D}$ which issues at most q queries, it holds:*

$$\left| Pr[\mathbf{D}^{\Sigma_3(\mathrm{IDEM}_{15}, \mathbf{P})} = 1] - Pr[\mathbf{D}^{\Sigma_2(\widetilde{E}, \widetilde{S})} = 1] \right| \leq \frac{2^{14} \cdot q^6}{2^n}.$$

4   Seq-indifferentiability for 7-Round IDEM

According to [LS13, ABD+13], there is a seq-distinguisher for SEM_3. Consider IDEM_6. If we fix the key k_2 to an arbitrary value, then the construction is reduced to a 3-round single-key Even-Mansour. By this, a seq-distinguisher for IDEM_6 is easily obtained.

It is natural to ask whether the additional n-bit key offers more freedom to the adversary and enable to attack more than this trivial 2×3 rounds. The second main result – also the main theorem of this section – provides a negative answer, and is as follows:

Theorem 2. *The 7-round Even-Mansour cipher IDEM_7 from seven independent random permutations $\mathbf{P} = (\mathbf{P}_1, \ldots, \mathbf{P}_7)$ and two n-bit keys (k_1, k_2) alternatively xored is strongly and statistically $(q, \sigma, t, \varepsilon)$-seq-indifferentiable from $\mathbf{E}$, where $\sigma = q^3$, $t = O(q^3)$, and $\varepsilon \leq \frac{27q^6}{2^n} = O(\frac{q^6}{2^n})$.*

Proof. The proof is much simpler than that of Theorem 1, since there is no recursive chain completion. In the following, we first present the simulator, then sketch the proof. The full proof is deferred to the full version [GL15b].

Simulator for IDEM 7. To make a distinction from the notations used in Sect. 3, we denote by $\mathcal{S}^{\mathbf{E}, \mathbf{P}}$ the simulator for IDEM_7 with access to $\mathbf{E}$ and $\mathbf{P}$. Similarly to $\mathbf{S}^{\mathbf{E}, \mathbf{P}}$, $\mathcal{S}^{\mathbf{E}, \mathbf{P}}$ also offers an interface $P(i, \delta, z)$ where $(i, \delta, z) \in \{1, \ldots, 7\} \times \{+, -\} \times \{0, 1\}^n$ and maintains a set P_i for each i to keep the already defined pairs of IO. The other notations P_i^+, P_i^-, and $|P_i|$ are all similar to those introduced in the context of IDEM_{15}. $\mathcal{S}^{\mathbf{E}, \mathbf{P}}$ uses an additional set ES to maintain the history of its queries to $\mathbf{E}$, which is similar to the set of $\widetilde{E}^{\mathbf{E}}$ introduced in Sect. 3. We also use the notations ES^+, ES^-, and $|ES|$ similar to Sect. 3.

Upon a query to $\mathcal{S}^{\mathbf{E},\mathbf{P}}.\mathrm{P}(i,\delta,z)$, $\mathcal{S}^{\mathbf{E},\mathbf{P}}$ calls an inner procedure $\mathcal{S}^{\mathbf{E},\mathbf{P}}.\mathrm{P}^{in}$, and $\mathcal{S}^{\mathbf{E},\mathbf{P}}.\mathrm{P}^{in}$ answers with $P_i^\delta(z)$ if $x \in P_i^\delta$, or queries $\mathbf{P}.\mathrm{P}(i,\delta,z)$ to obtain the answer z' and adds z and z' to P_i if $z' \notin P_i^{\bar{\delta}}$ while aborts otherwise.

The chain completing mechanism of $\mathcal{S}^{\mathbf{E},\mathbf{P}}$ is much simpler than that of $\mathbf{S}^{\mathbf{E},\mathbf{P}}$, and is somehow close to that appeared in [CS15]: $\mathcal{S}^{\mathbf{E},\mathbf{P}}$ completes the potential partial chains upon receiving a new query $\mathcal{S}^{\mathbf{E},\mathbf{P}}.\mathrm{P}(i,\delta,x)$ with $i \in \{3,4,5\}$. More clearly, when the query is of the form $\mathcal{S}^{\mathbf{E},\mathbf{P}}.\mathrm{P}(3,+,x)$, $\mathcal{S}^{\mathbf{E},\mathbf{P}}.\mathrm{P}(4,-,y)$, or $\mathcal{S}^{\mathbf{E},\mathbf{P}}.\mathrm{P}(5,+,x)$, $\mathcal{S}^{\mathbf{E},\mathbf{P}}$ considers all newly created tuples $(x_3, x_4, x_5) \in P_3^+ \times P_4^+ \times P_5^+$, and computes $k_1 := P_4^+(x_4) \oplus x_5$, $k_2 := P_3^+(x_3) \oplus x_4$. $\mathcal{S}^{\mathbf{E},\mathbf{P}}$ then evaluates in IDEM_7 both backward and forward until obtaining the corresponding y_7 and x_7, that is, computing the following values by calling $\mathcal{S}^{\mathbf{E},\mathbf{P}}.\mathrm{P}^{in}$ and querying $\mathbf{E}$, in the order: (1) $y_2 := x_3 \oplus k_1$; (2) $y_1 := \mathcal{S}^{\mathbf{E},\mathbf{P}}.\mathrm{P}^{in}(2,-,y_2) \oplus k_2$; (3) $y_0 := \mathcal{S}^{\mathbf{E},\mathbf{P}}.\mathrm{P}^{in}(1,-,y_1) \oplus k_1$; (4) $y_7 := \mathbf{E}.\mathrm{E}(+,(k_1,k_2),y_0) \oplus k_2$; (5) $x_6 := P_5^+(x_5) \oplus k_2$; (6) $x_7 := \mathcal{S}^{\mathbf{E},\mathbf{P}}.\mathrm{P}^{in}(6,+,x_6) \oplus k_1$. $\mathcal{S}^{\mathbf{E},\mathbf{P}}$ finally aborts if $x_7 \in P_7^+$ or $y_7 \in P_7^-$, otherwise adds (x_7, y_7) to P_7 as a newly defined pair of IO.

When the query is $\mathcal{S}^{\mathbf{E},\mathbf{P}}.\mathrm{P}(3,-,y)$, $\mathcal{S}^{\mathbf{E},\mathbf{P}}.\mathrm{P}(4,+,x)$, or $\mathcal{S}^{\mathbf{E},\mathbf{P}}.\mathrm{P}(5,-,y)$, $\mathcal{S}^{\mathbf{E},\mathbf{P}}$ considers all newly created tuples $(x_3, x_4, x_5) \in P_3^+ \times P_4^+ \times P_5^+$, computes k_1 and k_2, evaluates in IDEM_7 both forward and backward until obtaining the corresponding x_1 and y_1, and finally adds (x_1, y_1) to P_1 or aborts if $x_1 \in P_1^+$ or $y_1 \in P_1^-$. The strategy is illustrated in Fig. 1 (right).

To simplify the reasoning, we introduce a modified simulator $\mathcal{T}^{\mathbf{E},\mathbf{P}}$, which is obtained by embedding two early abort conditions into $\mathcal{S}^{\mathbf{E},\mathbf{P}}$:

(i) when a chain C is to be adapted at P_1 (P_7, resp.), right after the assignment (lines 13 or 16 in the code below) inside the call to P^{in} which led to C being detected, if the value y_2 (x_6, resp.) corresponding to C has been in P_2^- (P_6^+, resp.), then $\mathcal{T}$ aborts. This is captured by the procedure CHECKFREEBUFFER;

(ii) right after an assignment in P_3, P_4, or P_5 (lines 13/16), $\mathcal{T}$ aborts if the assignment creates a "lock" in the middle three rounds: for $(i,j) \in \{(3,4),(4,5)\}$, if $\exists (x_i, y_i), (x_i', y_i') \in P_i$ and $(x_j, y_j), (x_j', y_j') \in P_j$ such that $x_i \oplus y_j = x_i' \oplus y_j'$ and $y_i \oplus x_j = y_i' \oplus x_j'$. This is captured by the procedure CHECKLOCK. This situation is harmful for the procedure COMPCHAIN in some cases.

With all the above in mind, we have the pseudocode of $\mathcal{S}$ and $\mathcal{T}$ as follows. Note that the *underlined lines* only exist in $\mathcal{T}$ (say, $\mathcal{S}$ does not early abort).

Intermediate System Σ_2'. Denote by $\Sigma_1'(\mathbf{E}, \mathcal{S}^{\mathbf{E},\mathbf{P}})$ the simulated system, and by $\Sigma_3'(\mathrm{IDEM}_7^{\mathbf{P}}, \mathbf{P})$ the real system. As a quite standard first step, we introduce an intermediate system $\Sigma_2'(\mathrm{IDEM}_7^{\mathcal{T}^{\mathbf{E},\mathbf{P}}}, \mathcal{T}^{\mathbf{E},\mathbf{P}})$, in which the cipher IDEM_7 calls the interfaces of $\mathcal{T}$ to compute (as done in [MPS12, CS15]). The three systems involved in this proof are depicted in Fig. 3.

Complexity of $\mathcal{S}/\mathcal{T}$. By construction, for $i \in \{3,4,5\}$, $|P_i|$ can be enlarged by at most 1 only if $\mathcal{S}/\mathcal{T}$ receives a query $\mathrm{P}(i,\delta,\cdot)$. Hence for any seq-distinguisher D of total oracle query cost at most q, $\mathcal{S}/\mathcal{T}$ completes at most $|P_3| \cdot |P_4| \cdot |P_5| \leq q^3$ chains, and queries $\mathbf{E}$ at most q^3 times (say, $|ES| \leq q^3$).

1: **Simulator** $\mathcal{S}^{\mathbf{E,P}}$: | **Simulator** $\mathcal{T}^{\mathbf{E,P}}$:
2: **Variables:** Sets $\{P\} = \{P_1, \ldots, P_7\}$ and ES, initially empty
3: **public procedure** $\mathrm{P}(i, \delta, z)$
4:   **return** $\mathrm{P}^{in}(i, \delta, z)$
5: **private procedure** $\mathrm{P}^{in}(i, \delta, z)$ 20:   **else if** $i = 4 \wedge \delta = +$ **then**
6:   **if** $z \notin P_i^\delta$ **then**   21:   **forall** $(x_3, x_5) \in P_3^+ \times P_5^+$ **do**
7:     $z' := \mathbf{P}.\mathrm{P}(i, \delta, z)$ 22: $\textsc{CompChain}(x_3, z, x_5, 4, 1)$
8:     **if** $z' \in P_i^{\overline{\delta}}$ **then** // when $i = 1, 7$ 23:   **else if** $i = 5 \wedge \delta = +$ **then**
9:       **abort**   24:   **forall** $(x_3, x_4) \in P_3^+ \times P_4^+$ **do**
10:    $\textsc{CheckFreeBuffer}(i, \delta, z')$ 25: $\textsc{CompChain}(x_3, x_4, z, 5, 7)$
11:    **if** $\delta = +$ **then**   26:   **else if** $i = 3 \wedge \delta = -$ **then**
12:      $\textsc{CheckLock}(i, z, z')$ 27:   **forall** $(x_4, x_5) \in P_4^+ \times P_5^+$ **do**
13:      $P_i := P_i \cup \{(z, z')\}$ 28: $\textsc{CompChain}(z', x_4, x_5, 3, 1)$
14:    **else** // $\delta = -$ 29:   **else if** $i = 4 \wedge \delta = -$ **then**
15:      $\textsc{CheckLock}(i, z', z)$ 30:   **forall** $(x_3, x_5) \in P_3^+ \times P_5^+$ **do**
16:      $P_i := P_i \cup \{(z', z)\}$ 31: $\textsc{CompChain}(x_3, z', x_5, 4, 7)$
17:    **if** $i = 3 \wedge \delta = +$ **then**   32:   **else if** $i = 5 \wedge \delta = -$ **then**
18:      **forall** $(x_4, x_5) \in P_4^+ \times P_5^+$ **do** 33:   **forall** $(x_3, x_4) \in P_3^+ \times P_4^+$ **do**
19:        $\textsc{CompChain}(z, x_4, x_5, 3, 7)$ 34: $\textsc{CompChain}(x_3, x_4, z', 5, 1)$
                                                            35:   **return** $P_i^\delta(z)$

36: **private procedure** $\textsc{CompChain}(x_3, x_4, x_5, i, l)$
37:   $k_1 := P_4^+(x_4) \oplus x_5$ 50: $P_1 := P_1 \cup \{(x_1, y_1)\}$
38:   $k_2 := P_3^+(x_3) \oplus x_4$ 51:   **else** // $l = 7$
39:   **if** $l = 1$ **then**   52: $y_2 := x_3 \oplus k_1$
40:     $x_6 := P_5^+(x_5) \oplus k_2$ 53: $y_1 := \mathrm{P}^{in}(2, -, y_2) \oplus k_2$
41:     $x_7 := \mathrm{P}^{in}(6, +, x_6) \oplus k_1$ 54: $y_0 := \mathrm{P}^{in}(1, -, y_1) \oplus k_1$
42:     $x_8 := \mathrm{P}^{in}(7, +, x_7) \oplus k_2$ 55: $x_8 := \mathbf{E}.\mathrm{E}(+, (k_1, k_2), y_0)$
43:     $y_0 := \mathbf{E}.\mathrm{E}(-, (k_1, k_2), x_8)$ 56: $ES := ES \cup \{(y_0, x_8, (k_1, k_2))\}$
44:     $ES := ES \cup \{(y_0, x_8, (k_1, k_2))\}$ 57: $y_7 := x_8 \oplus k_2$
45:     $x_1 := y_0 \oplus k_1$ 58: $x_6 := P_5^+(x_5) \oplus k_2$
46:     $y_2 := x_3 \oplus k_1$ 59: $x_7 := \mathrm{P}^{in}(6, +, x_6) \oplus k_1$
47:     $y_1 := \mathrm{P}^{in}(2, -, y_2) \oplus k_2$ 60:   **if** $x_7 \in P_7^+ \vee y_7 \in P_7^-$ **then**
48:     **if** $x_1 \in P_1^+ \vee y_1 \in P_1^-$ **then**   61:   **abort**
49:       **abort**   62: $P_7 := P_7 \cup \{(x_7, y_7)\}$

63: **private procedure** $\textsc{CheckFreeBuffer}(i, \delta, z')$
64:   **if** $(i, \delta) = (3, +) \wedge \exists (x_4, y_5) \in P_4^+ \times P_5^-$ s.t. $z' \oplus x_4 \oplus y_5 \in P_6^+$ **then**
65:     **abort**
66:   **else if** $(i, \delta) = (4, +) \wedge \exists (x_3, x_5) \in P_3^+ \times P_5^+$ s.t. $x_3 \oplus z' \oplus x_5 \in P_2^-$ **then**
67:     **abort**
68:   **else if** $(i, \delta) = (5, +) \wedge \exists (y_3, x_4) \in P_3^- \times P_4^+$ s.t. $y_3 \oplus x_4 \oplus z' \in P_6^+$ **then**
69:     **abort**
70:   **else if** $(i, \delta) = (3, -) \wedge \exists (y_4, x_5) \in P_4^- \times P_5^+$ s.t. $z' \oplus y_4 \oplus x_5 \in P_2^-$ **then**
71:     **abort**
72:   **else if** $(i, \delta) = (4, -) \wedge \exists (y_3, y_5) \in P_3^- \times P_5^-$ s.t. $y_3 \oplus z' \oplus y_5 \in P_6^+$ **then**
73:     **abort**
74:   **else if** $(i, \delta) = (5, -) \wedge \exists (x_3, y_4) \in P_3^+ \times P_4^-$ s.t. $x_3 \oplus y_4 \oplus z' \in P_2^-$ **then**
75:     **abort**

76: **private procedure** $\textsc{CheckLock}(i, x, y)$
77:   **if** $i = 3 \wedge \exists ((x_3, y_3), (x_4, y_4), (x_4', y_4')) \in P_3 \times P_4 \times P_4$

78:      s.t. $x \oplus y_4' = x_3 \oplus y_4 \wedge y \oplus x_4' = y_3 \oplus x_4$ **then abort**
79:   **if** $i = 5 \wedge \exists((x_5, y_5), (x_4, y_4), (x_4', y_4')) \in P_5 \times P_4 \times P_4$
80:      s.t. $x_4 \oplus y_5 = x_4' \oplus y \wedge y_4 \oplus x_5 = y_4' \oplus x$ **then abort**
81:   **if** $i = 4 \wedge \exists((x_3, y_3), (x_3', y_3'), (x_4, y_4)) \in P_3 \times P_3 \times P_4$
82:      s.t. $x_3 \oplus y_4 = x_3' \oplus y \wedge y_3 \oplus x_4 = y_3' \oplus x$ **then abort**
83:   **if** $i = 4 \wedge \exists((x_5, y_5), (x_5', y_5'), (x_4, y_4)) \in P_5 \times P_5 \times P_4$
84:      s.t. $x_4 \oplus y_5 = x \oplus y_5' \wedge y_4 \oplus x_5 = y \oplus x_5'$ **then abort**

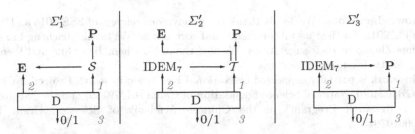

Fig. 3. Systems used in the seq-indifferentiability proof for $IDEM_7$. The number in red and *italic* illustrates the order of the queries/actions (of the sequential distinguisher) (Color figure online).

The running time of S is clearly dominated by the executions of COMPCHAIN, the number of which is $O(q^3)$. Therefore, S runs in time $O(q^3)$.

Indistinguishability of Outputs. We first upper bound the abort probability of T. Consider the two early abort conditions first:

(i) The overall probability that T aborts during CHECKFREEBUFFER is at most $\frac{2q^6}{2^n - q}$;

(ii) The overall probability that T aborts during CHECKLOCK is at most $\frac{2q^4}{2^n - q}$;

Then the two types of main abortions of T are as follows:

(i) a random answer from P_1 or P_7 collides with a previously added adapted value. The overall probability is at most $\frac{4q^6}{2^n - 2q^3}$;

(ii) T aborts due to adaptations. The overall probability is at most $\frac{5q^6}{2^n - 2q^3}$; this is obtained by carefully analyzing each case. A key point is that the buffer rounds ensure that any two chains completed during the same call to P^{in} will diverge at the adaptation round – the case is slightly similar to $IDEM_{15}$.

These cumulate to $\frac{26q^6}{2^n}$ (assuming $q^3 < \frac{2^n}{4}$). For a tuple $(\mathbf{E}, \mathbf{P})$, if T does not abort in $\Sigma_2'(IDEM_7^{T^{\mathbf{E},\mathbf{P}}}, T^{\mathbf{E},\mathbf{P}})$, then S does not abort in $\Sigma_1'(\mathbf{E}, S^{\mathbf{E},\mathbf{P}})$; and then the final bound $\frac{27q^6}{2^n} = \frac{26q^6}{2^n} + \frac{q^6}{2^n}$ is obtained by a randomness mapping argument, where the statistical distance $\frac{q^6}{2^n}$ is due to $|ES| \leq q^3$ random values from $\mathbf{E}$. $\square$

5   Conclusion

As a first step towards understanding the security of iterated Even-Mansour with key-length larger than the block-size, this work analyzes (seq-)indifferentiability of Even-Mansour with two independent round-keys alternatively xored, and proves that 7 rounds is necessary and sufficient to achieve sequential indifferentiability while 15 rounds is sufficient to achieve full indifferentiability.

Acknowledgements. We deeply thank the anonymous referees of FSE 2015 and ASIACRYPT 2015, for their useful comments and corrections. We thank Meicheng Liu and Jianghua Zhong for their suggestions. We also thank Yu Chen, Jian Guo, and Wentao Zhang, for their encouragements.

This work is partially supported by National Key Basic Research Project of China (2011CB302400), National Science Foundation of China (61379139) and the "Strategic Priority Research Program" of the Chinese Academy of Sciences, Grant No. XDA06010701.

References

[ABD+13] Andreeva, E., Bogdanov, A., Dodis, Y., Mennink, B., Steinberger, J.P.: On the indifferentiability of key-alternating ciphers. In: Canetti, R., Garay, J.A. (eds.) CRYPTO 2013, Part I. LNCS, vol. 8042, pp. 531–550. Springer, Heidelberg (2013). http://eprint.iacr.org/2013/061.pdf

[ABK98] Anderson, R., Biham, E., Knudsen, L.: Serpent: A proposal for the advanced encryption standard. NIST AES Proposal **174**, 1–23 (1998)

[BDK+10] Biryukov, A., Dunkelman, O., Keller, N., Khovratovich, D., Shamir, A.: Key recovery attacks of practical complexity on AES-256 variants with up to 10 rounds. In: Gilbert, H. (ed.) EUROCRYPT 2010. LNCS, vol. 6110, pp. 299–319. Springer, Heidelberg (2010)

[BDPVA08] Bertoni, G., Daemen, J., Peeters, M., Van Assche, G.: On the indifferentiability of the sponge construction. In: Smart, N.P. (ed.) EUROCRYPT 2008. LNCS, vol. 4965, pp. 181–197. Springer, Heidelberg (2008)

[BK09] Biryukov, A., Khovratovich, D.: Related-key cryptanalysis of the full AES-192 and AES-256. In: Matsui, M. (ed.) ASIACRYPT 2009. LNCS, vol. 5912, pp. 1–18. Springer, Heidelberg (2009)

[BKL+12] Bogdanov, A., Knudsen, L.R., Leander, G., Standaert, F.-X., Steinberger, J., Tischhauser, E.: Key-alternating ciphers in a provable setting: encryption using a small number of public permutations. In: Pointcheval, D., Johansson, T. (eds.) EUROCRYPT 2012. LNCS, vol. 7237, pp. 45–62. Springer, Heidelberg (2012)

[CDMS10] Coron, J.-S., Dodis, Y., Mandal, A., Seurin, Y.: A domain extender for the ideal cipher. In: Micciancio, D. (ed.) TCC 2010. LNCS, vol. 5978, pp. 273–289. Springer, Heidelberg (2010)

[CGH04] Canetti, R., Goldreich, O., Halevi, S.: The random oracle methodology, revisited. J. ACM **51**(4), 557–594 (2004)

[CHK+14] Coron, J.-S., Holenstein, T., Künzler, R., Patarin, J., Seurin, Y., Tessaro, S.: How to build an ideal cipher: the indifferentiability of the feistel construction. J. Cryptology, 1–54 (2014). http://link.springer.com/article/10.1007/s00145-014-9189-6

[CLL+14] Chen, S., Lampe, R., Lee, J., Seurin, Y., Steinberger, J.: Minimizing the Two-Round Even-Mansour Cipher. In: Garay, J.A., Gennaro, R. (eds.) CRYPTO 2014, Part I. LNCS, vol. 8616, pp. 39–56. Springer, Heidelberg (2014)

[CS14] Chen, S., Steinberger, J.: Tight security bounds for key-alternating ciphers. In: Nguyen, P.Q., Oswald, E. (eds.) EUROCRYPT 2014. LNCS, vol. 8441, pp. 327–350. Springer, Heidelberg (2014)

[CS15] Cogliati, B., Seurin, Y.: On the provable security of the iterated even-mansour cipher against related-key and chosen-key attacks. In: Oswald, E., Fischlin, M. (eds.) EUROCRYPT 2015. LNCS, vol. 9056, pp. 584–613. Springer, Heidelberg (2015). http://eprint.iacr.org/2015/069.pdf

[DDKS14] Dinur, I., Dunkelman, O., Keller, N., Shamir, A.: Cryptanalysis of iterated even-mansour schemes with two keys. In: Sarkar, P., Iwata, T. (eds.) ASIACRYPT 2014. LNCS, vol. 8873, pp. 439–457. Springer, Heidelberg (2014)

[DDKS15] Dinur, I., Dunkelman, O., Keller, N., Shamir, A.: Key recovery attacks on iterated even-mansour encryption schemes. J. Cryptology, 1–32 (2015). http://link.springer.com/article/10.1007/s00145-015-9207-3

[DGHM13] Demay, G., Gaži, P., Hirt, M., Maurer, U.: Resource-restricted indifferentiability. In: Johansson, T., Nguyen, P.Q. (eds.) EUROCRYPT 2013. LNCS, vol. 7881, pp. 664–683. Springer, Heidelberg (2013)

[DKS13] Dunkelman, O., Keller, N., Shamir, A.: Slidex attacks on the even-mansour encryption scheme. J. Cryptology 28, 1–28 (2013)

[DR02] Daemen, J., Rijmen, V.: The Design of Rijndael: AES-The Advanced Encryption Standard. Springer, Heidelberg (2002)

[EM93] Even, S., Mansour, Y.: A construction of a cipher from a single pseudorandom permutation. In: Imai, H., Rivest, R.L., Matsumoto, T. (eds.) ASIACRYPT 1991. LNCS, vol. 739, pp. 210–224. Springer, Heidelberg (1993)

[FP15] Farshim, P., Procter, G.: The related-key security of iterated even–mansour ciphers. In: Leander, G. (ed.) FSE 2015. LNCS, vol. 9054, pp. 342–363. Springer, Heidelberg (2015). http://eprint.iacr.org/2014/953.pdf

[GL15a] Guo, C., Lin, D.: On the indifferentiability of key-alternating feistel ciphers with no key derivation. In: Dodis, Y., Nielsen, J.B. (eds.) TCC 2015, Part I. LNCS, vol. 9014, pp. 110–133. Springer, Heidelberg (2015). http://eprint.iacr.org/

[GL15b] Guo, C., Lin, D.: A synthetic indifferentiability analysis of interleaved double-key even-mansour ciphers. Cryptology ePrint Archive, Report 2015/861 (2015). http://eprint.iacr.org/

[GPPR11] Guo, J., Peyrin, T., Poschmann, A., Robshaw, M.: The LED block cipher. In: Preneel, B., Takagi, T. (eds.) CHES 2011. LNCS, vol. 6917, pp. 326–341. Springer, Heidelberg (2011)

[KHP07] Kim, J.-S., Hong, S.H., Preneel, B.: Related-key rectangle attacks on reduced AES-192 and AES-256. In: Biryukov, A. (ed.) FSE 2007. LNCS, vol. 4593, pp. 225–241. Springer, Heidelberg (2007)

[LPS12] Lampe, R., Patarin, J., Seurin, Y.: An asymptotically tight security analysis of the iterated even-mansour cipher. In: Wang, X., Sako, K. (eds.) ASIACRYPT 2012. LNCS, vol. 7658, pp. 278–295. Springer, Heidelberg (2012)

[LS13] Lampe, R., Seurin, Y.: How to construct an ideal cipher from a small set of public permutations. In: Sako, K., Sarkar, P. (eds.) ASIACRYPT 2013, Part I. LNCS, vol. 8269, pp. 444–463. Springer, Heidelberg (2013). http://eprint.iacr.org/2013/255.pdf

[MPS12] Mandal, A., Patarin, J., Seurin, Y.: On the public indifferentiability and correlation intractability of the 6-round feistel construction. In: Cramer, R. (ed.) TCC 2012. LNCS, vol. 7194, pp. 285–302. Springer, Heidelberg (2012)

[MRH04] Maurer, U.M., Renner, R.S., Holenstein, C.: Indifferentiability, impossibility results on reductions, and applications to the random oracle methodology. In: Naor, M. (ed.) TCC 2004. LNCS, vol. 2951, pp. 21–39. Springer, Heidelberg (2004)

[RSS11] Ristenpart, T., Shacham, H., Shrimpton, T.: Careful with composition: limitations of the indifferentiability framework. In: Paterson, K.G. (ed.) EUROCRYPT 2011. LNCS, vol. 6632, pp. 487–506. Springer, Heidelberg (2011)

[Seu09] Seurin, Y.: Primitives et protocoles cryptographiques àsécurité prouvée. Ph.D. thesis, Université de Versailles Saint-Quentin-en-Yvelines, France (2009)

[Ste12] Steinberger, J.: Improved security bounds for key-alternating ciphers via hellinger distance. Cryptology ePrint Archive, Report 2012/481 (2012). http:// eprint.iacr.org/

[Ste15] Steinberger, J.: Block ciphers: from practice back to theory. In: TCC 2015 Invited Talk (2015)

Midori: A Block Cipher for Low Energy

Subhadeep Banik[1]([✉]), Andrey Bogdanov[1], Takanori Isobe[2], Kyoji Shibutani[2],
Harunaga Hiwatari[2], Toru Akishita[2], and Francesco Regazzoni[3]

[1] Technical University of Denmark, Kongens Lyngby, Denmark
{subb,anbog}@dtu.dk
[2] Sony Corporation, Tokyo, Japan
{Takanori.Isobe,Kyoji.Shibutani,Harunaga.Hiwatari,
Toru.Akishita}@jp.sony.com
[3] University of Lugano, Lugano, Switzerland
regazzoni@alari.ch

Abstract. In the past few years, lightweight cryptography has become
a popular research discipline with a number of ciphers and hash functions proposed. The designers' focus has been predominantly to minimize the hardware area, while other goals such as low latency have been
addressed rather recently only. However, the optimization goal of low
energy for block cipher design has not been explicitly addressed so far.
At the same time, it is a crucial measure of goodness for an algorithm.
Indeed, a cipher optimized with respect to energy has wide applications,
especially in constrained environments running on a tight power/energy
budget such as medical implants.

This paper presents the block cipher Midori (The name of the cipher
is the Japanese translation for the word Green.) that is optimized with
respect to the energy consumed by the circuit per bt in encryption or
decryption operation. We deliberate on the design choices that lead to
low energy consumption in an electrical circuit, and try to optimize each
component of the circuit as well as its entire architecture for energy.
An added motivation is to make both encryption and decryption functionalities available by small tweak in the circuit that would not incur
significant area or energy overheads. We propose two energy-efficient
block ciphers Midori128 and Midori64 with block sizes equal to 128 and
64 bits respectively. These ciphers have the added property that a circuit
that provides both the functionalities of encryption and decryption can
be designed with very little overhead in terms of area and energy. We
compare our results with other ciphers with similar characteristics: it
was found that the energy consumptions of Midori64 and Midori128 are
by far better when compared ciphers like PRINCE and NOEKEON.

Keywords: Lightweight block cipher · Low energy circuits

1 Introduction

The field of lightweight cryptography has gone into overdrive as evident from the
number of cipher proposals that have emerged in the past few years, like CLEFIA

© International Association for Cryptologic Research 2015
T. Iwata and J.H. Cheon (Eds.): ASIACRYPT 2015, Part II, LNCS 9453, pp. 411–436, 2015.
DOI: 10.1007/978-3-662-48800-3_17

[32], KATAN [13], KLEIN [18], LED [19], PRESENT [11], Piccolo [31], PRINCE [12], SIMON/SPECK [6] to name a few. However, the Advanced Encryption Standard (AES) [16] still remains the de-facto standard when it comes to practical lightweight encryption. The past few years have seen several low-power/area architectures for AES being reported in literature [17,27,30]. However, there has been little work that goes on to determine the design choices that lead to the most energy-efficient architecture. There are many parameters that contribute to the efficiency of a given lightweight design, with area, power, throughput and energy being the foremost among them. Power and energy, are correlated parameters, as energy is essentially the time integral of power, and power is equivalent to the energy consumed per unit time or simply the rate of energy consumption. Energy consumption, thus, is a measure of the total work done by voltage source during the execution of an operation. Hence, in many ways, energy rather than power may be a more relevant parameter to measure the efficiency of a design. Serial architectures of any block cipher that reduce the width of the datapath and reuse components, have a smaller power footprint than round based implementations in which the data path is equal to the block length of the cipher. However, serial implementations usually have high latency, that is, they take much longer to compute the result of an encryption operation than their round based counterparts, and as a result may end up consuming more energy. Therefore, there is no guarantee that low power architectures would necessarily lead to low energy architectures and vice versa.

In [5,21], an evaluation of several lightweight block ciphers with respect to various hardware performance metrics, with a particular focus on the energy cost was done. A formal model for energy consumption in any r-round unrolled block cipher architecture was proposed in [3]. However these papers do not specifically outline design choices that lead to energy-efficient designs.

1.1 Our Contributions

In this paper, we at first try to identify design choices that are energy-efficient and the related tradeoffs that are involved as a result of it. We throw some light at the design considerations that govern low energy circuits, and look at several factors like clock frequency, architecture, loop unrolling and lay down some general thumb rules that help in optimizing for energy. Then, we choose components specifically tailored to meet the requirements of low energy design. In particular, we develop energy-efficient linear layers and non-linear layers.

We use 4×4 almost MDS binary matrices which are more efficient than 4×4 MDS matrices in the terms of area and signal-delay. Note that the branch numbers (the smallest nonzero sum of active inputs and outputs of the matrix) of MDS and almost MDS matrices are 5 and 4, respectively. However, due to a smaller branch number, ciphers employing almost MDS matrices are likely to require the more number of rounds to guarantee its security against several attacks. To address this issue, we propose *optimal cell-permutation layers* which are aimed at improving diffusion speed and increasing the numbers of active

S-boxes in each round with low implementation overheads. Our optimal cell-permutations drastically improve the minimum number of differentially/linearly active S-boxes in each round, and achieve faster diffusion compared to ShiftRow-type permutation. We construct a lightweight and small-delay 4-bit S-box by focusing on the dependency of the computation in S-boxes. The signal delay in our S-boxes is 1.5 times and twice faster than those of PRINCE and PRESENT, respectively. Since the S-box layer is one of the most critical and expensive operations of the cipher, our new S-boxes sufficiently contribute to low energy consumptions.

Combining those new constructions, we design a family of low energy block ciphers Midori which is composed of two variants: Midori64 and Midori128. These provide the functionality for both encryption and decryption with minimal area and energy overhead. The two variants support a 128-bit secret key and a 64/128-bit block, respectively. Security wise, Midori64 and Midori128 do not claim related, known and chosen-key security as it is not relevant in our target application. Using the STM 90 nm standard cell library, both these ciphers consume less than 1.89 pJ/bit encrypted, which is by far better when compared ciphers like PRINCE and NOEKEON [16]. These ciphers are particularly useful for applications that run on tight energy budget, e.g. active RFID tags, sensor nodes, medical implants and battery operated portable devices.

1.2 Organization of the Paper

In Sect. 2, we look at some design considerations that help to minimize energy consumption in block cipher circuits. In Sect. 3, we outline the algorithmic specifications of the Midori128 and Midori64 ciphers. In Sect. 4, we explain our design decisions vis-a-vis the observations of Sect. 2. In Sect. 5, we outline the security analysis of the ciphers. Section 6 contains implementation results of our cipher in hardware using the standard cell library of the STM 90 nm logic process. Section 7 concludes the paper.

2 Design Considerations for Low Energy

For any given block cipher, three factors are likely to play a dominant role in determining the quantity of energy dissipated in the circuit:

(a) Frequency of the Clock used to drive the circuit,
(b) Architecture of the individual components,
(c) Unrolling round functions in the circuit.

We will try to understand the significance of each of these parameters in the context of energy consumption. Let us start with clock frequency. Two components characterize the amount of energy dissipated in a CMOS circuit :

– Dynamic dissipation due to the charging and discharging of load capacitances and the short-circuit current,
– Static dissipation due to leakage current and other current drawn continuously from the power supply.

The total energy dissipation for a CMOS gate can be written as

$$E_{gate} = E_{load} + E_{sc} + E_{leakage}$$

The quantity E_{load} is the energy dissipated for charging and discharging the capacitive load C_L of a gate when output transitions occur. The energy dissipated per $0 \to 1/1 \to 0$ transition is given as

$$E = \int_0^t vi \, dt = \int_0^t vC_L \frac{dv}{dt} dt = C_L \int_0^{V_{DD}} v \, dv = \frac{1}{2} C_L V_{DD}^2.$$

The energy due to the short-circuit current, E_{sc} is dissipated in a CMOS gate, when during a transition both the n and the p-transistors are on for a short period of time. The energy due to leakage currents $E_{leakage}$ is rather small, and is mainly caused due to the sub-threshold leakage current, which is the drain-source current in a CMOS gate when the transistor is OFF. This figure is becoming increasingly important as the technology is scaling down making the sub-threshold leakage more significant. However as pointed out in [3,21], the effect of the leakage energy at high clock frequencies is minimal. As such, energy becomes a metric which is a measure of the total switching activity of a circuit during the process. For sufficiently high frequencies, the energy consumption required to compute an encryption/decryption operation is essentially independent of frequency of operation. In our experiments, for circuits implemented using the standard cell library based on the STM 90 nm low leakage process, at frequencies higher than 1 MHz, leakage energy is usually less than 1 % of the total energy dissipated in the circuit.

To understand the significance of the other parameters we performed the following experiments. Consider a case in which two Rijndael S-boxes are placed one after the other in a circuit as shown in Fig. 1. The signals to the input of the first S-box, the second S-box, and the output of the 2nd S-box are named S1xD, S2xD and S3xD respectively. Note that, analyzing this situation is particularly useful for understanding the energy consumption trends of unrolled designs where logic blocks are placed sequentially one after the other.

Let us assume that the signal S1xD comes from an 8-bit register, so that it "cleanly" switches between successive byte values, i.e. all the bits of S1xD make logic transitions at the same point of time which is usually the rising clock edge

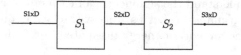

Fig. 1. S-boxes placed sequentially

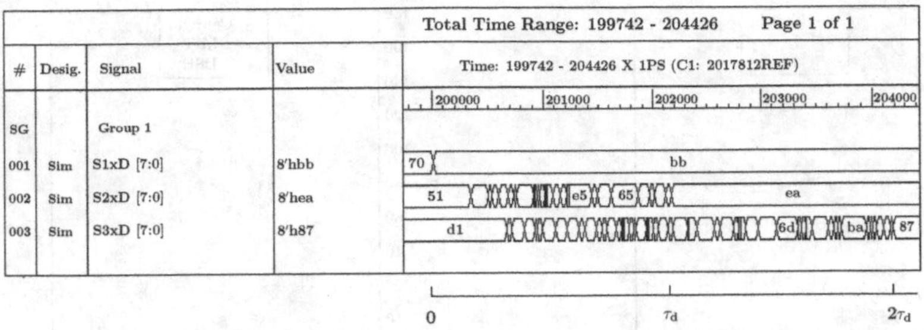

Fig. 2. The signals S1xD, S2xD, S3xD

for synchronous circuits. The signal S2xD will switch between various values in a given time interval $0 \to \tau_d$, before settling down to a stable value. The value τ_d which is the delay experienced by the signal S1xD usually depends on the cell library and the architecture adopted to implement the S-boxes. Another parameter dependent on the logic process and architecture of the S-box is the switching activity of S2xD which can be informally defined as the number of logic transitions made by this signal in the period $0 \to \tau_d$.

The second S-box S_2, sees this signal S2xD, which is switching between various values in the time interval $0 \to \tau_d$. Therefore, the switching activity of S_2 is actually at least double that of S_1, as it would continue switching for another τ_d before producing a stable signal. Figure 2 provides an example in which, the three signals for the pair of Rijndael S-boxes (implemented using the Canright [14] architecture in the standard cell library of the STM 90 nm logic process, at 10 MHz) are shown. The synthesis for each S-box was done separately, so that the synthesis tool would not group together gates from the first and the second S-box in order to save area. Since the energy consumption of a logic block depends on the switching activity of all its nodes, the S-box S_2 should naturally consume more energy than S_1. Again the exact energy consumed by S_2 relative to S_1 depends on factors like

(a) the logic process and hence the value of τ_d,
(b) the architecture of the S-box and hence the amount of "extra" switching experienced by S_2 and
(c) the algebraic structure of the S-box, i.e. its component Boolean functions.

The extra switching activity would be proportional to the average number of gates that undergo a $0 \to 1/1 \to 0$ transition during the period $\tau_d \to 2\tau_d$ (the average is typically taken over all possible transitions of the signal S1xD). Similarly if a third S-box S_3 were placed after S_2, then too it would experience an increase in switching activity relative to S_2 that would depend on the average number of gates switched in the period $2\tau_d \to 3\tau_d$. The increase in switching activity of S_3 over S_2 is likely to be roughly the same as that of S_2 over S_1, since the number of gates in S_2 that switch in $\tau_d \to 2\tau_d$ and those in S_3 between $2\tau_d \to 3\tau_d$ when averaged over $\binom{256}{2}$ transitions of S1xD, is likely to be same. And so if it so happens that S_1, S_2 and S_3 drive the same amount of capacitive

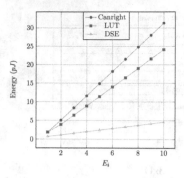
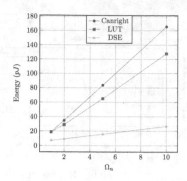

Fig. 3. Energy per cycle E_i in i^{th} S-box S_i

Fig. 4. Energy Ω_n required to compute $S^{10}(x)$ using n S-boxes

load, the difference between the energy consumed between S_2 and S_1 is likely to be the same as between S_3 and S_2.

Taking these ideas forward, if we connect a series of n S-boxes sequentially, the energy consumed by each S-box in a given period of time is likely to be more than the previous S-box, as the switching activity of the S-boxes are likely to increase from the first to the last. We tested three different architectures for the Rijndael S-box. The first is the Canright [14] architecture which is acknowledged to be smallest known implementation in terms of gate area. The second is the Look-up Table (LUT) based architecture as synthesized by the Synopsys Design Compiler. The LUT architecture, while larger than the Canright architecture in terms of area, is much faster in terms of signal delay from the input to output port. The third is a Decoder-Switch-Encoder (DSE) based architecture [7], which is optimal in terms of power/energy consumption. Over the years there has been much research on low power Rijndael S-boxes [28,34], but the DSE based architecture is widely believed to be most power/energy-efficient on account of its unique architecture. The 8-bit input is first decoded to a set of 256 wires. The S-box functionality is achieved by a shuffling of wires after which the output is produced by an encoding of the 256 shuffled wires (i.e. the inverse of the decoding process). The entire circuit can be constructed by AND/NAND gates, which have very low switching probability and since the S-box functionality is provided by wire shuffling, all 8-bit S-boxes can be constructed in this manner. The architecture offers very low switching per change of input bit: a maximum of 25 % of the gates switch when one of the input bits is flipped.

We connected 10 instances of the S-box constructed using the Canright architecture (using the standard cell library of the STM 90 nm logic process) sequentially and used the Synopsys Power Compiler to estimate the energy consumed per clock cycle E_i in each of the successive S-boxes S_i at a clock frequency of 10 MHz. We repeated the same experiment for the LUT and DSE based S-boxes. The results can be seen in Fig. 3. It can be seen that the successive instances of the LUT based S-box which has a delay of around 2.1 ns consumes much less energy as compared to the Canright S-box which has a delay of around 2.9 ns. In both the LUT and Canright architectures, the switching activity in the circuit is roughly proportional to the signal delay across the input and output ports. This is however not the case for DSE S-box, which although has a delay of around 2.3 ns, experiences much lower increase in successive values of E_i because the total switching activity in the delay period is much lower.

The above analysis is particularly relevant due to two reasons. The first pertains to the structure of especially SPN based ciphers, in which each round typically consists of a substitution, a linear layer and a key addition placed sequentially. A substitution layer with low switching activity and signal delay ensures that the linear layer consumes less energy. Similarly a linear layer with similar characteristics ensures that any circuit placed after it consumes less energy. The second pertains to the consideration of round unrolled circuits. An r-round unrolled circuit for a block cipher is one in which, the circuit computes the results of r successive round functions in a single clock cycle. So if the block cipher specification calls for N executions of the round function, an r-round unrolled circuit will compute the result of the encryption operation in $\left\lceil \frac{N}{r} \right\rceil$ cycles. An r-round unrolled architecture is constructed by placing the circuits for r round functions sequentially, followed by a register. The above analysis suggests that any multiple round unrolled circuit is unlikely to be efficient in terms of energy consumption. In the above example, using the LUT based S-box, computing the result of two S-box operations (i.e. $S(S(x))$) over 2 cycles costs $2 * 1.88 = 3.76$ pJ. Computing the same over one cycle by sequential placement of 2 S-boxes will cost $1.88 + 3.91 = 5.79$ pJ. Similarly computing three S-box operations over three cycles takes 5.64 pJ, whereas the same over one cycle would take $1.88 + 3.91 + 6.40 = 12.39$ pJ. Figure 4 shows the cumulative energy cost Ω_n of computing $S^{10}(x)$ using a sequence of n S-boxes (i.e. in $\frac{10}{n}$ cycles), for different values of n. It can be seen that, irrespective of the architecture of the S-box, the energy consumption is optimal for $n = 1$, i.e. computing the operation over 10 cycles using a single S-box, even if this involves updating the register 10 times in the process.

2.1   S-Box: 4-Bit Vs 8-Bit

In light of the above analysis, it is clear that a design using a 4-bit S-box is more efficient in terms of energy consumed per cycle than a design using an 8-bit S-box. This is primarily due to the fact that a 4-bit S-box will typically have a lower signal delay as compared to an 8-bit S-box. However 8-bit S-boxes offer higher non-linearity and lower values of the DP/LP co-efficient, and so in order to sustain similar security margins, a design using a 4-bit S-box will typically need more executions of the round function. To put things, in perspective we performed the energy evaluation of the circuit of the SPN round function (with blocksize equal to 128 bits) in which we experimented with two different substitution layers, one having sixteen 8-bit S-boxes and the other having thirty two 4-bit S-boxes. The Rijndael MixColumn was used in both cases, and the STM 90 nm cell library was used to synthesize the circuits. For this purpose

Table 1. A comparison of energy per cycle for round functions constructed with (A) 16 8-bit S-boxes, (B) 32 4-bit S-boxes.

	S-box	Delay in S	Energy per cycle
		(ns)	(pJ)
A	DSE (8-bit)	2.25	14.00
	Rijndael(LUT)	2.10	38.88
	mCrypton	1.59	13.20
	Whirlpool	1.33	16.38
B	DSE (4-bit)	0.81	7.92
	PRINCE	0.36	4.87
	PRESENT	0.45	6.18

four different 8-bit S-boxes were chosen. Apart from the LUT and DSE based Rijndael S-boxes, we chose the S-boxes used in mCrypton [24] and Whirlpool [4]. Unlike AES, these S-boxes can be functionally defined in terms of smaller 4-bit S-boxes, and so can be implemented efficiently in hardware. Additionally we chose three 4-bit S-boxes: the generic DSE based S-box (note that since the S-box functionality is provided by a wire shuffle, all DSE S-boxes will have same energy consumption), and the S-boxes used in PRINCE [12] and PRESENT [11].

Table 1 reports the energy per cycle figures at a frequency of 10 MHz. It can be seen that the DSE architecture is not as effective as energy saving measure for 4-bit S-boxes. It is also interesting to note that from the point of view of energy 4-bit S-boxes out performs their 8-bit counterparts by a ratio of around 2:1. Thus, the use of 4-bit S-boxes seems to be an efficient configuration even if the number of rounds in the encryption algorithm has to be increased in order to maintain security margins.

2.2   Feistel Vs SPN and Complex Vs Simple Round Function

As far as designing lightweight ciphers is concerned, both SPN and Feistel architectures have their respective advantages and disadvantages. Feistel structures (e.g. TWINE [33], Piccolo [31], SIMON [6]) usually apply a round function to only one half of the state and as such structures can be implemented in hardware with low average power. Also, implementing the inverse of Feistel constructions is not very difficult and hence a circuit that provides functionalities for both encryption and decryption can be designed with minimal overhead. However, given the fact that Feistel structures introduce non-linearity in only one half of the state in every round and hence, to maintain security margins, such constructions usually require more executions of the round functions as compared to SPN structures. As such Feistel, constructions are not suited for low latency implementations. Most SPN constructions, on the other hand, usually apply its transformation function to the entire state and so can be implemented using fewer rounds. In principle, if n rounds of SPN function and m rounds of Feistel function (where $m > n$) have the same security margin and similar energy expenditure, then using the n round SPN function makes more sense since lesser energy is consumed to update the state and key register for n rounds. A similar argument can be used to resolve the choice between (a) Simple round functions with more rounds (e.g. PRESENT [11]) and (b) Complex round functions with lesser rounds.

2.3 Effect of Key Schedule

Generating separate round keys in each round by means of a key schedule operation can eat into the energy budget as it incurs the added cost of updating the key register in every round. For example using the STM 90 nm standard cell library, in AES (with DSE S-box), the key schedule consumes a total of 25 % of the total energy consumed. For PRESENT, the key schedule consumes close to 32 % of the total energy. So designs meant primarily for low energy consumption, designers should look to avoid the key schedule operation. This would also be efficient in terms of area as it would not be necessary to include a key register in the design.

2.4 Main Conclusion: Low-Energy Design Choices

We can now state some conclusions that will serve as pointers for a good low energy block cipher design. From the point of view of energy, we know that a **round based architecture** is usually optimal. Thus we concentrate on an efficient round based construction that would with minimal overhead provide both the functionalities of encryption and decryption. A cipher like PRINCE, although provides both encryption/decryption functionalities with minimal tweak in the circuit, does not have an equally energy-efficient round based construction [12], as it needs to accommodate 3 different round functions in the same circuit. We have also seen that components with low switching and delay tend to perform better energy wise. So another requirement is choosing components with **low area and delay**. In this context, it makes sense to choose **4-bit S-boxes** over 8-bit S-boxes. We choose **SPN** architecture over Feistel to minimize the number of rounds in the design. And since providing the functionalities of both encryption and decryption is an added motivation, we try to include components which in addition to having low area/delay, are also **involutions**. Having such components would minimize any additional overhead required for providing the functionalities of both encryption and decryption. We will now present the specifications for the proposed block cipher and in Sect. 4 we will explain the design decisions in the context of the observations made in this Section.

3 Specification

Midori is a family of two block ciphers: Midori64 and Midori128. Both ciphers accept 128-bit keys, and have a different block size n ($n = 64$ for Midori64 and

Table 2. Parameters for Midori64 and Midori128

	block size(n)	key size	cell size(m)	number of rounds
Midori64	64	128	4	16
Midori128	128	128	8	20

$n = 128$ for Midori128). The basic parameters of Midori64 and Midori128 are shown in Table 2.

Midori is a variant of a Substitution Permutation Network (SPN), which consists of the S-layer and the P-layer, and uses the following 4×4 array called state as a data expression:

$$S = \begin{bmatrix} s_0 & s_4 & s_8 & s_{12} \\ s_1 & s_5 & s_9 & s_{13} \\ s_2 & s_6 & s_{10} & s_{14} \\ s_3 & s_7 & s_{11} & s_{15} \end{bmatrix},$$

where the sizes of each cell m are 4 and 8 bits for Midori64 and Midori128, respectively, i.e., $s_i \in \{0,1\}^m$, $m = 4$ for Midori64 and $m = 8$ for Midori128. A 64-bit or a 128-bit plaintext P is loaded into the state, and the i-th round output state is defined as S_i, namely $S_0 = P$.

3.1   S-Boxes and Matrices

S-box: Midori utilizes two types of bijective 4-bit S-boxes, Sb_0 and Sb_1, where $Sb_0, Sb_1 : \{0,1\}^4 \rightarrow \{0,1\}^4$ (see Table 3). Sb_0 and Sb_1 are used in Midori64 and Midori128, respectively. Note that Sb_0 and Sb_1 both have the involution property.

Midori128 utilizes four different 8-bit S-boxes SSb_0, SSb_1, SSb_2 and SSb_3, where $SSb_0, SSb_1, SSb_2, SSb_3 : \{0,1\}^8 \rightarrow \{0,1\}^8$ Mathematically, each SSb_i consists of input and output bit permutations and two Sb_1's as shown in Fig. 5. Each output bit permutation is taken as the inverse of the corresponding input bit permutation to keep the involution property. Let the input bit permutation of each SSb_i be referred to as p_i. Let $x_{[i]}$ denote the i-th bit of x, where $x_{[0]}$ is the most significant bit (MSB). Then denoting $p_i(x) = y^{(i)}$, we have

$$y^{(0)}_{[0,1,2,3,4,5,6,7]} = x_{[4,1,6,3,0,5,2,7]}, \quad y^{(1)}_{[0,1,2,3,4,5,6,7]} = x_{[1,6,7,0,5,2,3,4]}$$

$$y^{(2)}_{[0,1,2,3,4,5,6,7]} = x_{[2,3,4,1,6,7,0,5]}, \quad y^{(3)}_{[0,1,2,3,4,5,6,7]} = x_{[7,4,1,2,3,0,5,6]}$$

Table 3. 4-bit bijective S-boxes Sb_0 and Sb_1 in hexadecimal form

x	0	1	2	3	4	5	6	7	8	9	a	b	c	d	e	f
$Sb_0[x]$	c	a	d	3	e	b	f	7	8	9	1	5	0	2	4	6
$Sb_1[x]$	1	0	5	3	e	2	f	7	d	a	9	b	c	8	4	6

The output permutation used in each SSb_i is simply the inverse of the map p_i. Matrix: Midori utilizes an involutive binary matrix M defined as follows:

$$M = \begin{pmatrix} 0 & 1 & 1 & 1 \\ 1 & 0 & 1 & 1 \\ 1 & 1 & 0 & 1 \\ 1 & 1 & 1 & 0 \end{pmatrix}.$$

The matrix M updates four m-bit values (x_0, x_1, x_2, x_3) as follows:

$$^t(x_0, x_1, x_2, x_3) \leftarrow M \cdot {}^t(x_0, x_1, x_2, x_3),$$

where the operations between a matrix and a vector are performed over $\mathrm{GF}(2^m)$.

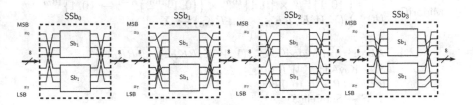

Fig. 5. SSb_0, SSb_1, SSb_2 and SSb_3

3.2 Round Function

The round function of Midori consists of an S-layer SubCell: $\{0, 1\}^n \rightarrow \{0, 1\}^n$, a P-layer ShuffleCell and MixColumn: $\{0, 1\}^n \rightarrow \{0, 1\}^n$ and a key-addition layer KeyAdd: $\{0, 1\}^n \times \{0, 1\}^n \rightarrow \{0, 1\}^n$. Each layer updates an n-bit state S as follows.

SubCell (S): Sb_0 and SSb_i are applied to every 4 and 8-bit cell of the state S of Midori64 and Midori128 in parallel, respectively. Namely, $s_i \leftarrow \mathsf{Sb}_0[s_i]$ for Midori64 and $s_i \leftarrow \mathsf{SSb}_{(i \bmod 4)}[s_i]$ for Midori128, where $0 \le i \le 15$.
ShuffleCell (S): Each cell of the state is permuted as follows:
$(s_0, s_1, ..., s_{15}) \leftarrow (s_0, s_{10}, s_5, s_{15}, s_{14}, s_4, s_{11}, s_1, s_9, s_3, s_{12}, s_6, s_7, s_{13}, s_2, s_8)$.
MixColumn (S): M is applied to every $4m$-bit column of the state S, i.e.,
$^t(s_i, s_{i+1}, s_{i+2}, s_{i+3}) \leftarrow M {}^t(s_i, s_{i+1}, s_{i+2}, s_{i+3})$ and $i = 0, 4, 8, 12$.
KeyAdd (S, RK_i): The i-th n-bit round key RK_i is XORed to a state S.

3.3 Data Processing Part

The data processing part of Midori for encryption $\mathsf{MidoriCore}_{(R)}$ performs as follows:

$$\mathsf{MidoriCore}_{(R)} : \begin{cases} \{0, 1\}^{16m} \times \{0, 1\}^{16m} \times \{\{0, 1\}^{16m}\}^{R-1} \rightarrow \{0, 1\}^{16m} \\ (X, WK, RK_0, ..., RK_{R-2}) \mapsto Y \end{cases}$$

$$
\boxed{
\begin{aligned}
&\textbf{Algorithm } \mathsf{MidoriCore}_{(R)}(X, WK, RK_0, ..., RK_{R-2}) : \\
&S \leftarrow \mathsf{KeyAdd}(X, WK) \\
&\textbf{for } i = 0 \textbf{ to } R - 2 \textbf{ do} \\
&\quad S \leftarrow \mathsf{SubCell}(S) \\
&\quad S \leftarrow \mathsf{ShuffleCell}(S) \\
&\quad S \leftarrow \mathsf{MixColumn}(S) \\
&\quad S \leftarrow \mathsf{KeyAdd}(S, RK_i) \\
&S \leftarrow \mathsf{SubCell}(S) \\
&Y \leftarrow \mathsf{KeyAdd}(S, WK)
\end{aligned}
}
$$

where $R = 16$ for Midori64 and $R = 20$ for Midori128. Similarly, the inverse data processing part $\mathsf{MidoriCore}_{(R)}^{-1}$ operates as follows:

$$
\mathsf{MidoriCore}_{(R)}^{-1} : \begin{cases} \{0,1\}^{16m} \times \{0,1\}^{16m} \times \{\{0,1\}^{16m}\}^{R-1} \to \{0,1\}^{16m} \\ (Y, WK, RK_{R-2}, ..., RK_0) \mapsto X \end{cases}
$$

$$
\boxed{
\begin{aligned}
&\textbf{Algorithm } \mathsf{MidoriCore}_{(R)}^{-1}(Y, WK, RK_{R-2}, ..., RK_0) : \\
&S \leftarrow \mathsf{KeyAdd}(Y, WK) \\
&\textbf{for } i = (R - 2) \textbf{ to } 0 \textbf{ do} \\
&\quad S \leftarrow \mathsf{SubCell}(S) \\
&\quad S \leftarrow \mathsf{MixColumn}(S) \\
&\quad S \leftarrow \mathsf{InvShuffleCell}(S) \\
&\quad S \leftarrow \mathsf{KeyAdd}(S, L^{-1}(RK_i)) \\
&S \leftarrow \mathsf{SubCell}(S) \\
&X \leftarrow \mathsf{KeyAdd}(S, WK)
\end{aligned}
}
$$

where L^{-1} (inverse of the linear layer) denotes the composition of the operations $\mathsf{InvShuffleCell} \circ \mathsf{MixColumn}$, and $\mathsf{InvShuffleCell}$ permutes each cell of the state as follows.

$$
(s_0, s_1, ..., s_{15}) \leftarrow (s_0, s_7, s_{14}, s_9, s_5, s_2, s_{11}, s_{12}, s_{15}, s_8, s_1, s_6, s_{10}, s_{13}, s_4, s_3).
$$

3.4   Round Key Generation

For Midori64, a 128-bit secret key K is denoted as two 64-bit keys K_0 and K_1 as $K = K_0 \| K_1$. Then, $WK = K_0 \oplus K_1$ and $RK_i = K_{(i \bmod 2)} \oplus \alpha_i$, where $0 \le i \le 14$. For Midori128, $WK = K$ and $RK_i = K \oplus \beta_i$, where $0 \le i \le 18$. The constants β_i are defined in Table 4. It can be seen that the constants are in the form of 4×4 binary matrices. They are added bitwise to the LSB of every round key byte in Midori128 and round key nibble in Midori64 respectively. Note that $\alpha_i = \beta_i$ for $0 \le i \le 14$.

3.5   Midori Ciphers

Midori block ciphers are composed of two variants: Midori64 and Midori128 consisting of $\mathsf{MidoriCore}_{(16)}$ with $m = 4$ and $\mathsf{MidoriCore}_{(20)}$ with $m = 8$, respectively. $\mathsf{MidoriCore}_{(16)}$ is depicted in Fig. 6 as an example.

Table 4. The Round Constants β_i

i	β_i	i	β_i	i	β_i	i	β_i	i	β_i	i	β_i	i	β_i
0	0010 0100 0011 1111	1	0110 1010 1000 1000	2	1000 0101 1010 0011	3	0000 1000 1101 0011	4	0001 0011 0001 1001	5	1000 1010 0010 1110	6	0000 0011 0111 0000
7	0111 0011 0100 0100	8	1010 0100 0000 1001	9	0011 1000 0010 0010	10	0010 1001 1001 1111	11	0011 0001 1101 0000	12	0000 1000 0010 1110	13	1111 1010 1001 1000
14	1110 1100 0100 1110	15	0110 1100 1000 1001	16	0100 0101 0010 1000	17	0010 0001 1110 0110	18	0011 1000 1101 0000				

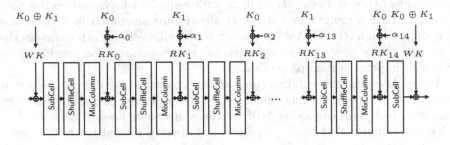

Fig. 6. Overview of Midori64

4 Design Decision

Here, we explain our design decisions vis-a-vis the observations of Sect. 2.

4.1 Linear Layer

Linear layers of the each variant consist of a cell-permutation (ShuffleCell) and four 4×4 matrix operations (MixColumn). Those operations are performed over $GF(2^4)$ and $GF(2^8)$ for the 64 and 128-bit variants, respectively.

MDS Vs Almost MDS. Using the NanGate 45 nm open cell library, Table 5 compares three types of 4×4 matrices, involutive MDS (M_A), non-involutive MDS (M_B) and involutive almost MDS matrices (M_C) from implementation aspects. These matrices are considered lightweight in each of the three aforementioned criteria [26, 31].

$$M_A = \begin{pmatrix} 1 & 2 & 6 & 4 \\ 2 & 1 & 4 & 6 \\ 6 & 4 & 1 & 2 \\ 4 & 6 & 2 & 1 \end{pmatrix}, M_B = \begin{pmatrix} 2 & 3 & 1 & 1 \\ 1 & 2 & 3 & 1 \\ 1 & 1 & 2 & 3 \\ 3 & 1 & 1 & 2 \end{pmatrix}, M_C = \begin{pmatrix} 0 & 1 & 1 & 1 \\ 1 & 0 & 1 & 1 \\ 1 & 1 & 0 & 1 \\ 1 & 1 & 1 & 0 \end{pmatrix}.$$

From Table 5, M_C is obviously preferable over the others in terms of the gate size and the path delay. In fact, circulant-type almost MDS matrices are adopted

Table 5. Comparison of three matrices

	M_A	M_B	M_C
Area [GE]	108	104	48
Delay [ns]	0.93	0.68	0.37
Diffusion	MDS	MDS	Almost MDS
Involution	yes	no	yes

Table 6. Comparison of S-boxes

	PRESENT	PRINCE	Sb_0	Sb_1
Area [GE]	24.33	16	13.3	15.33
Delay [ns]	0.47	0.36	0.24	0.32
Involution	No	No	Yes	Yes

in PRINCE [12], PRIDE [1], FIDES [8] and CLOC [20]. Moreover, Khoo et al. showed that, for a 64-bit block size employing the AES-like structure, the combination of 4×4 almost MDS matrices (M_C) with ShiftRow and 16 4-bit S-boxes is the most efficient in both a round-based and a serialized implementation by proposing a new comparison metric FOAM (figure of adversarial merit), which combines the inherent security provided by cryptographic structures and components along with their implementation properties [22].

While M_C has efficient implementation properties, its diffusion speed is slower and the minimum number of active S-boxes in each round is smaller than those of ciphers employing MDS matrices due to its lower branch number. It has been known that those properties are directly related to the immunity against several attacks including impossible differential, saturation, differential and linear attacks. To improve security of the almost MDS with low implementation overheads, we adopt *optimal cell-permutation layers* which are aimed at improving diffusion speed and increasing the number of active S-boxes in each round. The diffusion speed is measured by the number of rounds taken to attain full diffusion, which is the property that all output cells are affected by all input cells. Importantly, changing cell-permutation patterns generally does not require additional implementation costs in a round-based and an unrolled hardware implementation.

Approach to Find Optimal Cell-Permutation Layers for Almost MDS. Since it is computationally hard to exhaustively count the minimum number of active S-boxes for all possible permutations ($= 16! \approx 2^{44.25}$) by Matsui's search approach [9, 25], we take the following two-step approach to reduce the search space. In the fist step, we restrict the cell-permutations to row-based cell-permutations which permute four cells in each row, e.g. ShiftRow in AES. The number of possible row-based cell-permutations is estimated as $2^{18.3}$ ($= (4!)^4$). This step is based on the fact that the full diffusion property relies on only row-based property of the cell-permutation. As a result of our searches, we find that a class of row-based cell-permutations achieves full diffusion in 3 rounds and its necessary and sufficient condition is as follows.

Condition 1 *(3-round full diffusion). For a 4×4 cell-array, after applying a cell-permutation once and twice, each input cell in a column is mapped into a cell in the different column.*

From our search, **576** row-based cell-permutations satisfy Condition 1. Interestingly, ShiftRow-type permutation is not included in this class, i.e. it requires 4 rounds for full diffusion.

In the second step, we add a column-based cell-permutation, which permutes four cells in each column, after applying the class of permutations satisfying Condition 1. The target cell permutation consists of the combination of the row-based and column-based permutations. Note that adding a column-based cell-permutation to the row-based permutations satisfying Condition 1 does not affect the full diffusion property. The number of all possible cell-permutations of this class is estimated as $2^{27.51}$ ($= 576 \times (4!)^4$). Consequently, we find a class of cell-permutation achieving the largest number of active S-boxes in each round and the smallest number of rounds to attain full diffusion when satisfying Condition 1 and the following Condition 2 *or* 3.

Condition 2 *(The number of active S-box). For a 4 × 4 cell-array, after applying a cell-permutation twice and twice inversely, each input cell in a column is mapped into a cell in the same row.*

Condition 3 *(The number of active S-box). For a 4 × 4 cell-array, after applying a cell-permutation once and three times inversely, each input cell in a column is mapped into a cell in the same row.*

The numbers of cell-permutations satisfying Conditions 2 and 3 are both 576. We define such **1152** cell-permutation as *optimal cell-permutations*. Table 7 shows the minimum numbers of differentially/linearly active S-boxes of the optimal cell-permutations and the ShiftRow-type permutation. Our optimal cell-permutations drastically improve the minimum number of differentially/linearly active S-boxes in each round while keeping the 3-round full diffusion property. Thus, our optimal permutations achieve security against several attacks such as differential/linear and impossible attacks in the same number of rounds compared to ShiftRow-type permutation. Midori128 and Midori64 adopt one of optimal cell permutations satisfying both Conditions 1 and 2 as follows.

$$(s_0, s_1, ..., s_{15}) \leftarrow (s_0, s_{10}, s_5, s_{15}, s_{14}, s_4, s_{11}, s_1, s_9, s_3, s_{12}, s_6, s_7, s_{13}, s_2, s_8).$$

Starting from the state S_0, each cell of S_0 is mapped to S_1, S_2, S_1^{-1} and S_2^{-1} after applying the above cell-permutation once, twice, once inversely and twice inversely, respectively, as follows.

$$S_0 = \begin{bmatrix} s_0 & s_4 & s_8 & s_{12} \\ s_1 & s_5 & s_9 & s_{13} \\ s_2 & s_6 & s_{10} & s_{14} \\ s_3 & s_7 & s_{11} & s_{15} \end{bmatrix}, S_1 = \begin{bmatrix} s_0 & s_{14} & s_9 & s_7 \\ s_{10} & s_4 & s_3 & s_{13} \\ s_5 & s_{11} & s_{12} & s_2 \\ s_{15} & s_1 & s_6 & s_8 \end{bmatrix}, S_2 = \begin{bmatrix} s_0 & s_2 & s_3 & s_1 \\ s_{12} & s_{14} & s_{15} & s_{13} \\ s_4 & s_6 & s_7 & s_5 \\ s_8 & s_{10} & s_{11} & s_9 \end{bmatrix},$$

$$S_1^{-1} = \begin{bmatrix} s_0 & s_5 & s_{15} & s_{10} \\ s_7 & s_2 & s_8 & s_{13} \\ s_{14} & s_{11} & s_1 & s_4 \\ s_9 & s_{12} & s_6 & s_3 \end{bmatrix}, S_2^{-1} = \begin{bmatrix} s_0 & s_2 & s_3 & s_1 \\ s_{12} & s_{14} & s_{15} & s_{13} \\ s_4 & s_6 & s_7 & s_5 \\ s_8 & s_{10} & s_{11} & s_9 \end{bmatrix}.$$

Table 7. The number of minimum number of differentially/linearly active S-boxes (AS) of Midori64 and Midori128

Round number	4	5	6	7	8	9	10	11	12	13	14	15	16
Min. # of AS (Optimal Cell-Permutation)	16	23	30	**35**	38	41	50	57	62	**67**	72	75	84
Min. # of AS (ShiftRow-type Permutation)	16	18	20	26	**32**	34	36	42	48	50	52	58	**64**

From those mappings, it is clear that the relation among S_2^{-1}, S_0 and S_2 satisfies Condition 2. Similarly, all of the pairs (S_2^{-1}, S_1^{-1}), (S_1^{-1}, S_0), (S_0, S_1), (S_1, S_2) satisfy Condition 1.

4.2   S-Box Layer

According to analysis of Sect. 2.1, 4-bit S-boxes are usually more efficient than 8-bit S-boxes in terms of energy consumption per cycle. Also, the small path delay and the small gate area lead to low-energy implementation. To optimize S-layer regarding energy consumption, we aim to develop a small-delay and lightweight 4-bit S-box which fulfill the following requirements: (1) the maximal probability of a differential is 2^{-2}, (2) the maximal absolute bias of a linear approximation is 2^{-2} and (3) involution. The requirement (3) enables us to reduce the number of possible S-boxes from $2^{44.25}$ to $2^{25.5}$.

Approach to Find Small-Delay and Lightweight 4-Bit S-Box. Our approach starts with a key observation that *the path delay is highly related to the dependency of the computation.* We introduce a metric *depth* to estimate the path delay of S-boxes.

Definition 1 (depth): *The depth is defined as the sum of sequential path delays of basic operations AND, OR, NAND, NOR and NOT.*

Example. The depth of the computation of $(x \oplus y) \cdot z$ is estimated as the sum of path delays of XOR and AND, because "$\cdot z$" operation is feasible only after the computation of $(x \oplus y)$,

In our search, we assume that depths of XOR, AND/OR, NAND/NOR and NOT are weighted as 2, 1.5, 1 and 0.5, respectively, based on the number of the transistors to be sequentially proceeded in the operation. The required gates of NOT, NAND/NOR, AND/OR and XOR/XNOR are estimated as 0.5, 1, 1.5 and 2 [GEs], respectively. We search all S-boxes whose depth is $1, 1.5, 2, \ldots$, and check whether the S-boxes satisfy our security requirements. As a result, we can find Sb_0 (see Table 3) whose depth and gate size are the lowest and the smallest ones in our search. Sb_0 can be expressed as follows, where inputs and outputs are defined as $\{a, b, c, d\}$ and $\{a', b', c', d'\}$, and a and a' are the most significant bits.

Table 8. Input-output bit relations of each S-box

	SSb_0	SSb_1	SSb_2	SSb_3
A	(1, 3, 4, 6)	(0, 1, 6, 7)	(1, 2, 3, 4)	(1, 2, 4, 7)
B	(0, 2, 5, 7)	(2, 3, 4, 5)	(0, 5, 6, 7)	(0, 3, 5, 6)

$$a' = \left(\overline{c} \text{ NAND } (a \text{ NAND } b) \right) \text{ NAND } (a \text{ OR } d)$$

$$b' = \left((a \text{ NOR } d) \text{ NOR } (b \text{ AND } c) \right) \text{ NAND } \left((a \text{ AND } c) \text{ NAND } d \right)$$

$$c' = (b \text{ NAND } d) \text{ NAND } \left((b \text{ NOR } d) \text{ OR } a \right)$$

$$d' = \left(a \text{ NOR } (b \text{ OR } c) \right) \text{ NOR } \left((a \text{ NAND } b) \text{ NAND } (c \text{ OR } d) \right)$$

For instance, let us consider the computation of c'. In this computation, $(b \text{ NAND } d)$ and $(b \text{ NOR } d)$ can be done at first. After that, the computation of $(b \text{ NOR } d) \text{ OR } a$ is done. Then, the last operation of NAND is executable. Thus, the depth of c' is estimated as 3.5 ($= 1 + 1.5 + 1$). The depths of the remaining a', b' and d' are also estimated as 3.0 or 3.5.

Considering additional requirement *full diffusion property*, we find Sb_1 which has the lowest depth and the smallest gate area among 4-bit bijective S-boxes satisfying the requirements (1), (2), (3) and the full diffusion property. Sb_1 is expressed as follows :

$$a' = \left((b \text{ NAND } c) \text{ NAND } a \right) \text{ NAND } \left((a \text{ NOR } d) \text{ NAND } b \right)$$

$$b' = \left((a \text{ XOR } c) \text{ NOR } b \right) \text{ NOR } \left((b \text{ NAND } c) \text{ AND } d \right)$$

$$c' = (c \text{ NAND } d) \text{ NAND } \left((a \text{ XOR } b) \text{ NAND } (b \text{ OR } d) \right)$$

$$d' = \left((a \text{ NAND } b) \text{ NAND } c \right) \text{ NAND } (b \text{ OR } d)$$

Note that an S-box satisfies the full diffusion property if and only if any inputs $\{a, b, c, d\}$ of the S-box non-linearly affect all outputs $\{a', b', c', d'\}$. This full diffusion property enables us to ensure a 3-round property regarding the diffusion in Midori128 (we will explain it in the end of this section).

Evaluation. Table 6 shows the comparison of S-boxes of PRESENT, PRINCE, Sb_0 and Sb_1 using NanGate 45 nm open cell library. The path delay of Sb_0 is 1.5 times and twice smaller than PRINCE and PRESENT, respectively, and the gate size is also smaller than the others. Those of Sb_1 are comparable to PRINCE's S-box. Additionally Sb_0 and Sb_1 have the involution property.

8-Bit S-Boxes Based on 4-Bit S-Boxes. From the observation in Sect. 2.1, we adopt 8-bit S-boxes consisting of two 4-bit S-boxes processed in parallel to minimize the path delay in the round-based implementation. Moreover, in order to avoid having the unfavorable independent property exploited in the full-round attack on KLEIN [23], we add properly-chosen *bit-permutations* to the begin and the end of 8-bit S-boxes as shown in Fig. 5. As described in Sect. 3.1, each output bit-permutation is the inverse of the corresponding input bit-permutation to keep the involution property. With a property of our P-layer and those bit-permutations, we claim that no independent property is found after 3 rounds in Midori128. Since Sb_1 has the full diffusion property, any input bit of SSb_i affects the corresponding 4 bits output as shown in Table 8. For example, in SSb_1, any of the i-th input bit affects all of the i-th output bits, where $i \in \{0, 1, 6, 7\}$. We choose bit-permutations for SSb_0, SSb_1, SSb_2 and SSb_3 so that those satisfy the following property.

Property 1. *Affected 4-bit positions of outputs of an S-box are included in both of two different input groups of the other three S-boxes.*

For example, the group A of SSb_1 is $\{0, 1, 6, 7\}$. Then, those bit positions are found in the groups A and B of SSb_0. This implies that the $\{0, 1, 6, 7\}$-th input bits of SSb_0 affect all 8 bits output. For the matrix operation ${}^t(y_0, y_1, y_2, y_3) \leftarrow M^t(x_0, x_1, x_2, x_3)$, we have the following property.

Property 2. *Each input cell affects three cells in the different cell positions from the input.*

For instance, x_0 deterministically affects y_1, y_2 and y_3, and does not affect y_0. From Properties 1 and 2, we obtain the following theorem.

Theorem 1. *In Midori128, any input bit nonlinearly affect all 128 bits of the state after 3 rounds.*

Proof. An input bit affects 4 bits in the corresponding cell after the first S-layer due to the full diffusion property of Sb_1. From Property 2, the affected 4 bits in the cell are diffused to three cells in the same column but the different cell position after MixColumn. Note that, in the affected three cells, the affected bit positions are the same. From Property 1, in each affected three cells, the affected 4 bits are spreads over all 8 bits in the cell after the 2nd S-layer. Therefore, all bits are affected by any input after 3 rounds (see Fig. 7).   □

4.3   Key Scheduling Function

To save energy, Midori128 does not employ any key scheduling function. The same 128 bit key is used as the whitening key and to generate the round key. To make an efficient circuit for decryption, the i-th round key is defined as $L^{-1}(K) \oplus L^{-1}(\beta_{18-i})$, where L^{-1} denotes the inverse of the linear layer. Computation of $L^{-1}(K)$ involves a one-time computation with the key at the beginning at the

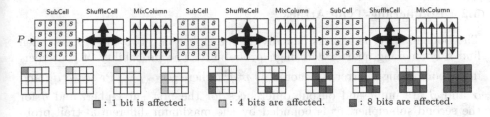

SubCell   ShuffleCell   MixColumn   SubCell   ShuffleCell   MixColumn   SubCell   ShuffleCell   MixColumn

■ : 1 bit is affected.   □ : 4 bits are affected.   ■ : 8 bits are affected.

Fig. 7. Theorem 1 : 3-round full diffusion property

decryption function and so does not consume any significant energy. The round key generation of Midori64, is slightly more complicated, as it involves selecting K_0 and K_1, i.e. the most significant and least significant halves of the 128 bit key in alternate rounds. This can be achieved by the use of a single multiplexer. For efficient decryption, a one-time computation of $L^{-1}(K_0)$ and $L^{-1}(K_1)$ can be done at the beginning of the algorithm, which again does not consume any significant energy.

4.4 Round Constant

Both Midori128 and Midori64 use 4×4 binary matrices as round constants. The constants have been derived from the hexadecimal encoding of the fractional part of $\pi = 3.243\text{f} \ 6\text{a}88 \ 85\text{a}3 \ \cdots$. For example, the 1st, 2nd, 3rd, 4th rows of β_0 when read as a 4-bit binary constant, are the encoding of the hex values 2,4,3,f respectively. Similarly for the other β_i's. These are added bitwise to the LSB of each round key byte in Midori128 and round key nibble in Midori64. The round constants were chosen in this manner with a view to have an energy-efficient decryption circuit. Both β_i and $L^{-1}(\beta_i)$ are 4×4 binary matrices, and so in both Midori128 and Midori64, the round constant addition requires a total of 16 XOR gates only. The constants β_i and $L^{-1}(\beta_i)$ can be stored in lookup tables and filtered accordingly in each round.

5   Security Evaluation

5.1   Differential/Linear Cryptanalysis

The minimum number of differentially and linearly active S-boxes of each round is estimated as shown in Table 7. The maximum differential and linear probabilities of Sb_0, SSb_0, SSb_1, SSb_2 and SSb_3 are 2^{-2}, respectively. Midori64 and Midori128 have more than 32 and 64 active S-boxes after 7 and 13 rounds. Thus, we expect that variants of Midori64 and Midori128 reduced to 7 rounds and 13 rounds do not have any differential and linear trails whose probabilities are higher than 2^{-64} and 2^{-128}.

5.2   Boomerang-Type Attack

The boomerang-type attacks first divide the cipher into two sub-ciphers, then find a boomerang quartet with high probability. The probability of constructing a boomerang quartet is denoted as $\hat{p}^2\hat{q}^2$, where $\hat{p} = \sqrt{\sum_\beta \Pr^2[\alpha \to \beta]}$, and α and β are input and output differences for the first sub-cipher, and $\hat{q}$ for the second sub-cipher. $\hat{p}^2$ is bounded by the maximum differential trail probability, i.e., $\hat{p}^2 \leq \max_\beta \Pr[\alpha \to \beta]$, and $\hat{q}^2$ as well. Let p, q be the maximum differential trail probability for the first and the second sub-ciphers. Then, p, q are bounded by multiplying the minimum number of active S-boxes in each sub-cipher. From Table 7, any combination of two sub-ciphers for consisting of Midori64 and Midori128 after 8 and 14 rounds has at least 32 and 64 active S-boxes in total. Note that these bounds of boomerang attacks are very conservative ones, i.e., it requires unrealistic assumptions of $\hat{p}^2 = p$ and $\hat{q}^2 = q$. Actually, in our active S-box search, we did not find such special events. Thus, we expect that much smaller rounds than 8 and 14 rounds are secure against boomerang-type attacks.

5.3   Impossible Differential Attacks

Midori64 and Midori128 achieve the 3-round full diffusion property. Thus, differences of all cells in a state becomes unknown after SubCell of 4 rounds, i.e., there is no any probability-one (truncated) differential characteristic. Following the miss-in-the-middle approach, the maximum number of rounds of impossible differential characteristics is estimated as 7 rounds.

In order to obtain the lower bound of rounds of impossible differential, we try to find actual impossible differential characteristics. We utilize several deterministic properties of four binary matrices M. This approach was also adopted in the security evaluation of FIDES [8]. As a result, we find 6-round impossible differentials such that *if only one active cell is input, 6-rounds of* Midori64 *and* Midori128 *never produces only one active cell.* We believe that full rounds of Midori64 and Midori128 have sufficient number of rounds as the security margin.

5.4   Meet-in-the-Middle Attacks

The 3-round full diffusion property with our S-boxes enable us to claim that any inserted key bit of $\{K_0, K_1\}$ or K non-linearly affects all bits of the state after 3 rounds in the forward and the backward directions in Midori64 and Midori128, respectively. Thus, the number of rounds used for the partial matching (PM) [2] is upper bounded by 5 $(= (3-1)+(3-1)+1)$. The condition for the initial structure (IS) [29], also called independent biclique [10], is that key differential trails in the forward direction and those in the backward direction do not share active non-linear components. For Midori64 and Midori128, since any key differential affects all 16 S-boxes after at least 4 rounds in the forward and the backward directions, there is no such differential which shares active S-box in more than 4 rounds. Thus, the number of rounds used for IS is upper bounded by 3. Assuming that the splice-and-cut technique allows an attacker to add more 3 rounds in the worst case, at most 11-round $(3 + 3 + 5)$ MitM attack may be feasible. However, because of white keys in the begin and the end and the actual constraint of key orders, we consider that it is difficult to construct 11-round attacks on Midori64 and Midori128.

5.5   Other Attacks

We also consider other-types of attacks including a integral differential, a truncated differential, a slide, a reflection, and an algebraic attack. Consequently, we expect that none of them work better than brute force attacks.

6   Implementation

The main design objectives of Midori were first to achieve efficiency in energy consumption and second to provide both the encryption and decryption functionalities with minimal overhead. In this context, it is essential to have a round based design optimal in terms of energy consumption, since unrolled designs are unlikely to be efficient in terms of energy consumption. The S-box and the Mix-Column layer were specifically chosen for their energy-efficiency and their involutive property. Both these layers have very small logic depth which makes the energy consumption per round figure as small as possible. Structurally Midori-Core and MidoriCore^{-1} differ only in the order of application of ShuffleCell, Mix-Column and InvShuffleCell operations. And so, the circuit for the round based implementation of the cipher, that accommodates both encryption and decryption can be realized in Fig. 8.

Since the ShuffleCell operation (Sh) and MixColumn (MC) do not commute, the linear layer which is basically the composition of MC∘Sh $(= L$ say), must be inverted during the decryption by $L^{-1} = $ Sh^{-1}∘MC. In hardware, this can be achieved in two ways. The first involves filtering the outputs of the L and L^{-1} operations through a single multiplexer. This requires two instances of the MixColumn logic in the circuit, and since this layer is the most expensive in

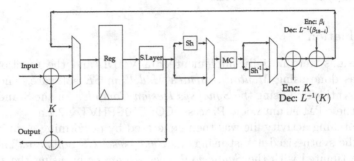

Fig. 8. The round based encryption/decryption architecture

Table 9. A comparison of energy consumption of Midori with selected ciphers for the STM 90 nm Logic Process. (Average Power reported at 10 MHz)

#	Cipher	Block Size	Architecture	Area (in GE)	Energy pJ	Energy/bit pJ	Average Power (μW)	Critical Path (ns)
1	AES	128	ED	21274	769.0	6.01	699.1	4.08
			E	12459	350.7	2.74	318.8	3.32
2	NOEKEON	128	ED	3439	331.5	2.59	184.2	3.79
			E	2284	338.0	2.64	187.8	3.38
3	SIMON 128/128	128	ED	3480	855.6	6.68	124.0	2.67
			E	2420	664.1	5.19	96.2	2.66
4	Midori128	128	ED	3661	228.3	1.78	108.7	2.44
			E	2522	187.3	1.46	89.2	2.25
5	PRESENT	64	ED	2186	250.2	3.91	75.8	2.32
			E	1440	172.3	2.69	52.2	2.09
6	PRINCE	64	ED	2650	146.3	2.29	112.5	4.09
			E	2286	144.7	2.26	111.3	4.06
7	Midori64	64	ED	2450	121.0	1.89	71.2	2.12
			E	1542	103.0	1.61	60.6	2.06

terms of area and energy consumed, it is not the most efficient way to achieve this functionality. The second method which is better in terms of both area and energy is the one shown in Fig. 8. This involves using two multiplexers for filtering the outputs of the Sh and Sh^{-1} operations and a single instance of the MixColumn logic. To perform the decryption operation using this circuit, the round key needs to be changed to $L^{-1}(K)$, and correspondingly the ith round constant to $L^{-1}(\beta_{18-i})$. The first involves a cheap one-time change to the master key, while keeping the whitening key constant. The round constant functionality can be achieved by employing two lookup tables, one each for encryption and decryption and filtering the appropriate round constant through a multiplexer. The round constants have been chosen in a manner so that both β_i and $L^{-1}(\beta_i)$ are 4×4 binary matrices, and so this layer requires a total of 16 XOR gates only. The circuit for the 64-bit variant is the same as in Fig. 8, except that it requires an extra filtering between K_0 and K_1 (the most and least significant halves of the secret key) in alternate rounds.

6.1   Evaluation

All the designs were initially implemented in VHDL and the functional verification was done using *Mentor Graphics ModelSim SE* software. The designs were then synthesized using the *Synopsys Design Compiler* for the Standard Cell library of the STM 90 nm Logic Process: CORE90GPHVT v 2.1.a.

The switching activity file was then generated by performing a timing simulation on the synthesized netlist using the *Synopsys VCS* Software. The energy was then estimated with the *Synopsys Power Compiler* by using the switching activity file. An operating frequency of 10 MHz was used in all the simulations since the effect of the leakage power is minimal at this frequency, and so the energy consumed is more or less independent of the clock frequency. The results of the simulation for the 90 nm logic process are presented in Table 9 along with similar evaluations for AES, NOEKEON, SIMON 128/128, PRESENT, PRINCE. It can be seen that Midori128/Midori64 performs better than NOEKEON/PRINCE which were also designed to make the combined functionalities of encryption and decryption easily available. In Fig. 9 we compare the energy/bit consumption of the ED architectures all the seven ciphers along with the cumulative latency figure (calculated as critical path × number of rounds). It can be seen that Midori128 and Midori64 fare optimally with respect to both parameters.

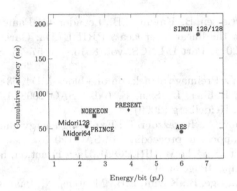

Fig. 9. Cumulative latency vs Energy/bit figures

7   Conclusion

In this paper we present the block ciphers Midori128 and Midori64, optimized with respect to energy consumption. We first identify design choices that make a given algorithm efficient in terms of energy. Thereafter we propose two design components i.e. MixColumn matrix and S-box, that help us achieve the objectives of low energy design. These components are additionally involutive, that makes it easier to design a circuit with functionalities for both encryption and decryption. The energy of the proposed design was then found to be optimal in comparison with state of the art block ciphers available in literature.

Appendix A: Test Vectors

A. Midori128

1.
Plaintext	:	00000000000000000000000000000000
Key	:	00000000000000000000000000000000
Ciphertext	:	c055cbb95996d14902b60574d5e728d6

```
        Plaintext   :  51084ce6e73a5ca2ec87d7babc297543
   2.   Key         :  687ded3b3c85b3f35b1009863e2a8cbf
        Ciphertext  :  1e0ac4fddff71b4c1801b73ee4afc83d
```

B. Midori64

```
        Plaintext   :  0000000000000000
   1.   Key         :  00000000000000000000000000000000
        Ciphertext  :  3c9cceda2bbd449a

        Plaintext   :  42c20fd3b586879e
   2.   Key         :  687ded3b3c85b3f35b1009863e2a8cbf
        Ciphertext  :  66bcdc6270d901cd
```

References

1. Albrecht, M.R., Driessen, B., Kavun, E.B., Leander, G., Paar, C., Yalçın, T.: Block ciphers – focus on the linear layer (feat. PRIDE). In: Garay, J.A., Gennaro, R. (eds.) CRYPTO 2014, Part I. LNCS, vol. 8616, pp. 57–76. Springer, Heidelberg (2014)

2. Aoki, K., Sasaki, Y.: Preimage attacks on one-block MD4, 63-step MD5 and more. In: Avanzi, R.M., Keliher, L., Sica, F. (eds.) SAC 2008. LNCS, vol. 5381, pp. 103–119. Springer, Heidelberg (2009)

3. Banik, S., Bogdanov, A., Regazzoni, F.: Exploring energy efficiency of lightweight block ciphers. To appear in proceedings of SAC (2015)

4. Barreto, P., Rijmen, V.: The WHIRLPOOL Hash Function. http://www.larc.usp.br/pbarreto/WhirlpoolPage.html

5. Batina, L., Das, A., Ege, B., Kavun, E.B., Mentens, N., Paar, C., Verbauwhede, I., Yalçın, T.: Dietary recommendations for lightweight block ciphers: power, energy and area analysis of recently developed architectures. In: Hutter, M., Schmidt, J.-M. (eds.) RFIDsec 2013. LNCS, vol. 8262, pp. 101–110. Springer, Heidelberg (2013)

6. Beaulieu, R., Shors, D., Smith, J., Treatman-Clark, S., Weeks, B., Wingers, L.: The SIMON and SPECK families of lightweight block ciphers. In: IACR eprint archive. https://eprint.iacr.org/2013/404.pdf

7. Bertoni, G., Macchetti, M., Negri, L., Fragneto, P.: Power-efficient ASIC synthesis of cryptographic S-boxes. In: 14th ACM Great Lakes Symposium on VLSI, pp. 277–281. ACM (2004)

8. Bilgin, B., Bogdanov, A., Knežević, M., Mendel, F., Wang, Q.: FIDES: Lightweight authenticated cipher with side-channel resistance for constrained hardware. In: Bertoni, G., Coron, J.-S. (eds.) CHES 2013. LNCS, vol. 8086, pp. 142–158. Springer, Heidelberg (2013)

9. Biryukov, A., Nikolić, I.: Automatic search for related-key differential characteristics in byte-oriented block ciphers: application to AES, Camellia, Khazad and others. In: Gilbert, H. (ed.) EUROCRYPT 2010. LNCS, vol. 6110, pp. 322–344. Springer, Heidelberg (2010)

10. Bogdanov, A., Khovratovich, D., Rechberger, C.: Biclique cryptanalysis of the full AES. In: Lee, D.H., Wang, X. (eds.) ASIACRYPT 2011. LNCS, vol. 7073, pp. 344–371. Springer, Heidelberg (2011)

11. Bogdanov, A., Knudsen, L.R., Leander, G., Paar, C., Poschmann, A., Robshaw, M., Seurin, Y., Vikkelsoe, C.: PRESENT: an ultra-lightweight block cipher. In: Paillier, P., Verbauwhede, I. (eds.) CHES 2007. LNCS, vol. 4727, pp. 450–466. Springer, Heidelberg (2007)

12. Borghoff, J., Canteaut, A., Güneysu, T., Kavun, E.B., Knezevic, M., Knudsen, L.R., Leander, G., Nikov, V., Paar, C., Rechberger, C., Rombouts, P., Thomsen, S.S., Yalçın, T.: PRINCE – a low-latency block cipher for pervasive computing applications. In: Wang, X., Sako, K. (eds.) ASIACRYPT 2012. LNCS, vol. 7658, pp. 208–225. Springer, Heidelberg (2012)
13. De Cannière, C., Dunkelman, O., Knežević, M.: KATAN and KTANTAN — a family of small and efficient hardware-oriented block ciphers. In: Clavier, C., Gaj, K. (eds.) CHES 2009. LNCS, vol. 5747, pp. 272–288. Springer, Heidelberg (2009)
14. Canright, D.: A very compact S-box for AES. In: Rao, J.R., Sunar, B. (eds.) CHES 2005. LNCS, vol. 3659, pp. 441–455. Springer, Heidelberg (2005)
15. Daemen, J., Peeters, M., Assche, G.V., Rijmen, V.: Nessie Proposal: NOEKEON. http://gro.noekeon.org/Noekeon-spec.pdf
16. Daemen, J., Rijmen, V.: The Design of Rijndael: AES - The Advanced Encryption Standard. Springer, Heidelberg (2002)
17. Feldhofer, M., Wolkerstorfer, J., Rijmen, V.: AES implementation on a grain of sand. IEEE Proc. Inf. Secur. **152**(1), 13–20 (2005)
18. Gong, Z., Nikova, S., Law, Y.W.: KLEIN: a new family of lightweight block ciphers. In: Juels, A., Paar, C. (eds.) RFIDSec 2011. LNCS, vol. 7055, pp. 1–18. Springer, Heidelberg (2012)
19. Guo, J., Peyrin, T., Poschmann, A., Robshaw, M.: The LED block cipher. In: Preneel, B., Takagi, T. (eds.) CHES 2011. LNCS, vol. 6917, pp. 326–341. Springer, Heidelberg (2011)
20. Iwata, T., Minematsu, K., Guo, J., Morioka, S.: CLOC: authenticated encryption for short input. In: Cid, C., Rechberger, C. (eds.) FSE 2014. LNCS, vol. 8540, pp. 149–167. Springer, Heidelberg (2015)
21. Kerckhof, S., Durvaux, F., Hocquet, C., Bol, D., Standaert, F.-X.: Towards green cryptography: a comparison of lightweight ciphers from the energy viewpoint. In: Prouff, E., Schaumont, P. (eds.) CHES 2012. LNCS, vol. 7428, pp. 390–407. Springer, Heidelberg (2012)
22. Khoo, K., Peyrin, T., Poschmann, A.Y., Yap, H.: FOAM: searching for hardware-optimal SPN structures and components with a fair comparison. In: Batina, L., Robshaw, M. (eds.) CHES 2014. LNCS, vol. 8731, pp. 433–450. Springer, Heidelberg (2014)
23. Lallemand, V., Naya-Plasencia, M.: Cryptanalysis of KLEIN. In: Cid, C., Rechberger, C. (eds.) FSE 2014. LNCS, vol. 8540, pp. 451–470. Springer, Heidelberg (2015)
24. Lim, C.H., Korkishko, T.: mCrypton – A lightweight block cipher for security of low-cost RFID tags and sensors. In: Song, J.-S., Kwon, T., Yung, M. (eds.) WISA 2005. LNCS, vol. 3786, pp. 243–258. Springer, Heidelberg (2006)
25. Matsui, M.: On correlation between the order of S-Boxes and the strength of DES. In: De Santis, A. (ed.) EUROCRYPT 1994. LNCS, vol. 950, pp. 366–375. Springer, Heidelberg (1995)
26. Sim, S.M., Khoo, K., Oggier, F., Peyrin, T.: Lightweight MDS involution matrices. In: Leander, G. (ed.) FSE 2015. LNCS, vol. 9054, pp. 471–493. Springer, Heidelberg (2015)
27. Moradi, A., Poschmann, A., Ling, S., Paar, C., Wang, H.: Pushing the limits: a very compact and a threshold implementation of AES. In: Paterson, K.G. (ed.) EUROCRYPT 2011. LNCS, vol. 6632, pp. 69–88. Springer, Heidelberg (2011)
28. Morioka, S., Satoh, A.: An optimized S-Box circuit architecture for low power AES design. In: Kaliski, B.S., Paar, C. (eds.) CHES 2002. LNCS, vol. 2523, pp. 172–186. Springer, Heidelberg (2003)

29. Sasaki, Y., Aoki, K.: Finding preimages in full MD5 faster than exhaustive search. In: Joux, A. (ed.) EUROCRYPT 2009. LNCS, vol. 5479, pp. 134–152. Springer, Heidelberg (2009)
30. Satoh, A., Morioka, S., Takano, K., Munetoh, S.: A compact Rijndael hardware architecture with S-Box optimization. In: Boyd, C. (ed.) ASIACRYPT 2001. LNCS, vol. 2248, pp. 239–254. Springer, Heidelberg (2001)
31. Shibutani, K., Isobe, T., Hiwatari, H., Mitsuda, A., Akishita, T., Shirai, T.: *Piccolo*: An ultra-lightweight blockcipher. In: Preneel, B., Takagi, T. (eds.) CHES 2011. LNCS, vol. 6917, pp. 342–357. Springer, Heidelberg (2011)
32. Shirai, T., Shibutani, K., Akishita, T., Moriai, S., Iwata, T.: The 128-bit blockcipher CLEFIA (Extended Abstract). In: Biryukov, A. (ed.) FSE 2007. LNCS, vol. 4593, pp. 181–195. Springer, Heidelberg (2007)
33. Suzaki, T., Minematsu, K., Morioka, S., Kobayashi, E.: *TWINE*: a lightweight block cipher for multiple platforms. In: Knudsen, L.R., Wu, H. (eds.) SAC 2012. LNCS, vol. 7707, pp. 339–354. Springer, Heidelberg (2013)
34. Tillich, S., Feldhofer, M., Großschädl, J.: Area, delay, and power characteristics of standard-cell implementations of the AES S-box. In: Vassiliadis, S., Wong, S., Hämäläinen, T.D. (eds.) SAMOS 2006. LNCS, vol. 4017, pp. 457–466. Springer, Heidelberg (2006)

Optimally Secure Block Ciphers
from Ideal Primitives

Stefano Tessaro[✉]

Department of Computer Science, University of California, Santa Barbara, USA
tessaro@cs.ucsb.edu
http://www.cs.ucsb.edu/~tessaro/

Abstract. Recent advances in block-cipher theory deliver security analyses in models where one or more underlying components (e.g., a function or a permutation) are *ideal* (i.e., randomly chosen). This paper addresses the question of finding *new* constructions achieving the highest possible security level under minimal assumptions in such ideal models.

We present a new block-cipher construction, derived from the Swap-or-Not construction by Hoang et al. (CRYPTO '12). With n-bit block length, our construction is a secure pseudorandom permutation (PRP) against attackers making $2^{n-O(\log n)}$ block-cipher queries, and $2^{n-O(1)}$ queries to the underlying component (which has itself domain size roughly n). This security level is nearly optimal. So far, only key-alternating ciphers have been known to achieve comparable security using $O(n)$ independent random permutations. In contrast, we only use a *single function* or *permutation*, and still achieve similar efficiency.

Our second contribution is a generic method to enhance a block cipher, initially only secure as a PRP, to additionally withstand related-key attacks without substantial loss in terms of concrete security.

Keywords: Block-cipher theory · Related-key security

1 Introduction

Several recent works provide ideal-model security proofs for key-alternating ciphers [2,14–17,19,23,25,26,31,50] and for Feistel-like ciphers [20,29,34,38,42]. In these proofs, the underlying components (wich are either permutations or functions) are chosen uniformly at random, and are *public*, i.e., the attacker can evaluate them. At the very least, these proofs target pseudorandom permutation (PRP) security: The block cipher, under a secret key, must be indistinguishable from a random permutation, provided the attacker makes at most q queries to the cipher, and at most q_F queries to the underlying component, for q and q_F as large as possible.

S. Tessaro—Partially supported by NSF grant CNS-1423566 and by the Glen and Susanne Culler chair.

ⓒ International Association for Cryptologic Research 2015
T. Iwata and J.H. Cheon (Eds.): ASIACRYPT 2015, Part II, LNCS 9453, pp. 437–462, 2015.
DOI: 10.1007/978-3-662-48800-3_18

Ideal-model proofs imply that the block cipher is secure against *generic* attacks (i.e., treating every component as a black box). Heuristically, however, one hopes for even more: Namely, that under a careful implementation of the underlying component, the construction retains the promised security level.

CONTRIBUTIONS. This paper contributes along two different axes:

– **Weaker Assumptions.** We present a new block-cipher design achieving near-optimal security, i.e., it remains secure even when q and q_F approach the sizes of the block-cipher and component domains, respectively. Our construction can be instantiated from a *function* or, alternatively, from a *single* permutation. This is the first construction from a function with such security level, and previous permutation-based constructions all relied on multiple permutations to achieve such high security.
– **Related-key Security.** We show how to enhance our construction to achieve related-key security without significantly impacting its efficiency and security. This is achieved via a *generic transformation* of independent interest.

This work should not be seen primarily as suggesting a new practical block-cipher construction, but rather as understanding the highest achievable security level in the model block ciphers are typically analyzed. The resulting technical questions are fairly involved, and resolving them is where we see our contributions.

Still, we hope that our approach may inspire designers. Our instantiation from a permutation gives a possible path for a first proof-of-concept implementation, where one simply takes a single-round of AES as the underlying permutation. (And in fact, even a simpler object may be sufficient).

1.1   First Contribution: Full-Domain Security

We start by explaining our construction from a (random) *function*. Concretely, we consider block-cipher constructions BC with block length n and key length κ using an underlying keyless *function* F with m-bit inputs. We say that BC is (q, q_F)-*secure* (as a PRP) if no attacker can distinguish with substantial advantage the *real world* – where it can query q_F times a randomly sampled function F and overall q times the block cipher BC_K^F (using the function F and a random secret key K) – from an *ideal world* where BC_K^F is replaced by an independent random permutation of the n-bit strings. (In fact, we typically also allow inverse queries to the block cipher and the permutation).

OUR GOAL. Let us first look at what can we expect for q and q_F when a cipher is (q, q_F)-secure. Clearly, $q_F \leq 2^m$ and $q \leq 2^n$, assuming queries are distinct. However, one can also prove that (roughly) $q_F < 2^\kappa$ is necessary, otherwise, the adversary can mount a brute-force key search attack. Moreover, $q \leq 2^m$ must also hold (cf. e.g. [28] for a precise statement of these bounds).

Here, we target (near) optimal security, i.e., we would like to achieve security for q and q_F as close as possible to 2^n and 2^m, respectively, whenever $m \geq n$. That is, the construction should remain secure even if the adversary can query most of its domain, and of that of the underlying function F. We note that the question is meaningful for every value of $m \geq n$, but we specifically target the case where $m \approx n$, e.g., $m = n$, or $m = n + O(\log n)$.

Previous constructions from functions fall short of achieving this: Gentry and Ramzan [29], and the recent generalization of their work by Lampe and Seurin [38], use a Feistel-based approach with $m = n/2$, and this hence yields (at best) $(2^{n/2}, 2^{n/2})$-security. (The work of [38] approaches that security level for increasing number of rounds). In contrast, key-alternating ciphers (KACs) have been studied in several works [2,14–16,19,23,26,31,50], and the tightest bounds show them to be $(2^{n(1-\epsilon)}, 2^{n(1-\epsilon)})$-secure, when using $O(1/\epsilon)$ rounds calling each an (independent) n-bit random *permutation*. However, there is no way of making direct use of KACs given only a non-invertible *function*.

THE WSN CONSTRUCTION. Our construction – which we call *Whitened Swap-or-Not* (WSN) – adds simple whitening steps to the Swap-or-Not construction by Hoang, Morris, and Rogaway [33], which was designed for the (different) setting where the component functions are secret-key primitives. Concretely, the WSN construction, on input $X = X_0$, iterates R times a very simple round structure of the form

$$X_{i+1} \leftarrow X_i \oplus (F_{b(i)}(W_i \oplus \max\{X_i, X_i \oplus K_i\}) \cdot K_i),$$

where W_i and K_i are round keys, max of two strings returns the largest with respect to lexicographic ordering, and $F_{b(i)}(x)$ returns the first bit of $F(x)$ in the first half of the rounds, and the second bit in the second half. (Moreover, $\cdot$ denotes simple scalar multiplication with a bit, i.e., $b \cdot X = X$ if $b = 1$, and 0^n else). *In particular, our construction requires F to only output 2 bits.* The round structure is very weak[1], and it differs from the construction of [33] in that the *same* round function is invoked over multiple rounds, and as this function is public, we use a key W_i to whiten the input. We prove the following:

Main Theorem. (Informal) The WSN construction for $R = O(n)$ rounds is $(2^{n-O(\log n)}, 2^{n-O(1)})$-secure.

Note that $O(n)$ rounds are clearly asymptotically optimal.[2] For some parameter cases, techniques from [47,49] can in fact be used to obtain a $(2^n, 2^{n(1-\epsilon)})$-secure PRP, at the cost of a higher number of rounds.

FUNCTIONS VS. PERMUTATIONS. It is beyond the scope of this paper to assess whether a function is a better starting point than a permutation *in practice*. Independently of this, we believe that studying constructions from *functions* is a fundamental theoretical problem for at least two reasons.

---

[1] A single round can easily be distinguished from a random permutation with a constant number of queries, as every input x is mapped to either x or $x \oplus K_i$.

[2] Even for one single query, every internal call to F can supply at most one bit of randomness, and the output must be (information theoretically) indistinguishable from a random n-bit string, and thus $\Omega(n)$ calls are necessary.

Foremost, functions are combinatorially simpler than permutations, and thus, providing constructions from them (and thus enabling a secure permutation structure) is an important theoretical question, akin to (and harder than) the problem of building PRPs from PRFs covered by a multitude of papers. Also, practical designs from keyless round functions have been considered (cf. e.g. [1]).

In addition, our construction *only* requires $c = 2$ output bits, and it is worth investigating whether such short-output functions are also harder to devise than permutations. We in fact provide some theoretical evidence that this may not be the case. We prove that an elegant construction by Hall, Wagner, Kelsey, and Schneier [32] can be used to transform *any* permutation from $n + c$ bits to $n + c$ bits into a function from n bits to c bits which is *perfectly indifferentiable* [44] from a random function. This property ensures that the *concrete security* of every cipher using a function $F : \{0,1\}^n \to \{0,1\}^c$ is preserved if we replace F with the construction from π, and allow the adversary access to π and its inverse π^{-1}. The construction makes 2^c permutation calls, and thus makes only sense for small c. In contrast, it should be noted that the only indifferentiable construction of a permutation from functions is complex and weakly secure [34], and that no suitable constant-complexity high-security constructions of large-range functions from permutations exist, the most secure construction being [41,46].

A SINGLE-PERMUTATION INSTANTIATION. With $c = 2$, combining the WSN construction with the HWKS construction yields a secure cipher with n-bit block length from a single permutation on $(n + 2)$-bit strings. In contrast, we are not aware of any trick to instantiate KACs from a single permutation retaining provable nearly-optimal security, even by enlarging the domain of the permutation. The only exception is the work of [15], which however only considers two rounds and hence falls short of achieving full-domain security.

The complexity of the resulting construction matches (asymptotically) that of KACs when targeting $(2^{n-O(\log n)}, 2^{n-O(1)})$-security. Nonetheless, a clear advantage of KACs is that their security degrades smoothly when reducing the amount of rounds, whereas here $O(n)$ rounds remain necessary even for $(1,0)$-security. We note that in the setting of functions constructions with such smooth security degradations are not known, even in the simpler setting of [33].

REDUCING THE KEY LENGTH. Arguably, an obvious drawback of our construction is that the key length grows with the number of rounds. We note that this is also true for key-alternating ciphers, and it is not unique to our construction.

It is worth noting that the key length *can* be reduced via standard techniques without affecting security, by deriving the round keys from a single $(n - d)$-bit master key K as $K_i \leftarrow H(K \parallel \langle i-1 \rangle)$ and $W_i \leftarrow H(K \parallel \langle R+i-1 \rangle)$ for all $i \in [R]$ and a function $H : \{0,1\}^n \to \{0,1\}^n$ (to be modeled as random in the proof), where $\langle \cdot \rangle$ denotes the $(d = \lceil \log(2R) + 1 \rceil)$-bit binary encoding of an integer in $[2R]$. (Note that $d = O(\log n)$.) The security proof is fairly straightforward, and omitted – it essentially accounts to excluding the event that H is queried on one of the values related to the key, and the reducing the analysis to the one with large keys. This adds an additional $q_H \cdot R/2^{n-d}$ term to the bound, where q_H is the number of queries to H. H can in fact be built from the very same function F, but this requires a slightly more involved analysis.

1.2 Second Contribution: Related-Key Security

In the second part, we show how to generically make *any* block-cipher construction secure against *related-key attacks* (or RKA secure, for short) while preserving full-domain security and *small* input length of the underlying function.

On RKA security. Several attacks over the last two decades (cf. e.g. [8–13, 35]) have motivated RKA security as the new golden standard for block-cipher security. As formalized by Bellare and Kohno [5], RKA security is parameterized by a class of key transformations Φ. Then, pseudorandomness security defined above is extended to allow the attacker for block-cipher queries of the form $(\phi, +, X)$ or $(\phi, -, Y)$ for $\phi \in \Phi$ and $X, Y \in \{0,1\}^n$, resulting in $\mathsf{BC}_{\phi(K)}(X)$ and $\mathsf{BC}_{\phi(K)}^{-1}(Y)$.

It is easy to see that WSN is *not* RKA secure if the class Φ allows for XORing chosen offsets to individual keys. Querying an input X (with the original key), and querying $X \oplus \Delta$ while adding Δ to K_1 results in the same output with probability $1/2$. In the random permutation model, two recent works [19, 26] have shown that KACs are RKA secure (for appropriate key scheduling), yet the resulting construction is only $(2^{n/2}, 2^{n/2})$-secure. Here, in contrast, we target full-domain security of the cipher.

Related-key secure key-derivation. We consider a generic approach to shield ciphers from related-key attacks using *related-key secure key-derivation functions* (or RKA-KDF, for short). These are functions $\mathsf{KDF} : \{0,1\}^\kappa \to \{0,1\}^\ell$ with the property that under a random secret key K, the outputs of $\mathsf{KDF}(\phi(K))$, for different $\phi \in \Phi$, look random and independent. A similar concept was proposed by Lucks [40], and further formalized by Barbosa and Farshim [3]. For any secure block cipher BC, the new block cipher computes, for key K and input X, the value $\mathsf{BC}_{\mathsf{KDF}(K)}(X)$, and is easily proved to be RKA-secure. Note that this approach is very different from the one used for standard-model RKA-secure PRF and PRP constructions (as in [4]), which leverage algebraic properties of PRF constructions.[3]

Building RKA-KDFs in *ideal* models may appear too easy: A hash function $H : \{0,1\}^\kappa \to \{0,1\}^\ell$, when modeled as a random oracle [6], *is* a secure RKA-KDF. However, such construction can be broken in $2^{\kappa/2}$ queries by a simple collision argument.[4] If our goal is to achieve security almost 2^n to preserve security of e.g. WSN above, then we need to set $\kappa \geq 2n$. But what if we are

---

[3] Also, our requirements are stronger than those for non-malleable codes and non-malleable key-derivation [24, 27].

[4] For example, for $Q := 2^{\kappa/2}$, and an additive RKA attack asking for random $\Delta_1, \ldots, \Delta_Q$, one of the values $H(K \oplus \Delta_i)$ is going to collide with constant probability with one of the values $H(X_i)$, for independent κ-bit strings $X_1, \ldots, X_Q$, allowing to distinguish.

building our block cipher from a primitive with n-bit inputs, like the very same primitive used to build the block cipher, as in the WSN setting above?

One approach is to use a domain extender in the sense of indifferentiability [44]. The only known construction with (near) optimal security is due to Maurer and Tessaro [45] (MT), and further abstracted by Dodis and Steinberger [22]. Unfortunately, instantiations of the MT construction are very inefficient, and make $O(n^c)$ calls to the underlying function for some undetermined (and fairly large) c.

MT-BASED RKA-KDFs. As our second contribution, we provide a highly parallelizable construction of a RKA-KDF from a keyless function with nearly optimal security, i.e., its outputs are pseudorandom even when evaluated on $q = 2^{n(1-\epsilon)}$ related keys, and the underlying function can be evaluated $q_F = 2^{n(1-\epsilon)}$ times, where $\epsilon > 0$. Our construction is a variant of the MT construction. However, while the latter is inefficient as it relies on a complex combinatorial object, called an input-restricting function family, here, we show that to achieve RKA-KDF security it is sufficient to use a much simpler *hitter* [30], which can for instance be built from suitable constant-degree expander graphs.

Overall, our construction needs $O(n)$ calls to *independent* n-to-n-bit functions. (It can also be reformulated to call a single n-to-n-bit function). We see it as a challenging open problem to improve the complexity, but we note that this already yields the most efficient known approach to ensure high related-key security for block ciphers built from ideal primitives.

INDIFFERENTIABILITY. The question of building a block cipher from a random function which is as secure as an ideal cipher (with respect to indifferentiability) was studied and solved by [20,34]. In the same vein, indifferentiable KAC-like cipher constructions from permutations have been given [2,31,37]. While these constructions are related-key secure, their concrete security is fairly weak.

2   Preliminaries

2.1   Notation

Throughout this paper, we let $[n] := \{1, \ldots, n\}$. Further, we denote by $\mathsf{Fcs}(m, n)$ the set of functions mapping m-bit strings to n-bit strings, and by $\mathsf{Fcs}(*, n)$ the set of functions $\{0,1\}^* \to \{0,1\}^n$. Similarly, we let $\mathsf{Perms}(n) \subset \mathsf{Fcs}(n, n)$ be the set of permutations on $\{0,1\}^n$. Given a string $X \in \{0,1\}^m$, we denote by $X[i \ldots j]$ (for $i < j$) the sub-string consisting of bits $i, i+1, \ldots, j-1, j$ of X. We also write $X^{\leq i}$ instead of $X[1 \ldots i]$. Further, given another string $X' \in \{0,1\}^n$, we denote by $X \parallel X'$ the $(m+n)$-bit concatenation of X and X'.

Algorithms, constructions, and adversaries in this paper are with respect to some (not further specified) RAM model of computation. We explicitly denote by $\mathsf{C}[F]$ the fact that a construction C (implementing a function) makes queries to another function F, and we denote by $\mathcal{A}^{\mathsf{O}}$ the fact that an adversary $\mathcal{A}$ accesses an oracle O. We denote by $x \xleftarrow{\$} \mathsf{S}$ the process of sampling x from the set S

uniformly at random, and by $y \xleftarrow{\$} \mathcal{A}^O$ the process of running the randomized algorithm $\mathcal{A}$ with access to a randomized oracle O, and sampling its output y. Also, we denote by $\mathcal{A}^O \Rightarrow y$ the event that the concrete value y is output in the same experiment. In general, we use a notation close to the one of Bellare and Rogaway's Game Playing framework [7], which we hope to be self evident.

Additionally, we denote by $\Pr[X = x]$ the probability that the random variable takes the value x, and by $\mathsf{E}[X]$ its expected value. Also, the *statistical distance* between two random variables X and X' is $\mathsf{SD}(X, X') = \frac{1}{2} \sum_x |\Pr[X = x] - \Pr[X' = x]|$, where the sum is over all values which can be taken by X or X'.

2.2 Ideal Models

Our analyses are in the *random function model*, where algorithms and adversaries are relative to a randomly chosen function $F \xleftarrow{\$} \mathsf{Fcs}(m, \ell)$ for parameters m and ℓ. A variant of the model grants access to *multiple* independent random functions $F_1, \ldots, F_t \xleftarrow{\$} \mathsf{Fcs}(m, \ell)$, but these can equivalently be implemented in the *single* random function model for $m' = m + \lceil \log t \rceil$, where the individual functions F_i are obtained as $F_i(X) = F(\langle i \rangle \parallel X)$, with $\langle i \rangle$ representing a $\lceil \log t \rceil$-bit encoding of i. We often denote $F = (F_1, \ldots, F_t)$ to stress this dual representation explicitly. Therefore, all upcoming definitions are in the single random function model without loss of generality.

We also recall that we can build a function F from m bits to ℓ bits by making ℓ calls to a function from $m + \lceil \log \ell \rceil$ bits to a single bit, i.e., $F(X) = F'(\langle 0 \rangle \parallel X) \parallel \cdots \parallel F'(\langle \ell - 1 \rangle \parallel X)$. The statement can be made precise via the notion of perfect indifferentiability [44], which we review in Appendix A.

The definitions of this section also naturally extend to the *random permutation model*, where adversaries and algorithms can query one or more random permutations sampled uniformly from $\mathsf{Perms}(n)$. In particular, adversaries are also allowed query the inverses of these permutations.

2.3 Block Ciphers and (related-Key) Pseudorandomness

Let $\mathsf{BC}[F] : \{0,1\}^\kappa \times \{0,1\}^n \to \{0,1\}^n$ be an efficient construction making calls to a function $F \in \mathsf{Fcs}(m, \ell)$. (We generally omit F whenever clear from the context). We say that $\mathsf{BC} = \mathsf{BC}[F]$ is a (κ, n)-block cipher if $\mathsf{BC}(K, \cdot)$ is a permutation for all κ-bit K and all $F \in \mathsf{Fcs}(m, \ell)$, and use the notation BC_K to refer to this permutation. Typically, we assume that BC_K and BC_K^{-1} are very efficient to compute given K, where efficiency in particular implies a small number of calls to F.

(MULTI-USER) PRPs. We require block ciphers to be secure *pseudorandom permutations (PRPs)* [39]. In particular, we consider a multi-user version of PRP security, which captures joint indistinguishability of an (a-priori unbounded) number of block-cipher instantiations under different independent keys. The traditional (single-user) PRP notion is recovered by considering adversaries making queries for one single key. While the single- and multi-user versions are related by a hybrid argument, sticking with the latter will allow potentially tighter bounds in the second part of this paper, as the standard hybrid argument cannot be made very tight given only an overall bound on the number of queries.

To this end, we consider two security games PRP-$b_{\mathsf{BC},F}^{\mathcal{A}}$ for $b \in \{0,1\}$. In both, $F \xleftarrow{\$} \mathsf{Fcs}(m, \ell)$ is initially sampled, as well as independent keys $K_1, K_2, \ldots \xleftarrow{\$} \{0,1\}^\kappa$, and permutations $P_1, P_2, \ldots \xleftarrow{\$} \mathsf{Perms}(n)$.[5] Then, the adversary $\mathcal{A}$ is executed, and is allowed to issue two types of queries:

- **Function queries** x, returning $F(x)$
- **Construction queries** (i, σ, z), where $i \in \mathbb{N}$, $\sigma \in \{-, +\}$, $z \in \{0,1\}^n$. For $b = 1$, the query returns $\mathsf{BC}_{K_i}(z)$ (if $\sigma = +$, this is a *forward query*) or $\mathsf{BC}_{K_i}^{-1}(z)$ (if $\sigma = -$, and this is a *backward query*). For $b = 0$, the query returns $P_i(z)$ or $P_i^{-1}(z)$, respectively.

Finally, $\mathcal{A}$ outputs a bit, which is also the game's output. Then, PRP-security of BC is defined via the following advantage metric

$$\mathsf{Adv}_{\mathsf{BC},F}^{\mathsf{PRP}}(\mathcal{A}) := \Pr\left[\mathsf{PRP}\text{-}1_{\mathsf{BC},F}^{\mathcal{A}} \Rightarrow 1\right] - \Pr\left[\mathsf{PRP}\text{-}0_{\mathsf{BC},F}^{\mathcal{A}} \Rightarrow 1\right] .$$

We also denote by $\mathsf{Adv}_{\mathsf{BC},F}^{\mathsf{PRP}}(q, q_F)$ the maximal advantage of an adversary $\mathcal{A}$ making at most q construction queries and q_F function queries. Informally, we say that BC is (q, q_F)-secure if $\mathsf{Adv}_{\mathsf{BC},F}^{\mathsf{PRP}}(q, q_F)$ is "small", i.e., negligible in κ.

RELATED-KEY SECURE PRPs. We target the traditional notion of a related-key secure (or RKA-secure) PRP introduced by Bellare and Kohno [5]. In particular, for a key length κ, we consider a family $\Phi \subseteq \mathsf{Fcs}(\kappa, \kappa)$ of *key transformations*. Given a (κ, n)-block cipher $\mathsf{BC} = \mathsf{BC}[F]$ as above, we define the following two games RKA-PRP-1 and RKA-PRP-0. The game RKA-PRP-$b_{\mathsf{BC},F,\Phi}^{\mathcal{A}}$ proceeds as follows: It first samples $F \xleftarrow{\$} \mathsf{Fcs}(m, \ell)$, a key $K \xleftarrow{\$} \{0,1\}^\kappa$, and 2^κ independent permutations $P_{k'} \xleftarrow{\$} \mathsf{Perms}(n)$ for all κ-bit k'. Then, $\mathcal{A}$ issues two types of queries:

- **Function queries** x, returning $F(x)$
- **Construction queries** (σ, ϕ, X), where $\sigma \in \{-, +\}$, $\phi \in \Phi$, $z \in \{0,1\}^n$. For $b = 1$, the query returns $\mathsf{BC}_{\phi(K)}(z)$ (if $\sigma = +$, this is a *forward query*) or $\mathsf{BC}_{\phi(K)}^{-1}(z)$ (if $\sigma = -$, and this is a *backward query*). For $b = 0$, the query returns $P_{\phi(K)}(z)$ or $P_{\phi(K)}^{-1}(z)$, respectively.

Finally, $\mathcal{A}$ outputs a bit, which is also the game's output. We define the RKA-PRP advantage as

$$\mathsf{Adv}_{\mathsf{BC},F,\Phi}^{\mathsf{RKA\text{-}PRP}}(\mathcal{A}) = \Pr\left[\mathsf{RKA}\text{-}\mathsf{PRP}\text{-}1_{\mathsf{BC},F,\Phi}^{\mathcal{A}} \Rightarrow 1\right] - \Pr\left[\mathsf{RKA}\text{-}\mathsf{PRP}\text{-}0_{\mathsf{BC},F,\Phi}^{\mathcal{A}} \Rightarrow 1\right] .$$

The advantage measure $\mathsf{Adv}_{\mathsf{BC},F,\Phi}^{\mathsf{RKA\text{-}PRP}}(q, q_F)$ is defined by taking the maximum.

---

[5] As we are sampling infinitely many objects, once can think of sampling these lazily the first time they are needed.

3   The Whitened Swap-or-Not Construction

3.1   The Construction

We present a construction of a block cipher using a function $F : \{0,1\}^n \rightarrow \{0,1\}^2$, which we refer to as the *Whitened Swap-or-Not construction*, or WSN for short. This construction naturally extends the Shuffle-or-Not construction by Hoang, Morris, and Rogaway [33] to the keyless-function setting.

For any even round number $R = 2r$, the construction WSN $=$ WSN$^{(R)}$ expects *round* keys $K_1, \ldots, K_R$ and *whitening* keys $W_1, \ldots, W_R$, which are all n-bit strings. Its computation proceeds as follows, where $j(i) = 1$ if $i \leq r$, and $j(i) = 2$ else, and we interpret F as two functions F_1 and F_2 such that $F_j(x)$ returns the j-th bit of $F(x)$ for $j \in \{1, 2\}$.

Construction WSN$^{(R)}_{K_1,\ldots,K_R,W_1,\ldots,W_R}(X)$:   // $X \in \{0,1\}^n$

$X_0 \leftarrow X$
For $i = 1, \ldots, R$ **do**
$\quad X'_{i-1} \leftarrow \max\{X_{i-1}, X_{i-1} \oplus K_i\}$
$\quad B_i \leftarrow F_{j(i)}(W_i \oplus X'_{i-1})$
$\quad$ **If** $B_i = 1$ **then** $X_i \leftarrow X_{i-1} \oplus K_i$ **else** $X_i \leftarrow X_{i-1}$
Return X_R

In the description, the max of two strings is with respect to the lexicographic order, and note that its purpose is to elect a unique representant for every pair $\{X, X \oplus K_i\}$. As in [33], the construction extends naturally to domains which are arbitrary abelian groups. However, we will stick with the special case of bit strings in the following.

It is easy to see that the construction can efficiently be inverted given the keys, simply by reversing the order of the rounds.

3.2   Security of the WSN Construction

Compared with the original Swap-or-Not construction, WSN adds at each round a whitening key W_i to the input of a (publicly evaluable) round function $F_{j(i)}$, as opposed to using a *secret* independent random function F_i (which in particular *cannot* be queried directly by the adversary). It is a well-known folklore fact that for a function $F : \{0,1\}^n \rightarrow \{0,1\}$, the construction mapping a key W and an input X to $F(W \oplus X)$ is indistinguishable from a random function under a random secret key W when F is random and publicly evaluable.

However, the high security of WSN does not follow by simply composing this folklore fact with the original analysis [33]. This is because the folklore construction can easily be distinguished from a random function via $\Theta(2^{n/2})$ queries to $F(W \oplus \cdot)$ (or a random function $f \xleftarrow{\$} \mathsf{Fcs}(n, 1)$), and $\Theta(2^{n/2})$ queries to

F.[6] To overcome this, a valid black-box instantiation would use a more complex construction mapping X to $F(W_1 \oplus X) \oplus \cdots \oplus F(W_k \oplus X)$ (analyzed in [28]) for the round functions within Swap-or-Not. This would however result in roughly $\Theta(n^2)$ calls to F, as opposed to $\Theta(n)$ achieved by WSN.

SECURITY OF WSN. The following theorem establishes the concrete security of the WSN construction with $R = 2r$ rounds.

Theorem 1 (Security of WSN). *For all $q, q_F > 0$ and for all $r \in \mathbb{N}$, we have*

$$\mathsf{Adv}^{\mathrm{PRP}}_{\mathsf{WSN}^{(2r)}, F}(q, q_F) \leq 2\sqrt{2}\sqrt{q}2^{n/4}\left(\frac{1}{2} + \frac{q \cdot r + q_F}{2 \cdot 2^n}\right)^{r/4}.$$

The proof of Theorem 1 is given in Sect. 3.3 below. Note that if $r \cdot q + q_F = (1 - \alpha)2^n$, then the above term can be made to be 2^{-n} for $r = O(n/\alpha)$. For example, this allows to infer security for $q = 2^{n - \log n - O(1)}$ and $q_F = 2^{n-2}$.

We also have no reason to believe that the construction would be insecure if we used a function with a single output bit throughout the evaluation, but we could not find a suitable proof and leave this analysis as an open problem.

SINGLE-PERMUTATION INSTANTIATION. The WSN construction can be instantiated from a single permutation if we are ready to enlarge the domain of the permutation to $n + 2$ bits. This follows from a result of independent interest, proved in Appendix B. Namely, we prove that a 2^c-call construction of a function $F^\pi \in \mathsf{Fcs}(n, c)$ from any permutation $\pi \in \mathsf{Perms}(n + c)$ due to Hall, Wagner, Kelsey, and Schneier [32] is *perfectly* indifferentiable [44] from a random function. This in particular implies (by the composition theorem in Appendix A) that we can replace the function F by our construction and still achieve the *same* security bound in the random permutation model.

FULL-DOMAIN SECURITY. Two recently published works [47,49] enhance swap-or-not to *full-domain* security (i.e., security against $q = 2^n$ queries) at the cost of making $O(n^2)$ calls to the construction in the worst-case. (The later work [47] shows how to reduce the complexity to $O(n)$ in the *average* case). One could hope to use their results generically to obtain $(2^n, 2^{n(1-\epsilon)})$-security in our setting.

Unfortunately, these results require security for $q = 2^{n-1}$, which is unattainable by the above bound. By inspecting the proof of Theorem 1, it is however not hard to verify that a version of the WSN construction with *independent* round functions $F_1, \ldots, F_R$ can be made to achieve $(2^{n-1}, 2^{n(1-\epsilon)})$-security (in essence, this is because one can easily reduce the exponential term in the bound to $(\frac{1}{2} + \frac{q_F + q}{2 \cdot 2^n})^{r/4}$) and the results from [47,49] can be used in a black-box way.

Nevertheless, we point out that in contrast to the *small-domain* setting of [47, 49], here we are mostly targeting a large n (e.g., $n = 128$), for which $2^{n(1-\epsilon)}$ security can be largely sufficient. The additional cost may thus not be necessary.

---

[6] Roughly, pick $X_1, \ldots, X_Q, X'_1, \ldots, X'_Q$ to be independent uniform n-bit strings of length $n - k$, for some $k = \lceil \log n \rceil$ and $Q \approx 2^{n/2}$. Then one just queries $Y_{z,i} \leftarrow F(W \oplus (X_i \| z))$ and $Y'_{z,j} \leftarrow F(X'_j \| z)$ for all $i, j \in [Q]$ and $z \in \{0, 1\}^k$. The distinguisher finally outputs one if and only if there exist i and j such that $Y_{i,z} = Y'_{j,z}$ for all $z \in \{0, 1\}^k$.

3.3   Proof of Theorem 1

Our proof shares similarities with the original analysis of Swap-or-Not [33], but dealing with the setting where the function F is public requires a careful extension and different techniques. To this end, we follow an approach used in previous works by Lampe, Patarin, and Seurin [36], and by Lampe and Seurin [38] to reduce security analyses for PRP constructions in ideal models to a non-adaptive analysis. (With some extra care due to the fact that we deal with the multi-user PRP security notion). In particular, we are first going to prove that the WSN construction, restricted to half of its rounds, satisfies a weaker non-adaptive security requirement, which we introduce in the following paragraph.

NON-ADAPTIVE SECURITY. Let $\mathsf{BC} = \mathsf{BC}[F]$ be a (κ, n)-block cipher construction based on some function $F : \{0, 1\}^m \to \{0, 1\}^\ell$. Now, let us fix a set of tuples $T_F = \{(x_i, y_i)\}_{i \in [q_F]}$ with $x_i \in \{0, 1\}^m$ and $y_i \in \{0, 1\}^\ell$ for all $i \in [q_F]$, and such that every x_i appears only in one pair in T_F. Moreover, let us fix a sequence $\mathbf{X}$ of q distinct inputs such that $\mathbf{X}[j] = (i_j, X_j)$ for all $j \in [q]$, where $i_j \in \mathbb{N}$ and $X_j \in \{0, 1\}^n$.

Then we consider two processes – sampling two sequences $\mathbf{Y}$ and $\mathbf{Y}'$ of q n-bit strings – defined as follows:

- $\mathbf{Y}$ (the *real world* distribution) is obtained by sampling random κ-bit strings $K_1, K_2, \ldots \xleftarrow{\$} \{0, 1\}^\kappa$, sampling a random $F \xleftarrow{\$} \mathsf{Fcs}(m, \ell)$ *conditioned on* satisfying $F(x_i) = y_i$ for all $i \in [q_F]$, and finally letting $\mathbf{Y}[j] \leftarrow \mathsf{BC}[F]_{K_{i_j}}(X_j)$ for all $j \in [q]$.
- $\mathbf{Y}'$ (the *ideal world* distribution) is obtained by sampling random permutations $P_1, P_2, \ldots \xleftarrow{\$} \mathsf{Perms}(n)$, and letting $\mathbf{Y}[j] \leftarrow P_{i_j}(X_j)$ for all $i \in [q]$.

Then, we define the advantage metric

$$\mathsf{Adv}_{\mathsf{BC},F}^{\mathsf{NCPAPRP}}(\mathbf{X}, T_F) := \mathsf{SD}(\mathbf{Y}, \mathbf{Y}') \,,$$

where SD denotes statistical distance. Moreover, let $\mathsf{Adv}_{\mathsf{BC},F}^{\mathsf{NCPAPRP}}(q, q_F)$ be the maximum of $\mathsf{Adv}_{\mathsf{BC},F}^{\mathsf{NCPAPRP}}(\mathbf{X}, T_F)$ taken over all q-sequences $\mathbf{X}$ and all sets T_F of size q_F.

FROM NON-ADAPTIVE TO ADAPTIVE SECURITY. We make use of the following lemma. The proof is very similar to previous works [36,38] and makes crucial use of Patarin's H-coefficient method [48]. The main difference is that our version deals with the multi-user PRP security notion. (A self-contained version of the proof is found in the full version).

Given a (κ, n)-block cipher $\mathsf{BC}[F]$ relying on a function $F : \{0, 1\}^m \to \{0, 1\}^\ell$, then let $\mathsf{BC}[F_1] \circ \mathsf{BC}^{-1}[F_2]$ be the $(2\kappa, n)$-block cipher which relies on two functions $F_1, F_2 : \{0, 1\}^m \to \{0, 1\}^\ell$, and which on input $X \in \{0, 1\}^n$ and given key $K_1 \parallel K_2 \in \{0, 1\}^{2\kappa}$, returns $\mathsf{BC}[F_2]_{K_2}^{-1}(\mathsf{BC}[F_1]_{K_1}(X))$. The following lemma tells us that if BC is non-adaptively secure (as in the above notion), then $\mathsf{BC} \circ \mathsf{BC}^{-1}$ is *adaptively* secure in the sense of being a secure PRP for attackers making both forward and backward queries.

Lemma 1 (Non-adaptive $\Rightarrow$ Adaptive Security). *For all q, q_F, we have*

$$\mathsf{Adv}^{\mathsf{PRP}}_{\mathsf{BC}[F_1] \circ \mathsf{BC}^{-1}[F_2], (F_1, F_2)}(q, q_F) \leq 4 \cdot \sqrt{\mathsf{Adv}^{\mathsf{NCPAPRP}}_{\mathsf{BC}[F], F}(q, q_F)} \ .$$

Note that a stronger version of this statement (essentially without the square root) can be proved [18, 43] in the setting where $q_F = 0$.

NON-ADAPTIVE ANALYSIS OF WSN. We first adopt a slightly different representation of the WSN construction. In particular, let $\overline{\mathsf{WSN}}^{(r)} = \overline{\mathsf{WSN}}^{(r)}[F]$ be the construction relying on a function $F : \{0, 1\}^n \to \{0, 1\}$ which operates as the original WSN construction for r rounds, but always uses the the function F (instead of using one function F_1 for the first half, and the function F_2 for the second half of the evaluation). Then, it is easy to see that

$$\mathsf{WSN}^{(2r)}[F_1, F_2] = \overline{\mathsf{WSN}}^{(r)}[F_1] \circ \left(\overline{\mathsf{WSN}}^{(r)}[F_2] \right)^{-1} , \tag{1}$$

where in particular we have used the fact that the inverse of $\overline{\mathsf{WSN}}$ is just the $\overline{\mathsf{WSN}}$ itself, with round and whitening keys scheduled in the opposite order.

The key element of our proof is the following lemma, which, combined with Lemma 1 and Eq. (1) immediately yields Theorem 1.

Lemma 2 (Non-adaptive Security of $\overline{\mathsf{WSN}}$). *For all q and q_F, and $N = 2^n$,*

$$\mathsf{Adv}^{\mathsf{NCPAPRP}}_{\overline{\mathsf{WSN}}^{(r)}[F], F}(q, q_F) \leq \frac{1}{2} q \sqrt{N} \left(\frac{1}{2} + \frac{q \cdot r + q_F}{2N} \right)^{r/2} \ .$$

Proof (Of Lemma 2). We fix a sequence of q distinct queries $\mathbf{X}$, as well as a set T_F of q_F input-output pairs. For now, we only consider the single-key setting, i.e., all queries $\mathbf{X}[j]$ are of the *same* index $i_j = 1$, and thus we omit these indices i_j. (We argue below how the multi-user case follows easily from our proof). Denote the randomly chosen round keys as $\mathbf{K} = (\mathbf{K}[1], \ldots, \mathbf{K}[r])$ and the corresponding whitening keys as $\mathbf{W} = (\mathbf{W}[1], \ldots, \mathbf{W}[r])$.

We are going to consider the evolution of the evaluation of $\overline{\mathsf{WSN}}$ on these inputs *simultaneously*, and denote the joint state after $t \in \{0\} \cup [r]$ rounds as $\mathbf{X}_t = (\mathbf{X}_t[1], \ldots, \mathbf{X}_t[q])$, with $\mathbf{X}_0 = \mathbf{X}$. With $\mathbf{U}$ uniformly distributed on the set of q distinct n-bit strings, we are going to upper bound

$$\mathsf{Adv}^{\mathsf{NCPAPRP}}_{\overline{\mathsf{WSN}}^{(r)}[F], F}(\mathbf{X}, T_F) = \mathsf{SD}(\mathbf{X}_r, \mathbf{U}) \ .$$

For any $i \in [q]$, denote by $\mathbf{Q}_t[i]$ the set of input-output pairs corresponding to the t F queries made to compute $\mathbf{X}_t[i]$ from $\mathbf{X}_0[i]$. Let now $U_{t,i}$ be a uniformly distributed value on the set $S_{t,i} := \{0, 1\}^n \setminus \{\mathbf{X}_t[1], \ldots, \mathbf{X}_t[i-1]\}$, and let $\mathbf{U}_{t,i}$ to be a uniform $(q-i)$-tuple of distinct strings from $S_{t,i+1}$. Then, for all $t \in \{0\} \cup [r]$,

$$\mathsf{SD}(\mathbf{X}_t, \mathbf{U}) \leq \sum_{i=1}^{q} \mathsf{SD}((\mathbf{X}_t^{\leq i-1}, \mathbf{U}_{t,i-1}), ((\mathbf{X}_t^{\leq i}, \mathbf{U}_{t,i}))$$

$$\leq \sum_{i=1}^{q} \mathsf{SD}((\mathbf{Q}_t^{\leq i-1}, \mathbf{X}_t^{\leq i-1}, \mathbf{U}_{t,i}, \mathbf{U}_{t,i}), (\mathbf{Q}_t^{\leq i-1}, \mathbf{X}_t^{\leq i}, \mathbf{U}_{t,i}))$$

$$= \sum_{i=1}^{q} \mathsf{SD}((\mathbf{Q}_t^{\leq i-1}, \mathbf{X}_t^{\leq i}), (\mathbf{Q}_t^{\leq i-1}, \mathbf{X}_t^{\leq i-1}, U_{t,i})) = \sum_{i=1}^{q} \mathsf{E}\left[\mathsf{SD}(\mathbf{X}_t[i], U_{t,i})\right]. \quad (2)$$

since $\mathsf{SD}(f(X), f(Y)) \leq \mathsf{SD}(X, Y)$ for all f, X, Y, and the i-th expectation in the sum is over $\mathbf{Q}_t^{\leq i-1}$, $\mathbf{X}_t^{\leq i-1}$, $\mathbf{W}^{\leq t}$, and $\mathbf{K}^{\leq t}$.

For all $a \in S_{t,i}$, we now we define the random variable $p_{t,i}(a)$ as the probability that $\mathbf{X}_t[i] = a$ conditioned on the actual values taken by the random variables $\mathbf{Q}_t^{\leq i-1}$, $\mathbf{X}_t^{\leq i-1}, \mathbf{W}^{\leq t}, \mathbf{K}^{\leq t}$. (In particular, $p_{t,i}(a)$ is a random variable itself, as it is a function of these random variables). Also, let $N_i := N - i + 1$. Then, by Cauchy-Schwarz and Jensen's inequalities, we obtain

$$
\begin{aligned}
\mathsf{E}\left[\mathsf{SD}(\mathbf{X}_t[i], U_{t,i})\right] &= \frac{1}{2} \cdot \mathsf{E}\left[\sum_{a \in S_{t,i}} \left|p_{t,i}(a) - \frac{1}{N_i}\right|\right] \\
&\leq \frac{1}{2} \cdot \sqrt{N} \sqrt{\mathsf{E}\left[\sum_{a \in S_{t,i}} \left(p_{t,i}(a) - \frac{1}{N_i}\right)^2\right]}.
\end{aligned}
\quad (3)
$$

We are going to give a recursive formula for $\mathsf{E}[\Delta_{t,i}]$, where

$$\Delta_{t,i} := \sum_{a \in S_{t,i}} \left(p_{t,i}(a) - \frac{1}{N_i}\right)^2.$$

Note that $\Delta_{0,i} = \mathsf{E}[\Delta_{0,i}] = 1 - \frac{1}{N_i}$. It is now convenient to assume that $\mathbf{Q}_t^{\leq i-1}$, $\mathbf{X}_t^{\leq i-1}, \mathbf{K}^{\leq t}, \mathbf{W}^{\leq t}$ are fixed to some values (and thus so are $\Delta_{t,i}$ and $p_{t,i}(a)$), and we are going to study $\mathsf{E}[\Delta_{t+1,i}]$, where the expectation is now over $\mathbf{X}_{t+1}^{\leq i-1}$, $\mathbf{K}[t+1]$, $\mathbf{W}[t+1]$ and $\mathbf{Q}_{t+1}^{\leq i-1}$. In particular, define $\mathcal{Q}_b$ (for $b \in \{0,1\}$) to be the set of all inputs of queries to F for which we know the corresponding output, i.e., $x \in \mathcal{Q}_b$ if $(x, b) \in T_F$ or $(x, b) \in \mathbf{Q}_t[j]$ for some $j \in [i-1]$. Moreover let $\mathcal{Q} := \mathcal{Q}_0 \cup \mathcal{Q}_1$ and $Q := |\mathcal{Q}|$, and note that $Q \leq t \cdot (i-1) + q_F$.

With the above being fixed, we are now considering the random experiment where we sample $\mathbf{K}[t+1]$ and $\mathbf{W}[t+1]$, and we are going to compute the expectation of $\Delta_{t+1,i}$ in this experiment. More concretely, we define a function $\varphi : S_{t,i} \to S_{t+1,i}$ (which is also a random variable, as it depends on $S_{t+1,i}$, $\mathbf{K}[t+1]$ and $\mathbf{W}[t+1]$) as follows:

$$\varphi(a) = \begin{cases} a & \text{if (1) } \max\{a \oplus \mathbf{K}[t+1], a\} \oplus \mathbf{W}[t+1] \in \mathcal{Q}_0, \text{ or} \\ & \text{(2) } a \oplus \mathbf{K}[t+1] \notin S_{t+1,i} \\ & \text{and } \max\{a \oplus \mathbf{K}[t+1], a\} \oplus \mathbf{W}[t+1] \notin \mathcal{Q}, \text{ or} \\ & \text{(3) } a \oplus \mathbf{K}[t+1] \in S_{t,i} \text{and } \max\{a \oplus \mathbf{K}[t+1], a\} \oplus \mathbf{W}[t+1] \notin \mathcal{Q}, \\ a \oplus \mathbf{K}[t+1] & \text{if (4) } \max\{a \oplus \mathbf{K}[t+1], a\} \oplus \mathbf{W}[t+1] \in \mathcal{Q}_1, \text{ or} \\ & \text{(5) } a \notin S_{t+1,i} \text{and } \max\{a \oplus \mathbf{K}[t+1], a\} \oplus \mathbf{W}[t+1] \notin \mathcal{Q}. \end{cases}$$

Note that φ is a bijection. Indeed, if $\mathbf{X}_t[i] = a$ implies $\mathbf{X}_{t+1}[i] = a'$ (where $a' \in \{a, a \oplus \mathbf{K}[t+1]\}$), then $\varphi(a) = a'$ (this corresponds to exactly one of the first four cases), and otherwise we let $\varphi(a) = a$. Also note that φ does not depend (directly) on $\mathbf{Q}_{t+1}^{\leq i-1}$, only on $S_{t+1,i}$, $\mathbf{K}[t+1]$, $\mathbf{W}[t+1]$, and $\mathbf{Q}_t^{\leq i-1}$. Using both the bijectivity of φ as well as the linearity of expectation,

$$\mathsf{E}\left[\Delta_{t+1,i}\right] = \sum_{a \in S_{t,i}} \mathsf{E}\left[\left(p_{t+1,i}(\varphi(a)) - \frac{1}{N_i}\right)^2\right].$$

Recall that the expectation here is over the choice of $\mathbf{Q}_{t+1}^{\leq i-1}$, $\mathbf{K}[t+1]$ and $\mathbf{W}[t+1]$. We prove the following lemma in Appendix C.

Lemma 3. *For all $a \in S_{t,i}$,*

$$\mathsf{E}\left[\left(p_{t+1,i}(\varphi(a)) - \frac{1}{N_i}\right)^2\right] = \left(1 - \frac{3}{4}\frac{N_i(N-Q)}{4 \cdot N^2}\right)\left(p_{t,i}(a) - \frac{1}{N_i}\right)^2 + \frac{1}{4}\frac{N-Q}{N^2}\Delta_{t,i}.$$

We can thus replace $\mathsf{E}\left[\left(p_{t+1,i}(\varphi(a)) - \frac{1}{N_i}\right)^2\right]$ in the above, and using the fact that $\Delta_{t,i} = \sum_{a \in S_{t,i}}(p_{t,i}(a) - \frac{1}{N_i})^2$, this simplifies to

$$\mathsf{E}\left[\Delta_{t+1,i}\right] = \sum_{a \in S_{t,i}} \mathsf{E}\left[\left(p_{t+1,i}(\varphi(a)) - \frac{1}{N_i}\right)^2\right] \leq \left(1 - \frac{N_i \cdot (N - Q_t)}{2 \cdot N^2}\right)\Delta_{t,i},$$

where $Q_t = t(i-1) + q_F$. Now, we come back to thinking of $\mathbf{X}_t^{\leq i-1}$, $\mathbf{K}^{\leq t}$ and $\mathbf{W}^{\leq t}$ as being randomly chosen (rather than fixed), and evaluate $\mathsf{E}[\Delta_{t,i}]$ recursively. The above in particular implies that $\mathsf{E}[\Delta_{t,i}] \leq \left(1 - \frac{N_i \cdot (N-Q)}{2 \cdot N^2}\right)\mathsf{E}[\Delta_{t-1,i}]$, and thus

$$\mathsf{E}[\Delta_{r,i}] \leq \left(1 - \frac{N_i \cdot (N - Q_{r-1})}{2 \cdot N^2}\right)^r \leq \left(\frac{1}{2} + \frac{r \cdot q + q_F}{2N}\right)^r.$$

Now, we can put this together with (2) and (3), and see that

$$\mathsf{SD}(\mathbf{X}_r, \mathbf{U}_q) \leq \frac{1}{2}q \cdot \sqrt{N} \cdot \left(\frac{1}{2} + \frac{r \cdot q + q_F}{2N}\right)^{r/2}.$$

Note that for the multi-user case, the proof is essentially the same, with slightly more complex notation. The only difference is that we define $S_{t,i}$ and all related quantities only with respect to the previous queries for the same key / user. The upper bounds are the same however, as they only depend on N, q and q_F. This concludes the proof of Lemma 2. □

4 Related-Key Security

4.1 Related-Key Secure Key Derivation

We consider the general notion of a related-key secure key-derivation function, or RKA-KDF for short. Informally, for a class of key-transformation functions $\Phi \subseteq \mathsf{Fcs}(\kappa, \kappa)$, this is a function $\mathsf{KDF} : \{0,1\}^\kappa \to \{0,1\}^\ell$ such that $\mathsf{KDF}(\phi(K))$ gives independent, pseudorandom values for every $\phi \in \Phi$. A similar notion was considered by Lucks [40] and by Barbosa and Farshim [3].

FORMAL DEFINITION. Let $\mathsf{KDF}[F] : \{0,1\}^\kappa \to \{0,1\}^\ell$ be a construction that calls a function $F : \{0,1\}^m \to \{0,1\}^n$. In Fig. 1, we define the security games RKA-KDF-0 and RKA-KDF-1 involving an adversary $\mathcal{A}$ and a class of key-transformations $\Phi \subseteq \mathsf{Fcs}(\kappa, \kappa)$. In the real world (Game RKA-KDF-0), the adversary $\mathcal{A}$ makes queries to a random function F via the F oracle and can obtain evaluations of $\mathsf{KDF}[F](\phi(K))$ for multiple $\phi \in \Phi$ of its choice via the Eval oracle, and these values should be indistinguishable from random values, which are returned by the Eval oracle in the ideal world (i.e., in Game RKA-KDF-1). The RKA-KDF-advantage is then defined as

$$\mathsf{Adv}^{\mathsf{RKA\text{-}KDF}}_{\mathsf{KDF},F,\Phi}(\mathcal{A}) = \Pr\left[\mathsf{RKA\text{-}KDF\text{-}0}^{\mathcal{A}}_{\mathsf{KDF},F,\Phi} \Rightarrow 1\right] - \Pr\left[\mathsf{RKA\text{-}KDF\text{-}1}^{\mathcal{A}}_{\mathsf{KDF},F,\Phi} \Rightarrow 1\right],$$

and $\mathsf{Adv}^{\mathsf{RKA\text{-}KDF}}_{\mathsf{KDF},F,\Phi}(q, q_F)$ is obtained by maximizing the above over all adversaries making q queries to Eval and making q_F queries to F via the F oracle.

Procedure MAIN:	**Procedure** Eval(ϕ):
// Game RKA-KDF-b, $b \in \{0,1\}$	// Game RKA-KDF-b, $b \in \{0,1\}$
$F \overset{\$}{\leftarrow} \mathsf{Fcs}(m,n)$, $G \overset{\$}{\leftarrow} \mathsf{Fcs}(*,\ell)$	**If** $b = 0$ **then**
$K \overset{\$}{\leftarrow} \{0,1\}^\kappa$	$\qquad$ **Return** $\mathsf{KDF}[F](\phi(K))$
$b' \overset{\$}{\leftarrow} \mathcal{A}^{\mathsf{F,Eval}}$	**Else return** $G(\phi)$
Return b'	
	Procedure F(x):
	Return $F(x)$

Fig. 1. RKA-KDF security. The procedure Eval, in both games, takes as input a function $\phi \in \Phi$. Also, the notation $G(\phi)$ denotes G applied to some unique bit-encoding of the function ϕ.

Remark 1. An alternative definition has the Eval oracle return $G(\phi(K))$ for a random function $G \overset{\$}{\leftarrow} \mathsf{Fcs}(\kappa, \ell)$. Our choice is better suited to the composition theorem below, and shifts the burden of dealing with the combinatorics of Φ to the RKA-KDF security proof.

THE COMPOSITION THEOREM. We can compose an arbitrary (ℓ, n)-block cipher construction $\mathsf{BC}[F]$ and a key-derivation function $\mathsf{KDF} : \{0,1\}^\kappa \to \{0,1\}^\ell$ using the same function F, into a new (κ, n) block cipher $\overline{\mathsf{BC}} = \overline{\mathsf{BC}}[F, \mathsf{KDF}]$ such that

$$\overline{\mathsf{BC}}_K(X) = \mathsf{BC}_{\mathsf{KDF}(K)}(X) . \tag{4}$$

for every $K \in \{0,1\}^\kappa$ and $X \in \{0,1\}^n$. The following theorem shows that if BC is a secure PRP and KDF is RKA-KDF secure, then the composition $\overline{\mathsf{BC}}$ is a related-key secure PRP. Note that the fact that we consider multi-user PRP security is central in allowing us a tight reduction.

Theorem 2 (The Composition Theorem). *Let* $\overline{\mathsf{BC}} = \overline{\mathsf{BC}}[F, \mathsf{KDF}]$ *be the* (κ, n)-*block cipher defined above, and assume that* BC *makes at most* t *calls to* F *upon each invocation. Let* $\Phi \subset \mathsf{Fcs}(\kappa, \kappa)$ *be a class of key transformations. Then, for all* q, q_F,

$$\mathsf{Adv}_{\overline{\mathsf{BC}}, F, \Phi}^{\mathsf{RKA\text{-}PRP}}(q, q_F) \leq 2 \cdot \mathsf{Adv}_{\mathsf{KDF}, F, \Phi}^{\mathsf{RKA\text{-}KDF}}(q, q_F + q \cdot t) + \mathsf{Adv}_{\mathsf{BC}, F}^{\mathsf{PRP}}(q, q_F).$$

Proof (Sketch). One uses RKA-KDF security to transition from RKA-PRP-1 to a setting where each query (ϕ, x) to the block cipher is replied with an independent key K_ϕ as $\mathsf{BC}_{K_\phi}(x)$, i.e., we map every ϕ with an independent κ-bit key K_ϕ. This is exactly PRP-1 (except that users are now identified by elements of Φ) and results in the additive term $\mathsf{Adv}_{\mathsf{KDF}}^{\mathsf{RKA\text{-}KDF}}(q, q_F + q \cdot t)$ in the bound by a standard reduction. Similarly, one uses RKA-KDF security to transition from RKA-PRP-0 to a setting where each query (ϕ, x) to the block cipher is replied with an independent permutation P_ϕ, and this exactly maps to PRP-0, and results in another additive term $\mathsf{Adv}_{\mathsf{KDF}}^{\mathsf{RKA\text{-}KDF}}(q, q_F + q \cdot t)$. The final bound follows by the triangle inequality.   □

Note that in a similar way, if KDF and BC use *different* functions F and F', then we can reduce $\mathsf{Adv}_{\mathsf{KDF}, F, \Phi}^{\mathsf{RKA\text{-}KDF}}(q, q_F + q \cdot t)$ to $\mathsf{Adv}_{\mathsf{KDF}, F, \Phi}^{\mathsf{RKA\text{-}KDF}}(q, q_F)$.

4.2   Efficient RKA-KDF-secure Construction

This section presents an RKA-KDF-secure construction from a (small number of) random functions with n-bit domain approaching $(2^{n(1-\epsilon)}, 2^{n(1-\epsilon)})$-security. (As we argue below, this can be turned into a construction from a single function $F : \{0,1\}^n \to \{0,1\}$ with standard tricks). Our construction will guarantee Φ-RKA-KDF-security for every class $\Phi \subseteq \mathsf{Fcs}(\kappa, \kappa)$ with the following two properties for (small) parameters $\gamma, \lambda \in [0, 1]$:

γ-collision resistance. $\Pr\left[K \xleftarrow{\$} \{0,1\}^\kappa : \phi(K) = \phi'(K)\right] \leq \gamma$ for any two distinct $\phi, \phi' \in \Phi$.

λ-uniformity. For any $\phi \in \Phi$, we have that $\mathsf{SD}(K, \phi(K)) \leq \lambda$ for $K \xleftarrow{\$} \{0,1\}^\kappa$, i.e., $\phi(K)$ is λ-close to uniform for a random key K.

For example, $\Phi^\oplus = \{K \mapsto K \oplus \Delta : \Delta \in \{0,1\}^\kappa\}$ is both 0-collision-resistant and 0-uniform.

COMBINATORIAL HITTERS. Our construction makes use of the standard combinatorial notion of a hitter [30], which we introduce with a slightly different parameterization than what used in the literature. Consider a family of functions $\mathsf{E} = (\mathsf{E}_1, \ldots, \mathsf{E}_t)$ such that $\mathsf{E}_i : \{0,1\}^\kappa \to \{0,1\}^n$.

Definition 1 (Hitters). *The functions* $\mathsf{E} = (\mathsf{E}_1, \ldots, \mathsf{E}_t)$ *with* $\mathsf{E}_i : \{0,1\}^\kappa \to \{0,1\}^n$ *are an* (α, β)*-hitter if for all subsets* $\mathcal{Q} \subseteq \{0,1\}^n$ *with* $|\mathcal{Q}| \leq \beta \cdot 2^n$, $\Pr[K \leftarrow \{0,1\}^\kappa : \forall i \in [t] : \mathsf{E}_i(K) \in \mathcal{Q}] \leq \alpha$.

In our setting, we are going to have $\beta = 2^{-n\epsilon}$ (for some (small) $\epsilon > 0$, and in particular $1 - \beta \geq \frac{1}{2}$) and $\alpha = 2^{-n}$. There are polynomially-computable *explicit* constructions of hitters (cf. e.g. [30] for an overview) with sufficiently good parameters for our purposes, where

$$\kappa = 2n + O(\log(1/\alpha)) = O(n) , \quad t = O(\log(1/\alpha)) = O(n) . \tag{5}$$

The full version gives further details about a concrete example of a "reasonably" cheap construction relying on random walks on constant-degree expander graphs. We will require our hitters to be *injective*, i.e., for any two inputs X and X', there must exist i such that $\mathsf{E}_i(X) \neq \mathsf{E}_i(X')$. It is easy to enforce injectivity for any hitter by just adding $O(\kappa/n)$ functions to the family.

THE MT CONSTRUCTION. We now present our construction of an RKA-KDF-secure function, which follows the framework of Maurer and Tessaro [45]. Let $\mathsf{E} = (\mathsf{E}_1, \ldots, \mathsf{E}_t)$ be such that $\mathsf{E}_i : \{0,1\}^\kappa \to \{0,1\}^n$. Moreover, let $F_{i,j} : \{0,1\}^n \to \{0,1\}^{2\kappa+n}$ for $i \in [t]$ and $j \in [r]$, $G_j : \{0,1\}^n \to \{0,1\}^\ell$ for $j \in [r]$. For simplicity, denote $F = (F_{i,j})_{i \in [t], j \in [r]}$ and $G = (G_i)_{i \in [t]}$.

The MT$[\mathsf{E}, F, G]$ construction operates as follow. (Here, $\odot$ denotes multiplication of $(2\kappa+n)$-bit-strings interpreted as elements of the corresponding extension field $\mathbb{F}_{2^{2\kappa+n}}$).

Construction MT$[F, G](K)$: $// K \in \{0,1\}^\kappa$

(1) For all $j \in [r]$, compute

$$S[j] \leftarrow \left(\bigodot_{i=1}^{t} F_{i,j}(\mathsf{E}_i(K)) \right) [1 \ldots n] .$$

(2) Compute $K' \leftarrow \bigoplus_{j=1}^{r} G_i(S[i])$.
(3) Return K'.

RKA-KDF SECURITY. The above construction is indifferentiable from a random oracle [22, 45] whenever E is a so-called *input-restricting function family*. While this combinatorial property would also imply RKA-KDF security, explicit constructions of such function families require a very large $t = O(n^c)$ for a large constant c, as discussed in [22].

Here, in contrast, we show that for RKA-KDF security it is sufficient if E is a good hitter. The following theorem summarizes the concrete parameters of our result. The complete proof is deferred to the full version for lack of space. We give some intuition further below.

Theorem 3 (RKA-KDF-Security of MT). *Let* E *be an* $(\alpha, \beta = q_F/2^n)$-*injective hitter. Moreover, let* $\Phi \subseteq \mathsf{Fcs}(\kappa, \kappa)$ *be a* (γ, λ)-*well behaved set of key transformations. Then, for all adversaries* $\mathcal{A}$ *making* q *queries to* Eval, q_F *queries to the* F-*functions, and* q_G *queries to the* G-*functions,*

$$\mathsf{Adv}^{\mathsf{RKA\text{-}KDF}}_{\mathsf{MT},(F,G),\Phi}(\mathcal{A}) \leq \frac{4rt}{2^n} + q(\alpha + \lambda) + q^2\gamma + q \cdot \left(\frac{q_G + q}{2^n}\right)^r.$$

<u>INSTANTIATIONS.</u> Let us target security for $q_F = q = 2^{n(1-\epsilon)}$ (e.g., $\epsilon = O(1/n)$), $\ell = n$, and additive attacks $\Phi = \Phi^{\oplus}$ with $\gamma = \lambda = 0$. First note that because we want $\alpha \approx 2^{-n}$ and $\beta = 2^{-\epsilon n}$, then we can use E with $\kappa = O(n)$ and $t = O(n)$ by (5). Moreover, we need to ensure that $2^{r(1-n)} \cdot 2^{n(1-\epsilon)(r+1)} < 1$ or alternatively $r(n\epsilon - 1) > n(1 - \epsilon)$, which is true for $r = r(\epsilon) = \Omega(\frac{1-\epsilon}{\epsilon})$, and $r = O(n)$ for $\epsilon = O(1/n)$.

Therefore, the construction evaluates a linear number of functions with linear output $O(n)$, or alternatively, $O(n^2)$ single-bit functions $\{0,1\}^n \to \{0,1\}$. This can be turned into evaluating $O(n^2)$ one single function $\{0,1\}^{n+2\log n + O(1)} \to \{0,1\}$.[7] Improving upon this appears to be a significant barrier.

The MT construction can be combined with the WSN construction above to obtain an RKA-secure block cipher with $(2^{n(1-\epsilon)}, 2^{n(1-\epsilon)})$-security via Theorem 2 for any class Φ with small λ, γ.

<u>OVERVIEW OF THE PROOF OF THEOREM</u> 3. We explain here the basic ideas behind the proof of Theorem 3.

To start with, it is convenient to first consider a toy construction, using only t functions $F = (F_i)_{i \in [t]}$ with $F_i \in \mathsf{Fcs}(n, \ell)$, in conjunction with a hitter $\mathsf{E} = (\mathsf{E}_1, \ldots, \mathsf{E}_t)$ as above. On input $K \in \{0,1\}^\kappa$, it outputs $\bigoplus_{i=1}^t F_i(\mathsf{E}_i(K))$. Also, let us only consider RKA-KDF attackers which make all q_F of their F queries *beforehand*, and only then query Eval on inputs $\phi_1, \ldots, \phi_q$, where the ϕ_i functions are such that $\phi_i(K)$ is uniform for a uniform K.

Assume without loss of generality the uniform key K is sampled after the F-queries have been made. Since E is an $(\alpha, \beta = q_F/2^n)$-hitter, then by the union bound, for every $k \in [q]$ there exists some $i^*(k)$ such that $\mathsf{E}_{i^*(k)}(\phi_k(K))$ was *not* queried to $F_{i^*(k)}$ in the first phase, except with probability $q \cdot \alpha$. Therefore, for all $k \in [q]$, the value $\bigoplus_{i=1}^t F_i(\mathsf{E}_i(\phi_k(K)))$ is *individually* uniform, even given the transcript of the F queries, but unfortunately, this does not guarantee independence of these outputs. Indeed, for two k and k', we may well have $i^*(k) = i^*(k')$, and we cannot exclude that for all $i \neq i^*(k)$ both values $F_i(\mathsf{E}_i(\phi_k(K)))$ and $F_i(\mathsf{E}_i(\phi_{k'}(K)))$ are known as part of the F-queries made in the first phase. Then, the output values for k and k' are clearly correlated.

Instead, by using two rounds with functions $(F_{i,j})_{i \in [t], j \in [r]}$ and $(G_j)_{j \in [r]}$ (where $F_{i,j} \in \mathsf{Fcs}(n, n)$ and $G_j \in \mathsf{Fcs}(n, \ell)$), we would generate values $S_k[j] \leftarrow$

---

[7] Note that we can play a bit with parameters, and given a function $F : \{0,1\}^n \to \{0,1\}$, interpret it as a function $\{0,1\}^{n'+2\log(n')} \to \{0,1\}$ for a suitable n' only marginally smaller than n, and obtain an instantiation of our construction with respect to n' still making roughly $O(n^2)$ calls to F.

$\bigoplus_{i=1}^{t} F_{i,j}(\mathsf{E}_i(\phi_k(K)))$ hoping that, in addition to being individually uniform as above, $S_k[j]$ and $S_{k'}[j]$ are unlikely to collide for any $k \neq k'$.

If the final output of the construction is $\bigoplus_{j=1}^{r} G_j(S_k[j])$, the above would imply security: Indeed, with very high probability, we can show that for every k, there is going to always exist some j^* such that $S_k[j^*]$ was never queried to G_{j^*} previously directly by the attacker (because of the individual uniformity of the value) and that no other $k' \neq k$ is such that $S_{k'}[j^*] = S_k[j^*]$. (Exploiting independence of the $S_k[j]$'s, the probability that such j^* does not exist can be made very small, of the order $\left(\frac{q_G+q}{2^n}\right)^r$).

There is a final catch. Imagine we are in the above "unfortunate" setting, i.e., for two k and k' and $j \in [r]$, we have $i^*(k) = i^*(k')$, and for all $i \neq i^*(k)$, $F_{i,j}(\mathsf{E}_i(\phi_k(K)))$ and $F_{i,j}(\mathsf{E}_i(\phi_{k'}(K)))$ are known. Then, the fact that $S_k[j]$ and $S_{k'}[j]$ collided is already determined by the transcript of the F queries, independent of $F_{i^*(k),j}(\mathsf{E}_{i^*(k)}(\phi_k(K)))$. Our approach to address this problem is to make the output of the F-values larger (roughly $2\kappa + n$ bits) and to use multiplication. This will make sure that given that any two partial product defined by the F queries as above will not collide (over $2\kappa + n$ bits), and thus (by the fact that multiplication with truncation gives a universal hash function), the final products, truncated at n bits, will also be unlikely to collide.

A   Indifferentiability

We briefly review the notion of indifferentiability by Maurer et al. [44] as needed in this paper.

Let $\mathsf{C}[G] : \{0,1\}^m \to \{0,1\}^\ell$ be a construction from a function $G : \{0,1\}^a \to \{0,1\}^b$. We say that C is *indifferentiable* from a random function if $\mathsf{C}[G]$, for $G \xleftarrow{\$} \mathsf{Fcs}(a,b)$, is "as good as" a randomly chosen function $F \xleftarrow{\$} \mathsf{Fcs}(m,\ell)$ in a setting where an adversary is given access to *both* $\mathsf{C}[G]$ and the underlying function G. This is formalized by requiring the existence of a simulator S, accessing F, which mimics the behavior of G in a way that makes real and ideal worlds indistinguishable.

<u>FORMAL DEFINITION.</u> For an adversary $\mathcal{A}$ and a simulator S, the *indifferentiability* advantage is

$$\mathsf{Adv}^{\mathsf{indiff}}_{\mathsf{C}[G],G,S}(\mathcal{A}) = \Pr\left[G \xleftarrow{\$} \mathsf{Fcs}(a,b) : \mathcal{A}^{\mathsf{C}[G],G} \Rightarrow 1\right] - $$
$$- \Pr\left[F \xleftarrow{\$} \mathsf{Fcs}(m,\ell) : \mathcal{A}^{F,S^F} \Rightarrow 1\right].$$

Similarly, for a construction $\mathsf{C}[\pi]$ from a permutation $\pi \in \mathsf{Perms}(a)$, we define

$$\mathsf{Adv}^{\mathsf{indiff}}_{\mathsf{C}[\pi],\pi,S}(\mathcal{A}) = \Pr\left[\pi \xleftarrow{\$} \mathsf{Perms}(a) : \mathcal{A}^{\mathsf{C}[\pi],\pi,\pi^{-1}} \Rightarrow 1\right] - $$
$$- \Pr\left[F \xleftarrow{\$} \mathsf{Fcs}(m,\ell) : \mathcal{A}^{F,S^F} \Rightarrow 1\right].$$

Note that in the latter case, the simulator S simulates both the behavior of π and π^{-1} queries. We are going to call queries to the first oracle (i.e., either $C[G]$, $C[\pi]$ or F) *construction queries*, and queries to the second oracle (either G, π, π^{-1}, or S^F) *primitive queries*.

In this paper, we are going to only consider an information-theoretic version of indifferentiability.

Definition 2 (Indifferentiability). *A construction* $C[\Sigma]$ *(where Σ is either a permutation or a function) is* (ϵ, s)*-indifferentiable from a random function if there exists a simulator S such that for all adversary A making q construction queries, and q_Σ primitive queries,* $\mathsf{Adv}^{\mathsf{indiff}}_{C[\Sigma],\Sigma,S}(A) \leq \epsilon(q, q_\Sigma)$*, and where additionally, upon each invocation via a primitive queries, the simulator Σ makes at most s queries. Moreover, the simulator answers each query in time polynomial in q_Σ.*

We say that $C[\Sigma]$ is *perfectly indifferentiable* if it is $(0, 1)$-indifferentiable.

COMPOSITION THEOREM. We use the following fact below, which follows from general composition theorems [21,44] adapted to the specific case of block ciphers considered in this paper.

Theorem 4 (Composition Theorem for Block Ciphers). *Let* $\mathsf{BC} = \mathsf{BC}[F]$ *be a (κ, n)-block cipher making at most t calls to a function $F : \{0,1\}^m \to \{0,1\}^\ell$, and let $C[\Sigma]$ be a construction using a primitive Σ which is (ϵ, s)-indifferentiable from a random function. Consider the (κ, n)-block cipher $\mathsf{BC}' = \mathsf{BC}'[\Sigma] = \mathsf{BC}[C[\Sigma]]$, i.e., calls to F are replaced by calls to $C[\Sigma]$. Then,*

$$\mathsf{Adv}^{\mathsf{PRP}}_{\mathsf{BC}'[\Sigma],\Sigma}(q, q_\Sigma) \leq \mathsf{Adv}^{\mathsf{PRP}}_{\mathsf{BC}[F],F}(q, s \cdot q_\Sigma) + 2 \cdot \epsilon(t \cdot q, q_\Sigma) .$$

B   From Permutations to Functions

In this section, we revisit the security of a construction by Hall, Wagner, Kelsey, and Schneier [32] to build a random function $F : \{0,1\}^n \to \{0,1\}^c$ from a permutation $\pi : \{0,1\}^{n+c} \to \{0,1\}^{n+c}$. In particular, here we show that their construction achieves the stronger notion of *perfect indifferentiability* defined above in Appendix A, and thus can be used to replace (in a black-box way) the function F in the WSN construction. Note that in [32], only indistinguishability was shown. We believe that this result is of interest beyond the scope of this paper.

THE CONSTRUCTION. Let $\pi : \{0,1\}^{n+c} \to \{0,1\}^{n+c}$ be a permutation. The 2^c-query construction $\mathsf{F}_C[\pi] : \{0,1\}^n \to \{0,1\}^c$ proceeds as follows, on input $X \in \{0,1\}^n$: It outputs the c-bit value Z^* such that $\pi(X \parallel Z^*)$ is the smallest element in $\{\pi(X \parallel Z) : Z \in \{0,1\}^c\}$, where smallest is according to lexicographic order. (Or any other total order on strings).

SECURITY. The following theorem establishes security of F in terms of indifferentiability.[8]

Theorem 5 (Indifferentiability of F). *The construction* $F_c = F_c[\pi]$ *is perfectly indifferentiable from a random function.*

Proof. We need to prove that there exists a simulator S such that $\mathsf{Adv}^{\mathsf{indiff}}_{F,\pi,S}(\mathcal{A}) = 0$ for all adversaries $\mathcal{A}$, and moreover, S simulates a permutation from $\mathsf{Perms}(n+c)$, together with its inverse, and makes at most one single query to a given function $F \xleftarrow{\$} \mathsf{Fcs}(n,c)$ upon each invocation.

To help with the definition of the simulator, for a function $f \in \mathsf{Fcs}(n,c)$ and a permutation $\tau \in \mathsf{Perms}(n+c)$, we define a new permutation $\pi[\tau, f] \in \mathsf{Perms}(n+c)$. To this end, for every $x \in \{0,1\}^n$, we define

$$y_x^* = \min\{\tau(x \parallel z) : z \in \{0,1\}^c\}$$

and $y_x = \tau(x \parallel f(x))$. Note that y_x^* is the output of τ on input $x \parallel F_c[\tau](x)$ and thus if $f = F_c[\tau]$, $y_x = y_x^*$. The permutation $\pi[\tau, f]$ is such that

$$\pi[\tau, f](x \parallel z) = \begin{cases} y_x^* & \text{if } \tau(x \parallel z) = y_x, \text{i.e., } f(x) = z \\ y_x & \text{if } \tau(x \parallel z) = y_x^* \\ \tau(x \parallel z) & \text{else.} \end{cases}$$

In other words, $\pi[\tau, f]$ re-arranges τ to assign $\pi[\tau, f](x \parallel f(x))$ the *smallest* value among $\tau(x \parallel z')$ for $z' \in \{0,1\}^c$. Clearly, given τ, $\pi[\tau, f](x \parallel z)$ can be computed with a single query to f and 2^c queries to τ. Moreover, note that the inverse $\pi^{-1}[\tau, f]$ is

$$\pi^{-1}[\tau, f](y) = \begin{cases} \tau^{-1}(y_x) \text{ if } y = y_x^* \\ \tau^{-1}(y_x^*) \text{ if } y = y_x \\ \tau^{-1}(y) \text{ else.} \end{cases}$$

Note that the check $y = y_x^*$ and $y = y_x$ can be implemented by first computing $\tau^{-1}(y)$, which returns $x \parallel z$, and then querying $\tau(x \parallel z')$ for all $z' \neq z$, as well as $f(x)$. In particular, $\pi^{-1}[\tau, f]$ can also be evaluated with one query to f, given τ.

The simulator S now simply does the following when given oracle access to f: It maintains a random permutation $\tau \xleftarrow{\$} \mathsf{Perms}(n+c)$ (implemented via lazy sampling), and on a forward query $x \parallel z$, replies as $\pi[\tau, f](x \parallel z)$, and on inverse query y it replies as $\pi^{-1}[\tau, f](y)$. By the above, this requires one f query per evaluation.

Therefore, to prove perfect indifferentiability, it is enough to prove that $(F_c[\pi], \pi)$ (for $\pi \xleftarrow{\$} \mathsf{Perms}(n+c)$) and $(f, \pi[\tau, f])$ (for $f \xleftarrow{\$} \mathsf{Fcs}(n,c)$ and $\tau \xleftarrow{\$} \mathsf{Perms}(n+c)$) are identically distributed. This can be done in two steps:

---

[8] A previous version of this paper provided a somewhat more cumbersome yet equivalent description of the simulator. The far more elegant description using $\pi[\tau, f]$ was suggested by an anonymous reviewer we wish to thank.

1. First, note that $\mathsf{F}_c[\pi[\tau, f]] = f$. This is because on input x, F_c outputs z such that $\pi[\tau, f](x \parallel z)$ is smallest. This must be $z = f(x)$, because $\pi[\tau, f]$ is such that $\pi[\tau, f](x \parallel f(x)) = y_x^*$, which is the smallest value among $\tau(x \parallel z')$, and thus also among $\pi[\tau, f](x \parallel z')$.

2. Therefore, it suffices to show that the permutation $\pi[\tau, f]$ is uniformly distributed. This is because $\pi[\tau, f]$ is obtained by sampling a random permutation τ, and then for all x, swapping y_x^* with the output of $x \parallel z$ for a randomly chosen $z = f(x)$. This gives a uniform random permutation.

This concludes the proof. $\qquad\qquad\qquad\qquad\qquad\qquad\qquad\qquad\qquad\qquad\qquad\qquad\quad$ □

C   Proof of Lemma 3

For every $a \in S_{t,i}$, we now define now two subsets partitioning $\{0,1\}^n \times \{0,1\}^n$, i.e., the key space for round $t + 1$:

$$\mathcal{WK}_a^+ := \{(w, k) : a \oplus k \in S_{t,i} \wedge \max\{a \oplus k, a\} \oplus w \notin \mathcal{Q}\}$$

$$\mathcal{WK}_a^- := \{(w, k) : a \oplus k \notin S_{t,i} \vee \max\{a \oplus k, a\} \oplus w \in \mathcal{Q}\}$$

It is easy to see that

$$\left|\mathcal{WK}_a^+\right| = N_i \cdot (N - Q), \quad \left|\mathcal{WK}_a^-\right| = N^2 - N_i \cdot (N - Q)$$

because for every a we have exactly $|S_{t,i}| = N_i$ values of k such that $a \oplus k \in S_{t,i}$, and moreover, we have (for each such value k) exactly $N - Q$ possible values of w with $\max\{a, a \oplus k\} \oplus w \notin \mathcal{Q}$. Also, note that for $(w, k) \in \mathcal{WK}_a^-$,

$$\mathsf{E}\left[(p_{t+1,i}(\varphi(a)) - 1/N_i)^2 \,\Big|\, \mathbf{K}[t+1] = k, \mathbf{W}[t+1] = w\right] = p_{t,i}(a)^2,$$

whereas for $(w, k) \in \mathcal{WK}_a^+$,

$$\mathsf{E}\left[\left(p_{t+1,i}(\varphi(a)) - \frac{1}{N_i}\right)^2 \,\Big|\, \mathbf{K}[t+1] = k, \mathbf{W}[t+1] = w\right] =$$
$$= \left(\frac{p_{t,i}(a) + p_{t,i}(a \oplus k)}{2} - \frac{1}{N_i}\right)^2.$$

Putting all of this together, we obtain

$$\mathsf{E}\left[\left(p_{t+1,i}(\varphi(a)) - \tfrac{1}{N_i}\right)^2\right] =$$

$$= \tfrac{1}{N^2} \sum_{k,w} \mathsf{E}\left[\left(p_{t+1,i}(\varphi(a)) - \tfrac{1}{N_i}\right)^2 \,\Big|\, \mathbf{K}[t+1] = k, \mathbf{W}[t+1] = w\right]$$

$$= \frac{1}{N^2}\left[\sum_{(w,k)\in\mathcal{WK}_a^-} \left(p_{t,i}(a) - \tfrac{1}{N_i}\right)^2 + \sum_{(w,k)\in\mathcal{WK}_a^+} \left(\tfrac{p_{t,i}(a)+p_{t,i}(a\oplus k)}{2} - \tfrac{1}{N_i}\right)^2\right]$$

$$= \left(1 - \frac{N_i(N-Q)}{N^2}\right)\left(p_{t,i}(a) - \tfrac{1}{N_i}\right)^2 + \frac{N-Q}{N^2}\sum_{y\in S_{t,i}} \left(\tfrac{p_{t,i}(a)+p_{t,i}(y)}{2} - \tfrac{1}{N_i}\right)^2,$$

where we have used the structure of $\mathcal{WK}_a^\dagger$, and the fact that for every $y \in S_{t,i}$ there exists k such that $a \oplus k = y$, and corresponding $N - Q$ values of w. In particular, we can expand

$$\sum_{y \in S_{t,i}} \left(\tfrac{p_{t,i}(a) + p_{t,i}(y)}{2} - \tfrac{1}{N_i} \right)^2 = \tfrac{1}{4} \sum_{y \in S_{t,i}} \left(\left(p_{t,i}(a) - \tfrac{1}{N_i} \right) + \left(p_{t,i}(y) - \tfrac{1}{N_i} \right) \right)^2$$

$$= \tfrac{N_i}{4} \cdot \left(p_{t,i}(a) - \tfrac{1}{N_i} \right)^2 + \tfrac{1}{4} \Delta_{t,i} \, ,$$

where we have used in passing the fact that $\sum_{y \in S_{t,i}} (p_{t,i}(a) - \tfrac{1}{N_i}) = 0$. When we plug this back into the above, we then get

$$\mathsf{E}\left[\left(p_{t+1,i}(\varphi(a)) - \tfrac{1}{N_i} \right)^2 \right] = \left(1 - \tfrac{3}{4} \tfrac{N_i(N-Q)}{4 \cdot N^2} \right) \left(p_{t,i}(a) - \tfrac{1}{N_i} \right)^2 + \tfrac{1}{4} \tfrac{N-Q}{N^2} \Delta_{t,i} \, .$$

This concludes the proof of Lemma 3.

References

1. Adams, C.: RFC 2144 - The CAST-128 Encryption Algorithm. Internet Activities Board, May 1997
2. Andreeva, E., Bogdanov, A., Dodis, Y., Mennink, B., Steinberger, J.P.: On the indifferentiability of key-alternating ciphers. In: Canetti, R., Garay, J.A. (eds.) CRYPTO 2013, Part I. LNCS, vol. 8042, pp. 531–550. Springer, Heidelberg (2013)
3. Barbosa, M., Farshim, P.: The related-key analysis of feistel constructions. In: Cid, C., Rechberger, C. (eds.) FSE 2014. LNCS, vol. 8540, pp. 265–284. Springer, Heidelberg (2015)
4. Bellare, M., Cash, D.: Pseudorandom functions and permutations provably secure against related-key attacks. In: Rabin, T. (ed.) CRYPTO 2010. LNCS, vol. 6223, pp. 666–684. Springer, Heidelberg (2010)
5. Bellare, M., Kohno, T.: A theoretical treatment of related-key attacks: RKA-PRPs, and applications. In: Biham, E. (ed.) EUROCRYPT 2003. LNCS, vol. 2656, pp. 491–506. Springer, Heidelberg (2003)
6. Bellare, M., Rogaway, P.: Random oracles are practical: a paradigm for designing efficient protocols. In: ACM CCS, vol. 93, pp. 62–73 (1993)
7. Bellare, M., Rogaway, P.: The security of triple encryption and a framework for code-based game-playing proofs. In: Vaudenay, S. (ed.) EUROCRYPT 2006. LNCS, vol. 4004, pp. 409–426. Springer, Heidelberg (2006)
8. Biham, E.: New types of cryptanalytic attacks using related keys. J. Cryptol. 7(4), 229–246 (1994)
9. Biham, E., Dunkelman, O., Keller, N.: A related-key rectangle attack on the full KASUMI. In: Roy, B. (ed.) ASIACRYPT 2005. LNCS, vol. 3788, pp. 443–461. Springer, Heidelberg (2005)
10. Biham, E., Dunkelman, O., Keller, N.: Related-key impossible differential attacks on 8-round AES-192. In: Pointcheval, D. (ed.) CT-RSA 2006. LNCS, vol. 3860, pp. 21–33. Springer, Heidelberg (2006)

11. Biryukov, A., Dunkelman, O., Keller, N., Khovratovich, D., Shamir, A.: Key recovery attacks of practical complexity on AES-256 variants with up to 10 rounds. In: Gilbert, H. (ed.) EUROCRYPT 2010. LNCS, vol. 6110, pp. 299–319. Springer, Heidelberg (2010)

12. Biryukov, A., Khovratovich, D.: Related-key cryptanalysis of the full AES-192 and AES-256. In: Matsui, M. (ed.) ASIACRYPT 2009. LNCS, vol. 5912, pp. 1–18. Springer, Heidelberg (2009)

13. Biryukov, A., Khovratovich, D., Nikolić, I.: Distinguisher and related-key attack on the full AES-256. In: Halevi, S. (ed.) CRYPTO 2009. LNCS, vol. 5677, pp. 231–249. Springer, Heidelberg (2009)

14. Bogdanov, A., Knudsen, L.R., Leander, G., Standaert, F.-X., Steinberger, J., Tischhauser, E.: Key-alternating ciphers in a provable setting: encryption using a small number of public permutations. In: Pointcheval, D., Johansson, T. (eds.) EUROCRYPT 2012. LNCS, vol. 7237, pp. 45–62. Springer, Heidelberg (2012)

15. Chen, S., Lampe, R., Lee, J., Seurin, Y., Steinberger, J.: Minimizing the two-round even-mansour cipher. In: Garay, J.A., Gennaro, R. (eds.) CRYPTO 2014, Part I. LNCS, vol. 8616, pp. 39–56. Springer, Heidelberg (2014)

16. Chen, S., Steinberger, J.: Tight security bounds for key-alternating ciphers. In: Nguyen, P.Q., Oswald, E. (eds.) EUROCRYPT 2014. LNCS, vol. 8441, pp. 327–350. Springer, Heidelberg (2014)

17. Cogliati, B., Lampe, R., Seurin, Y.: Tweaking even-mansour ciphers. In: Gennaro, R., Robshaw, M. (eds.) CRYPTO 2015. LNCS, vol. 9215. Springer, Heidelberg (2015)

18. Cogliati, B., Patarin, J., Seurin, Y.: Security amplification for the composition of block ciphers: simpler proofs and new results. In: Joux, A., Youssef, A. (eds.) SAC 2014. LNCS, vol. 8781, pp. 129–146. Springer, Heidelberg (2014)

19. Cogliati, B., Seurin, Y.: On the provable security of the iterated even-mansour cipher against related-key and chosen-key attacks. In: Oswald, E., Fischlin, M. (eds.) EUROCRYPT 2015. LNCS, vol. 9056, pp. 584–613. Springer, Heidelberg (2015)

20. Coron, J.-S., Patarin, J., Seurin, Y.: The random oracle model and the ideal cipher model are equivalent. In: Wagner, D. (ed.) CRYPTO 2008. LNCS, vol. 5157, pp. 1–20. Springer, Heidelberg (2008)

21. Dodis, Y., Ristenpart, T., Shrimpton, T.: Salvaging Merkle-Damgård for practical applications. In: Joux, A. (ed.) EUROCRYPT 2009. LNCS, vol. 5479, pp. 371–388. Springer, Heidelberg (2009)

22. Dodis, Y., Steinberger, J.: Domain extension for MACs beyond the birthday barrier. In: Paterson, K.G. (ed.) EUROCRYPT 2011. LNCS, vol. 6632, pp. 323–342. Springer, Heidelberg (2011)

23. Dunkelman, O., Keller, N., Shamir, A.: Minimalism in cryptography: the even-mansour scheme revisited. In: Pointcheval, D., Johansson, T. (eds.) EUROCRYPT 2012. LNCS, vol. 7237, pp. 336–354. Springer, Heidelberg (2012)

24. Dziembowski, S., Pietrzak, K., Wichs, D.: Non-malleable codes. In: ICS 2010, pp. 434–452 (2010)

25. Even, S., Mansour, Y.: A construction of a cipher from a single pseudorandom permutation. J. Cryptol. 10(3), 151–162 (1997)

26. Farshim, P., Procter, G.: The related-key security of iterated even–mansour ciphers. In: Leander, G. (ed.) FSE 2015. LNCS, vol. 9054, pp. 342–363. Springer, Heidelberg (2015)

27. Faust, S., Mukherjee, P., Venturi, D., Wichs, D.: Efficient non-malleable codes and key-derivation for poly-size tampering circuits. In: Nguyen, P.Q., Oswald, E. (eds.) EUROCRYPT 2014. LNCS, vol. 8441, pp. 111–128. Springer, Heidelberg (2014)

28. Gaži, P., Tessaro, S.: Secret-key cryptography from ideal primitives: asystematic overview. In: IEEE Information Theory Workshop - ITW (2015)

29. Gentry, C., Ramzan, Z.: Eliminating random permutation oracles in the even-mansour cipher. In: Lee, P.J. (ed.) ASIACRYPT 2004. LNCS, vol. 3329, pp. 32–47. Springer, Heidelberg (2004)

30. Goldreich, O.: A sample of samplers - a computational perspective on sampling (survey). Electron. Colloquium Comput. Complex. (ECCC), 4(20) (1997)

31. Guo, C., Lin, D.: On the indifferentiability of key-alternating feistel ciphers with no key derivation. In: Dodis, Y., Nielsen, J.B. (eds.) TCC 2015, Part I. LNCS, vol. 9014, pp. 110–133. Springer, Heidelberg (2015)

32. Hall, C., Wagner, D., Kelsey, J., Schneier, B.: Building PRFs from PRPs. In: Krawczyk, H. (ed.) CRYPTO 1998. LNCS, vol. 1462, pp. 370–389. Springer, Heidelberg (1998)

33. Hoang, V.T., Morris, B., Rogaway, P.: An enciphering scheme based on a card shuffle. In: Safavi-Naini, R., Canetti, R. (eds.) CRYPTO 2012. LNCS, vol. 7417, pp. 1–13. Springer, Heidelberg (2012)

34. Holenstein, T., Künzler, R., Tessaro, S.: The equivalence of the random oracle model and the ideal cipher model, revisited. In: 43rd ACM STOC, pp. 89–98, June 2011

35. Kelsey, J., Schneier, B., Wagner, D.: Key-schedule cryptanalysis of IDEA, G-DES, GOST, SAFER, and triple-DES. In: Koblitz, N. (ed.) CRYPTO 1996. LNCS, vol. 1109, pp. 237–251. Springer, Heidelberg (1996)

36. Lampe, R., Patarin, J., Seurin, Y.: An asymptotically tight security analysis of the iterated even-mansour cipher. In: Wang, X., Sako, K. (eds.) ASIACRYPT 2012. LNCS, vol. 7658, pp. 278–295. Springer, Heidelberg (2012)

37. Lampe, R., Seurin, Y.: How to construct an ideal cipher from a small set of public permutations. In: Sako, K., Sarkar, P. (eds.) ASIACRYPT 2013, Part I. LNCS, vol. 8269, pp. 444–463. Springer, Heidelberg (2013)

38. Lampe, R., Seurin, Y.: Security analysis of key-alternating feistel ciphers. In: Cid, C., Rechberger, C. (eds.) FSE 2014. LNCS, vol. 8540, pp. 243–264. Springer, Heidelberg (2015)

39. Luby, M., Rackoff, C.: How to construct pseudo-random permutations from pseudo-random functions. In: Williams, H.C. (ed.) CRYPTO 1985. LNCS, vol. 218, pp. 447–447. Springer, Heidelberg (1986)

40. Lucks, S.: Ciphers secure against related-key attacks. In: Roy, B., Meier, W. (eds.) FSE 2004. LNCS, vol. 3017, pp. 359–370. Springer, Heidelberg (2004)

41. Mandal, A., Patarin, J., Nachef, V.: Indifferentiability beyond the birthday bound for the Xor of two public random permutations. In: Gong, G., Gupta, K.C. (eds.) INDOCRYPT 2010. LNCS, vol. 6498, pp. 69–81. Springer, Heidelberg (2010)

42. Mandal, A., Patarin, J., Seurin, Y.: On the public indifferentiability and correlation intractability of the 6-round feistel construction. In: Cramer, R. (ed.) TCC 2012. LNCS, vol. 7194, pp. 285–302. Springer, Heidelberg (2012)

43. Maurer, U.M., Pietrzak, K., Renner, R.S.: Indistinguishability amplification. In: Menezes, A. (ed.) CRYPTO 2007. LNCS, vol. 4622, pp. 130–149. Springer, Heidelberg (2007)

44. Maurer, U.M., Renner, R.S., Holenstein, C.: Indifferentiability, impossibility results on reductions, and applications to the random oracle methodology. In: Naor, M. (ed.) TCC 2004. LNCS, vol. 2951, pp. 21–39. Springer, Heidelberg (2004)

45. Maurer, U.M., Tessaro, S.: Domain extension of public random functions: beyond the birthday barrier. In: Menezes, A. (ed.) CRYPTO 2007. LNCS, vol. 4622, pp. 187–204. Springer, Heidelberg (2007)

46. Mennink, B., Preneel, B.: On the xor of multiple random permutations. In: Applied Cryptography and Network Security - ACNS (2015)

47. Morris, B., Rogaway, P.: Sometimes-recurse shuffle. In: Nguyen, P.Q., Oswald, E. (eds.) EUROCRYPT 2014. LNCS, vol. 8441, pp. 311–326. Springer, Heidelberg (2014)

48. Patarin, J.: The "coefficients H" technique. In: Avanzi, R.M., Keliher, L., Sica, F. (eds.) SAC 2008. LNCS, vol. 5381, pp. 328–345. Springer, Heidelberg (2009)

49. Ristenpart, T., Yilek, S.: The mix-and-cut shuffle: small-domain encryption secure against N queries. In: Canetti, R., Garay, J.A. (eds.) CRYPTO 2013, Part I. LNCS, vol. 8042, pp. 392–409. Springer, Heidelberg (2013)

50. Steinberger, J.: Improved security bounds for key-alternating ciphers via hellinger distance. Cryptology ePrint Archive, Report 2012/481 (2012). http://eprint.iacr.org/2012/481

Authenticated Encryption

Security of Full-State Keyed Sponge and Duplex: Applications to Authenticated Encryption

Bart Mennink[1]([⊠]), Reza Reyhanitabar[2], and Damian Vizár[2]

[1] Department of Electrical Engineering,
ESAT/COSIC, KU Leuven, IMinds, Leuven, Belgium
bart.mennink@esat.kuleuven.be
[2] EPFL, Lausanne, Switzerland
{reza.reyhanitabar,damian.vizar}@epfl.ch

Abstract. We provide a security analysis for *full-state* keyed Sponge and *full-state* Duplex constructions. Our results can be used for making a large class of Sponge-based authenticated encryption schemes more efficient by concurrent absorption of associated data and message blocks. In particular, we introduce and analyze a new variant of SpongeWrap with almost free authentication of associated data. The idea of using *full-state* message absorption for higher efficiency was first made explicit in the Donkey Sponge MAC construction, but without any formal security proof. Recently, Gaži, Pietrzak and Tessaro (CRYPTO 2015) have provided a proof for the *fixed-output-length* variant of Donkey Sponge. Yasuda and Sasaki (CT-RSA 2015) have considered *partially* full-state Sponge-based authenticated encryption schemes for efficient incorporation of associated data. In this work, we unify, simplify, and generalize these results about the security and applicability of full-state keyed Sponge and Duplex constructions; in particular, for designing more efficient authenticated encryption schemes. Compared to the proof of Gaži et al., our analysis directly targets the original Donkey Sponge construction as an *arbitrary-output-length* function. Our treatment is also more general than that of Yasuda and Sasaki, while yielding a more efficient authenticated encryption mode for the case that associated data might be longer than messages.

Keywords: Sponge construction · Duplex construction · Full-state absorption · Authenticated encryption · Associated data

1 Introduction

Since its introduction, the Sponge construction by Bertoni, Daemen, Peeters and Van Assche [4] has faced an immense increase in popularity. As "simple" hash function mode, it is the fundament of the SHA-3 standard Keccak [5], but also its keyed variants have become very popular modes of operation for a permutation to build a wide spectrum of symmetric-key primitives: reseedable pseudorandom number generators [7], pseudorandom functions and message authentication codes (PRFs/MACs) [9,11], Extendable-Output Functions ("XOFs") [24]

© International Association for Cryptologic Research 2015
T. Iwata and J.H. Cheon (Eds.): ASIACRYPT 2015, Part II, LNCS 9453, pp. 465–489, 2015.
DOI: 10.1007/978-3-662-48800-3_19

and authenticated encryption (AE) modes [10,11]. The keyed Sponge principle also got adopted in Spritz, a new RC4-like stream cipher [26], and in 10 out of 57 submissions to the currently running CAESAR competition on authenticated encryption [1,3]. These use cases reinforce the fact that Sponge-based constructions will continue to play an important role, not only in the new hashing standard SHA-3, but in various next-generation cryptographic algorithms.

The classical Sponge construction consists of a sequential application of a permutation p on a state of b bits. This state is partitioned into an r-bit rate or outer part and a c-bit capacity or inner part, where $b = r + c$. In the absorption phase, message blocks of size r bits are absorbed by the outer part and the state is transformed using p, while in the squeezing phase, digests are extracted from the outer part r bits at a time. In the indifferentiability framework of Maurer, Renner and Holenstein [20], Bertoni et al. [6] proved that the Sponge construction is secure up to the $O(2^{c/2})$ birthday-type bound. The capacity part is left untouched throughout the evaluation of the Sponge construction: a violation of this paradigm would make the indifferentiability security result void.

In this work, we strive for optimality, and investigate the most efficient ways of using Sponges for message authentication and authenticated encryption in a provably secure manner. In both directions, we consider a generalization of the currently known schemes to *full-state absorption*, the most efficient usage of the underlying permutation, and we show that these schemes are secure. Due to the full-state absorption, we cannot anymore rely on the classical indifferentiability result of the Sponge (as was for instance done in [2,10]), and a new security analysis is required. We will elaborate on both directions in the following.

Message Authentication. Bertoni et al. [9] introduced the keyed Sponge as a simple evaluation of the Sponge function on the key and the message, $\text{Sponge}(K\|M)$, and proved security beyond $O(2^{c/2})$. Chang et al. considered a slight variant of the keyed Sponge where the key is processed in the inner part of the Sponge, and observed that it can be seen as the Sponge based on an Even-Mansour blockcipher. At FSE 2015, Andreeva, Daemen, Mennink and Van Assche [2] considered a generic and improved analysis of both the outer- and inner-keyed Sponge. So far, however, these constructions have only been considered with the classical r-bit absorption.

The idea of using *full-state* message absorption for achieving higher efficiency was first made explicit in the Donkey Sponge MAC construction [11],[1] but without any formal security proof. The recently introduced Donkey-inspired MAC function Chaskey [22] did get a formal security analysis, but its proof is thwarted towards Chaskey and does not apply to the Donkey Sponge.

A thorough analysis of the full-state message absorption keyed Sponge had to wait for Gaži, Pietrzak and Tessaro [17], who prove nearly tight security up to $O(\ell q(q + N)/2^b + q(q + \ell + N)/2^c)$, where the adversary makes q queries of maximal length ℓ, and makes N primitive calls. However, their analysis only applies to the *fixed-output-length* variant, and the proof does not directly seem

---

[1] We note that apart from full-state absorption, the Donkey Sponge also uses less rounds in the underlying permutation during the absorbing phase.

to extend to the original *arbitrary-output-length* keyed Sponge. In this work, we provide a direct proof for this more general case.

In more detail, we present a generalized scheme, dubbed *Full-state Keyed Sponge (FKS)*, whose security implies the security of Donkey Sponge in the ideal permutation setting, and prove that it is secure up to approximately $\frac{2(q\ell)^2}{2^b} + \frac{2q^2\ell}{2^c} + \frac{\mu N}{2^k}$, where k is the size of the key, and μ is a parameter called the "multiplicity". We note that usage of the outer-keyed Sponge makes no longer any difference from the usage of the inner-keyed variant in the presence of full-state absorption (see also Sect. 8). Our proof of FKS follows the modular approach of Andreeva et al., but due to the full-state absorption, we cannot rely on the indifferentiability result of [6], and present a new and more detailed analysis.

Authenticated Encryption. Encryption via the Sponge can be done (and is typically done) via the Duplex construction [10], a stateful construction consisting of an initialization interface and a duplexing interface. The initialization interface can be called to initialize an all-zero state; the duplexing interface absorbs a message of size $< r$ bits and squeezes $\leq r$ bits of the outer part. The security of the Duplex traces back to the indifferentiability of the classical Sponge, yielding a $O(2^{c/2})$ security bound.

Bertoni et al. [10] showed that the Duplex, in turn, allows for authenticated encryption in the form of SpongeWrap. This mode is, de facto, the basis of the majority of Sponge-based submissions to the CAESAR competition. Jovanovic et al. [18] re-investigated Sponge-based authenticated encryption schemes, starring NORX, and derived beyond birthday-bound security. These results are, however, all for the usual r-bit absorption. Yasuda and Sasaki [27] have considered several full-state and *partially* full-state Sponge-based authenticated encryption schemes for efficient incorporation of associated data, directly lifting Jovanovic et al.'s security proofs. The concurrent absorption mode proposed by Yasuda and Sasaki (Fig. 3 in [27]) fails to utilize the full-state absorption when the associated data becomes longer than the message, forcing the mode switch from a full-state mode to the classical r-bit absorbing Sponge mode; hence, we refer to this as a *partially* full-state AE mode. Full-state data absorption was also proposed by Reyhanitabar, Vaudenay and Vizár [25] in their compression function based AE mode p-OMD.

We generically aim to optimize the efficiency in Sponge-based authenticated encryption. To this end, we first formalize the *Full-state Keyed Duplex (FKD)* construction. It differs from the original Duplex in the fact that (i) the key is explicitly used to initialize the state (In this, the FKD is similar to the Monkey Duplex [11]) and (ii) the absorption is performed on the entire state. Note that the possibility to absorb in the entire state enforces the explicit usage of the key. Next, we prove that FKD is provably secure, i.e., indistinguishable from a random oracle with the same interfaces. As before, we cannot rely on the classical indifferentiability proof due to the full-state absorption; however, we show how to adapt the FKS proof to a special case directly related to the security of FKD.

We exemplify the better absorption capabilities of FKD by the introduction of a *Full-state SpongeWrap (FSW)*. The FSW construction is more general than that of Yasuda and Sasaki, who only considered specific AE constructions, and interestingly, our approach also yields a more efficient (truly full-state) authenticated encryption mode irrespective of the relative lengths of messages and their associated data.

Organization of the Paper. Notations and preliminary concepts are presented in Sect. 2. We present the Full-state Keyed Sponge and Full-state Keyed Duplex in Sect. 3. The security model is discussed in Sect. 4. In Sect. 5 we prove security of FKS and in Sect. 6 of FKD. The introduction of the Full-state SpongeWrap, and the application of FKD to this construction is given in Sect. 7. Section 8 provides a brief discussion on related-key security and our security models.

2   Notations and Conventions

The set of all strings of length b is denoted as $\{0,1\}^b$ for any $b \geq 1$ and the set of all finite strings of arbitrary length is denoted as $\{0,1\}^*$. We will denote the empty string of length 0 as ε. For any positive b, we let $\{0,1\}^{<b} = \bigcup_{i=0}^{b-1} \{0,1\}^i$ denote set of all strings of length less than b including ε. For two strings $X, Y \in \{0,1\}^*$ we let $X \parallel Y$ denote the string obtained by concatenation of X and Y. For a string $X \in \{0,1\}^x$ we let $\mathsf{left}_\ell(X)$ denote the ℓ leftmost bits of X and $\mathsf{right}_r(X)$ the r rightmost bits of X such that $X = \mathsf{left}_\chi(X) \parallel \mathsf{right}_{x-\chi}(X)$ for any $0 \leq \chi \leq x$. For integral b, r, c such that $b = r + c$, and for $t \in \{0,1\}^b$, we let $\mathsf{outer}(t) = \mathsf{left}_r(t)$ and $\mathsf{inner}(t) = \mathsf{right}_c(t)$.

For a non-empty finite set S let $a \xleftarrow{\$} S$ denote sampling an element a from S uniformly at random. We let $|Z|$ denote the cardinality if Z is a set and the length if Z is a string. We let $\mathrm{Perm}(b)$ denote the set of all permutations of b-bit strings and $\mathrm{Func}(b)$ the set of all functions over b-bit strings.

Given two strings X, Y, let

$$\mathsf{llcp}_b(X,Y) = \max_{i \geq 0} \{i \ : \ \mathsf{left}_{ib}(X) = \mathsf{left}_{ib}(Y)\}$$

denote the length of the longest common prefix between X and Y in b-bit blocks. For a string X and a non-empty set of strings $\{Y_1, \ldots, Y_n\}$ let

$$\mathsf{llcp}_b(X; Y_1, \ldots, Y_n) = \max\{\mathsf{llcp}_b(X, Y_1), \ldots, \mathsf{llcp}_b(X, Y_n)\}.$$

For any two pairs of integers $(i,j), (i',j')$, we say that $(i',j') < (i,j)$ if either $i' < i$ or if $i' = i$ and $j' < j$. We say that $(i',j') \leq (i,j)$ if $(i',j') < (i,j)$ or if $(i',j') = (i,j)$. In other words, we use lexicographical ordering to determine ordering of integer-tuples.

3   Sponge Constructions

3.1   Full-State Keyed Sponge

We consider the Full-State Keyed Sponge (FKS) construction that is using a public permutation $p : \{0,1\}^b \to \{0,1\}^b$. It is furthermore parameterized with

r, k, which are required to satisfy $r < b$ and $k \le b - r =: c$. The parametrization is sometimes left implicit if it is clear from the context. FKS gets as input a key $K \in \{0,1\}^k$, a message $M \in \{0,1\}^*$, and a natural number z, and it outputs a string $Z \in \{0,1\}^z$:

$$\text{FKS}^p(K, M, z) = \text{FKS}^p_K(M, z) = Z.$$

It operates on a state $t \in \{0,1\}^b$, which is initialized using the key K. The message M is first padded to a length a multiple of b bits, using $\text{pad}_b(M) = M \| 10^{b-1-|M| \bmod b}$, which is then viewed as m b-bit message blocks $M^1 \| ... \| M^m$.[2] These message blocks are processed one-by-one, interleaved with evaluations of p. After the absorption of M, the outer r bits of the state are output and the state is processed via p until a sufficient amount of output bits are obtained. FKS is depicted in Fig. 1, and Algorithm 1 provides a formal specification of FKS.

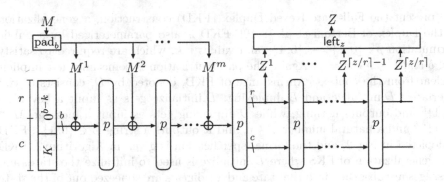

Fig. 1. The FKS construction.

Algorithm 1. $\text{FKS}[p, r, k](K, M, z)$	**Algorithm 2.** $\text{FKD}[p, r, k]$
1: $t \leftarrow 0^{b-k} \| K$	1: **Interface** FKD.initialize(K)
2: $M^1 \| \cdots \| M^m \xleftarrow{b} \text{pad}_b(M)$	2: $t \leftarrow 0^{b-k} \| K$
3: **for** $i = 1, \ldots, m$ **do**	
4: $s \leftarrow t \oplus M^i$	1: **Interface** FKD.duplexing(M, z)
5: $t \leftarrow p(s)$	2: **if** $z > r$ **or** $|M| \ge b$ **then**
6: $Z \leftarrow \text{left}_r(t)$	3: **return** $\perp$
7: **while** $|Z| < z$ **do**	4: $s \leftarrow t \oplus \text{pad}_b(M)$
8: $t \leftarrow p(t)$	5: $t \leftarrow p(s)$
9: $Z \leftarrow Z \| \text{left}_r(t)$	6: **return** $\text{left}_z(t)$
10: **return** $\text{left}_z(Z)$	

---

[2] In fact, any injective padding function works, as long as the last block is always non-zero.

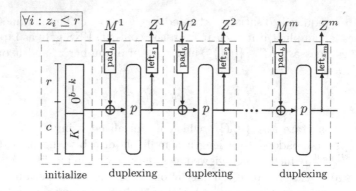

Fig. 2. The FKD construction.

3.2 Full-State Keyed Duplex

We present the Full-state Keyed Duplex (FKD) construction, a generalization of the Duplex of Bertoni et al. [8,10]. FKD is also parameterized by a public permutation $p : \{0,1\}^b \to \{0,1\}^b$ and values r, k, which are required to satisfy $r < b$ and $k \leq b - r =: c$. Again, the parametrization is sometimes left implicit if clear from the context. An instance of FKD, denoted by D, consists of two interfaces: D.initialize and D.duplexing. D.initialize gets as input a key $K \in \{0,1\}^k$ and outputs nothing, while D.duplexing gets as input a message $M \in \{0,1\}^{<b}$ and a natural number $z \leq r$, and it outputs a string $Z \in \{0,1\}^z$. FKD is depicted in Fig. 2, and the formal specification is given in Algorithm 2. FKD is a generalization of FKS where D.initialize is used to initialize the state, and messages are absorbed into the state and/or digests are squeezed out of the state using D.duplexing calls.

4   Security Models and Tools

Multiplicity. Let $\{(x_i, y_i)\}_{i=1}^{\sigma}$ be a set of σ evaluations of a permutation p. Following Andreeva et al. [2], we define the total maximal multiplicity as $\mu = \mu_{\text{fwd}} + \mu_{\text{bwd}}$, where

$$\mu_{\text{fwd}} = \max_a |\{i \in \{1, \dots, \sigma\} : \text{outer}(x_i) = a\}|,$$

$$\mu_{\text{bwd}} = \max_a |\{i \in \{1, \dots, \sigma\} : \text{outer}(y_i) = a\}|.$$

The multiplicity is a quantity that characterises the data that are available to the adversary during the attack. We have $2 \leq \mu \leq 2\sigma$ per definition, however the upper bound 2σ is never reached in practical applications of sponge-based constructions. Being a sum of forward and backward multiplicities, the total multiplicity can be seen as a measure of adversary's ability to control the outer part of the permutation inputs and outputs respectively. In case of sponge-based designs, the backward multiplicity can be expected to be approximately $\sigma 2^{-r}$ while the forward multiplicity varies with concrete applications [2].

4.1   Adversaries and Patarin's Coefficient-H Technique

We consider an information-theoretic adversary A that has access to one or more oracles X; this is denoted by A^X and the notation $A^X \Rightarrow 1$ means that A, after interaction with X, returns 1. It is a classical fact (for a simple proof see [14]) that in the information-theoretic setting, adversaries can be assumed to be deterministic without loss of generality.

We use Patarin's Coefficient-H technique [23]; more precisely, a revisited formulation of it by Chen and Steinberger [14]. Consider a deterministic information-theoretic adversary A whose goal is to distinguish two oracles X and Y:

$$\Delta_A (X;Y) = \left| \Pr \left[A^X \Rightarrow 1 \right] - \Pr \left[A^Y \Rightarrow 1 \right] \right|.$$

Here, X and Y are randomized algorithms; the randomization depends on the specific scenario and for now is left implicit. The interaction with any of the two systems X or Y is summarized in a transcript τ. Denote by D_X the probability distribution of transcripts when interacting with X, and similarly, D_Y the distribution of transcripts when interacting with Y. A transcript τ is called *attainable* if $\Pr [D_Y = \tau] > 0$, meaning that it can occur during interaction with Y. Denote by $\mathcal{T}$ the set of all attainable transcripts. The Coefficient-H technique states the following, for the proof of which we refer to [14].

Lemma 1 (Coefficient-H Technique [14,23]). *Consider a fixed deterministic adversary* A. *Let* $\mathcal{T} = \mathcal{T}_{\text{good}} \cup \mathcal{T}_{\text{bad}}$ *be a partition into* good *transcripts* $\mathcal{T}_{\text{good}}$ *and* bad *transcripts* $\mathcal{T}_{\text{bad}}$. *If there exists an* ε *such that for all* $\tau \in \mathcal{T}_{\text{good}}$,

$$\frac{\Pr [D_X = \tau]}{\Pr [D_Y = \tau]} \geq 1 - \varepsilon,$$

then, $\Delta_A (X;Y) \leq \varepsilon + \Pr [D_Y \in \mathcal{T}_{\text{bad}}]$.

The two partitions of $\mathcal{T}$ are labeled as $\mathcal{T}_{\text{good}}$ and $\mathcal{T}_{\text{bad}}$ to aid the intuitiveness of the proof. The transcripts in $\mathcal{T}_{\text{good}}$ are "good" in the sense that they give us a high value of $\Pr [D_X = \tau]/\Pr [D_Y = \tau]$ and thus small ε while the "bad" transcripts from $\mathcal{T}_{\text{bad}}$ fail to do so.

4.2   Security Models for FKS and FKD

Let $RO^\infty : \{0,1\}^* \to \{0,1\}^\infty$ be a random oracle which takes inputs of arbitrary but finite length and returns random infinite strings, where each output bit is selected uniformly and independently for every input M.

Let F be either FKS or FKD, which is based on a permutation $p : \{0,1\}^b \to \{0,1\}^b$ and a key $K \in \{0,1\}^k$. We will define the security of F in two settings: the public permutation setting, where the adversary has query access to the permutation (security comes from the secrecy of K), and the secret permutation setting (with no explicit key K), where the adversary has no access to the underlying permutation and the security comes from the secrecy of the permutation.

We use the notations F_K^p and F_0^π to refer to the public permutation and secret permutation based schemes, respectively; where, π is a secret random permutation.

In both settings, we consider an adversary that aims to distinguish the *real F* from an *ideal* (reference) primitive—an oracle RO with the same interface. For $F = \text{FKS}$ the corresponding ideal primitive RO is defined by $RO_{\text{FKS}}(M, z) = \text{left}_z(RO^\infty(M))$. For $F = \text{FKD}$ the corresponding reference primitive RO_{FKD} is a stateful oracle with two interfaces: (1) RO_{FKD}^r.initialize() that initializes the state of the oracle, St, to the empty string, and (2) RO_{FKD}^r.duplexing(M, z) that, on input $M \in \{0, 1\}^{<b}$ and a natural number z, first updates the state as $\text{St} \leftarrow \text{St} \| pad_b(M)$ and then outputs $\text{left}_z(RO^\infty(\text{St}))$.

We define the distinguishing advantage of any adversary A against F based on a *public* permutation by

$$\mathbf{Adv}_{F_K^p, p}^{\text{ind}}(A) = \left| \Pr\left[K \xleftarrow{\$} \{0, 1\}^k, p \xleftarrow{\$} \text{Perm}(b) : A^{F_K^p, p, p^{-1}} \Rightarrow 1 \right] - \Pr\left[p \xleftarrow{\$} \text{Perm}(b) : A^{RO, p, p^{-1}} \Rightarrow 1 \right] \right|.$$

The distinguishing advantage of A against F based on a *secret* permutation is defined by

$$\mathbf{Adv}_{F_0^\pi}^{\text{ind}}(A) = \left| \Pr\left[\pi \xleftarrow{\$} \text{Perm}(b) : A^{F_0^\pi} \Rightarrow 1 \right] - \Pr\left[A^{RO} \Rightarrow 1 \right] \right|.$$

The resource parameterized advantage functions are defined as usual. Let $\mathbf{Adv}_{F_K^p, p}^{\text{ind}}(q, \ell, \mu, N) = \max_A \mathbf{Adv}_{F_K^p, p}^{\text{ind}}(A)$ be the maximum advantage over all adversaries that make q queries to the left oracle, all of maximal length ℓ permutation calls if $F = \text{FKS}$ or that make at most q initialize() calls to the left oracle and issue at most ℓ duplexing queries after each initialization if $F = \text{FKD}$ with total maximal multiplicity μ in both cases, and that make N direct queries to the public permutation. To simplify the analysis, we assume that each of the q oracle queries in fact consists of *exactly* ℓ permutation (or that the adversary indeed makes ℓ duplexing calls after each initialization). This is without loss of generality, it can simply be achieved by giving extra squeezing outputs to the adversary. Similarly, we define $\mathbf{Adv}_{F_0^\pi}^{\text{ind}}(q, \ell, \mu) = \max_A \mathbf{Adv}_{F_0^\pi}^{\text{ind}}(A)$, noticing that in this case $N = 0$, thus it is omitted from the resources.

4.3   Security Model for Even-Mansour

Our proof relies on a reduction to the security of a low-entropy single-key Even-Mansour construction [15,16]. In more detail, let $p : \{0, 1\}^b \rightarrow \{0, 1\}^b$ be a permutation and $K \in \{0, 1\}^k$ be a key. The Even-Mansour blockcipher is defined as

$$E_K^p(M) = p(M \oplus (0^{b-k} \| K)) \oplus (0^{b-k} \| K).$$

We define the distinguishing advantage of any adversary A against E based on a public permutation p as

$$\mathbf{Adv}^{\mathrm{prp}}_{E^p_K,p}(A) = \left| \Pr\left[K \xleftarrow{\$} \{0,1\}^k, p \xleftarrow{\$} \mathrm{Perm}\,(b) \;:\; A^{E^p_K,p,p^{-1}} \Rightarrow 1 \right] - \right.$$
$$\left. \Pr\left[\pi, p \xleftarrow{\$} \mathrm{Perm}\,(b) \;:\; A^{\pi,p,p^{-1}} \Rightarrow 1 \right] \right|.$$

Let $\mathbf{Adv}^{\mathrm{prp}}_{E^p_K,p}(q,\mu,N) = \max_A \mathbf{Adv}^{\mathrm{prp}}_{E^p_K,p}(A)$ be the maximum advantage over all adversaries that make q queries to the left oracle, with total maximal multiplicity μ, and that make N direct queries to the public permutation.

5   Security Analysis of FKS

We prove the following result for FKS:

Theorem 1. *Let $b, r, c, k > 0$ be such that $b = r + c$ and $k \le c$. Let FKS be the scheme of Sect. 3.1. Then,*

$$\mathbf{Adv}^{\mathrm{ind}}_{\mathrm{FKS}^p_K,p}(q,\ell,\mu,N) \le \frac{2(q\ell)^2}{2^b} + \frac{2q^2\ell}{2^c} + \frac{\mu N}{2^k}.$$

The proof follows to a certain extent the modular approach of [2], and in particular also uses the observation that FKS^p_K can alternatively be considered as $\mathrm{FKS}^{E^p_K}_0$, a clever observation used before by Chang et al. [13]. Note that this observation only works for $k \le c$: it consists of xoring two dummy keys $K \oplus K$ in-between every two adjacent permutation calls, and if $k > c$ this would entail a difference in the squeezing blocks of FKS. This trick splits the security of FKS^p_K into the security of the Even-Mansour blockcipher and the security of FKS with secret primitive. Looking back at [2], the security of Inner-keyed Sponge/Outer-keyed Sponge [2] with secret permutations was simply reverted to the classical indifferentiability result of [6]. Because this is a rather loose approach, and additionally because the indifferentiability bound cannot be used for FKS due to its full-state absorption, we consider the security of FKS with secret primitive in more detail and derive an improved bound.

Proof (Proof of Theorem 1). Consider any adversary A with resources (q, ℓ, μ, N). Note that $\mathrm{FKS}^p_K = \mathrm{FKS}^{E^p_K}_0$. Therefore, by a modular argument,

$$\mathbf{Adv}^{\mathrm{ind}}_{\mathrm{FKS}^p_K,p}(A) = \Delta_A\left(\mathrm{FKS}^{E^p_K}_0, p; RO_{\mathrm{FKS}}, p\right)$$
$$\le \Delta_B\left(\mathrm{FKS}^{\pi}_0, p; RO_{\mathrm{FKS}}, p\right) + \Delta_C\left(E^p_K, p; \pi, p\right)$$
$$= \mathbf{Adv}^{\mathrm{ind}}_{\mathrm{FKS}^{\pi}_0}(B) + \mathbf{Adv}^{\mathrm{prp}}_{E^p_K,p}(C)$$

for some adversary B with resources (q, ℓ, μ) and adversary C with resources $(q\ell, \mu, N)$. Note that B also has access to p, but queries to this oracle are meaningless as its left oracle (FKS^{π}_0 or RO_{FKS}) is independent of p.

In [2], it is proven that $\mathbf{Adv}^{\mathrm{prp}}_{E^p_K,p}(C) \le \frac{\mu N}{2^k}$ for any C. In Lemma 2, we prove that $\mathbf{Adv}^{\mathrm{ind}}_{\mathrm{FKS}^{\pi}_0}(B) \le \frac{2(q\ell)^2}{2^b} + \frac{2q^2\ell}{2^c}$ for any adversary B. $\square$

Lemma 2. *Let $b, r, c > 0$ be such that $b = r + c$. Let* FKS *be the scheme of Sect. 3.1. Then,*

$$\mathbf{Adv}^{\mathrm{ind}}_{\mathrm{FKS}_0^\pi}(q, \ell, \mu) \leq \frac{2(q\ell)^2}{2^b} + \frac{2q^2\ell}{2^c}.$$

Proof. Given that the padding is publicly known and injective, we can generalize the setting, and assume that the ith query M_i has length divisible by b and that $M_i^{m_i} \neq 0^b$, i.e. we assume that all the queries are already padded. More detailed, for $1 \leq i \leq q$, we let $m_i = |M_i|/b$ and $M_i = M_i^1 \parallel M_i^2 \parallel \ldots \parallel M_i^{m_i}$ s.t. $|M_i^j| = b$ for $1 \leq j \leq m_i$. We further assume, that the adversary always asks for output of length divisible by r and that every query induces exactly ℓ primitive calls. This is without loss of generality: we can simply output "free bits" to the adversary. We will denote the b-bit state of FKS just *before* the jth application of π is made when processing the ith query as s_i^j for $1 \leq j \leq \ell$. Similarly, we will denote the b-bit state of FKS just *after* the jth application of π in i^{th} query as t_i^j for $1 \leq j \leq \ell$. We will call the former in-states and the latter out-states. Note that every in-state s_i^j is determined by the out-state t_i^{j-1} and the block of query M_i^j as $s_i^j = t_i^{j-1} \oplus M_i^j$ in the absorbing phase or just by t_i^j in the squeezing phase as depicted in Fig. 3.

To aid the simplicity of further analysis we additionally define initial dummy out-states $t_i^0 = 0^b$ and extended queries $\bar{M}_i = M_i \parallel 0^{(\ell - m_i)b}$ for $1 \leq i \leq q$. Now we can express every in-state, be it absorbing or squeezing, as $s_i^j = t_i^{j-1} \oplus \bar{M}_i^j$. We will group the out-states of ith query as $T_i = \{t_i^0, t_i^1, \ldots, t_i^\ell\}$. Because each query induces exactly ℓ calls to π, we know that a query M_i will be answered by a string $Z_i = Z_i^1 \parallel \ldots \parallel Z_i^{z_i}$ with $z_i = \ell - m_i + 1$ and $|Z_i^j| = r$ for $1 \leq j \leq z_i$. In particular, we have that $Z_i^j = \mathsf{outer}\left(t_i^{m_i+j-1}\right)$.

The RP-RF Switch. We start by replacing the random permutation $\pi \xleftarrow{\$} \mathrm{Perm}(b)$ by a random function $f \xleftarrow{\$} \mathrm{Func}(b)$ in the experiment. This will contribute the term $(q\ell)^2/2^b$ to the final bound by a standard hybrid argument so we have $\mathbf{Adv}^{\mathrm{ind}}_{\mathrm{FKS}_0^\pi}(q, \ell, \mu) \leq \mathbf{Adv}^{\mathrm{ind}}_{\mathrm{FKS}_0^f}(q, \ell, \mu) + (q\ell)^2/2^b$.

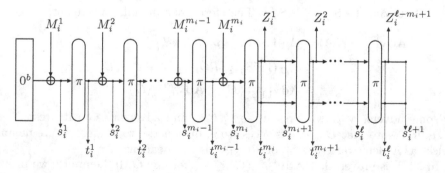

Fig. 3. Processing the ith query.

Patarin's Coefficient-H Technique. We will use the coefficient-H technique to show that $\mathbf{Adv}^{\mathrm{ind}}_{\mathrm{FKS}_0^f}(q, \ell, \mu) \leq (q\ell)^2/2^b + 2q^2\ell/2^c$. The two systems an adversary is trying to distinguish are FKS_0^f and RO_{FKS}. We will refer to the former as X and to the latter as Y. In either of the worlds, the adversary makes q queries $M_1, \ldots, M_q$ and learns the responses $Z_1, \ldots, Z_q$. The transition from queries M_i to $\bar{M}_i$ is injective, and additionally the length m_i of M_i is implicit from $\bar{M}_i$. Therefore, we can summarize the interaction of the adversary with its oracle (X or Y) with a transcript $(\bar{M}_1, \ldots, \bar{M}_q, Z_1, \ldots, Z_q)$.

To facilitate the analysis, we will disclose additional information $T_1, \ldots, T_q$ to the adversary at the end of the experiment. In the real world, these are the out-states $T_i = \{t_i^0, t_i^1, \ldots, t_i^\ell\}$ as discussed in the beginning of the proof. In the ideal world, these are dummy variables that satisfy the following intrinsic properties of the Sponge construction:

1. $t_i^0 = 0^b$ for $1 \leq i \leq q$,
2. if $\mathsf{llcp}_b(\bar{M}_i, \bar{M}_{i'}) = n$ for $1 \leq i, i' \leq q$ then $t_i^j = t_{i'}^j$ for $1 \leq j \leq n$,
3. $\mathrm{outer}\left(t_i^{j+m_i-1}\right) = Z_i^j$ for $1 \leq i \leq q$ and $1 \leq j \leq z_i$,

but are perfectly random otherwise. Note that in both worlds, $Z_1, \ldots, Z_q$ are fully determined by $T_1, \ldots, T_q$, so we can drop them from the transcript. Thus a transcript of adversary's interaction with FKS will be $\tau = (\bar{M}_1, \ldots, \bar{M}_q, T_1, \ldots, T_q)$.

With respect to Lemma 1, we will show that there exists a definition of bad transcripts $\mathcal{T}_{\mathrm{bad}}$, such that $\Pr[D_X = \tau]/\Pr[D_Y = \tau] = 1$ for any $\tau \in \mathcal{T}_{\mathrm{good}} = \mathcal{T} \backslash \mathcal{T}_{\mathrm{bad}}$, and thus $\mathbf{Adv}^{\mathrm{ind}}_{\mathrm{FKS}_0^f}(q, \ell, \mu) \leq \Pr[D_Y \in \mathcal{T}_{\mathrm{bad}}]$.

Definition of a Bad Transcript. Stated formally, a transcript τ is labeled as *bad* if
$$\exists (1,1) \leq (i,j), (i',j') \leq (q, \ell) \text{ such that:}$$
$$j \neq j' \lor \mathsf{llcp}_b(\bar{M}_i, \bar{M}_{i'}) < j = j' \leq \ell, \tag{1}$$
$$t_i^{j-1} \oplus \bar{M}_i^j = t_{i'}^{j'-1} \oplus \bar{M}_{i'}^{j'}.$$

This formalization of a bad transcript comes with an intuitive, informal interpretation; as long as all relevant *inputs* $s_i^j = t_i^{j-1} \oplus \bar{M}_i^j$ to the random function f induced by the Sponge function are distinct the output of the Sponge will be distributed uniformly. We do not require uniqueness of all in-states because the adversary can trivially force their repetition by issuing queries with common prefixes, as we have argued earlier. However these collisions are not a problem because uniqueness of the queries implies that $\mathsf{llcp}_b(\bar{M}_i, \bar{M}_{i'}) < \max\{m_i, m_{i'}\}$ for any two queries $\bar{M}_i, \bar{M}_{i'}$. Even if the adversary truncates an old query and thus forces an old absorbing in-state s to be squeezed for output, it is still not a problem because the adversary *has not seen* the image $f(s)$ before. Note that albeit in-states do not exist in the ideal world, they can be defined by the same relation as in the real world, i.e. $s_i^j = t_i^{j-1} \oplus \bar{M}_i^j$.

Bounding the Ratio of Probabilities of Good Transcripts. In the ideal world, the out-states $\{t_i^0\}_{i=0}^q$ are always assigned a value trivially. Beside that, we will also trivially assign a single randomly sampled value to multiple state variables, that are affected by the common prefixes of the queries. The remaining out-states are sampled uniformly at random. It follows that there are exactly $\eta(\tau) = \sum_{i=1}^{q} \ell - \text{llcp}_b (M_i; M_1, \ldots, M_{i-1})$ b-bit values in any transcript τ, that are sampled independently and uniformly. We thus have $\Pr[D_Y = \tau] = 2^{-\eta(\tau)b}$ for any τ.

Let Ω_X be the set of all possible real-world oracles. We have that $|\Omega_X| = 2^{b2^b}$. Let $\text{comp}_X(\tau) \subseteq \Omega_X$ be the set of all oracles compatible with the transcript τ, i.e. the set of the real-world oracles that are capable of producing τ in an experiment. We will compute the probability of seeing τ in the real world as $\Pr[D_X = \tau] = |\text{comp}_X(\tau)|/|\Omega_X|$. Note that a real-world oracle is completely determined by the underlying function f.

If $\tau \in \mathcal{T}_{\text{good}}$, then every in-state $s_i^j = t_i^{j-1} \oplus \bar{M}_i^j$ that does not trivially collide with some other in-state $s_{i'}^{j'}$ due to common prefix of $\bar{M}_i^j$ and $\bar{M}_{i'}^{j'}$ must be distinct. The number of domain points of f that have an image assigned by τ is easily seen to be $\eta(\tau) = \sum_{i=1}^{q} \ell - \text{llcp}_b(M_i; M_1, \ldots, M_{i-1})$. A compatible function f can therefore have arbitrary image values on the remaining $2^b - \eta(\tau)$ domain points. Thus we compute $|\text{comp}_X(\tau)| = 2^{b(2^b - \eta(\tau))}$ and

$$\Pr[D_X = \tau] = \frac{|\text{comp}_X(\tau)|}{|\Omega_X|} = \frac{2^{b(2^b - \eta(\tau))}}{2^{b2^b}} = 2^{-\eta(\tau)b} = \Pr[D_Y = \tau].$$

Bounding the Probability of a Bad Transcript in the Ideal World. We can bound the probability of τ being bad (cf. (1)) by first bounding the collision probability of an arbitrary but fixed pair of in-states $s_i^j, s_{i'}^{j'}$ (i.e. the event $s_i^j = s_{i'}^{j'}$ occurs) and then summing this probability for all possible values of $(i, j), (i', j')$ with $(i', j') \neq (i, j)$. Because this probability varies significantly, we will split all in-states into three classes and bound probabilities of individual collisions between these classes.

We will associate to each in-state s_i^j a label stamp_i^j. We set $\text{stamp}_i^j = \texttt{free}$ if $1 < j = \text{llcp}_b(\bar{M}_i; \bar{M}_1, \ldots, \bar{M}_{i-1}) + 1 \leq m_i$ such that $m_{i^*} < j$ for some $i^* < i$. We will set $\text{stamp}_i^1 = \texttt{initial}$ for $1 \leq i \leq q$ and $\text{stamp}_i^j = \texttt{fixed}$ in the remaining cases. Informally, we have $\text{stamp}_i^j = \texttt{free}$ whenever the adversary forces $\text{outer}\left(t_i^{j-1}\right) = Z_{i^*}^{j - m_{i^*} - 1}$ by reusing exactly first $j - 1$ blocks of a previous query $\bar{M}_{i^*}$ in $\bar{M}_i$ and sets $\bar{M}_i^j \neq \bar{M}_{i^*}^j = 0^b$. By doing this, it freely but non-trivially chooses $\text{outer}\left(s_i^j\right) = \text{outer}\left(s_{i^*}^j \oplus \bar{M}_{i^*}^j \oplus \bar{M}_i^j\right)$. Note that if the adversary puts $\bar{M}_i^j = \bar{M}_{i^*}^j$, this is not counted as a free state (the states will in fact be the same). We have $\text{stamp}_i^j = \texttt{initial}$ for the initial in-state of every query.

As the condition (1) is symmetrical w.r.t. (i, j) and (i', j'), and as it cannot be satisfied if $(i, j) = (i', j')$, it can be rephrased as

$$\exists (1, 1) \leq (i', j') < (i, j) \leq (q, \ell) \text{ such that:}$$

$$\text{llcp}_b(\bar{M}_i; \bar{M}_1, \ldots, \bar{M}_{i-1}) < j \leq \ell, \ s_i^j = s_{i'}^{j'}. \tag{2}$$

Doing so is without loss of generality, as each s_i^j with $j \leq \mathsf{llcp}_b (\bar{M}_i; \bar{M}_1, \ldots, \bar{M}_{i-1})$ is identical with some previous state that has already been checked for collisions with $s_{i'}^{j'}$ for every possible (i', j'). In the further analysis, we will be working with (2) rather than with (1).

We will now bound the probability of collision of an arbitrary pair of in-states $(s_i^j, s_{i'}^{j'}) = (t_i^{j-1} \oplus \bar{M}_i^j, t_{i'}^{j'-1} \oplus \bar{M}_{i'}^{j'})$ with $\mathsf{stamp}_i^j = \mathtt{fixed}$. We fix arbitrary i and investigate the following three cases for j. In each case we treat every $(i', j') < (i, j)$.

Case 1: $\mathsf{llcp}_b (\bar{M}_i; \bar{M}_1, \ldots, \bar{M}_{i-1}) + 1 < j \leq m_i$. In this case, t_i^{j-1} is undetermined when the adversary issues the query $\bar{M}_i$. This implies that it will be independent from all $t_{i'}^{j'-1}$ for any $(i', j') < (i, j)$. The probability of the collision $t_i^{j-1} \oplus \bar{M}_i^j = t_{i'}^{j'-1} \oplus \bar{M}_{i'}^{j'}$ is easily seen to be 2^{-b}.

Case 2: $\max \left\{ \mathsf{llcp}_b (\bar{M}_i; \bar{M}_1, \ldots, \bar{M}_{i-1}) + 1, m_i \right\} < j \leq \ell$. Here $t_i^{j-1} = Z_i^{j-m_i} \| \mathsf{inner} \left(t_i^{j-1} \right)$ and $\bar{M}_i^j = 0^b$. Although the adversary learns the value of $Z_i^{j-m_i}$ during the experiment, this is independent of all $s_{i'}^{j'}$ with $(i', j') < (i, j)$ (because $j + 1 > \mathsf{llcp}_b (\bar{M}_i; \bar{M}_1, \ldots, \bar{M}_{i-1})$). Even if $\mathsf{stamp}_{i'}^{j'} \in \{\mathtt{free}, \mathtt{initial}\}$ and $\mathsf{outer} \left(s_{i'}^{j'} \right) = \alpha$ for some value α chosen by the adversary, the collision $Z_i^{j-m_i} \| \mathsf{inner} \left(t_i^{j-1} \right) = \alpha \| \mathsf{inner} \left(s_{i'}^{j'} \right)$ happens with probability 2^{-b}.

Case 3: $j = \mathsf{llcp}_b (\bar{M}_i; \bar{M}_1, \ldots, \bar{M}_{i-1}) + 1$. If $j = \mathsf{llcp}_b (\bar{M}_i, \bar{M}_{i'}) + 1$, the in-state $s_{i'}^{j'=j}$, call it a twin-state of s_i^j, cannot collide with s_i^j, as by the second trivial property $t_i^{j-1} = t_{i'}^{j-1}$ and by $j - 1 = \mathsf{llcp}_b (\bar{M}_i, \bar{M}_{i'})$ we have $\bar{M}_i^j \neq \bar{M}_{i'}^j$. Note that if there was an $i^* < i$ with $m_{i^*} \leq \mathsf{llcp}_b (\bar{M}_i, \bar{M}_{i^*}) = j-1$ and $j \leq m_i$ then we would have $\mathsf{stamp}_i^j = \mathtt{free}$. However if we had the same situation but with $j > m_i$ then $\bar{M}_i$ and $\bar{M}_{i^*}$ would be identical. So $\mathsf{outer} \left(t_i^{j-1} \right)$ has not been set and revealed to the adversary by any previous output value and for any non-twin, in-state $s_{i'}^{j'}$, the probability of collision is at most 2^{-b} by a similar argument as in **Case 1**.

There are no more than $q\ell$ choices for (i, j) and no more than $q\ell$ possible (i', j') for every (i, j) so the overall probability that the condition (2) will be evaluated due to a pair of in-states with $\mathsf{stamp}_i^j = \mathtt{fixed}$ is at most $(q\ell)^2 / 2^b$.

If $\mathsf{stamp}_i^j = \mathtt{free}$ then $\mathsf{outer} \left(s_i^j \right)$ is under adversary's control. However the value of $\mathsf{inner} \left(t_i^{j-1} \right)$ is always generated at the end of the experiment. By a case analysis similar to the previous one we can verify that the probability of a collision due to a pair of in-states with $\mathsf{stamp}_i^j = \mathtt{free}$ is not bigger than 2^{-c}. It is apparent from the definition of a $\mathtt{free}$ in-state that there is at most one such in-state for each query. Having $q\ell$ in-states in total, there are at most $q(q\ell)$ pairs with $\mathsf{stamp}_i^j = \mathtt{free}$ and the probability of $\tau \in \mathcal{T}_{\mathrm{bad}}$ due to such a pair is at most $q^2 \ell / 2^c$.

If $\mathsf{stamp}_i^j = \mathsf{initial}$ then s_i^j cannot non-trivially collide with any other initial in-state. A collision with a non-initial state $s_{i'}^{j'}$ implies that $t_{i'}^{j'-1} = \bar{M}_{i'}^{j'} \oplus \bar{M}_i^1$. If $j' > m_{i'}$ or if there is some M_{i*} with $m_{i*} < j' <= \mathsf{llcp}_b(M_{i'}, \bar{M}_{i*}) + 1$, then outer $\left(t_{i'}^{j'-1}\right)$ is known to the adversary. However inner $\left(t_{i'}^{j'-1}\right)$ is always generated at the end of the experiment. By a case analysis similar to the one we carried out earlier, it can be verified that the collision $s_i^1 = s_{i'}^{j'}$ occurs with probability no bigger than 2^{-c}. There is exactly one initial in-state in each query, so similarly as with the free in-states, the overall probability of a transcript being bad due to a pair with an initial in-state is at most $q^2\ell/2^c$. By summing all the partial collision probabilities we obtain that $\Pr[D_Y \in \mathcal{T}_{bad}] \leq (q\ell)^2/2^b + 2q^2\ell/2^c$. $\square$

6   Security Analysis of FKD

For FKD, we prove the following result:

Theorem 2. *Let $b, r, c, k > 0$ be such that $b = r + c$ and $k \leq c$. Let FKD be the scheme of Sect. 3.2. Then,*

$$\mathbf{Adv}_{\mathrm{FKD}_K^p, p}^{\mathrm{ind}}(q, \ell, \mu, N) \leq \frac{(q\ell)^2}{2^b} + \frac{(q\ell)^2}{2^c} + \frac{\mu N}{2^k}.$$

The proof uses Lemma 3 to transform a FKD adversary into an FKS adversary, similarly to [8,10]. While this would be sufficient to prove the security of the Duplex construction, the bound induced solely by Lemma 3 suffers from a quantitative degradation: we have that $\mathbf{Adv}_{\mathrm{FKD}_K^p, p}^{\mathrm{ind}}(q, \ell, \mu, N) \leq \mathbf{Adv}_{\mathrm{FKS}_K^p, p}^{\mathrm{ind}}(q\ell, \ell, \mu, N)$, resulting in a bound $\frac{2q^2\ell^4}{2^b} + \frac{2q^2\ell^3}{2^c} + \frac{\mu N}{2^k}$ according to Theorem 1. In reality, there *will be* a quantitative gap between the security of FKD construction and that of FKS present, but it will be smaller. This is because an FKS adversary constructed from an FKD adversary issues queries of a specific structure which is far from general. In below proof for FKD, we use this property. In more detail, we derive a specific class of "constrained adversaries" and generalize the proof of Lemma 2 to these adversaries.

Proof (Proof of Theorem 2). Consider any adversary $\boldsymbol{A}$ with resources (q, ℓ, μ, N). We have that $\mathrm{FKD}_K^p = \mathrm{FKD}_0^{E_K^p}$. Therefore, by a modular argument,

$$\mathbf{Adv}_{\mathrm{FKD}_K^p, p}^{\mathrm{ind}}(\boldsymbol{A}) = \Delta_{\boldsymbol{A}}\left(\mathrm{FKD}_0^{E_K^p}, p; RO_{\mathrm{FKD}}, p\right)$$

$$\leq \Delta_{\boldsymbol{B}}\left(\mathrm{FKD}_0^\pi, p; RO_{\mathrm{FKD}}, p\right) + \Delta_{\boldsymbol{C}}\left(E_K^p, p; \pi, p\right)$$

$$\leq \mathbf{Adv}_{\mathrm{FKD}_0^\pi}^{\mathrm{ind}}(\boldsymbol{B}) + \mathbf{Adv}_{E_K^p, p}^{\mathrm{sprp}}(\boldsymbol{C})$$

for some adversary $\boldsymbol{B}$ with resources (q, ℓ, μ) and adversary $\boldsymbol{C}$ with resources (q, ℓ, μ, N). Note that $\boldsymbol{B}$ also has access to p, but these queries are meaningless as its left oracle (FKD_0^π or RO_{FKD}) is independent of p.

In [2], it is proven that $\mathbf{Adv}^{\mathrm{sprp}}_{E^P_{K},p}(C) \leq \mu N/2^k$. In Corollory 3 we show that any FKD adversary B can be turned into a special "constrained" adversary B' against FKS with resources $(q\ell, \ell, \mu)$:

$$\mathbf{Adv}^{\mathrm{ind}}_{\mathrm{FKD}^{\pi}_0}(B) \leq \mathbf{Adv}^{\mathrm{ind}}_{\mathrm{FKS}^{\pi}_0}(B').$$

In Lemma 4, we prove that $\mathbf{Adv}^{\mathrm{ind}}_{\mathrm{FKS}^{\pi}_0}(B') \leq (q\ell)^2/2^b + (q\ell)^2/2^c$ for any such adversary B'. □

For the remainder of the proof, we introduce the mapping $Q_{\mathrm{FKS}} : (\{0,1\}^{<b})^+ \rightarrow \{0,1\}^*$. For any $b > 0$ and for all $X_1, \ldots, X_n \in \{0,1\}^{<b}$ we let

$$Q_{\mathrm{FKS}}(X_1, \ldots, X_n) = \mathrm{pad}_b(X_1) \parallel \ldots \parallel \mathrm{pad}_b(X_{n-1}) \parallel X_n.$$

Lemma 3 (Duplexing Lemma [10]). *Let $b, r, c, k > 0$ be such that $b = r + c$ and $k \leq c$. Let $D = \mathrm{FKD}^p$ as defined in Sect. 3.2. Then for the ith duplexing query (M_i, z_i) made after the last $D.\mathrm{initialize}(K)$ we have*

$$Z_i = D.\mathrm{duplexing}\,(M_i, z_i) = \mathrm{FKS}^p(K, Q_{\mathrm{FKS}}(M_1, \ldots, M_i), z_i).$$

Moreover, the mapping $Q_{\mathrm{FKS}} : (\{0,1\}^{<b})^+ \rightarrow \{0,1\}^*$ is injective.

The proof of the lemma uses similar arguments as that of Bertoni et al. [10]. A complete proof can be found in the full version of this paper [21].

The result of Lemma 3 can be used to reduce any FKD adversary to a constrained FKS adversary. More specifically, any adversary A against FKD that makes q initialize calls and duplexes ℓ blocks after each initialization can be reduced to a constrained FKS adversary $A' = R_{\mathrm{FKS}}(A)$. To answer the jth duplexing query (M^j_i, z^j_i) made by A after the ith initialize call, A' queries its own oracle with $(Q_{\mathrm{FKS}}(M^1_i, \ldots, M^j_i), z^j_i)$. A' copies the output of A at the end of the experiment.

Corollary 3. *Let A be an adversary against FKD that makes q initialize calls and duplexes ℓ blocks after each initialization and $R_{\mathrm{FKS}}(A)$ the constrained FKS adversary as defined above. It follows from Lemma 3, that $\mathbf{Adv}^{\mathrm{ind}}_{\mathrm{FKD}^{\pi}_0}(A) \leq \mathbf{Adv}^{\mathrm{ind}}_{\mathrm{FKS}^{\pi}_0}(R_{\mathrm{FKS}}(A))$.*

We denote by $\mathcal{A}'_{q,\ell}$ the set of constrained adversaries against FKS, that were induced by some FKD adversary that makes q initialize calls and duplexes ℓ blocks after each initialization:

$$\mathcal{A}'_{q,\ell} = \{R_{\mathrm{FKS}}(A) \,:\, A \text{ an FKD adversary with resources } (q, \ell)\}.$$

Lemma 4. *Let $b, r, c > 0$ be such that $b = r + c$. Let FKS be the scheme of Sect. 3.1. Then,*

$$\mathbf{Adv}^{\mathrm{ind}}_{\mathrm{FKS}^{\pi}_0}(A') \leq \frac{(q\ell)^2}{2^b} + \frac{(q\ell)^2}{2^c},$$

for any constrained adversary $A' \in \mathcal{A}'_{q,\ell}$.

The proof follows to large extent the framework of the proof of Lemma 2. We show in particular, that although the constrained adversary makes $q\ell$ queries, each query induces only a single `free` or `initial` state; the remaining internal in-states, if any, are always identical to the in-states of a previous query and they thus do not contribute to the probability of observing a bad transcript. This gives us at most $q\ell$ `free` or `initial` in-states and the bound follows. A complete proof can be found in the full version of this paper [21].

7   Full-State SpongeWrap and its Security

Our results from Sect. 6 can be used to prove security of modified, more efficient versions of existing Sponge-based AE schemes. As an interesting instance, we introduce Full-state SpongeWrap, a variant of the authenticated encryption mode SpongeWrap [8,10], offering improved efficiency with respect to processing of associated data (AD).

7.1   Authenticated Encryption for Sequences of Messages

We will focus on authenticated encryption schemes that act on sequences of AD-message pairs. Following Bertoni et al.[3] [8,10] we will think of an authenticated encryption scheme as an object W surfacing three APIs:

- W.initialize(K, N): calling this function will initialize W with a secret key from the set of keys $\mathcal{K}$ and a nonce from the set of nonces $\mathcal{N}$.
- W.wrap(A, M): this function inputs an AD-message pair (A, M) and outputs a ciphertext-tag pair (C, T), where $|C| = |M|$ and T is a τ-bit tag authenticating (A, M) and all the queries processed by W so far (i.e. since the last initialization call).
- W.unwrap(A, C, T): this function accepts a triple of AD, ciphertext and tag, and outputs a message M if C is an encryption of M and T is a valid tag for (A, M), and all the previous queries processed by W so far; otherwise it outputs an error symbol $\perp$.

Here, the AD, messages and ciphertexts are finite strings and we have $|C| = |M|$. τ is a positive integer and we call it the expansion of W. We require that W is initialized before making the first wrapping or unwrapping call. For a given key K, we will use W_K to refer to the corresponding keyed instance, omitting K from the list of inputs; that is, W.initialize(K, N) = W_K.initialize(N).

Security of Authenticated Encryption. We follow Bertoni et al. [8,10] for defining the security of AE. We split the twofold security goal of AE into two separate requirements: privacy and authenticity.

Let W be a scheme for authenticated encryption, as described above, that internally makes calls to a public random permutation p. We formalize the privacy of W by an experiment in which an adversary A is given access to p, p^{-1}

---

[3] Bertoni et al. do not consider an explicit nonce as we do; they rather require the header of the first wrapping call to be unique.

and an oracle O that provides two interfaces: O.initialize(N) and O.wrap(A, M). We have $O \in \{W_K, RO_W\}$, where W_K is an instance of the real scheme with the key K, and RO_W is an ideal primitive that acts as follows: it keeps a *list* of strings $St \in (\{0,1\}^*)^*$ as its internal state. On calling RO_W.initialize(N) the list St is set to the empty list and then the nonce N is added to the list (denote this operation by $St \leftarrow St\|N$); now each call RO_W.wrap(A, M) will first update the list as $St \leftarrow St\|(A, M)$ and then will output $\mathsf{left}_{|M|+\tau}(RO^\infty(\langle St\rangle))$, where $\langle St\rangle$ denotes an *injective* encoding of the list St into a *string* in $\{0,1\}^*$. (Note that the list St preserves the boundaries between N and all the queried AD-message pairs).

The adversary must distinguish between the two worlds: the real world where it is interacting with W_K and the ideal world where it is interacting with RO_W. The advantage of the adversary in doing so is defined as

$$\mathbf{Adv}_{W[p]}^{\mathrm{priv}}(A) = \left| \Pr\left[K \xleftarrow{\$} \mathcal{K} \ : \ A^{W_K, p, p^{-1}} \Rightarrow 1 \right] - \Pr\left[A^{RO_W, p, p^{-1}} \Rightarrow 1 \right] \right|.$$

It is assumed that the adversary meets the *nonce-requirement*, i.e. that every initialize() it makes is done with a fresh nonce.

For the definition of authenticity property, consider an experiment where an adversary A is given access to the oracle W_K and is allowed to ask the queries W_K.initialize(N) and W_K.wrap(A, M). It is assumed that A respects the *nonce-requirement* in the wrapping queries. A is again allowed to query p. The adversary can also attempt forgeries at any time during the experiment; we say that the adversary forges if it outputs a sequence $(N, (A_1, C_1, T_1), \ldots, (A_n, C_n, T_n))$ such that after calling W.initialize(K, N) and then W.unwrap(A_i, C_i, T_i) for $1 \leq i \leq n-1$, W.unwrap(A_n, C_n, T_n) does not return $\perp$. The sequence $(N, (A_1, C_1, T_1), \ldots, (A_n, C_n, T_n))$ must be such that the adversary has not obtained (C_n, T_n) from a wrapping query that followed an initialization with N and a series of wrapping queries $(A_1, M_1), \ldots, (A_n, M_n)$ with some $M_1, \ldots, M_n$. The adversary does not have to use a unique nonce in the forgery. Note that it can be assumed w.l.o.g. that every forgery attempt is either a fresh nonce followed by a single AD-ciphertext-tag triplet or of the form $(N, (A_1, C_1, T_1), \ldots, (A_n, C_n, T_n))$ with $(N, (A_1, C_1, T_1), \ldots, (A_{n-1}, C_{n-1}, T_{n-1}))$ being learned by the adversary from a sequence of previous wrapping queries. We define the advantage of A as

$$\mathbf{Adv}_{W[p]}^{\mathrm{auth}}(A) = \Pr\left[K \xleftarrow{\$} \mathcal{K} \ : \ A^{W_K, p, p^{-1}} \text{ forges} \right].$$

We let $\mathbf{Adv}_{W[p]}^{\mathrm{priv}}(q_v, q, \ell, \mu, N) = \max_A \mathbf{Adv}_{W[p]}^{\mathrm{priv}}(A)$ be the maximum advantage over all adversaries that make q initialize queries to the left oracle, and after each initialization do wrapping queries that induce at most ℓ permutation calls (including the initialization) and with total maximal multiplicity μ, and that make N direct queries to the public permutation, and that make at most q_v forgery attempts. We similarly let $\mathbf{Adv}_{W[p]}^{\mathrm{auth}}(q, \ell, \mu, N) = \max_A \mathbf{Adv}_{W[p]}^{\mathrm{auth}}(A)$.

Algorithm 3. Outline of an FSW$[p, r, k, n, \tau]$ wrap/unwrap(A, M) query

---

1: **while** there are both AD and message bits to process **do**
2:     take $\leq r$ bit block of M and $\leq c - 5$ bit block of A
3:     wrap/unwrap the message block
4:     **if** both A and M end **then**
5:         produce tag using frame bits $\bar{F}_{\text{AM}}$
6:     **else if** only A ends **or** only M ends **then**
7:         process the blocks using frame bits $F_{\text{AM|}}$
8:     **else**
9:         process the blocks using frame bits F_{AM}
10: **while** there are message bits to process **do**
11:     take $\leq r$ bit block of M
12:     wrap/unwrap the message block
13:     **if** M ends **then**
14:         produce tag using frame bits $\bar{F}_{\text{M}}$
15:     **else**
16:         process the blocks using frame bits F_{M}
17: **while** there are AD bits to process **do**
18:     take $\leq r + c - 5$ bit block of A, split it into r bit and $c - 5$ bit parts
19:     **if** A ends **then**
20:         produce tag using frame bits $\bar{F}_{\text{A}}$
21:     **else**
22:         process the parts using frame bits F_{A}
23: prepare r random bits for next query using frame bits F_{N}

7.2 Full-State SpongeWrap

The Full-State SpongeWrap (FSW) is a permutation mode for authenticated encryption of AD-message sequences as described in Sect. 7.1. It is parametrized by a b-bit permutation p, the maximal message block size r, the key size k, the nonce size n, and the tag size $\tau > 0$. We require that $k \leq b - r =: c$ and $n < r$. The set of keys is $\mathcal{K} = \{0, 1\}^k$ and the set of nonces is $\mathcal{N} = \{0, 1\}^n$. The FSW construction uses an instance of FKD internally to process the inputs block by block. To ensure domain separation of different stages of processing a query, we use three *frame bits* placed at the same position in each duplexing call to FKD as explained in Table 1.

The main motivation of the FSW is concurrent absorption of message and AD to achieve maximal efficiency in terms of minimizing the number of permutation calls made. Since we can only process r bits of a message input at a time, we can use the remainder of the state for the frame bits and a block of AD. This implies the lengths of message and AD blocks processed with each permutation call; $r + 1$ bits for padded message block, 3 frame bits and (having in mind that the input to FKD is always padded) this leaves us at most $(b-1)-(r+1)-3 = c-5$ bits for a block of AD. To minimize the number of permutation calls made in all possible situations, we further specify special treatment for the wrap/unwrap

Table 1. Labeling and usage of the frame bits within FSW.

Label	Value	Usage
F_N	000	process nonce, derive initial mask of a query
F_{AM}	001	block of A and M inside query
F_M	010	block of M inside query
F_A	011	block of A inside query
$F_{AM|}$	100	last block of A and M inside query
$\bar{F}_{AM}$	101	last block of A and M, query ends, produces tag
$\bar{F}_M$	110	last block of M, query ends, produces tag
$\bar{F}_A$	111	last block of A, query ends, produces tag

queries with more AD blocks than message blocks. An informal outline of a wrap/unwrap query is given in Algorithm 3. This outline nicely illustrates how the frame bits are used for domain separation.

We next give a complete algorithmic description of the FSW. To keep it compact, we introduce the following notations. For any $L \in \{0,1\}^{\leq r}$, $R \in \{0,1\}^{\leq c-5}$ and $F \in \{0,1\}^3$, we let

$$Q(L, F, R) = \mathrm{pad}_{r+1}(L) \parallel F \parallel R. \tag{3}$$

Note that $r+4 \leq |Q(L,F,R)| \leq b-1$ for any L, F, R. We let $(L, R) = \mathsf{lsplit}(X, n)$ for any $X \in \{0,1\}^*$ such that $L = \mathsf{left}_{\min(|X|,n)}(X)$ and $\mathsf{right}_{|X|-|L|}(X)$. We let $X_1 \parallel X_2 \parallel \ldots \parallel X_m \xleftarrow{r} X$ denote partitioning a string X in such a way that $X = X_1 \parallel X_2 \parallel \ldots \parallel X_m$, $|X_i| = r$ for $1 \leq i < m$ and $0 < |X_m| \leq r$. Note that $m = \lceil |X|/r \rceil$. We will use the abbreviation $D.\mathrm{dpx}(M, z)$ for the interface $D.\mathrm{duplexing}\,(M, z)$ of an FKD D. The interfaces of $\mathrm{FSW}[p, r, k, n, \tau]$ are defined in Algorithm 4. A schematic depiction of how the wrap interface processes various types of inputs can be found in the full version of this paper [21].

7.3 Security of FSW

The security of FSW is relatively easy to analyze, thanks to the result from Sect. 6.

Lemma 5. Let $W = \mathrm{FSW}[p, r, k, n, \tau]$ be an instance of FSW as described in Sect. 7.2. Denote any query to W.initialize and a list of subsequent queries to W.wrap by $(N, (A_1, M_1), \ldots, (A_n, M_n))$. Then, FSW injectively maps this sequence to a sequence of corresponding FKD duplexing queries $(Q_1, \ldots, Q_d)$.

We prove the injectivity of the mapping by showing how it can be inverted. Thanks to the way the frame bits are used (Fig. 4), it is possible to determine which duplexing calls belong to a single wrap query. More than that, we can also determine the boundaries of message and AD using the frame bits and then we can reconstruct them thanks to the use of the padding. The full proof can be found in the full version of this paper [21].

Algorithm 4. $\text{FSW}[p, r, k, n, \tau]$

1: **Interface** $W.\text{initialize}(K, N)$
2: $\quad D.\text{initialize}(K)$
3: $\quad S \leftarrow \text{pad}_r(N) \| 0 \| F_N \| 0^{c-5}$
4: $\quad Z \leftarrow D.\text{dpx}(S, r)$

1: **Interface** $W.\text{wrap}(A, M)$
2: $\quad M_1 \| \ldots \| M_m \overset{r}{\leftarrow} M$
3: $\quad (A', A^*) \leftarrow \text{lsplit}(A, m(c - 5))$
4: $\quad A'_1 \| \ldots \| A'_{a'} \overset{c-5}{\leftarrow} A'$
5: $\quad A^*_1 \| \ldots \| A^*_{a*} \overset{b-5}{\leftarrow} A^*$
6: $\quad$ **if** $m = a' = a^* = 0$ **then**
7: $\qquad T \leftarrow \varepsilon$
8: $\qquad F \leftarrow \bar{F}_A$
9: $\quad$ **for** $i \leftarrow 1$ **to** $a' - 1$ **do**
10: $\qquad C_i \leftarrow M_i \oplus Z$
11: $\qquad Z \leftarrow D.\text{dpx}(Q(M_i, F_{\text{AM}}, A'_i), r)$
12: $\quad$ **if** $0 < a' < m$ **or** $0 < a', a^*$ **then**
13: $\qquad C_{a'} \leftarrow M_{a'} \oplus \text{left}_{|M_{a'}|}(Z)$
14: $\qquad Z \leftarrow D.\text{dpx}(Q(M_{a'}, F_{\text{AM}|}, A'_{a'}), r)$
15: $\quad$ **else if** $0 < m = a'$ **and** $a^* = 0$ **then**
16: $\qquad C_{a'} \leftarrow M_{a'} \oplus \text{left}_{|M_{a'}|}(Z)$
17: $\qquad T \leftarrow D.\text{dpx}(Q(M_{a'}, \bar{F}_{\text{AM}}, A'_{a'}), r)$
18: $\qquad F \leftarrow \bar{F}_{\text{AM}}$
19: $\quad$ **for** $i \leftarrow a' + 1$ **to** $m - 1$ **do**
20: $\qquad C_i \leftarrow M_i \oplus Z$
21: $\qquad Z \leftarrow D.\text{dpx}(Q(M_i, F_{\text{M}}, \varepsilon), r)$
22: $\quad$ **if** $a' < m$ **then**
23: $\qquad C_m \leftarrow M_m \oplus \text{left}_{|M_m|}(Z)$
24: $\qquad T \leftarrow D.\text{dpx}(Q(M_m, \bar{F}_{\text{M}}, \varepsilon), r)$
25: $\qquad F \leftarrow \bar{F}_{\text{M}}$
26: $\quad$ **for** $i \leftarrow 1$ **to** $a^* - 1$ **do**
27: $\qquad (L, R) \leftarrow \text{lsplit}(A^*_i, r)$
28: $\qquad D.\text{dpx}(Q(L, F_A, R), 0)$
29: $\quad$ **if** $a^* > 0$ **then**
30: $\qquad (L, R) \leftarrow \text{lsplit}(A^*_{a*}, r)$
31: $\qquad T \leftarrow D.\text{dpx}(Q(L, \bar{F}_A, R), r)$
32: $\qquad F \leftarrow \bar{F}_A$
33: $\quad$ **while** $|T| < \tau$ **do**
34: $\qquad T \leftarrow T \| D.\text{dpx}(Q(\varepsilon, F, \varepsilon), r)$
35: $\quad Z \leftarrow D.\text{dpx}(Q(\varepsilon, F_N, \varepsilon), r)$
36: $\quad C \leftarrow C_1 \| \ldots \| C_m$
37: $\quad$ **return** $C, \text{left}_\tau(T)$

1: **Interface** $W.\text{unwrap}(A, C, T)$
2: $\quad C_1 \| \ldots \| C_m \overset{r}{\leftarrow} C$
3: $\quad (A', A^*) \leftarrow \text{lsplit}(A, m(c - 5))$
4: $\quad A'_1 \| \ldots \| A'_{a'} \overset{c-5}{\leftarrow} A'$
5: $\quad A^*_1 \| \ldots \| A^*_{a*} \overset{b-5}{\leftarrow} A^*$
6: $\quad$ **if** $m = a' = a^* = 0$ **then**
7: $\qquad T' \leftarrow \varepsilon$
8: $\qquad F \leftarrow \bar{F}_A$
9: $\quad$ **for** $i \leftarrow 1$ **to** $a' - 1$ **do**
10: $\qquad M_i \leftarrow C_i \oplus Z$
11: $\qquad Z \leftarrow D.\text{dpx}(Q(M_i, F_{\text{AM}}, A'_i), r)$
12: $\quad$ **if** $0 < a' < m$ **or** $0 < a', a^*$ **then**
13: $\qquad M_{a'} \leftarrow C_{a'} \oplus \text{left}_{|C_{a'}|}(Z)$
14: $\qquad Z \leftarrow D.\text{dpx}(Q(M_{a'}, F_{\text{AM}|}, A'_{a'}), r)$
15: $\quad$ **else if** $0 < m = a'$ **and** $a^* = 0$ **then**
16: $\qquad M_{a'} \leftarrow C_{a'} \oplus \text{left}_{|C_{a'}|}(Z)$
17: $\qquad T' \leftarrow D.\text{dpx}(Q(M_{a'}, \bar{F}_{\text{AM}}, A'_{a'}), r)$
18: $\qquad F \leftarrow \bar{F}_{\text{AM}}$
19: $\quad$ **for** $i \leftarrow a' + 1$ **to** $m - 1$ **do**
20: $\qquad M_i \leftarrow C_i \oplus Z$
21: $\quad \cdot \qquad Z \leftarrow D.\text{dpx}(Q(M_i, F_{\text{M}}, \varepsilon), r)$
22: $\quad$ **if** $a' < m$ **then**
23: $\qquad M_m \leftarrow C_m \oplus \text{left}_{|C_m|}(Z)$
24: $\qquad T' \leftarrow D.\text{dpx}(Q(M_m, \bar{F}_{\text{M}}, \varepsilon), r)$
25: $\qquad F \leftarrow \bar{F}_{\text{M}}$
26: $\quad$ **for** $i \leftarrow 1$ **to** $a^* - 1$ **do**
27: $\qquad (L, R) \leftarrow \text{lsplit}(A^*_i, r)$
28: $\qquad D.\text{dpx}(Q(L, F_A, R), 0)$
29: $\quad$ **if** $a^* > 0$ **then**
30: $\qquad (L, R) \leftarrow \text{lsplit}(A^*_{a*}, r)$
31: $\qquad T' \leftarrow D.\text{dpx}(Q(L, \bar{F}_A, R), r)$
32: $\qquad F \leftarrow \bar{F}_A$
33: $\quad$ **while** $|T'| < \tau$ **do**
34: $\qquad T' \leftarrow T' \| D.\text{dpx}(Q(\varepsilon, F, \varepsilon), r)$
35: $\quad Z \leftarrow D.\text{dpx}(Q(\varepsilon, F_N, \varepsilon), r)$
36: $\quad M \leftarrow M_1 \| \ldots \| M_m$
37: $\quad$ **if** $T = \text{left}_\tau(T')$ **then**
38: $\qquad$ **return** M
39: $\quad$ **else**
40: $\qquad$ **return** $\bot$

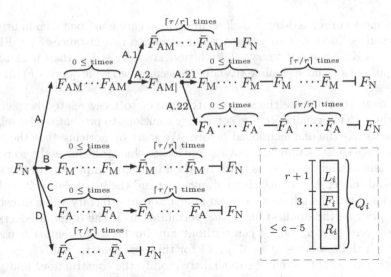

Fig. 4. The tree of all possible frame bits sequences for a single AD-message pair (top-left). The composition of an FKD query Q_i (bottom-right).

Theorem 3. *Let $b, r, c, k, n, \tau > 0$ be such that $b = r + c$, $k \leq c$ and $n < r$. Let FSW be the scheme of Sect. 7.2. Then,*

$$\mathbf{Adv}^{\mathrm{priv}}_{\mathrm{FSW}}(q, \ell, \mu, N) \leq \frac{(q\ell)^2}{2^b} + \frac{(q\ell)^2}{2^c} + \frac{\mu N}{2^k},$$

$$\mathbf{Adv}^{\mathrm{auth}}_{\mathrm{FSW}}(q, \ell, \mu, N) \leq \frac{(q\ell)^2}{2^b} + \frac{(q\ell)^2}{2^c} + \frac{\mu N}{2^k} + \frac{q_v}{2^\tau}.$$

We start by defining the *ROFSW*—an idealized FSW that internally uses the RO^r_{FKD} instead of FKD (and thus does not use p at all). By Thm. 2 we have that

$$\mathbf{Adv}^{\mathrm{priv}}_{\mathrm{FSW}}(q, \ell, \mu, N) \leq \mathbf{Adv}^{\mathrm{priv}}_{\mathrm{ROFSW}}(q, \ell, \mu) + \frac{(q\ell)^2}{2^b} + \frac{(q\ell)^2}{2^c} + \frac{\mu N}{2^k},$$

$$\mathbf{Adv}^{\mathrm{auth}}_{\mathrm{FSW}}(q, \ell, \mu, N) \leq \mathbf{Adv}^{\mathrm{auth}}_{\mathrm{ROFSW}}(q, \ell, \mu) + \frac{(q\ell)^2}{2^b} + \frac{(q\ell)^2}{2^c} + \frac{\mu N}{2^k}.$$

We consequently analyse the security of *ROFSW*, which is a relatively straightforward task because it internally uses a RO^r_{FKD}. We obtain $\mathbf{Adv}^{\mathrm{priv}}_{\mathrm{ROFSW}}(q, \ell, \mu) = 0$ and $\mathbf{Adv}^{\mathrm{auth}}_{\mathrm{ROFSW}}(q_v, q, \ell, \mu) \leq q_v/2^\tau$. A complete proof can be found in the full version of this paper [21].

8 Discussion

Related-Key Security. Our treatment of the security of the full-state constructions is in the traditional model where the adversary has no control over selection of the secret keys or relations among different keys. If one considers the stronger model of related-key attack security then care must be taken in utilizing these schemes. Indeed, if an adversary has access to two instances $F_1 = \text{FKS}^p_{K_1}$ and $F_2 = \text{FKS}^p_{K_2}$, and it knows the relation $\Delta = K_1 \oplus K_2$, then it can make the outputs of F_1 and F_2 collide trivially by asking two b-bit queries $F_1(M)$ and $F_2(M \oplus \Delta)$.

Although it is outside the scope of this paper to treat related-key security thoroughly, we informally propose some easy solutions to prevent trivial related-key attacks like the one mentioned before. We start by noticing that the inner-keyed Sponge construction [2] is not susceptible to this problem, as the secret key and the adversarial data blocks never overlap; hence, a simple way of thwarting such trivial related-key attacks is to always prepend the input data with a block of b zeroes. Thus the adversary can no longer xor an arbitrary value directly to the key prior to the application of the permutation. If the original adversarial resources were (q, ℓ, μ, N), we can without any further argumentation use the bound with the resources $(q, \ell + 1, \mu, N)$ for this new construction.

Another possibility would be to slightly modify the constructions and partition the input data into an r-bit starting block and b-bit blocks afterward. The initial block would be xored to the outer r bits of the initial state. Our security analysis would carry over to this construction with minimal modifications.

Generalized Security Model. The security analyses of FKS and FKD cover those of the original Sponge and Duplex constructions as special cases. Beyond that, for the security analysis of FKD itself, we have generalized the security model of the original Duplex construction from Bertoni et al. [9,10]. While in the analysis of Bertoni et al. the analysis of the multiple-initializations scenario is left rather implicit, we include it explicitly in our model.

This generalized setting seems more closely matching the use of the Duplex construction in several AE schemes which do not require sessions and new session keys, where one would initialize the Duplex (or FKD) construction for every query. This is well demonstrated by the example of FSW. More precisely, the way we design and analyze the security of FSW allows for a very versatile use. FSW can be used to secure AD-message pairs in a single *session* [12], i.e. using a single initialize call during the lifetime of the key or alternatively every AD-message pair can be preceded by an initialize call with a unique nonce. In fact, FSW can be used for anything between these two extremes; for example, a setting where every AD-message pair is processed with a unique nonce, but can get fragmented into smaller sub-pairs. The security analysis of FSW covers each of these use cases.

On the Keying of the Sponge. As we have claimed in the introduction, the difference in the security of the outer-keyed and inner-keyed Sponges vanishes in presence of the full state absorption. On one hand, using a key of more than c bits does not increase the security level, as the extra bits cannot be used by the low-entropy Even-Mansour construction. On the other hand, absorbing several b-bit blocks of the key only results into a derived key of effective length of c bits. We remark that both the outer- and inner-keyed Sponges can be seen as special cases of FKS, by using more restrictive padding rules that only place the message blocks in the outer part of the state.

Boosting Sponge-based AE. Out of 57 CAESAR candidates, 10 are using a Sponge-based design. The method we used to enhance SpongeWrap can be straightforwardly adjusted to boost the performance of five of these 10 schemes: Keyak, Ketje, STRIBOB, CBEAM and ICEPOLE [3]. This is because all the said designs are using frame bits for domain separation. The other designs cannot benefit from our modifications, either due to a domain separation method relying on intangibility of the inner part of the state (NORX), or due to producing tag from the inner part of the state (Ascon, Primates), or because they are already using the inner part of the state (Artemia) or because the designs do not follow the general structure of the Sponge Wrap (Pi Cipher) [3]. We note that if Ketje was to benefit from the technique we have introduced, it would be necessary to increase the number of rounds of the underlying permutation.

Acknowledgments. Bart Mennink is a Postdoctoral Fellow of the Research Foundation – Flanders (FWO). He is supported in part by the Research Council KU Leuven: GOA TENSE (GOA/11/007). Reza Reyhanitabar and Damian Vizár are supported in part by Microsoft Research under MRL Contract No. 2014-006 (DP1061305). We would like to thank the ASIACRYPT reviewers for their constructive comments. We would also like to thank Joan Daemen and Gilles Van Assche for an insightful discussion.

References

1. Abed, F., Forler, C., Lucks, S.: Classification of the CAESAR candidates. In: IACR Cryptology ePrint Archive 2014, vol. 792 (2014). http://eprint.iacr.org/2014/792
2. Andreeva, E., Daemen, J., Mennink, B., Van Assche, G.V.: Security of keyed sponge constructions using a modular proof approach. In: Leander [19], pp. 364–384. http://dx.doi.org/10.1007/978-3-662-48116-5_18
3. Bernstein, D.J.: Cryptographic competitions: CAESAR. http://competitions.cr.yp.to
4. Bertoni, G., Daemen, J., Peeters, M., Van Assche, G.: Sponge functions. In: ECRYPT Hash Workshop (2007). http://csrc.nist.gov/groups/ST/hash/documents/JoanDaemen.pdf
5. Bertoni, G., Daemen, J., Peeters, M., Van Assche, G.: Keccak Specifications. In: NIST SHA-3 Submission (2008). http://keccak.noekeon.org/
6. Bertoni, G., Daemen, J., Peeters, M., Van Assche, G.: On the indifferentiability of the sponge construction. In: Smart, N.P. (ed.) EUROCRYPT 2008. LNCS, vol. 4965, pp. 181–197. Springer, Heidelberg (2008)
7. Bertoni, G., Daemen, J., Peeters, M., Van Assche, G.: Sponge-based pseudo-random number generators. In: Mangard, S., Standaert, F.-X. (eds.) CHES 2010. LNCS, vol. 6225, pp. 33–47. Springer, Heidelberg (2010)
8. Bertoni, G., Daemen, J., Peeters, M., Van Assche, G.: Duplexing the sponge: single-pass authenticated encryption and other applications. In: IACR Cryptology ePrint Archive 2011, vol. 499 (2011). http://eprint.iacr.org/2011/499

9. Bertoni, G., Daemen, J., Peeters, M., Van Assche, G.: On the security of the keyed sponge construction. In: Symmetric Key Encryption Workshop 2011 (2011)

10. Bertoni, G., Daemen, J., Peeters, M., Van Assche, G.: Duplexing the sponge: single-pass authenticated encryption and other applications. In: Miri, A., Vaudenay, S. (eds.) SAC 2011. LNCS, vol. 7118, pp. 320–337. Springer, Heidelberg (2012)

11. Bertoni, G., Daemen, J., Peeters, M., Van Assche, G.: Permutation-based encryption, authentication and authenticated encryption. In: Workshop Records of DIAC 2012, pp. 159–170 (2012)

12. Bertoni, G., Daemen, J., Peeters, M., Van Assche, G., Van Keer, R.: CAESAR submission: Keyak v1, March 2014. http://competitions.cr.yp.to/round1/keyakv1.pdf

13. Chang, D., Dworkin, M., Hong, S., Kelsey, J., Nandi, M.: A keyed sponge construction with pseudorandomness in the standard model. In: NIST SHA-3 2012 Workshop. http://csrc.nist.gov/groups/ST/hash/sha-3/Round3/March2012/documents/papers/CHANG_paper.pdf

14. Chen, S., Steinberger, J.: Tight security bounds for key-alternating ciphers. In: Nguyen, P.Q., Oswald, E. (eds.) EUROCRYPT 2014. LNCS, vol. 8441, pp. 327–350. Springer, Heidelberg (2014)

15. Even, S., Mansour, Y.: A construction of a cipher from a single pseudorandom permutation. In: Matsumoto, T., Imai, H., Rivest, R.L. (eds.) ASIACRYPT 1991. LNCS, vol. 739. Springer, Heidelberg (1993)

16. Even, S., Mansour, Y.: A Construction of a cipher from a single pseudorandom permutation. J. Cryptol. **10**(3), 151–162 (1997)

17. Gazi, P., Pietrzak, K., Tessaro, S.: The exact PRF security of truncation: tight bounds for keyed sponges and truncated CBC. In: Gennaro, R., Robshaw, M. (eds.) CRYPTO 2015. LNCS, vol. 9215, pp. 368–387. Springer, Heidelberg (2015). http://dx.doi.org/10.1007/978-3-662-47989-6_18

18. Jovanovic, P., Luykx, A., Mennink, B.: Beyond $2^c/2$ security in sponge-based authenticated encryption modes. In: Sarkar, P., Iwata, T. (eds.) ASIACRYPT 2014. LNCS, vol. 8873, pp. 85–104. Springer, Heidelberg (2014)

19. Leander, G.: FSE 2015. Security and Cryptology, vol. 9054. Springer, Heidelberg (2015). http://dx.doi.org/10.1007/978-3-662-48116-5

20. Maurer, U.M., Renner, R.S., Holenstein, C.: Indifferentiability, impossibility results on reductions, and applications to the random oracle methodology. In: Naor, M. (ed.) TCC 2004. LNCS, vol. 2951, pp. 21–39. Springer, Heidelberg (2004)

21. Mennink, B., Reyhanitabar, R., Vizár, D.: Security of full-state keyed and duplex sponge: Applications to authenticated encryption. In: IACR Cryptology ePrint Archive 2015, vol. 541 (2015). http://eprint.iacr.org/2015/541

22. Mouha, N., Mennink, B., Van Herrewege, A., Watanabe, D., Preneel, B., Verbauwhede, I.: Chaskey: an efficient MAC algorithm for 32-bit microcontrollers. In: Joux, A., Youssef, A. (eds.) SAC 2014. LNCS, vol. 8781, pp. 306–323. Springer, Heidelberg (2014)

23. Patarin, J.: The "Coefficients H" technique. In: Avanzi, R.M., Keliher, L., Sica, F. (eds.) SAC 2008. LNCS, vol. 5381, pp. 328–345. Springer, Heidelberg (2009)

24. Perlner, R.: Extendable-output functions (XOFs). In: NIST SHA-3 2014 Workshop (2014). http://csrc.nist.gov/groups/ST/hash/sha-3/Aug2014/documents/perlner_XOFs.pdf

25. Reyhanitabar, R., Vaudenay, S., Vizár, D.: Boosting OMD for almost free authentication of associated data. In: Leander [19], pp. 411–427. http://dx.doi.org/10.1007/978-3-662-48116-5_20
26. Rivest, R.L., Schuldt, J.C.N.: Spritz - a Spongy RC4-like Stream Cipher and Hash Function (2014). https://people.csail.mit.edu/rivest/pubs/RS14.pdf
27. Sasaki, Y., Yasuda, K.: How to incorporate associated data in sponge-based authenticated encryption. In: Nyberg, K. (ed.) CT-RSA 2015. LNCS, vol. 9048, pp. 353–370. Springer, Heidelberg (2015)

Heuristic Tool for Linear Cryptanalysis with Applications to CAESAR Candidates

Christoph Dobraunig[(✉)], Maria Eichlseder, and Florian Mendel

Graz University of Technology, Graz, Austria
christoph.dobraunig@iaik.tugraz.at

Abstract. Differential and linear cryptanalysis are the general purpose tools to analyze various cryptographic primitives. Both techniques have in common that they rely on the existence of good differential or linear characteristics. The difficulty of finding such characteristics depends on the primitive. For instance, AES is designed to be resistant against differential and linear attacks and therefore, provides upper bounds on the probability of possible linear characteristics. On the other hand, we have primitives like SHA-1, SHA-2, and KECCAK, where finding good and useful characteristics is an open problem. This becomes particularly interesting when considering, for example, competitions like CAESAR. In such competitions, many cryptographic primitives are waiting for analysis. Without suitable automatic tools, this is a virtually infeasible job. In recent years, various tools have been introduced to search for characteristics. The majority of these only deal with differential characteristics. In this work, we present a heuristic search tool which is capable of finding linear characteristics even for primitives with a relatively large state, and without a strongly aligned structure. As a proof of concept, we apply the presented tool on the underlying permutations of the first round CAESAR candidates ASCON, ICEPOLE, KEYAK, Minalpher and PRØST.

Keywords: Linear cryptanalysis · Authenticated encryption · Automated tools · Guess-and-determine · CAESAR competition

1 Introduction

Research in symmetric cryptography in the last few years is mainly driven by dedicated high-profile open competitions such as NIST's AES and SHA-3 selection procedures, or ECRYPT's eSTREAM project. While these focused competitions in symmetric cryptography are generally viewed as having provided a tremendous increase in the understanding and confidence in the security of these cryptographic primitives, the impressive increase of submissions to such competitions reveal major problems related to the analytical effort for the cryptographic community. To better evaluate the security margin of the various submissions, automatic tools are needed to assist cryptanalysts with their work.

One important class of attacks is linear cryptanalysis [15,25]. The success of these attacks relies on the existence of suitable linear characteristics. The difficulty of finding such characteristics depends on the primitive. For example,

© International Association for Cryptologic Research 2015
T. Iwata and J.H. Cheon (Eds.): ASIACRYPT 2015, Part II, LNCS 9453, pp. 490–509, 2015.
DOI: 10.1007/978-3-662-48800-3_20

the wide-trail design strategy [7] incorporated by AES provides lower bounds on the minimum number of active S-boxes in a linear characteristic and therefore, gives an upper bound on the highest possible bias. On the other hand, we have primitives with weak alignment [1], such as the winner of the SHA-3 competition KECCAK, where finding good characteristics is an open problem, and heuristic search results are required to evaluate the security margin of the primitive. This is particularly interesting in the context of the CAESAR competition [26]. We noticed that many first round submissions focus their analysis on differential cryptanalysis, but provide only few results for linear cryptanalysis.

Our Contribution. The main contribution of this paper is a dedicated automatic tool for linear cryptanalysis, which is available at github[1]. The tool performs heuristic searches for good linear characteristics in cryptographic primitives. It was designed for primitives based on substitution-permutation networks (SP networks).

The modular design of the tool allows easy extension to other cryptographic primitives. It also allows to easily develop and test new dedicated search strategies. To facilitate further improvements and analysis, the tool is publicly available and its source code is published together with this paper. Such a tool is particularly useful when designing new cryptographic primitives. It allows to easily explore the effects of, for instance, different S-boxes and linear layers on linear characteristics and reveals possible bad decisions in an early stage of the design process. Even in wide-trail designs with provable bounds, it can be useful to evaluate different choices for building blocks with respect to their long-term behaviour over a larger number of rounds, where the quality of the best characteristics can deviate significantly from the derived bounds (i.e., two algorithms with the same bounds may behave quite differently in a heuristic search, which can be a basis for the decision of choosing one design over the other).

As a proof of concept and to demonstrate the advantages of the tool, we have chosen the first round CAESAR candidates ASCON [9], ICEPOLE [19], KEYAK [4], Minalpher [22] and PRØST [13] as analysis targets. ASCON, ICEPOLE, and KEYAK are sponge-based authenticated encryption schemes. All three primitives use permutations that are not strongly aligned, making it hard to find good linear characteristics. We demonstrate the capability of our automated search tool by giving linear characteristics suitable for different attack scenarios. In comparison, the permutations used in Minalpher and PRØST provide more "structure" by incorporating an "AES-like" design strategy. Hence, the designers of these two primitives are able to give computer-aided bounds on the minimum number of active S-boxes by using mixed-integer linear programming (MILP) for a number of rounds sufficient to thwart attacks. For Minalpher and PRØST, we show that our tool is capable of finding linear characteristics which match the provided bounds. Our results are summarized in Table 1 (Sect. 4).

---

[1] https://github.com/iaikkrypto/lineartrails.

Related Work. While several automatic tools for differential cryptanalysis have been published in the last few years [5, 6, 8, 12, 14, 16, 20, 23], in particular for hash functions, the work on automatic tools dedicated to linear cryptanalysis is very limited. One example is a tool designed by Sun et al. [24], extending previous work of Mouha et al. [21]. They model the differential and linear behavior of a block cipher as a mixed-integer linear program (MILP) and use general-purpose MILP tools to solve the optimization problem (i.e., find the optimal characteristics for the – often simplified – model of the cipher). This approach works well for lightweight ciphers like Simon or Present, but faces problems when it comes to large-state and less structured ciphers such as ASCON, ICEPOLE, and KEYAK. Hence, a dedicated search tool for linear characteristics will complement the existing tools.

Outline. This paper is divided into two main parts: the description of our new automated search tool for linear characteristics in Sect. 3, and its application to the CAESAR candidates in Sect. 4. However, first, we start with a short introduction to linear cryptanalysis and our notation in Sect. 2. Then, we deal with the propagation of linear masks in Sect. 3.2 and discuss the proposed search strategy for linear characteristics in Sect. 3.3. The application of the tool (Sect. 4) is first discussed in detail for KEYAK in Sect. 4.1. Then, our results for the other ciphers are summarized and briefly discussed in Sect. 4.2 to 4.5. Finally, we conclude in Sect. 5.

2   Linear Cryptanalysis

The goal of linear cryptanalysis [15, 25] is to identify good affine linear approximations for the target cipher. More specifically, we want to find linear equations between the plaintext bits, ciphertext bits and key bits that hold with probability significantly different from $\frac{1}{2}$ (bias). Then, in the actual attack phase, these equations can be used to derive information on the key bits from known plaintext-ciphertext pairs.

For linear cryptanalysis, the operation of the cipher, or building blocks of the cipher, is considered as a vectorial boolean function $f : \mathbb{F}_2^m \to \mathbb{F}_2^n$ (where the key bits might be part of $\mathbb{F}_2^m$). A (probabilistic) linear relation between input and output bits of f is then characterized by two linear masks $\alpha \in \mathbb{F}_2^m, \beta \in \mathbb{F}_2^n$. For $x \in \mathbb{F}_2^m$, $y \in \mathbb{F}_2^n$ with $y = f(x)$, the masks represent the relation

$$\alpha^t \cdot x = \beta^t \cdot y,$$

where $v^t \cdot w$ denotes the natural inner product of vectors. The quality of a linear approximation α, β is measured by the probability that the corresponding relation holds; or more precisely, by how far this probability deviates from the average $\frac{1}{2}$. This deviation is referred to as the bias of the masks α, β:

$$\varepsilon_{\alpha,\beta} = \text{bias}_f(\alpha,\beta) = \left| \mathbb{P}\left[\alpha^t \cdot x = \beta^t \cdot y \mid y = f(x) \right] - \frac{1}{2} \right|$$

$$= \frac{1}{2^m} \cdot \left| \left| \{ x \in \mathbb{F}_2^m \mid \alpha^t \cdot x = \beta^t \cdot f(x) \} \right| - 2^{m-1} \right|.$$

If m is very small, the expression for $\varepsilon_{\alpha,\beta}$ can easily be evaluated explicitly for all masks α,β to determine the best masks. This information is summarized in the linear distribution table (LDT), where non-zero entries mark masks α,β with non-zero bias.

However, this is obviously infeasible for the complete cipher at once. To obtain an approximation of the complete cipher, it is split into smaller parts that are easier to analyze. Matsui's piling-up lemma [15] is used to combine the individual biases of multiple building blocks to derive the overall bias (under the assumption that the validity of the partial approximations is independent). If ε denotes the bias of the overall approximation of the block cipher, Matsui [15] showed that the necessary number of plaintext-ciphertext pairs to derive the bit of key information from the approximation is proportional to $\frac{1}{\varepsilon^2}$.

The difficult part is to find a network or "trail" of partial approximations that are compatible with each other and give a good overall bias. In particular, each involved approximation must have non-zero bias, otherwise the overall bias becomes zero. For this reason, we refer to non-zero entries in the individual LDTs as "valid transitions" of masks for this building block. In the the following, such a "trail" of partial linear approximations is called linear characteristic.

Several algorithms and improvements thereof have been proposed for finding characteristics with the highest overall bias, typically by a sort of branch-and-bound algorithms. For more complex, modern ciphers, such a complete search is not feasible. Two possible approaches to handle this situation are (a) to design ciphers in a way to allow to prove bounds on the best possible bias, and (b) to use heuristic search methods to find stronger biases (for reduced versions of the cipher) to make better predictions on the security margin of the complete cipher.

In the following, we will focus on the second approach, and heuristically search for good characteristics. Unlike the original, complete search algorithms, our search will not proceed in a "linear", round-by-round manner. Instead, we will take inspiration from similar searching tools for differential cryptanalysis [8], and randomize the search order. This naturally implies that we will often start building inconsistent characteristics, which will need to be fixed or discarded.

3   An Automated Tool for Linear Cryptanalysis

The proposed automated tool can be roughly split into two main parts. The first part is described in Sect. 3.2 and deals with the description of cryptographic primitives within the search tool, including the representation of linear approximations and, most importantly, their propagation. The other part of the tool is the choice of the search algorithm to find good linear characteristics (see Sect. 3.3). Before we start with the description of the tool, we take a look at the requirements we have for the design and implementation of such a heuristic search tool.

3.1   Implementation Requirements for the Search Tool

In order for any automatic cryptanalysis tool to be useful for general application, for example to analyze the 57 first round CAESAR submissions, there are a number of flexibility and usability requirements:

- **Easy to Add New Primitives.** This is one of the main goals for the creation of this tool. To fulfill this requirement, we have decided to put the focus on primitives based on SP networks, i.e., with alternating S-box and linear layers. This simplifies the design process of the tool, since we did not have to consider every possible specialty, while still having a large group of applicable primitives. The programming interface should be designed to require as little effort as possible for converting, for example, a CAESAR reference implementation to a suitable cipher definition for the tool – ideally, it should possible to just copy the corresponding code fragments for the round transformation steps.
- **An Easily Adaptable, Parameterized Search Algorithm.** The linear tool implements a heuristic guess-and-determine search algorithm. This algorithm delivers good results for various primitives. However, the success of the search is highly dependent on various different parameters, such as the configuration of the searching order and conflict-handling behavior. Therefore, it is crucial that these parameters can be adjusted easily. For this reason, our standard guess-and-determine algorithm is parameterizable via an XML-file. This XML-file specifies the search starting point and allows to configure various other parameters.
- **Easy to Add other Search Algorithms.** The currently implemented, stack-based guess-and-determine algorithm is certainly not the only possible way to search for linear characteristics. To be open for new ideas and evaluate other algorithms, we have designed the tool in a way that the search algorithm is clearly separated from the description of the cipher and thus, can be replaced easily. This opens the door for experiments with various alternative search algorithms and will hopefully lead to new insights in this direction.
- **Portability of the Code.** We do not want the tool to require a specific operating system or platform to run. Therefore, we have reduced the dependence from external libraries whenever possible, and omitted the use of platform-specific instructions.

3.2   Propagation of Linear Masks

Our overall search strategy is based on the "guess-and-determine" approach. We want to build a consistent linear characteristic with high bias step by step, starting from a "mostly unknown" (undetermined) characteristic of masks, and progressively deciding which bits should be selected (activated) by the final mask.

For this purpose, we repeatedly "guess" the value of small parts of the masks, and then "determine" the consequences of this guess (in particular, whether this updated partial characteristic can still be completed to a "valid" characteristic). We refer to the "determining" step as propagation of information.

Representation of Partial Linear Masks. The tool represents the linear masks on bit-level. During the search, we work with partially-determined search masks. We represent an active bit in the linear mask with 1 and an inactive bit of a linear mask with 0. Mask bits that are not yet determined are represented by ?.

Propagation in SP Networks. We want to find linear characteristics for SP networks. Such a network consists of iterative applications of a substitution layer (consisting of relatively small S-boxes) and an (affine) linear layer (which typically covers larger parts of the state at once). We use different techniques for the propagation of information in these two layer types. The goal of the propagation step is to investigate whether the guess allows to derive explicit values for other ("neighbouring") bits, and in particular whether this explicit information is contradictory. The constraints that allow this propagation can be derived from the linear distribution table of the involved functions, since the characteristic must not contain any mask transitions with bias 0.

Propagation in the Non-linear Layer. We only deal with non-linear layers which can be represented by parallel applications of S-boxes. So the propagation of the linear masks at the input and the output of the S-boxes can be treated individually, since the parallel applications are considered independent of each other (any dependencies induced by the linking linear layers are treated separately). Therefore, we can do the propagation separately per S-box.

Many state-of-the-art ciphers use relatively small S-boxes. In many recent cipher proposals, the S-boxes map 4- to 5-bit inputs to outputs of the same size. Even the largest S-boxes hardly ever exceed a size of 8 bits. Therefore, the propagation of the linear masks can be done in a brute-force manner, based on the linear distribution table (LDT) of the S-box. The LDT is an exhaustive list of all valid (biased) mask transitions from mask α to mask β. Our current "knowledge" of the values of some input and output mask bits limits the set of available transitions. Depending on the concrete values of α and β and the remaining transition options, we have one of the following outcomes of the propagation:

1. **Contradiction:** The LDT reveals that no valid, biased transitions remain that satisfy the fixed mask bits; i.e., there is no linear relationship involving the bits currently marked by α and β as 1 (and optionally the ? bits). In other words, we have a contradiction. This means that the current, partially determined linear characteristic is in fact invalid. This situation has to be handled by the search algorithm by, e.g., backtracking and resolving the contradiction.

2. **Updated Bits:** The LDT reveals that one or more biased transitions respecting the partially determined α and β remain. In addition, all remaining transitions share the same value (0 or 1) for one or more of the current ? bits. Thus, we can refine some previously undefined bits in the masks to active or inactive bits by using information from the LDT. Before taking any further guesses, this newly-won information must in turn be propagated in all connected function components.

3. **No Updates:** The LDT reveals that α and β are possible, but no additional explicit bit-wise information can be won. Nothing else happens.

Propagation in the Linear Layer. There are two main differences between the linear and non-linear layers from the propagation perspective: On the one hand, the linear layer typically involves significantly more variables than individual S-boxes. On other hand, propagating partial linear masks for linear functions can be achieved easily using basic linear algebra.

Assume that the function $f : \mathbb{F}_2^m \to \mathbb{F}_2^n$ is linear, i.e., $f(x) = A \cdot x = y$ for some $A \in \mathbb{F}_2^{n \times m}$. Note that we can include affine linear functions in the same model, since the affine (constant) part is irrelevant for the bias of the linear model if we do not consider the sign of the bias. Then, for a fully determined mask α, β, the bias $\varepsilon_{\alpha,\beta}$ is either 0 (wrong model) or $\frac{1}{2}$ (exact, correct model). More specifically, α, β is a valid input-output mask if

$$
\begin{aligned}
\forall x : \; \alpha^t \cdot x = \beta^t \cdot f(x) \quad &\Leftrightarrow \quad \forall x : \; \alpha^t \cdot x = \beta^t \cdot (A \cdot x) \\
&\Leftrightarrow \quad \forall x : \; \alpha^t \cdot x = (A^t \cdot \beta)^t \cdot x \\
&\Leftrightarrow \quad \alpha = A^t \cdot \beta.
\end{aligned}
$$

If α and β are only partially determined, all propagation can be performed by propagating the information in the linear system $\alpha = A^t \cdot \beta$. For this purpose, we always keep the half-solved system in reduced row echelon form for all linear layers. Whenever mask values in α or β are updated, we perform partial Gaussian elimination to retain reduced row echelon form. If in the process, other bits of α or β are determined (case 1 or 2 from above), this information is extracted from the system and instead stored in the regular representation α, β of the mask bits that is also used for S-box propagation.

Update Process. Every time the propagation step leads to new, explicit information in the linear masks (i.e., mask bits that were previously undetermined are now fixed, case 2), this information has to be propagated over the connected linear or non-linear layers, which share those updated mask bits. In other words, the propagation step needs to be iterated to update the neighbouring layers. Since we require that every linear layer is only connected to non-linear layers and vice versa, we can use a very simple update process scheduling: After each guess or update, we first perform propagation on all non-linear layers (with updated bits), then on all linear layers (with updated bits). This process is iterated until the propagation process has converged and no additional explicit information can be learned anymore, or a contradiction was detected.

3.3   Search for Linear Characteristics

In this section, we discuss our proposed search strategy. The search strategy guides the guessing behavior (choice of bits or bit sets to guess, and their values), as well as the backtracking behavior after detecting contradictions. We currently implement a simple stack-based search algorithm, similar to the strategy used in recent tools for differential characteristic search [16,17]. We first give an algorithmic overview, before detailing the choices made for individual ingredients.

Basic Search Algorithm. We start from a mostly-undetermined characteristic A_0 as a starting point, and incrementally guess more and more of its mask bits. We refer to the current characteristic as A, and keep a history of the guesses that led from A_0 to A in the stack S. For each guess, we select a guessable item X in the current characteristic A. Depending on the search strategy, this can be a single bit, or all bits of an S-box input-output mask (unlike some tools for differential characteristic search which only consider individual bits). The choice of X is guided by the search and backtracking strategy. The characteristics stored in S are used for backtracking, where some of the most recent guesses are undone to resolve conflicts, i.e., we return to an older status stored in S. The basic search algorithm is summarized in Algorithm 1.

Algorithm 1. Guess-and-determine search algorithm

---

choose characteristic A_0 as starting point
loop
  push A_0 to empty stack S
  repeat
    get the topmost characteristic A from S
    select a guessable item (bit or S-box) X in A
    for all most preferable possible values x of X **do**
      guess X to x
      propagate information
      if contradiction detected **then**
        undo guess x and all resulting updates
      else
        push A to S and **break**
    if no valid assignment x was found **then**
      backtrack by popping characteristics from S
      mark X critical
  until exhausted or solution characteristic found

Choice of the Starting Point. The starting point is a linear characteristic, in which most mask bits are still undetermined. The appearance of the starting point depends highly on the scenario in which the linear characteristic will be used, since it can be used to define which bits of the resulting characteristic must definitely be active or inactive.

For instance, consider the search for a linear characteristic for a block cipher or a permutation. In principle, every bit of the input or the respective output mask can be active in such a scenario. So, we can use a starting point where nearly every bit of the respective input and output linear masks is free for guessing during the search. On the other hand, if we consider sponge-like modes, we have more restrictions on the characteristic. Here, the attacker can only observe or control a fraction of the state on the input and the output. Depending on the actual attack, it can be necessary that bits belonging to unknown parts of the state remain inactive, and that only observable or controllable bits are active.

Besides defining the possible use-cases of the linear characteristic, the choice of the starting point also greatly influences the expected success of the search. By fixing parts of the starting point, it is possible to reduce the search space significantly, and thus accelerate the search to quickly find results that would otherwise be out of range. However, reducing the search space also has the potential to exclude classes of good characteristics. Thus, the starting point is usually not too much restricted at the beginning of the analysis of a certain cipher. Instead, the choice of the starting point is an adaptive process based on the cryptographer's intuition and the cipher's structure, using information from previous searches.

Guessing Strategy. The guessing strategy specifies which undetermined bit or S-box is picked next for guessing, and how it will be refined. In S-box-based designs, the search success can profit significantly from guessing in an S-box-oriented manner; that is, by guessing the value of all bits in an S-box input-output mask at once. We refer to this as "guessing the S-box". Even if guesses are made S-box after S-box, the propagation procedure can produce half-guessed S-boxes with some bits fixed and others undetermined. It is also possible to mix S-box-wise and bit-wise guessing.

We refer to an S-box as "guessable" if the linear input and output masks contain at least one remaining undetermined ?-bit, and "fixed" or "not guessable" otherwise. In addition, the search configuration may limit the selection of S-boxes currently eligible for guessing, depending on the guessing progress. The most important example for this is the "critical" status that is assigned to an S-box after a failed attempt to find any valid assignment for this S-box, and assigns top priority to this S-box. Additionally, it can be useful to impose cipher-specific rules; for example, to demand that all S-boxes of the first few rounds must be fixed before we start guessing values in the following rounds.

To guide the guessing procedure, each guessable S-box is assigned a probability for being selected as the next guessing target, for example based on the criteria described above. In addition, all possible assignments for a guessable S-box are ranked by how promising they are estimated to be for high-bias characteristics. Of course, the primary guess for potentially inactive S-boxes (i.e., only with bits 0 and ? so far) is to keep them inactive (i.e., all 0). If this is not possible, the S-box is marked as active. If the selected guessable S-box is already marked active, we rank all possible assignments of the linear masks according to their linear bias and the number of active bits. We pick a random optimally ranked assignment as primary guess. If the following propagation reveals that this assignment is in fact impossible, we try other assignments until no alternative is left, or we have reached a predefined threshold on the number of trials.

Backtracking. If all alternative assignments fail (or a predefined threshold of trials is reached), we need to backtrack. To resolve this conflict, we return to an earlier version of the linear characteristic as stored on the stack S. Again, we try to guess the same critical S-box that caused the conflict. If we cannot resolve the conflict here, we jump further back, until it can be resolved.

Restarts. To better randomize the search process and avoid being "stuck" with a few unhappy first guesses, it is helpful to occasionally restart the complete search. For this purpose, we define a limit of "credits" or resources for one search run. When this limit is exhausted before finding a valid, fully determined characteristic, we clear the stack S and restart from scratch with the starting point A_0. Additionally, the search is also restarted after completing a successful run, with the hope of finding new, better characteristics. If the cryptographer detects promising patterns in the preliminary results, these can serve as a basis for improved starting points for the next run.

4 Application to CAESAR Candidates

In this section, we demonstrate the advantages of our tool for linear cryptanalysis by applying it to several first round CAESAR candidates: KEYAK, ASCON, ICEPOLE, PRØST, and Minalpher. All the analyzed candidates are permutation-based (rather than based on block ciphers). This is, however, not a constraint of the linear tool, which works just as well for block ciphers, since the typical round-key additions do not influence the linear characteristics. Rather, it is representative of the trend that a significant portion of CAESAR candidates with new, dedicated SPN primitives are permutation-based, since most block-cipher modes employ AES.

For each candidate, we first consider linear characteristics for the (round-reduced) permutation. However, for many modes (in particular for sponges), an attacker cannot influence the complete input to the permutation, or cannot observe its complete output. For this reason, we also investigate characteristics with additional constraints, where parts of the linear masks are fixed beforehand. We define the following three types of linear characteristics:

- **Type I (Permutation):** For this type of characteristics, we do not require any additional restrictions regarding the positions of active bits in the linear characteristic. Hence, a characteristic of this type might not be usable in a concrete attack on the duplex-like constructions of KEYAK, ASCON, and

ICEPOLE. Nevertheless, even for modes where Type-I characteristics allow no direct attacks, they still give insights in the resistance of the cryptographic primitive against linear attacks.

- **Type II (Output Constrained):** Linear characteristics of this type have the restriction that all active bits at the end of the characteristic have to be "observable". For duplex-like constructions, this means that all active mask bits have to be in the outer (rate) part of the state. Such linear characteristics can be used to create key-stream distinguishers in known-plaintext scenarios for duplex-like constructions, or even to perform key-recovery attacks.
- **Type III (Input and Output Constrained):** In addition to Type-II characteristics, also all active bits of the input have to be in the outer (rate) part of the state. This type of linear characteristic can act as a key-stream distinguisher in known-plaintext scenarios for duplex-like constructions, targeting the encryption of the plaintext. A similar type of linear relations has been used for instance by Minaud [18] to detect linear biases in the key-stream of the CAESAR candidate AEGIS.

We first discuss our approach and our findings for KEYAK in more detail, and then briefly present our results for ASCON, ICEPOLE, PRØST, and Minalpher.

4.1 Keyak

Brief Description of Keyak. KEYAK is a family of authenticated encryption algorithms designed by Bertoni et al. [4] and is one of the 57 submissions to the first round of the CAESAR competition. It is based on the round-reduced KECCAK-f permutation and follows the duplex construction [2]. The designers have defined four instances of KEYAK; all instances share the same capacity $c = 252$ and use 12 rounds of the KECCAK-f permutation, but differ in their state size b and the degree of parallelism p:

- **River Keyak:** $b = 800$, $p = 1$ (serial),
- **Lake Keyak:** $b = 1600$, $p = 1$ (serial),
- **Sea Keyak:** $b = 1600$, $p = 2$ (parallel),
- **Ocean Keyak:** $b = 1600$, $p = 4$, (parallel).

The KEYAK *Duplex Mode.* Figure. 1 sketches the encryption of serial KEYAK without associated data: The initialization takes as input the secret key K and public nonce N, and applies the permutation f once. This ensures that one always starts with a random-looking state at the beginning of the encryption of the plaintext. Afterwards, the plaintext is processed by xoring it block-wise to the internal state, separated by invocations of the permutation f. The ciphertext blocks are extracted from the state after adding the plaintext. After all data is processed, the finalization applies the permutation f once more and returns the tag. For more details on KEYAK, including the rules for processing associated data, we refer to the specification [4].

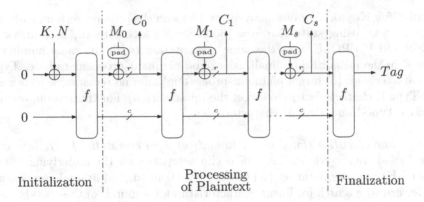

Fig. 1. Simplified sketch of LAKE KEYAK encryption (without associated data).

The KEYAK *Permutation.* The KEYAK permutation is a round-reduced version of the KECCAK-f permutation, reduced to 12 rounds. It operates on the $5 \times 5 = 25$ w-bit words ("lanes") $S[x][y][*]$ of the state S, with $w = 32$ or 64. Each round applies the five steps $R = \iota \circ \chi \circ \pi \circ \rho \circ \theta$, where all steps except ι are equivalent for each round.

- Step θ adds to every bit of the state $S[x][y][z]$ the bitwise sum of the neighbouring columns $S[x-1][*][z]$ and $S[x+1][*][z-1]$. This procedure can be described by the following equation:

$$\theta: \quad S[x][y][z] \leftarrow S[x][y][z] + \sum_{y'=0}^{4} S[x-1][y'][z] + \sum_{y'=0}^{4} S[x+1][y'][z-1].$$

- Step ρ rotates the bits in every lane by a constant value,

$$\rho: \quad S[x][y][z] \leftarrow S[x][y][z + C(x,y)],$$

where $C(x,y)$ is a constant value.
- Step π permutes the lanes using the following function:

$$\pi: \quad S[x][y][*] \leftarrow S[x'][y'][*], \quad \text{where} \quad \begin{pmatrix} x \\ y \end{pmatrix} = \begin{pmatrix} 0 & 1 \\ 2 & 3 \end{pmatrix} \cdot \begin{pmatrix} x' \\ y' \end{pmatrix}.$$

- Step χ is the only non-linear step in KECCAK and operates on each row of 5 bits:

$$\chi: \quad S[x][y][z] \leftarrow S[x][y][z] \oplus ((\neg S[x+1][y][z]) \wedge S[x+2][y][z]),$$

which can be seen as applying a 5-bit S-box in parallel to all rows.
- Step ι adds a round-dependent constant to the state. For the actual values of the constants, we refer to the design document [4].

The designers provide some results on the linear properties of this permutation online, as part of the KECCAKTOOLS package [3].

Results for Keyak. For our analysis, we focus on the primary recommendation LAKE KEYAK using state size $b = 1600$. Since LAKE KEYAK, in contrast to ASCON and ICEPOLE, uses the same permutation (with the same number of rounds) in the initialization, finalization, and plaintext-processing phase, Type-III characteristics (to target plaintext-processing) offer no remarkable advantage over Type-II characteristics (to target the initialization). For this reason, we only consider Type-I and Type-II characteristics.

Type-I Characteristics (for 3 and 4 Rounds of the Permutation). We first consider Type-I characteristics, i.e., linear characteristics for the underlying round-reduced KEYAK permutation (KECCAK-f) without any additional restrictions. We performed a search for linear characteristics for 4 rounds of the 1600-bit permutation. The best linear characteristic we found has 33 active S-boxes, which results in a bias of 2^{-34}. The best linear characteristic for 3 rounds with 13 active S-boxes and a bias of 2^{-14} can be obtained by omitting the first round of the 4-round linear characteristic. Our results are very similar to the characteristic given in the KECCAKTOOLS package [3].

Type-II Characteristics (for 3 and 4 Rounds of the Initialization). The previous 3 and 4-round characteristics have active bits in the inner part (last four 64-bit words) of the state after round 4. Therefore, we cannot use this characteristic in an actual attack. For an attack on the initialization of round-reduced KEYAK, we have to apply additional restrictions on the linear characteristics. Since we can only observe the outer (rate) part of the state at the output of the permutation after the initialization, we apply the restriction that only this part is allowed to contain active bits. Note that the input of the first permutation call is either known or constant. Therefore, we have no problems with active bits there.

For the initialization reduced to 3, or 4 rounds, we found characteristics which only have active S-boxes in the rate part of the state. Thus, considering a known-plaintext attack, we know all the output bits of these S-boxes and can invert them. This leads to the fact that the last round does not influence the bias. So we have an expected bias of 2^{-13} for the 3-round version, and 2^{-49} for the 4-round version of these characteristics. Taking the last S-box layer also into account, the bias of those characteristics would be 2^{-26} and 2^{-70}, respectively. When inverting the last S-boxes, both characteristics result in trivial key-stream distinguishers for round-reduced versions of KEYAK with complexity 2^{26} and 2^{98}, respectively. Moreover, these distinguishers could also be used in a key-recovery attack on round-reduced KEYAK, resulting in similar complexities.

Configuration of the Search. As already mentioned, the proposed search tool is a heuristic one and thus, the quality of the results heavily depends on the applied heuristic search criteria, as well as on the definition of the starting points. For the search process that led to the Type-II characteristics for 3 and 4 rounds of KEYAK, we used a quite natural starting point: For both starting points, the only restriction is that the S-boxes of the last round which "belong" to the inner part of the state must be inactive. In addition, one S-box in the second round is marked as active (to exclude the trivial, entirely inactive solution).

We split the search into two stages. In the first stage, we only pick potentially inactive guessable S-boxes, and set them to the best possible assignment (typically a completely inactive input and output linear mask). Which S-box is picked and refined is determined by a heuristic that picks the S-boxes according to a previously configured weight distribution. These weights can be manually assigned in the search configuration file (the same file in which the starting point is defined). In case of the search for the 3-round Type-II characteristic, the weights were assigned so that S-boxes of the first and second round have a 50 times higher chance to be picked compared to an S-box of the last round. The intention behind this distribution is that the majority of the active S-boxes of the resulting linear characteristic should be located in the last round, because their output is known in an attack. Hence, these active S-boxes can be inverted and do not contribute to the bias. Our heuristic for the 4-round Type-II characteristic prefers S-boxes from rounds 2 and 3 over S-boxes from rounds 1 and 4. Additionally, round 1 is favored over the last round 4. In the second stage, after every guessable and potentially inactive S-box is already determined, we continue by guessing active and yet not fully determined S-boxes until the linear characteristic is fully determined.

As can be seen above, the choice of the starting point and search heuristic depend on the structure of the target primitive, the planned use for the linear characteristic, and on the intuition of the cryptographer. Thus, better search strategies and starting points might exist, which may lead to better linear characteristics than those given in this paper.

4.2 Ascon

Brief Description of Ascon. ASCON is a family of sponge-based candidates, designed by Dobraunig et al. [9]. Compared to KEYAK, it features a significantly smaller state of 320 bits, and the linear layer is applied to each of the 5 64-bit words independently. The 5-bit S-box, on the other hand, is closely related (affine equivalent) to that of KEYAK. The primary proposal ASCON-128 has a rate of 64 bits and hence, a capacity of 256 bits.

Results for Ascon. For the linear tool, the simple design of the linear layer means that its linear model can be split into 5 separate, independent matrices. Combined with a small state size, this property greatly reduces the cost for linear algebra needed to perform the propagation compared to KEYAK.

Our findings for ASCON are summarized in Table 1. The number of active S-boxes of Type-I characteristics found with the help of this tool have already been included in work presented at CT-RSA 2015 [10]. Note that the characteristics given here are optimized for a minimum number of active S-boxes, rather than minimal bias. For ASCON-128, we additionally search for Type-II and Type-III characteristics. However, regarding Type-III characteristics, no meaningful results were obtained.

4.3 ICEPOLE

Brief Description of ICEPOLE. ICEPOLE is a family of authenticated encryption schemes designed by Morawiecki et al. [19]. It consists of the three proposals ICEPOLE-128, ICEPOLE-128a, and ICEPOLE-256a, which all use the same underlying 1280-bit permutation. All variants use 12 rounds of the permutation for initialization, and 6 rounds for processing of plaintext and finalization. However, they differ in details like size of the rate, key, nonce and tag.

The 1280-bit state of ICEPOLE is stored in $5 \times 4 = 20$ 64-bit words. For the linear layer, an MDS matrix over $\mathbb{F}_{2^5}$ is first applied 64 times in parallel (to each 20-bit slice of the state). Then, each word is rotated, and the words swap positions. The S-box layer applies 5-bit S-boxes (4 parallel row-wise applications for each 20-bit slice).

ICEPOLE's designers perform no dedicated linear analysis, but compare the cipher's resistance to linear cryptanalysis to its well-studied resistance against differential cryptanalysis. They conclude that the attack complexity after 5–6 rounds should be "completely intractable" [19]. At FSE 2015, Huang et al. [11] presented 3-round linear characteristics that they use in a differential-linear attack on ICEPOLE.

Results for ICEPOLE. The Type-II and Type-III characteristics given in Table 1 are constrained with respect to a capacity of 254 bits (due to padding, 256 bits are not observable), as defined for ICEPOLE-128 and ICEPOLE-128a. In the case of ICEPOLE, we do not have an immediate output of a ciphertext block right after the 12 rounds of the initialization. Before this happens, there is the option to process a secret message number and at least an empty associated data block is processed. Hence, 6 or even another 12 additional rounds have to be passed before an output suitable for our Type-II characteristic is accessible. Thus—in the worst case—our key-stream distinguisher using Type-II characteristics works for round-reduced versions of ICEPOLE-128, where the initialization plus the following processing is reduced to 5 out of 24 rounds with a complexity of about 2^{120}.

Type-III characteristics can be used to create distinguishers that target the processing of the plaintext. Here, every version of ICEPOLE uses the 6 round version of the ICEPOLE permutation. Thus, by using the Type-III characteristic in Table 1, the key-stream produced by round-reduced variants of ICEPOLE-128, where the permutation used in the plaintext processing is reduced to 4 (out of 6), rounds can be distinguished from a perfect randomly generated key-stream with a complexity of about 2^{88}. The bias of the 5-round Type-III characteristic is $2^{-87.08}$ and hence, the complexity of a resulting key-stream distinguisher cannot harm the 128-bit security of ICEPOLE-128. ICEPOLE-256a, on the other hand, claims a security level of 256 bits regarding the confidentiality. However, it has a higher capacity of 318 bits and therefore, the characteristics given in Table 1 cannot be used directly. Taking the higher capacity of ICEPOLE-256a into account, we get a Type-III characteristic with a bias of $2^{-89.49}$, which can be used to distinguish the key-stream of a round-reduced variant of ICEPOLE-256a, where the permutation used during the encryption is reduced to 5 (out of 6 rounds). Note that ICEPOLE-256a limits the number of blocks encrypted under a single key by 2^{62}. However, this type of key-stream distinguishers exploit relations between ciphertext block C_i and the key-stream used to generate the following ciphertext block C_{i+1}. Thus, distinguishers using Type-III characteristics in this way do not rely on the fact that always the same key is used.

Table 1 contains the results with the best bias, but not necessarily the minimal number of active S-boxes we found. For example, for 6 rounds, we also found a Type-I characteristic with only 103 active S-boxes, but a bias of $2^{-133.49}$ (compared to 104 active S-boxes with bias $2^{-126.32}$ as in the table).

4.4 Minalpher

Brief Description of Minalpher. Minalpher is designed by Sasaki et al. [22]. In contrast to the previous 3 candidates, Minalpher is no sponge-based design. Instead, the permutation is applied in a new tweakable block cipher construction, called tweakable Even-Mansour. For this construction, the permutation size only needs to be twice the security level, so for 128-bit security, Minalpher has the smallest of all investigated permutation sizes with only 256 bits. This small state is further divided into two halves, whose only interaction in each of the 17.5 rounds is that one half is once xored to the other half, and the two halves swap places. Besides the interaction between the halves and some nibble reordering, the linear layer features a near-MDS matrix multiplication over $\mathbb{F}_{2^4}$. The S-box size of 4 bits is also nibble-oriented.

For Minalpher's construction, only Type-I characteristics are useful. We understand our results as an analysis of the underlying permutation. However, since Minalpher claims security in nonce misuse settings and under unverified plaintext release, the Type-I characteristics could also be useful for attacks on the cipher. In particular, for a fixed nonce, the construction allows to control the entire permutation input (at least differentially, due to the Even-Mansour construction, which xors a key- and nonce-dependent value before and after the permutation) and observe the entire output (again, differentially).

The designers analyze the minimum number of active S-boxes (for differential cryptanalysis) theoretically, and prove a minimum number of 22 S-boxes for 4 rounds. For up to 7 rounds, they extend the bounds with the help of mixed integer linear programming (MILP). The bounds obtained this way for the numbers of rounds r also covered by this work are 22 ($r = 4$), 41 ($r = 5$), and 58 ($r = 6$). The designers claim that the same bounds apply for linear cryptanalysis.

Results for Minalpher. The existing bounds serve as a kind of benchmark for our tool to check its capabilities. As shown in Table 1, we were able to match the given bounds for up to 6 rounds. For better comparability, we included our results with the minimal number of active S-boxes, but not necessarily the best bias, in the table. For example, for 6 rounds, we also found a Type-I characteristic with a smaller bias of 2^{-61}, but with 60 active S-boxes (compared to 58 active S-boxes with bias 2^{-62} in the table).

4.5 Prøst

Brief Description of Prøst. PRØST, designed by Kavun et al. [13], offers both a sponge-based mode and two block-cipher-based modes, where the latter use the PRØST permutation in a single-key Even-Mansour construction. Each of the three modes offers two security levels: one based on the 256-bit PRØST-128 permutation, and one based on the 512-bit PRØST-256 permutation. The state is stored as 4×4 words of 16 or 32 bits, respectively. Both the 4-bit S-box (row-wise) and the 16-bit linear mixing function (MDS over $\mathbb{F}_{2^4}$ are applied in a bit-sliced way (using 1 bit of each word). Then, each word is rotated. The number of rounds per permutation call is $r = 16$ (PRØST-128) or $r = 18$ (PRØST-256), respectively.

Table 1. Results for KEYAK, ASCON, ICEPOLE, Minalpher, and PRØST. The corresponding linear characteristics can be found in the full version of this paper.

Cipher	Type	Rounds	Active S-boxes	Bias
KEYAK	I	3	13	2^{-14}
		4	33	2^{-34}
	II	3^*	12	2^{-13}
		4^*	43	2^{-49}
ASCON	I	3	13	2^{-15}
		4	43	2^{-50}
		5	67	2^{-94}
	II	2	6	2^{-8}
		3	23	2^{-30}
		4	61	2^{-83}
ICEPOLE	I	5	38	$2^{-55.08}$
		6	104	$2^{-126.32}$
	II	4	22	$2^{-30.42}$
		5	38	$2^{-59.49}$
	III	3	10	$2^{-16.66}$
		4	22	$2^{-43.25}$
		5	42	$2^{-87.08}$
Minalpher	I	4	22	2^{-23}
		5	41	2^{-42}
		6	58	2^{-62}
PRØST-256	I	4	25	2^{-26}
		5	41	2^{-42}
		6	105	2^{-107}
		7	169	2^{-175}

*Last S-box layer inverted.

We focus our analysis on PRØST-256 (formally offering 128-bit security). Like Minalpher, PRØST comes with a MILP-based proof for the minimum number of active S-boxes for differential and linear characteristics. For PRØST-256, the bounds for different round numbers are 25 ($r = 4$), 41 ($r = 5$), 105 ($r = 6$), and 169 ($r = 7$).

Results for PRØst. Again, we used the existing bounds as benchmarks for our linear tool. The tool is able to match each bound, mostly with optimal or near-optimal bias (2^{-26} for $r = 4$, 2^{-42} for $r = 5$, 2^{-107} for $r = 6$, and 2^{-175} for $r = 7$).

5   Conclusion

We presented a dedicated tool for the automatic linear cryptanalysis of substitution-permutation networks. The goal of the tool is to identify linear characteristics for a cryptographic function, which can subsequently be used by the cryptanalyst to mount key-recovery or distinguishing attacks. The heuristic search is based on an efficient guess-and-determine approach, which has previously been proven successful for searching differential characteristics. We described how to perform efficient propagation of linear masks in linear and non-linear building blocks of a cipher.

From the cryptanalyst's perspective, the tool is simple to use, flexible, and easy to extend with regard to search strategies and target ciphers. The open-source tool will be freely available to help analyze CAESAR candidates and other symmetric cryptographic primitives. We hope that our work will be a valuable contribution to get a better understanding of the security of these ciphers regarding linear cryptanalysis. In particular, we hope to encourage experiments with alternative, sophisticated search strategies optimized for different target ciphers.

We demonstrated the efficiency of our tool by applying it to several CAESAR candidates. The results obtained by searching for linear characteristics for the Minalpher and PRØST-256 permutation show that the presented heuristic search tool can keep pace with MILP-based approaches. However, due to the heuristic nature, we are not capable of providing bounds on the minimum number of active S-boxes.

On the other side, when looking at the results obtained for ASCON, ICEPOLE and KEYAK– all designs with weak alignment – we have been able to find new linear characteristics with a good bias that might be used in a key-recovery or distinguishing attack on round-reduced versions of the ciphers in the future. One highlight are the Type-III characteristics for round-reduced versions of ICEPOLE, which can be used to distinguish the key-stream of ICEPOLE in a nonce-respecting scenario.

Our results show that the existence of a publicly available analysis tool for linear characteristics will be of great help in the design of symmetric cryptographic primitives, in order to evaluate the resistance against linear attacks already in an early stage of the design. Thus, we think that this tool will facilitate new designs which are more balanced in their resistance against linear and differential attacks than some of today's designs.

Acknowledgments. The work has been supported in part by the Austrian Science Fund (project P26494-N15) and by the Austrian Research Promotion Agency (FFG) and the Styrian Business Promotion Agency (SFG) under grant number 836628 (SeCoS).

References

1. Bertoni, G., Daemen, J., Peeters, M., Van Assche, G.: On alignment in Keccak. http://keccak.noekeon.org/KeccakAlignment.pdf
2. Bertoni, G., Daemen, J., Peeters, M., Van Assche, G.: Duplexing the sponge: single-pass authenticated encryption and other applications. In: Miri, A., Vaudenay, S. (eds.) SAC 2011. LNCS, vol. 7118, pp. 320–337. Springer, Heidelberg (2012)
3. Bertoni, G., Daemen, J., Peeters, M., Van Assche, G.: KeccakTools software (2014). http://keccak.noekeon.org/
4. Bertoni, G., Daemen, J., Peeters, M., Van Assche, G., Van Keer, R.: Keyak. Submission to the CAESAR competition (2014). http://competitions.cr.yp.to/round1/keyakv1.pdf
5. Biryukov, A., Velichkov, V.: Automatic search for differential trails in ARX ciphers. In: Benaloh, J. (ed.) CT-RSA 2014. LNCS, vol. 8366, pp. 227–250. Springer, Heidelberg (2014)
6. Brier, E., Khazaei, S., Meier, W., Peyrin, T.: Linearization framework for collision attacks: application to CubeHash and MD6. In: Matsui, M. (ed.) ASIACRYPT 2009. LNCS, vol. 5912, pp. 560–577. Springer, Heidelberg (2009)
7. Daemen, J., Rijmen, V.: AES and the wide trail design strategy. In: Knudsen, L.R. (ed.) EUROCRYPT 2002. LNCS, vol. 2332, pp. 108–109. Springer, Heidelberg (2002)
8. De Cannière, C., Rechberger, C.: Finding SHA-1 characteristics: general results and applications. In: Lai, X., Chen, K. (eds.) ASIACRYPT 2006. LNCS, vol. 4284, pp. 1–20. Springer, Heidelberg (2006)
9. Dobraunig, C., Eichlseder, M., Mendel, F., Schläffer, M.: Ascon. Submission to the CAESAR competition (2014). http://competitions.cr.yp.to/round1/asconv1.pdf
10. Dobraunig, C., Eichlseder, M., Mendel, F., Schläffer, M.: Cryptanalysis of ascon. In: Nyberg, K. (ed.) CT-RSA 2015. LNCS, vol. 9048, pp. 371–387. Springer, Heidelberg (2015)
11. Huang, T., Tjuawinata, I., Wu, H.: Differential-linear cryptanalysis of ICEPOLE. In: Leander, G. (ed.) FSE 2015. LNCS, vol. 9054, pp. 243–263. Springer, Heidelberg (2015)
12. Indesteege, S., Preneel, B.: Practical collisions for EnRUPT. In: Dunkelman, O. (ed.) FSE 2009. LNCS, vol. 5665, pp. 246–259. Springer, Heidelberg (2009)
13. Kavun, E.B., Lauridsen, M.M., Leander, G., Rechberger, C., Schwabe, P., Yalçin, T.: Prøst. Submission to the CAESAR competition (2014). http://competitions.cr.yp.to/round1/proestv11.pdf
14. Leurent, G.: Construction of differential characteristics in ARX designs application to skein. In: Canetti, R., Garay, J.A. (eds.) CRYPTO 2013, Part I. LNCS, vol. 8042, pp. 241–258. Springer, Heidelberg (2013)

15. Matsui, M.: Linear cryptanalysis method for DES cipher. In: Helleseth, T. (ed.) EUROCRYPT 1993. LNCS, vol. 765, pp. 386–397. Springer, Heidelberg (1994)
16. Mendel, F., Nad, T., Schläffer, M.: Finding SHA-2 characteristics: searching through a minefield of contradictions. In: Lee, D.H., Wang, X. (eds.) ASIACRYPT 2011. LNCS, vol. 7073, pp. 288–307. Springer, Heidelberg (2011)
17. Mendel, F., Nad, T., Schläffer, M.: Improving local collisions: new attacks on reduced SHA-256. In: Johansson, T., Nguyen, P.Q. (eds.) EUROCRYPT 2013. LNCS, vol. 7881, pp. 262–278. Springer, Heidelberg (2013)
18. Minaud, B.: Linear biases in AEGIS keystream. In: Joux, A., Youssef, A. (eds.) SAC 2014. LNCS, vol. 8781, pp. 290–305. Springer, Heidelberg (2014)
19. Morawiecki, P., Gaj, K., Homsirikamol, E., Matusiewicz, K., Pieprzyk, J., Rogawski, M., Srebrny, M., Wójcik, M.: ICEPOLE. Submission to the CAESAR competition (2014). http://competitions.cr.yp.to/round1/icepolev1.pdf
20. Mouha, N., Preneel, B.: Towards finding optimal differential characteristics for ARX: Application to Salsa20. IACR Cryptology ePrint Archive, Report 2013/328 (2013). http://eprint.iacr.org/2013/328
21. Mouha, N., Wang, Q., Gu, D., Preneel, B.: Differential and linear cryptanalysis using mixed-integer linear programming. In: Wu, C.-K., Yung, M., Lin, D. (eds.) Inscrypt 2011. LNCS, vol. 7537, pp. 57–76. Springer, Heidelberg (2012)
22. Sasaki, Y., Todo, Y., Aoki, K., Naito, Y., Sugawara, T., Murakami, Y., Matsui, M., Hirose, S.: Minalpher. Submission to the CAESAR competition (2014). http://competitions.cr.yp.to/round1/minalpherv1.pdf
23. Schläffer, M., Oswald, E.: Searching for differential paths in MD4. In: Robshaw, M. (ed.) FSE 2006. LNCS, vol. 4047, pp. 242–261. Springer, Heidelberg (2006)
24. Sun, S., Hu, L., Wang, M., Wang, P., Qiao, K., Ma, X., Shi, D., Song, L.: Automatic enumeration of (related-key) differential and linear characteristics with predefined properties and its applications. IACR Cryptology ePrint Archive, Report 2014/747 (2014). http://eprint.iacr.org/2014/747
25. Tardy-Corfdir, A., Gilbert, H.: A known plaintext attack of FEAL-4 and FEAL-6. In: Feigenbaum, J. (ed.) CRYPTO 1991. LNCS, vol. 576, pp. 172–182. Springer, Heidelberg (1992)
26. The CAESAR committee: CAESAR: Competition for authenticated encryption: Security, applicability, and robustness (2014). http://competitions.cr.yp.to/caesar.html

Collision Attacks Against CAESAR Candidates

Forgery and Key-Recovery Against AEZ and Marble

Thomas Fuhr[1]([⊠]), Gaëtan Leurent[2], and Valentin Suder[3]

[1] ANSSI, Paris, France
thomas.fuhr@ssi.gouv.fr
[2] Inria, Paris, France
[3] University of Waterloo, Waterloo, Canada

Abstract. In this paper we study authenticated encryption algorithms inspired by the OCB mode (Offset Codebook). These algorithms use secret offsets (masks derived from a whitening key) to turn a block cipher into a tweakable block cipher, following the XE or XEX construction.

OCB has a security proof up to $2^{n/2}$ queries, and a matching forgery attack was described by Ferguson, where the main step of the attack recovers the whitening key. In this work we study recent authenticated encryption algorithms inspired by OCB, such as Marble, AEZ, and COPA. While Ferguson's attack is not applicable to those algorithms, we show that it is still possible to recover the secret mask with birthday complexity. Recovering the secret mask easily leads to a forgery attack, but it also leads to more devastating attacks, with a key-recovery attack against Marble and AEZ v2 and v3 with birthday complexity.

For Marble, this clearly violates the security claims of full n-bit security. For AEZ, this matches the security proof, but we believe it is nonetheless a quite undesirable property that collision attacks allow to recover the master key, and more robust designs would be desirable.

Our attack against AEZ is generic and independent of the internal permutation (in particular, it still works with the full AES), but the key-recovery is specific to the key derivation used in AEZ v2 and v3. Against Marble, the forgery attack is generic, but the key-recovery exploits the structure of the E permutation (4 AES rounds). In particular, we introduce a novel cryptanalytic method to attack 3 AES rounds followed by 3 inverse AES rounds, which can be of independent interest.

Keywords: CAESAR competition · Authenticated encryption · Cryptanalysis · Marble · AEZ · PMAC · Forgery · Key-recovery

1 Introduction

The purpose of an *Authenticated Encryption* scheme is to provide both privacy and integrity with a single cryptographic algorithm. In 2014, the CAESAR competition was launched with the goal to identify good Authenticated Encryption schemes as better alternatives to current options such as AES-GCM [14]. 57

© International Association for Cryptologic Research 2015
T. Iwata and J.H. Cheon (Eds.): ASIACRYPT 2015, Part II, LNCS 9453, pp. 510–532, 2015.
DOI: 10.1007/978-3-662-48800-3_21

candidates have been submitted to the CAESAR competition, and they must now be analyzed carefully.

In this paper, we provide a security analysis of the AES-based candidates Marble [5] and AEZ v3 [7]. Both designs are inspired by OCB [16], designed in 2001 by Rogaway, Bellare, Black, and Krovetz. They are built as modes of operation of a block cipher[1], using secret offsets at the input and/or output of the block cipher calls.

OCB. In OCB, a whitening key L is derived from the master key K, and the i-th message block M_i is enciphered to $C_i = E_K(M_i \oplus \gamma_i \cdot L) \oplus \gamma_i \cdot L$, where γ_i is a (Gray) counter, $\cdot$ is a finite field multiplication, and $\gamma_i \cdot L$ is the i-th offset. This design principle was later formalized as the XE and XEX construction [15], and proved to turn efficiently a secure block cipher into a secure tweakable block cipher [12]. OCB with a an n-bit block cipher is proven secure up to $2^{n/2}$ queries, and Ferguson showed a collision attack matching the bound [3]. The attack uses a long message M, so that there is a collision between two block cipher inputs:

$$M_i \oplus \gamma_i \cdot L = M_j \oplus \gamma_j \cdot L$$

The collision can be detected because $M_i \oplus C_i = M_j \oplus C_j$, and the value of L is recovered as $(\gamma_i \oplus \gamma_j)^{-1} \cdot (M_i \oplus M_j)$. When L is known, it is easy to forge messages.

Marble. Marble [5] is a CAESAR candidate by Jian Guo inspired by COPA [1]. COPA was designed in 2013, and combines OCB's offsets with an internal dependency chain in order to achieve some security in the case of nonce repetition. Marble uses two internal chains in order to prevent birthday attacks on the internal chain, and uses reduced-round AES as building blocks. Marble claims security against nonce-repetition, and against release of unverified plaintexts, but cannot hide common prefixes in case of nonce reuse (Marble is online). As opposed to most CAESAR candidates, Marble claims full 128-bit security (beyond the birthday bound). The structure of Marble can be seen in Fig. 2.

Results presented so far on Marble include a cipher-text distinguisher with complexity 2^{64}, similar to the distinguisher against the counter mode [17].

AEZ. AEZ is a CAESAR candidate designed by Hoang, Krovetz, and Rogaway. The authors define the security notion of *Robust AE*, which is the optimal security achievable when nonces are repeated, and unverified plaintexts are released. AEZ is claimed to achieve this security notion. In this paper, we focus on the current version of AEZ, AEZ v3, as proposed on the `crypto-competition` mailing list, and presented at DIAC [7]. AEZ v3 has also been accepted at Eurocrypt 2015, and presented as one of the honorable mentions for the best papers

---

[1] For efficiency reasons, Marble and AEZ actually use 4-rounds of AES rather than a full block cipher.

award [8]. Our result can also be applied to AEZ v2, but not to AEZ v1, because of a different key expansion.

As far as we are aware, no cryptanalysis of AEZ as been published so far.

Our Results. In this paper, we describe generic collision attacks against Marble and AEZ, allowing to recover the whitening key with about $2^{n/2}$ chosen message queries. When the whitening key is known the security offered by Marble and AEZ crumbles and we show a forgery attack using a single extra encryption query. Moreover, we extend this result to key-recovery attacks using properties of the internal permutations and/or the key scheduling.

Our results are summarized in Tables 1 and 2. The data complexity is listed in number of message blocks (16 bytes). We now detail our results on each Authenticated-Encryption with Associated-Data (AEAD) scheme.

Marble. Our attack against Marble uses queries with repeated nonces, which should be secure according to the security claims of Marble. Since Marble claims security beyond the birthday bound (allowing up to 2^n block of data), the forgery attack using collisions clearly violates the security claims. In addition, we show that if unverified plaintexts are released, *i.e.* if we can obtain plaintexts from ciphertexts without a valid tag, then we can further recover the master key K. For this attack, we build special queries so that only 3 forward AES rounds and 3 backwards AES rounds are active, and we develop a novel method to attack such a reduced cipher with only known plaintext/ciphertexts. Our attack can be build upon two different distinguishers. the first one is based on the detection of collision events, and the second one on a statistical property. In both cases, our attack requires about 2^{33} queries and its time complexity is 2^{64}; we believe this result is also of independent interest.

Following the disclosure of this attack, Guo proposed a minor modification of the specifications of Marble as version 1.2 [6]. However, our attack is still applicable to the modified version, as shown independently by ourselves and Lu [13]. Guo later decided to withdraw Marble from the CAESAR competition.

AEZ. Our analysis of AEZ v3 focuses on the processing of Associated Data. In particular, if AEZ is used with an empty message and no nonce, it turns into a variant of PMAC, and the security notion of Robust AE becomes the usual MAC security notion. We show how to recover the whitening key of this variant of PMAC with a collision attack (a collision attack is also possible against the standard PMAC, *e.g.* following [11]). More importantly, the key derivation of PMAC allows to recover the master key K from the whitening key.

This attack does not violate the security proof, but matches the security bound. However, collision attacks usually have a more limited impact (*e.g.* only affecting authenticity), and it seems quite unfortunate that a collision attack leads to a key-recovery. This property should probably be avoided when possible[2].

---

[2] In AEZ v4, for the second round of the competition, the designers took into account our result and modified the key derivation in order to prevent this property.

Table 1. Our results against Marble.

Attack (Sec. claim)	Data	Time
Recover L	$2^{65} \times 2$ CP	2^{64}
Forgery (2^{128})	$2^{65} \times 2$ CP	2^{64}
Key-recovery (2^{128}):		
Collision-based[a]	$2^{65} \times 2$ CP + $2^{32.6} \times 130$ CC	2^{64}
Collision-based	$2^{65} \times 2$ CP + $2^{33} \times 130$ CC	2^{64}

[a] The chosen ciphertexts use the decryption-misuse model.

Table 2. Our results against AEZ.

Attack	Data[a]	Time	Success probability
Key-recovery	$2^{66.6}$	1	1
Key-recovery	2^{44}	1	$2^{-45.2}$

[a] The AEZ specification requires to rotate the key after 2^{44} blocks

COPA. After the release of an early version of this paper, Lu applied the same techniques to COPA, and described an attack to recover the whitening key [13]. The main attack in this paper actually targets the associated data processing, which uses PMAC, and can be applied to PMAC. However, the impact of this result is unclear because COPA and PMAC do not claim security beyond the birthday bound, and this attack cannot be turned into a key-recovery attack.

Outline of the Paper. Since our collision attack on AEZ is much simpler than the attack against Marble, we first explain it in Sect. 2. Then we give a short description of the Marble authenticated encryption algorithm in Sect. 3. In Sect. 4, we show how to recover the whitening key L and describe our forgery attack. Finally, we demonstrate in Sect. 5 how to recover the encryption key K from decryption-misuse queries.

2   Collision Attack Against AEZ

We first explain the collision attack on AEZ and the resulting key-recovering attack.

2.1   Short Description of AEZ

For simplicity, we consider AEZ used with only associated data, without any nonce or message (the attack can easily be applied with a fixed nonce and message if required). In this case, AEZ turns into a variant of PMAC, and the security claim becomes the usual MAC security definition. A particularity of

AEZ is that it allows a vector-valued input, *i.e.* it can authenticate a sequence of strings rather than a single string.

More precisely, the MAC is computed as follow:

- The key derivation algorithm generates keys k_0, k_1 and whitening keys J, L
- Full data blocks A_i^j of the j-th string (indexed from 1) are processed as:

$$X_i^j = E_{k_0}\left(A_i^j \oplus (i \bmod 8) \cdot J \oplus 2^{\lfloor (i-1)/8 \rfloor} \cdot L \oplus 8j \cdot J\right)$$

- If the last block is partial, it is enciphered as:

$$X_i^j = E_{k_0}(A_i^j \oplus 8j \cdot J)$$

- The first block to be processed is the ciphertext extension τ (corresponding to the tag length). It is $\tau = 128$ by default.
- The tag is computed as $E_{k_1}(\bigoplus_{i,j} X_i^j)$

where E is a full or reduced-round AES. This is illustrated by Fig. 1.

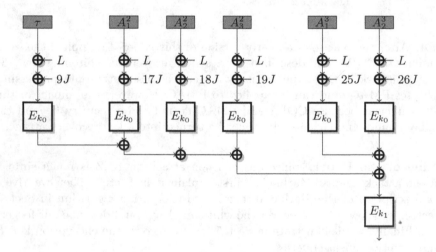

Fig. 1. AEZ used as a MAC (no message, no nonce, two AD strings).

2.2   Collision Attack on AEZ

In order to mount a collision attack against AEZ, we consider two sets of messages, with C a fixed block:

- $\mathcal{A} = \{A_x \mid x \in \{0 \dots 2^{64} - 1\}\}$, with $A_x = \left(\tau; C; (C \parallel [x] \parallel 0^{64})\right)$
- $\mathcal{B} = \{B_y \mid y \in \{0 \dots 2^{64} - 1\}\}$, with $B_y = \left(\tau; (C \parallel 0^{64} \parallel [y]); C\right)$

All message are made of two separate strings; message in $\mathcal{A}$ have a string of one block and a string of two blocks, while messages in $\mathcal{B}$ have a string of two blocks and a string of one block. In particular, we have:

$$- \text{MAC}(A_x) = E_{k1}\Big(E_{k_0}(\tau \oplus L \oplus 9J) \oplus E_{k_0}(C \oplus L \oplus 17J)$$
$$\oplus E_{k_0}(C \oplus L \oplus 25J) \oplus E_{k_0}(([x] \parallel 0^{64}) \oplus L \oplus 26J)\Big)$$
$$- \text{MAC}(B_y) = E_{k1}\Big(E_{k_0}(\tau \oplus L \oplus 9J) \oplus E_{k_0}(C \oplus L \oplus 17J)$$
$$\oplus E_{k_0}((0^{64} \parallel [y]) \oplus L \oplus 18J) \oplus E_{k_0}(C \oplus L \oplus 25J)\Big)$$

This leads to a simple collision attack: $\text{MAC}(A_x) = \text{MAC}(B_y)$ if and only if $[x] \parallel [y] = 8 \cdot J$ (where $8 = 18 \oplus 26$). With $\mathcal{A}$ and $\mathcal{B}$ of size 2^{64} as defined above, there is exactly one collision, and the collision immediately reveals the value of $J = 8^{-1} \cdot ([x] \parallel [y])$.

Key Recovery. Surprisingly, the key derivation of AEZ allows to recover the master key K from the whitening key J. More precisely, if the master key K is of length 128 bits or smaller, J is an encryption of K under a known constant C: $J = \text{AES4}_C(K)$. This can easily be inverted: $K = \text{AES4}_C^{-1}(J)$. We note that this is not the case in PMAC, where the whitening key is an encryption of 0 under the secret key K: $L = \text{AES}_K(0)$.

This attack matches the security proof of AEZ and does not violate the security claims. However, a complete break of AEZ after the birthday bound is not expected: most schemes with birthday-bound security are more resilient and collision attacks don't allow key-recovery.

It should be mentioned that the Eurocrypt version of AEZ does not explicitly specify a key derivation algorithm, and leaves it as an open problem:

"The key $K \in \text{BYTE}^*$ is mapped to three 16-byte subkeys (I, J, L) using the key-derivation function (KDF) named Extract that is called at line 401. The definition of Extract is omitted from the figures and regarded as orthogonal to the rest of AEZ. See the AEZ spec for the current Extract: $\text{BYTE}^* \rightarrow \text{BYTE}^{48}$. In our view, it is an unresolved matter what the security properties (and even what signature) of a good KDF should be. Work has gone off in very different directions, and the area is currently the subject of a Password Hashing Competition (PHC) running concurrently with CAESAR."

Clearly, the key derivation of the AEZ v3 specification does not have the security properties of a good KDF.

Data Limit. The AEZ specification requires users to change the key after encrypting 2^{48} bytes, *i.e.* 2^{44} blocks. This prevents the attack as described above. However, we can perform the attack with smaller sets $\mathcal{A}$ and $\mathcal{B}$ of size $2^{41.4}$, with a success probability of $2^{-45.2}$. This is still much more efficient than generic attacks: with a time complexity of 2^{44}, a brute-force key search only succeeds with a probability of 2^{-84}.

3   Description of Marble

Marble is an authenticated encryption algorithm designed by Guo [5] with key-length and tag-length of both 128 bits. A plaintext and its associated data are divided into blocks of 128 bits and are then proceeded consecutively. Its internal permutation is based on a modified version of the AES block cipher. Unlike other authenticated encryption algorithms, Marble does not require a nonce.

Marble has very strong security claims: it claims to offer full 128-bit security, *i.e.* an attack should take time $T = 2^{128}$ even after large amount of data have been encrypted with the same key (up to $D = 2^{128}$). This is in contrast to many CAESAR candidates and classical modes of operations for block ciphers (*e.g.* GCM), which only offer a birthday level of security, *i.e.* the ciphers are secure as long as $T \cdot D < 2^{128}$.

In addition, Marble does not use nonces, and the security claim even holds if unverified plaintexts are released, *i.e.* the adversary can request the decryption of a ciphertext C without knowing a valid tag corresponding to C (decryption-misuse oracle). A few other CAESAR candidate allow the release of unverified plaintext (AEZ, POET, APE, Minalpher), but they only claim birthday security.

An overview of Marble is depicted in Fig. 2. The Marble mode of operation makes use of two 128-bit chaining variables s_1 and s_2, initialized with constants $const_1$ and $const_2$. The associated data and the plaintext are padded independently, so both resulting fields A and P can be divided into 128 bit blocks. We do not describe the padding function here, as it does not affect our attacks. We will denote a message to encrypt by $(AD \parallel M)$, where AD is a vector containing l_A 128-bit blocks of associated data and M is a vector containing l_M 128-bit blocks of plaintexts.

The internal primitive used is a modified block cipher, as intermediate values of the block are combined with the incoming chaining variables. Formally, the primitive uses 3 internal keyed permutations E_1, E_2 and E_3 and processes 128-bit blocks as follows. On input (P, s_1, s_2), (C, s_1', s_2') is defined as

$$X = E_{1K}(P)$$
$$(X', s_1') = (3X + s_1, X + s_1)$$
$$Y = E_{2K}(X')$$
$$(Y', s_2') = (3Y + s_2, Y + s_2)$$
$$C = E_{3K}(Y')$$

Note that additions and multiplications are performed in the binary Galois Field $\mathbb{F}_{2^{128}}$ defined by the primitive polynomial $x^{128} + x^7 + x^2 + x + 1$. Furthermore, polynomials $\Sigma a_i X^i$ are denoted by the integers $\Sigma a_i 2^i$. Therefore, please note that additions and multiplications on such objects have to be interpreted as operations in the binary field (and not on the integer ring) and have to be handled carefully.

In the case of Marble, each one of the three permutations E_1, E_2 and E_3, is composed with 4 full-round of AES (*i.e.* SubByte, ShiftRow, MixColumn and

AddRoundKey). One can notice that no key addition is performed at the beginning of those permutations. 12 subkeys are therefore required. A 128-bit master key K is derived into 11 subkeys using the AES-128 key-schedule algorithm. The master key itself is used as the 12th subkey. For more information about the AES block cipher, we refer to [2].

The Marble encryption then works as follows. First, a mask L is derived from the key K by encrypting a constant $const_3$ (which also sets key-dependent $s_1[0], s_2[0]$). For each associated data block A_i ($i \geq 1$), a pre-whitening key is defined as $2^{i-1}3^2L$. For each plaintext block M_i, a pre-whitening key and a post-whitening key are defined as 2^iL and $2^{i-1}3L$. These blocks are processed iteratively, starting with associated data, as follows:

1. Addition (*i.e.* xor) of the pre-whitening key;
2. Application of the internal primitive;
3. For plaintext blocks, application of the post whitening key, which results in ciphertext blocks.

Finally, the tag is computed by encrypting an extra block defined as the sum of all plaintext blocks and all encrypted additional data blocks, with pre-whitening key $2^{\ell_M}7L$ and post-whitening key $2^{\ell_M-1}3L$.

4   A Universal Forgery Attack on Marble

In this section, we first describe a method to find the mask L using about 2^{65} chosen plaintext queries. Then, we use this knowledge to compute forgeries. Our attack enables to modify the associated data field in a way that affects neither the ciphertext nor the authentication tag. It can therefore be used to compute universal forgeries in a chosen plaintext setting.

4.1   Recover the Mask L

The main idea of the attack is to build a pair of message $M \neq M'$ so that the inputs to the E_1 functions are the same for both messages. This is possible if M and M' have the same total length, but the associated data and message parts have different lengths. When the inputs to E_1 collide, all the intermediate computations collide, and we can detect this event on the resulting ciphertexts. Please note that as different multiples of L are used for post-whitening, this operation is more tricky than detecting a collision on ciphertexts. In the following we use 2 blocks of AD and 1 block of message for M, but 1 block of AD and 2 blocks of message for M'.

More precisely, we encrypt messages M_α and M'_β, for different values $\alpha, \beta \in \mathbb{F}_{2^{128}}$, defined as follows (where $A \in \mathbb{F}_{2^{128}}$ is a constant value):

- $M_\alpha = (AD[1], AD[2] \parallel M[1]) = (A, 8\alpha \parallel 6\alpha)$;
- $M'_\beta = (AD[1] \parallel M'[1], M'[2]) = (A \parallel 8\beta, 6\beta)$.

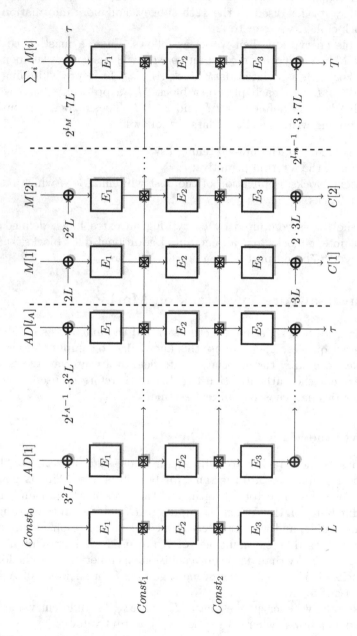

Fig. 2. General design of Marble.

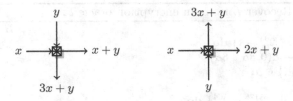

Fig. 3. The TRANS operation.

In the following, we consider sets of 2^{64} values α and β so that $\alpha \oplus \beta$ covers all values in $\mathbb{F}_{2^{128}}$. The inputs to the E_1 layer will be respectively (we note that $3^2 = 5$ in $\mathbb{F}_{2^{128}}$):

$$x_1 = A \oplus 5L \qquad x_2 = 8\alpha \oplus 10L \qquad x_3 = 6\alpha \oplus 2L \qquad \text{for } M_\alpha$$
$$x_1' = A \oplus 5L \qquad x_2' = 8\beta \oplus 2L \qquad x_3' = 6\beta \oplus 4L \qquad \text{for } M_\beta'$$

In particular, we have:

$$x_1 \oplus x_1' = 0 \qquad x_2 \oplus x_2' = 8(\alpha \oplus \beta \oplus L) \quad x_3 \oplus x_3' = 6(\alpha \oplus \beta \oplus L)$$

Therefore, the inputs to E_1 collide when $\alpha \oplus \beta = L$.

We denote the output of the E_3 layer as y_i (respectively y_i'), and the corresponding ciphertexts as $C_\alpha[1]$ (respectively $(C_\beta'[1], C_\beta'[2])$). We have:

$$C_\alpha[1] = y_3 \oplus 3L \qquad C_\beta'[1] = y_2' \oplus 3L \qquad C_\beta'[2] = y_3' \oplus 6L$$

In particular, if $\alpha \oplus \beta = L$, we have $x_i = x_i'$ for $i \leq 3$, therefore $y_i = y_i'$ for $i \leq 3$, and $C_\alpha[1] \oplus C_\beta'[2] = 5L$ (since $3 \oplus 6 = 5$). In order to detect this event efficiently we match the set of values $\{C_\alpha[1] \oplus 5\alpha\}$ and $\{C_\beta'[2] \oplus 5\beta\}$. When $\alpha \oplus \beta = L$, we have a match, and we can easily filter false positives using a new message pair with a different value of the constant A. The full algorithm is given by Algorithm 1, using 2^{65} short encryption queries.

4.2 An Attack Against Marble 1.2

After the first release of our attack, Guo made a minor modification to the specification of Marble [6]. Namely, the input mask for the last block of associated data is changed from $2^{i-1}3^2L$ to $2^{i-1}3^3L$. Our attacks can be adapted as follows.

The adversary needs to query an encryption oracle for messages M_α and M_β', defined as

- $M_\alpha = (AD[1], AD[2] \parallel M[1]) = (10\alpha, 28\alpha \parallel 6\alpha)$;
- $M_\beta' = (AD[1] \parallel M'[1], M'[2]) = (10\beta \parallel 28\beta, 6\beta)$.

Using the notations of Sect. 4.1, the inputs to the E_1 layer will be :

$$x_1 = 10\alpha \oplus 5L \qquad x_2 = 28\alpha \oplus 30L \qquad x_3 = 6\alpha \oplus 2L \qquad \text{for } M_\alpha$$
$$x_1' = 10\beta \oplus 15L \qquad x_2' = 28\beta \oplus 2L \qquad x_3' = 6\beta \oplus 4L \qquad \text{for } M_\beta'$$

Algorithm 1. Recover L from an encryption oracle $\mathcal{E}$.

$H \leftarrow \varnothing$ ▷ H is a hash table
for $\alpha \in \{0, 1, \ldots, 2^{64} - 1\}$ **do**
　　$(C[1] \parallel T) \leftarrow \mathcal{E}(0, 8\alpha \parallel 6\alpha)$
　　$H\{C[1] \oplus 5\alpha\} \leftarrow \alpha$
end for
for $\beta \in \{0, 2^{64}, \ldots, 2^{128} - 2^{64}\}$ **do**
　　$(C'[1], C'[2] \parallel T) \leftarrow \mathcal{E}(0 \parallel 8\beta, 6\beta)$
　　if $H\{C'[2] \oplus 5\beta\}$ exists **then**
　　　　$\alpha \leftarrow H\{C'[2] \oplus 5\beta\}$
　　　　$(D[1] \parallel T) \leftarrow \mathcal{E}(1, 8\alpha \parallel 6\alpha)$
　　　　$(D'[1], D'[2] \parallel T) \leftarrow \mathcal{E}(1 \parallel 8\beta, 6\beta)$
　　　　if $D[1] \oplus 5\alpha = D'[2] \oplus 5\beta$ **then**
　　　　　　return $\alpha \oplus \beta$
　　　　end if
　　end if
end for

In particular, we have:

$$x_1 \oplus x_1' = 10 \cdot (\alpha \oplus \beta \oplus L),$$
$$x_2 \oplus x_2' = 28 \cdot (\alpha \oplus \beta \oplus L),$$
$$x_3 \oplus x_3' = 6 \cdot (\alpha \oplus \beta \oplus L).$$

If for some (α, β), $\alpha \oplus \beta = L$, then $x_i = x_i'$ for $i = 1, 2, 3$. Then, the outputs of E_3 verify $y_1 = y_1'$, $y_2 = y_2'$ and $y_3 = y_3'$ and therefore, $C_\alpha[1] \oplus 3L = C_\beta'[2] \oplus 6L$. As $3 \oplus 6 = 5$, This can also be expressed as:

$$C_\alpha[1] \oplus 5\alpha = C_\beta'[2] \oplus 5\beta.$$

Therefore, L has to be searched among the values $(\alpha \oplus \beta)$ for which this relation holds. As for our attack on the previous version of Marble, about 2^{64} different values of both α and β are required.

4.3 Computing Forgeries on Marble Without Whitening Keys

Once we have retrieved L, we can consider a simplified description of Marble where the masks are removed, as depicted in Fig. 4. In its mask-less description, Marble possesses an interesting property as described in Fig. 5: a series of identical input blocks X has a periodic effect on the internal state.

Indeed, if we let $E_1(X) = u$, $E_2(3S_1 \oplus u) = v$ and $E_2(3S_1 \oplus 2u) = w$, it is easy to see that after encrypting 4 blocks X, the internal states S_1 and S_2 remain unchanged. Furthermore, if we use a series of 8 consecutive identical associated data blocks X, the effect on τ also cancels out. This leads to a universal forgery attack: for any associated data AD and plaintext M, the adversary computes the masked value B of a chunk of 8 identical blocks of associated data after AD

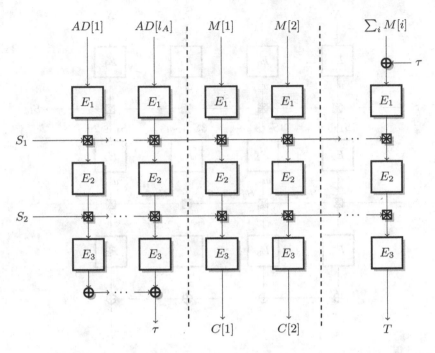

Fig. 4. Mask-less description of Marble. S_1 and S_2 are unknown key-dependent values.

and queries the encryption oracle on $((A, B) \parallel M)$. The answer $(C \parallel T)$ to that query is also a valid ciphertext for $(AD \parallel M)$, therefore the adversary can return $(C \parallel T)$ as a forgery. The attack is given as Algorithm 2.

Algorithm 2 . Compute the ciphertext $(C \parallel T)$ from $(AD \parallel M)$ using known L.

$B \leftarrow (2^l \cdot 3^2 \cdot L)_{l=l_A}^{l_A+7}$
$(C \parallel T) \leftarrow \mathcal{AE}_K((AD, B) \parallel M)$ ▷ Encryption oracle call
return $(C \parallel T)$

---

5   Key-Recovery Attack

We now show how to recover the master key once the mask L has been determined. In order to simplify the description of the attack, we now focus on the mask-less variant of Marble; however the full attack can easily be adapted to the full version of Marble with a known mask.

The main idea is to collect pairs of messages with a fixed difference in some internal state variables. This will allow us to attack a reduced cipher composed

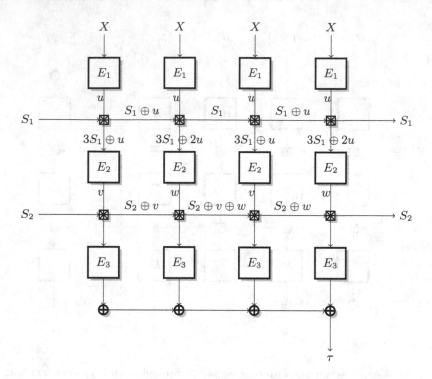

Fig. 5. Collision on the internal state of the associated data.

by 4 AES rounds followed by 4 inverse AES rounds rather than a 12-round AES (see details below). Moreover, we apply this strategy to E_1 rather that to E_3 because the whitening key of E_1 is directly derived from L. Since L is known, the first AES round of E_1 is key-independent. Therefore we can peel it off, and attack only 3 forward rounds and 3 inverse rounds. However, this requires us to use decryption queries, but we can't forge valid tags for an arbitrary ciphertext yet, so we use a decryption-misuse oracle.

5.1 Gathering Pairs

The first step is to collect pairs of plaintext blocks that have the same difference in the S_1 lane (after the permutation E_1). In order to construct such plaintexts, we build pairs of ciphertexts with specific differences and values. More precisely, we consider pairs of messages as follows (with the same associated data AD):

$$\widetilde{C}_x = (AD \parallel 0, 0^{120}, 0, 0, 0, 0, 0, 0, 0, 0, x) \qquad C_x[i] = \widetilde{C}_x[i] \oplus 2^{i-1} \cdot 3 \cdot L$$

$$\widetilde{C}'_x = (AD \parallel 1, 0^{120}, 1, 0, 0, 0, 0, 1, 1, 1, x) \qquad C'_x[i] = \widetilde{C}'_x[i] \oplus 2^{i-1} \cdot 3 \cdot L$$

where 0 and 1 are constant one-block values and x takes a different value for each pair. We decrypt these pairs and we collect the final plaintext blocks.

We now study the differences in the S_2 lane (before the permutation E_3). Using the definition of the TRANS operation as given in Fig. 3, S_2 is updated as follows during decryption:

$$S_2[i+1] = 2 \cdot S_2[i] \oplus E_3^{-1}(\widetilde{C}[i])$$

With the messages C_x and C'_x, we have

$$S_2[129] = 2^{129} S_2[0] \oplus (1 \oplus 2 \oplus \cdots \oplus 2^{128})A,$$
$$S'_2[129] = 2^{129} S_2[0] \oplus (1 \oplus 2 \oplus \cdots \oplus 2^{128})A \oplus (2^0 \oplus 2^1 \oplus 2^2 \oplus 2^7 \oplus 2^{128})(A \oplus B),$$

where $A = E_3^{-1}(0)$ and $B = E_3^{-1}(1)$. Since $2^{128} = 2^0 \oplus 2^1 \oplus 2^2 \oplus 2^7$, we have $S_2[129] = S'_2[129]$. This is shown in Fig. 6, where $\delta = A \oplus B$.

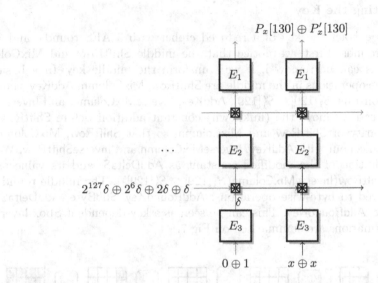

Fig. 6. Difference propagation in decryption. A red arrow means that there is a fixed unknown difference. A black arrow means that the difference is null.

We now consider the final plaintext block given by the decryption oracle.

$$\widetilde{P}_x[130] = P_x[130] \oplus 2^{130} \cdot L$$
$$= E_1^{-1}\left(E_2^{-1}\left(E_3^{-1}(x) \oplus 3 \cdot S_2[129]\right) \oplus 3 \cdot S_1[129]\right)$$
$$\widetilde{P}'_x[130] = P'_x[130] \oplus 2^{130} \cdot L$$
$$= E_1^{-1}\left(E_2^{-1}\left(E_3^{-1}(x) \oplus 3 \cdot S_2[129]\right) \oplus 3 \cdot S'_1[129]\right)$$

With $U_x = E_2^{-1}\left(E_3^{-1}(x) \oplus 3 \cdot S_2[129]\right)$, we have

$$\widetilde{P}_x[130] = E_1^{-1}(U_x) \oplus 3 \cdot S_1[129]$$
$$\widetilde{P}'_x[130] = E_1^{-1}(U_x) \oplus 3 \cdot S'_1[129]$$

Therefore, the pair $\widetilde{P}_x[130], \widetilde{P}'_x[130]$ can be seen as a plaintext/ciphertext pair for a cipher with 4 AES rounds, a middle key $S_1[129] \oplus S'_1[129]$, and 4 inverse AES rounds:

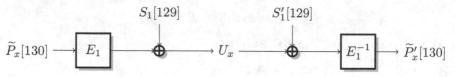

In addition, we can peel off the outer rounds since there is no whitening key in E_1.

5.2 Extracting the Key

We must now extract the key of a reduced cipher with 3 AES rounds, and 3 inverse AES rounds. First, we notice that the middle ShiftRow and MixColumn operations can be removed, if we transform the middle key. In a basic description, the operations in the middle are ShiftRow, MixColumn, AddKey, then XORing the constant $S_1[129] \oplus S'_1[129]$, AddKey, InverseMixColumn, and Inverse-ShiftRow. Instead we move the (unknown) constant addition before ShiftRow, using the linearity of ShiftRow and MixColumn, so that ShiftRow, MixColumn and AddKey cancel out with AddKey, InverseMixColumn and InverseShiftRow. We denote the addition of the modified constant as AddDeltaS, and its value as $\delta_S =$ InverseShiftRow(InverseMixColumn($S_1[129] \oplus S'_1[129]$)). The middle rounds are then reduced to byte-wise operations: AddRoundKey, SubByte, AddDeltaS, InverseSubByte, AddRoundKey. This can be seen as a key-dependent Sbox layer. These transformations are summarised on Fig.7.

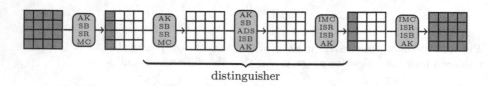

distinguisher

Fig. 7. Reduced cipher composed of 3 AES rounds, the addition of δ_S and 3 inverse AES rounds. The distinguisher covers the middle part of this cipher.

The first step of our attack is to guess a diagonal of the first round key, which allows to compute a column after the first round and before the last round. Next we focus on the middle rounds. The middle rounds have only one MixColumn operation, and one InverseMixColumn without byte shuffling in between. Therefore it can be seen as four parallel 32-bit functions, acting on each diagonal (similar to the Super-SBox technique [4]). Note that if the key guess is wrong, the resulting function can not be decomposed into 4 parallel functions. For each function, 1 input byte and 1 output byte are known. We describe below two different distinguishers for the middle rounds, that lead to key recovery attack with similar complexities. The first one is based on a rare event that can easily be detected, the second one relies on the detection of a statistical bias in the generic case.

Collision-based Distinguisher. For our first distinguisher, we focus on the constant δ_S. We only know that δ_S is non-zero on the full state. Considering that it is distributed uniformly on non-zero constants, it cancels one of the diagonals[3] with probability $(2^{96} - 1)/(2^{128} - 1) \approx 2^{-32}$. Then, an average of 2^{30} different choices of AD are necessary to reach a value of δ_S that cancels on one of the diagonals. Let us consider that it occurs on the first diagonal (w.l.o.g.), which contains bytes $0, 7, 10, 13$. Then, the value of these bytes collide before and after the AddDeltaS operation. Then, the values of the first column of the block (bytes $0, 1, 2, 3$) are not affected by the middle rounds. If we continue the decryption process towards both ends of the modified version of the AES, the collision passes through the InverseMixColumn operation. After undoing the ShiftRow, SubByte and textsfAddKey operation, we notice that the values of bytes $0, 5, 10, 15$ are equal at the beginning and at the end of the middle part of the cipher.

For each choice of AD, we then generate 3 (plaintext-ciphertext) pairs $(\widetilde{P}_x, \widetilde{P}'_x)$ for 3 values of x. Then, we proceed as follows.

In each of the 2^{30} sets, we guess separately the 32 bits on each of the 4 anti-diagonals[4] of the first round key. This enables to compute one full column of the state before and after the middle rounds, for each value of x. For each byte b_i in this column, we store a list L_i of the key values such that the input byte and the output byte of the middle rounds are equal for each x.

Then we consider the first diagonal before and after the middle rounds. It contains bytes $0, 5, 10$ and 15 of the block. Remember that the diagonals contain the inputs and outputs of 4 independent functions F_i. From the 4 lists of partial keys $L_j, j = 0, 5, 10, 15$, we can build all the keys such that the input of F_i collides with the output for each value of x. Using the known plaintexts and ciphertexts for the full cipher, we can try all these keys as candidates. Then, we repeat the whole process with the other three diagonals.

We now explain why this attack works.

Filtering Keys. Following the analysis above, the right key can be decomposed into 4 partial keys covering each diagonal of the block. If δ_S cancels on one of the columns, then the partial values of the right key will appear on the four corresponding lists L_i, and the full key will be among the combination of elements of the four lists. Therefore, the right key will be detected by our algorithm.

False Positives. For each wrong partial 32-bit key, it is stored in the corresponding list L_j if the input and output of F_i collide on byte j, for each of the 3 values of x. This occurs with probability 2^{-24}, if we consider the input and output

---

[3] defined as the images of columns by the ShiftRow operation.

[4] defined as the pre-images of columns by the ShiftRow operation.

byte computed for F_i as independent. Therefore, we have on average one false positive of each of the 4 diagonals of the key. Considering that the number of false positives are independent for each diagonal of the key, there are on average $(2^8)^4 = 2^{32}$ keys to try, for each of the 4 diagonals and each of the 2^{30} sets of values. The expected number of key candidates is marginally increased to $(2^8 + 1)^4 \approx 2^{32}$ when the difference δ_S cancels on the diagonal, as each set of partial keys at least contains the right key.

$\delta_S \neq 0$ *on column i.* As above, each wrong key guess is stored with probability 2^{-24}, which leads to $(2^8)^4$ false positives on average, that are discarded by exhaustive search.

Summary of the Attack. We focus on the key recovery attack on the mask-less version of Marble. In the decryption-misuse setting, it requires the decryption of 6×2^{30} ciphertexts composed of 130 blocks of plaintext and 1 block of associated data, which correspond to 2^{30} sets of 3 pairs. To build the lists of partial keys, one has to perform $6 \times 1/4$ of an AES round for each partial key guess, leading to a total of 3×2^{31} AES rounds, for each set and each diagonal. The average complexity of this step for the full attack is then 3×2^{63} AES rounds. The most time-consuming part of the attack is the exhaustive search among the remaining candidates, which requires 2^{64} AES encryptions on average (2^{32} per column and per set).

Linear Cryptanalysis. The method described in Sect. 5.1 leads to the knowledge of plaintext-ciphertext pairs for a cipher that consists of 3 AES rounds, a key addition and 3 inverse AES rounds. The adversary therefore targets a cipher with a reduced number of rounds, in a known plaintext setting. Using linear cryptanalysis therefore seems a natural idea. As shown above, one can guess 4 key bytes, which leads to the knowledge of 4 input and 4 output bytes of the inner 4 rounds of this cipher.

In [9], Keliher and Sui demonstrate that the maximum expected linear probability over 2 AES rounds is about $LP \approx 1.638 \times 2^{-28}$. In our case, we can concatenate a linear trail with its inverse. When averaging over the possible values of the key and of the intermediate difference δ_S, the maximum expected probability for a 4-round characteristic would be about $LP^2 \approx 1.342 \times 2^{-55}$. This number also gives an estimation of the amount of data required for the attack to work. Even by taking into account the possible bias due to the linear hull effect, the complexity of the linear attack is expected to be far higher than the one suggested by the experiments below.

A refinement of the linear attack consists in noticing that between the two middle rounds, each byte of the block is affected only by a key byte and a byte of δ_S, but not by other bytes of the block. Therefore, the two middle Sbox layers could be expressed as one layer of 8-bit key-dependent Sboxes, leading to trails with 6 active Sboxes instead of 10. Nevertheless, the best linear trail would then depend on the unknown value of δ_S, which would make it hard to exploit. Instead, we use the following statistical distinguisher.

Statistical Distinguisher. Intuitively, if we have many partial input/output pairs, we should detect some correlation between the inputs and output. Indeed, when the key guess is wrong, the function composing the distinguisher behaves as a 128-bit permutation instead of the parallel application of four 32-bit functions. Hence, the input and output bytes are less correlated. We focus on a property that does not require to know in advance which values are correlated, and works for any function based on (four 32-bit) parallel permutations.

For each known plaintext/ciphertext, we partially encrypt/decrypt one round on a specific diagonal and we denote one known input/output byte of the distinguisher by (α, β) respectively. It is possible to take into account the four known input/output pairs, but the distinguisher presented below works with only one position and is easier to explain. We use 2^{16} counters $c_{\alpha,\beta}$ to count how many times each pair (α, β) occurs with D available data. If the key guess is correct, there should be some correlation between α and β, which results in a higher value for some counters (and lower values for the other counters). In order to detect this effect, we compute the sample variance s^2 of the 2^{16} counters:

$$s^2 = 2^{-16} \sum_{\alpha,\beta} (c_{\alpha,\beta} - \overline{c})^2, \quad \text{where } \overline{c} = 2^{-16} \sum_{\alpha,\beta} c_{\alpha,\beta}.$$

We expect that s^2 is higher when the key guess is correct, because of the correlation between α and β. For a wrong key guess, the computation between α and β can not be decomposed into 4 parallel functions, and this correlation should vanish. The resulting attack is described by Algorithm 3.

Algorithm 3. Recover the key of a reduced AES (3 direct rounds and 3 inverse rounds)

Input: Plaintext/ciphertext pairs (P, C)
   for $0 \leq K < 2^{32}$ **do**   ▷ Partial key guess
      Initialize $c_{\alpha,\beta} = 0$
      for all (P, C) **do**
         Compute α, β
         $c_{\alpha,\beta} \leftarrow c_{\alpha,\beta} + 1$
      end for
      $\overline{c} \leftarrow 2^{-16} \sum_{\alpha,\beta} c_{\alpha,\beta}$
      $s^2[K] \leftarrow 2^{-16} \sum_{\alpha,\beta} (c_{\alpha,\beta} - \overline{c})^2$
   end for
   return $\arg\max_K s^2[K]$

In order to analyze this algorithm, we model the counters using random variables $C_{\alpha,\beta}$, and the sample variance as S^2 for a wrong key guess, and S_*^2 for the right key. Our goal is to show that when D is large enough, we have $\Pr[S^2 > S_*^2]$ negligible, i.e. the correct key is ranked first.

Wrong key Guess. We know that starting from α, if we revert the initial round with the wrong key, then compute three rounds forward with the correct keys, add δ_S, compute three round backwards with the correct keys, and finally one round forward with the wrong key, we reach a state with β. Therefore, α and β are partial inputs/outputs of a 128-bit permutation.

If we model this function by a random 128-bit permutation, the number of data matching a pair (α, β), in images and pre-images of this 128-bit function, follows an hypergeometric distribution. Indeed, for each input which first byte has value α, the output is drawn uniformly without replacement among all the possible outputs of the function. The success is determined by whether the first byte of the output equals β. The number of draws is 2^{120}, and there are 2^{120} success cases among 2^{128} possible values.

(α, β) occurs with D data, knowing that the probability of success is $p = 2^{120}/2^{128} = 2^{-8}$. Let us call this variable $X_{\alpha,\beta}$. Hence we have

$$
\begin{aligned}
\mathrm{E}[X_{\alpha,\beta}] &= (2^{120})^2/2^{128} = 2^{112} \\
\mathrm{Var}[X_{\alpha,\beta}] &= (2^{120})^2/2^{128} \times (1 - 2^{-8})^2/(1 - 2^{-128}) \approx 2^{112} - 2^{105}.
\end{aligned}
$$

Next we study $Y_{\alpha,u}$ the number of times each value α, u is reached with D samples, for each possible value u of the remaining 15 bytes of the input of F. The $Y_{\alpha,u}$ follow a multinomial distribution, with:

$$
\begin{aligned}
\mathrm{E}[Y_{\alpha,u}] &= 2^{-128}D, \\
\mathrm{Var}[Y_{\alpha,u}] &= 2^{-128}(1 - 2^{-128})D, \\
\mathrm{Var}[Y_{\alpha,u}, Y_{\alpha',u'}] &= -2^{-256}D.
\end{aligned}
$$

Let us denote by $S_{\alpha,\beta}$ the set of values u such that $F(\alpha, u) = (\beta, v)$ for some v. It contains exactly $X_{\alpha,\beta}$ elements. The counters $C_{\alpha,\beta}$ can then be expressed as

$$
C_{\alpha,\beta} = \sum_{u \in S_{\alpha,\beta}} Y_{\alpha,u}.
$$

The variables $Y_{\alpha,u}$ all follow the same distribution. From the law of total variance, we have:

$$
\mathrm{Var}[C_{\alpha,\beta}] = \mathrm{E}_{X_{\alpha,\beta}} \left(\mathrm{Var}\left[\sum_{u \in S_{\alpha,\beta}} Y_{\alpha,u} | X_{\alpha,\beta} \right] \right) + \mathrm{Var}_{X_{\alpha,\beta}} \left(\mathrm{E}\left[\sum_{u \in S_{\alpha,\beta}} Y_{\alpha,u} | X_{\alpha,\beta} \right] \right)
$$

After expanding the sums and reordering the terms to make variances and covariances of the $Y_{\alpha,u}$ appear, we get:

$$
\begin{aligned}
\mathrm{Var}[C_{\alpha,\beta}] &= \mathrm{E}\left[X_{\alpha,\beta} \mathrm{Var}[Y_{\alpha,u}] + (X_{\alpha,\beta}^2 - X_{\alpha,\beta}) \mathrm{Var}[Y_{\alpha,u}, Y_{\alpha,u'}] \right] + \mathrm{Var}\left[X_{\alpha,\beta} \mathrm{E}[Y_{\alpha,u}] \right] \\
&= \mathrm{E}(X_{\alpha,\beta}) \mathrm{Var}(Y_{\alpha,u}) + \mathrm{E}[X_{\alpha,\beta}^2 - X_{\alpha,\beta}] \mathrm{Var}(Y_{\alpha,u}, Y_{\alpha,u'}) + \mathrm{E}[Y_{\alpha,u}]^2 \mathrm{Var}[X_{\alpha,\beta}] \\
&= \mathrm{E}(X_{\alpha,\beta}) \mathrm{Var}(Y_{\alpha,u}) + \left(\mathrm{Var}[X_{\alpha,\beta}] + \mathrm{E}[X_{\alpha,\beta}]^2 - \mathrm{E}[X_{\alpha,\beta}] \right) \mathrm{Var}(Y_{\alpha,u}, Y_{\alpha,u'}) \\
&\quad + \mathrm{E}[Y_{\alpha,u}]^2 \mathrm{Var}[X_{\alpha,\beta}]
\end{aligned}
$$

We have numeric expressions for each term of this expression, therefore we can compute the following approximation:

$$\mathrm{Var}[C_{\alpha,\beta}] \approx 2^{-16}D + 2^{-144}D^2.$$

Correct Key. Let us now assume that the key guess is correct, *i.e.* the pairs (α, β) are valid partial input/output but of a 32-bit function this time. We can then re-apply the above analysis by adjusting the parameters to fit the 32-bit function. In this case, $X_{\alpha,\beta}$ denotes the number of partial values of the data matching the pair (α, β) in the right column. In that case, we have an hypergeometric distribution with 2^{24} draws without replacement from a set of 2^{32} values, among which 2^{24} define a success event.

Therefore, we have

$$\begin{aligned}
\mathrm{E}[X_{\alpha,\beta}] &= (2^{24})^2/2^{32} = 2^{16} \\
\mathrm{Var}[X_{\alpha,\beta}] &= (2^{24})^2/2^{32} \times (1-2^{-8})^2/(1-2^{-32}) \approx 2^{16} - 2^9.
\end{aligned}$$

Similarly, we can define variables $Y_{\alpha,u}$ as the number of times a given input of the 32-bit function F is reached among D samples, drawn uniformly. As in the previous case, the $Y_{\alpha,u}$ follow a multinomial distribution, with:

$$\begin{aligned}
\mathrm{E}[Y_{\alpha,u}] &= 2^{-32}D, \\
\mathrm{Var}[Y_{\alpha,u}] &= 2^{-32}(1-2^{-32})D, \\
\mathrm{Var}[Y_{\alpha,u}, Y_{\alpha',u'}] &= -2^{-64}D.
\end{aligned}$$

The same formula can be used to compute the variance of the counters $C_{\alpha,\beta}$. We get approximately:

$$\mathrm{Var}[C_{\alpha,\beta}] \approx 2^{-16}D + 2^{-48}D^2.$$

Distinguisher. We obtain an efficient distinguisher with $D = 2^{32}$: for a wrong key guess, the variance of the counter is about 2^{16}, but it is about 2^{17} for the right key. In order to evaluate the probability that the correct key is ranked first, we must evaluate how far the sample variance s^2 is from the true variance $\mathrm{Var}[C_{\alpha,\beta}]$.

For a wrong key guess, if we use a single counter and repeat the experiment with 2^{16} independent sets of D plaintexts, each counter $C_{\alpha,\beta}$ can be approximated by a binomial distribution with parameters D and $p = 2^{-16}$. If we approximate them as a normal distribution with parameters $\mu = 2^{-16}D$ and $\sigma = \sqrt{2^{-16}(1-2^{16})D}$, we know that the distribution of the sample variance S^2 for a wrong key guess follows a χ^2 distribution [10, Proposition 2.11]:

$$S^2 \sim \frac{\sigma^2}{(n-1)}\chi^2_{n-1} \sim 2^{-32}D\chi^2_{2^{16}-1}$$

In particular, we have $\Pr[S^2 > 2^{16} + 2^{12}] < 2^{-90}$, therefore we don't expect that the sample variance of a wrong key is above $2^{16} + 2^{12}$. In practice, we use a single set of D plaintexts, and we evaluate the sample variance of the 2^{16} counters; our experiments show that the distribution is close to a χ^2 distribution (the maximum value of s^2 with 2^{16} samples was $2^{16} + 1420$).

For the right key, we don't have an analytic expression of the distribution of the sample variance, but we can perform experiments. Our experiments show that with very high probability $S_*^2 > 2^{16} + 2^{12}$, as seen in Fig. 8. Of our 2^{16} experiments, the minimum value of s_*^2 was $102795 \approx 2^{16} + 2^{15}$. Using $D = 2^{32}$, we have a large margin and we expect the attack to work with significantly less data, but recovering L will be the bottleneck of the attack.

While this attack does not use any property of the parallel 32-bit function, we expect that it can be improved in the specific case of AES rounds. In particular, we notice a small peak around 3×2^{16} in Fig. 8, which is due to zero bytes in δ_s, and it should be possible to take advantage of this.

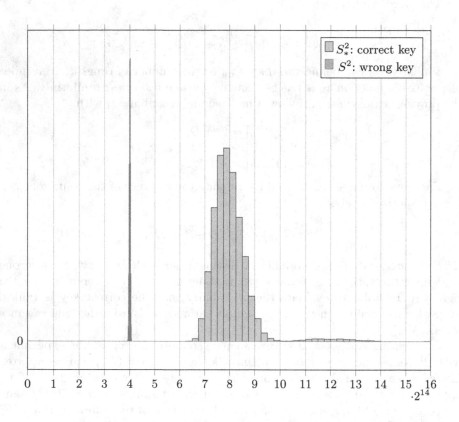

Fig. 8. Experimental results: distribution of the sample variance S^2 and S_*^2 with $D = 2^{32}$ (2^{16} experiments with random keys).

6   Conclusion

Our results show that collision attacks can have a strong impact on the security of authenticated encryption schemes. It seems that extracting the whitening key using collisions is possible in many OCB-based designs, and this can sometimes be extended into a full key-recovery attack.

On AEZ, we show how to recover the whitening key, and we point out that the key derivation of AEZ v2 and v3 has the unfortunate property that the master key can easily be recovered from the whitening key. This allows a complete break after the birthday bound. Even with a limit on the amount of data processed with a single key, this still gives an attack with a higher success probability than generic attacks. While this does not violate the security proof of AEZ, we believe it would be better to avoid this property.

Our results on Marble show that it does not provide the security features claimed by the author, *i.e.* beyond birthday bound security and decryption-misuse resistance. We note that Marble still offers a stronger security than many CAESAR candidates and classical modes of operations when using nonces (or unique AD). Once usage requirements are relaxed, our results also show that the security of Marble is similar to the security of other misuse resistant CAESAR candidates (*e.g.* APE-128, POET, AEZ, Minalpher) but it collapses badly after the birthday bound under a decryption-misuse setting, even leading to a full key recovery.

It seems that adding one extra operation on the state between the processing of the associated data and of the message would avoid our attacks, but a thorough analysis would be necessary to ensure that the resulting construction is secure. As our attack heavily relies on the fact that S_1 and S_2 keep the same values on two different plaintexts, one could xor a constant block (for example, 16 bytes encoding the byte positions in the block, $(0, 1, \ldots, 15))$ on S_1 and S_2 after processing the associated data.

In addition, our key-recovery attack of Marble suggests that 4 AES rounds in the E functions are insufficient if the adversary can find a shortcut to target two E functions instead of three. In particular, this suggest that a deeper investigation of the security of AEZ with a modified key schedule would be interesting.

Up to our knowledge, the statistical distinguisher presented to recover the encryption key of a reduced-round AES, has never been used before. Although it permits to attack few rounds, it seems that it is more efficient than a classical linear attack. We believe that it is sufficient enough for this kind of distinguisher to benefit from further research.

References

1. Andreeva, E., Bogdanov, A., Luykx, A., Mennink, B., Tischhauser, E., Yasuda, K.: Parallelizable and authenticated online ciphers. In: Sako, K., Sarkar, P. (eds.) ASIACRYPT 2013, Part I. LNCS, vol. 8269, pp. 424–443. Springer, Heidelberg (2013)
2. Daemen, J., Rijmen, V.: The Design of Rijndael: AES - The Advanced Encryption Standard. Information Security and Cryptography, Springer (2002)
3. Ferguson, N.: Collision attacks on OCB. Comments to NIST (2002). http://csrc.nist.gov/groups/ST/toolkit/BCM/documents/comments/General_Comments/papers/Ferguson.pdf

4. Gilbert, H., Peyrin, T.: Super-Sbox cryptanalysis: improved attacks for AES-like permutations. In: Hong, S., Iwata, T. (eds.) FSE 2010. LNCS, vol. 6147, pp. 365–383. Springer, Heidelberg (2010)

5. Guo, J.: Marble Specification Version 1.0. Submission to the CAESAR competition, March 2014

6. Guo, J.: Marble Specification version 1.2 (January 2015), posted on the `crypto-competition` mailing list

7. Hoang, V.T., Krovetz, T., Rogaway, P.: AEZ v3: Authenticated-Encryption by Enciphering. In: DIAC 2014: Directions in Authenticated Ciphers, Santa Barbara, August 2014

8. Hoang, V.T., Krovetz, T., Rogaway, P.: Robust authenticated-encryption AEZ and the problem that it solves. In: Oswald, E., Fischlin, M. (eds.) EUROCRYPT 2015. LNCS, vol. 9056, pp. 15–44. Springer, Heidelberg (2015)

9. Keliher, L., Sui, J.: Exact maximum expected differential and linear probability for two-round advanced encryption standard. IET Inf. Secur. $1(2)$, 53–57 (2007)

10. Knight, K.: Mathematical Statistics. Chapman & Hall, New York (1999)

11. Lee, C.-H., Kim, J.-S., Sung, J., Hong, S.H., Lee, S.-J.: Forgery and key recovery attacks on PMAC and mitchell's TMAC variant. In: Batten, L.M., Safavi-Naini, R. (eds.) ACISP 2006. LNCS, vol. 4058, pp. 421–431. Springer, Heidelberg (2006)

12. Liskov, M., Rivest, R.L., Wagner, D.: Tweakable block ciphers. J. Cryptology $24(3)$, 588–613 (2011)

13. Lu, J.: On the Security of the COPA and Marble Authenticated Encryption Algorithms against (Almost) Universal Forgery Attack, February 2015

14. NIST: Advanced Encryption Standard (AES) (November 2001), federal Information Processing Standards Publication FIPS 197

15. Rogaway, P.: Efficient instantiations of tweakable blockciphers and refinements to modes OCB and PMAC. In: Lee, P.J. (ed.) ASIACRYPT 2004. LNCS, vol. 3329, pp. 16–31. Springer, Heidelberg (2004)

16. Rogaway, P., Bellare, M., Black, J., Krovetz, T.: OCB: a block-cipher mode of operation for efficient authenticated encryption. In: Reiter, M.K., Samarati, P. (eds.) CCS 2001, Proceedings of the 8th ACM Conference on Computer and Communications Security, Philadelphia, Pennsylvania, USA, November 6–8, 2001. pp. 196–205. ACM (2001)

17. Sasaki, Y.: Universal Forgery on POET and Ciphertext Distinguisher on POET and Marble (July 2014), posted on the `crypto-competition` mailing list

Symmetric Analysis

Optimized Interpolation Attacks on LowMC

Itai Dinur[1]([⊠]), Yunwen Liu[2], Willi Meier[3], and Qingju Wang[2,4]

[1] Département d'Informatique, École Normale Supérieure, Paris, France
dinur@di.ens.fr
[2] Department of Electrical Engineering, ESAT/COSIC, KU Leuven and iMinds,
Leuven, Belgium
[3] FHNW, Windisch, Switzerland
[4] Department of Computer Science and Engineering, Shanghai Jiao Tong University,
Shanghai, China

Abstract. LowMC is a collection of block cipher families introduced at
Eurocrypt 2015 by Albrecht et al. Its design is optimized for instantiations of multi-party computation, fully homomorphic encryption, and
zero-knowledge proofs. A unique feature of LowMC is that its internal
affine layers are chosen at random, and thus each block cipher family
contains a huge number of instances. The Eurocrypt paper proposed
two specific block cipher families of LowMC, having 80-bit and 128-bit
keys.

In this paper, we mount interpolation attacks (algebraic attacks introduced by Jakobsen and Knudsen) on LowMC, and show that a practically
significant fraction of 2^{-38} of its 80-bit key instances could be broken 2^{23}
times faster than exhaustive search. Moreover, essentially all instances
that are claimed to provide 128-bit security could be broken about 1000
times faster. In order to obtain these results we optimize the interpolation attack using several new techniques. In particular, we present an
algorithm that combines two main variants of the interpolation attack,
and results in an attack which is more efficient than each one.

Keywords: Block cipher · LowMC · High-order differential
cryptanalysis · Interpolation attack

1 Introduction

LowMC is a collection of block cipher families designed by Albrecht et al. and
presented at Eurocrypt 2015. The cipher is specifically optimized for practical
instantiations of multi-party computation, fully homomorphic encryption, and
zero-knowledge proofs. In such applications, non-linear operations result in a
heavy computational penalty compared to linear ones. The designers of LowMC

Q.Wang—The fourth author is in part supported by the National Natural Science
Foundation of China (no. 61472250), Major State Basic Research Development Program (973 Plan, no. 2013CB338004).

T. Iwata and J.H. Cheon (Eds.): ASIACRYPT 2015, Part II, LNCS 9453, pp. 535–560, 2015.
DOI: 10.1007/978-3-662-48800-3_22

took an extreme approach, combining very dense affine layers with simple non-linear layers that have algebraic degree of 2.

Perhaps the most distinctive feature of LowMC is that its affine layers are chosen at random, and thus each block cipher family contains a huge number of instances. As this may enable a malicious party to instantiate LowMC with a hidden backdoor, its designers propose to use the Grain stream cipher [3] as a source of pseudo-random bits in order to restrict the freedom available in the LowMC instantiation. The designers also mention that it is possible to use any sufficiently random source to generate the affine layers, and this source does not necessarily need to be cryptographically secure.

The Eurocrypt paper proposed two specific block cipher families of LowMC, having 80-bit and 128-bit keys. The internal number of rounds in each family was set in order to guarantee a security level that corresponds to its key size. For this purpose, the resistance of LowMC was evaluated against a variety of well-known cryptanalytic attacks. One of the main considerations in setting the internal number of rounds was to provide resistance against algebraic attacks (such as high-order differential cryptanalysis [7]). Indeed, LowMC is potentially susceptible to algebraic attacks due to the low algebraic degree of its internal round, but the designers argue that LowMC has sufficiently many rounds to resist such attacks.

In this paper, we evaluate the resistance of LowMC against algebraic attacks and refute the designers' claims regarding its security level. Our results are given in Table 1, and show that a fraction of 2^{-38} of the LowMC 80-bit key instances could be broken in about 2^{57} time, using 2^{39} chosen plaintexts. The probability of 2^{-38} is practically significant, namely, a malicious party can easily find weak instances of LowMC by running its source of pseudo-random bits with sufficiently many seeds, and checking whether the resultant instance is weak (which can be done efficiently using basic linear algebra).

For LowMC with 128-bit keys, we describe an attack that breaks a fraction of 2^{-122} of its instances in time 2^{86} using 2^{70} chosen plaintexts. We note that this specific attack does not violate the formal security claims of the LowMC designers, as they do not consider attacks that apply to less than 2^{-100} of the instances as valid. Nevertheless, the designers of LowMC allow to instantiate it using a pseudo-random source that is not cryptographically secure. Our result shows that this is risky, as using an over-simplified source for pseudo-randomness may give a malicious party additional control over the LowMC instantiation, and allow finding weak instances much faster than exhaustively searching for them in 2^{122} time.

Finally, we describe an attack that can break essentially all LowMC instances with 128-bit keys. Although the attack is significantly slower than the weak-instance attack, it is still about 1000 times faster than exhaustive search, and uses 2^{73} chosen plaintexts.

All of our results were obtained using the interpolation attack, which is an algebraic attack introduced by Jakobsen and Knudsen in 1997 [4]. In an inter-polation attack, the attacker considers some intermediate encryption value b as

Table 1. Attacks on LowMC

Instance Family	Number of Rounds	Section	Rounds Attacked	Fraction of Instances	Data[†]	Time[††]	Memory[†††]
LowMC-80	11	6.1	9	1	2^{35}	2^{38}	2^{35}
		6.2	10	1	2^{39}	2^{57}	2^{39}
		6.3	all (11)	2^{-38}	2^{39}	2^{57}	2^{39}
LowMC-128	12	7.1	11	1	2^{70}	2^{86}	2^{70}
		7.1	all (12)	2^{-122}	2^{70}	2^{86}	2^{70}
		7.2	all (12)	1	2^{73}	2^{118}	2^{80}

[†] Given in chosen plaintexts.
[††] Given in LowMC encryptions.
[†††] Given in 256-bit words.

a polynomial in the ciphertext bits. The aim of the attacker is to interpolate the algebraic normal form (ANF) of b by recovering its unknown coefficients, and this typically allows to recover the secret key using ad-hoc techniques.

In order to recover the unknown coefficients, the attacker allocates a variable for each one of them. Assuming that b has a low-degree representation in terms of the plaintext bits, the attacker collects linear equations on the variables, typically by using high-order differentials in a chosen plaintext attack. After obtaining sufficiently many equations, the unknown variables are recovered by solving the resultant linear equation system. The efficiency of the attack depends on the algebraic degree of b in terms of the plaintext, but also on the number of allocated variables which is determined by the number of unknown coefficients in the ANF representation of b in terms of the ciphertext.

Although our results were obtained using the well-known interpolation attack, its straightforward application does not seem to threaten the security of LowMC. Therefore, we had to develop new techniques such as using carefully chosen plaintext structures which allow to efficiently derive the linear system of equations. However, our main new contribution is described next by considering two variants of the interpolation attack.

In the original variant of the interpolation attack over $GF(2)$ (which we refer to as variant 1), the attacker views the ANF of some intermediate encryption bit b as an initially unknown polynomial $F_K(C)$ in the ciphertext bits $C = c_1, \ldots, c_n$, where $K = x_1, \ldots, x_\kappa$ is the unknown (fixed) secret key. In a dual approach to the interpolation attack, which we refer to as variant 2 (used, for example, in [8]), the attacker interpolates the full polynomial $F(K, C)$ by considering each monomial in the key bits $x_1, \ldots, x_\kappa$ with a non-zero coefficient as a separate (linearized) variable. For example, consider the polynomial

$$F(c_1, c_2, x_1, x_2, x_3) = c_1 c_2 x_1 + c_1 c_2 x_2 + c_1 x_1 + c_1 x_2 + c_2 x_1 + x_1 x_2 + x_3 + 1.$$

We can write

$$F_{(x_1, x_2, x_3)}(c_1, c_2) = \alpha_1 c_1 c_2 + \alpha_2 c_1 + \alpha_3 c_2 + \alpha_4,$$

and thus in the first variant we have 4 variables: $\alpha_1, \alpha_2, \alpha_3, \alpha_4$. In this variant, the actual representation of the variables in terms of the key is not considered. In the dual variant, we write

$$F(c_1, c_2, x_1, x_2, x_3) = x_1 x_2(1) + x_1(c_1 c_2 + c_1 + c_2) + x_2(c_1 c_2 + c_1) + x_3(1) + 1,$$

and we have 4 variables: $x_1 x_2, x_1, x_2, x_3$.

The advantage of variant 2 over the first variant is that it directly recovers the secret key, and furthermore, in some cases it may result in a smaller number of variables in the equation system. At the same time, in order to derive the actual equation system the attacker has to evaluate the polynomial F for each ciphertext. This process is less efficient in variant 2, since each evaluation of $F(K, C)$ is expensive (it requires evaluating all the complex ciphertext expressions that are multiplied with the variables), whereas in variant 1 each evaluation of $F_K(C)$ is relatively simple (it requires evaluating simple monomials in the ciphertext). Therefore, the choice of which variant to use in order to optimize the attack depends on the underlying cryptosystem.

Our main idea is to combine the two dual variants of interpolations attacks: we first derive the equation system efficiently using the original variant of [4]. Then, we transform a carefully chosen variable subset to variables which are linearized monomials in the key bits, as in variant 2. This results in a mixed variable set that is smaller than the variable sets of each variant. Consequently, we obtain an attack which is more efficient than each one of the two variants.

In our example above, we can express $\alpha_1 = x_1 + x_2$, $\alpha_2 = x_1 + x_2$ and $\alpha_3 = x_1$, resulting in only 3 variables: x_1, x_2, α_4. Obviously, our toy example merely demonstrates the idea at a very high level, and the actual choice of which variables to transform as well as the analysis of the resultant algorithm are more involved.

The paper is organized as follows. In Sect. 2 we give some preliminaries, while in Sect. 3 we give a brief description of LowMC. Our basic attack on 9-round LowMC with an 80-bit key is described in Sect. 4, while our generic framework for optimized interpolation attacks is described in Sect. 5. In Sects. 6 and 7 we apply our optimized attack to LowMC with 80 and 128-bit keys, respectively. Finally, we conclude the paper in Sect. 8.

2 Preliminaries

In this section, we describe preliminaries that are used in the rest of the paper.

2.1 Boolean Algebra

For a finite set S, denote by $|S|$ its size. Given a vector $u = (u_1, \ldots, u_n) \in GF(2^n)$, let $wt(u)$ denote its Hamming weight.

Any function F from $GF(2^n)$ to $GF(2)$ can be described as a multivariate polynomial, whose algebraic normal form (ANF) is unique and given as

$$F(x_1, \ldots, x_n) = \sum_{u=(u_1,\ldots,u_n) \in GF(2^n)} \alpha_u M_u, \text{ where } \alpha_u \in \{0,1\} \text{ is the coefficient of}$$

the monomial $M_u = \prod_{i=1}^{n} x_i^{u_i}$, and the sum is over $GF(2)$. The algebraic degree of

the function F is defined as $deg(F) \triangleq max\{wt(u)|\alpha_u \neq 0\}$. Therefore, a function

F with a degree bounded by $d \leq n$ can be described using $\sum_{i=0}^{d} \binom{n}{i}$ coefficients.

To simplify our notations, we define $\binom{n}{\leq d} \triangleq \sum_{i=0}^{d} \binom{n}{i}$.

The ANF coefficient α_u of F can be interpolated by summing (over $GF(2)$) over $2^{wt(u)}$ evaluations of F: define the set of inputs S to contain all the $2^{wt(u)}$ n-bit vectors whose bits set to 1 is a subset of the bits set to 1 in $u_1, \ldots, u_n$. More formally, let $S = \{x = (x_1, \ldots, x_n)|\bar{u} \wedge x = 0\}$ (where $\bar{u}$ is bitwise NOT applied to u, and $\wedge$ is bitwise AND), then $\alpha_u = \sum_{(x_1,\ldots,x_n) \in S} F(x_1, \ldots, x_n)$. Note that this implies that a function F with a degree bounded by $d \leq n$ can be fully interpolated given its evaluations on the set of $\binom{n}{\leq d}$ inputs whose Hamming weight is at most d, namely $\{x = (x_1, \ldots, x_n)|wt(x) \leq d\}$.

Given the truth table of an arbitrary function F (as a bit vector of 2^n entries), the ANF of F can be represented as a bit vector of 2^n entries, corresponding to its 2^n coefficients α_u. This ANF representation can be efficiently computed using the *Moebius transform*, which is an FFT-like algorithm. The Moebius transform performs n iterations on its input vector (the truth table of F), where in each iteration, half of the array entries are XORed into the other half. In total, its complexity is about $n \cdot 2^n$ bit operations. For more details on the Moebius transform, refer to [5].

2.2 High-Order Differential Cryptanalysis and Interpolation Attacks

In this section, we give a brief summary of high-order differential cryptanalysis and interpolation attacks.

High-Order Differential Cryptanalysis. High-order differential cryptanalysis was introduced in [7] as an algebraic attack that is particularly efficient against ciphers of low algebraic degree. The basic variant of high-order differential cryptanalysis over $GF(2)$ considers some target bit b (which can be either a ciphertext or an intermediate encryption value) and analyzes its ANF representation in terms of the plaintext P, denoted by $F_K(P)$ (where K is the unknown secret key). Given that $deg(F_K(P)) \leq dg$ independently of K for dg (relatively) small, then the attacker chooses an arbitrary linear subspace S of dimension $dg + 1$, and evaluates the cipher (in a chosen plaintext attack) over its 2^{dg+1} inputs. Since every differentiation reduces the algebraic degree of the target bit by 1 and $deg(F_K(P)) \leq dg$, the value of the high-order differential over S for the target bit b (namely, the sum of evaluations of b over $GF(2)$) is equal to zero (refer to [7] for details). High-order differential properties may be used in key recovery attacks, depending on the specification of the cipher (refer to [6]). However, such key recovery methods are not part of the framework described in this section.

Interpolation Attacks. The interpolation attack was introduced in 1997 by Jakobsen and Knudsen as an algebraic attack on block ciphers [4]. The attack is closely related to high-order differential cryptanalysis[1] and (similarly to high-order differential cryptanalysis) is particularly efficient against block ciphers whose round function is of low algebraic degree. The interpolation attack has several variants, and can be applied over a general finite field, exploiting known or chosen plaintexts. Here, we give a high-level description of the chosen plaintext interpolation attack over $GF(2)$, as this is the variant we apply to LowMC.

The attack considers some intermediate encryption target bit b of the block cipher, whose ANF representation can be expressed from the decryption side in terms of the ciphertext and key as $F(C, K)$. The key K is viewed as an unknown constant, and thus we can write $F_K(C) = F_K(c_1, \ldots, c_n) = \sum_{u=(u_1,\ldots,u_n) \in GF(2^n)} \alpha_u M_u$, where $\alpha_u \in \{0, 1\}$ is the coefficient of the monomial $M_u = \prod_{i=1}^{n} c_i^{u_i}$. Therefore, the coefficients α_u of $F_K(C)$ generally depend on the secret key and are unknown in advance. The goal of the interpolation attack is to recover (interpolate) the unknown coefficients of $F_K(C)$, and then use various ad-hoc techniques (which are not part of the framework described in this section) in order to recover the actual secret key.

In order to deduce the unknown coefficients of $F_K(C)$, they are considered as variables (i.e., linearized), and recovered by solving a linear equation system. For the purpose of constructing the equation system, the attacker assumes that the algebraic degree dg of the bit b in terms of the bits of the plaintext is relatively small, which allows to use high-order differential cryptanalysis (as described above). More specifically, a high-order differential property is devised by encrypting a subspace S of plaintexts of dimension $dg + 1$, and performing high-order differentiation with respect to this subspace, whose outcome is zero on the bit b.

When expressed in terms of the ciphertexts $C_1, \ldots, C_{2^{dg+1}}$ (obtained by encrypting the plaintexts of S), this gives the equation $\sum_{t=1}^{2^{d+1}} F_K(C_t) = 0$. For each ciphertext C_t, $F_K(C_t)$ is merely a linear expression in the variables α_u (the coefficient of α_u in this expression is easily deduced by evaluating M_u on C_t), and thus the subspace S gives rise to one linear equation in the variables α_u. In order to solve for the unknown variables α_u, the attacker considers several such subspaces, each giving one equation. In total, the number of equations (and

---

[1] In fact, some of its variants directly exploit high-order differential properties, as we describe next.

subspaces considered) needs to be roughly equal to the number of the unknown α_u variables, assuming the equations are sufficiently "random".

From the high-level description above, it is easy to conclude that the data and time complexities of the attack depend on the value of the degree dg and the number of unknown variables α_u. Therefore, in order to mount efficient interpolation attacks, the attacker tries to minimize these parameters, as we demonstrate in our attacks on LowMC.

2.3 Model of Computation

Since an exhaustive key search attack (which evaluates the LowMC encryption function) and our attacks use different bitwise operations, comparing these attacks cannot be done simply by counting the number of encryption function evaluations. Instead, we compare the complexity of straight-line implementations of the algorithms, counting the number of bit operations (such as XOR, AND, OR) on pairs of bits. This computation model ignores operations such as moving a bit from one position to another (which only requires renaming variables in straight-line programs). As calculated in Sect. 3, the straight-line implementation of one encryption function evaluation of LowMC requires about 2^{19} bit operations. Consequently, a straight-line implementation of exhaustive search for 80-bit and 128-bit keys requires about 2^{99} and 2^{147} bit operations, respectively, and these are quantities of reference for our attacks.

3   Description of LowMC

LowMC is a collection of SP-network instances, proposed at Eurocrypt 2015 [1] by Albrecht et al. The specification defined two specific instance families which are analyzed in this paper, both having a block size of $n = 256$ bits, and are characterized by their key size κ, which is either 80 or 128 bits. In this paper, we refer to these instance families as LowMC-80 and LowMC-128. The encryption function of LowMC applies a sequence of rounds to the plaintext, where each round contains a (bitwise) round-key addition layer, an Sbox layer, and an affine layer (over $GF(2)$). LowMC was designed with distinct features (as detailed in the pseudocode below): it has a linear key schedule and its affine layers are selected at random, where each selection defines a separate instance of the family. The Sbox layer of LowMC is composed of 3-bit Sboxes with degree 2 over $GF(2)$ (the actual specification of the Sboxes is irrelevant for our analysis and is omitted from this paper). Furthermore, the Sbox layers are only partial, namely, in each Sbox layer, only $3m < n$ bits go through an Sbox (where m is a parameter), while the rest of the $n - 3m$ bits remain unchanged.

Each family instance of LowMC is also defined with a data limit lim, which determines the maximal (recommended) data complexity before changing the key. In other words, the cipher is guaranteed to offer security according to its key size as long as the adversary cannot obtain more than 2^{lim} plaintext-ciphertext pairs. The parameters of the two instance families are given in Table 2.

Table 2. LowMC instance families

Instance Family	key size κ	Block Size n	Sboxes m	Data lim	Rounds r
LowMC-80	80	256	49	64	11
LowMC-128	128	256	63	128	12

The pseudocode of the encryption function (taken from [1]) is given below.

```
ciphertext = encrypt (plaintext,key)
  //initial whitening
  state = plaintext + MultiplyWithGF2Matrix(KMatrix(0),key)
  for (i = 1 to r)
    //m computations of 3-bit Sbox, n-3m bits remain the same
    state = Sboxlayer (state)
    //affine layer
    state = MultiplyWithGF2Matrix(LMatrix(i),state)
    state = state + Constants(i)
    //generate round key and add to the state
    state = state + MultiplyWithGF2Matrix(KMatrix(i),state)
  end
  ciphertext = state
```

The matrices $LMatrix(i)$ are chosen at random from all invertible binary $n \times n$ matrices, while the matrices $KMatrix(i)$ are chosen independently and uniformly at random from all binary $n \times \kappa$ matrices of rank $min(n, \kappa)$. The constants $Constants(i)$ are chosen independently and uniformly at random from all binary vectors of length n.

In this paper, we denote the 256-bit state at the input to the i'th key addition layer by X_{i-1} (e.g., the plaintext is denoted X_0), the input to the i'th Sbox layer by Y_{i-1} and the input to the i'th affine layer by Z_{i-1}. We refer to the $3m$ bits of the state that go through Sboxes in the Sbox layer as the S-part, while the remaining $n - 3m$ bits are referred to as the I-part. Given a state W, denote by $W|SP$ and $W|IP$ the S-part and I-parts of the state, respectively (e.g., $Y_5|IP$ is the I-part of the input state to the 6'th Sbox layer).

It is common practice in cryptanalysis of block ciphers to exchange the order of the final two affine operations over $GF(2)$ (namely, the keyless affine transformation and key addition). This allows the attacker to "peel off" the last affine transformation at a negligible cost by working with an equivalent last-round key (obtained by an affine transformation on the original last-round key). For the sake of simplicity, we assume in the following that we have already "peeled off" the last affine transformation of the cipher. Therefore, the final states of the last round r are denoted by X_{r-1}, Y_{r-1}, Z_{r-1} and Y_r, which denotes the ciphertext (after "peeling off" the final affine transformation).

Each affine layer of LowMC involves multiplication of the 256 state with a 256×256 matrix. This multiplication requires roughly 2^{16} bit operations, and therefore a single encryption of LowMC (that contains more than 8 rounds) requires more than $2^{16} \cdot 8 = 2^{19}$ bit operations (as already noted in Sect. 2.3).

4   A Basic 9-Round Attack on LowMC-80

In this section we describe our basic interpolation attack on 9-round LowMC, which is given first without optimizations for the sake of clarity. We begin by considering the elements that are required for the attack.

4.1   The High-Order Differential Property

We construct the high-order differential property used in the interpolation attack. A similar property was described by the LowMC designers [1], but we reiterate it here for the sake of completeness.

The algebraic degree of a single round of LowMC-80 over $GF(2)$ is 2, and therefore the algebraic degree of any bit at the input to the 6'th Sbox layer of LowMC-80, Y_5, in the input bits, X_0, is at most 32. Moreover, as the bits of the I-part of LowMC do not go through Sboxes in the first round, then the degree at the input to the 7'th Sbox layer, Y_6, in the bits of the I-part, $X_0|IP$, (given that the input bits of the S-part, $X_0|SP$, are constant) is at most 32. Furthermore, since the bits of the I-part of the 7'th Sbox layer do not go through an Sbox, the degree of any bit of $Z_6|IP$ in the input bits of the I-part, $X_0|IP$, is at most 32 (given that $X_0|SP$ is constant).

The last property implies that the value of a 33-order differential over any 33-dimensional subspace selected from $X_0|IP$, (keeping $X_0|SP$ constant) is zero for any bit of $Z_6|IP$. Moreover, as we selected a subspace whose bits do not go through an Sbox in the first round, the value of a 32-order differential for any bit of $Z_6|IP$ over any 32-dimensional subspace from $X_0|IP$, is a constant (independent of the key). This observation implies that we can select several 32-dimensional subspaces, and compute in a preprocessing phase the constants obtained by summing (over $GF(2)$) over a target bit of $Z_6|IP$ (for an arbitrary fixed value of the key). Each such constant (derived from a 32-dimensional subspace) gives one bit of information that we will exploit as the constant value of an equation in the interpolation attack.

4.2   Bounding the Number of Variables

In the interpolation attack on 9-round LowMC-80, we select a target bit from $Z_6|IP$ and denote its ANF representation in the 256-bit ciphertext (obtained after inverting the final affine transformation) and 80-bit key by $F(C, K)$. We consider K as an unknown constant, and write $F_K(C) = F_K(c_1, \ldots, c_{256}) = \sum_{u=(u_1,\ldots,u_{256})\in GF(2^{256})} \alpha_u M_u$, where $\alpha_u \in \{0, 1\}$ is the coefficient of the monomial $M_u = \prod_{i=1}^{256} c_i^{u_i}$. As the complexity of the attack depends on the number of variables α_u, it is important to estimate their number with good accuracy. An initial estimation can be made by observing that the algebraic degree of the (inverse) round of LowMC-80 is 2,[2] and thus $deg(F_K(C)) \leq 4$. This implies that $\alpha_u = 0$

---

[2] The algebraic degree of any invertible 3-bit Sbox is (at most) 2.

in case $wt(u) > 4$, and therefore the number of unknown variables is upper bounded by $\binom{256}{<4} \approx 2^{27}$.

The initial upper bound on the number of variables can be significantly improved by considering the specific round function of LowMC-80. For this purpose, it will be convenient to use additional notation to describe the variables α_u according to the degree of M_u, by defining the set of variables U_i for a positive integer i as $U_i = \{\alpha_u$ that is not identically zero as a function of the key$|wt(u) = i \wedge u \in GF(2^{256})\}$. We have already seen that U_i is empty for $i > 4$ (as these variables are identically zero independently of the key), and we now derive tighter bounds on $|U_i|$ for $i \leq 4$. Thus, we analyze the symbolic representation of the state variables in the decryption direction, starting from the ciphertext Y_9, up to Z_6, as polynomials in the ciphertext bits $c_1, \ldots, c_{256}$.

The ciphertext Y_9 contains 256 bits of $c_1, \ldots, c_{256}$, while in order to compute Z_8 we merely add (unknown) constants to these bits (recall that we "peeled off" the last affine layer). Then, the inverse Sbox layer is applied to Z_8 to obtain the state Y_8. Each 3-bit Sbox may contribute (up to) 3 quadratic monomials to Y_8, and 6 monomials in total, e.g., an Sbox corresponding to ciphertext bits c_1, c_2, c_3 may contribute the monomials $c_1, c_2, c_3, c_1c_2, c_1c_3, c_2c_3$. Note that these monomials may appear in the ANF of different bits of Y_8 with different unknown coefficients (e.g., c_1x_1 and c_1x_2 may appear in the ANF of two different bits of Y_8). However, in interpolation attacks, we consider the ANF of the target bit, in which the coefficient α_u of every monomial M_u in the ciphertext is linearized and considered as a single variable. Therefore, the important quantity is the number of possibilities to create the monomials M_u (for this reason, the monomial c_1 is counted only once even if it appears in the ANF of different bits of Y_8 with different unknown coefficients).

Since there are 49 Sboxes, the total number of monomials M_u in the ANF of the state bits of Y_8 is bounded by $|U_2| \leq 3 \cdot 49 = 147$, $|U_1| \leq 256$ (which is the trivial bound) and $|U_i| = 0$ for $i \geq 3$. As the affine and key addition mappings do not influence the number of monomials M_u, this bound applies also to X_8 and Z_7.

Next, the inverse Sbox layer is applied to Z_7 to obtain the state Y_7, for which we already know that $|U_i| = 0$ for $i > 4$. Since the Sbox layer is of degree 2, a trivial upper bound on the number of variables α_u in Y_7 is obtained by multiplying the $147 + 256 = 403$ monomials in unordered pairs, giving $|\bigcup_{i=1}^{4} U_i| \leq \binom{403}{2} + 403 < 2^{16.5}$. Since the key addition and affine layers do not influence the number of monomials, the upper bound of $2^{16.5}$ also applies to X_7 and Z_6, and it is much smaller than our initial bound of about 2^{27}.

We denote the set of variables $\bigcup_{i=1}^{4} U_i$ by U, and note that the explicit set $\{u|\alpha_u \in U\}$ (which gives the relevant monomials M_u) can be easily derived during preprocessing (which involves a more explicit computation of the monomial set $\{M_u|\alpha_u \in U\}$, whose size is bounded above).

4.3   Obtaining the Data

After deducing that the number of variables in the system of equations is $|U| \approx 2^{16.5}$, we conclude that we need to differentiate over about $2^{16.5}$ 32-dimensional subspaces in order to obtain sufficiently many equations to solve the system. A trivial way to do this is to select about $2^{16.5}$ arbitrary linearly independent 32-dimensional subspaces from the $256 - 3 \cdot 49 = 109$ bits of $X_0|IP$. This results in an attack with data complexity of $2^{32+16.5} = 2^{48.5}$, and is rather wasteful. A more efficient approach (which was previously used in various papers such as [2]), is to select a large 37-dimensional subspace S from $X_0|IP$, containing $\binom{37}{32} > 2^{18}$ linearly independent 32-dimensional subspaces, which should suffice for the attack (assuming that the constructed system of equations is sufficiently random). The subspaces are indexed according to $37 - 32 = 5$ constant indexes that are set to zero in S.

4.4   The Basic Interpolation Attack

We now describe a basic interpolation attack on 9-round LowMC-80. We note that this attack is incomplete, as it only computes the $|U|$ variables α_u using $e \approx |U|$ equations, without recovering the actual secret key. The details of this final step will be given in the optimized attack in Sect. 5.2. For the sake of convenience, we describe the attack in two phases: the preprocessing phase (which is independent of the data and secret key) and online phase. However, we take into account both phases in the total complexity evaluation.

Assume we selected a target bit b from $Z_6|IP$, a subspace S of dimension 37 from $X_0|IP$, and $e \approx |U|$ 32-dimensional subspaces $S_1, \ldots, S_e$ in S. The detailed attack is described below.

Preprocessing:

1. Compute an e-bit array of free coefficients for $e \approx |U|$ equations, denoted by a_0: evaluate b on the subset of inputs of S (with the key set to zero), and obtain a bit array of size 2^{37}. Finally, calculate the free coefficients by summing on b for the e 32-dimensional subspaces $S_1, \ldots, S_e$ in S, and store the result in a_0.
2. Calculate the $|U|$ vectors $\{u|\alpha_u \in U\}$: This can be done by first calculating the 403 monomials M_u past the first Sbox layer, and multiplying them in pairs (as described in Sect. 4.2).

Online:

1. Ask for the encryptions of the 2^{37} plaintexts in S and store the ciphertexts in a table.

---

2. Allocate a $2^{37} \times |U|$ matrix A, where row $A[t]$ is a bit array that represents the evaluation $F_K(C_t)$ (namely, $\sum_{\{u|\alpha_u \in U\}} \alpha_u M_u(C_t)$).

3. For each ciphertext C_t, calculate $A[t]$ by evaluating $F_K(C_t)$:
   (a) For each $\{u|\alpha_u \in U\}$, evaluate the monomial $M_u(C_t)$ (the coefficient of α_u) and set the corresponding bit entry in $A[t]$ according to the result.

4. Allocate an $e \times |U|$ matrix E over $GF(2)$, representing the equation system on U.

5. For each 32-dimensional subspace S_j in S, namely $S_1, \ldots, S_e$ (that match the subspaces considered in preprocessing Step 1):
   (a) Populate the row (equation) $E[j]$ by summing over the 2^{32} rows of A corresponding to S_j.

6. Solve the equation system $E\boldsymbol{x} = \boldsymbol{a_0}$, where $\boldsymbol{x}$ represents the vector of variables of U and $\boldsymbol{a_0}$ is the vector of free coefficients calculated in preprocessing Step 1.

The data complexity of the attack is 2^{37} chosen plaintexts. The total time complexity of the attack is about 2^{65} bit operations, dominated by online Step 5 (for each of the e subspaces, we sum over 2^{32} bit vectors of size $|U|$, requiring about $e \cdot 2^{32} \cdot |U| \approx 2^{65}$ bit operations). The memory complexity of the attack is about $2^{37} \cdot |U| \approx 2^{53.5}$ bits, dominated by the storage of the matrix A in online Step 2.

We note that in the complexity evaluation of the attack we ignore indexing issues that arise (for example) in Step 3.a (that maps between a variable $\alpha_u \in U$ and its corresponding column index in $A[t]$), and in Step 5 (that maps between a subspace S_j in S and the corresponding 5 constant indexes of S). The reason that we can ignore these mappings in the complexity evaluation is that they are independent of the secret key and data, and therefore, they can be precomputed and integrated into the straight-line implementation of the program.

5   The Optimized Interpolation Attack

In this section, we introduce three optimizations of the basic 9-round attack above. The first optimization reorders the steps of the algorithm in order to reduce the memory complexity, while the second optimization further exploits the structure of chosen plaintexts to reduce the time complexity of the attack. Finally the third optimization is based on a novel technique in interpolation attacks, and allows to (further) reduce the data and time complexities. We first describe informally how to apply the optimizations to the basic 9-round attack on LowMC-80 above, and then devise a more formal and generic framework that can be applied to other LowMC variants.

The first two optimizations focus on online steps 2–5, which compute the equation system E from the 2^{37} ciphertexts. First, we reduce the memory complexity by noticing that we do not need to allocate the matrix A. Instead, we work column-wise and focus on a single column $A[*][\ell]$ at a time, corresponding to some $\{u | \alpha_u \in U\}$. We evaluate $M_u(C_t)$ for all ciphertexts (which gives an array of 2^{37} bits, $\boldsymbol{a_\ell}$) and then populate the corresponding column $E[*][\ell]$ by summing over the 32-dimensional subspaces $S_1, \ldots, S_e$ on $\boldsymbol{a_\ell}$.

Next, we reduce the time complexity by optimizing the summation process: given a bit array $\boldsymbol{a_\ell}$ of 2^{37} entries, the goal is to sum over many 32-dimensional subspaces (indexed according to 5 bits which are set to zero). This can be done efficiently using the Moebius transform (refer to Sect. 2.1). For this purpose, we can view $\boldsymbol{a_\ell}$ as evaluating a 37-variable polynomial over $GF(2)$, and the summation over a 32-dimensional subspace of $\boldsymbol{a_\ell}$ is equal to the coefficient of its corresponding 32-degree monomial. All these coefficients are computed by the Moebius transform in about $37 \cdot 2^{37}$ bit operations. We stress that the reason that we can use the Moebius transform in this case is purely combinatorial and is due to the way that we selected the structure of subspaces for the interpolation attack. Indeed, there does not seem to be any obvious algebraic interpretation to $\boldsymbol{a_\ell}$ when viewed as a polynomial.

Finally, we optimize the data complexity (and further reduce the time complexity): In order to achieve this, examine the polynomial $F(K, C)$ (as a function of both the key and ciphertext) for the target bit b selected in $Z_6 | IP$. Due to the linear key schedule of LowMC, this polynomial is of degree 4, similarly to $F_K(C)$ (in which the key is treated as a constant). We consider a variable $\alpha_u \in U$ and analyze its ANF in terms of the 80 key bit variables. Since α_u is multiplied with M_u in $F(K, C)$, then $deg(\alpha_u) + deg(M_u) \leq 4$, implying that if $deg(M_u) \geq 2$, then $deg(\alpha_u) \leq 2$. This simple observation is borrowed from cube attacks [2] and can be used to significantly reduce the number of variables U, as described next.

Consider all the variables in $U_2 \bigcup U_3 \bigcup U_4$, and recall that their number was upper-bounded in Sect. 4.2 by roughly $2^{16.5}$. However, since all of these variables are polynomials of degree (at most) 2 in the 80 key bits, they reside in a linear subspace of monomials of dimension $\binom{80}{2} + 80 = 3240$. This implies that we can significantly reduce the total number of variables from $\approx 2^{16.5}$ to $3240 + 256 = 3496 < 2^{12}$ (including the 256 variables of U_1) by considering linear relations between the variables $U_2 \bigcup U_3 \bigcup U_4$. An immediate consequence of the reduction of variables is that we need less equations to solve the equation system, and therefore, we require less subspaces (or data) to obtain these equations. More specifically, a subspace of dimension 35 contains $\binom{35}{32} = 6545 > 2^{12}$ subspaces of dimension 32, which should suffice for the attack.

Assuming that we interpolate the variables of $U_2 \bigcup U_3 \bigcup U_4$ in terms of the key and recover their values, then the key itself should be very easy to deduce, as the variables of U_3 are merely key bits.

We note that while the idea above exploits the linear key schedule of LowMC, the technique is general and can be applied to block ciphers with arbitrary key schedules. In this case, it would consider each round key as independent. This increases the number of variables in the (linearized) key, but not necessarily by a significant factor. For example, if LowMC-80 had a non-linear key schedule, the optimization above would interpolate $U_2 \bigcup U_3 \bigcup U_4$ in terms of $\binom{80}{2} + 80 = 3240$ monomials in the key of round 9, and only 80 additional linear monomials and $3 \cdot 49 = 294$ quadratic monomials in the key of round 8 that are created by the inverse Sbox layer of round 8 (we can assume that the key of round 8 is added right after the 8'th Sbox layer, as the key addition and affine layer are interchangeable).

5.1   Transformation of Variables

In this section, we begin to describe our generic framework for interpolation attacks on LowMC by formalizing the last optimization described above.

Given an instance of LowMC with a 256-bit block, a key size of κ, and m Sboxes per layer, we assume that we want to interpolate a target bit b through the final r_1 rounds of the cipher. We first describe in a more generic way how to calculate the initial set of variables U, and bound its size. As in the 9-round attack, the number of monomials in the 256 ciphertext bits at Y_{r-1} (after inverting the final Sbox layer) is bounded by $256 + 3m$. The target bit b is a polynomial of degree 2^{r_1-1} in the state Y_{r-1}, and thus it contains at most $\binom{256+3m}{\leq 2^{r_1-1}}$ monomials. Therefore, the set of monomials with (apriori) unknown coefficients can be computed by multiplying the $256 + 3m$ monomials in unordered tuples (with no repetition) of size up to 2^{r_1-1}. Thus,

$$|U| \leq \binom{256 + 3m}{\leq 2^{r_1-1}},$$

and this set can be computed with $|U|$ multiplications of tuples. Note again that this bound is generally better than the trivial bound of $|U| \leq \binom{256}{\leq 2^{r_1}}$, which is obtained due to the fact that b is a polynomial of degree 2^{r_1} in the 256 ciphertext bits.

We consider the target bit b as a polynomial in both the ciphertext and the key, namely, $F(K, C) = F(x_1, \ldots, x_\kappa, c_1, \ldots, c_{256}) = \displaystyle\sum_{u=(u_1,\ldots,u_n) \in GF(2^n)} \alpha_u M_u,$

where $M_u = \displaystyle\prod_{i=1}^{n} c_i^{u_i}$ and $\alpha_u(x_1, \ldots, x_\kappa)$ is a polynomial from $GF(2^\kappa)$ to $GF(2)$. We partition the variables of $|U|$ into subsets according to the degree of their monomials in the ciphertext, which is bounded by $deg(F_K(C)) = 2^{r_1}$. Denote $d = 2^{r_1}$ and write $U = \displaystyle\bigcup_{i=1}^{d} U_i$, where $U_i = \{\alpha_u \in U | deg(M_u) = i\}$. Due to the linear key schedule of LowMC, we have $deg(F(K, C)) = deg(F_K(C)) = d$, and therefore $deg(\alpha_u) + deg(M_u) \leq d$. This allows us to transform the variable set U into a smaller variable set, considering internal linear relations due to the fact that $deg(\alpha_u) \leq d - deg(M_u)$. We stress again that the variable transformation technique can be applied to block ciphers with arbitrary key schedules by considering each round key as independent.

We choose an integral *splitting index* $1 \leq sp \leq d+1$, and write $U = U' \bigcup U''$, where $U' = \bigcup_{i=1}^{sp-1} U_i$ and $U'' = \bigcup_{i=sp}^{d} U_i$. The observation above implies that the algebraic degree of the variables in U'' (in terms of the key) is bounded by $d - sp$, namely, $deg(\alpha_u) \leq d - sp$, for each $\alpha_u \in U''$. Therefore, we can interpolate each variable of U'' in terms of the key, and express it as $\alpha_u = \sum_{\{v=(v_1,\ldots,v_\kappa)|wt(v) \leq d-sp\}} \beta_u M_v$, where $\beta_v \in \{0,1\}$ is the coefficient of the monomial $M_v = \prod_{i=1}^{\kappa} x_i^{v_i}$. Note that the coefficients β_v are independent of the key and can be computed during preprocessing. This interpolation transforms the set of variables U'' into the set of variables V, which are low degree monomials in the key bits $V = \{M_v = \prod_{i=1}^{\kappa} x_i^{v_i} | v = (v_1,\ldots,v_\kappa) \wedge wt(v) \leq d - sp\}$. Similarly to the partition of U, we partition the variables of V into subsets according to the degree of their monomials in the key, namely $V_i = \{M_v \in V | deg(M_v) = i\}$. In addition, we define $V_{\leq i} = \bigcup_{j=1}^{i} V_i$. Note that $\alpha_u \in U_i$ is a linear combination of variables in $V_{\leq (d-i)}$.

Recall that our initial set of variables is expressed as $U = U' \bigcup U''$, where $U' = \bigcup_{i=1}^{sp-1} U_i$ and $U'' = \bigcup_{i=sp}^{d} U_i$. This set of variables is transformed via interpolation into a new set of variables $W = U' \bigcup V$.

We compute bounds on sizes of the variables sets as follows:

$$|U'| \leq \binom{256}{\leq sp-1}, \quad |V| \leq \binom{\kappa}{\leq d-sp},$$

$$|W| = |U'| + |V| \leq \binom{256}{\leq sp-1} + \binom{\kappa}{\leq d-sp}.$$

The Variable Transformation Algorithm. We now describe the algorithm which interpolates a variable $\alpha_u \in U_i$ in terms of the variable set $V_{\leq (d-i)}$. For the sake of efficiency, the algorithm is performed in two phases, where in the first phase, we evaluate the polynomial α_u in terms of the key for all relevant keys of low Hamming weight and store the results. Note that each evaluation of α_u requires summing on 2^i evaluations of the target bit b. In the second phase, we use the evaluations to interpolate α_u in terms of $V_{\leq (d-i)}$.

---

1. Allocate a bit array a_1 of size $|V_{\leq (d-i)}|$ for the evaluations of α_u.
2. Evaluate α_u for each key with Hamming weight at most $d - i$. Namely, for each key in the set $\{K | wt(K) \leq d - i\}$:

(a) Evaluate $F(K, C)$ (the target bit) on the subset of 2^i inputs (with the fixed key K) $\{K, C | \bar{u} \wedge C = 0\}$, sum the result over $GF(2)$, and store it in a_1.

3. Allocate a bit array a_2 of size $|V_{\leq(d-i)}|$ for interpolation of α_u in terms of $V_{\leq(d-i)}$.

4. For each $M_v \in V_{\leq(d-i)}$ (with index ℓ), the coefficient β_v of M_v in α_u is calculated as follows:

(a) Sum the $2^{wt(v)}$ values of a_1 calculated for the subset of keys $\{K | \bar{v} \wedge K = 0\}$, and store the result in $a_2[\ell]$.

The total number of evaluations of b in Step 2 is $2^i \cdot |V_{\leq(d-i)}|$, each requiring $r_1 \cdot 2^{16}$ bit operations. Therefore, the total complexity of this step is $r_1 \cdot 2^{16+i} \cdot |V_{\leq(d-i)}|$. Step 4 requires less than $|V_{\leq(d-i)}| \cdot 2^{d-i}$ bit operations. In total, the interpolation of $\alpha_u \in U_i$ requires $|V_{\leq(d-i)}| \cdot (r_1 \cdot 2^{16+i} + 2^{d-i})$ bit operations.

Since $U'' = \bigcup_{i=sp}^{d} U_i$, we can write the complexity of interpolating all the variables as $\sum_{i=sp}^{d} |U_i| \cdot |V_{\leq(d-i)}| \cdot (r_1 \cdot 2^{16+i} + 2^{d-i})$. A simple way to bound this complexity is

$$|U''| \cdot |V| \cdot (r_1 \cdot 2^{16+d} + 2^{d-sp}) \approx |U''| \cdot |V| \cdot r_1 \cdot 2^{16+d}.$$

In some cases, we can obtain a refined bound by writing the complexity as

$$|U_{sp}| \cdot |V_{\leq(d-sp)}| \cdot (r_1 \cdot 2^{16+sp} + 2^{d-sp}) + \sum_{i=sp+1}^{d} |U_i| \cdot |V_{\leq(d-i)}| \cdot (r_1 \cdot 2^{16+i} + 2^{d-i}) \leq$$

$$|U_{sp}| \cdot |V_{\leq(d-sp)}| \cdot (r_1 \cdot 2^{16+sp} + 2^{d-sp}) + |U''| \cdot |V_{\leq(d-sp-1)}| \cdot (r_1 \cdot 2^{16+d} + 2^{d-sp+1}) \approx$$

$$|U_{sp}| \cdot |V| \cdot (r_1 \cdot 2^{16+sp} + 2^{d-sp}) + |U''| \cdot |V_{\leq(d-sp-1)}| \cdot r_1 \cdot 2^{16+d}.$$

Note that the bound is potentially better than the trivial one of $|U''| \cdot |V| \cdot r_1 \cdot 2^{16+d}$ as $|U_{sp}| \leq \binom{256}{sp}$, which may be smaller than $|U''|$. Moreover $|V_{\leq(d-sp-1)}| \leq \binom{\kappa}{\leq d-sp-1}$, which is smaller than $|V|$.

Transformation of Equations. After computing the transformation of variables from U'' to V, we need to apply the actual transformation to every equation over U that we calculated. Namely, we are interested in transforming an equation over the variable set $U = U' \bigcup U''$, into an equation over variable set $W = U' \bigcup V$. Obviously, the coefficients of the variables of U' remain the same, and we need to apply the transformation for every variable $\alpha_u \in U''$.

The complexity of transforming a single variable $\alpha_u \in U_i$ in a single equation is simply equal to its number of coefficients over V, namely $|V_{\leq(d-i)}|$. Therefore, the complexity of transforming all the variables $\alpha_u \in U''$ in an equation is $\sum_{i=sp}^{d} |U_i| \cdot |V_{\leq(d-i)}|$. A simple upper bound on this complexity is

$$|U''| \cdot |V|.$$

Similarly to the variable transformation algorithm, a refined upper bound can be calculated as

$$|U_{sp}| \cdot |V| + |U''| \cdot |V_{\leq(d-sp-1)}|.$$

In total, if we transform e equations, the complexity calculations above are multiplied by e.

Finally, we observe that the splitting index determines the complexity of the variable and equation transformation algorithms. Furthermore, the splitting index also determines $|W|$, which in turn determines the number of equations e. In general, we will choose sp in order to minimize $|W|$, which in turn minimizes the data and time complexity of the attack.

5.2   Details of the Optimized Interpolation Attack

Given an instance of LowMC with a 256-bit block, a key size of κ, and m Sboxes per layer, we interpolate a target bit b through the final r_1 rounds of the cipher. Let U, U', U'', V and W be as defined above, and let $e \approx |W|$ denote the number of equations. Assume S is a sufficiently large subspace of plaintexts, such that it contains e smaller subspaces $S_1, \ldots, S_e$ whose high-order differential on b is a constant value (independent of the key).

The preprocessing phase of the optimized attack in described below.

Preprocessing:

1. Compute an e-bit array of free coefficients for $e \approx |U'|$ equations, denoted by a_0: evaluate b on the subset of inputs (plaintexts) of S (with the key set to zero), and obtain a bit array of size $|S|$. Then, calculate the free coefficients by applying the Moebius transform to the bit array, and copy the values of sums over $S_1, \ldots, S_e$ to a_0.
2. Calculate the $|U|$ vectors $\{u|\alpha_u \in U\}$: This is done by first calculating the $256 + 3m$ monomials past the first Sbox layer, and multiplying them in unordered tuples (with no repetition) of size up to 2^{r_1-1} (as described in Sect. 5.1).

Step 1 involves $|S|$ evaluations of the encryption scheme and one application of the Moebius transform on a vector of size S. Altogether, it requires $|S| \cdot 2^{19} + \log(|S|) \cdot |S| \approx |S| \cdot 2^{19}$ bit operations (as $\log(|S|) \ll 2^{19}$). Step 2 requires

$|U|$ monomial multiplications, each monomial can be represented with a 256-bit array, and therefore this step requires $2^8 \cdot |U|$ bit operations.

A summary of the complexity analysis of the preprocessing phase is as follows.

Step 1: $2^{19} \cdot |S|$
Step 2: $2^8 \cdot |U|$

In terms of memory, Step 1 requires $|S|$ bits, while Step 2 requires $2^8 \cdot |U|$ bits.

Online:

1. Ask for the encryptions of the plaintexts in S and store the ciphertexts in a table.
2. Allocate a bit vector of size $|S|$ for the storage of the vectors a_ℓ (the ℓ'th column of the matrix A in the basic attack).
3. Allocate an $e \times |W|$ matrix E over $GF(2)$, representing the (reduced) equation system on W. The matrix is vertically decomposed into two smaller matrices: E_1 of size $e \times |U'|$ and E_2 of size $e \times |V|$.
4. For each $\{M_u | \alpha_u \in U\}$ with an index ℓ:
   (a) For each ciphertext C_t, calculate $a_\ell[t]$ by evaluating $M_u(C_t)$.
   (b) Use the Moebius transform to sum over all subspaces of a_ℓ.
   (c) If $\alpha_u \in U'$, populate column ℓ of E_1: For each subspace S_j in S, namely $S_1, \ldots, S_e$, obtain its corresponding sum from a_ℓ and copy it to $E_1[j][\ell]$.
   (d) Otherwise, $\alpha_u \in U''$:
       i. Given that $\alpha_u \in U_i$, interpolate the coefficients of $V_{\leq (d-i)}$ in α_u as described in Sect. 5.1.
       ii. For each subspace S_j in S, obtain its corresponding boolean sum from a_ℓ (the coefficient of α_u over U). If the sum is 1, then add (over $GF(2)$) the interpolated coefficients into their indexes in $E_2[j]$ (as described in Sect. 5.1).
5. Solve the equation system $Ex = a_0$, where x represents the vector of variables of $W = U' \bigcup V$ and a_0 is the vector of free coefficients calculated in preprocessing Step 1.
6. Deduce the κ-bit secret key, which is simply given by the monomials V_1 (namely, the monomials of degree 1 in V).

The complexity of Step 1 is $|S|$ encryptions, or $|S| \cdot 2^{19}$ bit operations. In Step 4, we iterate over $|U|$ monomials, where for each one we first evaluate $M_u(C_t)$ for each ciphertext in Step 4.a. Each such evaluation can be performed with d bit operations (as $deg(M_u) \leq d$), and thus monomial evaluations require about $d \cdot |S| \cdot |U|$ bit operations. Next, we apply the Moebius transform in Step 4.b, requiring about $\log(|S|) \cdot |S|$ bit operation, and therefore the complexity of all the transforms is about $\log(|S|) \cdot |S| \cdot |U|$. The complexity of interpolating all the variables in Step 4.d.i, is bounded in Sect. 5.1 by $|U''| \cdot |V| \cdot r_1 \cdot 2^{16+d}$. The complexity of Step 4.d.ii (over all $\alpha_u \in U''$) is bounded in Sect. 5.1 by $e \cdot |U''| \cdot |V| \approx |W| \cdot |U''| \cdot |V|$.

The complexity of Step 5 is $|W|^3$ bit operations using Gaussian elimination. A summary of the complexity analysis of the online phase is as follows. Since we generally do not have a good bound for $|U''|$, we simply replace it with $|U|$ (as $|U''| \leq |U|$), and further assume that $e \approx |W|$.

Step 1: $|S| \cdot 2^{19}$
Step 2: $|S|$
Step 3: $|W| \cdot |W|$
Step 4.a: $d \cdot |S| \cdot |U|$
Step 4.b: $\log(|S|) \cdot |S| \cdot |U|$
Step 4.c: $|U'| \cdot |W|$
Step 4.d.i: $|U| \cdot |V| \cdot r_1 \cdot 2^{16+d}$
Step 4.d.ii: $|W| \cdot |U| \cdot |V|$
Step 5: $|W|^3$
Step 6: negligible

Alternatively, we can use the refined complexity bounds for steps 4.d.i and 4.d.ii, as calculated in Sect. 5.1.

Step 4.d.i: $|U_{sp}| \cdot |V| \cdot (r_1 \cdot 2^{16+sp} + 2^{d-sp}) + |U| \cdot |V_{\leq(d-sp-1)}| \cdot r_1 \cdot 2^{16+d}$
Step 4.d.ii: $|W| \cdot (|U_{sp}| \cdot |V| + |U| \cdot |V_{\leq(d-sp-1)}|)$

The total data complexity of the algorithm is $|S|$ chosen plaintexts. The total time complexity is dominated by steps 4 and 5, as calculated above. The memory complexity is potentially dominated by a few steps: the storage of variables in preprocessing that requires $2^8 \cdot |U|$ bits, the storage of ciphertexts in Step 1 that requires $2^8 \cdot |S|$ bits, and the storage of E in Step 3 that requires $|W| \cdot |W|$ bits.

6 Optimized Interpolation Attacks on LowMC-80

In this section we apply the optimized interpolation attack on LowMC-80, for which $\kappa = 80$ and $m = 49$.

6.1 A 9-Round Attack

As in the basic attack described in Sect. 4.4, we select the target bit b in $Z_6|IP$, using subspaces of dimension 32 to obtain the equations. We interpolate through $r_1 = 2$ rounds, implying that $d = 2^{r_1} = 4$. Therefore $|U| = \binom{256+3m}{\leq 2^{r_1}-1} = \binom{403}{\leq 2} \approx 2^{16.5}$.

As described at the beginning of Sect. 5, we use $sp = 2$. We compute the size of the relevant variable sets $|U'| \leq \binom{256}{\leq sp-1} = \binom{256}{\leq 1} \approx 2^8$, $|V| \leq \binom{\kappa}{\leq d-sp} = \binom{80}{\leq 2} < 2^{12}$, $|W| = |U'| + |V| < 2^{12}$.

We choose a subspace S of dimension 35 from $X_0|IP$, containing $\binom{35}{32} > 2^{12} > |W|$ 32-dimensional subspaces, which should suffice for the attack.

In terms of time complexity, the analysis of the critical steps of the attack is as follows:

Step 4.a: $d \cdot |S| \cdot |U| \approx 4 \cdot 2^{35} \cdot 2^{16.5} = 2^{53.5}$
Step 4.b: $\log(|S|) \cdot |S| \cdot |U| \approx 35 \cdot 2^{35} \cdot 2^{16.5} = 2^{56.5}$
Step 4.c: $|U'| \cdot |W| \approx 2^8 \cdot 2^{12} = 2^{20}$
Step 4.d.i: $|U| \cdot |V| \cdot r_1 \cdot 2^{16+d} \approx 2^{16.5} \cdot 2^{12} \cdot 2 \cdot 2^{20} = 2^{49.5}$
Step 4.d.ii: $|W| \cdot |U| \cdot |V| \approx 2^{12} \cdot 2^{16.5} \cdot 2^{12} = 2^{40.5}$
Step 5: $|W|^3 \approx 2^{12 \cdot 3} = 2^{36}$

In total, the time complexity of the optimized 9-round attack is about 2^{57} bit operations (or $2^{57-19} = 2^{38}$ encryptions), mostly dominated by Step 4.b. The data complexity is 2^{35} chosen plaintexts. The memory complexity is dominated by the storage of ciphertexts in Step 1, and is about $|S| \cdot 2^8 = 2^{43}$ bits.

We note that while the improvement of the optimized attack compared to the basic one is rather moderate for the 9-round attack, the effect of our optimizations is more pronounced in the attacks described next, as the reduction in the number of variables becomes more significant (a comparison for the attack on full LowMC-128 is at the end of Sect. 7.2).

6.2 A 10-Round Attack

Similarly to the 9-round attack, in order to attack 10 rounds of LowMC-80, we select the target bit b in $Z_6|IP$, using subspaces of dimension 32 to obtain the equations. We interpolate through $r_1 = 3$ rounds, implying that $d = 2^{r_1} = 8$. Therefore $|U| = \binom{256+3m}{\leq 2^{r_1}-1} = \binom{403}{\leq 4} < 2^{30.5}$.

In this attack we use $sp = 4$, and compute the size of the relevant variable sets $|U'| \leq \binom{256}{\leq sp-1} = \binom{256}{\leq 3} \approx 2^{21.5}$, $|V| \leq \binom{\kappa}{\leq d-sp} = \binom{80}{\leq 4} < 2^{21}$, $|W| = |U'| + |V| < 2^{22.5}$. We use the refined analysis for steps 4.d.i and 4.d.ii, and thus we also calculate $|U_{sp}| = |U_4| = \binom{256}{4} < 2^{27.5}$ and $|V_{\leq(d-sp-1)}| = \binom{80}{\leq 3} < 2^{16.5}$.

We choose a subspace S of dimension 39 from $X_0|IP$, containing $\binom{39}{32} > 2^{23} > |W|$ 32-dimensional subspaces.

In terms of time complexity, the analysis of the critical steps of the attack is as follows (using the refined analysis for steps 4.d.i and 4.d.ii):

Step 4.a: $d \cdot |S| \cdot |U| \approx 8 \cdot 2^{39} \cdot 2^{30.5} = 2^{72.5}$
Step 4.b: $\log(|S|) \cdot |S| \cdot |U| \approx 39 \cdot 2^{39} \cdot 2^{30.5} \approx 2^{75}$
Step 4.c: $|U'| \cdot |W| \approx 2^{21.5} \cdot 2^{22.5} = 2^{44}$
Step 4.d.i: $|U_{sp}| \cdot |V| \cdot (r_1 \cdot 2^{16+sp} + 2^{d-sp}) + |U| \cdot |V_{\leq(d-sp-1)}| \cdot r_1 \cdot 2^{16+d} \approx 2^{27.5} \cdot 2^{21} \cdot (3 \cdot 2^{20} + 2^4) + 2^{30.5} \cdot 2^{16.5} \cdot 3 \cdot 2^{24} \approx 2^{70} + 2^{72.5} \approx 2^{73}$
Step 4.d.ii: $|W| \cdot (|U_{sp}| \cdot |V| + |U| \cdot |V_{\leq(d-sp-1)}|) \approx 2^{22.5} \cdot (2^{27.5} \cdot 2^{21} + 2^{30.5} \cdot 2^{16.5}) \approx 2^{22.5} \cdot (2^{48.5} + 2^{47}) \approx 2^{71.5}$
Step 5: $|W|^3 \approx 2^{22.5 \cdot 3} = 2^{67.5}$

In total, the time complexity of the optimized 10-round attack is about 2^{76} bit operations (or 2^{57} encryptions), mostly dominated by Step 4.b. The data complexity is 2^{39} chosen plaintexts. The memory complexity is dominated by the storage of ciphertexts in Step 1, and is about $2^8 \cdot |S| = 2^{47}$ bits (note that the storage of E requires $2^{22.5 \cdot 2} = 2^{45}$ bits).

6.3 An Attack on Full LowMC-80 for Weak Instances

The 9 and 10-round attacks described above can be extended by an additional round with negligible cost for a subset of weak instances containing a fraction of about 2^{-38} of all instances. In particular, this implies that about 2^{-38} of the instances of full 11-round LowMC-80 can be attacked significantly faster than exhaustive search.

Consider the 10-round attack: as shown above, we can construct an efficient high-order differential property for any choice of target bit of $Z_6|IP$, and also for any linear combination of the bits of $Z_6|IP$. When considering interpolation from the decryption side on a full 11-round instance, we can efficiently interpolate the polynomial $F_K(C)$ for any bit of $Z_7|IP$, or any linear combination of the bits of $Z_7|IP$. Assume that there exists a linear dependency between the 109 bits of $Z_6|IP$ and the 109 bits of $Z_7|IP$. In this case, the linear combination in terms of $Z_6|IP$ does not go through an Sbox in round 8. Therefore, it is possible to extend the high-order differential property on this linear combination by another round with essentially no extra cost, and choose the target bit for interpolation to be the corresponding linear combination on the bits of $Z_7|IP$. The existence of this linear dependency is determined by the affine layer of round 7 (the transformation between Z_6 and X_7), and assuming that random invertible matrices behave roughly the same (with respect to the event considered) as random matrices, the probability of this event is about $2^{109+109-256} = 2^{-38}$ (over the choice of the 7'th affine layer).

We note that there exists an additional subset of weak instances of about the same size since the described attacks can also be mounted using chosen ciphertexts (where interpolation is performed on the decrypted plaintexts). In this case, the weakness of a given instance is determined by the choice of the third affine layer.

7 Optimized Interpolation Attacks on LowMC-128

In this section we apply the optimized interpolation attack on LowMC-128, for which $\kappa = 128$ and $m = 63$.

7.1 An 11-Round Attack and Weak Instances of LowMC-128

We describe our attack on 11-round LowMC-128 and then extend it to full LowMC-128 for weak instances. We select the target bit b in $Z_7|IP$, and interpolate through $r_1 = 3$ rounds, implying that $d = 2^{r_1} = 8$. Therefore $|U| = \binom{256+3m}{<2^{r_1}-1} = \binom{445}{\leq 4} < 2^{31}$.

In this attack we use $sp = 4$, and compute the size of the relevant variable sets $|U'| \leq \binom{256}{\leq sp-1} = \binom{256}{\leq 3} \approx 2^{21.5}$, $|V| \leq \binom{\kappa}{\leq d-sp} = \binom{128}{\leq 4} \approx 2^{23.5}$, $|W| = |U'| + |V| \approx 2^{24}$.

For the high-order differential property, we use subspaces of dimension $2^6 = 64$ whose bits are not multiplied together in the first round. The outcome of such a high-order differential is a constant (independent of the key) for $1 + 6 = 7$ rounds, and this property can be extended beyond the 8'th Sbox layer when selecting the target bit from $Z_7|IP$.

Since $|W| \approx 2^{24}$, we require roughly the same number of 64-dimensional subspaces to construct the equation system and mount the attack. Therefore, we take a larger subspace of dimension 70, containing $\binom{70}{64} > 2^{24} \approx |W|$ 64-dimensional subspaces. As $X_0|IP$ contains only 67 bits, we choose the subspace from these 67 bits and additional 3 bits in $X_0|SP$, contained in 1 active Sbox. Since the active Sbox is non-linear, we guess the 3 linear key expressions that are added to its input, which allow us to construct the required $\approx 2^{24}$ 64-dimensional subspaces from a 70-dimensional subspace after the first Sbox layer.

The guess of the 3 key bits can be avoided by selecting the $70 - 64 = 6$ constant bits of the 64-dimensional subspaces from the 67 bits of $X_0|IP$ in the 70-dimensional subspace. This restriction keeps the selected Sbox fully active in all subspaces, and thus the linear subspace after the first Sbox layer (at Z_0) is independent of the key bits. The number of such restricted 64-dimensional subspaces is $\binom{67}{6} > 2^{24} \approx |W|$, and hence they should suffice for the attack.

Finally, we notice that the Moebius transforms (Step 4.b) can be optimized due to the way that we chose the subspaces in S, as for all of them, 3 specific bits of $X_0|SP$ are active. In order to exploit this, we perform the Moebius transform on a 2^{70} bit vector in two phases: in the first phase, we partition the 2^{70} big subspace into 2^{67} 3-dimensional subspaces according to the 67 bits of $X_0|IP$, and sum on all of them in time 2^{70}, obtaining a vector of size 2^{67}. In the second phase, we perform the Moebius transform on the 2^{67} vectors computed in the first phase. Therefore, the complexity of a single Moebius transform is reduced from $70 \cdot 2^{70} \approx 2^{76}$ to $2^{70} + 67 \cdot 2^{67} \approx 2^{73}$. The complexity of online Step 4.b now becomes $|U| \cdot 2^{73} \approx 2^{104}$ bit operations.

The time complexity analysis of the critical steps of the attack is as follows:

Step 4.a: $d \cdot |S| \cdot |U| \approx 8 \cdot 2^{70} \cdot 2^{31} = 2^{104}$
Step 4.b: 2^{104} (as noted above)
Step 4.c: $|U'| \cdot |W| \approx 2^{21.5} \cdot 2^{24} = 2^{45.5}$
Step 4.d.i: $|U| \cdot |V| \cdot r_1 \cdot 2^{16+d} \approx 2^{31} \cdot 2^{23.5} \cdot 3 \cdot 2^{24} \approx 2^{80.5}$
Step 4.d.ii: $|W| \cdot |U| \cdot |V| \approx 2^{24} \cdot 2^{31} \cdot 2^{23.5} = 2^{78.5}$
Step 5: $|W|^3 \approx 2^{24 \cdot 3} = 2^{72}$

In total, the time complexity of the attack is about 2^{105} bit operations, dominated by steps 4.a and 4.b. The data complexity is 2^{70} chosen plaintexts.

The memory complexity is dominated by the storage of ciphertexts in Step 1, and is about $|S| \cdot 2^8 = 2^{78}$ bits.

Extending the Attack to Full LowMC-128 for Weak Instances. Similarly to the attacks on LowMC-80, the 11-round attack on LowMC-128 can be extended by an additional round with no increase in complexity for a subset of weak instances. However, the fraction of these instances is much smaller, as the I-part of LowMC-128 contains only 67 bits, and is smaller than the one of LowMC-80. A similar analysis to the one of Sect. 6.3 shows that the fraction of such weak instances for LowMC-128 is roughly $2^{67+67-256} = 2^{-122}$. As noted in the Introduction, this attack does not violate the formal security claims of the LowMC designers.

7.2   An Attack on Full LowMC-128

We now describe our attack on full (12-round) LowMC-128. This attack is more marginal than the previous attacks, and we have to use essentially all of our previously described optimizations, as well as new ones in order to obtain an attack which is faster than exhaustive search.

In order to attack 12 rounds of LowMC-128, we extend the interpolation of the 11-round attack past another round, interpolating $Z_7|IP$ through $r_1 = 4$ Sbox layers, and hence $d = 2^4 = 16$, $|U| = \binom{256+3m}{<2^{r_1}-1} = \binom{445}{\leq 8} \approx 2^{55}$.

In this attack we use $sp = 8$, and compute the size of the relevant variable sets $|U'| \leq \binom{256}{\leq sp-1} = \binom{256}{\leq 7} \approx 2^{43.5}$, $|V| \leq \binom{\kappa}{\leq d-sp} = \binom{128}{\leq 8} \approx 2^{40.5}$, $|W| = |U'| + |V| \approx 2^{44}$. We use the refined analysis for steps 4.d.i and 4.d.ii, and thus we also calculate $|U_{sp}| = |U_8| = \binom{256}{8} < 2^{48.5}$ and $|V_{\leq(d-sp-1)}| = \binom{128}{\leq 7} < 2^{36.5}$.

The High-Order Differential Property. We can try to mount the attack with high-order differentials on subspaces of dimension 64 for the target bit in $Z_7|IP$, but this results in an attack which is at best very marginally faster than exhaustive search. The main new optimization introduced in this attack is the use of reduced subspaces of dimension 60. Obviously, the result of a high-order differentiation over such a subspace is not a constant, but (as we show next) its algebraic degree in the key bits is bounded by 8. Consequently, the resultant function (polynomial) of each high-order differentiation can be expressed in terms of our reduced variable set $V = |V_{\leq(8)}|$. This polynomial can be interpolated during preprocessing and does not contribute additional variables to the equation system.

We select a big subspace S of dimension 73 that contains all the 67 bits of $X_0|IP$ and 6 additional bits of 2 active Sboxes in $X_0|SP$, and (similarly to the 11-round attack) define the 60-dimensional subspaces according to their $73-60 = 13$ constant bits in $X_0|IP$. The number of such subspaces is $\binom{67}{13} > 2^{44} \approx |W|$, and therefore they should suffice for the attack.

In order to show that the result of a high-order differentiation of the target bit in $Z_7|IP$ over a selected 60-dimensional is of degree 8 in the key bits, consider

· the state Z_0 obtained after the first Sbox layer. The algebraic degree of the target bit b (selected from $Z_7|IP$) in Z_0 is bounded by $2^6 = 64$. As the linear subspace undergoes a one-to-one transformation in the first Sbox layer (through the fully active 2 Sboxes), it remains a linear subspace in Z_0. Therefore, the algebraic degree of the high-order differentiation in the bits of Z_0 and the key is upper-bounded by $64 - 60 = 4$. Since each bit of Z_0 is a polynomial in the key of degree (at most) 2, the algebraic degree of the high-order differentiation in the bits of the key is upper-bounded by $4 \cdot 2 = 8$, as claimed.

The Preprocessing Phase. The main change in this attack compared to the one of Sect. 5.2 is in preprocessing Step 1, where in addition to interpolating the $e \approx |W|$ free coefficients, we interpolate the $e \cdot |V| \approx |W| \cdot |V|$ coefficients of V (since we selected 60-dimensional subspaces instead of 64-dimensional subspaces). The modified preprocessing step is described below. It is similar to the variable transformation algorithm of Sect. 5.1, interpolating first over the plaintexts and then over the keys. Note that the matrix E of linear equations is allocated and initialized already at this stage.

---

1. Allocate an $e \times |W|$ matrix E over $GF(2)$, representing the (reduced) equation system on W. The matrix is vertically decomposed into two smaller matrices: E_1 of size $e \times |U'|$ and E_2 of size $e \times |V|$.
2. Allocated an $e \cdot |V|$ evaluation matrix EV.
3. Allocate a $|S| = 2^{73}$ bit array a_1 for the evaluations of the target bit b.
4. For each key in the set $\{K|wt(K) \leq 8\}$ (with index ℓ):
    (a) Evaluate b (the target bit) on the set S of 2^{73} inputs (with the fixed key K) and store the result in a_1.
    (b) Apply the Moebius transform on a_1.
    (c) Populate column ℓ of EV: For each subspace S_j in S, namely $S_1, \ldots, S_e$, obtain its corresponding sum from a_1 and copy it to $E_1[j][\ell]$.
5. For each equation $1, \ldots, e$ (with index j):
    (a) For each $M_v \in V_{\leq 8} = V$ (with index ℓ):
        i Sum the $2^{wt(\bar{v})}$ values of $EV[j]$ calculated for the subset of keys $\{K|\bar{v} \wedge K = 0\}$, and store the result in $E_2[j][\ell]$.

We first note that similarly to the 11-round attack, the complexity of the Moebius transform can be optimized (due to the way that we selected the subspaces) in a 2-step process from $73 \cdot 2^{73}$ to $2^{73} + 67 \cdot 2^{67} \approx 2^{74}$.

We analyze the complexity of the computationally heavy steps 4 and 5. The complexity of Step 4.a (for all $\{K|wt(K) \leq 8\}$) is $|V| \cdot |S| \cdot 2^{19} \approx 2^{40.5} \cdot 2^{73} \cdot 2^{19} = 2^{132.5}$. The complexity of Step 4.b (using the optimized Moebius transform) is

$|V| \cdot 2^{74} \approx 2^{114.5}$. The complexity of Step 4.c is $e \cdot |V| \approx |W| \cdot |V| \approx 2^{44} \cdot 2^{40.5} = 2^{84.5}$. The complexity of Step 5.a.i is bounded by $e \cdot |V| \cdot 2^{8} \approx 2^{44} \cdot 2^{40.5} \cdot 2^{8} = 2^{92.5}$. In total, Step 4.a dominates the time complexity, which is about $2^{132.5}$ bit operations.

Analysis of the Full Attack. In terms of time complexity, the analysis of the critical steps of the online attack is as follows (using the optimized Moebius transform and the refined analysis for steps 4.d.i and 4.d.ii):

Step 4.a: $d \cdot |S| \cdot |U| \approx 16 \cdot 2^{73} \cdot 2^{55} = 2^{132}$
Step 4.b: $|U| \cdot 2^{74} \approx 2^{129}$
Step 4.c: $|U'| \cdot |W| \approx 2^{43.5} \cdot 2^{44} = 2^{87.5}$
Step 4.d.i: $|U_{sp}| \cdot |V| \cdot (r_1 \cdot 2^{16+sp} + 2^{d-sp}) + |U| \cdot |V_{\leq(d-sp-1)}| \cdot r_1 \cdot 2^{16+d} \approx 2^{48.5} \cdot 2^{40.5} \cdot (4 \cdot 2^{24} + 2^{8}) + 2^{55} \cdot 2^{36.5} \cdot 4 \cdot 2^{32} \approx 2^{115} + 2^{125.5} \approx 2^{125.5}$
Step 4.d.ii: $|W| \cdot (|U_{sp}| \cdot |V| + |U| \cdot |V_{\leq(d-sp-1)}|) \approx 2^{44} \cdot (2^{48.5} \cdot 2^{40.5} + 2^{55} \cdot 2^{36.5}) \approx 2^{44} \cdot (2^{89} + 2^{91.5}) \approx 2^{136}$
Step 5: $|W|^{3} \approx 2^{44 \cdot 3} = 2^{132}$

The online phase complexity is about 2^{136} dominated by[3] Step 4.d.ii. The total complexity of the attack is less than 2^{137} bit operations, which is about $2^{128+19-137} = 2^{10}$ times faster than exhaustive search (including the preprocessing phase, whose complexity is about $2^{132.5}$). The data complexity of the attack is 2^{73} chosen plaintexts. The memory complexity is dominated by the storage of E, whose size is about $|W| \cdot |W| \approx 2^{88}$ bits.

Note that without the variable transformation, merely Step 5 (Gaussian elimination) would require about $2^{55 \cdot 3} = 2^{165}$ bit operations, which is much slower than exhaustive search.[4]

8   Conclusions

In this paper, we introduced new techniques for interpolation attacks, including a new variable transformation algorithm that can lead to savings in their data and time complexities. We applied the optimized interpolation attack to LowMC, and refuted the claims of the designers regarding the security level of both the 80 and 128-bit key variants. As a future work item, it will be interesting to optimize our techniques further and apply them to additional block ciphers.

---

[3] We note that the analysis of Step 4.d.ii can be refined further, and its actual complexity is lower by a factor between 2 and 4. Moreover, the actual algorithm of this step can be optimized, but we do not consider such low-level optimizations here for the sake of simplicity.

[4] Solving the equation system remains slower than exhaustive search even when using more advanced algorithms which are based on Strassen's algorithm [9], requiring about $2^{55 \cdot 2.8} = 2^{154}$ bit operations. While there are known algorithms that perform better in theory, most of them are very complex and inefficient in practice.

References

1. Albrecht, M.R., Rechberger, C., Schneider, T., Tiessen, T., Zohner, M.: Ciphers for MPC and FHE. In: Oswald, E., Fischlin, M. (eds.) EUROCRYPT 2015. LNCS, vol. 9056, pp. 430–454. Springer, Heidelberg (2015)
2. Dinur, I., Shamir, A.: Cube attacks on tweakable black box polynomials. In: Joux, A. (ed.) EUROCRYPT 2009. LNCS, vol. 5479, pp. 278–299. Springer, Heidelberg (2009)
3. Hell, M., Johansson, T., Maximov, A., Meier, W.: The grain family of stream ciphers. In: Robshaw, M., Billet, O. (eds.) New Stream Cipher Designs. LNCS, vol. 4986, pp. 179–190. Springer, Heidelberg (2008)
4. Jakobsen, T., Knudsen, L.R.: The interpolation attack on block ciphers. In: Biham, E. (ed.) FSE 1997. LNCS, vol. 1267, pp. 28–40. Springer, Heidelberg (1997)
5. Joux, A.: Algorithmic Cryptanalysis, 1st edn. Chapman & Hall/CRC, Boca Raton (2009)
6. Knudsen, L.R.: Truncated and higher order differentials. In: Preneel, B. (ed.) Fast Software Encryption. LNCS, vol. 1008, pp. 196–211. Springer, Heidelberg (1994)
7. Lai, X.: Higher order derivatives and differential cryptanalysis. In: Blahut, R.E., Costello, D.J., Maurer, U., Mittelholzer, T. (eds.) Communications and Cryptography. SLSECS, vol. 276, pp. 227–233. Springer, Heidelberg (1994)
8. Shimoyama, T., Moriai, S., Kaneko, T.: Improving the higher order differential attack and cryptanalysis of the KN cipher. In: Okamoto, E., Davida, G., Mambo, M. (eds.) Information Security. LNCS, vol. 1396, pp. 32–42. Springer, Heidelberg (1997)
9. Strassen, V.: Gaussian elimination is not optimal. Numerische Mathematik **13**, 354–356 (1969)

Another Tradeoff Attack
on Sprout-Like Stream Ciphers

Bin Zhang[1,2(✉)] and Xinxin Gong[1(✉)]

[1] TCA Laboratory, SKLCS, Institute of Software,
Chinese Academy of Sciences, Beijing, China
[2] State Key Laboratory of Cryptology, P.O.Box 5159, Beijing 100878, China
{zhangbin,gongxinxin}@tca.iscas.ac.cn

Abstract. Sprout is a new lightweight stream cipher with shorter internal state proposed at FSE 2015, using key-dependent state updating in the keystream generation phase. Some analyses have been available on eprint so far. In this paper, we extend the design paradigm in general and study the security of Sprout-like ciphers in a unified framework. Our new penetration is to investigate the k-normality of the augmented function, a vectorial Boolean function derived from the primitive. Based on it, a dedicated time/memory/data tradeoff attack is developed for such designs. It is shown that Sprout can be broken in 2^{79-x-y} time, given $[c \cdot (2x + 2y - 58) \cdot 2^{71-x-y}]$-bit memory and 2^{9+x+y}-bit keystream, where x/y is the number of forward/backward steps and c is a small constant. Our attack is highly flexible and compares favorably to all the previous results. With carefully chosen parameters, the new attack is at least 2^{20} times faster than Lallemand/Naya-Plasencia attack at Crypto 2015, Maitra et al. attack and Banik attack, 2^{10} times faster than Esgin/Kara attack with much less memory.

Keywords: Cryptanalysis · Stream ciphers · Sprout · Tradeoff

1 Introduction

Design of secure lightweight stream ciphers for constrained hardware environments is important both in theory and practice. The most area/power consuming component in a lightweight design is the number of memory gates, which corresponds to the internal state size of the primitive. On the other hand, a common rule of thumb for stream cipher design is that the internal state size should be at least twice as long as the key size to resist against time/memory/data (TMD) tradeoff attacks [4].

This design principal indeed works, and security analysis of the eSTREAM finalists, e.g., Grain v1, Mickey v2 and Trivium [7] evolves rather slowly. At FSE

This work is supported by the program of the National Natural Science Foundation of China (Grant No. 61572482) and the National Grand Fundamental Research 973 Program of China (Grant No. 2013CB338002).

T. Iwata and J.H. Cheon (Eds.): ASIACRYPT 2015, Part II, LNCS 9453, pp. 561–585, 2015.
DOI: 10.1007/978-3-662-48800-3_23

2015, another design paradigm for stream ciphers is proposed and instantiated by a new design, called Sprout, aiming to reduce the internal state size, thus the hardware area size by using key-dependent state updating in the keystream generation phase [2]. It is expected that the immunity against TMD tradeoff attacks will not be compromised.

Surprisingly, there have been some cryptanalysis of Sprout appearing on the IACR eprint monthly after ESC 2015 and FSE 2015. In the time order of the open literature, a related key chosen IV attack on Sprout is presented in [9], but the designers have already ruled out the related key model in [2]. Then the first attack in the single key model is found in [12] by using a list merging technique with a time complexity around 2^{69} Sprout encryptions at Crypto 2015. In [13], another attack based on a SAT solver is given with a complexity of 2^{54} attempts, where each attempt takes a time equivalent to $6.6 \cdot 2^{54} \cdot 2^e$ encryptions which is more than 2^{80} if $e > 23$. Thus, it is questionable whether this work in [13] translates into a feasible attack on Sprout or not. To directly challenging the design rationale, Esgin and Kara presented a TMD tradeoff attack in [8] with an online time complexity of 2^{33} Sprout encryptions and 770 TB of memory after a pre-computation around 2^{53} basic operations. Finally in [3], a key recovery attack is launched against Sprout with a complexity of $2^{66.7}$ Sprout encryptions together with some other analysis results.

In this paper, we extend the design paradigm in general and study the security of Sprout-like ciphers in a unified framework. The model involves the secret key not only in the initialization process but also in the non-linear state updating in a Sprout-like manner during the keystream generation phase. Then based on the notion of normality first introduced by Dobbertin in [6], we investigate the k-normality of the augmented function [5], a vectorial Boolean function derived from the underlying primitive. This property is relevant for the design and analysis of cryptosystems. In [14] and [15], security implications of k-normal Boolean functions are considered when they are employed in certain stream ciphers. We make a systematic security analysis based on this property for Sprout-like stream ciphers and develop a dedicated TMD tradeoff attack framework for such designs. In particular, it is shown that Sprout can be broken in 2^{79-x-y} time, given $\left[c \cdot (2x + 2y - 58) \cdot 2^{71-x-y}\right]$-bit memory and 2^{9+x+y}-bit keystream, where x is the number of forward steps, y is the number of backward steps and c is a small constant. Our attack is highly flexible and compares favorably to all the previous attacks on Sprout. With carefully chosen attack parameters, our method is at least 2^{20} times faster than Lallemand/Naya-Plasencia attack at Crypto 2015, Maitra et al. attack and Banik attack, 2^{10} times faster than Esgin/Kara attack with much less memory. Practical simulations confirmed our analysis.

This paper is structured as follows. In Sect. 2, the stream cipher Sprout is described and generalized to a generic Sprout-like model. In Sect. 3, based on a natural extension of normality from Boolean functions to vectorial Boolean functions, a generic TMD cryptanalysis framework of such ciphers is formalized with complexity analysis. In Sect. 4, the framework is applied to Sprout with comparisons to other attacks. Section 5 provides the experimental results. Finally, some conclusions are given in Sect. 6.

2   Sprout-Like Stream Ciphers

In this section, a brief description of Sprout that is relevant to our work and a generic Sprout-like model that inherits the design spirit are presented. The following notations will be used throughout the paper.

- $L^t = [l_t, l_{t+1}, ..., l_{t+39}]$, the internal state of the LFSR at time t.
- $N^t = [n_t, n_{t+1}, ..., n_{t+39}]$, the internal state of the NFSR at time t.
- $[a, b] \triangleq \{a, a + 1, ..., b\}$, for two positive integers a, b $(a < b)$.
- $N^t_{[a,b]} \triangleq \{n_{t+a}, n_{t+a+1}, ..., n_{t+b}\}$ and $L^t_{[a,b]} \triangleq \{l_{t+a}, l_{t+a+1}, ..., l_{t+b}\}$, for two positive integers a, b $(a < b)$.
- $IV = (iv_0, iv_1, ..., iv_{69})$, the 70-bit initialization vector.
- $K = (k_0, k_1, ..., k_{79})$, the 80-bit secret key.
- k_t^*, the round key bit generated at time t.
- z_t, the keystream bit generated at time t.
- c_t^4, the round constant at time t, generated by a counter.

2.1   Description of Sprout

Sprout adopts a structure similar to the Grain family of stream ciphers [1,10,11], which consists of four parts, an 80-bit fixed key register, a 40-bit NFSR with a linked 40-bit LFSR, and a counter register, depicted in Fig. 1. Since storing a fixed key requires less area size than realizing a register of the same length, it is reported in [2] that the hardware area of Sprout is significantly less compared to the existing lightweight stream ciphers.

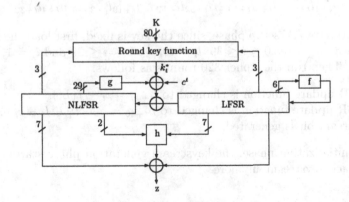

Fig. 1. Keystream generation of sprout

Denote the feedback functions of the NFSR, the LFSR and the nonlinear filter function by g, f and h respectively. There is a 9-bit counter register in Sprout, of which the lower 7 bits are a modulo 80 counter, denoted by $(c_t^6, c_t^5, c_t^4, c_t^3, c_t^2, c_t^1, c_t^0)$

at time t. The 4-th LSB c_t^4 of the counter is employed in the keystream genera-tion. It should be noted that, c_t^4 has a cycle of length 80, i.e., in each cycle, this bit takes the values $\underbrace{0, 0, ..., 0}_{16} \underbrace{1, 1, ..., 1}_{16} \underbrace{0, 0, ..., 0}_{16} \underbrace{1, 1, ..., 1}_{16} \underbrace{0, 0, ..., 0}_{16}$.

The 40-bit LFSR is updated recursively by f as $l_{t+40} = l_t \oplus l_{t+5} \oplus l_{t+15} \oplus l_{t+20} \oplus l_{t+25} \oplus l_{t+34}$. The NFSR is updated recursively by a non-linear feedback function g as

$$
\begin{aligned}
n_{t+40} &= k_t^* \oplus l_t \oplus c_t^4 \oplus g(N^t) \\
&= k_t^* \oplus l_t \oplus c_t^4 \oplus n_t \oplus n_{t+13} \oplus n_{t+19} \oplus n_{t+35} \oplus n_{t+39} \\
&\quad \oplus n_{t+2}n_{t+25} \oplus n_{t+3}n_{t+5} \oplus n_{t+7}n_{t+8} \oplus n_{t+14}n_{t+21} \oplus n_{t+16}n_{t+18} \\
&\quad \oplus n_{t+22}n_{t+24} \oplus n_{t+26}n_{t+32} \oplus n_{t+33}n_{t+36}n_{t+37}n_{t+38} \\
&\quad \oplus n_{t+10}n_{t+11}n_{t+12} \oplus n_{t+27}n_{t+30}n_{t+31}.
\end{aligned}
$$

Let $u_t = l_{t+4} \oplus l_{t+21} \oplus l_{t+37} \oplus n_{t+9} \oplus n_{t+20} \oplus n_{t+29}$, then

$$
k_t^* = \begin{cases} k_t, & 0 \le t \le 79 \\ k_{t(\bmod 80)} \cdot u_t, & \text{otherwise.} \end{cases}
$$

Given the internal state at time t, the keystream bit is generated as

$$
z_t = h(n_{t+4}, l_{t+6}, l_{t+8}, l_{t+10}, l_{t+32}, l_{t+17}, l_{t+19}, l_{t+23}, n_{t+38}) \oplus l_{t+30} \oplus \left(\bigoplus_{i \in A} n_{t+i} \right),
$$

where $A = \{1, 6, 15, 17, 23, 28, 34\}$, and the filter function is

$$
h(\cdot) = n_{t+4}l_{t+6} \oplus l_{t+8}l_{t+10} \oplus l_{t+32}l_{t+17} \oplus l_{t+19}l_{t+23} \oplus n_{t+4}l_{t+32}n_{t+38}.
$$

During the key/IV setup phase, since the key is fixed, first load the IV in the following way: $n_i = iv_i, 0 \le i \le 39$; $l_i = iv_{i+40}, 0 \le i \le 29$ and $l_i = 1, 30 \le i \le 38$, $l_{39} = 0$. Then run the cipher 320 rounds as follows.

- the LFSR update function is changed to $l_{t+40} = z_t \oplus f(L^t)$.
- the NFSR update function is changed to $n_{t+40} = z_t \oplus k_t^* \oplus l_t \oplus c_t^4 \oplus g(N_t)$.
- no keystream bit is generated.

After the initialization phase, the keystream generation phase starts and there is no feedback keystream anymore.

2.2   A Model for Sprout-Like Stream Ciphers

There are three functions involved in the model: a non-linear function $G(x)$, a linear function $F(x)$ and a non-linear filter function $h(\cdot)$.

The internal state of the model consists of the non-linear state N and the linear state L. At each step, the function $G(\cdot)$ is applied to N and $F(\cdot)$ to L, respectively. Besides, there may also be some other mixing procedure that xoring some bits of N into L, and vice versa. Further, the secret key is involved in the non-linear state updating selectively by a function $u(\cdot)$. The output of the current state is also computed as the xoring of the bits from both N and L and a non-linear filter function $h(\cdot)$, which takes some input values from both N and L, respectively. Some notations that will be used in the description are listed here.

- $L^t = [L_0^t, L_1^t, ..., L_{l_1-1}^t]$, the internal state of the linear component.
- $N^t = [N_0^t, N_1^t, ..., N_{l_2-1}^t]$, the internal state of the non-linear component.
- $rL^t = \{L_{\gamma_1}^t, L_{\gamma_2}^t, ..., L_{\gamma_{a_1}}^t\}$, a subset of L^t and the linear part of $u(\cdot)$.
- $rN^t = \{N_{\delta_1}^t, N_{\delta_2}^t, ..., N_{\delta_{a_2}}^t\}$, a subset of N^t and the non-linear part of $u(\cdot)$.
- $pL^t = \{L_{\alpha_1}^t, L_{\alpha_2}^t, ..., L_{\alpha_{n_1}}^t\}$, a subset of L^t with the variables of the filter function $h(\cdot)$ coming from the LFSR.
- $pN^t = \{N_{\beta_1}^t, N_{\beta_2}^t, ..., N_{\beta_{n_2}}^t\}$, a subset of N^t with the variables of the filter function $h(\cdot)$ coming from the NFSR.
- $qL^t = \{L_{\sigma_1}^t, L_{\sigma_2}^t, ..., L_{\sigma_{m_1}}^t\}$, a subset of L^t and the linear masking in the keystream generation function.
- $qN^t = \{N_{\tau_1}^t, N_{\tau_2}^t, ..., N_{\tau_{m_2}}^t\}$, a subset of N^t and the non-linear masking in the keystream generation function.
- $pqN^t = pN^t \cup qN^t$, the variables used in the keystream generation coming from the NFSR.

The general framework is specified by the following items (we only focus on the keystream generation phase).

1. Components

- The linear component is $L^t = [L_0^t, L_1^t, ..., L_{l_1-1}^t] \in F_2^{l_1}$, whose initial state is denoted by L^0. It is updated recursively as $L^{t+1} = F(L^t)$. Without loss of generality, we assume this process is invertible, and the inverse process is $L^{t-1} = F'(L^t)$.
- The non-linear component is $N^t = [N_0^t, N_1^t, ..., N_{l_2-1}^t] \in F_2^{l_2}$, whose initial state is denoted by N^0. It is updated recursively as

$$N^{t+1} = G(N^t \oplus L_1(L^t)) \oplus L_2(L^t) \oplus u(rL^t, rN^t) \cdot R(t, K) \oplus C_t,$$

where $G(\cdot)$ is a (l_2, l_2)-vectorial Boolean function, C_t is a counter related vector of length l_2. Note that whether the key is involved in the state updating is dependent on the value of $u(\cdot)$. If $u(rL^t, rN^t) = 1$, the key will be involved. Similarly, we assume this non-linear process is invertible, and the inverse process is computed as

$$N^{t-1} = G'(N^t \oplus L_1'(L^{t-1})) \oplus L_2'(L^{t-1}) \oplus u(rL^{t-1}, rN^{t-1}) \cdot R(t-1, K) \oplus C_{t-1}.$$

- A filter function $h(\cdot)$ from $F_2^{n_1+n_2}$ into F_2 is used as part of the output function in the form $h(pL^t, pN^t)$, which takes n_1 input values $\{L_{\alpha_1}^t, L_{\alpha_2}^t, ..., L_{\alpha_{n_1}}^t\}$ from L^t and n_2 input values $\{N_{\beta_1}^t, N_{\beta_2}^t, ..., N_{\beta_{n_2}}^t\}$ from N^t, respectively.

- A linear Boolean function $l(\cdot)$ from $F_2^{m_1+m_2}$ into F_2 is used as part of the output function in the form $l(qL^t, qN^t)$, which takes m_1 input values $\{L_{\sigma_1}^t, L_{\sigma_2}^t, ..., L_{\sigma_{m_1}}^t\}$ from L^t and m_2 input values $\{N_{\tau_1}^t, N_{\tau_2}^t, ..., N_{\tau_{m_2}}^t\}$ from N^t, respectively.
- An output function $\phi(\cdot) = l(\cdot) \oplus h(\cdot)$, which generates the keystream $\{z_t\}_{t\geq 0}$ based on the inputs taken from both L^t and N^t, $t = 0, 1, ...$

2. Keystream Generation

The keystream $\{z_t\}_{t\geq 0}$ is recursively generated as

$$z_t = h(pL^t, pN^t) \oplus l(qL^t, qN^t), \; t = 0, 1, ...$$

Let U be the subspace of F_2^m and denote the dimension as $dim(U)$, define $\overline{U} := \{a \in F_2^m : a \notin U\} \cup \{0\}$ as the complementary space of U. Now a coset of the subspace U is represented by $U_a := a \oplus U, a \in \overline{U}$, also called a *flat*. The following definitions are needed in our model.

Definition 1. *An m-variable Boolean function f is k-normal (resp. k-weakly normal) if there exists a flat $V \subseteq F_2^m$ of dimension k such that f is constant (resp. affine) on V.*

For example, the 5-variable Boolean function $h(\cdot)$ in Grain-v1 is 2-normal and 3-weakly normal, and the 9-variable Boolean function $h(\cdot)$ in Sprout and Grain-128a is 5-normal.

Next, we study a natural generalization of the above definition for vectorial Boolean functions [5].

Definition 2. *An (m, n)-function $F: F_2^m \to F_2^n$ is called k-normal if there exists a flat $V \subseteq F_2^m$ of dimension k such that F is constant on V.*

In our analysis, we investigate the k-normality of the augmented function defined as follows.

Definition 3. *For a $(n_1 + n_2)$-variable Boolean function $h(pL^t, pN^t)$, the $(b+f+1)$-th augmented function of h, $H^{(b,f)} : F_2^{M_1+M_2} \to F_2^{b+f+1}$ is defined as*

$$H^{(b,f)}(PL^t, PN^t) = \left(h(pL^{t-b}, pN^{t-b}), ..., h(pL^t, pN^t), ..., h(pL^{t+f}, pN^{t+f})\right),$$

where b, f are two positive integers, and

$$PL^t \triangleq \bigcup_{i=-b}^{f} pL^{t+i}, \; M_1 \triangleq |PL^t| \leq \sum_{i=-b}^{f} |pL^{t+i}| = n_1(b+f+1),$$

$$PN^t \triangleq \bigcup_{i=-b}^{f} pN^{t+i}, \; M_2 \triangleq |PN^t| \leq \sum_{i=-b}^{f} |pN^{t+i}| = n_2(b+f+1).$$

3. Assumptions

- *3.1*: there exists two positive integers b, f such that $\bigcup_{i=-b}^{f} pqN^{t+i} \subseteq N^t$ for any $t \geq b$. In this case, the output segment $z_{t-b}, ..., z_t, ..., z_{t+f}$ can be computed from the complete state (L^t, N^t) at time t.

- *3.2*: $H^{(b,f)}$, the $(b + f + 1)$-th augmented function of the filter function h, is a k-normal Boolean function such that $H^{(b,f)}(x_1, ..., x_n) = 0^{b+f+1}$ when x_j is fixed for all $j \in \Omega$, where Ω is a subset of $[1, n]$ and $|\Omega| = n - k$.
- *3.3*: there exists two positive integers d, e such that $\bigcup_{i=-d}^{e} rN^{t+i} \subseteq N^t$ for any $t \geq d$. In this case, $u(rL^{t+i}, rN^{t+i}), i = -d, ..., -1, 0, 1, ..., e$ can be computed from the complete state (L^t, N^t) at time t.
- *3.4*: assume $pqN^{t+f+1} \not\subset N^t$ and $pqN^{t+f+1} \subset N^{t+1}$ for any $t \geq b$, meaning that we cannot get pqN^{t+f+1} from the state (L^t, N^t). Note that the secret key is incorporated in the non-linear state updating selectively, if we assume a special state (L^t, N^t) such that $u(rL^t, rN^t) = 0$, N^{t+1} can be computed from (L^t, N^t), thus we further get the output bit z_{t+f+1}. Repeat this process for x steps, i.e., we assume a special state (L^t, N^t) such that $u(rL^{t+i}, rN^{t+i}) = 0$ for $i = 0, 1, ..., x - 1$, then we get the output bits $z_{t+f+1}, ..., z_{t+f+x}$.
- *3.5*: assume $rN^{t+e+1} \not\subset N^t$ and $rN^{t+e+1} \subset N^{t+1}$ for any $t \geq d$. For the above special state (L^t, N^t) such that $u(rL^{t+i}, rN^{t+i}) = 0$ for $i = 0, 1, ..., x - 1$, if $x - 1 \leq e$, we have only unknowns from (L^t, N^t); if $x - 1 > e$, then the unknowns from N^{t+1}, N^{t+2},... will appear with some nonlinear equations $N^{t+j+1} = G(N^{t+j} \oplus L_1(L^{t+j})) \oplus L_2(L^{t+j}) \oplus C_{t+j}, j = 0, 1, ..., x - e - 2$.
- *3.6*: assume $pqN^{t-b-1} \not\subset N^t$ and $pqN^{t-b-1} \subset N^{t-1}$ for any $t \geq b$, which means we cannot get pqN^{t-b-1} from the state (L^t, N^t). If we assume a special state (L^t, N^t) such that $u(rL^{t-1}, rN^{t-1}) = 0$, N^{t-1} can be computed from (L^t, N^t), thus we further get the output bit z_{t-b-1}. Repeat this process for y steps, i.e., we assume a special state (L^t, N^t) such that $u(rN^{t-j}, rL^{t-j}) = 0$ for $j = 1, ..., y$, then we get the output bits $z_{t-b-1}, ..., z_{t-b-y}$.
- *3.7*: assume $rN^{t-d-1} \not\subset N^t$ and $rN^{t-d-1} \subset N^{t-1}$ for any $t \geq d$. For the above special state (L^t, N^t) such that $u(rL^{t-j}, rN^{t-j}) = 0$ for $j = 1, ..., y$, if $y \leq d$, we have only unknowns from (L^t, N^t); if $y > d$, then the unknowns from N^{t-1}, N^{t-2},... will appear with some nonlinear equations $N^{t-j-1} = G'(N^{t-j} \oplus L_1'(L^{t-j-1})) \oplus L_2'(L^{t-j-1}) \oplus C_{t-j-1}, j = 0, 1, ..., y - d - 1$.

It is easy to check that the proposed model includes a number of primitives, e.g., Sprout and the Grain family. For Grain family, the term $u(rL^t, rN^t) = 0$ for any time t. For Sprout, $N^t = [n_t, n_{t+1}, ..., n_{t+39}]$, $L^t = [l_t, l_{t+1}, ..., l_{t+39}]$, and for any t, $u(rL^t, rN^t) = l_{t+4} \oplus l_{t+21} \oplus l_{t+37} \oplus n_{t+9} \oplus n_{t+20} \oplus n_{t+29}$. The positive integers b, f, d, e are $b = 1$, $f = 1$, $d = 9$, $e = 10$ respectively.

3   A TMD Tradeoff Attack Framework

In this section, we provide a systematic security analysis for Sprout-like stream ciphers. A dedicated TMD tradeoff attack framework is developed for such designs based on the k-normality of the augmented function.

The goal of cryptanalysis is to recover the internal state which has generated a sample segment, and if possible, given the internal state, to further restore the secret key. There are two phases in the framework: the pre-processing phase and the processing phase. The offline pre-processing phase is performed only once and is independent of the employed secret key and the keystream sample.

3.1   Pre-Processing Phase

In the offline pre-processing phase, some tables are prepared which will be used later in the processing phase. Given the parameters l_1, l_2 and b, f, d, e, x, y, define a two-dimensional counter array $\bar{\mathbf{C}} = [C_{t-y}, ..., C_{t-1}, C_t, C_{t+1}, ..., C_{t+(x-1)}]$, we construct the State-Keystream pair tables as follows.

1. Under the assumptions in the model, construct a system of equations which implies a "special" state (L^t, N^t) satisfying the following conditions.

   - (1.1) $H^{(b,f)}(PL^t, PN^t) = 0^{b+f+1}$ and $l(qL^{t+i}, qN^{t+i}) = 0$, for $i = -b, ..., -1, 0, 1, ..., f$.
   - (1.2) $u(rL^{t+i}, rN^{t+i}) = 0$ for $i = 0, 1, ..., x - 1$, from which we can get the output bits $z_{t+f+1}, ..., z_{t+f+x}$.
   - (1.3) $u(rL^{t-j}, rN^{t-j}) = 0$ for $j = 1, ..., y$, from which we can get the output bits $z_{t-b-1}, ..., z_{t-b-y}$.

2. Suppose Assumptions 3.2, 3.5 and 3.7 hold,

   - if $x - 1 \le e$ and $y \le d$, the above system of equations has only unknowns from the state (L^t, N^t).
   - if $x - 1 > e$ and $y \le d$, the unknowns from N^{t+1}, N^{t+2},... will appear with some non-linear equations:

   $$N^{t+j+1} = G(N^{t+j} \oplus L_1(L^{t+j})) \oplus L_2(L^{t+j}) \oplus C_{t+j}, \; j = 0, 1, ..., x - e - 2.$$

   Define another counter array $\bar{\mathbf{C}}' = [C_t, C_{t+1}, ..., C_{t+(x-e-2)}]$, note that the round constant vectors in $\bar{\mathbf{C}}'$ are involved in these equations.

   - if $x - 1 \le e$ and $y > d$, the unknowns from N^{t-1}, N^{t-2},... will appear with some nonlinear equations:

   $$N^{t-j-1} = G'(N^{t-j} \oplus L_1'(L^{t-j-1})) \oplus L_2'(L^{t-j-1}) \oplus C_{t-j-1}, \; j = 0, 1, ..., y-d-1.$$

   Define counter array $\bar{\mathbf{C}}' = [C_{t-(y-d)}, ..., C_{t-2}, C_{t-1}]$, the round constant vectors in $\bar{\mathbf{C}}'$ are involved in these equations.

   - if $x - 1 > e$ and $y > d$, the unknowns from N^{t+1}, N^{t+2},... and N^{t-1}, N^{t-2},... will appear with some nonlinear equations:

   $$N^{t+j+1} = G(N^{t+j} \oplus L_1(L^{t+j})) \oplus L_2(L^{t+j}) \oplus C_{t+j}, j = 0, 1, ..., x - e - 2,$$
   $$N^{t-j-1} = G'(N^{t-j} \oplus L_{1'}(L^{t-j-1})) \oplus L_{2'}(L^{t-j-1}) \oplus C_{t-j-1}, j = 0, 1, ..., y - d - 1.$$

   Define counter array $\bar{\mathbf{C}}' = [C_{t-(y-d)}, ..., C_{t-1}, C_t, C_{t+1}, ..., C_{t+(x-e-2)}]$, the round constant vectors in $\bar{\mathbf{C}}'$ are involved in these equations.

3. For each possible counter array $\bar{\mathbf{C}}'$, solve the constructed system of equations and get the special states (L^t, N^t) satisfying 1 and 2. Memorize the special state (L^t, N^t) in the first column of a row in table $T_{\bar{\mathbf{C}}'}$, further for this state and for each possible counter array $\bar{\mathbf{C}}^* = \bar{\mathbf{C}} \backslash \bar{\mathbf{C}}'$, get the corresponding $(x+y)$ output bits $\underbrace{z_{t-b-1}, ..., z_{t-b-y}}_{y}, \underbrace{z_{t+f+1}, ..., z_{t+f+x}}_{x}$ and store them in the second column as a sub-row in table $T_{\bar{\mathbf{C}}'}$.

Remarks. Denote the number of rows (in the first column) of table $T_{\bar{C}'}$ as 2^r, if $r < x + y$, we only need to store $(x + y - r)$ output bits in the second column, indexed by r-bit of the output. Next, let $Z_t^{(b+f+1)} = [z_{t-b}, ..., z_t, ..., z_{t+f}] \in F_2^{b+f+1}$, then an internal state satisfying the condition (1.1) implies $Z_t^{(b+f+1)} = 0^{b+f+1}$. Further, for each counter array $\bar{C}'$, $N^{t+1}, ..., N^{t+x}$ and $N^{t-1}, ..., N^{t-y}$ can be computed directly from a "special" state (L^t, N^t) according to the non-linear state updating function without involving the secret key.

3.2   Processing Phase

Now we discuss how to recover the internal state which has generated a sample segment, and if possible, given the internal state, to further restore the secret key. The following two propositions have provided us a direct way of key recovery from an internal state candidate and some keystream bits.

Proposition 1. *For a special state (L^t, N^t) satisfying the conditions (1.1) and (1.2), $N^{t+1}, ..., N^{t+x}$ can be computed directly from the complete state (L^t, N^t) and the non-linear state updating function without involving the secret key. Besides, if $u(rL^{t+x}, rN^{t+x}) = 1$, we may get some secret key information $R(t+x, K)$ when the keystream bit $z_{t+f+x+1}$ is known. Further, more key information $R(t+x+j, K)$, $j = 0, 1, ...$ will probably be obtained when more keystream bits $z_{t+f+x+j+1}$, $j = 0, 1, ...$ are known.*

Proof. The first half is clear from the condition (1.2).

For a special state (L^t, N^t), if $u(rL^{t+x}, rN^{t+x}) = 1$, the secret key information $R(t + x, K)$ is incorporated into the updating of the non-linear part from N^{t+x} to N^{t+x+1}. One can check that the keystream bit $z_{t+f+x+1}$ is dependent on N^{t+x+1}. In a word, $R(t + x, K)$ is likely to affect (if $u(rL^{t+x}, rN^{t+x}) = 1$) the keystream bit $z_{t+f+x+1}$. Accordingly, we may obtain some key information $R(t + x, K)$ from $z_{t+f+x+1}$. This procedure can be repeated many times.   □

Similar to the proof of Proposition 1, we have the following proposition.

Proposition 2. *For a special state (L^t, N^t) satisfying the conditions (1.1) and (1.3), $N^{t-1}, ..., N^{t-y}$ can be computed directly from the complete state (L^t, N^t) and the non-linear state updating function without involving the secret key. Besides, if $u(rL^{t-y-1}, rN^{t-y-1}) = 1$, we may get some key information $R(t-y-1, K)$ when the keystream bit $z_{t-b-y-1}$ is known. Further, more key information $R(t - y - j, K)$, $j = 1, 2, ...$ will probably be obtained when more keystream bits $z_{t-b-y-j}$, $j = 1, 2, ...$ are known.*

By utilizing the pre-computed tables and the given keystream sample, the processing phase is carried out as follows.

The Internal State Recovery Algorithm. Given the parameters b, f, x, y, the tables $T_{\bar{C}'}$, and the keystream sample $\{z_t\}_{t \geq 0}$, the processing steps are as follows.

1. Search the keystream sequence $\{z_t\}_t$ for the next non-considered block of $(b+f+1)$ zeros. If there are no more blocks, output a flag that the algorithm has failed.
2. For each detected block, compute the corresponding counter array $\bar{C}$, $\bar{C}'$ and $\bar{C}^*$ from the time t, compare the x-bit segment of the keystream subsequent to the block and y-bit segment prior to the block with the memorized $(x+y)$-bit segments in the second column (sub-row is indexed by $\bar{C}^*$) of the table $T_{\bar{C}'}$, and do the following:
   - If the matching does not exist, go the processing Step 1.
   - If the $(x+y)$-bit output segment matches with a segment in table $T_{\bar{C}'}$, go to Step 3.
3. Read the corresponding state, and if appropriate, recover (part of) the secret key according to Propositions 1 and 2 from this state and more keystream bits.

3.3   Complexity Analysis

In the Sprout-like model, the keystream bit is generated as $z_t = h(pL^t, pN^t) \oplus l(qL^t, qN^t)$. For the (b, f) derived from Assumption 3.3, we define a flat $V^{(b,f)}$ of dimension $dim(V^{(b,f)})$ such that $H^{(b,f)} = 0^{b+f+1}$ over it. i.e.,

$$V^{(b,f)} = \{(L^t, N^t) : H^{(b,f)}(PL^t, PN^t) = 0^{b+f+1}\},$$

We have the following lemma which is closely related to the time complexity of processing (table look-ups) of our proposed algorithm.

Lemma 1. *Suppose $Pr[u(\cdot) = 0] = p$, assume all the events in (1.1),(1.2) and (1.3) are independent, then the probability that an internal state (L^t, N^t) is a special state satisfying the conditions (1.1), (1.2) and (1.3) simultaneously when the keystream segment $Z_t^{(b+f+1)} = 0^{b+f+1}$ is*

$$Pr\left[(L^t, N^t) \text{ is a special state}\big|\, Z_t^{(b+f+1)} = 0^{b+f+1}\right] = \frac{1}{2^{l_1+l_2-dim(V^{(b,f)})}} \cdot p^{x+y},$$

where $V^{(b,f)}$ is a flat such that $H^{(b,f)} = 0^{b+f+1}$ over it.

Proof. For any internal state (L^t, N^t) and keystream segment $Z_t^{(b+f+1)}$, the underlying assumptions directly imply the following:

$$Pr\left[(L^t, N^t) \text{ is a special state}\right] = \frac{1}{2^{l_1+l_2-dim(V^{(b,f)})}} \cdot \frac{1}{2^{b+f+1}} \cdot p^{x+y},$$

and $Pr\left[Z_t^{(b+f+1)} = 0^{b+f+1}\right] = 2^{-(b+f+1)}$, and

$$Pr\left[Z_t^{(b+f+1)} = 0^{b+f+1}\big|\, (L^t, N^t) \text{ is a special state}\right] = 1.$$

On the other hand,

$$\Pr\left[(L^t, N^t) \text{ is a special state}\Big| Z_t^{(b+f+1)} = 0^{b+f+1}\right]$$

$$= \frac{\Pr[(L^t, N^t) \text{ is a special state}] \cdot \Pr\left[Z_t^{(b+f+1)} = 0^{b+f+1}\Big|(L^t, N^t) \text{ is a special state}\right]}{\Pr\left[Z_t^{(b+f+1)} = 0^{b+f+1}\right]}$$

$$= \frac{1}{2^{l_1+l_2-\dim(V^{(b,f)})}} \cdot p^{x+y}$$

which yields the statement of the lemma. $\qquad\square$

Theorem 1. *Suppose $V^{(b,f)}$ is a flat such that $H^{(b,f)} = 0^{b+f+1}$ over it, then the complexities of the proposed generic algorithm for cryptanalysis are as follows:*

(1) The processing data complexity is $D = 2^{l_1+l_2-\dim(V^{(b,f)})} \cdot 2^{b+f+1} \cdot p^{-(x+y)}$.

(2) The expected space complexity in the pre-processing phase is proportional to the sum of number of rows in each table $T_{\bar{C}'}$.

(3) The processing (table look-ups) time complexity is proportional to $2^{l_1+l_2-\dim(V^{(b,f)})} \cdot p^{-(x+y)}$.

(4) The pre-processing time complexity is equivalent to the workload for solving the system of equations constructed.

Proof. The data complexity is determined by the probability that an internal state is a special state satisfying conditions (1.1), (1.2) and (1.3) simultaneously in the pre-processing phase, which is given in the proof of Lemma 1 as $2^{-(l_1+l_2-dim(V^{(b,f)}))} \cdot 2^{-(b+f+1)} \cdot p^{x+y}$. Thus we have $D = 2^{l_1+l_2-dim(V^{(b,f)})} \cdot 2^{b+f+1} \cdot p^{-(x+y)}$.

For each possible counter array $\bar{C}'$, we have constructed the corresponding table $T_{\bar{C}'}$, thus the estimated space complexity is proportional to the sum of number of rows in each table $T_{\bar{C}'}$.

In the processing phase, the expected number of table look-ups depends on the probability that an internal state (L^t, N^t) is a special state satisfying the conditions (1.1), (1.2) and (1.3) simultaneously when the keystream segment $Z_t^{(b+f+1)} = 0^{b+f+1}$, which is given in Lemma 1 as $2^{-(l_1+l_2-dim(V^{(b,f)}))} \cdot p^{x+y}$. Thus the number of table look-ups is $2^{l_1+l_2-dim(V^{(b,f)})} \cdot p^{-(x+y)}$.

The pre-processing time complexity is determined by the workload for solving the system of equations constructed. $\qquad\square$

4   Cryptanalysis of Sprout

In this section, we apply the framework proposed in Sect. 3 to Sprout with the comparisons to the previous relevant attacks.

4.1   Fitting into the Model

Sprout fits into the model with the parameters $l_1 = l_2 = 40$, which are the length of LFSR and NFSR respectively. The keystream bit z_t at time t is generated as

$$z_t = h(n_{t+4}, l_{t+6}, l_{t+8}, l_{t+10}, l_{t+32}, l_{t+17}, l_{t+19}, l_{t+23}, n_{t+38})$$
$$\oplus l_{t+30} \oplus n_{t+1} \oplus n_{t+6} \oplus n_{t+15} \oplus n_{t+17} \oplus n_{t+23} \oplus n_{t+28} \oplus n_{t+34},$$

where $h(\cdot) = n_{t+4}l_{t+6} \oplus l_{t+8}l_{t+10} \oplus l_{t+32}l_{t+17} \oplus l_{t+19}l_{t+23} \oplus n_{t+4}l_{t+32}n_{t+38}$.

As described in Sect. 2, whether the secret key is involved in the NFSR state updating is determined by the value $u_t = l_{t+4} \oplus l_{t+21} \oplus l_{t+37} \oplus n_{t+9} \oplus n_{t+20} \oplus n_{t+29}$, to fit in the model, we have $rL^t = \{l_{t+4}, l_{t+21}, l_{t+37}\}$, $rN^t = \{n_{t+9}, n_{t+20}, n_{t+29}\}$ and the two parameters $d = 9$, $e = 10$ such that $\bigcup_{i=-9}^{10} rN^{t+i} \subseteq N^t$.

Let $pL^t = \{l_{t+6}, l_{t+8}, l_{t+10}, l_{t+17}, l_{t+19}, l_{t+23}, l_{t+32}\}$, $pN^t = \{n_{t+4}, n_{t+38}\}$, $qL^t = \{l_{t+30}\}$, and $qN^t = \{n_{t+1}, n_{t+6}, n_{t+15}, n_{t+17}, n_{t+23}, n_{t+28}, n_{t+34}\}$. From this we have $b = 1$, $f = 1$ to fit into the Sprout-like model. Given $(b, f) = (1, 1)$,

$$H^{(1,1)}(PL^t, PN^t) = \left(h(pL^{t-1}, pN^{t-1}), h(pL^t, pN^t), h(pL^{t+1}, pN^{t+1}) \right),$$

where $PL^t = L^t_{[5,11]} \cup L^t_{[16,20]} \cup L^t_{[22,24]} \cup L^t_{[31,33]}$ and $PN^t = N^t_{[3,5]} \cup N^t_{[37,39]}$, thus $H^{(1,1)}(\cdot)$ is a $(24, 3)$-vectorial Boolean function.

Suppose $n_{t+3} = n_{t+4} = n_{t+5} = 0$, by a computer computation, there are $12096 (> 2^{13})$ possible values for the following 16-bit of LFSR such that $H^{(1,1)}(\cdot) = 0^3$:

$$P^t = [l_{t+7}l_{t+8}l_{t+9}l_{t+10} \| l_{t+11}l_{t+16}l_{t+17}l_{t+18}$$
$$\| l_{t+19}l_{t+20}l_{t+22}l_{t+23} \| l_{t+24}l_{t+31}l_{t+32}l_{t+33}] \subseteq L^t.$$

For example, $P^t = 0x0000, 0x8000, 0x4000, 0xc000, \ldots$ when denoted by hexadecimal digits. We denote all the $12096 (> 2^{13})$ values of P^t as $a_1, a_2, a_3, a_4, \ldots, a_{12096}$ such that $a_1 = 0x0000$, $a_2 = 0x8000$, $a_3 = 0x4000$, $a_4 = 0xc000, \ldots$ respectively.

For Sprout, we will use 2^{13} flats defined as follows:

$$V_i = \{(L^t, N^t) : P^t = a_i \text{ and } n_{t+j} = 0, j = 3, 4, 5\}, \ i = 1, \ldots, 2^{13}$$

Note that in each V_i, 19 bits of (L^t, N^t) are fixed, then each V_i has a dimension of $dim(V_i) = 61$, and $H^{(1,1)}(\cdot) = 0^3$ over V_i. Further we define a flat V as $V = \bigcup_{i=1}^{2^{13}} V_i$. Thus the dimension of V is $dim(V) = 74$.

4.2   Cryptanalysis

We first discuss how to construct tables that will be used in the processing phase.

Pre-processing Phase. Given the parameters x, y and do the following:

1. Define a counter array as $C = [c^4_{t-y}, \ldots, c^4_{t-1}, c^4_t, c^4_{t+1}, \ldots, c^4_{t+(x-1)}]$ of size $|C| = x + y$. For an internal state (L^t, N^t) such that $n_{t+3} = n_{t+4} = n_{t+5} = 0$ (thus there are 77 unknowns), construct a system of equations which implies a state (L^t, N^t) satisfying the following conditions.
   - (a). $l(qL^{t+i}, qN^{t+i}) = 0$, for $i = -1, 0, 1$.
   - (b). $u_{t+i} = 0$ for $i = 0, 1, \ldots, x - 1$, from which we can get the output bits $z_{t+2}, \ldots, z_{t+x+1}$ (suppose the round constants $c^4_t, c^4_{t+1}, \ldots, c^4_{t+(x-1)}$ are known).

- (c). $u_{t-j} = 0$ for $j = 1, ..., y$, from which we can get the output bits $z_{t-2}, ..., z_{t-y-1}$ (suppose the round constants $c_{t-1}^4, c_{t-2}^4, ..., c_{t-y}^4$ are known).

2. We discuss it in the following situations:
   - **Case 1:** If $x \leq 11$ and $y \leq 9$, we have the corresponding system of $(3 + x + y)$ linear equations with only 77 unknowns from the state (L^t, N^t): 40 unknowns from L^t and 37 unknowns from N^t.

$$\begin{cases} l_{t+30+k} \oplus (\bigoplus_{i \in A} n_{t+i+k}) = 0, k = -1, 0, 1 \\ l_{t+4+i} \oplus l_{t+21+i} \oplus l_{t+37+i} \oplus n_{t+9+i} \oplus n_{t+20+i} \oplus n_{t+29+i} = 0, i = 0, 1, ..., x-1 \\ l_{t+4-j} \oplus l_{t+21-j} \oplus l_{t+37-j} \oplus n_{t+9-j} \oplus n_{t+20-j} \oplus n_{t+29-j} = 0, j = 1, ..., y \end{cases}$$

   - **Case 2:** If $x \geq 12$ and $y \leq 9$, in addition to the 77 unknowns from the state (L^t, N^t), the unknowns $n_{t+40}, n_{t+41}, ..., n_{t+40+(x-12)}$ will appear with some non-linear equations. Thus we obtain a system of equations with $(66 + x)$ unknowns, and $(2x + y - 8)$ equations $((3 + x + y)$ linear equations and $(x - 11)$ non-linear equations). Define another counter array $C' = [c_t^4, c_{t+1}^4, ..., c_{t+(x-12)}^4]$ of size $|C'| = x - 11$, note that the round constants in C' are involved in this system.

$$\begin{cases} l_{t+30+k} \oplus (\bigoplus_{i \in A} n_{t+i+k}) = 0, k = -1, 0, 1 \\ l_{t+4+i} \oplus l_{t+21+i} \oplus l_{t+37+i} \oplus n_{t+9+i} \oplus n_{t+20+i} \oplus n_{t+29+i} = 0, i = 0, 1, ..., x-1 \\ l_{t+4-j} \oplus l_{t+21-j} \oplus l_{t+37-j} \oplus n_{t+9-j} \oplus n_{t+20-j} \oplus n_{t+29-j} = 0, j = 1, ..., y \\ n_{t+40+m} \oplus l_{t+m} \oplus c_{t+m}^4 \oplus g(N^{t+m}) = 0, m = 0, 1, ..., x - 12 \, (non-linear) \end{cases}$$

   - **Case 3:** If $x \leq 11$ and $y \geq 10$, in addition to the 77 unknowns from the state (L^t, N^t), the unknowns $n_{t-1}, n_{t-2}, ..., n_{t-(y-9)}$ will appear with some non-linear equations. Thus we obtain a system of equations with $(68 + y)$ unknowns, and $(x + 2y - 6)$ equations $((3 + x + y)$ linear equations and $(y - 9)$ non-linear equations). Define $C' = [c_{t-(y-9)}^4, ..., c_{t-1}^4]$ of size $|C'| = y - 9$, the round constants in C' are involved in this system.

$$\begin{cases} l_{t+30+k} \oplus (\bigoplus_{i \in A} n_{t+i+k}) = 0, k = -1, 0, 1 \\ l_{t+4+i} \oplus l_{t+21+i} \oplus l_{t+37+i} \oplus n_{t+9+i} \oplus n_{t+20+i} \oplus n_{t+29+i} = 0, i = 0, 1, ..., x-1 \\ l_{t+4-j} \oplus l_{t+21-j} \oplus l_{t+37-j} \oplus n_{t+9-j} \oplus n_{t+20-j} \oplus n_{t+29-j} = 0, j = 1, ..., y \\ n_{t-n} \oplus l_{t-n} \oplus c_{t-n}^4 \oplus g'(N^{t-n+1}) = 0, n = 1, ..., y - 9 \, (non-linear) \end{cases}$$

   - **Case 4:** If $x \geq 12$ and $y \geq 10$, in addition to the 77 unknowns from the state (L^t, N^t), the unknowns $n_{t+40}, n_{t+41}, ..., n_{t+40+(x-12)}$ and $n_{t-1}, n_{t-2}, ..., n_{t-(y-9)}$ will appear with some non-linear equations. Thus we obtain a system of equations with $(57 + x + y)$ unknowns, and $(2x + 2y - 17)$ equations $((3 + x + y)$ linear equations and $(x + y - 20)$ non-linear equations). Define $C' = [c_{t-(y-9)}^4, ..., c_{t-1}^4, c_t^4, c_{t+1}^4, ..., c_{t+(x-12)}^4]$ of size $|C'| = x + y - 20$, the round constants in C' are involved in the system.

$$\begin{cases} l_{t+30+k} \oplus (\bigoplus_{i \in A} n_{t+i+k}) = 0, k = -1, 0, 1 \\ l_{t+4+i} \oplus l_{t+21+i} \oplus l_{t+37+i} \oplus n_{t+9+i} \oplus n_{t+20+i} \oplus n_{t+29+i} = 0, i = 0, 1, ..., x-1 \\ l_{t+4-j} \oplus l_{t+21-j} \oplus l_{t+37-j} \oplus n_{t+9-j} \oplus n_{t+20-j} \oplus n_{t+29-j} = 0, j = 1, ..., y \\ n_{t+40+m} \oplus l_{t+m} \oplus c_{t+m}^4 \oplus g(N^{t+m}) = 0, m = 0, 1, ..., x - 12 \, (non-linear) \\ n_{t-n} \oplus l_{t-n} \oplus c_{t-n}^4 \oplus g'(N^{t-n+1}) = 0, n = 1, ..., y - 9 \, (non-linear) \end{cases}$$

3. For each possible counter array C', solve the constructed system of equations. Observe that all the round constants in C' are added to the system linearly, by guessing at most $2^{74-(x+y)}$ appropriate unknowns we can solve the system and get $2^{74-(x+y)}$ solutions (L^t, N^t) for each possible counter array C'.

4. For each possible counter array C', check each of the $2^{74-(x+y)}$ solutions (L^t, N^t). If $(L^t, N^t) \in V_i$, i.e., $P^t = a_i$ for any $i = 1, 2, ..., 2^{13}$, store the 61-bit (L^{*t}, N^{*t}) in the first column of a row in table $T_{C',i}$, where $L^{*t} = L^t \backslash P^t$ and $N^{*t} = N^t \backslash \{n_{t+3}, n_{t+4}, n_{t+5}\}$. Further for this state and for each possible round constants of $C^* = C \backslash C'$, get the corresponding $(x + y)$ output bits $(z_{t-y-1}, ..., z_{t-2}, z_{t+2}, ..., z_{t+x+1})$ and put them in the second column as a sub-row in table $T_{C',i}$. Thus there are expected 2^{58-x-y} rows in the first column and $2^{58-x-y} \times \frac{Count(|C|)}{Count(|C'|)}$ rows in the second column, where $Count(n)$ represents the number of all the possible counter arrays of size n.

We list in Table 1 the number of all the possible counter arrays $Count(n)$ of size n.

Table 1. The size of the counter array n and the number $Count(n)$ for all the possible counter arrays

n	6	7	8	9	10	11	12	13	14	15	16	17
$Count(n)$	12	14	16	18	20	22	24	26	28	30	32	33
n		18	19	20	21	22	23	24	25	26	27	28	29
$Count(n)$		35	37	39	41	43	45	47	49	51	53	55	57
n		30	31	32	33	34	35	36	37	38	39	40	41
$Count(n)$		59	61	63	64	65	66	67	68	69	70	71	72
n		42	43	44	45	46	47	48	49	50	51	52	53
$Count(n)$		73	74	75	76	77	78	79	80	80	80	80	80

Remarks. First, it can be seen that, the necessary and sufficient condition for a state (L^t, N^t) to be a "special" state is that $(L^t, N^t) \in V$ and the conditions (a), (b) and (c) hold. Second, for each possible counter array C, $n_{t+40,...,}n_{t+40+(x-1)}$ and $n_{t-1,...,}n_{t-y}$ can be computed directly from a special state (L^t, N^t) according to the state updating of NFSR without involving the key information. Third, denote the number of rows (in the first column) of table $T_{C',i}$ as 2^{r_i}, if $r_i < x+y$, we only need to store $(x + y - r_i)$ output bits in the second column, indexed by r_i-bit of the output. Finally, in the pre-processing phase, we have obtained $Count(|C'|) \times 2^{13}$ tables $T_{C',i}$, each having 2^{58-x-y} rows in the first column to store "special" states and $2^{58-x-y} \times \frac{Count(|C|)}{Count(|C'|)}$ rows in the second column to store the corresponding output bits.

Lemma 2. *The probability that an internal state (L^t, N^t) is a special state (such that $(L^t, N^t) \in V$ and the conditions (a), (b) and (c) hold) when the keystream segment $Z_t^{(3)} = 0^3$ is given by the following:*

$$\Pr\left[(L^t, N^t) \text{ is a special state} \middle| Z_t^{(3)} = 0^3\right] = 2^{-(6+x+y)}.$$

Proof. For any internal state (L^t, N^t) and keystream segment $Z_t^{(3)}$, the underlying assumptions directly imply the following:

$$\Pr\left[(L^t, N^t) \text{ is a special state}\right] = \frac{1}{2^{40+40-\dim(V)}} \times \frac{1}{2^{3+x+y}} = 2^{-(9+x+y)},$$

and $\Pr\left[Z_t^{(3)} = 0^3\right] = 2^{-3}$, and

$$\Pr\left[Z_t^{(3)} = 0^3 \Big| (L^t, N^t) \text{ is a special state}\right] = 1.$$

On the other hand,

$$\Pr\left[(L^t, N^t) \text{ is a special state} \Big| Z_t^{(3)} = 0^3\right]$$
$$= \frac{\Pr\left[(L^t,N^t) \text{ is a special state}\right] \times \Pr\left[Z_t^{(3)}=0^3 \Big| (L^t,N^t) \text{ is a special state}\right]}{\Pr\left[Z_t^{(3)}=0^3\right]}$$
$$= 2^{-(6+x+y)}.$$

which yields the statement of the lemma. $\qquad\qquad\qquad\qquad\qquad\qquad\square$

Next, we will present a *State Checking and Key Recovery Mechanism* specified for Sprout, by which we have the opportunity to check whether a state candidate is correct, and if so, further recover the key for a correct guess.

State Checking and Key Recovery Mechanism. For a state candidate at time t, $L^t = [l_t, l_{t+1}, ..., l_{t+39}]$, $N^t = [n_t, n_{t+1}, ..., n_{t+39}]$, create an 80-bit vector K for the possible values associated with it:

1. Compute the value of n_{t-1} given by the keystream bit z_{t-2} as $n_{t-1} = z_{t-2} \oplus h(n_{t+2}, l_{t+4}, l_{t+6}, l_{t+8}, l_{t+30}, l_{t+15}, l_{t+17}, l_{t+21}, n_{t+36}) \oplus l_{t+28} \oplus \left(\bigoplus_{i \in A'} n_{t+i-2}\right)$ where $A' = \{6, 15, 17, 23, 28, 34\}$. And compute l_{t-1} by the LFSR updating equation as $l_{t-1} = l_{t+39} \oplus l_{t+33} \oplus l_{t+24} \oplus l_{t+19} \oplus l_{t+14} \oplus l_{t+4}$, and deduce from n_{t-1}, l_{t-1} the value k_{t-1}^* by the NFSR updating equation as $k_{t-1}^* = n_{t+39} \oplus c_{t-1}^4 \oplus l_{t-1} \oplus g(N^{t-1})$.
2. Compute the value of $u_{t-1} = l_{t+3} \oplus l_{t+20} \oplus l_{t+36} \oplus n_{t+8} \oplus n_{t+19} \oplus n_{t+28}$ and combine it with the value of k_{t-1}^* obtained in Step 1:
   - if $u_{t-1} = 0$ and $k_{t-1}^* = 0$, set $t \to t-1$ and go back to Step 1.
   - if $u_{t-1} = 0$ and $k_{t-1}^* = 1$, there is a contradiction, conclude that this guess for state is not correct and stop.
   - if $u_{t-1} = 1$ and $k_{t-1}^* = 0$, check if $k_{(t-1) \bmod 80}$ has already been set in K. If no, set it to 0. Set $t \to t-1$ and go back to Step 1. Else, if there is a contradiction, conclude that this guess for state is not correct and stop.
   - if $u_{t-1} = 1$ and $k_{t-1}^* = 1$, check if $k_{(t-1) \bmod 80}$ has already been set in K. If no, set it to 1. Set $t \to t-1$ and go back to Step 1. Else, if there is a contradiction, conclude that this guess for state is not correct and stop.

Similar to the statements in [8], the probability that a state candidate survives for $2r$ clocks is 2^{-r}. On average for each 2 clocks, half of the possible guesses will be eliminated. For 2^s candidate states, the average number of clocks for each elimination is

$$\sum_{i=0}^{s} \frac{2 \times 2^{s-i}}{2^s} = \sum_{i=0}^{s} \frac{1}{2^{i-1}} \approx 4$$

We can conclude that 4 clocks of output is enough for checking the validity of a candidate state and the recovery of the key bits for each candidate.

Next we illustrate the algorithm for the internal state and key recovery in the processing phase.

Processing Phase. Given the parameter x, y, the corresponding $Count(|C'|) \times 2^{13}$ tables $T_{C',i}$ and the given keystream sample $\{z_t\}_{t \geq 0}$, the processing steps are as follows:

1. Search the keystream sequence $\{z_t\}_t$ for the next non-considered block of 3 zeros. If there are no more blocks, output a flag that the algorithm has failed to recover the key.

2. For each detected block, compute the corresponding counter array C, C' and C^* from the time t. For $i = 1, ..., 2^{13}$, compare the x-bit segment of the keystream subsequent to the block and y-bit segment prior to the block with the memorized $(x+y)$-bit segments in the second column (sub-row is indexed by C^*) of the table $T_{C',i}$, and do the following:
   - If the matching does not exist, go to the processing Step 1.
   - If the $(x+y)$-bit sample segments match with a segment in table $T_{C',i}$, go to Step 3.
3. Read the corresponding state, check whether it is a correct state or not and recover the secret key by the *State Checking and Key Recovery Mechanism* stated above. If this state survives, recover and output the key, else go to Step 1.

Theorem 2. *For two positive integers x, y, the dedicated TMD tradeoff on Sprout has complexities as follows: (1) The data complexity for the processing is $D = 2^{9+x+y}$; (2) The expected memory M (-bit) of pre-processing is computed as follows:*

$$M = \begin{cases} Count(|C'|) \times 2^{71-x-y} \times \left[61 + \frac{Count(x+y)}{Count(|C'|)} \cdot (x+y)\right], & \text{if } x+y < 30, \\ Count(|C'|) \times 2^{71-x-y} \times \left[61 + \frac{Count(x+y)}{Count(|C'|)} \cdot (2x+2y-58)\right], & \text{if } x+y \geq 30. \end{cases}$$

(3) The time complexity of processing is $2^{70.66-x-y}$ Sprout encryptions along with 2^{6+x+y} table look-ups. (4) The time complexity of pre-processing is proportional to 2^{74-x-y}.

Proof. The data complexity is determined by the probability that an internal state (L^t, N^t) is a special state (such that $(L^t, N^t) \in V$ and the conditions (a), (b) and (c) hold), which is given in the proof of Lemma 2 as $2^{-(9+x+y)}$. Thus we have $D = 2^{9+x+y}$.

As for the memory, we need $Count(|C'|) \times 2^{13}$ tables $T_{C',i}$, each having 2^{58-x-y} rows in the first column to store 61-bit "special" states (3+16=19-bit are fixed for each table) and $2^{58-x-y} \times \frac{Count(|C|)}{Count(|C'|)}$ rows in the second column to store the corresponding output bits. If $x+y < 30$, each row in the second column contains $(x+y)$ output bits; if $x+y \geq 30$, each row in the second column contains $2(x + y) - 58$ output bits, indexed by $58 - x - y$ bits of the output. Hence the memory M(-bit) is computed as follows:

$$M = \begin{cases} Count(|C'|) \times 2^{13} \times 2^{58-x-y} \times \left[61 + \frac{Count(x+y)}{Count(|C'|)} \cdot (x+y)\right], & \text{if } x + y < 30, \\ Count(|C'|) \times 2^{13} \times 2^{58-x-y} \times \left[61 + \frac{Count(x+y)}{Count(|C'|)} \cdot (2x+2y-58)\right], & otherwise. \end{cases}$$

In the processing phase, the expected number of table look-ups is determined by the probability that an internal state (L^t, N^t) is a special state when the keystream segment $Z_t^{(3)} = 0$, which is given in Lemma 2 as $2^{-(6+x+y)}$. Thus the number of table look-ups is 2^{6+x+y}. For each $(x + y)$-bit keystream bits, we have $2^{71-2(x+y)}$ state candidates producing the output. As stated before, 4 more clocks of output is enough for checking the validity of the state and the recovery of the key bits for each candidate. In total, the time complexity is $2^{6+x+y} \times 2^{71-2(x+y)} \times 4 = 2^{79-x-y}$, which is equivalent to $\frac{2^{79-x-y}}{324} = 2^{70.66-x-y}$ Sprout encryptions.

The pre-processing time complexity is equivalent to solving the constructed system of equations. We see that by guessing at most $2^{74-(x+y)}$ appropriate unknowns we can solve the system for each possible counter array C'. As the counter values are added to the systems linearly, we can do the Gauss elimination only once to store separate tables for each of the $Count(|C'|)$ counter arrays. $\square$

4.3 Detailed Workload for $x = 16$, $y = 15$

We now focus on the workload to solve the system of equations for $x = 16$, $y = 15$. For a state (L^t, N^t), let $n_{t+j} = 0, j = 3, 4, 5$ and define $N^{*t} = N^t \backslash \{n_{t+3}, n_{t+4}, n_{t+5}\}$. We need to solve the following systems of equations, which amounts to 34 linear equations, 11 non-linear equations and 88 unknowns $L^t, N^{*t}, n_{t+40}, n_{t+41}, ..., n_{t+44}, n_{t-1}, n_{t-2}, ..., n_{t-6}$.

$$\begin{cases} 1: l_{t+29} \oplus n_t \oplus n_{t+5} \oplus n_{t+14} \oplus n_{t+16} \oplus n_{t+22} \oplus n_{t+27} \oplus n_{t+33} = 0 \\ 2: l_{t+30} \oplus n_{t+1} \oplus n_{t+6} \oplus n_{t+15} \oplus n_{t+17} \oplus n_{t+23} \oplus n_{t+28} \oplus n_{t+34} = 0 \\ 3: l_{t+31} \oplus n_{t+2} \oplus n_{t+7} \oplus n_{t+16} \oplus n_{t+18} \oplus n_{t+24} \oplus n_{t+29} \oplus n_{t+35} = 0 \\ 4: u_{t-15} = l_{t-11} \oplus l_{t+6} \oplus l_{t+22} \oplus n_{t-6} \oplus n_{t+5} \oplus n_{t+14} = 0 \\ 5: u_{t-14} = l_{t-10} \oplus l_{t+7} \oplus l_{t+23} \oplus n_{t-5} \oplus n_{t+6} \oplus n_{t+15} = 0 \\ \quad\cdots\cdots \\ 18: u_{t-1} = l_{t+3} \oplus l_{t+20} \oplus l_{t+36} \oplus n_{t+8} \oplus n_{t+19} \oplus n_{t+28} = 0 \\ 19: u_t = l_{t+4} \oplus l_{t+21} \oplus l_{t+37} \oplus n_{t+9} \oplus n_{t+20} \oplus n_{t+29} = 0 \\ 20: u_{t+1} = l_{t+5} \oplus l_{t+22} \oplus l_{t+38} \oplus n_{t+10} \oplus n_{t+21} \oplus n_{t+30} = 0 \\ \quad\cdots\cdots \\ 34: u_{t+15} = l_{t+19} \oplus l_{t+36} \oplus l_{t+52} \oplus n_{t+24} \oplus n_{t+35} \oplus n_{t+44} = 0 \end{cases}$$

$$
\begin{cases}
35: n_{t+40} \oplus l_t \oplus c_t^4 \oplus g(N^t) = 0 \\
36: n_{t+41} \oplus l_{t+1} \oplus c_{t+1}^4 \oplus g(N^{t+1}) = 0 \\
\cdots\cdots \\
39: n_{t+44} \oplus l_{t+4} \oplus c_{t+4}^4 \oplus g(N^{t+4}) = 0 \\
40: n_{t-1} \oplus l_{t-1} \oplus c_{t-1}^4 \oplus g'(N^t) = 0 \\
41: n_{t-2} \oplus l_{t-2} \oplus c_{t-2}^4 \oplus g'(N^{t-1}) = 0 \\
\cdots\cdots \\
45: n_{t-6} \oplus l_{t-6} \oplus c_{t-6}^4 \oplus g'(N^{t-5}) = 0
\end{cases}
$$

In the following part, L^t is treated as a column vector of size 40. First of all, we choose the 40 equations numbered by 1,2,...,34 and 40,41,...,45 from the above systems to represent L^t by the unknowns $N^{*t}, n_{t+40}, n_{t+41}, ..., n_{t+44}, n_{t-1}, n_{t-2}, ..., n_{t-6}$ as $\mathrm{M} \cdot L^t = v$, where M is the 40×40 coefficient matrix of L^t, and v is a column vector of size 40, and

$$
\mathrm{M} \cdot L^t = [l_{t+29}, l_{t+30}, l_{t+31}, l_{t-11} \oplus l_{t+6} \oplus l_{t+22}, ...,
$$
$$
l_{t+19} \oplus l_{t+36} \oplus l_{t+52}, l_{t-1}, ..., l_{t-6}]^T,
$$

and

$$
v = [\bigoplus_{i \in B} n_{t+i-1}, \bigoplus_{i \in B} n_{t+i}, \bigoplus_{i \in B} n_{t+i+1}, n_{t-6} \oplus n_{t+5} \oplus n_{t+14}, ...,
$$
$$
n_{t+24} \oplus n_{t+35} \oplus n_{t+44}, n_{t-1} \oplus c_{t-1}^4 \oplus g'(N^t), ..., n_{t-6} \oplus c_{t-6}^4 \oplus g'(N^{t-5})]^T.
$$

We have checked that rank(M) = 39. Take l_t as a free variable, we obtain an invertible coefficient matrix of size 39×39. Let $L'^t = L^t \setminus \{l_t\}$, then each variable in L'^t can be uniquely represented as linear combinations of $N^{*t}, n_{t+40}, n_{t+41}, ..., n_{t+44}, n_{t-1}, n_{t-2}, ..., n_{t-6}$ and l_t, together with 1 non-linear equation with these unknowns. Plugging in the values $l_{t+1}, l_{t+2}, l_{t+3}, l_{t+4}$ in equations numbered by 36,...,39, we get a system with 6 non-linear equations and 49 unknowns $N^{*t}, n_{t+40}, n_{t+41}, ..., n_{t+44}, n_{t-1}, n_{t-2}, ..., n_{t-5}$ and l_t. Define a set GUESS = $\{n_{t+j} : j \in S\}$ of size 33, where

$$
S = \{-1, 0, 1, 3, 4, 6, 7, 9, 11, 12, 13, 15, 16, 17, 19, 20, 21,
$$
$$
23, 24, 25, 26, 28, 30, 31, 32, 33, 34, 35, 36, 37, 39, 40, 41\}.
$$

By guessing the 33 unknowns in the set GUESS,

– If $n_{t+9} = 0$, we come up with 2^{32} systems with 6 linear equations and 16 unknowns; For each of these systems, we do the Gauss elimination once by choosing an invertible coefficient matrix of 6×6. The systems can be solved with $2^{32} \times (6^3 + 2^{10}) = 2^{42.27}$ basic operations.

– If $n_{t+9} = 1$, we further guess n_{t+8}, thus we get 2^{33} systems with 6 linear equations and 15 unknowns. Similarly, the systems can be solved with $2^{33} \times (6^3 + 2^9) = 2^{42.51}$ basic operations.

In total, the pre-computation is approximately $2^{43.39}$ basic operations.

We list in Table 2 more instances that illustrate the complexities of the TMD tradeoff attacks on Sprout. The Comparison of our TMD tradeoff attacks with the previous ones in [8] and [12] are presented in Table 3. With carefully chosen attack parameters, our method is at least 2^{20} times faster than the attack in [12], 2^{10} times faster than the attack in [8] with much less memory.

Table 2. The complexity issues of the attack on Sprout

x, y	$Count(x + y)$	Data	Memory(-bit),(TB)	Time	Pre-computation
16,14	59	2^{39}	$2^{51.39}$-bit, 336 TB	$2^{40.66}$	$2^{44.03}$
16,15	61	2^{40}	$2^{50.63}$-bit, 198 TB	$2^{39.66}$	$2^{43.39}$
17,15	63	2^{41}	$2^{49.85}$-bit, 115 TB	$2^{38.66}$	$2^{43.81}$
17,16	64	2^{42}	$2^{49.03}$-bit, 65 TB	$2^{37.66}$	$2^{45.36}$
18,16	65	2^{43}	$2^{48.20}$-bit, 36 TB	$2^{36.66}$	$2^{47.09}$

Table 3. Comparison of our time/memory/data Tradeoff attacks with the previous ones

Attack	Data	Memory(-bit),(TB)	Time	Pre-computation
[12]	112	$\geq 2^{52.32}$-bit, $\geq$ 639 TB	$2^{66.80}$	$2^{68.87}$
[8]	2^{40}	$2^{52.58}$-bit, 770 TB	$2^{30.66}$	$2^{54.29}$
[8]	2^{41}	$2^{52.64}$-bit, 399 TB	$2^{29.66}$	$\approx 2^{56.70}$
[8]	2^{42}	$2^{50.69}$-bit, 207 TB	$2^{28.66}$	$\approx 2^{59.07}$
[8]	2^{43}	$2^{49.74}$-bit, 108 TB	$2^{27.66}$	$\approx 2^{61.42}$
ours	2^{39}	$2^{51.39}$-bit, 336 TB	$2^{40.66}$	$2^{44.03}$
ours	2^{40}	$2^{50.63}$-bit, 198 TB	$2^{39.66}$	$2^{43.39}$
ours	2^{41}	$2^{49.85}$-bit, 115 TB	$2^{38.66}$	$2^{43.81}$
ours	2^{42}	$2^{49.03}$-bit, 65 TB	$2^{37.66}$	$2^{45.36}$
ours	2^{43}	$2^{48.20}$-bit, 36 TB	$2^{36.66}$	$2^{47.09}$

5 Practical Implementation

To verify the validity of our attack, we experimentally test it on a reduced cipher with similar structure and properties as Sprout. In general, the simulation results match well with the theoretical estimates.

5.1 The Reduced Version of Sprout

Similarly, there is an 8-bit counter register, of which the lower 6 bits are a modulo 40 counter, denoted by $(c_t^5, c_t^4, c_t^3, c_t^2, c_t^1, c_t^0)$ at a given round t. The 3-th LSB c_t^3 of the counter is employed in the keystream generation. It should be noted that, c_t^3 has a cycle of length 40, i.e., in each cycle, this bit takes the values $\underbrace{0, 0, ..., 0}_{8}\underbrace{1, 1, ..., 1}_{8}\underbrace{0, 0, ..., 0}_{8}\underbrace{1, 1, ..., 1}_{8}\underbrace{0, 0, ..., 0}_{8}$.

The reduced version of Sprout uses a 20-bit LFSR and a 20-bit NFSR. At time t, the LFSR state is $L^t = [l_t, l_{t+1}, ..., l_{t+19}]$, and it is updated recursively by f as $l_{t+20} = l_t \oplus l_{t+1} \oplus l_{t+14} \oplus l_{t+15} \oplus l_{t+16} \oplus l_{t+19}$. The NFSR state $N^t = [n_t, n_{t+1}, ..., n_{t+19}]$ is updated recursively by a nonlinear feedback function g as

$$n_{t+20} = k_t^* \oplus c_t^3 \oplus l_t \oplus g(N^t)$$
$$= k_t^* \oplus c_t^3 \oplus l_t \oplus n_t \oplus n_{t+13} \oplus n_{t+15} \oplus n_{t+17} \oplus n_{t+19}$$
$$\oplus n_{t+2}n_{t+5} \oplus n_{t+3}n_{t+7} \oplus n_{t+8}n_{t+9} \oplus n_{t+1}n_{t+14} \oplus n_{t+16}n_{t+18} \oplus n_{t+6}n_{t+12}$$
$$\oplus n_{t+13}n_{t+16}n_{t+17}n_{t+18} \oplus n_{t+10}n_{t+11}n_{t+12} \oplus n_{t+4}n_{t+7}n_{t+11}.$$

Let u_t be $u_t = l_{t+1} \oplus l_{t+4} \oplus l_{t+17} \oplus n_{t+4} \oplus n_{t+10} \oplus n_{t+14}$, then

$$k_t^* = \begin{cases} k_t, \ 0 \leq t \leq 39 \\ k_{t(\text{mod } 40)} \cdot u_t, \ \text{otherwise} \end{cases}$$

Given the internal state at time t, the keystream bit z_t is generated as

$$z_t = h(n_{t+4}, l_{t+6}, l_{t+8}, l_{t+10}, l_{t+12}, l_{t+17}, l_{t+19}, l_{t+3}, n_{t+18}) \oplus l_{t+10} \oplus \left(\bigoplus_{i \in A} n_{t+i} \right),$$

where $A = \{1, 3, 6, 15, 17\}$, and the filter function $h(\cdot)$ is defined as

$$h(\cdot) = n_{t+4}l_{t+6} \oplus l_{t+8}l_{t+10} \oplus l_{t+12}l_{t+17} \oplus l_{t+19}l_{t+3} \oplus n_{t+4}l_{t+12}n_{t+18}.$$

During the key/IV setup phase, since the key is fixed, first load the IV in the following way: $n_i = iv_i, 0 \leq i \leq 19$; $l_i = iv_{i+20}, 0 \leq i \leq 14$ and $l_i = 1, 15 \leq i \leq 18$, $l_{19} = 0$. Then run the cipher 160 rounds as follows.

- the LFSR update function is changed to $l_{t+20} = z_t \oplus f(L^t)$.
- the NFSR update function is changed to $n_{t+20} = z_t \oplus k_t^* \oplus l_t \oplus c_t^3 \oplus g(N_t)$.
- no keystream bit is generated.

After the initialization phase, the keystream generation phase starts and there is no feedback keystream anymore.

5.2   Attack Process

Suppose $l_{t+4} = 0$, $n_{t+3} = n_{t+4} = n_{t+5} = 0$, by a computer computation, there are $1728(> 2^{10})$ possible values for the following 13-bit of LFSR such that $H^{(1,1)}(\cdot)$ is 0 ($\in F_2^3$).

$$P^t = [l_{t+2}||l_{t+3}l_{t+7}l_{t+8}l_{t+9}||l_{t+10}l_{t+11}l_{t+12}l_{t+13}||l_{t+16}l_{t+17}l_{t+18}l_{t+19}] \subseteq L^t.$$

For example, $P^t = $ 0x0000, 0x0001, ... We denote all the 1728 values of P^t as $a_1, a_2, ..., a_{1024}, ..., a_{1728}$, where the first 1024 values are $a_1 = $ 0x0000, $a_2 = $ 0x0001,..., $a_{1024} = $ 0x1ba1. For convenience, several notations are defined as follows:

- $L^{*t} = L^t \backslash (\{l_{t+4}\} \cup P^t)$ of 6-bit.
- $N^{*t} = N^t \backslash \{n_{t+3}, n_{t+4}, n_{t+5}\}$ of 17-bit.

- Define $C = [c_{t-5}^3, ..., c_{t-1}^3, c_t^3, c_{t+1}^3, ..., c_{t+5}^3]$ of length 11, the employed counter array. There are 21 different counter arrays, denoted by hexadecimal numbers, they are

$$0x007, 0x00f, 0x01f, 0x03f, 0x07f, 0x0ff, 0x1fe,$$
$$0x3fc, 0x7f8, 0x7f0, 0x7e0, 0x7c0, 0x780, 0x700,$$
$$0x601, 0x403, 0x600, 0x400, 0x000, 0x001, 0x003,$$

- Define $C' = [c_{t-1}^3]$ of length 1, and $C^* = [c_{t-5}^3, ..., c_{t-2}^3, c_t^3, c_{t+1}^3, ..., c_{t+5}^3]$ of length 10. There are 2 different values for C', denoted as $c_1' = 0x0$, $c_2' = 0x1$. If $c_{t-1}^3 = 0x0$, there are 13 different values for C^*, they are $0x007, 0x00f,$ $0x01f, 0x03f, 0x3c0, 0x380, 0x301, 0x203, 0x300, 0x200, 0x000, 0x001, 0x003$; If $c_{t-1}^3 = 0x1$, there are 8 different values for C^*, they are $0x03f, 0x07f, 0x0fe,$ $0x1fc, 0x3f8, 0x3f0, 0x3e0, 0x3c0$.

Pre-processing Phase. For any state (L^t, N^t), suppose $l_{t+4} = 0$, $n_{t+3} = n_{t+4} = n_{t+5} = 0$. In the pre-processing phase, we construct 2×2^{10} tables T_{c_j', a_i} indexed with c_j' and a_i, for $c_1' = 0x0$, $c_2' = 0x1$ and $a_1 = 0x0000$, $a_2 = 0x0001, ..., a_{1024} = 0x1ba1$. In Table T_{c_j', a_i}, 23-bit (L^{*t}, N^{*t}) are stored in the first column of a row such that

$$\begin{cases} P^t = [l_{t+2}||l_{t+3}l_{t+7}l_{t+8}l_{t+9}||l_{t+10}l_{t+11}l_{t+12}l_{t+13}||l_{t+16}l_{t+17}l_{t+18}l_{t+19}] = a_i \\ l_{t+10+i} \oplus n_{t+1+i} \oplus n_{t+3+i} \oplus n_{t+6+i} \oplus n_{t+15+i} \oplus n_{t+17+i} = 0, i = -1, 0, 1 \\ u_{t+j} = l_{t+1+j} \oplus l_{t+4+j} \oplus l_{t+17+j} \oplus n_{t+4+j} \oplus n_{t+10+j} \oplus n_{t+14+j} = 0, j = 0, 1, ..., 5 \\ u_{t-k} = l_{t+1-k} \oplus l_{t+4-k} \oplus l_{t+17-k} \oplus n_{t+4-k} \oplus n_{t+10-k} \oplus n_{t+14-k} = 0, k = 1, 2, ..., 5 \\ n_{t-1} \oplus l_{t-1} \oplus c_{t-1}^3 \oplus g'(N^t) = 0 \ (non - linear) \end{cases}$$

Similarly, we can solve all the systems by choosing a set of unknowns as $GUESS = \{n_t, n_{t+1}, n_{t+6}, n_{t+7}, n_{t+10}, n_{t+11}, n_{t+15}, n_{t+16}, n_{t+17}\}$ with approximately 2^{23} basic operations. Besides, for each (c_j', a_i) pair, there are expected 2^9 solutions, we store the 23-bit (L^{*t}, N^{*t}) of the internal state $(4+13=17-$ bit are fixed for each table) in the first column of a row in table T_{c_j', a_i}. Further for this state and for each possible round constants C^*, get the corresponding 11-bit output $(z_{t-6}, ..., z_{t-2}, z_{t+2}, ..., z_{t+7})$ and put them in the second column as a sub-row indexed by C^*. The number of sub-row is 13 for $c_1' = 0x0$, while the number is 8 for $c_2' = 0x1$. In total, the memory needed is $M = 2^{10} \times 2^9 \times (23 + 11 \times 13) + 2^{10} \times 2^9 \times (23 + 11 \times 8) \approx 2^{27.11}$-bit, i.e., 17.3 MB[1].

Next, we present the *State Checking and Key Recovery Mechanism* specified for the reduced version of Sprout, which is similar to the one stated for Sprout.

State Checking and Key Recovery Mechanism. For a candidate state at time t, $L^t = [l_t, l_{t+1}, ..., l_{t+19}]$, $N^t = [n_t, n_{t+1}, ..., n_{t+19}]$, create a 40-bit vector K for the possible values associated with it:

---

[1] Since each table is expected to have 2^9 rows, we can only store 2 output bits in the second column of each row, indexed by 9 bits of the output. Thus, the memory can be reduced to $2^{10} \times 2^9 \times (23 + 2 \times 13) + 2^{10} \times 2^9 \times (23 + 2 \times 8) \approx 2^{25.46}$-bit, i.e., 5.5 MB.

1. Compute the value of n_{t-1} given by the keystream bit z_{t-2} as $n_{t-1} = z_{t-2} \oplus h(n_{t+2}, l_{t+4}, l_{t+6}, l_{t+8}, l_{t+10}, l_{t+15}, l_{t+17}, l_{t+1}, n_{t+16}) \oplus l_{t+8} \oplus (\bigoplus_{i \in A'} n_{t+i-2})$ where $A' = \{3, 6, 15, 17\}$. And compute l_{t-1} by the LFSR updating equation as $l_{t-1} = l_{t+19} \oplus l_{t+18} \oplus l_{t+15} \oplus l_{t+14} \oplus l_{t+13} \oplus l_t$, and deduce from n_{t-1}, l_{t-1} the value k_{t-1}^* by the NFSR updating equation as $k_{t-1}^* = n_{t+19} \oplus c_{t-1}^3 \oplus l_{t-1} \oplus g(N^{t-1})$.

2. Compute the value of $u_{t-1} = l_t \oplus l_{t+3} \oplus l_{t+16} \oplus n_{t+3} \oplus n_{t+9} \oplus n_{t+13}$ and combine it with the value of k_{t-1}^* obtained in Step 1:
   - if $u_{t-1} = 0$ and $k_{t-1}^* = 0$, set $t \to t - 1$ and go back to Step 1.
   - if $u_{t-1} = 0$ and $k_{t-1}^* = 1$, there is a contradiction, conclude that this guess for state is not correct and stop.
   - if $u_{t-1} = 1$ and $k_{t-1}^* = 0$, check if $k_{(t-1) \mod 40}$ has already been set in K. If no, set it to 0. Set $t \to t - 1$ and go back to Step 1. Else, if there is a contradiction, conclude that this guess for state is not correct and stop.
   - if $u_{t-1} = 1$ and $k_{t-1}^* = 1$, check if $k_{(t-1) \mod 40}$ has already been set in K. If no, set it to 1. Set $t \to t - 1$ and go back to Step 1. Else, if there is a contradiction, conclude that this guess for state is not correct and stop.

By utilizing the pre-computed tables and the given keystream sample, the processing phase is carried out as follows.

The Internal State Recovery Algorithm. Given the 2×2^{10} tables $T_{c'_j, a_i}$, and the keystream sample $\{z_t\}_{t \geq 0}$ having at least 2^{21} sample segments, the processing steps are as follows:

1. Search the keystream sequence $\{z_t\}_{t \geq 6}$ for the next non-considered block of 3 zeros, i.e., $z_{t-1} z_t z_{t+1} = 000$. If there are no more blocks, output a flag that the algorithm has failed.

2. For each detected block, compute the corresponding $C' = [c_{t-1}^3] \triangleq c'$ and $C^* = [c_{t-5}^3, ..., c_{t-2}^3, c_t^3, c_{t+1}^3, ..., c_{t+5}^3] \triangleq c^*$ from the time t. For $a_1 = \text{0x0000}$, $a_2 = \text{0x0001}, ..., a_{1024} = \text{0x1ba1}$, compare $(z_{t+2} z_{t+3} ... z_{t+7})$ after the zero-segment and $(z_{t-6} z_{t-5} ... z_{t-2})$ before the zero-segment with the memorized 11-bit segments in the second column of a sub-row indexed by c^* from the tables T_{c', a_i}, and do the following:
   - If the matching does not exist, go to the processing Step 1.
   - If the 11-bit sample segments match with a segment in table T_{c', a_i}, go to Step 3.

3. Read the corresponding state, check whether it is a correct state or not and recover the secret key by the *State Checking and Key Recovery Mechanism* stated before. If this state survives, recover and output the key, else go to Step 1.

5.3   Simulation Results

Our attacks have been fully implemented on one core of a single PC, running with Windows 7, Intel Core i3-2120 CPU @ 3.30 GHz and 4.00 GB RAM. In general, the experimental results match the theoretical analysis quite well. We present the details as follows.

In our experiment, first of all, we constructed 2×2^{10} tables indexed by (c'_j, a_i) pairs for $c'_1 = \texttt{0x0}$, $c'_2 = \texttt{0x1}$ and $a_1 = \texttt{0x0000}$, $a_2 = \texttt{0x0001}$,...,$a_{1024} = \texttt{0x1ba1}$, storing the special internal states. We used 2×2^{10} text files to store the $(State, Keystream_1, keystream_2, ..., keystream_{count(|C^*|)})$ tuples named with the corresponding c'_j and a_i. Note that $count(|C^*|) = 13$ for $c'_1 = \texttt{0x0}$ and $count(|C^*|) = 8$ for $c'_2 = \texttt{0x1}$. Experimental results show that there are 496 or 504 or 520 or 528 rows in each table, and totally $524448(\approx 2^{19})$ rows for $c'_1 = \texttt{0x0}$, $524128(\approx 2^{19})$ rows for $c'_2 = \texttt{0x1}$. Thus the memory needed in the simulation is $524448 \times (23 + 11 \times 13) + 524128 \times (23 + 11 \times 8) \approx 2^{27.11}$-bit, i.e., 17.3 MB, which matches the theoretical estimate quite well.

For the key recovery algorithm illustrated above, the data complexity is estimated by the probability that an internal state (L^t, N^t) is a special state satisfying:

(1) $l_{t+4} = 0$, $n_{t+3} = n_{t+4} = n_{t+5} = 0$,
(2) $P^t = a_1$ or $P^t = a_2$ or ... $P^t = a_{1024}$,
(3) $l_{t+10+d} \oplus \left(\bigoplus_{i \in A} n_{t+i+d} \right) = 0$ for $d = -1, 0, 1$,
(4) $u_{t+j} = 0$, for $j = 0, 1, ..., 5$,
(5) $u_{t-k} = 0$, for $k = 1, 2, ..., 5$,

Thus the theoretical estimate is $D = 2^{21}$. In the experiment, we used the RC4 cipher to randomly generate 2^{15} (K, IV) pairs and for each randomly chosen (K, IV) pair, we ran the cipher and generated 2^{21} keystream bits. Results show that we can get a special state at time $t \le 2^{21}$ for $20423(\approx 2^{14.32})$ (K, IV) pairs. For example, suppose (K, IV) pair be

$$K = 101010010101100110101011001001100011000110110$$
$$IV = 1101010110100100111010011001011011011$$

where the left-most bit represents the value for index 0. At time $t = 580697(\approx 2^{19.14})$, a special state arises in Table $T_{c', a_{140}}$, where $c' = \texttt{0x0}$ and $a_{140} = \texttt{0x0191}$, such that $(1)(3)(4)(5)$ hold and $P^t = \texttt{0x0191}$. This internal state is

$$L^t = 11110000010001110000$$
$$N^t = 00100000011011010110$$

In the internal state and key recovery algorithm, we search the keystream sequence for the 3 zeros blocks, and for each block, we try to find matching pairs, and further recover the key. In the experiment, we first searched the given keystream sequence and collected the time instances t implying 3 zeros. The expected number of such instances is $2^{21} \times 2^{-3} = 2^{18}$. Besides, for each 11-bit output, the expected number of candidate states is $\frac{2^{19}}{2^{11}} = 2^8$ producing this output. Thus we go through all the time instances, and for each time instance, we go through all the candidate states. We have also verified by experiments that 4 more clocks of output is enough for checking the validity of the state and the recovery of the key bits for each candidate. In total, the estimate of the time complexity is $2^{18} \times 2^8 \times 4 = 2^{28}$. In the simulation, for the (K, IV) pair above, we have recovered all the key bits within 1 hour.

6   Conclusion

In this paper, we have studied the security of Sprout-like stream ciphers in a unified framework from the viewpoint of k-normality of the augmented function. We made a systematic security analysis based on this property and developed a dedicated TMD tradeoff attack framework for such designs. In particular, it is shown that Sprout can be broken by various TMD tradeoffs. Our attack is highly flexible and compares favorably to all the previous attacks on Sprout, which demonstrates the superiority of the new method. We believe that stream ciphers with shorter internal state may suffer from the time/memory/data tradeoff attacks and the k-normality of the augmented function should be taken into account for new stream cipher designs.

References

1. Ågren, M., Hell, M., Johansson, T., Meier, W.: A New Version of Grain-128 with Authentication, Symmetric Key Encryption Workshop 2011, DTU, Denmark
2. Armknecht, F., Mikhalev, V.: On lightweight stream ciphers with shorter internal states. In: Leander, G. (ed.) FSE 2015. LNCS, vol. 9054, pp. 451–470. Springer, Heidelberg (2015)
3. Banik Subhadeep., Some Results on Sprout. http://eprint.iacr.org/2015/327.pdf
4. Biryukov, A., Shamir, A.: Cryptanalytic time/memory/data tradeoffs for stream ciphers. In: Okamoto, T. (ed.) ASIACRYPT 2000. LNCS, vol. 1976, pp. 1–13. Springer, Heidelberg (2000)
5. Braeken, A., Wolf, C., Preneel, B.: Normality of vectorial functions. In: Smart, N.P. (ed.) Cryptography and Coding 2005. LNCS, vol. 3796, pp. 186–200. Springer, Heidelberg (2005)
6. Dobbertin, H.: Construction of bent functions and balanced Boolean functions with high nonlinearity. In: Preneel, B. (ed.) Fast Software Encryption. LNCS, vol. 1008, pp. 61–74. Springer, Heidelberg (1995)
7. http://www.ecrypt.eu.org/stream/
8. Esgin, M.F., Kara, O.: Practical Cryptanalysis of Full Sprout with TMD Tradeoff Attacks. http://eprint.iacr.org/2015/289.pdf
9. Hao, Y.: A Related-Key Chosen-IV Distinguishing Attack on Full Sprout Stream Cipher. http://eprint.iacr.org/2015/231.pdf
10. Hell, M., Johansson, T., Meier, W.: Grain - a Stream Cipher for Constrained Environments. Int. J. Wirel. Mobile Comput. 2(1), 86–93 (2007)
11. Hell, M., Johansson, T., Maximov, A., Meier, W.: A stream cipher proposal: grain-128. In: IEEE International Symposium on Information Theory - ISIT 2006 (2006)
12. Lallemand, V., Naya-Plasencia, M.: Cryptanalysis of full sprout. In: Gennaro, R., Robshaw, M. (eds.) Advances in Cryptology – CRYPTO 2015. LNCS, vol. 9215, pp. 663–682. Springer, Heidelberg (2015)
13. Maitra, S., Sarkar, S., Baksi, A., Dey, P.: Key Recovery from State Information of Sprout: Application to Cryptanalysis and Fault Attack. http://eprint.iacr.org/2015/236.pdf

14. Mihaljevic, M.J., Gangopadhyay, S., Paul, G., Imai, H.: Internal state recovery of grain-v1 employing normality order of the filter function. Inf. Secur. IET **6**(2), 55–64 (2012)
15. Mihaljevic, M.J., Gangopadhyay, S., Paul, G., Imai, H.: Generic cryptographic weakness of k-normal boolean functions in certain stream ciphers and cryptanalysis of Grain-128. Periodica Math. Hung. **65**(2), 205–227 (2012)

Reverse-Engineering of the Cryptanalytic Attack Used in the Flame Super-Malware

Max Fillinger[(⊠)] and Marc Stevens

CWI, Amsterdam, The Netherlands
max.fillinger@cwi.nl, marc@marc-stevens.nl

Abstract. In May 2012, a highly advanced malware for espionage dubbed Flame was found targeting the Middle-East. As it turned out, it used a forged signature to infect Windows machines by MITM-ing Windows Update. Using counter-cryptanalysis, Stevens found that the forged signature was made possible by a chosen-prefix attack on MD5 [25]. He uncovered some details that prove that this attack differs from collision attacks in the public literature, yet many questions about techniques and complexity remained unanswered.

In this paper, we demonstrate that significantly more information can be deduced from the example collision. Namely, that these details are actually sufficient to reconstruct the collision attack to a great extent using some weak logical assumptions. In particular, we contribute an analysis of the differential path family for each of the four near-collision blocks, the chaining value differences elimination procedure and a complexity analysis of the near-collision block attacks and the associated birthday search for various parameter choices. Furthermore, we were able to prove a lower-bound for the attack's complexity.

This reverse-engineering of a non-academic cryptanalytic attack exploited in the real world seems to be without precedent. As it allegedly was developed by some nation-state(s) [11,12,19], we discuss potential insights to their cryptanalytic knowledge and capabilities.

Keywords: MD5 · Hash function · Cryptanalysis · Reverse engineering · Signature forgery

1 Introduction

1.1 End-of-Life of a Cryptographic Primitive

The end-of-life of a widely-used cryptographic primitive is an uncommon event, preferably orchestrated in an organized fashion by replacing it with a next generation primitive as a precaution as soon as any kind of weakness has been exposed. Occasionally such idealistic precautions are thrown to the wind for various reasons. Unfortunately, the sudden introduction of practical attacks may then seriously reduce the security of systems protected by the cryptographic primitive. The ensuing forced mitigation efforts need to overcome important hurdles in a

© International Association for Cryptologic Research 2015
T. Iwata and J.H. Cheon (Eds.): ASIACRYPT 2015, Part II, LNCS 9453, pp. 586–611, 2015.
DOI: 10.1007/978-3-662-48800-3_24

short amount of time and thus prove to be less successful than precautionary mitigation efforts. The topic of this paper, namely an exposed cryptanalytic attack on the hash function MD5 exploited in the real-world eight years after the first practical break of MD5, is a recent example of the above.

1.2   Collisions for MD5

The cryptographic hash function MD5 found widespread use for many years since its inception in 1991 by Ron Rivest [21]. It became the de facto industry standard in combination with RSA to generate digital signatures upon which our Internet's Public Key Infrastructure (PKI) for TLS/SSL has been build. This despite early collision attacks on the compression function of MD5 by den Boer and Bosselaers [2] and Dobbertin [6].

That changed after in 2004 the first real MD5 collision attack, as well as example collisions, were presented by Wang et al. in a major breakthrough in hash function cryptanalysis [28,29]. Improvements to their attack were published in a series of papers (e.g., see [9,10,13,22,24,27,30,31]). Unfortunately, no convincing threatening scenarios arose due to the important restriction that colliding message pairs $M = P||C||S$, $M' = P||C'||S$ can only differ in the random-looking C, C'.

This restriction was lifted with the introduction of the first chosen-prefix collision attack on MD5 [26] that for any two equal-length prefixes P and P' constructs short random-looking C and C' such that $P||C||S$ and $P'||C'||S$ collide for any common suffix S. Chosen-prefix collisions make it significantly easier to construct collisions with meaningful differences, i.e., often it suffices to choose M and M' appropriately and to hide C and C' somewhere within the messages. It enabled the first truly convincing attack scenario using MD5 collisions, namely the construction of a rogue Certificate Authority (CA) certificate presented in 2009 [27]. As it turned out, many CAs had voluntarily stopped using MD5. Nevertheless, the remaining few MD5-using CAs endangered the entire PKI as any PKI is only as strong as its weakest link, i.e., CA.

Based on these developments, various authorities explicitly disallowed MD5 in digital signatures (e.g., The CA/Browser Forum adopted Baseline Requirements for CAs in 2011[1], Microsoft updated its Root CA Program in 2009[2]).

1.2.1   Counter-Cryptanalysis

Due to its widespread and pervasive use, MD5 remains supported to accommodate old signatures even up to the time of this writing. Any party world-wide still signing with an MD5-based digital signature scheme – against all advice – may be attacked using a chosen-prefix collision attack. Furthermore, a resulting digital signature forgery can be exploited against nearly everyone due to the near-ubiquitous support of MD5-based signatures. Stevens recently proposed

---

[1] https://cabforum.org/wp-content/uploads/Baseline_Requirements_V1.pdf
[2] http://technet.microsoft.com/en-us/library/cc751157.aspx

to counter these threats using *counter-cryptanalysis* [25], specifically a collision detection algorithm, i.e., an algorithm that asserts whether any given *single* message belongs to a colliding message pair that was constructed using a MD5 and/or SHA-1 collision attack. The main idea is to guess the colliding part (i.e., the C') of the assumed sibling colliding message and to verify whether an internal collision occurs. Once a collision has been verified, one knows the near-collision blocks for both messages, however, one cannot reconstruct earlier parts of the missing message with counter-cryptanalysis.

Collision detection can strengthen digital signatures by invalidating forged digital signatures, thereby allowing the continued secure use of MD5-based signatures. However, collision detection is clearly not a permanent solution and cannot replace proper migration to the more secure SHA-2 and SHA-3.

1.3   The Super-Malware 'Flame'

1.3.1   Background

Flame is a highly advanced malware for espionage and was discovered in May 2012 by the Iranian CERT, Kaspersky Lab and CrySyS Lab [11,12]. It seemed to have targeted the Middle-East, with the most infections in Iran. Among the targets were government-related organizations, private companies, educational institutions as well as specific individuals. According to these reports by malware experts Flame was developed by some nation-state(s) with near-certainty. It seems the best report so far on the origin is a Washington Post article reporting that – according to unnamed officials and experts – Flame was a joint U.S.-Israel classified effort [19].

For espionage, Flame collected keyboard inputs, Skype conversations and local documents of potential interest. It could also record screen contents, microphone audio, webcam video as well as network traffic, sometimes triggered by the use of specific applications of interest like Instant Messaging applications.

According to Kaspersky [12], Flame was active since at least 2010. However, CrySyS Lab reports Flame or a preliminary version thereof may have been active since 2007 due to an observed file in the security enterprise webroot in 2007. Infections seem to have occurred with surgical precision with each target carefully selected instead of wildly spreading, which may be one of the reasons why it has evaded discovery for several years.

We refer to the analyses by Kaspersky Lab and CrySyS Lab [11,12] for more details on the functionality, purpose and origin of Flame. Here we focus on the variant chosen-prefix collision attack that enabled its propagation.

1.3.2   Propagation

As described by Sotirov [23], Flame used WPAD (Web Proxy Auto-Discovery Protocol) to register itself as a proxy for the domain update.windows.com to launch Man-In-The-Middle attacks for Windows Update on other computers on the local network. By forcing a fall-back from the secure HTTPS protocol to the insecure HTTP protocol, Flame was able to push validly signed Windows

Update patches of its choice. This included a properly, but illegitimately, signed Windows Update patch by which Flame could spread to other machines. Flame's code-signing certificate for this security patch was obtained by fooling a certain part of Microsoft's PKI into signing a colliding – innocuous-looking – sibling certificate using an MD5-based signature algorithm. As the to-be-signed parts of both certificates were carefully crafted to result in the same MD5-hash using a chosen-prefix collision attack, the MD5-based signature was valid for both certificates.

Even though Microsoft was fully aware of the severe weaknesses of MD5 and spent great effort on migrating to more secure hash functions for *new* digital signatures at least since 2008, their software continued to accept (old) MD5-based digital signatures. Unfortunately, the use of MD5-based signatures for licensing purposes in their Terminal Server Licensing Service was overlooked in their efforts.[3] This, together with other unforeseen circumstances, allowed the creation of Flame's properly, but illegitimately, signed security patch that was trusted by *all* versions of Windows [16].[4]

1.3.3 Unknown Variant Chosen-Prefix Collision Attack

On the 3rd of June 2012, Microsoft blogged that in their initial analysis of Flame they *"identified that an older cryptography algorithm could be exploited and then be used to sign code as if it originated from Microsoft"* [17]. An immediate guess was that this cryptically worded statement refers to the construction of a rogue code-signing certificate using a chosen-prefix collision attack on MD5 similar to [27]. Only the certificates in the chain leading to the forged signature on Flame's executable were circulating on the Internet [20], its sibling colliding certificate remains lost. Using his collision detection technique, Stevens was able to reconstruct the collision part of the missing sibling colliding certificate [25].

Having both colliding parts one can observe the differential paths used for this attack which Stevens uses to provide a preliminary analysis of Flame's attack:

Flame's differential paths clearly show a chosen-prefix collision attack that starts with a chaining-value difference containing many bit differences that is gradually reduced to zero by the four sequential "near-collision" block pairs. However, these differential paths do not match any family of published chosen-prefix collision attacks [27], but instead were variants based on the first differential paths for MD5 by Wang et al.[29]. Also, they show characteristics that do not match those from known differential path construction methods for MD5. The author provides arguments indicating an unnecessary costly differential path construction method was used. Furthermore, experimental results were given constructing replacement paths with significantly fewer bitconditions in only about 15 s on average on a single Intel i7-2600 CPU (equivalent to about 2^{29} MD5 compressions).

---

[3] Microsoft invalidated this part of their PKI after the discovery of Flame in 2012.

[4] Any license certificate produced by the Terminal Server Licensing Service could directly be used to attack Windows Vista and earlier versions, but not later versions.

Based on the differential paths and the observation that the best known message modification technique was used, for each block a lower bound for the average complexity to find the near-collision blocks is given. Note the implicit assumption that the differential path including the target output chaining value difference is fixed before the near-collision block search.

Based on the weight of the observed chaining value difference after the birthday search that need to be eliminated by the four near-collision attacks, an indicative complexity estimate of about 2^{42} MD5 compressions is given. Although further constraints make it more likely to be even higher instead of lower. Lacking a more detailed analysis of the chaining value difference elimination strategy, no more accurate prediction could be given.

Although Stevens was able to show a yet unknown variant attack was used, so far, no reconstruction of Flame's attack has been presented and many questions regarding techniques and complexities remained unanswered. Specifically there is no analysis so far for the possible differential path family for each block, and therefore for the chaining value reduction procedure that selects which chaining value differences (the tail of the differential path) to eliminate in each block. This in turn makes it hard to provide accurate complexity estimates for each of the four near-collision attacks as well as for the associated birthday attack. Furthermore, the work in this paper makes it clear that Stevens' assumption that each near-collision block targets a specific chaining value difference is inaccurate, making his preliminary comments on the attack complexity incorrect.

2   Our Contributions

In [25] Stevens presented proof that Flame uses a yet unknown chosen-prefix collision attack and made indications of the complexity to find solutions for the recovered differential paths. No attack reconstruction or more accurate complexity estimates were given.

Our paper is entirely based on the four near-collision block pairs shown in Appendix B that can be recovered from the single available certificate in Flame's attack using counter-cryptanalysis. This paper significantly improves upon Stevens' preliminary reconstruction and we demonstrate for the first time that a single example of a collision pair is actually sufficient to reconstruct the used collision attack to a great extent under some weak logical assumptions. Furthermore, the high level of detail of our reconstruction even admits concrete conclusions under a complexity analysis, specifically we prove a lower-bound for the estimated attack complexity and provide a cost figure for the closest fit of attack parameters. Our work shows that Stevens' indications of the near-collision costs are not the real expected costs. In particular the attack does not use fixed differential paths, but allows some random chaining value differences to occur in the first two blocks that can be efficiently negated in the last two blocks. Lacking more information about the near-collision attack procedures, Stevens was also unable to give real indications of the birthday search complexity of Flame's chosen-prefix collision attack. However, our reconstruction as well as our complexity analysis includes the birthday search and shows there is a trade-off between the birthday search cost and the total cost of the near-collision attacks.

At a high level we can draw some insights from our analysis into the cryptanalytic capabilities and the available resources of Flame's creators. In particular, the complexity for the closest fit of attack parameters is equivalent to $2^{49.3}$ MD5 compressions which takes roughly 40,000 CPUcore hours. That means for say 3-day attempts to succeed in reasonable time given the large number of required attempts, one needs about 560 CPUcores, which is large but not unreasonable even for academic research groups. With an estimated complexity of $2^{44.55}$ MD5 compressions from [27], this seems to be suboptimal. Not only the overall complexity seems to be suboptimal, also the differential path construction method and the near-collision speed-up techniques seem to be suboptimal. Overall we can report that it is clear that significant expertize in cryptanalysis was required, yet there are no indications at all of superior techniques, but instead that various parts are sub-optimal. It seems a working attack that succeeds in reasonable time was more important than optimizing the overall attack using all of the state of the art techniques[5].

Noteworthy is the following thought by an anonymous reviewer: developing a new variant attack required significant human effort which would have been unnecessary if its creators had enough computational power to do a general birthday search of complexity $2^{64.85}$ MD5 compressions in reasonable time. This may indicate a reasonable upper bound on available resources. Although, given the public availability of the Hashclash tools [8] since mid 2009, it might have been unnecessary in the first place which would imply they explicitly chose to build their attack or use their already built attack for Flame for some reasons.

At a more detailed level, our analysis revealed that a central idea behind the attack seems to be that the near-collision blocks operate in *pairs*: The first two blocks together eliminate one part of the intermediate hash value differential, allowing mostly random changes to other parts. The remaining differences (including the random changes from the first pair) are eliminated by the second two blocks. This idea allows a significant reduction in the expected complexity compared to the previous estimate by Stevens [25], where each near-collision pair was assumed to target specific intermediate hash value differences.

We have deduced the most likely parametrized family of differential paths for each near-collision block from which one is selected to eliminate specific intermediate hash value differences, as well as the complementary parametrized birthday search procedure that results in an intermediate hash value difference that can be eliminated using the 4 families of differential paths. We provide a complexity analysis for plausible parameter choices. Furthermore, we prove Theorem 6 stating a lower-bound complexity independent of parameter choices to be $2^{46.6}$ calls to the compression function in Sect. 4.3.3, and provide parameter

---

[5] At the time there seems to have been no reason to hold back more advanced techniques, given that counter-cryptanalysis was not publicly known then. Also, if there was a concern about revealing their knowledge then they could have easily used the publicly available Hashclash tools [8] instead.

choices that achieve this cost. Sotirov estimated that obtaining their forgery was significantly more difficult than the original Rogue CA construction, thus requiring many collisions in order to succeed [23]. This indicates that significant computational resources need to have been brought to bear to execute each chosen-prefix collision attack in a relatively short amount of time in order to succeed in their overall aim to obtain a forgery.

Lacking more examples or other hints about the actual attack procedure, it seems to be very hard to determine more specifics of Flame's chosen-prefix collision attack with any significant level of certainty. This includes the differential path construction algorithm and the collision search algorithm. For more details and analysis of less important aspects to our complexity analysis we refer to the full version of this paper.

The remainder of this paper is as follows. We start in Sect. 3 with an exposition of the main known techniques for chosen-prefix collision attacks. In Sect. 4.1, we break down the data from the recovered near-collision block pairs. We present our reconstruction in Sect. 4.2 and its complexity analysis in Sect. 4.3.

3   MD5 Chosen-Prefix Collision Attacks

3.1   MD5

The hash function MD5 maps an arbitrarily long input message M to a 128-bit output string. Its design follows the Merkle-Damgård construction [5,15], using a *compression function* which we call MD5Compress and a chaining value denoted IHV.

1. Unambiguously pad M to a length that is a multiple of 512.
2. For $i = 0, \ldots, N - 1$, let M_i denote the ith 512-bit block of M. Let

$$IHV_0 = IV = (67452301_{16}, \text{efcdab89}_{16}, 98\text{badcfe}_{16}, 10325476_{16})$$

3. For $i = 1, \ldots, N$, let $IHV_i = \text{MD5Compress}(IHV_{i-1}, M_{i-1})$.
4. Output IHV_N converted back from little-endian representation.

The description of MD5Compress we give here is not the standard one but an equivalent "unrolled" formulation [7] that is better suited for cryptanalysis. The compression function has 64 steps and computes a sequence of *working states* Q_t for inputs $IHV_{\text{in}} \in \{0,1\}^{128}$, $M \in \{0,1\}^{512}$:

1. Split IHV_{in} and M into 32-bit words; $IHV_{\text{in}} = a\|b\|c\|d$, $M = m_0\|\ldots\|m_{15}$
2. Let $Q_{-3} = a$, $Q_{-2} = d$, $Q_{-1} = c$ and $Q_0 = b$.
3. For $t = 0, \ldots, 63$, compute

$$F_t = f_t(Q_t, Q_{t-1}, Q_{t-2}); \quad T_t = F_t + Q_{t-3} + AC_t + W_t;$$
$$R_t = RL(T_t, RC_t); \quad Q_{t+1} = Q_t + R_t;$$

4. Output $IHV_{\text{out}} = (Q_{61} + a, Q_{64} + b, Q_{63} + c, Q_{62} + d)$.

where $AC_t = \lfloor 2^{32} \cdot |\sin(t+1)| \rfloor$ and W_t, $f_t(X, Y, Z)$ and RC_t are given by:

Step	W_t	$f_t(X, Y, Z)$	RC_t
$0 \le t < 16$	m_t	$(X \wedge Y) \oplus (\overline{X} \wedge Z)$	$(7, 12, 17, 22)_{[t \bmod 4]}$
$16 \le t < 32$	$m_{(1+5t) \bmod 16}$	$(Z \wedge X) \oplus (\overline{Z} \wedge Y)$	$(5, 9, 14, 20)_{[t \bmod 4]}$
$32 \le t < 48$	$m_{(5+3t) \bmod 16}$	$X \oplus Y \oplus Z$	$(4, 11, 16, 23)_{[t \bmod 4]}$
$48 \le t < 64$	$m_{(7t) \bmod 16}$	$Y \oplus (X \vee \overline{Z})$	$(6, 10, 15, 21)_{[t \bmod 4]}$

3.2   General Approach

When constructing a chosen-prefix collision pair $P\|C\|S_{any}$ and $P'\|C'\|S_{any}$ for given prefixes P and P' and arbitrary suffix S_{any}, we may assume without loss of generality that P and P' are of equal length and that their length is a multiple of the MD5 message block size. (Otherwise, one can just add padding.) A chosen-prefix collision attack consists of two stages. The first is the Birthday Search where one searches for equal-length suffixes S_b and S'_b such that the difference in the intermediate hash value after processing $P\|S_b$ and $P'\|S'_b$ has a particular form necessary for the second stage. In the second stage, one constructs near-collision block pairs $(S_1, S'_1), (S_2, S'_2), \ldots, (S_n, S'_n)$ such that after processing $P\|S_b\|S_1\|\ldots\|S_n$ and $P'\|S'_b\|S'_1\|\ldots\|S'_n$ the intermediate hash values are equal. Thus one has found the desired $C = S_b\|S_1\|\ldots\|S_n$ and $C' = S'_b\|S'_1\|\ldots\|S'_n$ for which the pair $P\|C\|S_{any}$ and $P'\|C'\|S_{any}$ form a collision for any suffix S_{any}. We explain the construction of the near-collision block pairs below.

3.3   Differential Cryptanalysis

Differential cryptanalysis is based on the analysis of the propagation of input differences throughout a cryptosystem. This technique was publicly introduced in 1993 by Eli Biham and Adi Shamir who first applied it to block ciphers [1]. Differential cryptanalysis of hash functions has been very successful. One of the key techniques introduced by Wang et al. against MD5 was the simultaneous use of the difference modulo 2^{32} and the bitwise XOR difference resulting in a bitwise signed difference.

Let I and I' be two different inputs, for any variable X involved in the computation for input I, we denote the respective variable for input I' as X'. For $X, X' \in \mathbb{Z}_{2^{32}}$, we denote by $\delta X = X' - X \bmod 2^{32}$ the *arithmetic* differential. When it is necessary to keep track of the bitwise differences as well, we use the *Binary Signed Digit Representation* (BSDR). The BSDR differential is $(\Delta X[i])_{i=0,\ldots,31}$ where $\Delta X[i] = X'[i] - X[i] \in \{-1, 0, 1\}$. We can easily calculate the arithmetic difference from the BSDR: $\delta X = \sum_i \Delta X[i] \cdot 2^i \bmod 2^{32}$.

For a BSDR ΔX, we define the *weight* $w(\Delta X)$ as the number of indices i where $\Delta X[i] \ne 0$. For $\delta X \ne 0$, there are multiple BSDRs ΔX such that $\delta X = \sum_{i=0}^{31} \Delta X[i] \cdot 2^i$. However, there is a normal form, called the non-adjacent form (NAF). The non-adjacent form of δX is the *unique* BSDR ΔX such that $\sum_{i=0}^{31} \Delta X[i] \cdot 2^i = \delta X$, $\Delta X[31] \ge 0$ and ΔX has no adjacent non-zero entries. The NAF is a minimal-weight BSDR of δX. We define the *NAF-weight* $w(\delta X)$ as the weight of the NAF of δX.

3.4   Differential Paths

A *differential path* is an exact description of how differences propagate through two related evaluations of MD5Compress. In particular, a differential path describes for every step t the differences δQ_{t-3}, ΔQ_{t-2}, ΔQ_{t-1}, ΔF_t, δW_t, δT_t, δR_t and δQ_{t+1} such that:

- $\delta T_t = \delta Q_{t-3} + \sigma(\Delta F_t) + \delta W_t$;
- $\delta Q_{t+1} = \sigma(\Delta Q_t) + \delta R_t$;
- $Pr[\Delta F_t | \Delta Q_{t-2}, \Delta Q_{t-1}, \Delta Q_t] > 0$;
- $Pr[\delta R_t | \delta T_t] > 0$.

We say that an input pair $(IHV, m_0 \| \ldots \| m_{15})$, $(IHV', m_0' \| \ldots \| m_{15}')$ for MD5Compress *solves* the differential path up to step t if differences for the message block and the intermediate variables are as specified in the differential path up to step t.

Although the first differential paths for MD5 were constructed entirely by hand [29], two quite different ways to construct differential paths have since been introduced: Stevens' meet-in-the-middle approach [26] and De Cannière and Rechberger's coding-theory based technique [4,14].

Suppose a pair of inputs solves a differential path up to some step. This pair of inputs might fail to solve the next step because of the Boolean function or because of the bit rotation. To handle the Boolean functions, bit conditions are used that allow efficient checks whether our inputs have the correct values for ΔF_t. The rotations are taken care of probabilistically.

3.5   Tunnels

Message modification, specifically Tunneling [10], is an important technique that can drastically speed up collision attacks. Under some preconditions, a tunnel allows us to change a certain working state bit $Q_t[i]$ and corresponding message bits without affecting $Q_{t+1}, \ldots, Q_{t'}$ for some $t' > t$. As an example, consider the most important known tunnel T_8 with the following requirements:

- $Q_9[i]$ is free, i.e., no difference and no boolean function bitcondition
- $Q_{10}'[i] = Q_{10}[i] = 0$, and $Q_{11}'[i] = Q_{11}[i] = 1$

Under these conditions, we can flip bits $Q_9[i] = Q_9'[i]$ and adjust m_8, m_9 and m_{12} without affecting $Q_{10}, \ldots, Q_{24}$ and $Q_{10}', \ldots, Q_{24}'$.

To see why T_8 is useful, suppose that we have a differential path and a partial solution thereof up to and including Q_{24}. We say that a bit-position $i \in \{0, \ldots, 31\}$ is *active* for T_8 if it satisfies the requirements. We call the number k of active bit-positions the *strength* of T_8. The tunnel allows us to generate 2^k different partial solution up to Q_{24} — one for each possible value of the active bit-positions. Since the probability that a partial solution can be extended to a full solution is rather small, cheaply generating *many* partial solutions reduces the cost of the collision attack significantly. In Table 1, we describe the three tunnels that are the most relevant for the Flame collision attack. In Sect. 4.1.3, we discuss how these tunnels might have been used in the collision attack.

Table 1. Most important tunnels for `MD5Compress`

Tunnel	Flip bit	Aux. bitconditions	Affected states	Affected message words
T_4	$Q_9[b]$	$Q_{10}[b] = Q_{11}[b] = 1$	$Q_{22}, \ldots, Q_{64}$	m_8, m_9, m_{10}, m_{12}
T_5	$Q_{10}[b]$	$Q_{11}[b] = 0$	$Q_{22}, \ldots, Q_{64}$	$m_9, m_{10}, m_{12}, m_{13}$
T_8	$Q_9[b]$	$Q_{10}[b] = 0, Q_{11}[b] = 1$	$Q_{25}, \ldots, Q_{64}$	m_8, m_9, m_{12}

4 Reverse-Engineering Flame's Attack

4.1 Breakdown of Data

In Appendix B we list the chaining values and near-collision blocks from the available Flame certificate and the ones for the associated 'legitimate' certificate that can be recovered using counter-cryptanalysis. The differential path for each near-collision block pair can directly be observed by comparing the two compression function computations. In this section we first list several specific observations about these (reconstructed) Flame near-collision blocks and the observed differential paths that are relevant to our reconstruction.

4.1.1 Some Features of the Near-Collision Blocks

Observation 1 ([23]). *Due the constrained space where the near-collision blocks were to be hidden in the certificate, the collision attack could only use four near-collision blocks.*

Observation 2. *Blocks 1 and 3 of the Flame collision attack use the message block differences from the first differential path of Wang et al.'s identical-prefix attack, $\delta m_4 = \delta m_{14} = 2^{31}$, $\delta m_{11} = 2^{15}$ and $\delta m_i = 0$ for $i \neq 4, 11, 14$. Blocks 2 and 4 use the differences from the second differential path of the identical prefix attack, $\delta m_4 = \delta m_{14} = 2^{31}$, $\delta m_{11} = -2^{15}$ and $\delta m_i = 0$ for $i \neq 4, 11, 14$.*

Observation 3. *The working state differences ΔQ_6 are maximal in all four near-collision blocks, i.e., for every $i = 0, \ldots, 31$, we have $\Delta Q_6[i] \neq 0$. The ΔQ_6 of Blocks 1 and 3 are equal, likewise for Blocks 2 and 4.*

Observation 4. *The four blocks all have a common structure: Up to and including step 5, the differences δQ_t vary among all four blocks. Then, there is a maximal difference in step 6. After that, the values for ΔQ_t and ΔF_t are mostly identical in the first and third and in Blocks 2 and 4, leading up to long sequences of trivial steps. The final five steps again differ greatly among all four blocks.*

4.1.2 Notes on Differential Path Construction

From this last observation, we conclude that a differential path beginning based on the input IHVs and a differential path ending were generated separately and then combined. Such differential path construction can be done for MD5 using

Stevens' meet-in-the-middle approach [26] or De Cannière and Rechberger's coding-theory based technique [4,14]. The latter technique is less likely to have been used, since all observed differential paths don't show its characteristic very long carry chains over the non-predetermined part $Q_1, \ldots, Q_5$. Stevens already showed that suitable differential paths can be constructed in about 15 s on an Intel i7-2600 CPU, so in time equivalent to approximately 2^{29} MD5 compressions [25]. As this shows that differential path construction can be done very fast and does not have to cost a significant fraction of the overall attack complexity and lacking more example collisions for analysis, our paper will focus on the complexity-wise more costly parts of the attack.

4.1.3   Tunnel Strengths in the Near-Collision Blocks

In order to estimate the complexity of the Flame collision attack, it is important to know whether and to what extent the attackers used *tunnels*. The tunnels T_4, T_5 and T_8 are the most important in speeding up the attack. See Table 1 for a description of the three relevant tunnels.

Observation 5 ([25, Sect. 3.3]). *The table below lists per near-collision block the observed strength of tunnel T_8, the maximal strength possible given the respective differential path and the average strength that would have been observed if the tunnel was not used.*

Near-collision Block	Observed strength	Maximal strength	Average strength
1	7	17	4.25
2	13	18	4.5
3	10	17	4.25
4	9	18	4.5

It is clear that the tunnel T_8 has been used, since the observed tunnel strengths are much larger than one expects to see if T_8 was *not* used. Although not presented here, we'd like to note that the tunnel strengths for T_4 and T_5 are smaller than average, but one cannot conclude that T_4 and T_5 were not used since for each bit only one of T_4, T_5 and T_8 can be active due to conflicting preconditions.

For our complexity estimates, we will assume the strengths of these three tunnels to be the average over all four blocks. That is, we assume that tunnel T_4 has strength 3, T_5 has strength 7.5 and T_8 has strength 9.75. Reconstructing the exact tunnel-exploitation method would be interesting and could lead to more precise complexity estimates. We discuss some possible methods in Sect. 4.2.3.

4.2   Our Reconstruction of the Chosen-Prefix Collision Attack

In this section, we describe our reconstruction of the collision attack, in particular the differential path construction, the *families* of differential paths endings that were used, the cost of the Birthday Search and of the message block construction.

Central to our reconstruction attempt is the idea that the first two blocks eliminate δc from $\delta IHV = (\delta a, \delta b, \delta c, \delta d)$ up to a constant term while allowing random changes in parts of δb. The second two blocks then eliminate δb and the constant term in δc. This allows for the first two blocks to be constructed faster than estimated in the preliminary analysis in [25].

It seems that the four near-collision attacks can be grouped into two *pairs*: Blocks 1 and 2 form a pair, and, likewise, Blocks 3 and 4. In each of the pairs, the first block uses the message block differences of the first near-collision block in the identical-prefix attack by Wang et al. and the second block uses the difference of the second near-collision block in that attack. That is, in Blocks 1 and 3, the only differences in the message block are $\delta m_4 = \delta m_{14} = 2^{31}$ and $\delta m_{11} = 2^{15}$. In Blocks 2 and 4, the differences are negated, i.e., $\delta m_{11} = -2^{15}$.

To determine the complexity of the Birthday Search and of the message block construction algorithm, we describe a family of end-segments of the differential path for each of the four near-collision blocks. We compute the complexity of the Birthday Search and the complexity of the algorithm for generating near-collision blocks on the basis of our reconstruction of the end-segments.

Table 2. Chaining value difference corrections ($\delta IHV_{\text{out}} - \delta IHV_{\text{in}}$) for each block

	Block 1	Block 2
δa	$[31]$	$[31, 5]$
δd	$[31, 25]$	$[31, -25, -9, 5]$
δc	$[31, 25, -14, -12, 9]$	$[31, 26, 24, 20, -9, 5]$
δb	$[31, 25, -18, -15, -12, 9, 1]$	$[-26, 24, 21, -14, -9, 5, 0]$

	Block 3	Block 4
δa	$[31]$	$[31]$
δd	$[31, 25, 9]$	$[31, -25, -9]$
δc	$[31, 26, -24, -14, 9]$	$[31, -25, 14, -9]$
δb	$[30, 26, -24, 20, -17, 15, 9, -3]$	$[-25, 14, -9, -5, 3, 0]$

4.2.1   Differential Path Family

Four near-collision blocks are used to eliminate the chaining value differences after the birthday search of the chosen-prefix collision attack. In this section we reconstruct the family of differential paths used for each of the four near-collision blocks based on the observed chaining value differences, the observed differential paths and the possible variations thereof that are compatible with the overall attack.

In particular, each of the four near-collision attacks uses a carry expansion of a particular bit difference in the last few steps of Wang's original differential paths for MD5 to allow for some controlled additional differences to affect the chaining value differences. This can be seen in the recovered paths shown in Appendix A: for each block there is a primary carry chain either in δQ_{62} or δQ_{63} starting at bit position either 5 or 25 used for controlled differences, other small carry chains are random artifacts and not actively used. Our reconstruction is based on these primary carry chains and we will parametrize the amount of allowed carries. Using other carry chains significantly complicates the overall attack strategy, does not lead to significant benefits and does not fit the observed paths, hence we apply Ockham's razor principle and keep to the most straightforward explanation.

The differences that are added to $\delta IHV = (\delta a, \delta b, \delta c, \delta d)$ using each near-collision block are summarized in Table 2. We begin with an outline of what we assume to be the elimination strategy. The differences in δc are eliminated by the first two blocks using carry chains in δQ_{62} starting at bit positions 25 and 5 respectively, but a difference of -2^{24} is introduced which is then eliminated in the final two blocks. For δb, matters are more complicated. Given the following observations:

– deliberate changes to δb possible in blocks 1,2 can be deferred to blocks 3,4;
– random changes to δb possible in blocks 1,2 can be handled in blocks 3,4;
– blocks 1 and 2 actually increase the NAF-weight of δb;

we found that the best explanation is that the changes to δb in the first two blocks are mostly random and that the elimination of differences in δb is done in Block 3 and Block 4 using carry chains in δQ_{63} starting at bit positions 25 and 5 respectively. This explanation in fact reduces the complexity for Blocks 1 and 2 as they only need to control δQ_{64} (that affects δb) to a very small extent.

Table 3. End segment of block 1.

Steps	Bitconditions
60	+BBBB1B. ....
61	+BBBB1B. ....
62	X+-----. ....
63	X.....+. .... .DDDDD+D ....
64	***...+. ...***** ***AAA+A ....****

$$\delta Q_{63} = 2^{31} + 2^{25} + 2^9 + C_{14}2^{14} + \textstyle\sum_{i=8}^{8+w_1} C_i 2^i, 1 \leq w_1 \leq 5$$

$$\delta Q_{64} = \delta Q_{63} + \textstyle\sum_{i=29}^{31} X_i 2^i + \sum_{i=14}^{20} X_i 2^i + \sum_{i=0}^{v_1} X_i 2^i, -1 \leq v_1 \leq 3$$

We have generalized the observed differential path endings to a reasonable extent, i.e., making our reconstructed path families more general would make matters significantly more complex, while similar benefits might also be obtained by simply choosing larger parameters for our families below. The four differential path families are described as follows. For Block i we use a parameter w_i that specifies the length of the carry chain and thus over how many bits one can fully control the differences. Block 4 uses an additional carry chain whose length is determined by u_4. For Blocks 1 and 2 we use an additional parameter v_1 and v_2 that control an amount of bit positions in which random differences are allowed as they can be handled in Blocks 3 and 4, these parameters v_1 and v_2 depend on the value of u_4.

Table 4. End segment of block 2.

Steps	Bitconditions
60	+............ .... BBBBBB1B BBB.....
61	-....... 0000.... BBBBBB1B BB+.....
62	+.....-. .... -++++++ ---.....
63	XDDDD-D+ DDDD.... ....-B B+-.....
64	**DDD-A+ AAAD**** ***...-. ..+*****

$$\delta Q_{63} = 2^{31} - 2^{26} + 2^{24} - 2^9 + 2^5 + \sum_{i=20}^{24+w_2} C_i 2^i,$$
$$\delta Q_{64} = \delta Q_{63} + \sum_{i=27}^{29} B_i 2^i + B_{20} 2^{20} + \sum_{i=0}^{v_2} X_i 2^i +$$
$$\sum_{i=13}^{19} X_i 2^i + \sum_{i=30}^{31} X_i 2^i$$
$$0 \le w_2 \le 6, \quad -1 \le v_2 \le 4$$

1. Block 1 uses a carry chain starting in δQ_{62} at bit position 25 up to $25 + w_1$ to control differences in δQ_{63} over bit positions 8 up to $8 + w_1$. Given the differences that can be covered in Blocks 3 and 4, we can allow arbitrary differences in δQ_{64} at bit position ranges $[0, v_1]$, $[14, 20]$, and $[29, 31]$.
2. Block 2 uses a carry chain starting in δQ_{62} at bit position 5 up to bit position $9 + w_2$ to control differences in δQ_{63} over bit positions 20 up to $24 + w_2$. Given the differences that can be covered in Blocks 3 and 4, we can allow arbitrary differences in δQ_{64} at bit position ranges $[0, v_2]$, $[13, 19]$, and $[30, 31]$.
3. Block 3 uses a carry chain starting in δQ_{63} at bit position 24 up to $26 + w_3$ to control differences in δQ_{64} over bit positions 13 up to $15 + w_3$.
4. Block 4 uses a carry chain starting in δQ_{63} at bit position 14 up to bit position $14 + w_4$ to control differences in δQ_{64} over bit positions 13 up to $15 + w_3$, and a second carry chain at bit positions 9 up to $9 + u_4$ to control differences over bit positions 30 up to $(30 + u_4 \bmod 32)$ that wrap around to the lower bit positions.

Note that in Block 4, the parameter u_4 must be large enough to eliminate the random changes to δb that are made in Blocks 1 and 2. That is, if $\max(v_1, v_2) \le 2$, we need $u_4 \ge \max(v_1, v_2) + 2$ and otherwise, we need $u_4 = 4$. Also, in Sect. 4.3.2 we will estimate the complexity of solving these differential paths.

Table 5. End segment of block 3.

Steps	Bitconditions
61	+BBBBBBB .... .1....0. ....
62	+BBBBB+B .... .0....+. ....
63	X+----B- ....-....+. ....
64	.+....+.- ....DDDD D-D...+. ....-...

$$\delta Q_{64} = 2^{30} + 2^{26} - 2^{24} + 2^9 - 2^3 + \sum_{i=13}^{15+w_3} B_i 2^i, 0 \le$$
$$w_3 \le 4$$

We now describe each differential path family more fully in Tables 3, 4, 5 and 6 by giving a template and specifying equations that the values of $\delta Q_{61}, \ldots, \delta Q_{64}$

Table 6. End segment of block 4.

Steps	Bitconditions
61	`+.....1. ...BBBBB 11BBBBB. ........`
62	`-.....-. ...BBBBB 00BBBB-. ........`
63	`X.....-. ...+---- ---++++. ........`
64	`DD....-. ........ .+....-D DD-D+DDD`

$$\delta Q_{64} = -2^{25} + 2^{14} - 2^9 - 2^5 + 2^3 + \sum_{i=30}^{\min(u_4,1)} B_i 2^i +$$
$$\sum_{i=0}^{u_4-2} B_i 2^i + \sum_{i=3}^{3+w_4} B_i 2^i$$
$$1 \le w_4 \le 6, \quad 0 \le u_4 \le 4$$

must satisfy. In the templates, a symbol $q_t[i]$ at step (row) i and bit position (column) i can be any of the following:

- '.': represents $Q_t[i] = Q'_t[i]$;
- '+': represents $Q_t[i] = 0$, $Q'_t[i] = 1$;
- '-': represents $Q_t[i] = 1$, $Q'_t[i] = 0$;
- '0': represents $Q_t[i] = Q'_t[i] = 0$;
- '1': represents $Q_t[i] = Q'_t[i] = 1$;
- '~': represents $Q_t[i] = Q'_t[i] = Q_{t-1}[i]$;
- '?': represents $(Q_t[i] = Q'_t[i]) \wedge (Q_t[i] = 1 \vee Q_{t-2}[i] = 0)$;
- 'D': a variable differential bitcondition, i.e., $q_t[i] \in \{.,+,-\}$;
- 'B': a variable Boolean function bitcondition, i.e., $q_t[i] \in \{.,0,1,?\}$;
- 'X': a non-constant bitcondition, i.e., $q_t[i] \in \{+,-\}$;
- '*': a (for now) irrelevant differential bitcondition;
- 'A': the same differential as above, $q_t[i] = q_{t-1}[i]$.

The equations may contain the following terms:

- w_i, v_i, u_i: Parameters of the differential path family. Higher values for the w_i mean that the differential path family can cancel more differences but is, on average, harder to solve.
- C_i, B_i: These terms can take on values in $\{-1,0,1\}$ and correspond to the variable differential bitconditions ('D's) in step 63 or 64, respectively. A member of the differential path family is determined by the C_i and B_i.
- X_i: These terms take on values in $\{-1,0,1\}$ and correspond to the irrelevant bitconditions ('*'s). While the B_i and C_i fix a differential path in the family, the X_i are determined only after a successful near-collision block search.

We say that a pair of inputs (IHV, B), (IHV', B') to `MD5Compress` solves the last four steps of the differential path if there is some setting for the X_i such that $\delta Q_{61}, \ldots, \delta Q_{64}$ satisfy the given equations. This is a more lax definition than what we use elsewhere, i.e., we do *not* require a solution to solve the exact bitconditions but use bitconditions as a tool to show which δQ_i are possible.

4.2.2 Birthday Search

We calculate the Birthday Search complexity for the maximal parameter values. It is easy to compute the Birthday Search complexity for lower values, namely for each carry that is dropped, the complexity increases by a factor of $2^{0.5}$.

Given the elimination strategy, we can now specify the Birthday Search target. We require that there are fixed differences in δa and δd and that for those bit positions i that we can not manipulate in our four near-collision blocks, we need $c[i] = c'[i]$ or $b[i] = b'[i]$ (after subtracting the constant bit differences). Given these constraints, the Birthday Search looks for a collision of the function

$$f(x) = (a, \tilde{b}_{10}, \ldots, \tilde{b}_{13}, \tilde{b}_{21}, \ldots, \tilde{b}_{26}, c_0, \ldots, c_7, c_{15}, \ldots, c_{19}, c_{31}, d)$$

$$\text{where } (a, b, c, d) = \begin{cases} \texttt{MD5Compress}(IHV, B\|x) + \left(-2^5, 0, -2^5, 2^9 - 2^5\right) & x \text{ even} \\ \texttt{MD5Compress}\left(IHV', B'\|x\right) & x \text{ odd} \end{cases}$$

$$\text{and } \tilde{b} = b - c$$

with IHV and IHV' the intermediate hash values after processing the two chosen prefixes. Not every collision of f is useful. The probability p that a collision is useful is at most $1/2$ since we require that the two parts use different prefixes. Therefore the expected number of compression function calls required to find a useful collision is $\sqrt{\pi \cdot 2^{88}/(2 \cdot p)} \approx \sqrt{\pi} \cdot 2^{44} \approx 2^{44.8}$ [18].

As we already mentioned, we use parameters to make trade-offs between message block construction and Birthday Search cost. For every carry we do not rely on, we introduce another bit position where b and b' or c and c' may not differ, increasing the Birthday Search complexity by a factor of $2^{0.5}$. This allows us to trade off Birthday Search complexity against complexity in the message block construction.

4.2.3 Tunnel Exploitation Analysis

As explained in Sect. 4.1.3, the tunnel strength in the Flame differential paths was neither average nor maximized. We now derive a formula for the expected tunnel strength when each tunnel bit is active with probability α. Let m be the number of bits that could be active for $\mathcal{T}_8$. For a random solution up to step 24, let $\mathbf{S}$ be the random variable measuring the strength of $\mathcal{T}_8$ and $\mathbf{solve}$ the event that the partial solution can be extended to a full solution using $\mathcal{T}_8$. Assuming $\Pr[\mathbf{solve} \mid \mathbf{S} = k] \approx 2^k \cdot p$ for some p independent of k, we can calculate

$$\mathbb{E}[\mathbf{S} \mid \mathbf{solve}] \approx \sum_{k=0}^{m} \frac{k \cdot \binom{m}{k}}{\sum_{i=0}^{m} \binom{m}{i} \cdot (2\alpha)^{i-k} \cdot (1 - \alpha)^{k-i}}$$

An explanation for the observed tunnel strengths (Observation 5) proposed in [25] is that the Flame authors did not try to maximize the tunnel strength but used tunnels in their message block construction algorithm to the extent that they were available. This corresponds to setting $\alpha = 1/4$. On the other hand, we consider the alternative hypothesis that many bits in working state

Q_{10} were fixed to '0' to bring the probability closer to $\alpha = 1/2$. In Table 7, we list the expectation and variance of the tunnel strength for both values of α. These results show that the initial explanation by Stevens with $\alpha = 1/4$ is rather unlikely, while the explanation with $\alpha = 1/2$ is more probable.

Table 7. Summary of the observed (s), maximal (m) and expected (μ) tunnel strength, and the standard deviation (σ).

Block	s	m	$\alpha = 1/4$		$\alpha = 1/2$	
			μ	σ	μ	σ
1	7	17	6.80	2.02	8.67	1.70
2	13	18	7.20	2.08	10.00	1.83
3	10	17	6.80	2.02	9.33	1.76
4	9	18	7.20	2.08	10.67	1.89

4.3   Cost Estimation

4.3.1   A Formula for the Expected Cost

We now estimate the cost of generating a near-collision block. Since the bitconditions are concentrated on the first 16 working states and the tunnel T_8 is used, we assume that the algorithm can be broken down into the following steps:

1. Generate a full differential path/generate a set of initial working states that connects to the lower differential path.
2. Select $Q_1, \ldots, Q_{16}$ according to the path and tunnel requirements.
3. Try to obtain a solution[6] up to step 24 with the help of tunnels T_4 and T_5. Go back to step 2 and choose different Q_i if it is not possible to obtain a solution and use early abort to reduce the cost of this step.
4. Attempt to generate a solution for the whole path from our solution up to step 24 using tunnel T_8. We use early abort to some extent.
5. Check if the values for $\delta Q_{61}, \ldots, \delta Q_{64}$ are correct. If yes, we have a solution.

The expected cost of this algorithm is as follows: C_{path} is the differential path construction cost; let the random variable Z be the number of input pairs with $\delta Q_{57} = \cdots = \delta Q_{60} = 2^{31}$ that we need to evaluate until we find an input pair where $\delta Q_{61}, \ldots, \delta Q_{64}$ are as specified by the differential path. The expected cost of finding a solution with the correct values for $\delta Q_{61}, \ldots, \delta Q_{64}$ is then

$$C_{\mathrm{block}} = C_{\mathrm{path}} + \mathbb{E}[Z] \cdot 2^{13.6}.$$

The factor $2^{13.6}$ represents the measured average complexity of finding input pairs with $\delta Q_{57} = \cdots = \delta Q_{60} = 2^{31}$ for Flame's differential paths. This complexity is very stable for all near-collision blocks as the differential paths are

---

[6] We say that a pair of inputs solves a path up to step t if it agrees with the bitconditions $q_{-3}, \ldots, q_t$ and with the δQ_{t+1} from the differential path.

only varied in the first 16 steps which don't affect complexity and the last few steps which are instead covered by $\mathcal{Z}$. Hence, the expected complexity of finding a near-collision block is $\mathbb{E}[\mathcal{Z}] \cdot 2^{13.6}$.

We give estimates for $\mathbb{E}[\mathcal{Z}]$ in the next section. As discussed in Sect. 4.1.2, $\mathcal{C}_{\text{path}}$ can be as low as 2^{29} MD5 compressions, which will be negligible compared to the other parts of the attack.

4.3.2   Estimating the Expected Number of Attempts

In this section, we want to estimate the expected number of input pairs with $\delta Q_{57} = \cdots = \delta Q_{60} = 2^{31}$ we have to generate until a solution for the differential path is found. We call input pairs with $\delta Q_{57} = \cdots = \delta Q_{60} = 2^{31}$ *admissible* input pairs and we call the values for $\delta Q_{61}, \ldots, \delta Q_{64}$ that we want the *target*.

Let $\mathcal{T}_{i,u_i,v_i,w_i}$ be the random variable that gives the target for Block i with parameters u_i, v_i and w_i. Selecting a target is done by selecting the values for $B_k, C_k \in \{-1, 0, 1\}$ as in Sect. 4.2.1. Let $\mathcal{Z}_\tau$ be the random variable that counts the admissible inputs we have to try until τ is solved and let $\mathcal{Z}_{i,u_i,v_i,w_i}$ be the random variable obtained by first sampling $\tau \leftarrow \mathcal{T}_{i,u_i,v_i,w_i}$ and then sampling $\mathcal{Z}_\tau$. To compute the total expected cost, we need $\mathbb{E}[\mathcal{Z}_{i,u_i,v_i,w_i}]$. To obtain an empirical estimate λ_{i,u_i,v_i,w_i}, we repeat the following process until a fixed number of targets is solved: We first sample $\tau \leftarrow \mathcal{T}_{i,u_i,v_i,w_i}$ and then select random admissible inputs and message blocks until we find one that solves the target.[7] When the chosen number of targets is solved, we let the average number of attempts to solve a target be our estimate λ_{i,u_i,v_i,w_i}. We then obtain $\mathcal{C}_{i,u_i,v_i,w_i} = \lambda_{i,u_i,v_i,w_i} \cdot 2^{13.6}$ as an estimate for the cost of solving the differential path for Block i with parameters u_i, v_i and w_i. The simulation outcomes for the four blocks are given in Tables 8, 9, 10 and 11.

To save time, we do *not* generate admissible inputs as in Sect. 4.3.1. Instead, we select working states $Q_{57}, \ldots, Q_{60}$ and message words $m_0, \ldots, m_{15}$ at random and compute $Q'_{57}, \ldots, Q'_{60}$ and $m'_0, \ldots, m'_{15}$ by applying the appropriate arithmetic differentials. This procedure requires the assumption that the probability for hitting the target does not change when we select $Q_{57}, \ldots, Q_{60}$ and message words at random, which is justified by the pseudo-randomness of MD5.

Our estimate of the Birthday Search cost in Sect. 4.2.2 assumes that the parameters w_i and u_4 are maximal. For smaller parameter values, the cost must be multiplied by a "Birthday Factor" μ_i which we give in Table 12.

4.3.3   Total Cost.

Let us now combine our estimates for the cost of solving the paths for different parameter setting with the Birthday Search complexity. We will calculate the following costs:

---

[7] Recall that we say that an input solves a differential path if there exists a setting for the X_k such that $\delta Q_{61}, \ldots, \delta Q_{64}$ are as described by the path.

Table 8. Estimated complexities for the first near-collision block.

$\log_2 \mathcal{C}_{1,v_1,w_1}$	$v_1 = -1$	$v_1 = 0$	$v_1 = 1$	$v_1 = 2$	$v_1 = 3$
$w_1 = 1$	24.1	23.7	23.6	23.4	22.9
$w_1 = 2$	25.8	25.2	25.0	24.8	24.3
$w_1 = 3$	27.9	27.2	26.8	26.5	26.1
$w_1 = 4$	29.8	29.1	28.6	28.2	27.8
$w_1 = 5$	30.4	29.7	29.2	28.7	28.3

Table 9. Estimated complexities for the second near-collision block.

$\log_2 \mathcal{C}_{2,v_2,w_2}$	$v_2 = -1$	$v_2 = 0$	$v_2 = 1$	$v_2 = 2$	$v_2 = 3$	$v_2 = 4$
$w_2 = 0$	35.4	34.8	34.7	34.6	34.6	34.6
$w_2 = 1$	37.0	36.2	36.1	36.0	36.0	36.0
$w_2 = 2$	39.2	38.2	37.9	37.8	37.8	37.8
$w_2 = 3$	41.6	40.5	40.0	39.8	39.7	39.7
$w_2 = 4$	44.0	42.9	42.4	42.0	41.8	41.8
$w_2 = 5$	46.7	45.5	45.0	44.6	44.2	44.0
$w_2 = 6$	49.3	48.1	47.6	47.2	46.8	46.4

Table 10. Estimated complexities for the third near-collision block.

	$\log_2 \mathcal{C}_{3,w_3}$
$w_3 = 0$	32.3
$w_3 = 1$	34.3
$w_3 = 2$	36.4
$w_3 = 3$	38.1
$w_3 = 4$	38.3

- $\mathcal{C}_{\mathrm{msg}}$: expected cost when minimizing the message block construction cost.
- $\mathcal{C}_{\mathrm{flame}}$: expected cost when minimizing the message block construction cost while keeping the parameters consistent with the observed paths.
- $\mathcal{C}_{\mathrm{search}}$: expected cost when minimizing the Birthday Search cost.
- $\mathcal{C}_{\mathrm{min}}$: minimal expected cost.

Firstly, for $\mathcal{C}_{\mathrm{msg}}$, we choose $w_1, \ldots, w_4$ to be as small as possible. We have to balance the parameters v_1 and v_2 against u_4. Increasing v_1 and v_2 does not speed up the message block construction by much, so we pick $v_1, v_2 = -1$ which allows us to pick $u_4 = 1$. The combined Birthday Factor for these parameters is $2^{11.0}$. We therefore have

$$\mathcal{C}_{\mathrm{msg}} = 4 \cdot \mathcal{C}_{\mathrm{path}} + 2^{11.0} \cdot 2^{44.3} \cdot p^{-1/2} + 2^{24.1} + 2^{35.4} + 2^{32.3} + 2^{34.1} \approx 2^{55.8}$$

Table 11. Estimated complexities for the fourth near-collision block.

$\log_2 C_{4,u_4,w_4}$	$u_4 = 0$	$u_4 = 1$	$u_4 = 2$	$u_4 = 3$	$u_4 = 4$
$w_4 = 1$	34.1	33.8	35.6	38.2	38.7
$w_4 = 2$	35.2	35.0	36.7	39.4	39.8
$w_4 = 3$	37.0	36.5	38.4	41.0	41.4
$w_4 = 4$	38.8	38.4	40.2	42.7	43.8
$w_4 = 5$	40.8	40.6	42.3	44.6	44.8
$w_4 = 6$	43.0	42.4	43.6	46.9	47.8

Table 12. "Birthday Factors" for the four near-collision blocks.

Block i	1	2	3	4
$\log_2 \mu_i$	$(5 - w_1)/2$	$(6 - w_2)/2$	$(4 - w_3)/2$	$(10 - w_4 - u_4)/2$

where C_{path} is the cost of constructing a full differential path and p is the probability that a collision is useful. We assume that $p \approx 1/2$ and use the fact that $4 \cdot C_{\text{path}}$ can be negligible compared to the other costs (see Sect. 4.1.2).

For C_{flame}, we must choose minimal values for the w_i that are compatible with the differential paths. That is, we must take $w_1 = 4$, $w_2 = 3$, $w_3 = 4$ and $w_4 = 1$. We have $v_1 \geq 1$ and $v_2 \geq 0$, therefore, we must have $u_4 \geq 3$. We can minimize the cost by choosing $v_1 = v_2 = u_4 = 4$. Then, we have a Birthday Factor of $2^{4.5}$. With the same assumptions as before, this gives us

$$C_{\text{flame}} = 4 \cdot C_{\text{path}} + 2^{48.8} \cdot p^{-1/2} + 2^{27.8} + 2^{39.7} + 2^{38.3} + 2^{38.7} \approx 2^{49.3}.$$

For C_{search}, we have a Birthday Factor of 1 and

$$C_{\text{search}} = 4 \cdot C_{\text{path}} + 2^{44.3} \cdot p^{-1/2} + 2^{28.3} + 2^{46.4} + 2^{38.3} + 2^{47.8} \approx 2^{48.4}$$

To minimize the total expected cost, we take $w_1 = 5$, $v_1 = 3$, $w_2 = 5$, $v_2 = 4$, $w_3 = 4$, $w_4 = 5$ and $u_4 = 4$. Then, we have a Birthday Factor of $2^{1.0}$ and

$$C_{\text{min}} = 4 \cdot C_{\text{path}} + 2^{45.8} \cdot p^{-1/2} + 2^{28.3} + 2^{44.0} + 2^{38.3} + 2^{44.8} \approx 2^{46.6}$$

We now show that this cost is indeed minimal:

Theorem 6 . *Given the values for $\mathbb{E}[\mathcal{Z}]$ from Sect. 4.3.2 and assuming that the probability p for a useful collision in the Birthday Search is 1/2, the expected cost of the collision attack is equivalent to at least $C_{\text{min}} = 2^{46.6}$ executions of* MD5Compress. *For suitably chosen parameters, this cost can be achieved.*

Proof. We have already given parameters which show that the second part of the theorem holds. To see that this parameter choice indeed gives us the minimal cost, let us try to improve upon it: It is easy to see that the Birthday Factor μ must satisfy $1 < \mu \leq 2^{1.5}$ for if $\mu = 1$, the attack complexity is $C_{\text{search}} > C_{\text{min}}$ and if $\mu = 2^{2.0}$, the Birthday Search cost is already larger than $\mathcal{C}_{\min}$. If $\mu = 2^{0.5 \cdot k}$, we can reduce the w_i or u_i parameters by k. Since Blocks 2 and 4 have the highest complexity, this is where these reductions should be spent.

For $\mu = 2^{0.5}$, in order to improve upon $\mathcal{C}_{\min}$, we need to construct the near-collision blocks with a cost $\leq 2^{45.8}$, for $\mu = 2^{1.0}$, the cost needs to be $\leq 2^{45.4}$ and for $\mu = 2^{1.5}$ it needs to be $\leq 2^{44.2}$. It turns out that the only way these constraints can be satisfied is by setting $\mu = 2^{1.0}$ and reducing the parameters w_2 and w_4. But these are precisely the parameters that give us $\mathcal{C}_{\min}$.   □

The parameters for $\mathcal{C}_{\min}$ are consistent with the observed differential paths. Assuming that our reconstruction is correct, we can conclude that the expected cost of the collision attack used by the Flame authors is lower-bounded by $2^{46.6}$ calls to MD5Compress. However, it seems likely that the cost of the actual attack was higher than $\mathcal{C}_{\min}$ since the observed number of carries is always lower than the $\mathcal{C}_{\min}$-parameters. Nevertheless, the actual collision attack might have been faster in practice: Since Birthday Search can be executed very cost-effectively on massively parallel architecture (e.g., GPUs), it might be advantageous to shift a larger part of the workload to the Birthday Search step.

The expected cost of the [27]-attack with four near-collision blocks is roughly $1/4$ of the lower bound of the Flame attack; its expected cost is equivalent to $2^{44.55}$ calls to MD5Compress (see [25, Sect. 3.7]). The cost of the Birthday Search dominates the total cost.

5   Conclusion

In this paper we have demonstrated for the first time that a cryptanalytic attack can be reconstructed from a single output example, specifically, a single example half of a collision pair. We have provided a complexity analysis proving a lower-bound for its cost. Furthermore, we showed that in terms of theoretical cost, the Flame attack is less efficient than the [27]-attack, although it might achieve a better real-world performance when the Birthday Search is performed on a massively parallel architecture.

Our reverse-engineering of a yet unknown cryptanalytic attack seems to be without precedent. As allegedly Flame was developed by some nation-state(s), the example collision and its analysis in this work provide some insights to their cryptanalytic knowledge and capabilities. With respect to the complexity, the closest fit of attack parameters is equivalent to $2^{49.3}$ MD5 compressions which takes roughly 40,000 CPUcore hours. That means for say 3-day attempts to succeed in reasonable time given the large number of required attempts, one needs about 560 CPUcores, which is large but not unreasonable even for academic research groups. With the respect to cryptanalytic knowledge there are no indications at all of superior techniques, rather various parts seem to be sub-optimal compared to the state-of-the-art in the literature. In particular it is clear that one could do better using the state-of-the-art in the literature, i.e., lower theoretical complexity to craft a 4-block chosen-prefix collision (see Theorem 6), and generate differential paths with significantly lower density of bitconditions in negligible time (as previously observed [25]). Nevertheless, the apparent signifi-cant resources more than make up for that and it seems a working attack that succeeds in reasonable time was more important than optimizing the overall attack using all state of the art techniques.

A   Flame Differential Paths

Here, we show the differential paths for all four Flame near-collision blocks, see also Sect. 4.2.1. The column 'Probability' lists the theoretical unconditional rotation probabilities from δT_t to δR_t. If this rotation probability for this δT_t is not maximal, we list the maximal possible rotation probability for this δT_t albeit for a different δR_t between braces. In the next column 'Cond. Est.', we give *empirical estimates* for the probabilities of the rotations *conditioned* on that the Q_i satisfy their bitconditions (Tables 13, 14, 15 and 16).

Table 13. Differential path sections of the 1st near-collision block of Flame's attack

t	Bitconditions $\mathsf{q}_t[31]\dots\mathsf{q}_t[0]$	Probability	Cond. est.
2	+0-0-.00 .-++00+- 0-1-+.1+ 1+-0++^.	0.247 (0.628)	0.166
3	+010-000 .-+++0+1 +--.+^1+ -+-+++-.	0.911	1
4	-00-10+. .11-+-0+ +++11--0 -101-+0.	0.381 (0.561)	1
5	0-+-++-^ ^0110+1- -110+0-0 -0001+1^	0.229 (0.435)	1
6	++-----+- ---+---- -----+++ ++++++++	0.425 (0.514)	1
7	111.-111 1101011. 110-1001 +0100.00	0.838	1
8	00+0.111 10111101 -1101100 .1110011	0.063 (0.444)	0.171
9	..0.1.. .....-.. 0.10+... 0-....0.	0.516	0.563
59	+.......	1	1
60	+.11110.	1	1
61	+.11000. .... .001.00.	0.992	1
62	-.+----. ....0....	0.391 (0.609)	0.427
63	+.?0??+. ....--+.+-.	0.867	0.855
64	+......+ ++++++.. -..-.+-. .....+-.		

Table 14. Differential path sections of the 2nd near-collision block of Flame's attack

t	Bitconditions $q_t[31]\ldots q_t[0]$				Probability	Cond. est.
2	.01.-011	00+-++0+	0--+.--0	++10+0+0	0.849	0.492
3	..1.-+11	+001++^+	01-+0110	0+1++0++	0.623	0.833
4	..-.1-11	++1-++-+	-1111--+	++0+-+-1	0.100 (0.547)	1
5	^^1^+1--	10-01011	0+10-1-+	0-+++000	0.399 (0.431)	0.499
6	+-++++++	++++----	------+-	--+-----	0.458 (0.518)	1
7	0010-000	01111011	1011-111	10.10010	0.961	1
8	00000100	1111111+	-1001111	1-010111	0.468	0.673
9	...-1...	.-.....1	0..1+...	.1....^.	0.468 (0.469)	0.495
58	+.......	....	.......'.	....	1	1
59	+.......	....	....	..0.....	1	1
60	+.....0.	....	...1001.	110.....	0.5	0.507
61	-....100	...0....	...1..1.	00+.....	0.496	0.749
62	+....1-.	....	...-+++.	+--.....	0.972	0.948
63	+....++-	...+....	...???-.	?+-.....	0.238 (0.270)	0.262
64	......--	..+.....	.-....-.	.+-....+		

Table 15. Differential path sections of the 3rd near-collision block of Flame's attack

t	Bitconditions $q_t[31]\ldots q_t[0]$				Probability	Cond. est.
2	10-01110	+++1---+	+10+....	0-0++++1	0.404(0.408)	0.374
3	-0-01^1+	+0+1--10	0-++^^.0	01+0+00.	0.941	1
4	--0++-00	0-0+11++	++-1-+10	-+00+-1.	0.085(0.593)	1
5	-1++-0-1	+1-00+1-	+0++110-	-1--1+^^	0.776	1
6	++----+-	---+----	-----+++	++++++++	0.514	1
7	1000-010	00.1010.	101-0101	+0001.00	0.838	1
8	11+1.101	01011100	-1000101	.1000011	0.437	0.0566
9	..0.1...	.....-..	0.10+...	0-....0.	0.516	0.573
58	+.......	....	....	....'....	1	1
59	+.......	....	....	....	1	1
60	-.....0.	....	......1.	....	1	1
61	-.0110.0	....	.1....0.	....	0.496	0.515
62	+..01.+.	....	.0....+.	....	0.498	0.492
63	+.+---?-	....	.-....+.	....	0.404	0.396
64	.+...+.-	...++++	-.....+.	....-...		

Table 16. Differential path sections of the 4th near-collision block of Flame's attack

t	Bitconditions $q_t[31] \ldots q_t[0]$				Probability	Cond. est.
2	+--.-0-.	-+1+0--0	1+1-1-++	-1-00+--	0.691	0.757
3	+--1-^1.	.+100--+	10---1+0	---0++-1	0.309	1
4	-010+-1.	10-1-01+	0-000-1-	0+-10-1-	0.574	1
5	+00-+00^	0++-11-0	+++0-111	01-+-100	0.749	1
6	+-++++++	++++----	------+-	--+-----	0.518	0.507
7	.111-110	01.010.0	0101-110	1101.011	0.961	0.735
8	11110110	0101000+	-0101111	0-100111	0.032(0.476)	0.0508
9	...-1...	.-.....1	0..1+...	.1....^.	0.468(0.469)	0.522
58	-.......	....	....	....	1	1
59	-.......	....	....	....	1	1
60	+.....0.	....	....00.	....	1	1
61	+.....1.	....	11.....1.	....	0.496	0.525
62	-.....-.	....	10...-+.	....	0.5	0.493
63	+.....-.	....	+-...?-.	....	0.500	0.503
64	...-++.	....	+-....-.	..-.+..+		

B   Message Blocks and *IHV*s

	Flame certificate	Legitimate certificate
IHV_7	a262d0136907c960bb84d9d73b74732e	8262d01365179fa09bd4c9cf1b76732e
B_8	7f7b4b7bc6beeb3f9f983da38487547e	7f7b4b7bc6beeb3f9f983da38487547e
	728771254b6835ae65bd6c8fdc8dacc4	728771a54b6835ae65bd6c8fdc8dacc4
	e89892dedc5362f5726a2527a31246eb	e89892dedc5362f5726a2527a39246eb
	7f6d58cd3083d77a85b848e60e011168	7f6d58cd3083d77a85b848660e011168
IHV_8	63fc3d453bdacbc8826faa39cc7df2cc	43fc3dc5395c9d8a62719ab3ac7ff24e
B_9	657d53380b40f43b684359c13c05c340	657d53380b40f43b684359c13c05c340
	269d5197e2eb2eb8c2196e4e94463bd8	269d5117e2eb2eb8c2196e4e94463bd8
	d4fd0d00d168fadff3fa188a7c659bda	d4fd0d00d168fadff3fa188a7ce59ada
	23119f16a68b23248887226919c211ea	23119f16a68b2324888722e919c211ea
IHV_9	7aeea241ddd49e30b9ce4dab4b8e0ff4	7aeea241fc1490efb9ce4daa4b8e0ff4
B_{10}	9d3681adfbe88bd2d0eb06f21a868dc6	9d3681adfbe88bd2d0eb06f21a868dc6
	84f388c5e0d964c64895d4bed3544891	84f38845e0d964c64895d4bed3544891
	e66ce91e33971542eeb46d1f150b27dd	e66ce91e33971542eeb46d1f158b27dd
	08bb81deb6961639d926446a5fd16b3f	08bb81deb6961639d92644ea5fd16b3f
IHV_{10}	ac3aa31bd79e7f3a9b34ec0a850e3940	ac3aa39bee607f3c9bf6eb8c851039c2
B_{11}	1271dcf09962d2431458f86ef82235d2	1271dcf09962d2431458f86ef82235d2
	90f7fd936ac449b8cb0ce965a8f722b5	90f7fd136ac449b8cb0ce965a8f722b5
	f2051920ef2563c7b3974a823eb2e3ee	f2051920ef2563c7b3974a823e32e3ee
	b45ecb1db3598f8df47901b1b6688914	b45ecb1db3598f8df4790131b6688914

References

1. Biham, E., Shamir, A.: Differential Cryptanalysis of the Data Encryption Standard. Springer-Verlag, London (1993)
2. den Boer, B., Bosselaers, A.: Collisions for the compression function of MD-5. In: Helleseth, T. (ed.) EUROCRYPT 1993. LNCS, vol. 765, pp. 293–304. Springer, Heidelberg (1994)
3. Brassard, G. (ed.): CRYPTO 1989. LNCS, vol. 435. Springer, Heidelberg (1990)
4. De Cannière, C., Rechberger, C.: Finding SHA-1 characteristics: general results and applications. In: Lai, X., Chen, K. (eds.) ASIACRYPT 2006. LNCS, vol. 4284, pp. 1–20. Springer, Heidelberg (2006)
5. Damgård, I.: A design principle for hash functions. In: Brassard [3], pp. 416–427
6. Dobbertin, H.: The Status of MD5 After a Recent Attack. RSA CryptoBytes, 2(2) (1996)
7. Hawkes, P., Paddon, M., Rose, G.G.: Musings on the Wang et al. MD5 Collision. Cryptology ePrint Archive, Report 2004/264 (2004)
8. Hashclash project webpage. http://code.google.com/p/hashclash
9. Klima, V.: Finding MD5 Collisions on a Notebook PC Using Multi-message Modifications. Cryptology ePrint Archive, Report 2005/102 (2005)
10. Klima, V.: Tunnels in Hash Functions: MD5 Collisions Within a Minute. Cryptology ePrint Archive, Report 2006/105 (2006)
11. CrySyS Lab: sKyWIper (a.k.a. Flame a.k.a. Flamer): A complex malware for targeted attacks. Laboratory of Cryptography and System Security, Budapest University of Technology and Economics, 31 May 2012
12. Kaspersky Lab: The Flame: Questions and Answers. Securelist blog, 28 May 2012
13. Liang, J., Lai, X.: Improved Collision Attack on Hash Function MD5. Cryptology ePrint Archive, Report 2005/425 (2005)
14. Mendel, F., Rechberger, C., Schläffer, M.: MD5 is weaker than weak: attacks on concatenated combiners. In: Matsui, M. (ed.) ASIACRYPT 2009. LNCS, vol. 5912, pp. 144–161. Springer, Heidelberg (2009)
15. Merkle, R.C.: One Way Hash Functions and DES. In: Brassard [3], pp. 428–446
16. Microsoft: Flame malware collision attack explained. Security Research and Defense, Microsoft TechNet Blog, 6 June 2012
17. Microsoft: Microsoft certification authority signing certificates added to the Untrusted Certificate Store. Security Research and Defense, Microsoft TechNet Blog, 3 June 2012
18. van Oorschot, P.C., Wiener, M.J.: Parallel collision search with cryptanalytic applications. J. Cryptol. 12(1), 1–28 (1999)
19. Nakashima, E., Miller, G., Tate, J.: U.S., Israel developed Flame computer virus to slow Iranian nuclear efforts, officials say. The Washington Post, June 2012
20. Ray, M.: Flame's Windows Update Certificate Chain. Randombit Cryptography Mailing List, June 2012. http://lists.randombit.net/pipermail/cryptography/2012-June/002969.html
21. Rivest, R.L.: The MD5 Message-Digest Algorithm. Internet Request for Comments, RFC 1321, April 1992
22. Sasaki, Y., Naito, Y., Kunihiro, N., Ohta, K.: Improved Collision Attack on MD5. Cryptology ePrint Archive, Report 2005/400 (2005)
23. Sotirov, A.: Analyzing the MD5 collision in Flame, June 2012
24. Stevens, M.: Fast Collision Attack on MD5. Cryptology ePrint Archive, Report 2006/104 (2006)

25. Stevens, M.: Counter-cryptanalysis. In: Canetti, R., Garay, J.A. (eds.) CRYPTO 2013, Part I. LNCS, vol. 8042, pp. 129–146. Springer, Heidelberg (2013)
26. Stevens, M., Lenstra, A.K., de Weger, B.: Chosen-prefix collisions for MD5 and colliding X.509 certificates for different identities. In: Naor, M. (ed.) EUROCRYPT 2007. LNCS, vol. 4515, pp. 1–22. Springer, Heidelberg (2007)
27. Stevens, M., Sotirov, A., Appelbaum, J., Lenstra, A., Molnar, D., Osvik, D.A., de Weger, B.: Short chosen-prefix collisions for MD5 and the creation of a rogue CA certificate. In: Halevi, S. (ed.) CRYPTO 2009. LNCS, vol. 5677, pp. 55–69. Springer, Heidelberg (2009)
28. Wang, X., Feng, D., Lai, X., Yu, H.: Collisions for Hash Functions MD4, MD5, HAVAL-128 and RIPEMD. Cryptology ePrint Archive, Report 2004/199 (2004)
29. Wang, X., Yu, H.: How to break MD5 and other Hash functions. In: Cramer, R. (ed.) EUROCRYPT 2005. LNCS, vol. 3494, pp. 19–35. Springer, Heidelberg (2005)
30. Xie, T., Feng, D.: How To Find Weak Input Differences for MD5 Collision Attacks. Cryptology ePrint Archive, Report 2009/223 (2009)
31. Yajima, J., Shimoyama, T.: Wang's sufficient conditions of MD5 are not sufficient. Cryptology ePrint Archive, Report 2005/263 (2005)

Analysis of SHA-512/224 and SHA-512/256

Christoph Dobraunig, Maria Eichlseder$^{(\boxtimes)}$, and Florian Mendel

Graz University of Technology, Graz, Austria
maria.eichlseder@iaik.tugraz.at

Abstract. In 2012, NIST standardized SHA-512/224 and SHA-512/256, two truncated variants of SHA-512, in FIPS 180-4. These two hash functions are faster than SHA-224 and SHA-256 on 64-bit platforms, while maintaining the same hash size and claimed security level. So far, no third-party analysis of SHA-512/224 or SHA-512/256 has been published. In this work, we examine the collision resistance of step-reduced versions of SHA-512/224 and SHA-512/256 by using differential cryptanalysis in combination with sophisticated search tools. We are able to generate practical examples of free-start collisions for 44-step SHA-512/224 and 43-step SHA-512/256. Thus, the truncation performed by these variants on their larger state allows us to attack several more rounds compared to the untruncated family members. In addition, we improve upon the best published collisions for 24-step SHA-512 and present practical collisions for 27 steps of SHA-512/224, SHA-512/256, and SHA-512.

Keywords: Hash functions · Cryptanalysis · Collisions · Free-start collisions · SHA-512/224 · SHA-512/256 · SHA-512 · SHA-2

1 Introduction

The SHA-2 family of hash functions is standardized by NIST as part of the Secure Hash Standard in FIPS 180-4 [21]. This standard is not superseded by the upcoming SHA-3 standard. Rather, the SHA-3 hash functions supplement the SHA-2 family. Thus, it is likely that the SHA-2 family will remain as ubiquitously deployed in the foreseeable future as it is now. Therefore, the continuous application of state-of-the-art cryptanalytic techniques for quantifying the security margin of hash functions of the SHA-2 family is of significant practical importance.

In this work, we focus on the two most recent members of the SHA-2 family, SHA-512/224 and SHA-512/256. As already observed by Gueron et al. [10], using truncated SHA-512 variants like SHA-512/256 gives a significant performance advantage over SHA-256 on 64-bit platforms due to the doubled input block size. At the same time, the shorter 256-bit hash values are more economic, compatible with existing applications, and offer the same security level as SHA-256. In addition, the resulting chop-MD [5] structure of SHA-512/224 and SHA-512/256 with is wide-pipe structure provides cryptographic benefits over the standard

© International Association for Cryptologic Research 2015
T. Iwata and J.H. Cheon (Eds.): ASIACRYPT 2015, Part II, LNCS 9453, pp. 612–630, 2015.
DOI: 10.1007/978-3-662-48800-3_25

Merkle-Damgård [7,20] structure by prohibiting generic attacks like Joux' multicollision attack [12], Kelsey and Kohno's herding and Nostradamus attacks [13], and Kelsey and Schneier's second preimages for long messages [14].

However, no cryptanalysis dedicated to SHA-512/224 and SHA-512/256 has been published so far. Therefore, we examine the effects of truncating the hash value of SHA-512. We show that due to this truncation, practical free-start collision for 43-step SHA-512/256 and 44-step SHA-512/224 are possible. Moreover, we improve upon the previous best collisions for 24-step SHA-512 [11,23] and show collisions for 27 steps of SHA-512, SHA-512/224, and SHA-512/256. Since all of our results are practical, we provide examples of colliding message pairs for every attack. Our results are summarized in Table 1 together with previously published collision attacks.

Table 1. Best published collision attacks on the SHA-512 family.

Hash size	Type	Steps	Complexity	Reference
all	collision	24/80	practical	[11,23]
	collision	**27**/80	**practical**	**Sect.** 4.3
	semi-free-start collision	38/80	practical	[9]
	semi-free-start collision	**39**/80	**practical**	**Sect.** 4.1
512	free-start collision	57/80	$2^{255.5}$	[17]
384	free-start collision	40/80	2^{183}	[17]
256	**free-start collision***	**43**/80	**practical**	**Sect.** 4.2
224	**free-start collision***	**44**/80	**practical**	**Sect.** 4.2

* without padding.

Related Work. No dedicated cryptanalysis of SHA-512/224 or SHA-512/256 has been published so far. However, there is a number of results targeting SHA-512. The security of SHA-512 against preimage attacks was first studied by Aoki et al. [1]. They presented MITM preimage attacks on 46 steps of the hash function. This was later extended to 50 steps by Khovratovich et al. [15]. However, due to the wide-pipe structure of SHA-512/224 and SHA-512/256, these attacks do not carry over to SHA-512/224 and SHA-512/256.

The currently best known practical collision attack on the SHA-512 hash function is for 24 steps. It was published independently by Indesteege et al. [11] and by Sanadhya and Sarkar [23]. Both attacks are trivial extensions of the attack strategy of Nikolić and Biryukov [22] which applies to both SHA-256 and SHA-512. Recently, Eichlseder et al. [9] demonstrated how to extend these attacks to get semi-free-start collisions for SHA-512 reduced to 38 steps with practical complexity. Furthermore, second-order differential collisions for SHA-512 up to 48 steps with practical complexity have been shown by Yu et al. [27]. We want to note that all these practical collision attacks on SHA-512 are also applicable to its truncated variants.

Additionally, Li et al. showed in [17] that particular preimage attacks on SHA-512 can also be used to construct free-start collision attacks for the step-reduced hash function and its truncated variants. They show a free-start collision for 57-step SHA-512 and 40-step SHA-384. Both attacks are only slightly faster than the respective generic attacks.

Outline. The remainder of the paper is organized as follows. We describe the design of the SHA-2 family in Sect. 2. Then, we briefly explain our attack strategy and discuss the choice of suitable starting points for our attacks in Sect. 3. The actual attacks on step-reduced SHA-512/224 and SHA-512/256 are presented in Sect. 4.

2   Description of SHA-512 and Other SHA-2 Variants

The SHA-2 family of hash functions is specified by NIST as part of the Secure Hash Standard (SHS) [21]. The standard defines two main algorithms, SHA-256 and SHA-512, with truncated variants SHA-224 (based on SHA-256) and SHA-512/224, SHA-512/256, and SHA-384 (based on SHA-512). In addition, NIST defines a general truncation procedure for arbitrary output lengths up to 512 bits. Below, we first describe SHA-512, followed by its truncated variants SHA-512/224 and SHA-512/256 that this paper is focused on. Finally, the main differences to SHA-256 and SHA-224 are briefly discussed.

SHA-512. SHA-512 is an iterated hash function that pads and processes the input message using t 1024-bit message blocks m_j. The 512-bit hash value is computed using the compression function f:

$$h_0 = \text{IV},$$
$$h_{j+1} = f(h_j, m_j) \qquad \text{for } 0 \le j < t.$$

The hash output is the final 512-bit chaining value h_t.

In the following, we briefly describe the compression function f of SHA-512. It basically consists of two parts: the message expansion and the state update transformation. A more detailed description of SHA-512 is given by NIST [21].

We use $+$ (or $-$) to denote addition (or subtraction) modulo 2^{64}; $\oplus$ (or $\wedge$) is bitwise exclusive-or (or bitwise and) of 64-bit words, and $\ggg n$ (or $\gg n$) denotes rotate-right (or shift-right) by n bits.

Padding and Message Expansion. The message expansion of SHA-512 splits each 1024-bit message block into 16 64-bit words M_i, $i = 0, \ldots, 15$, and expands these into 80 expanded message words W_i as follows:

$$W_i = \begin{cases} M_i & 0 \le i < 16, \\ \sigma_1(W_{i-2}) + W_{i-7} + \sigma_0(W_{i-15}) + W_{i-16} & 16 \le i < 80. \end{cases} \quad (1)$$

The functions $\sigma_0(x)$ and $\sigma_1(x)$ are given by

$$\sigma_0(x) = (x \ggg 1) \oplus (x \ggg 8) \oplus (x \gg 7),$$
$$\sigma_1(x) = (x \ggg 19) \oplus (x \ggg 61) \oplus (x \gg 6).$$

State Update Transformation. We use the alternative description of the SHA-512 state update by Mendel et al. [18], which is illustrated in Fig. 1.

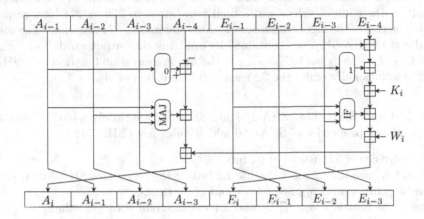

Fig. 1. The state update transformation of SHA-512.

The state update transformation starts from the previous 512-bit chaining value $h_j = (A_{-1}, \ldots, A_{-4}, E_{-1}, \ldots, E_{-4})$ and updates it by applying the step functions 80 times. In each step $i = 0, \ldots, 79$, one 64-bit expanded message word W_i is used to compute the two state variables E_i and A_i as follows:

$$E_i = A_{i-4} + E_{i-4} + \Sigma_1(E_{i-1}) + \text{IF}(E_{i-1}, E_{i-2}, E_{i-3}) + K_i + W_i, \quad (2)$$
$$A_i = E_i - A_{i-4} + \Sigma_0(A_{i-1}) + \text{MAJ}(A_{i-1}, A_{i-2}, A_{i-3}). \quad (3)$$

For the definition of the step constants K_i, we refer to the standard document [21]. The bitwise Boolean functions IF and MAJ used in each step are defined by

$$\text{IF}(x, y, z) = (x \wedge y) \oplus (x \wedge z) \oplus z,$$
$$\text{MAJ}(x, y, z) = (x \wedge y) \oplus (y \wedge z) \oplus (x \wedge z),$$

and the linear functions Σ_0 and Σ_1 are defined as follows:

$$\Sigma_0(x) = (x \ggg 28) \oplus (x \ggg 34) \oplus (x \ggg 39),$$
$$\Sigma_1(x) = (x \ggg 14) \oplus (x \ggg 18) \oplus (x \ggg 41).$$

After the last step of the state update transformation, the previous chaining value is added to the output of the state update (Davies-Meyer construction). The result of this feed-forward sum is the chaining value h_{j+1} for the next message block m_{j+1} (or the final hash value h_t):

$$h_{j+1} = (A_{79} + A_{-1}, \ldots, A_{76} + A_{-4}, E_{79} + E_{-1}, \ldots, E_{76} + E_{-4}). \tag{4}$$

SHA-512/256 and SHA-512/224. These truncated variants of SHA-512 differ only in their initial values and a final truncation to 256 or 224 bits, respectively. The rest of the algorithmic description remains exactly the same. The message digest of SHA-512/256 is obtained by omitting the output words $E_{79} + E_{-1}$, $E_{78} + E_{-2}$, $E_{77} + E_{-3}$, and $E_{76} + E_{-4}$ of the last compression function call. SHA-512/224 additionally omits the 32 least significant bits of $A_{76} + A_{-4}$.

SHA-256 and SHA-224. SHA-256 and SHA-512 are closely related. Thus, we only point out properties of SHA-256 which differ from SHA-512:

– The wordsize is 32 instead of 64 bits.
– IV and K_i are the 32 most significant bits of the respective SHA-512 value.
– The step function is applied 64 instead of 80 times.
– The linear functions $\sigma_0, \sigma_1, \Sigma_0$ and Σ_1 use different rotation values.

SHA-224 is a truncated variant of SHA-256 with different IV, in which the output word $E_{60} + E_{-4}$ is omitted.

3   Attack Strategy

Starting from the ground-breaking results of Wang et al. [25,26], the search techniques used for practical collisions have been significantly improved, hitting their current peak in the attacks on SHA-256 [2,19] and SHA-512 [9,27]. In spite of all achieved improvements, the top-level attack strategy has remained essentially the same. At first, a suitable starting point for the search must be determined to define the search space and hopefully make the ensuing search process feasible. The search itself usually involves two phases: The search for a suitable differential characteristic, and the message modification phase to determine a collision-producing message pair for this characteristic. The search for this characteristic and message pair can either be done by hand or, for more complex functions like SHA-2, using an automatic search tool. We use a heuristic search tool based on a guess-and-determine strategy, which we briefly describe in Sect. 3.1. Afterwards, we discuss the choice of suitable starting points in Sect. 3.2.

3.1 Guess-and-Determine Search Tool

To search for differential characteristics and colliding message pairs, we use an automatic search tool, which implements a configurable heuristic guess-and-determine search strategy. Roughly, the tool is partitioned into two separate, but closely interacting parts: The representation of the analyzed cryptographic primitive and the search procedure.

Representation. The tool internally represents differences at bit level, allowing to store all possible stages from a completely unrestricted bit over signed differences down to exact values. Thus, the same tool can be used in the search for a characteristic and in the search for a message pair. The conditions are grouped in words representing the internal variables of the hash function. These words can then be connected with any operations (typically bitwise functions or modular additions) to define the hash function.

Search. The search procedure uses the bitwise conditions as variables, and attempts to find a solving assignment with the help of a heuristic guess-and-determine strategy [8], similar to SAT solvers. The following steps are repeated until a solution is found:

- **Guess**: Pick a bit and guess its value (e.g., no difference, or a specific assignment).
- **Determine**: The previous guess influences other connected bit conditions. Determine these effects, which might result in further refinement of other bit conditions, or a contradiction.
- **Backtrack**: If a contradiction is detected, resolve this conflict by undoing previous guesses and replacing them with other choices.

This simple approach alone is not sufficient to go through the whole search space, so numerous refinements have been proposed to fine-tune this method. These include the detection of two-bit conditions [18], backtracking strategies, and a look-ahead approach to guide the search [9]. Additionally, SHA-2-specific heuristics and strategies [18,19] have been proposed, deciding which parts of the state to guess with higher priority.

3.2 Finding Starting Points for SHA-2

To model SHA-2 as a satisfiability problem for the search tool, we need to introduce suitable intermediate variables. Based on the alternative description from Sect. 2, we only use the words A_i and E_i of the state, plus the words W_i of the message expansion. Figure 2 illustrates the update rules for A, E and W by highlighting the input words for updating each word: Each row represents one of the 80 step iterations, with its three state words A_i, E_i, and W_i.

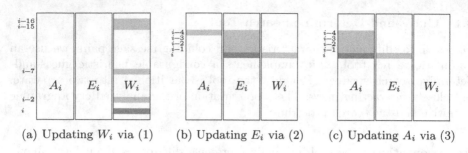

(a) Updating W_i via (1)   (b) Updating E_i via (2)   (c) Updating A_i via (3)

Fig. 2. Update rules to compute $A_i, E_i,$ and W_i (■) from other state words (■).

Local Collisions. All our results are based on "local collisions" in the message expansion: by carefully selecting (expanded) message words in the middle steps so that the differences can cancel out in as many consecutive steps as possible in the forward and backward expansion, i.e., the first and last few expanded message words contain no differences. The t middle steps with differences can induce differences in the A_i and E_i words. However, the W_i words can be used to achieve zero difference in the last 4 of the t words E_i, and in the last 8 of the t words A_i. This is necessary to obtain words with zero difference in the very last 4 steps of the state update and thus in the output chaining value.

As an example, the starting point for the 27-step collisions for SHA-256 [18] allows differences in expanded message words $W_7, W_8, W_{12}, W_{15},$ and W_{17}, as well as state words $E_7, \ldots, E_{13}$ and $A_7, \ldots, A_{10}$. The exact bitwise signed differences are chosen during the search such that any potential differences in $W_{19}, W_{22}, W_{23}, W_{24}$, as well as $E_{14}, \ldots, E_{17}$ and $A_{10}, \ldots, A_{13}$ cancel out. The resulting starting point is illustrated in Fig. 3a. We show in Sect. 4.3 how the same starting point can be used for SHA-512.

The semi-free-start collision starting point covering the most steps so far is for 38 steps of SHA-256 [19] and SHA-512 [9], with a local collision spanning $t = 18$ steps. Considering the large number of steps, the number of expanded message words with differences and cancellations is remarkably low: only 6 words with differences, and 6 words imposing cancellation conditions.

To find candidates for a higher number of steps, we enumerated all possible selections of active message words (more precisely, of some $t \leq 20$ intermediate expanded message words, the "core words" of the local collision) and investigated the forward and backward expansion under certain assumptions: the t core words are chosen freely, according to the message expansion rule; in the forward and backward expansion, if at least 2 of the input words have differences, they are assumed to cancel out, while a single input word with difference never cancels out. Criteria for selecting suitable candidates then include a low number t of spanned steps and a low number of required cancellation constraints. The best (consistent) result for 39 steps, spanning $t = 19$ steps with 9 cancellations, is given in Fig. 3b.

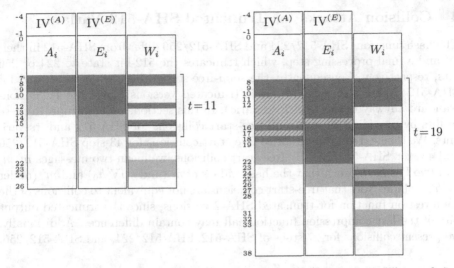

(a) 27-step collision of SHA-256 [18] and SHA-512 (Sect. 4.3).

(b) 39-step semi-free-start collision of SHA-512 (Sect. 4.1).

Fig. 3. SHA-2 starting points: Words with differences ■ and cancellations ■, ▨.

Semi-Free-Start Collisions and Collisions. The discussed starting points are targeted to find semi-free-start collisions, that is, different messages m, m' and an IV h_0 such that $f(h_0, m) = f(h_0, m')$. However, they can also be used for hash function collisions with the original IV h_0 by trading the freedom of the IV for freedom in the message words.

In order to find hash function collisions, the first few message words W_i must retain sufficient freedom (i.e., they should not be constrained by conditions from the message expansion for cancelling differences) to allow to match the correct IV value. Ideally, this means that the first 8 message words $W_0, \ldots, W_7$ are free of any conditions (no differences, but also not constrained by conditions from other message words connected via the message expansion). If the W_i differences are sparse enough overall, it can also be sufficient to have at least 5 words $W_0, \ldots, W_4$ free of conditions by providing the remaining freedom with a two-block approach [19].

The starting points of Fig. 3a and 3b both have at least 7 message words free of differences in the beginning. However, the local collision shown in Fig. 3b spans over $t = 19$ steps. Thus, the first message words are constrained by many conditions, leaving not enough freedom to match the correct IV. In contrast, the 11-step local collision shown in Fig. 3a provides enough freedom in the first 7 message words to be used in a single-block collision attack [18].

4   Collision Attacks for Truncated SHA-512 Variants

The hash functions SHA-512/224 and SHA-512/256 differ from SHA-512 in their IV and a final processing step, which truncates the 512-bit state to 224 or 256 bits, respectively. Consequently, the semi-free-start collisions demonstrated for SHA-512 [9] are also valid for these truncated versions (since the IV is non-standard anyway in this attack scenario). In this section, we first improve these results by providing 39-step semi-free-start collisions for SHA-512 and its variants. We then extend this result to free-start collisions for 43-step SHA-512/256 and 44-step SHA-512/224. By free-start collisions, we mean two messages m, m' and two IVs h_0, h_0' such that the hash values of m (under IV h_0) and m' (under IV h_0') collide. Note that free-start collisions are not equivalent to collisions of the compression function for truncated SHA-2 versions, since the truncated output bits of the last compression function call may contain differences. Additionally, we present collisions for 27 steps of SHA-512, SHA-512/224, and SHA-512/256.

4.1   Semi-free-start Collisions

We use the 39-step starting point from Fig. 3b. Previous work showed that sparse differences particularly in the A_i words are essential for the success probability of the message modification phase. For this reason, we additionally require that in 6 words between A_8 and A_{18}, namely $A_{11}, A_{12}, A_{13}, A_{14}, A_{15}$, and A_{17}, differences also cancel out. The five consecutive zero-difference words in A_i also force E_{15} to zero difference. These additional requirements are already marked in Fig. 3b (hatched area).

   The first task for the search procedure with the solving tool is to fix a suitable signed characteristic. Compared to the previously published 38-step SHA-512 semi-free-start collision [9], the local collision for our starting point spans 19 steps (compared to previously 18) and has 9 (previously 6) active expanded message words. Cancellations are also required in 9 (previously 6) expanded message words. This increases the necessity for very sparse differences in A_i and W_i in steps 16–26. For this reason, we require a single-bit difference in W_{26}, W_{17} and A_{18}, and very low Hamming weights for the other words. We finally found a characteristic with at most two active bits in almost all words of A_i and W_i (except $A_9, A_{10}, W_{11}, W_{12}$).

   After the characteristic is fixed, we need to find a complying message pair. We start by guessing the dense parts in A_i and E_i, hoping that the sparser conditions in the later steps are fulfilled probabilistically. Since the dense parts are already almost fully determined by the characteristics and the sparse parts pose only so few conditions, a message pair is easily found. The result is a semi-free-start collision valid for all SHA-512 variants. We give an example in Appendix A in Table 4a.

4.2   Free-Start Collisions

Free-start collisions are a generalization of semi-free-start collisions, so the 39-step results obtained in the previous section give a first result for SHA-512/224 and

SHA-512/256. However, we can take advantage of the truncated output bits to add several more steps. If we add another step in the beginning or in the end, the existing difference pattern remains unchanged, but there will be differences in the word W_0 (computable via backward expansion, which includes $W_{i+9} = W_9$, the previous W_8 from Fig. 3b) or in the new word W_{39} (via the normal forward expansion, which includes $W_{39-15} = W_{24}$), respectively. These, in turn, can imply differences in E_{-4} or in A_{39} and E_{39}, which translates to differences in the IV (turning semi-free-start into free-start results, and included in the hash value via the feed-forward) or directly in the compression function output, respectively.

The advantage of adding steps in the beginning is that it is possible to limit the additional differences in the state update words to E, and keep A free of new differences. Any differences in $E_{-1}, \ldots, E_{-4}$ will be added to the compression function output with the final feed-forward, but the corresponding words of the result are truncated, so the hash outputs still collide.

Free-Start Collisions for 43-Step SHA-512/256. Since SHA-512/256 truncates the last 4 output words of the compression function call ($E_{79} + E_{-1}$, $E_{78} + E_{-2}$, $E_{77} + E_{-3}$, and $E_{76} + E_{-4}$), differences in $E_{-1}, \ldots, E_{-4}$ are acceptable for a free-start collision. This observation allows us to add 4 additional steps in the beginning of the 39-step starting point from Fig. 3b. Shifting the characteristic "downwards" by 4 steps causes the previous message words $W_{12}, \ldots, W_{15}$ to turn into new expanded message words $W_{16}, \ldots, W_{19}$; in particular, this affects the difference in the previous word W_{12}. To determine a compatible difference pattern for the new first 4 words, the message expansion can be computed backwards from the new words $W_4, \ldots, W_{19}$ via

$$W_i = W_{i+16} - \sigma_1(W_{i+14}) - W_{i+9} - \sigma_0(W_{i+1}).$$

It turns out that all 4 new words will contain differences (W_3 from $W_{3+9} = W_{12}$; W_2 from $W_{2+1} = W_3$ and $W_{2+14} = W_{16}$; W_1 from $W_{1+1} = W_2$ and $W_{1+14} = W_{15}$; and W_0 from $W_{0+1} = W_1$, $W_{0+14} = W_{14}$ and $W_{0+16} = W_{16}$). However, similar to steps 27–30, the state words A_i and E_i can be kept free of differences for 4 steps. To achieve this, the search tool needs to find differences in the IV words $E_{-4}, \ldots, E_{-1}$ to cancel out those in $W_0, \ldots, W_3$ when computing $E_0, \ldots, E_3$. The resulting starting point is given in Fig. 4a.

For the search procedure with the solving tool, we fixed the signed differences of steps 12–30 to the same values as the 39-step SHA-512 semi-free-start collision of Sect. 4.1. Then, to complete the characteristic, we first search for a valid solution for the dense part of the middle steps (A_i and E_i in steps 13–16, and E_i in steps 17–27), and finally fix the corresponding message words W_i in steps 13–17, which determines the complete state, including the dense differences in the prepended steps and IV.

The search only takes seconds on a standard computer; an example for a free-start collision is given in Appendix A in Table 3a.

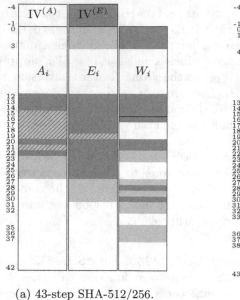
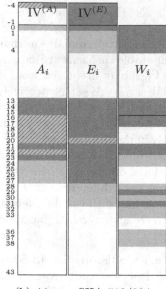

(a) 43-step SHA-512/256.   (b) 44-step SHA-512/224.

Fig. 4. Potential free-start starting points (differences ■ and cancellations ▨, ▨).

Free-Start Collisions for 44-Step SHA-512/224. A very similar strategy can be employed to extend the previous 43-step free-start collision by another step for SHA-512/224. Prepending an additional step shifts the difference of previous word E_{-1} to E_0, which in turn requires a cancellation in A_0 and a difference in A_{-4}, as illustrated in Fig. 4b. However, only the least significant 32 bits of the corresponding compression function output word are truncated. Furthermore, this output word is computed from A_{-4} via modular addition, so even differences only in the lower 32 bits can possibly cause differences in the untruncated output bits.

Fortunately, the underlying characteristic of signed differences as used for the 39-step SHA-512 semi-free-start collision is well compatible with our constraints: The difference in A_{-4} needs to cancel that in W_4 in a modular addition (via E_0, by Eqs. (3) and (2) or Fig. 2, since all other involved words have zero difference). This difference of W_4, in turn, is dictated by that in W_{13} (by the update rule for W_{20}, where again all other involved words have zero difference). None of these equalities involves any of the bitwise functions σ, Σ, MAJ or IF. Thus, the modular difference in A_{-4} must be the same as that in W_{13}, which is already fixed by the underlying characteristic to a modular difference of $+32$. Written as bitwise differences, this will translate to a single-bit difference (in the sixth least significant bit) with probability $\frac{1}{2}$ (which does not carry over to the untruncated bits of the final output with overwhelming probability). Indeed, the example for a free-start collision given in Appendix A in Table 2a only displays this single-bit difference in A_{-4} (and no carries in the output bits).

4.3   Collisions

So far, the best practical collisions found for SHA-512 are those for 24 steps, proposed independently by Sanadhya and Sarkar [23] and Indesteege et al. [11], together with 24-step collisions for SHA-256. While the results for SHA-256 have since been improved to 27 [18], 28 [19] (both practical), and finally 31 steps [19] (theoretical attack with almost practical complexity), no such improvements have been proposed for SHA-512 so far. The main reason for this seems to be the doubling in state size from SHA-256 to SHA-512; this larger search space increases the difficulty of the problem for the search tools.

Starting Point for SHA-512. Since the message expansion is essentially the same for all SHA-2 variants (except for different word sizes and rotation values, of course), the SHA-256 starting points can theoretically also be used for SHA-512. However, the resulting search complexity is different. For our results, we used the 27-step starting point (based on a local collision over the $t = 11$ steps 7–17), as illustrated in Fig. 3a. Just as the 39-step semi-free-start starting point (Fig. 3b), it requires that differences cancel in E in 4 of the t steps $(E_{14}, \ldots, E_{17})$ and in A in the 4 previous steps $(A_{10}, \ldots, A_{13})$, as well as in several steps of the message expansion.

Finding a solution from this starting point requires significantly more effort than for SHA-256. Of course, we also tried to expand our search to the closely related 28-step starting point, which adds an additional step in the beginning of the 27-step version. However, with the additional constraints imposed on the message expansion by this added step we could not find any suitable (reasonably sparse) characteristics.

In contrast to the results from Sect. 4.2, since the IV needs to exactly match the original IV, we were not able to take advantage of the final truncation to simplify the search process, or add additional steps. We first search a characteristic for SHA-512, and then try to use it to match the different IVs for SHA-512/224 and SHA-512/256.

Search Strategy. The search progresses in several stages, as illustrated in Fig. 5:

1. **Fix Signed Characteristic:**
   (a) **Find Candidate Characteristic** (Fig. 5a): First fix the signed differences of the message expansion W (5 words) and state update A (3 words). Since the word W_{17} poses conditions on the first few message words, whose freedom we will later need to match the IV, we focus on keeping its signed difference as sparse as possible, with only few difference bits. With much lower priority, also determine the differences in the state update words E (7 words) to complete the signed characteristic. The characteristic is very dense in E, but this only has limited influence on the success of the IV matching phase.

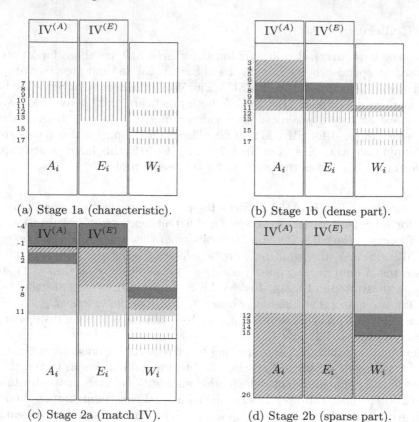

(a) Stage 1a (characteristic).   (b) Stage 1b (dense part).

(c) Stage 2a (match IV).   (d) Stage 2b (sparse part).

Fig. 5. Stages of the 27-step collision search (guessed values ■ and differences ▫, derived values ▨, and previously fixed values ▩ and differences ▫).

    (b) **Verify Dense Parts** (Fig. 5b): Fully determine the values of A and E in the densest steps 7–9 to verify the validity of the candidate characteristic. If necessary, fix any remaining free bits of A and E in steps 10–11. This fully determines $A_3, \ldots, A_{11}$, $E_7, \ldots, E_{11}$ and W_{11}.

To maneuver the search process in the large search space and detect contradictions as soon as possible, we need to apply the look-ahead strategies previously employed for semi-free-start collisions on SHA-512 [9] in this stage (with 16 look-ahead candidates per guess).

2. **Message Modification to Match IV:** Starting from the best signed characteristics of the previous stage, with the correct IV inserted, find a solution message pair step by step:

    (a) **Match IV** (Fig. 5c): Fix the values in the more difficult, heavily constrained words first (W_{10}, W_9, W_8, W_7). Choosing W_{10} and W_9 also determines A_2 and A_1 (via E_6 and E_5). Together with W_7, W_8, and the IV, this determines all values in steps 0–11.

(b) **Finalize Message for Sparse Parts** (Fig. 5d): choosing the 4 remaining message words $W_{12}, \ldots, W_{15}$ allows to satisfy the remaining, sparse parts of the characteristic in steps 12–26 with high probability.

Unlike the other stages, guesses are not made randomly here, but systematically word-by-word. Since most conditions are from modular additions, we always start from the least significant bits and proceed towards the more significant bits. This last stage needs to be repeated for each IV separately, which takes some hours on a single CPU per target IV.

Results. Our results for collisions for 27-step SHA-512/224, SHA-512/256, and SHA-512 are given in Appendix A in Tables 2b, 3b, and 4b, respectively.

Acknowledgments. This research (or a part of this research) is supported by Cryptography Research and Evaluation Committee (CRYPTREC) and by the Austrian Research Promotion Agency (FFG) and the Styrian Business Promotion Agency (SFG) under grant number 836628 (SeCoS).

A   Examples

An example for the semi-free-start collisions of Sect. 4.1 is given in Table 4a. Results for the free-start collisions of Sect. 4.2 are given in Tables 2a and 3a, and for the collisions of Sect. 4.3 in Tables 2b, 3b, and 4b.

Table 2. Results for SHA-512/224.

(a) Example of a free-start collision for 44 steps of SHA-512/224.

h_0	fef65b64d3694995 959fbfb82ed84eb1 1d9e855642e62ef2 335cc6d027695d91
	921d197e5cfa2803 e26c6eb26163a692 9ff3cf4d26f1de78 5323942861d9139a
h_0^*	fef65b64d3694995 959fbfb82ed84eb1 1d9e855642e62ef2 335cc6d027695db1
	a712860cdcfa1ff8 470749bbf7628f44 20cdfd694df67216 8e07b5fa2c7fedf0
Δh_0	0000000000000000 0000000000000000 0000000000000000 0000000000000020
	350f9f72800037fb a56b2709960129d6 bf3e32246b07ac6e dd2421d24da6fe6a
m	7a19df6089d00684 03ed2a0d0c29e00e 36c91e35f681fbb8 bb2b47428aeff294
	dce94ccc981d39a3 44230f73cf56d9ef e9d46b26b44950c8 550bed4b9419741c
	58a98894206e00de f3448a6f761d384d 9ae59f3a3bcc5bba ece85d5c77be431b
	6e3cf817e9376cc7 b74a2a43c0b96c93 7c5b51d6fe2a0c26 5a9868e5bf2e422d
m^*	5e031bbe28b2d027 ded424ef85255cc3 ad2f514be0830c1f a635dab40aeffa9f
	dce94ccc981d3983 44230f73cf56d9ef e9d46b26b44950c8 550bed4b9419741c
	58a98894206e00de f3448a6f761d384d 9ae59f3a3bcc5bba ece85d5c77be431b
	6e3cf817e9376cc7 b74a2a43c0b96cb3 5c5b51d6fe2a0c36 5a9868e5bf2e420d
Δm	241ac4dea162d6a3 dd390ee2890cbccd 9be64f7e1602f7a7 1d1e9df68000080b
	0000000000000020 0000000000000000 0000000000000000 0000000000000000
	0000000000000000 0000000000000000 0000000000000000 0000000000000000
	0000000000000000 0000000000000020 2000000000000010 0000000000000020
h_1	e309edf68f4d89b8 5c356e0359eb0dab 76b4a45ec3c2cd25 8bd0955d

(b) Example of a collision for 27 steps of SHA-512/224.

m	20dbf13a352116a9 295506e205afd435 abfe4826742c1a1a 279f07c7813dd9be
	47da77c701a98858 25aec1349d486501 37a992a15616ea31 e2b122ecf19e90d3
	2fff6025dc03dd67 032c261d740f459e 2e2599bd6e7e74df d490bd22815eb494
	72fedf1f607df6e3 87fc91fcfb7397fd e647b1b499eee17f 2dff8e493cbc8a4c
m^*	20dbf13a352116a9 295506e205afd435 abfe4826742c1a1a 279f07c7813dd9be
	47da77c701a98858 25aec1349d486501 37a992a15616ea31 5cc1250cb19e90d3
	203fdfe5dc03dd66 032c261d740f459e 2e2599bd6e7e74df d490bd22815eb494
	f0bc01167075f6eb 87fc91fcfb7397fd e647b1b499eee17f d3f8fe713d7c8a4c
Δm	0000000000000000 0000000000000000 0000000000000000 0000000000000000
	0000000000000000 0000000000000000 0000000000000000 be7007e040000000
	0fc0bfc000000001 0000000000000000 0000000000000000 0000000000000000
	8242de0910080008 0000000000000000 0000000000000000 fe07703801c00000
h_1	65b11e66e48da563 1b70d12da92e2dba 8f338768bb95601b 60b995bb

Table 3. Results for SHA-512/256.

(a) Example of a free-start collision for 43 steps of SHA-512/256.

h_0	159b52516f10f30d 546b2042f240afee f25339b24c441edf d62c698666558242 e5a9e39861fbd81d d2138eacc20d5224 a332c16df23609fb 73f78341dfd7a4e5
h_0^*	159b52516f10f30d 546b2042f240afee f25339b24c441edf d62c698666558242 e5a9e39861fbd83d 72e259ce420d5a0f 4db37906cc361264 ae579d9e0275b446
Δh_0	0000000000000000 0000000000000000 0000000000000000 0000000000000000 0000000000000020 a0f1d7628000082b ee81b86b3e001b9f dda01edfdda210a3
m	cfbec86f1cf6821e dd3343c25aad835a 2a08612b753f3d6b b328d40d2c624ef7 b3e51f8a3a63bd6f 4abdf96375bbf609 a8c5c1f784672e86 a78e2aa625830d4b 169dcb5039bf3d9f fbcc43ffebd8ae47 1b3eaefccf5c6a46 f668a2a728851b4e 374601ea44422bdb 2ca290d26a23a02f 6685babbfdcb5e22 e000111457201fd4
m^*	ee37d77210586a56 b2a4122800ad72cf 89399609f53f3560 b328d40d2c624ed7 b3e51f8a3a63bd6f 4abdf96375bbf609 a8c5c1f784672e86 a78e2aa625830d4b 169dcb5039bf3d9f fbcc43ffebd8ae47 1b3eaefccf5c6a46 f668a2a728851b4e 374601ea44422bfb 0ca290d26a23a03f 6685babbfdcb5e02 e07e151457202055
Δm	21891f1d0caee848 6f9751ea5a00f195 a331f7228000080b 0000000000000020 0000000000000000 0000000000000000 0000000000000000 0000000000000000 0000000000000000 0000000000000000 0000000000000000 0000000000000000 0000000000000020 2000000000000010 0000000000000020 007e040000003f81
h_1	1d7041bbbffa676a 03d8c440d9246b9d 20ce2d17c5b0b2c4 7e6e4d33a7f54afd

(b) Example of a collision for 27 steps of SHA-512/256.

m	306b0c2ebe7c1341 c8b55d4df1c5f4fe b91a173aeceb818a 33b5977f9b46e58b 6c6d5a4f87f1364f 1b7e33249d4acf4f b7f784ecdcaefc1f a33edafe7afc0452 dfc0200932c2b9df faec7d05e3518e56 ec2e19a7ee867396 d490bd22815eb494 72fedf1f887df303 f95891f08483da25 c327d0afa2c4f902 2c5f0c0806a4e298
m^*	306b0c2ebe7c1341 c8b55d4df1c5f4fe b91a173aeceb818a 33b5977f9b46e58b 6c6d5a4f87f1364f 1b7e33249d4acf4f b7f784ecdcaefc1f 1d4edd1e3afc0452 d0009fc932c2b9de faec7d05e3518e56 ec2e19a7ee867396 d490bd22815eb494 f0bc01169875f30b f95891f08483da25 c327d0afa2c4f902 d2587c300764e298
Δm	0000000000000000 0000000000000000 0000000000000000 0000000000000000 0000000000000000 0000000000000000 0000000000000000 be7007e040000000 0fc0bfc000000001 0000000000000000 0000000000000000 0000000000000000 8242de0910080008 0000000000000000 0000000000000000 fe07703801c00000
h_1	fcba5c8faf05fd68 c676b8f17b5daae3 6233801174b7fd01 0ff72ab4a869c54f

Table 4. Results for SHA-512.

(a) Example of a semi-free-start collision for 39 steps of SHA-512.

h_0	eccf3da189dd9668 b1ec21a4fd53b8d8 609ce4465f772770 adf4e7738e2978f6 8edd237ea50eebc9 231b3af0102a926d db45e613e8d2fd52 ad384433420073f6
m	a0ec9872cfffe63c df5c6a2b59f4c453 f2bea3763fc8fa7a 6a47e8ff0a995116 fa59232e8b617048 4c9690984c084498 28bee8f5701eab16 8d57686ecbdce623 3879318f901ff782 72644b0ca55a6142 6cb281dab11480b4 4a8198441f401ff2 5ffd956ed11a2b5f 9a640988d68287d3 74942df792f2637f b2819dc61f772d4f
m^*	a0ec9872cfffe63c df5c6a2b59f4c453 f2bea3763fc8fa7a 6a47e8ff0a995116 fa59232e8b617048 4c9690984c084498 28bee8f5701eab16 8d57686ecbdce623 3879318f901ff7a2 52644b0ca55a6152 6cb281dab1148094 4aff9c441f402073 6001956ed11a2a5f 9a640988d68287d3 74942df792f2637f b2819dc61f772d4f
Δm	0000000000000000 0000000000000000 0000000000000000 0000000000000000 0000000000000000 0000000000000000 0000000000000000 0000000000000000 0000000000000020 2000000000000010 0000000000000020 007e040000003f81 3ffc000000000100 0000000000000000 0000000000000000 0000000000000000
h_1	3aa73bfae7b82789 711f2024cf0f636e 0c6965f707279a53 8227fba8617aa955 fdd9e2ca8c4d0038 57db244560d7b70b 08ec5698343353c0 9e9b739ee307ea92

(b) Example of a collision for 27 steps of SHA-512.

m	537e7a4986aa2fce 11206ad0306c752b 90124a9e1c9b0ce2 8c14e0356fd26f5f fd3ef90ea3e4366f 35d8c2ba58abd92f b23e476632eca1fd e2b122ef46649b73 dfc020070e628f37 7acf74d1d1007558 6c6359a6fe7fe2f0 d490bd22815eb494 72fedf1f807df6f3 a8585af19b6dd9d1 3d2053b0c295522b 2d970e0e52a49081
m^*	537e7a4986aa2fce 11206ad0306c752b 90124a9e1c9b0ce2 8c14e0356fd26f5f fd3ef90ea3e4366f 35d8c2ba58abd92f b23e476632eca1fd 5cc1250f06649b73 d0009fc70e628f36 7acf74d1d1007558 6c6359a6fe7fe2f0 d490bd22815eb494 f0bc01169075f6fb a8585af19b6dd9d1 3d2053b0c295522b d3907e3653649081
Δm	0000000000000000 0000000000000000 0000000000000000 0000000000000000 0000000000000000 0000000000000000 0000000000000000 be7007e040000000 0fc0bfc000000001 0000000000000000 0000000000000000 0000000000000000 8242de0910080008 0000000000000000 0000000000000000 fe07703801c00000
h_1	d838f1d2ae4bf185 3fc837ae9bbc28d4 6b2f2977f58a9697 99c48839f0e8bdca c9c0a86fed1d921a 2f823b1fa1913751 3ba170b902c6da30 9c4e5807be51a7e7

References

1. Aoki, K., Guo, J., Matusiewicz, K., Sasaki, Y., Wang, L.: Preimages for step-reduced SHA-2. In: Matsui, M. (ed.) ASIACRYPT 2009. LNCS, vol. 5912, pp. 578–597. Springer, Heidelberg (2009)
2. Biryukov, A., Lamberger, M., Mendel, F., Nikolic, I.: Second-order differential collisions for reduced SHA-256. In: Lee and Wang [16], pp. 270–287
3. Brassard, G. (ed.): CRYPTO 1989. LNCS, vol. 435. Springer, Heidelberg (1990)
4. Canteaut, A. (ed.): FSE 2012. LNCS, vol. 7549. Springer, Heidelberg (2012)
5. Coron, J., Dodis, Y., Malinaud, C., Puniya, P.: Merkle-Damgård revisited: How to construct a hash function. In: Shoup [24], pp. 430–448
6. Cramer, R. (ed.): EUROCRYPT 2005. LNCS, vol. 3494. Springer, Heidelberg (2005)
7. Damgård, I.: A design principle for hash functions. In: Brassard [3], pp. 416–427
8. De Cannière, C., Rechberger, C.: Finding SHA-1 characteristics: general results and applications. In: Lai, X., Chen, K. (eds.) ASIACRYPT 2006. LNCS, vol. 4284, pp. 1–20. Springer, Heidelberg (2006)
9. Eichlseder, M., Mendel, F., Schläffer, M.: Branching heuristics in differential collision search with applications to SHA-512. In: Cid, C., Rechberger, C. (eds.) FSE 2014. LNCS, vol. 8540, pp. 473–488. Springer, Heidelberg (2015)
10. Gueron, S., Johnson, S., Walker, J.: SHA-512/256. In: Latifi, S. (ed.) Information Technology: New Generations – ITNG 2011, pp. 354–358. IEEE Computer Society (2011)
11. Indesteege, S., Mendel, F., Preneel, B., Rechberger, C.: Collisions and other non-random properties for step-reduced SHA-256. In: Avanzi, R.M., Keliher, L., Sica, F. (eds.) SAC 2008. LNCS, vol. 5381, pp. 276–293. Springer, Heidelberg (2009)
12. Joux, A.: Multicollisions in iterated hash functions. Application to cascaded constructions. In: Franklin, M. (ed.) CRYPTO 2004. LNCS, vol. 3152, pp. 306–316. Springer, Heidelberg (2004)
13. Kelsey, J., Kohno, T.: Herding hash functions and the nostradamus attack. In: Vaudenay, S. (ed.) EUROCRYPT 2006. LNCS, vol. 4004, pp. 183–200. Springer, Heidelberg (2006)
14. Kelsey, J., Schneier, B.: Second preimages on n-bit hash functions for much less than 2^n work. In: Cramer [6], pp. 474–490
15. Khovratovich, D., Rechberger, C., Savelieva, A.: Bicliques for preimages: Attacks on Skein-512 and the SHA-2 family. In: Canteaut [4], pp. 244–263
16. Lee, D.H., Wang, X. (eds.): ASIACRYPT 2011. LNCS, vol. 7073. Springer, Heidelberg (2011)
17. Li, J., Isobe, T., Shibutani, K.: Converting meet-in-the-middle preimage attack into pseudo collision attack: Application to SHA-2. In: Canteaut [4], pp. 264–286
18. Mendel, F., Nad, T., Schläffer, M.: Finding SHA-2 characteristics: Searching through a minefield of contradictions. In: Lee and Wang [16], pp. 288–307
19. Mendel, F., Nad, T., Schläffer, M.: Improving local collisions: new attacks on reduced SHA-256. In: Johansson, T., Nguyen, P.Q. (eds.) EUROCRYPT 2013. LNCS, vol. 7881, pp. 262–278. Springer, Heidelberg (2013)
20. Merkle, R.C.: One way hash functions and DES. In: Brassard [3], pp. 428–446
21. National Institute of Standards and Technology: FIPS PUB 180–4: Secure Hash Standard. Federal Information Processing Standards Publication 180-4, U.S. Department of Commerce, March 2012. http://csrc.nist.gov/publications/fips/fips180-4/fips-180-4.pdf

22. Nikolić, I., Biryukov, A.: Collisions for step-reduced SHA-256. In: Nyberg, K. (ed.) FSE 2008. LNCS, vol. 5086, pp. 1–15. Springer, Heidelberg (2008)
23. Sanadhya, S.K., Sarkar, P.: New collision attacks against up to 24-step SHA-2. In: Chowdhury, D.R., Rijmen, V., Das, A. (eds.) INDOCRYPT 2008. LNCS, vol. 5365, pp. 91–103. Springer, Heidelberg (2008)
24. Shoup, V. (ed.): CRYPTO 2005. LNCS, vol. 3621. Springer, Heidelberg (2005)
25. Wang, X., Yin, Y.L., Yu, H.: Finding collisions in the full SHA-1. In: Shoup [24], pp. 17–36
26. Wang, X., Yu, H.: How to break MD5 and other hash functions. In: Cramer [6], pp. 19–35
27. Yu, H., Bai, D.: Boomerang attack on step-reduced SHA-512. IACR Cryptology ePrint Archive, Report 2014/945 (2014). http://eprint.iacr.org/2014/945

Cryptanalysis

Tradeoff Cryptanalysis
of Memory-Hard Functions

Alex Biryukov[✉] and Dmitry Khovratovich

University of Luxembourg, Luxembourg City, Luxembourg
alex.biryukov@uni.lu, khovratovich@gmail.com

Abstract. We explore time-memory and other tradeoffs for memory-hard functions, which are supposed to impose significant computational and time penalties if less memory is used than intended. We analyze three finalists of the Password Hashing Competition: Catena, which was presented at Asiacrypt 2014, yescrypt and Lyra2.

We demonstrate that Catena's proof of tradeoff resilience is flawed, and attack it with a novel *precomputation tradeoff*. We show that using $M^{4/5}$ memory instead of M we have no time penalties and reduce the AT cost by the factor of 25. We further generalize our method for a wide class of schemes with predictable memory access. For a wide class of data-dependent schemes, which addresses memory unpredictably, we develop a novel *ranking tradeoff* and show how to decrease the time-memory and the time-area product by significant factors. We then apply our method to yescrypt and Lyra2 also exploiting the iterative structure of their internal compression functions.

The designers confirmed our attacks and responded by adding a new mode for Catena and tweaking Lyra2.

Keywords: Password hashing · Memory-hard · Catena · Tradeoff · Cryptocurrency · Proof-of-work

1 Introduction

Memory-hard functions are a fast emerging trend which has become a popular remedy to the hardware-equipped adversaries in various applications: cryptocurrencies, password hashing, key derivation, and more generic Proof-of-Work constructions. It was motivated by the rise of various attack techniques, which can be commonly described as optimized exhaustive search. In cryptocurrencies, the hardware arms race made the Bitcoin mining [29] on regular desktops tremendously inefficient, as the best mining rigs spend 30,000 times less energy per hash than x86-desktops/laptops[1]. This causes major centralization of the mining efforts which goes against the democratic philosophy behind the Bitcoin design. This in turn prevents wide adoption and use of such cryptocurrency in

---

[1] The estimate comes from the numbers given in [6]: the best ASICs make 2^{32} hashes per joule, whereas the most efficient laptops can do 2^{17} hashes per joule.

© International Association for Cryptologic Research 2015
T. Iwata and J.H. Cheon (Eds.): ASIACRYPT 2015, Part II, LNCS 9453, pp. 633–657, 2015.
DOI: 10.1007/978-3-662-48800-3_26

economy, limiting the current activities in this area to mining and hoarding, whith negative effects on the price. Restoring the ability of CPU or GPU mining by the use of memory-hard proof-of-work functions may have dramatic effect on cryptocurrency adoption and use in economy, for example as a form of decentralized micropayments [15]. In password hashing, numerous leaks of hash databases triggered the wide use of GPUs [3,34], FPGAs [27] for password cracking with a dictionary. In this context, constructions that intensively use a lot of memory seem to be a countermeasure. The reasons are that memory operations have very high latency on GPU and that the memory chips are quite large and thus expensive on FPGA and ASIC environments compared to a logic core, which computes, e.g. a regular hash function.

Memory-intensive schemes, which bound the memory bandwidth only, were suggested earlier by Burrows et al. [8] and Dwork et al. [17] in the context of spam countermeasures. It was quickly realized that to be a real countermeasure, the amount of memory shall also be bounded [18], so that memory must not be easily traded for computations, time, or other resources that are cheaper on certain architecture. Schemes that are resilient to such tradeoffs are called *memory-hard* [21,30]. In fact, the constructions in [18] are so strong that even tiny memory reduction results in a huge computational penalty.

Disadvantage of Classical Constructions and New Schemes. The provably tradeoff-resilient superconcentrators [32] and their applications in [18,19] have serious performance problems. They are terribly slow for modern memory sizes. A superconcentrator requiring N blocks of memory makes $O(N \log N)$ calls to F. As a result, filling, e.g., 1 GB of RAM with 256-bit blocks would require dozens of calls to F per block ($C \log N$ calls for some constant C). This would take several minutes even with lightweight F and is thus intolerable for most applications like web authentication or cryptocurrencies. Using less memory, e.g., several megabytes, does not effectively prohibit hardware adversaries.

This has been an open challenge to construct a reasonably fast and tradeoff-resilient scheme. Since the seminal paper by Dwork et al. [18] the first important step was made by Percival, who suggested scrypt [30]. The idea of scrypt was quite simple: fill the memory by an iterative hash function and then make a pseudo-random walk on the blocks using the block value as an address for the next step. However, the entire design is somewhat sophisticated, as it employs a stack of subfunctions and a number of different crypto primitives. Under certain assumptions, Percival proved that the time-memory product is lower bounded by some constant. The scrypt function is used inside cryptocurrency Litecoin [4] with 128 KB memory parameter and is now adapted as an IETF standard for key-derivation [5]. scrypt is a notable example of *data-dependent* schemes where the memory access pattern depends on the input, and this property enabled Percival to prove some lower bound on adversary's costs. However, the performance and/or the tradeoff resilience of scrypt are apparently not sufficient to discourage hardware mining: the Litecoin ASIC miners are more efficient than CPU miners by the factor of 100 [1].

The need for even faster, simpler, and possibly more tradeoff-resilient constructions was further emphasized by the ongoing Password Hashing Competition [2], which has recently selected 9 finalists out of the 24 original submissions. Notable entries are Catena [20], just presented at Asiacrypt 2014 with a security proof based on [26], and yescrypt and Lyra2 [25], which both claim performance up to 1 GB/sec and which were quickly adapted within a cryptocurrency proof-of-work [7]. The tradeoff resilience of these constructions has not been challenged so far. It is also unclear how possible tradeoffs would translate to the cost

Our Contributions. We present a rigorous approach and a reference model to estimate the amortized costs of password brute-force on special hardware using full-memory algorithms or time-space tradeoffs. We show how to evaluate the adversary's gains in terms of area-time and time-memory products via computational complexity and latency of the algorithm.

Then we present our tradeoff attacks on the last versions of Catena and yescrypt, and the original version of Lyra2. Then we generalize them to wide classes of data-dependent and data-independent schemes. For Catena we analyze the faster Dragonfly mode and show that the original security proof for it is flawed and the computation-memory product can be kept constant while reducing the memory. For ASIC-equipped adversaries we show how to reduce the area-time product (abbreviated further by AT) by the factor of 25 under reasonable assumptions on the architecture. The attack algorithm is then generalized for a wide class of data-independent schemes as a *precomputation method*.

Then we consider data-dependent schemes and present the first generic tradeoff strategy for them, which we call the *ranking method*. Our method easily applies to yescrypt and then to the second phase of Lyra2, both taken with minimally secure time parameters. We further exploit the incomplete diffusion in the core primitives of these designs, which reduces the time-memory and time-area products for both designs.

Altogether, we show how to decrease the time-memory product by the factor of 2 for yescrypt and the factor of 8 for Lyra2. Our results are summarized in Table 1. To the best of our knowledge, our methods are the first generic attacks so far on data-dependent or data-independent schemes[2].

Related Work. So far there have been only a few attempts to develop tradeoff attacks on memory-hard functions. A simple tradeoff for scrypt has been known in folklore and was recently formalized in [20]. Alwen and Serbinenko analyzed a simplified version of Catena in [9]. Designers of Lyra2 and Catena attempted to attack their own designs in the original submissions [20, 25]. Simple analysis of Catena has been made in [16].

Paper Outline. We introduce necessary definitions and metrics in Sect. 2. We attack Catena-Dragonfly in Sect. 3 and generalize this method in Sect. 4. Then we present a generic ranking algorithm for data-dependent schemes in Sect. 5

---

[2] The full version of this paper is available at [14].

Table 1. Our tradeoff gains on Catena, yescrypt and Lyra2 with minimal secure parameters, 2^{30} memory bytes and reference hardware implementations (Sect. 2). TM loss is the maximal factor by which we can reduce the time-memory product compared to the full-memory implementation. AT loss is the maximal factor for time-area product reduction. Compactness of TM and AT is the maximal memory reduction factor which does not increase the TM or AT, resp., compared to the default implementation.

	Catena-Dragonfly	Generic 1-pass	yescrypt	Lyra2 v1
Time	$T = 3M$	$T = M$	$T = 4/3M$	$T = 2M$
	Section 3	Section 5	Section 6	Appendix A
TM loss	200	1.28	2.1	8
AT loss	25	1.28	2.1	3
TM compactness	64	4	5.8	16
AT compactness	64	4	4.5	5

and attack yescrypt with this method in Sect. 6. The attack on Lyra2 is quite sophisticated and we leave it for Appendix A.

2   Preliminaries

2.1   Syntax

Let $\mathcal{G}$ be a hash function that takes a fixed-length string I as input and outputs tag H. We consider functions that iteratively fill and overwrite memory blocks $X[1], X[2], \ldots, X[M]$ using a compression function F:

$$X[i_j] = f_j(I), \ 1 \leq j \leq s; \tag{1}$$
$$X[i_j] = F(X[\phi_1(j)], X[\phi_2(j)], \ldots, X[\phi_k(j)]), \ s < j \leq T, \tag{2}$$

where ϕ_i are some indexing functions referring to some already filled blocks and f_j are auxiliary hash functions (similar to F) filling the initial s blocks for some positive s.

We say that the function *makes p passes* over the memory, if $T = pM$. Usually p and M are tunable parameters which are responsible for the total running time and the memory requirements, respectively.

2.2   Time-Space Tradeoff

Let $\mathcal{A}$ be an algorithm that computes $\mathcal{G}$. The *computational complexity* $C(\mathcal{A})$ is the total number of calls to F and f_i by $\mathcal{A}$, averaged over all inputs to $\mathcal{G}$. We do not consider possible complexity amortization over successive calls to $\mathcal{A}$. The *space complexity* $S(\mathcal{A})$ is the peak number of blocks (or their equivalents) stored by $\mathcal{A}$, again averaged over all inputs to $\mathcal{G}$. Suppose that $\mathcal{A}$ can be represented as a directed acyclic graph with vertices being calls to F. Then the *latency* $L(\mathcal{A})$ is the length of the longest chain the graph from the input to the output. Therefore, $L(\mathcal{A})$ is the minimum time needed to run $\mathcal{A}$ assuming unlimited parallelism and instant memory access.

A straightforward implementation of the scheme (1) results in an algorithm with computational complexity T and latency $L = T$ and space complexity M. However, it might be possible to compute $\mathcal{G}$ using less memory. According to [24], any function, that is described by Eq. (1) and whose reference block indices $\phi_j(i)$ are known in advance, can be computed using $c_k \frac{T}{\log T}$ memory blocks for some constant c_k depending on the number k of input blocks for F. Therefore, any p-pass function can be computed using less than $M = T/p$ memory for sufficiently large M.

Let us fix some *default algorithm* $\mathcal{A}$ of $\mathcal{G}$ with (C_1, M_1, L_1) being computational and space complexities and latency of $\mathcal{A}$, respectively. Suppose that there is a *time-space tradeoff* given by the family of algorithms[3] $\mathcal{B} = \{B_q\}$ that compute $\mathcal{G}$ using $\frac{M_1}{q}$ space for different q. The idea is to store only one of q memory blocks on average and recompute the missing blocks whenever they are needed. Then we define the *computational penalty* $CP_{\mathcal{B}}(q)$ as

$$CP_{\mathcal{B}}(q) = \frac{C(B_q)}{C_1}$$

and *latency penalty* $LP_{\mathcal{B}}(q)$.

$$LP_{\mathcal{B}}(q) = \frac{L(B_q)}{L_1},$$

2.3   Attackers and Cost Estimates

We consider the following attack. Suppose that $\mathcal{G}$ with time and memory parameters (T, M) is used as a password hashing function with $I = (P, S)$, where P is a secret password and S is a public salt. An attacker gets H and S (e.g., from a database leak) and tries to recover P. He attempts a dictionary attack: given a list L of most probable passwords, he runs $\mathcal{G}$ on every $P \in L$ and checks the output.

Definition 1. *Let Φ be a cost function defined over a space of algorithms. Let also $\mathcal{G}_{T,M}$ be a hash function with fixed algorithm $\mathcal{A}_0$ (default algorithm). Then $\mathcal{G}_{T,M}$ is called (α, Φ)-secure if for every algorithm $\mathcal{B}$ for $\mathcal{G}_{T,M}$*

$$\Phi(\mathcal{B}) > \alpha\Phi(\mathcal{A}).$$

In other words, $\mathcal{G}_{T,M}$ can not be computed cheaper than by the factor of $\frac{1}{\alpha}$.

The cost function is more difficult to determine. We suggest evaluating amortized computing costs for a single password trial. Depending on the architecture, the costs vary significantly for the same algorithm $\mathcal{A}$. For the ASIC-equipped attackers, who can use parallel computing cores, it is widely suggested that the

---

[3] As well as $\mathcal{A}$, the family $\mathcal{B}$ admits parallel implementations.

costs can be approximated by the *time-area product* AT [9,11,28,35]. Here T is the time complexity of the used algorithm and A is the sum of areas needed to implement the memory cells and the area needed to implement the cores. Let the area needed to implement one block of memory be the unit of area measurement. Then in order to know the total area, we need *core-memory ratio* R_c, which is how many memory blocks we can place on the area taken by one core.

Suppose that the adversary runs algorithm B_q using M/q memory and l computing cores, thus having computational complexity $C_q = C(B_q)$. The running time is lower bounded by the latency $L_q = L(B_q)$ of the algorithm. If $L_q < C_q/l$, i.e. the computing cores can not finish the work in minimum time, then the time T can be approximated by C_q/l, and the costs are estimated as follows:

$$\text{AT}_{B_q}(l) = \left(lR_c + \frac{M}{q} \right) \frac{C_q}{l} = C_q \left(R_c + \frac{M}{ql} \right)$$

We see that the costs drop as l increases. Therefore, the adversary would be motivated to push it to the maximum limit C_q/L_q. Thus we obtain the final approximation of costs:

$$\text{AT}_{B_q} = C_q R_c + L_q \frac{M}{q}. \tag{3}$$

Here we assume unlimited memory bandwidth. Taking the bandwidth restrictions into account is even more difficult, as they depends on the relative frequency of the computing core and the memory as well as on the architecture of the memory bus. Moreover, the memory bandwidth of the algorithm depends on the implementation and is not easy to evaluate. We leave rigorous memory bandwidth evaluation and restrictions for the future work.

We recall that the value R_c is depends on the architecture, the function F, and the block size. To give a concrete example, suppose that the block is 64 bytes and F is the Blake-512 hash function. We use the following reference implementations[4]:

- The 50-nm DRAM [22], which takes 550 mm^2 per GByte;
- The 65-nm Blake-512 [23], which takes about 0.1 mm^2.

Then the core-memory ratio is $\frac{2^2 4 \cdot 0.1}{550} \approx 3000$. For more lightweight hash functions this ratio will be smaller.

The actual functions F in the designs that we attack are often ad-hoc and have not implemented yet in hardware. Moreover, the numbers may change when going to smaller feature size. To make our estimates of the attack costs architecture-independent, we introduce a simpler metric — the *time-memory product* TM:

$$\text{TM}_{B_q} = L_q \frac{M}{q}, \tag{4}$$

which for not so high computational penalties gives a good approximation of AT.

---

[4] We take low-area implementations, as possible parallelism is already taken into account.

In our tradeoff attacks, we are mainly interested to compare the AT and TM costs of B_q with that of the default algorithm $\mathcal{A}$. Thus we define the *AT ratio* of B_q as $\frac{\text{AT}_{B_q}}{\text{AT}_{\mathcal{A}}}$, and the *TM ratio* of B_q as $\frac{\text{TM}_{B_q}}{\text{TM}_{\mathcal{A}}}$.

We note that for the same TM value the implementation with less memory is preferable, as its design and production will be cheaper. Thus we explore how much the memory can be reduced keeping the AT or TM costs below those of the default algorithm.

Definition 2. *Tradeoff algorithms $\mathcal{B}$ have AT compactness q if it is the maximal q such that*

$$\text{AT}_{B_q} \leq \text{AT}_{\mathcal{A}}.$$

Tradeoff algorithms $\mathcal{B}$ have TM compactness q if it is the maximal q such that

$$\text{TM}_{B_q} \leq \text{TM}_{\mathcal{A}}.$$

For the concrete schemes we take "minimally secure" values of T, i.e. those that supposed to have (α, Φ)-security for reasonably high α. Unfortunately, no explicit security claim of this kind is present in the design documents of the functions we consider.

Data-Dependent and Data-Independent Schemes. The existing schemes can be categorized according to the way they access memory. The *data-independent schemes* Catena [20], Pomelo [36], Argon2i [13] computes $\phi(j)$ independently of the actual password in order to avoid timing attacks like in [33]. Then the algorithm $\mathcal{B}$ that uses less memory can recompute the missing blocks just by the time they are requested. Therefore, it has the same latency as the full-memory algorithm, i.e. $L(\mathcal{B}) = L_0$. For these algorithms the time-memory product can be arbitrarily small, and the minimum AT value is determined by the core-memory ratio.

The *data-dependent* schemes scrypt [30] yescrypt [31], Argon2d [13] compute $\phi(j)$ using the just computed block: $\phi(j) = \phi(j, X_{i_{j-1}})$. Then precomputation is impossible, and for each recomputing block the latency is increased by the latency of the recomputation algorithm, so $L_q > L_0$. There exist hybrid schemes [25], which first run a data-independent phase and then a data-dependent one.

3   Cryptanalysis of Catena-Dragonfly

3.1   Description

Short History. Catena was first published on ePrint [20] and then submitted to the Password Hashing Competition. Eventually the paper was accepted to Asiacrypt 2014 [21]. In the middle of the reviewing process, we discovered and communicated the first attack on Catena to the authors. The authors have introduced a new mode for Catena in the camera-ready version of the Asiacrypt paper, which is resistant to the first attack. The final version of Catena, which is the finalist of the Password Hashing Competition, contains two modes: Catena-Dragonfly (which we abbreviate to Catena-D), which is an extension to the original Catena, and Catena-Butterfly, which is a new mode advertised as tradeoff-resistant. In this paper we present the attack on Catena-Dragonfly, which is very similar to the first attack on Catena.

Specification. Catena-D is essentially a mode of operation over the hash function F, which is be instantiated by Blake2b [10] in the full or reduced-round version. The functional graph of Catena-D is determined by the time parameter λ (values $\lambda = 1, 2$ are recommended) and the memory parameter n, and can be viewed as $(\lambda + 1)$-layer graph with 2^n vertices in each layer (denoted by Catena-D-λ). We denote the X-th vertex in layer l (both count from 0) by $[X]^l$. With each vertex we associate the corresponding output of the hash function F and denote it by $[X^l]$ as well. The outputs are stored in the memory, and due to the memory access pattern it is sufficient to store only 2^n blocks at each moment. The hash function H has 512-bit output, so the total memory requirements are 2^{n+6} bytes.

First layer is filled as follows

- $[0]^0 = G_1(P, S)$, where G_1 invokes 3 calls to F;
- $[1]^0 = G_2(P, S)$, where G_2 invokes 3 calls to F
- $[i]^0 \leftarrow F([i-1]^0, [i-2]^0)$, $2 \leq i \leq 2^n - 1$.

Then $2^{3n/4}$ nodes of the first layer are modified by function Γ. The details of Γ are irrelevant to our attack.

The memory access pattern at the next layers is determined by the *bit-reversal permutation* ν. Each index is viewed as an n-bit string and is transformed as follows:

$$\nu(x_1 x_2 \ldots x_n) = x_n x_{n-1} \ldots x_1, \text{ where } x_i \in \{0, 1\}.$$

The layers are then computed as

- $[0]^j = F([0]^{j-1} \,||\, [2^n - 1]^{j-1})$;
- $[i]^j = F([i-1]^j \,||\, [\nu(i)]^{j-1})$.

Thus to compute $[X]^l$ we need $[\nu(X)^{l-1}]$. The latter can be then overwritten[5]. An example of Catena-D with $\lambda = 2$ and $n = 3$ is shown at Fig. 1.

The bit-reversal permutation is supposed to provide memory-hardness. The intuition is that it maps any segment to a set of blocks that are evenly distributed at the upper layer.

Original Tradeoff Analysis. The authors of Catena-D originally provided two types of security bounds against tradeoff attacks. Recall that Catena-D-λ can be computed with $\lambda 2^n$ calls to F using 2^n memory blocks. The Catena-D designers

---

[5] In terms of Eq. (1) we could enumerate all blocks as $[i]^j = j||\underbrace{i}_{n \text{ bits}}$ so that $\phi(j||i) = (j-1)||\nu(i)$.

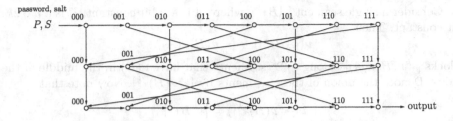

Fig. 1. Catena-D-2 with $n = 3$. 3 layers, 8 vertices per layer.

demonstrated that Catena-D-λ can be computed using λS memory blocks with time complexity[6]

$$T \leq 2^n + 2^n \left(\frac{2^n}{2S}\right)^{\lambda-1} + 2^n \left(\frac{2^n}{2S}\right)^{\lambda}$$

Therefore, if we reduce the memory by the factor of q, i.e. use only $\frac{2^n}{q}$ blocks, we get the following penalty:

$$P_\lambda(q) \approx \left(\frac{q}{2}\right)^{\lambda}. \tag{5}$$

The second result is the lower bound for tradeoff attacks with memory reduction by q:

$$P_\lambda(q) \geq \Omega\left(q^{\lambda}\right). \tag{6}$$

However the constant in $\Omega()$ is too small (2^{-18} for $\lambda = 3$) to be helpful in bounding tradeoff attacks for small q. More importantly, the proof is flawed: the result for $\lambda = 1$ is incorrectly generalized for larger λ. The reason seems to be that the authors assumed some independence between the layers, which is apparently not the case (and is somewhat exploited in our attack).

In the further text we demonstrate a tradeoff attack yielding much smaller penalties than Eq. (5) and thus asymptotically violating Eq. (6).

3.2 Our Tradeoff Attack on Catena-D

The idea of our method is based on the simple fact that

$$\nu(\nu(X)) = X,$$

where X can be a single index or a set of indices. We exploit it as follows. We partition layers into *segments* of length 2^k for some integer k, and store the first block of every segment (first two blocks at layer 0). As the index of such a block ends with k zeros, we denote the set of these blocks as $[*^{n-k}0^k]$. We also store all $2^{3n/4}$ blocks modified by Γ, which we denote by $[\Gamma]$.

---

[6] This result is a part of Theorem 6.3 in [20].

Consider a single segment $[AB*^k]$, where A is a k-bit constant, B is a $n-2k$-bit constant. Then

$$\nu([AB*^k]) = [*^k \nu(B)\nu(A)].$$

Blocks $[*^k \nu(B)\nu(A)]$ belong to 2^k segments that have $\nu(B)$ in the middle of the index. Denote the union of these segments by $[*^k \nu(B)*^k]$. Now note that

$$\nu([*^k \nu(B)*^k]) = [*^k B*^k],$$

and

$$\nu(\nu([*^k B*^k])) = [*^k B*^k].$$

Therefore, when we iterate the permutation ν, we are always within some 2^k segments. We suggest the computing strategy in Algorithm 1. At layer t we recompute 2^k full segments from layers 0 to $t-2$ and 2^k subsegments of length $\nu(A)$ (interpreted as a number in the binary form) at layer $t-1$. Therefore, the total cost of computing layer t is

$$C(t) = \sum_A \sum_B ((t-1)2^{2k} + \nu(A)2^k + 2^k) = \tag{7}$$

$$= \sum_A ((t-1)2^n + \nu(A)2^{n-k} + 2^{n-k}) =$$

$$(t-1)2^{n+k} + 2^{n+k-1} + 2^n = (t - \frac{1}{2})2^{n+k} + 2^n.$$

The total cost of computing Catena-D-λ is

$$2^n \left(\frac{\lambda^2}{2} 2^k + \lambda + 1 \right).$$

We store $(t+1)2^{n-k}$ blocks as segment starting points, $2^{3n/4}$ blocks $[\Gamma]$ and 2^{2k} blocks for intermediate computations. For $k = \log q + \log(\lambda+1)$ and $q < 2^{n/4}$

Algorithm 1. Tradeoff for Catena-Dragonfly.

1. Compute layer 0 storing $[*^{n-k}0^k]^0$ and $[*^{n-k}0^{k-1}1]^0$, i.e. the first two blocks of every segment.
2. Compute Γ and store all the updated blocks $[\Gamma]$ in the memory.
3. Compute layer 1 segmentwise: for each segment $[AB*^k]^1$ recompute blocks $[*^k \nu(B)\nu(A)]^0$ using stored blocks from layer 0 and $[\Gamma]$. Store blocks $[*^{n-k}0^k]^1$.
4. Compute layer 2 segmentwise: for each segment $[AB*^k]^2$ recompute 2^k segments $[*^k B^k]^0$ using stored blocks from layer 0, then use them to recompute blocks $[*^k \nu(B)\nu(A)]^1$ using $[\Gamma]$, then compute $[AB*^k]^2$. Store blocks $[*^{n-k}0^k]^2$.
5. Compute layer 3 segmentwise: for each segment $[AB*^k]^3$ recompute 2^k segments $[*^k \nu(B)^k]^0$ using stored blocks from layer 0, then recompute 2^k segments $[*^k B^k]^1$ using stored blocks from layer 1 and $[\Gamma]$, then recompute blocks $[*^k \nu(B)\nu(A)]^2$, then compute $[AB*^k]^3$. Store blocks $[*^{n-k}0^k]^3$.
6. Compute other layers in the similar fashion.

we store about $2^n/q$ blocks, so the memory is reduced by the factor of q. This value of k yields the total computational complexity of

$$C_q = 2^n \left(\frac{q\lambda^2(\lambda+1)}{2} + \lambda + 1 \right) \qquad (8)$$

Since the computational complexity of the memory-full algorithm is $(\lambda+1)2^n$, our tradeoff method gives the computational penalty

$$\frac{q\lambda^2}{2} + 1.$$

Since Catena is a data-independent scheme, the latency of our method does not increase. Therefore, the time-memory product (Eq. (4)) can be reduced by the factor of $2^{n/4}$. We can estimate how AT costs evolves assuming the reference implementation in Sect. 2.3:

$$\text{AT}_{B_q} = 2^n \left(\frac{q\lambda^2(\lambda+1)}{2} + \lambda + 1 \right) \cdot 3000 + (\lambda+1)2^n \frac{2^n}{q}.$$

For $q = 2^{n/5}$ and $\lambda = 2$ we get

$$\text{AT}_{B_{2^{n/5}}} = 2^n \left(6 \cdot 2^{n/5} \right) \cdot 2^{11.5} + 3 \cdot 2^{9n/5}.$$

For $n = 24$ (1 GB of RAM) we get

$$\text{AT}_{B_{24.8}} \approx 2^{24+2.5+4.8+11.5} + 2^{43.2+1.5} \approx 2^{44}.$$

whereas

$$\text{AT}_{B_1} = 2^{49.5}.$$

Therefore, we expect the time-area product dropped by the factor of about 25 if the memory is reduced by the factor of 30. In the terms of Definition 1, Catena-D-2 is not $(1/25, \text{AT})$-secure. Our tradeoff method also have AT and TM compactness at least $2^{n/5} = 64$.

On other architectures the AT may drop even further, and we expect that an adversary would choose the one that maximizes the tradeoff effect, so the actual impact of our attack can be even higher.

Violation of Catena-D Lower Bound. Our method shows that the Catena-D lower bound is wrong. If we summarize the computational costs for λ layers, we obtain the following computational penalty for the memory reduction by the factor of q:

$$CP_\lambda(q) = O(\lambda^3 q),$$

which is asymptotically smaller than the lower bound $\Omega(q^\lambda)$ (Eq. (6)) from the original Catena submission [20].

Table 2. Computation-memory tradeoff for Catena-D-3 and Catena-D-4.

Memory fraction	Catena-D-3		Catena-D-4	
	Computational penalty			
	Our	[20]	Our	[20]
$\frac{1}{2}$	7.4	36.2	13.8	512
$\frac{1}{4}$	15.5	252	26.6	7373
$\frac{1}{8}$	30.1	1872	52	2^{17}
$\frac{1}{2^l}$	$2^{l+1.9}$	2^{3l}	$2^{l+2.8}$	$2^{4l+1.5}$

3.3 Other Results for Catena

Our attack on Catena can be further scrutinized and generalized to non-even segments. More details are provided in [14] with the summary given in Table 2.

4 Generic Precomputation Tradeoff Attack

Now we try to generalize the tradeoff method used in the attack on Catena for a class of data-independent schemes. We consider schemes $\mathcal{G}$ where each memory block is a function of the previous block and some earlier block:

$$X[i] \leftarrow F(X[i-1], X[\phi(i)]), 0 \leq i < T$$

where ϕ is a deterministic function such that $\phi(i) < i$. A group of existing password hashing schemes falls into this category: Catena [20], Pomelo [36], Lyra2 [25] (first phase). Multiple iterations of such a scheme are equivalent to a single iteration with larger T and an additional restriction

$$x - M \leq \phi(x),$$

so that the memory requirements are M blocks.

The crucial property of the data-independent attacks is that they can be tested and tuned offline, without hashing any real password. An attacker may spend significant time to search for an optimal tradeoff strategy, since it would then apply to the whole set of passwords hashed with this scheme.

Precomputation Method. Our tradeoff method generalizes as follows. We divide memory into segments and store only the first block of each segment. For every segment I we calculate its image $\phi(I)$. Let $\overline{\phi(I)}$ be the union of segments that contain $\phi(I)$. We repeat this process until we get an invariant set $U_k = U(I)$:

$$\underbrace{I}_{U_0} \rightarrow \underbrace{\overline{\phi(I)}}_{U_1} \rightarrow \underbrace{\overline{\phi(\phi(I))}}_{U_2} \cdots \rightarrow U_k.$$

The scheme $\mathcal{G}$ is then computed according to Algorithm 2.

Algorithm 2. Precomputation method

- For all segments precompute the block indices in the union chains $I \rightarrow U(I)$
- Compute $\mathcal{G}$ by segment. For each segment I:
    1. Compute blocks $U(I) = U_k$;
    2. Compute blocks in U_{k-1}, then in U_{k-2}, up to $U_0 = I$.
    3. Store the first block of I in the memory.

The total amount of calls to F is $\sum_{i \geq 0} |U_i|$, and the penalty to compute I is

$$CP(I) = \frac{\sum_{i \geq 0} |U_i|}{|I|}.$$

How efficient the tradeoff is depends on the properties of ϕ and the segment partition, i.e. how fast U_i expands. As we have seen, Catena uses a bit permutation for ϕ, whereas Lyra2 uses a simple arithmetic function or a bit permutation [20,25]. In both cases U_i stabilizes in size after two iterations. If ϕ is a more sophisticated function, the following heuristics (borrowed from our attacks on data-dependent schemes) might be helpful:

- Store the first T_1 computed blocks and the last T_2 computed blocks for some T_1, T_2 (usually about N/q).
- Keep the list $\mathcal{L}$ of the most expensive blocks to recompute and store $M[i]$ if $\phi(i) \in \mathcal{L}$ (Fig. 2).

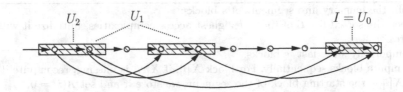

Fig. 2. Segment unions in the precomputation method.

5   Generic Ranking Tradeoff Attack

Now we present a generic attack on a wide class of schemes with data-dependent memory addressing. Such schemes include scrypt [30] and the PHC finalists yescrypt [31], Argon2d [13], and Lyra2 [25]. We consider the schemes described by Eq. (1) with $k = 2$ and the following addressing (cf. also Fig. 3):

$$
\begin{aligned}
X[1] &= f(I) \\
&\text{for } 1 < i < T \\
r_i &= g(X[i-1]); \\
X[i] &= F(X[i-1], X[r_i]).
\end{aligned}
\tag{9}
$$

Here g is some indexing function. This construction and our tradeoff method can be easily generalized to multiple functions F, to stateful functions (like in Lyra2), to multiple inputs, outputs, and passes, etc. However, for the sake of simplicity we restrict to the construction above.

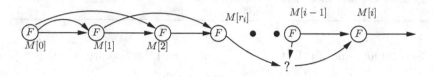

Fig. 3. Data-dependent schemes.

Our tradeoff strategy is following: we compute the blocks sequentially and for each block $X[i]$ decide if we store it or not. If we do not store it, we calculate its *access complexity* $A(i)$ – the number of calls needed to recompute it as a sum of access complexities of $X[i-1]$ and $X[r_i]$ plus one. If we store $X[i]$, its access complexity is 0.

The storing heuristic rule is the crucial element of our strategy. The idea is to store the block if $A(r_i)$ is too high.

Our *ranking tradeoff method* works according to Algorithm 3 (Fig. 4).

Algorithm 3. Ranking method
---
1. Split the memory into segments of s blocks;
2. Keep the sorted list $\mathcal{L}$ of the T/l highest access complexities, initialize it with all zeros;
3. Temporarily store last w blocks.
4. Compute blocks sequentially. For block $X[i]$, if $X[r_i]$ is missing, recompute it.
5. If $X[i]$ is the starting block of the segment, we store it and set $A(i) = 0$;
6. If $X[i]$ is not the starting block of the segment, but $A(r_i) \in \mathcal{L}$, we store $X[i]$ and set $A(i) = 0$;
7. If $X[i]$ is not the starting block of the segment, and $A(r_i) \notin \mathcal{L}$, we do not store $X[i]$ and set $A(i) = A(r_i) + A(i-1) + 1$.

Here w, s and l are parameters, and we usually set $l = 3s$. The computational complexity is computed as

$$C = \sum_i A(r_i).$$

We also compute the latency $L(i)$ of each block as $L(i) = \max(L(r_i), L(i-1)) + 1$ if we do not store $X[i]$ and $L(i) = 0$ if we store it. Then the total latency is

$$L = \sum_i L_i.$$

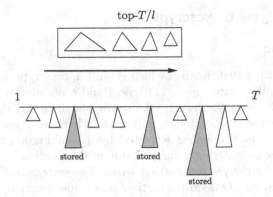

Fig. 4. Outline of the ranking tradeoff method.

We implemented our attack and tested it on the class of functions described by Eq. (9). For fixed w and s the total number of calls to F and the number of stored blocks is entirely determined by indices $\{r_i\}$. Thus we do not have to implement a real hash function, and it is sufficient to generate r_i according to some distribution, model the computation as a directed acyclic graph, and compute C and L for this graph. We made a number of tests with uniformly random r_i (within the segment $[0; i]$ and $T = 2^{12}$) and different values of w and s. Then we grouped C and L values by the memory complexity and figured the lowest complexities for each memory reduction factor. These values are given in Table 3.

Table 3. Computational, latency, AT (for $R_c = 3000$ and $M = 2^{24}$), and TM penalties for the ranking tradeoff attack on generic data-dependent schemes.

Memory fraction ($1/q$)	$\frac{1}{2}$	$\frac{1}{3}$	$\frac{1}{4}$	$\frac{1}{5}$	$\frac{1}{6}$	$\frac{1}{7}$	$\frac{1}{8}$	$\frac{1}{9}$	$\frac{1}{10}$
Computation penalty $CP(q)$	1.59	2.98	7.3	16.6	57.5	180	635	3340	$2^{13.2}$
Latency penalty $LP(q)$	1.56	2.55	4	5.8	8.7	11.6	15.4	21.1	24.8
AT ratio	0.78	0.85	1.02	1.16	1.45	1.69	2.04	2.97	4.24
TM ratio	0.78	0.85	1.02	1.16	1.45	1.65	1.9	2.34	2.48
Segment length s	3	5	8	10	13	16	18	21	23
Window size $\frac{w}{M}$	0.06	0.01	0.01	0	0	0	0	0	0

We conclude that generic 1-pass data-dependent schemes with random addressing are $(0.75, AT)$- and $(0.75, TM)$-secure using our ranking method. Both AT and TM ratios exceed 1 when $q \geq 4$, so both the AT- and the TM-compactness is about 4.

6   Cryptanalysis of yescrypt

6.1   Description

yescrypt [31] is another PHC finalist, which is built upon scrypt and is notable for its high memory filling rate (up to $2\,\mathrm{GB/sec}$) and a number of features, which includes custom S-boxes to thwart exhaustive search on GPU, multiplicative chains to increase the ASIC latency, and some others. yescrypt is essentially a family of functions, each member activated by a combination of flags. Due to the page limits, we consider only one function of the family.

Here we consider the yescrypt setting where flag yescrypt_RW is set, there is no parallelism, and no ROM (in the further text – just yescrypt). It operates on 1024-byte memory blocks $X[1], X[2], \ldots, X[M]$. The scheme works as follows:

$$X[1] \leftarrow F'(I);$$
$$X[i] \leftarrow F(X[i-1] \oplus X[\phi(i)]),\ 1 < i \le M;$$
$$Y \leftarrow X[M];$$
$$Y \leftarrow X[Y \mod M]) \leftarrow F(Y \oplus X[Y \pmod{M}]), M < i \le T.$$

Here F and F' are compression functions (the details of F' are irrelevant for the attack). Therefore, the memory is filled in the first M steps and then $(T - M)$ blocks are updated using the state variable Y. Here $\phi(i)$ is the data-dependent indexing function: it takes 32 bits of $X[i-1]$ and interprets it as a random block index among the last 2^k blocks, where 2^k is the largest power of 2 that is smaller than i.

Transformation F operates on 1024-byte blocks as follows:

- Blocks are partitioned into 16 64-byte subblocks $B_0, B_1, \ldots, B_{15}$.
- New blocks are produced sequentially:

$$B_0^{new} \leftarrow f(B_0^{old} \oplus B_{15}^{old});$$
$$B_i^{new} \leftarrow f(B_{i-1}^{new} \oplus B_i^{old}),\ 0 < i < 16.$$

The details of f are irrelevant to our attack.

6.2   Tradeoff Attack on yescrypt

Our crucial observation is that there is no diffusion from the last subblocks to the first ones. Thus if we store all B_0, we break the dependencies between consecutive blocks and the subblocks can be recomputed from B_1 to B_{15} with pipelining (Fig. 5). Suppose that the block $X[i]$ is computed with latency $L(i)$, i.e. its computation tree has $L(i)$ levels if measured in F. However, if we consider the tree of f, then the actual latency of $X[i]$ is $L(i) + 15$ instead of expected $16L(i)$ if measured in calls to f.

The tradeoff strategy is given in Algorithm 4.

Algorithm 4. Tradeoff attack on yescrypt.

1. Start the ranking tradeoff method with some parameters w, s;
2. Store B_0 of each block;
3. If $X[i]$ needs the missing block $X[r_i]$:
   (a) Compute B_0 of $X[i]$ using one call to f, as all previous B_0 are stored;
   (b) Compute B_1.
   (c) While B_1 is recomputed, start recomputing B_2, as it needs exactly the same subblocks used in the recomputation of B_1. This adds latency of one call to f.
   (d) Compute B_i for all other i.

If the missing block is recomputed by a tree of depth D, then the latency of the new block is $D + 16$ measured in calls to f, or $\frac{D}{16} + 1$ if measured in calls to F. This number should be compared to the latency $D+1$ if we had not exploited the iterative structure of F. Thus if the ranking method gives the total latency L (measured in F), the actual latency should be $\frac{L+15}{16}$.

For the smallest secure parameter ($T = 4M/3$) we get the final computational and latency penalties as well as AT and TM penalties are given in Table 4 (1/16-th of each block is added to the attacker's memory). We conclude yescrypt is only $(0.45, AT)$- and $(0.45, TM)$-secure, whereas the AT compactness is 4 and the TM compactness is 6. Since this numbers are worse than for generic 1-pass schemes, our attack clearly signals of a vulnerability in the design of BlockMix. We expect that our attack becomes inefficient for $T = 2M$ and higher.

Table 4. Computational, latency, AT (for $R_c = 3000$ and $M = 2^{24}$), and TM penalties for the ranking tradeoff attack on yescrypt mode of operation with 4/3 passes, using the iterative structure of F.

Memory	1	$\frac{1}{2}$	$\frac{1}{3}$	$\frac{1}{4}$	$\frac{1}{5}$	$\frac{1}{6}$	$\frac{1}{7}$	$\frac{1}{8}$
Computation penalty $CP(q)$	1	2.9	26	1135	2^{19}	-	-	-
Latency penalty $LP(q)$	1	1.1	1.4	2	3.5	6.3	11.1	17.5
TM ratio	1	0.55	0.47	0.5	0.75	1.05	1.59	2.19
AT ratio	1	0.55	0.46	0.7	95	-	-	-

7 Future Work

Our tradeoff methods apply to a wide class of memory-hard functions, so our research can be continued in the following directions:

- Application of our methods to other PHC candidates and finalists: Pomelo [36] and the modified Lyra2.

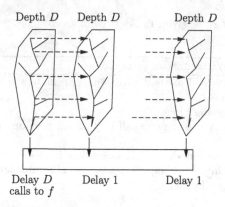

Fig. 5. Pipelining the block computation in yescrypt: only the first subblock is computed with delay D.

- Set of design criteria for the indexing functions that would withstand our attacks.
- New methods that directly target schemes that make multiple passes over memory or use parallel cores.
- Design a set of tools that helps to choose a proof-of-work instance in various applications: cryptocurrencies, proofs of space, etc.

8   Conclusion

Tradeoff cryptanalysis of memory hard functions is a young, relatively unexplored and complex area of research combining cryptanalytic techniques with understanding of implementation aspects and hardware constraints. It has direct real-world impact since its results can be immediately used in the on-going arms race of mining hardware for the cryptocurrencies.

In this paper we have analyzed memory-hard functions Catena-Dragonfly and yescrypt. We show that Catena-Dragonfly is not memory-hard despite original claims and the security proof by the designers', since a hardware-equipped adversary can reduce the attack costs significantly using our tradeoffs. We also show that yescrypt is more tradeoff-resilient than Catena, though we can still exploit several design decisions to reduce the time-memory and the time-area product by the factor of 2.

We generalize our ideas to the generic precomputation method for data-independent schemes and the generic ranking method for the data-dependent schemes. Our techniques may be used to estimate the attack cost in various applications from the fast emerging area of memory-hard cryptocurrencies to the password-based key derivation.

Acknowledgement. We would like to thank the authors of Catena for verifying and confirming our attack.

A   Cryptanalysis of Lyra2 V1

A.1   Description of Lyra2 V1

Lyra2 [25] is a PHC finalist, notable for its high memory filling rate (up to 1 GB/sec). Very recently, Lyra2 has been significantly changed for the second round of the competition. This section describes the original submission to PHC [25], Lyra2 v1 (just Lyra2 in the further text).

Lyra2 is a hybrid hashing scheme, which uses data-independent addressing in the first phase and data-dependent addressing in the second phase. Lyra2 operates on blocks of 768 bits (96 bytes) each, and fills the memory with $2^n \cdot C$ such blocks, where n and C are parameters, and C is by default set to 128 [25, p. 39]. In this paper we use $C = 128$. The entire memory is represented as a $(2^n \times C)$-matrix M, and we refer to its components as rows and columns. Rows are denoted by $M[i]$.

Lyra2 has two main phases: the single-iteration **Setup phase**, where the memory is addressed data-independently, and the multiple-iteration **Wandering phase**, where the memory is addressed data-dependently. The number T of iterations in the Wandering phase can be as low as 1, and we take this value in our analysis.

Setup Phase. The first phase fills rows sequentially from $M[0]$ to $M[2^n - 1]$ as follows:

$$M[0], M[1] \leftarrow f(\text{Password},\text{Salt});$$
$$\text{for } i \text{ from 2 to } 2^n - 1$$
$$M[i] \leftarrow F(M[i-1], M[\phi(i)]);$$
$$M[\phi(i)] \leftarrow M[\phi(i)] \oplus \overleftarrow{M[i]}.$$

Here $\phi(i) = 2^k - i$, where 2^k is the smallest power of 2 that is not smaller than i, $\overleftarrow{M[]}$ stands for the left rotation of each 768-bit word by 32 bits, and G is a cryptographic hash function.

The following details of F are relevant to our attack:

– Function F is stateful: it operates on the 1024-bit state S, which is preserved between rows.
– Function $F(X,Y)$ processes columns X_i, Y_i of X and Y sequentially. The internal state undergoes C rounds (similarly to the duplex-sponge construction [12]), where in round i column Z_i of the output Z is produced as follows:

$$S \leftarrow S \oplus X_i \oplus Y_i;$$
$$S \leftarrow P(S);$$
$$Z_i \leftarrow 768 \text{ least sign. bits of } (S).$$

Here P is a single round of the Blake2b internal permutation [10]. We do not exploit any specific property of P. Thus F can be seen as a duplex-sponge instantiated with a Blake2b round function.

We remind the reader that Z is used not only to produce a new row $M[i]$ but also to overwrite the row $M[2^k - i]$.

Wandering Phase. The Wandering phase transforms the blocks produced in the Setup phase. First, it reverses the ordering. Then it operates similarly to the Setup phase, but the second input block to F is taken pseudo-randomly:

$$\text{for } i \text{ from } 1 \text{ to } 2^n - 1$$
$$r_i \leftarrow g(M[i-1], i-1);$$
$$M[i] \leftarrow M[i] \oplus F(M[i-1], M[r_i]);$$
$$M[r_i] \leftarrow M[r_i] \oplus \overleftarrow{F}(M[i-1], M[r_i]).$$

Here g truncates the first input to the least significant 32 bits and xores with the second input. All indices are computed modulo 2^n.

A.2   Tradeoff Attack on the Setup and Wandering Phases of Lyra2

Our strategy for the Setup phase is similar to the one for Catena. Again, we exploit the properties of the indexing function ϕ.

Let us denote a segment of rows $\{M[i], M[i+1], \dots, M[j]\}$ by $M[i : j]$. Consider a, b such that $2^{k-1} < a < b < 2^k$. Then

$$\phi([a : b]) = [(2^k - b) : (2^k - a)].$$

Thus to construct a single segment we need another segment of the same length. This suggests the following strategy for computing 2^n rows in the Setup phase.

1. First 2^{n-l} rows $M[0], \dots, M[2^{n-l} - 1]$ for some $l > 0$ (parameter of the attack).
2. We split rows from $M[2^{n-l}]$ to $M[2^n - 1]$ into segments of length q for some $q < 2^{n-l}$. Store the entire state S at the start of each segment.

Then to compute segment $M[a : a + q - 1], 2^{k-1} < a < 2^k$ we have to compute $M[\phi(a : a+q-1)]$, which has been updated when computing segments between 2^{k-2} and 2^{k-1}. Eventually we reach the stored 2^{n-l} rows. To compute $M[a : a + q - 1], 2^{k-1} < a < 2^k$ we need to compute a segment in the interval $[2^i : 2^{i+1}]$ for each $n - l < i < k$ (Fig. 6).

Let us figure out the memory reduction and the computational overhead of this procedure. We store 2^{n-l} first rows and $\frac{2^n}{96q}$ rows for starting state in each segment, then a segment of length q during recomputation. For segments between rows $M[2^{n-l}]$ and $M[2^{n-l+1}]$ we need 1 call to F per row, as there is no recomputation. For segments between rows $M[2^{n-l+1}]$ and $M[2^{n-l+2}]$ we need 2 calls to F per row, and so on. In general, we make

$$(k - n + l)q \tag{10}$$

calls to F to compute a segment of length q between row indices 2^k and 2^{k+1}. For the entire Setup phase we spend

$$\underbrace{2^{n-l}}_{M[0:2^{n-l}-1]} + \underbrace{2^{n-l}}_{M[2^{n-l}:2^{n-l+1}-1]} + 2 \cdot 2^{n-l+1} + 3 \cdot 2^{n-l+2} + \cdots + l2^{n-1} \le (l-0.5)2^n$$

calls to F. The memory requirements are $2^{n-l} + q + \frac{2^n}{96q}$, which reaches the minimum of $2^{n-l} + 2^{n/2-4.5}$ for $q = 2^{n/2-5.5}$.

To summarize, our tradeoff algorithm B has computational penalty $(l-0.5)$ if the memory is reduced by the factor of 2^l (Table 5).

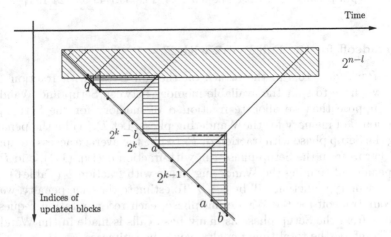

Fig. 6. Computing segment of length q with precomputation method in the Setup phase of Lyra2.

Table 5. Computational-memory tradeoff for the Setup phase of Lyra2: our method and designers' analysis.

Memory fraction	Computational penalty	[25]
1/4	1.5	2
1/8	2.5	4
2^{-l}	$l-0.5$	2^{l-1}

Access Complexity of a Single Row. In the next phase we will need to calculate the cost of recomputing a single row rather than a segment. To compute a single row, we need to recompute $(l-0.5)$ segments on average, so the average recomputation complexity is:

$$A = q(l-0.5). \tag{11}$$

Tradeoff Attack on the Wandering Phase of Lyra2 with $T = 1$. We apply the ranking method to the Wandering phase of Lyra2. Since Lyra2 updates two rows at once, its penalties are higher than in generic data-dependent schemes and are given in Table 6.

Table 6. Average computational and depth penalties for the ranking method on the Wandering phase of Lyra2, without exploiting the row pipelining.

Memory fraction	$\frac{1}{2}$	$\frac{1}{3}$	$\frac{1}{4}$	$\frac{1}{5}$	$\frac{1}{6}$	$\frac{1}{7}$	$\frac{1}{8}$	$\frac{1}{10}$	$\frac{1}{12}$	$\frac{1}{16}$
Computation penalty	2.7	10.4	75	1071	2^{14}	2^n	2^n	2^n	2^n	2^n
Depth penalty		2.4	4.8	8.9	15.4	23.8	35.7	49	83	124	193

A.3   Tradeoff for the Full Lyra2 with $T = 1$

Memory Partition. To run the attack on the full Lyra2 with fraction $1/l$ of memory, we have to split the available memory between Setup and Wandering phases. Suppose that we allocate fraction α of memory for the Setup phase and fraction β of memory for the Wandering phase. Let $P_S(\alpha)$ be the penalty of running the Setup phase with fraction α, $P_R(\alpha)$ be the average access complexity of a single row from the Setup phase run with fraction α (Eq. (11)), and $P_W(\beta)$ be the penalty of running the Wandering phase with fraction β (Table 6). Then the total memory reduction will be $\alpha + \beta$. To estimate the time penalty, we note that in our tradeoff for the Wandering phase, each recomputation requests as many rows from the Setup phase as many hash calls is made in the Wandering phase. Therefore, the total time penalty would be estimated as

$$P(\alpha + \beta) = \frac{P_S(\alpha)2^n + P_R(\alpha)P_W(\beta)2^n}{2^{n+1}},$$

as we construct $2 \cdot 2^n$ blocks in two phases.

Exploiting Iterative Compression Function. Similarly to the attack on yescrypt we can exploit the fact that Lyra2 produces blocks of a row columnwise. Therefore, we have to make D calls to P to compute the first column of the block, whereas computation of other columns can be pipelined: the second column of the deepest tree level can be computed simultaneously with the first column of one level higher. To compute all 128 columns, we spend time needed to compute $D + 128$ columns only, so the actual latency penalty is $1 + D/128$. Therefore, the total latency penalty can be calculated as follows:

$$D(\alpha + \beta) = \frac{D_S(\alpha) + \frac{D_R(\alpha) + D_W(\beta)}{128} + 1}{2},$$

where $D_S(\alpha) = 1 - (\log \alpha)/256$ is the average latency penalty in the Setup phase, $D_R(\alpha) = -\log \alpha - 0.5$ is the average latency penalty for accessing the row from the Setup phase, and $D_W(\beta)$ is the depth penalty for the Wandering phase given in Table 6.

The results are given in Table 7. We conclude that Lyra2 is only $(0.33, AT)$-secure and $(0.1, TM)$-secure. The AT compactness is 4 and the TM compactness is 16. Thus Lyra2 v1 is more susceptible to tradeoff attacks compared to yescrypt.

Table 7. Computational, latency, AT (assuming $R_c = 3000$ and $M = 2^{24}$) and TM penalties for the ranking tradeoff attack on Lyra2 v1 with $T = 1$. Memory fraction is given as a sum of Wandering and Setup allocations

Memory	1	0.45	0.31	0.26	0.12	0.06
Wandering+Setup		1/3+1/8	1/4+ 1/16	1/5+1/16	1/10+1/64	1/17+1/256
Comp. penalty $CP(q)$	1	14.2	133	1876	2^{19}	-
Lat. penalty $LP(q)$	1	1.03	1.05	1.08	1.37	1.93
TM penalty	1	0.47	0.33	0.28	0.16	0.12
AT penalty	1	0.47	0.35	0.63	94	-

References

1. Litecoin: Mining hardware comparison. https://litecoin.info/Mining_hardware_comparison
2. Password Hashing Competition. https://password-hashing.net/
3. Software tool: John the Ripper password cracker. http://www.openwall.com/john/
4. Litecoin - Open source P2P digital currency (2011). https://litecoin.org/
5. IETF Draft: The scrypt Password-Based Key Derivation Function (2012). https://tools.ietf.org/html/draft-josefsson-scrypt-kdf-02
6. Bitcoin: Mining hardware comparison (2014). https://en.bitcoin.it/wiki/Mining_hardware_comparison
7. Vertcoin: Lyra2RE reference guide (2014). https://vertcoin.org/downloads/Vertcoin_Lyra2RE_Paper_11292014.pdf
8. Abadi, M., Burrows, M., Manasse, M.S., Wobber, T.: Moderately hard, memory-bound functions. ACM Trans. Internet Techn. **5**(2), 299–327 (2005)
9. Alwen, J., Serbinenko, V.: High parallel complexity graphs and memory-hard functions. IACR Cryptology ePrint Archive 2014/238 (2014)
10. Aumasson, J.-P., Neves, S., Wilcox-O'Hearn, Z., Winnerlein, C.: BLAKE2: simpler, smaller, fast as MD5. In: Jacobson, M., Locasto, M., Mohassel, P., Safavi-Naini, R. (eds.) ACNS 2013. LNCS, vol. 7954, pp. 119–135. Springer, Heidelberg (2013)
11. Bernstein, D.J., Lange, T.: Non-uniform cracks in the concrete: the power of free precomputation. In: Sako, K., Sarkar, P. (eds.) ASIACRYPT 2013, Part II. LNCS, vol. 8270, pp. 321–340. Springer, Heidelberg (2013)
12. Bertoni, G., Daemen, J., Peeters, M., Van Assche, G.: Duplexing the sponge: single-pass authenticated encryption and other applications. In: Miri, A., Vaudenay, S. (eds.) SAC 2011. LNCS, vol. 7118, pp. 320–337. Springer, Heidelberg (2012)

13. Biryukov, A., Dinu, D., Khovratovich, D.: Argon and argon2: password hashing scheme. Technical report (2015). https://password-hashing.net/submissions/specs/Argon-v2.pdf
14. Biryukov, A., Khovratovich, D.: Tradeoff cryptanalysis of memory-hard functions. Cryptology ePrint Archive, Report 2015/227 (2015). http://eprint.iacr.org/
15. Biryukov, A., Pustogarov, I.: Proof-of-work as anonymous micropayment: rewarding a Tor relay. IACR Cryptology ePrint Archive 2014/1011 (2014). To appear at Financial Cryptography 2015
16. Chang, D., Jati, A., Mishra, S., Sanadhya, S.K.: Time memory tradeoff analysis of graphs in password hashing constructions. In: Preproceedings of PASSWORDS 2014, pp. 256–266 (2014). http://passwords14.item.ntnu.no/Preproceedings_Passwords14.pdf
17. Dwork, C., Goldberg, A.V., Naor, M.: On memory-bound functions for fighting spam. In: Boneh, D. (ed.) CRYPTO 2003. LNCS, vol. 2729, pp. 426–444. Springer, Heidelberg (2003)
18. Dwork, C., Naor, M., Wee, H.M.: Pebbling and proofs of work. In: Shoup, V. (ed.) CRYPTO 2005. LNCS, vol. 3621, pp. 37–54. Springer, Heidelberg (2005)
19. Dziembowski, S., Faust, S., Kolmogorov, V., Pietrzak, K.: Proofs of space. IACR Cryptology ePrint Archive 2013/796 (2013). To appear at Crypto 2015
20. Forler, C., Lucks, S., Wenzel, J.: Catena: a memory-consuming password scrambler. IACR Cryptology ePrint Archive, Report 2013/525 (2013). Version of 5 January 2014. http://eprint.iacr.org/eprint-bin/getfile.pl?entry=2013/525&version=20140105:194859&file=525.pdf
21. Forler, C., Lucks, S., Wenzel, J.: Memory-demanding password scrambling. In: Sarkar, P., Iwata, T. (eds.) ASIACRYPT 2014, Part II. LNCS, vol. 8874, pp. 289–305. Springer, Heidelberg (2014)
22. Giridhar, B., Cieslak, M., Duggal, D., Dreslinski, R.G., Chen, H.M., Patti, R., Hold, B., Chakrabarti, C., Mudge, T.N., Blaauw, D.: Exploring DRAM organizations for energy-efficient and resilient exascale memories. In: International Conference for High Performance Computing, Networking, Storage and Analysis (SC 2013), pp. 23–35. ACM (2013)
23. Gürkaynak, F., Gaj, K., Muheim, B., Homsirikamol, E., Keller, C., Rogawski, M., Kaeslin, H., Kaps, J.-P.: Lessons learned from designing a 65nm ASIC for evaluating third round SHA-3 candidates. In: Third SHA-3 Candidate Conference, March 2012
24. Hopcroft, J.E., Paul, W.J., Valiant, L.G.: On time versus space. J. ACM **24**(2), 332–337 (1977)
25. Simplicio Jr., M.A., Almeida, L.C., Andrade, E.R., dos Santos, P.C.F., Barreto, P.S.L.M.: The Lyra2 reference guide, version 2.3.2. Technical report, April 2014
26. Thomas Lengauer and Robert Endre Tarjan: Asymptotically tight bounds on timespace trade-offs in a pebble game. J. ACM **29**(4), 1087–1130 (1982)
27. Malvoni, K.: Energy-efficient bcrypt cracking. In: Passwords 2014 Conference (2014). http://www.openwall.com/presentations/Passwords14_Energ_Efficient_Cracking/
28. Mukhopadhyay, S., Sarkar, P.: On the effectiveness of TMTO and exhaustive search attacks. In: Yoshiura, H., Sakurai, K., Rannenberg, K., Murayama, Y., Kawamura, S. (eds.) IWSEC 2006. LNCS, vol. 4266, pp. 337–352. Springer, Heidelberg (2006)
29. Nakamoto, S.: Bitcoin: a peer-to-peer electronic cash system (2009). http://www.bitcoin.org/bitcoin.pdf
30. Percival, C.: Stronger key derivation via sequential memory-hard functions (2009). http://www.tarsnap.com/scrypt/scrypt.pdf

31. Peslyak, A.: Yescrypt - a password hashing competition submission. Technical report (2014). http://password-hashing.net/submissions/specs/yescrypt-v0.pdf
32. Pippenger, N.: Superconcentrators. SIAM J. Comput. **6**(2), 298–304 (1977)
33. Ristenpart, T., Tromer, E., Shacham, H., Savage, S.: Hey, you, get off of my cloud: exploring information leakage in third-party compute clouds. In: Proceedings of the 2009 ACM Conference on Computer and Communications Security, CCS 2009, Chicago, Illinois, USA, 9–13 November 2009, pp. 199–212 (2009)
34. Sprengers, M., Batina, L.: Speeding up GPU-based password cracking. In: SHARCS 2012 (2012). http://2012.sharcs.org/record.pdf
35. Thompson, C.D.: Area-time complexity for VLSI. In: STOC 1979, pp. 81–88. ACM (1979)
36. Wu, H.: POMELO: a password hashing algorithm. Technical report (2014). https://password-hashing.net/submissions/specs/POMELO-v1.pdf

Property Preserving Symmetric Encryption Revisited

Sanjit Chatterjee[1]([⊠]) and M. Prem Laxman Das[2]

[1] Department of Computer Science and Automation,
Indian Institute of Science, Bengaluru, India
sanjit@csa.iisc.ernet.in
[2] Society for Electronic Transactions and Security, Chennai, India
prem.lax@gmail.com

Abstract. At EUROCRYPT 2012 Pandey and Rouselakis introduced the notion of property preserving symmetric encryption which enables checking for a property on plaintexts by running a public test on the corresponding ciphertexts. Their primary contributions are: (i) a separation between 'find-then-guess' and 'left-or-right' security notions; (ii) a concrete construction for left-or-right secure orthogonality testing in composite order bilinear groups.

This work undertakes a comprehensive (crypt)analysis of property preserving symmetric encryption on both these fronts. We observe that the quadratic residue based property used in their separation result is a special case of testing equality of one-bit messages, suggest a very simple and efficient deterministic encryption scheme for testing equality and show that the two security notions, find-then-guess and left-or-right, are tightly equivalent in this setting. On the other hand, the separation result easily generalizes for the equality property. So contextualized, we posit that the question of separation between security notions is property specific and subtler than what the authors envisaged; mandating further critical investigation. Next, we show that given a find-then-guess secure orthogonality preserving encryption of vectors of length 2n, there exists left-or-right secure orthogonality preserving encryption of vectors of length n, giving further evidence that find-then-guess is indeed a meaningful notion of security for property preserving encryption. Finally, we cryptanalyze the scheme for testing orthogonality. A simple distinguishing attack establishes that it is not even the weakest selective find-then-guess secure. Our main attack extracts out the subgroup elements used to mask the message vector and indicates greater vulnerabilities in the construction beyond indistinguishability. Overall, our work underlines the importance of cryptanalysis in provable security.

Keywords: Bilinear pairings · Property preserving encryption · Predicate private encryption · Symmetric key

© International Association for Cryptologic Research 2015
T. Iwata and J.H. Cheon (Eds.): ASIACRYPT 2015, Part II, LNCS 9453, pp. 658–682, 2015.
DOI: 10.1007/978-3-662-48800-3_27

1   Introduction

The question of constructing practical cryptographic schemes for securing data in the cloud has attracted a lot of research during the last decade. Notions like order preserving encryption [8,10], attribute-based encryption [21,24,26], functional encryption [1,6,14–16,25] and format preserving encryption [7] are useful for this purpose. The notions of IBE [11,12,19] and public key encryption with keyword search [13,17,33,34] deal with testing of equality. Homomorphic encryption too [22,23,35] plays an important role in cloud security. These schemes aim to achieve data privacy, user privacy, secure computation on encrypted data, etc., on the cloud.

At EUROCRYPT 2012 Pandey and Rouselakis [29] defined the notion of *property preserving symmetric encryption* (PPEnc) which can be used for data clustering [27]. This notion, the authors claim, is most useful in the symmetric key setting. A PPEnc scheme is a collection of four algorithms, namely, Setup, Encrypt, Decrypt and Test where Test is used to check whether the underlying messages satisfy a particular property or not. The authors claim that it is sufficient to consider a simpler notion called *property preserving tag* (PPTag), obtained by dropping the decryption algorithm. The standard approach is to use a semantic secure symmetric key encryption scheme to encrypt the "payload" message while the encryption algorithm of PPTag is used to create a "tag" that is used as one of the inputs to Test to publicly check whether the message satisfies the property or not. In fact a similar approach was taken in [28,32]. Following the Bellare et al. approach for standard encryption [4,5], they define several security notions for property preserving encryption such as find-then-guess (FtG) and left-or-right (LoR) security. However, unlike Bellare et al. [4] who showed FtG implies LoR in the ordinary symmetric key setting, [29] claims that there is a separation between FtG and LoR notions and a hierarchy among the FtG classes that does not collapse. While the notion of property preserving encryption and its security are defined in the abstract setting of a general k-ary property, the separation results are conditioned on the assumed existence of a PPEnc for a concrete binary property based on quadratic residuosity, called P_{qr}. Finally, the paper proposes a scheme for achieving orthogonality, which is claimed to be LoR secure in the generic bilinear group model.

Property preserving encryption has a direct connection with predicate private encryption [32]. In such a scheme, given a token one can check whether a ciphertext satisfies a certain predicate or not. A PPTag scheme may be easily constructed from a predicate private encryption scheme by concatenating ciphertext and token for a given message. If one starts from a full secure predicate-private scheme, one obtains an LoR secure PPTag scheme [1,29]. In [29], the authors also claim that property preserving encryption is a generalization of order preserving encryption of Boldyreva et al. [8–10].

Our Motivation. Property preserving symmetric encryption is an interesting new concept, with a potential practical application for outsourcing computation and it is related to several other primitives like order preserving encryption and predicate encryption. Hence it is imperative that this notion be critically evaluated from the definitional perspective. Because of the separation, designers working on the problem of constructing property preserving encryption for various concrete properties may tend to disregard the FtG notion and only aim at the strongest LoR notion, which is likely to take considerably more resources, see, for example, [1]. Thus it is natural to ask whether the separation indicates any real gap between the two notions and generalizes to any concrete property of interest or is it an artifact related to the peculiarities of the property considered in [29]. The importance of cryptanalyzing the proposed provably secure construction requires no further emphasis.

Our Contributions. In Sect. 3, we revisit the separation results of [29]. As no concrete construction of FtG-secure scheme for P_{qr} was suggested to validate the separation results, we first attempt to build such a scheme. The first observation is that the quadratic residuosity property used in the separation results of [29], can be generalized to a property preserving test of equality. Hence we focus on equality property and show that one-time pad is sufficient to achieve FtG security for equality preserving encryption of one-bit messages. Furthermore, the two notions of FtG and LoR security in fact collapse in such a *deterministic* setting. This result is further generalized for equality testing of n-bit messages where we show a pseudo-random permutation is sufficient to achieve the strongest LoR security. Thus, on one hand we can easily generalize the separation results of [29] for the equality property, on the other we show that in concrete terms the two notions of FtG and LoR effectively collapse for this property. This points to the inherent ambiguity with respect to the actual implication of the separation results for concrete properties of interest. Thus contextualized, we note that the question of whether the separation results of [29] actually indicate any real world difference between the two notions of FtG and LoR security for property preserving encryption still remains open.

In Sect. 4, we look at the relation of FtG and LoR in the context of orthogonality property. We show that given an FtG secure orthogonality preserving encryption of vectors of length $2n$, there exists LoR secure orthogonality preserving encryption of vectors of length n. This result gives further credence to our already established evidence that FtG is indeed a meaningful notion of security for property preserving encryption. We also show that in the property preserving scenario orthogonality implies equality.

In Sect. 5, we cryptanalyze the scheme for testing orthogonality from [29]. We show that the PPEnc scheme given in [29, Sect. 5] is not even weakest selective find-then-guess secure, which falsifies the claim [29, Theorem 5.1] that it is LoR secure. Going beyond indistinguishability, we show that if an adversary is allowed just one query and then given a ciphertext for some unknown message vector $x = (x_1, \ldots, x_n)$, s/he can extract significant non-trivial information about x including whether x is orthogonal to any message of adversary's choice. Thus the attack defeats the very purpose of having property preserving encryption in the symmetric key setting and may be of independent interest in understanding the security of cryptographic schemes in the composite order pairing setting.

We draw our conclusion in Sect. 6. Some of the detailed proofs are provided in Appendix A.

2 Definitions

We recall the basic definition of property preserving encryption and notions of its security from [29]. The paper claims that the idea makes most sense in the symmetric key setting – in the public key setting an adversary can gain non-trivial information about a target ciphertext by encrypting messages of her own choice and then testing for the property on the target message.

As in [29], we too model any k-ary property on $\mathcal{M}$ as a Boolean function on $\mathcal{M}^k$. One of the main properties considered is orthogonality, which depends on computing inner products in finite dimensional vector spaces over finite fields. Let $v = (v_1, \ldots, v_n)$ and $w = (w_1, \ldots, w_n)$ be vectors over a finite field $\mathbb{F}_q$. The inner product between them is defined as $v \cdot w = v_1 w_1 + \ldots + v_n w_n \pmod{q}$. These vectors are *orthogonal* if $v \cdot w = 0$.

Definition 1. *A property preserving encryption scheme (PPEnc) for the k-ary property P is a collection of four probabilistic polynomial time (PPT) algorithms, which are defined as follows:*

1. Setup(1^λ): *This takes as input the security parameter and outputs the message space ($\mathcal{M}$), public parameters (PP) and the secret key (SK).*
2. Encrypt(PP, SK, m): *This algorithm outputs the ciphertext CT corresponding to the message m, using the secret key SK and public parameter PP.*
3. Decrypt(PP, SK, CT): *This algorithm outputs the plaintext message m.*
4. Test($CT_1, \ldots, CT_k, PP$): *This is a public algorithm that takes as inputs ciphertexts $CT_1, \ldots, CT_k$ corresponding to messages $m_1, \ldots, m_k$, respectively and outputs a bit.*

These set of four algorithms must satisfy the standard correctness requirement. In addition, if the Test algorithm outputs $b \in \{0, 1\}$ then, except with negligible probability, one has $P(m_1, \ldots, m_k) = b$.

A related notion of PPTag scheme was also defined. Informally, such a scheme does not have any decrypt module.

Definition 2. *A property preserving tag scheme (PPTag) for the k-ary property P is a collection of three probabilistic polynomial time (PPT) algorithms, which are defined as follows:*

1. Setup(1^λ): *This takes as input the security parameter and outputs the message space ($\mathcal{M}$), public parameters (PP) and the secret key (SK).*
2. Encrypt(PP, SK, m): *This algorithm outputs the ciphertext CT corresponding to the message m, using the secret key SK and public parameter PP.*
3. Test($CT_1, \ldots, CT_k, PP$): *This is a public algorithm that takes as inputs ciphertexts $CT_1, \ldots, CT_k$ corresponding to messages $m_1, \ldots, m_k$, respectively and outputs a bit.*

This set of algorithms must satisfy the standard correctness requirement. If the Test lgorithm outputs $b \in \{0, 1\}$ then, except with negligible probability, one has $P(m_1, \ldots, m_k) = b$.

Remark 1. In [29], the authors suggest the following strategy while designing a secure property preserving encryption scheme. The actual "payload" message is encrypted using an IND-CPA secure symmetric encryption scheme. For testing the property, a tag is constructed for each message using a PPTag scheme.

2.1   Security Notions

Inspired by the study of security notions of symmetric key encryption by Bellare et al. [4], Pandey and Rouselakis [29] propose several notions of security for property preserving symmetric encryption. These notions are defined by taking into account the specific nature of PPEnc. Here we informally describe the two notions of security for such schemes which are most relevant to our work. For more details refer to [29].

Definition 3. *For a k-ary property P, any two sequences $X = (x_1, \ldots, x_n)$ and $Y = (y_1, \ldots, y_n)$ of inputs are said to have the same equality pattern if*

$$P(x_{i_1}, \ldots, x_{i_k}) = P(y_{i_1}, \ldots, y_{i_k}), \ \forall (i_1, \ldots, i_k) \in [n]^k.$$

Find-then-Guess Security (FtG). Challenger and adversary $\mathcal{A} = (\mathcal{A}_1, \mathcal{A}_2)$ plays the following game $\mathsf{Game}^{\mathsf{FtG}}_{\Pi, \mathcal{A}, \lambda}(b)$ which is formally defined in [29, Sect. 3]. After the Setup phase, in $\mathcal{A}_1$, the adversary first adaptively queries the encryption oracle for messages $(m_1, \ldots, m_t)$. Then the adversary outputs the challenge messages (m_0^*, m_1^*). In $\mathcal{A}_2$, after the challenger returns the ciphertext of m_b^* for a random $b \in \{0, 1\}$, the adversary again adaptively queries $(m_{t+1}, \ldots, m_q)$. The adversary wins the game if s/he can correctly predict the bit b. Adversarial queries must satisfy the *extra* condition that the equality patterns of $(m_1, \ldots, m_t, m_0^*, m_{t+1}, \ldots, m_q)$ and $(m_1, \ldots, m_t, m_1^*, m_{t+1}, \ldots, m_q)$ are the same. Otherwise $\mathcal{A}$ can trivially win the game.

Definition 4. *Let $\Pi =$ Setup, Encrypt, Decrypt, Test be a symmetric key property preserving encryption scheme. Then Π is said to be FtG secure if there exists a negligible function $n(\cdot)$ such that for all PPT FtG adversaries $\mathcal{A}$ as above and for all $\lambda \in \mathbb{N}$ sufficiently large, the advantage of $\mathcal{A}$ in the FtG game is negligible:*

$$\mathsf{Adv}^{\mathsf{FtG}}_{\Pi, \mathcal{A}, \lambda} = \left| \Pr\left[\mathsf{Game}^{\mathsf{FtG}}_{\Pi, \mathcal{A}, \lambda}(1) = 1 \right] - \Pr\left[\mathsf{Game}^{\mathsf{FtG}}_{\Pi, \mathcal{A}, \lambda}(0) = 1 \right] \right| \leq n(\lambda).$$

They [29] further introduce a hierarchy in the FtG notion depending on the number of challenge queries. In particular, any adversary playing the FtG^η game, for $\eta \in \mathbb{N}$, is allowed to make η many challenge queries interleaved between encryption oracle queries. A *selective* FtG notion may be defined in the usual way, following [11], where the adversary outputs the challenge messages even before receiving the public parameters.

Left-or-Right Security (LoR). Challenger and adversary $\mathcal{A}$ plays the following game $\mathsf{Game}^{\mathsf{LoR}}_{\Pi, \mathcal{A}, \lambda}(b)$. After setup, $\mathcal{A}$ makes q encryption queries, where each query is of the form $(m_0^{(i)}, m_1^{(i)})$. The queries are such that the tuples $(m_0^{(1)}, \ldots, m_0^{(q)})$

and $(m_1^{(1)}, \ldots, m_1^{(q)})$ have the same equality pattern. The challenger returns the encryption of $m_b^{(i)}$ for each i where the random bit b is chosen at the beginning of game. At the end, the adversary has to output a guess b' of b and wins if $b' = b$. The game is formally defined in [29, Sect. 3]. The definition of adversarial advantage is as follows.

Definition 5. *Let* Π =Setup,Encrypt,Decrypt,Test *be a symmetric key property preserving encryption scheme. Then* Π *is said to be* LoR *secure if there exists a negligible function* $n(\cdot)$ *such that for all PPT* LoR *adversaries* $\mathcal{A}$ *as above and for all* $\lambda \in \mathbb{N}$ *sufficiently large, the advantage of* $\mathcal{A}$ *in the* LoR *game is negligible:*

$$\mathsf{Adv}_{\Pi,\mathcal{A},\lambda}^{\mathsf{LoR}} = \left| \Pr\left[\mathsf{Game}_{\Pi,\mathcal{A},\lambda}^{\mathsf{LoR}}(1) = 1\right] - \Pr\left[\mathsf{Game}_{\Pi,\mathcal{A},\lambda}^{\mathsf{LoR}}(0) = 1\right] \right| \leq n(\lambda).$$

3   Separation Results: A Closer Look

Let $\mathcal{QR}_p$ (resp. $\mathcal{QNR}_p$) be the set of quadratic residues (resp. quadratic non-residues) in $\mathbb{Z}_p^*$ for some prime p. Consider the quadratic residuosity property P_{qr} defined as follows:

$$P_{qr}(m_1, m_2) = \begin{cases} 1 \text{ if } m_1 \cdot m_2 \in \mathcal{QR}_p \\ 0 \text{ if } m_1 \cdot m_2 \in \mathcal{QNR}_p \end{cases} \tag{1}$$

Assuming there *exists* an FtG secure property preserving encryption scheme Π for P_{qr}; Pandey and Rouselakis construct an artificial scheme Π' which is FtG but not LoR secure [29, Theorem 4.1]. In a similar fashion they establish that FtG$^\eta \nrightarrow$ FtG$^{\eta+1}$ [29, Theorem 4.4]. Note that (i) the separation results are *specific* to property P_{qr} and (ii) *conditioned* on the existence of FtG secure scheme for P_{qr} and no such construction was known or suggested in [29].

Property preserving encryption is a rather broad category and a separation based on the specificity of a particular property does not necessarily provide enough insight about the relationship between different security notions for another concrete property or how two notions are related in general. For example, the separation result for P_{qr} in [29] does not give any clue whether the same will hold for another property, say orthogonality. Another crucial question is whether the separation is real or merely an artifact – is there any 'natural' construction for a 'natural' property that is FtG but not LoR secure.

Clearly, a thorough investigation of these questions requires identifying natural properties that encompass other properties and then analysing the real difference between security notions of property preserving encryption in the context of these natural properties. For example, consider the set of all unary properties. It is suggested [29] that for any unary property P, a PPTag scheme can be trivially obtained by providing $P(m)$ in the clear as part of the ciphertext. We note that in such a scenario, the two notions FtG and LoR actually collapse. The case for binary properties, however, is more subtle as we see next.

3.1   Equivalence Testing via Equality

We demonstrate that certain equivalence relations can be tested via equality property – P_{qr} property used in [29] is one such relation.

Claim 1. *To construct a* PPTag *scheme for* P_{qr}*; it suffices to construct a* PPTag *scheme for equality where the message space is* $\mathcal{M} = \{0, 1\}$.[1]

Proof. The argument is quite straightforward. A "sign" function S was used by [29] to define P_{qr} where $S(m) = 0$ if $m \in \mathcal{QR}_p$; else $S(m) = 1$. In other words, P_{qr} divides the message space $\mathcal{M} = \mathbb{Z}_p^*$ into 2 equivalence classes. Given any message in $\mathbb{Z}_p^*$ one can efficiently determine $S(m)$ and then use the PPTag scheme for equality over the message space $\{0, 1\}$ to encrypt $S(m)$. Product of two messages x and y belongs to $\mathcal{QR}_p$ if and only if both x and y belong to same equivalence class. Thus testing whether the product of x and y is a quadratic residue or not is now reduced to the task of testing whether $S(x)$ and $S(y)$ are equal or not.   □

The property P_{qr} used in [29] is a particular instance of a larger class of property $\mathcal{P}$. In particular, the property $\mathcal{P}$ induces an equivalence relation on a set $\mathcal{M}$ such that there exists an efficient algorithm to determine the class in which a given element lies. Another example of such property is to test, given two integers m and n, whether their difference is divisible by a fixed prime p. It is easy to see that a PPTag scheme for such a property $\mathcal{P}$ can be realized by any PPTag scheme for equality. Note, however, that there do exist equivalence relations for which the question of membership testing is not known to be easy.

3.2   Natural LoR Secure Equality Testing

We describe a property preserving encryption scheme for testing equality over message space $\{0, 1\}$.

1. Setup(1^λ): Set $SK = t$, where $t \in_R \{0, 1\}$.
2. Encrypt(SK, m): $CT = t \oplus m$.
3. Decrypt(SK, CT): $m' = CT \oplus t$.
4. Test(CT_1, CT_2): Return 1 if and only if $CT_1 = CT_2$.

It is well-known that as a symmetric key encryption scheme the above construction (or any deterministic encryption scheme) is not FtG secure in the sense of [4] but it is as a PPEnc as the following claim shows.

Claim 2. *The above construction is an* FtG *secure* PPEnc *for one-bit messages.*

---

[1] Here and afterwards we often focus on PPTag schemes as the problem of constructing a PPEnc is essentially reduced to the problem of constructing a PPTag scheme (see Remark 1).

Proof. The key idea is that an FtG adversary $\mathcal{A}$ is restricted by the equality pattern. If $\mathcal{A}$ makes the challenge query as $(0,1)$ or $(1,0)$ then s/he cannot make any encryption oracle query. Hence, the one-time pad ensures the challenge bit is information theoretically hidden from $\mathcal{A}$. On the other hand, if the challenge query is of the form $(0,0)$ or $(1,1)$ then there is no non-trivial information for $\mathcal{A}$ to learn either from the encryption queries or from the challenge.   □

The above result further leads us to the following interesting consequence. Let $E : \mathcal{K} \times \{0,1\} \longrightarrow \{C_0, C_1\}$ be a deterministic encryption scheme.

Claim 3. *If E is FtG secure PPEnc scheme for equality then it is LoR secure.*

Proof. Let $\mathcal{A}$ be a valid LoR adversary for E. We will construct a valid FtG adversary $\mathcal{B}$ for E, which is playing the FtG game with its own challenger $\mathcal{C}$ by internally running $\mathcal{A}$.

Observe that $\mathcal{A}$ has to respect the equality pattern and hence can only make queries from the following disjoint sets: $S_1 = \{(0,0),(1,1)\}$ and $S_2 = \{(0,1),(1,0)\}$. If $\mathcal{A}$ makes queries from the set S_1, then FtG $\longrightarrow$ LoR holds trivially.

Now let us analyze the case when $\mathcal{A}$ makes queries from $S_2 = \{(0,1),(1,0)\}$. Let us, without loss of generality, assume that $\mathcal{A}$'s first query is $(0,1)$. $\mathcal{B}$ sets the same message $(0,1)$ as its own FtG challenge query, forwards it to $\mathcal{C}$. In response $\mathcal{C}$ provides a challenge ciphertext C_b to $\mathcal{B}$, $b \in \{0,1\}$ by encrypting $\beta \in_R \{0,1\}$ using the encryption function E as per the rule of the FtG game. $\mathcal{B}$ forwards the same C_b to $\mathcal{A}$. Note that by the definition of FtG security, $\mathcal{B}$ cannot make any other query to $\mathcal{C}$. However, if $\mathcal{A}$ repeats the same query $(0,1)$, then $\mathcal{B}$ simply forwards the same ciphertext C_b. If $\mathcal{A}$ queries the other valid message pair $(1,0)$, then $\mathcal{B}$ returns ciphertext C_{1-b}. When $\mathcal{A}$ outputs a bit as its guess and halts, then $\mathcal{B}$ outputs the same bit to $\mathcal{C}$ and halts.

The simulation of $\mathcal{A}$'s environment by $\mathcal{B}$ is perfect. In fact, after the first query, $\mathcal{A}$ can on its own generate the response for all other queries it is going to make. Now the FtG security of E ensures that the encryption of 1 is indistinguishable from the encryption of 0. Hence, the advantage of $\mathcal{B}$ is same as that of $\mathcal{A}$ and the two notions actually collapse.   □

As a consequence we note that the one-time pad construction of PPEnc achieves LoR security. However, it is well-known that the same is not even FtG secure as standard symmetric key encryption scheme. Thus there exists binary property preserving encryption scheme secure in the strong LoR sense of property preserving encryption but does not even achieve FtG security as a standard symmetric key encryption scheme.

Based on our previous observations we suggest the following direct construction of LoR secure PPEnc for equality testing on $\mathcal{M} = \{0,1\}^n$. A PPTag can be obtained by dropping the Decrypt algorithm from the description.[2]

---

[2] Similar construction for testing equality in the context of authenticated encryption and searchable encryption schemes was suggested earlier by Rogaway-Shrimpton [31] and Amanatidis et al. [2]. Their constructions used deterministic MAC which is modeled as a PRF.

Property Preserving Encryption for Equality. We describe a scheme Π to test for equality of strings of length n.[3] Let $\{\mathcal{F}\}_n$ be a pseudo-random permutation (PRP) family and an element $F \in \{\mathcal{F}\}_n$ is defined as $F : \{0,1\}^n \times \{0,1\}^n \longrightarrow \{0,1\}^n$.

1. Setup(1^λ): Set a random n-bit binary string K as the secret key SK.
2. Encrypt(SK, m): $CT = F_K(m)$.
3. Decrypt(SK, CT): Return $F_K^{-1}(CT)$.
4. Test(CT_1, CT_2): Return 1 if and only if $CT_1 = CT_2$.

Claim 4. *If the underlying PRP family is secure, then Π is* LoR *secure.*

Proof. (Sketch) The claim is established through a simple hybrid argument. Let the adversary $\mathcal{A}$ for the LoR game set $(m_{0,1}^*, m_{1,1}^*), \ldots, (m_{0,t}^*, m_{1,t}^*)$ as challenges. We claim that the games $\mathsf{Game}_0 : m_{0,1}^*, \ldots, m_{0,t}^*$ and $\mathsf{Game}_1 : m_{1,1}^*, \ldots, m_{1,t}^*$ are indistinguishable. We note that, by the security of the PRP, the Game_0 is indistinguishable from a game where the challenger computes the response from a random permutation. Similarly, challenges output in Game_1 will be indistinguishable from the output of a random permutation. $\square$

3.3   Separation Between FtG and LoR Notions for Equality

After establishing the existence of natural PPEnc/PPTag scheme for equality testing satisfying LoR security (and, hence, FtG security), we now generalize the result of [29, Theorem 4.1] to show that the separation holds for the equality property and need not necessarily be restricted to small number of equivalence classes. Let $\mathcal{M}$ be the message space. Suppose $z = \lceil \log_2 |\mathcal{M}| \rceil$ so that every element $m \in \mathcal{M}$ can be represented by a bit string of length z. Note that z (and not $|\mathcal{M}|$) is a polynomial in the security parameter. Let $\Pi = (\mathsf{Setup}, \mathsf{Encrypt}, \mathsf{Test})$ be any FtG secure PPTag scheme for equality. From this scheme we construct another scheme $\Pi' = (\mathsf{Setup}', \mathsf{Encrypt}', \mathsf{Test}')$ for realizing the same property. The construction uses a PRF family $\mathcal{F} : \{0,1\}^\kappa \times \{0,1\}^z \longrightarrow \{0,1\}^z$.[4]

1. Setup$'$(1^λ): Calls Setup of Π to obtain (PP, SK) and chooses $k \in_R \{0,1\}^\kappa$ (as the key for the PRF). The algorithm outputs PP as the public parameters for Π' and sets the secret key as $SK' = (SK, k)$.
2. Encrypt$'$(PP, SK', m): While encrypting $m \in \mathcal{M}$, the encryption algorithm of Π is used to obtain $ct = \mathsf{Encrypt}(PP, SK, m)$. Then choose a bit $b \in_R \{0,1\}$. The ciphertext of Π' is computed as

$$CT = \begin{cases} (ct, b, F_k(m)), & \text{if } b = 0, \\ (ct, b, F_k(m) \oplus m), & \text{otherwise.} \end{cases}$$

---

[3] For the case of PPTag there is no need to decrypt and hence the construction can be extended to arbitrary length messages by the use of a CRHF H with n-bit digests.

[4] The PRF can be replaced by a set of $|\mathcal{M}|$ random bit strings when $|\mathcal{M}|$ is *small* (i.e., polynomial in the security parameter). On the other hand, for arbitrary length messages one can use a collision resistant hash function (CRHF) H to first map the message to a digest of z-bit and then apply the PRF on the digest.

3. $\mathsf{Test}'(CT_1, CT_2, PP)$: Given $CT_1 = (ct_1, b_1, t_1)$ and $CT_2 = (ct_2, b_2, t_2)$, the algorithm outputs $\mathsf{Test}(ct_1, ct_2, PP)$.

The following two lemma generalize the result of [29] and together establish that the separation result for FtG and LoR holds for equality property. We provide the proofs in Appendix A.

Lemma 1. *If the scheme Π is FtG secure and $\mathcal{F}$ is a secure PRF then Π' constructed as above is also FtG secure. In particular, $\epsilon_{\Pi'} \leq \epsilon_{\Pi} + 2\epsilon_{\mathcal{F}}$ where ϵ_X denotes the advantage in the corresponding security game for the primitive $X \in \{\Pi, \mathcal{F}, \Pi'\}$.*

Lemma 2. *There is an LoR adversary for the scheme Π' with non-negligible advantage.*

Remark 2. We point out an interesting consequence of the above separation result. Shen-Shi-Waters [32] proposed two security notions, the single challenge and full challenge security for predicate private symmetric encryption (see [32] for the definitions of security). The strategy outlined in Lemmas 1 and 2 in the context of PPTag can be adapted to establish a similar separation between single challenge and full challenge security of predicate private encryption. Suppose we are given a single challenge secure predicate private scheme for equality, called Ψ. From that we construct another scheme Ψ' where the only changes are in the Setup and Encrypt as described in the context of Π' above. In particular, the encryption algorithm of Ψ' outputs a ciphertext of Ψ together with either $(b, F_k(m))$ or $(b, F_k(m) \oplus m)$ depending upon whether $b = 0$ or $b = 1$. A similar argument as in the case of PPTag above shows that Ψ' is single challenge secure but not full secure.

Hierarchy Among FtG Classes. We briefly comment on the separation result for the hierarchy among FtG classes given in [29]. The reader may refer to the full version [20] for further details. The equality property over small message space is used to establish the result. We start with a scheme Π which is FtG^η secure and derive a scheme Π' which is not $\mathsf{FtG}^{\eta+1}$ secure. For each message m the Setup algorithm of Π' stores a set of random bit strings $\{t_{m,1}, \ldots, t_{m,\eta}\}$ as part of secret key. Encryption algorithm of Π' chooses $b \in_R \{1, \ldots, \eta + 1\}$ and returns

$$\Pi'.\mathsf{Encrypt}(PP, SK, m) = (\Pi.\mathsf{Encrypt}(PP, SK, m), b, val),$$

where

$$val = \begin{cases} t_{m,b}, & \text{if } 1 \leq b \leq \eta \\ t_{m,1} \oplus \ldots \oplus t_{m,\eta}) \oplus m, & \text{if } b = \eta + 1. \end{cases}$$

The derived scheme Π' is not $\mathsf{FtG}^{\eta+1}$ secure, but FtG^η secure.

3.4   The Bottom Line

At this point a reader may wonder what could be a plausible conclusion of our analysis. On one hand, a PRP is sufficient to construct LoR secure PPEnc for equality and the two notions of FtG and LoR collapse in such a setting. On the other, for the same property there is a *theoretical* gap between FtG and LoR notions of security which may or may not be the case for other properties of interest. In fact, in the next section we show that for orthogonality any FtG secure PPEnc for vectors of length $2n$ gives an LoR secure PPEnc for vectors of length n.

It seems the only reasonable conclusion is that no conclusive evidence exists indicating any real world difference between the two notions of security for PPEnc in general. This leads us to the following open question: is there a 'natural' construction of a scheme for testing equality or, for that matter, any other 'natural' property, which is FtG secure but not LoR secure. Resolving this question will shed further light into the usefulness of the hierarchy of security notions introduced in [29].

4   Orthogonality: Relation Between FtG and LoR and with Equality

We show that it is possible to construct an LoR secure scheme from FtG secure scheme for orthogonality which provides evidence that FtG is a meaningful notion for property preserving encryption. Next, we show that orthogonality implies equality in the property preserving scenario.

4.1   FtG$_{2n}$ implies LoR$_n$

Shen, Shi and Waters showed [32, Theorem 2.8] that a single challenge secure symmetric key predicate-only encryption scheme for testing orthogonality of vectors of length $2n$ may be used to construct one achieving full security for n length vectors. Inspired by their technique we derive a similar result for property preserving orthogonality testing.

Let Θ_{2n} be an FtG secure PPTag encryption scheme for testing orthogonality of vectors of length $2n$. We construct a PPTag scheme Θ_n for testing orthogonality of vectors of length n as follows. In the following we assume that the underlying field on which the vectors are defined does not have characteristic 2 (this is a technical requirement in the security argument). For $x = (x_1, \ldots, x_n)$ and $y = (y_1, \ldots, y_n)$, as usual $x \| y := (x_1, \ldots, x_n, y_1, \ldots, y_n)$.

1. $\Theta_n \cdot \mathsf{Setup}(1^\lambda)$: The public parameters and the secret key are the same as the corresponding ones of Θ_{2n}.
2. $\Theta_n \cdot \mathsf{Encrypt}(PP, SK, x)$: The ciphertext is $\Theta_{2n} \cdot \mathsf{Encrypt}(PP, SK, x\|x)$.
3. $\Theta_n \cdot \mathsf{Test}(CT_1, CT_2, PP)$: The test is carried out using that of the Θ_{2n} scheme as $\Theta_n \cdot \mathsf{Test}(CT_1, CT_2, PP) = 1$ if and only if $\Theta_{2n} \cdot \mathsf{Test}(CT_1, CT_2, PP) = 1$.

Next, we show that Θ_n is LoR secure. The proof proceeds via a sequence of hybrids. Any adversary who can distinguish two adjacent games can break the FtG security of Θ_{2n}.

Theorem 1. *The scheme Θ_{2n} is* FtG *secure implies the derived scheme Θ_n is* LoR *secure.*

Proof. (Sketch) Recall that we have assumed that the underlying field on which the vectors are defined does not have characteristic 2. We observe that $x \cdot y = 0$ if and only if $(x||x) \cdot (y||y) = 0$. The encoding which maps x to $x||x$ is used for proving LoR security via a hybrid argument.

Let $\mathcal{A}$ be a valid LoR adversary for Θ_n. The adversary $\mathcal{A}$ sets as challenges the pairs $(x_0^{(1)}, x_1^{(1)}), \ldots, (x_0^{(q)}, x_1^{(q)})$ to the challenger $\mathcal{C}$. The challenger fixes a random bit b and returns encryption of $x_b^{(i)}$, $1 \leq i \leq q$. The adversary outputs a bit b' at the end of the game and wins if $b = b'$.

We prove that the distributions of the ciphertexts of the sequence of messages $(x_0^{(1)}, x_0^{(2)}, \ldots, x_0^{(q)})$ and $(x_1^{(1)}, x_1^{(2)}, \ldots, x_1^{(q)})$ are indistinguishable. That is, the adversary $\mathcal{A}$ cannot distinguish the games $\mathcal{G}_0$ and $\mathcal{G}_1$ of Table 1. The proof proceeds via a sequence of hybrid games. We tabulate the sequence of hybrids in Table 1. In $\mathcal{G}_B$, the value α is chosen at random from the underlying field. We mention that a sequence of intermediate games is defined between two consecutive games for proving indistinguishability, where only one ciphertext is changed. One such sequence between $\mathcal{G}_A$ and $\mathcal{G}_B$ is given in Table 1.

Table 1. Left: sequence of hybrids $\mathcal{G}_0$ through $\mathcal{G}_1$; right: intermediate games between $\mathcal{G}_A$ and $\mathcal{G}_B$

$\mathcal{G}_0 : x_0^{(1)}||x_0^{(1)}, \ldots, x_0^{(q)}||x_0^{(q)}$
$\mathcal{G}_A : x_0^{(1)}||0, \ldots, x_0^{(q)}||0$
$\mathcal{G}_B : x_0^{(1)}||\alpha x_1^{(1)}, \ldots, x_0^{(q)}||\alpha x_1^{(q)}$
$\mathcal{G}_C : 0||\alpha x_1^{(1)}, \ldots, 0||\alpha x_1^{(q)}$
$\mathcal{G}_D : x_1^{(1)}||\alpha x_1^{(1)}, \ldots, x_1^{(q)}||\alpha x_1^{(q)}$
$\mathcal{G}_1 : x_1^{(1)}||x_1^{(1)}, \ldots, x_1^{(q)}||x_1^{(q)}$

$\mathcal{G}_A : x_0^{(1)}||0, x_0^{(2)}||0, \ldots, x_0^{(q)}||0$
$\mathcal{G}_{A,1} : x_0^{(1)}||\alpha x_1^{(1)}, x_0^{(2)}||0, \ldots, x_0^{(q)}||0$
$\mathcal{G}_{A,2} : x_0^{(1)}||\alpha x_1^{(1)}, x_0^{(2)}||\alpha x_1^{(2)}, x_0^{(3)}||0, \ldots, x_0^{(q)}||0$
$\vdots \qquad \vdots$
$\mathcal{G}_B : x_0^{(1)}||\alpha x_1^{(1)}, \ldots, x_0^{(q)}||\alpha x_1^{(q)}$

We first argue that $\mathcal{G}_0$ and $\mathcal{G}_A$ are indistinguishable. Consider an intermediate game, called $\mathcal{G}_{0,1}$, defined as $x_0^{(1)}||0, x_0^{(2)}||x_0^{(2)}, \ldots, x_0^{(q)}||x_0^{(q)}$.

Notice that this game differs from $\mathcal{G}_0$ only in the first component. We claim that $\mathcal{G}_0$ and $\mathcal{G}_{0,1}$ are indistinguishable. For, suppose $\mathcal{A}$ can distinguish them. Setting $(x_0^{(1)}||x_0^{(1)}, x_0^{(1)}||0)$ as challenge messages and querying the rest of the elements, $\mathcal{A}$ can be used to construct a valid FtG adversary for Θ_{2n}. We proceed by defining a sequence of games where any two consecutive games vary exactly at one component. Similar argument would show that $\mathcal{G}_B$ and $\mathcal{G}_C$ are indistinguishable. The games $\mathcal{G}_C$ and $\mathcal{G}_D$ too may similarly be shown to be indistinguishable.

Recall that $\mathcal{G}_B$ was defined using a random parameter α. Even though, say for example $(x_0^{(1)}\|0) \cdot (x_0^{(2)}\|0) \neq 0$ holds, it may so happen that $(x_0^{(1)}\|x_1^{(1)}) \cdot (x_0^{(2)}\|x_1^{(2)}) = 0$. Thus, a random choice of α ensures that setting as the challenge $(x_0^{(1)}\|0, x_0^{(1)}\|\alpha x_1^{(1)})$ and the rest of the elements as queries one gets a valid FtG adversary for Θ_{2n}. This argument shows that $\mathcal{G}_A$ and $\mathcal{G}_B$ are indistinguishable. Similar argument shows that $\mathcal{G}_D$ and $\mathcal{G}_1$ are indistinguishable. $\qquad\square$

4.2   A Direct Test for Equality from Orthogonality

Katz et al. [28] suggested a simple encoding to test for equality using inner product: create a ciphertext for $\mathcal{I} = (1, I)$ and a token for $\mathcal{J} = (-J, 1)$. Now the inner product of $\mathcal{I}$ and $\mathcal{J}$ is 0 if and only if $I = J$. This encoding does not directly work for property preserving encryption as there is no separate token and the Test is performed only on the ciphertexts. Nevertheless, we show that one can construct a scheme for testing equality property, given a scheme for testing orthogonality of vectors. The new scheme inherits the same security as that of the underlying orthogonality testing scheme. Note that this result is of theoretical interest, but of little practical value as we already have much more efficient scheme for testing equality.

The setting is as follows. Let the message space be $\mathbb{F}_q$, where the finite field is assumed to contain $i = \sqrt{-1}$. Examples of fields which contain i are $\mathbb{F}_{2^n}$; $\mathbb{F}_p$, where $p \equiv 1 \pmod 4$; or extensions of the form $\mathbb{F}_q$ which contain i. The square root of -1 may be given explicitly or may be computed using Tonelli-Shanks algorithm [3, Chapter 7].

We encode any $x \in \mathbb{F}_q$ as a vector in $\mathbb{F}_q^5$, where the encoding is given by $x \mapsto v_x = (x^2+1, ix^2, ix, ix, i)$ (in characteristic 2 fields $m \mapsto v_m = (m+1, m, 1)$). The mapping $m \mapsto v_m$ is one-to-one. Observe that, elements x and y are equal if and only if $v_x \cdot v_y = 0$. We now describe a scheme Π' for testing equality, given a scheme Π for testing orthogonality of vectors of length 5 over $\mathbb{F}_q$.

1. Setup(1^λ): The public parameters and secret key for Π' are those of Π.
2. Encrypt(PP, SK, m): While encrypting $m \in \mathbb{F}_q$, the encryption algorithm first computes the encoding v_m corresponding to m. Then the ciphertext corresponding to m is $CT = \Pi.\text{Encrypt}(PP, SK, v_m)$.
3. Test(CT_1, CT_2, PP): Same as that of Π.

Lemma 3. *If Π is FtG (respectively LoR) secure then so is Π', correspondingly.*

Proof. We describe the FtG case as the LoR case may be similarly handled. Suppose Π' is not FtG secure, with $\mathcal{A}_{\Pi'}$ a valid adversary. We construct $\mathcal{A}_\Pi$, an FtG adversary for scheme Π, which internally runs $\mathcal{A}_{\Pi'}$. Whenever $\mathcal{A}_{\Pi'}$ makes an encryption query m, the adversary $\mathcal{A}_\Pi$ forwards v_m to the challenger $\mathcal{B}_\Pi$. On receiving the ciphertext, it forwards it to $\mathcal{A}_{\Pi'}$. When $\mathcal{A}_{\Pi'}$ sets (m_0^*, m_1^*) as challenge, the adversary $\mathcal{A}_\Pi$ forwards $(v_{m_0^*}, v_{m_1^*})$ to the challenger. On receiving the encryption of one of the two vectors $\mathcal{A}_\Pi$ forwards it to $\mathcal{A}_{\Pi'}$. The other queries made by $\mathcal{A}_{\Pi'}$ may be handled similarly. When $\mathcal{A}_{\Pi'}$ outputs a bit b' and halts, so does $\mathcal{A}_\Pi$. This is a perfect simulation and $\mathcal{A}_\Pi$ wins with the same advantage as that of $\mathcal{A}_{\Pi'}$. $\qquad\square$

5 Cryptanalysis of Pandey and Rouselakis Construction

The only construction proposed in [29] is a PPTag scheme for testing orthogonality of two vectors over a finite field. The proposed scheme works in the composite order bilinear pairing setting. It is claimed without proof in [29, Theorem 5.1] that the scheme achieves LoR security in the generic group model with a precise bound on the adversarial advantage.

We identify an inherent symmetry in the construction that is required for the public Test algorithm. The same symmetry allows the adversary to construct 'pseudo-ciphertext' for many messages from a valid ciphertext of a known message. Suitably manipulated pseudo-ciphertext can be exploited by the adversary to win the indistinguishability game with overwhelming probability. Thus the scheme is not even selective FtG secure. However, the properties of pseudo-ciphertexts allow an adversary to go even further. We show that, after making a single query, an adversary can gain non-trivial information about the underlying message vector given any valid ciphertext. In particular, the adversary can choose any vector and then check whether the unknown message is orthogonal to it or not. This effectively negates the main motivation of using the symmetric key setting for property preserving encryption.

5.1 Pandey and Rouselakis Construction

We recall the scheme of [29] for testing orthogonality of two vectors defined over a prime field $\mathbb{F}_p$, referred to as PR scheme hereafter.

1. Setup($1^\lambda, n$): Pick two distinct primes p and q uniformly at random in the range $(2^{\lambda-1}, 2^\lambda)$ where λ is the security parameter. Let $\mathbb{G}$ and $\mathbb{G}_T$ be two groups of order $N = pq$ such that there is an efficiently computable bilinear map $e : \mathbb{G} \times \mathbb{G} \longrightarrow \mathbb{G}_T$. Select a vector $(\gamma_1, \ldots, \gamma_n) \in \mathbb{Z}_q$ such that $\sum_{i=1}^n \gamma_i^2 = \delta^2 \pmod{q}$. Let g_0 (resp. g_1) be a generator of a subgroup of order p (resp. q) of $\mathbb{G}$. Set the message space as $\mathcal{M} = (\mathbb{Z}_N^* \cup \{0\})^n$. Set

$$PP = \langle n, N, \mathbb{G}, \mathbb{G}_T, e \rangle, \quad SK = \langle g_0, g_1, \{\gamma_i\}_{i=1}^n, \delta \rangle.$$

2. Encrypt(PP, SK, M): On input a message $M = (m_1, \ldots, m_n)$, select two random elements ϕ and ψ from $\mathbb{Z}_N$. The ciphertext is computed as

$$CT = (ct_0, \{ct_i\}_{i=1}^n) = \left(g_1^{\psi\delta}, \{g_0^{\phi m_i} \cdot g_1^{\psi\gamma_i}\}_{i=1}^n \right).$$

3. Test($CT^{(1)}, CT^{(2)}, PP$): When two ciphertexts $CT^{(1)} = (ct_0^{(1)}, \{ct_i^{(1)}\}_{i=1}^n)$ and $CT^{(2)} = (ct_0^{(2)}, \{ct_i^{(2)}\}_{i=1}^n)$ are input, the algorithm outputs 1 if and only if:

$$\prod_{i=1}^n e(ct_i^{(1)}, ct_i^{(2)}) = e(ct_0^{(1)}, ct_0^{(2)}).$$

Correctness ensures that Test outputs 1 only when the underlying messages are orthogonal, except with a negligible probability.

5.2   A Valid FtG Adversary

Notice that the construction ensures that the quadratic form relation $\gamma_1^2 + \gamma_2^2 + \ldots + \gamma_n^2 = \delta^2 \pmod{q}$ is formed in the exponent for one subgroup element of $\mathbb{G}_T$ while the inner product of the two message vectors is computed in the exponent of the other. However, the above equality implies that $\gamma_1(\gamma_1+\gamma_2)+\gamma_2(\gamma_2-\gamma_1)+ \gamma_3^2 + \ldots + \gamma_n^2 = \delta^2 \bmod q$ also holds.

Given a ciphertext for some message $x = (x_1, \ldots, x_n)$, say $(c_0, c_1, c_2, \ldots, c_n)$, the tuple $W = (c_0, c_1 \cdot c_2, c_2/c_1, c_3, \ldots, c_n)$ may be computed. We can hence easily see that the tuple W may be used in the Test algorithm in place of a valid ciphertext of $x' = (x_1 + x_2, x_2 - x_1, x_3, \ldots, x_n)$. The advantage is that, even though the adversary is forbidden to query x' in the security game, s/he may still obtain a ciphertext of x if it is a valid query, and then, compute and use W for testing for orthogonality to x'.

Many such relations among the secret key tuple $(\gamma_1, \ldots, \gamma_n)$ exist that are equal to δ^2. We give more such examples in Lemma 4. But, this observation motivates us to define the notion of *pseudo-ciphertext*.

Definition 6. *A pseudo-ciphertext for PR scheme, associated with a valid message z, is an element $W_z \in \mathbb{G}^{n+1}$ such that $\mathsf{Test}(CT_x, W_z, PP) = 1$ if and only if $\mathsf{Test}(CT_x, CT_z, PP) = 1$, except with negligible probability, where CT_x and CT_z are properly formed ciphertexts for x and z respectively.*

Next, we prove that [29] scheme is not FtG secure.

Proposition 1. *The PPTag scheme proposed in [29] for testing orthogonality is not even **selective FtG** secure.*

Proof. One can construct a valid selective FtG adversary for the $n = 2$ case as follows. The adversary sets $(0, 1)$ and $(1, 0)$ as challenges. Then s/he queries $(1, 1)$ and forms a pseudo-ciphertext for $(2, 0)$. Using that pseudo-ciphertext adversary can trivially win the indistinguishability game.

Now consider the case where $n \geq 3$. The claim is established in terms of the following attack game between the adversary ($\mathcal{A}$) and the challenger ($\mathcal{S}$).

(i) $\mathcal{A}$ outputs a pair of n-dimensional vectors (μ_0^*, μ_1^*) as the challenge messages where $n \ll N$. The challenges are of the form $\mu_0^* = (m_1, m_0, 1, \ldots, 1)$ and $\mu_1^* = (m_1, m_1, 1, \ldots, 1)$, where $m_1 \neq m_0$ are from $\mathbb{Z}_N^*$.
(ii) $\mathcal{A}$ receives the public parameter PP from challenger.
(iii) $\mathcal{A}$ queries $Q = ((m_1+m_0)/2, (m_0-m_1)/2, 1, \ldots, 1, -(n-3))$. Observe that Q is not orthogonal to either of the challenge messages μ_0^* and μ_1^* and hence, is a valid query. $\mathcal{S}$ responds with CT_Q, which is equal to

$$\left(g_1^{\psi\delta}, g_0^{\phi(m_1+m_0)/2} g_1^{\psi\gamma_1}, g_0^{\phi(m_0-m_1)/2} g_1^{\psi\gamma_2}, g_0^{\phi} g_1^{\psi\gamma_3}, \ldots, g_0^{\phi} g_1^{\psi\gamma_{n-1}}, g_0^{-(n-3)\phi} g_1^{\psi\gamma_n} \right)$$

for some $\psi, \phi \in_R \mathbb{Z}_N$. Given CT_Q, $\mathcal{A}$ takes the product and ratio of the third and second components of the ciphertext to obtain respectively $g_0^{m_0\phi} g_1^{\psi(\gamma_1+\gamma_2)}$ and $g_0^{-m_1\phi} g_1^{\psi(\gamma_2-\gamma_1)}$. $\mathcal{A}$ now computes the *pseudo-ciphertext* (Definition 6) $W_{Q'}$ for $Q' = (m_0, -m_1, 1, \ldots, 1, -(n-3))$ as

$$\left(g_1^{\psi\delta}, g_0^{m_0\phi} g_1^{\psi(\gamma_1+\gamma_2)}, g_0^{-m_1\phi} g_1^{\psi(\gamma_2-\gamma_1)}, g_0^{\phi} g_1^{\psi\gamma_3}, \ldots, g_0^{\phi} g_1^{\psi\gamma_{n-1}}, g_0^{-(n-3)\phi} g_1^{\psi\gamma_n}\right).$$

Note that the message vector Q' is orthogonal to μ_0^* but not to μ_1^*.
(**iv**) $\mathcal{A}$ now asks for the challenge ciphertext. Suppose that $\mathcal{S}$ responds with an encryption for μ_b^*

$$CT_b = \left(g_1^{\tilde\psi\tilde\delta}, g_0^{m_1\tilde\phi} g_1^{\gamma_1\tilde\psi}, g_0^{m_b\tilde\phi} g_1^{\gamma_2\tilde\psi}, g_0^{\tilde\phi} g_1^{\gamma_3\tilde\psi}, \ldots, g_0^{\tilde\phi} g_1^{\gamma_n\tilde\psi}\right),$$

where $b \in_R \{0,1\}$ and $\tilde\phi, \tilde\psi \in_R \mathbb{Z}_N$ are as chosen by $\mathcal{S}$.
(**v**) $\mathcal{A}$ runs the Test algorithm on $(CT_b, W_{Q'}, PP)$. This amounts to computing the following quantities:

$$A = e(g_1^{\psi\delta}, g_1^{\tilde\psi\tilde\delta}) \quad \text{and}$$

$$B = e(g_0^{m_0\phi} g_1^{\psi(\gamma_1+\gamma_2)}, g_0^{m_1\tilde\phi} g_1^{\gamma_1\tilde\psi}) \cdot e(g_0^{-m_1\phi} g_1^{\psi(\gamma_2-\gamma_1)}, g_0^{m_b\tilde\phi} g_1^{\gamma_2\tilde\psi}).$$

$$\prod_{i=3}^{n-1} e(g_0^{\phi} g_1^{\psi\gamma_i}, g_0^{\tilde\phi} g_1^{\gamma_i\tilde\psi}) \cdot e(g_0^{-(n-3)\phi} g_1^{\psi\gamma_n}, g_0^{\tilde\phi} g_1^{\gamma_n\tilde\psi}).$$

If $A = B$ then $\mathcal{A}$ outputs $b' = 0$, otherwise $\mathcal{A}$ outputs $b' = 1$.

We see that $A = B$ implies $b = 0$, except with negligible probability. Hence, the adversary wins the selective FtG game with overwhelming probability of success. $\square$

Remark 3. We give yet another attack on the scheme for even n. Let $x = (x_1, \ldots, x_n)$ be any valid message. Observe that both

$$\delta^2 = \gamma_1(\gamma_1 + \gamma_2) + \gamma_2(\gamma_2 - \gamma_1) + \ldots + \gamma_{n-1}(\gamma_{n-1} + \gamma_n) + \gamma_n(\gamma_n - \gamma_{n-1}),$$
$$\delta^2 = \gamma_1(\gamma_1 - \gamma_2) + \gamma_2(\gamma_2 + \gamma_1) + \ldots + \gamma_{n-1}(\gamma_{n-1} - \gamma_n) + \gamma_n(\gamma_n + \gamma_{n-1})$$

hold modulo q. Hence, from the ciphertext for x, pseudo-ciphertexts for both

$$\xi_1 = (x_1 + x_2, x_2 - x_1, \ldots, x_{n-1} + x_n, x_n - x_{n-1}) \text{ and}$$
$$\xi_2 = (x_1 - x_2, x_2 + x_1, \ldots, x_{n-1} - x_n, x_n + x_{n-1})$$

can be formed. Note that neither ξ_1 nor ξ_2 is orthogonal to x, while ξ_1 is orthogonal to ξ_2. Thus, for example, after setting (ξ_1, x) as the challenge pair, querying x and computing pseudo-ciphertext for ξ_2, the adversary can win the FtG game. A similar attack may also be worked out for odd n.

Remark 4. It would have been illustrating to see where exactly the proof of [29, Theorem 5.1] fails. Unfortunately no such proof is provided by the authors.

5.3   Insecurity Beyond Indistinguishability

Recall that in the ciphertext of PR scheme described in Sect. 5.1, the message components reside in the exponent and even the party who possesses the secret key does not have the ability to decrypt. Thus it is not reasonable to expect that one can attack the scheme in the sense of message recovery for high min-entropy messages. Our next attack demonstrates that an adversary is still capable of extracting significant amount of information. This will lead to a total break of the scheme when the messages come from a smaller domain, which could be the case in applications dealing with, for example, certain types of streaming data as envisaged in [29].

We assume that the adversary is allowed to make just one query and is given a valid ciphertext as response. We show how the adversary can process the given ciphertext and then utilize pairing to unmask the subgroup elements containing the message vector of any ciphertext, by working in the target group.

Attack for $n = 2$ Case. Suppose the adversary makes a query $(1/2, 1/2)$ and gets the ciphertext $(c_0, c_1, c_2) = (g_1^{\psi\delta}, g_0^{\phi/2} g_1^{\psi\gamma_1}, g_0^{\phi/2} g_1^{\psi\gamma_2})$. Observe that

$$(c_0, c_1 \cdot c_2, c_2/c_1) = (g_1^{\psi\delta}, g_0^{\phi} g_1^{\psi(\gamma_1+\gamma_2)}, g_1^{\psi(\gamma_2-\gamma_1)})$$

$$(c_0, c_1/c_2, c_1 \cdot c_2) = (g_1^{\psi\delta}, g_1^{\psi(\gamma_1-\gamma_2)}, g_0^{\phi} g_1^{\psi(\gamma_1+\gamma_2)})$$

are pseudo-ciphertexts (see Definition 6) for $(1, 0)$ and $(0, 1)$, respectively, which can be computed by the adversary. We represent the formation of the two pseudo-ciphertexts, respectively, via the following two matrices with the obvious interpretation:

$$M_1 = \begin{bmatrix} 1 & 1 \\ -1 & 1 \end{bmatrix} \text{ and } M_2 = \begin{bmatrix} 1 & -1 \\ 1 & 1 \end{bmatrix}.$$

Suppose now the adversary gets a ciphertext for some unknown message $x = (x_1, x_2)$ as $(C_0, C_1, C_2) = (g_1^{\tilde\psi\delta}, g_0^{\tilde\phi x_1} g_1^{\tilde\psi\gamma_1}, g_0^{\tilde\phi x_2} g_1^{\tilde\psi\gamma_2})$. With the pseudo-ciphertext for $(1, 0)$, the adversary computes

$$\frac{e(C_1, c_1 \cdot c_2) e(C_2, c_2/c_1)}{e(C_0, c_0)} = \frac{e(g_0^{\tilde\phi x_1} g_1^{\tilde\psi\gamma_1}, g_0^{\phi} g_1^{\psi(\gamma_1+\gamma_2)}) \cdot e(g_0^{\tilde\phi x_2} g_1^{\tilde\psi\gamma_2}, g_1^{\psi(\gamma_2-\gamma_1)})}{e(g_1^{\tilde\psi\delta}, g_1^{\psi\delta})}$$

$$= e(g_0, g_0)^{\phi\tilde\phi x_1}.$$

Thus the adversary now possesses $(e(g_0, g_0)^{\phi\tilde\phi x_1}, e(g_0, g_0)^{\phi\tilde\phi x_2})$, after processing the pseudo-ciphertext for $(0, 1)$ similarly.

This trivially breaks the FtG security of PR scheme. Moreover, the adversary can test if x is orthogonal to any $y = (y_1, y_2)$ of his choice by checking whether

$$\left(e(g_0, g_0)^{\phi\tilde\phi x_1}\right)^{y_1} \cdot \left(e(g_0, g_0)^{\phi\tilde\phi x_2}\right)^{y_2} = 1.$$

The adversary may also test for relations among the message coordinates, like whether $x_1 = \alpha x_2$ for some α in a testable range. If x comes from a small domain then one can exhaustively try for all candidate y to check whether x and y are orthogonal and thereby recover x with non-negligible probability.

Attack for General n. Before describing the attack, we show that many a pseudo-ciphertexts can be formed from a valid ciphertext.

Lemma 4. *For $1 \le i \le n$, let $M_i = ((m_{st}^{(i)}))$ be an $n \times n$ matrix defined as follows. Define $m_{it}^{(i)} = 1$, $1 \le t \le n$. For $1 \le s \le n$, but $s \ne i$*

$$m_{st}^{(i)} = \begin{cases} 1, & t = s \\ -1, & t = i \\ 0, & otherwise. \end{cases}$$

Let $CT = (c_0, c_1, \ldots, c_n)$ be a valid ciphertext for $x = (x_1, \ldots, x_n)$. Define $\xi_i = M_i x^T$. Define $W_i = (d_0^{(i)}, d_1^{(i)}, \ldots, d_n^{(i)})$ as follows. For all j, define

$$d_j^{(i)} = \begin{cases} c_0, & if \ j = 0 \\ \prod_{k=1}^{n} c_k^{m_{jk}^{(i)}}, & otherwise. \end{cases}$$

Then W_i is a pseudo-ciphertext for ξ_i.

Proof. We provide details for $i = 1$ – the general case is similar. Observe that by applying M_1 to x^T one obtains $\xi_1 = (\sum_{l=1}^{n} x_l, x_2 - x_1, \ldots, x_n - x_1)$. We also note that $M_1(\gamma_1, \ldots, \gamma_n)^T = (\sum_{l=1}^{n} \gamma_l, \gamma_2 - \gamma_1, \ldots, \gamma_n - \gamma_1)$. By an easy computation:

$$\gamma_1 \sum \gamma_l + \gamma_2(\gamma_2 - \gamma_1) + \ldots + \gamma_n(\gamma_n - \gamma_1) = \delta^2 \pmod{q}.$$

Let $(g_1^{\psi\delta}, g_0^{\phi x_1} g_1^{\psi\gamma_1}, \ldots, g_0^{\phi x_n} g_1^{\psi\gamma_n})$ be a valid ciphertext for x. From this, we compute a pseudo-ciphertext for ξ_1 as

$$W_1 = (g_1^{\psi\delta}, g_0^{\phi \sum x_l} g_1^{\psi \sum \gamma_l}, g_0^{\phi(x_2 - x_1)} g_1^{\psi(\gamma_2 - \gamma_1)}, \ldots, g_0^{\phi(x_n - x_1)} g_1^{\psi(\gamma_n - \gamma_1)}).$$

Let a ciphertext for $y = (y_1, \ldots, y_n)$ be given as

$$CT_y = (c_0, c_1, \ldots, c_n) = \left(g_1^{\tilde{\psi}\delta}, g_0^{\tilde{\phi} y_1} g_1^{\tilde{\psi}\gamma_1}, \ldots, g_0^{\tilde{\phi} y_n} g_1^{\tilde{\psi}\gamma_n} \right).$$

Suppose we run Test with CT_y and W_1. It is easy to see that:

$$\frac{e(c_0, g_0^{\phi \sum x_l} g_1^{\psi \sum \gamma_l}) \prod_{l=2}^{n} e(c_l, g_0^{\phi(x_l - x_1)} g_1^{\psi(\gamma_l - \gamma_1)})}{e(g_1^{\psi\delta}, g_1^{\psi\delta})} = e(g_0, g_0)^{\phi\tilde{\phi}(y \cdot \xi_1)}$$

$$= 1$$

if and only if y is orthogonal to ξ_1, except with negligible probability. $\qquad\square$

Corollary 1. *By querying the vector $x = (1/n, \ldots, 1/n)$, one can obtain the pseudo-ciphertexts for each of the unit vectors $e_i = (0, \ldots, 0, 1, 0 \ldots, 0)$ (1 in the ith place), $1 \leq i \leq n$.*

In the following theorem we describe the attack for general n.

Theorem 2. *Suppose in the proposed PR scheme of [29] the adversary is allowed to make one query for any message of its choice. Then, given a valid ciphertext for any unknown message $(x_1, \ldots, x_n)$, the adversary can extract the tuple of elements $(\eta; \eta^{\phi' x_1}, \ldots, \eta^{\phi' x_n})$ for some η belonging to the order-p subgroup of $\mathbb{G}_T$ and $\phi' \in \mathbb{Z}_N$.*

Proof. Let $(d_0, d_1, \ldots, d_n) = (g_1^{\psi\delta}, g_0^{\phi/n} g_1^{\psi\gamma_1}, \ldots, g_0^{\phi/n} g_1^{\psi\gamma_n})$ be the ciphertext for the queried message $(1/n, \ldots, 1/n)$. A ciphertext CT_x for some unknown $x = (x_1, \ldots, x_n)$ is given to the adversary, where $CT_x = (c_0, c_1, \ldots, c_n) = \left(g_1^{\tilde\psi\delta}, g_0^{\tilde\phi x_1} g_1^{\tilde\psi\gamma_1}, \ldots, g_0^{\tilde\phi x_n} g_1^{\tilde\psi\gamma_n} \right)$.

Notice that the unit vector e_i can be written as $e_i = M_i(1/n, \ldots, 1/n)^T$. From Lemma 4, the adversary can compute $W_i = (w_0^{(i)}, w_1^{(i)}, \ldots, w_n^{(i)})$, a pseudo-ciphertext for e_i as

$$W_i = \left(g_1^{\psi\delta}, g_1^{\psi(\gamma_1 - \gamma_i)}, \ldots, g_1^{\psi(\gamma_{i-1} - \gamma_i)}, g_0^{\phi} g_1^{\psi(\sum \gamma_j)}, g_1^{\psi(\gamma_{i+1} - \gamma_i)}, \ldots, g_1^{\psi(\gamma_n - \gamma_i)} \right).$$

The adversary further computes $\left(\prod_{l=1}^{n} e(c_l, w_l^{(i)}) \right) / e(c_0, w_0^{(i)}) = e(g_0, g_0)^{\phi\tilde\phi x_i}$. In a similar fashion, the adversary obtains a tuple over the order-p subgroup of the target group $\mathbb{G}_T$ as $\Omega = \left(e(g_0, g_0)^{\phi\tilde\phi x_1}, \ldots, e(g_0, g_0)^{\phi\tilde\phi x_n} \right)$. The adversary now computes $\eta := \left(\prod_{i=1}^{n} e(d_i, d_i) \right) / e(d_0, d_0) = e(g_0, g_0)^{\phi^2/n}$. Rewriting Ω as powers of η, s/he gets $\Omega = (\eta^{\phi' x_1}, \ldots, \eta^{\phi' x_n})$. Hence the result. $\square$

As already pointed out for the $n = 2$ case, the above argument shows that the adversary is capable of extracting a lot of information from the ciphertext of any unknown message vector x. Recall that the fundamental reason for having PPTag in symmetric setting is to prevent the adversary from being able to test whether a ciphertext of some unknown message satisfies a certain property and thereby learn some non-trivial information about the message. Given Ω the adversary can precisely do that and thus the scheme in [29] defeats the very purpose of symmetric key property preserving encryption.

6   Concluding Remarks

In this work we perform a comprehensive (crypt)analysis of property preserving symmetric encryption. On the definitional front, we revisit the FtG and LoR separation result in [29]. To do that we show equality property captures property P_{qr} used in the separation results and provide a simple construction for equality property to demonstrate that the separation results are non-vacuous.

Based on the security attributes of our construction and its generalization we raised the pertinent question of whether the separation results actually indicate any real world difference between the two notions of security and argue for a property specific study of the security notions. Continuing further in this direction, we see that an LoR-secure scheme may be constructed from a so-called weaker FtG-secure one for orthogonality. We demonstrate several attacks on the PPTag scheme for testing orthogonality from [29] refuting the claim that the scheme is provably secure. Our main attack successfully unmasks the subgroup elements where the message vector is mapped to and thereby points to greater vulnerability beyond the notion of indistinguishability.

Acknowledgements. The authors wish to thank the anonymous reviewers for their valuable comments. The authors also thank Chethan Kamath, Neal Koblitz, Alfred Menezes, Omkant Pandey, Yannis Rouselakis and Palash Sarkar for their comments on a preliminary version of this work.

A   Appendix

We first argue the separation result for polynomial size message space and use it to prove the general case.

A.1   Separation Result for Polynomial Size Message Space

Let $\mathcal{M} = \{\alpha_i \mid 1 \leq i \leq l\}$ be the message space and each α_i can be represented by a z-bit string where $z = \lceil \log_2 l \rceil$. We argue the separation result FtG $\not\rightarrow$ LoR for equality property in the case where l is polynomial in security parameter. Let Π be an FtG secure PPTag scheme for equality over $\mathcal{M}$. From this scheme we construct another scheme Π' for realizing the same property as follows.

1. $\Pi' \cdot \mathsf{Setup}(1^\lambda)$: The public parameters for Π' are exactly those of Π. The secret key SK' of Π' comprises of that of Π and a set of binary strings $\{t_i \mid 1 \leq i \leq l\}$, where each t_i is of length z and chosen independently and uniformly at random.
2. $\Pi' \cdot \mathsf{Encrypt}(PP, SK, m)$: Suppose $m = \alpha_i$; the algorithm chooses a random bit b and the output is defined as

$$\Pi'.\mathsf{Encrypt}(PP, SK', m) = \begin{cases} (\Pi.\mathsf{Encrypt}(PP, SK, m), b, t_i), & \text{if } b = 0, \\ (\Pi.\mathsf{Encrypt}(PP, SK, m), b, t_i \oplus \alpha_i), & \text{o.w.} \end{cases}$$

3. $\Pi'.\mathsf{Test}(CT_1, CT_2, PP)$: Same as that of Π, where only the relevant parts of the ciphertexts are used.

Lemma 5. *The scheme Π' is not* FtG *secure implies Π is not* FtG *secure.*

Proof. Consider a valid FtG adversary for Π', denoted by $\mathcal{A}$. We describe how an FtG adversary $\mathcal{B}$ for Π, with same advantage as that of $\mathcal{A}$ and which internally uses $\mathcal{A}$, can be constructed.

(**i**). $\mathcal{B}$ forwards to $\mathcal{A}$ whatever is received from its own challenger as public parameters of Π and initializes an empty table T.

(**ii**). Whenever $\mathcal{A}$ makes an encryption query for $m = \alpha_i$, $1 \le i \le l$, $\mathcal{B}$ forwards it to the simulator of Π. On receiving ct from the simulator, $\mathcal{B}$ checks whether the same query was made earlier or not. If the query is made for the first time, then it chooses $t \in_R \{0,1\}^z$, sets $t_i = t$ and updates the table T with $\{(i, t_i)\}$. Else, $\mathcal{B}$ reuses corresponding t_i from T. Finally $\mathcal{B}$ chooses a random bit b and forwards to $\mathcal{A}$

$$CT = \begin{cases} (ct, b, t_i), & \text{if } b = 0, \\ (ct, b, t_i \oplus \alpha_i), & \text{if } b = 1. \end{cases}$$

(**iii**). After a certain number of encryption queries $\mathcal{A}$ outputs the challenge (m_0^*, m_1^*). Two cases arise with respect to the challenges, which we describe below.

Case 1: The challenge messages m_0^* and m_1^* are equal.

Case 2: The challenge messages m_0^* and m_1^* are different. In this case, the adversary cannot make encryption query for these two messages.

$\mathcal{B}$ forwards (m_0^*, m_1^*) to the simulator of Π and gets ct^*. If the challenge messages are equal (**Case 1**), then (ct^*, b, val) may be computed by $\mathcal{B}$ in the same way as it responses to the encryption queries. If the challenge messages are different (**Case 2**), then none of m_0^* and m_1^* have been queried previously. $\mathcal{B}$ returns (ct^*, b, t^*), where $b \in_R \{0,1\}$ and $t^* \in_R \{0,1\}^z$. Let $\alpha_j \in \{m_0^*, m_1^*\}$ be the unknown message chosen by the simulator of Π. The strategy adopted by $\mathcal{B}$ gives a perfect simulation. This is because if $b = 0$ then t^* can be set as t_j whereas for $b = 1$, t^* can be set as $t_j \oplus \alpha_j$.

(**iv**). $\mathcal{B}$ follows the same strategy of step (ii) above to answer all the subsequent encryption queries of $\mathcal{A}$.

(**v**). When $\mathcal{A}$ outputs a bit b' and halts, so does $\mathcal{B}$.

Notice that all the ciphertexts which $\mathcal{B}$ computes for forwarding to $\mathcal{A}$ are properly distributed. $\mathcal{B}$ is a polynomial time algorithm and provides a perfect simulation. Hence, advantage of $\mathcal{B}$ is equal to that of $\mathcal{A}$. $\square$

Lemma 6. *There is an* LoR *adversary for the scheme Π' with non-negligible advantage.*

Proof. A valid LoR adversary sets as u challenges the same pair of the form (m_0, m_1), with $m_0 \ne m_1$. Equality pattern is clearly preserved between the left and right sequences. If the challenger outputs two ciphertexts for which the b-values are distinct, then the adversary can immediately distinguish the two sequences. The advantage will be $1 - 2^{-u+1}$. $\square$

The strategy outlined in the above proof can be used to prove Lemma 2.

A.2   Proof of Lemma 1

Recall that in the FtG game $\mathcal{A}$ makes a polynomial number of encryption oracle query m_i, $1 \leq i \leq q$, and a single challenge query (m_0^*, m_1^*) maintaining the equality pattern. Two cases arise depending upon whether the challenge messages m_0^* and m_1^* are equal or not. If $m_0^* = m_1^*$ then it is easy to see that any advantage of $\mathcal{A}$ against Π' translates into the same advantage against Π. Hence, we consider the case when $m_0^* \neq m_1^*$. Note that in this case none of the queries to the encryption oracle m_i is equal to m_b^*, for $b \in \{0, 1\}$. Otherwise, the equality pattern of the two sequences will be different allowing $\mathcal{A}$ to trivially distinguish.

Let Game_0 correspond to the queries $(m_1, \ldots, m_i, m_0^*, m_{i+1}, \ldots, m_q)$ while Game_1 to queries $(m_1, \ldots, m_i, m_1^*, m_{i+1}, \ldots, m_q)$ made by the adversary. Suppose $\mathcal{A}$ can distinguish whether it is playing Game_0 or Game_1 with a non-negligible advantage $\epsilon_{\Pi'}$. The proof will proceed through a hybrid argument. Given an adversary $\mathcal{A}$ against Π' we construct a series of four games and then show that if $\mathcal{A}$ can distinguish between any two consecutive games then we can construct either a PRF adversary against $\mathcal{F}$ or an FtG adversary against Π.

Game_0 The challenger runs the Setup algorithm of Π' and gives the PP to $\mathcal{A}$ and keeps the secret key $SK' = (SK, k)$ to itself. The challenger computes the ciphertext corresponding to $(m_1, \ldots, m_i, m_0^*, m_{i+1}, \ldots, m_q)$ using SK' as per the encryption algorithm of Π' and give them to $\mathcal{A}$.

Game_A The challenger runs the Setup algorithm of Π and gives the PP to $\mathcal{A}$ and keeps the secret key SK of Π to itself. Note that the challenger does not generate the PRF key k; instead it will maintain a table $\mathbb{T} = \langle x_i, y_i \rangle$ where x_i and y_i are two z-bit strings. The first entry in each row of $\mathbb{T}$ corresponds to the messages queried by $\mathcal{A}$ while the second entry is a random bit-string. The table is initially empty. Whenever $\mathcal{A}$ makes an encryption query for a message x, the challenger first checks whether there is a corresponding entry in $\mathbb{T}$. If not, it chooses a random z-bit string y and enters (x, y) in the table $\mathbb{T}$ sorted according to the first entry. $\mathcal{A}$ makes encryption queries for $(m_1, \ldots, m_i, m_0^*, m_{i+1}, \ldots, m_q)$. To answer the query of $\mathcal{A}$ for a message, say x, the challenger computes the ciphertext of Π on x and then uses the corresponding random string y from the entry (x, y) in $\mathbb{T}$ to create a ciphertext of Π'. Note that $\mathcal{A}$ makes at most q encryption oracle queries and a single challenge query. So the size of $\mathbb{T}$ is $O(q)$ and hence the challenger can consistently respond to all the queries of $\mathcal{A}$.

Claim 5. *If $\mathcal{A}$ can decide with a non-negligible advantage whether it is playing Game_0 or Game_A then we can construct a PRF distinguisher with the same advantage.*

Recall that in the PRF security game we are provided with an oracle which is either a function from the PRF family or a random function. In the former case the challenger will be playing Game_0 while in the latter case it'll be playing Game_A. Hence, any advantage of $\mathcal{A}$ in distinguishing between the two games translate into the same advantage of the challenger in breaking the PRF security.

Game_1 (resp. Game_B) will be identical to Game_0 (resp. Game_A) except the fact that $\mathcal{A}$ now queries with $(m_1, \ldots, m_i, m_1^*, m_{i+1}, \ldots, m_q)$. An identical argument as in the claim above establishes that any advantage of $\mathcal{A}$ in deciding whether it is playing Game_1 or Game_B translates into the same PRF advantage for the challenger.

Note that the only difference in Game_A and Game_B is in the challenge ciphertext (corresponding to m_0^* and m_1^*). The challenge is computed by calling the encryption algorithm of Π and appending either a random bit string or a one-time encryption of m_b^* (using that random string). Hence, an adversary distinguishing between Game_A and Game_B can be converted into an adversary breaking the FtG security of Π. As there are only polynomial many queries, this case is the same as the one where there are only small (polynomial in λ) number of messages. This case can be easily handled by using random strings. We have already given the analysis in the proof of Lemma 5.

References

1. Agrawal, S., Agrawal, S., Badrinarayanan, S., Kumarasubramanian, A., Prabhakaran, M., Sahai, A.: Function Private Functional Encryption and Property Preserving Encryption : New Definitions and Positive Results. Cryptology ePrint Archive, Report 2013/744 (2013). http://eprint.iacr.org/
2. Amanatidis, G., Boldyreva, A., O'Neill, A.: Provably-secure schemes for basic query support in outsourced databases. In: Barker, S., Ahn, G.-J. (eds.) Data and Applications Security 2007. LNCS, vol. 4602, pp. 14–30. Springer, Heidelberg (2007)
3. Bach, E., Shallit, J.O.: Algorithmic Number Theory. Foundations of Computing. MIT Press, Cambridge (1996)
4. Bellare, M., Desai, A., Jokipii, E., Rogaway, P.: A concrete security treatment of symmetric encryption. In: FOCS, pp. 394–403. IEEE Computer Society (1997)
5. Bellare, M., Desai, A., Pointcheval, D., Rogaway, P.: Relations among notions of security for public-key encryption schemes. In: Krawczyk, H. (ed.) CRYPTO 1998. LNCS, vol. 1462, p. 26. Springer, Heidelberg (1998)
6. Bellare, M., O'Neill, A.: Semantically-secure functional encryption: possibility results, impossibility results and the quest for a general definition. In: Abdalla, M., Nita-Rotaru, C., Dahab, R. (eds.) CANS 2013. LNCS, vol. 8257, pp. 218–234. Springer, Heidelberg (2013)
7. Bellare, M., Ristenpart, T., Rogaway, P., Stegers, T.: Format-preserving encryption. In: Jacobson Jr, M.J., Rijmen, V., Safavi-Naini, R. (eds.) SAC 2009. LNCS, vol. 5867, pp. 295–312. Springer, Heidelberg (2009)
8. Boldyreva, A., Chenette, N., Lee, Y., O'Neill, A.: Order-preserving symmetric encryption. In: Joux, A. (ed.) EUROCRYPT 2009. LNCS, vol. 5479, pp. 224–241. Springer, Heidelberg (2009)
9. Boldyreva, A., Chenette, N., Lee, Y., ONeill, A.: Order-Preserving Symmetric Encryption. Cryptology ePrint Archive, Report 2012/624 (2012). http://eprint.iacr.org/
10. Boldyreva, A., Chenette, N., O'Neill, A.: Order-preserving encryption revisited: improved security analysis and alternative solutions. In: Rogaway, P. (ed.) CRYPTO 2011. LNCS, vol. 6841, pp. 578–595. Springer, Heidelberg (2011)
11. Boneh, D., Boyen, X.: Efficient selective identity-based encryption without random oracles. J. Cryptol. 24(4), 659–693 (2011)

12. Boneh, D., Boyen, X., Goh, E.-J.: Hierarchical identity based encryption with constant size ciphertext. In: Cramer, R. (ed.) EUROCRYPT 2005. LNCS, vol. 3494, pp. 440–456. Springer, Heidelberg (2005)

13. Boneh, D., Di Crescenzo, G., Ostrovsky, R., Persiano, G.: Public key encryption with keyword search. In: Cachin, C., Camenisch, J.L. (eds.) EUROCRYPT 2004. LNCS, vol. 3027, pp. 506–522. Springer, Heidelberg (2004)

14. Boneh, D., Raghunathan, A., Segev, G.: Function-private identity-based encryption: hiding the function in functional encryption. In: Canetti, R., Garay , J.A. (eds.) [18], pp. 461–478 (2013)

15. Boneh, D., Raghunathan, A., Segev, G.: Function-private subspace-membership encryption and its applications. In: Sako, K., Sarkar, P. (eds.) ASIACRYPT 2013, Part I. LNCS, vol. 8269, pp. 255–275. Springer, Heidelberg (2013)

16. Boneh, D., Sahai, A., Waters, B.: Functional encryption: definitions and challenges. In: Ishai, Y. (ed.) TCC 2011. LNCS, vol. 6597, pp. 253–273. Springer, Heidelberg (2011)

17. Boneh, D., Waters, B.: Conjunctive, subset, and range queries on encrypted data. In: Vadhan, S.P. (ed.) TCC 2007. LNCS, vol. 4392, pp. 535–554. Springer, Heidelberg (2007)

18. Canetti, R., Garay, J.A. (eds.): CRYPTO 2013, Part II. LNCS, vol. 8043. Springer, Heidelberg (2013)

19. De Caro, A., Iovino, V., Persiano, G.: Fully secure anonymous HIBE and secret-key anonymous IBE with short ciphertexts. In: Joye, M., Miyaji, A., Otsuka, A. (eds.) Pairing 2010. LNCS, vol. 6487, pp. 347–366. Springer, Heidelberg (2010)

20. Chatterjee, S.,Das, M.P.L.: Property Preserving Symmetric Encryption Revisited. Cryptology ePrint Archive, Report 2013/830 (2013). http://eprint.iacr.org/

21. Garg, S., Gentry, C., Halevi, S., Sahai, A., Waters, B.: Attribute-based encryption for circuits from multilinear maps. In: Canetti, R., Garay, J.A. (eds.) [18], pp. 479–499 (2013)

22. Gentry, C.: Fully Homomorphic encryption using ideal lattices. In: Mitzenmacher, M. (ed.) STOC, pp. 169–178. ACM (2009)

23. Gentry, C., Halevi, S., Smart, N.P.: Fully homomorphic encryption with polylog overhead. In: Pointcheval, D., Johansson, T. (eds.) [30], pp. 465–482 (2012)

24. Gentry, C., Sahai, A., Waters, B.: Homomorphic encryption from learning with errors: conceptually-simpler, asymptotically-faster, attribute-based. In: Canetti, R., Garay, J.A. (eds.) CRYPTO 2013, Part I. LNCS, vol. 8042, pp. 75–92. Springer, Heidelberg (2013)

25. Goldwasser, S., Gordon, S.D., Goyal, V., Jain, A., Katz, J., Liu, F.-H., Sahai, A., Shi, E., Zhou, H.-S.: Multi-input functional encryption. In: Nguyen, P.Q., Oswald, E. (eds.) EUROCRYPT 2014. LNCS, vol. 8441, pp. 578–602. Springer, Heidelberg (2014)

26. Goyal, V., Pandey, O., Sahai, A., Waters, B.: Attribute-based encryption for fine-grained access control of encrypted data. In: Juels, A., Wright, R.N., De Capitani di Vimercati, S. (eds.) ACM Conference on Computer and Communications Security, pp. 89–98. ACM (2006)

27. Guha, S., Meyerson, A., Mishra, N., Motwani, R., O'Callaghan, L.: Clustering data streams: theory and practice. IEEE Trans. Knowl. Data Eng. 15(3), 515–528 (2003)

28. Katz, J., Sahai, A., Waters, B.: Predicate encryption supporting disjunctions, polynomial equations, and inner products. In: Smart, N.P. (ed.) EUROCRYPT 2008. LNCS, vol. 4965, pp. 146–162. Springer, Heidelberg (2008)

29. Pandey, O., Rouselakis, Y.: Property preserving symmetric encryption. In: Pointcheval, D., Johansson, T. (eds.) [30], pp. 375–391 (2012)
30. David, P., Johansson, T.: EUROCRYPT 2012. LNCS, vol. 7237. Springer, Heidelberg (2012)
31. Rogaway, P., Shrimpton, T.: A provable-security treatment of the key-wrap problem. In: Vaudenay, S. (ed.) EUROCRYPT 2006. LNCS, vol. 4004, pp. 373–390. Springer, Heidelberg (2006)
32. Shen, E., Shi, E., Waters, B.: Predicate privacy in encryption systems. In: Reingold, O. (ed.) TCC 2009. LNCS, vol. 5444, pp. 457–473. Springer, Heidelberg (2009)
33. Shi, E., Bethencourt, J., Chan, H.T.-H., Song, D.X., Perrig, A.: Multi-dimensional range query over encrypted data. In: IEEE Symposium on Security and Privacy, pp. 350–364. IEEE Computer Society (2007)
34. Song, D.X., Wagner, D., Perrig, A.: Practical techniques for searches on encrypted data. In: IEEE Symposium on Security and Privacy, pp. 44–55. IEEE Computer Society (2000)
35. van Dijk, M., Gentry, C., Halevi, S., Vaikuntanathan, V.: Fully homomorphic encryption over the integers. In: Gilbert, H. (ed.) EUROCRYPT 2010. LNCS, vol. 6110, pp. 24–43. Springer, Heidelberg (2010)

Refinements of the k-tree Algorithm for the Generalized Birthday Problem

Ivica Nikolić[1](✉) and Yu Sasaki[1,2]

[1] Nanyang Technological University, Singapore, Singapore
inikolic@ntu.edu.sg, sasaki.yu@lab.ntt.co.jp
[2] NTT Secure Platform Laboratories, Tokyo, Japan

Abstract. We study two open problems proposed by Wagner in his seminal work on the generalized birthday problem. First, with the use of multicollisions, we improve Wagner's k-tree algorithm that solves the generalized birthday problem for the cases when k is not a power of two. The new k-tree only slightly outperforms Wagner's k-tree. However, in some applications this suffices, and as a proof of concept, we apply the new 3-tree algorithm to slightly reduce the security of two CAESAR proposals. Next, with the use of multiple collisions based on Hellman's table, we give improvements to the best known time-memory tradeoffs for the k-tree. As a result, we obtain the a new tradeoff curve $T^2 \cdot M^{\lg k-1} = k \cdot N$. For instance, when $k = 4$, the tradeoff has the form $T^2 M = 4 \cdot N$.

Keywords: Generalized birthday problem · k-list problem · k-tree algorithm · Time-memory tradeoff

1 Introduction

Arguably, the most popular problem in private key cryptography is the collision search problem. It appears frequently not only in its classical usage, e.g. finding collisions for hash functions, but also as an intermediate subproblem of a wider cryptographic problem. The collision search has been widely studied and well understood. Besides this problem, and along with the search of multicollisions and multiple collisions, perhaps the next most popular is the generalized birthday problem (GBP).

The GBP is defined as follows: given k lists of random elements, choose a single element in each list, such that all the chosen elements sum up to a predefined value. Wagner is the first to investigate the GBP for all values of k and as an independent problem. In his seminal paper [31], he proposes an algorithm to solve GBP for all values of k and shows wide variety of applications ranging from blind signatures, to incremental hashing, low weight parity checks, and cryptanalysis of various hash functions.

Prior to Wagner, GBP problem has been mostly studied in the context of its application and only for a concrete number of lists (usually four lists, i.e. $k = 4$). Schroeppel and Shamir [28] find all solutions to the 4-list problem.

© International Association for Cryptologic Research 2015
T. Iwata and J.H. Cheon (Eds.): ASIACRYPT 2015, Part II, LNCS 9453, pp. 683–703, 2015.
DOI: 10.1007/978-3-662-48800-3_28

Bernstein [4] uses similar algorithm to enumerate all solutions to a particular equation. Boneh, Joux and Nguyen [10] use Schroeppel and Shamir's algorithm for solving integer knapsacks as well as Bleichenbacher [8] in his attack on DSA. Chose, Joux, and Mitton [11] use it to speed up search for parity checks for stream cipher cryptanalysis. Joux and Lercier [19] use related ideas in point-counting algorithms for elliptic curves. Blum, Kalai, and Wasserman [9] apply it to find the first known subexponential algorithm for the learning parity with noise problem. Ajtai, Kumar, and Sivakumar findings [1] base on Blum, Kalai, and Wasserman's algorithm as a subroutine to speed up the shortest lattice vector problem.

To solve the k-list problem, Wagner proposes a so-called k-tree algorithm. In a nutshell, the k-tree is a divide and conquer approach and at each step it operates on only two lists. The step operations are based on a simple collision search. When the k lists are composed of n-bit words, Wagner's k-tree algorithm solves the GBP in $\mathcal{O}(k \cdot 2^{\frac{n}{\lceil \lg k \rceil + 1}})$ time and memory and requires lists of around $2^{\frac{n}{\lceil \lg k \rceil + 1}}$ elements[1].

Even though the GBP has been shown to be very important to many problems in cryptography, more than a decade after its publication neither significant improvement to the k-tree algorithm nor other dedicated algorithms have emerged. However, moderate improvements and refinements have been published. As one of the most important, we single out the extended k-tree algorithm by Minder and Sinclair [21] that provides solution to GBP when the lists have smaller sizes and the time-memory tradeoffs by Bernstein et al. [5,6].

Our Contribution. Wagner points out a few open problems of the GBP and of the k-tree algorithm. Two of these problems, namely, improving the efficiency of k-tree when k is not a power of two and memory reduction of the k-tree, are in fact the main research topics of our paper.

The k-tree algorithm discards part of the lists when k is not a power of two (note how the complexity of k-tree takes lower bound of $\lg k$). For instance, 7-tree works only with 4 lists, while the remaining 3 lists are not processed. Our first improvement to the k-tree is to work with the discarded lists (we call them *passive* lists) by creating multicollisions from the lists. From each of the passive lists we create a multicollision set of values that coincide on certain l bits, where $l < n$. Then, we produce several solutions with the k-tree from the other (active) lists, and for the same l bits. Finally, the remaining $n - l$ bits are absorbed by combining the multicollisions from the passive lists, and the solutions from the active lists. The advantage of our approach over the classical k-tree is limited by the size of the multicollision sets, which in turn is bounded by the value of n. The speed-up factor can be approximated as $a \cdot n^c / \lg(b \cdot n)$, where a, b, c are constants that depend on k. The speed-up is sufficient to break the $\mathcal{O}(2^{\frac{n}{2}})$ complexity bound for the 3-list problem and to show that in applications this can matter. As an example, we show a security reduction of two authenticated

---

[1] Note, we use lg for $\log_2$.

encryption CAESAR [3] proposals, Deoxys [16] and KIASU [18], based on the latest results of Nandi [22]. He shows that a forgery attack for COPA based candidates can be reduced to the 3-list problem. We apply our improved 3-tree algorithm to this problem and reduce the security bound of the candidates by 2 bits.

Our second contribution are time-memory tradeoffs for the k-tree algorithm. This research topic has been investigated by Bernstein et al. Their best tradeoffs are described with the curves $TM^{\lg k} = k \cdot N$ and $T^2 \cdot M^{\lg k - 2} = \frac{k^2}{4} \cdot N$, where M and T, are the memory and time complexity, respectively, and N is the size of the space of elements. To achieve a better tradeoff, we play around with the idea of producing multiple collisions in a memory constrained environment with the use of Hellman's tables[2]. It allows us to significantly reduce the memory complexity of the first level of the k-tree algorithm and to achieve better tradeoffs. As a result, we obtain the tradeoff $T^2 M^{\lg k - 1} = k \cdot N$. This translates to $T^2 M = 4 \cdot N$ for $k = 4$, and $T^2 M^2 = 8 \cdot N$ for $k = 8$ (cf. $TM^2 = 4 \cdot N$ and $TM^3 = 8 \cdot N$ curves of Bernstein et al.). As illustrated further in Fig. 6, for a given amount of memory, the new tradeoff always leads to a lower time complexity than the previous tradeoffs. The improvement of the tradeoff can be seen on the case of generalized birthday problem for the hash function SHA-160 and $k = 8$. Our new tradeoff requires around 2^{50} SHA-1 computations and 2^{30} memory on 8 cores (with the use of van Oorschot and Wiener's parallel collisions search [30]), while with the same memory, the old tradeoff needs around 2^{65} SHA-1 computations.

2   The Generalized Birthday Problem

Wagner introduced the generalized birthday problem (GBP) as multidimensional generalization of the birthday problem. GBP is also called a k-list problem, and is formalized as follows:

Problem 1. *Given k lists $L_1, \ldots, L_k$ of elements drawn uniformly and independently at random from $\{0,1\}^n$, find $x_1 \in L_1, \ldots, x_k \in L_k$ such that $x_1 \oplus x_2 \oplus \ldots \oplus x_k = 0$.*

Obviously, if $|L_1| \times |L_2| \times \ldots \times |L_k| \geq 2^n$, then with a high probability the solution to the problem exists. The real challenge, however, is to find it efficiently.

When all the lists have a minimal size, i.e. $|L_i| = 2^{\frac{n}{k}}$, efficient algorithms to the k-list problem are known only for the cases when $k = 2$, and $k \geq n$. The former is due to the collisions search algorithm, i.e. 2-list problem is equivalent to finding collisions thus it can be solved in $2^{n/2}$. The latter is due to the Bellare and Micciancio result [2] which states that such problem can be solved by Gaussian elimination in $O(n^3 + kn)$. A trivial algorithm is known for the k-list when $2 < k < n$. The algorithm first creates two larger lists $\overline{L_1}, \overline{L_2}$, where $\overline{L_1} = \{X | X = x_1 \oplus \ldots \oplus x_{k/2}, x_i \in L_i\}$, $\overline{L_2} = \{Y | Y = x_{k/2+1} \oplus \ldots \oplus x_k, x_i \in L_i\}$ and

---

[2] Joux and Lucks [20] use this technique to generate multiple collisions, which later lead to multicollisions.

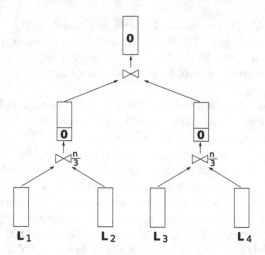

Fig. 1. Wagner's 4-tree algorithm.

subsequently it finds a collision between the two lists. The size of the lists is $2^{\frac{n}{2}}$ thus the time complexity of the algorithm is $\mathcal{O}(2^{\frac{n}{2}})$.

Wagner proposed the k-tree algorithm that solves GBP (k-list problem) faster under the assumption that the list sizes are larger. Further we describe the case when $k = 4$, refer to Fig. 1. Let us define $S \bowtie T$ as a list of elements common to both S and T, and let $low_l(x)$ stand for the l least significant bits of x. Furthermore, let us define $S \bowtie_l T$ as a set that contains all pairs from $S \times T$ that agree in their l least significant bits (the xor on the least significant bits is zero). Assume L_1, L_2, L_3, L_4 are four lists, each containing 2^l elements (l will be defined further). First we create a list L_{12} of values $x_1 \oplus x_2$, where $x_1 \in L_1, x_2 \in L_2$, such that $low_l(x_1 \oplus x_2) = 0$. Similarly, we create a list L_{34} of values $x_3 \oplus x_4$, where $x_3 \in L_3, x_4 \in L_4$, such that $low_l(x_3 \oplus x_4) = 0$. Finally, we search for a collisions between L_{12} and L_{34}. It is easy to see that such a collision reveals the required solution, i.e. $x_1 \oplus x_2 \oplus x_3 \oplus x_4 = 0$.

The main advantage of the k-tree algorithm lies in the way the solution is found – at each of the two levels, only a simple collision search algorithm is used, and only a certain number of bits is made to fulfill the final goal (the xor is zero on all bits). At the first level, the lists L_{12}, L_{34} contain words that have zeros on the l least significant bits, thus xor of any two words from the lists must have zeros on these bits. At the second level, the xor of the words from the two lists will result in zeros on the remaining $n - l$ bits, if there are enough pairs. To get the sufficient number of pairs, the value of l is defined as $l = n/3$. Then each of L_{12}, L_{34} will have $2^{n/3} \cdot 2^{n/3}/2^{n/3} = 2^{n/3}$ words, and thus at the second level there will be $2^{n/3} \cdot 2^{n/3} = 2^{2n/3}$ possible xors, one of which will have zeros on the remaining $n - n/3 = 2n/3$ bits. It is important to note that l is chosen as to balance the complexity of the two levels. Obviously, the total memory and the time complexities of the 4-tree algorithm are $\mathcal{O}(2^{n/3})$ each.

The very same idea is used to tackle any k-list problem, where k is a power of two. The only difference is in the choice of l, and in the number of levels. In general, the number of levels equals $\lg k$, and at each level except the final, additional l bits are set to zero. At the final level, the remaining $2l$ bits are zeroed. Hence, $l \cdot \lg k + l = n$, and thus $l = n/(\lg k + 1)$. The algorithm works in $\mathcal{O}(k 2^{\frac{n}{\lg k + 1}})$ time and memory and requires lists of sizes $2^{\frac{n}{\lg k + 1}}$. As an example, let us focus on 8-list problem, i.e. we have $L_1, \ldots, L_8$ lists, $\lg 8 = 3$ levels, and $l = n/4$. At the first level we build $L_{12}, L_{34}, L_{56}, L_{78}$, by combining two lists L_i, L_j, each with $2^l = 2^{n/4}$ elements that have zeros in the $n/4$ least significant bits. At the second, we build L_{1234} and L_{5678} that have again $2^{n/4}$ elements with zeros in the next $n/4$ bits. Finally, at the third level, we find the solution that will have zeros on the remaining $n/2$ bits.

Wagner's algorithm works similarly when k is not a power of two. The trick is to make some lists *passive*, i.e. to choose one element from each of the passive lists, and then continue with the algorithm as for the case of power of two and the remaining lists. For instance, to solve 6-list problem for lists $L_1, \ldots, L_6$, we take any element $v_5 \in L_5$ and $v_6 \in L_6$, and then solve the 4-list problem $x_1 \oplus x_2 \oplus x_3 \oplus x_4 = v_5 \oplus v_6$, for the lists $L_1, \ldots, L_4$. We can easily remove the non-zero condition $v_5 \oplus v_6$ in the right part, by adding this value to each element of the list L_1. Hence, the complexity of the k-list problem for the case when k is not a power of two equals the complexity to the closest (and smaller) power of two case. Thus, for any value of k, the k-tree algorithm works in $\mathcal{O}(k \cdot 2^{\frac{n}{\lfloor \lg k \rfloor + 1}})$ time and memory.

3   Improved Algorithm for the 3-List Problem

We focus on the 3-list problem and show how to improve the complexity of Wagner's 3-tree algorithm. Our improvement is based on the idea of multicollisions. The technique mimics the approach developed by Nikolić et al. [24] and further generalized by Dinur et al. [12]. We exploit the k-tree algorithm, but we also work with the passive lists and make them more active. Namely, instead of simply taking one element from the passive lists, we find in them partial multicollisions – sets of words that share the same value on particular bits. We then force the active lists on these bits to have a specific value (which is xor of all the values of the partial multicollisions), and at the final step, merge the results of the active and passive lists to obtain zero on the remaining bits. Let us take a closer look at this idea.

Definition 1. *The set of n-bit words $S = \{x_1, \ldots, x_p\}$ forms a p-partial multicollision on the s least significant bits, if $low_s(x_1) = low_s(x_2) = \ldots = low_s(x_p)$.*

This is to say that all p words are equal on the last s bits. Note, the choice to work with the least bits is not crucial but is introduced to simplify the presentation. Given an arbitrary set, we can create a p-partial multicollision from this set, i.e. we can find a subset that is p-partial multicollision. The maximal value of p depends on the size of the initial set and will be analyzed later in the section.

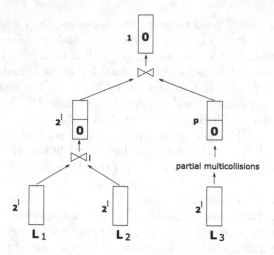

Fig. 2. Multicollision technique for $k = 3$. The values in blue denote the size of the lists.

Let us assume that we are given a 3-list problem with lists L_1, L_2, L_3, each of size 2^l. If we apply the k-tree algorithm, then l should equal $n/2$, the lists L_1, L_2 will be active, while L_3 will be passive. Instead of marking L_3 as passive, let us create a p-partial multicollision from L_3 on the l least significant bits (LSB) and denote this set as $\overline{L_3}$ (refer to Fig. 2). Without loss of generality we can assume that the colliding value of the l bits is zero (if not, we xor this value to all the elements of the list L_1). Furthermore, with the use of the join operator, from L_1, L_2 we create a list L_{12} of all values $x_1 \oplus x_2$, where $x_1 \in L_1, x_2 \in L_2$ and $low_l(x_1 \oplus x_2) = 0$; obviously $|L_{12}| \simeq 2^l$ with high probability. Finally, we use the join operator once again between L_{12} and $\overline{L_3}$, to find the required solution. As we have to cancel additional $n - l$ bits, the solution will exist with a high probability as long as $p|L_{12}| \geq 2^{n-l}$, that is, $p2^{2l-n} \geq 1$.

The complexity of our algorithm depends on the complexity of the two join operators and of producing multicollisions. The join operators (which are indeed simple collision searching algorithms) work in $\mathcal{O}(2^l)$ as in each of the cases, the sizes of the lists are not larger than 2^l. Furthermore, the partial multicollisions from $|L_3| = 2^l$ can be produced in $\mathcal{O}(2^l)$ time and memory[3]. Hence the multicollision technique solves the 3-list problem in $\mathcal{O}(2^l)$ time and memory.

Let us find the value of l. For this purpose we replace the inequality $p2^{2l-n} \geq 1$, with

$$p2^{2l-n} = 1, \tag{1}$$

---

[3] It is to initialize counters for each possible value of the colliding bits, then for each $x \in L_3$ increase the counter $low_l(x_3)$. After all elements have been processed, counter with the highest value corresponds to the largest multicollision set.

and obtain

$$l = \frac{n}{2} - \frac{1}{2} \lg p. \tag{2}$$

Therefore, the complexity of our algorithm is $\mathcal{O}(2^{\frac{n}{2}}/\sqrt{p})$, hence the speed-up factor is $\sqrt{p}$. Recall that p is the size of the multicollision set produced from the passive list L_3 – the larger the size, the greater the speed-up. Note, in the original Wagner's 3-tree algorithm, one element is chosen at random from L_3 and therefore the multicollision set consists of a single element. That is, for the classical 3-tree, $p = 1$ and the complexity is $\mathcal{O}(2^{\frac{n}{2}})$.

Let us examine the maximal possible value of p, i.e. the size of the p-partial multicollisions set on l bits produced from the set L_3 of size 2^l. Theorem 5 of [29] defines the number of elements in a set required to produce multicollision with a high probability, and by this theorem we obtain

$$(p!)^{1/p} 2^{\frac{p-1}{p} l} = 2^l. \tag{3}$$

A more straightforward way that we use to find the value of p is based on the so-called balls-into-bins problem: m balls are thrown into m bins (the bin for each ball is chosen uniformly at random), and the problem is to find the expected maximum load, i.e. the expected number of balls in a bin that contains the most balls. The solution to this problem is well known and the expected maximum load asymptotically is:

$$\Theta\left(\frac{\ln m}{\ln \ln m}\right). \tag{4}$$

Our multicollision problem is an instance of the balls-into-bins problem as the number of elements in the passive list L_3 (the number of balls), and the size of the multicollision space (the number of bins) are both 2^l. Therefore, the asymptotics of $p(l)$ can be evaluated as $\Theta(\frac{\ln 2^l}{\ln \ln 2^l}) = \Theta(\frac{l}{\ln l})$. Finally, as $l \approx \frac{n}{2}$, we obtain that the speed-up factor $\sqrt{p}$ of our improved 3-tree (over Wagner's 3-tree) is $\sqrt{\frac{n/2}{\ln n/2}}$, thus the complexity of our algorithm is

$$\mathcal{O}\left(2^{\frac{n}{2}}/\sqrt{\frac{n/2}{\ln n/2}}\right). \tag{5}$$

To find the actual speed-up for concrete values of n, we need to approximate the asymptotics of $p(l)$, i.e. need to find the approximate value of c in $p(l) = c\frac{l}{\ln l}$. For this purpose, we have run a series of experiments. For each value of $l = 10, \ldots, 28$, we have generated 2^l random values (of bit length l) and have checked the maximal number of multicollisions. For each l, the experiments have been repeated 20 times. The outcomes of the experiments are reported in Table 1.

Based on the experiments, the value of c can be approximated as $c \approx 1.3$. With such an assumption, we have computed the speed-up factor of our improved 3-tree for various values of n – we refer the reader to Table 2.

Table 1. Experimental search of number of multicollisions.

l	Average size	$\frac{l}{\ln l}$	c
10	5.80	4.34	1.34
11	5.85	4.59	1.27
12	6.10	4.83	1.26
13	6.45	5.07	1.27
14	7.00	5.30	1.32
15	7.15	5.54	1.29
16	7.55	5.77	1.31
17	7.90	6.00	1.32
18	8.15	6.23	1.31
19	8.50	6.45	1.32
20	8.70	6.68	1.30
21	9.05	6.89	1.31
22	9.50	7.12	1.33
23	9.65	7.34	1.31
24	10.30	7.55	1.36
25	10.40	7.77	1.34
26	10.60	7.98	1.33
27	11.05	8.19	1.35
28	11.15	8.40	1.33

Table 2. A comparison of the time complexities of Wagner's 3-tree with our new approach.

n	Speed-up ($\sqrt{p}$)	l
64	3.43	31
128	4.42	62
256	5.82	126
512	7.71	253

The above strategy is in line with the multicollision approach by Nikolić et al. used in the analysis of the lightweight block cipher LED [14]. The advanced approach by Dinur et al., however, cannot be used for further improvements. One of their main ideas is to work simultaneously with a few multicollisions, instead of only one. In the case of the k-tree algorithm, this would mean to produce from L_3 several p-partial multicollision sets. However, each such set will collide on s different value of the l LSBs, i.e. the elements of the first p-partial multicollision set will have the value v_1 on the l LSB, the elements of the second set will have the value v_2, etc. The different values will increase the complexity of the later stage of k-tree by a factor of s. When using the join operator on l bits of L_1 and L_2 there will be s targets (whereas previously we had only one), thus a simple collision search will have to be repeated s times. Therefore, in this particular case, Dinur et al. approach cannot be used.

Improvements for $k > 3$. Our technique can be applied as well to improve the k-tree algorithm for larger (and non-power of two) values of k. Again, we will start with the classical k-tree and assume that all the lists are of size 2^l (where $l < \frac{n}{\lg k + 1}$). Given k that is not a degree of 2, the number of active lists k_A is $2^{\lfloor \lg k \rfloor}$ and the number of passive lists k_P is $k - k_A$. For instance, for $k = 7$, it means that $k_A = 4, k_P = 3$ (see Fig. 3). Without loss of generality, assume that the first k_A lists are active, and the remaining lists are passive. First, we produce p-partial multicollision sets on $\lambda = l \cdot \lg k_A$ bits, independently for all of the k_P passive lists, and obtain $\overline{L_{k_A+1}}, \dots, \overline{L_k}$. Let $v_1, \dots, v_{k_P}$ be the common values of these sets, and $v = v_1 \oplus \dots \oplus v_{k_P}$. Obviously the set $L_P = \overline{L_{k_A+1}} \bowtie_l \dots \bowtie_l \overline{L_k}$ has cardinality $|L_P| = p^{k_P}$ and all the elements of the set have the value v on the λ LSBs. For the sake of simplicity, assume $v = 0$. Next, focus on the active lists and find 2^l solutions for k_A-problem on the same $\lambda = l \cdot \lg k_A$ bits by running the k-tree with initial lists of sizes 2^l. Note, this way at level $\lg k_A$ there will be one list with 2^l elements that have zeros on λ LSBs. If the number of solutions produced from the two independent steps satisfy $p^{k_P} \cdot 2^l \geq 2^{n-\lambda}$, then one of the elements of the list found by the k_A-tree algorithm can be matched with one of the elements of L_P, on the remaining $n - \lambda$ bits. As a result we will obtain one solution to the original k-list problem.

Let us focus on the complexity of the algorithm. The p-partial multicollisions produced from the passive lists $L_i, i = k_A + 1, \dots, k$, require around $k_P \cdot |L_i| = k_P \cdot 2^l$ operations. Under the assumption that k_P is small, the additional $k_P + p^{k_P}$ operations spent on producing v and L_P can be ignored as the whole complexity

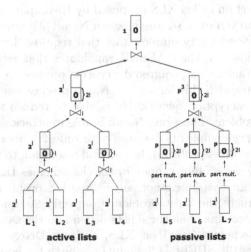

Fig. 3. Multicollision technique for $k = 7$. The values in blue denote the size of the lists.

will be dominated by the multicollisions. On the other hand, the production of 2^l elements with the k_A-tree requires around $k_A \cdot 2^{\frac{\lambda}{\lg k_A}} = k_A \cdot 2^l$ operations. As a result, the total complexity of the algorithm is around $k_P \cdot 2^l + k_A \cdot 2^l = (k_P + k_A) \cdot 2^l = k \cdot 2^l$. Let us find the value of l. For this reason we equate $p^{k_P} \cdot 2^l$ to $2^{n-\lambda}$ (specified in the inequality above), and obtain $k_P \lg p + l = n - l \cdot \lg k_A$, or equivalently

$$l = \frac{n}{\lg k_A + 1} - \frac{k_P \lg p}{\lg k_A + 1}. \tag{6}$$

Therefore the improved k-tree outperforms the classical k-tree by a factor of

$$\frac{k 2^{\frac{n}{\lg k_A + 1}}}{k 2^l} = 2^{\frac{k_P \lg p}{\lg k_A + 1}} = (2^{\lg p})^{\frac{k_P}{\lg k_A + 1}} = p^{\frac{k_P}{\lg k_A + 1}}. \tag{7}$$

The value of p can be approximated as follows. First note that we can no longer use the balls-into-bins problem, as the size of the lists (i.e. 2^l) not necessarily equals the size of the multicollision space (e.g., when $k = 7$, the space has 2^{2l} elements). Therefore, we use (3), to approximate the number of multicollisions. From (3), with a simple transformation we obtain that $\frac{l}{p} = \lg \frac{p}{e}$. The approximate solution of this equation is of the form $p = \frac{l}{\lg \frac{l}{e}}$. Therefore, the speed up factor of our improved k-tree algorithm can be evaluated as

$$\left(\frac{l}{\lg \frac{l}{e}}\right)^{\frac{k_P}{\lg k_A + 1}} \approx \frac{a \cdot n^c}{\lg(b \cdot n)}, \tag{8}$$

where a, b, c are constants that depend on the values of k_A and k_P.

Applications. The improvement of the 3-tree algorithm can be used for cryptanalysis of authenticated encryption schemes proposed to the ongoing CAESAR [3]. Some of these schemes, to process the final incomplete blocks of messages, use a construction called XLS proposed by Ristenpart and Rogaway [27]. Initially, XLS was proven to be secure, however Nandi [22] points out flaws in the security proof and shows a very simple attack that requires three queries to break the construction. However, the CAESAR candidates that rely on XLS, do not allow this trivial attack as the required decryption queries are not permitted by the schemes. To cope with this limitation, Nandi proposes another attack [23], that requires only encryption queries. He is able to reduce the design flaw of XLS to the 3-list problem. Therefore, Nandi is able to attack schemes that claim birthday bound *query* complexity because with only $2^{\frac{n}{3}}$ queries (equivalent to size of the lists in the 3-list problem), he can find a solution to the 3-list problem (in $2^{\frac{2n}{3}}$ time). However, Nandi cannot break the schemes that claim birthday bound *time* complexity, as he cannot solve the 3-list problem faster than $2^{\frac{n}{2}}$. Note, Nandi constructs the 3-list problem from only $2^{\frac{n}{3}}$ queries, rather than from $3 \cdot 2^{\frac{n}{3}}$, as the elements of all three lists depend on the same $2^{\frac{n}{3}}$ ciphertexts.

The CAESAR schemes based on XLS, such as Deoxys [16], Joltik [17], KIASU [18], Marble [13], SHELL [32], claim only birthday bound *time* complexity, thus Nandi's findings do not break the security claims of these candidates.

However, our improved 3-tree algorithm goes below the birthday bound and thus can be used to show a slight weakness in some of these candidates.

Let us focus on the 128-bit CAESAR candidates Deoxys and KIASU. The 3-list problem for XLS in these candidates has the parameter $n = 128$. According to Table 2, we can take $\sqrt{p} = 4.42$ and $l = 62$. Consequently, the complexity of a forgery based on the XLS weakness is $C \cdot 2^{62}$, where C is a constant factor. The value of C is 1 because: 1) as mentioned above, the 3 lists can be produced from the same 2^{62} ciphertexts, and 2) all of the operations required by the improved 3-tree algorithm are significantly less expensive than one encryption of the analyzed schemes. As a result, we obtain a forgery on the COPA modes of Deoxys and KIASU in 2^{62} encryptions and thus the security level of these schemes is reduced by 2 bits from the claimed 64 bits.

4 Improved Time-Memory Tradeoffs

In applications, usually the elements of the lists L_i are in fact outputs of functions f_i, thus GBP is often formulated as:

Problem 2. *Given non-invertible functions $f_1, \ldots, f_k : \{0,1\}^{n'} \to \{0,1\}^n$, $n' \geq n$, find $y_1, \ldots, y_k \in \{0,1\}^{n'}$ such that $f_1(y_1) \oplus f_2(y_2) \oplus \ldots \oplus f_k(y_k) = 0$.*

In some applications, all the functions are identical, and the problem is to find distinct inputs:

Problem 3. *Given a non-invertible function $f : \{0,1\}^{n'} \to \{0,1\}^n$, $n' \geq n$, find distinct $y_1, \ldots, y_k \in \{0,1\}^{n'}$ such that $f(y_1) \oplus f(y_2) \oplus \ldots \oplus f(y_k) = 0$.*

Both of the above problems give rise to the possibility of time-memory trade-offs, i.e., reducing the memory complexity of the k-tree algorithm at the expense of time. We will investigate time-memory tradeoffs for the GBP as defined in Problem 3. Recall that k-tree in its current form assumes that both time and memory are of equal magnitude, i.e. $T = M = O(k \cdot 2^{\frac{n}{\lg k + 1}})$.

Bernstein et al. [5,6] investigate k-tree in memory restricted environments and propose a few tradeoffs. Their main approach is to solve the k-list problem on less than n bits. Assume $M = 2^m$, where $M < 2^{\frac{n}{\lg k + 1}}$. Then, a k-list problem on $\bar{n} = m(\lg k + 1)$ bits (instead of n bits) can easily be solved with the k-tree algorithm. The first tradeoff idea is to perform a precomputation (or prefiltration) such that all the entries in each list have the value of 0 in the $n - \bar{n}$ most significant bits.[4] For the remaining $\bar{n}$ least significant bits, they apply the k-tree algorithm and thus find a solution for all n bits. The time complexity is the sum of the cost for precomputation and for solving the k-tree algorithm, which is $k \cdot (2^{n-\bar{n}} \cdot 2^m + 2^m) \approx k \cdot 2^{n-m \lg k}$. The tradeoff is therefore defined as

$$T \cdot M^{\lg k} = k \cdot N. \tag{9}$$

---

[4] It is pointed out in [6] that $n - \bar{n}$ bits can have an arbitrary value as long as the sum of all lists is zero. The technique is called *clamping through precomputations*.

Their second idea is similar but does not use precomputation. They apply the k-tree algorithm on $\bar{n} = m(\lg k + 1)$ bits until the value of the remaining $n - \bar{n}$ bits probabilistically becomes zero. Obviously, in total there will be $2^{n-\bar{n}}$ repetition of the k-tree, thus the time complexity becomes $T = k \cdot 2^m \cdot 2^{n-\bar{n}} = k \cdot 2^{n-m\lg k}$, which provides the same tradeoff as the previous one, i.e.,

$$T \cdot M^{\lg k} = k \cdot N. \tag{10}$$

The third idea also relies on reduction of n, but the technique is more advanced. Assume, $f_1 = f_2, f_3 = f_4, \ldots$, i.e. the functions are pairwise identical. The k-list problem is regarded as two separate $\frac{k}{2}$ problems, the first involving the functions $f_1, f_3, \ldots$, while the second $f_2, f_4, \ldots$. If the amount of available memory is 2^m, then it is possible to solve each of these $\frac{k}{2}$-list problems on up to $\bar{n} = m(\lg \frac{k}{2}+1) = m \cdot \lg k$ bits. By elevating the two $\frac{k}{2}$-lists to k-list, the remaining $n - \bar{n}$ bits can be zeroed with the use of memoryless collision search algorithm. Therefore the time complexity is $T = \frac{k}{2} \cdot 2^{\frac{n-\bar{n}}{2}} \cdot 2^m = \frac{k}{2} \cdot 2^{\frac{n}{2}-m(\frac{\lg k}{2}-1)}$ and their tradeoff curve is defined as

$$T \cdot M^{\frac{\lg k}{2}-1} = \frac{k}{2} \cdot N^{\frac{1}{2}},$$

which is converted to

$$T^2 \cdot M^{\lg k-2} = \frac{k^2}{4} \cdot N, \tag{11}$$

Because this method solves $\frac{k}{2}$-list problem, it is meaningful when $k > 4$. We note that when $M < 2^{\frac{n}{\lg k+2}}$, then (11) provides better tradeoff while for $M > 2^{\frac{n}{\lg k+2}}$, (9) is better.

The k-tree relies on producing multiple collisions. For instance, at the first level of 4-tree, $2^{\frac{n}{3}}$ colliding pairs on $\frac{n}{3}$ bits are produced. Producing these pairs is trivial when the amount of available memory is $2^{\frac{n}{3}}$. However, once the memory is reduced to $2^m, m < \frac{n}{3}$, the trivial collision search does not work.

The fact that k-tree requires multiple collisions, opens doors to the following technique based on Hellman's tables [15][5].

Fact 1. (Hellman's table) *Let $f : \{0,1\}^* \to \{0,1\}^n$ be an arbitrary-size input and n-bit output function, $N = 2^n$, and let $M = 2^m$ be the amount of available memory. Once the precomputation equivalent to MX evaluations of f is performed, the cost of generating new collisions for f is $\frac{N}{MX}$ per collision.*

The technique works as follows. Choose M distinct values $v_i^0 \in \{0,1\}^n$, where $i = 1, 2, \ldots, M$. For each of them, compute chains of length X with the target function f, i.e. compute $v_i^j \leftarrow f(v_i^{j-1})$ for $i = 1, \ldots, M, j = 1, \ldots, X$, and store only the first and last values of each chain, i.e. (v_i^0, v_i^X), in a precomputation

---

[5] Note, we could not exploit the more advanced Rivest's distinguished points and Oechslin's rainbow tables [25] to improve the analysis.

Fig. 4. Hellman's table T_{pre} when M memory is available.

table T_{pre}. The construction of T_{pre} is depicted in Fig. 4. Note, even though MX values exist in all the chains, only $2M$ values are stored in T_{pre}. Once T_{pre} is constructed, to generate a collision, start with a random point and construct a chain of length $\frac{N}{MX}$. As there are N possible values, and MX are in T_{pre}, one point of the new chain will collide with one point of the chains created during the construction of the table. The match can be detected by further extending the new chain at most X times, as eventually it will reach one of v_i^X stored in T_{pre}. Then, the exact colliding values can be detected by recalculating chains from v_i^0 and the starting value of the new chain. Obviously T_{pre} can be reused to find not only one, but multiple collisions.

Joux and Lucks [20] use this technique to produce 3-collisions. They set $M = X = 2^{\frac{n}{3}}$ to generate $2^{\frac{n}{3}}$ ordinary collisions with time $T = 2^{\frac{2n}{3}}$ and memory $M = 2^{\frac{n}{3}}$. Then, they find another collision between $2^{\frac{n}{3}}$ ordinary collisions and $2^{\frac{2n}{3}}$ single values. When they generate $2^{\frac{n}{3}}$ ordinary multiple collisions, Hellman's table has an important role to keep the memory M rather than MX.

Further, we will use Hellman's table to produce multiple collisions for the first level of k-tree, but only on certain l bits (where $l < n$).

4.1   Improved Time-Memory Tradeoffs for the 4-List Problem

We present a more efficient time-memory tradeoff for GBP. Our tradeoff curve depends on the number of available lists, which is parameterized by k. For a better understanding, first we explain our algorithm for $k = 4$.

The original 4-tree algorithm consists of two-level collision searches (the parameter l used below will be determined later).

Level 1. Construct two lists, L_{12} and L_{34}, each containing $2^{\frac{n-l}{2}}$ partial collisions on l bits.

Level 2. Find a collision between the elements of L_{12} and L_{34} on the remaining $n - l$ bits.

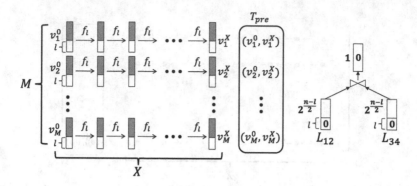

Fig. 5. Improved time-memory tradeoff for the 4-list problem with Hellman's table.

Our new 4-tree algorithm works similarly with the exception of Level 1. At this level, we first construct Hellman's table, and then we use it to find $2^{\frac{n-l}{2}}$ collisions. As a result, our algorithm decomposes Level 1 into two parts. Its complexity depends on the available memory M which in turn determines the length of the chains X. The updated 4-tree is illustrated in Fig. 5 and is specified as follows.

Level 1a. Construct Hellman's table containing M chains, each of length of X.
Level 1b. With the use of Hellman's table, find $2 \cdot 2^{\frac{n-l}{2}}$ partial collisions on l
   bits. Store a half ($2^{\frac{n-l}{2}}$) of them in a list L_{12} and the other half in L_{34}.
Level 2. Find a collision between the elements of L_{12} and L_{34} on the remaining
   $n - l$ bits.

Construction of Hellman's Table. For the Level 1a our algorithm first constructs Hellman's table which contains M chains of length X. However, unlike in [20], we have the following technical obstacle. The function f takes an n-bit input and produces an n-bit output and thus for such a function only full n-bit collisions can be identified. In other words, the classical Hellman's table cannot be used to find partial collisions.

To solve this problem, we define a reduction function $f_l : \{0,1\}^l \rightarrow \{0,1\}^l$ so that only the l bits are meaningful in the chain. For generating chains with f_l, $n - l$ bits of 0's are concatenated to the l-bit input, and this value is processed with $f : \{0,1\}^n \rightarrow \{0,1\}^n$. Finally, the n-bit output is truncated to l bits, and is used as the input to the next chain. That is, $f_l(x) = Trunc_l\big(f(0^{n-l}\|x)\big)$, where $Trunc_l(\cdot)$ truncates the n-bit input to the l least significant bits.

To summarize, we choose M distinct l-bit values v_i^0 for $i = 1, 2, \ldots, M$, for each of them generate a chain of length X by computing $v_i^{j+1} = f_l(v_i^j)$ where $j = 1, 2, \ldots, X$. In total, MX values are in all the chains and only the first and the last points of each chain are stored in T_{pre}. Thus Hellman's table requires around MX time and M memory.

Generation of l-bit Collisions. According to Fact 1, once Hellman's table T_{pre} is constructed, the complexity for producing l-bit collisions is reduced significantly. Considering that the size of the values in the chains is l bits and the length of each chain is X, Fact 1 shows that the cost is $\frac{2^l}{MX}$ per collision.

To generate an l-bit collision, we choose a random l-bit value and with the function f_l from it compute a chain of length $\frac{2^l}{MX} + X$. On average, one collision will occur before we reach the $\frac{2^l}{MX}$ value of this new chain against the MX values stored in T_{pre}. The computation of additional X values in the chain ensures that the corresponding v_i^X will appear as one of the ending points of T_{pre}. The exact colliding pairs are detected by recomputing the chains from v_i^0 and the chosen l-bit value.

From the definition of f_l, the two inputs colliding on f always have the form $(0^{n-l}\|l_1, 0^{n-l}\|l_2)$, where 0^{n-l} is a sequence of $n - l$ zero bits and l_1 and l_2 are some l-bit values. A collision of the two chains means that $Trunc_l(f(0^{n-l}\|l_1)) = Trunc_l(f(0^{n-l}\|l_2))$. Therefore, $f(0^{n-l}\|l_1)$ and $f(0^{n-l}\|l_2)$ only collide in the least significant l bits, while on the remaining $n - l$ bits behave randomly.

The collision generation process is iterated $2^{\frac{n-l}{2}}$ times and the input and output of each pair is stored in L_{12}. Similarly, the process is iterated additional $2^{\frac{n-l}{2}}$ times and the results are stored in L_{34}. Therefore the complexity of this step is around $2 \cdot 2^{\frac{n-l}{2}} \cdot \frac{2^l}{MX} = 2 \cdot \frac{N^{\frac{1}{2}} 2^{\frac{l}{2}}}{MX}$ time and $2 \cdot 2^{\frac{n-l}{2}} = 2 \cdot \frac{N^{\frac{1}{2}}}{2^{\frac{l}{2}}}$ memory.

Finding a Solution to the 4-list Problem. From the two lists L_{12} and L_{34} containing $2^{\frac{n-l}{2}}$ partial collisions on l bits, we find a collision on the remaining $n - l$ bits. This procedure is straightforward and it requires $2^{\frac{n-l}{2}} = \frac{N^{\frac{1}{2}}}{2^{\frac{l}{2}}}$ time and no additional memory.

Parameters and the Tradeoff. The complexities for each step are as follows:

Level 1a.	Time $= MX$,	Memory $= M$
Level 1b.	Time $= 2 \cdot \dfrac{N^{\frac{1}{2}} 2^{\frac{l}{2}}}{MX}$,	Memory $= 2 \cdot \dfrac{N^{\frac{1}{2}}}{2^{\frac{l}{2}}}$
Level 2.	Time $= \dfrac{N^{\frac{1}{2}}}{2^{\frac{l}{2}}}$,	Memory $=$ negligible

To balance the memory at Level 1a and Level 1b, M, N, l should satisfy the relation $M = 2 \cdot \frac{N^{\frac{1}{2}}}{2^{\frac{l}{2}}}$. From this relation, the time complexity of Level 2 becomes $\frac{M}{2}$, and thus is negligible compared to Level 1a when X is sufficiently large. To balance the time complexities of Level 1a and Level 1b, we need $MX = 2 \cdot \frac{N^{\frac{1}{2}} 2^{\frac{l}{2}}}{MX}$, which gives the relation $M^3 X^2 = 4 \cdot N$. Finally, as the time complexity T satisfies $T = MX$, we obtain the following tradeoff curve

$$T^2 \cdot M = 4 \cdot N. \tag{12}$$

For instance, when the available memory is $2^{\frac{n}{4}}$ (instead of $2^{\frac{n}{3}}$ as in the original 4-tree), then the updated 4-tree finds a solution in around $2^{\frac{3n}{8}}$ time. This is to be compared to Bernstein et al. tradeoffs given in (9) and (10) which require around $2^{\frac{n}{2}}$ time. Additional points of the tradeoff curve and comparison to previous results are given in Table 4.

During the analysis, we relied implicitly on several facts. First, we assumed that Hellman's table can contain an arbitrary number of points. In order to avoid collisions between the chains, however, the values of M and X cannot be arbitrary, but should depend on l. That is, during the construction of Hellman's table, the number of chains and their length is bounded by the value of l. Biryukov and Shamir in [7] call this a matrix stopping rule, and define it as $MX^2 \le 2^l$. It is trivial to see that this inequality holds in our case as $MX^2 = M\frac{4N}{M^3} = \frac{4N}{M^2} = \frac{4N}{(2N^{\frac{1}{2}}/2^{\frac{l}{2}})^2} = 2^l$. For instance, when $M = 2^{\frac{n}{4}}$, then $l \approx \frac{n}{2}$, $T = 2^{\frac{3n}{8}}$, $X = 2^{\frac{n}{8}}$. Therefore, obviously $MX^2 = 2^{\frac{n}{2}} = 2^l$. We assumed as well that the tradeoff applies only to Problem 3. However, a close inspections of our algorithm reveals that it can be applied to the case of pairwise identical functions, i.e., $f_1 = f_2, f_3 = f_4$. That is, the area of application of the tradeoff is wider, and is similar to the area of the tradeoff given by Bernstein et al. in (11). To deal with the extended case, we have to create two Hellman's tables at the initial stage, one for each pair of functions. Thus the time and memory complexities will increase by a factor of two at Level 1a, and will stay the same at Levels 1b and 2.

4.2 Improved Time-Memory Tradeoff for the k-list Problem

In this section, we generalize the time-memory tradeoff for the k-tree algorithm, where $k = 2^d$. Overall, we replace the collision generation at Level 1 of the k-tree algorithm with a generation based on Hellman's table. Hereafter, we call the bits whose sum is fixed to zero *clamped bits*.

The ordinary k-tree algorithm initially starts from 2^d lists containing $M = 2^m$ elements. At Level 1, 2^{d-1} lists containing M elements are generated with m bits clamped. At Level i for $i = 2, 3, \ldots, d-1$, 2^{d-i} lists containing M elements are generated with im bits clamped. At the last Level d there are two lists containing M elements with $(d-1)m$ bits clamped. As no longer M collisions are required, but rather only one, the sum on up to $(d+1)m$ bits can be 0, by setting $(d+1)m = n$, and thus the k-tree algorithm will find the solution to the k-list problem. However, if the memory size is restricted, i.e. $m \ll \frac{n}{d+1}$, the k-tree algorithm can enforce the sum of only $(d+1)m$ bits to zero.

Our algorithm replaces Level 1 with Hellman's table collision generation and performs the same procedure as the k-tree algorithm from Level 2 to Level d. To find the required solution after Level d, however, at Level 1 we clamp more

Table 3. A comparison of the number of clamping bits between the k-tree and our algorithm.

#lists		#Clamped bits	
		k-tree algorithm	Our algorithm
Level 1	2^{d-1}	m	l
Level i, $(i = 2, \ldots, d-1)$	2^{d-i}	im	$l + (i-1)m$
Level d	1	$(d+1)m$	$l + dm$

bits. Let the number of the clamped bits at Level 1 be l. After the first level we will have 2^{d-1} lists, each with $M = 2^m$ elements. Similarly, after Level i for $i = 2, 3, \ldots, d-1$, we will have 2^{d-i} lists containing M elements with $l + (i-1)m$ bits clamped. After the final Level d, we will have one element with $l + dm$ zero bits. Therefore, we set $l + dm = n$, i.e. $l = n - dm$, to get at least one solution on all n bits. In Table 3, we compare the number of clamped bits of the k-tree and our algorithm.

From the condition $l = n - dm$ and the parameters k and m, we can determine the reduction function f_l for Hellman's table. We create M chains of length X, and only store the first and last values of the chains in Hellman's table T_{pre}. Once T_{pre} is constructed, we can find an l-bit partial collision with a cost of $\frac{2^l}{MX}$ per a collision, which is equivalent to $\frac{N}{M^{d+1}X}$. At Level 1, we produce in total $(2^{d-1} \cdot M)$ l-bit collisions, and store them in 2^{d-1} lists each with M elements. The total cost for producing the partial collisions and thus the complexity of Level 1 is $2^{d-1} \cdot \frac{N}{M^d X}$.

Complexity Evaluation and the Tradeoff Curve. The complexity to generate T_{pre} is MX time and M memory. As mentioned above, Level 1 requires $2^{d-1} \cdot \frac{N}{M^d X}$ time and $2^{d-1} \cdot M$ memory. The time and memory complexities of the remaining Levels 2 to d are all M, thus negligibly small compared to the generation of T_{pre}. We balance the time complexity of Hellman's table generation and of Level 1, which gives the relation $T = MX = 2^{d-1} \cdot \frac{N}{M^d X}$, and can further be reduced to $(MX)^2 = 2^{d-1} \cdot \frac{N}{M^{d-1}}$ and approximately results in a tradeoff curve

$$T^2 \cdot M^{\lg k - 1} = k \cdot N \tag{13}$$

Note, the tradeoff given in Sect. 4.1 can be obtained from the above tradeoff by setting $k = 4$. In Table 4, we compare the previous tradeoffs given in (9), (11) to our new tradeoff for $k = 4, 8$ and for two particular memory amounts. Obviously, the time complexity of our algorithm is significantly smaller for the same amount of available memory.

The tradeoff curves of these three methods are also depicted in Fig. 6. The vertical axis and horizontal axis represent the logarithm of the time complexity t and memory complexity m, respectively. Curves for $k = 8$ and $k = 16$ are drawn in Fig. 6 with red lines and black lines, respectively. For $k = 8$ with $m \geq \frac{n}{4}$, the

Table 4. Comparison of tradeoffs. For simplicity, the constant multiplication for N is ignored.

Method		M	T	Other parameters
$k=4$	Bernstein et al. Eq.(9)	$2^{\frac{n}{4}}$	$2^{\frac{n}{2}}$	–
	$(T \cdot M^2 = N)$	$2^{\frac{n}{6}}$	$2^{\frac{2n}{3}}$	–
	Our	$2^{\frac{n}{4}}$	$2^{\frac{3n}{8}}$	$X = 2^{\frac{n}{8}}, l = \frac{n}{2}$
	$(T^2 \cdot M = N)$	$2^{\frac{n}{6}}$	$2^{\frac{5n}{12}}$	$X = 2^{\frac{n}{4}}, l = \frac{2n}{3}$
$k=8$	Bernstein et al. Eq.(9)	$2^{\frac{n}{5}}$	$2^{\frac{2n}{5}}$	–
	$(T \cdot M^3 = N)$	$2^{\frac{n}{6}}$	$2^{\frac{n}{2}}$	–
	Bernstein et al. Eq.(11)	$2^{\frac{n}{5}}$	$2^{\frac{2n}{5}}$	–
	$(T^2 \cdot M = N)$	$2^{\frac{n}{6}}$	$2^{\frac{5n}{12}}$	–
	Our	$2^{\frac{n}{5}}$	$2^{\frac{3n}{10}}$	$X = 2^{\frac{n}{10}}, l = \frac{2n}{5}$
	$(T^2 \cdot M^2 = N)$	$2^{\frac{n}{6}}$	$2^{\frac{n}{3}}$	$X = 2^{\frac{n}{6}}, l = \frac{n}{2}$

ordinary k-tree algorithm with $t = \frac{n}{4}$ can be performed. Thus, the time-memory tradeoffs are meaningful only when the memory amount is limited to $m < \frac{n}{4}$, and Fig. 6 only describes the curves in this range. Similarly, for $k = 16$ only $m < \frac{n}{5}$ is shown in the figure.

The previous curve given in (9) achieves the same time complexity as the k-tree algorithm when sufficient memory is available, while the time complexity is about 2^n when the available amount of memory is very limited. The previous curve given in (11) cannot reach the time complexity of the k-tree algorithm even if sufficient memory is available, while the time complexity is at most $2^{\frac{n}{2}}$ for very limited amount of memory. It is easy to see that our tradeoff takes advantages

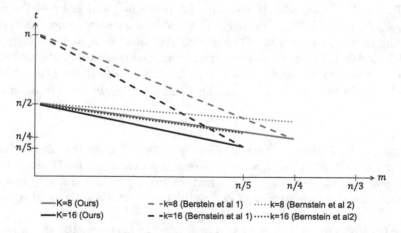

Fig. 6. Comparison of tradeoff curves. Our curve for $k = 8$ and Bernstein et al. 2 for $k = 16$ are overlapped in the range of $m < n/5$.

of those two curves, i.e. it requires the same complexity as the k-tree algorithm when sufficient memory is available and requires only $2^{\frac{n}{2}}$ time complexity when the available amount of memory is limited. Therefore, our tradeoff always allows a lower time complexity than both of the previous tradeoffs. It improves the time complexity and simplifies the situation, as it is the best for any value of m (unlike the previous two tradeoffs that outperformed each other for different values of m).

5   Conclusion

We have shown improvements to Wagner's k-tree algorithm for the case when k is not a power of two, and when the available memory is restricted. For the former case, our findings indicate that the passive lists can be used to reduce the complexity of the k-tree (in the case of 3-tree, by a factor of $\sqrt{\frac{n/2}{\ln n/2}}$). Rather than discarding the passive lists, we have produced multicollisions sets from them, and later, we have used the sets to decrease the size and thus the complexity of the k-tree algorithm. In the case of a memory restricted k-list problem, we have provided a new time-memory tradeoff based on the idea of Hellman's table. The precomputed table has allowed us to efficiently produce a large number of collisions at the very first level of the k-tree algorithm, and thus to reduce the memory requirement of the whole algorithm. As a result, we have achieved an improved tradeoff that follows the curve $T^2 M^{\lg k - 1} = k \cdot N$.

We point out that we have run series of experiments to confirm parts of the analysis. In particular, we have verified that the predicted number of multicollisions and we have completely implemented the tradeoff for $k = 4, n = 60$ and various sizes of available memory, e.g., $m = 8, 10, 14$. The outcome of the experiments has confirmed the tradeoff.

The 3-list problem appears frequently in the literature and as our improved 3-tree algorithm is the first that solves this problem with below the birthday bound complexity, we expect future applications of the algorithm. However, although our improved 3-tree asymptotically outperforms Wagner's 3-tree algorithm, the speed up factor is lower for smaller values of n. Thus we urge careful analysis when applying the improved 3-tree.

Bernstein [5] argues that the large memory requirement of Wagner's k-tree algorithm makes it impractical. He assumes that the memory access is far more expansive, thus the actual cost of the algorithm is miscalculated. He introduces tradeoffs (discussed in Sect. 4) to reduce the memory requirement, and to obtain algorithms of lower complexity (measured by the new metric). We note that as our tradeoffs are more memory effective, by the new metric they lead to better algorithms for the k-tree problem with pairwise identical functions.

There are several future research directions. One is to consider restrictions on the amount of available data. The functions f_i in the k-list problem are often assumed to be public, i.e. the attacker can evaluate them offline. When f_i are not public, the data needs to be collected by making online queries. Thus developing new time-memory-data tradeoffs for this scenario is an interesting open problem.

Another direction is to consider the weight of each function in the total cost of the algorithm, which leads to the case of an unbalanced GBP. This is based on the fact that in specific applications, it may occur that some of the functions are more costly to compute than other functions. The algorithm that solves an unbalanced GBP will be different than the one for the balanced GBP.

Acknowledgements. The authors would like to thank the anonymous reviewers of ASIACRYPT'15 for very helpful comments and suggestions (in particular, for pointing out the balls-into-bins problem). Ivica Nikolić is supported by the Singapore National Research Foundation Fellowship 2012 (NRF-NRFF2012-06).

References

1. Ajtai, M., Kumar, R., Sivakumar, D.: A sieve algorithm for the shortest lattice vector problem. In: Vitter, J.S., Spirakis, P.G., Yannakakis, M. (eds.) Proceedings on 33rd Annual ACM Symposium on Theory of Computing, 6–8 July 2001, pp. 601–610. ACM, Heraklion (2001)
2. Bellare, M., Micciancio, D.: A new paradigm for collision-free hashing: incrementality at reduced cost. In: Fumy, W. (ed.) EUROCRYPT 1997. LNCS, vol. 1233, pp. 163–192. Springer, Heidelberg (1997)
3. Bernstein, D.: CAESAR Competition (2013). http://competitions.cr.yp.to/caesar.html
4. Bernstein, D.J.: Enumerating solutions to $p(a) + q(b) = r(c) + s(d)$. Math. Comput. **70**(233), 389–394 (2001)
5. Bernstein, D.J.: Better price-performance ratios for generalized birthday attacks. In: Workshop Record of SHARCS 2007: Special-purpose Hardware for Attacking Cryptographic Systems (2007). http://cr.yp.to/rumba20/genbday-20070719.pdf
6. Bernstein, D.J., Lange, T., Niederhagen, R., Peters, C., Schwabe, P.: FSBday. In: Roy, B., Sendrier, N. (eds.) INDOCRYPT 2009. LNCS, vol. 5922, pp. 18–38. Springer, Heidelberg (2009)
7. Biryukov, A., Shamir, A.: Cryptanalytic time/memory/data tradeoffs for stream ciphers. In: Okamoto, T. (ed.) [26], vol. 1976, pp. 1–13. Springer, Heidelberg (2000)
8. Bleichenbacher, D.: On the generation of DSA one-time keys. In: The 6th Workshop on Elliptic Curve Cryptography (ECC 2002) (2002)
9. Blum, A., Kalai, A., Wasserman, H.: Noise-tolerant learning, the parity problem, and the statistical query model. In: Yao, F.F., Luks, E.M. (eds.) Proceedings of the Thirty-Second Annual ACM Symposium on Theory of Computing, 21–23 May 2000, pp. 435–440. ACM, Portland (2000)
10. Boneh, D., Joux, A., Nguyen, P.Q.: Why textbook ElGamal and RSA encryption are insecure. In: Okamoto, T. (ed.) [26], pp. 30–43 (2000)
11. Chose, P., Joux, A., Mitton, M.: Fast correlation attacks: an algorithmic point of view. In: Knudsen, L.R. (ed.) EUROCRYPT 2002. LNCS, vol. 2332, p. 209. Springer, Heidelberg (2002)
12. Dinur, I., Dunkelman, O., Keller, N., Shamir, A.: Key recovery attacks on 3-round Even-Mansour, 8-step LED-128, and Full AES2. In: Sako, K., Sarkar, P. (eds.) ASIACRYPT 2013, Part I. LNCS, vol. 8269, pp. 337–356. Springer, Heidelberg (2013)
13. Guo, J.: Marble v1. Submitted to CAESAR (2014)

14. Guo, J., Peyrin, T., Poschmann, A., Robshaw, M.: The LED block cipher. In: Preneel, B., Takagi, T. (eds.) CHES 2011. LNCS, vol. 6917, pp. 326–341. Springer, Heidelberg (2011)
15. Hellman, M.E.: A cryptanalytic time-memory trade-off. IEEE Trans. Inf. Theory **26**(4), 401–406 (1980)
16. Jean, J., Nikolić, I., Peyrin, T.: Deoxys v1. Submitted to CAESAR (2014)
17. Jean, J., Nikolić, I., Peyrin, T.: Joltik v1. Submitted to CAESAR (2014)
18. Jean, J., Nikolić, I., Peyrin, T.: KIASU v1. Submitted to CAESAR (2014)
19. Joux, A., Lercier, R.: "Chinese and Match", an alternative to Atkin "Match and Sort" method used in the sea algorithm. Math. Comput. **70**(234), 827–836 (2001)
20. Joux, A., Lucks, S.: Improved generic algorithms for 3-collisions. In: Matsui, M. (ed.) ASIACRYPT 2009. LNCS, vol. 5912, pp. 347–363. Springer, Heidelberg (2009)
21. Minder, L., Sinclair, A.: The extended k-tree algorithm. J. Cryptol. **25**(2), 349–382 (2012)
22. Nandi, M.: XLS is not a strong pseudorandom permutation. In: Sarkar, P., Iwata, T. (eds.) ASIACRYPT 2014. LNCS, vol. 8873, pp. 478–490. Springer, Heidelberg (2014)
23. Nandi, M.: Revisiting security claims of XLS and COPA. Cryptology ePrint Archive, Report 2015/444 (2015). http://eprint.iacr.org/
24. Nikolić, I., Wang, L., Wu, S.: Cryptanalysis of round-reduced LED. In: Moriai, S. (ed.) FSE 2013. LNCS, vol. 8424, pp. 112–130. Springer, Heidelberg (2014)
25. Oechslin, P.: Making a faster cryptanalytic time-memory trade-off. In: Boneh, D. (ed.) CRYPTO 2003. LNCS, vol. 2729, pp. 617–630. Springer, Heidelberg (2003)
26. Okamoto, T.: ASIACRYPT 2000. LNCS, vol. 1976. Springer, Heidelberg (2000)
27. Ristenpart, T., Rogaway, P.: How to enrich the message space of a cipher. In: Biryukov, A. (ed.) FSE 2007. LNCS, vol. 4593, pp. 101–118. Springer, Heidelberg (2007)
28. Schroeppel, R., Shamir, A.: A $T = O(2^{n/2}), S = O(2^{n/4})$ algorithm for certain NP-complete problems. SIAM J. Comput. **10**(3), 456–464 (1981)
29. Suzuki, K., Tonien, D., Kurosawa, K., Toyota, K.: Birthday paradox for multi-collisions. In: Rhee, M.S., Lee, B. (eds.) ICISC 2006. LNCS, vol. 4296, pp. 29–40. Springer, Heidelberg (2006)
30. van Oorschot, P.C., Wiener, M.J.: Parallel collision search with cryptanalytic applications. J. Cryptol. **12**(1), 1–28 (1999)
31. Wagner, D.: A generalized birthday problem. In: Yung, M. (ed.) CRYPTO 2002. LNCS, vol. 2442, p. 288. Springer, Heidelberg (2002)
32. Wang, L.: SHELL v1. Submitted to CAESAR (2014)

How to Sequentialize Independent Parallel Attacks?

Biased Distributions Have a Phase Transition

Sonia Bogos[✉] and Serge Vaudenay

EPFL, 1015 Lausanne, Switzerland
{soniamihaela.bogos,serge.vaudenay}@epfl.ch

Abstract. We assume a scenario where an attacker can mount several independent attacks on a single CPU. Each attack can be run several times in independent ways. Each attack can succeed after a given number of steps with some given and known probability. A natural question is to wonder what is the optimal strategy to run steps of the attacks in a sequence. In this paper, we develop a formalism to tackle this problem. When the number of attacks is infinite, we show that there is a magic number of steps m such that the optimal strategy is to run an attack for m steps and to try again with another attack until one succeeds. We also study the case of a finite number of attacks.

We describe this problem when the attacks are exhaustive key searches, but the result is more general. We apply our result to the learning parity with noise (LPN) problem and the password search problem. Although the optimal m decreases as the distribution is more biased, we observe a phase transition in all cases: the decrease is very abrupt from m corresponding to exhaustive search on a single target to $m = 1$ corresponding to running a single step of the attack on each target. For all practical biased examples, we show that the best strategy is to use $m = 1$. For LPN, this means to guess that the noise vector is 0 and to solve the secret by Gaussian elimination. This is actually better than all variants of the Blum-Kalai-Wasserman (BKW) algorithm.

1 Introduction

We assume that there are an infinite number of independent keys $K_1, K_2, \ldots$ and that we want to find at least one of these keys by trials with minimal complexity. Each key search can be stopped and resumed. The problem is to find the optimal strategy to run several partial key searches in a sequence. In this optimization problem, we assume that the distributions D_i for each K_i are known. We denote $D = (D_1, D_2, \ldots)$. Consider the problem of guessing a key K_i, drawn following D_i, which is not necessarily uniform. We assume that we try all key values exhaustively from the first to the last following a fixed ordering. If we stop the

S. Bogos—Supported by a grant of the Swiss National Science Foundation, 200021_143899/1.

T. Iwata and J.H. Cheon (Eds.): ASIACRYPT 2015, Part II, LNCS 9453, pp. 704–731, 2015.
DOI: 10.1007/978-3-662-48800-3_29

key search on K_i after m trials, the sequence of trials is denoted by $ii \cdots i = i^m$. It has a worst-case complexity m and a probability of success which we denote by $\text{Pr}_D(i^m)$.

Instead of running parallel key searches in sequence, we could consider any other attack which decomposes in *steps* of the same complexity and in which each step has a specific probability to be the succeeding one. We assume that the ith attack has a probability $\text{Pr}_D(i^m)$ to succeed within m steps and that each step has a complexity 1. The fundamental problem is to wonder how to run steps of these attacks in a sequence so that we minimize the complexity until one attack succeeds. For instance, we could run attack 1 for up to m steps and decide to give up and try again with attack 2 if it fails for attack 1, and so on. We denote by $s = 1^m 2^m 3^m \cdots$ this strategy. Unsurprisingly, when the D_i's are the same, the average complexity of s is the ratio $\frac{C_D(1^m)}{\text{Pr}_D(1^m)}$ where $C_D(1^m)$ is the expected complexity of the strategy 1^m which only runs attack 1 for m steps[1] and $\text{Pr}_D(1^m)$ is its probability of success.

Traditionally, when we want to compare single-target attacks with different complexity C and probability of success p, we use as a rule of the thumb to compare the ratio $\frac{C}{p}$. Quite often, we have a continuum of attacks $C(m)$ with a number of steps limited to a variable m and we tune m so that $p(m)$ is a constant such as $\frac{1}{2}$. Indeed, the curve of $m \mapsto \frac{C(m)}{p(m)}$ is often decreasing (so has an L shape) or decreasing then increasing (with a U shape) and it is optimal to target $p(m) = \frac{1}{2}$. But sometimes, the curve can be increasing with a Γ shape. In this case, it is better to run an attack with very low probability of success and to try again until this succeeds. In some papers, e.g. [14], we consider $\min \frac{C(m)}{p(m)}$ as a complexity metric to compare attacks. Our framework justifies this choice.

LPN and Learning with Errors (LWE) [21] are two appealing problems in cryptography. In both cases, the adversary receives a matrix V and a vector $C = Vs + D$ where s is a secret vector and D is a noise vector. For LPN, the best solving algorithm was presented in Asiacrypt 2014 [12]. It brings an improvement over the well-known BKW [5] and its variants [11,15]. The best algorithm has a sub-exponential complexity.

Assuming that V is invertible, by guessing D we can solve s and check it with extra equations. So, this problem can be expressed as the one of guessing a correct vector D of small weight, which defines a biased distribution. Here, the distribution of D corresponds to the weighted concatenation of uniform distributions among vectors of the same weight. We can thus study this problem in our formalism. This was used in [8]. This algorithm is also cited in [6] and by Lyubashevsky[2].

Both LPN and LWE fall in the aforementioned scenario of guessing a k-bit biased noise vector by a simple transformation. Work on breaking cryptosystems with biased keys was also done in [18].

---

[1] $C_D(1^m)$ can be lower than m since there is a probability to succeed before reaching the mth step.

[2] http://www.di.ens.fr/~lyubash/talks/LPN.pdf.

The guessing game that we describe in our paper also matches well the password guessing scenario where an attacker tries to gain access to a system by hacking an account of an employee. There exists an extensive work on the cryptanalytic time-memory tradeoffs for password guessing [2–4,13,19,20], but the game we analyse here requires no pre-computation done by the attacker.

Our Results. We develop a formalism to compare strategies and derive some useful lemmas. We show that when we can run an infinite number of independent attacks of the same distribution, an optimal strategy is of the form $1^m 2^m 3^m \cdots$ and it has complexity

$$\min_m \frac{C_D(1^m)}{\Pr_D(1^m)}$$

for some "magic" value m. This justifies the rule of the thumb to compare attacks with different probabilities of success.

When the probability that an attack succeeds at each new step decreases (e.g., because we try possible key values in decreasing order of likelihood), there are two remarkable extreme cases: $m = n$ (where n is the maximal number of steps) corresponds to the normal single-target exhaustive search with a complexity equal to the *guesswork entropy* [17] of the distribution; $m = 1$ corresponds to trying attacks for a single step until it works, with complexity 2^{-H_∞}, where H_∞ is the *min-entropy* of the distribution.

When looking at the "magic" value m in terms of the distribution D, we observe that in many cases there is a phase transition: when D is very close to uniform, we have $m = n$. As soon as it becomes slightly biased, we have $m = 1$. There is no graceful decrease from $m = n$ to $m = 1$.

We also treat the case where we have a finite number $|D|$ of independent attacks to run. We show that there is an optimal "magic" sequence $m_1, m_2, \ldots$ such that an optimal strategy has form

$$1^{m_1} 2^{m_1} \cdots |D|^{m_1} 1^{m_2} 2^{m_2} \cdots |D|^{m_2} \cdots$$

The best strategy is first to run all attacks for m_1 steps in a sequence then to continue to run them for m_2 steps in a sequence, and so on.

Although our results look pretty natural, we show that there are distributions making the analysis counter-intuitive. Proving these results is actually non trivial.

We apply this formalism to LPN by guessing the noise vector then performing a Gaussian elimination to extract the secret. The optimal m decreases as the probability τ to have an error in a parity bit decreases from $\frac{1}{2}$. For $\tau = \frac{1}{2}$, the optimal m corresponds to a normal exhaustive search. For $\tau < \frac{1}{2} - \frac{\ln 2}{2k}$, where k is the length of the secret, the optimal m is 1: this corresponds to guessing that we have no noise at all. So, there is a phase transition.

Furthermore, for LPN with $\tau = k^{-\frac{1}{2}}$, which is what is used in many cryptographic constructions, the obtained complexity is $\mathsf{poly} \cdot e^{\sqrt{k}}$ which is much better than the usual $\mathsf{poly} \cdot 2^{\frac{k}{\log_2 k}}$ that we obtain for variants of the BKW algorithm [6].

More generally, we obtain a complexity of $\mathsf{poly} \cdot e^{-k\ln(1-\tau)}$. It is not better than the BKW variants for constant τ but becomes interesting when $\tau < \frac{\ln 2}{\log_2 k}$.

When the number of samples is limited in the LPN problem with $\tau = k^{-\frac{1}{2}}$, we can still solve it with complexity $e^{\mathcal{O}(\sqrt{k}(\ln k)^2)}$ which is better than $e^{\mathcal{O}(\frac{k}{\ln \ln k})}$ with the BKW variants [16].

For LWE, the phase transition is similar, but the algorithm for $m = 1$ is not better than the BKW variants. This is due to the 0 noise having a much lower probability in LWE (which is $1 - \tau$ for LPN) in the discrete Gaussian distribution in $\mathbb{Z}_q$.

For password search, we tried several empirical distributions of passwords and obtained again that the optimal m is $m = 1$. So, the complexity is 2^{-H_∞}.

Besides the 3 problems we study here, we believe that our results can prove to be useful in other cryptographic applications.

Structure of the Paper. Section 2 formalizes the problem and presents a few useful results. In Sect. 3 we characterize the optimal strategies and show they can be given a special regular structure. We then apply this in Sect. 4 with LPN and password recovery. Due to lack of space, we do the same for LWE in the full version of this paper. We study the phase transition of the "magic" number m in Sect. 5 and conclude in Sect. 6.

2 The STEP Game

In this section we introduce our framework through which we address the fundamental question of what is the best strategy to succeed in at least one attack when we can step several independent attacks. Let $D = (D_1, D_2, \ldots)$ be a tuple of independent distributions. If it is finite, $|D|$ denotes the number of distributions. We formalize our framework as a game where we have a ppt adversary $\mathcal{A}$ and an oracle that has a sequence of keys $(K_1, K_2, \ldots)$ where $K_i \leftarrow D_i$. At the beginning, the oracle assigns the keys according to their distribution. These distributions are known to the adversary $\mathcal{A}$. The adversary will test each key K_i by exhaustive search following a given ordering of possible values. We can assume that values are sorted by decreasing order of likelihood to obtain a minimal complexity but this is not necessary in our analysis. We only assume a fixed order. So, our framework generalizes to other types of attacks in which we cannot choose the order of the steps. Each test on K_i corresponds to a step in the exhaustive search for K_i. In general, we write "i" in a sequence to denote that we run one new step of the ith attack. The sequence of "i"s defines a strategy s. It can be finite or not. The sequence of steps we follow is thus a sequence of indices. For instance, i^m means "run the K_i search for m steps". The oracle is an algorithm that has a special command: $\mathsf{STEP}(i)$. When queried with the command $\mathsf{STEP}(i)$, the oracle runs one more step of the ith attack (so, it increments a counter t_i and tests if $K_i = t_i$, assuming that possible key values are numbered from 1). If this happens then the adversary wins. The adversary wins as soon as one attack succeeds (i.e., he guesses one of the keys from the sequence $K_1, K_2, \ldots$).

Definition 1 (Strategies). *Let D be a sequence of distributions $D = (D_1, \ldots, D_{|D|})$ (where $|D|$ can be infinite or not). A strategy for D is a sequence s of indices between 1 and $|D|$. It corresponds to Algorithm 1. We let $\mathrm{Pr}_D(s)$ be the probability that the strategy succeeds and $C_D(s)$ be the expected number of STEP when running the algorithm until it stops. We say that the strategy is full if $\mathrm{Pr}_D(s) = 1$ and that it is partial otherwise.*

Algorithm 1. Strategy s in the STEP game

1: initialize attacks $1, \ldots, |D|$
2: **for** $j = 1$ to $|s|$ **do**
3:     STEP(s_j): run one more step of the attack s_j and stop if succeeded
4: **end for**
5: stop (the algorithm fails)

For example for $s = 11223344\cdots$, Algorithm 1 tests the first two values for each key.

Definition 2 (Distributions). *A distribution D_i over a set of size n is a sequence of probabilities $D_i = (p_1, \ldots, p_n)$ of sum 1 such that $p_j \geq 0$ for $j = 1, \ldots, n$. We assume without loss of generality that $p_n \neq 0$ (Otherwise, we decrease n). We can equivalently specify the distribution D_i in an incremental way by a sequence $D_i = [p'_1, \ldots, p'_n]$ (denoted with square brackets) such that*

$$p'_j = \frac{p_j}{p_j + \cdots + p_n} \qquad p_j = p'_j(1 - p'_1)\cdots(1 - p'_{j-1})$$

for $j = 1, \ldots, n$.

We have $\mathrm{Pr}_D(i^j) = p_1 + \cdots + p_j = 1 - (1 - p'_1)\cdots(1 - p'_j)$, the probability of the j first values under D_i.

When considering the key search, it may be useful to assume that distributions are sorted by decreasing likelihood. We note that the equivalent condition to $p_j \geq p_{j+1}$ with the incremental description is $\frac{1}{p'_j} + j \leq \frac{1}{p'_{j+1}} + j + 1$, for $j = 1, \ldots, n - 1$.

We define the distribution that the keys are not among the already tested ones.

Definition 3 (Residual Distribution). *Let $D = (D_1, \ldots, D_{|D|})$ be a sequence of distributions and let s be a strictly partial strategy for D (i.e., $\mathrm{Pr}_D(s) < 1$). We denote by "$\neg s$" the residual distribution in the case where the strategy s does not succeed, i.e., the event $\neg s$ occurs.*

We let $\#\mathrm{occ}_s(i)$ denote the number of occurrences of i in s. We have

$$D|\neg s = \left(D_1|\neg 1^{\#\mathrm{occ}_s(1)}, \ldots, D_{|D|}|\neg|D|^{\#\mathrm{occ}_s(|D|)}\right)$$

where $D_i|\neg i^{t_i} = [p'_{i,t_i+1}, \ldots, p'_{i,n_i}]$ if $D_i = [p'_{i,1}, \ldots, p'_{i,n_i}]$. Hence, defining distributions in the incremental way makes the residual distribution being just a shift of the original one.

We write $\Pr_D(s'|\neg s) = \Pr_{D|\neg s}(s')$ and $C_D(s'|\neg s) = C_{D|\neg s}(s')$.

Next, we prove a list of useful lemmas in order to compute complexities, compare strategies, etc.

Lemma 4 (Success Probability). *Let s be a strategy for D. The success probability is computed by*

$$\Pr_D(s) = 1 - \prod_{i=1}^{|D|} \Pr_{D_i}(\neg i^{\#occ_s(i)})$$

Proof. The failure corresponds to the case where for all i, K_i is not in $\{1, \ldots, \#occ_s(i)\}$. The independence of the K_i implies the result. $\square$

Lemma 5 (Complexity of Concatenated Strategies). *Let ss' be a strategy for D obtained by concatenating the sequences s and s'. If $\Pr_D(s) = 1$, we have $\Pr_D(ss') = \Pr_D(s)$ and $C_D(ss') = C_D(s)$. Otherwise, we have*

$$\Pr_D(ss') = \Pr_D(s) + \left(1 - \Pr_D(s)\right) \Pr_D(s'|\neg s)$$

$$C_D(ss') = C_D(s) + \left(1 - \Pr_D(s)\right) C_D(s'|\neg s)$$

Proof. The first equation is trivial from the definition of residual distributions and conditional probabilities.

The prefix strategy s succeeds with probability $\Pr_D(s)$. Let c be the complexity of s conditioned to the event that s succeeds. Clearly, the complexity of ss' conditioned to this event is equal to c. The complexity of ss' conditioned to the opposite event is equal to $|s| + C_D(s'|\neg s)$. So, $C_D(ss') = \Pr_D(s)c + (1 - \Pr_D(s))(|s| + C_D(s'|\neg s))$. The complexity of s conditioned to that s fails is equal to $|s|$. So, $C_D(s) = \Pr_D(s)c + (1 - \Pr_D(s))|s|$. From these two equations, we obtain the result. $\square$

Lemma 6 (Complexity with Incremental Distributions). *Let $D_i = [p'_{i,1}, \ldots, p'_{i,n_i}]$ and let s be a strategy for $D = (D_1, D_2, \ldots)$. We have*

$$\Pr_D(s) = 1 - \prod_{t'=1}^{|s|} (1 - p'_{s_{t'}, \#occ_{s_1 \cdots s_{t'}}(s_{t'})})$$

$$C_D(s) = \sum_{t=1}^{|s|} \prod_{t'=1}^{t-1} (1 - p'_{s_{t'}, \#occ_{s_1 \cdots s_{t'}}(s_{t'})})$$

Proof. By induction, the probability that the strategy fails on the first $t-1$ steps is $q_t = \prod_{t'=1}^{t-1}(1 - p'_{s_{t'}, \#occ_{s_1 \cdots s_{t'}}(s_{t'})})$. We can express $C_D(s) = \sum_{t=1}^{|s|} q_t$. So, we can deduce $\Pr_D(s)$ and $C_D(s)$. $\square$

Example 7. For $D_1 = (p_1, \ldots, p_n) = [p'_1, \ldots, p'_n]$ and $m \leq n$, due to Lemma 6 we have

$$\Pr_D(1^m) = p_1 + \cdots + p_m = 1 - (1 - p'_1) \cdots (1 - p'_m)$$

and

$$C_D(1^m) = \sum_{t=1}^{m} \prod_{j=1}^{t-1} (1 - p'_j)$$

$$= \sum_{t=1}^{m} (p_t + \cdots + p_n) = p_1 + 2p_2 + \cdots + mp_m + mp_{m+1} + \cdots + mp_n$$

The second equality uses the relations from Definition 2.

We want to concatenate an isomorphic copy w of a strategy v to another strategy u. For this, we make sure that w and u have no index in common.

Definition 8 (Disjoint Copy of a Strategy). *Two strategies v and w are isomorphic if there exists an injective mapping φ such that $w_t = \varphi(v_t)$ for all t and $D_{\varphi(i)} = D_i$ for all i. So, $C_D(v) = C_D(w)$. Let u and v be two strategies for D. Whenever possible, we define a new strategy $w = \mathsf{new}_u(v)$ such that v and w are isomorphic and w has no index in common with u.*

We can define it by recursion: if $w_1 = \varphi(v_1), \ldots, w_{t-1} = \varphi(v_{t-1})$ are already defined and $\varphi(v_t)$ is not, we set it to the smallest index i (if exists) which does not appear in u nor in $w_1, \ldots, w_{t-1}$ and such that $D_i = D_{v_t}$.

For instance, if $v = 1^m$, all D_i are equal, and i is the minimal index which does not appear in u, we have $\mathsf{new}_u(v) = i^m$.

Lemma 9 (Complexity of a Repetition of Disjoint Copies). *Let s be a non-empty strategy for D. We define new strategies $s_{+1}, s_{+2}, \ldots$, disjoint copies of s, by recursion as follows: $s_{+r} = \mathsf{new}_{ss_{+1}\cdots s_{+(r-1)}}(s)$. We assume that $s_{+1}, s_{+2}, \ldots, s_{+(r-1)}$ can be constructed. If $\Pr_D(s) = 0$, then*

$$C_D(ss_{+1}s_{+2} \cdots s_{+(r-1)}) = r \cdot C_D(s).$$

Otherwise, we have

$$C_D(ss_{+1}s_{+2} \cdots s_{+(r-1)}) = \frac{1 - (1 - \Pr_D(s))^r}{\Pr_D(s)} C_D(s)$$

For r going to ∞, we respectively obtain $C_D(ss_{+1}s_{+2} \cdots) = +\infty$ and

$$C_D(ss_{+1}s_{+2} \cdots) = \frac{C_D(s)}{\Pr_D(s)}$$

For instance, for $s = 1^m$ and D_i all equal, the disjoint isomorphic copies of s are $s_{+r} = (1+r)^m$. I.e., we run m steps the $(1+r)$th attack. So, $ss_{+1}s_{+2} \cdots s_{+(r-1)} = 1^m 2^m \cdots r^m$.

Proof. We prove it by induction on r. This is trivial for $r = 1$. Let $\bar{s}_r = ss_{+1}s_{+2}\cdots s_{+r}$. If it is true for $r - 2$, then

$$C_D(\bar{s}_{r-1}) = C_D(\bar{s}_{r-2}) + (1 - \Pr_D(\bar{s}_{r-2}))C_D(s_{+(r-1)}|\neg\bar{s}_{r-2})$$

$$= \begin{cases} \frac{1-(1-\Pr_D(s))^{r-1}}{\Pr_D(s)}C_D(s) + (1 - \Pr_D(\bar{s}_{r-2}))C_D(s_{+(r-1)}|\neg\bar{s}_{r-2}) & \text{if } \Pr_S(s) > 0 \\ (r-1) \cdot C_D(s) + (1 - \Pr_D(\bar{s}_{r-2}))C_D(s_{+(r-1)}|\neg\bar{s}_{r-2}) & \text{if } \Pr_S(s) = 0 \end{cases}$$

Clearly, we have $1 - \Pr_D(\bar{s}_{r-2}) = (1 - \Pr_D(s))^{r-1}$ and $C_D(s_{+(r-1)}|\neg\bar{s}_{r-2}) = C_D(s)$. So, we obtain the result. □

Example 10. For all D_i equal, if we let $s = 1^m$, we can compute

$$C_D(1^m2^m\cdots r^m) = \frac{1 - (1 - \Pr_D(1^m))^r}{\Pr_D(1^m)}C_D(1^m)$$

$$= \frac{1 - (p_{m+1} + \cdots + p_n)^r}{p_1 + \cdots + p_m}(p_1 + 2p_2 + \cdots + mp_m + mp_{m+1} + \cdots + mp_n)$$

We now consider $r = \infty$. For an infinite number of i.i.d distributions we have

$$C_D(1^m2^m\cdots) = \frac{C_D(1^m)}{\Pr_D(1^m)}$$

$$= \frac{p_1 + 2p_2 + \cdots + mp_m + mp_{m+1} + \cdots, mp_n}{p_1 + \cdots + p_m}$$

$$= \frac{\sum_{i=1}^{m} ip_i + m(1 - p_1 + \cdots + p_m)}{p_1 + \cdots + p_m}$$

$$= G_m + m\left(\frac{1}{\Pr_{D_i}(1^m)} - 1\right)$$

where $G_m = C_{D_1|1^m}(1^m)$ and $D_1|1^m = (\frac{p_1}{\Pr_{D_1}(1^m)}, \ldots, \frac{p_m}{\Pr_{D_1}(1^m)})$. If D_1 is ordered, G_m corresponds to the guesswork entropy of the key with distribution $D_1|1^m$.

We can see two extreme cases for $s = 1^m2^m\cdots$. On one end we have a strategy of exhaustively searching the key until it is found, i.e. take $m = n$. On the other extreme we have a strategy where the adversary tests just one key before switching to another key, i.e. $m = 1$. For the sequences $s = 12\cdots$ and $s = 1^n2^n\cdots$, i.e. $m = 1$ and $m = n$, when D_1 is ordered by decreasing likelihood, we obtain the following expected complexity:

$$m = 1 \Rightarrow \qquad C_D(12\cdots) = \frac{1}{p_1} = 2^{-H_\infty(D_1)}$$

$$m = n \Rightarrow \qquad C_D(1^n2^n\cdots) = C_D(1^n) = G_n,$$

where $H_\infty(D_1)$ and G_n denote the min-entropy and the guesswork entropy of the distribution D_1, respectively.

We now define a way to compare partial strategies.

Definition 11 (Strategy Comparison). *We define*

$$\mathsf{minC}_D(s) = \inf_{s';\mathrm{Pr}_D(ss')=1} C_D(ss')$$

the infimum of $C_D(ss')$, i.e. the greatest of its lower bounds. We write $s \leq_D s'$ if and only if $\mathsf{minC}_D(s) \leq \mathsf{minC}_D(s')$. A strategy s is optimal if $\mathsf{minC}_D(s) = \mathsf{minC}_D(\emptyset)$, where $\emptyset$ is the empty strategy (i.e. the strategy running no step at all).

So, s is better than s' if we can reach lower complexities by starting with s instead of s'. The partial strategy s is optimal if we can still reach the optimal complexity when we start by s.

Lemma 12 (Best Prefixes are Best Strategies). *If u and v are permutations of each other, we have $u \leq_D v$ if and only if $C_D(u) \leq C_D(v)$.*

Proof. Note that $\mathrm{Pr}_D(u) = 1$ is equivalent to $\mathrm{Pr}_D(v) = 1$. If $\mathrm{Pr}_D(u) = 1$, it holds that $\mathsf{minC}_D(u) = C_D(u)$ and $\mathsf{minC}_D(v) = C_D(v)$. So, the result is trivial in this case. Let us now assume that $\mathrm{Pr}_D(u) < 1$ and $\mathrm{Pr}_D(v) < 1$. For any s', by using Lemma 5 we have

$$C_D(us') = C_D(u) + \left(1 - \mathrm{Pr}_D(u)\right) C_D(s'|\neg u)$$

So,

$$\inf_{s';\mathrm{Pr}_D(us')=1} C_D(us') = C_D(u) + \left(1 - \mathrm{Pr}_D(u)\right) \inf_{s';\mathrm{Pr}_D(us')=1} C_D(s'|\neg u)$$

The same holds for v. Since u and v are permutations of each other, we have $D|\neg u = D|\neg v$. So, $\mathrm{Pr}_D(us') = \mathrm{Pr}_D(vs')$ and $C_D(s'|\neg u) = C_D(s'|\neg v)$. Hence, $\inf C_D(s'|\neg u) = \inf C_D(s'|\neg v)$. Furthermore, we have $\mathrm{Pr}_D(u) = \mathrm{Pr}_D(v)$. So, $\mathsf{minC}_D(u) \leq \mathsf{minC}_D(v)$ is equivalent to $C_D(u) \leq C_D(v)$. □

3   Optimal Strategy

The question we address in this paper is: what is the optimal strategy for the adversary so that he obtains the best complexity in our STEP formalism? That is, we try to find the optimal sequence s for Algorithm 1. At a first glance, we may think that a *greedy* strategy always making a step which is the most likely to succeed is an optimal strategy. We show below that this is wrong. Sometimes, it is better to run a series of unlikely steps in one given attack because we can then run a much more likely one of the same attack after these steps are complete. However, criteria to find this strategy are not trivial at all.

   The greedy algorithm is based on looking at the i for which the next applicable p'_j in D_i is the largest. With our formalism, this defines as follows.

Definition 13 (Greedy Strategy). *Let s be a strategy for D. We say that s is greedy if*

$$\mathrm{Pr}_D(s_t|\neg s_1 \cdots s_{t-1}) = \max_i \mathrm{Pr}_D(i|\neg s_1 \cdots s_{t-1})$$

for $t = 1, \ldots, |s|$.

The following example shows that the greedy strategy is not always optimal.

Example 14. We take $|D| = \infty$ and all D_i equal to $D_i = (\frac{2}{3}, \frac{7}{36}, \frac{5}{36}) = [\frac{2}{3}, \frac{7}{12}, 1]$. After testing the first key, we have $D|\neg 1 = (D', D_2, D_3, \ldots)$ with $D' = (\frac{7}{12}, \frac{5}{12}) = [\frac{7}{12}, 1]$. Since $\frac{2}{3} > \frac{7}{12}$, the greedy algorithm would then test a new key and continue testing new keys. I.e., we would have $s = 1234\cdots$ as a greedy strategy. By applying Lemma 5, the complexity is solution to $c = 1 + \frac{1}{3}c$, i.e., $c = \frac{3}{2}$. However, the one-key strategy $s = 111$ has complexity

$$\frac{2}{3} + 2\frac{7}{36} + 3\frac{5}{36} = \frac{53}{36} < \frac{3}{2}$$

so the greedy strategy is not the best one.

Remark: The above counterexample works even when $|D|$ is finite. If we take $D = (D_1, D_2)$ with $D_i = (\frac{2}{3}, \frac{7}{36}, \frac{5}{36}) = [\frac{2}{3}, \frac{7}{12}, 1]$, the greedy approach would test the strategy $s = 1211$ that has a complexity of

$$1 + \frac{1}{3}\left(1 + \frac{1}{3}\left(1 + \frac{5}{12} \cdot 1\right)\right) = \frac{161}{108}.$$

This is greater than $\frac{53}{36}$, the complexity of the strategy 111.

Next, we note that we may have no optimal strategy as the following example shows.

Example 15 (Distribution with No Optimal Strategy). Let q_i be an increasing sequence of probabilities which tends towards 1 without reaching it. Let $D_i = [q_i, q_i, \ldots, q_i, 1]$ of support n. We have $C(i^n) = \frac{1}{q_i}(1 - (1 - q_i)^n)$ which tends towards 1 as i grows. So, 1 is the best lower bound of the complexity of full strategies. But there is no full strategy of complexity 1.

When the number of different distributions is finite, optimal strategies exist.

Lemma 16 (Existence of an Optimal Full Strategy). *Let $D = (D_1, D_2, \ldots)$ be a sequence of distributions. We assume that we have in D a finite number of different distributions. There exists a full strategy s such that $C_D(s)$ is minimal.*

Proof. Clearly, $c = \inf C_D(s)$ over all full strategies s is well defined. Essentially, we want to prove that c is reached by one strategy, i.e. that the infimum is a minimum. First, if $c = \infty$, all full strategies have infinite complexity, and the result is trivial. So, we now assume that $c < +\infty$ and we prove the result by a diagonal argument.

We now construct $s = s_1 s_2 \cdots$ by recursion. We assume that $s_1 s_2 \cdots s_r$ is constructed such that $\min C(s_1 s_2 \cdots s_r) = c$. We concatenate $s_1, \ldots, s_r$ to i^m where m is such that $\Pr_D[i^{m-1}|\neg s_1 \cdots s_r] = 0$ and $\Pr_D[i^m|\neg s_1 \cdots s_r] > 0$. The values of i to try are the ones such that i appears in $s_1, \ldots, s_r$ (we have a finite number of them), and the ones which do not appear, but we can try only one for each different D_i. We take the choice minimizing $\min C(s_1 s_2 \cdots s_r i^m)$ and set $s_{r+1} = i^m$. So, we construct a strategy s.

If one key K_i is tested until exhaustion, we have $\Pr_D(s) = 1$. If no key is tested until exhaustion, there is an infinite number of keys with same distribution D_i which are tested. If $p = \Pr_D[i^m]$ is the nonzero probability with the smallest m of this distribution, there is an infinite number of tests which succeed with probability p. So, $\Pr_D(s) \geq 1 - (1-p)^\infty = 1$. In all cases, as s has a probability to succeed of 1, s is a full strategy.

What remains to be proven is that $C_D(s) = c$. We now denote by s_i the ith step of s.

Let q_t be the probability that s fails on the first $t-1$ steps. We have $C_D(s) = \sum_{t=1}^{|s|} q_t$. Let $\varepsilon > 0$. For each r, by construction, there exists a tail strategy v such that $C_D(s_1 \cdots s_{r-1} v) \leq c + \varepsilon$. Since q_t is also the probability that $s_1 \cdots s_{r-1} v$ fails on the first $t-1$ steps for $t \leq r$, we have $\sum_{t=1}^r q_t \leq C_D(s_1 \cdots s_{r-1} v) \leq c + \varepsilon$. This holds for all r. So, we have $C_D(s) \leq c + \varepsilon$. Since this holds for all $\varepsilon > 0$, we have $C_D(s) \leq c$. Consequently, $C_D(s) = c$: s is an optimal and full strategy. □

The following two results show what is the structure of an optimal strategy.

Theorem 17. *Let $D = (D_1, D_2, \ldots)$ be a sequence of distributions. We assume that we have in D a finite number of pairwise different distributions but an infinite number of copies of each of them in D. There exists a sequence of indices $i_1 < i_2 < \cdots$ and an integer m such that $D_{i_1} = D_{i_2} = \cdots$ and $s = i_1^m i_2^m \cdots$ is an optimal strategy of complexity $\frac{C_D(i_1^m)}{\Pr_D(i_1^m)}$.*

Here are examples of optimal m for different distributions.

Example 18 (Uniform Distribution). For the uniform distribution $p_i = \frac{1}{n}$, with $1 \leq i \leq n$. We get $\Pr_D(1^m) = \frac{m}{n}$ and $G_m = \frac{m+1}{2}$. With this we obtain $C_D(1^m 2^m \cdots) = n - \frac{m-1}{2}$. Thus, the value of m that minimizes the complexity is $m = n$ and $C_D(1^m 2^m \cdots) = \frac{n-1}{2}$. The best strategy is to exhaustively search the key until it is found.

Example 19 (Geometric Distribution). For the geometric distribution with parameter p, we have $p_i = (1-p)^{i-1} p$, with $i = 1, 2, \ldots$ or $D_i = [p, p, \ldots]$. Due to Lemma 5, we can see that for every infinite strategy s, $C_D(s) = \frac{1}{p}$.

In Appendix A we study concatenations of uniform distributions.

We note that Theorem 17 does not extend if some distribution has a finite number of copies as the following example shows.

Example 20 (Distribution with No Optimal Strategy of the Form $i_1^m i_2^m \cdots$). Let $D_1 = [1 - \varepsilon, \varepsilon, \varepsilon, \ldots, \varepsilon, 1]$ of support n and $D_2 = D_3 = \cdots = [p, \ldots, p, 1]$ for $\varepsilon < p \leq \frac{1}{2}$ and n large enough. Given a full strategy s, the formula in Lemma 5 defines a sequence $q_t(s) = p'_{s_t, \#occ_{s_1 \cdots s_t}(s_t)}$. We can see that for all full strategies s and s', if $|s| \leq |s'|$ and $q_t(s) \geq q_t(s')$ for $t = 1, \ldots, |s|$, then $C_D(s) \leq C_D(s')$.

With this, we can see that $s = 12^n$ is better than all full strategies with length at least $n + 1$. There are only two full strategies with smaller length: 1^n and 2^n. We have $C_D(2^n) = \frac{1-(1-p)^n}{p} \approx \frac{1}{p} \geq 2$ as n grows. We have $C_D(12^n) = 1 + \varepsilon\frac{1-(1-p)^n}{p} \approx 1 + \frac{\varepsilon}{p}$ as n grows, so $C_D(12^n) < C_D(2^n)$ for n large enough. We have $C_D(1^n) = 1 + \varepsilon\frac{1-(1-\varepsilon)^{n-1}}{\varepsilon} = 2 - (1-\varepsilon)^{n-1} \approx 2$ so $C_D(12^n) < C_D(1^n)$ for n large enough. For all strategies of length at least $n + 1$, $s = 12^n$ collected the largest possible p' values. So, the best strategy is $s = 12^n$. It is better than any strategy of form $i_1^m i_2^m \cdots$.

When we have a finite number of distributions, we may have no optimal strategy of the form in Theorem 17. We may have multiple layers of repetition of i^m as the following result shows.

Theorem 21. *Let D_1 be a distribution of finite support n. Let $D = (D_1, D_2, \ldots, D_{|D|})$ be a finite sequence of length $|D|$ in which $D_1 = D_2 = \cdots = D_{|D|}$. There exists a sequence $m_1, \ldots, m_r$ such that the strategy*

$$s = 1^{m_1} 2^{m_1} \cdots |D|^{m_1} 1^{m_2} 2^{m_2} \cdots |D|^{m_2} \cdots 1^{m_r}$$

is optimal.

We provide toy examples below.

Example 22. We take $D = (D_1, D_2)$ with $D_1 = D_2 = (\frac{3}{5}, \frac{9}{25}, \frac{1}{50}, \frac{1}{50}) = [\frac{3}{5}, \frac{18}{20}, \frac{1}{2}, 1]$. Here are the complexities of some full strategies.

$$C_D(1111) = \frac{146}{100} = 1.46$$

$$C_D(12111) = \frac{792}{500} = 1.584$$

$$C_D(11211) = \frac{732}{500} = 1.464$$

$$C_D(121211) = \frac{7892}{5000} = 1.5784$$

$$C_D(112211) = \frac{7292}{5000} = 1.4584$$

so the last strategy is the best one. Notice that this is also a greedy strategy.

Example 23. We take $D = (D_1, D_2)$ with $D_1 = D_2 = (\frac{70}{100}, \frac{20}{100}, \frac{5}{100}, \frac{3}{100}, \frac{1}{100}, \frac{1}{100}) = [\frac{70}{100}, \frac{2}{3}, \frac{1}{2}, \frac{3}{5}, \frac{1}{2}, 1]$. Here are the complexities of some full strategies.

$$C_D(111111) = 1.48$$
$$C_D(1211111) = 1.44$$
$$C_D(12121111) = 1.438$$
$$C_D(121212111) = 1.439$$
$$C_D(121122111) = 1.444$$

so $s = 12121111$ is the best one. For this example we have that the optimal strategy requires $m_1 = 1$, $m_2 = 1$ and $m_3 = 4$. It is also greedy.

3.1   Proof of Theorem 17

To prove the result, we first state a useful lemma.

Lemma 24 (Is It Better to Do s or s' First?). *If s and s' are non-empty and have no index in common (i.e., if $s_t \neq s'_{t'}$ for all t and t'), then $ss' \leq_D s's$ if and only if $\frac{C_D(s)}{\Pr_D(s)} \leq \frac{C_D(s')}{\Pr_D(s')}$ in $[0, +\infty]$, with the convension that $\frac{c}{p} = +\infty$ for $c > 0$ and $p = 0$.*

Proof. Due to Lemma 5, when $\Pr_D(s) < 1$ we have

$$C_D(ss') = C_D(s) + \left(1 - \Pr_D(s)\right) C_D(s'|\neg s)$$

Since s' does not make use of the distributions which are dropped in $D|\neg s$, we have $C_D(s'|\neg s) = C_D(s')$. So,

$$C_D(ss') = C_D(s) + \left(1 - \Pr_D(s)\right) C_D(s')$$

This is also clearly the case when $\Pr_D(s) = 1$. Similarly,

$$C_D(s's) = C_D(s') + \left(1 - \Pr_D(s')\right) C_D(s)$$

So, $C_D(ss') \leq C_D(s's)$ is equivalent to

$$C_D(s) + \left(1 - \Pr_D(s)\right) C_D(s') \leq C_D(s') + \left(1 - \Pr_D(s')\right) C_D(s)$$

So, this inequality is equivalent to $\frac{C_D(s)}{\Pr_D(s)} \leq \frac{C_D(s')}{\Pr_D(s')}$.   □

We can now prove Theorem 17.

Proof (of Theorem 17). Due to Lemma 16, we know that optimal full strategies exist. Let s be one of these. We let i be the index of an arbitrary key which is tested in s. We can write $s = u_0 i^{m_1} u_1 i^{m_2} \cdots i^{m_r} u_r$ where i appears in no u_j and $m_j > 0$ for all j, and $u_1, \ldots, u_{r-1}$ are non-empty.

Since s is optimal, by permuting i^{m_j} and either u_{j-1} or u_j, we obtain larger complexities. So, by applying Lemma 24, we obtain

$$\frac{C_D(i^{m_1})}{\Pr_D(i^{m_1})} \leq \frac{C_D(u_1|\neg u_0)}{\Pr_D(u_1|\neg u_0)} \leq \frac{C_D(i^{m_2}|\neg i^{m_1})}{\Pr_D(i^{m_1}|\neg i^{m_1})} \leq \cdots \leq C_D(u_r|\neg u_0 \cdots u_{r-1})$$

We now want to replace u_r in s by some isomorphic copy of s which is not overlapping with $u_0 i^{m_1} u_1 i^{m_2} \cdots i^{m_r}$. Due to the optimality of s, we would deduce

$$C_D(u_r|\neg u_0 \cdots u_{r-1}) \leq C_D(s|\neg u_0 \cdots u_{r-1}) = C_D(s)$$

so $\frac{C_D(i^{m_1})}{\Pr_D(i^{m_1})} \leq C_D(s)$ which would imply that the repetition of isomorphic copies of i^{m_1} are at least as good as s, so $\frac{C_D(i^{m_1})}{\Pr_D(i^{m_1})} = C_D(s)$ due to the optimality of s.

But to replace u_r in s by the isomorphic copy of s, we need to rewrite the original s containing u_r by some isomorphic copy in which indices are left free to implement another isomorphic copy of s.

For that, we split the sequence $(1, 2, 3, \ldots)$ into two subsequences v and v' which are non-overlapping (i.e. $v_t \neq v_{t'}$ for all t and t'), complete (i.e. for every integer j, v contains j or v' contains j), and representing each distribution with infinite number of occurrences (i.e. for all j, there exist infinite sequences $t_1 < t_2 < \cdots$ and $t'_1 < t'_2 < \cdots$ such that $D_j = D_{v_{t_\ell}} = D_{v'_{t'_\ell}}$ for all ℓ). For that, we can just construct v and v' iteratively: for each j, if the number of $j' < j$ such that $D_{j'} = D_j$ in v or v' is the same, we put j in v, otherwise (we may have only one more instance in v), we put j in v' (to balance again). For instance, if all D_i are equal, this construction puts all odd j in v and all even j in v'. Hence, we can define $s' = \mathsf{new}_v(s)$ and $s'' = \mathsf{new}_{v'}(s)$. s' will thus only use indices in v' while s'' will only use indices in v. Therefore, s' and s'' will be isomorphic, with no index in common. So, $C_D(s) = C_D(s') = C_D(s'')$.

Following the split of s, the strategy s' can be written $s' = u'_0 i'^{m_1} u'_1 i'^{m_2} \cdots i'^{m_r} u'_r$ with

$$\frac{C_D(i^{m_1})}{\Pr_D(i^{m_1})} = \frac{C_D(i'^{m_1})}{\Pr_D(i'^{m_1})} \leq C_D(u'_r | \neg u'_0 \cdots u'_{r-1}) = C_D(u'_r | \neg u'_0 i'^{m_1} u'_1 i'^{m_2} \cdots i'^{m_r})$$

If we replace u'_r in s' by s'', since s' is optimal, we obtain a larger complexity. So,

$$C_D(u'_0 i'^{m_1} u'_1 i'^{m_2} \cdots i'^{m_r} u'_r) \leq C_D(u'_0 i'^{m_1} u'_1 i'^{m_2} \cdots i'^{m_r} s'')$$

These two strategies have the prefix $u'_0 i'^{m_1} u'_1 i'^{m_2} \cdots i'^{m_r}$ in common. We can write their complexities by splitting this common prefix using Lemma 5. By eliminating the common terms, we deduce

$$C_D(u'_r | \neg u'_0 i'^{m_1} u'_1 i'^{m_2} \cdots i'^{m_r}) \leq C_D(s'' | \neg u'_0 i'^{m_1} u'_1 i'^{m_2} \cdots i'^{m_r}) = C_D(s'') = C_D(s)$$

We deduce

$$\frac{C_D(i^{m_1})}{\Pr_D(i^{m_1})} \leq C_D(s)$$

Let $i_1 < i_2 < \cdots$ be a sequence of keys using the distribution D_i. By Lemma 9, the strategy $i_1^m i_2^m \cdots$ has complexity $\frac{C_D(i^{m_1})}{\Pr_D(i^{m_1})}$. Since s is optimal, we have $\frac{C_D(i^{m_1})}{\Pr_D(i^{m_1})} \geq C_D(s)$. Therefore, $\frac{C_D(i^{m_1})}{\Pr_D(i^{m_1})} = C_D(s)$.   □

3.2 Proof of Theorem 21

For the proof of Theorem 21 we need the result of the following lemma.

Lemma 25. *Let* $s = ui^a v j^b w$ *be an optimal strategy with* n *occurrences of each key. We assume that* $i \neq j$, $a < b$, u *does not end with* i, v *has no occurrence of either* i *or* j, *and* w *has equal number of occurrences for* i *and* j. *Furthermore, we assume that either* $a \neq 0$, *or* v *is nonempty and starts with some* k *such that* u *does not end with* k. *Then,* $C_D(s) = C_D(uj^{b-a}i^a v j^a w)$.

Lemma 25 will be used in two ways.

1. For $s = u'j^c v j^b w$ with $c > 0$, $b > 0$, v with no i or j, and balanced occurrences of i and j in w, which has the same complexity as $s' = u'j^{b+c}vw$ (so, to apply the lemma we define $a = 0$, $u = u'j^c$, $k = j$, and $s = u'j^c i^0 v j^b w$; all hypotheses are verified except v non-empty, but the result is trivial for empty v). This means that we can regroup j^c and j^b when there are separated by a v with no i and followed by a balanced tail w.

2. For $s = ui^a v j^b w$ with $0 < a < b$, v with no i or j, and balanced occurrences of i and j in w, which has the same complexity as $s' = uj^{b-a}i^a v j^a w$. This means that we can balance i^a and j^b when there are separated by a v with no i or j and followed by a balanced tail w.

The proof of Lemma 25 is given in Appendix B.

In what follows, we say that a strategy is in a *normal form* if for all t, $i \mapsto$ #occ$_{s_1 \cdots s_t}(i)$ is a non-increasing function, i.e. #occ$_{s_1 \cdots s_t}(i) \geq$ #occ$_{s_1 \cdots s_t}(i+1)$ for all i. For instance, 1112322133 is normal as the number of STEP(1) is at no time lower than the number of STEP(2) and the same for the number of STEP(2) and STEP(3).

Since all distributions are the same, all strategies can be rewritten into an equivalent one in a normal form: for this, for the smallest t such that there exists i such that #occ$_{s_1 \cdots s_t}(i) <$ #occ$_{s_1 \cdots s_t}(i+1)$, it must be that $s_t = i+1$ and #occ$_{s_1 \cdots s_{t-1}}(i) =$ #occ$_{s_1 \cdots s_{t-1}}(i+1)$. We can permute all values i and $i+1$ in the tail $s_t s_{t+1} \cdots$ and obtain an equivalent strategy on which the function becomes non-increasing at step t and is unchanged before. By performing enough such rewriting, we obtain an equivalent strategy in normal form. For instance, 12231332 is not normal. The smallest t is $t = 3$ when we make a second STEP(2) while we only did a single STEP(1). So, we permute 1 and 2 at this time and obtain 12132331. Then, we have $t = 7$ and permute 2 and 3 to obtain 12132321. Then, again $t = 7$ to permute 1 and 2 to obtain 12132312 which is normal.

We now prove Theorem 21.

Proof (of Theorem 21). Let s be an optimal strategy. Due to the assumptions, it must be finite. We assume w.l.o.g. that s is in normal form. We note that we can always complete s in a form $s2^{a_2}3^{a_3} \cdots$ so that the final strategy has exactly n occurrences of each i. So, we assume w.l.o.g. that s has equal number of occurrences. We write $s = 1^{m_1}x_1 1^{m_2}x_2 \cdots 1^{m_r}x_r$ where the x_t's are non-empty and with no 1 inside.

As detailed below, we rewrite x_r (and push some steps earlier in x_{r-1}) so that we obtain a permutation of the blocks $2^{m_r}, \ldots, |D|^{m_r}$. The rewriting is done by preserving the probability of success (which is 1) and the complexity (which is the optimal complexity). Then, we do the same operation in x_{r-1}

and continue until x_1. When we are done, each x_t becomes a permutation of the blocks $2^{m_t}, \ldots, |D|^{m_t}$. Finally, we normalize the obtained rewriting of s and obtain the result.

We assume that s has already been rewritten so that for each $t' = t+1, \ldots, r$, the $x_{t'}$ sub-strategy is a permutation of the blocks $2^{m_{t'}}, \ldots, |D|^{m_{t'}}$. Then, we explain how to rewrite x_t. We make a loop for $j = 2$ to $|D|$. In the loop, we first regroup all blocks of j's by using Lemma 25 with $i = 1$: while we can write $x_t = u'j^c v j^b w'$ where $c > 0$, $b > 0$, v is non-empty with no j, and w' has no j, we write $u = 1^{m_1}x_1 1^{m_2}x_2 \cdots 1^{m_t}u'$ and $w = w'1^{m_{t+1}}x_{t+1} \cdots 1^{m_r}x_r$, and set $a = 0$ and $i = 1$. This rewrites $x_t = u'j^{b+c}v w'$ by preserving the complexity and making a permutation. When this while loop is complete, we can only find a single block of j's in x_t and write $x_t = vj^b w'$, where v and w' have no j. So, we apply again Lemma 25 to balance 1^{m_t} and j^b: we write $u = 1^{m_1}x_1 1^{m_2}x_2 \cdots x_{t-1}$ and $w = w'1^{m_{t+1}}x_{t+1} \cdots 1^{m_r}x_r$, and set $a = m_t$ and $i = 1$. This rewrites $1^{m_t}x_t$ to $j^{b-m_t}1^{m_t}vj^{m_t}w'$ by preserving the complexity and making a permutation. So, this rewrites x_t to $vj^{m_t}w'$ and x_{t-1} to $x_{t-1}j^{b-m_t}$. When the loop of j is complete, x_t is a permutation of the blocks $2^{m_t}, \ldots, |D|^{m_t}$.

Interestingly, the sequence $m_1, \ldots, m_r$ is unchanged from our starting optimal normal full strategy s. If we rather start from an optimal full strategy s which is not in normal form, we can still see how to obtain this sequence: for each t, $m_1 + \cdots + m_t$ is the next record number of steps for an attack i after the $m_1 + \cdots + m_{t-1}$ record. That is the number of steps for the attack i when s decides to move to another attack. $\square$

3.3   Finding the Optimal m

We provide here a simple criterion for the optimal m of Theorem 17.

Lemma 26. We let $D_1 = (p_1, \ldots, p_n) = [p'_1, \ldots, p'_n]$ be a distribution and define $D = (D_1, D_1, \ldots)$. Let m be such that $s = 1^m 2^m \cdots$ is an optimal strategy based on Theorem 17. We have $\frac{1}{p'_m} \le C_D(1^m 2^m \cdots) \le \frac{1}{p'_{m+1}}$.

Proof. We let $s = 2^m 3^m \cdots$ We know that $C_D(1^{m+1}s) \ge C_D(1^m s)$ since $1^m s$ is optimal. So,

$$0 \le C_D(1^{m+1}s) - C_D(1^m s)$$
$$= (1 - \Pr_D(1^m))(C_D(1s|\neg 1^m) - C_D(s))$$
$$= (1 - \Pr_D(1^m))(1 - p'_{m+1} \cdot C_D(s))$$

from which we deduce $\frac{1}{p'_{m+1}} \ge C_D(s)$. Similarly, we have

$$0 \ge C_D(1^m s) - C_D(1^{m-1}s)$$
$$= (1 - \Pr_D(1^{m-1}))(C_D(1s|\neg 1^{m-1}) - C_D(s))$$
$$= (1 - \Pr_D(1^{m-1}))(1 - p'_m \cdot C_D(s))$$

from which we deduce $\frac{1}{p'_m} \le C_D(s)$. $\square$

We note that if $p_m = p_{m+1}$, then

$$p'_{m+1} = \frac{p_{m+1}}{p_{m+1} + \cdots + p_n} = \frac{p_m}{p_{m+1} + \cdots + p_n} > \frac{p_m}{p_m + p_{m+1} + \cdots + p_n} = p'_m$$

which is impossible (given the result from Lemma 26). Consequently, we must have $p_m \neq p_{m+1}$. So, in distributions when we have sequences of equal probabilities p_t, we can just look at the largest index t in the sequence as a possible candidate for being the value m.

Lemma 26 has an equivalent for Theorem 21 (given in the full version of this paper due to lack of space).

4 Applications

4.1 Solving Sparse LPN

We will model the Learning Parity with Noise (LPN) problem in our STEP game. As we will see, we use the noise bits as the keys the adversary $\mathcal{A}$ is trying to guess. First of all, we formally give the definition of the LPN problem.

Definition 27 (Search LPN). *Let* $s \xleftarrow{U} \mathbb{Z}_2^k$, *let* $\tau \in]0, \frac{1}{2}[$ *be a constant noise parameter and let* Ber_τ *be the Bernoulli distribution with parameter* τ. *Denote by* $D_{s,\tau}$ *the distribution defined as*

$$\{(v, c) \mid v \xleftarrow{U} \mathbb{Z}_2^k, c = \langle v, s \rangle \oplus d, d \leftarrow \mathsf{Ber}_\tau\} \in \mathbb{Z}_2^{k+1}.$$

An LPN *oracle* $\mathcal{O}_{s,\tau}^{\mathsf{LPN}}$ *is an oracle which outputs independent random samples according to* $D_{s,\tau}$.

Given queries from the oracle $\mathcal{O}_{s,\tau}^{\mathsf{LPN}}$, *the search* LPN *problem is to find the secret* s.

As studied in [6], the LPN-solving algorithms which are based on BKW [5] have a complexity $\mathsf{poly} \cdot 2^{\frac{k}{\log_2 k}}$. The naive algorithm guessing that the noise is 0 and running a Gaussian elimination until this finds the correct solution works with complexity $\mathsf{poly} \cdot (1 - \tau)^{-k}$. So, the latter is much better as soon as $\tau < \frac{\ln 2}{\log_2 k}$, and in particular for $\tau = k^{-\frac{1}{2}}$ which is the case for some applications [1,9]. Experiments reported in [6] also show that for $\tau = k^{-\frac{1}{2}}$, the Gaussian elimination outperforms the BKW variants for $k > 500$.

The Gaussian elimination algorithm just reduces to finding a k-bit noise vector. It guesses that this vector is 0. If this does not work, the algorithm tries again with new LPN queries. We can see this as guessing at least one k-bit biased vector K_i which follows the distribution $D_i = \mathsf{Ber}_\tau^k$ defined by $\Pr[K_i = v] = \tau^{\mathsf{HW}(v)}(1 - \tau)^{k - \mathsf{HW}(v)}$ in our framework. The most probable vector is $v = 0$ which has probability $\Pr[K_i = 0] = (1 - \tau)^k$. The above algorithm corresponds to trying $K_1 = 0$ then $K_2 = 0$, ... i.e., the strategy $123\cdots$ in our framework. We can wonder if there is a better $1^m 2^m 3^m \cdots$. This is the problem we study below. We will see that the answer is no: using $m = 1$ is the best option as soon as τ is less than $\frac{1}{2} - \varepsilon$ for $\varepsilon = \frac{\ln 2}{2k}$ which is pretty small.

For instance, for $\mathsf{LPN}_{768, \frac{1}{\sqrt{768}}}$ we obtain $C_D(12\cdots) = 2^{41}$. I.e., 2^{41} calls to the STEP command which corresponds to collecting k LPN queries and making a Gaussian elimination to recover the secret based on the assumption that the error bits are all 0. If we add up the cost of running Gaussian elimination in order to recover the secret, we obtain a complexity of 2^{70}. This outperforms all the BKW variants and proves that $\mathsf{LPN}_{768, \frac{1}{\sqrt{768}}}$ is not a secure instance for a 80-bit security. Furthermore, this algorithm outperforms even the covering code algorithm [12]. Our results are strengthened by the results from [6] where we see that there is a big difference between the performance of $C_D(12\cdots)$ and the one of the covering code algorithm.

$D_{i\text{-}}$ is a composite distribution of uniform ones in the sense defined in Appendix A. Namely, $D_i = \sum_{w=0}^{k} \tau^k (1-\tau)^{k-w} U_w$ where U_w is uniform of support $\binom{k}{w}$. By Theorem 17, we know that there exists a magic m for which the strategy $s = 1^m 2^m \cdots$ is optimal. The analysis of composite distributions further says that m must be of form $m = B_w = \sum_{i=0}^{w} \binom{k}{i}$ for some magic w. Let c_m be the complexity of $1^m 2^m \cdots$. A value $w = k$, i.e. $m = n$ corresponds to the exhaustive search of the noise bits. For $w = 0$, i.e. $m = 1$, the adversary assumes that the noise is 0 every time he receives k queries from the LPN oracle.

We first computed experimentally the optimal m for the $\mathsf{LPN}_{100,\tau}$ instance where we take $0 < \tau < \frac{1}{2}$. The magic m takes the value 1 for a τ which is not close to $\frac{1}{2}$. As shown on Fig. 1, it changes to $n = 2^{100}$ around the value $\tau = 0.4965$. This boundary between two different strategies corresponds to the value $\tau = \frac{1}{2} - \frac{\ln 2}{2k}$ computed in our analysis below. Interestingly, there is no intermediate optimal m between 1 and n.

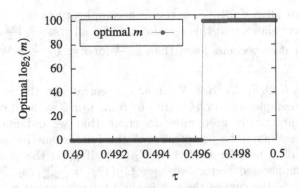

Fig. 1. The change of optimal m for solving $\mathsf{LPN}_{100,\tau}$

For Cryptographic Parameters, c_1 is Optimal. The optimal w depends on τ. The case when τ is lower than $\frac{1}{k}$ is not interesting as it is likely that no error occurs so all w lead to a complexity which is very close to 1. Conversely, for $\tau = \frac{1}{2}$, the exhaustive search has a complexity of $c_n = \frac{1}{2}(2^k+1)$ and $w = 0$ has a complexity of $c_1 = 2^k$. Actually, D_i is uniform in this case and we know that the optimal m completes batches of equal consecutive probabilities. So, the optimal strategy is the exhaustive search.

We now show that for $\tau < 0.16$, the best strategy is obtained for $w = 0$.

Below, we use $p_{B_w} = \tau^w (1 - \tau)^{k-w}$ and $c_1 = (1 - \tau)^{-k}$.

Let w_c be a threshold weight and let $\alpha = \Pr(1^{B_{w_c}})$. For $0 < w \leq w_c$, due to Lemma 26, if c_{B_w} is optimal we have

$$c_{B_w} \geq \frac{1}{p'_{B_w}} = \frac{\Pr_D(\neg 1^{B_w-1})}{p_{B_w}} \geq \frac{\Pr_D(\neg 1^{B_{w_c}})}{p_{B_w}} = \frac{1-\alpha}{p_{B_w}} = \frac{1-\alpha}{\left(\frac{\tau}{1-\tau}\right)^w} c_1 \geq \frac{1-\alpha}{\frac{\tau}{1-\tau}} c_1$$

For $\tau < 0.16$, we have $\frac{\tau}{1-\tau} < 0.20$. So, if $\alpha \leq \frac{4}{5}$ we obtain $c_{B_w} > c_1$. This contradicts that w is optimal. For $w_c = \tau k$, the Central Limit Theorem gives us that $\alpha \approx \frac{1}{2}$ which is less than $\frac{4}{5}$. So, no w such that $0 < w \leq \tau k$ is optimal.

Now, for $w \geq w_c$, we have

$$c_w = \frac{C_D(1^{B_w})}{\Pr_D(1^{B_w})} \geq C_D(1^{B_w}) = \sum_{i=1}^{B_w} ip_i + B_w \Pr_D(\neg 1^{B_w}) \geq B_{w_c} \Pr_D(\neg 1^{B_{w_c}}) = (1-\alpha)B_{w_c}$$

By using the bound $B_{w_c} \geq \left(\frac{k}{w_c}\right)^{w_c}$, for $w_c = \tau k$ we have $\alpha \approx \frac{1}{2}$ and we obtain $c_w \geq \frac{1}{2}\tau^{-\tau k}$. We want to compare this to $c_1 = (1 - \tau)^{-k}$. We look at the variations of the function $\tau \mapsto -k\tau \ln \tau - \ln 2 + k \ln(1 - \tau)$. We can see by derivating twice that for $\tau \in [0, \frac{1}{2}]$, this function increases then decreases. For $\tau = 0.16$, it is positive. For $\tau = \frac{1}{k}$, it is also positive. So, for $\tau \in [\frac{1}{k}, 0.16]$, we have $c_{B_w} \geq c_1$.

Therefore, for all $\tau < 0.16$, c_1 is the best complexity so $m = 0$ is the magic value. Experiment shows that this remains true for all $\tau < \frac{1}{2} - \frac{\ln 2}{2k}$. Actually, we can easily see that c_1 becomes lower than $\frac{2^k+1}{2}$ for $\tau \approx \frac{1}{2} - \frac{\ln 2}{2k}$. We will discuss this in Sect. 5.

Solving LPN *with* $\mathcal{O}(k)$ *Queries.* We now concentrate on the $m = n$ case to limit the query complexity to $\mathcal{O}(k)$. (In our framework, we need only k queries but we would practically need more to check that we did find the correct value). So, we estimate the complexity of the full exhaustive search on one error vector x of k bits for LPN, i.e., $C_D(1^n)$. If p_t is the probability that x is the t-th enumerated vector, we have $C_D(1^n) = \sum_{t=1}^n t p_t$. For t between $B_{w-1} + 1$ and B_w, the sum of the p_t's is the probability that we have exactly w errors. So, $C_D(1^n) \leq \sum_{w=0}^k B_w \Pr[w \text{ errors}]$. We approximate $\Pr[w \text{ errors}]$ to the continuous distribution. So, the Hamming weight has a normal distribution, with mean $k\tau$ and standard deviation $\sigma = \sqrt{k\tau(1-\tau)}$. We do the same for

$B_w \approx \frac{2^k}{\sqrt{2\pi}} \int_{-\infty}^{\frac{2w-k}{\sqrt{k}}} e^{-\frac{v^2}{2}} \, dv$. With the change of variables $w = k\tau + t\sigma$, we have

$$C_D(1^n) \leq \sum_{w=0}^{k} B_w \Pr[w \text{ errors}]$$

$$\approx \frac{2^k}{2\pi} \int_{-\infty}^{+\infty} \left(\int_{-\infty}^{\frac{2w-k}{\sqrt{k}}} e^{-\frac{v^2}{2}} \, dv \right) \frac{1}{\sigma} e^{-\frac{(w-k\tau)^2}{2\sigma^2}} \, dw$$

$$= \frac{2^k}{2\pi} \iint_{v \leq \frac{2k\tau - k + 2t\sigma}{\sqrt{k}}} e^{-\frac{t^2 + v^2}{2}} \, dv \, dt$$

The distance between the origin $(t, v) = (0, 0)$ and the line $v = \frac{2k\tau - k + 2t\sigma}{\sqrt{k}}$ is

$$d = \sqrt{k} \frac{1 - 2\tau}{\sqrt{1 + 4\tau(1 - \tau)}}$$

By rotating the region on which we sum, we obtain

$$C_D(1^n) \approx \frac{2^k}{2\pi} \iint_{x \geq d} e^{-\frac{x^2 + y^2}{2}} \, dx \, dy = \frac{2^k}{\sqrt{2\pi}} \int_{d}^{+\infty} e^{-\frac{x^2}{2}} \, dx \sim \frac{2^k}{d\sqrt{2\pi}} e^{-\frac{d^2}{2}}$$

On Fig. 2 we can see that this approximation of $C_D(1^n)$ is very good for $\tau = k^{-\frac{1}{2}}$.

So, the complexity $C_D(1^n)$ is asymptotically $2^{k\left(1 - \frac{1}{2\ln 2}\right) + \mathcal{O}(\sqrt{k})}$. Interestingly, the dominant part of $\log_2 C_D(1^n)$ is $0.2788 \times k$ and does not depend on τ as long as $\frac{1}{k} \ll \tau \ll \frac{1}{2}$. Although very good for the low k that we consider, this approximation of $C_D(1^n)$ deviates, probably because of the imprecise approximation of the B_w's. Next, we derive a bound which is much higher but asymptotically better (the curves crossing for $k \approx 50\,000$). We now use the bound $B_w \leq k^w$ and do the same computation as before. We have

$$C_D(1^n) \leq \sum_{w=0}^{k} k^w \Pr[w \text{ errors}]$$

$$\approx \frac{1}{\sqrt{2\pi}} \int_{-\infty}^{+\infty} k^{k\tau + t\sigma} e^{-\frac{t^2}{2}} \, dw$$

$$= \frac{e^{\frac{1}{2}(\sigma \ln k)^2 + k\tau \ln k}}{\sqrt{2\pi}} \int_{-\infty}^{+\infty} e^{-\frac{(t - \sigma \ln k)^2}{2}} \, dw$$

$$= e^{\frac{1}{2}(\sigma \ln k)^2 + k\tau \ln k}$$

So, $C_D(1^n) = e^{\frac{1}{2}\sqrt{k}(\ln k)^2 + \mathcal{O}(\sqrt{k}\ln k)}$ for $\tau = k^{-\frac{1}{2}}$. It is better than the $e^{\mathcal{O}\left(\frac{k}{\ln \ln k}\right)}$ of Lyubashevsky [16] in the sense that it is asymptotically better and that we use $\mathcal{O}(k)$ queries instead of $k^{1+\varepsilon}$. However, this new bound for $C_D(1^n)$ is very loose.

Outside the scenario of a sparse LPN, we display in Fig. 3 the logarithmic complexity to solve LPN in our STEP game when the noise parameter is constant.

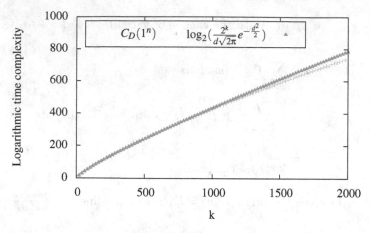

Fig. 2. $\log_2(C_D(1^n))$ vs. $\log_2\left(\frac{2^k}{d\sqrt{2\pi}}e^{-\frac{d^2}{2}}\right)$ for $\tau = k^{-\frac{1}{2}}$

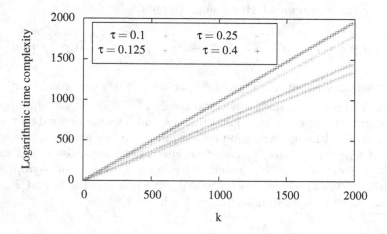

Fig. 3. $\log_2(C_D(1^n))$ for constant τ

Table 1. $\log_2(C_D(1^n))$ vs. $\log_2\left(\frac{2^k}{d\sqrt{2\pi}}e^{-\frac{d^2}{2}}\right)$ for $k = 2000$

τ	$\log_2(C_D(1^n))$	$\log_2\left(\frac{2^k}{d\sqrt{2\pi}}e^{-\frac{d^2}{2}}\right)$
0.1	1350.04	1314.81
0.125	1458.86	1429.33
0.25	1794.57	1788.49
0.4	1966.67	1966.55

Comparing $\log_2(C_D(1^n))$ with the approximation we obtained, i.e. $\log_2\left(\frac{2^k}{d\sqrt{2\pi}}e^{-\frac{d^2}{2}}\right)$, we obtain the following results which validate our approximations (See Table 1).

4.2   Password Recovery

There are many news nowadays with attacks and leaks of passwords from different famous companies. From these leaks the community has studied what are the worst passwords used by the users. Having in mind these statistics, we are interested to see what is the best strategy of an outsider that tries to get access to a system having access to a list of users. The goal of the attacker is to hack one account. He can try to hack several accounts in parallel. Within our framework, we compute to see what is the optimal m for the strategy $1^m 2^m \cdots$. In this given scenario, the strategy corresponds to making m guesses for each user until it reaches the end of the list and starting again with new guesses.

We consider the statistics that we have found for the 10 000 Top Passwords[3] and the one done for the database with passwords in clear from the RockYou hack[4]. Studies on the distribution of user's passwords were also done in [7,10, 22,23]. The first case-study analyses what are the top 10 000 passwords from a total 6.5 million username-passwords leaked. The most frequent passwords are the following:

$$
\begin{array}{ll}
\text{password} & p_1 = 0.00493 \\
\text{123456} & p_2 = 0.00400 \\
\text{12345678} & p_3 = 0.00133 \\
\text{1234} & p_4 = 0.00089
\end{array}
$$

In the case of the RockYou hack, where 32 million of passwords were leaked, we have that the most frequent passwords and their probability of usage is:

$$
\begin{array}{ll}
\text{123456} & p_1 = 0.009085 \\
\text{12345} & p_2 = 0.002471 \\
\text{123456789} & p_3 = 0.002400 \\
\text{Password} & p_4 = 0.000194
\end{array}
$$

Moreover, approximately 20 % of the users used the most frequent 5 000 passwords. What these statistics show is that users frequently choose poor and predictable passwords. While dictionary attacks are very efficient, we study here the case where the attacker wants to minimize the number of trials until he gets access to the system, with no pre-computation done. By using our formulas of computing $C_D(1^m 2^m \cdots)$, we obtain in both of the above distributions that $m = 1$ is the optimal one. This means that the attacker tries for each username the most probable password and in average after couple of hundred of users (for the two studies we obtain C_D to be ≈ 203 and ≈ 110), he will manage to access

---

[3] https://xato.net/passwords/more-top-worst-passwords/#.VNiORvnF-xW.
[4] http://www.imperva.com/docs/WP_Consumer_Password_Worst_Practices.pdf.

the system. We note that having $m = 1$ is very nice as for the typical password guessing scenario, we need to have a small m to avoid complications of blocking accounts and triggering an alarm that the system is under an attack.

5   On the Phase Transition

Given the experience of the previous applications, we can see that for "regular" distributions, the optimal m falls from $m = n$ to the minimal m as the bias of the distribution increases. We let n_1 be such that $p_1 = p_2 = \cdots = p_{n_1} \neq p_{n_1+1}$ and n_2 be such that $p_{n_1+1} = \cdots = p_{n_1+n_2} \neq p_{n_1+n_2+1}$. Due to Lemma 26, the magic value m can only be n_1, $n_1 + n_2$, or more. We study here when the curves of $C_D(1^{n_1}2^{n_1} \cdots)$, $C_D(1^{n_1+n_2}2^{n_1+n_2} \cdots)$, and $C_U(1^n) = \frac{n+1}{2}$ cross each other.

Lemma 28. *We consider a composite distribution* $D_1 = \alpha U_1 + \beta U_2 + (1 - \alpha - \beta)D'$, *where* U_1 *and* U_2 *are uniform of support* n_1 *and* n_2. *For* U *uniform, we have*

$$C_D(1^{n_1}2^{n_1} \cdots) \leq C_D(1^{n_1+n_2}2^{n_1+n_2} \cdots) \iff \alpha - \beta\frac{n_1}{n_2} \geq \alpha\left(\alpha + \beta\frac{1 - n_1/n_2}{2}\right)$$

$$C_D(1^{n_1}2^{n_1} \cdots) \leq C_U(1^n) \iff \frac{n/n_1 + 1}{2} \geq \frac{1}{\alpha}$$

Note that for $2^{-H_\infty} \geq \frac{2}{n}$, we have $\frac{\alpha}{n_1} \geq \frac{2}{n}$ so the second property is satisfied.

As an example, for $n_1 = n_2 = 1$, the first condition becomes $\alpha - \beta \geq \alpha^2$ which is the case of all the distribution we tried for password recovery. The second condition becomes $2^{-H_\infty} \geq \frac{2}{n+1}$, which is also always satisfied.

For LPN, we have $n_1 = 1$, $n_2 = k$, $\alpha = (1 - \tau)^k$, and $\beta = n_2\tau(1 - \tau)^{k-1}$. The first and second conditions become

$$(1 - \tau)^k \leq \frac{1 - 2\tau}{1 + \frac{k-3}{2}\tau} \quad \text{and} \quad (1 - \tau)^k \geq \frac{2}{2^k + 1}$$

respectively. They are always satisfied unless τ is very close to $\frac{1}{2}$: by letting $\tau = \frac{1}{2} - \varepsilon$ with $\varepsilon \to 0$, the right-hand term of the first condition is asymptotically equivalent to $\frac{8\varepsilon}{k+1}$ and the left-hand term tends towards 2^{-k}. The balance is thus for $\tau \approx \frac{1}{2} - \frac{k+1}{8}2^{-k}$. The second condition gives

$$\tau \leq 1 - \left(\frac{2^k + 1}{2}\right)^{-\frac{1}{k}} = \frac{1}{2} - \frac{\ln 2}{2k} - o\left(\frac{1}{k}\right)$$

So, we can explain the phase transition in $\mathsf{LPN}_{k,\tau}$ as follows: if we make τ decrease from $\frac{1}{2}$, for each fixed m, the complexity of all possible $C_D(1^m)$ smoothly decrease. The function for $m = n_1$ crosses the one of $m = n_1 + n_2$ before it crosses $\frac{n+1}{2}$ which is close to the value of the one for $m = n$. So, the curve for $m = n_1$ becomes interesting *after* having beaten the curve for $m = n_1 + n_2$. This proves that we never have a magic m equal to $n_1 + n_2$. Presumably, it is the case for all other curves as well. This explains the abrupt fall from $m = n$ to $m = 1$ which we observed on Fig. 1.

Proof. We have

$$C_D(1^{n_1}2^{n_1}\cdots) = \frac{C_D(1^{n_1})}{\Pr_D(1^{n_1})} = \frac{\alpha\frac{n_1+1}{2} + (1-\alpha)n_1}{\alpha}$$

and

$$C_D(1^{n_1+n_2}2^{n_1+n_2}\cdots) = \frac{C_D(1^{n_1+n_2})}{\Pr_D(1^{n_1+n_2})} = \frac{\alpha\frac{n_1+1}{2} + \beta\left(n_1 + \frac{n_2+1}{2}\right) + (1-\alpha-\beta)(n_1+n_2)}{\alpha+\beta}$$

so

$$\frac{C_D(1^{n_1})}{\Pr_D(1^{n_1})} \le \frac{C_D(1^{n_1+n_2})}{\Pr_D(1^{n_1+n_2})} \Longleftrightarrow$$

$$\frac{\alpha\frac{n_1+1}{2} + (1-\alpha)n_1}{\alpha} \le \frac{\alpha\frac{n_1+1}{2} + \beta\left(n_1 + \frac{n_2+1}{2}\right) + (1-\alpha-\beta)(n_1+n_2)}{\alpha+\beta} \Longleftrightarrow$$

$$\alpha - \beta\frac{n_1}{n_2} \ge \alpha\left(\alpha + \beta\frac{1 - n_1/n_2}{2}\right)$$

For the second property, we have

$$C_D(1^{n_1}2^{n_1}\cdots) \le C_U(1^n) \Longleftrightarrow \frac{C_D(1^{n_1})}{\Pr_D(1^{n_1})} \le C_U(1^n)$$

$$\Longleftrightarrow \frac{\alpha\frac{n_1+1}{2} + (1-\alpha)n_1}{\alpha} \le \frac{n+1}{2}$$

$$\Longleftrightarrow \frac{n/n_1 + 1}{2} \ge \frac{1}{\alpha}$$

$\square$

6   Conclusions

Our framework enables the analysis of different strategies to sequentialize algorithms when the objective is to make one succeed as soon as possible.

When the algorithms have the same distribution and are unlimited in number, the optimal strategy is of form $1^m 2^m \cdots$ for some magic m. As the distribution becomes biased, we observe a phase transition from the regular single-algorithm run 1^n (i.e., $m = n$) to the single-step multiple algorithms $123 \cdots$ (i.e., $m = 1$) which is very abrupt in the application we considered: LPN and password recovery.

The phase transition phenomenon is further studied. In particular, we show that the fall from $m = n$ to $m = 1$ does not go through any $m \in \{2, \ldots, \frac{k(k+1)}{2}\}$.

For LPN, the solving algorithm we obtain outperforms the classical ones.

When we have a limited number of algorithms, the optimal strategy has the form $1^{m_1} \cdots |D|^{m_1} 1^{m_2} \cdots |D|^{m_2} \cdots$. For LPN, this simple algorithm outperforms the classical ones, even the one from Asiacrypt 2014 [12] for the relevant parameters using $\tau \sim k^{-\frac{1}{2}}$.

A   Composite Distributions

We give a formula to compute the optimal strategies for distributions obtained by composing several distributions. The formula is useful when we want to regroup equal consecutive p_j's in a distribution D_1 so that D_1 appears as a composition of uniform distributions.

Lemma 29. *Let $U_1, \ldots, U_k$ be independent distributions of support $n_1, \ldots, n_k$, respectively. Let $U_i = (p_{i,1}, \ldots, p_{i,n_i})$. Given a distribution $(\alpha_1, \ldots, \alpha_k)$ of support k, we define $D_1 = \alpha_1 U_1 + \alpha_2 U_2 + \ldots + \alpha_k U_k$ by $D_1 = (\alpha_1 p_{1,1}, \ldots, \alpha_1 p_{1,n_1}, \alpha_2 p_{2,1}, \ldots, \alpha_k p_{k,n_k})$.*
Let $m = \sum_{j=1}^{i} n_j$. We have

$$\Pr_{D_1}(1^{n_1} 1^{n_2} \cdots 1^{n_i}) = \alpha_1 + \cdots + \alpha_i$$

$$C_{D_1}(1^{n_1} 1^{n_2} \cdots 1^{n_i}) = \sum_{j=1}^{i} \alpha_j C_{U_j}(1^{n_j}) + \sum_{j=1}^{i} n_j \left(1 - \sum_{k=1}^{j} \alpha_k \right)$$

We note that if all U_i are ordered and if $\alpha_i p_{i,n_i} \geq \alpha_{i+1} p_{i+1,1}$ for all $1 \leq i < k$, then D_1 is ordered as well.

We let $D = (D_1, D_1, \ldots)$. If we assume that U_i are uniform distributions, we can use the observation following Lemma 26 to deduce from Theorem 17 that the optimal strategy is $1^m 2^m \cdots$ for $m = \sum_{j=1}^{i} n_j$ and i minimizing

$$\mathsf{min}C_D(\emptyset) = \min_i \left(\frac{\sum_{j=1}^{i} \alpha_j C_{U_j}(1^{n_j}) + \sum_{j=1}^{i} n_j \left(1 - \sum_{k=1}^{j} \alpha_k \right)}{\sum_{j=1}^{i} \alpha_j} \right)$$

Proof. We prove it by induction on i. It is trivial for $i = 0$. We assume the result holds for $i - 1$. By induction, we have

$$C_{D_1}(1^{n_1} \cdots 1^{n_i}) = C_{D_1}(1^{n_1} \cdots 1^{n_i-1}) + (1 - \Pr_{D_1}(1^{n_1} \cdots 1^{n_i-1}))C_{D_1}(1^{n_i} | \neg(1^{n_1} \cdots 1^{n_i-1}))$$

$$= \sum_{j=1}^{i-1} \alpha_j C_{U_j}(1^{n_j}) + \sum_{j=1}^{i-1} n_j \left(1 - \sum_{k=1}^{j} \alpha_k \right) + \alpha_i C_{U_i}(1^{n_i}) + n_i \left(1 - \sum_{k=1}^{i} \alpha_k \right)$$

$$= \sum_{j=1}^{i} \alpha_j C_{U_j}(1^{n_j}) + \sum_{j=1}^{i} n_j \left(1 - \sum_{k=1}^{j} \alpha_k \right)$$

The second equality is obtained from the fact that

$$C_{D_1}(1^{n_i} | \neg(1^{n_1} \cdots 1^{n_i-1})) = \frac{\alpha_i}{\alpha_i + \cdots + \alpha_k}(p_{i,1} + 2p_{i,2} + \ldots + n_i p_{i,n_i}) + n_i \left(\frac{\alpha_{i+1} + \cdots + \alpha_k}{\alpha_i + \cdots + \alpha_k} \right)$$

$$= \frac{\alpha_i}{1 - \Pr_{D_1}(1^{n_1} \cdots 1^{n_i-1})} C_{U_i}(1^{n_i}) + n_i \left(\frac{1 - \Pr_{D_1}(1^{n_1} \cdots 1^{n_i-1}) - \alpha_i}{1 - \Pr_{D_1}(1^{n_1} \cdots 1^{n_i-1})} \right)$$

$\square$

B　Proof of Lemma 25

Proof. We will show below that there exists $d > 0$ such that $a \leq b - d$ and $C_D(s) = C_D(uj^d i^a v j^{b-d} w)$. Hence, we can rewrite s by replacing u by uj^d and b by $b - d$. Since $d > 0$ and $a \leq b - d$, we can just apply this rewriting rule enough time until b is lowered down to a. Hence, we obtain the result.

To find d, we first write $s = u_0 i^{m_1} u_1 i^{m_2} \cdots i^{m_r} u_r i^a v j^b w$ where i appears in no u_t, the m_t are nonzero, and $u_1, \ldots, u_r$ are non-empty. (Note that since $a < b$, we must have $m_1 + \cdots + m_r > 0$ so $r \geq 1$.) Let n' be the equal number of occurrences of i and j in $u i^a v j^b$. Let t be the smallest index such that $m_1 + \cdots + m_t > n' - b$. (For $t = 0$, the left-hand term is 0 but $n' \geq b$; for $t = r$, the left-hand term is $n' - a$ and we know that $a < b$; so, t exists and $t > 0$.) We write $m_t = m' + d$ such that $m_1 + \cdots + m_{t-1} + m' = n' - b$. So, $d > 0$. Note that $b - d = b - m_t + m' = n' - m_1 - \cdots - m_t = m_{t+1} + \cdots + m_r + a$. So, $b - d \geq a$. Clearly, $d \leq b$. We write $s = H i^d B i^a v j^d T$ with head $H = u_0 i^{m_1} u_1 i^{m_2} \cdots u_{t-1} i^{m'}$, body $B = u_t i^{m_{t+1}} \cdots i^{m_r} u_r$, and tail $T = j^{b-d} w$. Clearly, H has $n' - b$ occurrences of i and $H i^d B i^a v$ has $n' - b$ occurrences of j. Since s is optimal for D, $i^d B i^a v j^d$ is optimal for $D | \neg H$. We note that B does not start with i (t is between 1 and r and u_t is nonempty and with no i) and that $i^a v$ is non-empty and with no j (either $a \neq 0$ or v is nonempty and with no j). We split $i^d B i^a v j^d = i^d x_1 \cdots x_\ell i^a y_1 \cdots y_{\ell'} j^d$ where two consecutive blocks in the list $i^d, x_1, \ldots, x_\ell, i^a, y_1, \ldots, y_{\ell'}, j^d$ have no key in common. (For $a = 0$, we can always split so that x_ℓ and y_1 have no key in common by using the first term k of v which is not the last of u: we just take y_1 as a block of k's and x_ℓ as a block with no k.) We can apply Lemma 24 and obtain

$$\frac{C_D(i^d | \neg i^{n'-b})}{\Pr_D(i^d | \neg i^{n'-b})} \leq \frac{C_D(i^a | \neg i^{n'-a})}{\Pr_D(i^a | \neg i^{n'-a})} \leq \frac{C_D(y_1 | \neg \cdots)}{\Pr_D(y_1 | \neg \cdots)} \leq \frac{C_D(y_{\ell'} | \neg \cdots)}{\Pr_D(y_{\ell'} | \neg \cdots)} \leq \frac{C_D(j^d | \neg j^{n'-b})}{\Pr_D(j^d | \neg j^{n'-b})}$$

Since the first and the last terms are equal, all of them are equal. So, we can permute two consecutive blocks which have no index in common. Hence, we can propagate j^d earlier until it is stepped before i^a, since we know there is no other occurrence of j in the exchanged blocks. We obtain that

$$C_D(H i^d B i^a v j^d T) = C_D(H i^d B j^d i^a v T)$$

as announced.　　□

References

1. Alekhnovich, M.: More on average case vs approximation complexity. In: Proceedings of the 44th Symposium on Foundations of Computer Science (FOCS 2003), 11–14 October 2003, pp. 298–307. IEEE Computer Society, Cambridge (2003)
2. Avoine, G., Bourgeois, A., Carpent, X.: Analysis of rainbow tables with fingerprints. In: Foo, E., Stebila, D. (eds.) ACISP 2015. LNCS, vol. 9144, pp. 356–374. Springer, Heidelberg (2015)

3. Avoine, G., Carpent, X.: Optimal storage for rainbow tables. In: Lee, H.-S., Han, D.-G. (eds.) ICISC 2013. LNCS, vol. 8565, pp. 144–157. Springer, Heidelberg (2014)

4. Avoine, G., Junod, P., Oechslin, P.: Time-memory trade-offs: false alarm detection using checkpoints. In: Maitra, S., Veni Madhavan, C.E., Venkatesan, R. (eds.) INDOCRYPT 2005. LNCS, vol. 3797, pp. 183–196. Springer, Heidelberg (2005)

5. Blum, A., Kalai, A., Wasserman, H.: Noise-tolerant learning, the parity problem, and the statistical query model. In: Frances Yao, F., Luks, E.M. (eds.) Proceedings of the Thirty-Second Annual ACM Symposium on Theory of Computing, Portland, OR, USA, 21–23 May 2000, pp. 435–440. ACM (2000)

6. Bogos, S., Tramèr, F., Vaudenay, S.: On solving LPN using BKW and variants. Crypt. Commun. (2015, to appear)

7. Bonneau, J.: The science of guessing: analyzing an anonymized corpus of 70 million passwords. In: IEEE Symposium on Security and Privacy, SP 2012, San Francisco, California, USA, 21–23 May 2012, pp. 538–552. IEEE Computer Society (2012)

8. Carrijo, J., Tonicelli, R., Imai, H., Nascimento, A.C.A.: A novel probabilistic passive attack on the protocols HB and HB$^+$. IEICE Trans. **92-A**(2), 658–662 (2009)

9. Damgård, I., Park, S.: Is public-key encryption based on LPN practical? IACR Cryptology ePrint Archive, 2012:699 (2012)

10. Dell'Amico, M., Michiardi, P., Roudier, Y.: Password strength: an empirical analysis. In: INFOCOM 2010, 29th IEEE International Conference on Computer Communications, Joint Conference of the IEEE Computer and Communications Societies, San Diego, CA, USA, 15–19 March 2010, pp. 983–991. IEEE (2010)

11. Fossorier, M.P.C., Mihaljević, M.J., Imai, H., Cui, Y., Matsuura, K.: An algorithm for solving the LPN problem and its application to security evaluation of the HB protocols for RFID authentication. In: Barua, R., Lange, T. (eds.) INDOCRYPT 2006. LNCS, vol. 4329, pp. 48–62. Springer, Heidelberg (2006)

12. Guo, Q., Johansson, T., Löndahl, C.: Solving LPN using covering codes. In: Sarkar, P., Iwata, T. (eds.) ASIACRYPT 2014. LNCS, vol. 8873, pp. 1–20. Springer, Heidelberg (2014)

13. Hellman, M.E.: A cryptanalytic time-memory trade-off. IEEE Trans. Inf. Theory **26**(4), 401–406 (1980)

14. Huang, J., Vaudenay, S., Lai, X., Nyberg, K.: Capacity and data complexity in multidimensional linear attack. In: Gennaro, R., Robshaw, M. (eds.) CRYPTO 2015, Part I. LNCS, vol. 9215, pp. 141–160. Springer, Heidelberg (2015)

15. Levieil, É., Fouque, P.-A.: An improved LPN algorithm. In: De Prisco, R., Yung, M. (eds.) SCN 2006. LNCS, vol. 4116, pp. 348–359. Springer, Heidelberg (2006)

16. Lyubashevsky, V.: The parity problem in the presence of noise, decoding random linear codes, and the subset sum problem. In: Chekuri, C., Jansen, K., Rolim, J.D.P., Trevisan, L. (eds.) APPROX and RANDOM 2005. LNCS, vol. 3624, pp. 378–389. Springer, Heidelberg (2005)

17. Massey, J.L.: Guessing and entropy. In: Proceedings of the 1994 IEEE International Symposium on Information Theory, p. 204, June 1994

18. Meier, W., Staffelbach, O.: Analysis of pseudo random sequences generated by cellular automata. In: Davies, D.W. (ed.) EUROCRYPT 1991. LNCS, vol. 547, pp. 186–199. Springer, Heidelberg (1991)

19. Narayanan, A., Shmatikov, V.: Fast dictionary attacks on passwords using time-space tradeoff. In: Atluri, V., Meadows, C., Juels, A. (eds.) Proceedings of the 12th ACM Conference on Computer and Communications Security, CCS 2005, Alexandria, VA, USA, 7–11 November 2005, pp. 364–372. ACM (2005)

20. Oechslin, P.: Making a faster cryptanalytic time-memory trade-off. In: Boneh, D. (ed.) CRYPTO 2003. LNCS, vol. 2729, pp. 617–630. Springer, Heidelberg (2003)

21. Regev, O.: On lattices, learning with errors, random linear codes, and cryptography. J. ACM **56**(6), 1–40 (2009)
22. Schneier, B.: Real-world passwords, December 2006. https://www.schneier.com/blog/archives/2006/12/realworld_passw.html
23. Weir, M., Aggarwal, S., Collins, M.P., Stern, H.: Testing metrics for password creation policies by attacking large sets of revealed passwords. In: Al-Shaer, E., Keromytis, A.D., Shmatikov, V. (eds.) Proceedings of the 17th ACM Conference on Computer and Communications Security, CCS 2010, Chicago, Illinois, USA, 4–8 October 2010, pp. 162–175. ACM (2010)

Privacy and Lattices

Pure Differential Privacy for Rectangle Queries via Private Partitions

Cynthia Dwork[1], Moni Naor[2], Omer Reingold[3], and Guy N. Rothblum[3(✉)]

[1] Microsoft Research, Mountain View, USA
dwork@microsoft.com
[2] The Weizmann Institute, Rehovot, Israel
moni.naor@weizmann.ac.il
[3] Samsung Research America, Mountain View, USA
omer.reingold@gmail.com, rothblum@alum.mit.edu

Abstract. We consider the task of data analysis with pure differential privacy. We construct new and improved mechanisms for statistical release of interval and rectangle queries. We also obtain a new algorithm for counting over a data stream under continual observation, whose error has optimal dependence on the data stream's length.

A central ingredient in all of these result is a differentially private partition mechanism. Given set of data items drawn from a large universe, this mechanism outputs a partition of the universe into a small number of segments, each of which contain only a few of the data items.

1 Introduction

Differential privacy is a recent privacy guarantee tailored to the problem of statistical disclosure control: how to publicly release statistical information about a set of people without compromising the privacy of any individual [DMNS06] (see the book [DR14] for an extensive treatment). In a nutshell, differential privacy requires that the probability distribution on the published results of an analysis is "essentially the same," independent of whether any individual opts in to, or opts out of, the data set. (The probabilities are over the coin flips of the privacy mechanism.) Statistical databases are frequently created to achieve a social goal, and increased participation in the databases permits more accurate analyses. The differential privacy guarantee supports the social goal by assuring each individual that she incurs little risk by joining the database: anything that can happen is essentially equally likely to do so whether she joins or abstains.

In the differential privacy literature, privacy is achieved by the introduction of randomized noise into the output of an analysis. Moreover, sophisticated mechanisms for differentially private data analysis can incur a significant efficiency

M. Naor—Incumbent of the Judith Kleeman Professorial Chair. Research supported in part by grants from the Israel Science Foundation, BSF and Israeli Ministry of Science and Technology and from the I-CORE Program of the Planning and Budgeting Committee and the Israel Science Foundation (grant No. 4/11).

© International Association for Cryptologic Research 2015
T. Iwata and J.H. Cheon (Eds.): ASIACRYPT 2015, Part II, LNCS 9453, pp. 735–751, 2015.
DOI: 10.1007/978-3-662-48800-3_30

overhead. A rich and growing literature aims to minimize the "cost of privacy" in terms of the error and also in terms of computational efficiency. In this work we present new algorithms with improved error for several natural data analysis tasks.

There are several variants of differential privacy that have been studied. Most notably, these include the stronger (in terms of privacy-protection) notion of *pure* differential privacy, and its relaxation to *approximate* differential privacy. Our work focuses on mechanisms that guarantee *pure* differential privacy for the tasks of answering statistical queries, maintaining an online count of significant events in a data stream, and partitioning a large universe into a small number of contiguous segments, none of which contains too many input items (a type of "dimension reduction").

Before proceeding to outline our contributions, we recall the definition of differential privacy:

Definition 1.1 (Differential Privacy [DMNS06, DKM+06]). *A randomized algorithm* $M : \mathcal{U}^n \to Y$ *is* (ε, δ)-*differentially private if for every pair adjacent databases* x, x' *that differ only in one row, and for every* $S \subset Y$:

$$\Pr[M(x) \in S] \le e^{\varepsilon} \cdot \Pr[M(x') \in S] + \delta.$$

When $\delta = 0$, *we say the algorithm provides* (pure) ε -*differential privacy. When* $\delta > 0$, *we say that the algorithm provides* (approximate) differential privacy.

As discussed above, we focus on the stronger guarantee of *pure* differential privacy throughout this work.

1.1   Differentially Private Query Release: Interval and Rectangle Queries

Differentially private query release is a central problem in the literature. The goal is releasing the answers to a set of statistical queries while maintaining both differential privacy and low error. We focus on the case of *counting queries* (sometimes referred to as statistical queries). Let $\mathcal{U}$ be the set of possible data items (the data universe). A counting query q is specified by a predicate $q : \mathcal{U} \to \{0, 1\}$. For an n-element database $x \in \mathcal{U}^n$, the query output $q(x) \in [0, n]$ counts how many items in the database satisfy the query. The goal, given a set Q of queries and a database x, is to approximate $q(x)$ for each $q \in Q$, while (*i*) guaranteeing differential privacy (for the collection of all answers), and (*ii*) minimizing error in the answers.

We focus on the (challenging) setting where the query set Q is large. To avoid running in time proportional to $|Q|$ (which is too large), we will produce a differentially private *data synopsis*. Given the database x, the mechanism produces a synopsis: a data structure that can later be used to answer any query $q \in Q$. Thus, the synopsis is a small implicit representation for the answers to all queries in Q.

Differentially private query release, especially for counting queries, has been the focus of a rich literature. Starting with the works of Dinur, Dwork and Nissim [DN03, DN04], showed how to answer k queries (counting queries or general low-sensitivity queries) using computationally efficient mechanisms, with noise that grew with k for pure ε-DP [DMNS06], or $\sqrt{k}$ for approximate (ε, δ)-DP. Starting with the work of Blum, Ligett and Roth [BLR08], later works improved the dependence on the number of queries k to logarithmic. The running time for these mechanisms, however, can be prohibitive in many settings. Even the state-of-the-art mechanisms for answering general counting queries [HR10] require running time that is at least linear in the *size of the data universe* $|\mathcal{U}|$ (whereas the running time of earlier mechanisms was logarithmic in $|\mathcal{U}|$). Indeed, for many query sets Q, the best differentially private query release mechanisms that are known require either large error (as a function of $|Q|$), or large running time (as a function of $|\mathcal{U}|$ and $|Q|$). Indeed, under cryptographic assumptions, there are inherent limits on the computational efficiency and the accuracy of differentially private query release algorithms for specific sets of counting queries [DNR+09, Ull13, BUV14]. Thus, a significant research effort has aimed to design efficient and accurate DP mechanisms for specific natural sets of counting queries.

Our work continues this effort. We construct new and improved mechanisms for answering interval or threshold queries. We further extend these results to multi-dimensional rectangle queries, and for these queries we are able to increase the data dimensionality with relatively mild loss in accuracy and efficiency.

Interval Queries. We consider the natural class of *interval queries*. Here the data universe is the integers from 1 to D (i.e. $\mathcal{U} = [1, D]$, and $|\mathcal{U}| = D$).[1] Each query q is specified by an *interval* $I = [i, j] \subseteq [1, D]$, and associated with the predicate that outputs 1 on data elements that fall in that interval. Usually we think of D as being very large, much larger than (even exponential in) the database size n. For example, the data universe could represent a company's salary information, and interval queries approximate the number of employees whose salaries fall in a certain bracket.

In prior work, Dwork et al. [DNPR10] showed that this class could be answered with pure ε-differential privacy and error roughly $O(\frac{\log^2 D}{\varepsilon})$ (see the analysis in [CSS11]). They also showed an $\Omega(\log D)$ error lower bound for obtaining pure differential privacy. Our first contribution is a new mechanism that obtains pure differential privacy with error roughly $O(\frac{\log D + (\log^2 n)}{\varepsilon})$. In particular, the error's dependence on D is optimal.

Theorem 1.2 (DP Intervals). *The mechanism in Sect. 3.2 answers interval queries over $[1, D]$. For any privacy and accuracy parameters $\varepsilon, \beta > 0$, it guarantees (pure) ε-differential privacy. For any database x of size n, with all but β probability over the mechanism's coins, it produces a synopsis that answers all*

---

[1] Throughout this work, for integers i, j s.t. $i \leq j$, we use the notation $[i, j]$ to denote the (closed) interval of integers $\{i, i+1, \ldots, j-1, j\}$.

interval queries (simultaneously) with error $O(\frac{\log D + ((\log^2 n)\cdot\log(1/\beta))}{\varepsilon})$. *The running time to produce the synopsis (and to then answer any interval query) is* $(n \cdot \text{poly}(\log D, \log(1/\varepsilon), \log(1/\beta)))$.

While the error's dependence on $\log D$ is optimal, we do not know whether the dependence on $\log^2 n$ is optimal (i.e., whether the error is tight for cases where D is not much larger than n). This remains a fascinating question for future work.

The main idea behind this mechanism is partitioning the data universe $[1, D]$ into at most n contiguous segments, where the number of items in each segment is not too large. We give a new differentially private mechanism for constructing such a partition, see Sects. 1.3 and 2. Given this partition, we treat the n segments as a new smaller data universe, and use the algorithm of [DNPR10] to answer interval queries on this smaller data universe (this is where we incur the $\log^2 n$ error term).

Related Work: Approximately Private Threshold Queries. The class of interval queries generalizes the class of threshold queries, where each query is specified by $i \in [1, D]$ and counts how many items in the input database are larger than i (i.e., how many items are in the interval $[i, D]$). In fact, since answers to threshold queries can also be used to answer interval queries, these two classes are equivalent. Answering threshold queries with *approximate* (ε, δ)-DP was considered in the work of Beimel, Nissim and Stenner [BNS13], who obtain an upper bound of $2^{O(\log^* D)}$. In a beautiful recent independent work, Bun, Nissim, Stemmer and Vadhan [BNSV15] show a lower bound of $\Omega(\log^* D)$ for approximate-DP mechanisms (as well as an improved upper bound of roughly $2^{\log^* D}$). The main difference with our work is that we focus on the stricter guarantee of *pure* differential privacy, which (provably) incurs a larger error.

Rectangle Queries. We further study a natural generalization of interval queries: *rectangle queries*. These queries consider multi-dimensional data (in particular, c-dimentional for an integer $c > 1$). The data universe is $\mathcal{U} = [1, D]^c$. A rectangle query q is specified by a *rectangle* $R = ([i_1, j_1] \times \ldots \times [i_c, j_c]) \subseteq [1, D]^c$, and associated with the predicate that outputs 1 on data items that fall inside the set R. As was the case for interval queries, we usually think of D as larger than n, and of c as being smaller than either of these quantities (sub-logarithmic in n, or even constant). Continuing the example above, a database could contain employees' salaries, ages, years of experience, rank, etc. Rectangle queries can be used to approximate the numbers of employees that fall into various conjunctions of brackets, e.g. the number of employees in given age, experience and salary brackets. More generally, these queries are useful for multi-dimensional data, where many (or all) of the data dimensions are associated with an ordering on data items in that dimension.

We generalize the intervals mechanism to answer rectangle queries. While in many settings known differentially private algorithms suffer from a "curse of dimensionality" that increases the error or running time as the dimension grows, we give an algorithm whose error and running time have a mild dependence on the data dimensionality. In particular, the error is roughly $O((c^2 \cdot \log D) + ((\log n)^{O(c)}))$. The running time is roughly $n \cdot \text{poly}(\log^c n, \log D)$, and does not grow with D^c. For the (reasonable) setting of parameters where we think of $n \approx \log D$, the running time is only polynomial in $(\log \log D)^c$.

Theorem 1.3 (DP Rectangles). *The mechanism described in Sect. 3.3 answers c-dimensional rectangle queries over $[1, D]^c$. For any privacy and accuracy parameters $\varepsilon, \beta > 0$, the mechanism guarantees (pure) ε-differential privacy. With all but β probability over its coins, all rectangle queries (simultaneously) are answered with error $O(\frac{(c^2 \cdot \log D) + ((\log n)^{O(c)} \cdot \log(1/\beta))}{\varepsilon})$. The running time to produce the synopsis (and to then answer any rectangle query) is $(n \cdot \text{poly}(\log^c n, \log D, \log(1/\varepsilon), \log(1/\beta)))$.*

In prior work, Chan Shi and Song [CSS11] considered rectangle queries and obtained an error bound of roughly $(\log D)^{O(c)}$. Theorem 1.3 roughly replaces this with a $(\log n)^{O(c)}$ term, as well as an additive $O(c^2 \cdot \log D)$ (recall that typically $n << D$). We emphasize that the error's dependence on $\log D$ *does not grow exponentially with the dimentionality c*.

Muthukrishnan and Nikolov [MN12] show an $\Omega((\log n)^{c-O(1)})$ error lower-bound when $n \approx D$, even for (the relaxed notion of) (ε, δ)-differentially private algorithms (they refer to this as "orthogonal range counting"). Thus, the dependance on $\log n$ in the mechanism of Theorem 1.3 is optimal up to a (small) polynomial factor (the exact term in our upper bound is $O((\log n)^{1.5c+1})$).

The rectangles mechanism is a multi-dimensional generalization of the intervals mechanism (see more above and below). Recall that the intervals algorithm utilized a differentially private partition of the data universe into n segments. It then used the "tree-counter" algorithm of [DNPR10] to answer interval queries over these n segments. This is done by building a binary tree of noisy counts, whose leaves are the n segments. For the rectangle mechanism, we use a (k, d)-tree-like data structure (see [Ben75] and see also the rectangle mechanism of [CSS11]), building a "tree of trees" of noisy counts along the c dimensions of the data universe (after reducing the size of each dimension using a differentially private partition). We judiciously prune this tree to avoid an exponential blowup in its size (the naive implementation requires time and memory n^c). A careful analysis guarantees that even while we extend to c dimensions, the error (as a function of D) only grows to $O(c^2 \cdot D)$.

1.2 Counting Under Continual Observation

Dwork, Naor, Pitassi and Rothblum [DNPR10] introduced the problem of *counting under continual observation*. The goal is to monitor a stream of D bits, and continually maintain an approximation of the number of 1's that have been observed so far. For privacy, the entire collection of D outputs (where the i-th output approximates the count after processing i elements) should maintain ε-differential privacy, masking the value of any single bit. The canonical application is monitoring events, such as the number of influenza patients arriving at medical office, or the number of users visiting a webpage (where privacy hides any single access). Since its introduction online counting has found many applications. In most settings, the data stream is *sparse*: the number of 1's (the stream's "weight") is much smaller than D.

The online counter proposed by [DNPR10] (we refer to this as the "tree counter") had error roughly $O(\log^2 D)$ (see the analysis in [CSS11]). As an additional contribution, we present an improved counter (with pure or $(\varepsilon, 0)$ differential privacy) for sparse streams. In particular, thinking of the input as a boolean string where the number of 1's is at most n (and $n \ll D$), the error is improved to roughly $(\log D + (\log^2 n))$ (compared with roughly $(\log^2 D)$ for the tree counter). We note that the dependence on D is optimal, and matches the $\Omega(\log D)$ lower bound in [DNPR10].

Theorem 1.4. *For any $\varepsilon, \beta > 0$, the online counter from Sect. 3.1 guarantees ε-differential privacy. Taking n to be an upper bound on the input stream's weight, with all but β probability over the counter's coins, the maximal error over all D items is at most $O(\frac{\log D + ((\log^2 n) \cdot \log(1/\beta))}{\varepsilon})$.*

Here again, we partition the data stream (of length D) into at most n segments, where the number of items in each segment is not too large. This is done using an *online* partition mechanism, which can process the items one-by-one, and after processing each item can decide whether a segment is large enough to be "sealed", or whether to keep accumulating the current segment (see Sects. 1.3 and 2). Given this *online* partition mechanism, we can run the tree counter of [DNPR10] (or any other counter) on its output. As we process data items, we don't update the count until the current segment is sealed. When a segment is sealed, we feed the count within this segment into the tree counter, and obtain an updated count (we use here the fact that the tree counter can also operate on integer inputs, not just on bits).

1.3   Differentially Private Online Partition

As mentioned above, one of the main tools we use is a (pure) ε-differentially private partition algorithm. Given an n-item database $x \subseteq [1, D]$, this algorithm partitions the data universe $\mathcal{U} = [1, D]$ into (at most) n contiguous segments $(S_1 = [1, s_1], S_2 = [s_1 + 1, s_2], \dots, S_n = [s_{n-1} + 1, D])$ (where the s_i's are all integers). The guarantee is that w.h.p. the number of data elements in each of these segments is small, and bounded by roughly $O(\log D)$. These partitions are pervasive in the applications mentioned above. In a nutshell, we treat the segments as a new and reduced data universe. This reduces the size of the data universe from D to n, an exponential improvement for some of the parameter regimes of interest. Beyond its applications in this work, we find the partition mechanism to be of independent interest, and hope that it will find further applications.

Theorem 1.5. *For any $\varepsilon, \beta > 0$, the Partition mechanism in Sect. 2 guarantees ε-differential privacy. When run on a database of size n, with all* *but β probability over the mechanism's coins, it outputs at most n segments, and each segment is of weight at most $\frac{5(\log D + \log(1/\beta))}{\varepsilon}$. The running time is $n \cdot \mathrm{poly}(\log D, \log(1/\varepsilon), \log(1/\beta))$.*

The Partition algorithm and its analysis are inspired by an algorithm from [DNPR10] for transforming a class of streaming algorithms into ones that are private even under continual observation.

Another important property of this algorithm is that it can be run in an *online* manner. In this setting, the input is treated as a bit-stream of length D. The i-th input $y_i \in \{0,1\}$ indicates whether item i is in the dataset. Thus, this is a *sparse* stream with total weight n. The partition mechanism can process these bits one-by-one, making an online decision about when to "seal" each segment. We use this online of the partition algorithm to obtain an improved online counter.

2   Differentially Private Online Partition

The Mechanism. The (online) partition algorithm processes the input as a stream $x_1, \ldots, x_D \in \{0,1\}$. We use n to denote the weight of the stream (the number of 1's).[2] The output is a partition of $[D]$ into (contiguous) segments $P = (S_1, \ldots, S_j)$, such that:

1. W.h.p. the number of segments j is smaller than n.
2. The weight of the items in each segments is $O((\log D + \log(1/\beta))/\varepsilon)$ (where ε is the privacy parameter).

This is an online algorithm, in the sense that after processing the i-th data item, the algorithm either "seals" a new segment, ending at i, or it keeps the current segment "open" and proceeds to the next data item. We emphasize that the algorithm is oblivious to the input stream's weight. The Partition algorithm and its analysis are inspired by an algorithm from [DNPR10] for transforming a class of streaming algorithms into ones that are private even under continual observation (see also the discussion of the "sparse vector" abstraction in [DR14]).

Theorem 2.1. *For any $\varepsilon, \beta > 0$, the Partition Algorithm of Fig. 1 guarantees ε-differential privacy. Let n be the total weight of the input stream. With all but β probability over the algorithm's coins, it outputs at most n segments, and each segment is of weight at most $\frac{5(\log D + \log(1/\beta))}{\varepsilon}$.*

Before proving the partition algorithm's privacy and accuracy, we remark that the dependence on $\log D$ is optimal by the lower bound of [DNPR10]. Moreover, for an *offline* implementation, where the input is given as an n-item database $x \subseteq [1, D]$, we can reduce the running time to polylog$D$:

---

[2] More generally, we could also work with a stream of integers, and the weight would be the L_1 norm.

Partition (D, ε, β)

Initialize the threshold $T \leftarrow (3(\log D + \log(1/\beta))/\varepsilon)$, and indices $i, j \leftarrow 0$

Repeat the following loop: (each iteration of the loop seals a new segment)

1. Initialize the j-th segment:
   $j \leftarrow j + 1$, $count_j \leftarrow 0$, $\widetilde{T}_j \leftarrow T + Lap(1/\varepsilon)$
2. Repeat the following loop, processing the i-th data item in each iteration:
   (a) $i \leftarrow i + 1$, $count_j \leftarrow count_j + x_i$
   (b) $\widetilde{count}_i \leftarrow count_j + Lap(1/\varepsilon)$

   Keep the j-the segment open until $(\widetilde{count}_i > \widetilde{T}_j)$ or $(i \geq D)$
3. Seal the j-th segment: $s_j \leftarrow i$

Until $(i \geq D)$. Take $m \leftarrow j$ to be the final number of segments
Output the partition $P = \{[1, s_1], [s_1 + 1, s_2], \ldots, [s_{m-1} + 1, D]\}$ and the number of segments m

---

Fig. 1. Online DP partition algorithm

Remark 2.2 [Efficient Offline Implementation]. For the offline settings, where the input is an n-item database $x \subseteq [1, D]$, we can compute the partition in $n \cdot \text{polylog}(D)$ time as follows. We sort the n items so that $x_1 < x_2 < \ldots < x_n$ (where each $x_k \in [1, D]$). We then process the items one by one. When processing the k-th item x_k, assume that the last sealed segment was sealed at s_j. We count the number of database items in $[s_j + 1, x_k]$. This gives a certain probability p that the $(j + 1)$-th segment will be sealed at x_k. Until the $(j + 1)$-th segment is sealed, for every $y \in [x_k, x_{k+1} - 1]$, the probability that the $(j + 1)$-th segment is sealed at y remains p_k (because there are no additional items processed). We can now sample in $\text{polylog}D$ time whether the segment is sealed in the range $[x_k, x_{k+1} - 1]$. If we sample that the segment is sealed at some y^* in this range, then we run the above process again starting at y (with a new probability computed from the updated true count, which becomes 0). If not, then we run it again starting at x_{k+1} (again from the updated the count, which is incremented).

Proof (Proof of Theorem 2.1). We argue privacy and accuracy:

Privacy. Fix databases x, x', which differ in the i-th data item (for $i \in [D]$). Consider a partition P. Take $S_j \in P$ s.t. $i \in S_j$. Since the data streams are identical up to S_j, the probabilities of generating the prefix $S_1, \ldots, S_{j-1}$ are identical on x and x' (for any choice of random coins made in the first $j - 1$ segments, the outcome on both databases is identical). Below, we bound the ratio between the probabilities of generating $S_j = [s_{j-1}, s_j]$ as the j-th segment in both runs. After generating S_j as the j-th segment, the probabilities of the partition's suffix when running on the two databases are again identical, because the data are identical and no state is carried over (beyond the boundary s_j of the j-th segment).

We show a bijection between noise values when running on x and on x', such that for any noise value producing S_j on x, the bijection gives a noise value of similar probability that produces the same output on x'. We conclude that the probability p' of producing S_j on x' is not much smaller than the probability p of producing S_j on x, which implies Differential Privacy.

Towards this, take $\widetilde{T}_j, \widetilde{T}'_j$ be the j-th noisy thresholds in a run on x and on x' (respectively), and similarly take $\widetilde{count}_{s_j}$ and $\widetilde{count}'_{s_j}$ to be the noisy counts in runs on x and on x'. The bijection is defined as follows:

– For the case $x_i = 0$ and $x'_i = 1$, take:

$$\widetilde{T}'_j = \widetilde{T}_j + 1, \ \widetilde{count}'_{s_j} = \widetilde{count}_{s_j}$$

All other noise values are unchanged in the two runs. This bijection guarantees that if no item before s_j sealed the j-th segment on x, then no item before s_j will seal the j-th segment on x' (whose $count$ can only be larger by at most 1 at any point in the segment). Moreover, if s_j seals the j-th segment on x, then it will also seal the j-th segment on x' (because the noisy threshold there is larger by 1, and $count$ at s_j is larger by 1 in x').

– For the case $x_i = 1$ and $x'_i = 0$, take:

$$\widetilde{T}'_j = \widetilde{T}_j, \ \widetilde{count}'_{s_j} = \widetilde{count}_{s_j} + 1$$

All other noise values are unchanged in the two runs. This bijection guarantees that if no item before s_j sealed the j-th segment on x, then no item before s_j will seal the j-th segment on x' (whose $count$ can only be smaller at any point in the segment). Moreover, if s_j seals the j-th segment on x, then it will also seal the j-th segment on x' (because the noisy threshold there is smaller by 1, and $count$ at s_j is also smaller by 1 in x').

Since the bijection changed the magnitude of a single draw from $Lap(1/\varepsilon)$ by at most 1, we conclude that $p' \geq e^{-\varepsilon} \cdot p$, and the algorithm is ε-differentially private.

Accuracy. By construction, the algorithm makes at most $2D$ draws from the $Lap(1/\varepsilon)$ distribution. By the properties of the Laplace distribution, with all but β probability, all of these draws will have magnitude at most $((\log D + \log 2 + \log(1/\beta))/\varepsilon)$. Condition on this event for the remainder of the proof. Under this conditioning, whenever $\widetilde{count} > \widetilde{T}$, we have that $count$ (the true count within the segment) is greater than 0, and so all the segments are non-empty, and there can be at most n segments (because there are only n items in the dataset). Moreover, under the above conditioning, as soon as we have $count \geq 5(\log D + \log(1/\beta))/\varepsilon)$, we also have $\widetilde{count} > \widetilde{T}$, and so no segment can have weight larger than $5(\log D + \log(1/\beta))/\varepsilon)$.

3   From Partitions to Counting, Intervals and Rectangles

In this section we apply the partition algorithm to obtain improved differentially private mechanisms for online counting, and for answering interval and rectangle queries.

3.1   Online Counting Under Continual Observation

Counting under continual observation was first studied by [DNPR10], and has emerged as an important primitive with many applications. Given a stream of D data items (integers or boolean values), the goal is to process the items one-by-one. After processing the i-th item, the counter outputs an approximation to the sum of items $(1 \ldots i)$. Taken together, the counter's D outputs should be differentially private, and mask a change of 1 in any particular data item (flipping a bit if the values are boolean, or adding/subtracting 1 if they are integers). A (D, α, β)-counter guarantees that with all but β probability over its own coins, all D estimates it outputs (simultaneously) have error bounded by α.

Recap: The "Tree Counter". For privacy and error parameters $\varepsilon, \beta > 0$, "tree counter" of [DNPR10] is an ε-differentially private $(D, O(((\log^2 D) \cdot \log(1/\beta))/\varepsilon), \beta)$ counter: W.h.p., for all D outputs simultaneously, the error is bounded by roughly $(\log^2 D)$. The counter works by building a binary tree over the interval $[1, D]$. Each data item is a leaf in the tree, and each internal node at height ℓ (where leaves are at height 0) "covers" a sub-segment of length 2^ℓ. The $(D/2^\ell)$ nodes in height ℓ partition the interval $[1, D]$ into sub-segments of length 2^ℓ. The online counter maintains a noisy sum for the items in each internal node (filling up these counts as the items $(1, \ldots, D)$ are processed). To estimate the number of items in some segment $[1, k]$, they observe that the segment is exactly covered by at most $\log D$ internal nodes of the tree. The counter outputs the sum of these internal nodes as its estimate. The noise for each internal node is drawn from a Laplace distribution with magnitude $O(\log D/\varepsilon)$, so the sum of noises from the $\log D$ noise values is $O(\log^2 D)$ w.h.p. (the error analysis in [DNPR10] is a bit more slack, see [CSS11]). Privacy follows because any "leaf" (i.e. input element) only affects the counts of the $\log D$ internal nodes that "cover" it.

Improved Online Counter via Partitions. We show that the (online) partition algorithm of Sect. 2 gives an improved online counter (with pure or $(\varepsilon, 0)$ Differential Privacy) for the case of *sparse* streams. In particular, thinking of the input as a boolean string where the number of 1's is at most n (and $n << D$), the error is improved to roughly $(\log D + (\log^2 n))$ (compared with roughly $(\log^2 D)$ for the tree counter). We note that the dependence on D is optimal, and matches the $\Omega(\log D)$ lower bound in [DNPR10]. We note that the counter was conceived for (and is usually applied to) scenarios where D is much larger than n.

The improved counter operates by running any online counter (and in partic-ular the tree counter) "on top of" a partition obtained from the (online) partition algorithm. Initializing the count to 0, we process each new data item using the partition algorithm. If the algorithm keeps the current segment open, then we simply maintain the current count. If the algorithm seals a segment, then we "feed" that segment into the (online) counter as a new data item (using the true number of 1's in the current segment). We then update the current count using the counter's output. I.e. the segments of the partition now form the "leaves" of the tree used in the [DNPR10] online counter.[3] By differential privacy of the partition algorithm and the counter, the output of this composed algorithm is also differentially private.

Theorem 3.1 *Composing the Partition algorithm from Fig. 1 with the online tree counter from [DNPR10] gives an online counter. For any $\varepsilon, \beta > 0$, the composed algorithm guarantees ε-differential privacy. Let n be an upper bound on the input stream's weight. With all but β probability over the counter's coins, the maximal error over all D items is at most $O(\frac{\log D + ((\log^2 n) \cdot \log(1/\beta))}{\varepsilon})$.*

Proof. We run the partition algorithm with privacy parameter $(\varepsilon/2)$ and error parameter $(\beta/2)$. By Theorem 2.1, with all but $(\beta/2)$ probability, the online partition algorithm seals at most n segments, where the (true) number of 1's in each segment is at most $\frac{10(\log D + \log(2/\beta))}{\varepsilon}$. We then run the tree counter on this "stream" of n segments, with privacy parameter $(\varepsilon/2)$ and error parameter $(\beta/2)$. The partition into segments is $(\varepsilon/2)$-DP, and the output of the tree counter on the "stream" of n segments (given the true count in each of these segments) is also $(\varepsilon/2)$-DP. By composition of DP mechanisms, the complete output of the composed mechanism is ε-DP. For accuracy:

1. By the error guarantee of the tree counter, the n counts obtained when segments are sealed have error at most $O(\frac{(\log^2 n) \cdot \log(1/\beta)}{\varepsilon})$ (with all but a $(\beta/2)$ probability of error).
2. By the segment-size guarantee of the partition algorithm, the true count in a "open" segment that hasn't been sealed yet is bounded. Thus, the fact that counts are not updated before a segment is sealed incurs only a $O(\frac{\log D + \log(1/\beta)}{\varepsilon})$ additional (additive) error for the $(D - n)$ items that do not "seal" a segment.

By a union bound, with all but β probability, the total error is $O(\frac{\log D + ((\log^2 n) \cdot \log(1/\beta))}{\varepsilon})$.

3.2 Interval Queries

To answer interval queries on a database $x \subseteq [1, D]$, we run the partition algorithm and obtain a privacy-preserving partition of $[1, D]$ into (at most) n disjoint segments $(S_1, \ldots, S_n)$, where w.h.p. the count of items in each segment is small. We then construct a binary tree "on top of" these n segments, as in the improved online counter (see Sect. 3.1). I.e., the n segments are the tree's leaves, and each internal node at height h "covers" 2^h segments. For each node in this tree, covering an interval $[i, j]$, we add independent Laplace noise of magnitude $(\log n/\varepsilon)$, and release the (noisy) size of the intersection $x \cap [i, j]$ (the number of 1's in the interval). Privacy follows because the partition is DP, and given the partition

---

[3] We note that, in general, we could compose *any* online counter with the partition algorithm. We are not using any specific properties of the tree counter.

any data item only changes the counts in $\log n$ nodes of the tree. Note that this offline algorithm can be implemented in time poly($n, \log D$) (see Remark 2.2).

Given the tree of noisy counts, we can answer any interval query $I = [i, j]$ as follows. First, observe that any such interval can be "covered" by at most $2 \log n$ nodes of the tree: a collection of nodes whose (disjoint) leaves form the (minimal) collection of segments whose union contains I. To find such a cover, consider the lowest node k in the tree such that the segments its sub-tree cover the interval I (but this is not true for either of k's children). Now the "left" and "right" parts of the interval I are the parts contained in the left or right sub-trees of k (respectively). The left part of I is covered by at most $\log n$ nodes in the left sub-tree, and the right part of I is covered by at most $\log n$ nodes in the right sub-tree. Note that we can also find this cover efficiently. Once the above cover is obtains, we answer the query by simply outputting the sum of (noisy) counts of the nodes that cover the interval. Accuracy follows by the fact that the counts in each segment are small, and the noise in the sum of noisy counts is also small.

Theorem 3.2 (Theorem 1.2, Restated). *The mechanism described above answers interval queries. For any privacy and accuracy parameters $\varepsilon, \beta > 0$, the mechanism guarantees ε-differential privacy. For any database x of size n, with all but β probability over the mechanism's coins, all interval queries (simultaneously) are answered with error $O(\frac{\log D + ((\log^2 n) \cdot \log(1/\beta))}{\varepsilon})$. The running time to produce the synopsis (which can later be used to answer any interval query) is* poly($n, \log D, \log(1/\varepsilon), \log(1/\beta)$).

Proof. We run the partition algorithm with privacy parameter $(\varepsilon/2)$ and error parameter $(\beta/2)$. By Theorem 2.1, with all but $(\beta/2)$ probability, the online partition algorithm outputs at most n segments, where the (true) number of 1's in each segment is at most $\frac{10(\log D + \log(2/\beta))}{\varepsilon}$. We then build a tree of noisy counts on top of these n segments, adding Laplace noise of magnitude $(2 \log n/\varepsilon)$ to each node's true count, and releasing all of these noisy counts. The partition itself is $(\varepsilon/2)$-DP, and since each data item affects exactly $\log n$ counts in the tree, these noisy counts (taken all together and as a function of the partition) are $(\varepsilon/2)$-DP. Thus, the algorithm's output is altogether ε-DP. For an interval query $I = [i, j]$, we argue accuracy as follows:

The algorithm finds a "minimal cover": A collection of segments $(S_k, \ldots, S_\ell)$ s.t. the union of these segments contains the interval I, and (by minimality) the union of $(S_{k+1}, \ldots, S_{\ell-1})$ is *contained in* I (we ignore the borderline cases where $\ell - k \leq 1$, which is handled similarly). Let us denote the union of $(S_k, \ldots, S_\ell)$ by I', so that $I \subseteq I'$. We have that:

1. The (true) sum of items in I' is well approximated by the sum of noisy counts computed by the algorithm. In particular, with all but $(\beta/2)$ probability, the error in computing this sum (a sum of $\log n$ Laplacian RVs) is $O(\frac{(\log^2 n) \cdot \log(1/\beta)}{\varepsilon})$.

2. The difference between the (true) counts in I' and in I is at most the sum of counts in S_k and S_ℓ. This is because the only items that are in I' but not in I are those in S_k or S_ℓ (recall that I contains the union of $(S_{k+1}, \ldots, S_{\ell-1})$). By the accuracy of the partition algorithm, with all but $(\beta/2)$ probability, this difference is at most $O(\frac{\log D + \log(1/\beta)}{\varepsilon})$

By a union bound, we conclude that with all but β probability, the total error in computing the count on interval I is $O(\frac{\log D + ((\log^2 n) \cdot \log(1/\beta))}{\varepsilon})$.

3.3 Rectangle Queries

To answer c-dimensional rectangle queries on a database $x \subseteq [1, D]^c$, we run the partition algorithm on each "axis" of the input space separately. For each dimension $a \in [1, c]$, we partition the line $[1, D]$ into (at most) n segments, where for each of these segments, the number of input elements whose a-th coordinate falls into that segment is bounded. I.e., we compute a privacy-preserving partition $(S_1^a, \ldots, S_n^a)$, where for all i, the number of database elements whose a-th coordinate falls into S_i^a is bounded. For the remainder of the construction, we will consider the partition of the multi-dimensional space $[1, D]^c$ into a collection of rectangles:

$$\{(S_{i_1}^1 \times \ldots \times S_{i_c}^c)\}_{i_1, \ldots, i_c \in [1, n]}.$$

By the properties of the partition algorithm, these rectangles are disjoint and cover the input space.

Multi-Dimensional Tree. We construct a "multi-dimensional tree" of counts over the above partition. The construction is iterative, proceeding one dimension at a time from 1 to c:

- The dimension-1 tree is a binary tree, whose leaves are the segments $\{S_i^1\}_{i \in [1,n]}$ (as in the intervals algorithm). Each node of this dimension-1 tree corresponds to an interval T^1, a union of some number (a power of 2) of segments $\{S_i^1\}$ from the dimension-1 partition. Each such node contains a noisy count for the number of items whose first coordinate falls in the interval T^1. The node also contains a dimension-2 tree, which we call its "successor".
- For $a \in [2, c]$ each dimension-a tree is a binary tree whose leaves are the segments $\{S_i^a\}_{i \in [1,n]}$. The dimension-a tree has a "predecessor", a dimension-$(a-1)$ tree, corresponding to intervals $(T^1, \ldots, T^{a-1})$ in the first $(a-1)$ dimensions.

  Each node in the dimension-a tree corresponds to an interval T^a, a union of some number (a power of 2) of segments $\{S_i^a\}$ from the dimension-a partition. Each such node contains a noisy count for the number of items s.t. for all $i \in [1, a]$, their i-th coordinate falls in T^i. For $a < c$ (until the "final" dimension), each node also contains a dimensional-$(a+1)$ tree, which we call its "successor".

For privacy parameter ε, the noise added to each count is drawn from a Laplace random variable with magnitude $(4 \log^c n / \varepsilon)$. We view each node in the above construction as specifying a c-dimensional rectangle $T = (T^1 \times \ldots \times T^c)$ (for nodes in dimension-a trees where $a < c$, the intervals $(T^{a+1}, \ldots, T^c)$ are "full" and equal $[1, D]$). Each such node contains a noisy count of the number of input elements that fall into this rectangle, i.e. of $|x \cap T|$. The size of this data structure is roughly n^c. By pruning this tree, removing nodes with small noisy counts (and their successors), we can obtain a data structure of size $O(n \cdot \log^c n)$ (whose construction also requires time $O(n \cdot \log^c n)$), see Remark 3.4 below.

The following two claims will be used in arguing privacy and accuracy:

Claim. Adding or removing an element to the dataset only changes the counts of at most $(2 \log^c n)$ nodes in the multi-dimensional tree.

Proof. Let $x_j \in [1, d]^c$ be a data item. Let $(S_{i_1}^1, \ldots, S_{i_n}^n)$ be the (unique) segments of the partition s.t. the a-th coordinate of x_j is in $S_{i_a}^a$ (for all $a \in [1, c]$). We bound the number of nodes in the tree for which their corresponding rectangle T includes x_j (adding or removing x_j will not affect the counts in any other nodes). In the dimension-1 tree there are only $\log n$ such nodes: the leaf corresponding to the segment $S_{i_1}^1$, and its ancestors in the tree. Now observe that for the other nodes in the dimension-1 tree, their successors (and their successors) will never correspond to rectangles that include x_j. For the $\log n$ nodes that do include x_j, their successors are dimension-2 trees, and they each have $\log n$ nodes that include x_j. Thus, we have $\log^2 n$ nodes in dimension-2 trees that include x_j. For all other nodes, their successors will not include x_j. Continuing as above, we have that in the dimension-a trees there are $\log^a n$ nodes that include x_j. We conclude that in total, the number of nodes in the multi-dimensional tree that include x_j is bounded by:

$$\sum_{a=1}^{c} \log^a n \leq 2 \log^c n$$

Claim. For any rectangle $R = (R^1 \times \ldots \times R^c) \subseteq [1, D]^c$, there exists a tight "covering" of that rectangle using a set of at most $m = (2 \log n)^c$ nodes $T = \{T_1, \ldots, T_m\}$ from the multi-dimensional tree. Taking $Q = \bigcup_i T_i$ we have:

1. R is no larger than Q, in particular $R \subseteq Q$.
2. Q is not "much" larger than R. In particular, for each dimension a there exist segments S_j^a, S_k^a (segments of the a-th partition) s.t. for any element in $y \in (Q \setminus R)$ for some $a \in [1, c]$ the a-th coordinate of y is in either S_j^a or S_k^a (and thus, by the properties of the partition algorithm, the size of $(Q \setminus R)$ is not too large).

Proof. Similarly to the intervals algorithm, we begin with a separate "cover" for the intervals that constitute each dimension of the rectangle R. As in the intervals algorithm, for each dimension $a \in [1, c]$, there exists a collection $\mathcal{T}^a$ of $2 \log n$ intervals corresponding to nodes in the dimension-a tree that "cover" the interval R^a as follows. Taking $Q^a = \bigcup_{T \in \mathcal{T}^a} T$:

1. $R^a \subseteq Q^a$.
2. There exist two segments (S_j^a, S_k^a) of the a-th partition, s.t. $(Q^a \setminus (S_i^a \cup S_j^a)) \subseteq R^a$

Now the claim follows by taking $\mathcal{T}$, the set of tree nodes, to be $\mathcal{T} = (T^1 \times \ldots \times T^c)$. This is a set of at most $(2 \log n)^c$ nodes, as required. Moreover, taking:

$$Q = \bigcup_{T \in \mathcal{T}} T = (Q^1 \times \ldots \times Q^c),$$

by the above properties of the cover on each dimension separately, we get that:

$$R = (R^1 \times \ldots \times R^c) \subseteq (Q^1 \times \ldots \times Q^c) = (\bigcup_{T \in \mathcal{T}} T) = Q.$$

Moreover, for each dimension a we denote $Q_a' = (Q^a \setminus (S_j^a \cup S_k^a))$. We have that $Q' = Q_1' \times \ldots \times Q_c'$ has the properties that $Q' \subseteq R$, and for every element $y \in (Q \setminus Q')$, for some $a \in [1, c]$, its a-th coordinate is in $(S_j^a \cup S_k^a)$.

Answering Rectangle Queries. We use the multi-dimensional tree of noisy counts described above to answer rectangle queries. Given a rectangle query $R = (R^1 \times \ldots \times R^c) \subseteq [1, D]^c$, we decompose it into a "cover" $\mathcal{T}$ of $(2 \log n)^c$ tree nodes as promised in Claim 3.3. We answer the query R by adding up the noisy counts for the these m nodes and outputting this noisy sum. This can be done in time poly($\log^c n$).

Theorem 3.3 (Theorem 1.3, Restated). *The mechanism described above answers c-dimensional rectangle queries. For any privacy and accuracy parameters $\varepsilon, \beta > 0$, the mechanism guarantees ε-differential privacy. With all but β probability over its coins, all rectangle queries (simultaneously) are answered with error $O(\frac{(c^2 \cdot \log D) + (c \cdot (2 \log n)^{1.5c+1} \cdot \log(1/\beta))}{\varepsilon})$.*

Proof. By composition of DP mechanisms, privacy follows directly from: (i) privacy of the Partition algorithm (for computing the c partitions), and (ii) from Claim 3.3 and the fact that we add Laplace noise of magnitude $(4 \log^c n/\varepsilon)$ to each count.

For accuracy, observe that after we partition the axis, there are n^{2c} possible rectangle queries (rectangles whose covers are identical are essentially equivalent). For each such query R, we release a noisy count for its cover $\mathcal{T}$. The noise is a sum of (at most) $(2 \log n)^c$ independent Laplace RVs, each of magnitude $2 \log^c n$. With all but $(\beta/2)$ probability, the maximal noise added to the count of any of these covers is of magnitude at most $O(\frac{c \cdot (2 \log n)^{1.5c+1} \cdot \log(1/\beta)}{\varepsilon})$ (see the analysis for the sum of Laplacian RVs in [CSS11]). So for each rectangle R with cover $\mathcal{T}$, the error in the noisy count for $\mathcal{T}$ is bounded.

We run the partition algorithm c times, each with privacy parameter $(\varepsilon/2c)$ and error parameter $(\beta/2c)$. With all but $(\beta/2c)$ probability, the size of each segment in each of the c partitions is at most $O(\frac{c \cdot (\log D + \log c + \log(1/\beta))}{\varepsilon})$. Every point that is in $\mathcal{T}$ but not in R must have one of its coordinates be in a (fixed) set of $2c$ such segments. Thus, by the second property of the cover $\mathcal{T}$ (see Claim 3.3), the difference between the true counts of R and of $\mathcal{T}$ is at most $O(\frac{c^2 \cdot (\log D + \log c + \log(1/\beta))}{\varepsilon})$. The error bound follows by a triangle inequality (and a union bound).

Remark 3.4. The naive construction of the multi-dimensional tree requires time (and size) n^c. We improve this running time dramatically by judiciously "pruning" the tree. We take a threshold $t = O((\log n)^{c+1} \cdot \log(1/\beta))$, and as we construct the multi-dimensional tree (starting with the dimension-1 tree), for any node whose noisy count is smaller than t, we set that node to be "empty" (noisy count 0), and do not continue to its children in the current tree, nor to its successor. By this choice of t, w.h.p. over the noise, any node that is *not* marked as empty corresponds to a rectangle that is not empty in the input database.

Now when using the noisy counts to reconstruct the answers to a given rectangle, because we might be under-counting for all $(2 \log n)^c$ of the nodes that we use to "cover" the query, we obtain a slightly-larger error of $((\log n)^{O(c)} \cdot \log(1/\beta))$. The advantage, however, is that the running time and the size of the multi-dimensional tree are improved to $O(n \cdot \log^c n)$. To see this, recall that any node that is not marked as "empty" must have at least 1 data item in its corresponding rectangle. The bound on the tree size follows by induction over c (as does the improved running time).

References

[Ben75] Bentley, J.L.: Multidimensional binary search trees used for associative searching. Commun. ACM **18**(9), 509–517 (1975)

[BLR08] Blum, A., Ligett, K., Roth, A.: A learning theory approach to non-interactive database privacy. In: STOC (2008)

[BNS13] Beimel, A., Nissim, K., Stemmer, U.: Private learning and sanitization: pure vs. approximate differential privacy. In: Raghavendra, P., Raskhodnikova, S., Jansen, K., Rolim, J.D.P. (eds.) APPROX/RANDOM 2013. LNCS, vol. 8096, pp. 363–378. Springer, Heidelberg (2013)

[BNSV15] Bun, M., Nissim, K., Stemmer, U., Vadhan, SP.: Differentially private release and learning of threshold functions. CoRR, abs/1504.07553 (2015)

[BUV14] Bun, M., Ullman, J., Vadhan, S.: Fingerprinting codes and the price of approximate differential privacy. In: Proceedings of the 46th Annual ACM Symposium on Theory of Computing, pp. 1–10. ACM (2014)

[CSS11] Hubert Chan, T.-H., Shi, E., Song, D.: Private and continual release of statistics. ACM Trans. Inf. Syst. Secur. **14**(3), 26 (2011)

[DKM+06] Dwork, C., Kenthapadi, K., McSherry, F., Mironov, I., Naor, M.: Our data, ourselves: privacy via distributed noise generation. In: Vaudenay, S. (ed.) EUROCRYPT 2006. LNCS, vol. 4004, pp. 486–503. Springer, Heidelberg (2006)

[DMNS06] Dwork, C., McSherry, F., Nissim, K., Smith, A.: Calibrating noise to sensitivity in private data analysis. In: Halevi, S., Rabin, T. (eds.) TCC 2006. LNCS, vol. 3876, pp. 265–284. Springer, Heidelberg (2006)

[DN03] Dinur, I., Nissim, K.: Revealing information while preserving privacy. In: PODS, pp. 202–210 (2003)

[DN04] Dwork, C., Nissim, K.: Privacy-preserving datamining on vertically partitioned databases. In: Franklin, M. (ed.) CRYPTO 2004. LNCS, vol. 3152, pp. 528–544. Springer, Heidelberg (2004)

[DNPR10] Dwork, C., Naor, M., Pitassi, T., Rothblum, G.N.: Differential privacy under continual observation. In: Proceedings of the 42nd ACM Symposium on Theory of Computing, STOC 2010, Cambridge, Massachusetts, USA, 5–8 June 2010, pp. 715–724 (2010)

[DNR+09] Dwork, C., Naor, M., Reingold, O., Rothblum, G.N., Vadhan, S.: On the complexity of differentially private data release: efficient algorithms and hardness results. In: STOC, pp. 381–390 (2009)

[DR14] Dwork, C., Roth, A.: The algorithmic foundations of differential privacy. Found. Trends Theor. Comput. Sci. 9(3–4), 211–407 (2014)

[HR10] Hardt, M., Rothblum, G.N.: A multiplicative weights mechanism for interactive privacy-preserving data analysis. In: FOCS (2010)

[MN12] Muthukrishnan, S., Nikolov, A.: Optimal private halfspace counting via discrepancy. In: Proceedings of the 44th Symposium on Theory of Computing Conference, STOC 2012, New York, NY, USA, 19–22 May 2012, pp. 1285–1292 (2012)

[Ull13] Ullman, J.: Answering n {2+ o (1)} counting queries with differential privacy is hard. In: Proceedings of the Forty-Fifth Annual ACM Symposium on Theory of Computing, pp. 361–370. ACM (2013)

Implementing Candidate Graded Encoding Schemes from Ideal Lattices

Martin R. Albrecht[1]([⊠]), Catalin Cocis[2], Fabien Laguillaumie[3], and Adeline Langlois[4,5]

[1] Information Security Group, Royal Holloway, University of London, Egham, UK
martin.albrecht@rhul.ac.uk
[2] Technical University of Cluj-Napoca, Cluj-Napoca, Romania
[3] LIP (U. Lyon, CNRS, ENS Lyon, INRIA, UCBL),
Université Claude Bernard Lyon 1, Villeurbanne, France
[4] EPFL, Lausanne, Switzerland
[5] CNRS/IRISA, Rennes, France

Abstract. Multilinear maps have become popular tools for designing cryptographic schemes since a first approximate realisation candidate was proposed by Garg, Gentry and Halevi (GGH). This construction was later improved by Langlois, Stehlé and Steinfeld who proposed GGHLite which offers smaller parameter sizes. In this work, we provide the first implementation of such approximate multilinear maps based on ideal lattices. Implementing GGH-like schemes naively would not allow instantiating it for non-trivial parameter sizes. We hence propose a strategy which reduces parameter sizes further and several technical improvements to allow for an efficient implementation. In particular, since finding a prime ideal when generating instances is an expensive operation, we show how we can drop this requirement. We also propose algorithms and implementations for sampling from discrete Gaussians, for inverting in some Cyclotomic number fields and for computing norms of ideals in some Cyclotomic number rings. Due to our improvements we were able to compute a multilinear jigsaw puzzle for $\kappa = 52$ (resp. $\kappa = 38$) and $\lambda = 52$ (resp. $\lambda = 80$).

Keywords: Algorithms · Implementation · Lattice-based cryptography · Cryptographic multilinear maps

1 Introduction

Multilinear maps, starting with bilinear ones, are popular tools for designing cryptosystems. When pairings were introduced to cryptography [Jou04], many previously unreachable cryptographic primitives, such as identity-based encryption [BF03], became possible to construct. Maps of higher degree of linearity were conjectured to be hard to find – at least in the "realm of algebraic geometry" [BS03]. But in 2013, Garg, Gentry and Halevi [GGH13a] proposed a construction, relying on ideal lattices, of a so-called "graded encoding scheme" that approximates the concept of a cryptographic multilinear map.

© International Association for Cryptologic Research 2015
T. Iwata and J.H. Cheon (Eds.): ASIACRYPT 2015, Part II, LNCS 9453, pp. 752–775, 2015.
DOI: 10.1007/978-3-662-48800-3_31

As expected, graded encoding schemes quickly found many applications in cryptography. Already in [GGH13a] the authors showed how to generalise the 3-partite Diffie-Hellman key exchange first constructed with cryptographic bilinear maps [BS03] to N parties: the protocol allows N users to share a secret key with only one broadcast message each. Furthermore, a graded encoding scheme also allows constructing very efficient broadcast encryption [BS03, BWZ14]: a broadcaster can encrypt a message and send it to a group where only a part of it (decided by the broadcaster before encrypting) will be able to read it. Moreover, [GGH+13b] introduced indistinguishability obfuscation (iO) and functional encryption based on a variant of multilinear maps — multilinear jigsaw puzzles — and some additional assumptions.

The GGH Scheme. For a multilinearity parameter κ, the principle of the symmetric GGH graded encoding scheme is as follows: given a ring R and a principal ideal $\mathcal{I}$ generated by a small secret element $g \in R$, a plaintext is a small element of $R/\mathcal{I}$ and is viewed as a level-0 encoding. Given a level-0 encoding, it is easy increase the level to a higher level $i \leqslant \kappa$, but it is assumed hard to come back to an inferior level. The encodings are additively homomorphic at the same level, and multiplicatively homomorphic up to κ operations. The multiplication of a level-i and a level-j encoding gives a level-$(i + j)$ encoding. Additionally, a zero-testing parameter p_{zt} allows testing if a level-κ element is an encoding of 0, and hence also allows testing if two level-κ encodings are encoding the same elements. Finally, the extraction procedure uses p_{zt} to extract ℓ bits which are a "canonical" representation of a ring element given its level-κ encoding.

More precisely, in GGH we are given $R = \mathbb{Z}[X]/(X^n + 1)$, where n is a power of 2, a secret element z uniformly sampled in $R_q = R/qR$ (for a certain prime number q), and a public element y which is a level-1 encoding of 1 of the form $[a/z]_q$ for some small a in the coset $1 + \mathcal{I}$. We are also given m level-i encodings of 0 named $x_j^{(i)}$, for all $1 \leqslant i \leqslant \kappa$, and a zero-testing parameter p_{zt}. To encode an element of $R/\mathcal{I}$ at level-i (for $i \leqslant \kappa$), we multiply it by y^i in R_q (which give an element of the form $\left[c/z^i\right]_q$, where c is an arbitrary small coset representative). Then, we add a linear combination of encodings of 0 at level-i of the form $\sum_j \rho_j x_j^{(i)}$ to it where the ρ_j are sampled from a certain discrete Gaussian. This last step is the re-randomisation process and ought to ensure that the analogue of the discrete logarithm problem is hard: going from level-i to level-0, for example by multiplying the encoding by y^{-i}. We will see later that the encodings of zero made public for this step are a problem for the security of the scheme.

The asymmetric variant of this scheme replaces levels by "groups" which are identified with subsets of $\{1, \ldots, \kappa\}$. Addition of two elements in the same group stays within the group, multiplying two elements of different groups with disjoint index sets produces an element in the group defined by the union of their index sets. These groups are realised by defining one z_i for each index $1 \leqslant i \leqslant \kappa$ and then dividing by the appropriate product of z_i. Given a group characterised by $S \subseteq \{1, \ldots, \kappa\}$ we call the cardinality of S its level.

We can distinguish between GGH instances where encodings of zero are made publicly available to allow anyone to encode elements and those where this is not the case. The latter are also called "Multilinear Jigsaw Puzzles" and were introduced in [GGH+13b] as a building block for indistinguishability obfuscation. Such instances can be thought of as secret-key graded encoding schemes. To distinguish the two cases, we denote those instances where no encodings of zero $x_j^{(i)}$ are published as GGH_s. In such instances the secret elements g and z_i are required to encode elements at levels above zero.

Security. Already in [GGH13a] it was shown that an attacker can recover the ideal (g) and the coset of (g) for any encoding at level $\leq \kappa$ if encodings of zero are made available. However, since these representatives of either (g) or the cosets are not small, it was believed that these "weak discrete log" attacks would not undermine the central security goal of GGH – the analogue of the BDDH assumption. However, in [HJ15] it was shown that these attacks can be extended to recover short representatives of the cosets. As a consequence, if encodings of zero are published, then [HJ15] breaks the GGH security goals in many scenarios and it is not clear, at present, if and how GGH-like graded encoding schemes can be defended against such attacks. A candidate proposal to prevent weak discrete logarithm attacks was proposed in [CLT15, Appendix G], where the strategy is to change zero testing to make it non-linear in the encodings such that the attack does not work anymore. However, no security analyses was provided in [CLT15] and revision 20150516:083005 of [CLT15] drops any mention of this candidate fix. Hence, the status of GGH-like schemes where encodings of zero are published is currently unclear. However, we note that GGH_s, where no encodings of zero are made available, does not appear to be vulnerable to weak discrete log attacks if the freedom of an attacker to produce encodings of zero at the higher levels is also severely restricted to prevent generalisations of "zeroizing" attacks such as [CGH+15]. Such variants are the central building block of indistinguishability obfuscation, i.e. this case has important applications despite being more limited in functionality. Indeed, at present no known attack threatens the security of indistinguishability obfuscation constructed from graded encoding schemes such as GGH.

Alternative Constructions. An alternative instantiation of graded encoding schemes over the integers promising practicality was proposed by Coron, Lepoint and Tibouchi [CLT13]. This first proposal was also broken in polynomial time using public encodings of zero in [CHL+15]. The attack was later generalised in [CGH+15] and a candidate defence against these attacks was proposed in [CLT15]. The authors of [CLT15] also provided a C++ implementation of a heuristic variant of this scheme. They report that the **Setup** phase of an 7-partite Diffie-Hellman key exchange takes 4528 s (parallelised on 16 cores), publishing a share (**Publish**) takes 7.8 s per party (single core) and the final key derivation (**KeyGen**) takes 23.9 s per party (single core) for a level of security $\lambda = 80$.

Instantiation. The implementation reported in [CLT15] is to date the only implementation of a candidate graded encoding scheme. This is partly because instantiating the original GGH construction is too costly in practice for anything but toy instances. In 2014, Langlois, Stehlé and Steinfeld [LSS14a] proposed a variant of GGH called GGHLite, improving the re-randomisation process of the original scheme. It reduces the number m of re-randomisers, public encodings of zero, needed from $\Omega(n \log n)$ to 2 and also the size of the parameter σ_i^* of the Gaussian used to sample multipliers ρ_j during the re-randomisation phase from $\widetilde{\mathcal{O}}(2^\lambda \lambda n^{4.5} \kappa)$ to $\widetilde{\mathcal{O}}(n^{5.5} \sqrt{\kappa})$. These improvements allow reducing the size of the public parameters and improving the overall efficiency of the scheme. But even though [LSS14a] made a step forward towards efficiency and in some cases no public re-randomisation is required at all (GGH$_s$), GGH-like schemes are still far from being practical.

Our contribution. Our main contribution is a first and efficient implementation of improved GGH-like schemes which we make publicly available under an open-source license. This implementation covers symmetric and asymmetric flavours and we allow encodings of zero to be published or not. However, since the security of GGH-like constructions is unclear when encodings of zero are published, we do not discuss this variant in this paper. We note, however, that our implementation provides a good basis for implementing any future fixes and improvements for GGH-based graded encoding schemes.

   Implementing GGH-like schemes efficiently such that non-trivial levels of multilinearity and security can be achieved is not straight forward and to obtain an implementation we had to address several issues. In particular, we contribute the following improvements to make GGH-like multilinear maps instantiable:

- We show that we do not require (g) to be a prime ideal for the existing proofs to go through. Indeed, sampling an element $g \in \mathbb{Z}[X]/(X^n + 1)$ such that the ideal it generates is prime, as required by GGH and GGHLite, is a prohibitively expensive operation. Avoiding this check is then a key step to allow us to go beyond toy instances.
- We give a strategy to choose practical parameters for the scheme and extend the analysis of [LSS14a] to ensure the correctness of all the procedures of the scheme. Our refined analysis reduces the *bitsize* of q by a factor of about 4, which in turn reduces the required dimension n.
- We apply the analyses from [CS97] to pick parameters to defend against lattice attacks.
- For all steps during the instance generation we provide implementations and algorithms which work in quasi-linear time and efficiently in practice. In particular, we provide algorithms and implementations for inverting in some Cyclotomic number fields, for computing norms of ideals in some Cyclotomic number rings, for producing short representatives of elements modulo (g) and for sampling from discrete Gaussians in $\widetilde{\mathcal{O}}(n)$. For the latter we use Ducas and Nguyen's strategy [Duc13] Our implementation of these operations might

Table 1. Computing a κ-level asymmetric multilinear maps with our implementation without encodings of zero. Column λ gives the minimum security level we accepted, column λ' gives the actually expected security level based on the best known attacks for the given parameter sizes. Timings produced on Intel Xeon CPU E5–2667 v2 3.30 GHz with 256 GB of RAM, parallelised on 16 cores, but not all operations took full advantage of all cores. Setup gives the time for generating the GGH instance. Encode lists the time it takes to reduce an element $\in \mathbb{Z}_p$ with $p = \mathcal{N}(\mathcal{I})$ to a small element in $\mathbb{Z}[X]/(X^n + 1)$ modulo (g). Mult lists the time to multiply κ elements. All times are wall times.

λ	κ	λ'	n	$\log q$	Setup	Encode	Mul	$\|\text{enc}\|$
52	6	64.4	2^{15}	2117	114 s	26 s	0.05 s	8.3 MB
52	9	53.5	2^{15}	3086	133 s	25 s	0.12 s	12.1 MB
52	14	56.6	2^{16}	4966	634 s	84 s	0.62 s	38.8 MB
52	19	56.6	2^{16}	6675	762 s	75 s	1.38 s	52.2 MB
52	25	59.6	2^{17}	9196	2781 s	243 s	5.78 s	143.7 MB
52	52	62.7	2^{18}	19898	26695 s	1016 s	84.1 s	621.8 MB
80	6	155.2	2^{16}	2289	415 s	74 s	0.13 s	17.9 MB
80	9	86.7	2^{16}	3314	445 s	72 s	0.27 s	25.9 MB
80	14	120.9	2^{17}	5288	1525 s	252 s	1.38 s	82.6 MB
80	19	80.4	2^{17}	7089	1821 s	268 s	3.07 s	110.8 MB
80	25	138.8	2^{18}	9721	9595 s	967 s	13.52 s	303.8 MB
80	38	80.3	2^{18}	14649	20381 s	947 s	16.21 s	457.8 MB

be of independent interest (cf. [LP15] for recent work on efficient sampling from a discrete Gaussian distribution), which is why they are available as a separate module in our code.

- We discuss our implementation and report on experimental results.

Our results (cf. Table 1) are promising, as we manage to compute up to multilinearity level $\kappa = 52$ (resp. $\kappa = 38$) at security level $\kappa = 52$ (resp. $\lambda = 80$) in the asymmetric GGH$_\text{s}$ case. We note that much smaller levels of multilinearity have been used to realise non-trivial functionality in the literature. For example, [BLR+15] reports on comparisons between 16-bit encrypted values using a 9-linear map (however, this result holds in a generic multilinear map model). We note that the results in Table 1, where no encodings of zero are made available, are not directly comparable with those reported in [CLT15].

Technical Overview. Our implementation relies on FLINT [HJP14]. However, we provide our own specialised implementations for operations in the ring of integers of Cyclotomic number fields where the degree is a power of two and related rings as listed above.

Our variant of GGH foregoes checking if g generates a prime ideal. During instance generation [GGH13a, LSS14a] specify to sample g such that (g) is a prime ideal. This condition is needed in [GGH13a, LSS14a] to ensure that no non-zero encoding passes the zero-testing test and to argue that the non-interactive N-partite key exchange produces a shared key with sufficient entropy. We show that for both arguments we can drop the requirement that g generates a prime ideal. This was already mentioned as a potential improvement in [Gar13, Section 6.3] but not shown there. As rejection sampling until a prime ideal (g) is found is prohibitively expensive due to the low density of prime ideals in $\mathbb{Z}[X]/(X^n + 1)$, this allows speeding-up instance generation such that non-trivial instances are possible. We also provide fast algorithms and implementations for checking if $(g) \subset \mathbb{Z}[X]/(X^n + 1)$ is prime for applications which still require prime (g).

We also improve the size of the two parameters q and ℓ compared to [LSS14a]. We first perform a finer analysis than [LSS14a], which allows us to reduce the *size* of the parameter q by a factor 2. Then, we introduce a new parameter ξ, which controls what fraction of q is considered "small", i.e. passes the zero-testing test, which reduces the size of q further. This also reduces the number of bits extracted from each coefficient ℓ. Indeed, instead of setting $\ell = 1/4 \log q - \lambda$ where λ is the security parameter, we set $\ell = \xi \log q - \lambda$ with $0 < \xi \leqslant 1/4$. We then show that for a good choice of ξ this is enough to ensure the correctness of the extraction procedure and the security of the scheme. Overall, our refined analysis allows us to reduce the size of $q \approx \left(3n^{\frac{3}{2}}\sigma_1^\star\sigma'\right)^{8\kappa}$ in [LSS14a] to $q \approx \left(3n^{\frac{3}{2}}\sigma_1^\star\sigma'\right)^{(2+\varepsilon)\kappa}$ which, in turn, allows reducing the dimension n. When no encodings of zero are published we simply set $\sigma_1^\star = 1$ and apply the same analysis.

Open Problems. The most pressing question at this point is whether GGH-like constructions are secure. There exist no security proofs for any variant and recent cryptanalysis results recommend caution. Even speculating that secure variants of GGH-like multilinear maps can be found, performance is still an issue. While we manage to compute approximate multilinear maps for relatively high levels of κ in this work, all known schemes are still at least quadratic in κ which presents a major obstacle to efficiency. Any improvement which would reduce this to something linear in κ would mean a significant step forward. Finally, establishing better estimates for lattice reduction and tuning the parameter choices of our schemes are areas of future work.

Roadmap. We give some preliminaries in Sect. 2. In Sect. 3 we describe the GGH-like asymmetric graded encoding schemes and the multilinear jigsaw puzzles used for iO. In Sect. 4, we explain our modifications to GGH-like schemes, especially concerning the parameter q. We also recall a lattice attack to derive the parameter n and show that we do not require (g) to be prime. In Sect. 5, we give the details of our implementation.

2   Preliminaries

Lattices and Ideal Lattices. An m-dimensional *lattice* L is an additive subgroup of $\mathbb{R}^m$. A lattice L can be described by its basis $B = \{b_1, b_2, \ldots, b_k\}$, with $b_i \in \mathbb{R}^m$, consisting in k linearly independent vectors, for some $k \leqslant m$, called the *rank* of the lattice. If $k = m$, we say that the lattice has *full-rank*. The lattice L spanned by B is given by $L = \{\sum_{i=1}^{k} c_i \cdot b_i, c_i \in \mathbb{Z}\}$. The volume of the lattice L, denoted by $\mathrm{vol}(L)$, is the volume of the parallelepiped defined by its basis vectors. We have $\mathrm{vol}(L) = \sqrt{\det(B^T B)}$, where B is any basis of L.

For n a power of two, let $f(X) \in \mathbb{Z}[X]$ be a monic polynomial of degree n (in our case, $f(X) = X^n + 1$). Then, the polynomial ring $R = \mathbb{Z}[X]/f(X)$ is isomorphic to the integer lattice $\mathbb{Z}^n$, i.e. we can identify an element $u(X) = \sum_{i=0}^{n-1} u_i \cdot X^i \in R$ with its corresponding coefficient vector $(u_0, u_1, \ldots, u_{n-1})$. We also define $R_q = R/qR = \mathbb{Z}_q[X]/(X^n + 1)$ (isomorphic to $\mathbb{Z}_q^n$) for a large prime q and $K = \mathbb{Q}[X]/(X^n + 1)$ (isomorphic to $\mathbb{Q}^n$).

Given an element $g \in R$, we denote by $\mathcal{I}$ the principal ideal in R generated by g: $(g) = \{g \cdot u : u \in R\}$. The ideal (g) is also called an *ideal lattice* and can be represented by its $\mathbb{Z}$-basis $(g, X \cdot g, \ldots, X^{n-1} \cdot g)$. We denote by $\mathcal{N}(g)$ its norm. For any $y \in R$, let $[y]_g$ be the reduction of y modulo $\mathcal{I}$. That is, $[y]_g$ is the unique element in R such that $y - [y]_g \in (g)$ and $[y]_g = \sum_{i=0}^{n-1} y_i X^i g$, with $y_i \in [-1/2, 1/2), \forall i, 0 \leq i \leq n - 1$. Following [LSS14a] we abuse notation and let $\sigma_n(b)$ denotes the last singular value of the matrix $\mathrm{rot}(b) \in \mathbb{Z}^{n \times n}$, for any $b \in \mathcal{I}$. For $z \in R$, we denote by $\mathrm{MSB}_\ell \in \{0,1\}^{\ell \cdot n}$ the ℓ most significant bits of each of the n coefficients of z in R.

Gaussian Distributions. For a vector $c \in \mathbb{R}^n$ and a positive parameter $\sigma \in \mathbb{R}$, we define the Gaussian distribution of centre c and width parameter σ as $\rho_{\sigma,c}(x) = \exp(-\pi \frac{\|x-c\|^2}{\sigma^2})$, for all $x \in \mathbb{R}^n$. This notion can be extended to ellipsoid Gaussian distribution by replacing the parameter σ with the square root of the covariance matrix $\Sigma = BB^t \in \mathbb{R}^{n \times n}$ with $\det(B) \neq 0$. We define it by $\rho_{\sqrt{\Sigma},c}(x) = \exp(-\pi \cdot (x - c)^t (B^t B)^{-1}(x - c))$, for all $x \in \mathbb{R}^n$. For L a subset of $\mathbb{Z}^n$, let $\rho_{\sigma,c}(L) = \sum_{x \in L} \rho_{\sigma,c}(x)$. Then, the *discrete Gaussian distribution* over L with centre c and standard deviation σ (resp. $\sqrt{\Sigma}$) is defined as $D_{L,\sigma,c}(y) = \frac{\rho_{\sigma,c}(y)}{\rho_{\sigma,c}(L)}$, for all $y \in L$. We use the notations ρ_σ (resp. $\rho_{\sqrt{\Sigma}}$) and $D_{L,\sigma}$ (resp. $D_{L,\sqrt{\Sigma}}$) when c is 0.

Finally, for a fixed $Y = (y_1, y_2) \in R^2$, we define: $\tilde{\mathcal{E}}_{Y,s} = y_1 D_{R,s} + y_2 D_{R,s}$ as the distribution induced by sampling $\boldsymbol{u} = (u_1, u_2) \in R^2$ from a discrete spherical Gaussian with parameter s, and outputting $y = y_1 u_1 + y_2 u_2$. It is shown in [LSS14a, Theorem 5.1] that if $Y \cdot R^2 = \mathcal{I}$ and $s \geq \max(\|g^{-1} y_1\|_\infty, \|g^{-1} y_2\|_\infty) \cdot n \cdot \sqrt{2 \log(2n(1 + 1/\varepsilon))/\pi}$ for $\varepsilon \in (0, 1/2)$, this distribution is statistically close to the Gaussian distribution $D_{\mathcal{I}, sY^T}$.

3   GGH-like Asymmetric Graded Encoding Scheme

We now recall the definitions given in [GGH+13b, Section 2.2] for the notions of Jigsaw specifier, Multilinear Form and Multilinear Jigsaw puzzle.

Definition 1 ([GGH+13b, **Definition 5**]). *A Jigsaw specifier is a tuple (κ, ℓ, A) where $\kappa, \ell \in \mathbb{Z}^+$ are parameters and A is a probabilistic circuit with the following behavior: On input a prime number q, A outputs the prime q and an ordered set of ℓ pairs $(S_1, a_1), \ldots, (S_\ell, a_\ell)$ where each $a_i \in \mathbb{Z}_q$ and each $S_i \subseteq [\kappa]$.*

Definition 2 ([GGH+13b, **Definition 6 and 7**]). *A Multilinear Form is a tuple* $\mathcal{F} = (\kappa, \ell, \Pi, F)$ *where* $\kappa, \ell \in \mathbb{Z}^+$ *are parameters and* Π *is a circuit with* ℓ *input wires, made out of binary and unary gates.* F *is an assignment of an index set* $I \subseteq [\kappa]$ *to every wire of* Π. *A multilinear form must satisfies constraints given in the original definition (on gates, and the output wire is assigned to* $[\kappa]$).

We say that a Multilinear Form $\mathcal{F} = (\kappa', \ell', \Pi, F)$ *is compatible with* $X = ((S_1, a_1), \ldots, (S_\ell, a_\ell))$ *if* $\kappa = \kappa'$, $\ell = \ell'$ *and the input wires of* Π *are assigned to the sets* $S_1, \ldots, S_\ell$. *The evaluation of* $\mathcal{F}$ *on* X *is then doing arithmetic operations on the inputs depending on the gates. We say that the evaluation succeeds if the final output is* $([\kappa], 0)$.

We now define the Multilinear Jigsaw Puzzles.

Jigsaw Generator: $\mathsf{JGen}(\lambda, \kappa, \ell, \boldsymbol{A}) \rightarrow (q, X, \mathsf{puzzle})$. This algorithm takes as input λ, and a Jigsaw specifier (κ, ℓ, A). It outputs a prime q, a private output X and a public output puzzle. The generator is using a pair of PPT algorithms $\mathsf{JGen} = (\mathsf{InstGen}, \mathsf{Encode})$.

  $\mathsf{InstGen}(\lambda, \kappa) \rightarrow (q, \mathsf{params}, s)$. This algorithm takes λ and κ as inputs and outputs (q, params, s), where q is a prime of size at least 2^λ, params is a description of public parameters, and s is a secret state to pass to the encoding algorithm.
  $\mathsf{Encode}(q, \mathsf{params}, s, (S, a)) \rightarrow (S, u)$. The encoding algorithm takes as inputs the prime q, the parameters params, the secret state s, and a pair (S, a) with $S \subseteq [\kappa]$ and $a \in \mathbb{Z}_q$ and outputs u, an encoding of a relative to S.

  More precisely, the algorithm runs the Jigsaw specifier on input q to get ℓ pairs $(S_1, a_1), \ldots, (S_\ell, a_\ell)$. Then encodes all the plaintext elements by using the Encode algorithm on each (S_i, a_i) which return (S_i, u_i). We have:

$$X = (q, (S_1, a_1), \ldots, (S_\ell, a_\ell)) \text{ and } \mathsf{puzzle} = (\mathsf{params}, (S_1, u_1), \ldots, (S_\ell, u_\ell)).$$

Jigsaw Verifier: $\mathsf{JVer}(\mathsf{puzzle}, \mathcal{F}) \rightarrow \{0, 1\}$. This algorithm takes as input the public output of a Jigsaw Generator puzzle, and a multilinear form $\mathcal{F}$. It outputs either accept (1) or reject (0).

Correctness. For an output (q, X, puzzle) and a form $\mathcal{F}$ compatible with X, we say that the verifier JVer is correct if either the evaluation of $\mathcal{F}$ on X succeeds and $\mathsf{JVer}(\mathsf{puzzle}, \mathcal{F}) = 1$ or the evaluation fails and $\mathsf{JVer}(\mathsf{puzzle}, \mathcal{F}) = 0$. We require that with high probability over the randomness of the generator, the verifier will be correct on all forms.

Security. The hardness assumptions for the Multilinear Jigsaw puzzle requires that for two different polynomial-size families of Jigsaw Specifier $\{(\kappa_\lambda, \ell_\lambda, A_\lambda)\}_{\lambda \in \mathbb{Z}^+}$ and $\{(\kappa_\lambda, \ell_\lambda, A'_\lambda)\}_{\lambda \in \mathbb{Z}^+}$ the public output of the Jigsaw Generator on $(\kappa_\lambda, \ell_\lambda, A_\lambda)$ will be computationally indistinguishable from the public output of the Jigsaw Generator on $(\kappa_\lambda, \ell_\lambda, A'_\lambda)$.

3.1   Using GGH to Construct Jigsaw Puzzles

In Fig. 1, we describe a GGH-like asymmetric graded encoding scheme without encodings of zero based on the definition of GGHLite from [LSS14a]. Below, we explain how to use those procedures to construct the Jigsaw Generator, described in [GGH+13b, Appendix A].

- **Instance generation.** $\mathsf{InstGen}(1^\lambda, 1^\kappa)$: Given security parameter λ and multilinearity parameter κ, determine scheme parameters n, q, σ, σ', ℓ_{g-1}, ℓ_b, ℓ as in [LSS14a]. Then proceed as follows:
  - Sample $g \hookleftarrow D_{R,\sigma}$ until $\|g^{-1}\| \leq \ell_{g-1}$ and $\mathcal{I} = (g)$ is a prime ideal. Define encoding domain $R_g = R/(g)$.
  - Sample $z_i \hookleftarrow U(R_q)$ for all $0 < i \leq \kappa$.
  - Sample $h \hookleftarrow D_{R,\sqrt{q}}$ and define the zero-testing parameter $p_{zt} = \left[\frac{h}{g} \prod_{i=1}^{\kappa} z_i\right]_q$.
  - Return public parameters $\mathsf{params} = (n, q, \ell)$ and p_{zt}.
- **Encode at level-0** $\mathsf{Enc0}(\mathsf{params}, g, e)$: Compute a small representative $e' = [e]_g$ and sample an element $e'' \hookleftarrow D_{e'+\mathcal{I},\sigma'}$. Output e''.
- **Encode in group** $\{i\}$. $\mathsf{Enc}(\mathsf{params}, z_i, e)$: Given parameters params, z_i, and a level-0 encoding $e \in R$, output $[e/z_i]_q$.
- **Adding encodings.** $\mathsf{Add}(\mathsf{params}, u_1, u_2)$: Given encodings $u_1 = \left[c_1/\left(\prod_{i \in S} z_i\right)\right]_q$ and $u_2 = \left[c_2/\left(\prod_{i \in S} z_i\right)\right]_q$ with $S \subseteq \{1, \ldots, \kappa\}$:
  - Return $u = [u_1 + u_2]_q$, an encoding of $[c_1 + c_2]_q$ in the group S.
- **Multiplying encodings.** $\mathsf{Mult}(\mathsf{params}, u_1, u_2)$: Let $S_1 \subset [\kappa]$, $S_2 \subset [\kappa]$ with $S_1 \cap S_2 = \varnothing$, given an encoding $u_1 = \left[c_1/\left(\prod_{i \in S_1} z_i\right)\right]_q$ and an encoding $u_2 = \left[c_2/\left(\prod_{i \in S_2} z_i\right)\right]_q$:
  - Return $u = [u_1 \cdot u_2]_q$, an encoding of $[c_1 \cdot c_2]_q$ in $S_1 \cup S_2$.
- **Zero testing at level** κ. $\mathsf{isZero}(\mathsf{params}, p_{zt}, u)$: Given parameters params, a zero-testing parameter p_{zt}, and an encoding $u = \left[c/\left(\prod_{i=0}^{\kappa-1} z_i\right)\right]_q$ in the group $[\kappa]$, return 1 if $\|[p_{zt}u]_q\|_\infty < q^{3/4}$ and 0 else.

Fig. 1. GGH-like asymmetric graded encoding scheme adapted from [LSS14a].

Jigsaw Generator. The Jigsaw Generator uses $\mathsf{InstGen}$ to generate all the public (params and p_{zt}) and secret parameters of the multilinear map. Each level of the multilinear map will be associated with a subset of the set $[\kappa]$. To create the puzzle pieces, which are encodings of some elements of R at different level, the Generator simply encodes some random elements at level $S \subset [1, \kappa]$, those are given as puzzle.

Jigsaw Verifier. The verifier is given the public parameters params and p_{zt}, a valid form Π (which is defined [GGH+13b, Def. 6] in as a circuit made of binary and unary gates) and puzzle, an input for Π (which are some encodings). The verifier is then evaluating Π on these input using Add for addition gates and Mult for multiplication gates. The verifier must succeeds if the evaluation of $\mathcal{F}$ on X succeeds, which means that the final output of the evaluation is an encoding of zero at level κ. The verifier is invoking the zero-testing procedure, and outputs 1 if the test passes, 0 otherwise.

4   Modifications to and Parameters for GGH-like Schemes

In this section, we first show that we do not require a prime (g) and then describe a method which allows to reduce the size of two parameters: the modulus q and the number ℓ of extracted bits. In Sect. 4.3 then we describe the lattice-attack against the scheme which we use to pick the dimension n. Finally, we describe our strategy to choose parameters that satisfy all these constraints.

4.1   Non-prime (g)

Both GGHLite and GGH-like jigsaw puzzles as specified in Fig. 1 require to sample a g such that (g) is a prime ideal. However, finding such a g is prohibitively expensive. While checking each individual g whether (g) is a prime ideal is asymptotically not slower than polynomial multiplication, finding such a g requires to run this check often. The probability that an element generates a prime ideal is assumed to be roughly $1/(n^c)$ for some constant $c > 1$ [Gar13, Conjecture 5.18], so we expect to run this check n^c times. Hence, the overall complexity is at least quadratic in n which is too expensive for anything but toy instances.

Primality of (g) is used in two proofs. Firstly, to ensure that after multiplying $\kappa + 1$ elements in R_g the product contains enough entropy. This is used to argue entropy of the N-partite non-interactive key exchange. Secondly, to prove that $c \cdot h/g$ is big if $c, h \notin g$ (cf. Lemma 2). Below, we show that we can relax the conditions on g for these two arguments to still go through, which then allows us to drop the condition that (g) should be prime. We note, though, that some other applications might still require g to be prime and that future attacks might find a way to exploit non-prime (g).

Entropy of the Product. The next lemma shows that excluding prime factors $\leqslant 2N$ and guaranteeing $\mathcal{N}(g) \geqslant 2^n$ is sufficient to ensure 2λ bits of entropy in a product of $\kappa + 1$ elements in R_g with overwhelming probability. We note that both conditions hold with high probability, are easy to check and are indeed checked in our implementation.

Lemma 1. *Let $\kappa \geqslant 2$, λ be the security parameter and $g \in \mathbb{Z}[X]/(X^n + 1)$ with norm $p = \mathcal{N}(g) \geqslant 2^n$ such that p has no prime factors $\leqslant 2\kappa + 2$, and such that $n \geqslant \kappa \cdot \lambda \cdot \log(\lambda)$. Then, with overwhelming probability, the product of $\kappa + 1$ uniformly random elements in R_g has at least $\kappa \cdot \lambda \cdot \log(\lambda)/4$ bits of entropy.*

Proof. Write $p = \prod_{i=1}^{r} p_i^{e_i}$ where p_i are distinct primes and $e_i \geq 1$ for all i. Let us consider the set $\mathcal{S} = \{i \in \{1, \dots, r\} : e_i = 1\}$. Then, following [CDKD14] we define $p_s = \prod_{i \in \mathcal{S}} p_i$ as the square-free part of p. Asymptotically, it holds that

$\#\{p \leqslant x : p/p_s > p_s\}$ is $cx^{3/4}$ for some computable constant c (cf. [CDKD14]). Since in our case we have $x \geqslant 2^n$, this implies that with overwhelming probability it holds that $p_s \geqslant \sqrt{p}$ and hence $\log(p_s) \geqslant n/2$.

By the Chinese Remainder Theorem, R_g is isomorphic to $F_1 \times \cdots \times F_r$ where each "slot" $F_i = \mathbb{Z}_{p_i^{e_i}}$. The set of F_i, for $i \in \mathcal{S}$ corresponds to the square-free part of p. Those F_i are fields, and each of them has order $p_i \geqslant 2N$ which means that a random element in such F_i is zero with probability $1/p_i$. In those slots, the product of N elements has E_s bits of entropy, where

$$E_s = \sum_{i \in \mathcal{S}} \left(1 - \frac{N}{p_i}\right) \log(p_i).$$

First, as $p_i \geqslant 2N$ for all $i \in \mathcal{S}$, the quotient $N/p_i \leqslant 1/2$ and then $\left(1 - \frac{N}{p_i}\right) \geqslant 1/2$ for all $i \in \mathcal{S}$. This implies that

$$E_s \geqslant 1/2 \sum_{i \in \mathcal{S}} \log(p_i) = 1/2 \log \left(\prod_{i \in \mathcal{S}} p_i\right) = 1/2 \log(p_s).$$

Because $\log(p_s) \geqslant n/2$, we conclude that $E_s \geqslant \frac{n}{4} \geqslant \frac{\kappa \cdot \lambda \cdot \log(\lambda)}{4}$. $\square$

Probability of False Positive. It remains to be shown that we can ensure that there are no false positives even if (g) is not prime. In [GGH13a, Lemma 3] false positives are ruled out as follows. Let $u = [c/z^\kappa]_q$ where c is a short element in some coset of $\mathcal{I}$, and let $w = [p_{zt} \cdot u]_q$, then we have $w = [c \cdot h/g]_q$. The first step in [GGH13a] is to suppose that $\|g \cdot w\|$ and $\|c \cdot h\|$ are each at most $q/2$, then, since $g \cdot w = c \cdot h \mod q$ we have that $g \cdot w = c \cdot h$ exactly. We also have an equality of ideals: $(g) \cdot (w) = (c) \cdot (h)$, and then several cases are possible. If (g) is prime as in [GGH13a, Lemma 3], then (g) divides either (c) or (h) and either c or h is in (g). As, by construction, none of them is in (g) if c is not in $\mathcal{I}$, either $\|g \cdot w\|$ or $\|c \cdot h\|$ is more than $q/2$. Using this, they conclude that there is no small c (not in $\mathcal{I}$) such that w is small enough to be accepted by the zero-test.

Our approach is to simply notice that all we require is that (g) and (h) are co-prime. Checking if (g) and (h) are co-prime can be done by checking $\gcd(\mathcal{N}(g), \mathcal{N}(h)) = 1$. However, computing $\mathcal{N}(h)$ is rather costly because h is sampled from $D_{\mathbb{Z}^n, \sqrt{q}}$ and hence has a large norm. To deal with this issue we notice that if $\gcd(\mathcal{N}(g), \mathcal{N}(h)) \neq 1$ then we also have $\gcd(\mathcal{N}(g), \mathcal{N}(h \mod g)) \neq 1$ which can be verified with a simple calculation. Now, interpreting $h \mod g$ as "a small representative of h modulo g", we can compute $h \mod g$ as $h - g \cdot \lfloor g^{-1} \cdot h \rceil$, which produces an element of size $\approx \sqrt{n} \cdot \|g\|$. We can use this observation to reduce the complexity of checking if (g) and (h) are co-prime to computing two norms for elements of size $\|g\|$ and $\approx \sqrt{n} \cdot \|g\|$ and taking their gcd. Furthermore, this condition holds with high probability, i.e. we only have to perform this test $\mathcal{O}(1)$ times. Indeed, by ruling out likely common prime factors first, we expect to run this test exactly once. Hence, checking co-primality of (g) and (h) is much cheaper than finding a prime (g) but still rules out false positives.

Finally, we note that recent proposals of indistinguishability obfuscation from multilinear maps [Zim15, AB15] requires composite order maps. These are not the maps we are concerned with here as in [Zim15, AB15] it is assumed that the factorisation of (g) is known. However, we note that our techniques and implementation easily extend to this case by considering $g = g_1 \cdot g_2$ for known co-prime g_1 and g_2.

4.2   Reducing the Size of q

In this section, we show how to reduce q for which we consider the case where re-randomisers are published for level-1 but no other levels. This matches the requirements of the N-partite Diffie-Hellman key exchange but not the Jigsaw puzzle case. However, when no re-randomisers are published we may simply set $\sigma_1^\star = 1$ and apply the same analysis. Hence, assuming that re-randomisers are published fits our framework in all cases and makes our analysis compatible with previous work. We note that the analysis can be easily generalised to accommodate re-randomisers at higher levels than one by increasing q to accommodate "numerator growth".

The size of q is driven from both correctness and security considerations. To ensure the correctness of the zero-testing procedure, [LSS14a] showed the two following lower bounds on q. Equation 1 implies that false negatives do not exist, and Eq. 2 implies that the probability of false positive occurrence is negligible:

$$q > \max \left((n\ell_{g-1})^8, (3n^{\frac{3}{2}}\sigma_1^\star \sigma')^{8\kappa} \right), \tag{1}$$

$$q > (2n\sigma)^4. \tag{2}$$

The strongest constraint for q is given by the inequality $q > (3n^{\frac{3}{2}}\sigma_1^\star \sigma')^{8\kappa}$. It comes from the fact that for any level-κ encoding u of 0, the inequality $\|p_{zt}u\|_\infty < q^{3/4}$ has to hold. The condition is needed for the correctness of zero-testing and extraction.

New parameter ξ. The choice suggested in [LSS14a] is to extract $\ell = \log(q)/4 - \lambda$ bits from each element of the level-κ encoding. We show that this supplies much more entropy than needed and that we can sample a smaller fraction, $\ell = \xi \log(q) - \lambda$ bits. The equation for q can be rewritten in terms of the variable ξ, by setting the initial condition $\|p_{zt} u\|_\infty < q^{1-\xi}$.

Lemma 2 (Adapted from Lemma A.1 in [LSS14b]**).** *Let* $g \in R$ *and* $\mathcal{I} = (g)$, *let* $c, h \in R$ *such that* $c \notin \mathcal{I}$, (g) *and* (h) *are co-prime,* $\|c \cdot h\| < q/2$ *and* $q > (2tn\sigma)^{1/\xi}$ *for some* $t \geqslant 1$ *and any* $0 < \xi \leqslant 1/4$. *Then* $\|[c \cdot h/g]_q\| > t \cdot q^{1-\xi}$.

Proof. From [GGH13a, Lemma 3] and the discussion in Sect. 4.1 we know that since $\|c \cdot h\| < q/2$ we must have $\left\|g \cdot [c \cdot h/g]_q\right\| > q/2$ if (g) and (h) are co-prime (note that $c \cdot h \neq g \cdot [c \cdot h/g]_q$ in $R/(X^n + 1)$). So we have $\left\|g \cdot [c \cdot h/g]_q\right\| > q/2 \implies \sqrt{n} \|g\| \cdot \left\|[c \cdot h/g]_q\right\| > q/2 \implies \left\|[c \cdot h/g]_q\right\| > q/(2n\sigma)$. We have $t \cdot q^{1-\xi} = t \cdot q/q^\xi < t \cdot q/(2tn\sigma) = q/(2n\sigma)$ and the claim follows. $\qquad\square$

Correctness of Zero-Testing. We can obtain a tighter bound on q by refining the analysis in [LSS14a]. Recall that $\|[p_{zt} u]_q\|_\infty = \|[hc/g]_q\|_\infty = \|h \cdot c/g\|_\infty \leqslant \|h\| \cdot \|c/g\| \leqslant \|h\| \cdot \|c\| \cdot \|g^{-1}\|\sqrt{n}$. The first inequality is a direct application of the inequalities between the infinity norm of a product and the product of the Euclidean norms, the second comes from [Gar13, Lemma 5.9].

Since $h \leftarrow D_{R,\sqrt{q}}$, we have $\|h\| \leqslant \sqrt{n}q^{1/2}$. Moreover, c can be written as a product of κ level-1 encodings u_i, for $i = 1, \dots, \kappa$, i.e., $c = \prod_{i=1}^{\kappa} u_i$. Thus, $\|c\| \leqslant (\sqrt{n})^{\kappa-1}(\max_{i=1,\dots,\kappa} \|u_i\|)^\kappa$ since each of the $\kappa - 1$ multiplications brings an extra $\sqrt{n}$ factor. Let $u_{\max}$ be one of the u_i of largest norm. It can be written as $u_{\max} = e \cdot a + \rho_1 \cdot b_1^{(1)} + \rho_2 \cdot b_2^{(1)}$. As we sampled the polynomial g such that $\|g^{-1}\| \leqslant l_{g^{-1}}$ the inequality $\|[p_{zt} u]_q\|_\infty < q^{1-\xi}$ holds if:

$$n l_{g^{-1}} (\sqrt{n})^{\kappa-1} \|(e \cdot a + \rho_1 \cdot b_1^{(1)} + \rho_2 \cdot b_2^{(1)})\|^\kappa < q^{1/2-\xi}. \tag{3}$$

Then, since

$$\|e \cdot a + \rho_1 \cdot b_1^{(1)} + \rho_2 \cdot b_2^{(1)}\|^\kappa \leqslant (\|e\| \cdot \|a\| \sqrt{n} + \|\rho_1\| \cdot \|b_1^{(1)}\| \sqrt{n} + \|\rho_2\| \cdot \|b_2^{(1)}\| \sqrt{n})^\kappa,$$

$e \leftarrow D_{R,\sigma'}, a \leftarrow D_{1+I,\sigma'}, b_1^{(1)}, b_2^{(1)} \leftarrow D_{I,\sigma'}$ and $\rho_1, \rho_2 \leftarrow D_{R,\sigma_1^*}$, we can bound each of these values as $\|e\|, \|a\|, \|b_1^{(1)}\|, \|b_2^{(1)}\| \leqslant \sigma'\sqrt{n}, \|\rho_1\|, \|\rho_2\| \leqslant \sigma_1^*\sqrt{n}$ to get:

$$n l_{g^{-1}} (\sqrt{n})^{\kappa-1} (\sigma'\sqrt{n} \cdot \sigma'\sqrt{n} \cdot \sqrt{n} + 2 \cdot \sigma_1^*\sqrt{n} \cdot \sigma'\sqrt{n} \cdot \sqrt{n})^\kappa < q^{1/2-\xi},$$

$$\left(n l_{g^{-1}} (\sqrt{n})^{\kappa-1} ((\sigma')^2 n^{\frac{3}{2}} + 2\sigma_1^*\sigma' n^{\frac{3}{2}})^\kappa\right)^{\frac{2}{1-2\xi}} < q. \tag{4}$$

In [LSS14a], we had $\xi = 1/4$ (which give $2/(1 - 2\xi) = 4$), we hence have that this analysis allows to save a factor of 2 in the size of q even for the same ξ. If we additionally consider $\xi < 1/4$ bigger improvements are possible. For practical parameter sizes we reduce the size of q by a factor of almost 4 because ξ tends towards zero as κ and λ grow.

Correctness of Extraction. As in [LSS14a], we need that two level-κ encodings u and u' of different elements have different extracted elements, which implies that we need: $\|[p_{zt}(u - u')]_q\|_\infty > 2^{L-\ell+1}$ with $L = \lfloor \log q \rfloor$. This condition follows from Lemma 2 with t satisfying $t \cdot q^{1-\xi} > 2^{L-\ell+1}$, which holds for $t = q^\xi \cdot 2^{-\ell+1}$. As a consequence, the condition $q > (2tn\sigma)^{1/x}$ is still satisfied if we have $\ell > \log_2(8n\sigma)$, and to ensure that $t > 1$ we need that $\ell < \xi \log q + 2$. Finally, to ensure that ε_{ext}, the probability of the extraction to be the same for two different elements, is negligible, we need that $\ell \leqslant \xi \log_2 q - \log_2(2n/\varepsilon_{ext})$.

Picking ξ and q. Putting all constraints together, we let $\ell = \log(8n\sigma)$ and

$$\tilde{q} = n l_{g^{-1}} (\sqrt{n})^{\kappa-1} \left((\sigma')^2 n^{\frac{3}{2}} + 2\sigma_1^*\sigma' n^{\frac{3}{2}}\right)^\kappa.$$

To find ξ we solve $\ell + \lambda = \frac{2\xi}{1-2\xi} \cdot \log \tilde{q}$ for ξ and set $q = \tilde{q}^{\frac{2}{1-2\xi}}$.

4.3   Lattice Attacks

To pick a dimension n we rely on lattice attacks. The most efficient lattice attacks described [GGH13a] rely on computing weak discrete logarithms and hence do not seem to be applicable to either the case where no encodings of zero are published or the case where such attacks are ruled out in some other way. However, we may mount the attack from [CS97] against GGH-like graded encoding schemes. We explain it in the symmetric setting. Assume two encodings of random elements: $u_1 = [e_1/z]_q$ and $u_2 = [e_2/z]_q$. We have

$$\begin{bmatrix} u_1 \\ u_2 \end{bmatrix}_q = \begin{bmatrix} e_1/z \\ e_2/z \end{bmatrix}_q = \begin{bmatrix} e_1 \\ e_2 \end{bmatrix}_q$$

with e_1 and e_2 small. We set up the lattice $\Lambda = \begin{pmatrix} qI & 0 \\ X & I \end{pmatrix}$ where I is the $n \times n$ identity matrix, 0 is the $n \times n$ zero matrix, and U a rotational basis for $[u_1/u_2]_q$. By construction Λ contains the vector (e_1, e_2) which is short. We have $\det(\Lambda) = q^n$ and $\|(e_1, e_2)\| \approx \sqrt{2n}\sigma'$. In contrast, a random lattice with determinant q^n and dimension $2n$ is expected to have a shortest vector of norm $\approx q^{n/2n} = \sqrt{q}$ which is much longer than $\|(e_1, e_2)\|$. While Λ does not constitute a Unique-SVP instance because there are many short elements of norm roughly $\sqrt{2n}\sigma'$ we may consider all of these "interesting". Clearly, there is a gap between those "interesting" vectors and the expected length of short vectors for random lattices. To hedge against potential attacks exploiting this gap, we may hence want to ensure that finding those "interesting" short vectors is hard. The hardness of Unique-SVP instances is determined by the ratio of the second shortest $\lambda_2(\Lambda)$ and the shortest vector $\lambda_1(\Lambda)$ of the lattice. We assume that the complexity of finding a short element in Λ depends on the gap between (e_1, e_2) and $\sqrt{q}$ in a similar way.

In order to succeed, an attacker needs to solve something akin of a Unique-SVP instance with gap $\lambda_2(\Lambda)/\lambda_1(\Lambda)$. We need to pick parameters such that this problem takes at least 2^λ operations. The most efficient technique known in the literature to produce short lattice vectors is to run lattice reduction. The quality of lattice reduction is typically expressed as the root-Hermite factor σ_0. An algorithm with root-Hermite factor σ_0 is expected to output a vector v in a lattice L such that $\|v\| = \sigma_0^n \operatorname{vol}(L)^{1/n}$. Hence, in our case we require $\tau \cdot \sigma_0^{2n} \leqslant \lambda_2(\Lambda)/\lambda_1(\Lambda)$ and thus

$$\sigma_0 \leqslant \left(\frac{\sqrt{q}}{\sqrt{2n} \cdot \sigma' \cdot \tau} \right)^{1/(2n)}, \tag{5}$$

where τ is a constant which depends on the lattice structure and on the reduction algorithm used. Typically $\tau \approx 0.3$ [APS15], which we will use as an approximation.

Currently, the most efficient algorithm for lattice reduction is a variant of the BKZ algorithm [SE94] referred to as BKZ 2.0 [CN11]. However, its running time and behaviour, especially in high dimensions, is not very well understood: there is no consensus in the literature as to how to relate a given σ_0 to computational cost. We estimate the cost of lattice reduction as in [APS15].

We stress, though, that these assumptions requires further scrutiny. Firstly, this attack does not use p_{zt} which means we expect that better lattice attacks can be found eventually. Secondly, we are assuming that the lattice reduction estimates in [APS15] are accurate. However, should these assumptions be falsified, then this part of the analysis can simply be replaced by refined estimates.

4.4 Putting Everything Together

Our overall strategy is as follows. Pick an n and compute parameters σ, σ', $\sigma_1^\star$ as in [LSS14a] and ℓ_g and q as in Sect. 4.2. Now, establish the root-Hermite factor required to carry out the attack in Sect. 4.3 using Equation (5). If this σ_0 is small enough to satisfy security level λ terminate, otherwise double n and restart the procedure.

We give choices of parameters in Table 2.

Table 2. Parameter choices for multilinear jigsaw puzzles.

λ	κ	n	q	$\| \text{ enc } \|$	$\| \text{ params } \|$	σ_0	BKZ Enum	BKZ Sieve
52	2	2^{14}	$\approx 2^{781.5}$	$\approx 2^{23.6}$	$\approx 2^{23.6}$	1.006855	$\approx 2^{112.2}$	$\approx 2^{101.8}$
52	4	2^{15}	$\approx 2^{1469.0}$	$\approx 2^{25.5}$	$\approx 2^{25.5}$	1.007031	$\approx 2^{110.4}$	$\approx 2^{102.3}$
52	6	2^{15}	$\approx 2^{2114.9}$	$\approx 2^{26.0}$	$\approx 2^{26.0}$	1.010477	$\approx 2^{64.4}$	$\approx 2^{83.3}$
52	10	2^{15}	$\approx 2^{3406.8}$	$\approx 2^{26.7}$	$\approx 2^{26.7}$	1.017404	$\approx 2^{53.5}$	$\approx 2^{68.6}$
52	20	2^{16}	$\approx 2^{7014.8}$	$\approx 2^{28.8}$	$\approx 2^{28.8}$	1.018311	$\approx 2^{56.6}$	$\approx 2^{71.7}$
52	40	2^{17}	$\approx 2^{14599.3}$	$\approx 2^{30.8}$	$\approx 2^{30.8}$	1.019272	$\approx 2^{59.6}$	$\approx 2^{74.8}$
52	80	2^{18}	$\approx 2^{30508.4}$	$\approx 2^{32.9}$	$\approx 2^{32.9}$	1.020258	$\approx 2^{62.7}$	$\approx 2^{77.8}$
52	160	2^{18}	$\approx 2^{60827.8}$	$\approx 2^{33.9}$	$\approx 2^{33.9}$	1.040912	$\approx 2^{54.0}$	$\approx 2^{54.0}$
80	2	2^{14}	$\approx 2^{837.5}$	$\approx 2^{23.7}$	$\approx 2^{23.7}$	1.007451	$\approx 2^{98.2}$	$\approx 2^{94.5}$
80	4	2^{15}	$\approx 2^{1525.0}$	$\approx 2^{25.6}$	$\approx 2^{25.6}$	1.007330	$\approx 2^{103.7}$	$\approx 2^{98.8}$
80	6	2^{16}	$\approx 2^{2287.2}$	$\approx 2^{27.2}$	$\approx 2^{27.2}$	1.005661	$\approx 2^{160.9}$	$\approx 2^{128.3}$
80	10	2^{17}	$\approx 2^{3844.7}$	$\approx 2^{28.9}$	$\approx 2^{28.9}$	1.004882	$\approx 2^{209.0}$	$\approx 2^{150.9}$
80	20	2^{18}	$\approx 2^{7824.9}$	$\approx 2^{30.9}$	$\approx 2^{30.9}$	1.005074	$\approx 2^{198.9}$	$\approx 2^{148.5}$
80	40	2^{19}	$\approx 2^{16152.9}$	$\approx 2^{33.0}$	$\approx 2^{33.0}$	1.005294	$\approx 2^{188.4}$	$\approx 2^{145.7}$
80	80	2^{20}	$\approx 2^{33546.4}$	$\approx 2^{35.0}$	$\approx 2^{35.0}$	1.005528	$\approx 2^{179.7}$	$\approx 2^{143.6}$
80	160	2^{21}	$\approx 2^{69810.9}$	$\approx 2^{37.1}$	$\approx 2^{37.1}$	1.005769	$\approx 2^{171.3}$	$\approx 2^{141.4}$

5 Implementation

Our implementation relies on FLINT [HJP14]. We use its data types to encode elements in $\mathbb{Z}[X]$, $\mathbb{Q}[X]$, and $\mathbb{Z}_q[X]$ but re-implement most non-trivial operations for the ring of integers of a Cyclotomic number field where the degree is a power of two. Other operations — such as Gaussian sampling or taking approximate inverses — are not readily available in FLINT and are hence provided by our implementation. For computation with elements in $\mathbb{R}$ we use MPFR's mpfr_t [The13] with precision 2λ if not stated otherwise. Our implementation is available under the GPLv2+ license at https://bitbucket.org/malb/gghlite-flint. We give experimental results for computing multilinear maps using our implementation in Table 1.

For all operations considered in this section naive algorithms are available in $\mathcal{O}\left(n^2 \log q\right)$ or $\mathcal{O}\left(n^3 \log n\right)$ bit operations. However, the smallest set of parameters we consider in Table 1 is $n = 2^{15}$ which implies that if implemented naively each operation would take 2^{49} bit operations for the smallest set of parameters we consider. Even quadratic algorithms can be prohibitively expensive. Hence, in order to be feasible, all algorithms should run in quasi-linear time in n, or more precisely in $\mathcal{O}\left(n \log n\right)$ or $\mathcal{O}\left(n \log^2 n\right)$. All algorithms discussed in this section run in quasi-linear time.

5.1 Polynomial Multiplication in $\mathbb{Z}_q[X]/(X^n + 1)$

During the evaluation of a GGH-style graded encoding scheme multiplications of polynomials in $\mathbb{Z}_q[X]/(X^n+1)$ are performed. Naive multiplication takes $\mathcal{O}\left(n^2\right)$ time in n, Asymptotically fast multiplication in this ring can be realised by first reducing to multiplication in $\mathbb{Z}[X]$ and then to the Sch?nehage-Strassen algorithm for multiplying large integers in $\mathcal{O}(n \log n \log \log n)$. This is the strategy implemented in FLINT, which has a highly optimised implementation of the Sch?nehage-Strassen algorithm. Alternatively, we can get an $\mathcal{O}(n \log n)$ algorithm by using the *Number-Theoretic Transform* (NTT). Furthermore, using a negative wrapped convolution we can avoid reductions modulo $(X^n + 1)$:

Theorem 1 (Adapted from [Win96]**).** *Let ω_n be a nth root of unity in $\mathbb{Z}_q$ and $\varphi^2 = \omega_n$. Let $a = \sum_{i=0}^{n-1} a_i X^i$ and $b = \sum_{i=0}^{n-1} b_i X^i \in \mathbb{Z}_q[X]/(X^n + 1)$. Let $c = a \cdot b \in \mathbb{Z}_q[X]/(X^n+1)$ and let $\overline{a} = (a_0, \varphi a_1, \ldots, \varphi^{n-1} a_{n-1})$ and define $\overline{b}$ and $\overline{c}$ analogously. Then $\overline{c} = 1/n \cdot NTT_{\omega_n}^{-1}(NTT_{\omega_n}(\overline{a}) \odot NTT_{\omega_n}(\overline{b}))$.*

The NTT with a negative wrapped convolution has been used in lattice-based cryptography before, e.g. [LMPR08]. We note that if we are doing many operations in $\mathbb{Z}_q[X]/(X^n + 1)$ we can avoid repeated conversions between coefficient and "evaluation" representations, $(f(1), f(\omega_n), \ldots, f(\omega_n^{n-1}))$, of our elements, which reduces the amortised cost from $\mathcal{O}(n \log n)$ to $\mathcal{O}(n)$. That is, we can convert encodings to their evaluation representation once on creation and back only when running extraction. We implemented this strategy. We observe a considerable overall speed-up with the strategy of avoiding the conversions where possible. We also note that operations on elements in their evaluation representation are embarrassingly parallel.

5.2 Computing Norms in $\mathbb{Z}[X]/(X^n + 1)$

During instance generation we have to compute several norms of elements in $\mathbb{Z}[X]/(X^n + 1)$. The norm $\mathcal{N}(f)$ of an element f in $\mathbb{Z}[X]/(X^n + 1)$ is equal to the resultant $\mathrm{res}(f, X^n + 1)$. The usual strategy for computing resultants over the integers is to use a multi-modular approach. That is, we compute resultants modulo many small primes q_i and then combine the results using the Chinese Remainder Theorem. Resultants modulo a prime q_i can be computed in $\mathcal{O}(M(n) \log n)$ operations where $M(n)$ is the cost of one multiplication in $\mathbb{Z}_{q_i}[X]/(X^n + 1)$. Hence, in our setting computing the norm costs $\mathcal{O}(n \log^2 n)$ operations without specialisation.

However, we can observe that $\mathrm{res}(f, X^n + 1) \mod q_i$ can be rewritten as $\prod_{(X^n+1)(x)=0} f(x) \mod q_i$ as $X^n + 1$ is monic, i.e. as evaluating f on all roots of $X^n + 1$. Picking q_i such that $q_i \equiv 1 \mod 2n$ this can be accomplished using the NTT reducing the cost mod q_i to $\mathcal{O}(M(n))$ saving a factor of $\log n$, which in our case is typically > 15.

5.3   Checking if (g) is a Prime Ideal

While we show in Sect. 4.1 that we do not necessarily require a prime (g), some applications might still rely on this property. We hence provide an implementation for sampling such g.

To check whether the ideal generated by g is prime in $\mathbb{Z}[X]/(X^n + 1)$ we compute the norm $\mathcal{N}(g)$ and check if it is prime which is a sufficient but not necessary condition. However, before computing full resultants, we first check if $\mathrm{res}(g, X^n + 1) = 0 \mod q_i$ for several "interesting" primes q_i. These primes are 2 and then all primes up to some bound with $q_i \equiv 1 \mod n$ because these occur with good probability as factors. We list timings in Table 3.

Table 3. Average time of checking primality of a single (g) on Intel Xeon CPU E5–2667 v2 3.30 GHz with 256 GB of RAM using 16 cores.

n	$\log \sigma$	wall time	n	$\log \sigma$	wall time	n	$\log \sigma$	wall time
1024	15.1	0.54 s	2048	16.2	3.03 s	4096	17.3	20.99 s

5.4   Verifying that $(b_1^{(1)} b_2^{(1)}) = (g)$

If re-randomisation elements are required, then it is necessary that they generate all of (g), i.e. $(b_1^{(1)}, b_2^{(1)}) = (g)$. If $b_i^{(1)} = \tilde{b}_i^{(1)} \cdot g$ for $0 < i \leqslant 2$ then this condition is equivalent to $(\tilde{b}_1^{(1)}) + (\tilde{b}_2^{(1)}) = R$. We check the sufficient but not necessary condition $\gcd(\mathrm{res}(\tilde{b}_1^{(1)}, X^n + 1), \mathrm{res}(\tilde{b}_2^{(1)}, X^n + 1)) = 1$, i.e. if the respective ideal norms are co-prime. This check, which we have to perform for every candidate pair $(\tilde{b}_1^{(1)}, \tilde{b}_2^{(1)})$, involves computing two resultants and their gcd which is quite expensive. However, we observe that $\gcd(\mathrm{res}(\tilde{b}_1^{(1)}, X^n + 1), \mathrm{res}(\tilde{b}_2^{(1)}, X^n + 1)) \neq 1$ when $\mathrm{res}(\tilde{b}_1^{(1)}, X^n + 1) = 0 = \mathrm{res}(\tilde{b}_2^{(1)}, X^n + 1) \mod q_i$ for any modulus q_i. Hence, we first check this condition for several "interesting" primes and resample if this condition holds. These "interesting" primes are the same as in the previous section. Only if these tests pass, we compute two full resultants and their gcd.

Indeed, after having ruled out small common prime factors it is quite unlikely that the gcd of the norms is not equal to one which means that with good probability we will perform this expensive step only once as a final verification. However, this step is still by far the most time consuming step during setup even with our optimisations applied. We note that a possible strategy for reducing setup time is to sample $m > 2$ re-randomisers $b_i^{(1)}$ and to apply some bounds on the probability of m elements $\tilde{b}_i^{(1)}$ sharing a prime factor (after excluding small prime factors).

5.5   Computing the Inverse of a Polynomial Modulo $X^n + 1$

Instance generation relies on inversion in $\mathbb{Q}[X]/(X^n + 1)$ in two places. Firstly, when sampling g we have to check that the norm of its inverse is bounded by ℓ_g. Secondly, to set up our discrete Gaussian samplers we need to run many inversions in an iterative process. We note that for computing the zero-testing parameter we only need to invert g in $\mathbb{Z}_q[X]/(X^n + 1)$ which can be realised in n inversions in $\mathbb{Z}_q$ in the NTT representation.

In both cases where inversion in $\mathbb{Q}[X]/(X^n + 1)$ is required approximate solutions are sufficient. In the first case we only need to estimate the size of g^{-1} and in the second case inversion is a subroutine of an approximation algorithm (see below). Hence, we implemented a variant of [BCMM98] to compute the approximate inverse of a polynomial in $\mathbb{Q}[X]/(X^n + 1)$, with n a power of two.

The core idea is similar to the FFT, i.e. to reduce the inversion of f to the inversion of an element of degree $n/2$. Indeed, since n is even, $f(X)$ is invertible modulo $X^n + 1$ if and only if $f(-X)$ is also invertible. By setting $F(X^2) = f(X)f(-X) \mod X^n + 1$, the inverse $f^{-1}(X)$ of $f(X)$ satisfies

$$F(X^2)\, f^{-1}(X) = f(-X) \quad (\mathrm{mod}\ X^n + 1). \tag{6}$$

Let $f^{-1}(X) = g(X) = G_e(X^2) + XG_o(X^2)$ and $f(-X) = F_e(X^2) + XF_o(X^2)$ be split into their even and odd parts respectively. From Eq. 6, we obtain $F(X^2)$ $(G_e(X^2)+XG_o(X^2)) = F_e(X^2)+XF_o(X^2) \pmod{X^n+1}$ which is equivalent to

$$\begin{cases} F(X^2)G_e(X^2) = F_e(X^2) \pmod{X^n + 1} \\ F(X^2)G_o(X^2) = F_o(X^2) \pmod{X^n + 1}. \end{cases}$$

From this, inverting $f(X)$ can be done by inverting $F(X^2)$ and multiplying polynomials of degree $n/2$. It remains to recursively call the inversion of $F(Y)$ modulo $(X^{n/2} + 1)$ (by setting $Y = X^2$). This leads to an algorithm for approximately inverting elements of $\mathbb{Q}[X]/(X^n + 1)$ when n is a power of 2 which can be performed in $\mathcal{O}(n\log^2(n))$ operations in $\mathbb{Q}$. We give experimental results in Table 4.

We give experimental results comparing Algorithm 1 with FLINT's extended GCD algorithm in Table 4 which highlights that computing approximate inverses instead of exact inverses is necessary for anything but toy instances.

Algorithm 1. Approximate inverse of $f(X) \mod X^n + 1$ using `prec` bits of precision

---

if $n = 1$ **then**
    $g_0 \leftarrow f_0^{-1}$
else
    $F(X^2) \leftarrow f(X)f(-X) \mod X^n + 1$
    $\tilde{F}(Y) = F(Y)$ truncated to `prec` bits of precision
    $G(Y) \leftarrow \text{InverseMod}(\tilde{F}(Y), q, n/2)$
    Set $F_e(X^2), F_o(X^2)$ such that $f(-X) = F_e(X^2) + XF_o(X^2)$
    $T_e(Y), T_o(Y) \leftarrow G(Y) \cdot F_e(Y), G(Y) \cdot F_o(Y)$
    $f^{-1}(X) \leftarrow T_e(X^2) + XT_o(X^2)$
    $\tilde{f}^{-1}(X) = f^{-1}(X)$ truncated to `prec` bits of precision
    return $\tilde{f}^{-1}(X)$
end if

Table 4. Inverting $g \hookleftarrow D_{\mathbb{Z}^n, \sigma}$ with FLINT's extended Euclidean algorithm ("xgcd"), our implementation with precision 160 ("160"), iterating our implementation until $\|\tilde{f}^{-1}(X) \cdot f(X)\| < 2^{-160}$ ("160iter") and our implementation without truncation ("∞") on Intel Core i7–4850HQ CPU at 2.30 GHz, single core.

n	$\log \sigma$	xgcd	160	160iter	∞
4096	17.2	234.1 s	0.067 s	0.073 s	121.8 s
8192	18.3	1476.8 s	0.195 s	0.200 s	755.8 s

5.6 Small Remainders

The Jigsaw Generator as defined in [GGH+13b, Definition 8] takes as input elements a_i in $\mathbb{Z}_p$ where $p = \mathcal{N}(\mathcal{I})$ and produces level encodings with respect to some source group S_i. In particular, this algorithm produces some small representative of the coset a_i modulo (g) from large integers of size $\approx (\sigma\sqrt{n})^n$ if we represents elements in $\mathbb{Z}_p$ as integers $0 \leqslant a_i < p$. This can be accomplished by using Babai's trick and that g is small, i.e. by computing $a_i - g \cdot \lfloor g^{-1} \cdot a_i \rceil$ in $\mathbb{Q}[X]/(X^n + 1)$. However, in order for this operation to produce sufficiently small elements, we need g^{-1} either exactly or with high precision. Computing such a high quality approximation of g^{-1} can be prohibitively expensive in terms of memory and time. Our strategy for computing with a lower precision is to rewrite a_i as

$$a_i = \sum_{j=0}^{\lceil \log_2(a_i)/B \rceil} 2^{B \cdot j} \cdot a_{ij}$$

where $a_{ij} < 2^B$ for some B. Then, we compute small representatives for all $2^{B \cdot j}$ and a_{ij} using an approximation of g^{-1} with precision B. Finally, we multiply the small representatives for $2^{B \cdot j}$ and a_{ij} and add up their products. This produces a somewhat short element which we then reduce using our approximation of g^{-1} with precision B until its size does not decrease any more.

5.7   Sampling from a Discrete Gaussian

While the strategy in Sect. 5.6 produces short elements it does not necessarily produce elements which follow a spherical Gaussian distribution and hence do not leak geometric information about g. To produce such samples we need to sample from the discrete Gaussian $D_{(g),\sigma',c}$ where c is a small representative of a coset of (g). Furthermore, if encodings of zero are published, we are required to sample from $D_{(g),\sigma',0}$ and $D_{(g),\sigma',1}$. For this, a fundamental building block is to sample from the integer lattice. We implemented a discrete Gaussian sampler over the integers both in arbitrary precision – using MPFR — and in double precision — using machine `doubles`. For both cases we implemented rejection sampling from a uniform distribution with and without table ("online") lookups [GPV08] and Ducas et al's sampler which samples from $D_{\mathbb{Z},k\sigma_2}$ where σ_2 is a constant [DDLL13, Algorithm 12]. Our implementation automatically chooses the best algorithm based on σ, c and τ (the tail cut). In our case σ is typically relatively large, so we call the latter whenever sampling with a centre $c \in \mathbb{Z}$ and the former when $c \notin \mathbb{Z}$. We list example timings of our discrete Gaussian sampler in Table 5. We note that in our implementation we — conservatively — only make use of the arbitrary precision implementation of this sampler with precision 2λ.

Table 5. Example timings for discrete Gaussian sampling over $\mathbb{Z}$ on Intel Core i7–4850HQ CPU at 2.30 GHz, single core.

Algorithm	σ	c	double		mpfr_t	
			prec	samp./s	prec	samp./s
Tabulated [GPV08, SampleZ]	10000	1.0	53	660.000	160	310.000
Tabulated [GPV08, SampleZ]	10000	0.5	53	650.000	160	260.000
Online [GPV08, SampleZ]	10000	1.0	53	414.000	160	9.000
Online [GPV08, SampleZ]	10000	0.5	53	414.000	160	9.000
[DDLL13, Algorithm 12]	10000	1.0	53	350.000	160	123.000

Using our discrete Gaussian sampler over the integers we implemented discrete Gaussian samplers over lattices. Implemented naively this takes $\mathcal{O}(n^3 \log n)$ operations even if we ignore issues of precision. Following [Duc13], we implemented a variant of [Pei10] which we reproduce in Algorithm 2. Namely, we first observe that $D_{(g),\sigma',0} = g \cdot D_{R,\sigma' \cdot g^{-T}}$ and then use [Pei10, Algorithm 1] to sample from $D_{R,\sigma' \cdot g^{-T}}$ where g^{-T} is the conjugate of g^{-1}. That is, $g_0^T = g_0$ and $g_{n-i}^T = -g_i$ for $1 \leqslant i < n$ for $\deg(g) = n - 1$. We then proceed as follows. We first compute an approximate square root (see below) of $\Sigma_2' = g^{-T} \cdot g^{-1}$ up to λ bits of precision. We perform operations with $\log(n) + 4\,(\log(\sqrt{n} \parallel \sigma \parallel))$ bits of precision. If the square root does not converge for this precision, we double it and start over. We then use this value, scaled appropriately, as the initial value from which to start computing a square-root of $\Sigma_2 = \sigma'^2 \cdot g^{-T} \cdot g^{-1} - r^2$ with $r = 2 \cdot \lceil \sqrt{\log n} \rceil$. We terminate when the square of the approximation

---

Algorithm 2. Computing an approximate square root of $\sigma'^2 \cdot g^{-T} \cdot g^{-1} - r^2$.

$p, \; s' \leftarrow \log n + 4 \log(\sqrt{n} \, \| \, \sigma \, \|), \; 1$

$\Sigma_2' \leftarrow g^{-T} \cdot g^{-1}$

while $\|s'^2 - \Sigma_2'\| > 2^{-\lambda}$ **do**

$\quad s' \leftarrow_\approx \sqrt{\Sigma_2'}$ computed at prec. p until $\|s'^2 - \Sigma_2\| < 2^{-\lambda}$ or no more convergence

$\quad p \leftarrow 2p$

end while

$p, \; r \leftarrow p + 2 \log \sigma', \; 2 \cdot \lceil \sqrt{\log n} \rceil$

$\Sigma_2 \leftarrow \sigma \cdot g^{-T} \cdot g^{-1} - r^2$

$s \leftarrow_\approx \sqrt{\Sigma_2}$ computed at precision p using s' as initial approximation until $\|s^2 - \Sigma_2\| < 2^{-2\lambda}$

return s

Algorithm 3. Sampling from $D_{(g),\sigma'}$

$\sqrt{\Sigma_2}' \leftarrow_\approx \sqrt{\sigma'^2 \cdot g^{-T} \cdot g^{-1} - r^2}$

$x' \in \mathbb{R}^n \leftarrow \rho_{1,0}$

$x \leftarrow x'$ considered as an element $\in \mathbb{Q}[X]/(X^n + 1)$

$y \leftarrow \sqrt{\Sigma_2}' \cdot x$

return $g \cdot (\lfloor y \rceil_r)$

---

is within distance $2^{-2\lambda}$ to Σ_2. This typically happens quickly because our initial candidate is already very close to the target value.

Given an approximation $\sqrt{\Sigma_2}'$ of $\sqrt{\Sigma_2}$ we then sample a vector $x \leftarrow \mathbb{R}^n$ from a standard normal distribution and interpret it as a polynomial in $\mathbb{Q}[X]/(X^n+1)$. We then compute $y = \sqrt{\Sigma_2}' \cdot x$ in $\mathbb{Q}[X]/(X^n + 1)$ and return $g \cdot (\lfloor y \rceil_r)$, where $\lfloor y \rceil_r$ denotes sampling a vector in $\mathbb{Z}^n$ where the i-th component follows $D_{\mathbb{Z},r,y_i}$. This algorithm is then easily extended to sample from arbitrary centres c. The whole algorithm is summarised in Algorithm 3 and we give experimental results in Table 6.

5.8 Approximate Square Roots

Our Gaussian sampler requires an (approximate) square root in $\mathbb{Q}[X]/(X^n + 1)$. That is, for some input element Σ we want to compute some element $\sqrt{\Sigma}' \in \mathbb{Q}[X]/(X^n + 1)$ such that $\|\sqrt{\Sigma}' \cdot \sqrt{\Sigma}' - \Sigma\| < 2^{-2\lambda}$. We use iterative methods as suggested in [Duc13, Section 6.5] which iteratively refine the approximation of the square root similar to Newton's method. Computing approximate square roots of matrices is a well studied research area with many algorithms known in the literature (cf. [Hig97]). All algorithms with global convergence invoke approximate inversions in $\mathbb{Q}[X]/(X^n + 1)$ for which we call our inversion algorithm.

We implemented the Babylonian method, the Denman-Beavers iteration [DB76] and the Padé iteration [Hig97]. Although the Babylonian method only involves one inversion which allows us to compute with lower precision, we used Denman-Beavers, since it converges faster in practice and can be parallelised

Table 6. Approximate square roots of $\Sigma_2 = \sigma'^2 \cdot g^{-T} \cdot g - r^2 \cdot I$ for discrete Gaussian sampling over g with parameter σ' on Intel Core i7–4850HQ CPU at 2.30 GHz, 2 cores for Denman-Beavers, 4 cores for estimating the scaling factor, one core for sampling. The last column lists the rate (samples per second) of sampling from $D_{(g),\sigma'}$.

| prec | n | $\log \sigma'$ | Square root | | | |
			Iterations	Wall time	$\log \|\|(\sqrt{\Sigma_2}')^2 - \Sigma_2\|\|$	$D_{(g),\sigma'}/s$
160	1024	45.8	9	0.4 s	-200	26.0
160	2048	49.6	9	0.9 s	-221	12.0
160	4096	53.3	10	2.5 s	-239	5.1
160	8192	57.0	10	8.6 s	-253	2.0
160	16384	60.7	10	35.4 s	-270	0.8

on two cores. While the Padé iteration can be parallelised on arbitrarily many cores, the workload on each core is much greater than in the Denman-Beavers iteration and in our experiments only improved on the latter when more than 8 cores were used.

Most algorithms have quadratic convergence but in practice this does not assure rapid convergence as error can take many iterations to become small enough for quadratic convergence to be observed. This effect can be mitigated, i.e. convergence improved, by scaling the operands appropriately in each loop iteration of the approximation [Hig97, Section 3]. A common scaling scheme is to scale by the determinant which in our case means computing $\mathrm{res}(f, X^n + 1)$ for some $f \in \mathbb{Q}[X]/(X^n + 1)$. Computing resultants in $\mathbb{Q}[X]/(X^n + 1)$ reduces to computing resultants in $\mathbb{Z}[X](X^n + 1)$. As discussed above, computing resultants in $\mathbb{Z}[X]/(X^n + 1)$ can be expensive. However, since we are only interested in an approximation of the determinant for scaling, we can compute with reduced precision. For this, we clear all but the most significant bit for each coefficient's numerator and denominator of f to produce f' and compute $\mathrm{res}(f', X^n + 1)$. The effect of clearing out the lower order bits of f is to reduce the size of the integer representation in order to speed up the resultant computation. With this optimisation scaling by an approximation of the determinant is both fast and precise enough to produce fast convergence. See Table 6 for timings.

Acknowledgement. We would like to thank Guilhem Castagnos, Guillaume Hanrot, Bill Hart, Claude-Pierre Jeannerod, Clément Pernet, Damien Stehlé, Gilles Villard and Martin Widmer for helpful discussions. We would like to thank Steven Galbraith for pointing out the NTRU-style attack to us and for helpful discussions. This work has been supported in part by ERC Starting Grant ERC-2013-StG-335086-LATTAC. The work of Albrecht was supported by EPSRC grant EP/L018543/1 "Multilinear Maps in Cryptography".

References

[AB15]   Applebaum, B., Brakerski, Z.: Obfuscating circuits via composite-order graded encoding. In: Dodis, Y., Nielsen, J.B. (eds.) TCC 2015, Part II. LNCS, vol. 9015, pp. 528–556. Springer, Heidelberg (2015)

[APS15]   Albrecht, M.R., Player, R., Scott, S.: On the concrete hardness of learning with errors. Cryptology ePrint Archive, Report 2015/046 (2015). http:// eprint.iacr.org/2015/046

[BCMM98]   Bini, D., Del Corso, G.M., Manzini, G., Margara, L.: Inversion of circulant matrices over Z_m. In: Larsen, K.G., Skyum, S., Winskel, G. (eds.) ICALP 1998. LNCS, vol. 1443, p. 719. Springer, Heidelberg (1998)

[BF03]   Boneh, D., Franklin, M.: Identity-based encryption from the Weil pairing. SIAM J. Comput. **32**(3), 586–615 (2003)

[BLR+15]   Boneh, D., Lewi, K., Raykova, M., Sahai, A., Zhandry, M., Zimmerman, J.: Semantically secure order-revealing encryption: Multi-input functional encryption without obfuscation. In: Oswald and Fischlin [OF15b], pp. 563–594

[BS03]   Boneh, D., Silverberg, A.: Applications of multilinear forms to cryptography. Contemp. Math. **324**, 71–90 (2003)

[BWZ14]   Boneh, D., Waters, B., Zhandry, M.: Low overhead broadcast encryption from multilinear maps. In: Garay, J.A., Gennaro, R. (eds.) CRYPTO 2014, Part I. LNCS, vol. 8616, pp. 206–223. Springer, Heidelberg (2014)

[CDKD14]   Cloutier, M.É., de Koninck, J.M., Doyon, N.: On the powerful and square-free parts of an integer. J. Integer Sequences **17**(2), 28 (2014)

[CG13]   Canetti, R., Garay, J.A. (eds.): CRYPTO 2013, Part I. LNCS, vol. 8042. Springer, Heidelberg (2013)

[CGH+15]   Coron, J.S., Gentry, C., Halevi, S., Lepoint, T., Maji, H.K., Miles, E., Raykova, M., Sahai, A., Tibouchi, M.: Zeroizing without low-level zeroes: New MMAP attacks and their limitations. In: Gennaro and Robshaw [GR15], pp. 247–266

[CHL+15]   Cheon, J.H., Han, K., Lee, C., Ryu, H., Stehlé, D.: Cryptanalysis of the multilinear map over the integers. In: Oswald and Fischlin [OF15a], pp. 3–12

[CLT13]   Jean-Sébastien Coron, Tancrède Lepoint, and Tibouchi, M.: Practical multilinear maps over the integers. In: Canetti and Garay [CG13], pp. 476–493

[CLT15]   Coron, J.S., Lepoint, T., Tibouchi, M.: New multilinear maps over the integers. In: Gennaro and Robshaw [GR15], pp. 267–286

[CN11]   Chen, Y., Nguyen, P.Q.: BKZ 2.0: better lattice security estimates. In: Lee, D.H., Wang, X. (eds.) ASIACRYPT 2011. LNCS, vol. 7073, pp. 1–20. Springer, Heidelberg (2011)

[CS97]   Coppersmith, D., Shamir, A.: Lattice attacks on NTRU. In: Fumy, W. (ed.) EUROCRYPT 1997. LNCS, vol. 1233, pp. 52–61. Springer, Heidelberg (1997)

[DB76]   Denman, E.D., Beavers, Jr., A.N.: The matrix sign function and computations in systems. Appl. Math. Comput., vol. 2, pp. 63–94 (1976)

[DDLL13]   Ducas, L., Durmus, A., Lepoint, T., Lyubashevsky, V.: Lattice signatures and bimodal gaussians. In: Canetti and Garay [CG13], pp. 40–56

[Duc13]   Ducas, L.: Signatures Fondées sur les Réseaux Euclidiens: Attaques, Analyse et Optimisations. Ph.D. thesis, Université Paris, Diderot (2013)

[Gar13] Garg, S.: Candidate Multilinear Maps. Ph.D. thesis, University of California, Los Angeles (2013)

[GGH13a] Garg, S., Gentry, C., Halevi, S.: Candidate multilinear maps from ideal lattices. In: Johansson, T., Nguyen, P.Q. (eds.) EUROCRYPT 2013. LNCS, vol. 7881, pp. 1–17. Springer, Heidelberg (2013)

[GGH+13b] Garg, S., Gentry, C., Halevi, S., Raykova, M., Sahai, A., Waters, B.: Candidate indistinguishability obfuscation and functional encryption for all circuits. In: 54th FOCS, pp. 40–49. IEEE Computer Society Press, October 2013

[GPV08] Gentry, C., Peikert, C., Vaikuntanathan, V.: Trapdoors for hard lattices and new cryptographic constructions. In: Ladner, R.E., Dwork, C. (eds.) 40th ACM STOC, pp. 197–206. ACM Press, May 2008

[GR15] Gennaro, R., Robshaw, M.J.B.: CRYPTO 2015, Part I, vol. 9215. Springer, Heidelberg (2015)

[Hig97] Higham, N.J.: Stable iterations for the matrix square root. Numer. Algorithms 15(2), 227–242 (1997)

[HJ15] Hu, Y., Jia, H.: Cryptanalysis of GGH map. Cryptology ePrint Archive, Report 2015/301 (2015). http://eprint.iacr.org/2015/301

[HJP14] Hart, W., Johansson, F., Pancratz, S.: FLINT: fast library for number theory (2014). Version 2.4.4. http://flintlib.org

[Jou04] Joux, A.: A one round protocol for tripartite Diffie-Hellman. J. Cryptol. 17(4), 263–276 (2004)

[LMPR08] Lyubashevsky, V., Micciancio, D., Peikert, C., Rosen, A.: SWIFFT: a modest proposal for FFT hashing. In: Nyberg, K. (ed.) FSE 2008. LNCS, vol. 5086, pp. 54–72. Springer, Heidelberg (2008)

[LP15] Lyubashevsky, V., Prest, T.: Quadratic time, linear space algorithms for gram-schmidt orthogonalization and gaussian sampling in structured lattices. In: Oswald and Fischlin [OF15a], pp. 789–815

[LSS14a] Langlois, A., Stehlé, D., Steinfeld, R.: GGHLite: more efficient multilinear maps from ideal lattices. In: Nguyen, P.Q., Oswald, E. (eds.) EUROCRYPT 2014. LNCS, vol. 8441, pp. 239–256. Springer, Heidelberg (2014)

[LSS14b] Langlois, A., Stehlé, D., Steinfeld, R.: GGHLite: More efficient multilinear maps from ideal lattices. Cryptology ePrint Archive, Report 2014/487 (2014). http://eprint.iacr.org/2014/487

[OF15a] Oswald, E., Fischlin, M. (eds.): EUROCRYPT 2015. LNCS, vol. 9056. Springer, Heidelberg (2015)

[OF15b] Oswald, E., Fischlin, M. (eds.): EUROCRYPT 2015. LNCS, vol. 9057. Springer, Heidelberg (2015)

[Pei10] Peikert, C.: An efficient and parallel gaussian sampler for lattices. In: Rabin, T. (ed.) CRYPTO 2010. LNCS, vol. 6223, pp. 80–97. Springer, Heidelberg (2010)

[SE94] Schnorr, C.P., Euchner, M.: Lattice basis reduction: Improved practical algorithms and solving subset sum problems. Math. Program. 66(1–3), 181–199 (1994)

[The13] The MPFR team. GNU MPFR: The Multiple Precision Floating-Point Reliable Library, 3.1.2 edition (2013). http://www.mpfr.org/

[Win96] Winkler, F.: Polynomial Algorithms in Computer Algebra. Texts and Monographs in Symbolic Computation. Springer, Heidelberg (1996)

[Zim15] Zimmerman, J.: How to obfuscate programs directly. In: Oswald and Fischlin [OF15b], pp. 439–467

New Circular Security Counterexamples
from Decision Linear and Learning with Errors

Allison Bishop[1][✉], Susan Hohenberger[2], and Brent Waters[3]

[1] Columbia University, New York, USA
allison@cs.columbia.edu
[2] Johns Hopkins University, Baltimore, USA
susan@cs.jhu.edu
[3] University of Texas at Austin, Austin, USA
bwaters@cs.utexas.edu

Abstract. We investigate new constructions of n-circular counterexamples with a focus on the case of $n = 2$. We have a particular interest in what qualities a cryptosystem must have to be able to separate such circular security from IND-CPA or IND-CCA security. To start, we ask whether there is something special about the asymmetry in bilinear groups that is inherent in the works of [1,18] or whether it is actually the bilinearity that matters. As a further question, we explore whether such counterexamples are derivable from other assumptions such as the Learning with Errors (LWE) problem. If it were difficult to find such counterexamples, this might bolster our confidence in using 2-circular encryption as a method of bootstrapping Fully Homomorphic Encryption systems that are based on lattice assumptions.

The results of this paper broadly expand the class of assumptions under which we can build 2-circular counterexamples. We first show for any constant $k \geq 2$ how to build counterexamples from a bilinear group under the decision k-linear assumption. Recall that the decision k-linear assumption becomes progressively weaker as k becomes larger. This means that we can instantiate counterexamples from symmetric bilinear groups and shows that asymmetric groups do not have any inherently special property needed for this problem. We then show how to create 2-circular counterexamples from the Learning with Errors problem. This extends the reach of these systems beyond bilinear groups and obfuscation.

A. Bishop — Supported by NSF CNS 1413971 and NSF CCF 1423306.
S. Hohenberger — Supported by NSF CNS-1228443 and CNS-1414023, the Office of Naval Research under contract N00014-14-1-0333, and a Microsoft Faculty Fellowship.
B. Waters — Supported by NSF CNS-1228599 and CNS-1414082, DARPA SafeWare, a Google Faculty Research Award, the Alfred P. Sloan Fellowship, Microsoft Faculty Fellowship, and Packard Foundation Fellowship.

T. Iwata and J.H. Cheon (Eds.): ASIACRYPT 2015, Part II, LNCS 9453, pp. 776–800, 2015.
DOI: 10.1007/978-3-662-48800-3_32

1 Introduction

The notion of key dependent message security [12] moves beyond our classical notion of encryption security [22]. It demands a system remain secure even if an attacker gains access to ciphertexts that encrypt messages that are, or depend on, the very private keys of the system it is trying to attack. As a concrete example, consider a special case of key-dependent security called n-circular security. Here an encryption scheme is said to be n-circular secure, if an adversary is unable to distinguish $\mathsf{Enc}(pk_1, sk_2), \mathsf{Enc}(pk_2, sk_3), \ldots, \mathsf{Enc}(pk_n, sk_1)$ from corresponding zero encryptions.

While the notion of key dependent or circular security might first appear to be just a technical exercise, this very problem arises in multiple contexts. Camenisch and Lysyanskaya [17] applied circular secure encryption to build an anonymous credentials scheme with certain properties. Other works used circular security in formal methods to prove the soundness of symbolic protocols [2,26]. Perhaps the most compelling example comes from Gentry [20], who showed that a fully homomorphic scheme for limited depth can be "bootstrapped" to work for arbitrary depth circuits if the original system is sufficient to compute its own decryption circuit and is 1-circular secure.

The first positive examples of key-dependent message security were given in the random oracle model by Black et al. [12] and Camenisch and Lysyanskaya [17]. It was a significant time later when Boneh, Hamburg, Halevi and Ostrovsky [14] gave an elegant construction of an n-circular secure encryption in the standard model under the decision Diffie-Hellman assumption. Subsequently, a sequence of further works [5,7–9,15,16] gave standard model constructions of key dependent security for functions that could be arbitrary circuits on the private key(s).

All the above constructions and proofs were based on encryption schemes with specific properties. A natural question is whether key-dependent message security is implied by IND-CPA (or IND-CCA) security. If this were true, we would get it for free, without needing such specific properties of the encryption scheme.

A cursory examination of the problem shows that in the broadest sense the answer is no. One can derive a simple counterexample for 1-circular security (i.e., a system that encrypts its own private key) by slightly modifying a public key encryption system. To do so, simply augment a standard private key K with a randomly chosen $K' \in \{0,1\}^\lambda$ and append $y = f(K')$ to the public key where f is a one way function. When encrypting a message $m = (m_1, m_2)$ the system will give out the message in the clear if $f(m_2) = y$) and encrypt normally otherwise. Clearly, an encryption of the private key will be detectable. Yet, if the function f is one way and the original system is IND-CPA secure, the resulting system will still be IND-CPA secure.

While it can be trivially shown (by the argument above) that IND-CPA security does not imply 1-circular security, the case for $n \geq 2$ becomes significantly more challenging. Intuitively, when multiple public keys are thrown into the mix, we need a system that is powerful enough to allow for different ciphertexts to "talk" to each other in a manner that allows for cycle detection, but does not compromise IND-CPA security. So far there have been two approaches to this. For the case of $n = 2$, Acar et al. [1] and Cash, Green and Hohenberger [18] showed how to construct a counterexample from a certain class of *asymmetric* bilinear groups.[1] Here there must exist a bilinear map $e : \mathbb{G}_1 \times \mathbb{G}_2 \to \mathbb{G}_T$ where the decision Diffie-Hellman problem is believed to remain hard respectively within $\mathbb{G}_1$ and within $\mathbb{G}_2$ (this is called the SXDH assumption). A second approach by Koppula, Ramchen and Waters [25] showed a counterexample under the assumption of indistinguishability obfuscation for poly-sized circuits. Independently and concurrently, Marcedone and Orlandi [27] showed this under the stronger assumption of virtual black box obfuscation.

Our Goals and Results. In this work, we investigate new constructions of n-circular counterexamples with a focus on the case of $n = 2$. We have a particular interest in what qualities a cryptosystem must have to be able to separate circular security from IND-CPA and IND-CCA security.

To start, we ask whether there is something special about the asymmetry in bilinear groups that is inherent in the works of [1,18,34] or whether it is actually more the bilinearity that matters. As a further question, we explore how to derive such counterexamples from other assumptions such as the Learning with Errors (LWE) problem. If it were difficult to find such counterexamples, this might bolster are confidence in using 2-circular encryption as a method of bootstrapping [20] fully homomorphic encryption systems that are based on lattice assumptions.

The results of this paper broadly expand the class of assumptions from which we can build 2-circular counterexamples. We first show for any constant $k \geq 2$ how to build 2-circular counterexamples from a bilinear group under the decision k-linear assumption. Recall that the decision k-linear assumption becomes progressively weaker as k becomes larger. This means that we can instantiate counterexamples from symmetric bilinear groups and shows that asymmetric groups do not have any inherently special property needed for this problem. We then show how to create 2-circular counterexamples from the Learning with Error (LWE) problem. This extends the reach of these systems beyond bilinear groups and obfuscation, giving us a much broader understanding of circular security and its challenges.

Our Approach. We begin by introducing a new abstraction called an n-Cycle Tester that will simplify the process of finding and describing counterexamples by focusing on the core problem. A *cycle tester* consists of four algorithms (Setup, KeyGen, Enc, Test). The algorithms of Setup, KeyGen, Enc behave as in a normal encryption scheme with a common trusted setup algorithm, while the Test algorithm will take in an n-tuple of public keys and ciphertexts and

---

[1] In a similar vein, Rothblum [34] presented an elegant counterexample for bit-encryption under a generalization of the SXDH assumption applied to multilinear groups.

detect (with some non-negligible probability) the presence of a cycle. Notably absent is the inclusion of a decryption algorithm. Thus, a tester does not require that ciphertexts be decryptable in the traditional sense — it only matters that the Test algorithm work with some non-negligible probability. We found that relieving the responsibility of providing a system with decryption simplifies our constructions and allows us to focus on the main ideas. The security property required is IND-CPA security (recall that the basic IND-CPA game does not involve a decryption algorithm).

Of course, to obtain a full-fledged counterexample of an encryption system we actually do need to provide an encryption system that decrypts. We show how to generically derive such a counterexample for n-circular encryption by combining a standard IND-CPA secure cryptosystem (of sufficient message length) with a n-cycle tester. The idea is fairly straightforward. The setup algorithm of the counterexample will run the respective setup algorithms of the encryption and cycle tester schemes. The public key is the pair of these public keys and the secret key is the pair of secret keys. To encrypt a message $m = (m_1, m_2)$, first encrypt $m = (m_1, m_2)$ under the regular encryption system, then encrypt just m_2 under the cycle tester. We can now see that: (1) the cycle tester will allow for any key cycle to be detected and (2) the standard encryption scheme can be used for decryption. A simple hybrid argument shows that the IND-CPA security of the standard encryption scheme and cycle tester imply IND-CPA security of the derived counterexample system.

We also show that it is possible to extend this transformation idea to chosen ciphertext security, where we can combine any IND-CCA secure encryption system (of appropriate message length) with the same IND-CPA secure cycle tester to get an encryption system that is IND-CCA secure, but where encryption of key cycles can be detected.

Again, the usefulness of this framework is its modularity. We show these basic transformations once in Sect. 4, and then for each construction we only need to focus on the basic cycle tester abstraction.

A Cycle Tester from Asymmetric Bilinear Groups. As a baseline for our exploration (see [11] for the full details), we first create a 2-cycle tester from asymmetric groups using the SXDH assumption. Our construction is extracted from Cash et al. [18] (also similar to [1,34]), but simpler in that we only aim for the tester abstraction.

In our construction, the Setup algorithm creates an asymmetric pairing description $PP = (p, \mathbb{G}_1, \mathbb{G}_2, \mathbb{G}_T, e)$ of prime order p. It also produces generators $g \in \mathbb{G}_1$ and $h \in \mathbb{G}_2$. The message space will be $\mathbb{Z}_p^*$.

A key can be of one of two types. The cycle detection algorithm Test will work on any cycle of keys of two different types. The key generation algorithm KeyGen will first flip a coin $\beta \in \{0, 1\}$ to determine its type. It then picks a random key $s \in \mathbb{Z}_p^*$. If $\beta = 0$, it sets its public key to be $K = g^s \in \mathbb{G}_1$; otherwise, its public key is $K = h^s \in \mathbb{G}_2$.

The encryption algorithm will choose a random exponent $t \in \mathbb{Z}_p$ and if the key is of type $\beta = 0$, it produces the ciphertext as $(C_1 = K^{tm} = g^{stm},$

$C_2 = g^t) \in \mathbb{G}_1^2$; otherwise if $\beta = 1$, it produces the ciphertext as $(C_1 = K^{tm} = h^{stm}, \; C_2 = h^t) \in \mathbb{G}_2^2$. With ciphertexts of this form, the test algorithm follows straightforwardly. Suppose we had a pair of ciphertexts $\mathbf{y} = (C = (C_1, C_2), C' = (C_1', C_2'))$ that encrypted a cycle for keys of different types. The algorithm can test this by simply computing $e(C_1, C_2') \overset{?}{=} e(C_2, C_1')$. Plugging in s, s' as the respective keys, t, t' as the encryption randomness, and m, m' as the messages, we see that the test computes:

$$e(g^{stm}, h^{t'}) \overset{?}{=} e(g^t, h^{s't'm'}).$$

This equality holds if $m = s'$ and $m' = s$ and will not hold with high probability for a message independent of the private key.

One thing we emphasize here is that IND-CPA is clearly broken if the SXDH assumption does not hold. Consider an encryption $(C_1 = K^{tm} = g^{stm}, \; C_2 = g^t) \in \mathbb{G}_1^2$ for the message m. The group elements $g, (g^s)^m = g^{sm}, C_2 = g^t, C_1 = g^{stm}$ clearly form a DDH tuple. So if DDH is easy in $\mathbb{G}_1$, any $\beta = 0$ type key is susceptible to attack. An analogous statement holds in $\mathbb{G}_2$ for any $\beta = 1$ key. This potential attack demonstrates that the above construction relies strongly on properties of asymmetric groups. We next show how to remove that reliance.

A Cycle Tester from the Decision k-Linear Assumption. We next move to constructing a cycle tester from the decision k-linear assumption for any constant $k \geq 2$. Recall that the k-linear assumption [24,35] is a parameterized family of assumptions on the source elements of bilinear groups. The assumption class becomes progressively weaker for larger values of k. Importantly, by moving to the decision k-linear assumption we remove our dependence on asymmetric groups.[2] See [11] for a review.

In our construction, the setup algorithm first generates a bilinear source group $\mathbb{G}$ of prime order p with generator g. Then it chooses a random invertible (rank k) matrix $\mathbf{A} \in \mathbb{Z}_p^{k \times k}$ and computes $g^{\mathbf{A}}$, which along with the group description forms the common public parameters. (We use the notation $g^{\mathbf{M}}$ as shorthand for the set of group elements resulting from raising g to each matrix entry in $\mathbf{M}$.) The message and key spaces are defined to be the set of rank k matrices in $\mathbb{Z}_p^{k \times k}$.[3]

Once again the key generation algorithm will flip a coin β to determine its type. Next it chooses a random $\mathbf{W}$ from the set of invertible matrices in $\mathbb{Z}_p^{k \times k}$. If $\beta = 0$ the key is $g^{\mathbf{AW}}$; otherwise it is $g^{\mathbf{WA}}$.

---

[2] We emphasize though that our constructions could use an asymmetric form of bilinear maps if desired, although we describe things in terms of symmetric groups. The main point is that there is no longer a reliance on asymmetry or that DDH is hard within each group.

[3] In our scheme, we actually let the message and key space be $\{0,1\}^\lambda$ for security parameter λ and define a pseudorandom generator from this to rank k matrices. That way the message space is defined before the common setup is executed. However, for simplicity we will just assume here that the message and key spaces are the set of invertible $k \times k$ matrices.

The encryption algorithm takes as input a message $\mathbf{M} \in \mathbb{Z}_p^{k \times k}$ and then computes its inverse $\mathbf{M}^{-1}$. (Recall the message space is the set of invertible matrices.) If the type bit $\beta = 0$, the algorithm chooses a random row vector $\mathbf{r}$ of length k in $\mathbb{Z}_p$ (i.e. a random matrix of dimension $1 \times k$). The ciphertext is computed and output as $C_1 = g^{\mathbf{rAW}}$, $C_2 = g^{\mathbf{rAM}^{-1}}$. Thus, the ciphertext will consist of two row vectors in the exponent. We observe all terms are computable from the public keys and public parameters. If the type bit $\beta = 1$ the algorithm chooses a random column vector $\mathbf{r}$ of length k in $\mathbb{Z}_p$ (i.e., a random matrix of dimension $k \times 1$). The ciphertext is computed and output as $C_1 = g^{\mathbf{WAr}}$, $C_2 = g^{\mathbf{M}^{-1}\mathbf{Ar}}$.

Now suppose we have two ciphertexts $\mathbf{y} = (C = (C_1, C_2), C' = (C'_1, C'_2))$ of different types (with the first being of $\beta = 0$). We can then test for a cycle by testing if $e(C_1, C'_2) \stackrel{?}{=} e(C'_1, C_2)$. To see why, suppose we had a cycle, so we have that $\mathbf{M}'^{-1} = \mathbf{W}^{-1}$ and $\mathbf{M}^{-1} = \mathbf{W}'^{-1}$. Then, in the exponent, it follows that:

$$\mathbf{rAWM}'^{-1}\mathbf{Ar}' \stackrel{?}{=} \mathbf{rAM}^{-1}\mathbf{W}'\mathbf{Ar}'$$
$$\mathbf{rAIAr}' \stackrel{?}{=} \mathbf{rAIAr}'$$
$$\mathbf{rA}^2\mathbf{r}' \stackrel{?}{=} \mathbf{rA}^2\mathbf{r}'.$$

So if there is a cycle, the test will output 1. In contrast, if the messages encrypted are independent of the key, the test will output 0 with high probability.

Finally, we can give a simple proof of IND-CPA security from the decision k-linear assumption. More specifically, we will use the matrix k-linear assumption, introduced by Naor and Segev [29], that was shown to be equivalent to the decision k-linear assumption. Informally, the assumption says that it is hard to distinguish $g^{\mathbf{X}}$ and $g^{\mathbf{Y}}$ where $\mathbf{X}$ is a random matrix of rank $i > k$ and $\mathbf{Y}$ is a random matrix (of the same dimension) of rank $j > k$. I.e., the rank of matrices in the exponent cannot be determined as long as it is greater than k. For our purposes, we will be interested in using the difficulty of distinguishing between rank k and rank $k + 1$ matrices.

Let us examine IND-CPA security for an encryption under a type $\beta = 0$ key. (The argument for $\beta = 1$ will follow analogously.) We will devise a reduction algorithm that receives a matrix k-linear assumption challenge $g^{\mathbf{M}}$, where $\mathbf{M}$ is selected as either a random rank k matrix or rank $k + 1$ matrix. In the case where it is a rank k matrix, our reduction algorithm will use it to derive the key and ciphertext values of

$$g^{\mathbf{A}}, g^{\mathbf{AW}}, g^{\mathbf{rAW}}, g^{\mathbf{rA}}.$$

These can be used to generate a well-formed ciphertext of a given message. However, if the reduction algorithm receives a random matrix of rank $k + 1$, it will create key and ciphertext values distributed as

$$g^{\mathbf{A}}, g^{\mathbf{AW}}, g^{\mathbf{rAW}}, g^{\mathbf{uA}}.$$

In this case the fact that $\mathbf{u}$ is fresh randomness will information-theoretically hide the message from the attacker. It then follows that any attacker with non-negligible advantage against our system must break the matrix k-linear assumption.

In the full version [11], we present a different 2-cycle tester from the Decision Linear assumption in symmetric pairing groups. This construction can be viewed as closer to an extension of the SXDH one (sketched above and detailed in [11]) to symmetric groups where new variables and equations are introduced to prevent the use of pairings to disrupt IND-CPA security. However, it does not seem to generalize to a system that is secure using the decision k-linear assumption for $k > 2$ or help move toward a Learning with Errors Assumption. At the same time, when compared to our more general construction just given for the $k = 2$ (decision linear assumption) case, it achieves smaller public keys. Public keys here are two group elements as opposed to four. Our techniques for this construction might be of future interest for other applications of transforming constructions proved under asymmetric group assumptions to those that do not rely on them. We defer further details of these techniques to the full version [11].

A Cycle Tester from Learning with Errors Assumption. While there are now many known examples of cryptographic functionalities that can be achieved in both the bilinear and lattice settings, it is not at all clear how to imitate the pairings-based approach above to obtain a cycle tester from the LWE assumption. Typically, encryption schemes proven secure under LWE have ciphertexts that are large, noisy vectors in $\mathbb{Z}_q^m$ and secret keys that are short vectors in $\mathbb{Z}^m$, with decryption computing a dot product and then removing the small effect of the noise multiplied by the short key vector. It seems unlikely that we could build a cycle tester using only this kind of structure, as the cycle effect would be obscured by the interactions of large ciphertext vectors with the embedded noise.

Intuitively, we then expect that a cycle tester may use ciphertexts that have two parts: a noisy vector and a short vector. The large, noisy vectors will help us prove IND-CPA security from LWE, while the short vectors will help us perform the cycle test. Naturally, the main challenge is designing the relationship between the noisy and short vectors such that the short vectors do not break security when there is no cycle.

The secret key for our scheme will generate a matrix B and a corresponding short trapdoor basis T_B. For IND-CPA security, it is important that B is hidden, so one should ignore the notational collision and not think of this as corresponding to the public matrix A in an LWE challenge, but rather the columns of B will play the role of different hidden s vectors in typical LWE notation. The public key will be formed by choosing several random vectors $c_1, \ldots, c_\ell$ and publishing noisy versions of $c_1 B, \ldots, c_\ell B$ as well as the (non-noisy) vectors $c_1, \ldots, c_\ell$ (so these c_i's can be thought of as playing the role of the public matrix A in an LWE challenge).

To encrypt a message, the message will first be used to generate a matrix Z and a corresponding short trapdoor basis T_Z. The encryptor will mimic typical LWE-style encryption by forming a noisy version of sB for some vector s, but since it does not know B, it will form s as a linear combination of $c_1, \ldots, c_\ell$ with coefficients chosen randomly from $\{-1, 1\}$. Note that the encryptor can then compute both s (without noise) and a noisy version of sB. The noisy version of sB becomes the noisy part of the ciphertext, and the other part of the ciphertext is a short vector v such that Zv equals the transpose of s. Note that such a vector v can be sampled appropriately using the trapdoor basis T_Z.

For full details of how the cycle test works, see Sect. 6. The main idea is that when there is a 2-cycle, the secret key matrix B for one ciphertext is the same as the message matrix Z for the other ciphertext and vice versa. This leads to a common relationship between the short vector of one ciphertext and the noisy vector of the other, while when the B, Z matrices of each are fresh and unrelated, this relationship does not appear. One convenient feature of this scheme as compared to the bilinear schemes is that there is no need for different types of ciphertexts. Intuitively, the pairing relationship has been replaced by a dot product relationship between a short vector and a noisy one.

Proving IND-CPA security for this scheme can be accomplished in a few steps. First, since B is hidden and its columns act like the hidden vector s in a typical LWE challenge and the c_i's act like rows of the public matrix A, we can argue that LWE implies the noisy public versions of $c_i B$ can be replaced by uniformly random vectors, independent of the c_i's and B. Next, using a convenient variant of the left over hash lemma from [3], we argue that the random coefficients in $\{-1, 1\}$ that form s from the c_i's and the noisy ciphertext vector from the public noisy vectors supply sufficient entropy to replace both of these with fresh uniformly random vectors as well. We are then left with an encryption that samples a uniformly random s (now independent of the noisy part of the ciphertext) and samples the short part of the ciphertext as a short vector v such that Zv is the transpose of s. Here we can argue that the distribution of such a v is statistically close to a distribution that is independent of Z: this follows from a result in [21] that ensures us that the image of a short, Gaussian distributed vector v under multiplication by Z is uniformly distributed in $\mathbb{Z}_q^n$. Thus, by employing LWE followed by a sequence of statistical arguments, we can arrive at a point where the ciphertext is independent of the message, and this implies IND-CPA security.

Other Related Work. Haitner and Holenstein [23] show black box impossibility results for proving key-dependent message security from different cryptographic assumptions. Their goal deviates from ours in two important ways. First, their work focuses on impossibility results for ciphertext encrypting functions of its own private keys, whereas we are concerned with the circular case where there is a cycle over multiple private keys. Second, we are interested in concrete counterexamples. In particular, it may be possible that IND-CPA security implies certain key-dependent security properties even if there does not exist any black box reduction. In contrast our counterexamples will show that this is impossible if certain specific number theoretic assumptions hold.

2   Preliminaries

Background on pairings can be found in the full version [11].

2.1   The k-LIN Assumption

Decision Linear and the k-LIN Family (k-LIN). We now present a family of assumptions called the k-LIN assumptions (where $k = 1$ is the standard DDH assumption and $k = 2$ is called Decision Linear [13]) [10,24]. Let $\mathbb{G}$ be a group of prime order $p \in \Theta(2^\lambda)$. For all p.p.t. adversaries $\mathcal{A}$ and $k \geq 1$, the following probability is $1/2$ plus an amount negligible in λ:

$$\Pr[g, g_1, \dots, g_k \leftarrow \mathbb{G}; r_1, \dots, r_k \leftarrow \mathbb{Z}_p; T_0 = g^{(r_1 + \dots + r_k)}; T_1 \leftarrow \mathbb{G}; d \leftarrow \{0, 1\};$$
$$d' \leftarrow \mathcal{A}(g, g_1, \dots, g_k, g_1^{r_1}, \dots, g_k^{r_k}, T_d) : d = d'].$$

In the generic group model, these k-LIN assumptions become progressively weaker for increasing k.

In our proof of security in Sect. 5 we will use a theorem due to Naor and Segev [29] that shows that under the decision k-linear assumption no attacker can distinguish between a random rank i matrix and a random rank j matrix (in the exponent and of the same dimensions) for $i, j \geq k$.

2.2   Lattices and LWE

We let q, n, and m denote positive integers. Given a matrix $A \in \mathbb{Z}_q^{n \times m}$, we let $\Lambda_q^\perp(A)$ denote the lattice $\{x \in \mathbb{Z}^m : Ax = 0 \mod q\}$. For $u \in \mathbb{Z}_q^n$, we let $\Lambda_q^u(A)$ denote the set $\{x \in \mathbb{Z}^m : Ax = u \mod q\}$.

For a matrix $A \in \mathbb{Z}^{n \times m}$, we let $||A||$ denote the ℓ_2 length of the longest column of A, and we let $||A||_{GS}$ denote $||\widetilde{A}||$, where $\widetilde{A}$ is the Gram-Schmidt orthogonalization of the columns of A. We let A^t denote the transpose of the matrix A.

Learning with Errors (LWE). Given integers n, m, a prime q, and a noise distribution χ over $\mathbb{Z}$, the (n, m, q, χ)-LWE problem is to distinguish the distributions $(A, A^t s + e)$ and (A, u), where A is chosen uniformly from $\mathbb{Z}_q^{n \times m}$, s is chosen uniformly from $\mathbb{Z}_q^n$, e is chosen from χ^m, and u is chosen uniformly from $\mathbb{Z}_q^m$.

Under a quantum reduction, Regev [33] showed that for certain noise distributions, the LWE problem is as hard as the worst-case SIVP and GapSVP. Peikert [31] gave a reduction in the classical setting. Our construction will admit a range of parameters where solving the LWE problem is as hard as approximating the worst-case GapSVP to polynomial (in n) factors, which is believed to be computationally hard.

Trapdoor Generation. We will rely on the polynomial time algorithm Trap-Gen(1^n, 1^m, q) (developed in [4,6,28]). This is a randomized algorithm that when given $m = \Theta(n \log q)$, outputs a full rank matrix $A \in \mathbb{Z}_q^{n \times m}$ and an accompanying basis $T_A \in \mathbb{Z}^{m \times m}$ for $\Lambda_q^\perp(A)$ such that the distribution of A is negligibly close (in n) to uniform over $\mathbb{Z}_q^{n \times m}$ and $||T_A||_{GS} = \mathcal{O}(\sqrt{n \log q})$ with all but negligible probability (as a function of n).

Discrete Gaussian Distributions. We employ the discrete Gaussian distribution $\mathcal{D}_\sigma(\Gamma_q^u(A))$ on $\Gamma_q^u(A)$, parameterized by $\sigma > 0$ (as defined e.g. in [33]). The salient fact we will use about this distribution is that for a random matrix $A \in Z_q^{n \times m}$ and $\sigma = \tilde{\Omega}(\sqrt{n})$, a vector sampled from $\mathcal{D}_\sigma(\Lambda_q^u(A))$ has ℓ_2 norm less than $\sigma\sqrt{m}$ with probability at least 1 minus a quantity that is negligible in m.

We will rely on a polynomial time algorithm $\mathsf{SampleD}(A, T_A, u, \sigma)$ [21]. This is a randomized algorithm that when $\sigma = \|T_A\|_{GS} \cdot \omega(\sqrt{\log m})$, produces a random vector x from a distribution that is statistically close to $\mathcal{D}_\sigma(\Lambda_q^u(A))$.

We also employ the following result from [21] (appears as Corollary 5.4 in that work):

Lemma 1. *Let n and q be positive integers with q prime, and let $m \geq 2n \log q$. Then for all but a $2q^{-n}$ fraction of all $A \in \mathbb{Z}_q^{n \times m}$ and for any $\sigma \geq \omega(\sqrt{\log m})$, the distribution of the syndrome $u = Ae \mod q$ is statistically close to uniform over $\mathbb{Z}_q^n$, where e is distributed according to $\mathcal{D}_{\mathbb{Z}^m, \sigma}$.*

Randomness Extraction. We will use the leftover hash lemma (see [3] e.g. for an even stronger statement):

Lemma 2. *Suppose that $\ell > (j+1)\log q + \omega(\log j)$ and $q > 2$ is prime (for integers q, j, ℓ). Let R be an $\ell \times 1$ vector chosen uniformly in $\{1, -1\}^\ell \mod q$. Let A and B be matrices chosen uniformly in $\mathbb{Z}_q^{j \times \ell}$ and $\mathbb{Z}_q^{j \times 1}$ respectively. Then, the distribution (A, AR) is statistically close to the distribution (A, B).*

3 Security Definitions

In this work, we will focus on public key encryption schemes that admit a global setup algorithm.

Definition 1 (Public Key Encryption). *A public key encryption scheme $\Pi = (\mathsf{Setup}, \mathsf{KeyGen}, \mathsf{Enc}, \mathsf{Dec})$ for a message space M and secret key space S[4] is a tuple of algorithms specified as follows:*

- $\mathsf{Setup}(1^\lambda) \to \mathrm{PP}$. *The Setup algorithm takes as input the security parameter λ and outputs common public parameters PP.*
- $\mathsf{KeyGen}(\mathrm{PP}) \to (pk, sk)$. *The Key Generation algorithm takes as input the public parameters PP and outputs a public pk and secret key $sk \in S$.*
- $\mathsf{Enc}(pk, m \in M) \to C$. *The Encryption algorithm takes as input a public key pk and a message $m \in M$ and outputs a ciphertext C.*
- $\mathsf{Dec}(sk, C) \to m$. *The Decryption algorithm takes as input a secret key sk and a ciphertext C and outputs either an error message $\perp$ or a value $m \in M$.*

By $\mathsf{negl}(k)$ we denote some *negligible* function, i.e., one such that, for all $c > 0$ and all sufficiently large k, $\mathsf{negl}(k) < 1/k^c$. We abbreviate probabilistic polynomial time as PPT.

---

[4] Technically, the output of the Setup algorithm may be required to establish the message and secret key spaces. For instance, the setup algorithm may output a prime p and the message space might be set as $\mathbb{Z}_p^*$. For simplicity, we provide a name for these sets at the scheme level, even though the elements in these sets may not be defined until after Setup.

Perfect Correctness. An encryption scheme $\Pi = (\mathsf{Setup}, \mathsf{KeyGen}, \mathsf{Enc}, \mathsf{Dec})$ for message space M is said to be perfectly correct if for all $\lambda \in \mathbb{N}$, $m \in M$, and $(pk, sk) \in \mathsf{KeyGen}(\mathsf{Setup}(1^\lambda))$, it holds that $\mathsf{Dec}(sk, \mathsf{Enc}(pk, m)) = m$.

Security. We recall the notion of indistinguishability of encryptions under a chosen-plaintext attack [22].

Definition 2 (IND-CPA Security). *Let $\Pi = (\mathsf{Setup}, \mathsf{KeyGen}, \mathsf{Enc}, \mathsf{Dec})$ be a public-key encryption scheme. For scheme Π, adversary $\mathcal{A}$, and $\lambda \in \mathbb{N}$, let the random variable $\mathsf{IND\text{-}CPA}(\Pi, \mathcal{A}, \lambda)$ be defined by the probabilistic algorithm described on the left side of Fig. 1. We denote the $\mathsf{IND\text{-}CPA}$ advantage of $\mathcal{A}$ by $\mathsf{Adv}^{cpa}_{\Pi, \mathcal{A}}(\lambda) = 2 \cdot \Pr[\mathsf{IND\text{-}CPA}(\Pi, \mathcal{A}, \lambda) = 1] - 1$. We say that Π is $\mathsf{IND\text{-}CPA}$ secure if $\mathsf{Adv}^{cpa}_{\Pi, \mathcal{A}}(\lambda)$ is negligible for all PPT $\mathcal{A}$.*

We also consider the indistinguishability of encryptions under a chosen-ciphertext attack [19,30,32].

Definition 3 (IND-CCA Security). *Let $\Pi = (\mathsf{Setup}, \mathsf{KeyGen}, \mathsf{Enc}, \mathsf{Dec})$ be a public-key encryption scheme. Let the random variable $\mathsf{IND\text{-}CCA}(\Pi, \mathcal{A}, \lambda)$ be defined by an algorithm identical to $\mathsf{IND\text{-}CPA}(\Pi, \mathcal{A}, \lambda)$ above, except that $\mathcal{A}$ has access to an oracle $\mathsf{Dec}(sk, \cdot)$ that returns the output of the decryption algorithm and $\mathcal{A}$ cannot query this oracle on input y. We denote the $\mathsf{IND\text{-}CCA}$ advantage of $\mathcal{A}$ by $\mathsf{Adv}^{cca}_{\Pi, \mathcal{A}}(\lambda) = 2 \cdot \Pr[\mathsf{IND\text{-}CCA}(\Pi, \mathcal{A}, \lambda) = 1] - 1$. We say that Π is $\mathsf{IND\text{-}CCA}$ secure if $\mathsf{Adv}^{cca}_{\Pi, \mathcal{A}}(\lambda)$ is negligible for all PPT $\mathcal{A}$.*

3.1 Circular Security

We next define circular security of public-key encryption. This definition is derived from the Key-Dependent Message (KDM) security notion of Black et al. [12]. We follow prior counterexample definitions [1,18] which restrict the adversary's power (e.g., cannot ask for any affine function of the secret keys). The adversary is asked to distinguish between an encryption cycle or encryptions of zero as in [14,18]. The bit string zero is not actually in the message spaces we consider, but this value can be encoded to be in the space; equivalently, one can follow the approach of Acar et al. [1] which instead of zero, encrypts a fresh random message.

Definition 4 (IND-CIRC-CPA^n). *Let $\Pi = (\mathsf{Setup}, \mathsf{KeyGen}, \mathsf{Enc}, \mathsf{Dec})$ be a public-key encryption scheme. For integer $n > 0$, scheme Π, adversary $\mathcal{A}$ and $\lambda \in \mathbb{N}$, let the random variable $\mathsf{IND\text{-}CIRC\text{-}CPA}^n(\Pi, \mathcal{A}, \lambda)$ be defined by the probabilistic algorithm in the middle of Fig. 1. We denote the $\mathsf{IND\text{-}CIRC\text{-}CPA}^n$ advantage of $\mathcal{A}$ by*

$$\mathsf{Adv}^{n\text{-}circ\text{-}cpa}_{\Pi, \mathcal{A}}(\lambda) = 2 \cdot \Pr[\mathsf{IND\text{-}CIRC\text{-}CPA}^n(\Pi, \mathcal{A}, \lambda) = 1] - 1.$$

We say that Π is $\mathsf{IND\text{-}CIRC\text{-}CPA}^n$ secure if $\mathsf{Adv}^{n\text{-}circ\text{-}cpa}_{\Pi, \mathcal{A}}(\lambda)$ is negligible for all PPT $\mathcal{A}$.

IND-CPA$(\Pi, \mathcal{A}, \lambda)$
$b \xleftarrow{r} \{0, 1\}$
PP $\leftarrow$ Setup(1^λ)
$(pk, sk) \leftarrow$ KeyGen(PP)
$(m_0, m_1) \leftarrow \mathcal{A}(pk)$
$y \leftarrow$ Enc(pk, m_b)
$\hat{b} \leftarrow \mathcal{A}(y)$
Output $(\hat{b} \stackrel{?}{=} b)$

IND-CIRC-CPA$^n(\Pi, \mathcal{A}, \lambda)$
$b \xleftarrow{r} \{0, 1\}$
PP $\leftarrow$ Setup(1^λ)
For $i = 1$ to n:
$\quad (pk_i, sk_i) \leftarrow$ KeyGen(PP)
If $b = 1$ then
$\quad \mathbf{y} \leftarrow$ EncCycle(pk, sk)
Else
$\quad \mathbf{y} \leftarrow$ EncZero(pk, sk)
$\hat{b} \leftarrow \mathcal{A}(\mathbf{pk}, \mathbf{y})$
Output $(\hat{b} \stackrel{?}{=} b)$

EncCycle(pk, sk)
For $i = 1$ to n
$\quad m_i \leftarrow sk_{(i \bmod n)+1}$
$\quad y_i \leftarrow$ Enc(pk_i, m_i)
Output $\mathbf{y}$

EncZero(pk, sk)
For $i = 1$ to n
$\quad m_i \leftarrow 0^{|sk_{(i \bmod n)+1}|}$
$\quad y_i \leftarrow$ Enc(pk_i, m_i)
Output $\mathbf{y}$

Fig. 1. Experiments for Definitions 2 and 4, each for a message space M, and we assume that $m_0, m_1, sk_i \in M$. We write $\mathbf{pk}$, $\mathbf{sk}$, and $\mathbf{y}$ for $(pk_1, \ldots, pk_n)$, $(sk_1, \ldots, sk_n)$ and $(y_1, \ldots, y_n)$ respectively.

Discussion. Cash et al. [18] made a distinction between whether an adversary could *distinguish* an encryption cycle from encryptions of zero (as in the standard game above), or whether an adversary could actually *recover* the secret keys (and provided the latter type of counterexample). Recently, Koppula et al. [25] showed that if there exists (an IND-CPA secure) scheme with a PPT adversary that can distinguish an encryption cycle (in the standard game), then it can be transformed into another scheme with a corresponding adversary that can extract the secret keys from the cycle. Thus, in this work, we can focus exclusively on the standard definition.

4   A Framework for Generating Circular Counterexamples

We now present a general framework for creating circular security counterexamples, which we will instantiate under a variety of different assumptions in the subsequent sections. At the center of our framework is an abstraction called a "cycle tester". Like an encryption scheme, a cycle tester must be able to encode a message in an IND-CPA secure manner. However, unlike an encryption scheme, the cycle tester need not support a decryption operation, instead it must support a testing operation which can detect the presence of an encryption cycle.

After formalizing this abstraction, we provide two results that use it. First, we show how our tester can be combined with any IND-CPA encryption scheme (of appropriate message length) to provide a full blown counterexample. Second, we extend this idea to show how to combine any tester with any IND-CCA encryption scheme to get an IND-CCA counterexample.

In addition to letting us focus on a narrower primitive for our counterexample, this separation avoids duplication of work and minimizes assumptions. In particular, we can design a single tester and then both the IND-CPA and IND-CCA counterexamples follow. Most prior works did not address IND-CCA counterexamples. While Cash et al. [18] did, their IND-CCA counterexample required the use of NIZKs, which is a stronger assumption than simply assuming the existence of IND-CCA encryption schemes as we do here. Our abstraction and transformation essentially show that designing IND-CCA counterexamples is no harder than designing IND-CPA ones.

We remark that Koppula et al. [25] have a IND-CPA counterexample with structure similar to our general transformation, however, no generic or IND-CCA theorems are proven.

Definition 5 (n-Cycle Tester). *A cycle tester $\Gamma = (\mathsf{Setup}, \mathsf{KeyGen}, \mathsf{Enc}, \mathsf{Test})$ for message space M and secret key space S is a tuple of algorithms specified as follows:*

- $\mathsf{Setup}(1^\lambda) \to \mathrm{PP}$. *The Setup algorithm takes as input the security parameter λ and outputs common public parameters PP.*
- $\mathsf{KeyGen}(\mathrm{PP}) \to (pk, sk)$. *The Key Generation algorithm takes as input the public parameters PP and outputs a public key pk and secret key $sk \in S$.*
- $\mathsf{Enc}(pk, m \in M) \to C$. *The Encryption algorithm takes as input a public key pk and a message $m \in M$ and outputs a ciphertext C.*
- $\mathsf{Test}(\mathbf{pk}, \mathbf{y}) \to \{0, 1\}$. *On input $\mathbf{pk} = (pk_1, \ldots, pk_n)$ and $\mathbf{y} = (C_1, \ldots, C_n)$, the Testing algorithm outputs a bit in $\{0, 1\}$.*

It also must possess the following properties. Let $\Pi = (\mathsf{Setup}, \mathsf{KeyGen}, \mathsf{Enc}, \cdot)$ be an encryption scheme formed from the first three algorithms of the tester with an empty decryption algorithm. Then, it must hold that:

1. *(IND-CPA security) Π is IND-CPA secure according to Definition 2.*
2. *(Testing Correctness) the Testing algorithm's advantage in distinguishing encryption cycles, denoted $\mathsf{Adv}_{\Pi, \mathsf{Test}}^{n\text{-circ-cpa}}(\lambda)$ from Definition 4, is non-negligible.*

We now prove two theorems.

Theorem 1 (CPA Counterexample from Cycle Testers). *If there exists an IND-CPA-secure encryption scheme Π for message space $M = (M_1 \times M_2)$ and secret key space $S_1 \subseteq M_1$ and an n-cycle tester Γ for message space M_2 and secret key space $S_2 \subseteq M_2$, then there exists an IND-CPA-secure encryption scheme Π' for message space $M = (M_1 \times M_2)$ and secret key space $S = (S_1 \times S_2)$ that is n-circular insecure.*

Proof. Let $\Pi = (\mathsf{Setup}_1, \mathsf{KeyGen}_1, \mathsf{Enc}_1, \mathsf{Dec}_1)$ and $\Gamma = (\mathsf{Setup}_2, \mathsf{KeyGen}_2, \mathsf{Enc}_2, \mathsf{Test}_2)$. We construct an IND-CPA $\Pi' = (\mathsf{Setup}, \mathsf{KeyGen}, \mathsf{Enc}, \mathsf{Dec})$, together with its IND-CIRC-CPA^2 test algorithm Test, as follows.

$\mathsf{Setup}(1^\lambda)$: On input 1^λ, run $\mathrm{PP}_1 \leftarrow \mathsf{Setup}_1(1^\lambda)$ and $\mathrm{PP}_2 \leftarrow \mathsf{Setup}_2(1^\lambda)$. Output $\mathrm{PP} = (\mathrm{PP}_1, \mathrm{PP}_2)$.

$\mathsf{KeyGen}(\mathrm{PP})$: On input $\mathrm{PP} = (\mathrm{PP}_1, \mathrm{PP}_2)$, run $(pk_1, sk_1) \leftarrow \mathsf{KeyGen}_1(\mathrm{PP}_1)$ and $(pk_2, sk_2) \leftarrow \mathsf{KeyGen}_2(\mathrm{PP}_2)$. Output $pk = (pk_1, pk_2)$ and $sk = (sk_1, sk_2)$.

$\mathsf{Enc}(pk, m)$: On input $pk = (pk_1, pk_2)$ and $m = (m_1, m_2) \in M$, run $c_1 \leftarrow \mathsf{Enc}_1(pk_1, (m_1, m_2))$ and $c_2 \leftarrow \mathsf{Enc}_2(pk_2, m_2)$. Output $C = (c_1, c_2)$.

$\mathsf{Dec}(sk, C)$: On input $sk = (sk_1, sk_2)$ and $C = (c_1, c_2)$, output $\mathsf{Dec}_1(sk_1, c_1)$.

$\mathsf{Test}(\mathbf{pk}, \mathbf{y})$: On input $\mathbf{pk} = (pk_1, \ldots, pk_n)$ and $\mathbf{y} = (C_1, \ldots, C_n)$, parse $pk_i = (a_i, b_i)$ and $C_i = (c_i, d_i)$ and output the bit $\mathsf{Test}_2((b_1, \ldots, b_n), (d_1, \ldots, d_n))$.

The correctness of Test follows directly from that of Test_2. If $(\mathbf{pk}, \mathbf{y})$ contains an encryption cycle (or encryptions of zero, respectively), then so will $((b_1, \ldots, b_n), (d_1, \ldots, d_n))$, and thus by definition of the cycle tester, the test will distinguish between these cases with non-negligible advantage.

It remains to argue that Π' is an IND-CPA secure encryption scheme. This follows by a simple hybrid argument based on the fact that an encryption in Π' is a pair of encryptions from two different IND-CPA-secure schemes, Γ and Π. We omit this proof as it is a simplified version of the IND-CCA proof that we provide next.

Theorem 2 (CCA Counterexample from Cycle Testers). *Let k, ℓ be security parameters and $p(\cdot)$ be a polynomial. If there exists an IND-CCA-secure encryption scheme Π (with k-bit secret keys and $(p(\ell) + 2k)$-bit messages) and an n-cycle tester Γ (with k-bit secret keys, k-bit messages, and $p(\ell)$-bit ciphertexts), then there exists an IND-CCA-secure encryption scheme Π' for $2k$-bit messages that is n-circular insecure.*

Proof. Let $\Pi = (\mathsf{Setup}_1, \mathsf{KeyGen}_1, \mathsf{Enc}_1, \mathsf{Dec}_1)$ and $\Gamma = (\mathsf{Setup}_2, \mathsf{KeyGen}_2, \mathsf{Enc}_2, \mathsf{Test}_2)$ with the length constraints above. We construct an IND-CCA $\Pi' = (\mathsf{Setup}, \mathsf{KeyGen}, \mathsf{Enc}, \mathsf{Dec})$, together with its IND-CIRC-CPA^2 test algorithm Test, as follows. We can no longer simply append the cycle-tester encryption to the regular encryption, because changes to the cycle-testing portion might be leveraged to obtain a decryption of a portion of the challenge ciphertext. Instead, we encrypt this cycle-testing portion using the regular CCA-secure scheme.

$\mathsf{Setup}(1^\lambda)$: On input 1^λ, run $\mathrm{PP}_1 \leftarrow \mathsf{Setup}_1(1^\lambda)$ and $\mathrm{PP}_2 \leftarrow \mathsf{Setup}_2(1^\lambda)$. Output $\mathrm{PP} = (\mathrm{PP}_1, \mathrm{PP}_2)$.

$\mathsf{KeyGen}(\mathrm{PP})$: On input $\mathrm{PP} = (\mathrm{PP}_1, \mathrm{PP}_2)$, run $(pk_1, sk_1) \leftarrow \mathsf{KeyGen}_1(\mathrm{PP}_1)$ and $(pk_2, sk_2) \leftarrow \mathsf{KeyGen}_2(\mathrm{PP}_2)$. Output $pk = (pk_1, pk_2)$ and $sk = (sk_1, sk_2)$.

$\mathsf{Enc}(pk, (m_a, m_b))$: On input $pk = (pk_1, pk_2)$ and message $(m_a, m_b) \in \{0,1\}^k \times \{0,1\}^k$, run $c_2 \leftarrow \mathsf{Enc}_2(pk_2, m_b)$ and $c_1 \leftarrow \mathsf{Enc}_1(pk_1, (m_a, m_b, c_2))$. Output $C = (c_1, c_2)$.

$\mathsf{Dec}(sk, C)$: On input $sk = (sk_1, sk_2)$ and $C = (c_1, c_2)$, run $\mathsf{Dec}_1(sk_1, c_1)$. If it does not return a message of the form $(m_a, m_b, m_c) \in \{0,1\}^k \times \{0,1\}^k \times \{0,1\}^{p(\lambda)}$ or if $m_c \neq c_2$, then output $\perp$ (invalid ciphertext). Otherwise, output the message $(m_a, m_b) \in \{0,1\}^k \times \{0,1\}^k$.

$\mathsf{Test}(\mathbf{pk}, \mathbf{y})$: On input $\mathbf{pk} = (pk_1, \ldots, pk_n)$ and $\mathbf{y} = (C_1, \ldots, C_n)$, parse $pk_i = (a_i, b_i)$ and $C_i = (c_i, d_i)$ and output the bit $\mathsf{Test}_2((b_1, \ldots, b_n), (d_1, \ldots, d_n))$. Same as before.

As before, the correctness of Test follows directly from that of Test_2. If $(\mathbf{pk}, \mathbf{y})$ contains an encryption cycle (or encryptions of zero, respectively), then so will $((b_1, \ldots, b_n), (d_1, \ldots, d_n))$, and thus by definition of the cycle tester, the test will distinguish between these cases with non-negligible advantage.

4.1   Proving IND-CCA Security via a Sequence of Games

It remains to argue that Π' is an IND-CCA secure encryption scheme. This proof is significantly more involved than the IND-CPA case. We prove this using a sequence of games from an encryption of a message M_0 to an encryption of M_1 (where these messages come from the IND-CCA game). The public and secret keys are always distributed as in the real scheme, but the structure of the challenge ciphertext changes in each hybrid. We underline these changes for the reader. Let the challenge messages be described as $M_0 = (m_{0,a}, m_{0,b})$ and $M_1 = (m_{1,a}, m_{1,b})$. Then the hybrids are as follows:

Game 1. This corresponds to the original security game $\mathsf{IND\text{-}CCA}(\Pi', \mathcal{A}, \lambda)$ in which the challenger interacts with adversary $\mathcal{A}$, except that the challenge ciphertext is always an encryption of message M_0.

1. Run $\mathsf{Setup}(1^\lambda)$ to produce PP and then $\mathsf{KeyGen}(\mathrm{PP})$ to produce (pk, sk).
2. On decryption query C_i from $\mathcal{A}$, output $\mathsf{Dec}(sk, C_i)$.
3. Provide the challenge ciphertext as $C^* = (c_1^*, c_2^*)$, where $c_1^* = \mathsf{Enc}_1(pk_1, (m_{0,a}, m_{0,b}, c_2^*))$ and $c_2^* = \mathsf{Enc}_2(pk_2, m_{0,b})$. This is a valid encryption of M_0.
4. On decryption query $C_i \neq C^*$ from $\mathcal{A}$, output $\mathsf{Dec}(sk, C_i)$.

Game 2. This is the same as Game 1, except that we change how the second decryption queries to reject *all* requests where the first portion of the query matches the first portion of the challenge.

1. Run $\mathsf{Setup}(1^\lambda)$ to produce PP and then $\mathsf{KeyGen}(\mathrm{PP})$ to produce (pk, sk).
2. On decryption query C_i from $\mathcal{A}$, output $\mathsf{Dec}(sk, C_i)$.
3. Provide the challenge ciphertext as $C^* = (c_1^*, c_2^*)$, where $c_1^* = \mathsf{Enc}_1(pk_1, (m_{0,a}, m_{0,b}, c_2^*))$ and $c_2^* = \mathsf{Enc}_2(pk_2, m_{0,b})$. This is a valid encryption of M_0.
4. On decryption query $C_i = (c_{i,1}, c_{i,2}) \neq C^*$ from $\mathcal{A}$, if $c_{i,1} = c_1^*$ output $\perp$, otherwise output $\mathsf{Dec}(sk, C_i)$.

Game 3. This is the same as Game 2, except that we now encrypt M_1 in the cycle tester portion and continue to encrypt M_0 in the regular encryption portion. We continue to reject all decryption queries where the regular encryption portion matches the challenge.

1. Run $\mathsf{Setup}(1^\lambda)$ to produce PP and then $\mathsf{KeyGen}(\mathrm{PP})$ to produce (pk, sk).
2. On decryption query C_i from $\mathcal{A}$, output $\mathsf{Dec}(sk, C_i)$.
3. Provide the challenge ciphertext as $C^* = (c_1^*, c_2^*)$, where $c_1^* = \mathsf{Enc}_1(pk_1, (m_{0,a}, m_{0,b}, c_2^*))$ and $\underline{c_2^* = \mathsf{Enc}_2(pk_2, m_{1,b})}$.
4. On decryption query $C_i = (c_{i,1}, c_{i,2}) \neq C^*$ from $\mathcal{A}$, if $c_{i,1} = c_1^*$ output $\perp$, otherwise output $\mathsf{Dec}(sk, C_i)$.

Game 4. This is the same as Game 3, except that now the entire challenge ciphertext is an encryption of M_1. As before, we continue to reject all decryption queries where the regular encryption portion matches the challenge.

1. Run $\mathsf{Setup}(1^\lambda)$ to produce PP and then $\mathsf{KeyGen}(\mathrm{PP})$ to produce (pk, sk).
2. On decryption query C_i from $\mathcal{A}$, output $\mathsf{Dec}(sk, C_i)$.
3. Provide the challenge ciphertext as $C^* = (c_1^*, c_2^*)$, where $\underline{c_1^* = \mathsf{Enc}_1(pk_1, (m_{1,a}, m_{1,b}, c_2^*))}$ and $c_2^* = \mathsf{Enc}_2(pk_2, m_{1,b})$.
4. On decryption query $C_i = (c_{i,1}, c_{i,2}) \neq C^*$ from $\mathcal{A}$, if $c_{i,1} = c_1^*$ output $\perp$, otherwise output $\mathsf{Dec}(sk, C_i)$.

Game 5. This is the same as Game 4, except now all decryption queries are answered as normal. The challenge ciphertext always contains an encryption of M_1.

1. Run $\mathsf{Setup}(1^\lambda)$ to produce PP and then $\mathsf{KeyGen}(\mathrm{PP})$ to produce (pk, sk).
2. On decryption query C_i from $\mathcal{A}$, output $\mathsf{Dec}(sk, C_i)$.
3. Provide the challenge ciphertext as $C^* = (c_1^*, c_2^*)$, where $c_1^* = \mathsf{Enc}_1(pk_1, (m_{1,a}, m_{1,b}, c_2^*))$ and $c_2^* = \mathsf{Enc}_2(pk_2, m_{1,b})$.
4. On decryption query $C_i \neq C^*$ from $\mathcal{A}$, output $\mathsf{Dec}(sk, C_i)$.

4.2   Adversary's Probability of Outputting 1 in These Games

Let $\mathsf{Prob}_{\mathcal{A}}^i$ denote the probability that adversary $\mathcal{A}$ outputs a 1 in Game i. We will now show, by a series of steps, that for any adversary $\mathcal{A}$ the difference in its probability of outputting 1 between Game 1 (encryption of M_0) and Game 5 (encryption of M_1) is negligible. Thus, it cannot distinguish between these two games.

Claim. For any adversary $\mathcal{A}$, $\mathsf{Prob}_{\mathcal{A}}^2 = \mathsf{Prob}_{\mathcal{A}}^1$.

Proof. These games are identical except that in Game 1 all decryption queries $C_i = C^*$ are rejected whereas in Game 2 all decryption queries $C_i = (c_{i,1}, c_{i,2})$ such that $c_{i,1} = c_1^*$ for $C^* = (c_1^*, c_2^*)$ are rejected. This results, however, in identical behavior on the decryption queries. Whenever $c_{i,1} \neq c_1^*$, both games answer the queries normally. Whenever $C_i = C^*$, neither game answers this illegal challenge query. On $c_{i,1} = c_1^*$ but $c_{i,2} \neq c_2^*$, Game 2 will output $\perp$. However, Game 1's response is also to reject this query with the message $\perp$ for being a non-valid ciphertext, since the decryption of c_1^* results in an intermediate tuple of the form $(m_{0,a}, m_{0,b}, c_2^*)$ and the decryption algorithm checks that $c_2^* = c_{i,2}$, which won't be true in this case. Thus, the adversary gets identical responses to its decryption queries (and everything else) in both games. Since the games are identical, from the adversary's viewpoint, it will output 1 with the same probability.

Claim. If Γ is an IND-CPA-secure n-cycle tester with security parameter λ, then for any adversary $\mathcal{A}$, $\mathsf{Prob}_{\mathcal{A}}^3 - \mathsf{Prob}_{\mathcal{A}}^2 \leq \mathsf{negl}(\lambda)$.

Proof. We show that an attacker's probability of outputting 1 cannot be non-negligibly different in Games 2 and 3, because that would imply an attack on the IND-CPA security of the cycle tester. More formally, suppose there exists an adversary $\mathcal{A}$ such that $\mathsf{Prob}_{\mathcal{A}}^3 - \mathsf{Prob}_{\mathcal{A}}^2 = \epsilon$. Then we can construct an adversary $\mathcal{B}$ that uses $\mathcal{A}$ to show that Γ is not an IND-CPA-secure n-cycle tester. $\mathcal{B}$ works as follows:

1. $\mathcal{B}$ runs $\mathsf{Setup}_1(1^\lambda) \to \mathrm{PP}_1$ and $\mathsf{KeyGen}_1(\mathrm{PP}_1) \to (pk_{\hat{1}}, sk_1)$.
2. $\mathcal{B}$ obtains the public key pk_2 from the IND-CPA encryption challenger.
3. $\mathcal{B}$ sends $pk = (pk_1, pk_2)$ to $\mathcal{A}$.
4. $\mathcal{A}$ returns two messages $M_0 = (m_{0,a}, m_{0,b})$ and $M_1 = (m_{1,a}, m_{1,b})$.
5. $\mathcal{B}$ sends $(m_{0,b}, m_{1,b})$ to the cycle tester encryption challenger and obtains the challenge c_2^*.
6. $\mathcal{B}$ forms the challenge ciphertext by computing $c_1^* = \mathsf{Enc}_1(pk_1, (m_{0,a}, m_{0,b}, c_2^*))$ and sending $C^* = (c_1^*, c_2^*)$ to $\mathcal{A}$.
7. Eventually, $\mathcal{A}$ returns a bit $\hat{b}$ and $\mathcal{B}$ outputs $\hat{b}$ to its challenger.

In the above, $\mathcal{B}$ perfectly simulates Game 2 for adversary $\mathcal{A}$ if the challenge ciphertext c_2^* contains an encryption of $m_{0,b}$ and, in the other case, $\mathcal{B}$ perfectly simulates Game 3 for adversary $\mathcal{A}$ when the challenge ciphertext c_2^* contains an encryption of $m_{1,b}$. Moreover, $\mathcal{B}$ succeeds if and only if $\mathcal{A}$ succeeds. Thus, if $\mathsf{Prob}_{\mathcal{A}}^3 - \mathsf{Prob}_{\mathcal{A}}^2 = \epsilon$, then we have $\Pr[\mathcal{B}$ is correct$] = \frac{1}{2}\Pr[\mathcal{B}$ is correct $\mid$ IND-CPA challenger chose $0] + \frac{1}{2}\Pr[\mathcal{B}$ is correct $\mid$ IND-CPA challenger chose $1] = \frac{1}{2}\Pr[\mathcal{A}$ is correct $\mid$ Game $2] + \frac{1}{2}\Pr[\mathcal{A}$ is correct $\mid$ Game $3] = \frac{1}{2}(1 - \mathsf{Prob}_{\mathcal{A}}^2) + \frac{1}{2}(\mathsf{Prob}_{\mathcal{A}}^3) = \frac{1}{2}(1 - \mathsf{Prob}_{\mathcal{A}}^2) + \frac{1}{2}(\mathsf{Prob}_{\mathcal{A}}^2 + \epsilon) = \frac{1}{2} + \frac{\epsilon}{2}$. Since we assumed the cycle tester was IND-CPA secure, it must hold that $\epsilon \leq \mathsf{negl}(\lambda)$.

Claim. If Π is an IND-CCA-secure encryption scheme with security parameter λ, then for any adversary $\mathcal{A}$, $\mathsf{Prob}_{\mathcal{A}}^4 - \mathsf{Prob}_{\mathcal{A}}^3 \leq \mathsf{negl}(\lambda)$.

Proof. Suppose there exists an adversary $\mathcal{A}$ such that $\mathsf{Prob}_{\mathcal{A}}^4 - \mathsf{Prob}_{\mathcal{A}}^3 = \epsilon$. Then we can construct an adversary $\mathcal{B}$ that uses $\mathcal{A}$ to show that Π is not an IND-CCA-secure encryption scheme. $\mathcal{B}$ works as follows:

1. $\mathcal{B}$ obtains the public key pk_1 from the IND-CCA encryption challenger.
2. $\mathcal{B}$ runs $\mathsf{Setup}_2(1^\lambda) \to \mathrm{PP}_2$ and $\mathsf{KeyGen}_2(\mathrm{PP}_2) \to (pk_2, sk_2)$.
3. $\mathcal{B}$ sends $pk = (pk_1, pk_2)$ to $\mathcal{A}$.
4. On receiving a decryption query for ciphertext $C_i = (c_{i,1}, c_{i,2})$ from $\mathcal{A}$, $\mathcal{B}$ sends $c_{i,1}$ to its IND-CCA encryption challenger to obtain a message M. $\mathcal{B}$ returns M to $\mathcal{A}$.
5. $\mathcal{A}$ returns two messages $M_0 = (m_{0,a}, m_{0,b})$ and $M_1 = (m_{1,a}, m_{1,b})$.
6. $\mathcal{B}$ computes $c_2^* = \mathsf{Enc}_2(pk_2, m_{1,b})$ and sends $M_0' = (M_0, c_2^*)$ and $M_1' = (M_1, c_2^*)$ to the IND-CCA challenger and obtains the challenge c_1^*.
7. $\mathcal{B}$ sends the challenge ciphertext $C^* = (c_1^*, c_2^*)$ to $\mathcal{A}$.
8. On receiving a decryption query for ciphertext $C_i = (c_{i,1}, c_{i,2})$ where $c_{i,1} \neq c_1^*$ from $\mathcal{A}$, $\mathcal{B}$ sends $c_{i,1}$ to its IND-CCA encryption challenger to obtain a message M. $\mathcal{B}$ returns M to $\mathcal{A}$
9. Eventually, $\mathcal{A}$ returns a bit $\hat{b}$ and $\mathcal{B}$ outputs $\hat{b}$ to its challenger.

In the above, $\mathcal{B}$ perfectly simulates Game 3 for adversary $\mathcal{A}$ if the challenge ciphertext c_1^* contains an encryption of M_0' and, in the other case, $\mathcal{B}$ perfectly simulates Game 4 for adversary $\mathcal{A}$ when the challenge ciphertext c_1^* contains an encryption of M_1'. Moreover, $\mathcal{B}$ succeeds if and only if $\mathcal{A}$ succeeds. Thus, if $\mathsf{Prob}_{\mathcal{A}}^4 - \mathsf{Prob}_{\mathcal{A}}^3 = \epsilon$, then $\mathcal{B}$'s probability of success in the IND-CCA security game is $\Pr[\mathcal{B}$ is correct$] = \frac{1}{2}\Pr[\mathcal{B}$ is correct | IND-CCA challenger chose 0$] + \frac{1}{2}\Pr[\mathcal{B}$ is correct | IND-CCA challenger chose 1$] = \frac{1}{2}\Pr[\mathcal{A}$ is correct | Game 3$] + \frac{1}{2}\Pr[\mathcal{A}$ is correct | Game 4$] = \frac{1}{2}(1 - \mathsf{Prob}_{\mathcal{A}}^3) + \frac{1}{2}(\mathsf{Prob}_{\mathcal{A}}^4) = \frac{1}{2}(1 - \mathsf{Prob}_{\mathcal{A}}^3) + \frac{1}{2}(\mathsf{Prob}_{\mathcal{A}}^3 + \epsilon)$ $= \frac{1}{2} + \frac{\epsilon}{2}$. Since we assumed that Π was IND-CCA secure, it must hold that $\epsilon \leq \mathsf{negl}(\lambda)$.

Claim. For any adversary $\mathcal{A}$, $\mathsf{Prob}_{\mathcal{A}}^5 = \mathsf{Prob}_{\mathcal{A}}^4$.

Proof. These games are identical except that in Game 4 all decryption queries $C_i = (c_{i,1}, c_{i,2})$ such that $c_{i,1} = c_1^*$ for $C^* = (c_1^*, c_2^*)$ are rejected in Game 5 whereas all decryption queries $C_i = C^*$ are rejected. This results, however, in identical behavior on the decryption queries. This case is the mirror image of the argument in the proof of Claim 4.2.

Conclusion of the Proof of Theorem 2. Given the above claims, we can conclude that if Γ is an IND-CPA-secure n-cycle tester and Π is an IND-CCA-secure encryption scheme (with the appropriate length constraints), then for any adversary $\mathcal{A}$, it holds that $\mathsf{Prob}_{\mathcal{A}}^5 - \mathsf{Prob}_{\mathcal{A}}^1$ is negligible, implying that Π' is an IND-CCA-secure encryption scheme.

5   A 2-Cycle Tester from the k-DLIN Assumption

We now present a 2-cycle tester from the decision k-Linear assumption in pairing groups for any constant k (where this assumption is believed to hold for $k \geq 2$ in this bilinear setting and the assumption grows weaker as k increases). We will use a message space of $\{0,1\}^\lambda$. In our exposition we will use boldface to denote a matrix such as $\mathbf{M}$. We also use $g^{\mathbf{M}}$ as shorthand to denote the group elements corresponding to the raising g to each individual element of $\mathbf{M}$.

$\mathsf{Setup}(1^\lambda) \to \mathsf{PP}$. The setup algorithm first runs $\mathcal{G}(1^\lambda)$ to generate a (Type-1) group $\mathbb{G}$ of prime order p with generator g. Next it defines a pseudorandom generator $\mathsf{PRG} : \{0,1\}^\lambda \to \mathbb{Z}_p^{k \times k}$, which maps strings from $\{0,1\}^\lambda$ to *invertible* $k \times k$ matrices over $\mathbb{Z}_p$. Finally, it chooses a random invertible matrix $\mathbf{A} \in \mathbb{Z}_p^{k \times k}$ and computes $g^{\mathbf{A}}$. The public parameters, PP consist of the group description $\mathbb{G}$, the description of PRG and $g^{\mathbf{A}}$.

$\mathsf{KeyGen}(\mathsf{PP}) \to (pk, sk)$. The key generation algorithm first chooses random $w \in \{0,1\}^\lambda$. The secret key $sk = w$. Next, it computes $\mathsf{PRG}(w) \to \mathbf{W} \in \mathbb{Z}_p^{k \times k}$ and chooses a bit $\beta \in \{0,1\}$. Finally, in addition to implicitly including PP, it defines the public key as

$$pk = \begin{cases} (0, K = g^{\mathbf{AW}}) \in \{0,1\} \times \mathbb{G}^{k \times k} & \text{if } \beta = 0; \\ (1, K = g^{\mathbf{WA}}) \in \{0,1\} \times \mathbb{G}^{k \times k} & \text{if } \beta = 1. \end{cases}$$

$\mathsf{Enc}(pk = (\beta, K), m \in \{0,1\}^\lambda) \to ct$.
    The encryption algorithm first computes computes $\mathsf{PRG}(m) \to \mathbf{M} \in \mathbb{Z}_p^{k \times k}$ and then computes $\mathbf{M}^{-1}$. Note that since PRG maps to invertible matrices, $\mathbf{M}$ will have an inverse.

If the type bit $\beta = 0$ the key $K = g^{\mathbf{AW}}$ for some $\mathbf{W}$. The algorithm chooses $\mathbf{r}$ as a random row vector of length k in $\mathbb{Z}_p$ (i.e. a random matrix of dimension $1 \times k$). The ciphertext is computed and output as

$$C_1 = g^{\mathbf{rAW}}, \quad C_2 = g^{\mathbf{rAM}^{-1}}.$$

Thus, the ciphertext will consist of two row vectors in the exponent. We observe all terms are computable from the public keys and public parameters.

If the type bit $\beta = 1$ the key $K = g^{\mathbf{WA}}$ for some $\mathbf{W}$. The algorithm chooses $\mathbf{r}$ as a random column vector of length k in $\mathbb{Z}_p$ (i.e. a random matrix of dimension $k \times 1$). The ciphertext ct is computed and output as

$$C_1 = g^{\mathbf{WAr}}, \quad C_2 = g^{\mathbf{M}^{-1}\mathbf{Ar}}.$$

Test$(\mathbf{pk}, \mathbf{y}) \to \{0, 1\}$. Since we are testing for 2-cycles, parse $\mathbf{y} = (C = (C_1, C_2), C' = (C'_1, C'_2))$. If the key types are identical i.e. $\beta = \beta'$ then just output a random bit as a guess.

Otherwise, presume that $\beta = 0, \beta' = 1$ (if it is the other way around just flip the order). Then compute $e(C_1, C'_2) \stackrel{?}{=} e(C'_1, C_2)$ and output the result. Note here we overload notation so that the pairing operator e is over a matrix of group elements and means matrix multiplication in the exponent. (Or in this case a dot product in the exponent.)

Analysis of Test Algorithm. We analyze the correctness of the test algorithm. Let's consider two secret keys w, w' where $\text{PRG}(w) = \mathbf{W}$ and $\text{PRG}(w') = \mathbf{W}'$. Again, presume that $\beta = 0, \beta' = 1$. The corresponding public keys will be $pk = g^{\mathbf{AW}}$ and $pk = g^{\mathbf{W}'\mathbf{A}}$. Now consider an encryption of m under pk and m' under pk' where $\text{PRG}(m) = \mathbf{M}$ and $\text{PRG}(M') = \mathbf{M}'$. Let $\mathbf{r}$ and $\mathbf{r}'$ be the respective randomness used for each encryption.

The test equations outputs 1 iff $e(C_1, C'_2) \stackrel{?}{=} e(C'_1, C_2)$ this is equivalent to testing

$$\mathbf{rAWM'}^{-1}\mathbf{Ar'} \stackrel{?}{=} \mathbf{rAM}^{-1}\mathbf{W'Ar'}. \tag{1}$$

Let's first consider the case where we have an encryption of a cycle. This means that $m' = w$ and $m = w'$ so we have that $\mathbf{M'}^{-1} = \mathbf{W}^{-1}$ and $\mathbf{M}^{-1} = \mathbf{W'}^{-1}$. Substituting these in we see that

$$\mathbf{rAWM'}^{-1}\mathbf{Ar'} \stackrel{?}{=} \mathbf{rAM}^{-1}\mathbf{W'Ar'}$$
$$\mathbf{rAIAr'} \stackrel{?}{=} \mathbf{rAIAr'}$$
$$\mathbf{rA}^2\mathbf{r'} \stackrel{?}{=} \mathbf{rA}^2\mathbf{r'}.$$

Thus, on a cycle the test will output 1.

We now turn to the case of showing that an encryption of 0's will output 0 (when the keys have different β types) with all but negligible probability.

First, we first let $\mathbf{Z} = \text{PRG}(0^\lambda)^{-1}$ which is the matrix used to encrypt the all 0's string. Second, we consider the probability of the tester outputting 1, when $\mathbf{W}$ and $\mathbf{W}'$ are chosen uniformly at random (and independently from $\mathbf{Z}$) from the set of full rank matrices, as opposed to being the output of a pseudo-random generator. If there, was more than a negligible difference of the test in outputting 1 in these two cases, it would lead to an attack on the security of the pseudorandom generator.

We can now observe that the matrices $\mathbf{X} = \mathbf{AWZA}$ and $\mathbf{X}' = \mathbf{AZW'A}$ are distributed independently and uniformly random from full rank matrices. Note we substituted $\mathbf{Z}$ for both $\mathbf{M}'^{-1}$ and $\mathbf{M}^{-1}$ in Eq. 1. Then $\mathbf{u} = \mathbf{rX}$ and $\mathbf{u}' = \mathbf{rX}'$ are independently distributed as uniformly at random row vectors of length k. Finally, it follows that the probability that

$$\mathbf{ur}' \overset{?}{=} \mathbf{u}'\mathbf{r}'$$

is negligible in the security parameter. Thus, with probability negligibly close to 1 the test algorithm will output 0 when given an encryption of all 0's.

IND-CPA Security of the Tester

Theorem 3. *The above encryption scheme $\Pi = (\mathsf{KeyGen}, \mathsf{Enc}, \mathsf{Test})$ (where the decryption algorithm is ignored) is* IND-CPA-*secure under the k-Linear Assumption in* $\mathbb{G}$.

The proof of this theorem can be found in the full version [11].

6   A 2-Cycle Tester from Learning with Errors

We now present a 2-Cycle Tester whose IND-CPA security follows from the Learning with Errors Assumption. We note that our construction is similar to multi-bit Regev encryption.

6.1   Construction

$\mathsf{Setup}(1^n) \to \mathsf{PP}$. The setup algorithm chooses $m, q, \ell, \sigma, r, \alpha$. These parameters are chosen to satisfy the following constraints: $m \geq 2n \log q$, $\sigma \geq L\omega(\sqrt{\log m})$, $q \geq 5\sigma(m+1)$, $\ell > (n+m+1)\log q + \omega(\log(n+m))$, $r := \sigma\ell$, $\alpha \leq 1/(r\sqrt{m+1} \cdot \omega(\sqrt{\log n}))$, and $q > 2$ is prime. Here, L is defined as follows. We let z denote the number of uniform random bits employed by TrapGen to generate a matrix B in $\mathbb{Z}_q^{n \times m}$ along with a trapdoor basis T_B. L is a bound such that $\|T_B\|_{GS} \leq L$ with overwhelming probability. (We note that this range of parameters allows us to set α so that n/α is polynomial, and LWE is believed to be hard in this parameter regime.) The public parameters are $\mathsf{PP} = (m, q, \ell, \sigma, r, \alpha, z)$.

$\mathsf{KeyGen}(\mathsf{PP}) \to (pk, sk)$. The key generation algorithm chooses a uniformly random secret key sk in $\{0,1\}^z$ and runs $\mathsf{TrapGen}(sk)$ to produce a matrix $B \in \mathbb{Z}_q^{n \times m}$ and a corresponding trapdoor basis T_B. It then chooses independent and uniformly random vectors $c_1, \ldots, c_\ell \in \mathbb{Z}_q^n$ and noise vectors $\gamma_1, \ldots, \gamma_\ell$ from χ^m, where χ is distributed as $\lfloor q\Psi_\alpha \rceil \bmod q$, where Ψ_α is a distribution on $\mathbb{T}$ of a normal variable with mean 0 and standard deviation $\alpha/\sqrt{2\pi}$ reduced modulo 1. (We think of these vectors as row vectors.) In addition to implicitly including the PP, it sets

$$pk = \{c_1, \ldots, c_\ell, y_1 := c_1 B + \gamma_1, \ldots, y_\ell := c_\ell B + \gamma_\ell\}.$$

$\mathsf{Enc}(pk, m \in \{0, 1\}^z) \to ct$. The encryption algorithm runs $\mathsf{TrapGen}(m)$ to produce a matrix $Z \in \mathbb{Z}_q^{n \times m}$ and a corresponding trapdoor basis T_Z. It chooses random signs $r_1, \ldots, r_\ell \in \{-1, 1\}$ and computes $s := \sum_{i=1}^{\ell} r_i c_i$. It then uses T_Z to sample a short (column) vector v such that $Zv = s^t$, by calling the algorithm $\mathsf{SampleD}$. It computes $C = \sum_{i=1}^{\ell} r_i y_i$, and sets the ciphertext as $ct = (C, v)$.

$\mathsf{Test}((pk_0, pk_1), ((C_0, v_0), (C_1, v_1))) \to \{0, 1\}$. The cycle test algorithm compares $C_0 v_1$ to $C_1 v_0$ and checks if there are close modulo q (if their distance is $\leq 2q/5$). If so, it outputs 1. If not, it outputs 0.

Analysis of Test Algorithm. We let B_0, Z_0, s_0 be the B, Z and s values corresponding to ciphertext (C_0, v_0) and B_1, Z_1, s_1 be the analogous values for (C_1, v_1). When there is a cycle, we then have $Z_0 = B_1$ and $Z_1 = B_0$. We then have $B_0 v_1 = s_1^t$ and $B_1 v_0 = s_0^t$. Noting that $C_0 = s_0 B_0 + \psi_0$ for some small vector ψ_0, we see that

$$C_0 v_1 = s_0 B_0 v_1 + \psi_0 v_1 = s_0 s_1^t + \psi_0 v_1.$$

Similarly, $C_1 = s_1 B_1 + \psi_1$ for some small vector ψ_1, so we have that

$$C_1 v_0 = s_1 B_1 v_0 + \psi_1 v_0 = s_1 s_0^t + \psi_1 v_0.$$

We consider the size of $\psi_0 v_1 - \psi_1 v_0$ modulo q. First, $|\psi_0 v_1|$ is at most ℓ times the maximal size of $|\gamma_j v_1|$. Using the same analysis as in the proof of Lemma 8.2 of [21], each of these is $\leq \frac{q}{5\ell}$ with high probability. Thus, $|\psi_0 v_1 - \psi_1 v_0| \leq \frac{2q}{5}$ with high probability.

Since all of v_0, v_1, ψ_0, ψ_1 are short, this will cause these values to be close modulo q, so the cycle test will output 1 with high probability.

When there is no cycle, the matrices B_0 and B_1 are (statistically close) to independent, uniformly random matrices. Thus the probability that $s_0 B_0 v_1$ and $s_1 B_1 v_0$ will be within $\frac{2}{5} q$ modulo q is negligibly close to $\frac{2}{5}$. Thus the cycle test wins the distinguishing game with probability negligibly close to $\frac{1}{2} + \frac{1}{2} \cdot \frac{3}{5} = \frac{4}{5}$.

6.2   IND-CPA Security of the Tester

To prove that this construction satisfies IND-CPA, we define a sequence of security games.

Game$_0$ This is the regular IND-CPA security game for our construction:

1. The challenger runs $\mathsf{Setup}(1^n) \to \mathrm{PP} = (m, q, \ell, \sigma, r, \alpha, z)$.

2. The challenger chooses a uniformly random secret key sk in $\{0,1\}^z$ and runs TrapGen(sk) to produce a matrix $B \in \mathbb{Z}_q^{n \times m}$ and a corresponding trapdoor basis T_B. It then chooses independent and uniformly random vectors $c_1, \ldots, c_\ell \in \mathbb{Z}_q^n$ and noise vectors $\gamma_1, \ldots, \gamma_\ell$ from χ^m. It sets

$$pk = \{c_1, \ldots, c_\ell, y_1 := c_1 B + \gamma_1, \ldots, y_\ell := c_\ell B + \gamma_\ell\}.$$

The challenger gives the parameters PP and key pk to the attacker.

3. $\mathcal{A}$ The attacker submits two messages m_0, m_1 to the challenger.

4. The challenger flips a coin $b \in \{0,1\}$. It runs TrapGen(m_b) to produce a matrix $Z \in \mathbb{Z}_q^{n \times m}$ and a corresponding trapdoor basis T_Z. It chooses random signs $r_1, \ldots, r_\ell \in \{-1, 1\}$ and computes $s := \sum_{i=1}^{\ell} r_i c_i$. It then uses T_Z to sample a short (column) vector v such that $Zv = s^t$, by calling the algorithm SampleD. It computes $C = \sum_{i=1}^{\ell} r_i y_i$, and sets the ciphertext as (C, v).

5. The attacker receives the challenge ciphertext. It then outputs a guess b' and wins if $b' = b$.

Game₁

2. The challenger chooses a uniformly random secret key sk in $\{0,1\}^z$ and runs TrapGen(sk) to produce a matrix $B \in \mathbb{Z}_q^{n \times m}$ and a corresponding trapdoor basis T_B. It then chooses independent and uniformly random vectors $c_1, \ldots, c_\ell \in \mathbb{Z}_q^n$ and <u>uniformly random vectors $y_1, \ldots, y_\ell \in \mathbb{Z}_q^m$</u>. It sets $pk = \{c_1, \ldots, c_\ell, y_1, \ldots, y_\ell\}$.

Game₂

4. The challenger flips a coin $b \in \{0,1\}$. It runs TrapGen(m_b) to produce a matrix $Z \in \mathbb{Z}_q^{n \times m}$ and a corresponding trapdoor basis T_Z. <u>It chooses s randomly in $\mathbb{Z}_q^n$.</u> It then uses T_Z to sample a short (column) vector v such that $Zv = s^t$, by calling the algorithm SampleD. <u>It chooses C randomly from $\mathbb{Z}_q^m$</u> and sets the ciphertext as (C, v).

Game₃

4. <u>The challenger samples the vector v from $\mathcal{D}_{\mathbb{Z}^m, \sigma}$. It chooses C randomly from $\mathbb{Z}_q^m$</u> and sets the ciphertext as (C, v).

At this point, the distribution of the ciphertext is independent of the message, and it is clear that no PPT adversary can obtain a non-zero advantage.

Lemma 3. *Under the LWE assumption for the noise distribution χ, no PPT attacker can obtain a non-negligible difference in advantage between Game₀ and Game₁.*

Proof. We can collect the column vectors $c_1^t, \ldots, c_\ell^t$ into a $n \times \ell$ matrix we call D. We can collect the row vectors $y_1, \ldots, y_m$ into a $\ell \times m$ matrix we call Y and the row vectors $\gamma_1, \ldots, \gamma_\ell$ into a $\ell \times m$ matrix we call Γ. We can then write the public key as $D, D^t B + \Gamma$. Since B is never published, each column of B is a fresh, uniform vector in $\mathbb{Z}_q^n$, and therefore each column of $D^t B + \Gamma$ is distributed as an LWE sample with D playing the role of the $n \times m$ matrix A and the column of B playing the role of the random vector s. By a hybrid argument over the columns, we can thus rely on LWE to change each y_i to be uniformly distributed in $\mathbb{Z}_q^m$.

Lemma 4. *No PPT attacker can obtain a non-negligible difference in advantage between Game$_1$ and Game$_2$.*

Proof. For this, we will argue that the distributions of s, C in Game$_1$ and Game$_2$ are statistically close. This is a direct application of Lemma 2 with j set to be $n+m$. To see this, we consider the random signs $r_1, \ldots, r_\ell \in \{-1, 1\}$ as a column vector R of length ℓ. We then consider the (vertical) concatenation of s^t and C^t into a $n+m$ length column vector. In Game$_1$, this is produced as MR, where M is a $(n+m) \times \ell$ matrix formed by vertically concatenating D and Y^t as defined in the proof of the previous lemma. Since the matrices D, Y are now uniformly chosen, replacing MR by a uniformly random $(n+m) \times 1$ matrix (as in Game$_2$) is a statistically close distribution by Lemma 2.

Lemma 5. *No PPT attacker can obtain a non-negligible difference in advantage between Game$_2$ and Game$_3$.*

Proof. We will argue that the distributions of v in Game$_2$ and Game$_3$ are statistically close. We first observe that in Game$_2$, v is chosen so that $Zv = s^t$ for a uniformly random s that is now independent of the rest of the ciphertext. The distribution of v here produced by SampleD is statistically close to $\mathcal{D}_{\Lambda_q^s(Z), \sigma}$. Now by Lemma 1, if we consider the distribution $\mathcal{D}_{\mathbb{Z}^m, \sigma}$, the probability mass on the preimages of s^t under the mapping $Zv = s^t$ is (up to a negligible statistical distance) the same for each s. Thus, the distribution of v in both Game$_2$ and in Game$_3$ is statistically close to $\mathcal{D}_{\mathbb{Z}^m, \sigma}$.

References

1. Acar, T., Belenkiy, M., Bellare, M., Cash, D.: Cryptographic agility and its relation to circular encryption. In: Gilbert, H. (ed.) EUROCRYPT 2010. LNCS, vol. 6110, pp. 403–422. Springer, Heidelberg (2010)
2. Adão, P., Bana, G., Herzog, J., Scedrov, A.: Soundness and completeness of formal encryption: the cases of key cycles and partial information leakage. J. Comput. Secur. **17**(5), 737–797 (2009)
3. Agrawal, S., Boneh, D., Boyen, X.: Efficient lattice (H)IBE in the standard model. In: Gilbert, H. (ed.) EUROCRYPT 2010. LNCS, vol. 6110, pp. 553–572. Springer, Heidelberg (2010)
4. Ajtai, M.: Generating hard instances of the short basis problem. In: Wiedermann, J., Van Emde Boas, P., Nielsen, M. (eds.) ICALP 1999. LNCS, vol. 1644, pp. 1–9. Springer, Heidelberg (1999)
5. Alperin-Sheriff, J., Peikert, C.: Circular and KDM security for identity-based encryption. In: Fischlin, M., Buchmann, J., Manulis, M. (eds.) PKC 2012. LNCS, vol. 7293, pp. 334–352. Springer, Heidelberg (2012)

6. Alwen, J., Peikert, C.: Generating shorter bases for hard random lattices. In: Proceedings of 26th International Symposium on Theoretical Aspects of Computer Science, STACS 2009, 26–28 February 2009, Freiburg, Germany, pp. 75–86 (2009)

7. Applebaum, B.: Key-dependent message security: generic amplification and completeness. In: Paterson, K.G. (ed.) EUROCRYPT 2011. LNCS, vol. 6632, pp. 527–546. Springer, Heidelberg (2011)

8. Applebaum, B., Cash, D., Peikert, C., Sahai, A.: Fast cryptographic primitives and circular-secure encryption based on hard learning problems. In: Halevi, S. (ed.) CRYPTO 2009. LNCS, vol. 5677, pp. 595–618. Springer, Heidelberg (2009)

9. Barak, B., Haitner, I., Hofheinz, D., Ishai, Y.: Bounded key-dependent message security. In: Gilbert, H. (ed.) EUROCRYPT 2010. LNCS, vol. 6110, pp. 423–444. Springer, Heidelberg (2010)

10. Benson, K., Shacham, H., Waters, B.: The k-BDH assumption family: bilinear map cryptography from progressively weaker assumptions. In: Dawson, E. (ed.) CT-RSA 2013. LNCS, vol. 7779, pp. 310–325. Springer, Heidelberg (2013)

11. Bishop, A., Hohenberger, S., Waters, B.: New circular security counterexamples from decision linear and learning with errors. In: Cryptology ePrint Archive, Report 2015/715 (2015)

12. Black, J., Rogaway, P., Shrimpton, T.: Encryption-scheme security in the presence of key-dependent messages. In: 9th Annual International Workshop Selected Areas in Cryptography, SAC 2002, St. John's, Newfoundland, Canada, 15–16 August 2002. Revised Papers, pp. 62–75 (2002)

13. Boneh, D., Boyen, X., Shacham, H.: Short group signatures. In: Franklin, M. (ed.) CRYPTO 2004. LNCS, vol. 3152, pp. 41–55. Springer, Heidelberg (2004)

14. Boneh, D., Halevi, S., Hamburg, M., Ostrovsky, R.: Circular-secure encryption from decision diffie-hellman. In: Wagner, D. (ed.) CRYPTO 2008. LNCS, vol. 5157, pp. 108–125. Springer, Heidelberg (2008)

15. Brakerski, Z., Goldwasser, S.: Circular and leakage resilient public-key encryption under subgroup indistinguishability. In: Rabin, T. (ed.) CRYPTO 2010. LNCS, vol. 6223, pp. 1–20. Springer, Heidelberg (2010)

16. Brakerski, Z., Goldwasser, S., Kalai, Y.T.: Black-box circular-secure encryption beyond affine functions. In: Ishai, Y. (ed.) TCC 2011. LNCS, vol. 6597, pp. 201–218. Springer, Heidelberg (2011)

17. Camenisch, J.L., Lysyanskaya, A.: An efficient system for non-transferable anonymous credentials with optional anonymity revocation. In: Pfitzmann, B. (ed.) EUROCRYPT 2001. LNCS, vol. 2045, p. 93. Springer, Heidelberg (2001)

18. Cash, D., Green, M., Hohenberger, S.: New definitions and separations for circular security. In: Fischlin, M., Buchmann, J., Manulis, M. (eds.) PKC 2012. LNCS, vol. 7293, pp. 540–557. Springer, Heidelberg (2012)

19. Dolev, D., Dwork, C., Naor, M.: Nonmalleable cryptography. SIAM J Comput. **30**(2), 391–437 (2000)

20. Gentry, C.: Fully homomorphic encryption using ideal lattices. In: Proceedings of the 41st Annual ACM Symposium on Theory of Computing, STOC 2009, Bethesda, MD, USA, 31 May - 2 June 2009, pp. 169–178 (2009)

21. Gentry, C., Peikert, C., Vaikuntanathan, V.: Trapdoors for hard lattices and new cryptographic constructions. In: Proceedings of the 40th Annual ACM Symposium on Theory of Computing, Victoria, British Columbia, Canada, 17–20 May 2008, pp. 197–206 (2008)

22. Goldwasser, S., Micali, S.: Probabilistic encryption. J. Comput. Syst. Sci. **28**(2), 270–299 (1984)

23. Haitner, I., Holenstein, T.: On the (Im)Possibility of key dependent encryption. In: Reingold, O. (ed.) TCC 2009. LNCS, vol. 5444, pp. 202–219. Springer, Heidelberg (2009)

24. Hofheinz, D., Kiltz, E.: Secure hybrid encryption from weakened key encapsulation. In: Menezes, A. (ed.) CRYPTO 2007. LNCS, vol. 4622, pp. 553–571. Springer, Heidelberg (2007)

25. Koppula, V., Ramchen, K., Waters, B.: Separations in circular security for arbitrary length key cycles. In: Dodis, Y., Nielsen, J.B. (eds.) TCC 2015, Part II. LNCS, vol. 9015, pp. 378–400. Springer, Heidelberg (2015)

26. Laud, P.: Encryption cycles and two views of cryptography. In: NORDSEC 2002 - Proceedings of the 7th Nordic Workshop on Secure IT Systems (Karlstad University Studies 2002:31), pp. 85–100 (2002)

27. Marcedone, A., Orlandi, C.: Obfuscation $\Rightarrow$ (IND-CPA Security $!\Rightarrow$ Circular Security). In: Abdalla, M., De Prisco, R. (eds.) SCN 2014. LNCS, vol. 8642, pp. 77–90. Springer, Heidelberg (2014)

28. Micciancio, D., Peikert, C.: Trapdoors for lattices: simpler, tighter, faster, smaller. In: Pointcheval, D., Johansson, T. (eds.) EUROCRYPT 2012. LNCS, vol. 7237, pp. 700–718. Springer, Heidelberg (2012)

29. Naor, M., Segev, G.: Public-key cryptosystems resilient to key leakage. SIAM J. Comput. 41(4), 772–814 (2012)

30. Naor, M., Yung, M.: Public-key cryptosystems provably secure against chosen ciphertext attacks. In: STOC 1990, pp. 427–437 (1990)

31. Peikert, C.: Public-key cryptosystems from the worst-case shortest vector problem: extended abstract. In: Proceedings of the 41st Annual ACM Symposium on Theory of Computing, STOC 2009, Bethesda, MD, USA, 31 May - 2 June 2009, pp. 333–342 (2009)

32. Rackoff, C., Simon, D.R.: Non-interactive zero-knowledge proof of knowledge and chosen ciphertext attack. In: Feigenbaum, J. (ed.) CRYPTO 1991. LNCS, vol. 576, pp. 433–444. Springer, Heidelberg (1992)

33. Regev, O.: On lattices, learning with errors, random linear codes, and cryptography. In: Proceedings of the 37th Annual ACM Symposium on Theory of Computing, Baltimore, MD, USA, 22–24 May 2005, pp. 84–93 (2005)

34. Rothblum, R.: On the circular security of bit-encryption. In: Cryptology ePrint Archive, Report 2012/102 (2012). http://eprint.iacr.org/

35. Shacham, H.: A cramer-shoup encryption scheme from the linear assumption and from progressively weaker linear variants. IACR Cryptol. ePrint Arch. **2007**, 74 (2007)

Author Index

Printed in the United States
By Bookmasters